BiCMOS Integrated Circuit Design

IEEE PRESS
445 Hoes Lane, PO Box 1331
Piscataway, NJ 08855-1331

BiCMOS Integrated Circuit Design

with Analog, Digital, and Smart Power Applications

Edited by

M. I. Elmasry

Professor of Electrical and Computer Engineering
University of Waterloo

IEEE
PRESS

A Selected Reprint Volume
IEEE Solid-State Circuits Council, *Sponsor*

The Institute of Electrical and Electronics Engineers, Inc., New York

This book may be purchased at a discount from the publisher
when ordered in bulk quantities. For more information contact:

IEEE PRESS Marketing
Attn: Special Sales
PO Box 1331
445 Hoes Lane
Piscataway, NJ 08855-1331
Fax: (908) 981-8062

Printed in the United States of America
10 9 8 7 6 5 4 3 2 1

ISBN 0-7803-0430-6
IEEE Order Number: PC0346-7

Library of Congress Cataloging-in-Publication Data

BiCMOS integrated circuit design : with analog, digital, and smart power applications / edited by M. I. Elmasry.
p. cm.
Includes bibliographical references and index.
ISBN 0-7803-0430-6
1. Metal oxide semiconductors, Complementary—Design. 2. Bipolar integrated circuits—Design. I Elmasry, Mohamed I.
TK7871.99.M44B5 1994
621.3815—dc20 94-2478
CIP

Contents

Preface

THE introduction of bipolar small-scale integrated circuits (ICs) in the 1960s established the foundation to one of the most impressive engineering areas, in terms of technical and social impact: microelectronics. These early ICs, with only tens of devices on the same chip, revolutionized the way electronic systems were built. It was the first time the integration of a complete circuit was possible, with all its active and passive components, on a single monolithic silicon chip. Since that time, a long series of dreams have been realized with impressive speed and usually to the surprise of the most optimistic workers in the field.

The introduction of bipolar small-scale ICs was followed by two equally impressive evolutionary steps: the use of small- and medium-scale MOS integrated circuits in the 1970s and medium- and large-scale CMOS integrated circuits in the 1980s. And if the 1990s has to be marked with an IC landmark, then it is the use of BiCMOS integrated circuits.

BiCMOS is a circuit designer's dream come true! It offers the power of bipolar devices in circuit design, along with CMOS. However, the ability to integrate high-performance bipolar transistors with CMOS offers a great challenge to process, device, and circuit designers. The optimum use of these devices at high levels of integration must be achieved in order to justify the higher cost of BiCMOS, compared to that of CMOS or bipolar. The field of BiCMOS circuit design can be treated as an extension of CMOS or bipolar circuit design. For example, the field of BiCMOS digital circuit design can be treated as CMOS digital circuit design with an add-on bipolar speed enhancer. It can also be treated as a bipolar digital circuit design with an add-on CMOS power saver. However, the BICMOS technology environment can only be optimally used when the circuit designer is equally fluent in CMOS, bipolar, and BiCMOS circuit techniques. As a result (s)he can successfully apply this rich circuit medium to the design of analog, digital, and mixed analog/digital VLSI systems.

This book contains a collection of the most recent key developments in the BiCMOS field. In addition to covering BiCMOS analog and digital circuit design, it reviews BiCMOS process technology, modeling, and application to the area of smart power. It is hoped that this book will contribute to the education and the research efforts of a new generation of BiCMOS IC designers.

M. I. Elmasry
Waterloo, Ontario, Canada

Acknowledgments

I would like to acknowledge the significant contributions of Dr. Kevin O'Connor of AT&T Bell Laboratories and the IEEE Solid-State Circuit Council reviewers to the final version of this book. The original contributions to this book by Drs. R. H. Havemann, R. H. Eklund, H. J. Shin, A. G. Eldin, B. Leung and J. S. T. Huang are greatly appreciated. My BiCMOS research team at the University of Waterloo have made the last 10 years of my research life enjoyable, and they have made many contributions to this book and to the field. My special thanks to Drs. A. Bellaouar, S. Embabi and S. Rofail. The support of the professional staff at IEEE Press and that of my assistant Joan Pache is also acknowledged. Finally, the research support of MICRONET, NSERC, and ITRC is appreciated.

Part 1
Analysis and Design of BiCMOS Integrated Circuits

AT the heart of recent progress in the area of BiCMOS ICs is the ability to optimally design integrated systems and subsystems. Understanding the basic issues of design trade-offs; the effect of processing technologies and process/device/circuit optimization, modeling, and design methodology; circuit design innovations, fault characterization, and design for testability (DFT); and scaling and performance limits, contributes to the success of applying BiCMOS technology to a wide range of applications.

This part consists of four sections: Section 1.1 contains an introductory tutorial to BiCMOS integrated circuits. Section 1.2 deals with processing technology and contains an invited overview paper. Section 1.3 is the main section of this part and addresses the topic of BiCMOS circuit analysis and circuit techniques and, finally, Section 1.4 deals with scaling and performance limits.

In Section 1.1, following the introductory tutorial by Elmasry, the paper by Alvarez examines the state-of-the-art development of BiCMOS ICs at a given time frame, while the papers, by Ikeda et al. and Kubo et al. explain in detail the development of basic BiCMOS circuits.

In Section 1.2, following the invited overview paper by Havemann and Eklund on BiCMOS processing technologies, four papers are selected to highlight some different aspects of BiCMOS processing technologies, including a BiNMOS technology without buried layers and a complementary BiCMOS technology.

In Section 1.3, the topic of BiCMOS circuit analysis and circuit techniques is addressed in some detail. The paper by Embabi et al. introduces the topic, while the papers that follow deal with different aspects of circuit design. The paper by Hiraki et al. examines how full-swing BiCMOS circuits can be designed. The four papers by Heimsch et al. and by Elrabaa and Elmasry and by Yano et al. and by Ritts et al. examine different novel circuit techniques that can be used in BiCMOS circuit design. The following three papers by Raje et al., Wissel and Gould, and Elrabaa and Elmasry discuss the optimal use of CMOS and BiCMOS in a BiCMOS environment. The modeling aspects of BiCMOS circuits, including Current Mode Logic (CML) and Emitter Coupled Logic (ECL), are addressed in the three papers by Raje et al., Yuan, and Ghannam et al. Finally, the topic of fault characterization, testing, and design for testability is addressed in the paper by Salama and Elmasry.

In Section 1.4, three papers deal with the topic of scaling BiCMOS circuits and their operation at low supply voltages.

Paper 1.1

Introduction to BiCMOS Integrated Circuits: A Tutorial

M. I. ELMASRY, FELLOW, IEEE
VLSI RESEARCH GROUP
DEPARTMENT OF ELECTRICAL AND COMPUTER ENGINEERING
UNIVERSITY OF WATERLOO
WATERLOO, ONTARIO, CANADA N2L 3G1

***Abstract*—This introductory tutorial covers basic concepts in BiCMOS integrated circuit design. First-order circuit analysis of typical gates is given. Some key design issues are reviewed and proposed solutions are briefly summarized. Basic digital inverters are used to highlight different circuit techniques.**

1. Introduction

The main advantages of CMOS circuits are low power dissipation and high noise immunity [1, 2]. However, they have a limited driving capability, which causes performance degradation, especially in VLSI chips where signals drive large loading capacitances. Alternatively, bipolar circuits can drive high capacitive loads at the expense of high power dissipation [3]. Combining the two circuit structures offers BiCMOS integrated circuits.

Combining MOSFETs and bipolar devices into integrated circuit design has been proposed as early as 1969 [4]. The bipolar device offers the following advantages: greater current drive per unit area, better matching, superior linear performance, and low sensitivity to process variation. The MOSFET device offers a smaller area because of its self-isolation feature, lower parasitic capacitance, high input resistance, lower power, high yield, bidirectional current switching, and inherent memory capability. The ability of the two devices to charge and discharge loading capacitances is related to their transconductances and is given by:

$$g_m, \text{ bipolar} = \frac{I_c}{V_t}$$

$$g_m, \text{ MOSFET} = \sqrt{I_D}\,\sqrt{2\beta},$$

where I_c = the bipolar collector current $= I_s \exp(V_{BE}/V_t)$,

V_t = the thermal voltage $= \dfrac{kT}{q}$,

I_s = the bipolar saturation current,

I_D = the MOSFET drain current $= \dfrac{\beta}{2}(V_G - V_T)^2$,

$\beta = \mu C_{ox}\dfrac{W}{L}$,

V_T = the threshold voltage of the MOSFET.

The transconductance of the two devices relate to their current driving capabilities per unit input voltage. The exponential dependency of the g_m of the bipolar transistor on the input voltage is a key advantage over the linear dependency of that of the MOS transistor.

A BiCMOS process is usually based on an existing bipolar or CMOS process [5]. This results in a suboptimal but practical process that can be used in digital and analog circuit design. Initially, BiCMOS was used in digital circuit design. Bipolar and BiCMOS circuits were used as an add-on CMOS speed enhancer. Alternatively, CMOS and BiCMOS circuits were used as an add-on bipolar power saver. Figure 1 shows the delay-power curves for digital CMOS, BiCMOS, NMOS, and ECL logic families. The key advantage of BiCMOS technology is its unique ability to integrate these forms of logic families on the same chip, offering a wide range of delay, power, and density trade-offs [6].

BiCMOS use in analog circuit design is more recent and it was, initially, used only occasionally when the main device did not offer the needed characteristics. For example, MOSFETs were used in mainly bipolar integrated circuits when high impedance switches were needed. However, as state-of-the-art BiCMOS integrated circuit design matures, analog circuit design techniques are being developed that make use of both bipolar and MOSFET devices early in the design phase [7].

2. From CMOS to BiCMOS

Figure 2(a) shows a CMOS inverter buffer with the complementary transistors P_1 and N_1 driving a capacitive load C_L. If

This work is supported in part by NSERC, ITRC, MICRONET, BNR, and NTE.

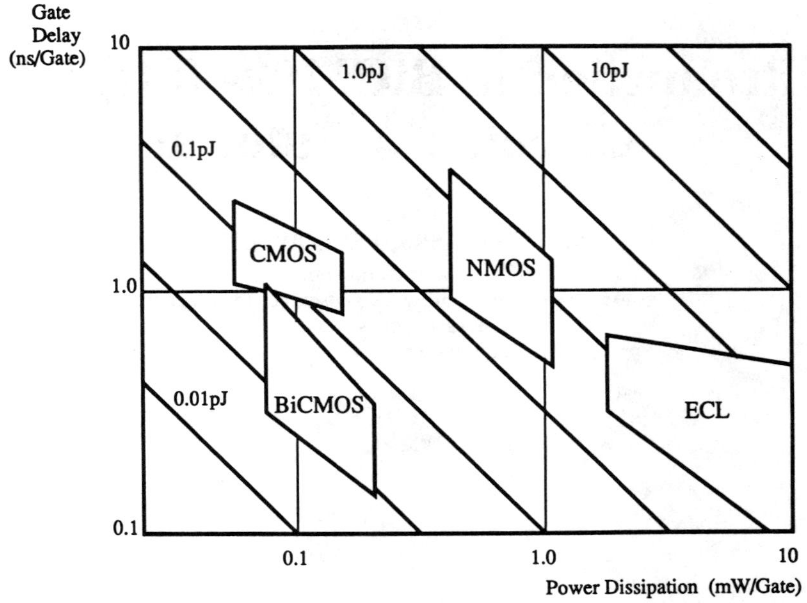

Fig. 1 Power-delay curves for CMOS, NMOS, ECL, and BiCMOS.

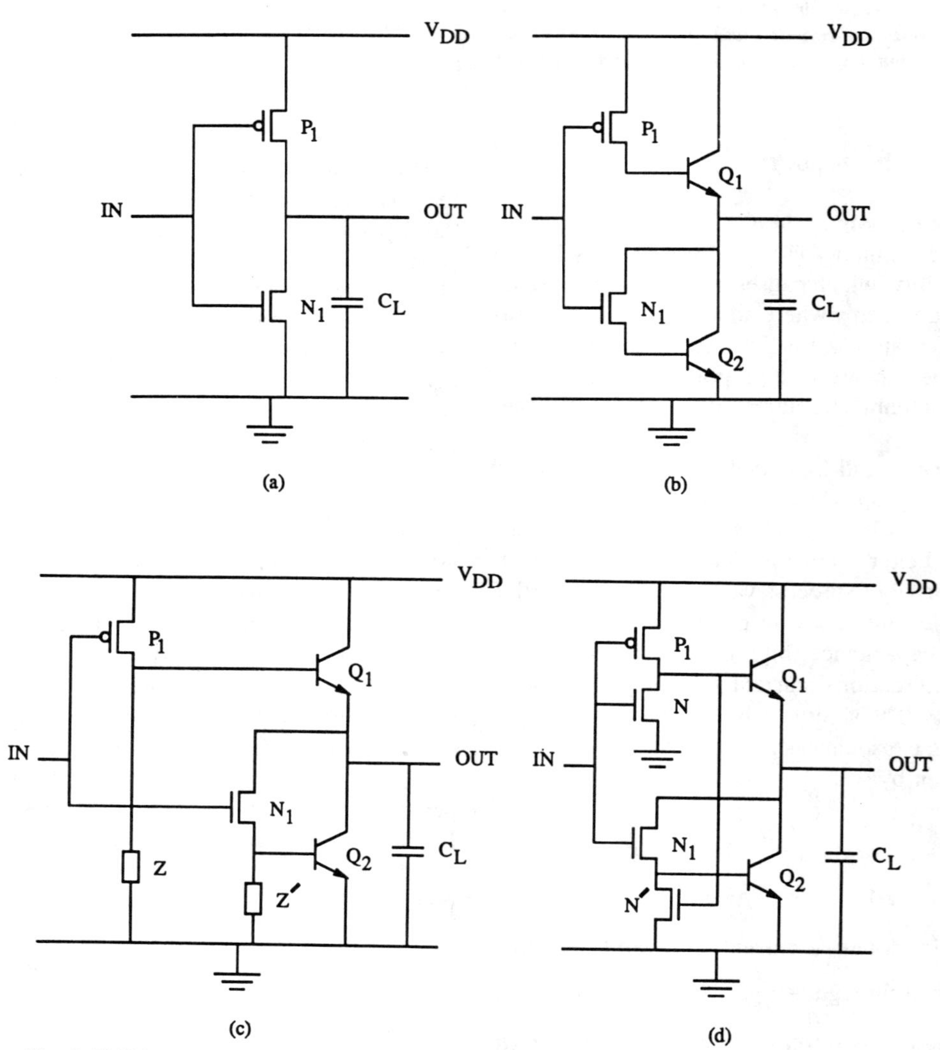

Fig. 2 CMOS and BiCMOS inverters.

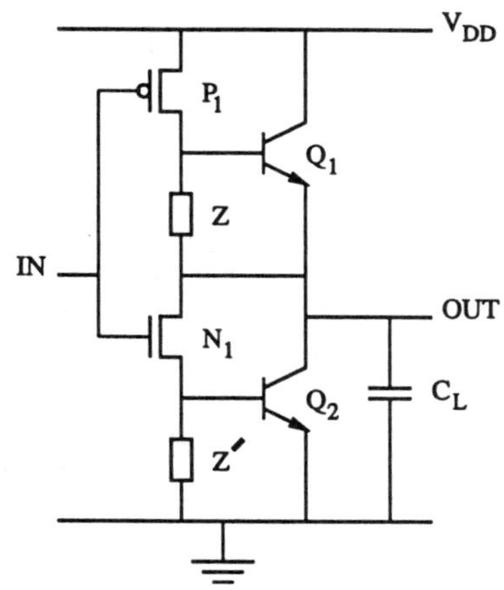

(e)

Fig. 2 CMOS and BiCMOS inverters (continued).

the circuit is used to drive a push-pull *NPN* bipolar transistor pair Q_1 and Q_2, which in turn drives the capacitive load C_L, a BiCMOS inverter buffer results, as shown in Fig. 2(b). When the input is low, P_1 is on and provides the base current to Q_1, turning it on. When the input is high, N_1 conducts and Q_2 provides a discharging path to C_L. In the circuit of Fig. 2(b), Q_1 and Q_2 have a high-resistance discharging path as they turn off. Thus, this circuit has a limited speed because of the long turn-off time of Q_1 and Q_2. This problem can be solved by adding bleeding elements that discharge the bases when Q_1 and Q_2 are turning off. Figure 2(c) shows one way of bleeding the base charge using resistors Z and Z' between the bases of Q_1 and Q_2 and ground. However, the path P_1–Z provides a dc path between V_{DD} and ground when the input is low. This drawback can be eliminated by using two techniques: Z can be replaced with an input-controlled NMOS transistor N, as shown in Fig. 2(d), or Z can be connected as shown in Fig. 2(e). In Fig. 2(d), Z' is also replaced with an NMOS transistor N', controlled by the output of the CMOS inverter, to speed up the discharging of Q_2.

3. The BiCMOS Inverter

In Fig. 2(d), which is referred to as the standard BiCMOS inverter, when the input voltage level is low, P_1 and, hence, Q_1 are on and C_L is charged with the emitter current of Q_1. The high output voltage level V_H is given by

$$V_H = V_{DD} - \text{drop across } P_1 - V_{BE}.$$

The drop across P_1 can be made sufficiently small by increasing the size of P_1. However, V_H still suffers from a V_{BE} voltage loss.

When the input voltage level is high N_1 and, hence, Q_2 are on and C_L is discharged mainly with the collector current of Q_2. The low output voltage V_L is given by

$$V_L = V_{BE} + \text{drop across } N_1.$$

Similarly, the drop across N_1 can be made sufficiently small by increasing the size of N_1. However, V_L still suffers from a V_{BE} voltage loss.

The V_{BE} voltage losses in BiCMOS can be reduced by circuit techniques. These V_{BE} voltage losses result from the fact that Q_1 and Q_2 do not enter the saturation region when they are on. This happens because sufficient base current is not supplied by P_1 and N_1 to Q_1 and Q_2 to drive them into saturation.

In the BiCMOS gates of Figs. 2(c), (d), and (e), the bipolar transistors operate in the forward active region. However, if the collector resistance is large (kΩ range), the bipolar transistors may saturate before the output reaches its final value. The saturation occurs because of the forward biasing of the intrinsic base-collector junction [6]. In this case, the output reaches its final value very slowly. Where the collector resiliance is low (20–50 Ω for 1-μm process) because of the use of the buried layer in state-of-the-art BiCMOS processes, the bipolar transistors do not saturate.

Z and Z' of Fig. 2(e) are replaced by two NMOS transistors N and N', which are controlled by the input, as shown in Fig. 2(d). The input controls N, while the complement of the input controls N'. When the input is low, P_1 is on, N is off, N_1 is off, and N' is on. This allows Q_1 to be on and Q_2 to be off; hence, the output is high. When the input is high, P_1 is off, N is on, N_1 is on, N' is off, and Q_2 is on; hence, the output is low.

The replacement of Z and Z' with N and N' speeds up the charging and discharging of Q_1 and Q_2. During the charging of C_L, the current supplied by P_1 is all used to charge up the base of Q_1, as N is turning off. Similarly, during the discharging of C_L, the current supplied by N_1 to the base of Q_2 is used to charge the base of Q_2 as N' is turning off.

The circuit of Fig. 2(d) suffers from an area penalty if NAND and NOR circuits are used. NAND and NOR gates derived from the inverter of Fig. 2(d) are achieved by replacing P_1, N_1, and N with an appropriate series/parallel device, as in conventional CMOS design, as shown in Figs. 3(a) and (b). As a result, this circuit version is used when the number of logic inputs is small.

The inverter of Fig. 2(e) is usually called R-type BiCMOS inverter. The resistors are implemented within a polysilicon layer. The gate occupies less area than that of Fig. 2(d) when complex gates are implemented. However, it has the drawback that the current supplied by the MOS transistor to the bipolar transistor is wasted through the resistors, which reduces the bipolar drive.

Figure 4 shows the charging path of C_L for the circuit of Fig. 2(d) when P_1 is turning on, charging the base of Q_1. For first-order calculations, the turn-on delay time of the circuit can be considered as the summation of three components [8]:

$$t_d = t_{d1} + t_{d2} + t_{d3}, \tag{1}$$

where t_{d1} = the time required to turn on Q_1 using the PMOS transistor current I_{sat},
$= C_J V_{BE} / I_{sat}$,

t_{d2} = the time required to establish the collector current I_c,
$= \tau_f I_c / I_{sat}$,

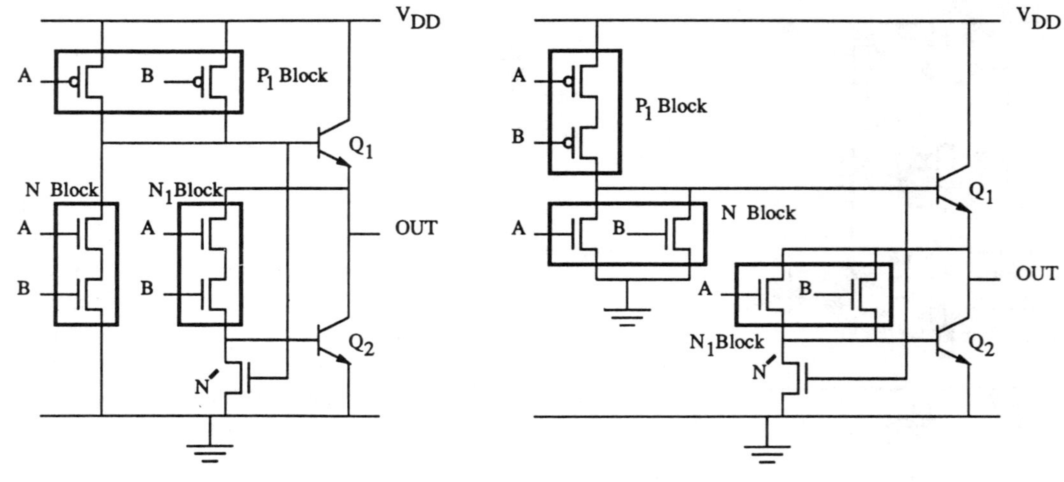

(a) (b)

Fig. 3 BiCMOS gates. (a) NAND; (b) NOR.

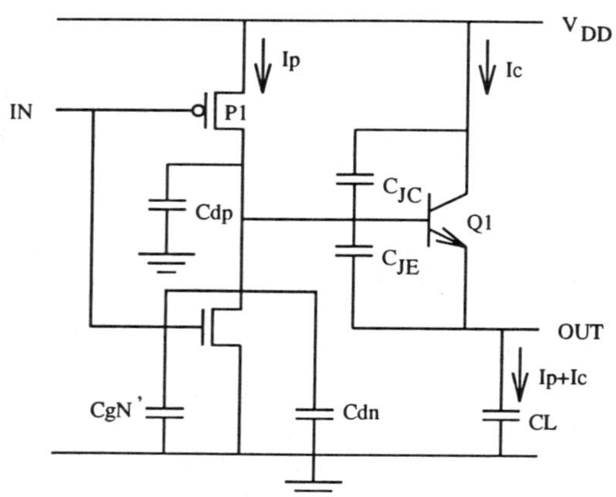

Fig. 4 Calculating BiCMOS delays.

t_{d3} = the time required to charge the output load capacitance C_L and the output parasitic capacitance C_{par} using the emitter current of Q_1 $(= I_{sat} + I_c)$,

$$= \frac{(C_L + C_{par})V_l}{(I_{sat} + I_c)} \approx \frac{(C_L + C_{par})V_l}{\beta I_{sat}}$$

V_{BE} = the base-emitter voltage of Q_1,

V_l = the logic swing across C_L,

$C_J = C_{DP} + C_{DN} + C_{JC} + C_{JE} + C_{GN}'$,

$C_{DP}, C_{DN}, C_{JC}, C_{JE}$ = junction capacitances of the drain of P_1, the drain of N, the collector, and emitter of Q_1, respectively; C_{GN}' is the gate-oxide capacitance of transistor N',

$\tau_f \approx 1/(2\pi f_T)$ forward transient time of Q_1,

f_T = the cutoff frequency of Q_1,

I_c = the collector current of Q_1,

I_{sat} = the drain saturation current of P_1,

$= K'/2(W/L)_p(|V_{GS}| - |V_{TP}|)^2$, for large-size devices ($L > 3$ μm),

$= KC_{ox}v_{sat}W_p(|V_{GS}| - |V_{TP}|)$, for small-size devices ($L \geq 0.5$ μm),

K' = a process-dependent parameter defined as $K' = \mu\, C_{ox}$, where μ is the carrier mobility and C_{ox} is the gate-oxide per unit area,

$K = \dfrac{1}{1 + E_c \dfrac{L}{V_{GS} - |V_{TP}|}}$, where E_c is a critical field where the carrier drift velocity is equal to the carrier saturation velocity [6],

v_{sat} = carrier saturation velocity.

A similar relation to that given in (1) can be used to calculate the discharging time of C_L.

For 1-μm typical BiCMOS technology, the averages of C_{DP}, C_{DN}, C_{GN}', C_{JE}, and C_{JC} are about 8, 2, 4, 8, and 25 fF, respectively, giving C_J = 47 fF. f_T is about 10 GHz, and τ_f is about 15 ps. For $\mu_n = 600$ cm^2/$V - s$, $t_{ox} = 40$ nm, $C_{ox} = 9 \times 10^{-8}$ F/cm^2, $|V_{TP}| = 1$V, $K'_n = 54\ \mu$A/V^2, then $I_{sat} = 0.9$ mA for long-channel devices when $|V_{GS}| = 4.2$V and $W_P = 5$ μm. If $t_{ox} = 25$ nm, $C_{ox} = 1.4 \times 10^{-7}$ F/cm^2 and v_{sat} are taken to be 10^7 cm/s and $K = 0.6$, then $I_{sat} = 1.4$ mA for a short-channel device when $|V_{GS}| = 4.2$V, $|V_{TP}| = 0.8$V, and $W_P = 5$ μm.

The calculation of I_c in (1) depends on whether the bipolar transistor is in low- or high-level injection.

$$I_c = \beta I_b \qquad \text{if } I_b = I_{sat} \leq J_{KF}A_e/\beta \quad \text{(low-level injection)}$$

$$I_c = \sqrt{J_{KF}A_e\beta I_b} \qquad \text{if } I_b = I_{sat} > J_{KF}A_e/\beta \quad \text{(high-level injection)},$$

where J_{KF} is the bipolar transistor forward knee current density and A_e is the emitter area. If $J_{KF} = 0.5$ mA/μm^2 and $\beta = 50$, then $J_{KF}A_e\beta = 50\ \mu$A, where $A_e = 5 \times 1$ μm^2. Hence, the transistor is in high level when $I_c > 2.5$ mA.

From the above values and substituting in (1) with $\tau_f = 15$ ps, $I_c = 14$ mA, $I_{sat} = 1.5$ mA, $C_J = 47$ fF, $C_L = 1$ pF, $V_{BE} = 0.8$ V, and $V_l = 3.5$ V (for a 5-V supply voltage), we

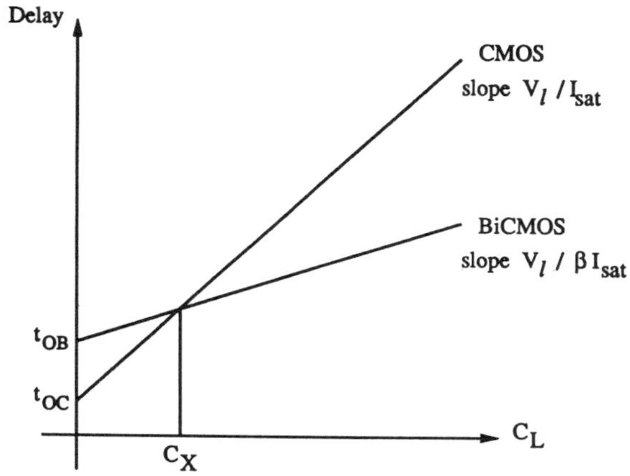

Fig. 5 Delay versus load capacitance for CMOS and BiCMOS.

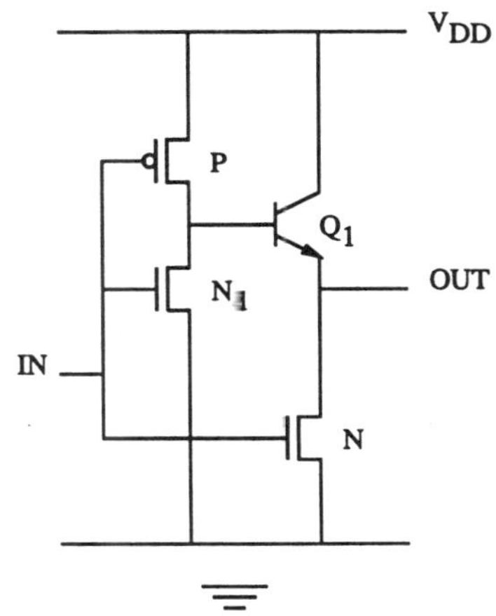

Fig. 6 The circuit schematic of the BiNMOS inverter [9].

have $t_{d1} = 25$ ps, $t_{d2} = 140$ ps, $t_{d3} = 226$ ps, and $t_d = 391$ ps. Thus, the contribution of t_{d1}, t_{d2}, and t_{d3} to t_d is 6%, 36%, and 58%, respectively. The contribution of t_{d3} to the total delay increases even further at higher values of C_L.

Figure 5 shows the typical delay versus C_L for a CMOS and a BiCMOS inverter buffer. At low values of C_L, the BiCMOS inverter has a higher delay because of its higher internal capacitances compared with that of the CMOS inverter. However, for $C_L > C_x$, the BiCMOS circuit offers less delay compared with the CMOS and becomes less sensitive to load capacitance. The value C_x, the crossover capacitance, determines the area of application where BiCMOS, rather than CMOS, circuits should be used.

Equation (1) reveals that when C_L is zero, $t_d = t_{d1} + t_{d2} + t_{d30}$, which corresponds to t_{OB} in Fig. 5. With comparison to CMOS

$$t_{dcmos} = \frac{(C_L + C_{par})}{I_{sat}} V_l.$$

When C_L is zero, $t_{dcmos} = \dfrac{C_{par}V_l}{I_{sat}}$, which corresponds to t_{OC} in Fig. 5 and is smaller than t_{OB}.

Figure 5 also shows the slope of the delay versus C_L; whereas in the case of CMOS it is V_l/I_{sat}, it is $V_l/\beta I_{sat}$ in the case of BiCMOS.

It should be noted that for a 5-V power supply, the R-type and the N-type BiCMOS circuits of Fig. 2 are widely used in BiCMOS digital circuit design. The bleeding resistors in the R-type BiCMOS are in the range of kΩ and are implemented within a polysilicon layer. This layer is used also in the BiCMOS process for implementing ECL circuits.

4. Full-Voltage Swing BiCMOS

One of the important design issues in BiCMOS is the reduction or elimination of V_{BE} voltage losses. V_{BE} voltage losses in basic BiCMOS circuits reduce the logic swing of the circuit. Hence, both noise margins and the overdrive voltage factor of the driven circuit are reduced. V_{BE} voltage losses do not scale down with reduced supply voltages, hence, their negative impact on circuit performance is increased at low V_{DD}. The major cause of this deterioration of performance is the pull-down section of the BiCMOS gate of Fig. 2(c). The V_{GS} voltage of the NMOS transistor N_1 is equal to $V_{DD} - 2V_{BE}$. As V_{DD} is reduced, V_{GS} is significantly reduced, resulting in degradation of drain current and, hence, the driving capability of the BiCMOS gate.

Three circuit techniques can be used to reduce the effect of the V_{BE} losses [6]:

1. Using the BiNMOS circuit configuration.
2. Modifying the basic BiCMOS circuit structure.
3. Using complementary and quasi-complementary BiCMOS.

4.1 BiNMOS

The basic BiNMOS inverter is shown in Fig. 6 [9]. The pull-down circuit of the basic BiCMOS inverter of Fig. 2(d) is replaced with a single NMOS transistor that is driven by the input. The pull-down section is kept as CMOS. The circuit can also be viewed as the basic CMOS inverter of Fig. 2(a) in which the slow pull-up PMOS transistor is replaced with a faster BiCMOS pull-up circuit. The BiNMOS inverter eliminates the pull-down V_{BE} voltage losses and reduces the circuit area. It offers lower delay than BiCMOS at reduced V_{DD}.

The BiNMOS gate of Fig. 6 suffers from two drawbacks. The first is the unsymmetrical rise and fall times. The second drawback is the fact that the high output reaches $V_{DD}-V_{BE}$ instead of V_{DD}. This increases the delay and the leakage current of the next driven gate and, hence, decreases the noise margins.

4.2. Modifying the Basic BiCMOS Circuit

Figure 7 shows four types of circuit structures that achieve full V_{DD} voltage swing. They are referred to as follows:

(a) the base-emitter shunt R-type BiCMOS circuit (Fig. 7a),
(b) the base-emitter shunt MOS-type BiCMOS circuit (Fig. 7b),

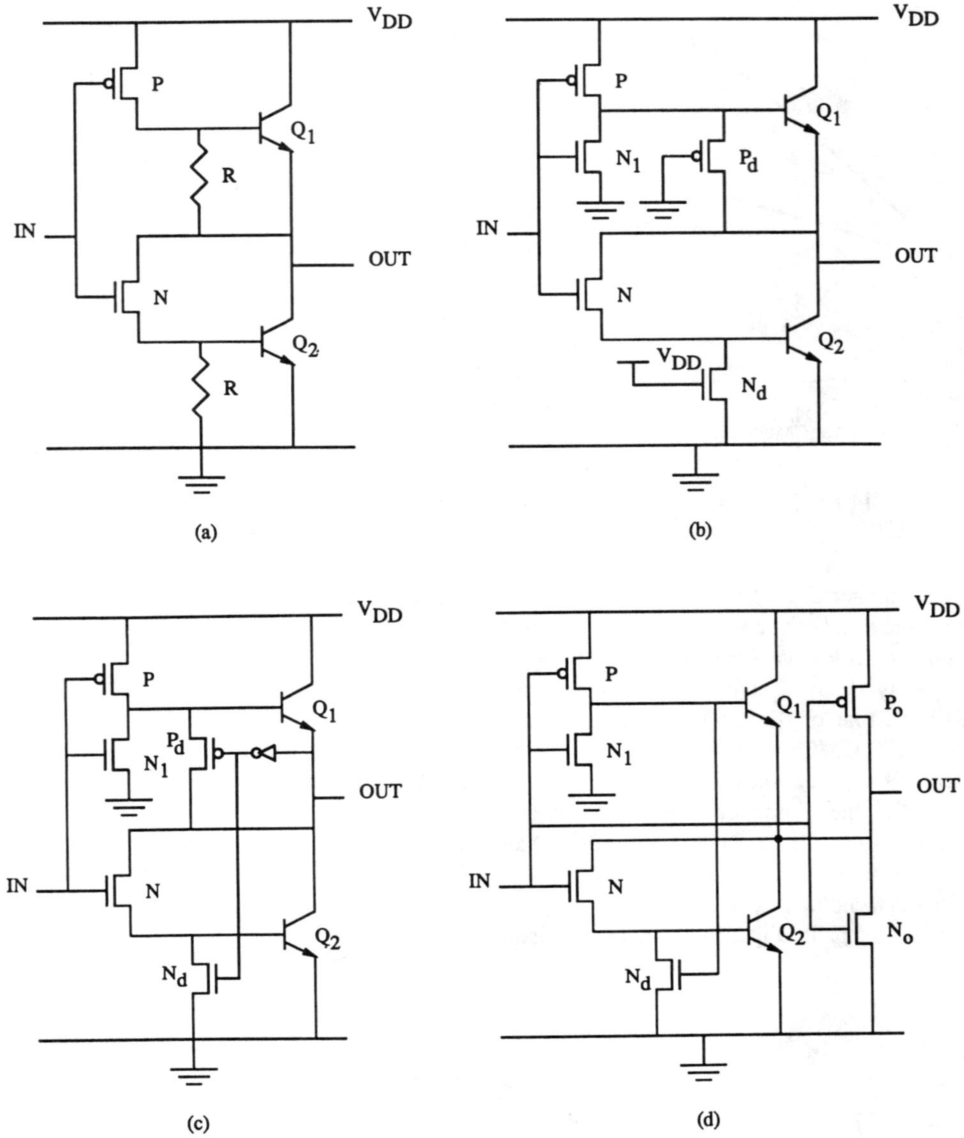

Fig. 7 Full-swing BiCMOS buffers [10].

(c) the base-emitter shunt feedback-type BiCMOS circuit (Fig. 7c), and
(d) the collector-emitter shunt BiCMOS circuit (Fig. 7d).

The R-type BiCMOS circuit of Fig. 7(a) [10] tries to achieve full V_{DD} voltage swing by shunting the base-emitter junctions of the two output bipolar transistors with resistors. The resistor and its associated MOSFET offer an alternative charging/discharging path to the output capacitance when the output voltage approaches V_{BE} of its final value. In charging the output capacitance, as Q_1 turns off at the end of the charging time, P and R provide some charging current, allowing the output to reach V_{DD}. Similarly, as Q_2 turns off at the end of the discharging time, N and R provide some discharging current allowing the output to reach ground.

To allow full-voltage swing, R must be low enough to allow significant charging/discharging current at the end of the charging/discharging time. However, at low values of R, the voltage drive of the bipolar transistors is reduced due to the current sharing at the base and, hence, the delay at the start of the charging/discharging time will increase. Also, the circuit cannot provide full voltage swing when the load capacitance is large and/or the switching frequency is high because of the difficulty of choosing an optimum value for R to minimize the overall delay.

The MOS-type BiCMOS circuit of Fig. 7(b) [10] is similar to the R-type except that the two resistors R are replaced with fixed-bias gate MOSFETs. The sizing of R is replaced with the sizing of P_d and N_d.

In the feedback-type BiCMOS circuit of Fig. 7(c) [10], the conduction of the two MOSFETs P_d and N_d are controlled by the output. This allows the two MOSFETs to act as high resistances at the early stages of the charging and discharging times and as low resistances at the later stages of the charging and discharging times. This results in a full-voltage swing without sacrificing the delay performance of the circuit.

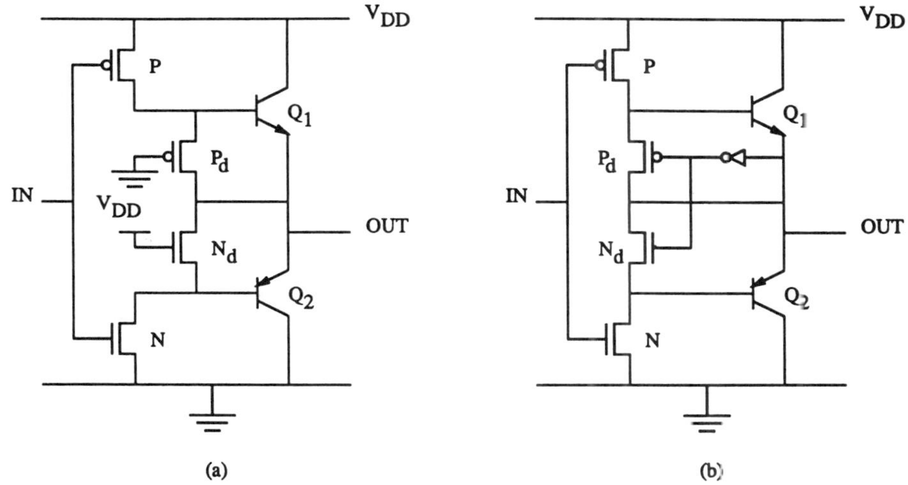

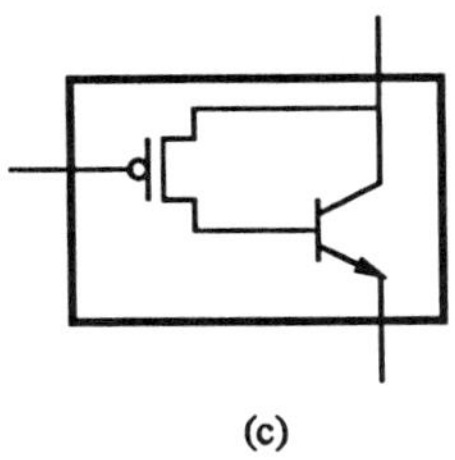

Fig. 8 (a), (b) Schematics of full-swing CBiCMOS inverters; (c) Quasi-pnp transistor.

The circuit in Fig. 7(d) [10] uses a different circuit principle to achieve full-voltage swing. A CMOS inverter is used in parallel at the output with the BiCMOS inverter; i.e., the CMOS's Po and No are used to shunt the collector-emitter of both Q_1 and Q_2. This allows the output voltage to reach V_{DD} and ground at the end of the charging and discharging times, respectively.

It should be noted that the full-swing achieved using the circuits of Fig. 7 is obtained at a slower rate as the output voltages approach $V_{DD} - V_{BE}$.

4.3. Complementary and Quasi-Complementary BiCMOS Circuits

Figure 8(a) shows the use of a complementary bipolar output stage. The bipolar transistors are operating in the emitter-follower mode. Their base-emitter junctions are shunted with fixed-bias MOS transistors P_d and N_d [11]. A trade-off must be applied in sizing P_d and N_d similar to the case of Fig. 7(a). The conduction of the shunting MOS transistors can be controlled by the output voltage in a feedback arrangement, as shown in Fig. 8(b). This results in an improved delay performance for the same reason given for the circuit of Fig. 7(c). The pnp bipolar transistor can be replaced with a quasi-pnp transistor consisting of a PMOS and an npn, as shown in Fig. 8(c), thus eliminating the need for a pnp device. The circuit is called a quasi-complementary BiCMOS (QCBiCMOS) circuit [12].

QCBiCMOS has two attractive features. The first one is that the drain current of the pull-down section does not suffer the 2 V_{BE} losses as in the case of conventional BiCMOS gate. The second advantage is that the pull-down waveform is steep due to the good charge retention capability of the bipolar transistor.

Alternatively, the output bipolar transistors can be used in the common-emitter mode of operation, as shown in Fig. 9 [13]. The complementary bipolar transistors Q_1 and Q_2 are driven into saturation with the input-controlled MOSFETs N and P, respectively. As a result, the output voltage levels are (V_{satn}) and ($V_{DD} - V_{satp}$), where V_{satn} and V_{satp} are the collector-emitter saturation voltages of the npn and the pnp bipolar transistors, respectively.

The basic circuit of Fig. 9 has two disadvantages: there is a dc current through the bases of Q_1 and Q_2, and the turn-off time is high because of the saturation of the bipolar transistors. These two problems can be solved by driving the bipolar transistors with base current pulses rather than dc base currents. During the base current pulses, the bipolar transistors operate in the active or soft-saturation region and at the end of the pulse, the base is isolated from the rest of the circuit and the bipolar transistors continue sourcing or sinking current to or from the output node. A circuit that uses this concept is shown in Fig. 10. It consists of a multidrain [14] CMOS input stage (P, N) [15]. The two discharging elements P_d and N_d are essential to discharge the bases of Q_1 and Q_2, respectively, to

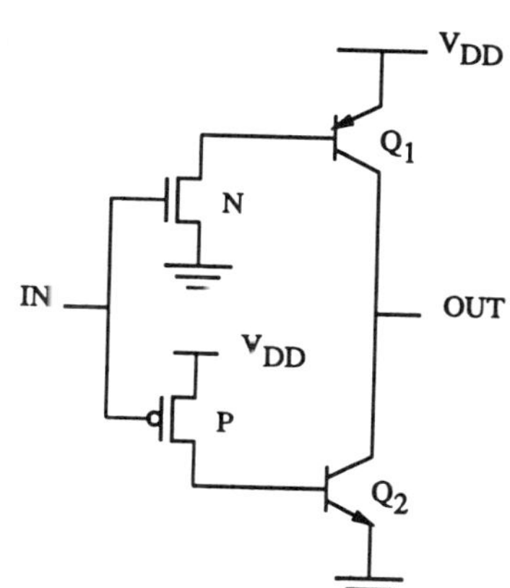

Fig. 9 The schematic of the common-emitter complementary BiCMOS buffer [13].

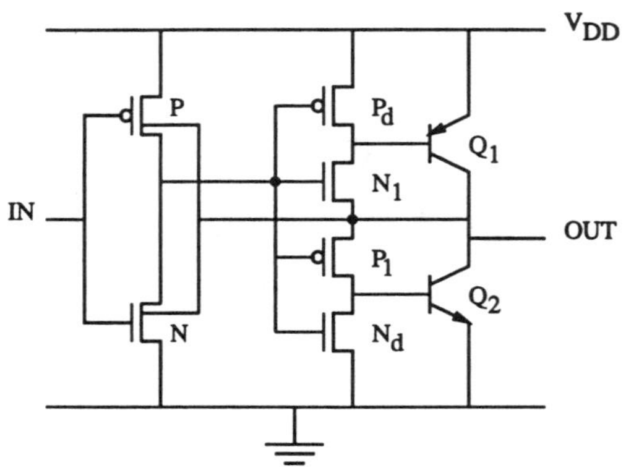

Fig. 10 The circuit schematic of the MOS-type CE CBiCMOS inverter.

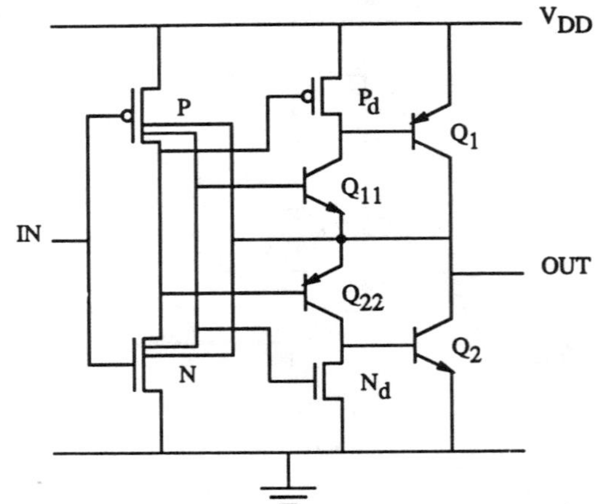

Fig. 11 The schematic of the semi-full-swing CE CBiCMOS inverter [15].

reduce their turn-off time. Transistors N_1 (P_1) act as an input-output controlled switch. The complementary pair Q_1 and Q_2 serves as an output driver. To avoid any dc level shifts due to leakage currents, the buffer has to be shunted by a minimum-size CMOS inverter for which the second drain of the multi-drain input CMOS is used.

The circuit operation can be explained as follows. When the input is low, transistor P turns on, causing N_1 to turn on. When the base of Q_1 is pulled down to $V_{DD} - V_{BE,on}$, Q_1 turns on, pulling up the output node. When the output voltage reaches V_{DD} - V_{TN}, transistor N_1 turns off, provided that the threshold of N_1 is greater than or equal to $V_{BE,on}$. The base of Q_1 is thus isolated and the base charge, if large enough, forces Q_1 into saturation. The pull-down operation can be similarly explained.

This implementation has several drawbacks. The base drive is poor, because the gate-source and drain-source voltages of N_1 (P_1) decrease as the output rises (falls). Thus, the base charge required for full-swing operation is hard to achieve. Moreover, the threshold voltages of N_1 and P_1 cannot be scaled down for low-supply voltage operation, because they have to be greater than or equal to V_{BE}. This results in degrading the base drive.

To improve the drive of the output drivers, N_1 and P_1 can be replaced by an npn (Q_{11}) and a pnp (Q_{22}) transistor, respectively, as shown in Fig. 11 [15]. The circuit is similar to the one shown in Fig. 10. In this case, Q_{11} and Q_{22} act as an input/output controlled switch. The bases of Q_{11} and Q_{22} should not be tied to the same drain node to avoid short-circuiting their base-emitter junctions. They are connected to separate drains of the multidrain P/N CMOS input stage. The circuit operation is similar to that of Fig. 10. If the input voltage is low, the transistor P turns on, causing Q_{11} and, in turn, Q_1 to enter the active region, pulling up the output node. When the output voltage reaches $V_{DD} - V_{BE,on}$, Q_{11} turns off. The base of Q_1 is hence isolated, preventing any leakage of its charge and thus allowing the output voltage to increase until Q_1 saturates. For the circuit of Fig. 11, it is difficult to satisfy the full-swing condition, especially at very low power-supply voltages (< 2V).

Figure 12 shows another implementation [15] of a complementary BiCMOS buffer. It consists of a multidrain CMOS in-

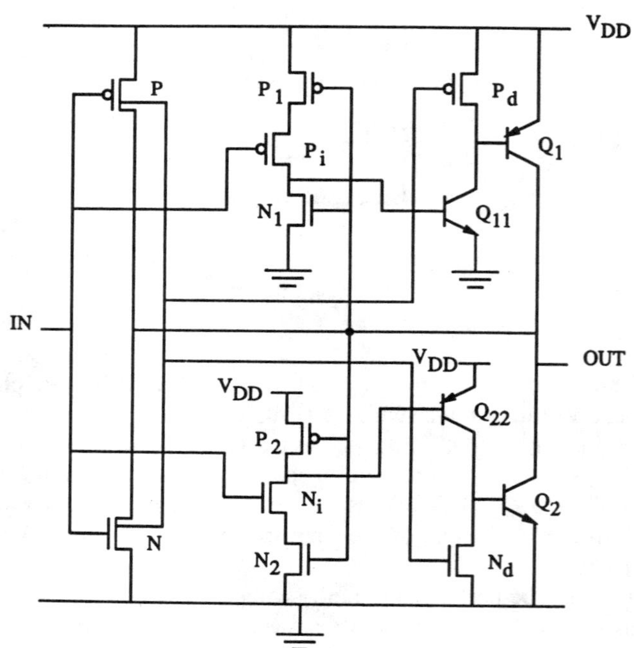

Fig. 12 The schematic of the full-swing CE CBiCMOS inverter [15].

put stage. The first drain is used to compensate for any leakage of the charge stored in the load capacitance during steady state. The second drain drives the discharging transistors P_d and N_d. The inverter composed of P_1 P_i, and N_1 controls the base current of Q_{11}. P_i is driven by the input, while P_1 and N_1 are driven by the output signal. The current amplifier Q_{11} serves as the base driver for Q_1. The pull-down section is the complement of the pull-up section. The operating principle of this circuit is similar to the one of Fig. 11. When the output is LOW and the input falls from HIGH to LOW, the switches P_i and P_1 are closed, turning on Q_{11} and Q_1, thus pulling up the output node. As the output voltage reaches the threshold voltage of the inverter P_1, P_i, and N_1, the voltage of the base of Q_{11} falls below $V_{BE,on}$, causing Q_{11} to turn off. The time period required to turn Q_{11} off can be tailored by sizing the transistors P_1, N_1, and P_i. As soon as Q_{11} cuts off, the base minority charge is trapped in the base of Q_1. Hence, Q_1 stays

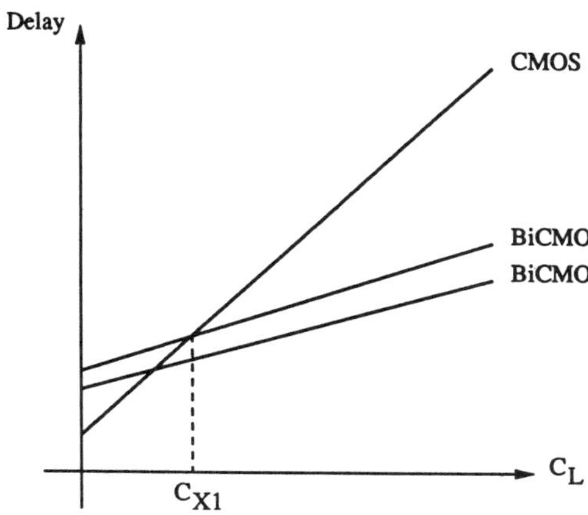

Fig. 13 The crossover capacitance for different types of BiCMOS.

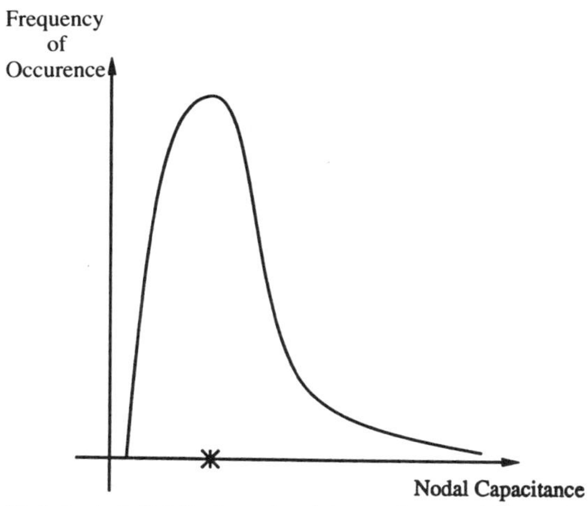

Fig. 14 A typical distribution of nodal capacitance within a given chip design.

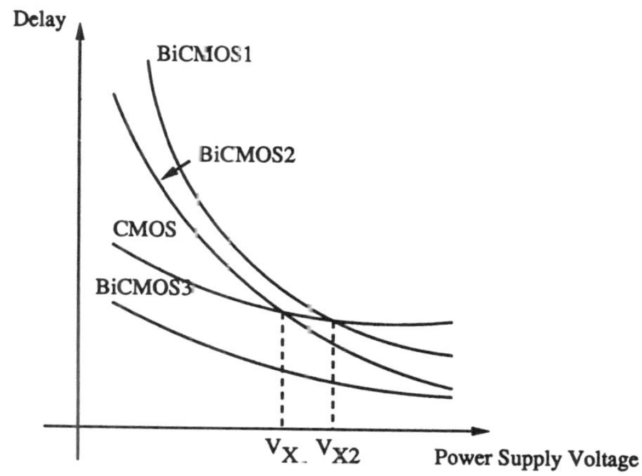

Fig. 15 The crossover power supply voltage for different types of BiCMOS.

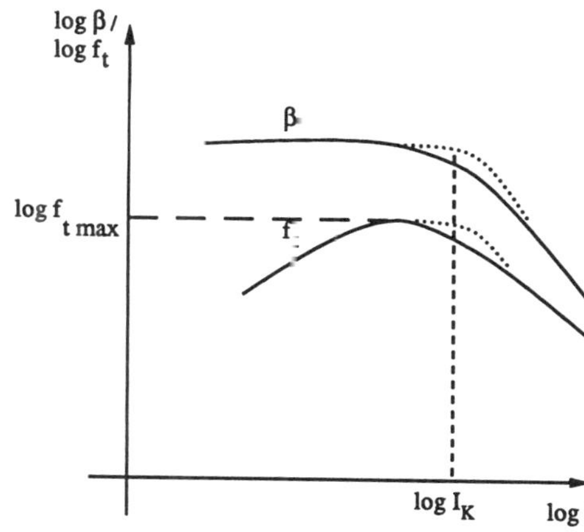

Fig. 16 β and f_t versus I_C for the bipolar transistor.

on, supplying the output load with the charging current until Q_1 saturates. In this implementation, the conduction periods of Q_{11} (Q_{22}) can be extended to allow the output voltage to get as close as possible to $V_{DD} - V_{CEsat}$ (V_{CEsat}) before the base current of Q_1 (Q_2) falls to zero.

5. BiCMOS Design Considerations

This section discusses some key issues regarding the design of BiCMOS integrated circuits. Detailed analysis and discussion of these issues are treated in the cited references, most are included in [16].

5.1. Crossover Capacitance and Crossover Power-Supply Voltage

One of the key BiCMOS design parameters for performance evaluation is the crossover capacitance C_x, defined as the load capacitance above which a BiCMOS buffer delay is less than that of a corresponding CMOS buffer for the same power-supply voltage, input capacitance, and design rules. Maintaining the same input capacitance of both CMOS and BiCMOS, in obtaining C_x, guarantees that the loading on the previous stage is the same. C_x is a function of the BiCMOS circuit type and the sizing of the devices. The lower its value, the higher the range of load capacitance for which the BiCMOS buffer can be utilized. C_x is affected by V_{DD} and the scaling design rules.

Figure 13 shows a typical delay versus C_L for CMOS and different types of BiCMOS circuits discussed in Section 4. The quantitative analysis is given in [17], where research development is reported to reduce C_x. An ideal value of C_x for a given BiCMOS chip design is equal to the value of nodal capacitance that occurs most frequently within a chip. Figure 14 shows a typical distribution of nodal capacitance within a given chip design [18].

Another key BiCMOS design parameter is the power-supply voltage V_x, above which a BiCMOS buffer delay is less than that of a corresponding CMOS buffer for the same design rules, input capacitance, and load capacitance. V_x is sensitive to the design rules and is a function of the BiCMOS circuit type. The lower the value of V_x, the more suitable the BiCMOS circuit is for scaled BiCMOS technologies. Figure 15 shows a typical delay versus power-supply voltage. Detailed quantitative analysis and discussion are given in [19], where research results are reported to reduce V_x. It is also possible to shift the t_d versus V_{DD} of BiCMOS, as in the case of BiCMOS3 [12], displaying a CMOS-like performance.

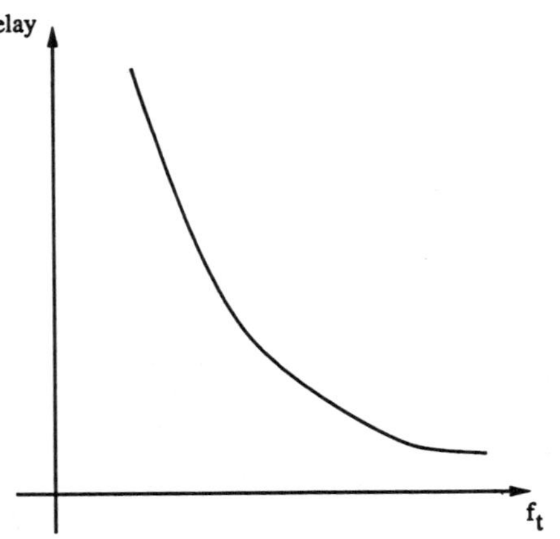

Fig. 17 BiCMOS delay versus f_t.

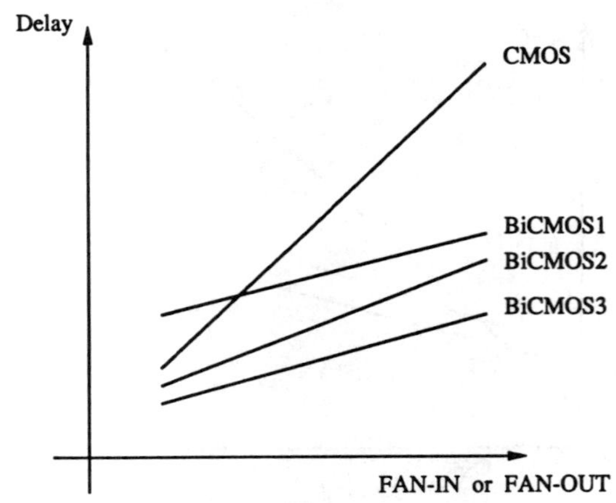

Fig. 18 Delay versus fan-in or fan-out for CMOS and BiCMOS.

5.2 High Current Effects in Bipolar Transistors

In the design and modeling of the bipolar transistors used in BiCMOS buffers, high current effects must be considered [20, 21]. Both the current gain (β) and the transition frequency (f_T) fall off at high collector currents, as shown in Fig. 16.

The rapid fall-off of both the bipolar gain and the transition frequency is due mainly to the very important phenomenon of base stretching, or Kirk effect, which occurs when the base is operating in high-level injection. The value of the knee current I_K is an important design parameter. It can be increased by increasing the emitter area.

As shown in (1), the contribution of t_{d2} to the total delay decreases as the load capacitance increases. Thus, as f_T increases, the total delay decreases. However, beyond a certain value of f_T, the decrease in the total delay is insignificant [22], as shown in Fig. 17.

5.3. Effect of Fan-In and Fan-Out

BiCMOS logic gates are used in digital chip design to complement the use of CMOS logic gates. However, both the fan

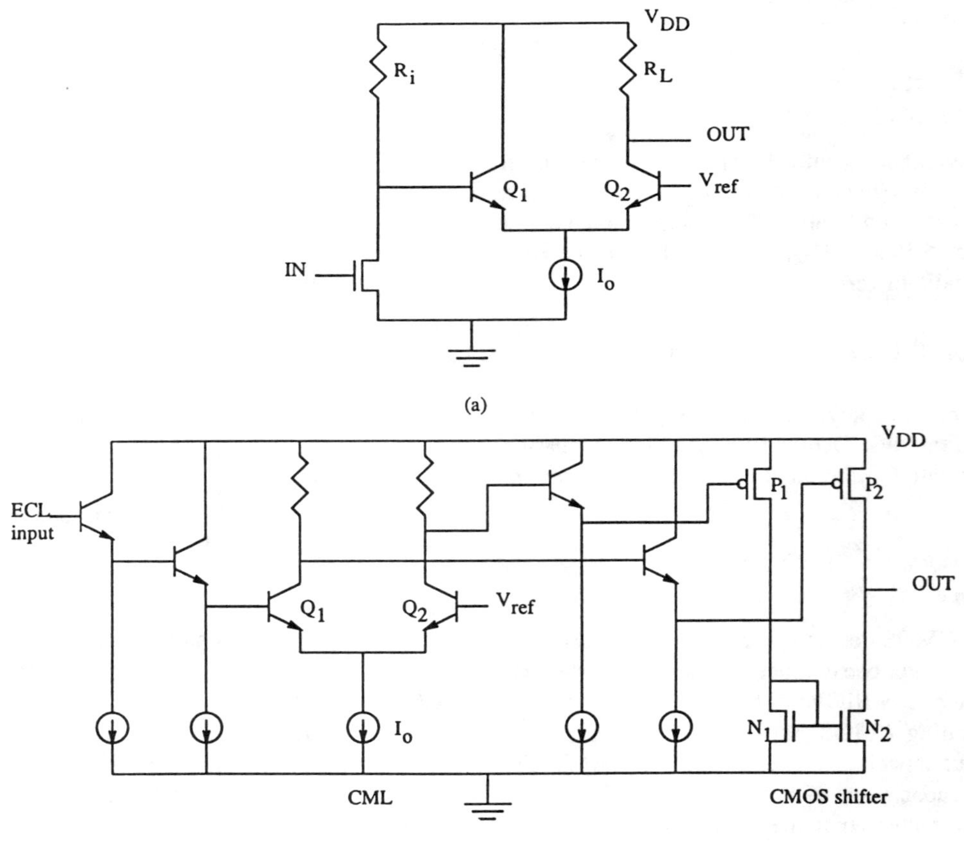

Fig. 19 NMOS/CML and ECL/CMOS interface circuits [31].

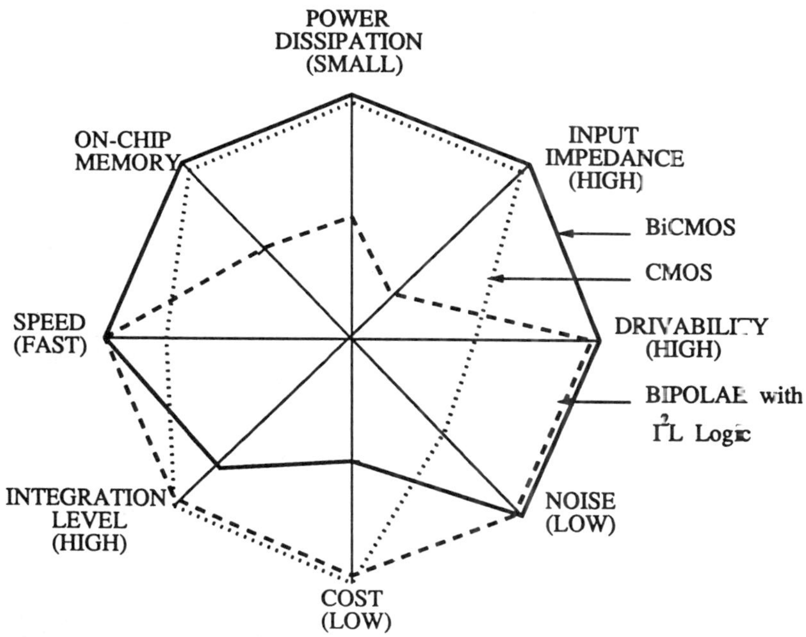

Fig. 20 Alternative devices for mixed A/D VLSIs [32].

in and fan-out affects the delay of the gate and the silicon area. Figure 18 shows a typical delay versus fan-in or fan-out for CMOS and three representative types of BiCMOS circuits. The designer must calibrate such performance using the available BiCMOS technology and a given area of applications. This will determine the design situation when CMOS or BiCMOS logic gates are used [23, 24].

5.4. Interface Circuits

One of the key design issues in the design of BiCMOS integrated circuits is the selection of interface circuits to provide level shifting between the different voltage levels of CMOS, BiCMOS, CML, and ECL [6, 25–30]. These level-shifting circuits add to the total delay, power, and silicon area. Their optimization is, in many cases, a critical step toward the overall optimization of BiCMOS chip performance.

Figure 19 shows two simple circuits: an NMOS/CML and an ECL/CMOS interface circuit [31]. In Fig. 19(a), the circuit is composed of a pull-up resistor R_i and pull-down NMOS driving a CML gate. In the circuit of Fig. 19(b), the level shifting is done in different stages. First, an input 2 V_{BE} level-shifting stage is used to drive a CML stage. The CML stage drives two emitter-followers, which drive a CMOS shifter stage.

5.5. BiCMOS in Mixed A/D VLSI

One of the important ongoing research areas regarding BiCMOS ICs is their application to the design of mixed analog/digital (A/D) VLSI systems. The area is still in its infancy and expected to grow in the coming years. The domain where bipolar, BiCMOS, and CMOS can be used in the design of mixed A/D VLSI systems is illustrated in Fig. 20 [32]. Eight parameters are chosen to compare the three technologies: CMOS, BiCMOS, and bipolar with I^2L. The figure demonstrates the superiority of BiCMOS over the other two technologies, except for the higher cost. However, it is believed such cost differential will be narrowed as the VLSI technology advances to the sub-0.5-micron range.

6. Conclusions

The integration of bipolar and CMOS devices on the same chip, into a BiCMOS technology, offers a unique opportunity for IC designers to design optimum circuits in terms of speed, power, and density. The paper introduces the topic of BiCMOS IC design and reviews some key design issues.

References

[1] M. I. Elmasry (ed.), *Digital MOS Integrated Circuits II*. New York: IEEE Press, 1992.

[2] J. P. Uyemura, *Circuit Design for CMOS VLSI*. Norwell, Massachussetts: Kluwer Academic Publishers 1992.

[3] M. I. Elmasry, *Digital Bipolar Integrated Circuits*. New York: Wiley-Interscience, 1983.

[4] H. C. Lin et al., "CMOS-bipolar transistor structure," *IEEE Trans. Electron Devices*, vol ED-16, no 11, pp. 945–951, Nov. 1969.

[5] R. H. Havemann and R. H. Eklund, "Overview of BiCMOS device and process integration," in M. I. Elmasry (ed.), *Analysis and Design of BiCMOS Integrated Circuits* New York: IEEE Press, 1993.

[6] S. H. K. Embabi, A. Bellaouar, and M. I. Elmasry, *Digital BiCMOS Integrated Circuits*. Norwell, Massachusetts: Kluwer Academic Publishers, 1993.

[7] B. H. Leung, "Analog circuit design in BiCMOS technology," in M. I. Elmasry (ed.), *Analysis and Design of BiCMOS Integrated Circuits*. New York: IEEE Press, 1994.

[8] M. I. Elmasry, "Digital MOS integrated circuits: A tutorial," in M. I. Elmasry (ed.), *Digital MOS Integrated Circuits II*. New York: IEEE Press, 1992.

[9] A. Watanabe et al. "Future BiCMOS technology for scaled supply voltage," in *Proc. IEDM Technical Digest*, pp. 429–433, 1989.

[10] Y. Nishio et al. "A feedback-type BiCMOS logic gate," *IEEE J. Solid-State Circuits*, vol 24, no. 5, pp. 1360–1362, Oct. 1989.

[11] H. J. Shin, "Full-swing logic circuits in a complementary BiCMOS technology," in *Proc. of the Symposium on VLSI*, pp. 89–90, 1990.

[12] K. Yano et al., "Quasi-complementary BiCMOS for sub-3-V digital circuits," *IEEE J. Solid-State Circuits*, vol 26, no. 11, pp. 1708–1719, Nov. 1991.

[13] H. J. Shin, "Performance comparison of driver configuration and full-swing techniques for BiCMOS logic circuits," *IEEE J. Solid-State Circuits*, vol 25, no. 3, pp. 863–865, June 1990.

[14] M. I. Elmasry, "Multidrain NMOS for VLSI logic design," *IEEE J. Solid-State Circuits*, vol 17, no. 2, pp. 409–411, April 1982.

[15] S. H. K. Embabi et al. "New full voltage swing BiCMOS buffers," *IEEE J. Solid-State Circuits*, vol 26, no. 2, pp. 150–153, Feb. 1991.

[16] M. I. Elmasry (ed.), *Analysis and Design of BiCMOS Integrated Circuits*. New York: IEEE Press, 1994.

[17] M. S. Elrabaa and M. I. Elmasry, "Design and optimization of buffer chains and logic circuits in a BiCMOS environment," *IEEE J. Solid-State Circuits*, vol. 27, no. 5, pp. 792–801, May 1992.

[18] A. R. Alvarez, J. Teplik, H. B. Liang, T. Hulseweh, D. W. Schucker, K. L. McLaughlin, K. A. Hansen, and B. Smith, "VLSI BiCMOS technology and applications," in *Proc. Int. Symp. VLSI Technology, Systems and Applications*, pp. 314–319, 1987.

[19] A. Bellaouar et al. "Scaling of digital BiCMOS circuits," *IEEE J. Solid-State Circuits*, vol. 25, no. 4, pp. 932–941, Aug. 1990.

[20] R. Gu, M. I. Elmasry, and D. J. Roulston, "A new model for bipolar transistor characeristics at high current," *IEEE J. Solid-State Circuits*, vol. 28, no. 2, pp. 173–175, Feb. 1993.

[21] D. J. Roulston, *Bipolar Semiconductor Devices*. New York: McGraw Hill, 1990.

[22] S. Nagano et al., "High performance BiCMOS circuits," in *Proc. BiCMOS Symposium*, pp. 105–110, 1987.

[23] T. Arnborg, "Performance predictions of scaled BiCMOS gates using physical simulation," *IEEE J. Solid-State Circuits*, vol 27, no. 5, pp. 754–760, May 1992.

[24] T. Akioka et al. "A 6ns 256-KBIT BiCMOS TTL SRAM," in *Proc. IEEE Custom Integrated Circuits Conf.* pp. 24.3.1–24.3.4, May 1990.

[25] A. Ohba et al. "A 7ns 1MB BiCMOS ECL SRAM with program-free redundancy," *Symp. VLSI Circuits Dig. Tech. Papers*, pp. 41–42, June 1990.

[26] K. Ogiue et al., "A 13ns/500mW 64Kb ECL RAM." in *Proc. Int. Solid-State Circuits Conf.* XXIX, pp. 212–213, Feb. 1986.

[27] Y. Maki et al. "A 6.5ns 1Mb BiCMOS ECL SRAM," in *Proc. Int. Solid-State Circuits Conf.* XXXIII, pp. 136–137, Feb. 1990.

[28] T. S. Yang et al., "A 4ns 4K × 1-bit two-port BiCMOS SRAM," *IEEE J. Solid-State Circuits*, vol. 23, pp. 1030–1040, Oct. 1988.

[29] R. Kertis et al. "A 12ns ECL I/O 256K × 1-bit SRAM using a 1μm BiCMOS technology," *IEEE J. Solid-State Circuits*, vol. 23, pp. 1048–1053, Oct. 1988.

[30] H. V. Tran et al., "A novel BiCMOS TTL input buffer: A merging of analog and digital circuit design technique," *Symp. VLSI Ciricuits Dig. Tech. Papers*, pp. 65–66, Aug. 1988.

[31] S. H. Embabi et al., "Analysis and optimization of BiCMOS digital circuit structures," *IEEE J. Solid-State Circuits*, vol. 26, no. 4, pp. 676–679, April 1991.

[32] M. Kubo et al., "Perspective on BiCMOS VLSI's," *IEEE J. Solid-State Circuits*, vol. 23, no. 1, pp. 5–11, Feb. 1988.

BiCMOS - Has The Promise Been Fulfilled?

A.R. Alvarez

Cypress Semiconductor, San Jose, CA 95134

Abstract

The era of modern BiCMOS technology started in the early 1980s [1-2] with the promise of significantly improving SRAM and logic (gate array) performance and greater integration of mixed-signal analog systems. Seven years latter what has actually been realized? The fastest commercially available SRAMs in the world, ranging in densities from 1K to 1M, are BiCMOS. BiCMOS 100K SOG arrays with 100 - 200MHz capability can be purchased from various vendors. BiCMOS microprocessors are starting to emerge from the lab and into production. Analog BiCMOS capabilities are being exploited in applications such as mobile communication. All in all, BiCMOS has been demonstrated to provide CMOS power and densities at bipolar speeds. At a given technology level, BiCMOS outperforms CMOS by a factor of 1.5 - 2.0X; 0.8μ BiCMOS exceeds the performance of 0.5μm CMOS. So why hasn't BiCMOS taken over? Even though every major semiconductor company now boasts some form of BiCMOS technology, its relative process complexity keeps cost stubbornly high. It has also taken the circuit and system designers the full seven years to learn how to best exploit BiCMOS, and we're still learning. Commercial CAD tools, while making significant progress, remain inadequate at the higher levels of abstraction. Finally, the role of BiCMOS in a reduced power supply environment is controversial.

Introduction

The market forces driving BiCMOS remain: speed, density, power, and cost. Failure to be competitive in any one of these areas or to match the CMOS learning curve will result in BiCMOS being relegated to an expensive niche technology. Given the above, where is BiCMOS going? Estimates of 1989 sales derived from BiCMOS technology range from \$140M to \$275M. Projections for 1994 are in the \$2.5 - \$3.8B [3-4]. If realized, this would be a 70 - 80% compounded annual growth rate over the 5 year period. Fueling this growth will continue to be SRAMs and Gate Arrays. Microprocessors and Mixed-Signal Analog will constitute a an increasingly larger (>25%) of the total. In design the real power of BiCMOS is in mixing and matching different design techniques (M-BiCMOS): ECL, CMOS, and BiCMOS circuits; effectively in one design. Speed critical paths are implemented in ECL for maximum performance, while the bulk of the chip is CMOS or BiCMOS thereby conserving power and area. From a technology viewpoint performance requirements continue to drive process complexity. In particular, the addition of double-poly self-aligned NPNs and vertical PNPs are two recent trends [5-6]. In manufacturing, BiCMOS's higher process complexity has necessitated strict adherence to SPC principles and particulate reduction programs. Due to the greater number of steps in a BiCMOS process, individual unit module Cpks must be greater than for CMOS (≥1.20 vs. ≥1.17). Greater complexity also increases testing, verification and database management. Finally, several approaches are being taken to resolve the speed degradation of the conventional BiCMOS buffer at 3.3V. The most promising approach replaces conventional BiCMOS buffers with new logic gates capable of operating below 3V.

Market Forces

Microprocessor trends, more than anything else, drive digital BiCMOS speed and density requirements (fig. 1). Clock rate sets the required SRAM access time (fig. 2) and PLD or ASIC propagation delay. The architecture (read MIPs) sets the required memory density and cost (fig. 3). Historically there has been a significant gap between the availability of CMOS and BiCMOS SRAMs at a given density level. This gap is narrowing quickly (fig. 4) and some companies are indicating that for sub-10ns 4M SRAMs and above BiCMOS will simply replace CMOS [4]. While the market factors for Analog/Mixed-Signal BiCMOS are more fragmented than the Digital market, not surprisingly the results are similar --- greater frequency performance than can be provided by CMOS and greater functionality than can be provided by Bipolar (fig. 5) [8].

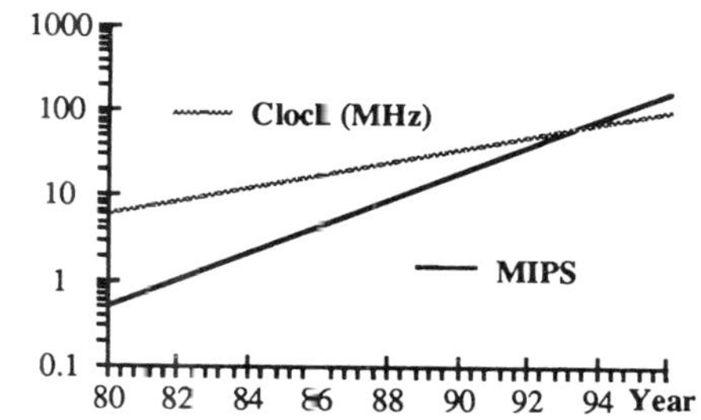

Fig. 1- Intel μP Trends [7]

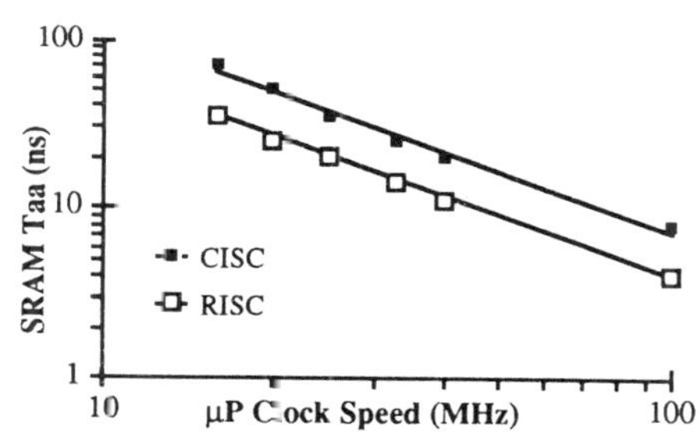

Fig. 2- Standard SRAM Taa Required For μPs

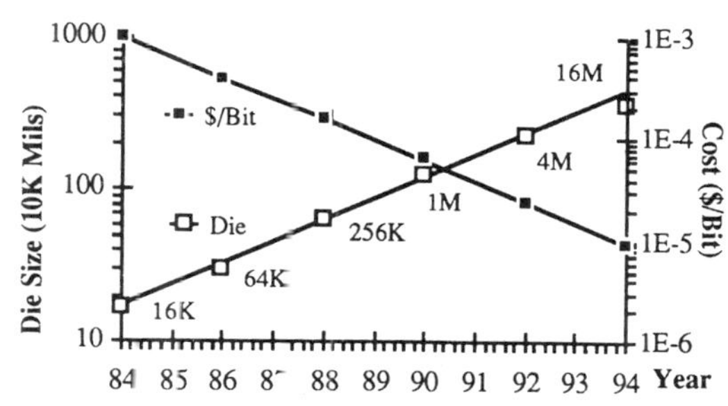

Fig. 3- Fast SRAM Die Size & Cost Trend

Reprinted from *IEDM 1991*, pp. 13.1.1–13.1.4, 1991.

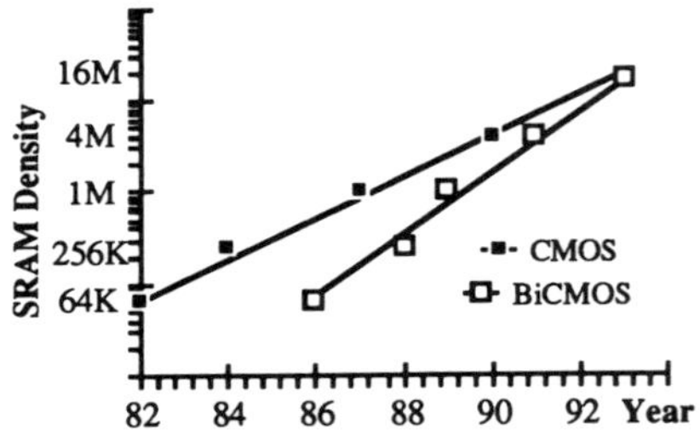

Fig. 4- Fast CMOS/BiCMOS SRAM Trend [ISSCC]

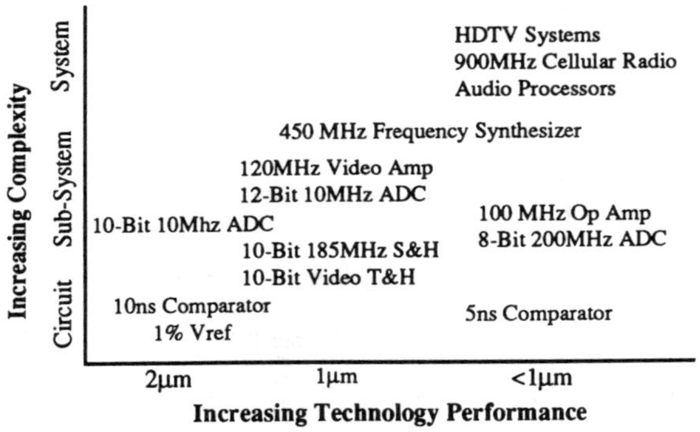

Fig. 5- Mixed-Signal/Analog BiCMOS Market [8]

Available Products

As for any technology, the measure of success is only through successful product applications. So, after 7 years of technology and circuit R&D what BiCMOS products can one really purchase in the marketplace? Table 1 provides a somewhat incomplete summary of digital product availability gleaned from trade journals, product announcements, datasheets, and market research. As noted earlier, sales of these and mixed-signal products is expected to grow at a compounded rate of 70 - 80%/year over the next 4 years yielding a total market on the order of $3 - 4B. It follows that the economic motivation justifying BiCMOS is clear.

	Speed	Density	# Vendors
SRAMs	3 - 5ns	≤ 16K	1 - 2
	7 - 15ns	64K -256K	5 - 8
	≥ 15ns	≥ 1M	3 - 5
Gate Arrays	≥ 200 MHz	5K - 150K	5 - 9
μPs	1992 Maybe??		2 - 3
PLD/Logic	7.5ns	22V10	1 - 2
	3ns	74ABT	3 - 4

Table I - Sampling of Available BiCMOS Products

Design & CAD Issues

At the product level the real power of BiCMOS is in mixing and matching different design techniques. Mixing BiCMOS, ECL and CMOS circuits (M-BiCMOS) effectively in one design allows speed critical paths to be implemented in ECL while the bulk of the chip is low power CMOS or BiCMOS. This approach is being taken in applications ranging from microprocessors [9], where NPNs can constitute less than 2% of the total transistor count, to communications [10] where NPNs are used in high data rate paths. Higher clock rates impose performance requirements at both the chip and package level, where BiCMOS offers speed and I/O flexibility advantages over CMOS. Yield and power dissipation limit bipolar integration densities (fig. 6) providing a second opportunity for BiCMOS products.

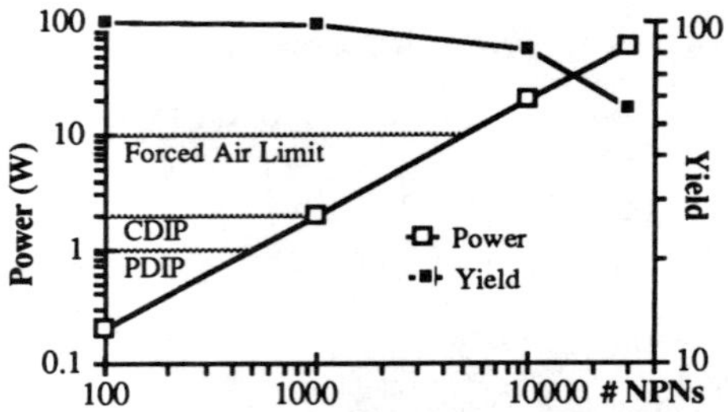

Fig. 6- Power (2mW/Gate) & Yield (0.99999) vs #NPNs

To date BiCMOS technology has had the greatest success in fast SRAM applications. At an equivalent lithography level and bit density BiCMOS provides better (1.5X) speed than CMOS (fig. 7). An additional advantage of BiCMOS is a smaller access time distribution (sigma) resulting from the reduced dependence of NPN performance on process variations. Since they share the same memory cell, die area for BiCMOS and CMOS SRAMs are approximately equal. Compared to Bipolar, BiCMOS SRAMs provide faster access time at the same power or the same access time at lower (typically 50%) power. In addition, the CMOS memory cells used in BiCMOS have superior stability, better alpha immunity, and smaller area than equivalent bipolar cells. M-BiCMOS design techniques have replaced early designs that kept Bipolar and CMOS blocks separate. Key considerations are when and how to perform level translations (before or after decode), sensing techniques, and the degree to which the output path is maintained ECL. At the 4M level the ability to provide a well regulated internal power supply to the memory array provides an additional edge.

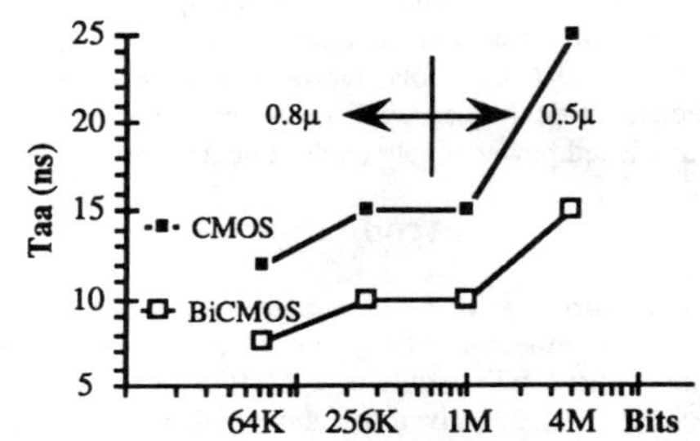

Fig. 7- CMOS & BiCMOS SRAM Taa

BiCMOS holds great potential in microprocessors (μP). μP performance is limited by three generic categories: 1) instructions per task, 2) cycles per instruction, and 3) time per cycle. These tasks can in turn be optimized by: 1) balancing the instruction set, 2) instruction pipelining and efficient hierarchical memory design, and 3) circuit and technology synergy [9]. While the first two categories are dominated by computer architecture, BiCMOS impacts time per cycle by speeding up the cache RAM and critical paths without sacrificing chip partitioning as in the case of Bipolar [9]. Performance is therefore enhanced by: reducing delays through critical paths, reducing clock skews, increasing I/O throughput, and allowing larger chip/memory sizes. Since RISC processors operate at higher clock rates than CISC, they are leading candidates for BiCMOS integration [11]. RISC's simpler instruction and need to minimize "wait states" implies that fast on- and off- chip cache is necessary. The problem is that small CPUs leave room only for an FPU and little on-chip cache. Since one-clock memory access requires one instruction access per clock and one data access every three clocks on average, use of separate instruction and data cache is required to eliminate contention [9,11]. These caches have to be 256K to 512K bits to minimize first-cache misses. The speeds required, ≤ 5ns, cannot be sustained in ≥0.5μ CMOS over worst-case temperature and voltage [11]. The inability to integrate the required amount of on-chip memory and high power dissipation make a bipolar approach impractical. Both CMOS-dominant (Hitachi/Matsushita) and ECL-dominant (Siemens) design philosophies have been demonstrated. Hitachi's 70MHz 32bit 1.0μ

500K transistor μP [12] and Matsushita's 100MFlop 64bit 0.8μ 1.3M transistor μP [13] deliver modest (25 - 35%) speed improvement over CMOS with little or no area or power penalty. Siemens's 1μ R3000 case study resulted in a clock rate improvement of 2.4X (33 to 80 MHz), at a 25% area penalty (57 vs 72 mm^2) and a 4X increase in power (16W vs 3.7W) [9]. The power dissipation, while significant, remains manageable and significantly below the 48W 2 chip bipolar version (R6000) [9]. In summary, BiCMOS can: 1) utilize CMOS methodologies used to design previous generations of μPs, 2) integrate large fast on-chip RAM, and 3) keep power manageable by focusing it where required to optimize the speed-power tradeoff.

While market forces in ASICs and Programmable Logic Devices (PLDs) are more varied than for SRAMs, higher μP clock rates have also impacted the need for high speed glue logic. BiCMOS gate arrays have been available since the mid 80s [14]. Array architectures have evolved from regular cells with NPN drivers either in the periphery only or in every cell to Sea-of-Gate (SOG) configurations with an optimized number of NPNs per macrocell [15-16]. Over the past six years gate delays (FO=3, L=3mm) have been reduced from 800ps to 200ps. In PLDs, CMOS has achieved high densities (≥3500 gates) at relatively low power (≤750mW) and modest (≥10ns) speeds, whereas Bipolar has produced the fastest parts. M-BiCMOS breaks traditional circuit compromises resulting in high speed and density at modest power (fig. 8). An example is a 7.5ns 22V10: the speed path is almost entirely ECL, while CMOS is used at the I/Os and programming control circuitry [17].

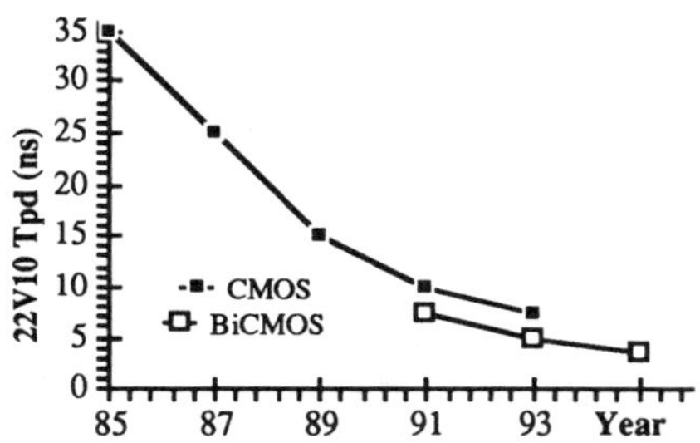

Fig. 8- CMOS & BiCMOS 22V10 Tpd vs Time

Faster systems require faster gates and higher throughput I/Os. Gate delays in BiCMOS technologies have been improving ~10%/year (fig. 9). For self-loaded ring oscillators, sub 100ps CMOS and ECL gates, and sub 200 ps BiCMOS gates have been realized. This extrapolation will not hold for the conventional BiCMOS gate as the channel length is further reduced because at some point a drop in Vcc is required. When Vcc is reduced the 2Vbe offset in the BiCMOS gate results in a severe performance penalty (fig. 10). It is important to realize that this is a BiCMOS gate issue, not a BiCMOS technology issue; an ECL gate operates with little or no speed degradation below 2 volts. This concern about degraded performance with reduced Vcc has resulted in 3 possible paths: 1) operate BiCMOS chips with dual Vcc (a reduced CMOS Vcc is generated from the external 5V supply), 2) introduce a high performance (≥3GHz) emitter follower PNP into the process [6], or 3) change the BiCMOS buffer design [18].

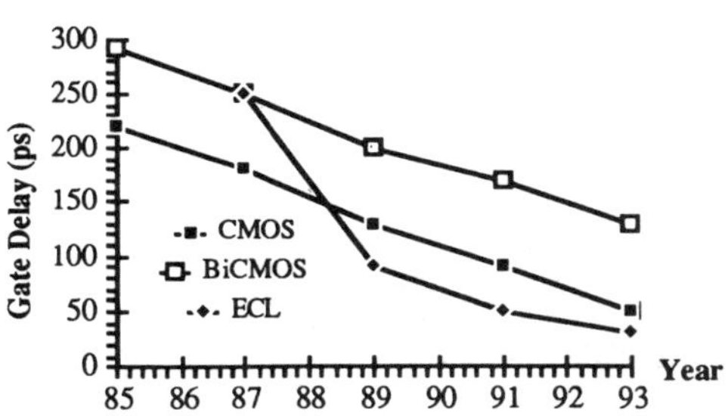

Fig. 9- CMOS, BiCMOS, ECL Gate Delays [IEDM]

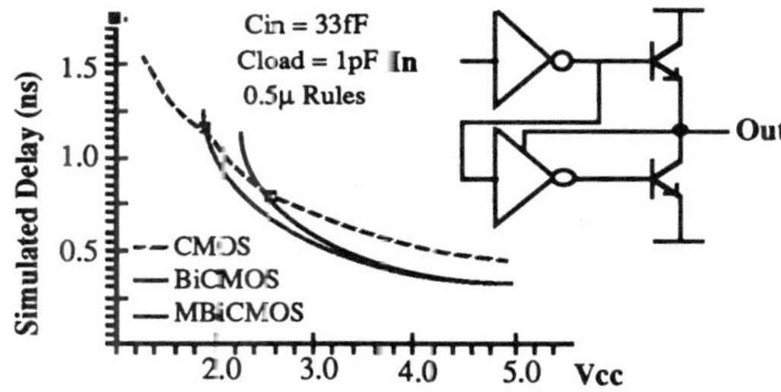

Fig. 10- CMOS, BiCMOS, MBiCMOS Gate Delay vs Vcc

The importance of I/O, TTL vs. ECL, is illustrated in fig. 11. In a TTL system a significant percentage of the total delay is taken up in getting on and off chip. The noisier TTL environment also requires a larger test guardband. For a given delay, 7.5ns in this example, an ECL chip could have twice as many 250ps gates, or conversely, could reduce the overall chip access time by almost 3ns. Because of the NPNs inherent in the process, BiCMOS offers considerable advantages in implementing true ECL and mixed TTL-ECL circuits.

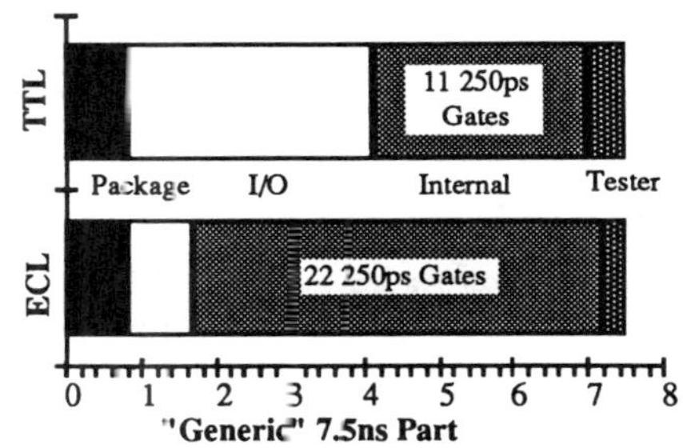

Fig. 11- TTL vs ECL Speed Breakdown

CAD tools for BiCMOS have progressed significantly over the past several years. Borrowing heavily from CMOS, physical tasks such as: schematic capture, netlisting, layout, layout extraction, verification, and DRCs, are now straightforward. Work is still required however to improve higher level tools for: logic simulation, cell generation, parasitic (especially resistance) extraction, and critical path optimization.

Technology & Manufacturing Issues

Static memories and gate arrays were early BiCMOS technology drivers. It follows that currently available BiCMOS processes evolved predominantly from a CMOS base. M-BiCMOS requires a greater emphasis on the bipolar technology. In either case, advanced process technology (≤0.8μm design rules, thin gate oxide and epitaxial layer, advanced isolation, and 2 - 3 layers of metallization) is required to retain the necessary advantage over CMOS. Two process trends have resulted in even higher performance: extensive use of advanced double poly NPNs [5], and high performance (≥3GHz) PNPs [6]. Both, however, result in greater process complexity.

The optimal combination of BiCMOS device parameters depends on the circuit techniques used. Since circuit applications drive the technology toward M-BiCMOS, both NPNs and MOSFETs must be high performance. This implies that high Ft, high drive, low Rb NPNs must be realized for the ECL portions of the design as well as high performance MOSFETs for the CMOS and BiCMOS circuits. In some instances this causes the NPN to be over-designed. For example, base resistance is not critical in a BiCMOS gate, but is a dominant factor in determining ECL gate delays. To achieve an optimal combination of parameters a sensitivity analysis of a critical circuit speed path can be performed (fig. 12). This insures a balanced tradeoff in technology direction. From fig. 13 it is seen that equivalent SRAM Taa can be obtained with different combinations of NPN and MOSFET performance.

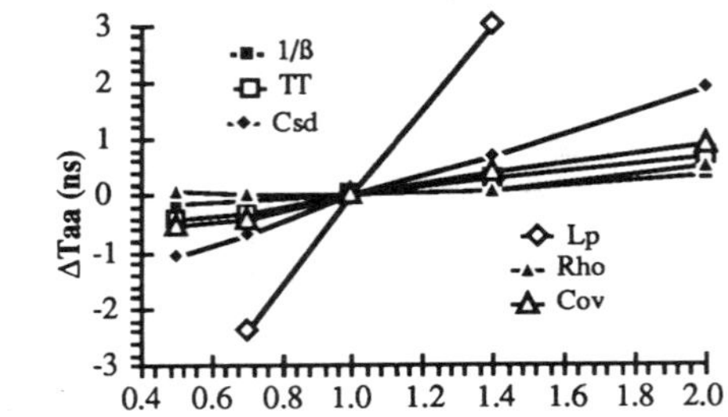

Fig. 12- 64K SRAM Taa Parameter Sensitivity

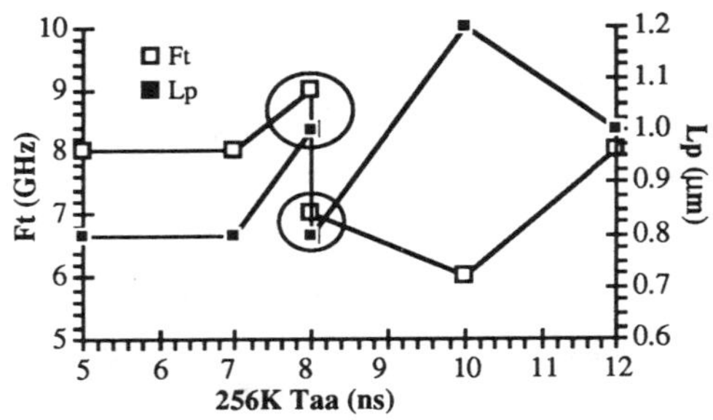

Fig. 13- 256K SRAM Taa vs Ft and Lp (ISSCC)

The Cypress process, typical of an advanced BiCMOS technology, features twin buried layers and wells, locos isolation, silicided TiSi2 poly, dual LDD FETs, and planarized double layer metal (fig. 14). The polysilicon emitter is the same as used for MOS gates. This decouples the NPN from the poly load used for 4T SRAM cells, making the process more modular. An Ft in excess of 10GHz, low Ccb, a self-aligned extrinsic base and 1.0µm NPN design rules produce 100ps ECL gate delays. NPN DC and AC knee current are optimized through tailoring of the well profile. This ensures high speed in the BiCMOS gate. DC and AC EB degradation have been evaluated and reliability results insure >10 year device operation. FETs have been optimized for high current drive (340/160 µA/µm Idsatn/p) while maintaining hot-carrier stress immunity through careful tailoring of the LDD profile. It is critical that FET performance match that of pure 0.8µm CMOS processes so that standard cell libraries can be exchanged between the two technologies.

The main disadvantage of BiCMOS is cost. Most of the extra cost is incurred because of greater manufacturing complexity. For high performance BiCMOS, mask count and total process steps increase by 25% compared to CMOS. Most of the extra process complexity is in the front-end --- buried layer(s) and epitaxy. Additional process steps: Deep N+ collector, local collector, intrinsic base, emitter, etc., are also required to realize the NPN. Depending on whether a single or double poly process is used or whether a PNP transistor is integrated in will determine the number of polysilicon/oxide depositions, implants, and diffusion cycles required. For very high performance processes it is not unusual to have over 20 masks and 120 operations. Because of the length and complexity of the technology, BiCMOS imposes more stringent requirements on cycle time, defect density, and process control. For example, in order to maintain the same level of "out of control" points in an SPC chart a 125 step BiCMOS technology must have Cpks ≥ 1.20 compared to 1.17 for a 100 step CMOS technology.

Conclusion

So, we return to the original question ---- "Has the promise been fulfilled?". It has taken ~7 years, but with the market for BiCMOS parts growing at ~70% per year, manufacturing technologies firmly entrenched in most major semiconductor companies, and a multitude of successful products on the market, one would have to answer the question positively. Problems still exist however. The extra process complexity requires chip manufacturers to command a premium for BiCMOS products. It follows that BiCMOS must provide sufficient improvement in speed to justify the higher prices. To date that has been possible in SRAMs, gate arrays, and PLDs; the µP market is yet to be tested. In the analog market the ability to integrate large mixed systems provides the compelling cost advantage of BiCMOS; this market is still emerging. The next issue is whether the BiCMOS advantage is sustainable at lower Vcc. Recent design break-throughs indicate that it is. This needs to be realized in products.

References

1. H. Higuchi et al, "Performance and Structures of Scaled-Down Bipolar Devices Merged with CMOSFETs," 1984 IEDM, pp. 694 - 697.
2. A.R. Alvarez et al, "2 Micron Merged Bipolar-CMOS Technology," 1984 IEDM, pp. 761 - 764.
3. S. Baker, "Will BiCMOS Enter Mainstream", Source: Forward Concepts, Electronic Engineering Times, Dec. 1990, pp. 1, 66-67.
4. S. Weber, "BiCMOS: The Next Great Wave in Silicon", Source: ICE, Electronics, Dec. 1990, pp. 40 - 46.
5. S. Konaka, et. al., "HSST/BiCMOS Technology with 26ps ECL and 45ps 2V CMOS Inverter," 1990 Int. Electron Dev. Mtg., pp. 493-496.
6. N. Rovedo, et. al., "Process Design for Merged Complementary BiCMOS," 1990 Int. Electron Dev. Mtg., pp. 485-488.
7. P. Gelsinger, P. Gargini, G. Parker, A. Yu, "Microprocessors circa 2000", Spectrum, pp. 43 - 47, Oct. 89.
8. M. Declercq, "Introduction to BiCMOS", Intensive Course on BiCMOS Technology & Circuit Design, Lausanne, Swit., Oct. 1990.
9. W. Heimsch, "BiCMOS Microprocessor Design," Ibid.
10. K. Hansen, "BiCMOS Analog Design," Ibid.
11. P. Bosshart, "Introduction to Microprocessor Design," Int. Electron Device Meeting Short Course, 1988.
12. T. Hotta, et. al., "A 70Mhz 32b Microprocessor with 1.0µm BiCMOS Macrocell Library," 1989 Int. Solid State Circuits Conf. Dig. Tech. Papers, pp. 124-125.
13. Y. Nakakura, et. al., "Design Methodology for Over 100MFlops 64bit MPU with 0.8µm BiCMOS," 1991 Symp. VLSI Tech, pp. 21 - 22.
14. S.C. Lee, et. al., "Bi-CMOS Circuits for High Performance VLSI", Symp. on VLSI Technology, pp.40-47, 1984.
15. J.D. Gallia et. al., "High-Performance BiCMOS 100K-Gate Array," J. Solid State Circuits, Vol 25., No. 1, Feb. 1990, pp. 142 - 148.
16. Y. Tanaka et. al., "A 0.5µm BiCMOS SOG with Selectable 5V/3.3V Operations," 1991 Custom Int. Circuit Conf, pp. 14.14.1 - 14.14.4.
17. T. DeLurio et al, "Programmable BiCMOS Technology Improves TTL PLDs", 1991 CICC.
18. P. Raje et al, "MBiCMOS: A Device and Circuit Technique Scalable to the Sub-micron, Sub-2V Regime," 1991 ISSCC, pp. 150-151.

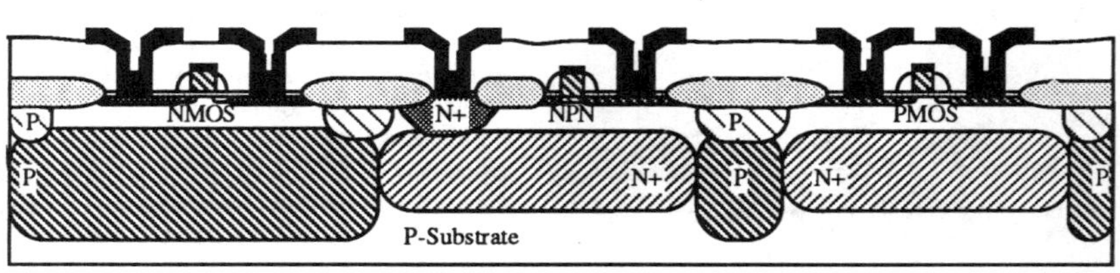

Fig. 14- BiCMOS Technology Cross-section

Paper 1.3

High-Speed BiCMOS Technology with a Buried Twin Well Structure

TAKAHIDE IKEDA, ATSUO WATANABE, YOJI NISHIO, IKURO MASUDA, NOBUO TAMBA, MASANORI ODAKA, AND KATSUMI OGIUE

***Abstract*—A buried twin well and polysilicon emitter structure is developed for high-speed BiCMOS VLSI's. A bipolar transistor of high cutoff frequency (f_T = 4 GHz) and small size (500 μm²) has been fabricated on the same chip with a standard 2-μm CMOS, without degrading the device characteristics of the MOSFET. Latchup immunity is improved due to the low well resistance of the buried layer. The well triggering current is a 0.5–1.0 order of magnitude higher than that of a standard n-well CMOS. To evaluate the utility of this technology, a 15-stage ring oscillator of the 2NAND BiCMOS gate is fabricated. The gate has a 0.71-ns propagation delay time and 0.25-mW power dissipation at 0.85-pF loading capacitance and 4-MHz operation. Drive ability is 0.24 ns/pF, which is 2.5 times larger than that of the equal-area CMOS gate.**

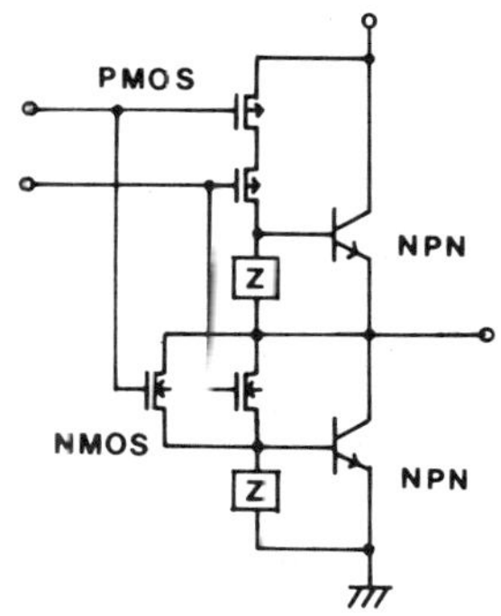

Fig. 1. A two-input NAND BiCMOS gate.

I. Introduction

RECENTLY, CMOS has become the main device of VLSI's due to its inherent low-power characteristics. However, system speed is restricted by the drive ability of the CMOS gate that drives large capacitive loads (~ 1– ~ 10 pF) for word lines of memory LSI's, and bus lines in or out of the logic LSI's, etc. For high-speed LSI's, the bipolar technology is superior to CMOS due to the area-efficient high transconductance of the bipolar transistor. However, power dissipation of bipolar gates such as TTL and ECL is about one order of magnitude larger than that of the CMOS gate. Therefore, the bipolar VLSI is difficult to realize without sacrificing switching speed. Another disadvantage of the bipolar LSI is the yield problem. Leakage current between the emitter and the collector is very sensitive to crystal defects.

Very-high-speed CMOS VLSI's are possible if the bipolar transistor is fabricated on the same chip with the CMOS and its high drive ability is used. One such recently developed BiCMOS gate suitable for driver part is shown in Fig. 1 [1]. The gate has high drive ability because bipolar transistors drive the load capacitance and has low-power characteristics because the current flows only during switching transient just like the CMOS gate. Impedance Z is the resistor or MOSFET that extracts the base charge from the bipolar transistor. To utilize the BiCMOS gate in CMOS VLSI's, high-performance and small-size bipolar transistors are required.

Manuscript received May 6, 1986; revised October 30, 1986.
T. Ikeda, A. Watanabe, Y. Nishio, and I. Masuda are with Hitachi Research Laboratory, Hitachi Ltd., Hitachi-shi, Ibaraki, 319-12 Japan.
N. Tamba, M. Odaka, and K. Ogiue are with the Device Development Center, Hitachi Ltd., Imai-cho, Oume-shi, Tokyo, 198 Japan.
IEEE Log Number 8613908.

Although several BiCMOS technologies that merge the bipolar transistor with the CMOS have already been presented [2]–[4], they are not enough to permit the above-mentioned application. One approach is the CMOS-based BiCMOS technology that uses an n-well as the collector of the bipolar transistor. It has the advantage of very few additional process steps for bipolar transistor fabrication. However, high current-drive ability can hardly be derived because of the high collector series resistance of the n-well (~ kΩ). The isolation width between n-wells cannot be reduced because of the low impurity concentration of the p-substrate. Another approach is a bipolar-based BiCMOS technology that uses buried and epitaxial layers. Usually, this approach uses the relatively thick epitaxial layer to admit deep well diffusion of the CMOS process, or it has not been considered to reduce isolation width. Therefore, both the cutoff frequency and the isolation width are not enough to satisfy the above-mentioned application. This paper describes the optimized device structure of the bipolar-based BiCMOS approach. One feature is the n^+ and p^+ buried twin well structure that uses the thin epitaxial layer to achieve high cutoff frequency and narrow isolation width of the bipolar transistor. The other is the polysilicon emitter that uses the second polysilicon layer of the CMOS process. The shallow junction and the small-size emitter can be formed without additional process steps. This technology, combined with the BiCMOS circuit technology, has been named high-performance BiCMOS (Hi-BiCMOS) technology, and has been applied successfully to high-speed gate arrays [1] and static memories [5], [6]. Further improvement of the device performance by scaling down this structure from 2 to 1.3 μm has also been presented [7].

Reprinted from *IEEE Trans. Electron Devices,* vol. ED-34, no. 6, pp. 1304–1310, June 1987.

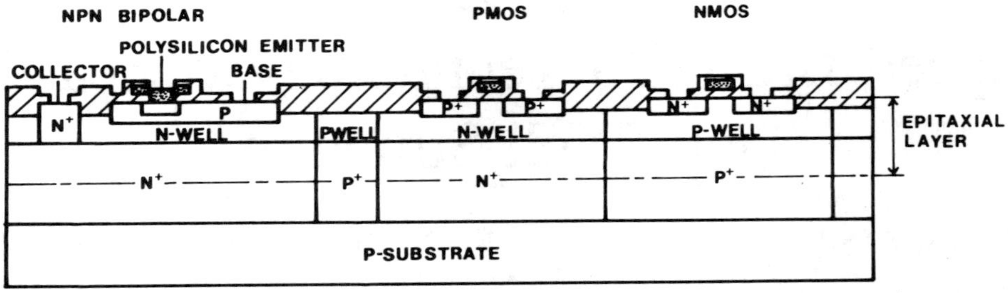

Fig. 2. Schematic cross section of the Hi-BiCMOS.

II. Device Structure and Fabrication Process

A. Device Structure

The schematic cross section of the Hi-BiCMOS device is shown in Fig. 2. Features of this structure are as follows. The first is the n^+ and p^+ buried twin well structure. The n^+ buried layer serves as the foundation for both the PMOS and the bipolar transistor, and the p^+ buried layer for both the NMOS and the isolation between the n^+ buried layer. The well is formed by combining the shallow surface twin well with the buried twin well. A thin epitaxial layer of 1.3–2 μm is used for high cutoff frequency and narrow isolation width of the bipolar transistor. Isolation between n^+ buried layers is acheived by combining the LOCOS and the p^+ buried layer.

The second feature is the polysilicon emitter formed using the second polysilicon layer of the CMOS process. The emitter-junction depth is 0.3 μm and the base width is 0.2 μm. The emitter width can be formed by the minimum feature size and is 2 μm in this study. As for the CMOS, the gate length is 2 μm and the gate-oxide thickness is 35 nm. The junction depth is 0.5 μm for both NMOS and PMOS, and a double-diffused drain is adopted for NMOS.

B. Fabrication Process

An outline of the fabrication process flow is shown in Fig. 3. The basic process is the 2-μm standard CMOS process with two polysilicon layers, which is usually used to fabricate CMOS SRAM with a polysilicon high-resistor memory cell.

The additional process steps necessary to fabricate the bipolar transistor are shown on the right-hand side. The p-substrate is essential for isolation of the bipolar transistor, instead of the n-substrate of the standard CMOS. The buried twin well is formed by the usual self-aligned technology that uses Si_3N_4 mask and local oxidation. The n^+ is formed by antimony diffusion and the p^+ is formed by boron-ion implantation. The n-epitaxial layer growth process is followed. The shallow-surface twin well is formed just above the buried twin well. The conventional high-temperature well diffusion step to form the deep well is not needed because the buried twin well forms the deep retrograde well. As a result, the upward diffusion distance of the buried layer is much reduced.

Collector and base ion-implantation steps are inserted before source and drain fabrication of the MOSFET. The

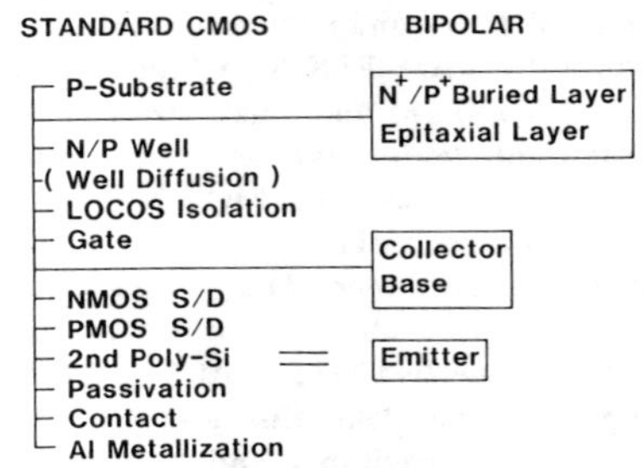

Fig. 3. Fabrication process flow of Hi-BiCMOS.

emitter is formed by phosphorus ion implantation into the second polysilicon layer at 50 keV, followed by annealing at 950°C. As can be seen from this flow chart, the bipolar process can be added without disturbing the CMOS process conditions.

III. Well Impurity Profile and Related Electrical Characteristics

Typical impurity profiles of the n- and p-well are shown in Fig. 4(a) and (b). The graphs are simulation data of BiCMOS and CMOS. Measured data by the MOS C–V method are also shown by the broken line. The remarkable feature of the profile is the existence of the low-impurity concentration region between the surface and the buried layer.

A. MOSFET's

One of the most sensitive electrical parameters of the MOSFET to well profile is the junction capacitance of the source and drain. The relation between the junction capacitance and epitaxial layer thickness is shown in Fig. 5. The voltage dependence of the junction capacitance varies depending on the well profile. Therefore, capacitance values are averaged by the junction voltage between 0 and 5 V. With a thicker epitaxial layer, capacitance values are smaller than those of the standard CMOS values due to the existence of the low-impurity-concentration region. The capacitance increases more rapidly in NMOS than in PMOS, depending on the epitaxial layer thickness and boron dose of the buried p-well. This is because the upward diffusion distance of boron is larger than that of antimony due to the large difference in the diffusion constant, as shown in Fig. 4. The body effect constant, which is also sensitive to the well profile, varies quite identically

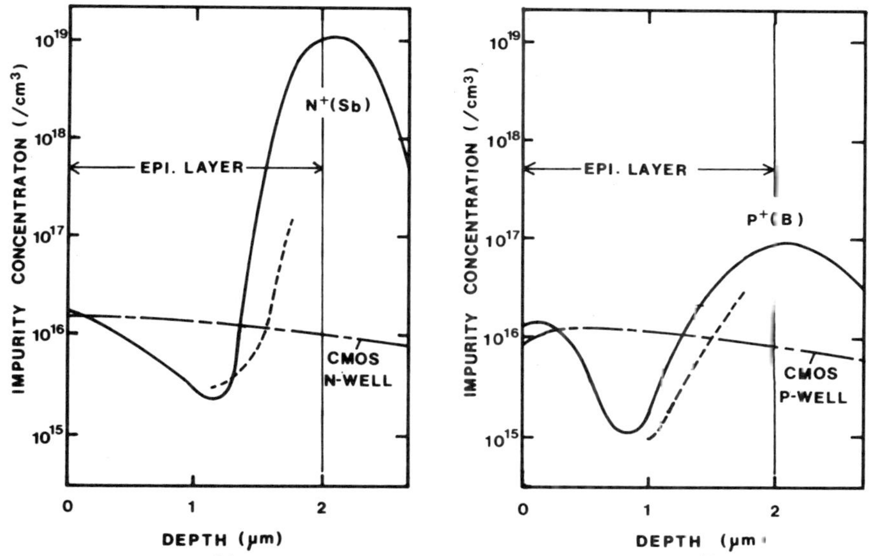

(a) (b)

Fig. 4. Impurity profiles of (a) n- and (b) p-wells of the Hi-BiCMOS and CMOS.——: Simulated profile of Hi-BiCMOS.— - — -: Simulated profile of CMOS. - - - - -: Measured profile of Hi-BiCMOS by MOS C–V method.

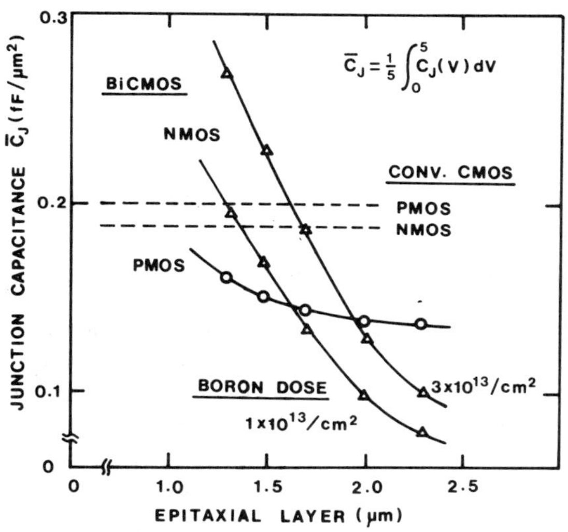

Fig. 5. Average junction capacitance $\overline{C}_J$ of MOSFET's as a function of epitaxial layer thickness.- - - -: Conventional CMOS.

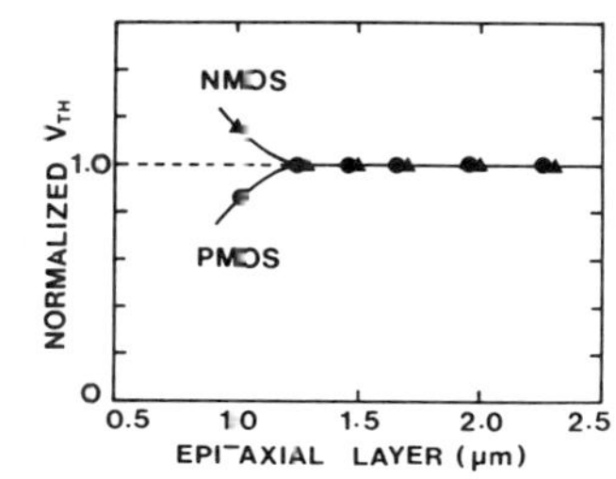

Fig. 6. Threshold voltage versus epitaxial layer thickness.

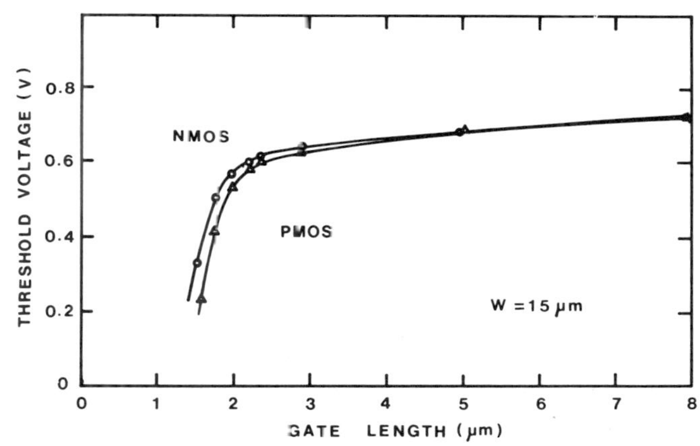

Fig. 7. Threshold voltage versus gate length of MOSFET's fabricated by Hi-BiCMOS technology.

to the junction capacitance, depending on the epitaxial layer thickness and boron dose.

The threshold voltage V_{TH} is one of the most important parameters of the MOSFET. The epitaxial layer thickness dependence of V_{TH} is shown in Fig. 6. It is determined by the surface impurity concentration of the well and is constant to an epitaxial layer thickness as thin as 1.3 μm. Curves of threshold voltage versus gate-channel length at an epitaxial layer thickness of 1.5 μm are shown in Fig. 7. They are quite the same as those for the standard CMOS. From these results, the epitaxial layer thickness can be reduced to 1.3 μm without degrading the electrical characteristics of the MOSFET. If the conventional high-temperature well-diffusion step of CMOS is applied to get a deep well, the upward-diffusion length of the buried layer becomes 4–5 μm. In this case, the effects of buried layers on MOSFET's cannot be eliminated without a very thick epitaxial layer (6–7 μm).

B. Bipolar Transistor

The bipolar transistor is formed in the n-well of Fig. 4(a). The epitaxial layer thickness is also one of the most important parameters affecting the electrical characteristics of bipolar transistors. The cutoff frequency f_T, the breakdown voltage between the collector and emitter BV_{CEO}, and the knee current I_K versus epitaxial layer thickness are shown in Fig. 8. I_K is defined as the collector current at which current gain h_{FE} falls off at high current to half the normal value. This value is an important

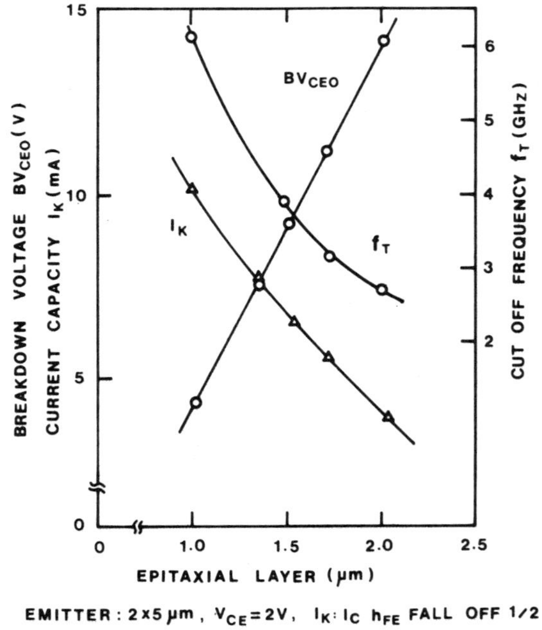

Fig. 8. Epitaxial layer thickness dependence of cutoff frequency f_T, breakdown voltage BV_{CEO}, and knee current I_K.

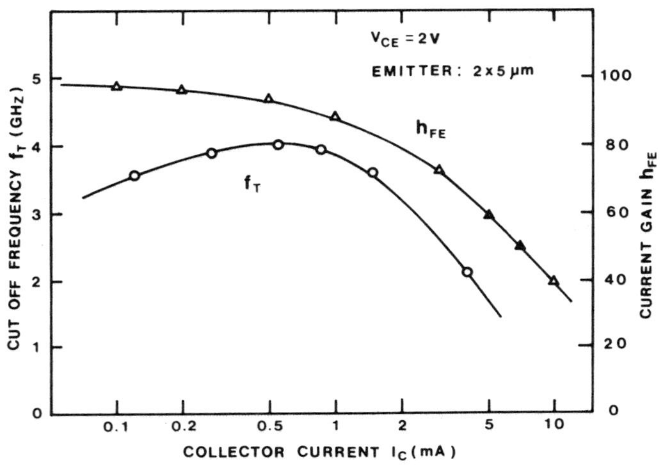

Fig. 9. Cutoff frequency f_T and current gain h_{FE} versus collector current.

measure of the drive ability of the BiCMOS gate because the peak the current of the BiCMOS gate is very high when a large capacitive load is driven and the bipolar transistor operates in the current gain fall-off region. At thinner epitaxial layers, f_T and I_K increase due to suppression of the Kirk effect, but the breakdown voltage decreases. The lower limit of the thickness is determined by the BV_{CEO} value required from the circuit design. Typical h_{FE} and f_T versus collector current are shown in Fig. 9. The epitaxial layer thickness of this sample is 1.5 μm and the maximum f_T is 4 GHz.

With respect to the area of the small-size bipolar transistor, more than 50 percent is occupied by the isolation area. Therefore, reduction of the isolation area is essential. A cross section of the bipolar-transistor isolation region is shown in Fig. 10. Isolation width C consists of parts A and B. A is the width between the n^+ buried layers

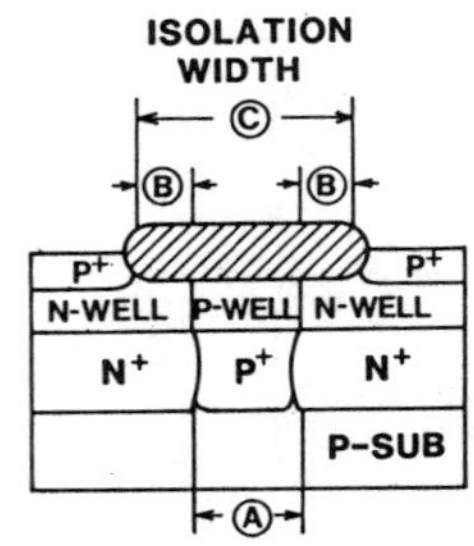

Fig. 10. Cross section of the isolated region of bipolar transistor.

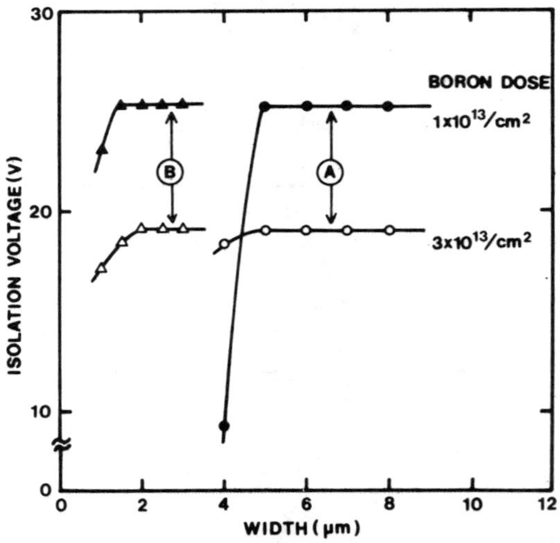

Fig. 11. Isolation voltage versus isolation width of A and B.

and the isolation voltage is determined by the impurity concentration of the p^+ buried layer. B is the width between the base of the bipolar transistor and the p-well, and the voltage is determined by the n-well concentration. The isolation voltage of A and B versus the corresponding isolation width is shown in Fig. 11. For a narrower width, the isolation voltage decreases by punchthrough. From the data of Fig. 11, total isolation width C can be fixed at, for example, 10 μm, including mask alignment margin. This isolation width is a little larger than that of the conventional 2-μm bipolar LSI that uses recessed oxide isolation technology (~6 μm). However, it is supposed to be much reduced as compared to the isolation width of heretofore-reported BiCMOS structures [2], [4], [8], [9]. As a result, the area of a bipolar transistor of 2 × 3 μm emitter size is as small as 500 μm^2 and is comparable to the area of the 2-μm standard MOSFET.

IV. Device Performance

A. Device Characteristics

The typical device characteristics of the bipolar transistor and MOSFET's are summarized in Table I. With the bipolar transistor, only isolation capacitance C_{cs} is larger than that of the 2-μm conventional bipolar transistor due to the larger isolation area. The device characteristics of MOSFET's such as threshold voltages, breakdown voltages, and transconductances are quite the same as those of the 2-μm standard CMOS.

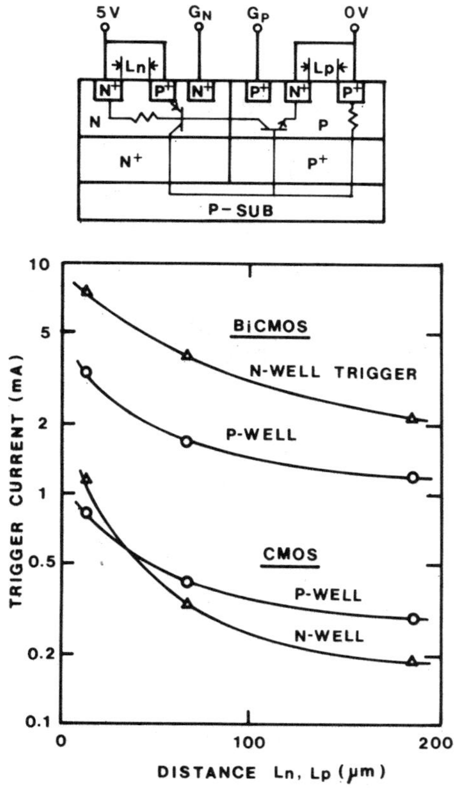

Fig. 12. Latchup test device structure and the well triggering current versus distance between the well electrode and the junction.

TABLE I
TYPICAL DEVICE CHARACTERISTICS OF THE BIPOLAR TRANSISTOR AND MOSFET'S

Device		Characteristics	Value	Unit
Bipolar	Emitter= 2x5 μm^2	h_{FE}	90	—
		BV_{CEO}	9	V
		BV_{CBO}	25	V
		R_{CS}	120	Ω
		f_T	4	GHz
		C_{EB}	40	fF
		C_{CB}	33	fF
		C_{CS}	145	fF
NMOS	W/L= 15/2 μm/μm	V_{TH}	0.55	V
		BV_{DS}	15	V
		β_{omax}	55	μS/V
PMOS		V_{TH}	−0.55	V
		BV_{DS}	14	V
		β_{omax}	18	μS/V

B. Latchup Characteristics

The buried layer is very effective in suppressing latchup due to low well resistance. A schematic cross section of the latchup test device structure and the well triggering current are shown in Fig. 12. The triggering current to initiate latchup is measured at different spaces between the well electrode and the n^+ or the p^+ junction. The same test structure of a conventional n-well CMOS is also evaluated to compare the triggering current. The sheet resistance of the n^+ buried layer is 76 Ω/□, whereas the sheet resistance of the conventional n-well in the p-substrate is 2–3 kΩ/□. The triggering current is 10 times larger for the n-well trigger and five times larger for the p-well trigger than that of the conventional n-well CMOS.

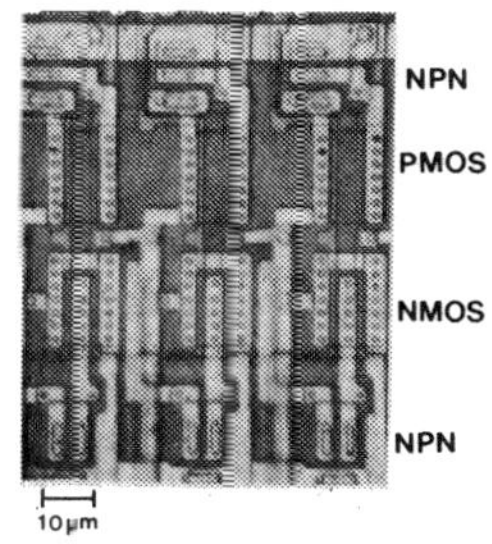

Fig. 13. Photograph of the BiCMOS gate.

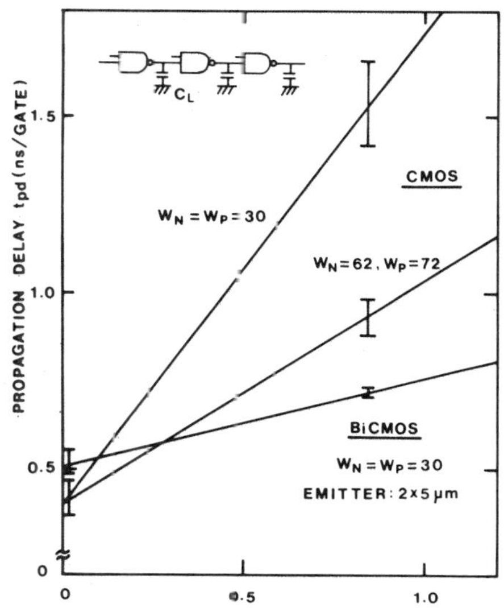

Fig. 14. Propagation delay time versus loading capacitance of the 2NAND BiCMOS gate and two kinds of CMOS gate. One has the same MOSFET size and the other has the same gate area as the BiCMOS gate.

C. BiCMOS Gate

To demonstrate the utility of this technology, 15-stage ring oscillators of the 2NAND BiCOMS gate shown in Fig. 1 are fabricated. Impedance Z is the 16-kΩ resistor. The emitter size of the bipolar transistor is 2 × 5 μm and the epitaxial layer is 1.5 μm thick. A photograph of part of the BiCMOS gate ring oscillator is shown in Fig. 13. The area of the bipolar transistor is almost equal to that of the MOSFET. The propagation delay time versus the load capacitance is shown in Fig. 14. Data on the CMOS gates are also plotted. One is designed to have the same MOSFET size as the MOSFET of the BiCMOS gate and the aspect ratio of the gate width to length W/L is 30/2 for both NMOS and PMOS. The other is designed to have the same gate area with the BiCMOS gate to compare the gate area efficiency of the drive ability. Gate size W/L is 62/2 for NMOS and 72/2 for PMOS. At a typical load capacitance of 0.85 pF, the delay time is 0.71 ns. Drive ability is represented by the slope of the graph and is 0.24 ns/pF. The value is 5.0 times larger than that of the CMOS gate of the same MOSFET size and 2.5 times larger than that of the same area CMOS gate. The power dissipation is 0.25 mW/gate at 0.85 pF and 4-MHz operation.

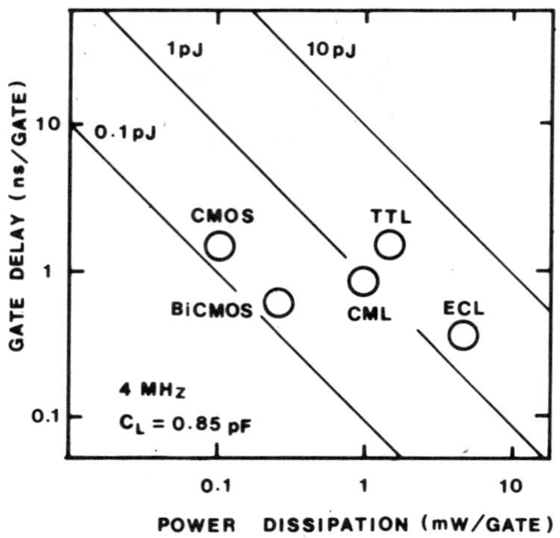

Fig. 15. Power-delay products of BiCMOS, CMOS, and bipolar gates with conventional 2-μm technology.

The power-delay products of the CMOS, BiCMOS, and bipolar gates of the same conventional 2-μm technology are shown in Fig. 15. The BiCMOS gate has the same product value as the CMOS gate, and twice the switching speed. As compared with bipolar gates, the gate delay is comparable to those of the TTL and CML, and power dissipation is 1/5–1/10.

V. Conclusions

A Hi-BiCMOS device structure featuring a buried twin well and a polysilicon emitter has been developed to produce high-performance small-size bipolar transistors on the same chip with standard CMOS. A thin epitaxial layer 1.3–2.0 μm thick can be used by adjusting the boron dose for the p^+ buried layer and by eliminating the high-temperature well diffusion process. The isolation width between bipolar transistors can be reduced to 10 μm by combining the p^+ buried layer and the LOCOS. A shallow-junction (0.3 μm) small-size (2 $\times$ 3-μm) emitter has been formed using the second polysilicon layer of the CMOS process. As a result, a bipolar transistor having a 4-GHz cutoff frequency and 500-μm^2 minimum area has been obtained by 2-μm technology. The fabrication process and the obtained device characteristics of MOSFET's are fully compatible with the standard 2-μm CMOS. Latchup immunity has been improved by the low well resistance of the buried layer. The well triggering current to initiate latchup has been increased 5–10 times in BiCMOS over the conventional n-well CMOS. The two-input NAND BiCMOS gate fabricated by this technology has a 0.71-ns/gate propagation delay time, 0.25-mW/gate power dissipation at 0.85 pF, and 4-MHz operation. The drive ability of the gate is 0.24 ns/pF and is 2.5 times superior to that of the same area CMOS gate.

The Hi-BiCMOS technology can be applied to various kinds of CMOS VLSI's such as memories, processors, and other digital analog-combined VLSI's to enchance system performance.

Acknowledgment

The authors would like to thank Dr. J. Suzuki, Dr. M. Kubo, T. Kamei, Y. Tanihara, M. Yoshimura, Dr. Y. Kawamoto, Dr. M. Okamura, Dr. M. Nishihara, K. Miyata, M. Tanaka, and T. Nitta for their encouragement and arrangement of this study. We would also like to thank S. Meguro, H. Higuchi, Dr. N. Momma, T. Nagano, and H. Uchida for their helpful discussions.

References

[1] Y. Nishio *et al.*, "A subnanosecond low power advanced bipolar-CMOS gate array," in *ICCD Dig. Tech. Papers*, p. 428, Oct. 1984.
[2] J. Miyamoto *et al.*, "A 1 μm n-well CMOS/bipolar technology for VLSI circuits," in *IEDM Tech. Dig.*, p. 63, Dec. 1983.
[3] F. Walczyk and J. Rubinstein, "A merged CMOS/bipolar VLSI process," in *IEDM Tech. Dig.*, p. 59, Dec. 1983.
[4] A. R. Alvarez, P. Meller, and B. Tien, "2 micron merged bipolar-CMOS technology," in *IEDM Tech. Dig.*, p. 761, Dec. 1984.
[5] K. Ogiue *et al.*, "A 15 ns / 250 mW 64 K static RAM," in *ICCD Dig. Tech. Papers*, p. 26, Oct. 1985.
[6] —, "A 13 ns/500 mW 64 Kb ECL RAM," in *ISSCC Dig. Tech. Papers*, p. 212, Feb. 1986.
[7] A. Watanabe *et al.*, "High speed BICMOS VLSI technology with buried twin well structure," in *IEDM Tech. Dig.*, p. 423, Dec. 1985.
[8] T. Matsuda *et al.*, A new BI-CMOS structure for analog/digital system," in *ESSDERC Tech. Dig.*, p. 174, Sept. 1981.
[9] H. Nakashiba *et al.*, "A subnanosecond BI-CMOS gate-array family," in *CICC Tech. Dig.*, p. 63, May 1986.

Paper 1.4

Perspective on BiCMOS VLSI's

MASAHARU KUBO, SENIOR MEMBER, IEEE, IKURO MASUDA, MEMBER, IEEE, KENJI MIYATA, AND KATSUMI OGIUE

***Abstract*—This paper describes a high-performance BiCMOS technology (Hi-BiCMOS) and its applications to VLSI's. By combining bipolar and CMOS devices in unit circuits of VLSI's, Hi-BiCMOS provides both speed performance competitive with bipolar LSI's and integration density close to that of CMOS LSI's. Hi-BiCMOS technology has been successfully used for static RAM's, dynamic RAM's, and gate arrays. The effectiveness of its applications to some type of processors has also been examined by evaluating test chips.**

I. Introduction

CMOS technology has become the mainstream fabrication technology for memories and microcomputers because of its high density and low power features. But some applications such as computer and communication systems require better speed performance than that obtained by CMOS technology so bipolar LSI's have been used in such fields. The target of high-performance BiCMOS (Hi-BiCMOS™) is to provide speed performance competitive with bipolar LSI's and integration density close to CMOS LSI's.

Early BiCMOS technology, which combined bipolar and CMOS devices on an LSI chip, e.g., combining bipolar drivers and CMOS logic circuits, has been tried in several types of LSI's. However, performance in this earlier BiCMOS was dominated by individual characteristics of the bipolar drivers and the CMOS logic circuits, so that its applications have been limited to specific fields. The Hi-BiCMOS opened a new field by introducing the new concepts listed below [1].

1) Bipolar and CMOS devices are formed on the same chip in an "uncompromised" form. It is essential that high-performance and small-size bipolar transistors are merged with the CMOS while maintaining compatibility with the standard CMOS process.

2) Bipolar and CMOS devices are combined in unit circuits composing an LSI, while, in earlier BiCMOS, these were placed on separate areas.

These points are discussed in detail below. Then some typical examples of Hi-BiCMOS applications are described which demonstrate its performance.

Manuscript received July 23, 1987; revised September 26, 1987.

M. Kubo is with Takasaki Works, Hitachi Ltd., 111 Nishi-Yokote-cho, Takasaki-shi, Gunma 370-11, Japan.

I. Masuda and K. Miyata are with the Hitachi Research Laboratory, Hitachi Ltd., 4026 Kuji-cho, Hitachi-shi, Ibaraki 319-12, Japan.

K. Ogiue is with the Device Development Center, Hitachi Ltd., 2326 Imai, Oumeshi, Tokyo 198, Japan.

IEEE Log Number 8718660.

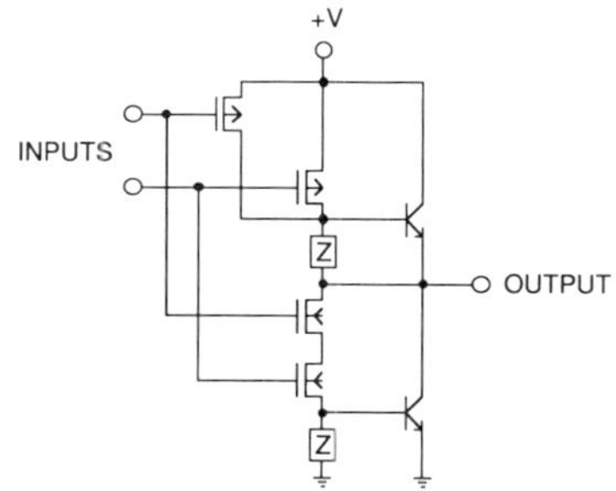

Fig. 1. Example of Hi-BiCMOS circuit (two-input NAND).

II. Hi-BiCMOS Technology

Fig. 1 shows the circuit configuration of a Hi-BiCMOS two-input NAND gate. Bipolar transistors and CMOS FET's are combined in the unit circuit, while keeping the features of each device. Namely, the Hi-BiCMOS gate provides both drivability from a bipolar output buffer and low power performance by a complementary operation similar to CMOS. Additionally, various practical methods to combine bipolar and CMOS devices have been introduced and some of them are shown in the following sections.

The Hi-BiCMOS logic gate achieves less gate delay than the CMOS logic gate owing to less dependence on load capacitance as shown in Fig. 2. The slope for the Hi-BiCMOS logic gate depends on current gain of the bipolar transistor determined by f_T and operation frequency f. So, f_T corresponding to the collector current must be sufficiently high as compared with the operating frequency. For ordinary cases, current gain should be more than 5. According to this consideration, f_T of more than 6 GHz is desired in 1.3-μm Hi-BiCMOS to achieve a gate delay competitive with ECL devices.

A cross-sectional view of the Hi-BiCMOS devices [2], [3] is shown in Fig. 3. It is based on closely packed N^+- and P^+-buried twin-well structures to minimize isolation space between PMOS and NMOS FET's. These wells are formed by a self-aligned technology. A thin epitaxial layer of around 1 μm is used to obtain a high-speed bipolar transistor. It also has a narrow isolation space between the N^+-buried layers which is achieved by the combination of LOCOS and the P^+-buried layer.

In these structures, CMOS FET's closely compatible with standard CMOS devices and bipolar transistors competi-

Reprinted from *IEEE J. Solid-State Circuits*, vol. 23, no. 1, pp. 5–11, Feb. 1988.

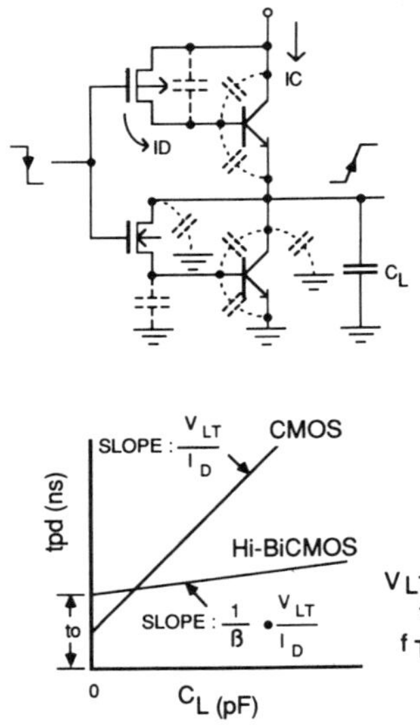

$$tpd = to + \frac{1}{\beta} \cdot \frac{V_{LT}}{I_D} \cdot C_L \quad ; \beta = \frac{f_T(I_C)}{f}$$

β : Current Amplification Factor
V_{LT} : Logical Threshold Voltage (V)
f : Operation Frequency (Hz)
f_T : Current Gain-Bandwidth Product (Hz)

Fig. 2. Analysis of gate delay.

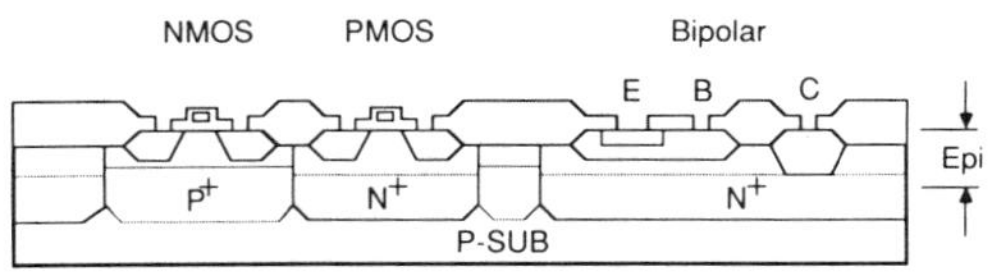

Fig. 3. Device structure of Hi-BiCMOS.

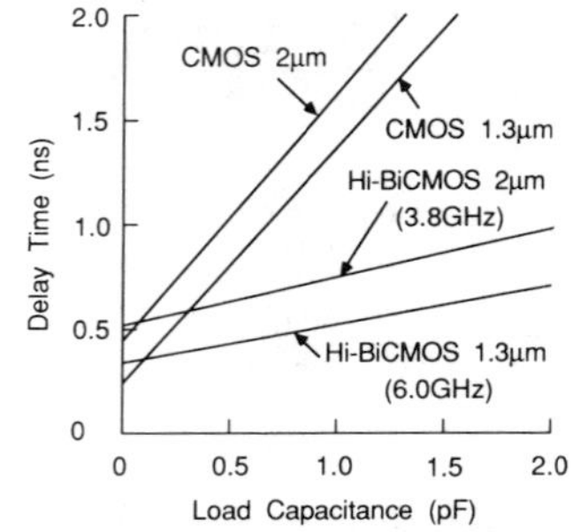

Fig. 4. Gate delay characteristics.

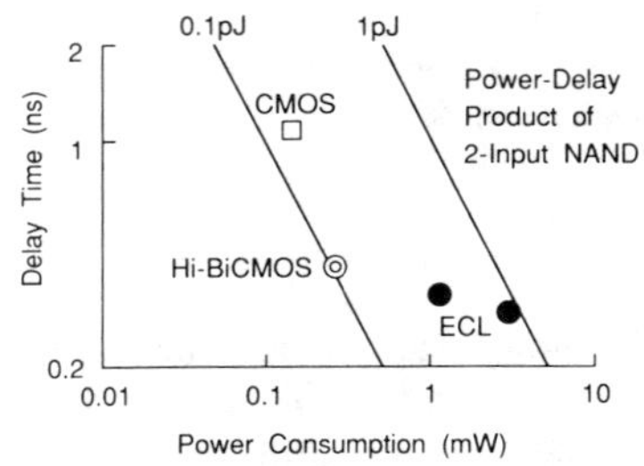

Fig. 5. Power–delay product of Hi-BiCMOS gate.

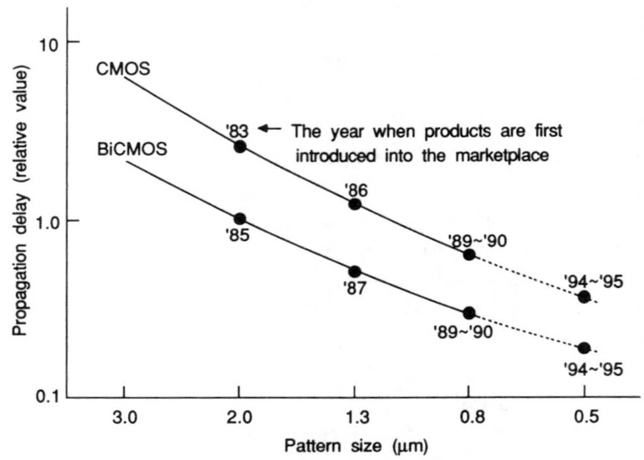

Fig. 6. Propagation delay versus pattern size.

tive with those for high-speed ECL devices can be formed on the same chip. This is the reason why this structure is called an "uncompromised" form.

As for the bipolar transistor, the current gain-bandwidth product f_T is sufficiently high, 3.8 GHz for 2 μm and 6.0 GHz for 1.3 μm, enough to realize the performance of Hi-BiCMOS. Other characteristics such as current gain, breakdown voltage, and collector resistance are comparable to those of conventional high-speed bipolar transistors.

As for MOSFET's, device characteristics such as threshold voltage, breakdown voltage, and transconductances are nearly the same as those of the conventional CMOS.

Fig. 4 plots measured gate delay characteristics for 2- and 1.3-μm technology. Their evaluations were done under a condition of the same MOS size. They clearly demonstrate that the Hi-BiCMOS gate achieves less dependence on load capacitance; its high f_T meets the operating frequency requirements. Similar behavior can be derived for other Hi-BiCMOS circuits besides the logic gate. But the advantages of the Hi-BiCMOS can be achieved only because high-speed bipolar transistors are combined with CMOS FET's.

Fig. 5 shows the power–delay product of the Hi-BiCMOS logic gate compared with those of ECL and CMOS. The Hi-BiCMOS logic gate fabricated with 1.3-μm technology is superior to any other conventional gate circuits based on equivalent technology. Hi-BiCMOS LSI's can have high performance anywhere on the chip using such circuits. However, such performance of the Hi-BiCMOS circuits is never obtained without high-speed bipolar transistors.

To summarize the versatility of Hi-BiCMOS, as shown above, Hi-BiCMOS has good drivability as compared with pure CMOS. It also has a small-signal handling capability, is less sensitive to process fluctuation, and provides an easy mixing of high-precision bipolar analog and low-power CMOS digital functions. On the other hand, as compared with pure bipolar, Hi-BiCMOS has higher density and less power consumption.

Fig. 6 shows the improvement of propagation delay by the introduction of a finer pattern technology. The propagation delay time of CMOS has decreased by a factor of 2 for every new generation technology according to the so-called "Scaling rule," under a constant supply voltage.

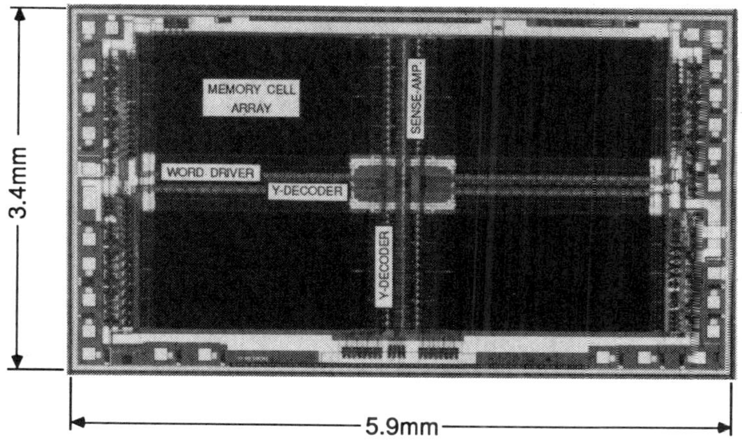

Fig. 7. Chip photomicrograph of 64-kbit SRAM.

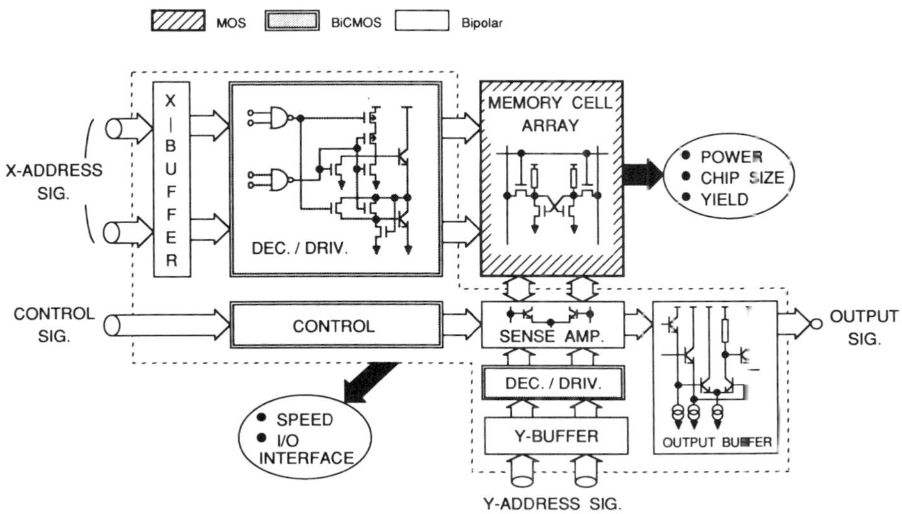

Fig. 8. Concept of Hi-BiCMOS ECL RAM.

The curves in the submicrometer range will become more gradual because of some device limiting factors associated with submicrometer CMOS.

A 0.8-μm BiCMOS will have a typical delay of 0.3 ns, which is better than equivalent CMOS by a factor of 1.5–2. Further performance improvements in BiCMOS are now known yet, however, 0.2 ns of delay time will be the target in 0.5-μm BiCMOS.

A supply voltage will be lowered after the year of 0.5-μm technology. In this case, the delay time of CMOS gates will not be improved, because the drain current of MOSFET's is reduced by a lowered supply voltage. On the other hand, the f_T of bipolar transistors can be improved in a finer pattern and not be reduced under the lowered supply voltage. Thus, in BiCMOS gates, the reduction of the MOS drain current can be compensated by the improvement of the f_T, so the BiCMOS will be superior to the CMOS even in the future.

The rate of introduction of new products to the marketplace will also become more gradual in the submicrometer generations. In BiCMOS, the first 2-μm BiCMOS products, with speed advantage of twice over CMOS, appeared on the marketplace in 1985. It was two years later than 2-μm CMOS but one year ahead over 1.3-μm CMOS products. As for 1.3-μm BiCMOS, the first products were introduced in 1987. CMOS products with the same performance will appear in the market in 1990. As a result, BiCMOS will be able to retain its higher performance for a longer period. It also has to be noted that the introduction year of both BiCMOS and CMOS products will merge in the submicrometer area because CMOS will need a more sophisticated process with longer process steps, and the additional processing required for BiCMOS will be relatively decreasing.

III Applications

A. Memories

1. Static RAM: Fig. 7 shows a chip photomicrograph of a 64-kbit Hi-BiCMOS static RAM with ECL I/O's and a 7-ns address access time [5]. This chip is fabricated with 1.3-μm Hi-BiCMOS technology, and its size is 3.4×5.9 mm^2. In this chip, the memory cell arrays are separated

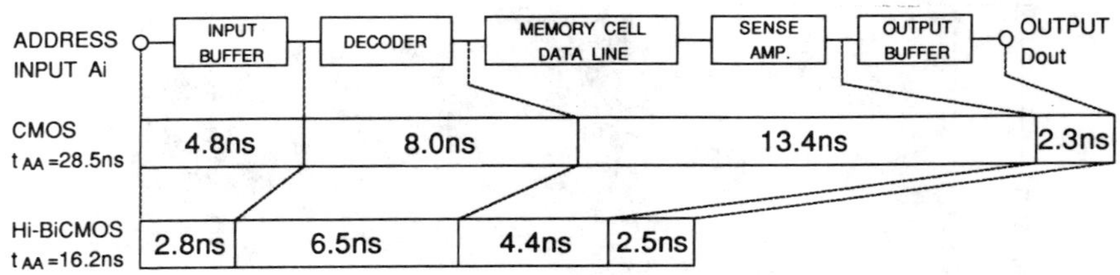

Fig. 9. Delay-time reduction of Hi-BiCMOS.

into four planes and row decoders are placed between them. In this section, structural features and performance of this RAM are discussed.

Fig. 8 shows a typical construction of static RAM's. The basic concept of the Hi-BiCMOS static RAM is to speed up its time-critical portions by utilizing bipolar and CMOS combining circuits. Several examples have been introduced so far for BiCMOS static RAM's. But, in these cases, use of bipolar transistors was limited mostly to output drivers or sense amplifiers [6].

The Hi-BiCMOS static RAM has achieved a greater reduction in access time by a complete mixing of bipolar and CMOS devices all over the chip. This construction is made possible owing to the speed performance of bipolar transistors. In particular, high-speed decoders using the Hi-BiCMOS logic gates and bipolar sense amplifiers for minimum data-line voltage swing are effective for improving the speed performance. Of course, high-density and low power features of CMOS static RAM's can be kept by using MOS memory cell arrays.

Each logic gate in the address decoding circuits has to drive long aluminum lines with a maximum of 32 inputs in the next stage, and the word drivers are also required to drive memory cells through word lines. The Hi-BiCMOS logic gates with high drivability are useful to shorten the delay time under such loading conditions. The voltage swing level of the common data line, which has a slow transition time due to parastic capacitances, is reduced to only 30 mV by the level control circuit. To amplify this small signal, a two-stage bipolar sense amplifier is effectively used. The voltage level of the common collector line is kept almost constant by bipolar transistors connected in series, so that the influence of the parasitic capacitance is minimized.

The difference in total delay time between 2-μm Hi-BiCMOS and 2-μm CMOS is compared in Fig. 9. The delay time of the Hi-BiCMOS static RAM is remarkably reduced at the decoding circuit and sense-amplifier blocks as compared with CMOS.

The Hi-BiCMOS static RAM's have two types of interface levels: TTL and ECL. Power dissipation of the TTL type is competitive with that of CMOS static RAM's in spite of its speed performance. The ECL type [7] uses the same memory cell as the TTL type. Its access time is less than that of the latter, though its power dissipation increases a little, because of the high-speed ECL output buffers. Fig. 10 shows that the relative performance of BiCMOS technology will be placed in static RAM as compared with pure CMOS and bipolar. Here, the horizontal axis represents memory bit capacity while the vertical axis shows address access time. BiCMOS will prevail over CMOS and bipolar in the range beyond 64 kbit and below 25-ns access time.

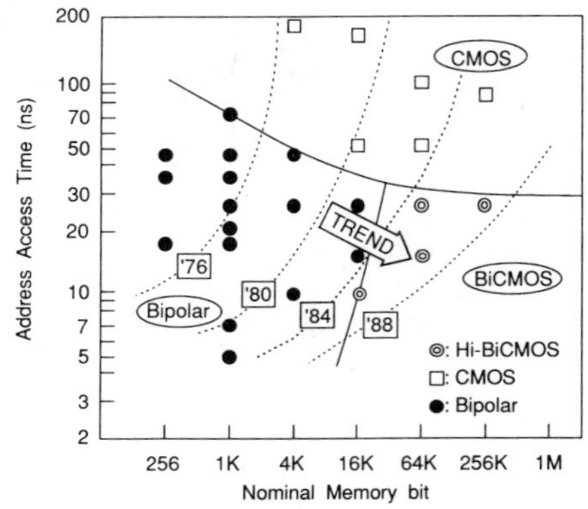

Fig. 10. Trend of Hi-BiCMOS memories.

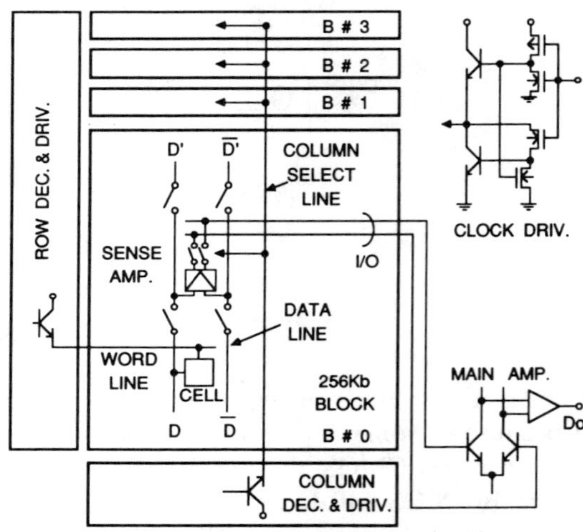

Fig. 11. Memory array configuration.

2. *Dynamic RAM:* A Hi-BiCMOS 1-Mbit dynamic RAM using 1.3-μm technology has also been developed. Fig. 11 shows the high-speed memory array configuration, well matched to the high-speed operation of peripheral circuitry due to BiCMOS circuitry. Bipolar transistors are used throughout word drivers, column drivers, output circuits, and clock drivers. Special attention is paid to reducing wiring resistance in memory arrays as well as in the peripheral circuitry. Polycide word lines are shunted with the second-layer aluminum to reduce the line delay. Data lines are composed of a first-layer aluminum. Further-

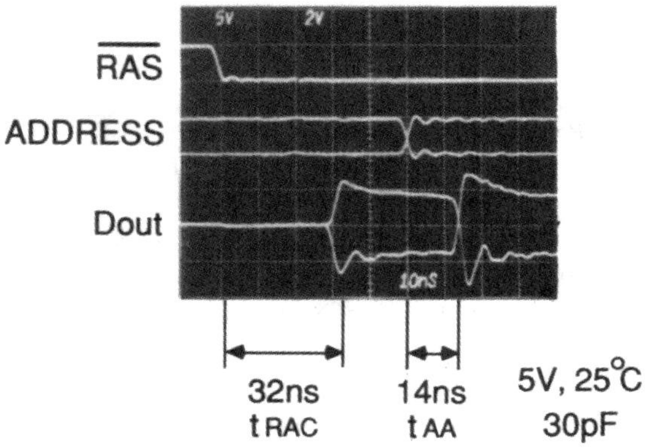

Fig. 12. Operating waveforms.

more, divided data lines and a shared sense-amplifier configuration to enhance speed and reduce power is adopted.

The chip size of the 1-Mbit Hi-BiCMOS DRAM is 5.0×14.9 mm^2, which is about 16 percent larger than a CMOS chip using the same memory cell and the same design rule. The increase in chip size is mainly due to widening Al wires to achieve low wiring resistance for high speed.

Fig. 12 shows operating waveforms of the 1-Mbit chip. A typical access time T_{RAC} from clock signal RAS of 32 ns and an address access time t_{AA} from column address ($ADDRESS$) of 14 ns, are realized with a cycle time of 60 ns and a low power of 450 mW.

B. Gate Arrays

A Hi-BiCMOS gate array is fabricated with 1.3-μm technology. It contains approximately 7560 basic cells, 4-kbit three-port RAM, and 182 input/output buffers in the 10.4×11.5-mm^2 chip. The basic cell in the logic circuit area is composed of a pair of bipolar transistors and three pairs of CMOS FET's so that it can be applied to the Hi-BiCMOS logic gates. This construction provides the 1.3-μm gate delay characteristics shown in Fig. 4 and also low power characteristics competitive with CMOS gate arrays. The Hi-BiCMOS gate array with a gate delay of 0.45 ns for typical loading conditions is more than twice as fast as a conventional CMOS gate array. Furthermore, the three-port RAM and the input/output buffers are composed by combining bipolar transistors and MOSFET's.

Another advantage of the Hi-BiCMOS is that its characteristics have less dependence on external conditions than those of CMOS. For example, when the four-input NOR gate and the two-input NAND gate are compared, the delay time is increased by only 50 percent for the Hi-BiCMOS, while it is 200 percent for the standard CMOS. Less dependence on fabrication process conditions, such as gates length and threshold voltage of MOSFET's, and ambient temperature are also observed. These are caused by the fact that the gate delay of the Hi-BiCMOS logic gates depends mainly on the characteristics of the bipolar transistor.

Fig. 13. Photomicrograph of processor test chip.

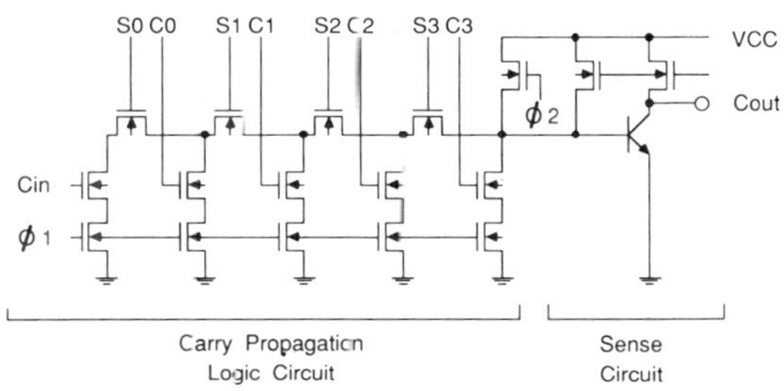

Fig. 14. Carry propagation circuit.

Hi-BiCMOS gate arrays have been successfully applied to many kinds of time-critical circuits for minicomputers, communication systems, and graphic systems. CMOS-type gate arrays, with their low-power requirements, and ECL-type gate arrays, with their high speed, have left a distinct gap in performance. The Hi-BiCMOS gate array with high-speed and low-power characteristics provides new features to fill this gap.

C. Processors

Recently, a standard cell approach has been introduced for processor design to improve design productivity, in which large logic circuit blocks are used as macro cells to increase integration density. When this approach is applied, load capacitance of individual logic circuits is not always large and drivability of a bipolar transistor is not always effective for improving speed performance. To overcome this problem, a new circuit technology utilizing the features of Hi-BiCMOS, differing from that of the gate array, must be introduced.

The design approach for the Hi-BiCMOS processor is similar to that of the static RAM insofar as most portions of the chip are composed of MOS devices and critical paths are accelerated using Hi-BiCMOS circuits. But new low-voltage swing circuits for macro cells are introduced [8].

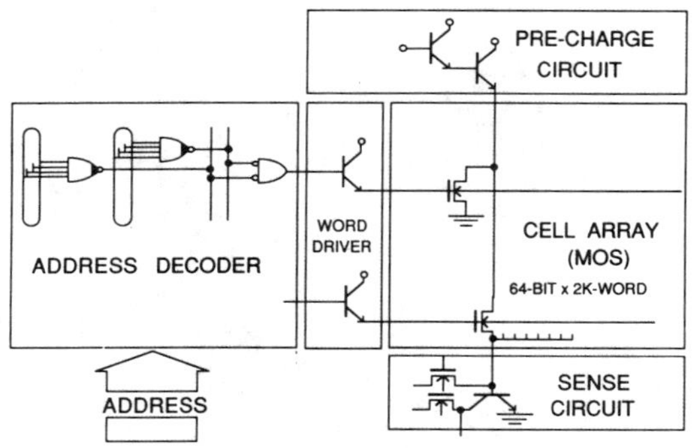

Fig. 15. Circuit diagram of ROM.

Basically, a typically processor consists of arithmetic units, registers, a ROM, and many kinds of macro cells and poly cells. Critical paths of this chip are a) the arithmetic operation cycle, and b) the ROM access cycle. Utilizing the Hi-BiCMOS technology speeds up these operations, as verified by evaluating a test chip. Fig. 13 shows the test chip fabricated with 2-μm Hi-BiCMOS technology. It includes a 64-bit$\times$2K step ROM, a 32-bit arithmetic unit, a barrel shifter, and a register file. By another test chip fabricated with 1.3-μm Hi-BiCMOS technology, a 64-bit multiplier and a cache memory were estimated.

Fig. 14 shows the carry propagation circuit in the arithmetic unit. In this circuit, MOS switches and a bipolar sensing circuit are combined. Voltage swing of this circuit can be limited within the base–emitter voltage of the bipolar transistor. This results in high-speed operation. The 32-bit carry propagation circuit can be obtained by linking eight 4-bit circuits in series. The measured propagation delay time for 32-bit operation is 5.6 ns, which is equivalent to a gate delay of approximately 0.15 ns.

The ROM circuit, shown in Fig. 15, can be designed in the same manner as the static RAM. The memory cell array is composed of MOS switches, and the Hi-BiCMOS logic gates are used in decoding circuits. The measured access cycle of the ROM is approximately 60 MHz under typical operating conditions.

The Hi-BiCMOS processor technology may possibly be extended to computers, some special-purpose processors, and ASIC's.

D. Others

BiCMOS technology is expected to extend to mixed analog/digital (A/D) VLSI's. Fig. 16 shows the three alternative device approaches: CMOS, bipolar with IIL, and BiCMOS. An additional favorable aspect of BiCMOS for this application is low-noise capability. CMOS A/D is good for digital-oriented applications and bipolar A/D is suitable for analog-oriented applications. BiCMOS A/D is the best approach for medium-range applications.

Another application of BiCMOS technology is telecommunication transmission. Various kinds of functions for transmitter and receiver are needed. As the data rate becomes higher, BiCMOS technology is getting important for covering those functions with wide bandwidth. In telecommunication and data communication, BiCMOS is expected to one of the key components for switcher and ISDN.

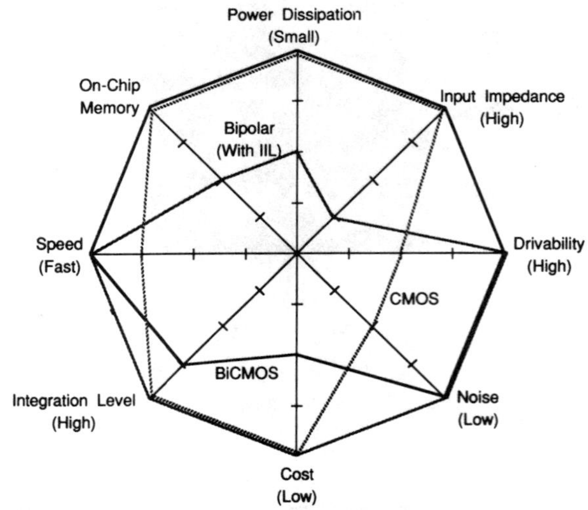

Fig. 16. Alternative devices for mixed A/D VLSI's.

BiCMOS devices are also expected to be used in consumer and automobile applications.

IV. Conclusions

The Hi-BiCMOS technology provides speed performance competitive with bipolar LSI's, while providing integration density and power dissipation close to CMOS LSI's. To keep these advantages in the future, more advanced devices and circuit technology will be required. Speed performance of the bipolar transistor represented by f_T must be improved to meet speed increases in short-channel MOSFET's. New ideas for circuits using both bipolar–CMOS combining circuits are also required to increase the variety of macro cell libraries.

A perspective on BiCMOS VLSI's is introduced and three major points result from the discussions in this paper:

1) BiCMOS will have a more significant role as the generations of fine pattern devices proceed;
2) the versatility inherent in BiCMOS will become more important in the era of system-on-chip VLSI's; and
3) diversification of BiCMOS applications to memory, gate array, processors, and mixed analog/digital systems will become more important.

Acknowledgment

The authors wish to thank Y. Kawamoto, M. Nishihara of Hitachi Research Laboratory, J. Suzuki of Tobu Semiconductor Inc., M. Tanaka of Hitachi Device Development Center, and T. Makimoto of Hitachi Takasaki Works for their encouragement and support during the Hi-BiCMOS program. The authors also wish to thank K. Itoh of Hitachi Central Research Laboratory, and H. Maejima and N. Momma of Hitachi Research Laboratory for their efforts on DRAM, processor, and device technologies, respectively.

References

[1] I. Masuda, Y. Nishio, and T. Ikeda, "High speed logic circuits combining bipolar and CMOS technology," *Trans. Inst. Electron. Commun. Eng. Japan*, vol. J67-C, no. 12, pp. 999–1005, Dec. 1984.

[2] T. Watanabe *et al.*, "High speed BiCMOS VLSI technology with buried twin well structure," in *IEDM Tech. Dig.*, 1985, pp. 423–426.

[3] T. Ikeda *et al.*, "Advanced BiCMOS technology for high speed VLSI," in *IEDM Tech. Dig.*, 1986, pp. 408–411.

[4] Y. Nishio, I. Masuda, T. Ikeda, M. Iwamura, K. Ogiue, and Y. Suzuki, "A subnanosecond low power advanced bipolar-CMOS gate array," in *Proc. ICCD*, Oct. 1984, pp. 428–433.

[5] S. Miyaoka *et al.*, "A 7ns/350mW 64K ECL compatible RAM," in *ISSCC Dig. Tech. Papers*, Feb. 1987, pp. 132–133.

[6] M. Odaka, S. Miyaoka, K. Ogiue, K. Tonomura, T. Ikeda, and T. Yasui, "A 12ns/350mW 16Kb ECL compatible RAM," in *Proc. ESSCIRC*, Sept. 1985, pp. 166–176.

[7] K. Ogiue, M. Odaka, S. Miyaoka, I. Masuda, T. Ikeda, and K. Tonomura, "13-ns, 500mW, 64-kbit ECL RAM using Hi-BiCMOS technology," *IEEE J. Solid-State Circuits*, vol. SC-21, no. 5, pp. 681–685, Oct. 1986.

[8] T. Hotta *et al.*, "CMOS/bipolar circuits for 60-MHz digital processing," *IEEE J. Solid-State Circuits*, vol. SC-21, no. 5, pp. 808–813, Oct. 1986.

Paper 1.5

Overview of BiCMOS Device and Process Integration

R. H. HAVEMANN AND R. H. EKLUND
SEMICONDUCTOR PROCESS AND DESIGN CENTER
TEXAS INSTRUMENTS INCORPORATED
DALLAS, TEXAS

1. Introduction

Over the past decade, BiCMOS technology has evolved from a tentative merger of stand-alone bipolar and MOS transistors [1–3] into sophisticated device structures and novel gate designs that are uniquely qualified to meet the high-performance demands of the nineties. BiCMOS has gained widespread acceptance in high-end digital and analog applications by offering a unique mix of design (CMOS, BiCMOS, BiNMOS, ECL) and device (NMOS, PMOS, bipolar) options to provide optimum circuit speed, power, and density. From a simple device perspective, the design trade-offs are clear—use bipolar transistors as drivers in critical speed paths or in ECL gates for high-speed switching and use CMOS devices where power and density are of utmost concern. In practice, BiCMOS designs have become much less straightforward due to the evolution of BiCMOS gate design and the emergence of new gate concepts [4–7].

From a system-level viewpoint, functional density and power constraints limit the use of bipolar devices in VLSI/ULSI circuits. BiCMOS circuits therefore tend to be CMOS-intensive, with bipolar transistors added where necessary to meet specific design or performance requirements. Additional process steps are required to integrate bipolar transistors into a CMOS core process flow; these steps become more numerous and complex if higher bipolar performance and device options (PNP and polysilicon resistor) are needed. The evolution of BiCMOS integration has been governed by the trade-off between circuit performance and processing cost. In general, this trade-off has led to the use of lower performance bipolar transistors in BiCMOS circuits as compared to pure bipolar designs. The reason for this compromise in bipolar performance is that for a given CMOS technology level, BiCMOS circuit performance tends to saturate as the level of bipolar performance increases. This performance saturation is illustrated in Fig. 1, which shows simulated access time plotted versus bipolar unity gain cutoff frequency for a 256-K BiCMOS ECL SRAM with 0.8-μm design rules. Note that access time improves dramatically over the lower range of f_T, but only incremental improvement is observed beyond an f_T of approximately 7–8 GHz. Since the fabrication of bipolar transistors becomes more difficult with increasing f_T due to shallower junction requirements and process control issues, a compromise value of f_T is typically chosen that meets the product cost and performance need. Additional emphasis may also be placed on harmonizing CMOS/bipolar processes and device structures to further reduce the cost of BiCMOS manufacturing.

However, as each successive generation of new products drives the need for higher bipolar performance, the BiCMOS process integration challenge becomes more difficult. Integration of state-of-the-art bipolar and CMOS devices may be required to meet future performance goals, and congruent

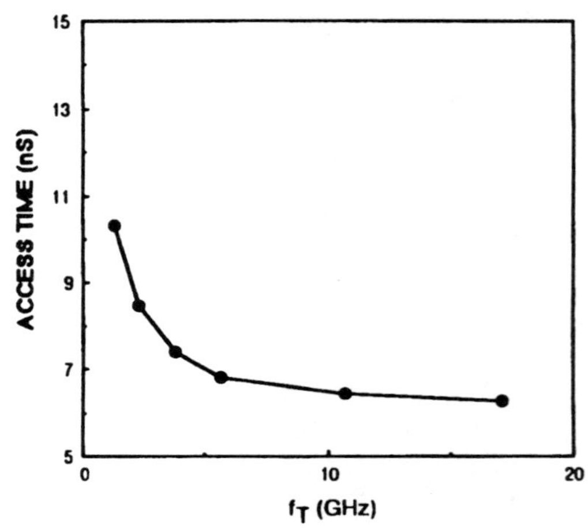

Fig. 1 Simulated access time of 256-K BiCMOS ECL SRAM as a function of bipolar f_T.

CMOS/bipolar device structure and processes will be necessary to reduce manufacturing complexity. The ensuing discussion addresses the BiCMOS process integration issue from the standpoint of merging high-performance bipolar transistors into a CMOS core process flow. Topics covered include the evolution of BiCMOS device structure, a CMOS and bipolar device structure comparison, bipolar device structure evolution, process integration issues for a typical BiCMOS process flow, and voltage scaling.

2. BiCMOS Process Evolution

The first rudimentary BiCMOS devices emphasized low cost and process simplicity over performance. The device cross section shown in Fig. 2(a) illustrates an early merged CMOS and bipolar device structure. The PMOS n-well was typically used as the collector of the NPN bipolar transistor, with the NMOS n^+ source/drain serving as both the emitter and collector diffusion. Only one extra mask (the bipolar base implant) was required to integrate this bipolar transistor into a CMOS core flow. Bipolar f_T was limited due to the lightly doped n-well and deep n^+ emitter junction, but bipolar performance was adequate considering the overall CMOS technology level and the limited use of bipolar transistors as stand-alone drivers.

Greater demand was placed on bipolar performance as BiCMOS circuit design began to mature and the core CMOS technology continued to scale. This led to the first significant departure from standard CMOS processing—the use of buried layers with overlying epitaxy. The need for epitaxy added considerable complexity and expense to BiCMOS manufacturing. However, due to the importance of the collector in determining bipolar performance [8], it became necessary to merge traditional bipolar processes into the CMOS flow. These processes included the formation of a buried n^+ layer or subcollector to provide low collector resistance and deposition of an thin overlying n-type epitaxial layer with doping and thickness tailored for optimum bipolar performance. The n^+ buried layer served a dual purpose by also reducing CMOS latchup susceptibility. A deep n^+ diffusion, or n^+ sinker, was typically used to provide a low-resistance path to the buried collector. These structural changes are illustrated in the BiCMOS device cross section shown in Fig. 2(b). A total of three extra masks (subcollector, n^+ sinker, and base) were required to add the collector enhancements to a CMOS core flow. An additional mask

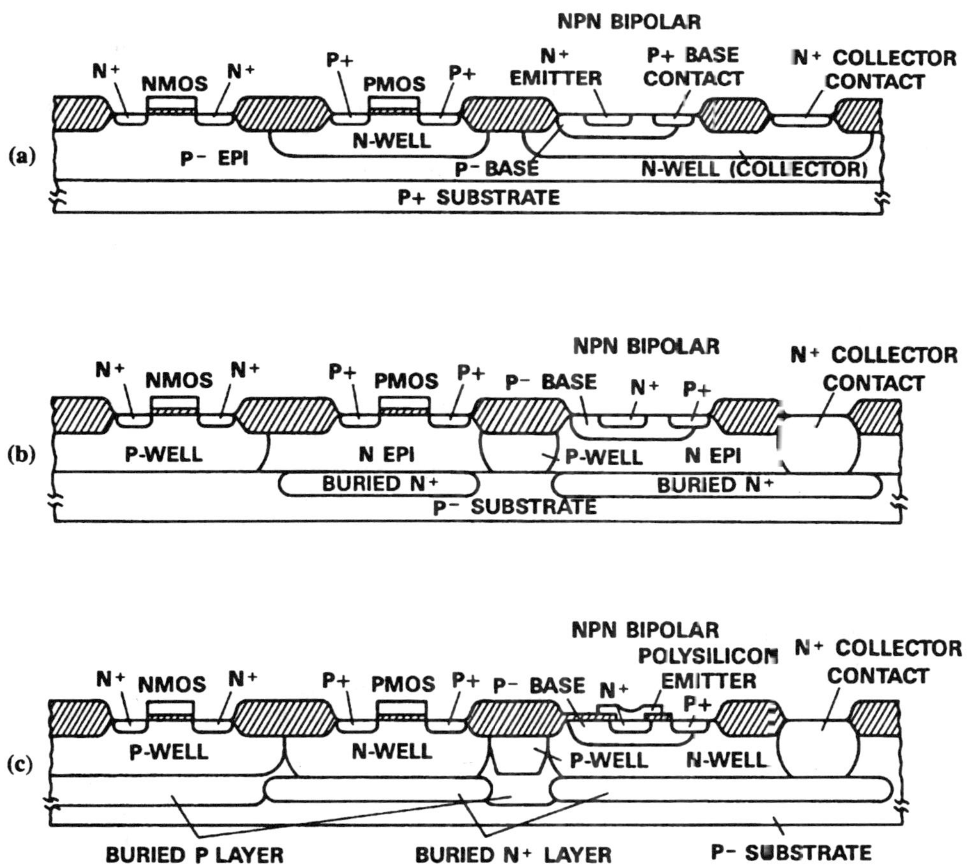

Fig. 2 Evolution of BiCMOS device structure: (a) simple addition of an NPN bipolar transistor to the basic CMOS; (b) addition of n^+ buried layer and n^+ sinker to reduce collector resistance; (c) use of self-aligned p/n^+ buried layers for improved packing density, separately optimized p/n-wells and polysilicon emitter for improved bipolar performance without compromising CMOS performance.

was also frequently used to separately optimize the emitter implant, providing further improvement in bipolar performance.

As emitter-junction-depth requirements continued to scale, direct emitter implants were eventually replaced by implanted polysilicon emitter diffusion sources [9] that could achieve even shallower depths. Bipolar polysilicon emitters provided dramatic improvements in performance but also added process complexity. Typically, two mask levels (the emitter diffusion window and polysilicon emitter) were required to form the emitter structure, in addition to extra process steps such as polysilicon deposition, n^+ emitter implant, and sequential polysilicon/oxide etch. Integration of the polysilicon emitter into the BiCMOS device structure is shown in Fig. 2(c); a detailed cross-sectional view of the most commonly used polysilicon emitter structure is also shown in Fig. 3(a).

The alternative self-aligned emitter/base structure [10] shown in Fig. 3(b) was developed for process simplification and scalability. Positive attributes of this structure included one less mask (the emitter diffusion window) as well as a self-alignment of the emitter-to-intrinsic base. The major disadvantage of this structure is the lack of a polysilicon etch stop over the single-crystal base region. For this reason, the structure shown in Fig. 3(a) is generally favored for current BiCMOS manufacturing processes. The process steps required to merge this structure into a BiCMOS flow can be simplified by using a split polysilicon approach [11]. This technique combines the CMOS gate and bipolar polysilicon emitter pattern so that only one extra mask level (the emitter diffusion window) is required to form the bipolar emitter.

The polysilicon emitter concept was eventually extended to include diffusion of both base and emitter dopants, either from a common polysilicon electrode [12, 13] or, more typically, from separate electrodes, as shown in Fig. 3(c). In the double electrode process, a first layer of polysilicon is normally used as the extrinsic base, while a second polysilicon layer forms the emitter of the transistor, self-aligned to the extrinsic base. The intrinsic base may be implanted prior to emitter polysilicon deposition or implanted/diffused from the emitter polysilicon as noted previously [12, 13]. Variations of the double polysilicon approach are numerous and have spawned an entire generation of "super self-aligned" emitter-base structures [4], [6], [14] that offer densities comparable to MOSFETs for common collector configurations. While the double polysilicon bipolar structure is more complex than single polysilicon structures, the use of p^+ polysilicon for the extrinsic base is compatible with surface-channel PMOS transistors, which similarly require a p^+ polysilicon gate. As discussed in the next section, such process commonalities are extremely important from the standpoint of reducing the integration cost required for future BiCMOS generations.

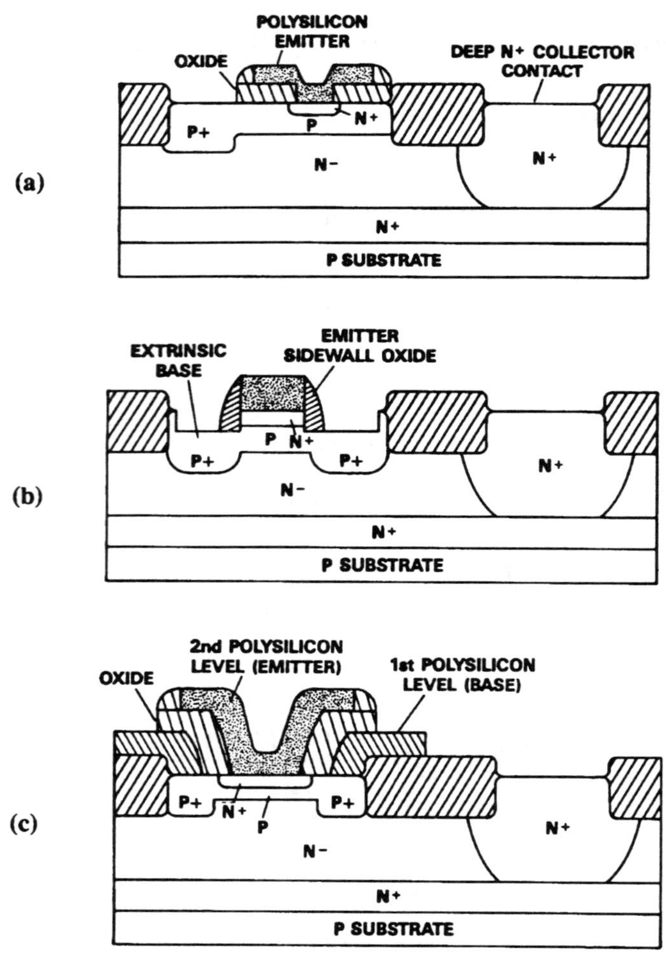

Fig. 3 Schematic cross sections of various types of bipolar polysilicon emitter structures: (a) single-level polysilicon emitter with an underlying protective oxide; (b) single-level polysilicon emitter self-aligned to extrinsic base; (c) double-level polysilicon bipolar transistor with p^+ polysilicon base contact and self-aligned n^+ polysilicon emitter.

3. CMOS and Bipolar Device Structure Commonality

The natural evolution of bipolar transistor structures into more complex designs has emphasized a need for congruent CMOS/bipolar processes and device structures to reduce BiCMOS manufacturing complexity and cost. Fortunately, the structural and process-control requirements for high-performance CMOS and bipolar devices have tended to converge in the submicrometer regime. This is not unexpected in view of the common need for three-dimensional scaling, but innovations such as the polysilicon emitter have added a surprising degree of similarity in device structure, as shown in Fig. 4.

In this example, the bipolar polysilicon emitter structure shows a strong resemblance to the MOS device. Both structures use an oxide to isolate the polysilicon electrode. For the MOSFET, the gate is isolated from the channel, whereas in the bipolar case, the polysilicon emitter is isolated from the intrinsic base extension. The base oxide is typically thicker than the gate oxide to provide lower emitter-base capacitance. Both structures also utilize a sidewall oxide to offset the p^+ implant that serves as the bipolar extrinsic base and the PMOS source/drain. Indeed, as will be described in the following section, identical process steps can be used to fabricate the CMOS and bipolar transistors following the polysilicon deposition.

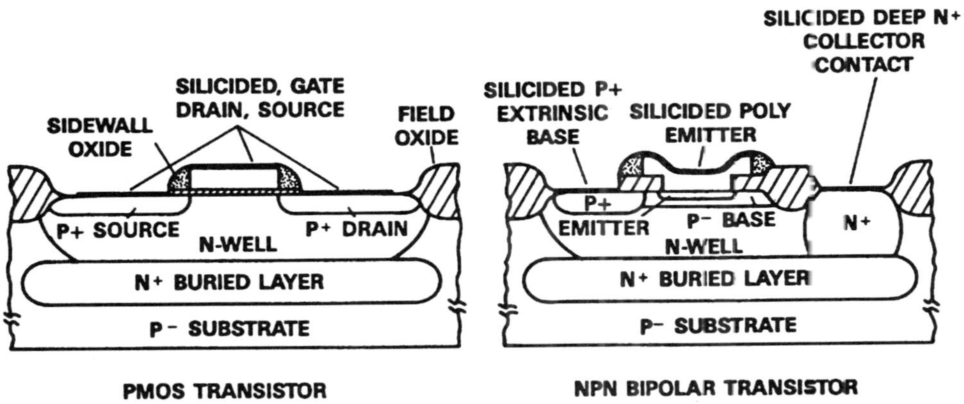

Fig. 4 Typical CMOS and bipolar transistor cross sections showing device structure commonality.

Another common PMOS/bipolar feature illustrated in Fig. 4 is the use of a self-aligned silicide, or ''salicide,'' to form a low-resistance cladding on exposed polysilicon and diffusion regions. The use of a salicide process is prevalent for high-performance microprocessor and gate-array applications where series resistance between metal contacts and active device regions must be minimized. A silicide cladding also reduces the emitter and series base resistance of bipolar devices as well as the source/drain series resistance of MOS devices. Frequently, emitter resistance is also lowered by placing metal contacts directly over the emitter diffusion. MOS devices have traditionally not allowed contacts over active device regions for reliability concerns, and this may also be an issue for bipolar emitters as polysilicon thickness and junction depths continue to scale.

4. Bipolar Process and Device Structure Scaling Issues

In spite of the progress made toward convergence of CMOS/bipolar device structure and process, several imitations remain. One major scaling issue is bipolar collector isolation. Unlike PMOS devices that are self-isolated and can share a common n-well, each bipolar transistor normally requires an isolated collector. The collector scaling problem stems from the conflicting requirement of a low-resistance subcollector (which requires a deep diffusion of the n^+ dopant) and tightly spaced collector regions with low capacitance. The collector scaling issue is highlighted in Fig. 5, where simulated collector-substrate capacitance and breakdown voltage are plotted as a function of substrate doping for two different collector spacings. For a given breakdown voltage specification, tighter collector spacing forces higher substrate doping or, more correctly, a higher boron concentration in the punch-through stopper that is normally implanted between collectors. Higher doping levels in turn lead to higher collector-substrate capacitance and degradation in performance, particularly at low current levels.

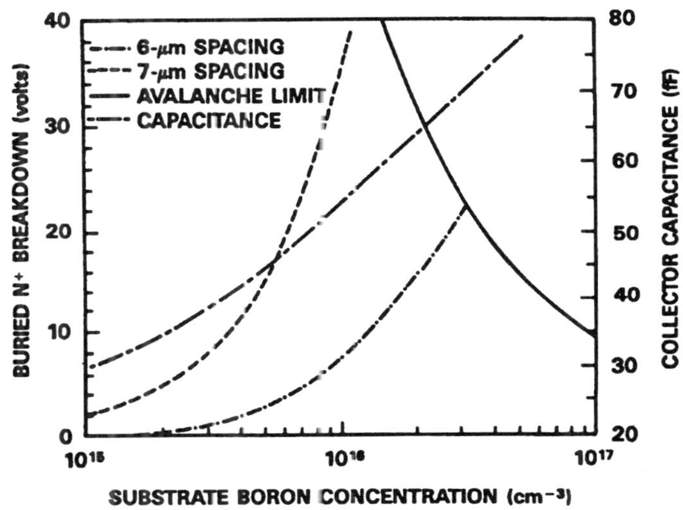

Fig. 5 Simulation of subcollector isolation characteristics for as-implanted buried n^+ spacings of 6 μm and 7 μm. Breakdown voltage is avalanche-limited for higher substrate doping and punchthrough-limited for lower substrate doping levels. Model assumes an NPN transistor with a perimeter of 33 μm and sidewall area of 100 μm^2 with a lateral diffusion of approximately 2 μm during the buried n^+ anneal.

As operating voltage is reduced, isolation requirements become more relaxed for a given collector spacing, as shown in Fig. 5. However, as collector spacing is scaled, doping levels must also increase, and it is not obvious that standard junction isolation techniques will suffice. For this reason, trench-isolated structures have been explored [15], but BiCMOS applications have been limited to isolation between n-wells [16] rather than n-well/p-well isolation due to leakage along the p-well sidewall [17]. Although p-well sidewall leakage may be reduced by use of diffused or angle-implanted boron channel stop, these techniques add complexity to an already overly expensive process. Alternatives to trench isolation include the use of selective epitaxy [18], but selectivity breakdown and sidewall leakage are still major issues.

Although lateral diffusion of the n^+ sinker must be accounted for in transistor layout design, scalability is not a serious issue since high-energy implants can be used to vertically distribute the n^+ dopant and provide good subcollector linkup with minimal thermal drive. As operating voltage is scaled, epitaxial layer thickness can be reduced and collector linkup will be enhanced. However, for high-voltage BiCMOS applications that require thick epitaxy, the n^+ sinker is a serious scaling issue. For these cases, a scalable, low-resistance path to the subcollector can be obtained by using an in-situ doped N^+ polysilicon plug, as shown in Fig. 6. This approach allows a doping level of greater than 10^{19} atoms/cm^3 to be maintained throughout the vertical plug, while a sidewall dielectric eliminates lateral diffusion and permits scaling of the bipolar structure.

An improvement in npn f_T and the current density for peak f_T can be obtained by the use of an additional implant to locally increase the collector doping under the emitter window [20]. This implant serves to increase the bipolar performance by cutting off the channeling tail of the base implant, thereby decreasing the base width, and by increasing the collector doping under the emitter to delay the onset of base pushout. By restricting the secondary collector implant to the region beneath the emitter window, only a minimal increase in the collector-base capacitance is incurred. Since the higher collector doping is localized under the bipolar emitter, this scheme can also be applied to a BiCMOS process flow without impacting the performance of the CMOS transistors through higher body effect.

Whereas base width plays a dominant role in determining the bipolar f_T, it is important to optimize the base profile for maximum performance within the constraint of supporting adequate emitter-base breakdown voltage and avoiding collector-emitter punchthrough. Tight process control is required for base and emitter formation to maintain predictable bipolar current gain and high-frequency performance. Critical variables include 1) base implant depth control, 2) base-to-polysilicon emitter interface quality, 3) polysilicon emitter grain size, and 4) thermal cycles. Of these, the base-to-polysilicon emitter interface is the most critical since interfacial oxides disrupt epitaxial regrowth of the polysilicon emitter and lead to high emitter resistance [9, 20]. Cleaning and annealing procedures must be optimized to break up this interfacial layer, particularly for the arsenic-doped polysilicon emitters prevalent in high-performance devices. Rapid thermal annealing (RTA) has been shown to be an effective emitter anneal [21], and will likely play a major role in meeting the anneal requirements of scaled BiCMOS devices.

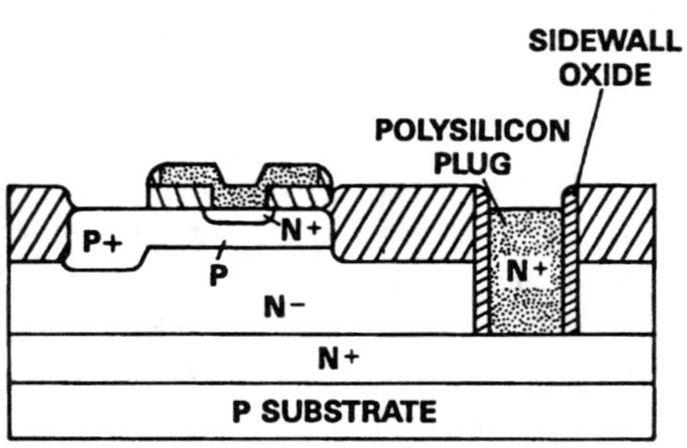

Fig. 6 The schematic device structure cross section of a bipolar transistor with an n^+ polysilicon plug used to contact the subcollector.

However, future bipolar scaling needs may require another revolutionary change in device structure. The requirement for vertical scaling leads to higher doping levels in the base to prevent punchthrough. This results in an increase in the emitter-base capacitance. To avoid a degradation in performance, operating current density must also be increased, supported by a rise in the collector doping level. These changes narrow the operating window of the bipolar transistor and also reduce the reverse diode breakdown voltages (BV_{ceo} and BV_{ebo}) required for compatibility with practical circuit applications. Several experimental approaches have been undertaken to overcome these limitations by altering the bandgap of the emitter or base material to form heterojunction bipolar transistors (HBTs). The most common approach is to alter the emitter material to increase the bandgap and thereby improve injection efficiency while relaxing the need for high emitter doping [22–26]. Alternative approaches have addressed a lower bandgap in the base material by growing a Si–Ge base [27], and bipolar f_T's as high as 75 GHz have been predicted for such devices [28]. Recently, a BiCMOS process with a Si–Ge HBT has been reported [29]. However, HBT technology is still in its infancy, and many issues must be resolved before HBTs can be fully integrated into BiCMOS processes.

5. BiCMOS Process Integration

As devices are scaled and demand for functional integration increases, the number of masking levels and process steps also increases. This dilemma is particularly challenging for BiCMOS processes and adds impetus to the need for CMOS and bipolar process congruence. While device-specific processes such as CMOS V_t and bipolar base implants may remain separate, other processes such as device isolation can be shared. As an example, Table 1 highlights the changes necessary to merge a bipolar transistor into a core CMOs process flow. This process was first introduced with 0.8-μm design rules [11] and later adapted for 0.6-μm BiCMOS [30]. A cross section of a BiCMOS device structure representative of the 0.6-μm process flow is illustrated in Fig. 7. Details of the 0.8-μm process flow are given elsewhere [31, 32]; the following discussion highlights the reasons for significant differences between the CMOS and BiCMOS processes.

The first major difference between the CMOS and BiCMOS flows in Table 1 is the starting material (substrate/epitaxy). In traditional CMOS technology, a p^+ substrate serves as a ground plane and low-resistance path for suppression of latchup. CMOS n/p-wells are formed in a p-type epitaxial layer grown on the p^+ substrate, and epitaxial layer thickness is chosen to allow for boron updiffusion from the p^+ substrate during n/p-well drive.

TABLE 1. Comparison of CMOS and BiCMOS Process Flows

CMOS Baseline Flow	Changes for Bipolar
P^+ substrate	P^- substrate
	Buried N^+ / P layer*
P-thick epitaxy	Intrinsic thin epitaxy
N/P-well drive	Additional deep N-well implant
Scaled LOCOS process	Reduced drive (thinner epitaxy)
Anti-Kooi / pregate oxidation	
	P base formation*
	P resistor pattern / implant*
	Deep N^+ sinker pattern / implant*
Blanket CMOS threshold implant	
Gate oxidation	
	Thin polysilicon deposition
	Emitter window pattern / etch*
Polysilicon deposition	
Doping from diffusion source	Doping from ion implant
Pattern / etch polysilicon	
LDD pattern / implant	
Sidewall oxide deposition / etch	
Pattern / implant N^+/P^+ S/Ds	
Source / Drain anneal	Anneal optimized for emitter
Silicide formation	
Low-temperature backend processing	

*Denotes extra mask required in comparison to CMOS baseline.

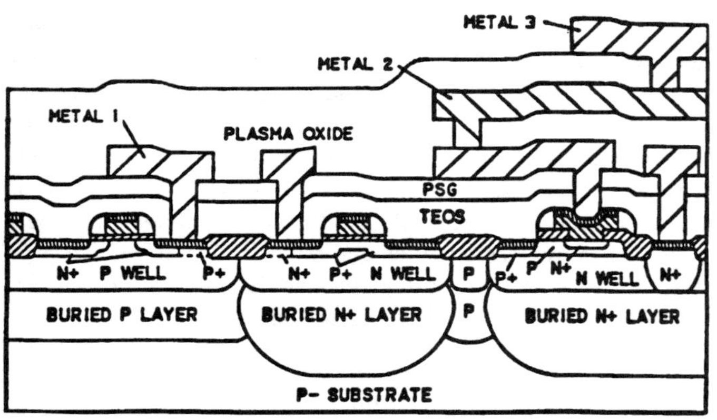

Fig. 7 The schematic device structure cross section for example 0.6–0.8-μm BiCMOS process.

As discussed previously, bipolar devices use a buried n^+ layer, or subcollector, to provide a low-resistance path for collector current. A lightly doped p-type substrate is chosen to minimize the capacitance between the n^+ subcollector and substrate. The buried n^+ layer is formed by pattern, implant, and anneal of a suitable n-type dopant (typically Sb or As). A self-aligned boron implant provides punchthrough protection between n^+ layers and prevents ground bounce; the combined n^+ p-buried layers are also very effective in suppressing CMOS latchup.

Following formation of the buried layers, an undoped epitaxial layer is grown. Epitaxial thickness is optimized for CMOS and bipolar performance as well as circuit/device specifications. In general, thinner epitaxy improves bipolar performance by lowering R_c and reducing the Kirk effect, while thicker epitaxy improves CMOS performance by reducing body effect and source/drain diode capacitance. In addition to these constraints, the epitaxy thickness and diffused well dopings must support diode breakdown voltages required by the circuit. For 5-V operation, epitaxy thickness is typically 1.2–1.5 μm, as compared with a ~5-μm thickness for 15-V analog applications. Epitaxial layer thickness and updiffusion of buried-layer dopants must be carefully controlled to maintain device parametrics such as bipolar BV_{cbo} and CMOS BV_{dss} as well as bipolar f_T.

The n-well and p-well doping concentrations are independently optimized along with CMOS source/drain and bipolar base diffusions to support the performance and operating conditions of the circuit. A traditional CMOS self-aligned twin-well process is used to implant and diffuse the wells for both the CMOS and BiCMOS flows listed in Table 1. However, a reduced well drive is possible in the BiCMOS case due to thinner epitaxy, and a dual low- and high-energy phosphorous implant is also used to achieve a more uniform n-well doping profile for the bipolar transistor. In subsequent processing, additional n-well doping can be implanted underneath the emitter-base region to reduce the Kirk effect and improve f_T, as discussed in the previous section.

After well formation, active device regions are isolated by using a scaled LOCOS process, which is common to both CMOS and BiCMOS processing. The bipolar base, resistor, and deep n^+ sinker are then patterned and implanted prior to a MOSFET-threshold-adjust implant. For high-current operation, it is important to minimize R_c by maintaining a uniformly low doping level throughout the deep n^+ sinker region. This is achieved by using low- and high-energy implants to appropriately distribute the n^+ dopant.

During gate oxidation, it is advisable to prevent oxidation of the base region to minimize oxidation-enhanced diffusion of the base along with segregation of the boron into the oxide at the silicon interface. These factors can lead to a degradation of the base profile, which results in lower performance, particularly for high f_T bipolar transistors with extremely shallow base implants. To avoid this problem, the base may be implanted after the gate oxidation or protected from its deleterious effects. In the BiCMOS device cross section shown in Fig. 7, a thick oxide limits oxidation over the base region and also reduces the polysilicon emitter-to-base capacitance, as discussed previously. Low-temperature gate oxidation and rapid thermal anneal may also be used to limit base diffusion and boron segregation.

The final step required for bipolar fabrication is emitter formation. As noted in the preceding section, there are several types of polysilicon emitters that support high-performance operation. One method that maximizes compatibility with CMOS processing is to use a split polysilicon deposition, with the first polysilicon layer passivating the gate oxide during emitter patterning and the second polysilicon deposition forming contact with the emitter interface. In all polysilicon emitter processes, predictable high-performance bipolar characteristics critically depend on control of the polysilicon-to-silicon interface. A clean, oxide-free silicon surface must be present to insure epitaxial regrowth at the polysilicon-silicon interface during polysilicon deposition and subsequent thermal processing.

After emitter polysilicon deposition, the emitter dopant is implanted and the polysilicon bipolar emitter and CMOS gate are patterned and etched. High-performance bipolar devices typically use an implanted arsenic dopant, whereas pure CMOS processes use a diffused phosphorus doping for the gate. From this point onward, the CMOS and BiCMOS process flows listed in Table 1 are identical, with the exception of source/drain anneal, which is tailored for optimum bipolar performance. Following source/drain anneal, a $TiSi_2$ salicide is used to clad polysilicon and silicon diffusions for low resistance. Low-temperature (<750°C) backend processing is required to minimize boron segregation into the $TiSi_2$; otherwise a standard multilevel interconnect can be used to complete processing.

In this example BiCMOS process, five extra mask levels are added to merge high-performance bipolar transistors into a baseline CMOS flow: 1) buried n^+, 2) p-base, 3) p-resistor, 4) deep n^+ sinker, and 5) emitter. Identical CMOS and BiCMOS process recipes are also used wherever possible to optimize process compatibility. As noted in Table 1, the CMOS and BiCMOS process flows are matched (with the exception of source/drain anneal time) after polysilicon gate formation, providing an efficient and cost-effective way of implementing the mixed BiCMOS technology.

6. Low-Voltage Applications and New Gate Concepts

At 0.5-μm design rules, the trend is to reduce the power supply to 3.3 V. A reduction in power-supply voltage allows lower breakdown voltages to be used in the bipolar transistor. As a result, thinner epitaxial layers and higher n-well doping levels can be applied, offering improved bipolar performance by lowering the collector resistance and also allowing the transistor to operate at a higher current density. The higher n-well doping required for bipolar performance improvement is also necessary to support the PMOS scaling requirements of tighter isolation spacing and shorter gate lengths. However, the higher tank doping levels do result in an increase in MOS body effective relative to 0.8-μm-generation CMOS devices.

As the power-supply voltage is reduced, it is necessary to reduce CMOS transistor threshold voltages in order to maintain adequate drive currents and also provide a design margin for stacked gates in NAND functions. Scaling of the threshold voltage is complicated for the case of the PMOS transistor. For most 5-V BiCMOS processes at the 0.8-μm node or larger, a buried-channel PMOS transistor with an n^+ poly gate is used for manufacturing simplicity. At the 3.3-V 0.5-μm node, the need for threshold voltages in the 0.5–0.6-V range will require a surface-channel PMOS transistor design in order to control the short-channel leakage characteristics. Source/drain implants are typically used to dope the n^+ and p^+ polysilicon gates, but an additional n^+ implant is required to provide a higher doping level for the bipolar polysilicon emitter.

In order to achieve the high speeds required for the telecommunications market, the bipolar transistor in the BiCMOS process must be optimized for state-of-the-art ECL performance. This requires not only controlling a narrow base width, but also minimizing the transistor parasitics. Double-polysilicon self-aligned transistor structures are typically used in order to minimize the base resistance and collector-base capacitance. As a result, state-of-the-art ECL performance can be realized in a BiCMOS process. Polysilicon resistors must also be added to the process instead of the simpler diffused resistor to provide lower parasitic capacitance.

In addition to process and transistor structure optimization, basic changes in gate design have also been proposed to enhance low-voltage BiCMOS performance. Some of the options being considered include BiNMOS, merged BiCMOS, and complementary BiCMOS [4–7]. The BiNMOS and merged BiCMOS gates can be built with the existing BiCMOS processes. However, the addition of a PNP transistor is required for the C-BiCMOS gate where the PNP is typically formed in a p-well with the substrate as the collector. Extra masking steps are required to form the PNP base and emitter doping. At the present time, it is not clear if the addition of a PNP to the BiCMOS process translates to a cost-effective improvement in circuit performance.

With proper optimization of the process and BiCMOS gate design, BiCMOS has been able to maintain a performance advantage over CMOS at reduced power-supply voltages. An example is presented in Figs. 8 and 9 where measured NAND gate delays are plotted for the case of a 0.6-μm BiCMOS technology optimized for operation below 5V. The BiCMOS advantage over CMOS is still evident, particularly for the case of heavily loaded gates. The BiCMOS process integration challenge at 0.5-μm design rules and beyond is to minimize process complexity and cost while achieving state-of-the-art bipolar and CMOS transistor performance.

7. Conclusions

While BiCMOS is maturing for high-performance circuit applications, it continues to evolve as a technology. A CMOS core flow remains the foundation for production processes, but the convergence of bipolar and CMOS device structures has encouraged the use of innovative high-performance bipolar transistors in BiCMOS development. The successful merger of these structures into future BiCMOS processes will depend largely on the degree of bipolar and CMOS process commonality that can be obtained. The performance versus cost trade-off will continue to drive the practical application of novel device structures and processes for new generations of BiCMOS products.

Acknowledgments

In addition to the many people at Texas Instruments who have contributed to the example 0.6–0.8-μm BiCMOS technology discussed in this paper, the authors wish to especially thank Ken Fung for the simulation shown in Fig. 1 and Jim Gallia for the simulations shown in Figs. 8 and 9.

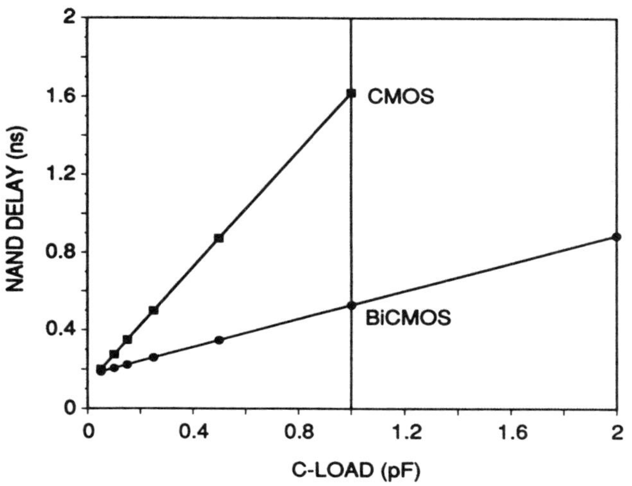

Fig. 8 A comparison of CMOS and BiCMOS NAND gate delays as a function of capacitive loading. The simulation assumes a CMOS gate length of 0.6 μm, an input capacitance of 0.05 pF, and a supply voltage of 3.3 V.

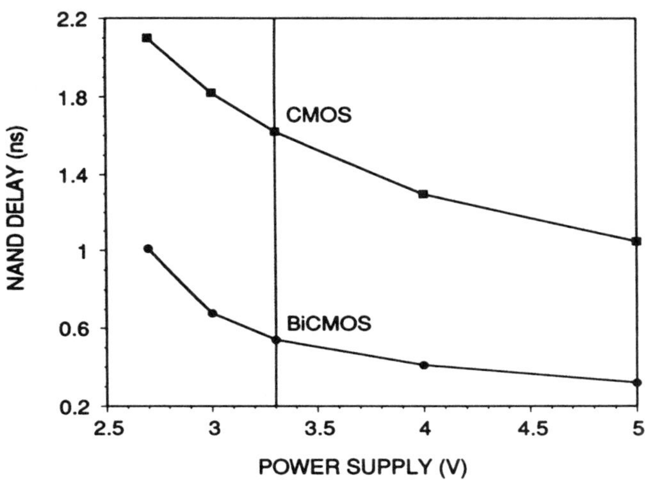

Fig. 9 A comparison of CMOS and BiCMOS NAND gate delays as a function of power-supply voltage. The simulation assumes a CMOS gate length of 0.6 μm, an input capacitance of 0.05 pF, and a load capacitance of 1 pF.

References

[1] J. Miyamoto et al., "A 1μm N-well/bipolar technology for VLSI circuits," in *Dig. Technical Papers, 1983 International Electron Devices Meeting*, pp. 63–66, Dec. 1983.

[2] H. Higuchi et al., "Performance and structures of scaled-down bipolar devices merged with CMOSFETs," in *Dig. Technical Papers, 1984 International Electron Devices Meeting*, pp. 694–697, Dec. 1984.

[3] A. Alvarez et al., "2 micron merged bipolar-CMOS technology," in *Dig. Technical Papers, International Electron Devices Meeting*, pp. 761–764, Dec. 1984.

[4] A. Watanabe et al., "Future BiCMOS technology for scaled supply voltage," in *Dig. Technical Papers, 1989 International Electron Devices Meeting*, p. 429, Dec. 1989.

[5] N. Rovedo et al., "Process design for merged complementary BiCMOS," in *Dig. Technical Papers, 1990 International Electron Devices Meeting*, pp. 485–488, Dec. 1990.

[6] P. Raje, et al., "MBiCMOS: A device and circuit technique scalable to the sub-micron, sub-2V regime," in *ISSCC Dig. Technical Papers*, p. 150, Feb. 1991.

[7] A. Shida et al., "3.3 V BiNMOS technology using NPN transistors without buried layers," in *Dig. Technical Papers, 1991 International Electron Devices Meeting*, pp. 93–96, Dec. 1991.

[8] E. Greenwich and K. McLaughlin, "Analysis and characterization of BiCMOS for high-speed digital logic," *IEEE J. Solid-State Circuits*, vol. 23, pp. 558–565, April 1988.

[9] G. Patton et al., "Physics, technology and modeling of polysilicon emitter contacts for VLSI devices," *IEEE Trans. Electron Devices*, p. 1754, Nov. 1986.

[10] M. Brassington et al., "An advanced submicron BiCMOS technology for VLSI applications," *Digest of Technical Papers, 1988 Symposium on VLSI Technology*, pp. 89–90, May 1988.

[11] R. Havemann et al., "An 0.8 μm256K BiCMOS SRAM technology," in *Dig. Technical Papers, 1987 International Electron Devices Meeting*, pp. 841–843, Dec. 1987.

[12] T. Yuzuriha et al., "Submicron bipolar-CMOS technology using 16GHz f_T double-polysilicon bipolar devices," in *Dig. Technical Papers, 1988 International Electron Devices Meeting*, pp. 748–751, Dec. 1988.

[13] H. Park et al., "High-speed polysilicon emitter-base bipolar transistor," *IEEE Electron Device Lett.* vol. EDL-7, pp. 658–660, Dec. 1986.

[14] T. Chiu et al., "The design and characterization of nonoverlapping super self-aligned BiCMOS technology," *IEEE Trans. Electron Devices*, vol. ED-38, pp. 141–149, Jan. 1991.

[15] J. Brighton et al., "Scaling issues in the evolutions of ExCL bipolar technology," in *Dig Technical Papers, 1988 Bipolar Circuits & Technology Meeting*, pp. 121–124, Sept. 1988.

[16] J. Kirchgessner et al., "An advanced 0.4μm BICMOS technology for high performance ASIC applications," in *Dig. Technical Papers, 1991 International Electron Devices Meeting*, pp. 97–100, Dec. 1991.

[17] K. Cham et al., "Characterization and modeling of the trench surface inversion problem for the trench isolated CMOS technology," in *Dig. Technical Papers, 1984 International Electron Devices Meeting*, pp. 23–26, Dec. 1984.

[18] N. Kasai et al., "1/4μm CMOS isolation technique using selective epitaxy," *IEEE Trans. Electron Devices*, vol. ED-34, pp. 1331–1336, June 1987.

[19] P. van Wijnen et al., "A localized collector implant to improve uniformity of polysilicon bipolar transistors," in *Dig. Technical Papers, 1989 Bipolar Circuits & Technology Meeting*, pp. 160–163, Sept. 1989.

[20] J. Stork and J. Cressier, "The impact of non-ohmic polysilicon emitter resistance on bipolar transistor performance," in *Dig. Technical Papers, 1986 Symposium on VLSI Technology*, p. 47, 1986.

[21] S. Sun et al., "A 0.4-micron fully complementary BiCMOS technology for advanced logic and microprocessor applications," in *Dig. Technical Papers, 1991 International Electron Devices Meeting*, pp. 85–88, Dec. 1991.

[22] P. Asbeck, "Heterojunction bipolar transistors: Status and directions," in *Dig. Technical Papers, 1989 Bipolar Circuits & Technology Meeting*, p. 65, Sept. 1989.

[23] S. Wright et al., "Molecular beam epitaxy growth of GaP on Si," *J. Appl. Phys.*, vol. 55, p. 2916. April 1984.

[24] Y. Kwark and R. Swanson, "N-type SIPOS and poly-silicon emitters," *Solid State Electronics*, vol. 30, p. 1121, Nov. 1987.

[25] M. Takahashi et al., "I-V characteristics of oxygen-doped Si epitaxial film (OXSEF)/Si heterojunctions," *IEEE Trans. Electron Devices*, vol. EDL-8, no. 10, pp. 475–4476, Oct. 1987.

[26] T. Sugii et al., "Beta-SiC/Si heterojunction biopolar transistors with high current gain," *IEEE Trans. Electron Devices*, vol. 9, no. 2, pp. 87–89, Feb. 1988.

[27] S. Iyer et al., "Heterojunction bipolar transistors using Si-Ge alloys," *IEEE Trans. Electron Devices*, vol. 36, no. 10, pp. 2043–2064, Oct. 1989.

[28] T. Won and H. Morkoc "High-speed performance of Si/Si(1-x)Ge(x) heterojunction bipolar transistors," *IEEE Electron Dev. Lett.*, vol. 10, no. 1, pp. 33–35, Jan. 1989.

[29] K. Imai et al., "A cryo-BiCMOS technology with Si/SiGe heterojunction bipolar transistors," in *Dig. Technical Papers, 1990 Bipolar Circuits and Technology Meeting*, pp. 90–93, Sept. 1990.

[30] R. Eklund et al., "A 0.5 μm BiCMOS technology for logic and 4Mbit-class SRAM's," in *Dig. Technical Papers, 1989 International Electron Devices Meeting*, pp. 425–428, Dec. 1989.
[31] R. Haken et al., "BiCMOS process technology," in *BiCMOS Technology and Applications*, Ed. A. Alvarez, Norwell, MA: Klewer Academic, 1989.
[32] S. Wolf, "Bipolar and BiCMOS process integration," in *Silicon Processing for the VLSI Era, Volume 2—Process Integration*, pp. 538–545. Sunset Beach, CA: Lattice Press, 1990.

The Design and Characterization of Nonoverlapping Super Self-Aligned BiCMOS Technology

Tzu-Yin Chiu, *Member, IEEE*, Gen M. Chin, *Member, IEEE*, Maureen Y. Lau, Ronald C. Hanson, Mark D. Morris, Kwing F. Lee, *Member, IEEE*, Mark T. Y. Liu, *Member, IEEE*, Alexander M. Voschenkov, *Senior Member, IEEE*, Robert G. Swartz, *Senior Member, IEEE*, Vance D. Archer, III, *Member, IEEE*, Sean N. Finegan, and Mark D. Feuer, *Member, IEEE*

***Abstract*—An optimal device structure for integrating bipolar and CMOS is described. Process design and device performance will be discussed. Both the vertical n-p-n and MOS devices have Non-OVer-lapping super self-Aligned (NOVA) structures. The base–collector and source/drain junction capacitances are significantly reduced. This structure allows complete silicidation of active polysilicon electrodes, cutting down the parasitic resistances of source, drain, and extrinsic base. The critical gate and emitter regions are protected from direct RIE exposure and damage. All shallow junctions are contacted by polysilicon electrodes which suppress silicide-induced leakage. An arsenic buried layer minimizes collector resistance and collector–substrate capacitance. A novel selective epitaxy capping technique suppresses lateral autodoping from the arsenic buried layer. Fully recessed oxide with polysilicon buffer layer is used to achieve a low defect density device isolation. CMOS with L_{eff} = 1.1 μm and W_n/W_p = 10 μm/10 μm exhibits averaged ring oscillator delay of 128 ps/stage. An n-p-n transistor with f_T, of 14 GHz and low power ECL ring oscillator with a delay of 87 ps/stage have been fabricated.**

I. Introduction

There is steadily growing interest in the area of new self-aligned device structures [1]–[11]. Double-polysilicon self-aligned devices have greatly increased bipolar circuit speed and reduced the power dissipation [1]–[2]. In the MOS area, various schemes to achieved a self-aligned structure have been demonstrated [4]–[6]. Processes that merge conventional CMOS with double poly bipolar structure have also been reported [7], [8]. Techniques to simultaneously achieve a self-aligned structure for bipolar and CMOS have also been explored [9]–[12]. The objective of this work is to develop a BiCMOS technology with an optimal device structure that is economical and manufacturable [14].

This paper will first introduce several unique features of the device design. The technologies used to achieve these features are then briefly described, followed by the electrical characterization of MOS and bipolar devices. The potential of this BiCMOS technology is demonstrated by ring oscillators and high-performance communication circuits.

Manuscript received March 12, 1990; revised July 3, 1990. The review of this paper was arranged by Associate Editor T. H. Ning.

T. Y. Chiu, G. M. Chin, M. D. Morris, K. F. Lee, M. T. Y. Liu, A. M. Voshchenkov, R. G. Swartz, V. D. Archer, S. N. Finegan, and M. D. Feuer are with AT&T Bell Laboratories, Holmdel, NJ 07733-1988.

M. Y. Lau was with AT&T Bell Laboratories, Holmdel, NJ 07733-1988. She is now with AT&T Bell Laboratories, Murray Hill, NJ 07974.

R. C. Hanson was with AT&T Bell Laboratories, Holmdel, NJ 07733-1988. He is now with Zilog Inc., Nampa, ID.

IEEE Log Number 9040192.

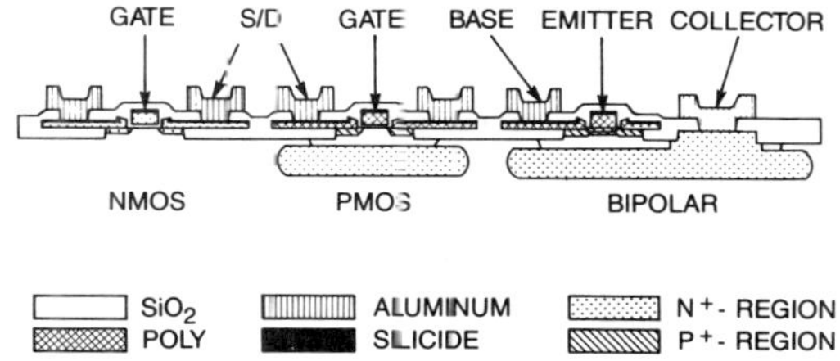

Fig. 1. Cross-sectional view of NOVA BiCMOS. Note that polysilicon electrodes are used to contact the base and source/drain. They are self-aligned to the emitter and gate poly, respectively. The two sets of poly electrodes are isolated by 200-nm-thick oxide spacer.

II. Device Concept and Design

Fig. 1 shows the cross section of NOVA BiCMOS devices. The active area of the bipolar transistor and MOS devices are virtually identical. The design rule is purposely conservative with the minimum feature and misalignment tolerance at 1.5 and 0.75 μm, respectively. The layout of the bipolar and MOS devices are shown in Fig. 2. The standard minimum bipolar device has base poly strapped around the emitter for two-sided contact. Based on the same design rule, the collector–base area of the NOVA device is the same as a double poly transistor. For bipolar devices requiring large emitter area, multi-emitter fingers are employed. Because the base electrode is silicided, emitter finger pitch can be as small as 3 μm. The MOS source/drain junctions have a nominal dimension of 1 μm, a significant compaction compared to conventional geometry that results in a dramatic reduction of junction parasitic capacitance. Edge capacitance is lowered by abutting all junctions against fully recessed oxide. The emitter and base contacts as well as the gate and source/drain electrodes do not overlap. This nonoverlapping structure allows complete salicidation of polysilicon electrodes, eliminating the major portion of the extrinsic base and source/drain resistances. Unlike the conventional double poly structure, the emitter and gate features are defined by lithography before base and S/D formation. Because of this fabrication sequence, shallow junctions, LDD and link-up implants, secure gate to source–drain overlaps can be incorporated without difficulties. Device scaling is therefore straightforward even though the benefits of emitter width and gate length reduction associated with a double poly structure are absent. To maintain a shallow emitter and a narrow base width, heat treatment is limited to below 900°C after the low-energy base implant until the final junction drive-in step.

Reprinted from *IEEE Trans. Electron Devices*, vol, 38, no. 1, pp. 141–150, Jan. 1991.

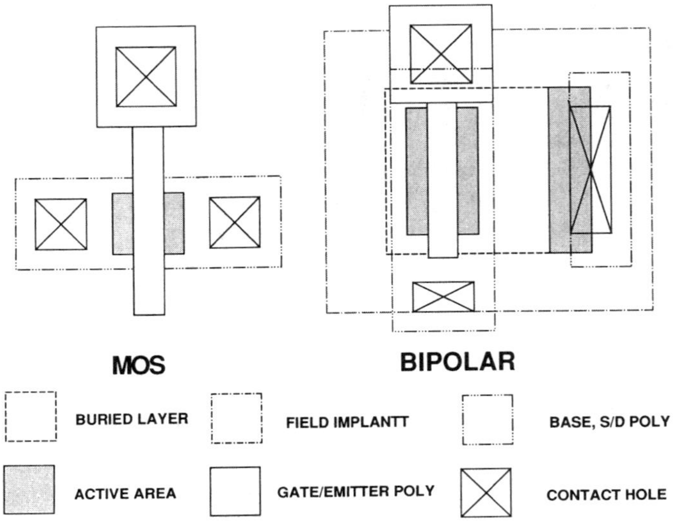

Fig. 2. Layout of the MOS and bipolar devices. All contact holes are placed on the field oxide to avoid aluminum spiking of shallow junctions. This figure also illustrates that the collector can be strapped with polysilicon layer.

III. Process Design

Fig. 3 shows the fabrication sequence. Either bipolar or CMOS devices alone requires 10 masks. Bipolar-CMOS merge takes 14 masks to first level metal. Additions to the CMOS process are indicated by the boxed item. Vacuum contact printing is used throughout. Besides vertical n-p-n and CMOS devices, a local interconnect level, lateral p-n-p, polysilicon resistor, and surface substrate contact are also included.

The starting p-type wafers have (100) orientation and 10-$\Omega \cdot$ cm resistivity. Buried layer is first defined. Its sheet resistance and capacitance must be low. Therefore, the subcollector should be as shallow and as high in dopant concentration as possible. Abutment against highly doped p regions must be avoided. Arsenic is chosen because its solid solubility is 30 times that of antimony. Lateral autodoping which plagued arsenic buried layer technology is suppressed by selective epitaxy growth capping shown in Fig. 4 [15]. The spreading resistance measurements over buried layer and off buried layer region is shown in Fig. 5. The blanket deposited layer is a 1-μm-thick p-epi with carrier concentration of 1×10^{15} cm^{-3}, matching the substrate concentration. In the off buried layer region, the doping control is uniform. The material over the buried layer remains n-type due to vertical autodoping. It has been determined that the n-type material remains confined largely within the boundary of the subcollector. Buried layer to buried layer isolation has been observed down to a separation of 3 μm.

After the epitaxial growth, n-well is implanted and driven in. Field oxidation for device isolation is carried out. Device isolation is closely intertwined with the epi process. With arsenic contamination suppressed, the buried layer is now isolated by a lightly doped p-type substrate and epi. A poly buffered SWAMI process has been developed for device isolation [12], [16], [17]. Fig. 6 shows the idealized drawing and the SEM cross section of the final structure. Field oxidation produces a 650-nm-thick oxide. Unlike conventional LOCOS, a boron field implant is introduced after the oxidation [18]. As a result, boron segregation is eliminated and the dose can be reduced by about one order of magnitude compared to the conventional method. Since the boron lateral diffusion is eliminated, field implant encroachment and its impact on narrow width effect as well as the edge capacitance of the NMOS S/D junctions are minimized. The boron implant penetrates the NMOS LOCOS mask into the active area and functions as the punchthrough stopper. It is displaced away from the n-well and buried layer to minimize collector capacitance. The inserted poly layer serves to both relieve nitride stress and position the punchthrough implant.

Boron threshold implants for CMOS and bipolar base are the next steps. The doping profile in the channel and field region for NMOS has been simulated by Supreme III (Fig. 7). The channel profile is the sum of two implants. The threshold implant is shared between NMOS and PMOS and must be very shallow. The second component is the field implant. Because of the limited thermal cycle, the field implant profile under the

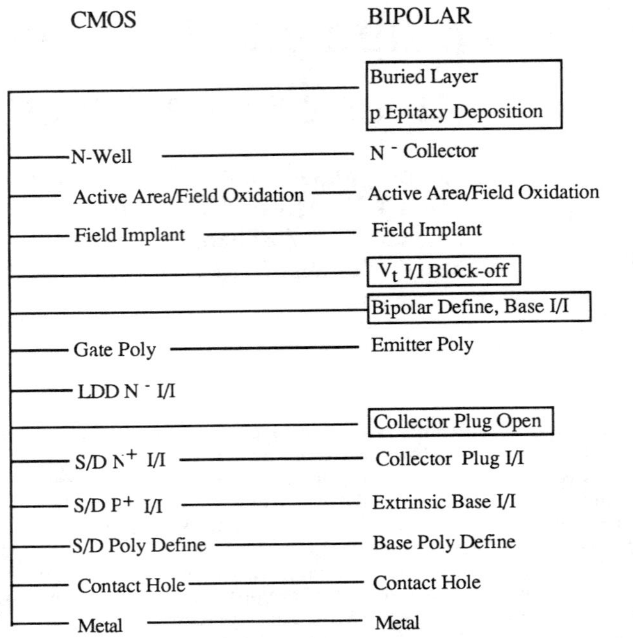

Fig. 3. Fabrication sequence of NOVA BiCMOS. Either CMOS or bipolar device alone can be fabricated with 10 masking steps. Four additional masks need to be added to the CMOS process to attain the merge. Those four masks are indicated by the boxes. For pure bipolar process, V_t implant block-off, bipolar define, and extrinsic base implant masks can be eliminated.

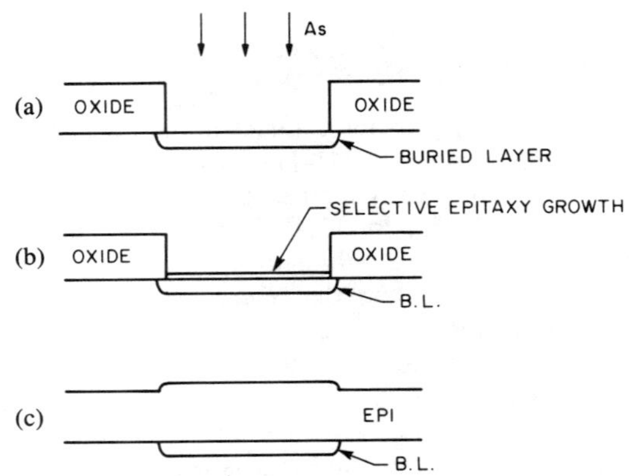

Fig. 4. SEG capping for lateral autodoping suppression. (a) Buried layer open and implant. (b) SEG layer is deposited over the buried layer for capping. The SEG layer is normally 300 nm thick. (c) Oxide is stripped and a blanket p-epi of desired doping is grown.

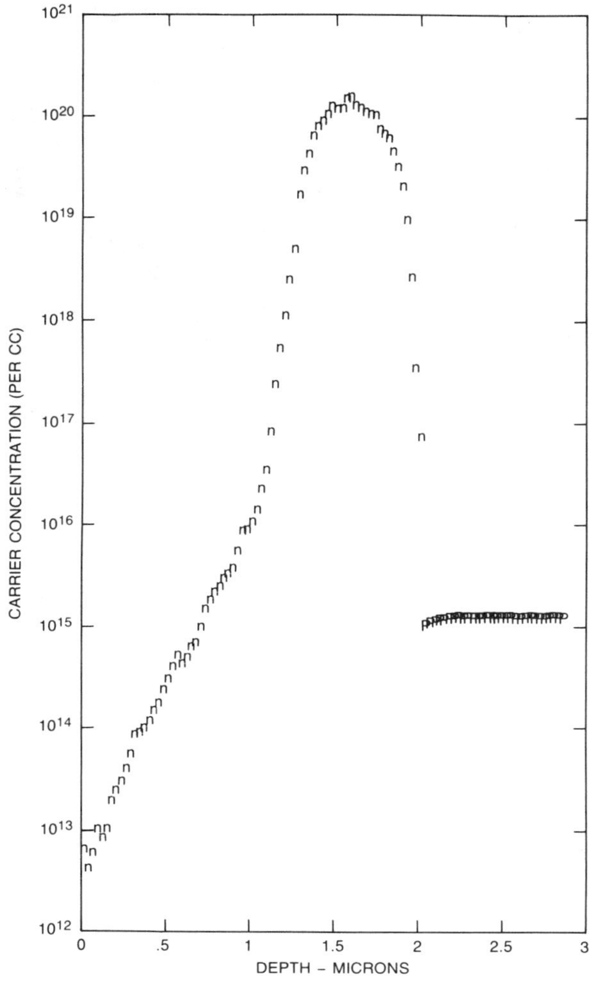

(a)

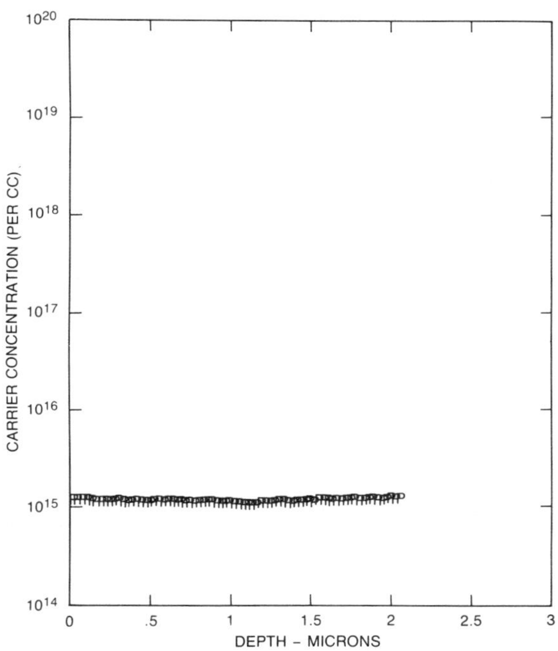

(b)

Fig. 5. Spreading resistance profiling of the epilayers. The letters in the profile indicate the conductivity type. Profiles (a) and (b) are measured on and off the buried layer region, respectively.

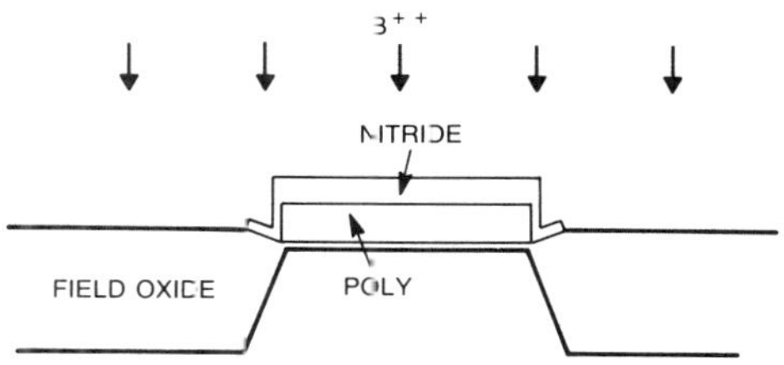

(a)

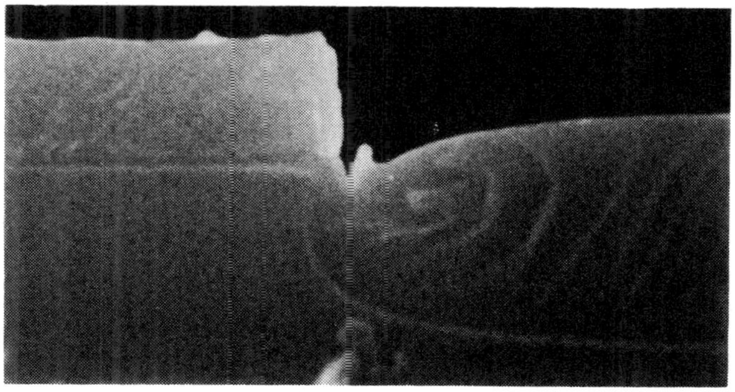

(b)

Fig. 6. (a) Idealized cross section of the field oxidation structure. The LOCOS mask has a nitride/poly and pad oxide stack with a thin nitride frame around the sidewall. Doubly ionized boron is used as the field implant dopant. Boron penetrates both the field oxide as well as the oxidation mask into the NMOS active area. The location in the active area is determined by the polysilicon thickness. (b) the SEM micrograph. Note that a small notch appears at the field oxide boundary.

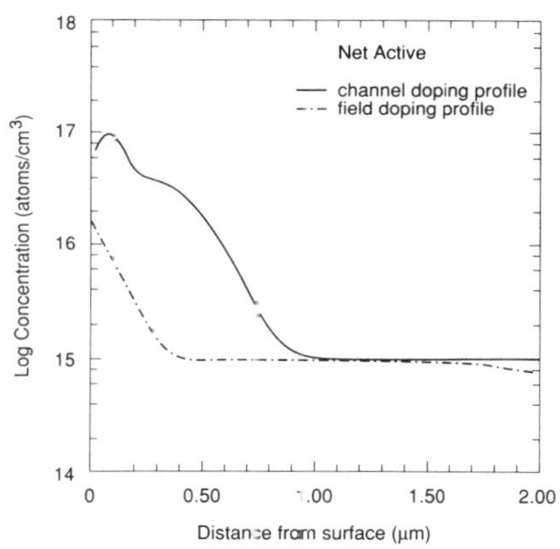

Fig. 7. The doping profile of the NMOS channel region and field isolation region as simulated by Supreme III. The channel doping profile is the composite of field implant and threshold implant.

isolation oxide is shallow and the surface concentration is maintained above 10^{16} cm^{-3}. The elimination of field implant lateral encroachment also has its drawback. Because of its absence, the NMOS subthreshold characteristic is very sensitive to the field oxide topography along the active area boundary. A small notch in the oxide edge gives rise to poor subthreshold slope (Table I). Further improvement of the oxide planarity is desirable.

After the threshold and base implants, fabrication of NOVA structures follows (Fig. 8). A sequential deposition of thin nitride and thick undoped polysilicon on top of arsenic implanted poly creats a 1-μm-thick stack. Next comes the emitter and gate stacks definition and etching. Link-up implant and LDD im-

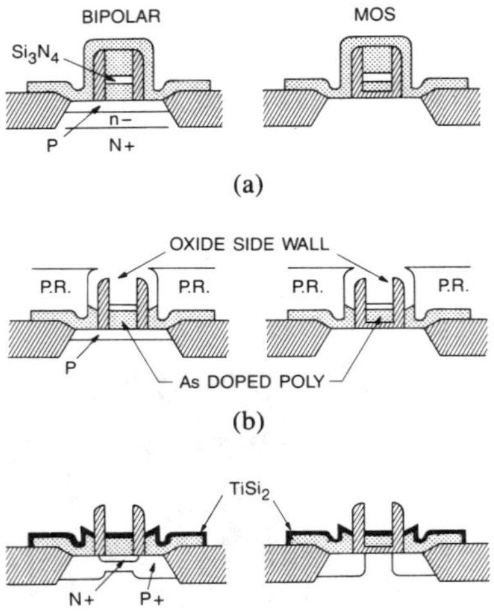

Fig. 8. (a) SDB poly is deposited over the emitter and gate stack. (b) Planarization resist is etched back, exposing the top of the emitter/gate stack. This figure depicts the structure after the isotropic polysilicon etch. (c) The devices after salicide process.

TABLE I

	NMOS	PMOS	Bipolar	
t_{ox}	230 Å	230 Å	A_e (μm^2)	1.5 × 3
X_{jsd}	300 nm	300 nm	X_{je}	150 nm
L_{eff}	1.1 μm	1.1 μm	W_b	300 nm
V_t	0.9 V	−0.88 V	X_{jbx}	300 nm
μ_{peak} ($cm^2/V \cdot s$)	460	200	h_{fe}	160
			F_t	14 GHz
g_m ($\mu A/V^2 \cdot \mu m$)	64	28	τ_f	11 ps
I_d ($V_d = V_{gs} = 5$ V);			r_e	17 Ω
mA/μm)	0.23	0.11	r_{bx}	14 Ω
k ($V^{0.5}$)	0.7	0.26	r_{bi}	416 Ω
C_{sd} ($fF/\mu m^2$)	0.526	0.26	C_{be}	13 fF
C_{gsd} ($fF/\mu m$)	0.12	0.12	C_{bc}	6.7 fF
S (mV/decade)	120	95	C_{cs}	30 fF

plant are then introduced into bipolar and MOS devices, respectively. A 200-nm-thick conformal TEOS oxide layer is deposited over the stack structure by means of LPCVD. After RIE etching, oxide spacer is formed along the stack sidewall. Source–drain and base (SDB poly) contact poly, 200 nm thick, is then deposited. SDB poly over the collector plug is removed next. Phosphorus is implanted into the collector and NMOS contact poly. Boron implant is shared by PMOS S/D poly and base contact poly. The SDB poly are then defined and etched (Fig. 8(a)). A thick planarization resist is then spun on the wafer. The resist is etched back exposing the top of the stack. Due to substrate topography, there is still about 1 μm of resist left on top of the field and SDB poly. A selective but isotropic polysilicon etch removes poly to the nitride etch stop. The base and S/D poly and emitter/gate electrodes are now isolated from each other (Fig. 8(b)). This is the first self-alignment. After the resist removal, a single furnace anneal cycle is used to drive in the dopants from the emitter, base, and S/D poly. The sheet resistances of poly layers on the field oxide depends strongly on the impurity species and dose. For our process, the poly layers doped with arsenic, phosphorus, and boron are respectively 250, 85, and 370 Ω/□. Following the selective etching of the nitride layer, $TiSi_2$ with a sheet resistance of 1–2 Ω/□

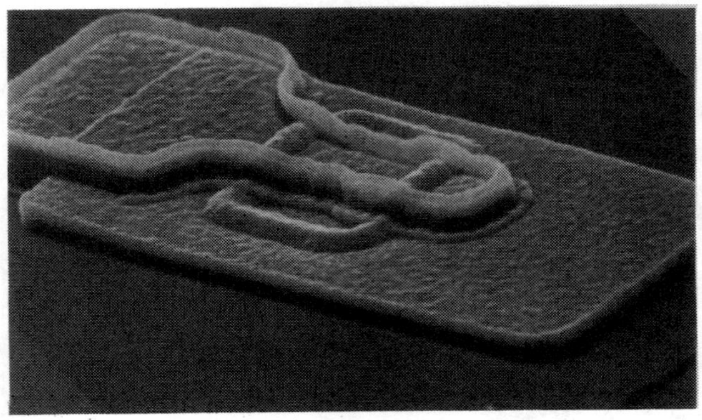

Fig. 9. SEM micrograph of a bipolar device after the planarization-etch process. Oxide spacer that isolates emitter and base electrodes is clearly visible. The active area is completelysealed by the poly.

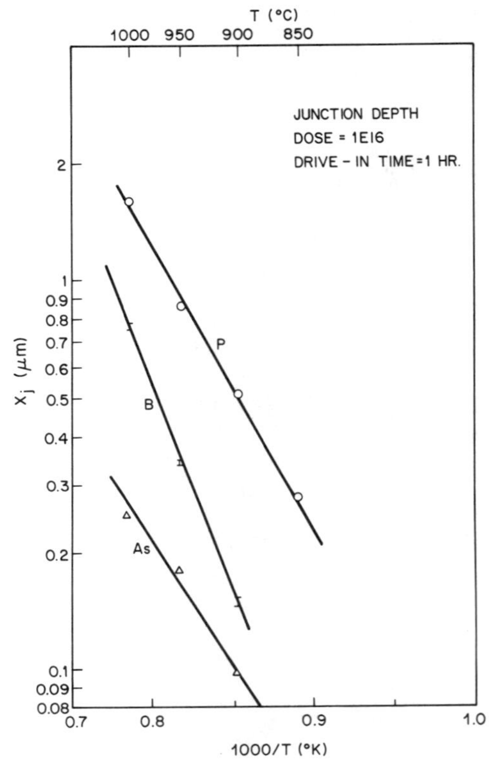

Fig. 10. Junction depth as a function of drive-in temperature. The dopants are driven-in from implanted polysilicon layer. BF_2 (25 keV) and P (50 keV) are implanted into 200-nm-thick poly and As (40 keV) is implanted into 250-nm poly. The dose is 1×10^{16} cm^{-2}. The drive-in time is 60 min in a dry O_2 ambient.

is formed over all polysilicon electrodes. The protruding sidewall ensures that no silicide bridging will occur (Fig. 8(c)). This is the second self-aligned process. A thick oxide dielectric is deposited followed by contact hole and metal definition. Fig. 9 is a SEM micrograph of a bipolar device just before the salicide process. The integrity of the oxide sidewall is clearly visible.

Extensive data have been collected for polysilicon diffusion sources. The junction depths as measured from the polysilicon/Si interface are determined by groove and stain method. The uncertainty of these measurements are ±20 nm. Each data point represents the average of three to four staining results. As the data in Fig. 10 indicate, the junction depths cover a large range

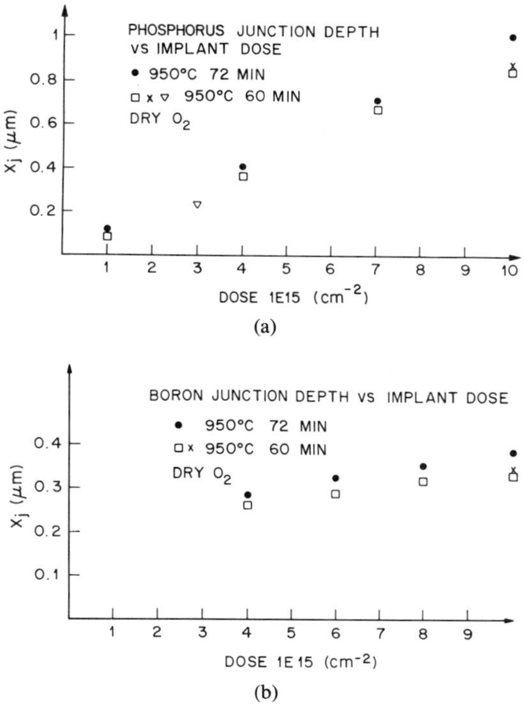

Fig. 11. The junction depth versus the implant dose for (a) phosphorus, (b) BF_2. The implant energies are the same as in Fig. 10.

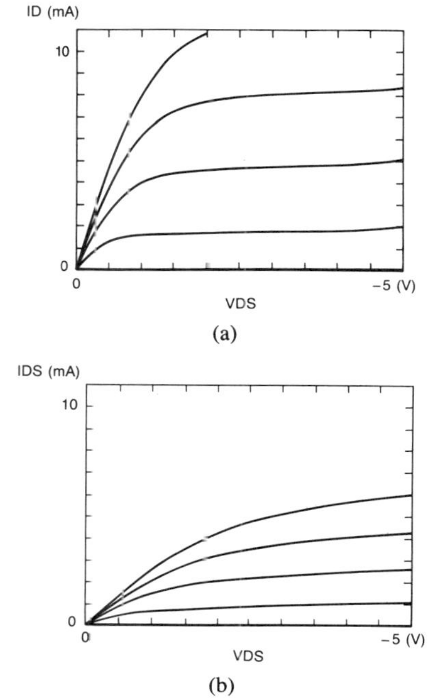

Fig. 13. The electrical characteristics of MOS device. The NMOS breaks down at about 7 V. The breakdown voltage of the PMOS is greater than 10 V. The devices are 50 μm wide with L_{eff} = 1.1 μm. The gate is swept from 0 V to $\pm$5 V at 1 V per step. No substrate bias is applied.

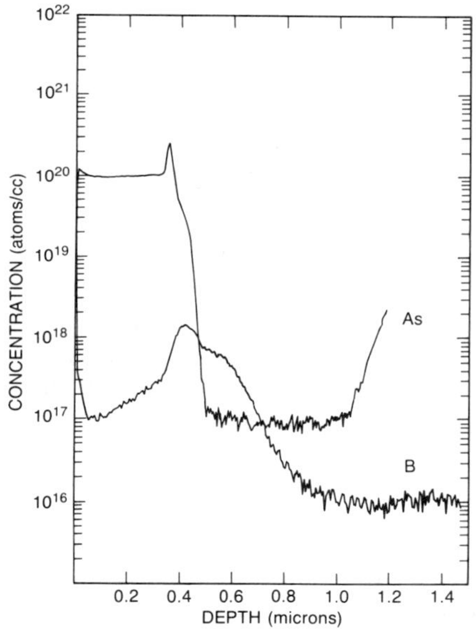

Fig. 12. SIMS measurement of the n-p-n vertical profile. The emitter is 150 nm deep and the base metallurgical junction width is about 300 nm wide.

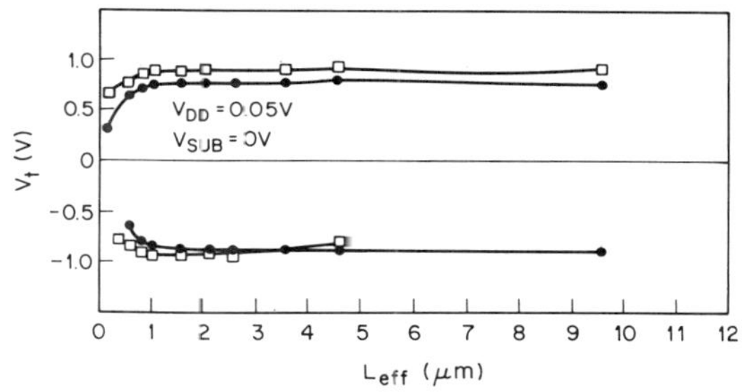

Fig. 14. The linear threshold voltage versus effective channel length. The threshold voltage is about $\pm$0.8–0.9 V and is fairly constant until L_{eff} approaches 0.6 μm.

as a function of drive-in temperature. We have chosen 950°C to set the various junctions. A second set of experiments, carried out at 950°C, determines the depth as a function of implant dose (Fig. 11). Boron and phosphorus are used to form the MOS source/drain and extrinsic base and arsenic doping is used to form the emitter. The source/drain and extrinsic base junctions are engineered to be 300 nm deep. The emitter junction depth is designed to be 150 nm deep. The collector and substrate contact regions are exposed to direct implant of the dopants. Due to high concentration enhanced diffusion, phosphorus atoms move substantially deeper and reach the buried layer. The bipolar active area has been profiled by SIMS as shown in Fig. 12. The emitter depth and base width are, respectively, 150 and 300 nm. These data have been incorporated into Supreme III for process modeling.

IV. Device Characteristics

The MOS transistors have gate oxide thickness of 23 nm. Fig. 13 shows the transistor I–V for NMOS and PMOS. The effective channel lengths are 1.1 μm for both devices. Fig. 14 shows the threshold variation as a function of gate length. A fairly constant threshold voltage is obtained at L_{eff} larger than 0.6 μm. Threshold voltage as a function of channel width is shown in Fig. 15. There is no increase in threshold voltage as the width is reduced. This observation is expected due to the elimination of field implant encroachment [17]. The field thresholds for p-substrate and N tub are greater than $\pm$10 V. V_t dependence on substrate bias is presented in Fig. 16. The PMOS

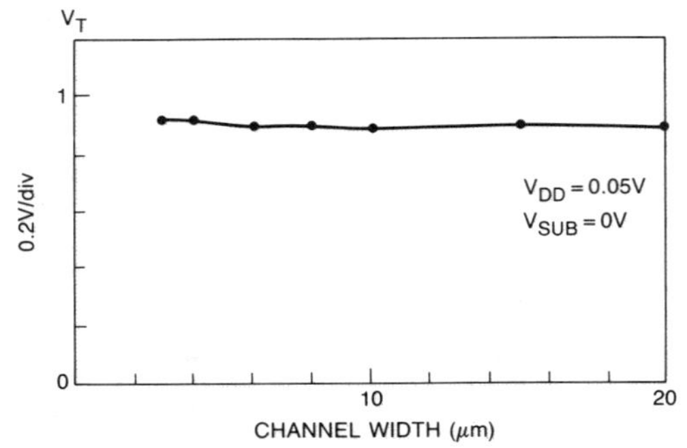

Fig. 15. The linear threshold voltage as a function of effective channel width, indicating that above W_{eff} = 1.5 μm, the threshold is about 0.9 V. This observation is consistent with the elimination of the field implant encroachment.

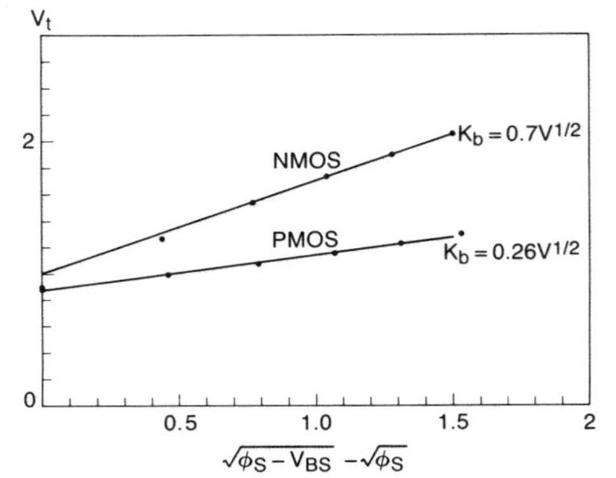

Fig. 16. The linear threshold voltage as a function of substrate bias. The substrate sensitivity is higher in NMOS than PMOS due to the difference in channel doping.

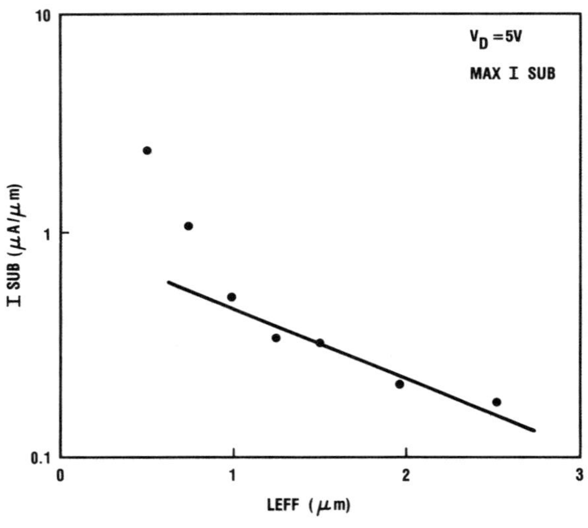

Fig. 17. The maximum substrate current as a function of L_{eff}. The drain voltage is fixed at 5 V while the gate voltage is swept from 0 to 5 V. The substrate is grounded. These devices do not have LDD implant.

substrate sensitivity is 0.26 $V^{0.5}$ and is quite acceptable. It is larger for NMOS because of the high channel concentration. Since the NMOS S/D are diffused phosphorus junctions and have fairly graded impurity profiles, we expect substrate current generation to be moderate even in the absence of LDD implant. The maximum substrate current is plotted as a function of gate

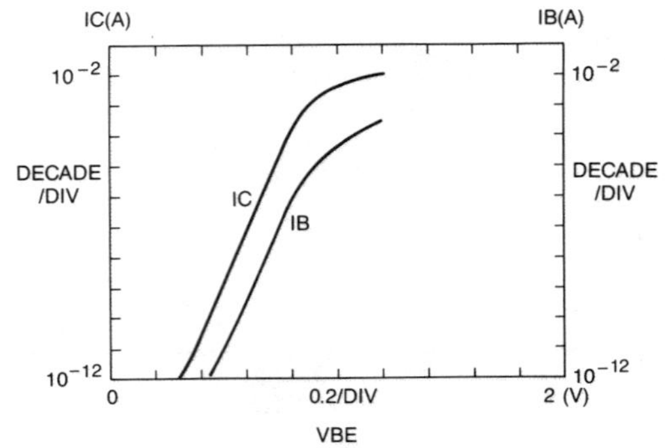

(a)

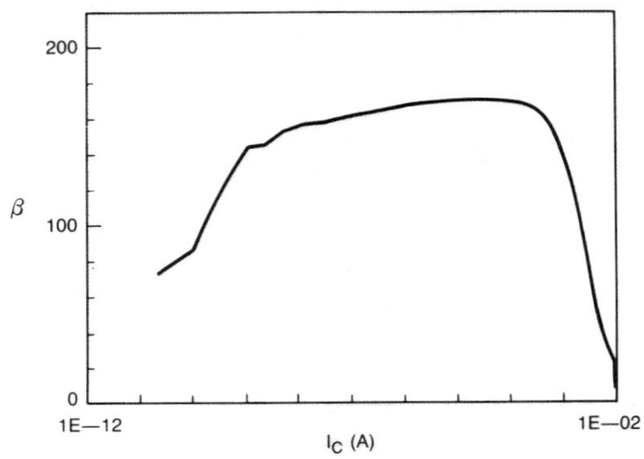

(b)

Fig. 18. (a) Gummel plot for n-p-n transistor. No reverse leakage is observed. (b) The differential current gain β versus I_c. The gain is about 160 and remains approximately constant for six decades. The emitter has a dimension of 1.5 × 9 μm^2.

length in Fig. 17. The amount of hot carriers generated is indeed comparable to LDD device structure [20].

The Gummel plot of an n-p-n transistor is shown in Fig. 18(a). Differential current gain β as function of I_c is plotted in Fig. 18(b). The β is about 160 and remains approximately constant for nearly six decades of I_c. This measurement indicates that the material quality of the epilayer with SEG cap is excellent. The base–emitter junction is ideal down to 1-pA current. We observe no residual RIE damages along the emitter–base perimeter. Link-up implant has been introduced in this device. Without the implant, excessive E/C leakage renders the device inoperable. Fig. 19 shows the dependence of f_t versus the collector current density. The collector current density at which f_t peaks indicates the onset of base push-out effect. As expected, the peak f_t current increases with increasing collector implant dose. The emitter and extrinsic base resistances have been characterized by several methods. Emitter resistance as obtained by Ning's measurement and the open collector method agree with each other only for very high current [21]. At the usual operating current, the open collector measurement produces a higher r_e. Extrinsic base resistance is usually between 10–20 Ω. This is negligible compared to the intrinsic base resistance. The device parameters are summarized in Table I.

V. Ring Oscillator and Circuit Performance

Fig. 20 shows the CMOS ring oscillator speed as a function of L_{eff}. Each data point is averaged over 20 to 30 ring oscillators. To study the impact of bipolar-CMOS integration, CMOS

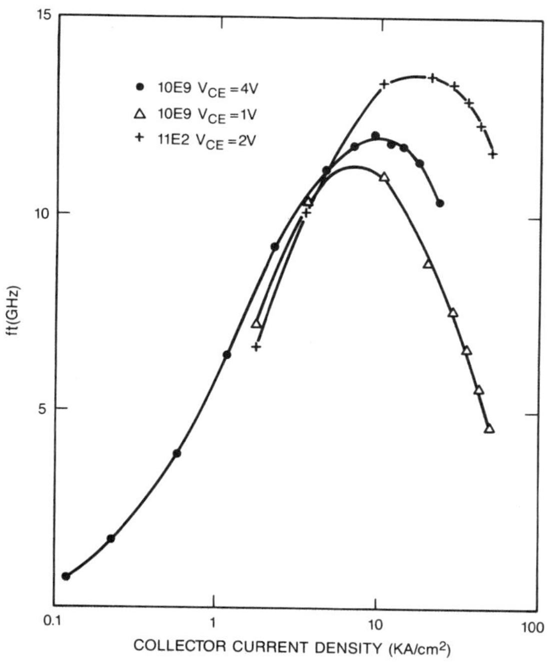

Fig. 19. The f_t versus the I_c. The n-well implant for sample 11E2 is 1.5 times that of sample 10E9. The peak f_t current density for 11E2 is indeed higher than 10E9.

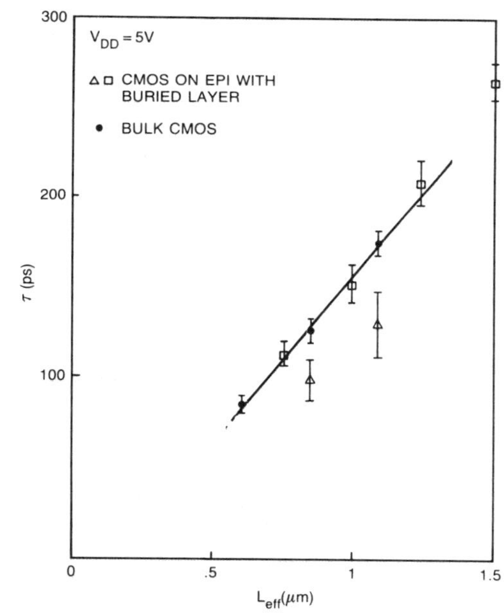

Fig. 20. Ring oscillator delay versus the effictive channel length. Twenty to thirty ring oscillators are probed for each channel length. The square and circular data have W_p/W_n = 10 μm/5 μm while the triangles have W_n/W_p = 10 μm/10 μm. "CMOS on epi with buried layer" are data from the full BiCMOS process, and "bulk CMOS" results are based on a pure CMOS process. The fan-in and fan-out are both one.

ring oscillators fabricated on buried layer with full BiCMOS process are compared to bulk Si, CMOS only, process. The circle and square data represent the results for ring oscillators with W_n/W_p = 5 μm/10 μm. The fact that both sets of ring oscillators fall on the same line strongly suggests that the integration results in no performance degradation in CMOS. An

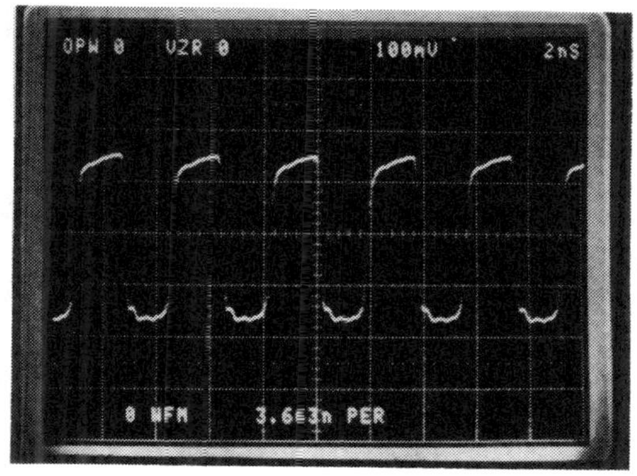

Fig. 21. The output of a 21-stage ECL ring oscillator. The period is 3.66 ns. Each stage has a single-ended output with a voltage swing of 0.5 V. The fan-in and fan-out are one. The sum of currents through the emitter coupled pair and the emitter follower are 0.3 and 0.1 mA, respectively. A −15-V substrate bias is applied.

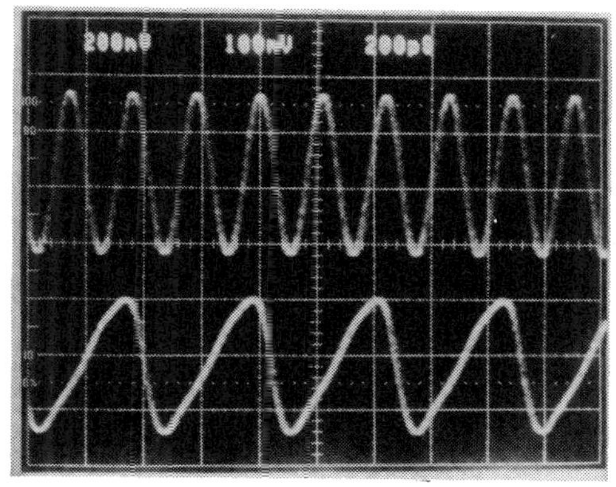

Fig. 22. Display of input and output waveforms of a divide by 2 circuit. The input frequency is 4.6 GHz. The substrate is connected to V_{ee} and is −5 V.

increase of device size to W_n/W_p = 10 μm/10 μm cuts down the delay time significantly. With an L_{eff} = 1.1 μm and 10-μm-wide devices, an average delay of 128 ps/stage is achieved. These data compare very favorably to the performance of conventional CMOS technologies.

Low-power ECL ring oscillators with a minimum delay of 93 ps/stage have been fabricated. The emitter dimension is 1.5 × 3 μm^2. The total gate current (including the follower) is 0.4 mA. By applying a reverse substrate bias of −15 V, the delay can be reduced to 87 ps/stage (Fig. 21). To demonstrate the high speed capability of the technology, static frequency dividers and two types of decision circuit have been designed with the same conservative 1.5-μm emitter width. Fig. 22 shows the divide-by-two circuit operating at 4.6 GHz with a total internal power dissipation of 30 mW. The test signal to the decision circuit is a non-return-to-zero (NRZ) $2^{15} - 1$ pseudo-random code. Conventional decision circuit functions up to 4 Gb/s with bit error rate (BER) of less than 10^{-9}. A biphase decision circuit demultiplexes the input into two parallel signal streams, passes each stream through its own decision circuit block, and then multiplexes the streams together again. This circuit functions to >5.1 Gb/s with BER < 10^{-9} (Fig. 23). This implies that the MUX and DEMUX circuitries are also fully functional at 5.1 Gb/s.

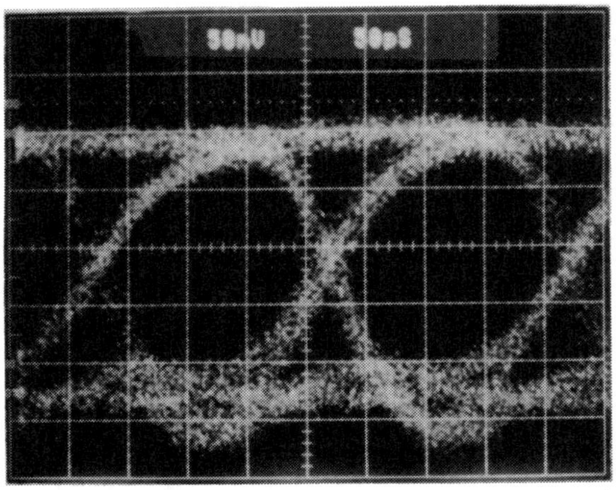

Fig. 23. The output eye diagram of a bi-phase decision circuit. The input is a pseudo-random NRZ signal at 5.1 Gb/s, which is the upper limit of our testing equipment. The bit error rate is less than 10^{-9}.

VI. Discussions and Conclusions

The NOVA device structure is highly optimized for bipolar-CMOS device integration. The bipolar and CMOS devices have compatible structures. The gate oxide and intrinsic base regions are protected from direct exposure to RIE during gate and emitter definition, and thus the gate oxide quality is equivalent to that of a conventional MOS technology. The nonoverlapping structure allows self-alignment without high gate to source-drain capacitance. Complete silicidation of all electrodes minimizes parasitic resistance. This reduction in extrinsic base resistance is especially important when the bipolar device is further scaled. Unlike the double poly structure, scaling is straightforward because the LDD and link-up implants are decoupled from the channel and the intrinsic base region, respectively. All active shallow junctions are formed and contacted by doped polysilicon electrodes. We have demonstrated that junction depth can be controlled by varying the implant dose. Under the same drive-in condition, impurities diffuse much deeper if the dopants are directly implanted into the substrate. We have taken advantage of these factors to obtain both shallow and deep junctions simultaneously. The polysilicon layers protect the shallow junctions from silicide-induced junction leakage and also function as local interconnect. The final topography is planar, a favorable characteristic for multi-level metal technology. The gate and emitter dimensions are defined and limited by the resolution of the lithography. Although their dimensional control will be excellent, the absence of minimum size reduction is a disadvantage as compared to double poly device structure.

In conclusion, NOVA BiCMOS with an advanced epi/isolation process has been successfully demonstrated. The device parameters are good, except for the high subthreshold slope of the NMOS device. Excellent circuit performance has been obtained. Based on a conservative design rule of 1.5 μm, ECL, and CMOS ring oscillator delays of 87 and 128 ps/stage have been obtained. The power dissipation for the ECL is 2 mW/gate with 5-V supply. A frequency divider operates up to 4.6 GHz at 30-mW power dissipation. Conventional decision circuit functions to 4 Gb/s, and multiplexer, demultiplexer, and bi-phase decision circuit are functional to 5.1 Gb/s. These results show that NOVA BiCMOS is suitable for gigabits per second digital VLSI application.

Acknowledgment

The authors wish to thank Dr. P. K. Tien for his steady support and invaluable inputs throughout this project. They also would like to thank K. Moerschel for providing the SEM micrographs.

References

[1] T. Sakai and M. Suzuki, "Super self-aligned bipolar technology," in *Proc. Symp. on VLSI Technology*, p. 16, 1983.

[2] D. D. Tang, P. M. Solomon, T. K. Ning, R. D. Isaac, and R. E. Burger, "1.25 μm deep-groove-isolated self-aligned bipolar circuits," *IEEE J. Solid State Circuit*, vol. SC-17, no. 3, p. 925, Oct. 1982.

[3] T. Nakamura, K. Nakazato, T. Miyazaki, M. Ogirima, T. Okabe, and M. Nagata, "290 psec I_2L circuits with five-fold self-alignment," in *IEDM Tech. Dig.*, p. 684, 1982.

[4] C. H. Dennison, A. T. Wu, P. K. Ko, C. I. Drowley, and D. Bradbury, "A novel MOS device structure with S/D contact over oxide," in *IEDM Tech. Dig.*, p. 204, 1985.

[5] H. Horie, T. Fukano, T. Ito, and H. Ishikawa, "Multiple self-alignment MOS technology (MUSA/MOST)," in *IEDM Tech. Dig.*, p. 638, 1984.

[6] C. K. Lau, H. Shibata, M. Saito, T. Matsuno, S. Samata, H. Ito, H. Sasaki, and K. Hashimoto, "A super self-aligned source/drain MOSFET," in *IEDM Tech. Dig.*, p. 358, 1987.

[7] K. Rajkanan, T. R. Gheewala, and J. Diedrick, "A high-performance BICMOS technology with double-polysilicon self-aligned bipolar devices," *IEEE Electron Device Lett.*, vol. EDL-8, no. 11, p. 509, 1987.

[8] T. Yamaguchi and T. H. Yuzuriha, "Process integration and device performance of a submicrometer BiCMOS with 16-GHz f_t double poly-bipolar devices," *IEEE Trans. Electron Devices*, vol. 36, no. 5, p. 890, 1989.

[9] F. Mieno, A. Shimizu, S. Nakamura, T. Deguchi, N. Haga, I. Matsumoto, Y. Furumura, Y. Yamauchi, K. Inayoshi, M. Maeda, and K. Yanagida, "Novel selective poly- and epitaxial-silicon growth (SPEG) technique for ULSI processing," in *IEDM Tech. Dig.*, p. 16, 1987.

[10] Y. Misawa, H. Homma, K. Sato, and N. Momma, "A self-aligning polysilicon electrode technology (SPEL) for future LSI," in *IEDM Tech. Dig.*, p. 32, 1987.

[11] T. Y. Chiu, G. M. Chin, M. Y. Lau, R. C. Hanson, M. D. Morris, K. F. Lee, A. M. Voshchenkov, R. G. Swartz, V. D. Archer, and S. N. Finegan, "A high speed super self-aligned bipolar-CMOS technology," in *IEDM Tech. Dig.*, p. 24, 1987.

[12] Y. Kobayashi, C. Yamaguchi, Y. Amemiya, and T. Sakai, "High performance LSI process technology: SST CBi-CMOS," in *IEDM Tech. Dig.*, p. 760, 1988.

[13] T. Y. Chiu, G. M. Chin, M. Y. Lau, R. C. Hanson, M. D. Morris, K. F. Lee, A. M. Voshchenkov, R. G. Swartz, V. D. Archer, M. T. Y. Liu, S. N. Finegan, M. D. Feuer, "Non-overlapping super self-aligned BiCMOS with 87 ps low power ECL," in *IEDM Tech. Dig.*, p. 752, 1988.

[14] K. G. Moerschel, T. Y. Chiu, W. A. Possanza, K. S. Lau, R. G. Swartz, R. A. Manz, T. Y. M. Liu, K. F. Lee, V. D. Archer, G. R. Hower, G. T. Mazsa, R. E. Carsia, J. A. Pavlo, M. P. Ling, J. L. Dolcin, F. M. Erceg, J. J. Eagan, C. J. Fassl, J. T. Glick, and M. A. Prozonic, "BEST: A BiCMOS-compatible super-self-aligned ECL technology," presented at CICC 1990.

[15] T. Y. Chiu, K. F. Lee, M. Y. Lau, S. N. Finegan, M. D. Morris, and A. M. Voshchenkov, "Lateral autodoping suppression by selective epitaxy capping and its application in high speed Bi-CMOS," in *Extended Abstracts 20th (1988 Int.) Conf. on Solid State Devices and Materials*, p. 45.

[16] K. Y. Chiu, J. L. Moll, K. M. Cham, J. Lin, C. Lage, S. Angelos, and R. L. Tillman, "The sloped-wall SWAMI-A defect free zero bird's beak local oxidation process for scaled VLSI technology," *IEEE Trans. Electron Devices*, vol. ED-30, p. 1506, 1983.

[17] Y. Han and B. Ma, "Isolation process using polysilicon buffer layer for scaled MOS/VLSI," *Extended Abstract (J. Electrochem. Soc.)*, p. 98, 1984.

[18] P. K. Ko, A. M. Voshchenkov, R. C. Hanson, P. Grabbe, D. M. Tennant, V. D. Archer, G. M. Chin, M. Lau, D. C. Soo,

and B. A. Wooley, "SiGMOS—A silicon gigabit/s NMOS technology," in *IEDM Tech. Dig.*, p. 751, 1983.
[19] K. M. Cham, S.-Y. Oh, D. Chin, J. L. Moll, K. Lee, and P. V. Voorde, in *Computer-Aided Design and VLSI Device Development*. Dordrecht, The Netherlands: Kluwer, 1988, ch. 11.
[20] Y. Matsumoto, T. Higuchi, T. Mizuno, S. Sawada, S. Shinozaki, and O. Ozawa, "An optimized and reliable LDD structure for 1-μm NMOSFET based on substrate current analysis," *IEEE J. Solid-State Circuits*, vol. SC-20, no. 1, p. 349, Feb. 1985.
[21] T. H. Ning and D. D. Tang, "Method for determining the emitter and base series resistance of bipolar transistors," *IEEE Trans. Electron Devices*, vol. ED-31, no. 4, p. 409, 1984.

A High-Performance 0.5-μm BiCMOS Technology for Fast 4-Mb SRAM's

James D. Hayden, *Senior Member, IEEE,* Thomas C. Mele, *Member, IEEE,* Asanga H. Perera, David Burnett, Fred W. Walczyk, Craig S. Lage, Frank K. Baker, Michael Woo, Wayne Paulson, Mitch Lien, Yee-Chaung See, *Senior Member, IEEE,* Dean Denning, and Stephen J. Cosentino, *Member, IEEE*

***Abstract*—A high-performance 0.5-μm BiCMOS technology has been developed for a fast 4-Mb SRAM class of circuits. Three layers of polysilicon are used to achieve a compact four transistor SRAM bit cell size of less than 20 μm^2 by creating self-aligned bit-sense and V_{ss} contacts. A WSi_x polycide emitter n-p-n transistor with an emitter area of 0.8 × 2.4 μm^2 provides a peak cutoff frequency (f_T) of 14 GHz with a collector–emitter breakdown voltage (BV_{CEO}) of 6.5 V. A selectively ion-implanted collector (SIC) is used to compensate the base channeling tail in order to increase f_T and knee current without significantly affecting collector–substrate capacitance. ECL gate delays as fast as 105 ps are obtained with this process.**

I. Introduction

A 0.5-μm BiCMOS technology designed to support a high-performance fast 4-Mb SRAM class of products must satisfy a number of requirements. The SRAM bitcell area must be less than 20 μm^2 to meet package size constraints. The active transistors, and particularly the gate oxide thickness, must be designed to support a 5-V power supply. The NMOS and PMOS transistors should exhibit acceptable MOSFET behavior for minimum effective channel lengths of 0.35 and 0.50 μm, respectively. To maintain the performance advantage of bipolar over CMOS, the n-p-n transistor should have a peak cutoff frequency f_T exceeding 10 GHz while the bipolar snapback voltage BV_{CEO} is maintained above 6.5 V. To satisfy all of these requirements, a 0.5-μm BiCMOS technology has been developed that combines a triple polysilicon process architecture with Framed Mask Poly-Buffered LOCOS (FMPBL) isolation [1], a disposable poly spacer module, a polycided emitter structure, and double-level metallization.

Addition of a bipolar transistor to a CMOS process provides the circuit designer with the capability to drive large capacitive loads with less cost in input capacitance and/or silicon area. Alternatively, because of the reduced sensitivity of gate delay to load capacitance, design can be simplified since the requirement for exact *a priori* estimates of circuit parasitics is eased. High-density BiCMOS SRAM's typically use a small number of bipolar transistors relative to the large number of CMOS devices comprising the memory array and periphery. Formation of the memory array and reduction of process complexity and defect density levels are of fundamental importance. The bipolar transistor must be added with minimal increase in process complexity while maintaining the technology as CMOS-like as possible.

An aggressive bit cell size of less than 20 μm^2 has been achieved with the introduction of an additional WSI_x polycide layer which forms self-aligned bit sense contact landing pads, a global interconnect to supply V_{ss} to the cell, and the emitter of the n-p-n bipolar transistors. The self-aligned contact landing pad relaxes the requirements for bit sense contact to word line and active edge spacing encountered in a conventional bit cell design.

II. Process Technology

The process that was developed is a CMOS-based 0.5-μm generation BiCMOS technology in which bipolar transistors were added to an existing 0.5-μm CMOS process. Growth of a thin epitaxial layer as well as addition of three masking steps; self-aligned buried layer, deep collector, and active base, were required to form the bipolar transistor. The original CMOS process featured self-aligned twin-well formation, framed mask poly-buffered LOCOS (FMPBL) isolation, a 150-Å gate, oxide thickness, surface-channel NMOS and buried-channel PMOS transistors, a reverse sequence, disposable polysilicon spacer module, three levels of polysilicon, and two layers of metallization. A large numerical aperture (NA = 0.54) G-line stepper was used for patterning all masking layers, resulting in pitches of 1.2 μm for gate polysilicon and 1.3 μm for active isolation.

The primary role of the bipolar transistor in this technology is to drive large capacitive loads in a BiCMOS gate configuration. To decrease the risk of bipolar saturation, a low collector resistance R_C is essential. Both an n^+ buried layer in a thin epitaxial region and a deep collector implant were added to the original CMOS process in order to minimize R_C. A self-aligned p^+ buried layer

Manuscript received August 15, 1991; revised November 26, 1991. The review of this paper was arranged by Associate Editor Y. Nishi.

J. D. Hayden, T. C. Mele, A. H. Perera, D. Burnett, F. W. Walczyk, C. S. Lage, F. K. Baker, M. Woo, W. Paulson, M. Lien, and D. Denning are with Advanced Products Research and Development Laboratory, Motorola Inc., Austin, TX 78721.

Y. C. See and S. J. Cosentino are with Advanced Custom Technology Center, Motorola Inc., Mesa, AZ 85210.

IEEE Log Number 9200234.

Reprinted from *IEEE Trans. Electron Devices*, vol. 39, no. 7, pp. 1669–1679, July 1992.

(or chan-stop) compensates the lateral diffusion and autodoping of the arsenic buried layer and reduces the n^+ to n^+ buried layer spacing at the expense of increased bipolar collector-substrate capacitance C_{CS} in the bipolar transistor. An offset between n^+ and p^+ buried layers will reduce C_{CS} but will increase process complexity, requiring an additional masking step and will increase the minimum required separation between n^+ buried layers.

A 1.6-μm thick lightly p-doped epitaxial layer was grown subsequent to buried-layer formation. The p doping as well as a heavier p-type cap layer are used to control arsenic autodoping. Thinner epitaxial layers, and well and field drives with reduced thermal cycles were developed to improve bipolar performance and reduce updiffusion of the buried layers. The final flat zone width is approximately 0.4 μm.

The self-aligned twin-well approach from the base CMOS process was retained; although, the temperature and time of subsequent thermal cycles was reduced to combat updiffusion of the buried layers.

To meet the needs of a high-density memory array for lower defect density levels and reduced process complexity, the isolation was maintained unchanged from the base CMOS process. This is at the cost of increased bipolar parasitic capacitances and reduced bipolar performance. A framed-mask poly-buffered LOCOS (FMPBL) isolation is used to achieve an active pitch of 1.3 μm and acceptable MOSFET narrow-width effects down to effective channel width of 0.25 μm [1], [2]. Gate oxide integrity and reliability were maintained relative to a standard LOCOS process, as indicated with time-zero breakdown and charge-to-breakdown measurements.

CMOS transistors are fabricated with a 150 Å gate oxide which provides a compromise between oxide reliability at 5-V operation and MOSFET short-channel behavior. Surface-channel NMOS and buried-channel PMOS transistors provide long-channel MOSFET characteristics down to effective channel lengths of 0.35 and 0.50 μm, respectively. Both transistor types use shallow threshold adjust and deeper punchthrough suppression implants. Moderately doped drain (MDD) regions are used for both NMOS and PMOS transistors to provide improved short-channel behavior. A reverse sequence, disposable polysilicon spacer module is used for MOSFET MDD formation to reduce the lateral encroachment of the source/drain regions and provide improved diode leakage [3].

Three levels of polysilicon are used in this process. The first layer is the gate electrode for the CMOS transistors. The second is a tungsten–polycide/polysilicon stack that performs three functions: creating self-aligned contact landing pads in the SRAM bit cell, forming the emitter of the n-p-n, and providing a global interconnect. The third polysilicon layer forms the teraohm resistor load for the bit cell.

The bipolar transistor is formed using the PMOS n well for the collector, the p^+ S/D implant as the extrinsic base contact, added implants for the active or intrinsic base region and the deep collector, and the second tungsten-polycide layer as the emitter. The extrinsic base and emitter regions are not self-aligned to each other. Allowing adequate margin for misalignment and dimensional variation, results in a large separation between extrinsic base and emitter. This translates into a relatively large base resistance R_B. The benefit of this approach is a simplified process in which lateral encroachment of the extrinsic base dopant is not a concern and design of a link base implant is not required. Reduced base-emitter breakdown voltage BV_{EBO} and enhanced bipolar hot carrier injection (HCI) effects are thus not an issue. The active base implant could be merged with the p-MDD implant without degrading PMOS characteristics. This implant, however, is also used to form the ECL load resistor and merging would reduce the freedom to adjust resistor values. The requirements of the PMOS transistor for a high n-well doping to suppress bulk punchthrough and the n-p-n bipolar for a low collector doping to ensure acceptable collector–base and collector–emitter breakdowns and reduced collector substrate capacitance were decoupled by the use of a PMOS punchthrough suppression implant. Placing this implant under the bipolar emitter–base junction also would serve to retard the onset of base pushout or the Kirk effect [4]. Alternatively, the selectively implanted collector or SIC [5] approach in which a deep n-type pocket is formed by an implant self-aligned to the emitter was investigated.

The polycide emitter and the self-aligned contact landing pad are formed simultaneously. Consequently, both are affected similarly by interfacial oxide and dopant loss during polycide formation. An HF dip is used prior to emitter polysilicon deposition to reduce the effect of any interfacial oxide. Additionally, the emitter is annealed with a rapid thermal anneal (RTA) which balls up interfacial oxide at the emitter polysilicon/silicon interface as well as activating the arsenic and driving it out of the emitter polysilicon.

Contacts are formed either using a conventional tapered etch process or with a straight-wall etch followed by a tungsten plug formation. The tungsten plugs are fabricated in the conventional fashion using a deposition plus etch-back process. The conventional TiN barrier is employed in either case. The patterned contact size is $0.6 \times 0.6\ \mu m^2$.

Two levels of metallization are required in this process. The RTA emitter anneal also flows the BPSG layer below Metal 1. A photoresist deposition plus etchback technique is used for planarization between metal layers.

A schematic cross section of this technology is shown in Fig. 1 and a SEM cross-sectional micrograph in Fig. 2. In Fig. 1, NMOS, PMOS, and n-p-n bipolar transistors are schematically depicted and all three polysilicon layers are represented. The first polysilicon layer is used for the MOSFET gates. The second layer forms the self-aligned contact landing pad in the SRAM array as well as the bipolar emitter. The third polysilicon layer forms the load resistor for the bit cell. Fig. 2 is an actual cross section of the SRAM bit cell. All three polysilicon layers can

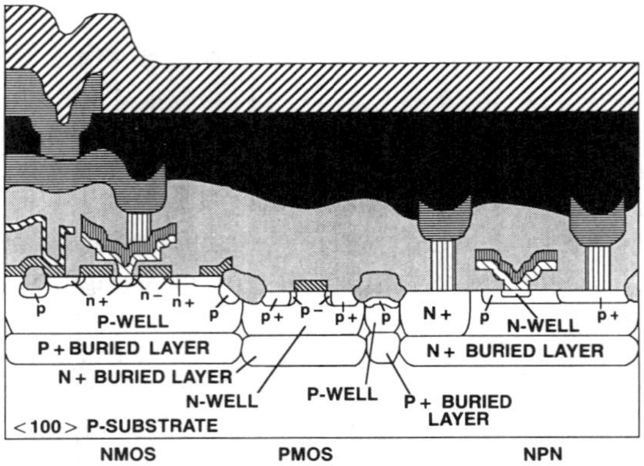

Fig. 1. Schematic cross section of the 0.5-μm BiCMOS technology.

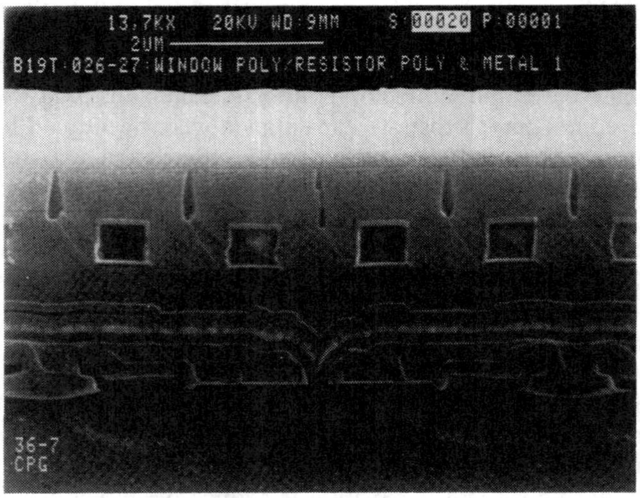

Fig. 2. SEM cross-sectional photomicrograph of the 0.5-μm BiCMOS technology.

again be seen. The second polysilicon layer acts here as both self-aligned contact and local interconnect. The first polysilicon layer forms the MOSFET gates and the third layer forms the bit cell load resistor.

III. CMOS Transistor Characteristics

Surface-channel NMOS and buried-channel PMOS devices are designed with nominal poly gate lengths of 0.70 and 0.90 μm and electrical or effective channel lengths L_{eff} of 0.50 and 0.65 μm, respectively. Long-channel behavior is maintained to minimum L_{eff} values of 0.35 and 0.50 μm. The NMOS and PMOS inverse subthreshold slopes ($|V_{DS}| = 5.0$ V) are 93 and 86 mV/decade, respectively, and off leakages ($|V_{DS}| = 5.0$ V) at minimum L_{eff} values are below 1 pA/μm as is illustrated in Fig. 3. Intrinsic gate delays of less than 100 ps/gate at a power supply voltage of 5.0 V have been obtained for unloaded CMOS ring oscillators. CMOS device characteristics are presented in Fig. 3 and summarized in Table I.

The triple-polysilicon technology allows a second polysilicon contact to the n^+ regions. The polysilicon contact adds little resistance, changing the drain saturation current by less than 2%, independent of whether the second polysilicon contact is placed at the source or drain terminal.

A reverse-sequence, disposable polysilicon spacer process has been used for MOSFET MDD formation [3] in order to reduce lateral encroachment of the MDD regions by approximately 0.05 μm per side.

IV. n-p-n Transistor Characteristics

The WSi_x polycide which forms the self-aligned contact landing pads in the bit cell is also used for the emitter of the n-p-n bipolar transistor. A schematic cross section of the bipolar transistor is shown in Fig. 4 and a SEM micrograph in Fig. 5. Fig. 6 is a high resolution TEM photomicrograph of the emitter polysilicon/silicon interface. The results of the emitter RTA anneal in which the interfacial oxide is "balled" up are apparent. Additionally, some degree of epitaxial realignment of the polysilicon film is typically observed.

A Gummel plot for the nominal n-p-n transistor with an emitter size of 0.8×2.4 μm^2 displays ideal base and collector I–V characteristics with a slope of 60 mV/decade down to current levels below 1 pA/μm^2 (Fig. 7). The bipolar transistor is designed for a current gain h_{FE} of 100, which is maintained over 7–8 orders of magnitude in collector current. The collector current at the onset of base push-out (knee current or I_K, defined at a 50% roll-off in h_{FE}) is approximately 500 μA/μm^2. The effects of bipolar saturation are reduced by maintaining a low collector resistance $R_{C\text{sat}}$ of 50 Ω. The peak cutoff frequency f_T is in excess of 12 GHz for the nominal transistor, where measurements are made at a constant collector-base voltage, V_{CB}, of 2.0 V and the peak in f_T occurs around 100–200 μA/μm^2. Emitter–base (BV_{EBO}), collector–base (BV_{CBO}), and collector–emitter breakdowns (BV_{CEO}) are maintained above 6.0, 18.0, and 6.5 V, respectively. Electrical parameters for the nominal WSi_x polycide emitter n-p-n device are listed in Table II.

Bipolar parasitic resistances and capacitances have in many instances been sacrificed for the sake of reduced process complexity. Non self-aligned emitter and extrinsic base regions result in larger base resistance R_B, and collector–base capacitance C_{CB}, but allow acceptable base–emitter breakdown voltage BV_{EBO}, and bipolar HCI resistance. Use of a LOCOS-type nonrecessed isolation degrades collector–substrate C_{CS} and collector–base C_{CB} capacitances. A self-aligned buried-layer process with a p^+ buried layer, while reducing n^+ to n^+ buried layer spacing, also degrades C_{CS}.

V. Epitaxial-Layer Thickness

The effect of epitaxial-layer thickness on bipolar transistor characteristics has been evaluated. A sensitivity analysis was performed in which the grown epitaxial layer thickness was varied from 1.4 to 1.95 μm while the implanted collector dose was maintained constant. The grown epitaxial layer thicknesses correlate to flat zone

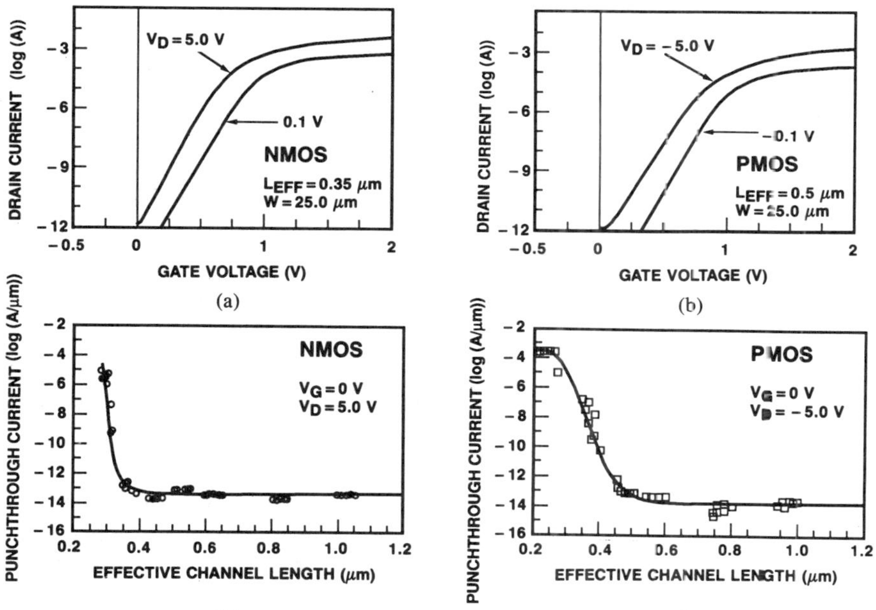

Fig. 3. CMOS transistor characteristics. (a), (c) are I_D versus V_{GS} curves at $|V_{DS}| = 0.1$ V, $|V_{DS}| = 5.0$ V for 0.35-μm NMOS and 0.50-μm PMOS transistors, respectively. (b), (d) are off-leakage currents (log $[I_D]$ at $V_G = 0$, $V_D = 5$ V) versus L_{eff}. The transistor widths are 25 μm in all figures.

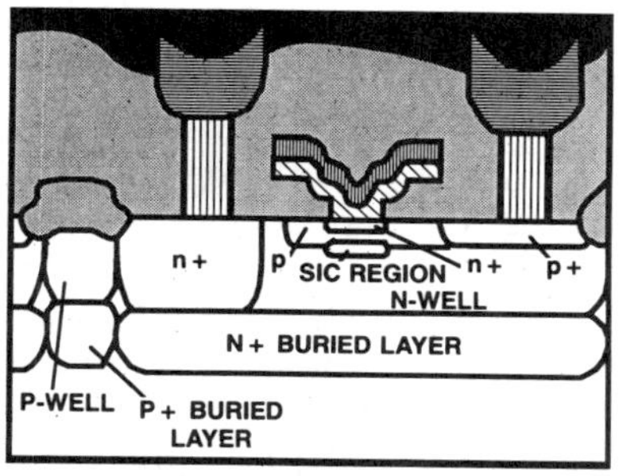

Fig. 4. Schematic cross section of the WSi_x polycide emitter n-p-n bipolar transistor. The location of the n-doping pocket created by the SIC implant is shown.

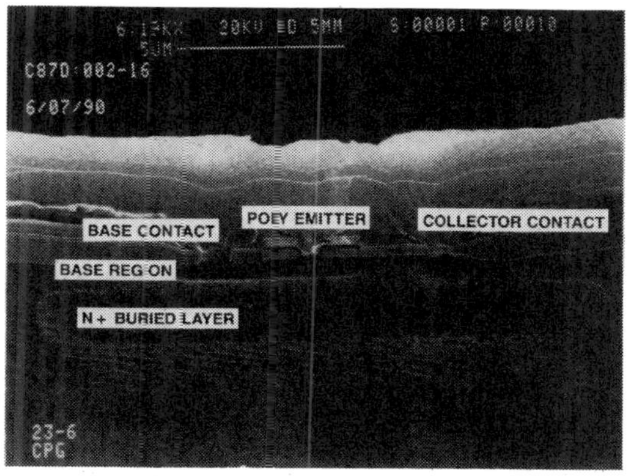

Fig. 5. SEM cross section of a WSi_x polycide emitter n-p-n transistor. The emitter area is 0.8 × 2.4 μm^2. The polycide is comprised of 2 kÅ WSi_x and 1 kÅ polysilicon.

TABLE I
CMOS Transistor Parameters

Parameter	NMOS	PMOS
L_{EFF} [nominal] (μm)	0.50	0.65
V_T (V)	0.90	−1.0
I_{DS} [sat, nominal] (mA/μm)	0.43	0.19
SS^{-1} [nominal] (mV/dec.)	93	86

widths from 0.20 to 0.75 μm, respectively. Fig. 8 demonstrates the improvement in knee current (defined as the collector current for a 50% reduction in transistor gain h_{FE}) and gate delay of a loaded BiCMOS ring oscillator. The delay in the onset of base push-out or the Kirk effect is predicted theoretically [4] and has been reported in the literature [10]. The thinner epi reduces the component of the collector resistance under the intrinsic device and increases the collector current at the onset of quasi-saturation. Retardation of the onset of base push-out and quasi-saturation both serve to increase the ability of the bipolar

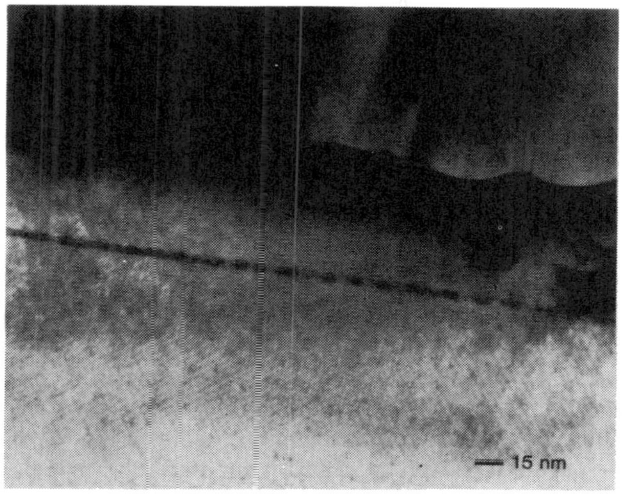

Fig. 6. TEM cross section of the emitter polysilicon/silicon interface showing balled up interfacial oxide as well as partial epitaxial regrowth.

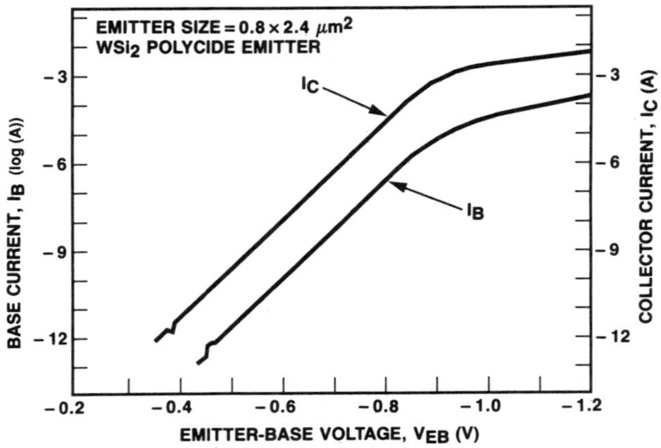

Fig. 7. Gummel plot for a nominal WSi_x polycide emitter n-p-n transistor. The emitter area is 0.8 × 2.4 μm^2. The polycide comprises 2 kÅ WSi_x and 1 kÅ polysilicon.

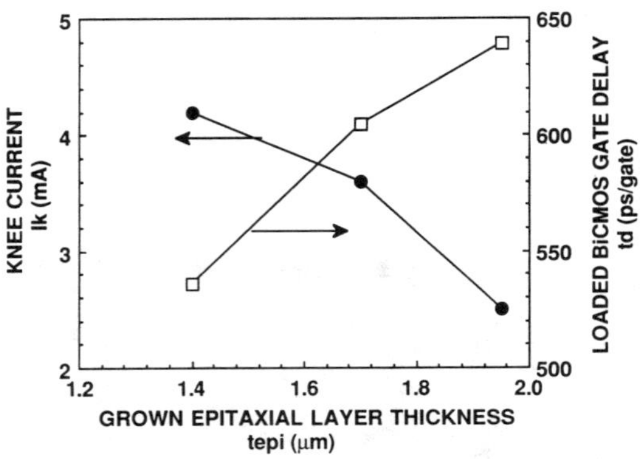

Fig. 8. Measured n-p-n bipolar knee current I_K and loaded BiCMOS gate delay as a function of epitaxial-layer thickness. The knee current is defined as the collector current for a 50% reduction in transistor gain.

TABLE II
n-p-n BIPOLAR TRANSISTOR CHARACTERISTICS
(The emitter area is 0.8 × 2.4 μm^2. The polycide comprises 2 kÅ WSi_x and 1 kÅ polysilicon.)

Parameter	Value
BV_{CBO}	18 V
BV_{CEO}	6.5 V
BV_{EBO}	6.0 V
C_{EB}	8 fF
C_{CB}	13 fF
C_{CS}	67 fF
R_E	25 Ω
f_T	14 GHz
h_{FE}	100

transistor to source large currents. This combination results in the observed reduction in BiCMOS gate delay. Fig. 9 indicates that there is a tradeoff between higher peak cutoff frequency and lower bipolar snapback voltage with thinner epitaxial layer thickness. The higher peak cutoff frequency f_T results from the delay in the onset of base push-out as well as a narrowing of the intrinsic base. The effective collector doping under the base is increased because a constant collector dose has been placed in a

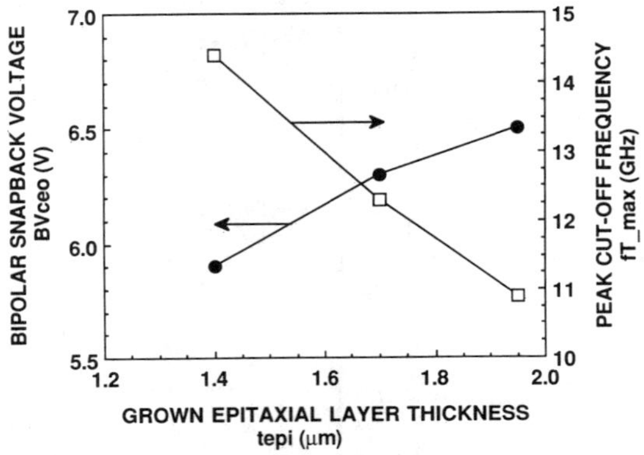

Fig. 9. Measured n-p-n bipolar snapback voltage BV_{CEO}, and peak cutoff frequency f_T, as a function of epitaxial layer thickness.

thinner epilayer and because of up-diffusion of the buried layer. The higher effective collector doping pinches the intrinsic base profile, increasing the peak cutoff frequency, f_T, the base Gummel number, and transistor gain h_{FE}. The higher effective collector doping and higher current gain both contribute to the observed reduction in the bipolar snapback voltage, BV_{CEO}.

VI. Optimization of the SIC Implant

The selectively ion implanted collector (SIC) [5] been proposed as a means of increasing cutoff frequency f_T and knee current I_K, without sacrificing collector–base capacitance C_{CB}. In the SIC approach, an n-type pocket, self-aligned to the emitter, is placed below the active base. This pocket cuts off the channeling tail from the boron implant, increasing f_T. Additionally, the higher n-type doping serves to delay the onset of base push-out or the Kirk effect [4]. The potential benefits of the SIC implant were investigated through extensive process and device simulation with PEPPER [6] and MEDUSA [7], respectively. Active base and SIC implant conditions were jointly optimized using a matrix of simulations based on an experimental design generated by the RS1 software package [8]. Fig. 10 shows simulated doping profiles through the intrinsic bipolar device with and without the SIC implant. Device simulations were performed to study current gain h_{FE}, peak cutoff frequency f_T, knee current (defined as the collector current for a 50% reduction in h_{FE}), I_K, and bipolar snapback voltage BV_{CEO}, as a function of active base and SIC implant dose and energy. Response surfaces were fit describing the dependence of these electrical parameters on the implant parameters and contour plots of the results of the device simulations were generated across the matrix of input parameters using RS1. These contour plots are shown in Figs. 11–14. Contour plots of cutoff frequency, knee current, and snapback voltage exhibit clear tradeoffs as implant conditions are varied. The primary dependence for f_T and I_K is on base implant energy. As the base implant energy is reduced, f_T increases while I_K decreases. A narrower base width decreases base transit time and improves peak cutoff frequency. A lower base implant energy results in more

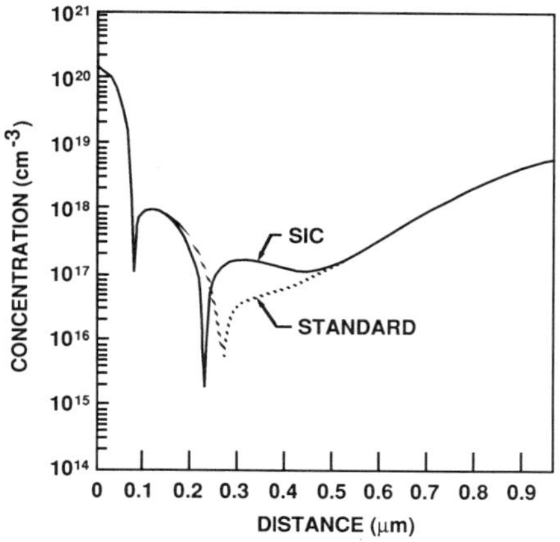

Fig. 10. Simulated doping profiles for the intrinsic bipolar transistor comparing the standard with the SIC devices.

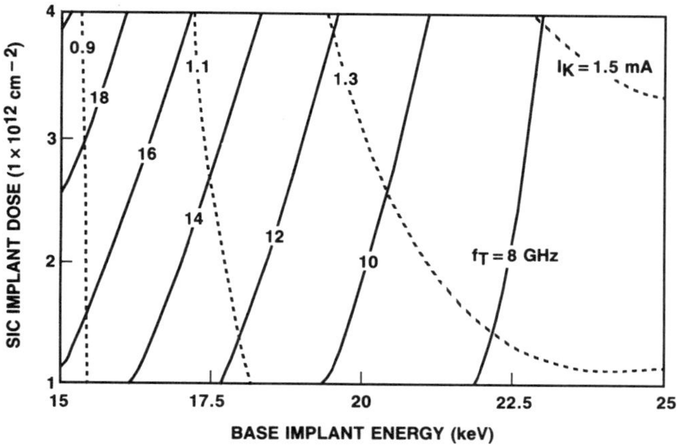

Fig. 11. Contour plots of simulated f_T and I_K characteristics for the nominal n-p-n device as functions of base implant energy and SIC dose. The base implant dose was 2×10^{13} cm^{-2} and SIC energy was 150 keV.

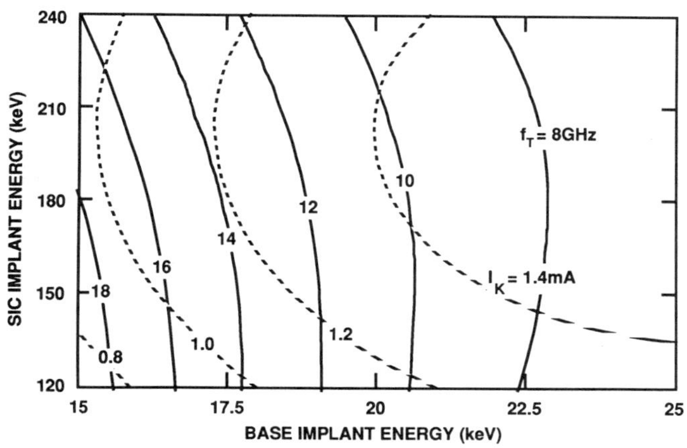

Fig. 12. Contour plots of simulated f_T and I_K characteristics for the nominal n-p-n device as functions of base implant energy and SIC implant energy. The base implant dose was 2×10^{13} cm^{-2} and SIC implant dose was 3×10^{12} cm^{-2}.

compensation of the base dopant by the emitter, lowering the peak base doping and lowering the threshold for the onset of high-level injection. Additionally, a shallower base profile implies a larger flat zone which lowers the

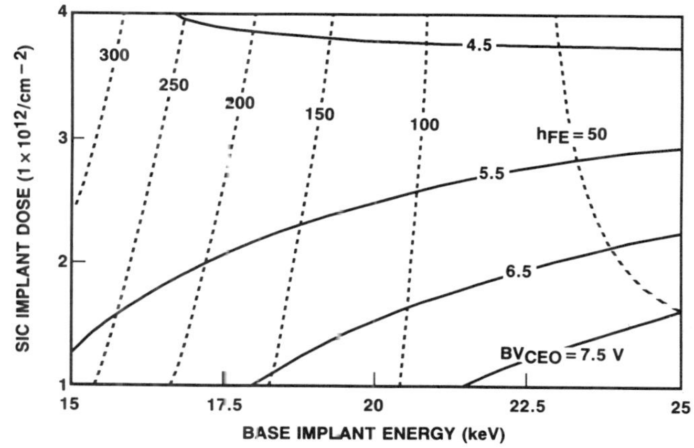

Fig. 13. Contour plots of simulated BV_{CEO} and h_{FE} characteristics for the nominal n-p-n device as functions of base implant energy and SIC implant dose, for a base dose of 2×10^{13} cm^{-2} and SIC energy of 150 keV.

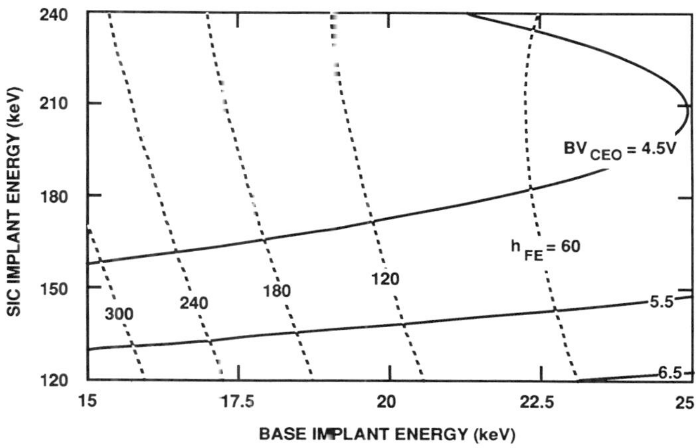

Fig. 14. Contour plots of simulated BV_{CEO} and h_{FE} characteristics for the nominal n-p-n device as functions of base implant energy and SIC implant energy. The base implant dose was 2×10^{13} cm^{-2} and SIC implant dose was 3×10^{12} cm^{-2}.

collector current required for the onset of base push-out or the Kirk effect. Higher SIC doses and lower SIC energies increase f_T by pinching off the base profile and reducing the base width. Higher SIC implant energies and doses improve the knee current by placing more n-type dopant below the intrinsic device and increasing the collector current required for the onset of the Kirk effect. By adjusting the SIC implant dose at fixed active base and SIC implant energies, I_K can be improved by approximately 20% as long as the base implant energy is above 20 keV. A lower SIC implant energy is required to improve I_K for lower energy base implants. Higher SIC implant doses degrade the bipolar snapback voltage because the heavier n-type doping lowers the collector–base breakdown voltage BV_{CBO} and, hence, BV_{CEO}. Lower SIC implant energies actually improve BV_{CEO} since more of the SIC implant is compensated by the base profile. Bipolar snapback tends to decrease with lower base implant energies. This correlates with the increase in transistor gain as the base Gummel number is reduced. Simulation

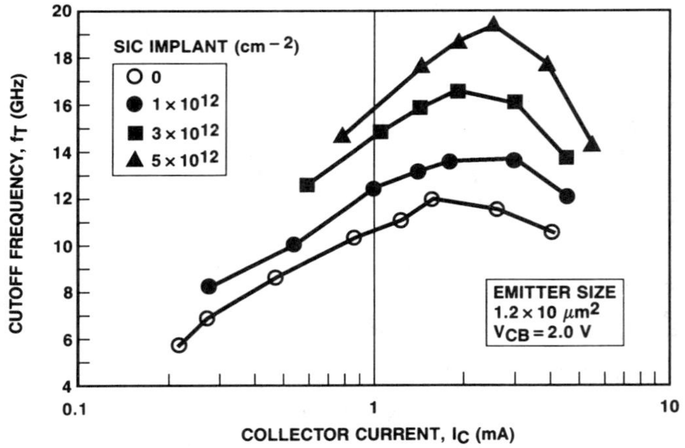

Fig. 15. Measured f_T versus I_C characteristics illustrating the effect of the SIC implant on peak cutoff frequency. The emitter area is $1.2 \times 10\ \mu m^2$ and $V_{CB} = 2.0$ V.

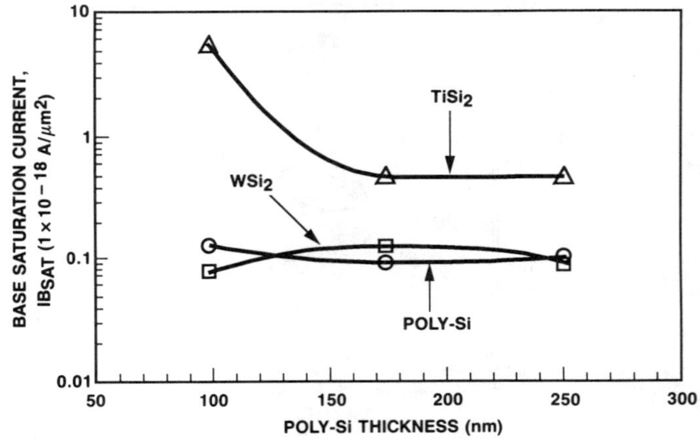

Fig. 17. n-p-n base saturation current I_{BSAT} versus the emitter polysilicon thickness for the three emitter structures. The WSi_x thickness was 2 kÅ and the $TiSi_2$ thickness was 500 Å.

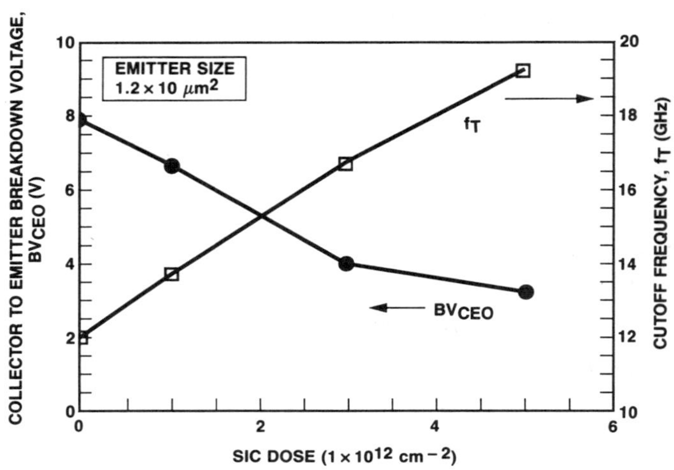

Fig. 16. Measurements of f_T and BV_{CEO} as function of the SIC implant dose. The emitter area is $1.2 \times 10\ \mu m^2$.

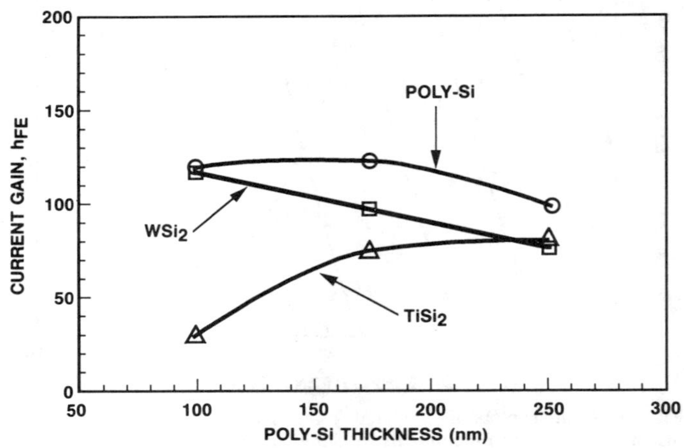

Fig. 18. n-p-n peak current gain h_{FE} versus emitter polysilicon thickness for the three emitter structures. The WSi_x thickness was 2 kÅ and the $TiSi_2$ thickness was 500 Å.

results indicate that an improvement in I_K is expected to be coupled with the reduced BV_{CEO}.

The selectively ion implanted collector (SIC) was also investigated experimentally. Fig. 15 is a plot of peak cutoff frequency versus collector current for a series of SIC implant doses. The SIC implant increases f_T from 12 GHz with no implant to a maximum of 19 GHz. Fig. 16 demonstrates that the improved cutoff frequency must be traded off against a lower bipolar snapback voltage BV_{CEO}. A peak cutoff frequency of 14 GHz can be achieved while maintaining a BV_{CEO} of 6.5 V.

VII. Tungsten Polycide

The emitter polysilicon layer must be strapped with a silicide to provide a low sheet resistance interconnect for the V_{SS} connection in the bit cell. A WSi_x polycide and a $TiSi_2$ salicide were compared experimentally. The impact of adding a $TiSi_2$ or WSi_x layer on top of the emitter polysilicon was evaluated by examining the n-p-n transistor characteristics as a function of emitter polysilicon thickness and emitter rapid thermal anneal (RTA) sequence. Formation of a $TiSi_2$ salicide on the emitter polysilicon caused a significant increase in base saturation current (IB_{SAT}) and decrease in transistor gain h_{FE} as the emitter polysilicon thickness was decreased. WSi_x polycide only slightly affected the n-p-n characteristics. This is illustrated in Figs. 17 and 18. When the $TiSi_2$ formation is performed prior to the emitter RTA anneal, the structure of the polysilicon emitter is dramatically affected. During salicidation, a large portion of the arsenic implanted into the emitter is lost due to arsenic segregation and consumption during the $TiSi_2$ reaction. This results in a reduction in the arsenic dopant segregated at the polysilicon/silicon interface which in turn reduces the effectiveness of the blocking layer or potential barrier at the polysilicon/silicon interface [9]. The lower potential barrier causes devices with $TiSi_2$ strapped emitters to become more susceptible to the effects of reducing the final emitter polysilicon thickness. As the polysilicon thickness approaches the hole minority-carrier diffusion length in the polysilicon emitter, an increase in IB_{SAT} is observed [9]. The thermal stability of the $TiSi_2$ layer is also an is-

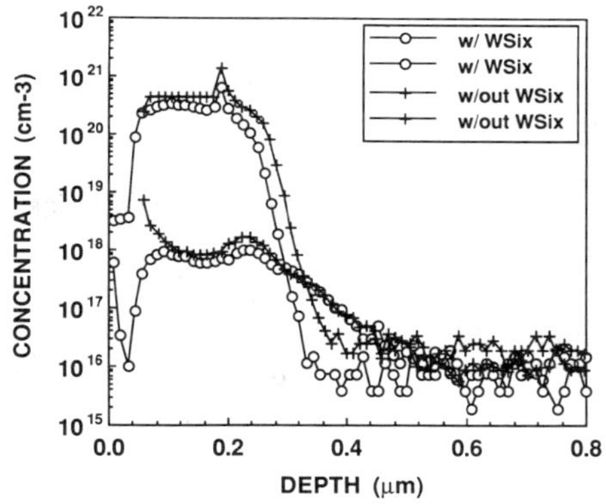

Fig. 19. SIMS profiles for emitter and base regions of n-p-n bipolar with polysilicon and WSi_x polycide emitters.

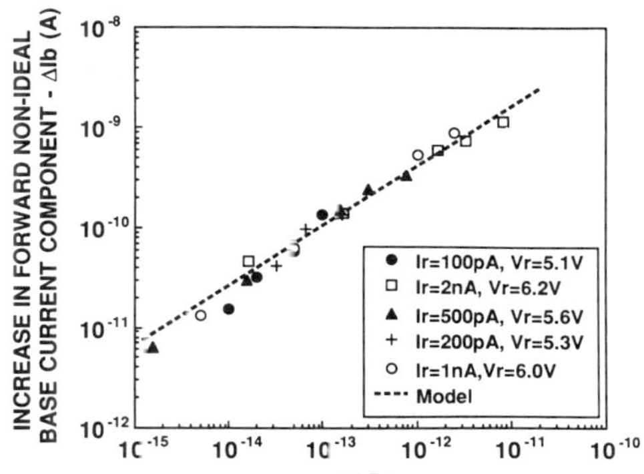

Fig. 20. Increase in nonideal base current component ΔI_B as a function of $I_R^{1.7} * t$ under band-to-band tunneling conditions.

sue. Performing the RTA emitter anneal after formation of the $TiSi_2$, will often result in agglomeration of $TiSi_2$. In WSi_x polycide devices, some arsenic segregation also occurs, but to a lesser extent. Fig. 19 compares the SIMS profiles for the polysilicon and the WSi_x polycide emitter n-p-n transistors. Because of arsenic segregation, the WSi_x device has a shallower emitter and thus a wider base and larger base Gummel number. The result is a decrease in transistor gain h_{FE} and emitter resistance R_E. The tungsten polycide also exhibited superior thermal stability, exhibiting negligible degradation after the RTA emitter anneal.

VIII. Bipolar HCI

It is well known that under reverse-bias emitter–base stress, bipolar transistors undergo HCI (hot carrier injection) degradation which is manifested as an increase in the nonideal base current component and a decrease in the low current gain [11]. Physically, charge is injected into the oxide above the emitter–base junction edge. This results in an increase in the density of interface states (DIT) above the junction edge, which contribute a recombination leakage component to the overall current. In standard operation of a BiCMOS gate, the emitter–base junctions of the bipolar transistors are transiently reverse-biased and bipolar HCI is hence a significant concern. A comprehensive study was carried out of HCI degradation in the bipolar transistors fabricated in this process. Earlier work was confirmed in which the generation of HCI damage is a function of the level of reverse bias stress the device is subjected to. In particular, when bipolar transistors experience a low-level stress corresponding to the band-to-band tunneling region of the reverse emitter–base breakdown characteristic, the nonideal base current component increases as $I_r^{1.7} * t$, as shown in Fig. 20. When the bipolar transistors are stressed under avalanche breakdown conditions, the nonideal base current component ΔI_B increases as a function of the total stress charge Q. Here, Q is equal to the product of the stress current and total stress time. This is illustrated in Fig. 21. In Fig. 22, the results of bipolar HCI stress under band-to-band tunneling and avalanche breakdown conditions are combined and the time for 50% beta degradation is plotted as a function of reverse stress current for different emitter–base forward

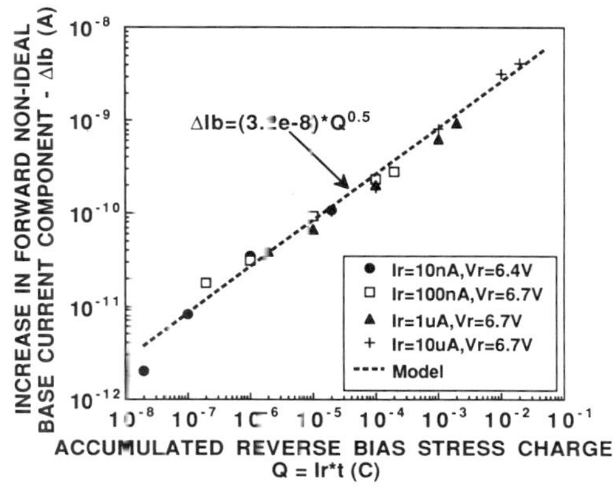

Fig. 21. Increase in nonideal base current component ΔI_B as a function of the total reverse stress charge $Q = I_R * t$ under reverse emitter–base junction avalanche breakdown conditions.

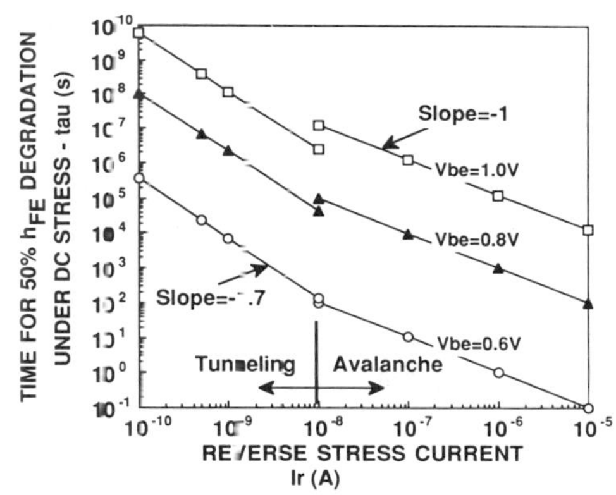

Fig. 22. The time to 50% reduction in transistor gain h_{FE} as a function of reverse bias stress current.

biases. It should be noted that the bipolar HCI stressing was carried out under dc conditions whereas the typical emitter–base reverse bias stress a bipolar transistor is subjected to is a transient waveform and the resulting duty cycle is quite low.

IX. ECL Ring Oscillator Measurements

ECL gate delay was measured as a means of characterizing bipolar performance. Fig. 23 compares gate delay to ECL ring oscillators with single- and double-base n-p-n bipolar transistors laid out with conservative design

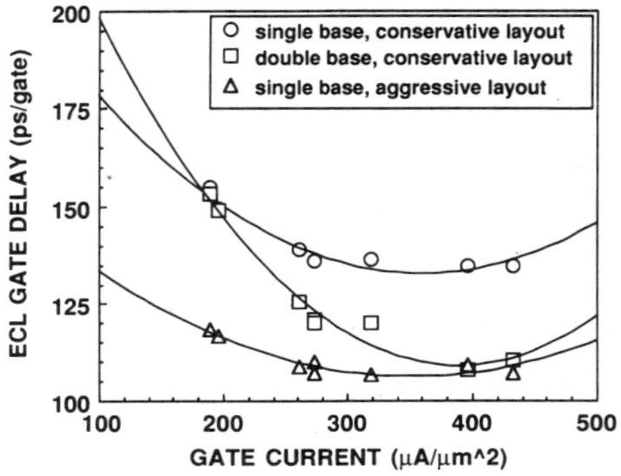

Fig. 23. Measured ECL gate delay as a function of gate current. The plot compares oscillators with single- and double-base bipolar transistors laid out with conservative design rules and single-base bipolar transistors laid out with aggressive design rules.

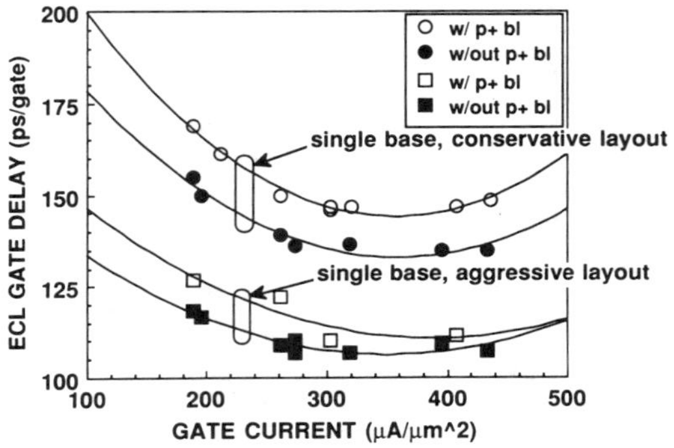

Fig. 24. Measured ECL gate delay as a function of gate current. The plot demonstrates the effect the p^+ buried layer has in increasing collector-substrate capacitance C_{CS} and degrading ECL gate delay.

rules and single-base n-p-n bipolar transistors laid out with aggressive design rules. The single-base n-p-n with aggressive design rules achieves an ECL gatedelay of 105 ps/gate at a gate current of 350 $\mu A/\mu m^2$. An ECL ring oscillator with double-base n-p-n bipolar transistors laid out with aggressive design rules would exhibit even faster gate delays. Increased process complexity in the form of a more recessed isolation, a lithographic offset between n^+ and p^+ buried layers, and a double-polysilicon, fully self-aligned bipolar structure, would provide improved ring oscillator performance, at the expense of reduced SRAM yield.

The impact of the p^+ buried layer in increasing the peripheral component of the collector–substrate capacitance C_{CS} was investigated. Fig. 24 demonstrates that leaving out the p^+ buried-layer implant improves ECL gate delay by approximately 10%. This is at the expense of increasing the minimum n^+ buried layer to n^+ buried spacing from 2.5 to 6.5 μm.

X. Conclusions

A high-performance 0.5-μm BiCMOS technology has been demonstrated which uses a triple-polysilicon process architecture for a fast 4-Mb SRAM class of products. Three layers of polysilicon were used to achieve a compact four transistor bit cell size that is less than 20 μm^2 by forming self-aligned bit-sense and V_{ss} contacts. A WSi_x polycide emitter n-p-n transistor was implemented with an emitter area of 0.8×2.4 μm^2 and peak cutoff frequency of 14 GHz. A selectively ion-implanted collector was used to compensate the base channeling tail as well as increase cutoff frequency and knee current while maintaining a collector-to-emitter breakdown voltage of 6.5 V. n-p-n base and SIC implant conditions were optimized through device simulations coupled with statistical experimental design. A minimum ECL gate delay of 105 ps was achieved at a gate current of 350 $\mu A/\mu m^2$.

Acknowledgment

Special thanks are extended to the APRDL CMOS SRAM group for their collaboration on the 0.5-μm CMOS array technology, to L. Parker and C. Gunderson for their assistance with last minute measurements, and to J. Teplik and V. de la Torre of ATC for technical discussions. The authors would also like to acknowledge R. Sivan and L. Parrillo for their guidance throughout this project. And finally, the authors would like to thank the APRDL manufacturing, process, and test engineering groups.

References

[1] B. Y. Nguyen, P. Tobin, M. Lien, M. Woo, J. Leiss, and J. D. Hayden, "Framed mask poly-buffered LOCOS isolation for submicron VLSI technology," in *ECS Spring 1990 Meeting Extended Abstracts*, vol. 90-1, p. 614.

[2] J. D. Hayden *et al.*, "A high-performance half-micrometer generation CMOS technology for fast SRAM's," *IEEE Trans. Electron Devices*, vol. 38, pp. 876–886, Apr. 1991.

[3] L. C. Parrillo, J. R. Pfiester, J.-H. Lin, E. O. Travis, and R. D. Sivan, "An advanced 0.5 μm CMOS disposable LDD spacer technology," in *1989 Symp. on VLSI Technology*, p. 31.

[4] D. J. Roulston, *Bipolar Semiconductor Devices*. New York: McGraw Hill, 1990, p. 254.

[5] S. Konaka, E. Yamamoto, K. Sakuma, Y. Amemiya, and T. Sakai, "A 20-ps Si bipolar lc using advanced super self-aligned process technology with collector ion implantation," *IEEE Trans. Electron Devices*, vol. 36, p. 1370, July 1989.

[6] B. J. Mulvaney, W. B. Richardson, G. Siebers, and T. Crandle, *PEPPER 1.2*. Austin, TX: Microelectronics and Computer Technology Corp., Jan. 1989.

[7] W. L. Engl, *MEDUSA (Version 4.1)*, Tech. Univ. of Aachen, West Germany.

[8] RS/1, Release 4, BBN Software Products Corporation, Cambridge MA, Oct. 1988.

[9] P. Ashburn, "Polysilicon emitter technology," in *BCTM Tech. Dig.*, Sept. 1989, p. 90.

[10] T. Ikeda, A. Watanabe, Y. Nishio, I. Masuda, N. Tamba, M. Odaka, and K. Ogiue, "High-speed BiCMOS technology with a buried twin well structure," *IEEE Trans. Electron Devices*, vol. ED-34, pp. 1304–1309, June 1987.

[11] J. D. Burnett and C. Hu, "Modeling hot-carrier effects in polysilicon emitter bipolar transistors," *IEEE Trans. Electron Devices*, vol. 35, pp. 2238–2244, Dec. 1988.

Paper 1.8

3.3V BINMOS TECHNOLOGY USING NPN TRANSISTORS WITHOUT BURIED LAYERS

A. SHIDA, M. KAGAMIHARA, M. KOMATSU, K. KUMAGAI* and M. HIRATA

SYSTEM LSI Development Division, NEC Corporation
1120 Shimokuzawa, Sagamihara, Kanagawa 229, Japan
*VLSI Development Division, NEC Corporation

ABSTRACT

A VLSI technology suitable for low voltage BiNMOS logic has been developed. A BiNMOS device structure, *N+* substrate based *E*mitter-followered *CMOS*(NECMOS) is proposed. NECMOS has a simple structure using a *s*ubstrate *NPN*(S-NPN) transistor which does not need buried layers. The use of an N+ substrate and a thin epitaxial layer is capable of realizing low collector resistance for S-NPN. A polysilicon emitter type S-NPN transistor, with a maximum cut-off frequency of 9GHz, has been integrated to a 0.8μm CMOS with three additional masks. The fan-out coefficient of the propagation delay time for a BiNMOS 2-NAND gate is 1/2.4 of that for a CMOS 2NAND under 3.3V operation.

INTRODUCTION

BiCMOS technology has been making steady progress because of its optimum use combining high speed bipolar and high density, low power CMOS. In logic application, there have been two major trends in the development of the BiCMOS process; the low-end BiCMOS and the high-end BiCMOS[1]. In the low-end approach, the major purpose is performance improvement utilizing BiCMOS circuits in CMOS logic. Therefore, the issues relating to trades-off between performance and cost are important.

According to the scaling requirements for CMOS transistors, power supply voltage tends to be reduced to ensure device reliability. Under low voltage operation, BiNMOS logic has better performance than totempole BiCMOS from the viewpoint of higher density and faster operation[2][3]. In conventional BiCMOS LSI, NPN transistors are used as pull up and down devices, and isolation for NPN transistors requires buried layers which demand many fabrication steps and results in higher cost. On the other hand, in BiNMOS logic, NPN transistors are used only for pull up devices, and collector isolation of NPN transistors is not necessary because the potential is common. A S-NPN transistor which does not need buried layers is the optimum pull up device for BiNMOS logic.

A S-NPN transistor formed in the conventional epi-CMOS process has poor performance due to its high collector resistance(over 1k-ohm), because the epitaxial layer thickness is large to allow impurity diffusion by the high temperature process. To overcome this shortcoming, we propose an NECMOS technology with a low temperature process, in which the N+ substrate and a thin epitaxial layer is used as the NPN collector. This paper describes submicron NECMOS technology for BiNMOS logic application.

DEVICE STRUCTURE

Figure 1 shows a cross section view of the completed device. NECMOS devices are fabricated on antimony doped N+ substrate(0.01ohm·cm) with an N type thin epitaxial layer. An N well formed in the epitaxial layer is used as the NPN collector, as well as a PMOS tub. The epitaxial layer thickness is an important consideration, because it relates to collector resistance for NPN transistor and vertical isolation for NMOS transistors. The epitaxial layer thickness is decided to 1.2μm.

To prevent punchthrough from the N+ SD of the NMOS transistor to the N+ substrate, the doping level of the P well should be above 1E17cm-2 at a depth below 0.5μm. Optimization of the P well doping profile was performed by process simulation(SUPREM-2). The 1.2μm epitaxial layer requires a low temperature(under 1000°C) well drive-in process to suppress impurity upward-diffusion from the N+ substrate. The simulated impurity profile for the P well is shown in Fig. 2. It has a peak impurity concentration of 3E17cm-2 at a depth of 0.7μm using high energy(300KeV) boron I/I. The surface flat region which affects NMOS characteristics is defined by a 50KeV boron I/I. Also, 130KeV boron I/I is introduced to connect these two regions. This retrograde P well also prevents punchthrough and allows submicron LOCOS isolation for neighboring NMOS transistors.

In the NECMOS structure, there is no need to form a VDD contact in the basic cell because substrate resistivity is low and the substrate potential is fixed at the VDD level using a back-side contact. Therefore NECMOS realizes high density BiNMOS logic.

In this work, epitaxial layer thickness and well impurity profile have been optimized to integrate high performance S-NPN to submicron CMOS.

PROCESS DESCRIPTION

The main fabrication steps are shown in Fig. 3. It is based on a 0.8μm CMOS process. Integration of the polysilicon emitter S-NPN has been accomplished

Reprinted from *IEDM 1991*, pp. 4.3.1–4.3.4, 1991.

by adding three photo-masks for direct contact(deep collector) and emitter formation. Base regions for NPN and LDD regions for PMOS are formed by simultaneous ion-implantation.

Beginning with a 1.2μm epitaxial layer on an N+ substrate, twin well regions are formed by a self-aligned process. After a conventional LOCOS formation, deep boron ion-implantation(300KeV) is carried out to form the retrograde P well. The gate oxide thickness is 15nm. A Policide gate(W-silicide/ N+ polysilicon) is used for both NMOS and PMOS. After the MOS gate electrode formation, LDD source/ drain are implanted. A polysilicon emitter is used to realize minimum emitter size of 0.8x3μm. LDD sidewall oxide is formed by an etch-back process after the emitter polysilicon patterning to keep the emitter-base junction damage-free.

DEVICE CHARACTERISTICS

Characteristics for the fabricated devices are summarized in Table.1. Current gain, BVceo, BVebo and collector resistance(Rcs) are 60, 7V, 5.3V and 90ohm, respectively. Collector resistance of less than 70 ohm is obtained for collector contact from the back-side. Figure 4 shows the relationship between cut-off frequency(fT) and collector current (Ic). Maximum fT(fTmax) is 9GHz at a Vce of 3V. The channel conductance of 0.8μm CMOS is 440μS/V for NMOS and 220μS/V for PMOS with a 10μm gate width, the same characteristics as in the case of pure 0.8μm CMOS.

Epitaxial layer thickness is most considerable parameter which affect to vertical isolation for NMOS transistors. Isolation characteristics are related to a parasitic NPN transistor which is composed of the N+ substrate as a collector, the P well as a base, and a N+ source as an emitter. Figure 5 shows the measured dependence of fTmax for S-NPN and the parasitic NPN transistor BVceo on the epitaxial layer thickness(tepi). The use of a thin epi results in fT improvement, but causes lower BVceo. Operating at the collector-emitter bias of the parasitic NPN over BVceo, trigger current injection into the P well causes a low impedance state for the parasitic PNPN structure with the P+SD region floating(Fig.6). This low impedance state is retained after the removal of the trigger current. This phenomena, with one parasitic NPN, is explained by reverse base hole current[4]. Measured holding voltage(Vhold) for this low impedance state is shown in Fig.5. BVceo of the parasitic NPN transistor is a prime parameter for this phenomena and tepi has strong dependence on it. Around the 1.2μm epitaxial layer thickness, Vhold is grater than 5V and there is no problem of operating at 3.3V.

The impurity profile for the P well is an important factor for NMOS isolation. Figure 7 shows the dependence of punchthrough voltage between the N+ substrate and the N+ S/D formed in the P well on deep boron I/I(300KeV) dose. Punchthrough voltage in excess of 10V is obtained for a deep boron dose grater than 5E12cm-2. Lateral isolation for NMOS transistors is carried out by field oxidation and a guard ring formed by an optimized channel stopper boron I/I(before LOCOS oxidation) and the deep boron I/I(after LOCOS oxidation). The lateral isolation characteristics were evaluated by measuring inversion and punchthrough voltage for parasitic MOS transistors as a function of LOCOS width. Submicron LOCOS isolation was obtained for both NMOS and PMOS transistors.

To confirm high speed performance under low voltage, BiNMOS logic speed was measured. Logic circuits are designed with an ECMOS basic cell which realizes high density BiNMOS SOG[5]. The ECMOS cell consists of a CMOS logic and EFU(Emitter Followered Unit), which is a bipolar transistor with a base discharge resistor. Figure 8 shows the tpd dependence for 2-NAND and INVERTER as a function of fan-outs. The fan-out coefficient for a 2-NAND gate is 1/2.4 of that for a CMOS 2NAND under 3.3V operation. Typical gate delay(2NAND double gate) is 350ps when typically loaded(F/O=2,Al=2mm). Figure 9 shows the gate delay dependence on supply voltage. BiCMOS gate delay(simulated) increases rapidly as the supply voltage becomes less than 3.3V. On the other hand, a BiNMOS gate has better performance than a CMOS gate above a supply voltage of 2.5V.

CONCLUSION

The NECMOS technology has been developed for realizing low voltage BiNMOS logic. A S-NPN transistor which does not need buried layers is integrated to 0.8μm CMOS with three additional photo masks. An N+ substrate(0.01ohm·cm) and a 1.2μm epitaxial layer realizes fTmax of 9GHz for S-NPN. The effectiveness of this technology has been demonstrated by evaluating BiNMOS circuits operating at 3.3V.

NECMOS looks to be a promising technology for realizing fewer process steps and higher density BiNMOS logic LSI operating at around 3V.

ACKNOWLEDGMENT

The authors would like to thank Dr. Yoshida, Dr. Suzuki, Mr. Hiki and Mr. Katsuta for helpful suggestions and constant encouragement.

REFERENCES

[1] L.Gal et al.,"Performance and Technology Trade-offs of BiCMOS Submicron Processes for ASIC Applications", Proceedings of the IEEE CICC,18.8.1, May 1991.

[2] A.E.Gamal et al., "BiNMOS :A Basic Cell for BiNMOS Sea-of-Gates", Proceedings of the IEEE CICC, 8.3.1, May 1989.

[3] H.Miyakawa et al., "A 3.3V, 0.5μm BiCMOS Technology for BiNMOS and ECL Gate", Proceedings of the IEEE CICC, 18.3.1,May 1991.

[4] Y.Okazaki et al., "New Well Structure for Deep Sub-μm CMOS/BiCMOS Using Thin Epitaxy Over Buried Layer and Trench Isolation", Symposium on VLSI Technology Dig., pp.83-84,1990.

[5] K.Kumagai et al., "A 150k Gate 250ps BiCMOS SOG with an Emitter-followered CMOS(ECMOS) Cell", Proceedings of the IEEE CICC, 4.3.1,May 1990.

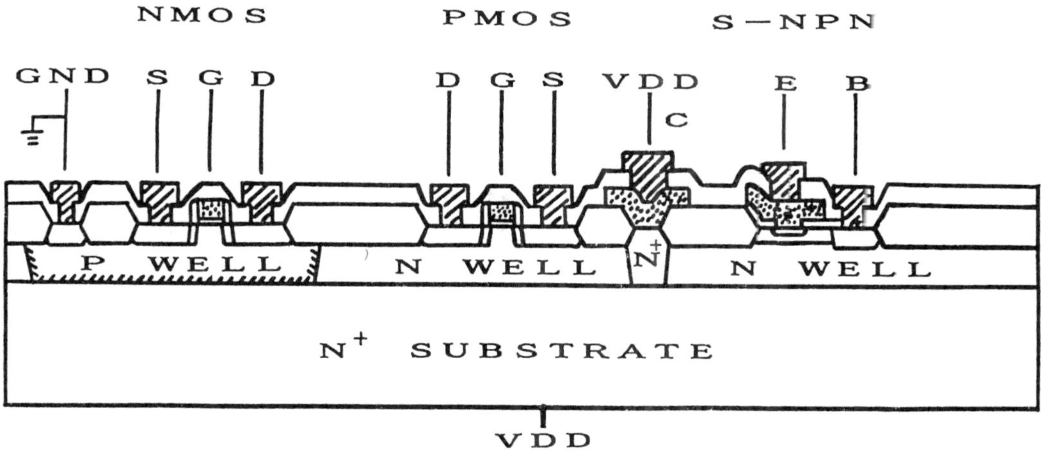

Figure 1 Schematic cross section of the NECMOS
N well on N+ substrate is used as the S-NPN collector. VDD contact is not formed in the basic cell.

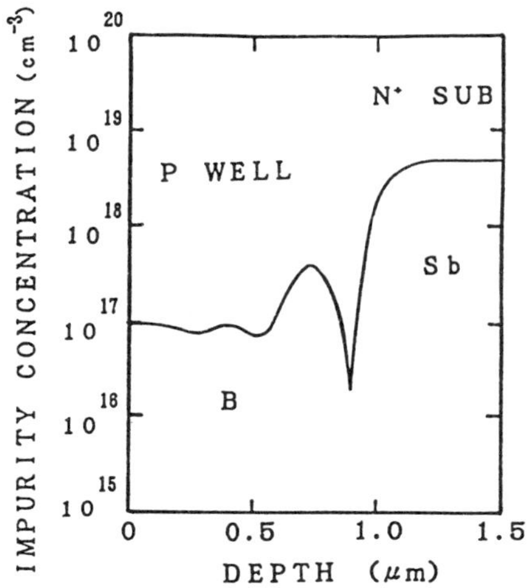

Figure 2 Simulated impurity profile for the P well
Retrograde P well is formed with multiple energy ion-implantation.

Table.1 Device characteristics

S-NPN	emitter size	0.8x3 μm
	current gain	60
	BVceo	7.0 V
	BVcbo	14 V
	BVebo	5.3 V
	Rsc	90 ohm
	fTmax	9 GHz
NMOS	L/W	0.8/10 μm
	tox	15 nm
	bata	440 μS/V
	snapback volt.	6.0 V
PMOS	L/W	1.0/10 μm
	tox	15 nm
	beta	220 μS/V

CMOS	NECMOS [added]
Twin well	
Field isolation	
Channel doping for NMOS(Retrograde P well) for PMOS	
Gate oxidation	
	Direct contact
Gate polyside	
LDD P	
LDD B (Base B)	
	Emitter contact
	Emitter Poly Si
LDD sidewall	
SD AS	
SD BF2	
Contact & Metallization	

Figure 3 NECMOS process flow

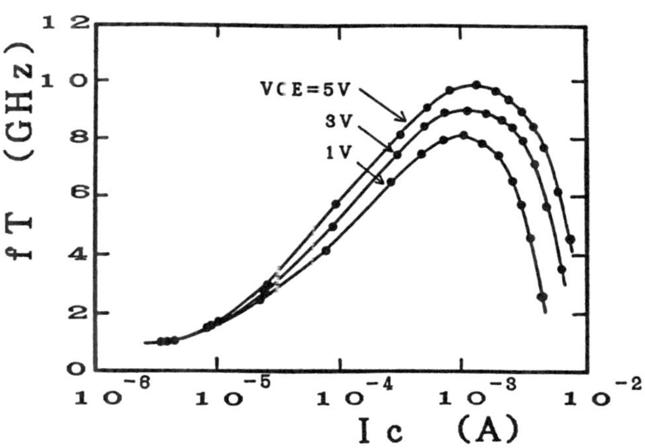

Figure 4 Dependence of fT on collector current
FTmax is 9GHz(at Vce=3V) with an emitter size of 0.8x3μm.

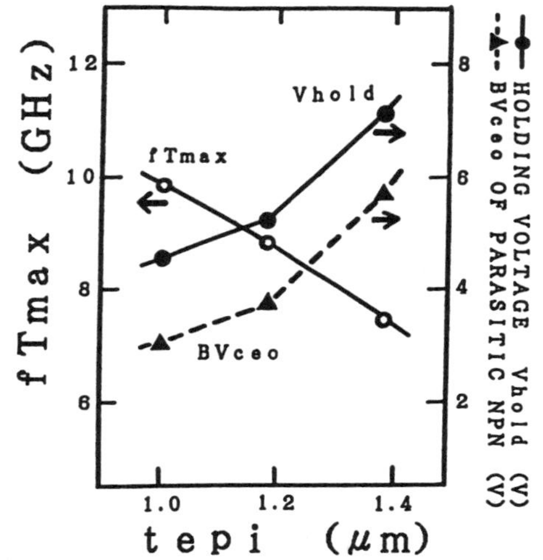

Figure 5 Dependence of fTmax for S-NPN and BVceo for parasitic NPN on epitaxial layer thickness
Deep boron dose is 1E13cm-2. Tepi is an important parameter in the NECMOS process.

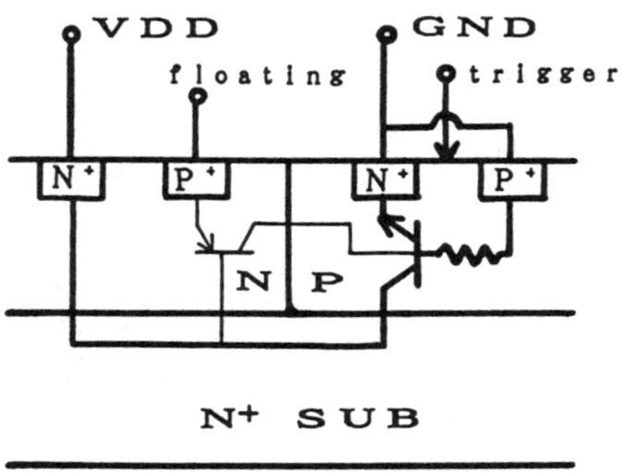

Figure 6 Parasitic bipolar portion of a 4-terminal PNPN structure
Low impedance state is retained after removal of trigger current with P+ S/D floating.

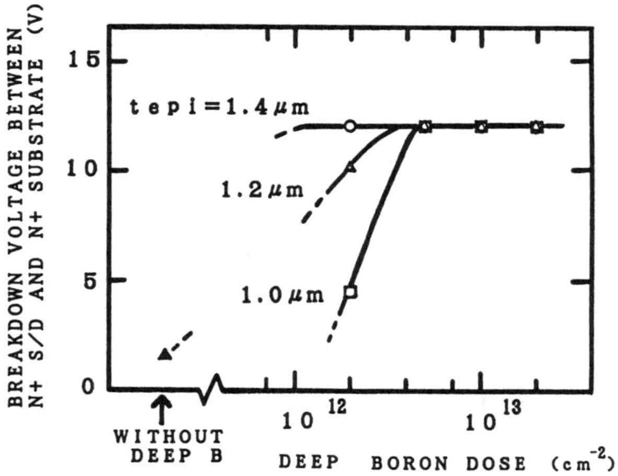

Figure 7 Punchthrough voltage between N+ S/D and N+ substrate

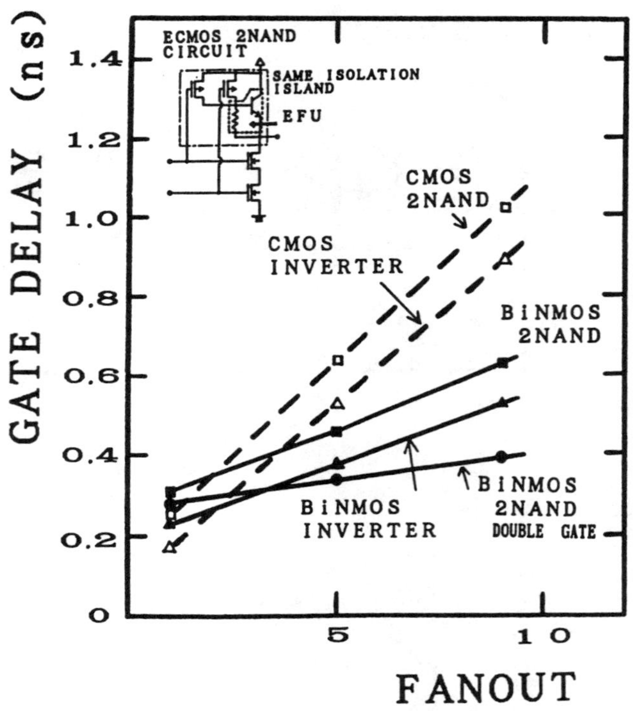

Figure 8 BiNMOS gate delay vs. FAN-OUT
Fan-out coefficient is 40ps/fanout for a 2NAND single gate.
(Wn/Wp=15/15μm, Ae=0.8x3μm, VDD=3.3V)

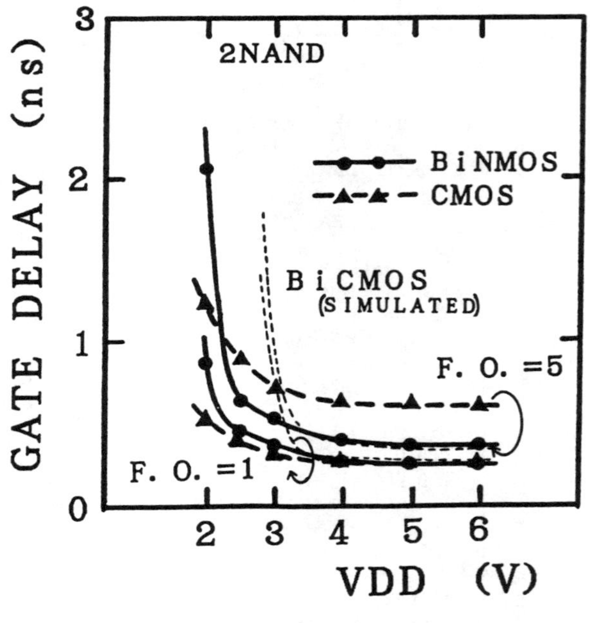

Figure 9 BiNMOS gate delay vs. supply voltage
NECMOS performance with S-NPN under low voltage is shown.

Paper 1.9

A 0.4-Micron Fully Complementary BiCMOS Technology for Advanced Logic and Microprocessor Applications

Shih Wei Sun, Paul G. Y. Tsui, Bradley M. Somero, Jeff Klein, Fabio Pintchovski, John R. Yeargain, and Bernie Pappert✿

Motorola Inc., Advanced Products Research and Development Laboratory, ✿High-End MPU Design
3501 Ed Bluestein Blvd., Austin, Texas 78721, (512) 928-5028

ABSTRACT

A modular process architecture has been adopted to develop a versatile yet manufacturable, single-poly, four-level metal, fully complementary BiCMOS (C-BiCMOS) technology for sub-0.5 μm logic and microprocessor products. Both the poly-emitter vertical NPN and PNP bipolar transistors are integrated into a dual-poly (n^+/p^+) gate CMOS process flow. Using a pedestal implant in the emitter window, the NPN performance has been enhanced to 26 GHz. Lateral-PNP, and $TiSi_2$ Schottky barrier diode formed during the titanium self-aligned silicide process are available for various circuit applications. Stacking of the tungsten-plug contacts and vias are allowed in the multi-level metallization module. A process window analysis has also been performed to derive the optimal device design targets. Comparing with the CMOS counterpart, approximately 40% speed improvement (at 3.3 V V_{cc}) in a 68030 critical path has been demonstrated using this logic BiCMOS technology.

INTRODUCTION

For the next-generation microprocessors (MPU) with a clock frequency above 100 MHz and the on-chip cache density surpassing 1Mb, high-performance BiCMOS technology is considered necessary. The BiCMOS design also considerably reduces the design cycle, since the BiCMOS buffer delay is much less sensitive to the loading capacitance as compared to that of the CMOS design. To develop an advanced sub-0.5 μm BiCMOS technology for logic and MPU products, the following points need to be considered. (1) The low voltage BiCMOS device and circuit performance are critical as the supply voltage is reduced to meet the device reliability and chip power limit requirements. (2) The process architecture has to be modular, and be able to provide a migration path for subsequent integration of the analog, EPROM, and EEPROM functions into the baseline process flow. (3) Most importantly, the technology has to be cost-effective and manufacturable to facilitate future transfer to volume production.

In this work, a versatile 0.4 μm single-poly, four-level metal, C-BiCMOS technology has been developed based on a CMOS logic process [1]. The modular process architecture allows the selection and expansion of the integration level for specific MPU and microcontroller product applications. Process simplicity and low voltage BiCMOS device and circuit performance have been emphasized throughout this development project. Process integration, device characteristics, and circuit performance of this logic BiCMOS technology will be described in this paper.

PROCESS INTEGRATION

The schematic cross-section of this C-BiCMOS technology is illustrated in Fig. 1, with the typical process and device parameters given in Table 1. The single layer of polysilicon is used as the n^+ and p^+ poly gate of n-channel and p-channel MOSFET's, as the n^+ and p^+ poly emitter of NPN and PNP bipolars, and as the ECL poly resistors in this technology. The n^+/p^+ poly and n^+/p^+ buried contact cross-coupled connection have also been employed in an embedded bulk 6T SRAM for stable operation at reduced supply voltages.

Self-aligned n^+ and p^+ buried layers and twin wells are formed before and after a 1.6 μm epi deposition process. Framed mask poly buffered LOCOS isolation provides a final encroachment of < 0.15 μm per side [2]. The bipolar active base is formed by implantation through a stack of thin polysilicon (500 Å) and base oxide layers. Emitter windows of the bipolar transistors are defined simultaneously with the CMOS buried contacts. Emitter polysilicon of the NPN and PNP bipolars are implanted with As^+ and BF_2^+ respectively.

A lightly-doped-drain module is used to fabricate the 100 Å T_{ox} MOSFET's with the n^+/p^+ dual-poly gate doped concurrently with the source/drain (S/D) implants. The S/D implants also serve as the bipolar extrinsic base implants, which are self-aligned to the edges of the polysilicon emitter. A 1050 C rapid-thermal-annealing (RTA) is then used to drive the emitter and to anneal the dopants. The two-step RTA, titanium self-aligned silicide (SALICIDE) process yields a sheet resistivity of < 4 ohm/sq. with junction leakage less than 1 nA/cm^2 and 10 pA/cm at 10 V, room temperature. $TiSi_2$ Schottky barrier diode is formed in the n-well without any additional processing steps. Stacked contacts and vias are plugged with an integrated tungsten deposition-and-etchback process for the backend four-level metallization.

DEVICE CHARACTERISTICS

Forward Gummel plots of the poly-emitter vertical NPN (0.45x1.35 μm^2) and PNP (0.6x1.8 μm^2) bipolar transistors are illustrated in Fig. 2. I_B tracks V_{BE} ideally down to 1 pA and below. The maximum current gain (beta) is approximately 140 and 70 for the NPN and PNP devices (Fig. 3). Beta plateau across six orders of magnitude of collector current has been obtained. Peak cut-off frequency (f_t) of NPN has been enhanced from 18 to 26 GHz using a pedestal implant process in the emitter window (Fig. 4). Compromise in BV_{CEO}, from 5.5 V to 4.0 V, has to be made as a result of the pedestal implant. The measured unity gain cut-off frequency for the PNP transistor is 4 GHz.

The MOSFET output device characteristics are shown in Fig. 5. The 3.3 V $I_{D,SAT}$ of the 100 Å T_{ox} NMOS (0.40 μm) and PMOS (0.38 μm) are 440 μA/μm and 210 μA/μm respectively. V_T roll-off and subthreshold device characteristics are well behaved, as illustrated in Fig. 6 and Fig. 7. Natural MOSFET's processed with both the V_T adjustment and punch-through implants blocked are also available in this technology. Fig. 8 shows the quasi-static capacitance-voltage characteristics of the implanted-gate MOS capacitors. By optimizing the process integration, neither boron penetration [3] nor depletion of the n^+/p^+ poly gate under inversion operating conditions [4] are noticeable. Hot carrier immunity of the MOSFET's have been evaluated under dc drain-avalanche stress conditions. The

Reprinted from *IEDM 1991*, pp. 4.1.1–4.1.4, 1991.

extrapolated lifetime at a supply voltage of 3.3 V is beyond 20 years for the 0.4 μm L_{eff} n-channel MOSFET.

A $TiSi_2$/n-well Schottky barrier diode (SBD) has been developed for termination of the transmission lines, clamping of the bipolar transistors, and other circuit applications. The room temperature current-voltage characteristics of the SBD is shown in Fig. 9. Forward ideality factor is 1.05. Junction leakage is low and tightly distributed by isolating the SBD from the field periphery with a p^+ guard ring. Barrier height of the $TiSi_2$ SBD is 0.58 eV, extracted from the well behaved temperature dependence of the reverse leakage current shown in Fig. 10. Gummel plot and output characteristics of the NPN bipolar transistor clamped with a $TiSi_2$ SBD is shown in Fig. 11. The background base current introduced by the SBD is less than 10 pA/μm^2. Collector-to-emitter saturation voltage (V_{CE_SAT}) is approximately 0.52 V for the SBD-clamped NPN transistor. Distribution of the $TiSi_2$ SBD device parameters are excellent, such as the junction leakage, ideality factor, and V_{CE_SAT} of the SBD-clamped NPN transistor (Fig. 12).

For cost-effective C-BiCMOS applications, a gated lateral-PNP device is available without any extra processing steps. A combination of MOSFET and bipolar device operations have been utilized for the lateral bipolar transistors [5]. The lateral-PNP performance improvement obtained in this work is achieved by realizing the gate-assisted device operation in combination with the sub-0.5 μm PMOS L_{eff}, thin gate oxide (100 Å), and low PMOS threshold voltage. As shown in Fig. 13, peak current gain above 100 can be obtained on the lateral-PNP transistors provided in this technology.

The four-level metallization module processed with the stacked and plugged contacts and vias is illustrated by the SEM photomicrograph of Fig. 14. Tungsten (W) contact and via plugs are formed in an integrated deposition-and-etchback process. TiN layers are deposited, both as the glue layer and as the electromigration barrier (between W and Al), before and after the W-plug formation process. Plasma-enhanced oxide and nitride have been used as the final passivation materials.

To derive the optimal device design targets, a process window analysis has been performed for the CMOS portion of this technology. Fig. 15 is an example of the experimentally determined NMOS T_{ox} and L_{eff} process boundaries for a supply voltage of 3 V. The device and circuit requirements selected for this analysis include: drain-induced-barrier-lowering (DIBL) < 40 mV/V, I_{SUB} < 750 nA/μm, $I_{D,SAT}$ > 350 μA/μm, gate oxide E_{max} < 4 MV/cm, and ring-oscillator (loaded with fan-out of 3) propagation delay, t_{PD} < 150 ps. It is clearly shown in Fig. 15 that the L_{eff} distribution is by far the most critical parameter in meeting these targeted device and circuit performances.

CIRCUIT PERFORMANCE

A 68030 speed path consisting of 11-inversion stages with 3 of the stages heavily loaded (~ 4 pF each) has been used for bench marking this advanced BiCMOS technology. As shown in Fig. 16 comparing the CMOS and BiCMOS implementations of this 68030 speed path at room temperature, the measured BiCMOS speed advantage can be sustained at a supply voltage of 2.5 V by employing a p-channel MOSFET to supply the base current of the pull-down NPN transistor. For 3.3 V product generation, BiCMOS is the apparent technology choice for high-performance microprocessor applications.

CONCLUSION

A versatile 0.4 μm, single-poly, four-level-metal, C-BiCMOS technology has been developed for advanced logic and MPU applications. Process integration, device characteristics, and circuit performance have been described for this logic BiCMOS technology. The modular C-BiCMOS process architecture developed in this work provides both high performance and manufacturability in the arena of sub-0.5 μm logic and MPU product families.

ACKNOWLEDGEMENTS

The authors would like to thank R. Cardwell for his technical support, Y.-S. Kim for ac measurements, R. Bertram and M. Conner for layout design of the test vehicle, P. Madrid for his work on the process window analysis, and the APRDL pilot line for wafer processing. L. Parrillo's management support is also greatly appreciated.

REFERENCES

[1] S. W. Sun et al., Proc. IEEE CICC, 18.7 (1989).
[2] B-Y. Nguyen et al., ECS Abstract, 90-1, p. 614 (1990).
[3] F. K. Baker et al., IEEE IEDM Tech. Dig. p. 443 (1989).
[4] C. Y. Wong et al., IEEE IEDM Tech. Dig. p. 238 (1988).
[5] K. K. Au et al., Proc. IEEE CICC, 18.5 (1990).

CMOS:	NMOS	PMOS
Gate Oxide, T_{ox} (Å)	100	100
Threshold Voltage, V_T (V)	0.60	- 0.60
Channel Length, L_{eff} (μm)	0.40	0.38
$I_{D,SAT}$ at 3.3 V (μA/μm)	440	210
S/D, Gate R_S (Ohm/Sq.)	4.0	4.0
BIPOLAR:	NPN	PNP
Emitter Size, A_E (μm^2)	0.45x1.35	0.6x1.8
Current Gain, Beta	140	70
Peak f_t (GHz)	18~26	4
BV_{CEO} (V)	5.5~4.0	- 6.5
BV_{EBO} (V)	6.0	- 7.0

Table 1. Typical Process and Device Parameters.

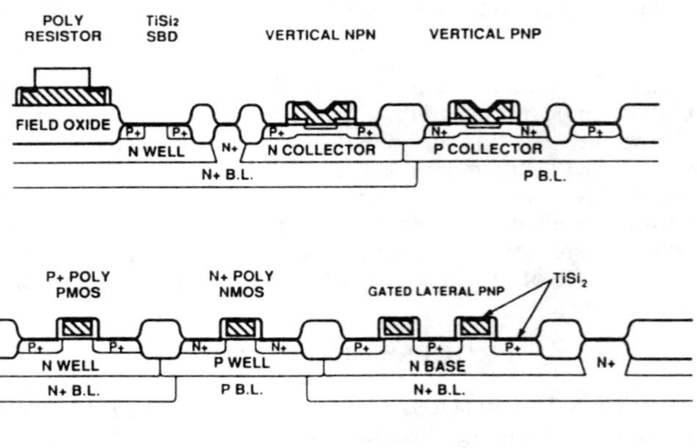

Fig. 1. Schematic cross-section of the 0.4-micron fully complementary BiCMOS technology.

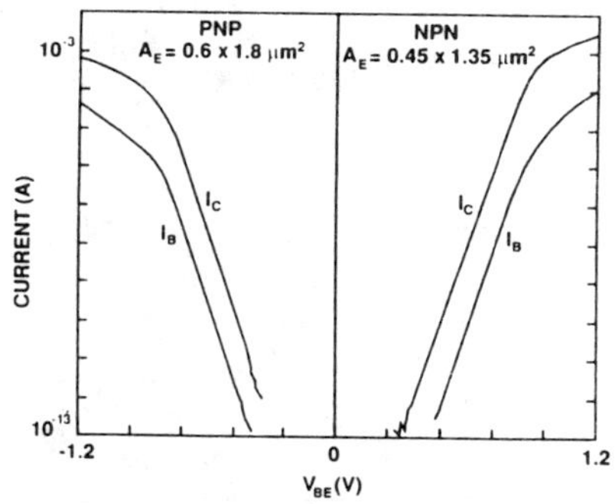

Fig. 2. Forward Gummel plots of the poly-emitter, vertical NPN and PNP bipolar transistors.

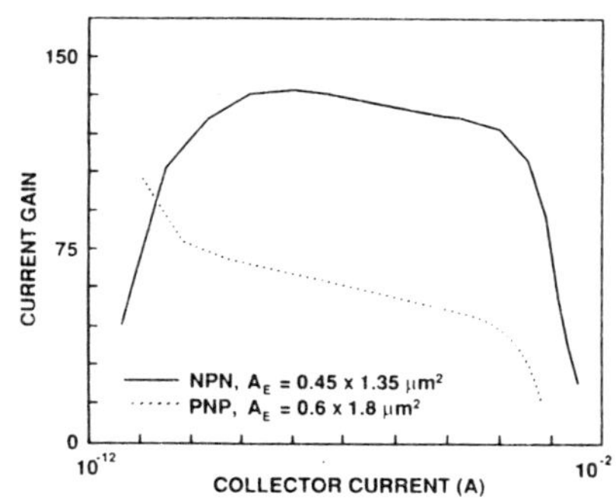

Fig. 3. NPN and PNP beta versus collector current.

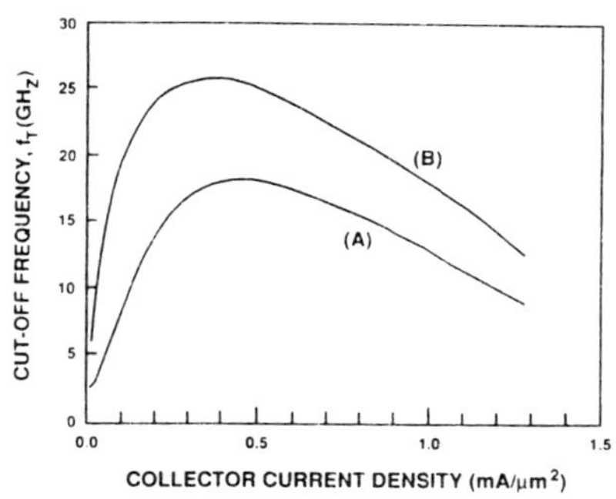

Fig. 4. f_t vs. J_C of the NPN transistors, with the pedestal implant doses of zero (A), and $2x10^{12}$ cm^{-2} (B) in the emitter window.

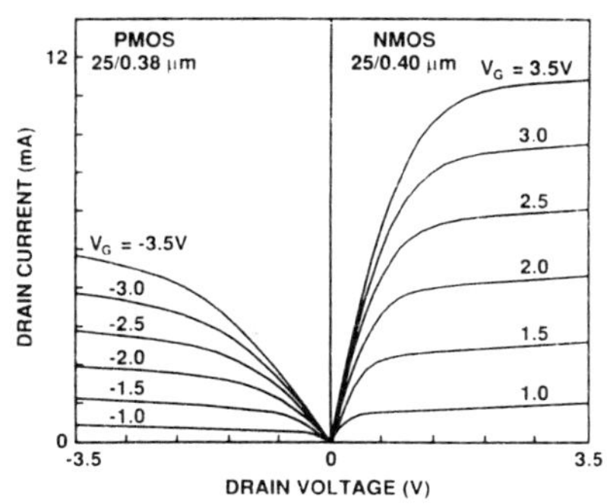

Fig. 5. Output device characteristics of the 100 Å T_{OX}, n-channel and p-channel MOSFET's.

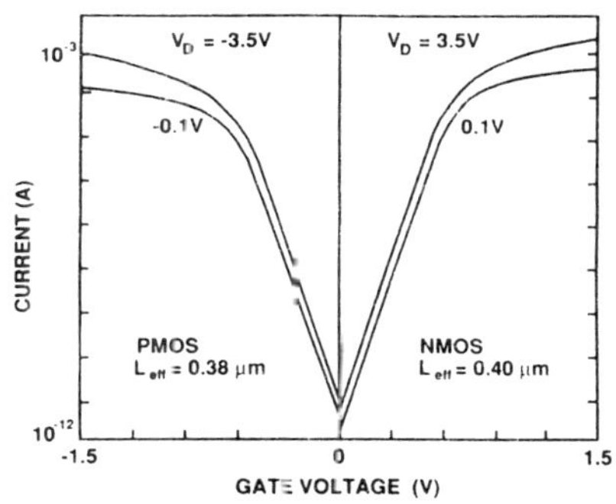

Fig. 6. Subthreshold I-V's of the NMOS and PMOS.

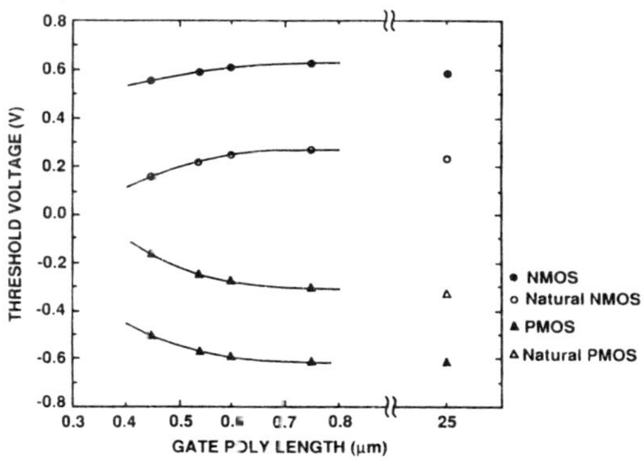

Fig. 7. V_T roll-off characteristics of the MOSFET's.

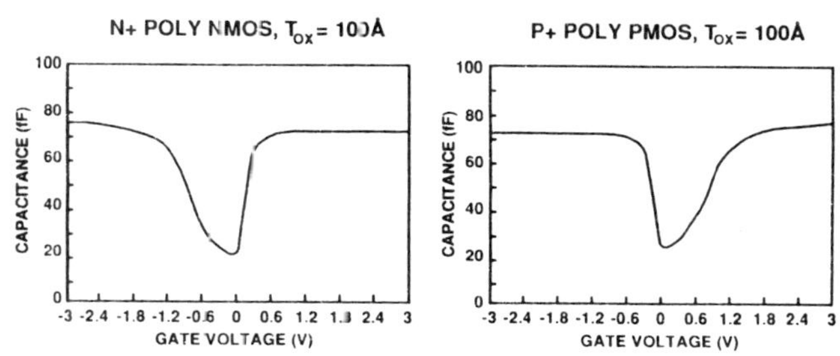

Fig. 8. Quasi-static CV characteristics of the MOS capacitors processed with the ion-implanted polysilicon gate.

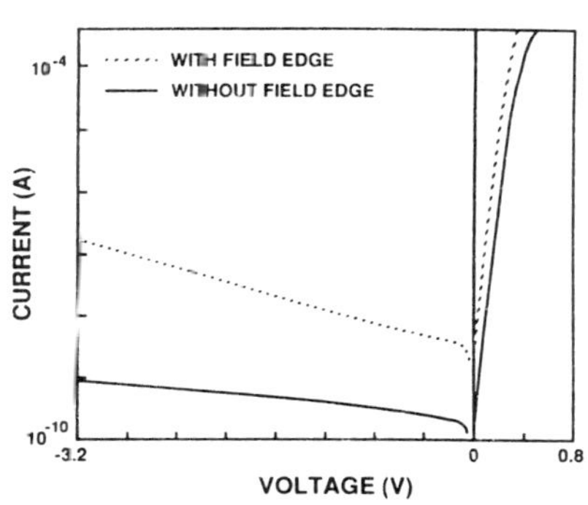

Fig. 9. $TiSi_2$ SBD I-V curves at room temperature.

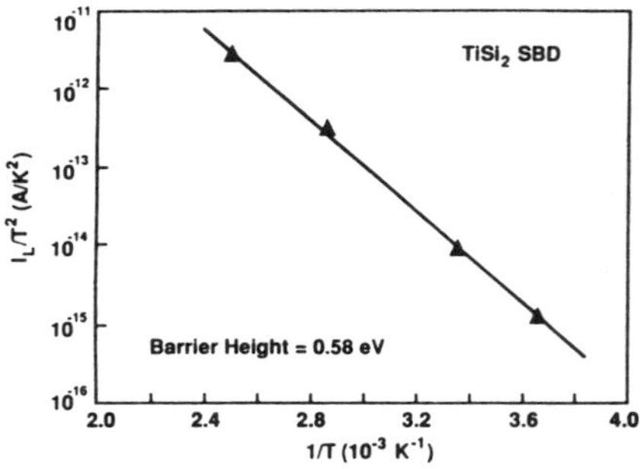

Fig. 10. Temperature dependence of the reverse leakage current of the $TiSi_2$/n-well SBD.

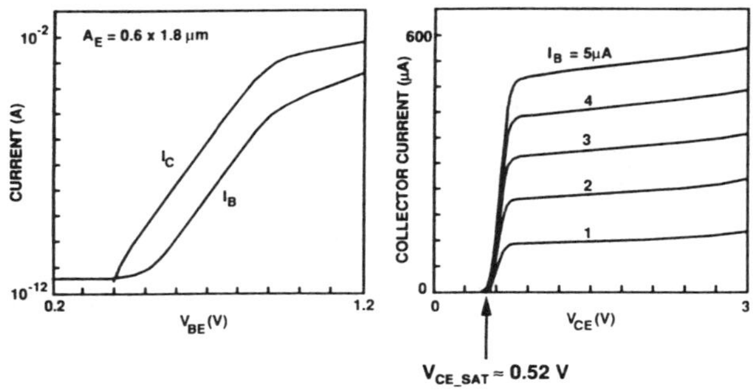

Fig. 11. Device characteristics of the NPN bipolar transistor clamped with a $TiSi_2$ SBD.

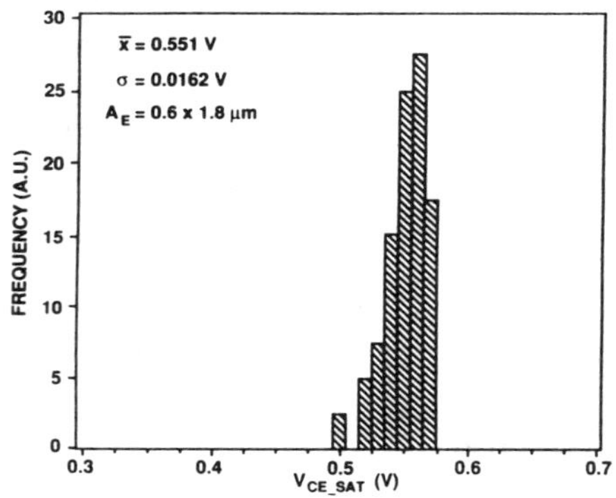

Fig. 12. V_{CE_SAT} histogram of the NPN bipolar transistor clamped with a $TiSi_2$/n-well SBD.

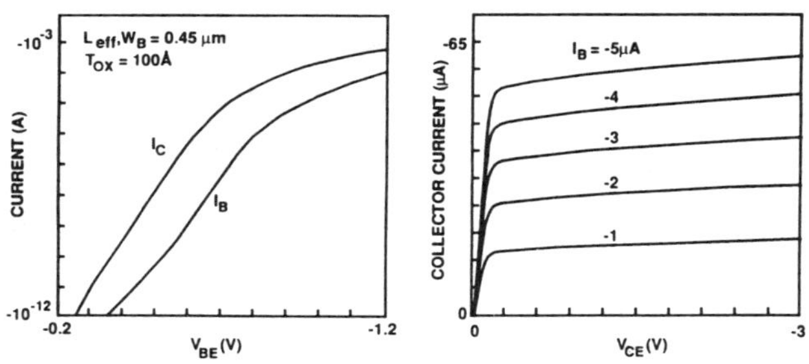

Fig. 13. Gummel plot and output transistor characteristics of the gated lateral-PNP device.

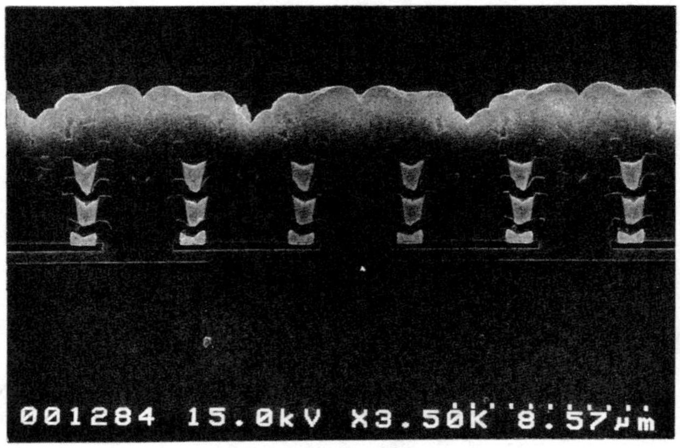

Fig. 14. SEM photomicrograph of the four-level metal module, with the stacked and plugged contacts and vias.

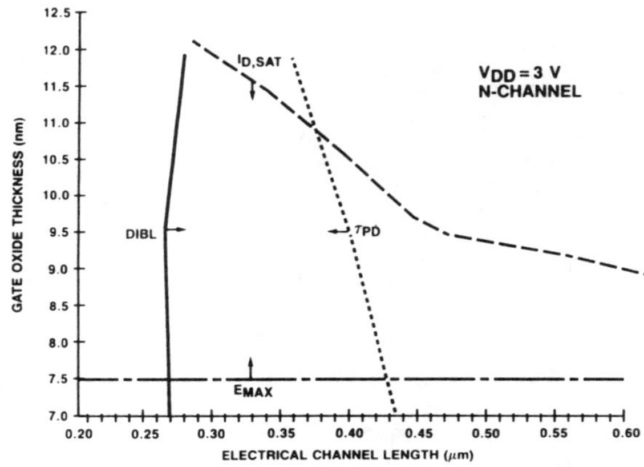

Fig. 15. Experimentally determined T_{OX} and L_{eff} process window of the n-channel MOSFET at a supply voltage of 3 V.

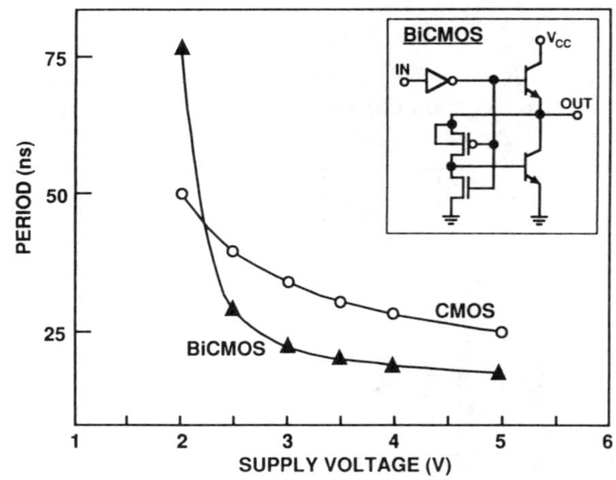

Fig. 16. 68030 speed path comparison between the CMOS and the BiCMOS implementations.

Paper 1.10

Analysis and Optimization of BiCMOS Digital Circuit Structures

S. H. K. Embabi, A. Bellaouar, and M. I. Elmasry

Abstract—**This brief paper presents analytical analysis and performance optimization for the transient behavior of some basic BiCMOS circuit structures. The analysis offers insight into the critical circuit and device parameters that affect the performance of these circuits. Techniques to improve the speed of each structure and the trade-off factors involved in designing such circuits are discussed.**

I. Introduction

IN THIS brief paper we present circuit analyses and a discussion for optimizing three basic BiCMOS digital circuit structures: BiCMOS buffer, NMOS/CML, and ECL/CMOS interface circuits. The analytical modeling of the transient behavior offers insight into the critical circuit and device parameters that affect the performance of these circuits and is used to develop a first-order scaling theory [1]. The derived delay expressions in this work can also be used in CAD tools for optimizing BiCMOS circuits and systems. To verify our models, we compared the results of the analyses with SPICE3B1 simulations using a generic 1-μm BiCMOS technology. In SPICE, we have used model level-4 for the MOS devices and the Gummel–Poon model for the bipolar transistors. The key bipolar model parameters are indicated in Fig. 1(c).

II. BiCMOS Pull-up/Pull-Down Buffer

The delay analysis of the BiCMOS buffer (Fig. 1(a)) presented in this work is simple and accounts for low- and high-level injection conditions in the bipolar transistors, which were not accounted for in other work [2]. Our analysis is fully analytical as opposed to previously reported semi-analytical analysis [3]. For the transient analysis we use the equivalent circuit of the pull-up section as shown in Fig. 1(b). For the delay calculation, we assume that the input MOS devices operate in the saturation region. This assumption is only valid for submicrometer MOSFET's, which have small saturation voltage [4]. The PMOS can thus be modeled by a current source $I_{D_{sat}}$ ($I_{D_{sat}} = Kv_{sat}C_{ox}W_p(V_{GS} - V_T)$). The current gain of the BJT is modeled [5] by $\beta = \beta_o/(1 + I_c/I_K)$ to account for high-level injection (HLI), where β_o is the current gain under low-level injection (LLI) conditions and I_K is the knee current.

In this work the 50% rise time is derived. However, the derived expressions are valid for the fall time. The rise time t_d is considered to be the summation of three components: t_1 is the time required for the pull-up BJT to be turned on; t_2 is the time elapsed between the end of t_1 and the time at which the collector current reaches its peak; and t_3 is the

Manuscript received October 26, 1989; revised November 12, 1990. This work was supported in part by NSERC and ITRC research grants.

The authors are with the VLSI Group, Department of Electrical and Computer Engineering, University of Waterloo, Waterloo, Ont., Canada N2L 3G1.

IEEE Log Number 9042355.

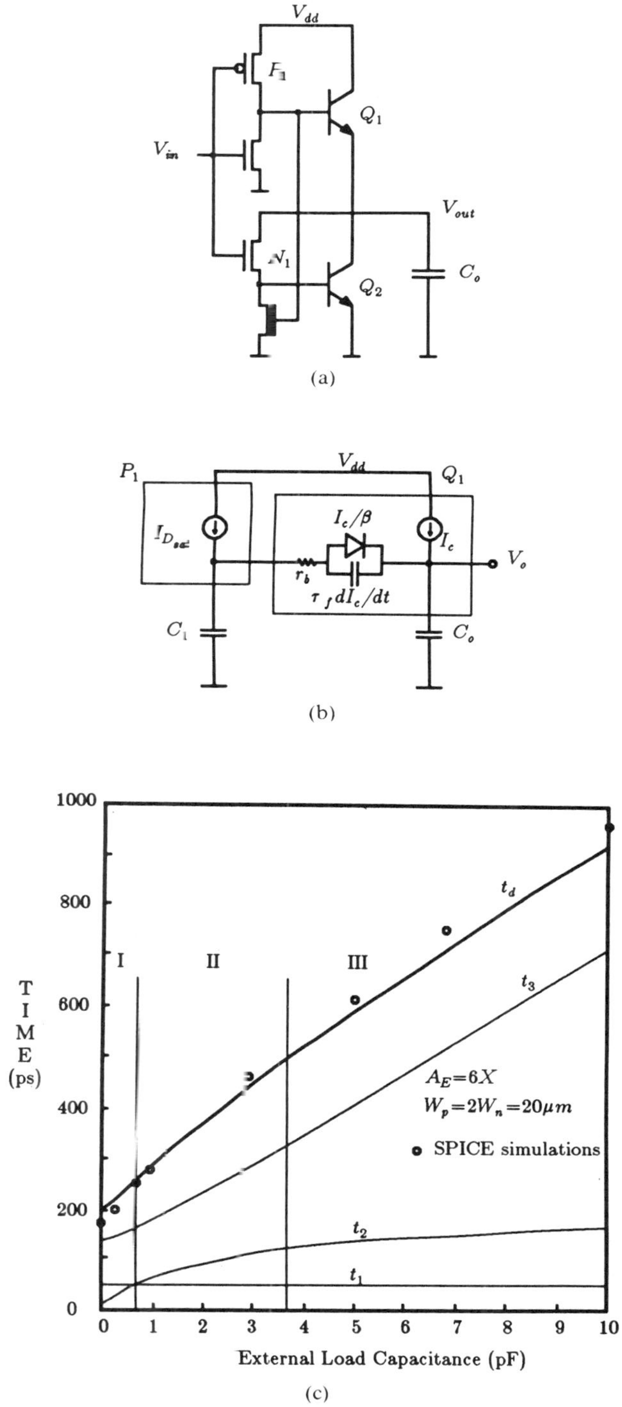

Fig. 1. (a) BiCMOS buffer. (b) Model of the BiCMOS buffer. (c) The effect of external loading on the delay components. Key SPICE model parameters of the BJT ($A_E = 1X = 0.8\times 4\ \mu m^2$): $\beta_f = 100$, $\tau_f = 6$ ps, $I_{Kf} = 2$ mA, $C_{js} = C_{jc} = 7$ fF, $C_{je} = 30$ fF, $R_c = 150\ \Omega$, $R_b = 200\ \Omega$.

Reprinted from *IEEE J. Solid-State Circuits,* vol. 26, no. 4, pp. 676–679, April 1991.

time elapsed between the end of t_2 and the time when the output voltage reaches $V_{dd}/2$. These three delay components are given by [6]

$$t_1 = \frac{C_1 V_{be_{on}}}{I_{D_{sat}}} \tag{1a}$$

$$t_2 = \frac{\tau_f}{I_{D_{sat}}}\left(\alpha(C_1+C_o) - I_{D_{sat}}\right) \approx \begin{cases} \tau_f \dfrac{C_o}{C_1 + C_o/\beta_o}, & C_o < C_{o_{crit}} \text{ (LLI)} \\ \tau_f \left(\left(\dfrac{\beta_o I_K}{I_{D_{sat}}} \right)^{1/2} - \dfrac{I_K}{I_{D_{sat}}} \dfrac{C_o + \beta_o C_1}{C_o + C_1} \right), & C_o \gg C_{o_{crit}} \text{ (HLI)} \end{cases} \tag{1b}$$

$$t_3 = \frac{V_s}{2\alpha} \approx \begin{cases} \dfrac{V_s}{2 I_{D_{sat}}} \left(\dfrac{C_o}{\beta_o} + C_1 \right), & C_o < C_{o_{crit}} \text{ (LLI)} \\ \dfrac{V_s}{2} \left(\dfrac{C_o + C_1}{\sqrt{\beta_o I_K I_{D_{sat}}}} + \dfrac{C_o + \beta_o C_1}{2} \dfrac{1}{\beta_o I_{D_{sat}}} \right), & C_o \gg C_{o_{crit}} \text{ (HLI)} \end{cases} \tag{1c}$$

where

$$\alpha = -0.5B + 0.5\sqrt{B^2 + C}$$

with

$$B = \frac{I_K}{C_1 + C_o} \left(\frac{C_o + C_1 \beta_o}{C_1 + C_o} - 2 \frac{I_{D_{sat}}}{I_K} \right)$$

and

$$C = \frac{4 \beta_o I_K I_{D_{sat}}}{(C_1 + C_o)^2}.$$

V_s is the output voltage swing, τ_f is the forward transit time, C_1 represents the parasitic capacitance at the base of the pull-up BJT, and C_o is the total output capacitance. The definition of $C_{o_{crit}}$ is given by

$$C_{o_{crit}} = \frac{\beta_o C_1}{2(\beta_o I_{D_{sat}}/I_K)^{1/2} - 1}. \tag{2}$$

Equations (1b) and (1c) show that the general expressions for t_2 and t_3 can be simplified for high- and low-level injection conditions. Fig. 1(c) shows t_1, t_2, t_3, and t_d versus the external load capacitance C_L along with SPICE results for t_d. In region I, where $C_o \ll C_{o_{crit}}$, the bipolar transistor operates under low-level injection. The t_3 component is the major contributor to the delay time. However, the sensitivity of the delay to load capacitance in this region is determined by t_2, therefore, the slope of the delay is a function of τ_f/C_1. In regions II and III, the bipolar transistor operates under high-level injection. In region III, the contribution of t_2 to the total delay and its effect on the delay sensitivity to C_L are not significant provided that τ_f is in the order of 10 to 20 ps. As C_L increases, the delay becomes less dependent on τ_f and more dependent on C_L, V_s, and I_K. The delay time depends on the emitter area of the output BJT's. A too small emitter area causes high-level injection in the output bipolar transistors while a too large emitter area means larger parasitic capacitances. A near-optimum emitter area can be found by differentiating the expression of t_3 under the high-level injection condition in (1c) with respect to emitter area and equating to zero.

III. NMOS/CML Interface Circuit

The NMOS/CML interface circuit shown in Fig. 2(a) is composed of a pull-up resistor R_i and a pull-down NMOS driving a CML gate and is modeled by the equivalent circuit shown in Fig. 2(b). C_i is the parasitic capacitance of the drain of the NMOS, C_1 is the sum of the bipolar base–emitter and base–collector capacitances, r_b is the base resistance of the BJT's (Q_1 and Q_2) of the CML, and I_o is the current source of the CML.

A. Rise Time

The rise time is expressed as the summation of two components.

1) t_{or} is the time required to turn transistor Q_1 on and is given by [6]

$$t_{or} = \tau_c \ln\left(\frac{2}{1 + 2(V_{be}(I_o) - V_{be_{on}})/V_s} \right) \tag{3}$$

where $\tau_c = R_i C_i + R_i C_1 + r_b C_1$ and V_s is the voltage swing at the input of the CML.

2) t_1 is the time required for the output to rise to the 50% point of swing. The nodal analysis for this circuit yields a transcendent expression for the output voltage, which can be simplified over certain domains for the ratio τ_s/τ_l, where τ_s represents the time required for the CML current to switch from one branch to the other and can be approximated by [6]

$$\tau_s \approx (R_i + 2r_b)(\tau_f/R_L)\left(1 + \left(1 - 2r_b R_L \tau_c/\tau_f(R_i + 2r_b)^2\right)^{1/2}\right).$$

The load time constant τ_l is given by $\tau_l = R_L C_o$. Using the simplified expressions of the output voltage, we can derive the following equation for t_1 [6]:

$$t_1 = \begin{cases} 0.7\tau_l, & \tau_l \gg \tau_s \\ 1.2\tau_l = 1.2\tau_s, & \tau_l = \tau_s \\ 0.4\tau_s, & \tau_l \ll \tau_s. \end{cases} \tag{4}$$

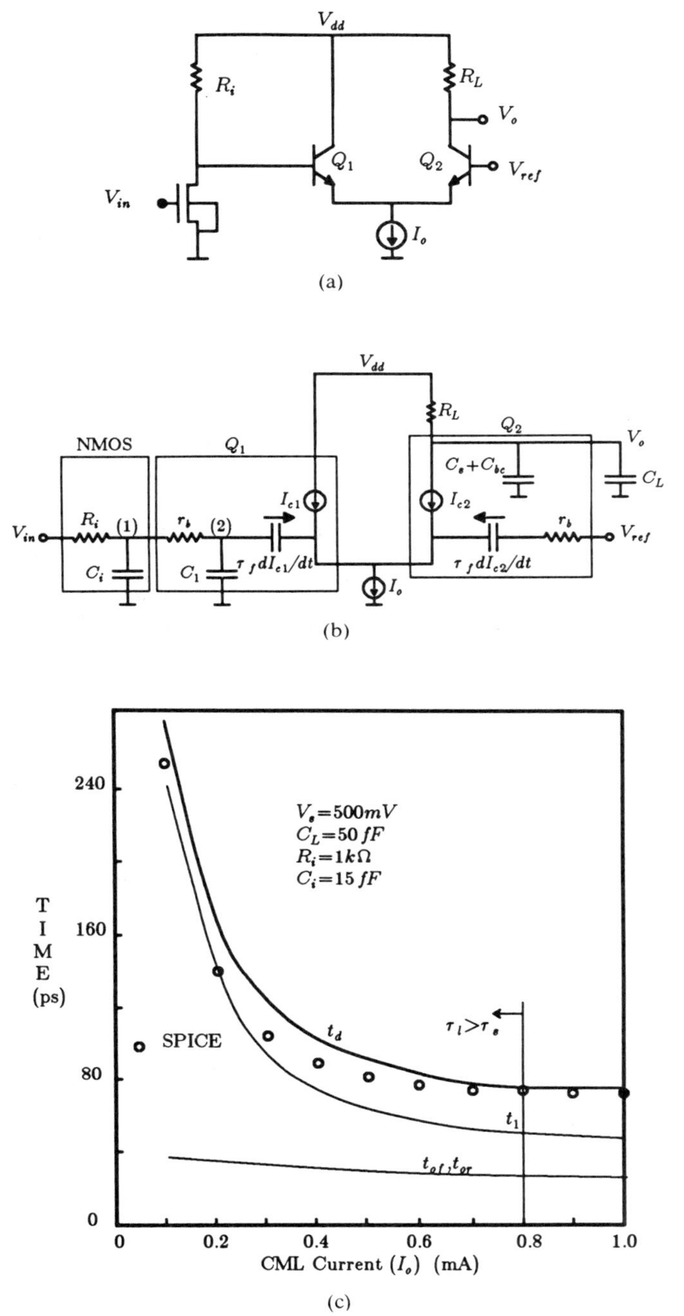

Fig. 2. (a) NMOS/CML interface circuit. (b) Model of NMOS/CML interface circuit. (c) Delay time and its components versus the CML current.

B. Fall Time

Using the model shown in Fig. 2(b) the fall-time analysis is treated in a similar manner as the rise-time analysis. The fall time is considered to be the summation of the following components.

1) t_{of}, which is the time required for Q_2 to turn on, is given by [6]

$$t_{of} = \tau_c \ln\left(\frac{2}{1+2\Delta V_{be}/V_s + 2(R_i + r_b)I_o/V_s\beta_o}\right) \quad (5)$$

where ΔV_{be} is the difference between the base–emitter voltage when the collector current reaches its maximum (I_o) and the turn-on base–emitter voltage.

2) t_1 is the time required for the output to reach the midpoint of the swing after Q_2 turns on and is given by (4).

Fig. 2(c) shows the dependence of t_{of}, t_{or}, t_1 and the total average delay on the current source of the CML. The total delay approaches its minimum as the ratio τ_l/τ_s approaches unity. Fig. 2(c) also shows that the average of the turn-on and turn-off times of the transistor Q_1 is about 30% of the total delay time at the minimum delay. The figure illustrates the good agreement between the analytical model and the results of SPICE simulations.

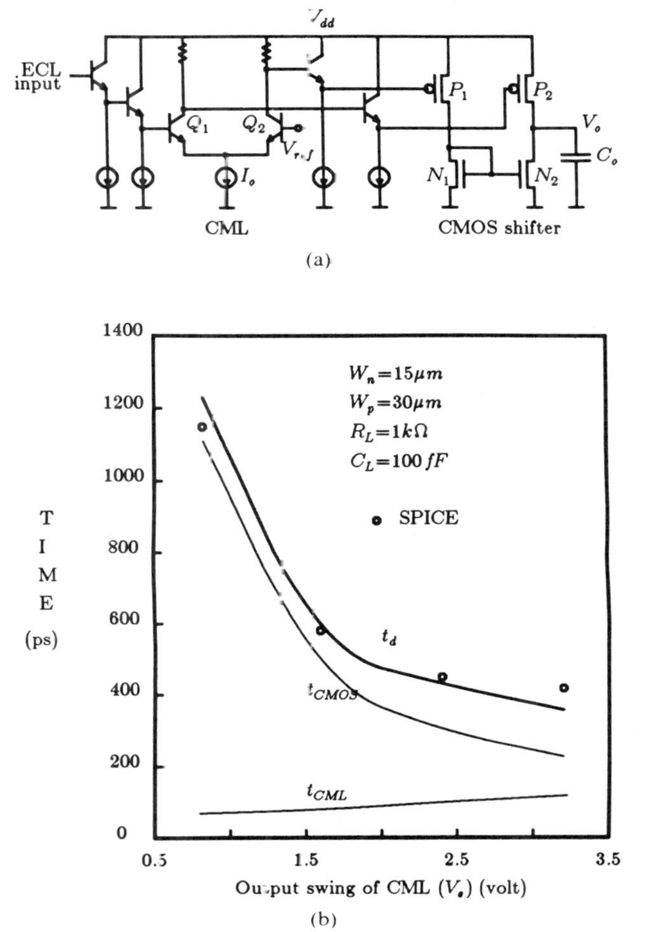

Fig. 3. (a) ECL/CMOS interface circuit. (b) The effect of the CML output swing on the delay components.

IV. ECL/CMOS Interface Circuit

The delay time of the ECL/CMOS interface circuit (Fig. 3(a)) is the sum of the delay of the CML (t_{CML}) and the delay of the CMOS shifter (t_{CMOS}). In our analysis the delay of the emitter followers is neglected. Based on the analysis of the NMOS/CML, it can be shown that the delay of the CML can be given by [6]

$$t_{CML} = 0.7\tau_l = 0.7\frac{V_s}{I_o}\left(C_{cs_{CML}} + C_{bc_{CML}} + C_{bc_{EF}} + \frac{C_{PMOS}}{1+\beta_o}\right) \quad (6)$$

where $C_{cs_{\mathrm{CML}}}$ and $C_{bc_{\mathrm{CML}}}$ are the collector–substrate and base–collector capacitances of the CML bipolar transistors, respectively. $C_{bc_{\mathrm{EF}}}$ is the base–collector capacitance of the emitter-follower bipolar transistor and C_{PMOS} is the input capacitance of the PMOS transistor.

The average 50% delay time of the CMOS shifter is given by [6]

$$t_{\mathrm{CMOS}} = 1.2\frac{C_{cg}}{K_n v_{\mathrm{sat}} C_{ox} W_n} + \frac{C_o}{I_p}\frac{V_{dd}}{2}. \tag{7}$$

I_p is the saturation current of P_1, C_o is the output capacitance, v_{sat} is the saturation velocity, and C_{cg} is the parasitic capacitance of the common gate node. C_{ox} and W_n are the oxide capacitance and the width of the transistor N_1. K_n is the saturation current factor of N_1.

The key issues of the design of the ECL/CMOS converter are the output swing of the CML stage and the sizing of the MOS devices of the CMOS shifter. Increasing the swing degrades the delay of the CML and improves that of the CMOS shifter. Fig. 3(b) shows that increasing the swing to 2.5 V improves the total delay. Beyond that point, the rate of increase in t_{CML} approaches the rate of the decrease of t_{CMOS}. The current source of the CML can be determined based on the condition $\tau_s = \tau_l$. The CMOS shifter can be designed separately because its loading on the CML can be neglected as shown by (6). Increasing the size of the PMOS transistors reduces the delay of the CMOS shifter as shown by (7) provided that the external load capacitance dominates over the parasitic capacitance at the output node. The size of the NMOS transistors does not affect the average delay as such, however, SPICE simulations have shown that taking W_n equal to half W_p gives symmetrical rise and fall times.

V. Conclusion

In this brief paper we have presented analytical analyses for the delay of some basic digital BiCMOS circuits, based on device models that account for many of the features of the state-of-the-art devices such as short-channel effects in MOS transistors. The delay expressions offer an insight into the dependencies of speed on the different circuit and device parameters and the trade-offs involved in the design and optimization of such circuits.

References

[1] A. Bellaouar, S. H. K. Embabi, and M. I. Elmasry, "Scaling of digital BiCMOS circuits," *IEEE J. Solid-State Circuits*, vol. 25, no. 4, pp. 932–941, Aug. 1990.

[2] E. W. Greenich and K. L. McLaughlin, "Analysis and characterization of BiCMOS for high-speed digital logic," *IEEE J. Solid-State Circuits*, vol. 23, no. 2, pp. 566–572, Apr. 1988.

[3] G. P. Rosseel and R. W. Dutton, "Influence of device parameters on the switching speed of BiCMOS buffers," *IEEE J. Solid-State Circuits*, vol. 24, no. 1, pp. 90–99 Feb. 1989.

[4] K.-Y. Toh *et al.*, "An engineering model for short-channel MOS devices," *IEEE J. Solid-State Circuits*, vol. 23, no. 4, pp. 950–958, Aug. 1988.

[5] K. Nakazato *et al.*, "Characteristics and scaling properties of n-p-n transistors with a sidewall base contact structure," *IEEE Trans. Electron Devices*, vol. ED-32, no. 2, pp. 328–332, Feb. 1985.

[6] S. H. K. Embabi, A. Bellaouar, and M. I. Elmasry, "Analysis of BiCMOS digital circuit structures," Dept. of Elec. and Computer Eng., Univ. of Waterloo, Waterloo, Ont., Canada, Internal Rep., 1989.

Paper 1.11

A 1.5-V Full-Swing BiCMOS Logic Circuit

Mitsuru Hiraki, *Member, IEEE*, Kazuo Yano, *Associate Member, IEEE*, Masataka Minami, Kazushige Sato, Nozomu Matsuzaki, Atsuo Watanabe, Takashi Nishida, Katsuro Sasaki, *Member, IEEE*, and Koichi Seki, *Member, IEEE*

***Abstract*—A BiCMOS logic circuit applicable to sub-2-V digital circuits has been developed for the first time. A transiently saturated full-swing BiCMOS (TS-FS-BiCMOS) logic circuit operates twice as fast as CMOS at 1.5-V supply. A newly developed transient-saturation technique, with which bipolar transistors saturate only during switching periods, is the key to sub-2-V operation because a high-speed full-swing operation is achieved to remove the voltage loss due to the base-emitter turn-on voltage. Furthermore, both small load dependence and small fan-in dependence of gate delay time are also attained with this technique. A two-input gate fabricated with 0.3-μm BiCMOS technology verifies the performance advantage of TS-FS-BiCMOS over other BiCMOS circuits and CMOS at sub-2-V supply.**

I. Introduction

THE switching speed of BiCMOS circuits is seriously deteriorated as the supply voltage is reduced. This difficulty limits the performance of low-voltage BiCMOS ULSI's. In conventional BiCMOS circuits (Fig. 1(a)), which are widely used at 5-V supply, the gate-source voltage of $M1$ is reduced by $2V_{BE}$. This voltage loss seriously degrades the switching speed of the circuit when the supply voltage is reduced to about 3 V [1]. This voltage loss is caused by two factors. One factor is that the fast output voltage transition obtained by the bipolar transistor $Q1$ is V_{BE} smaller than the supply voltage. The other factor is that the source node voltage of $M1$ rises up to V_{BE}.

To improve the low-voltage performance of BiCMOS circuits, several types of BiCMOS circuits have been proposed, such as complementary BiCMOS (C-BiCMOS) [2], quasi-complementary BiCMOS (QC-BiCMOS) [3]–[5], merged BiCMOS (MBiCMOS) [6], and BiNMOS [7], and the performance advantage of these BiCMOS circuits has been demonstrated at 3-V supply. The idea behind these BiCMOS circuits is to ground the source node of $M1$ to remove one V_{BE} voltage loss. As an example, Fig. 1(b) shows the configuration of C-BiCMOS. Although the voltage rising of the source node is prevented, these circuits also suffer from the one V_{BE} voltage loss in the output voltage transition. This is because these circuits have the same pull-up driver configuration as the conventional BiCMOS circuit. Because of this voltage loss, the switching speed is seriously degraded even with these circuits when the supply voltage is further reduced to about 2 V. For sub-2-V BiCMOS ULSI's, therefore, a high-speed full-swing circuit technique is essential.

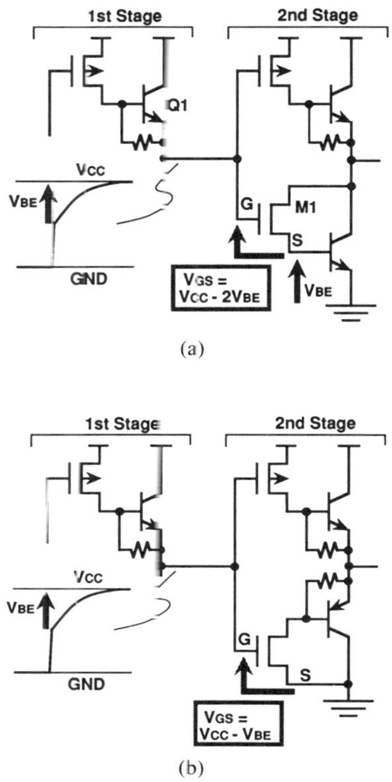

Fig. 1. Voltage loss in reported BiCMOS circuits: (a) a conventional BiCMOS circuit, and (b) C-BiCMOS.

This paper describes a novel full-swing BiCMOS logic circuit that breaks through the limit for sub-2-V operation [8]. This full-swing BiCMOS logic circuit operates twice as fast as CMOS at 1.5-V supply. The newly developed transient-saturation technique enables high-speed full-swing operation. First the concept of the full-swing BiCMOS circuit using this transient saturation technique is presented in Section II. Next, some simulated performance of the circuit are described in Section III, and design issues of the circuit are discussed in Section IV. Then, some experimental results are shown in Section V. Finally, Section VI summarizes this work.

Manuscript received April 6, 1992; revised June 18, 1992.

M. Hiraki, K. Yano, M. Minami, N. Matsuzaki, T. Nishida, K. Sasaki, and K. Seki are with the Central Research Laboratory, Hitachi Ltd., 1-280 Higashi-koigakubo, Kokubunji-shi, Tokyo 185, Japan.

K. Sato and A. Watanabe are with the Device Development Center, Hitachi Ltd., Tokyo 198, Japan.

IEEE Log Number 9202809.

Reprinted from *IEEE J. Solid-State Circuits*, vol. 27, no. 11, pp. 1568–1574, Nov. 1992.

II. Circuit Concept

To achieve a high-speed full-swing operation in BiCMOS circuits, full-swing operation of bipolar transistors must be realized. The saturation type circuit shown in Fig. 2(a) achieves full-swing operation of a bipolar transistor. But the switching speed of a saturated bipolar transistor is slow because once the excess minority carriers are charged into the base, they remain there until the next switching period [9], [10]. The full-swing BiCMOS circuit that we propose is based on the concept of saturating the bipolar transistor transiently. The idea of this transient-saturation technique is illustrated in Fig. 2(b). The transient-saturation technique enables the bipolar transistor to achieve a high-speed full-swing operation because the excess minority carriers are first charged into the base and then discharged out of the base immediately after the output voltage transition.

The configuration of the full-swing BiCMOS circuit using the transient-saturation technique is shown in Fig. 3(a). Simulated waveforms are also provided in Fig. 3(b) to explain the transient operation of this circuit. Since this is a noninverting circuit, the output is high when the input is high. When the input falls, the pull-down transition occurs as follows. $MP2$ turns on, and the base of $Q1$ is charged through the current path from V_{CC} to $B1$. $Q1$ continues driving the load until the output voltage nearly reaches zero. Thus, $Q1$ achieves full-swing operation. Although $Q1$ saturates, this does not slow the next pull-up transition because the excess minority carriers of $Q1$ are discharged immediately after the pull-down transition through $MN4$, which turns on as the feedback signal FB is pulled high. Note that the node voltage of $B1$ falls nearly to zero immediately after the pull-down transition. The pull-up transition is easily understood because of the symmetrical circuit configuration. Thus, bipolar transistors saturate transiently, only in the switching period. Therefore, the proposed BiCMOS circuit is called transiently-saturated full-swing BiCMOS (TS-FS-BiCMOS).

Static power dissipation is avoided because $MN3$ cuts off the dc current path after the pull-up transition and $MP3$ cuts it off after the pull-down transition. An arbitrary noninverting logic function is realized by replacing $MN1$ and $MN2$ with nMOS logic, and $MP1$ and $MP2$ with pMOS logic. To realize an inverting logic function in TS-FS-BiCMOS circuits, a CMOS gate should be placed as the stage before a TS-FS-BiCMOS buffer.

The static output voltage is held by the small-size CMOS latch composed of $INV1$ and $INV2$. Although both $Q1$ and $Q2$ are off in steady state, this circuit suffers little from a switching noise at the output node. The reason for this is as follows. If the switching noise is large enough to flip the CMOS latch, the bipolar transistor ($Q1$ or $Q2$) begins to drive the load again. As a result, the output voltage quickly goes back to the static level. If the switching noise is small enough not to flip the CMOS latch, it does not flip the next stage gate, either.

To verify the advantage of the transient-saturation technique, the circuit operation was simulated for the configuration without $MN4$. With this configuration, $Q1$ continues to saturate even after the pull-down transition. The simulated waveforms are shown with hatched lines in Fig. 3(b). These show that the configuration without the transient-saturation technique results in about 2.5 times longer delay time owing to the excess minority carriers.

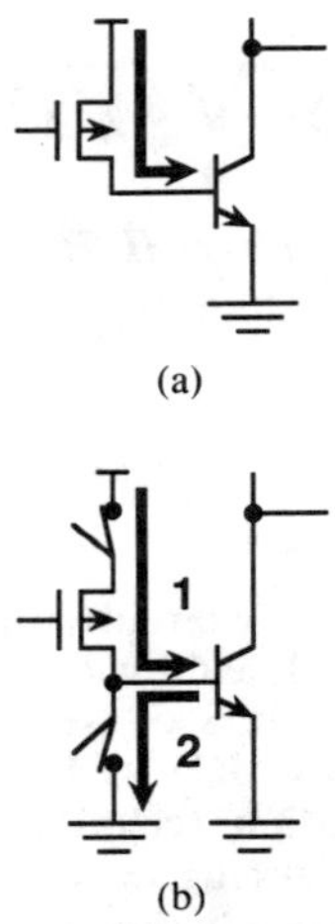

Fig. 2. Full-swing circuit technique: (a) saturation type (conventional), and (b) transient-saturation type (proposed).

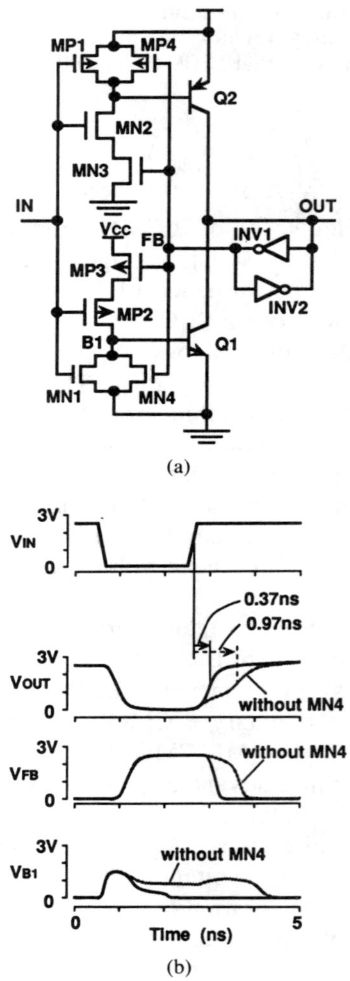

Fig. 3. (a) Configuration of TS-FS-BiCMOS buffer, and (b) simulated voltage waveforms.

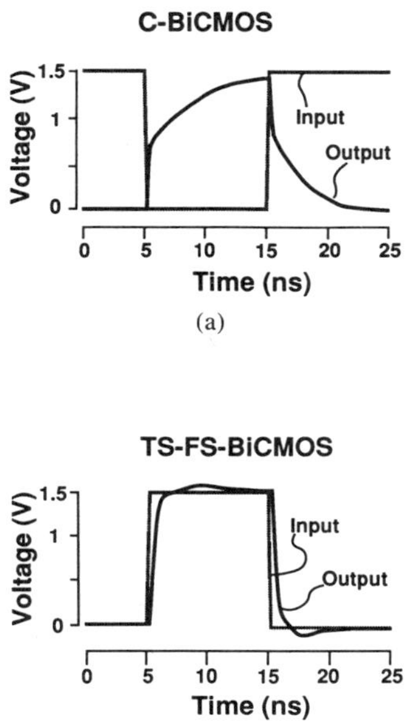

Fig. 4. Simulated input and output voltage waveforms of (a) C-BiCMOS, and (b) TS-FS-BiCMOS.

Fig. 4 shows the simulated output voltage waveforms of C-BiCMOS and TS-FS-BiCMOS operating at 1.5-V supply for the step function input. In C-BiCMOS, fast output voltage transitions obtained by the bipolar transistors are V_{BE} smaller than the supply voltage and the following transitions by additional elements are slow (Fig. 4(a)). This slow transition not only degrades the driving capability but also causes large leakage current in the next stage gate, especially at sub-2-V supply voltages. In contrast, high-speed full-swing operation is achieved in TS-FS-BiCMOS even at 1.5-V supply (Fig. 4(b)).

III. Circuit Performance

SPICE simulation was performed to compare the performance of TS-FS-BiCMOS with those of CMOS, conventional BiCMOS, C-BiCMOS, QC-BiCMOS, and BiNMOS. Configurations of the simulated circuits are shown in Fig. 5. Here, all the circuits have an input capacitance of 60 fF. QC-BiCMOS (Fig. 5(e)) is a circuit that uses a quasi-p-n-p connection instead of a p-n-p transistor of C-BiCMOS. BiNMOS (Fig. 5(f)) is a circuit whose pull-down driver is composed of nMOS logic. The device parameters used in the simulation are listed in Table I. In this simulation, 0.3-μm technology was assumed.

Fig. 6 shows the simulated dependence of the gate delay time on supply voltage. TS-FS-BiCMOS is faster than any other BiCMOS circuit below 2-V supply. Moreover, TS-FS-BiCMOS is twice as fast as CMOS at 1.5-V supply. This has been impossible in any other BiCMOS circuit. For example, the conventional BiCMOS circuit loses its performance advantage over CMOS at about 3 V because of $2V_{BE}$ voltage loss. C-BiCMOS, QC-BiCMOS, and BiNMOS lose their performance advantage over CMOS at about 2 V owing to one V_{BE} voltage loss. Fig.

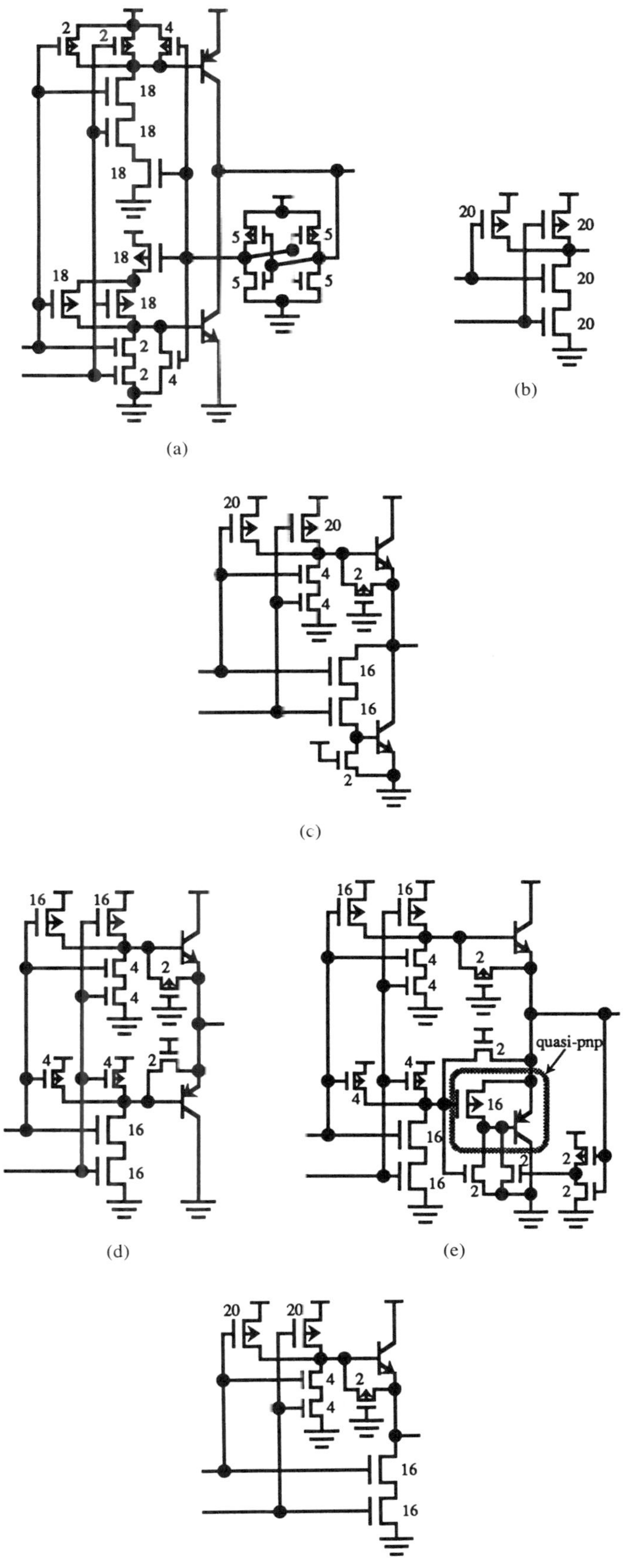

Fig. 5. Compared two-input gates. (a) TS-FS-BiCMOS, (b) CMOS, (c) conventional BiCMOS, (d) C-BiCMOS, (e) QC-BiCMOS, and (f) BiNMOS. The numbers indicate gate widths of MOSFET's.

TABLE I
DEVICE PARAMETERS USED FOR SPICE SIMULATION

MOS	L_g	0.35 μm (nMOS), 0.35 μm (pMOS)
	L_{eff}	0.3 μm (nMOS), 0.3 μm (pMOS)
	I_{DS}	4.8 mA (nMOS), 2.4 mA (pMOS) ($V_{DS} = V_{GS} = 3.3$ V, $W =$ 10 μm)
	T_{OX}	10 nm
	C_j	20 fF ($W = 10$ μm)
	V_{th}	0.3 V (nMOS), -0.3 V (pMOS)
Bipolar transistor	H_{FE}	89 (n-p-n), 89 (p-n-p)
	F_{TMAX}	15 GHz (n-p-n), 6 GHz (p-n-p)
	C_{JE}	12.4 fF (n-p-n), 23.8 fF (p-n-p)
	C_{JC}	10.6 fF (n-p-n), 16.8 fF (p-n-p)
	R_E	25 Ω (n-p-n), 25 Ω (p-n-p)
	$R_{bb'}$	300 Ω (n-p-n), 250 Ω (p-n-p)
	R_c	42 Ω (n-p-n), 42 Ω (p-n-p)

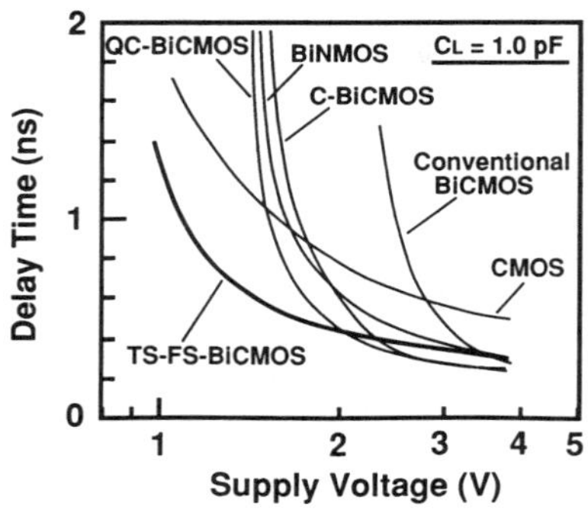

Fig. 6. Simulated dependence of gate delay time on supply voltage. TS-FS-BiCMOS is a two-AND gate. CMOS and BiCMOS circuits are two-NAND gates.

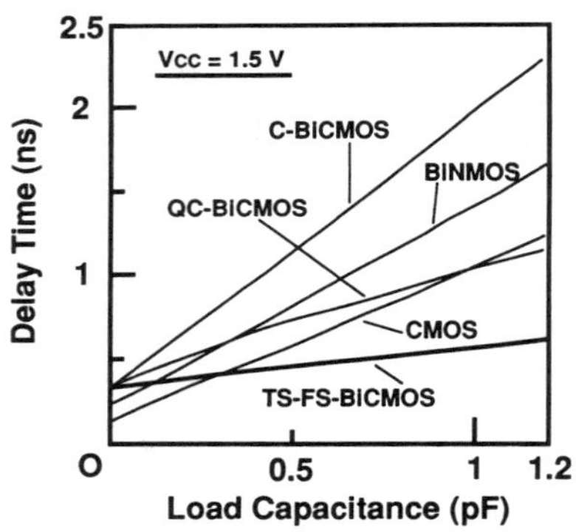

Fig. 7. Simulated dependence of gate delay time on load capacitance.

7 shows the simulated dependence of the gate delay time on load capacitance at the supply voltage of 1.5 V. The load dependence at this low supply is smaller for TS-FS-BiCMOS than for any other BiCMOS circuit because of the high-speed full-swing operation of TS-FS-BiCMOS.

These performances of TS-FS-BiCMOS were shown based on 0.3-μm technology. To predict circuit performance in the post-0.3-μm generation, performance improvement due to device scaling was also studied. Fig. 8 shows the simulated gate delay time for the technology of each generation. The device parameters were determined by assuming that dimensions in the devices were scaled proportionally. The supply voltage was scaled considering hot-carrier immunity [11] and is indicated along the upper horizontal axis of Fig. 8. The threshold voltages of the MOSFET's were also scaled. The simulation results shows that the performance of TS-FS-BiCMOS improves down to less than 0.1 μm, whereas the performance of C-BiCMOS cannot be improved at less than 0.2 μm because of its base–emitter voltage loss.

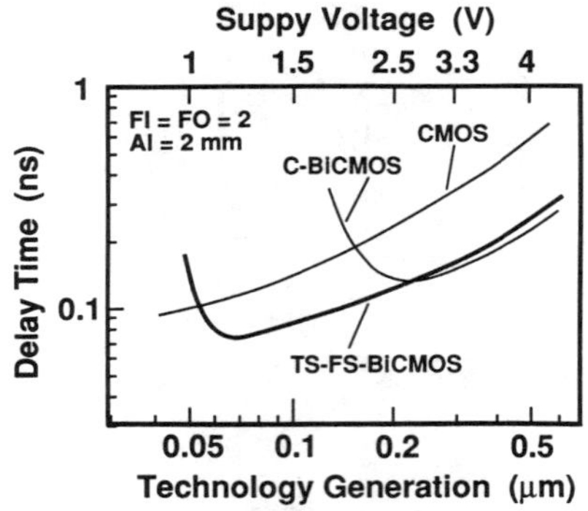

Fig. 8. Simulated delay time for each technology generation.

IV. DESIGN ISSUES

Since TS-FS-BiCMOS uses the feedback signal to discharge the excess minority carriers, the feedback CMOS inverter must be carefully designed. Fig. 9 shows the simulated gate delay time T_d as a function of the input transition interval T_p. Here, W_n denotes the gate width of nMOS in the feedback CMOS inverter. When the driving capability of the feedback CMOS inverter is too small (when $W_n = 1$ μm, for example), a fatal increase in gate delay time is observed for short cycle time operations. This is because the base-carrier discharging is not yet completed when the next transition starts. To suppress this delay-time increase, the CMOS inverter must be designed so that it has enough driving capability. For example, the delay-time increase can be reduced down to only 10% by increasing W_n to 5 μm, which corresponds to one-fourth the gate width of *MP*1 or *MP*2 driving the bipolar transistor.

In Table II, the device count and gate area of TS-FS-BiCMOS are compared with those of C-BiCMOS and CMOS for two-input gates. The gate area of TS-FS-BiCMOS is 1.4 times larger than that of C-BiCMOS and 3.5 times larger than that of CMOS, owing to the larger number of devices of TS-FS-BiCMOS. But this disadvantage of TS-FS-BiCMOS can be considerably relieved by the feature shown in Fig. 10. Fig. 10 shows the simulated dependence of gate delay time on fan-in for TS-FS-BiCMOS and CMOS. Fan-in dependence of TS-FS-BiCMOS is smaller than that of CMOS. It should be noted that this feature of TS-FS-BiCMOS allows a very complex logic function to be achieved with a single-stage gate in TS-FS-BiCMOS whereas several-stage gates are needed in CMOS for the same logic function. Therefore, the overall area of TS-FS-BiCMOS can be drastically reduced in spite of its gate area disadvantage. Fig. 11 is provided to explain why the fan-in dependence of TS-FS-BiCMOS

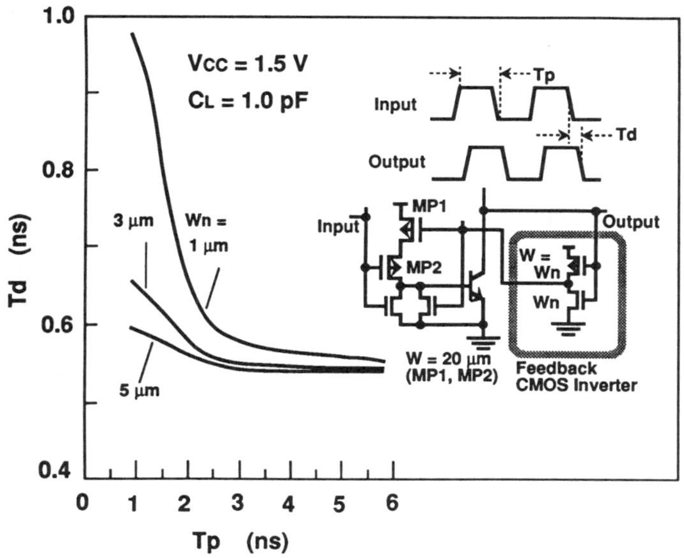

Fig. 9. Simulated delay time as a function of input transition interval.

TABLE II
DEVICE COUNT AND GATE AREA OF TWO-INPUT GATES

	Device Count		
	MOS	Bipolar	Relative Area
TS-FS-BiCMOS	16	2	3.5
C-BiCMOS	10	2	2.5
CMOS	4	—	1

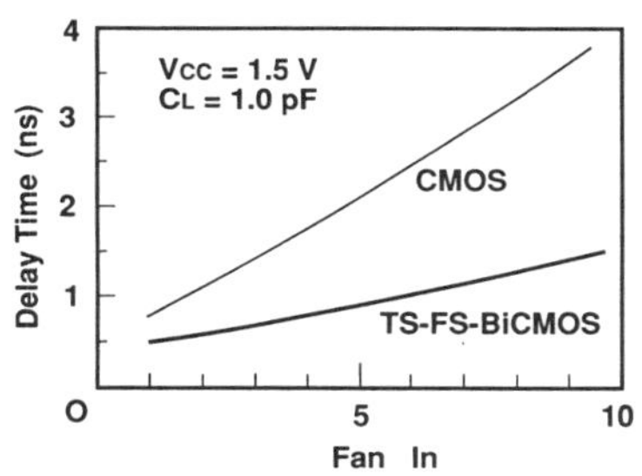

Fig. 10. Simulated dependence of gate delay time on fan-in.

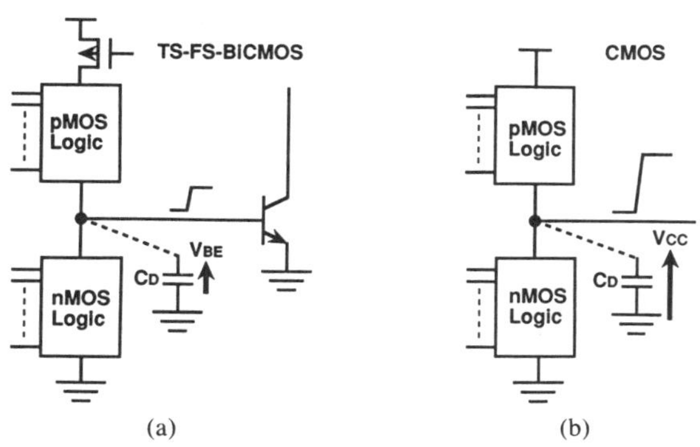

Fig. 11. Voltage swing of parasitic capacitance in (a) TS-FS-BiCMOS, and (b) CMOS. C_D denotes parasitic capacitance.

delay is smaller than that of CMOS delay. The difference of fan-in dependence comes from the difference of the voltage swing of the parasitic capacitances. As fan-in increases, the parasitic capacitance increases both in TS-FS-BiCMOS and CMOS, and this causes the increase of the gate delay time. But the voltage swing of the parasitic capacitance is only one V_{BE} in TS-FS-BiCMOS instead of as large as V_{CC} in CMOS. This results in smaller fan-in dependence for FS-FS-BiCMOS than for CMOS.

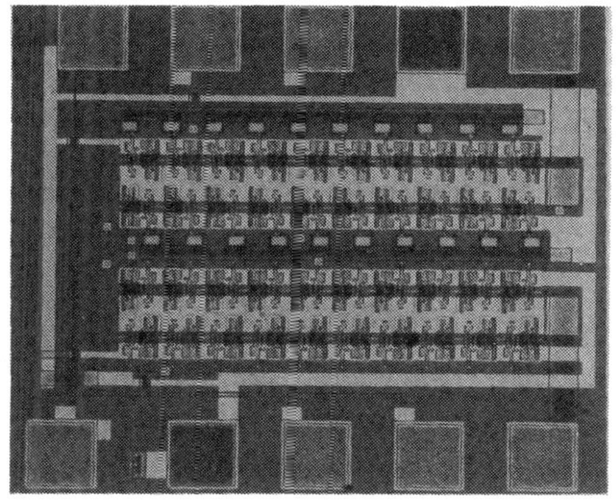

Fig. 12. Photomicrograph of a test chip.

V. EXPERIMENTAL RESULTS

TS-FS-BiCMOS, conventional BiCMOS, C-BiCMOS, QC-BiCMOS, BiNCMOS, and CMOS were fabricated using 0.3-μm BiCMOS technology with triple-poly-Si and double-metal process. The gate lengths of both nMOS and pMOS were 0.4 μm and the gate oxide thickness was 9 nm. The threshold voltages of nMOS and pMOS were 0.5 and -0.1 V, respectively. Bipolar transistors of both n-p-n type and p-n-p type were fabricated in vertical structure with poly-Si emitters for all the BiCMOS circuits except p-n-p transistors of TS-FS-BiCMOS. P-n-p transistors of TS-FS-BiCMOS were fabricated in a lateral structure to simplify the fabrication process. This is because the collector of the p-n-p transistor needs to be isolated from the p-type substrate in TS-FS-BiCMOS. Cutoff frequencies of the vertical n-p-n and p-n-p transistors were 8 and 6 GHz, respectively. The cutoff frequency of the lateral p-n-p transistor was 500 MHz.

Fig. 12 shows a photomicrograph of a test chip used for gate delay time measurement. The measurement was performed using a 17-stage gate chain. Fig. 13 shows the measured waveforms of the input and the output of a loaded TS-FS-BiCMOS gate chain operating at 1.5-V supply. Fig. 14 shows the measured dependence of the gate delay time on supply voltage for TS-FS-BiCMOS, conventional BiCMOS, C-BiCMOS, QC-BiCMOS, and BiNMOS. TS-FS-BiCMOS is faster than any other BiCMOS circuit below 2-V supply. These measured results demonstrate the performance advantage of TS-FS-BiCMOS over other BiCMOS circuits at sub-2-V supply. Here, the delay time of TS-FS-BiCMOS shown in Fig. 14 is pull-down delay time. The pull-up delay time of TS-FS-BiCMOS was about 2.5 times longer than the pull-down delay time because of the low switching speed of the lateral p-n-p transistor. In a practical chip, therefore, vertical p-n-p transistors should be used in TS-FS-BiCMOS

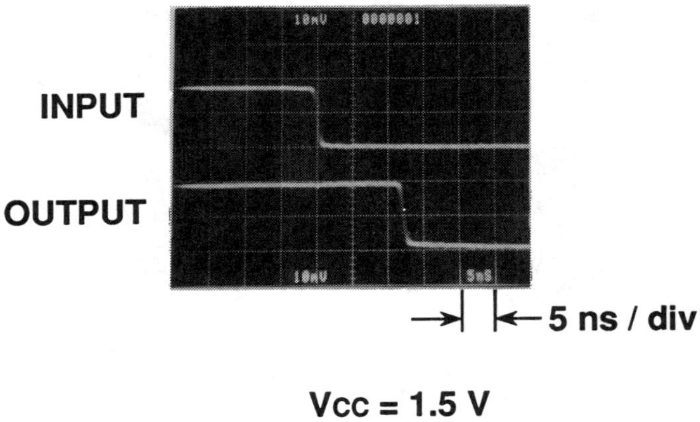

Fig. 13. Measured voltage waveforms of a TS-FS-BiCMOS gate chain.

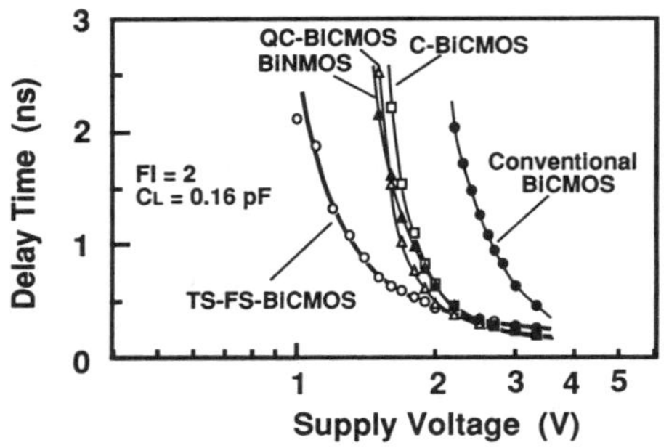

Fig. 14. Measured dependence of gate delay time on supply voltage.

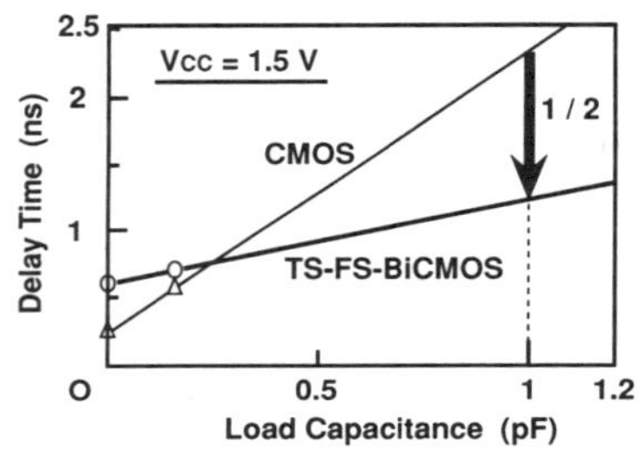

Fig. 15. Measured dependence of gate delay time on load capacitance.

for high-speed pull-up operation. Fig. 15 shows the measured dependence of the gate delay time on load capacitance for TS-FS-BiCMOS and CMOS operating at 1.5-V supply. As predicted in the circuit simulation, the fabricated TS-FS-BiCMOS is twice as fast as CMOS even at 1.5-V supply.

VI. Summary

A BiCMOS circuit applicable to 1.5-V digital circuits has been developed for the first time. The transient-saturation technique is the key to sub-2-V operation because high-speed full-swing operation is achieved with this technique. Both small load dependence and small fan-in dependence of gate delay time are also attained with this technique. The simulated and measured results show that TS-FS-BiCMOS is faster than any other BiCMOS circuit below 2-V supply and is twice as fast as CMOS even at 1.5-V supply. Therefore, TS-FS-BiCMOS is the most promising circuit for BiCMOS ULSI's in sub-2-V regime.

Acknowledgment

The authors would like to thank K. Shimohigashi and T. Nagano for encouragement, and H. Higuchi, M. Suzuki, and S. Tachibana for technical discussions.

References

[1] H. Momose, K. M. Cham, C. I. Drowley, H. R. Grinolds, and H. S. Fu, "0.5 micron BiCMOS technology," in *IEDM Tech. Dig.*, 1987, pp. 838–840.

[2] H. J. Shin, "Full-swing logic circuits in a complementary BiCMOS technology," in *Dig. Symp. VLSI Circuits*, 1990, pp. 89–90.

[3] K. Yano *et al.*, "Quasi-complementary BiCMOS for sub-3-V digital circuits," in *Dig. Symp. VLSI Circuits*, 1991, pp. 123–124.

[4] K. Yano *et al.*, "3.3-V BiCMOS circuit techniques for 250-MHz RISC arithmetic modules," in *Proc. CICC*, 1991, pp. 15.8.1–15.8.4.

[5] O. Nishii *et al.*, "A 1,000MIPS BiCMOS microprocessor with superscalar architecture," in *ISSCC Dig. Tech. Papers*, 1992, pp. 114–115.

[6] P. Raje, R. Ritts, K. Cham, J. Plummer, and K. Saraswat, "Merged BiCMOS: A device and circuit technique scalable to the sub-micron sub-2 V regime," in *ISSCC Dig. Tech. Papers*, 1991, pp. 150–151.

[7] A. E. Gamal, J. L. Kouloheris, D. How, and M. Morf, "BiNMOS: A basic cell for BiCMOS sea-of-gates," in *Proc. CICC*, 1989, pp. 8.3.1–8.3.4.

[8] M. Hiraki *et al.*, "A 1.5-V full-swing BiCMOS logic circuit," in *ISSCC Dig. Tech. Papers*, 1992, pp. 48–49.

[9] H. J. Shin, "Performance comparison of driver and full-swing techniques for BiCMOS logic circuits," *IEEE J. Solid-State Circuits*, vol. 25, no. 3, pp. 863–865, 1990.

[10] S. H. Embabi, A. Bellaouar, M. I. Elmasry, and R. A. Hadaway, "New full-voltage-swing BiCMOS buffers," *IEEE J. Solid-State Circuits*, vol. 26, no. 2, pp. 150–153, 1991.

[11] E. Takeda, Y. Ohji, and H. Kume, "High field effects in MOSFETs," in *IEDM Tech. Dig.*, 1985, pp. 60–63.

Paper 1.12

Merged CMOS/Bipolar Current Switch Logic (MCSL)

WOLFGANG HEIMSCH, ASSOCIATE MEMBER, IEEE, BIRGIT HOFFMANN, ROLAND KREBS, ERNST G. MÜLLNER, BRUNO PFÄFFEL, AND KLAUS ZIEMANN

Abstract—**A merged CMOS/bipolar current switch logic (MCSL) is presented. CMOS/ECL level conversion and logical operation are realized simultaneously. This circuit technique allows a supply voltage reduction to 3.3 V. A carry delay time of 150 ps/bit for a 4-bit BiCMOS full adder was measured. This is about five times faster than an optimized CMOS adder.**

I. Introduction

IN RECENT YEARS, combined bipolar/CMOS technologies were applied to shorten the delay times through critical paths of complex circuits, since BiCMOS combines the high switching speed of bipolar ECL circuits with the low area, low power consumption, and ease of design of CMOS. One remarkable application of BiCMOS technology was a carry propagation circuit in a 32-bit arithmetic unit of a digital processing circuit [1], in which the system speed was increased by more than a factor of 2. In that case bipolar transistors were used as sense and buffer devices. To obtain higher speed improvement factors (bipolar switching time t_{bip}/MOS switching time t_{MOS}) for the whole circuit, the voltage swing ΔV of the logic circuits has to be minimized as shown in (1):

$$t = \frac{C \cdot \Delta U}{I}$$

$$\frac{t_{bip}}{t_{MOS}} = \frac{(C_{wire} + C_{bip})\Delta V_{bip}}{(C_{wire} + C_{MOS})\Delta V_{MOS}}. \qquad (1)$$

It can be seen that for a given current, t_{bip}/t_{MOS} depends directly on the ratio ($\Delta V_{bip}/\Delta V_{MOS}$) of the voltage swings at any node capacitance C. C is composed of the wiring capacitance C_{wire} and the input capacitance of any transistor at this node, C_{bip} or C_{MOS}. Therefore, in complex structures, the time-critical path which determines the overall switching speed has to be implemented in a logic with a small voltage swing, e.g. ECL, using bipolar transistors (Fig. 1) while the remaining larger part can be realized in CMOS logic. But then the problem of level conversion between the CMOS environment and the ECL parts arises. To avoid time, area, and power consumption at the interface between CMOS and ECL, a logic which performs level conversion and logical operation at the same time is desired. This logic must operate with a large input voltage swing to communicate with a CMOS environment and a small output voltage swing for the ECL logic in the time-critical path. The new merged CMOS/bipolar current switch logic (MCSL) closes this gap between CMOS logic with its large input and large output voltage swing and ECL logic with its small input and small output voltage swing.

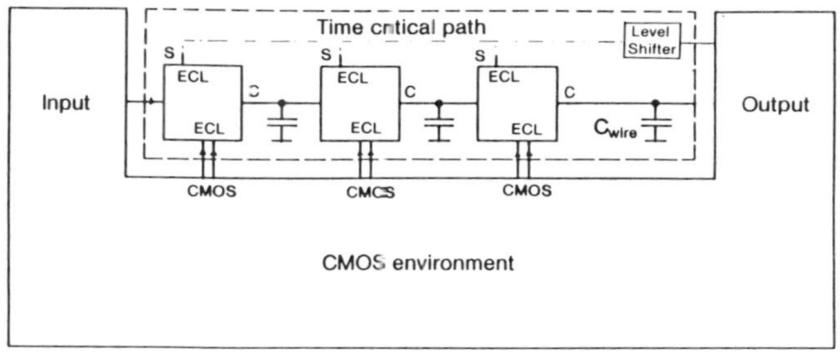

Fig. 1. Concept for using MCSL.

MCSL is introduced in Section II of this paper and is applied to a BiCMOS ripple adder in Section III. The adder exhibits bipolar performance without any additional circuits for level conversion at the input. In contrast to a pure bipolar solution, area and power consumption are reduced by 50 percent for each bit. The advantage in area consumption results from the smaller number of transistors and the smaller spacing of the MOS part. Only 28 transistors in comparison to 48 transistors (considering emitter followers and level shifters) are necessary for each bit. The advantage in power consumption results from the smaller number of current paths. Only two gate and four emitter-follower currents instead of four gate and eight emitter-follower currents are necessary. In comparison to a pure CMOS adder cell, the MCSL adder exhibits a speed improvement by a factor of 5 while the area consumption is only three times larger.

Manuscript received April 3, 1989; revised June 7, 1989.
The authors are with Corporate Research and Development, Siemens AG, 8000 Munich 83, West Germany.
IEEE Log Number 8929852.

Reprinted from *IEEE J. Solid-State Circuits*, vol, 24, no. 5, pp. 1307–1311, Oct. 1989.

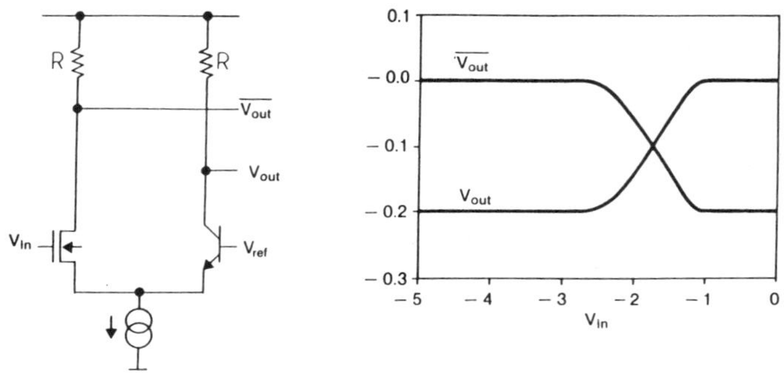

Fig. 2. Merged CMOS/bipolar current switch with transfer characteristic.

II. Merged CMOS/Bipolar Current Switch Logic (MCSL)

In contrast to an ECL differential pair, the basic gate is a merged CMOS/bipolar current switch (Fig. 2). In analogy to an ECL gate, a current source supplies the structure with a constant current. The MOS transistor determines whether the current flows through the left or the right branch of the structure. A small voltage drop across resistor R allows the connection to the ECL logic. The inputs are controlled directly by a CMOS voltage swing. Therefore the structure is able to operate in a full CMOS input environment. Hence a level conversion and a logical operation are performed at the same time, thus reducing propagation delay time, area consumption, and power consumption. Compared to ECL the reference voltage is not critical because it must be centered with respect to the large input signal. The reference voltage is chosen such that the supply voltage is distributed equally over the gates and the current source. Supply voltage fluctuations are not critical because the effects of these variations are compensated for by an appropriate bias driver. The transfer characteristic of the MCSL gate can be seen in Fig. 2. The output resistance R is 200 Ω. The current source provides a current of 1 mA to the MCSL gate. The MOS transistor has a W/L of 10/1.5 and the bipolar h_{fe} was 100. To demonstrate the advantages of this MCSL logic, a BiCMOS ripple adder is fabricated.

III. Comparison of MCSL, CMOS, and ECL Ripple Adders

A. MCSL BiCMOS Ripple Adder

The ripple adder is a typical example of a circuit with a time-critical path. This latter is the carry path, because in each full-adder cell the two input bits and the carry of the previous stage are combined to form the sum and the carry for the next stage. Therefore the sum of the last bit cannot be performed until the carries of the previous stages are determined. The n-bit addition time is

$$t_{add} = (n-1)t_{carry} + t_{sum}. \tag{2}$$

A BiCMOS adder cell using the new MCSL logic is shown in Fig. 3. The circuit for carry and sum is realized using three-level series gating. In the lower two levels the two input bits to be added (CMOS voltage swing) control the current path. The n-channel transistors used in the first level and the p-channel transistors used in the second level minimize the circuit area. In principle, it is possible to use either n-channel or p-channel transistors in both levels, but with the smaller gate–source voltage and body effect, this would lead to very large p-channel transistors in level 1 or n-channel transistors in level 2, thus increasing adding time. In the third level, which is controlled by the carry bit of the previous stage, only bipolar transistors are used as they provide a fast propagation of the carry bit to the next adder cell. The ECL sum output communicates directly with fast bipolar logic. An optional ECL/CMOS level converter [2] offers the possibility to work within a CMOS environment. The output voltage swing is adjusted by the output resistors and the corresponding current source.

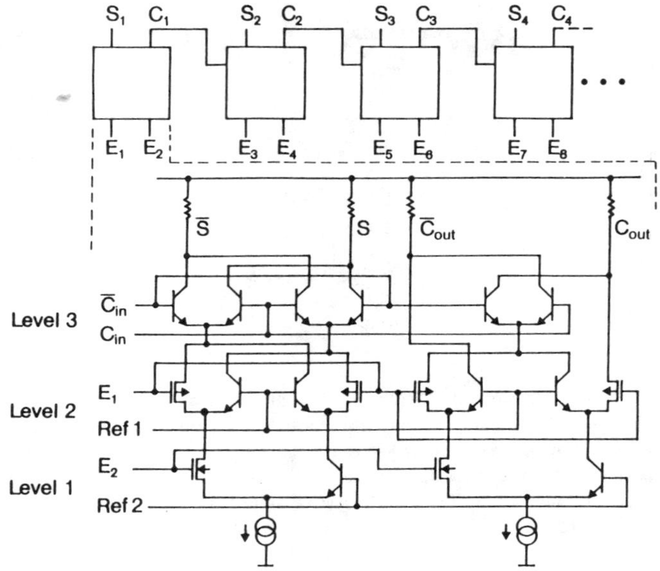

Fig. 3. MCSL ripple adder with full-adder cell.

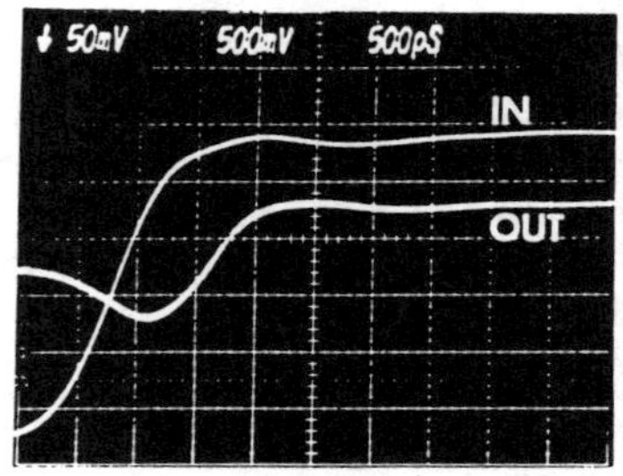

Fig. 4. BiCMOS adder: sum output first adder cell. Voltage scale: input 1 V/div; output 50 mV/div. Time scale: 500 ps/div.

The measurements of a 4-bit MCSL adder show a carry delay time of 150 ps/bit. The input and output signals of the MCSL and CMOS 4-bit adders are shown in Figs. 4 and 5. The power consumption per bit was 16 mW and is determined by two gate currents (Fig. 3) and four emitter-follower currents for sum s, $\overline{\text{sum}}$ $\bar{s}$, carry C_{out}, and $\overline{\text{carry}}$ $\overline{C_{out}}$. The area consumption per bit was 18 000

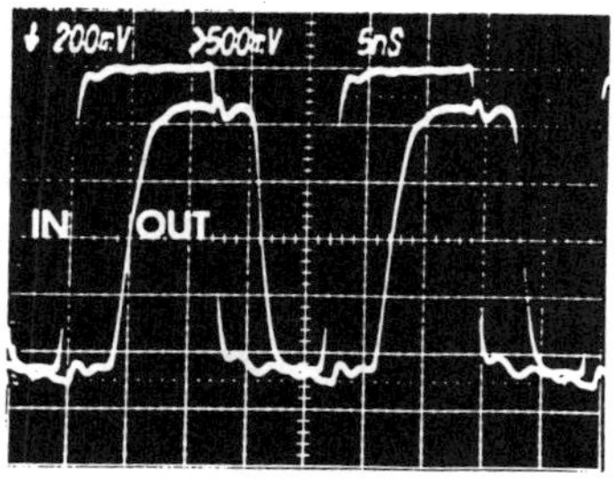

Fig. 5. CMOS adder: carry output fourth adder cell. Voltage scale: input 1 V/div; output 200 mV/div. Time scale: 2 ns/div.

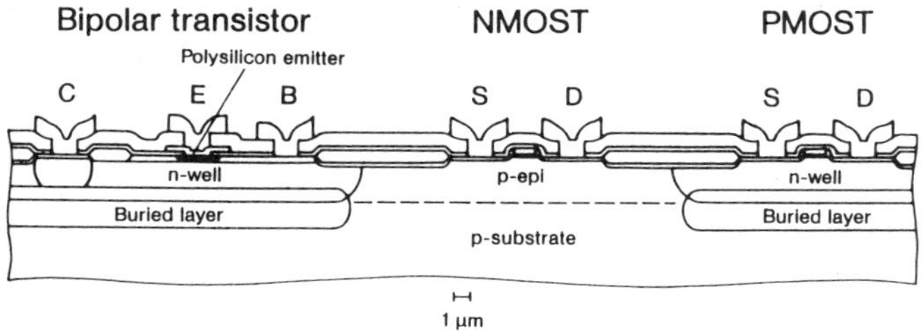

Bipolar transistor

Emitter size	$1.6 \times 10\ \mu m^2$
hfe	100
Transit time	18 ps
Re	11 Ω
Rc	40 Ω
Rb	325 Ω
C_{je}	60 fF
C_{jc}	78 fF
C_{js}	70 fF

MOS transistor

Gate oxide thickness	25 nm
N-channel L_{eff}	1.1 μm
V_{th}	0.87 V
P-channel L_{eff}	1.1 μm
V_{th}	−0.9 V

Fig. 6. BiCMOS process features.

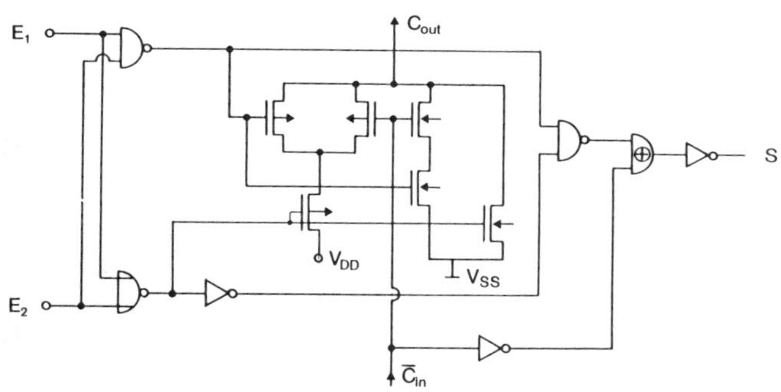

Fig. 7. Pure CMOS ripple adder cell.

μm^2. The MCSL adder results are based on a 1.5-μm BiCMOS process (Fig. 6) using double metal and triple polysilicon. The gate length of the MOS transistors is 1.5 μm. The bipolar transistors have a polysilicon emitter and a transit time of 18 ps.

To further evaluate the results of the MCSL adder, a pure CMOS and a pure ECL adder were fabricated.

B. CMOS Ripple Adder

In the carry path of the pure CMOS ripple adder (Fig. 7), an AND/NOR mixed gate is used. This results in a short delay time for the carry path at the expense of the sum path consisting of five gate delays. A carry delay time of

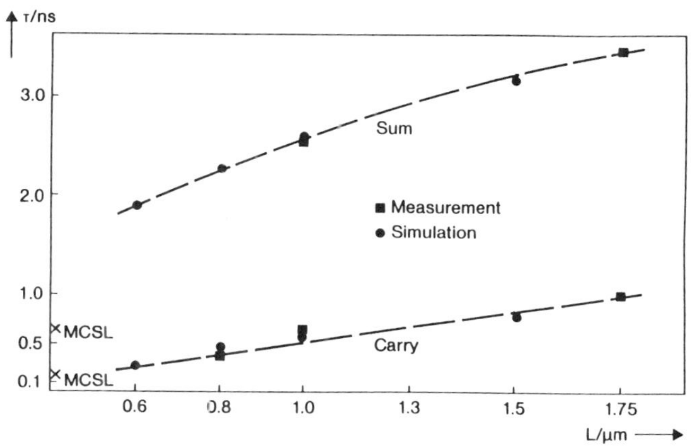

Fig. 8. Sum and carry delay times of different CMOS generations.

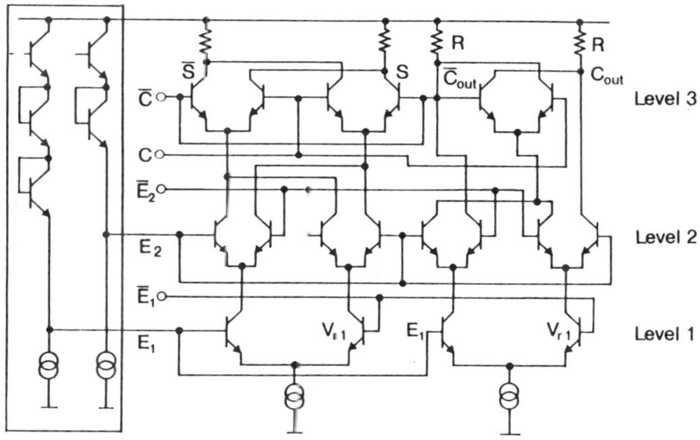

Fig. 9. Pure bipolar ripple adder cell.

TABLE I

COMPARISON OF MCSL, CMOS, AND ECL ADDERS

	ECL	MCSL	CMOS
Carry path	150 ps	150 ps	800 ps
Sum path	250 ps	700 ps	3.2 ns
16 bit addition	2.5 ns	2.95 ns	15.2 ns
Bipolar transistors per bit	48	28	–
Area per bit	32 000 μ^2	18 000 μ^2	5 900 μ^2
Gate currents	4	2	–
Emitter follower	8	4	–
Power per bit	32 mW	16 mW	0.2 mW at 10 MHz

800 ps, a factor of 5 slower than the MCSL version, was determined. In Fig. 3, the carry and sum delay times are plotted as a function of gate length. Fig. 8 shows that a 0.6-μm pure CMOS technology is necessary to obtain approximately the same result (200 ps) as for the conservative 1.5-μm BiCMOS MCSL adder. This clearly demonstrates the superiority of the MCSL BiCMOS adder over the pure CMOS version.

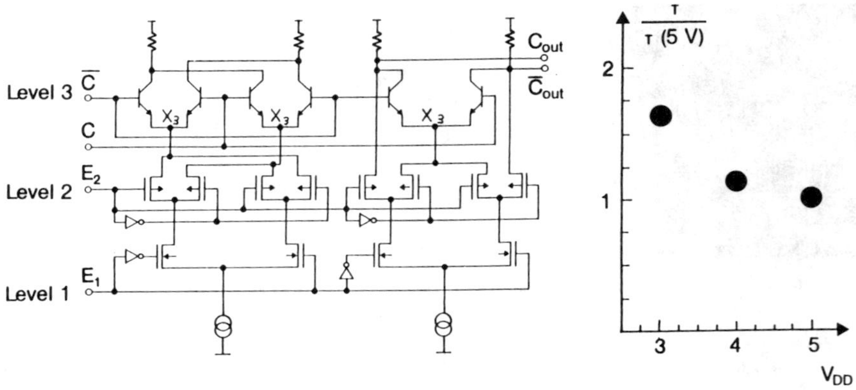

Fig. 10. MCSL ripple adder cell for reduced supply voltages.

C. ECL Ripple Adder

The pure ECL solution is shown in Fig. 9. This circuit also contains an emitter-coupled pair of bipolar transistors used in the first and second levels of the three-level series gating structure. An additional level conversion is therefore necessary, symbolized by the emitter-follower path in the surrounded box in Fig. 9. The measured carry delay time is 150 ps, the same value as in the MCSL version, since the carry path, represented by the third level of the series-gated structure, is identical to that of the MCSL version. The power consumption is determined by four gate current paths—two for the full-adder cell and two for level conversion—and eight emitter-follower paths. Therefore the power consumption increases by a factor of 2 compared with the MCSL version. The area consumption increases from 18 000 to 32 000 μm^2/bit. Metal 1 pitch was 3.5 μm.

D. Comparison of Adder Performances

The different adders are compared in Table I. The MCSL version shows the same carry delay times as the pure ECL version and is faster than the pure CMOS version by a factor of 5. The 16-bit addition also reflects the speed advantage in the carry path. The ECL/CMOS level conversion is not included in the results of Table I, because the ECL sum output should communicate directly with fast bipolar logic. An optional ECL/CMOS level converter offers the possibility of working directly within the CMOS environment. Then, for a load of 300 fF, 700 ps would have to be added to the addition time. MCSL shows a good compromise between the large area and power dissipation of the ECL version and the smaller values for CMOS. This clearly demonstrates the superiority of the MCSL version over the pure ECL or pure CMOS solutions.

IV. MCSL for Reduced Supply Voltages

For future technologies it is necessary to reduce the supply voltages to lessen the high electric field effects arising from the small dimensions. This leads to a problem for three-level series gating using bipolar transistors because of saturation effects and the consequent dramatically increasing delay time. Three-level series gating is essential to obtaining high switching speed with a reasonable power consumption. With a voltage drop of 1 V at the current source (Fig. 9), 200-mV voltage swing at the resistor, and 0.8 V for $V_{be,\text{on}}$, a minimal supply voltage of 3.6 V is necessary to avoid saturation. Fig. 10 shows the MCSL version which guarantees an even further reduction in supply voltage. By replacing the remaining bipolar transistors in levels 1 and 2 (Fig. 3) with MOS transistors, the three-level series gating can be retained. To ensure fast propagation of the carry bit, bipolar transistors have to remain in the third level. The voltage drop between source and drain of the MOS transistors can be reduced below 0.8 V ($V_{be,\text{on}}$) and the potential at nodes X_3 is lowered so that saturation of the emitter-coupled pair of bipolar transistors in the third level can be avoided. The switching speed of the critical carry path in the upper level is unaffected by this measure and a short carry delay time is thus guaranteed. In Table II the simulated delay time for the carry signal and the sum signal as well as the power distribution of the MCSL adder cell are shown. The results of $V2$, $V3$, and $V4$ are based on a bipolar f_t of 16 GHz. The adder cell with MOS and bipolar transistors in levels 1 and 2 is the fastest ($V2$). Using the adder cell in Fig. 10 at a supply voltage of 5 V with MOS transistors only on levels 1 and 2, the delay time increases. But it should be noted that the power consumption decreases because the emitter followers are gone ($V3$). At a reduced supply voltage of 3.3 V the delay time further increases and the power consumption decreases ($V4$). The increase of the adding time is due to the fact that the widths of the MOS transistors have to be adapted to the lower supply voltages. Therefore the capacitive loads are higher on internal nodes and at the coupling

TABLE II
SIMULATED RESULTS FOR THE DIFFERENT ADDER VERSIONS

Version	sum delay	carry delay	16 bit add.	power	circuit	Lgeo	supply voltage
V1	700ps	150ps	2.95ns	16 mW/bit	Fig.3	1.5um	5V
V2	500ps	150ps	2.75ns	2.7mW/bit	Fig.3	1.0um	5V
V3	550ps	185ps	3.32ns	1.5mW/bit	Fig.10	1.0um	5V
V4	780ps	210ps	3.93ns	1.0mW/bit	Fig.10	1.0um	3.3V

between MOS gates and drain and source increases. The current to discharge the internal nodes is the same for the different supply voltages. This leads to an increase of the signal propagation time from level 1 to level 3 and also to an increase of the adding time. Despite this, the carry delay time remains short.

V. Conclusion

The merged CMOS/bipolar current switch logic (MCSL) was presented. CMOS/ECL level conversion and logical operation are realized simultaneously with merged bipolar/CMOS gates. A supply voltage reduction down to 3.3 V is possible. As an example of this circuit technique, an adder with a carry delay time of 150 ps was designed and measured. This is about five times faster than an optimized CMOS design at comparable feature size.

Acknowledgment

The authors would like to thank M. Stegherr, M. Böhner, and B. Zehner for helpful discussions, M. Wurm for circuit design and verification, and H. Klose for assembly technology.

References

[1] T. Hotta *et al.*, "CMOS/bipolar circuits for 60-MHz digital processing," in *ISSCC Dig. Tech. Papers*, Feb. 1986, pp. 190–191.

[2] K. Ogiue *et al.* "13-ns, 500-mW, 64-kbit ECL RAM using HI-BICMOS technology," *IEEE J. Solid-State Circuits*, vol. SC-21, no. 5, pp. 681–684, Oct. 1986.

[3] W. Heimsch *et al.*, "Merged CMOS/bipolar current switch logic (MCSL)," in *ISSCC Dig. Tech. Papers*, Feb. 89, pp. 122–123.

Paper 1.13

Multiemitter BiCMOS CML Circuits

M. Elrabaa and M. I. Elmasry

***Abstract*—New BiCMOS current-mode logic (CML) circuits employing multiemitter devices are proposed. They perform logic functions in addition to conversion from CMOS to CML (or ECL). Their transient behavior was analyzed, and their delay expressions were obtained and verified using HSPICE [6]. These expressions were used to optimize their design. Their performance is compared with other BiCMOS CML circuits with similar functionality.**

Nomenclature

- β Bipolar gain.
- τ_F Bipolar transit time.
- C_{be} Base-emitter junction capacitance.
- C_{jc} Collector junction capacitance.
- C_{cm} Miller's capacitance at the collector.
- C_{js} Collector-substrate junction capacitance.
- C_{gs} Gate-source capacitance in the linear region in the MOS HSPICE transient model.
- C_{gd} Gate-drain capacitance in the linear region in the MOS HSPICE transient model.
- C_D Source/drain depletion capacitance.
- C_{ovl} MOS overlap capacitance.
- C_e Capacitance at the common emitter node, $C_e = C_D + C_{ovl} + C_{be}$.
- C_L Load capacitance.
- V_{th} MOS threshold voltage.
- V_s Output voltage swing.
- V_{ref} Reference voltage.
- V_{BE} Base-emitter junction voltage of a turned-on bipolar.
- $I_{D\text{sat}}$ MOS saturation current.
- ΔQ_{gd} Channel charge supplied by the drain node after the MOS enters the linear region.
- ΔQ_{sd} Channel charge supplied by the source node after the MOS enters the linear region.
- I_{SDf} MOS drain current when $V_{SG} = V_{DD}$ and the drain voltage is $V_{\text{ref}} - V_{BE}$.
- r Portion (percentage) of input transition where the MOS device at the input is in strong inversion.
- V_T Thermal voltage, $V_T \approx 26$ mV.

Manuscript received March 7, 1991; revised November 1, 1991. This work was supported in part by NSERC, ITRC, and MICRONET research grants.

The authors are with the VLSI Research Group, Department of Electrical and Computer Engineering, University of Waterloo, Waterloo, Ont., Canada N2L 3G1.

IEEE Log Number 9105978.

I. Introduction

OPTIMAL performance of large digital systems requires the use of CMOS, BiCMOS, and CML (or ECL) logic: CMOS for low-power densed logic and on-chip memories with BiCMOS gates as buffers and sense amplifiers, and the faster CML (or ECL) logic for implementing critical paths. Hence, it is necessary to have fast conversion logic to convert signals from CMOS (or BiCMOS) levels to CML level and vice versa. This paper deals with CMOS/CML conversion circuits.

Conversion circuits have been proposed and reported [1], [2] with speeds comparable to that of CML. Two of these circuits are considered for comparison. They can perform logic functions with mixed CMOS/CML inputs. However, they consume large static power and silicon area. Moreover, these circuits suffer in performance at reduced voltage supplies [1]. A new set of conversion circuits using merged BiCMOS structures with multiemitter bipolar transistors are proposed. Their operation, transient analysis, design, and a comparison between them and the existing circuits with similar functionality in terms of speed, power, area, and low-voltage operation are presented.

II. Conventional CMOS/CML Conversion Circuits

Two series-gated circuits are considered in this section—one with a conventional CMOS/CML translator (Fig. 1(a)) and one with a merged current switch logic (MCSL) [1] (Fig. 1(b)). The CMOS/CML translator operation was analyzed in [3], and analytical delay expressions were given. A drawback of this circuit is that, when the output of the CMOS gate at the CML input is rising, the CML part will only start to switch when this voltage approaches V_{ref}, which is less than V_{DD} by $V_s/2$. Thus, most of the rise time of the CMOS gate is added to the total gate delay. The MCSL has a better performance because it begins switching after a smaller change in the input; however, it suffers from performance degradation at low levels of supply voltage [1]. Moreover, the number of levels of series gated (stacked) CML gates for both circuits is limited by the supply voltage.

III. Multiemitter BiCMOS CML Circuits

Two new types of circuits are proposed to overcome the limitations of the above circuits. One is the low-power multiemitter merged MOS/bipolar CML (low-power M^3BiCML) of Fig. 1(c), and the second is the multiemitter merged MOS/bipolar CML (M^3BiCML) of Fig. 1(d).

Reprinted from *IEEE J. Solid-State Circuits,* vol., 27, no., 3, pp. 454–458, March 1992.

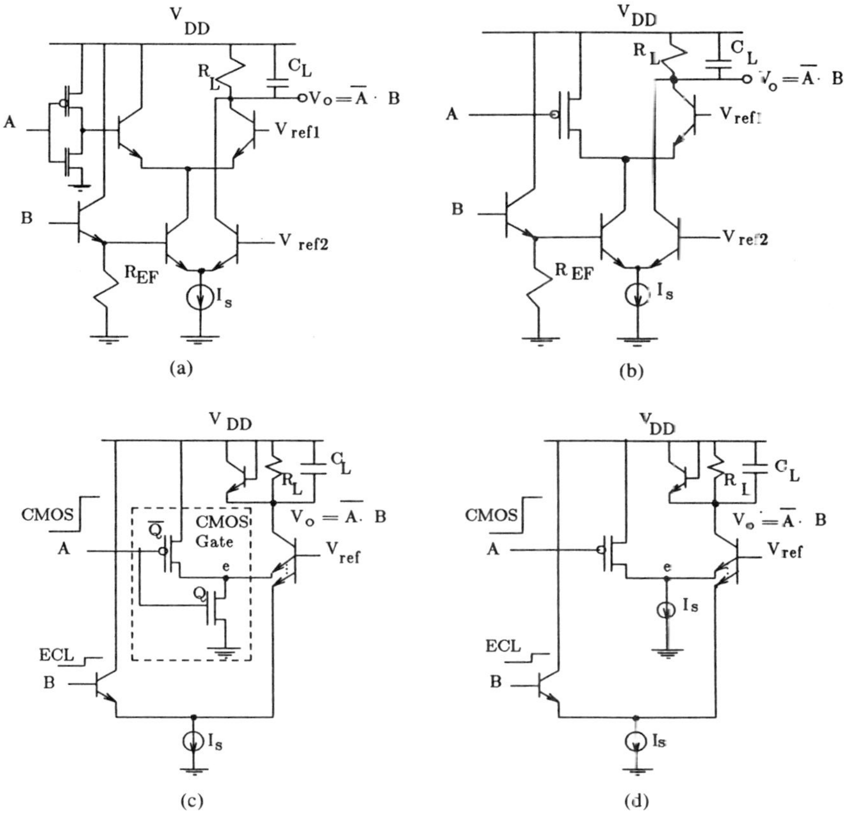

Fig. 1. Circuits diagrams of (a) conventional BiCMOS CMOS/CML translator, (b) the merged current switch gate, (c) low-power multiemitter merged MOS/bipolar CML, and (d) multiemitter merged MOS/bipolar CML. The diode-connected transistors are used to limit the output voltage swing.

A. Low-Power M^{3}BiCML

This circuit is composed of a conventional CML with a CML input, and a merged BiCMOS CML with a CMOS input. An NMOS transistor Q is acting as a current source when the A input is high. The channel width of Q is adjusted to the required value of current, which is the same as I_s in the CML part, and R_L is chosen to give the desired voltage swing. For equal probability of the two states (ZERO and ONE) of the input A, the merged part will draw power for only half the time. Hence, the average static power dissipated in this part is

$$P_{\text{ave}} = 0.5\, I_s V_{DD}.$$

The voltage swing at the emitter of the merged part (node e) is small ($V_{DD} - (V_{\text{ref}} - V_{BE})$). As it is shown later, a portion of this circuit delay is proportional to this voltage swing. A faster version of this circuit results if the positions of the NMOS (Q) and the PMOS ($\overline{Q}$) devices are interchanged since the voltage swing at node e would be reduced by V_{th}. The resulting circuit is called noninverting LPM^{3}BiCML.

Using multiemitter devices enables the implementation of more complex functions (or handling more inputs) at lower voltage supplies since no series gating is required [8]. The output capacitance is smaller than that of the conventional series-gated circuits since the collector area (and hence capacitance) of the multiemitter bipolar is smaller than the sum of the collector area of two bipolar transistors. The delay of the CML part was reported in many papers [4], [5]. The delay of the merged BiCMOS part is analyzed next. The input is assumed to be a voltage step rising (or falling) in a time $t_{\text{rise/fall}}$ (50 ps in this work), and the delay is measured from $t_{\text{rise/fall}}/2$ to the time the output reaches $V_{DD} - 0.5V_s$. The calculated results were compared to HSPICE simulations using the parameters in Table I.

1) Rise-Time Analysis: As the input decreases, the drain current of Q decreases, and by the end of t_{fall} this current and the emitter current of the bipolar transistor would have fallen to zero. Assuming t_{fall} is small enough, most of the base charge of the bipolar remains unchanged. This charge is then removed through diffusion in a time of the order of τ_F. Meanwhile, the collector current becomes zero and the output node capacitance C_o is charged to $V_{DD} - 0.5V_s$ in $0.7R_L C_o$, $C_o = C_L + C_{cm} + C_{js}$, and $C_{cm} = 2C_{jc}$ [4]. Thus, the rise time is given by

$$T_r = 0.5t_{\text{fall}} + \tau_F + 0.7R_L C_o. \tag{1}$$

2) Fall-Time Analysis: At the end of input change, the emitter node is still close to V_{DD}. Then it is discharged by $I_{D\text{sat}}$ of the NMOS transistor Q to about $V_{\text{ref}} - V_{BE}$, forcing the bipolar transistor to turn on in about τ_F. Finally, the output capacitance is discharged to $V_{DD} - 0.5V_s$ in $0.7R_L C_o$. Thus, the fall time is

TABLE I

	V_{th}	L_{eff}	$C_{S/D}$	C_{ovl}	Bipolar	
NMOS	0.065 V	1 μm	4E-4 F/m	2E-10 F/m	G_{jc} = 7.5 fF	C_{js} = 31 fF
PMOS	−0.95 V	1 μm	5E-4 F/m	2E-10 F/m	C_{be} = 7 fF	τ_F = 6 ps
	V_{DD} = 5 V	V_s = 0.8 V	$V_{BE(\text{on})}$ = 0.83 V		R_B = 200	β = 100

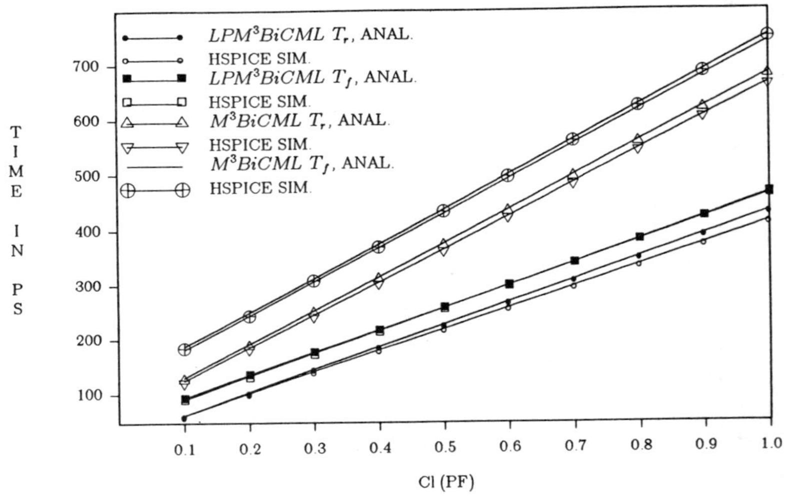

Fig. 2. The calculated and simulated rise and fall times of the new circuits.

$$T_F = 0.5t_{\text{rise}} + C_e[V_{DD} - (V_{\text{ref}} - V_{\text{BE}})]/I_{D\text{sat}} + \tau_F + 0.7R_L C_o. \tag{2}$$

A comparison between the above expressions and HSPICE is shown in Fig. 2. The total average delay is $0.5(T_R + T_F)$.

As can be seen from the above expressions for the rise and fall times, the PMOS width should be minimum (and satisfying the noise margins) to minimize C_e.

B. The M³BiCML

The M³BiCML circuit of Fig. 1(d) consists of a CML and a merged BiCMOS CML. The merged part is the same as the MCSL reported in [1]. The M³BiCML consumes more power than the LPM³BiCML. However, it has a superior ability to achieve more complex functions in smaller area at low power supplies since the current source is implemented by a bipolar transistor not a MOS transistor. Moreover, series gating is possible, giving the designer more freedom. The delay of the merged part is analyzed next.

1) Rise-Time Analysis: The rise time is divided into three times. The first time is the time required for the MOS device to turn on (during input transition) while the emitter current increases to $I_e(0)$, approximated as follows:

$$I_e(0) = I_s + 2(\Delta Q_{gd} - \Delta Q_{sd})/t_{\text{fall}} \tag{3}$$

where $\Delta Q_{gd} = C_{gd}\Delta V_{gd}$, ΔV_{gd} is the change in gate-to-drain voltage, and $\Delta Q_{sd} = 0.5rI_{SDf}\,t_{\text{fall}}$.

In the second time, V_e increases by about $4V_T$, and the emitter current becomes zero. These changes can be approximated to be linear and occur in a time t_{r1}:

$$t_{r1} = \frac{4V_T C_e + \tau_F I_e(0)}{I_{SD\text{ave}} - I_s} \tag{4}$$

where $I_{SD\text{ave}}$ is the average drain current during this time.

The final time is the *RC* charging of the output. Thus, the total rise time is

$$T_r = 0.5t_{\text{fall}} + t_{r1} + 0.7R_L C_o. \tag{5}$$

2) Fall-Time Analysis: By the end of the input transition, the PMOS is turned off. However, the channel charge causes the emitter node voltage to rise by ΔV_e. Since the drain current of the short-channel PMOS is approximately linear with V_{SG} (for $V_{SG} > V_{th}$) [7], it is linear with time. Hence, ΔV_e is approximated as

$$\Delta V_e = \frac{\Delta V_{gd} C_{gd} - t_{\text{rise}} I_s(1 - 0.5r)}{C_e}. \tag{6}$$

Next, the emitter node is discharged to $V_{\text{ref}} - V_{BE}$ in t_{dis}:

$$t_{dis} = C_e[V_{ei} - (V_{\text{ref}} - V_{BE})]/I_s \tag{7}$$

where V_{ei} is the voltage at the emitter node at the end of the input transition (including ΔV_e). Finally, the bipolar transistor turns on in about τ_F and the output capacitance is discharged in $0.7R_L C_o$. The total fall time is therefore

$$T_f = 0.5t_{\text{rise}} + t_{dis} + \tau_F + 0.7R_L C_o. \tag{8}$$

Comparisons between the above expressions of the rise

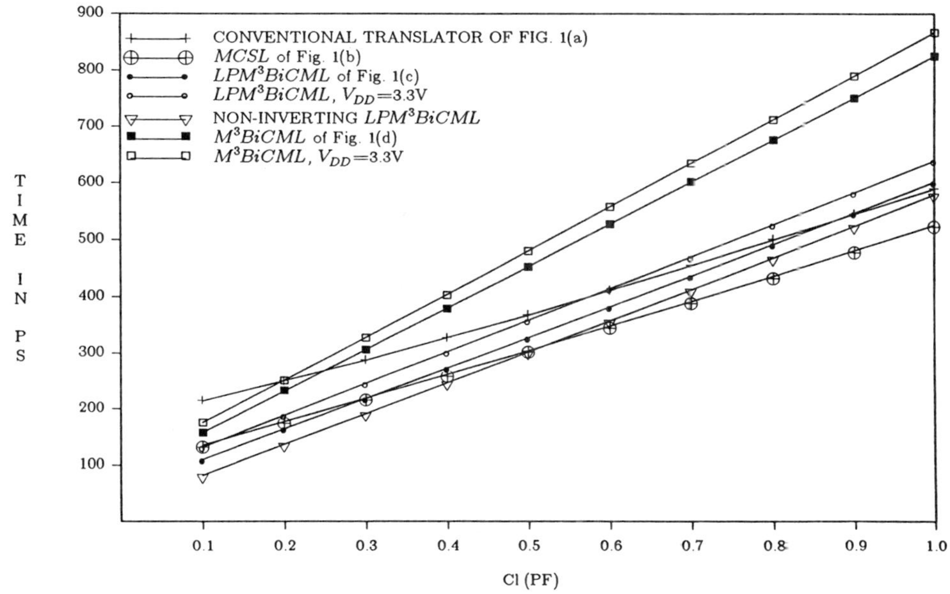

Fig. 3. The simulated average delays versus load capacitance for V_{DD} = 5 V except as indicated.

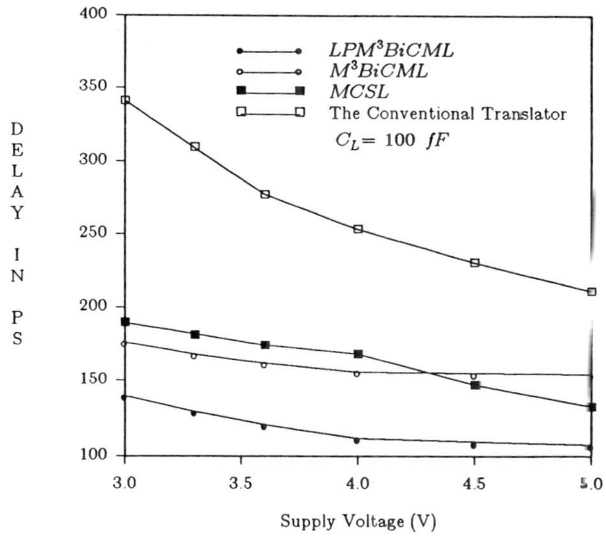

Fig. 4. The delays of the four circuits versus the supply voltage.

TABLE II

	LPM³BiCML	M³BiCML	Conv. Trans.	MCSL
Power (mW)	7.0	6.75	6 8	6.8
Area (normalized)	1	1.32	2 27	1.83
Delay (ps)	110	157.2	214	135.3
Delay-Power Product (PJ)	0.77	1.06	1 46	0.92

and fall times and HSPICE simulations are shown in Fig. 2. The total delay is $0.5(T_r + T_f)$.

The rise time is independent of the PMOS width (W_P), however, the fall time has a term (t_{dis}) that increases with W_P. Thus, W_P should be kept at a minimum value that satisfies the noise margin requirements. The emitter's areas (as in previous circuits) are also set to a minimum to avoid high-level injection effects and to minimize the parasitic capacitances [4].

IV. Results and Comparison

The average delays of the circuits described in this paper were measured as a function of load capacitance using HSPICE for the same level of power and are shown in

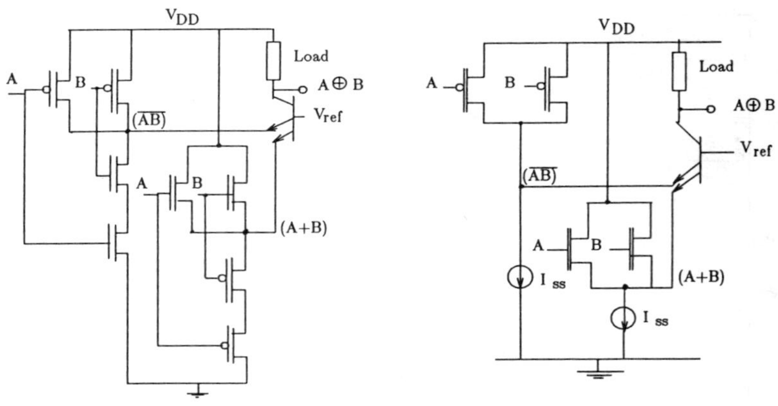

Fig. 5. XOR implementations using (a) LPM³BiCML structure, and (b) M³BiCML structure.

Fig. 3. The LPM³BiCML and the noninverting LPM³BiCML have the lowest delays for moderate values of load capacitance most likely to be encountered in a critical path. The delays of the new circuits under reduced supply voltage (3.3 V) do not increase significantly. Also, the delays of the four circuits as a function of the supply voltage are shown in Fig. 4, where it is clear that the new proposed circuits have a superior performance under lower supply voltages. Table II shows the relative area of each circuit, along with the delay, power, and delay–power product for a load capacitance of 0.1 pF. LPM³BiCML and M³BiCML circuits offer a powerful logic element and can implement complex logic functions at lower supply voltages, and an example of an XOR implementation is shown in Fig. 5.

V. Conclusion

New merged BiCMOS CML circuits are proposed. They utilize multiemitter BJT's instead of series gating to minimize the limitations imposed by lower supply voltage. Analytical expressions of their delays that can be used to optimize their design are obtained. They offer higher speed at the same power level for moderate load capacitances and up to 50% saving in area. Their performances do not degrade with the scaling down of the supply voltage. The use of multiemitter bipolar transistors instead of series gating enhances their logic functions' implementation capability and makes them adaptable to low supply voltage operation.

References

[1] W. Heimsch *et al.*, "Merged CMOS/bipolar current switch logic (MCSL)," *IEEE J. Solid-State Circuits*, vol. 24, pp. 1307–1311, 1989.

[2] A. R. Alvarez, *BiCMOS Technology and Applications*. Norwell, MA: Kluwer Academic, 1989.

[3] S. H. K. Embabi *et al.*, "Analysis and optimization of BiCMOS digital circuit structures," VLSI Res. Group, Dept. Elec. Comput. Eng., Univ. of Waterloo, Waterloo, Ont., Canada, Internal Rep., 1991.

[4] M. Y. Ghannam *et al.*, "An analytical model for the determination of the transient response of CML and ECL gates," *IEEE Trans. Electron Devices*, vol. 37, pp. 191–201, 1990.

[5] W. Fang, "Accurate analytical delay expressions for ECL and CML circuits and their applications to optimizing high-speed bipolar circuits," *IEEE J. Solid-State Circuits*, vol. 25, pp. 572–583, 1990.

[6] *HSPICE User's Manual*, Meta-Software, Inc., Campbell, CA, 1990.

[7] K.-Y. Toh *et al.*, "An engineering model for short-channel MOS devices," *IEEE J. Solid-State Circuits*, vol. 23, pp. 950–958, 1988.

[8] M. I. Elmasry, *Digital Bipolar Integrated Circuits*. New York: Wiley, 1983.

Quasi-Complementary BiCMOS for Sub-3-V Digital Circuits

Kazuo Yano, *Member, IEEE*, Mitsuru Hiraki, *Member, IEEE*, Shoji Shukuri, Yasuo Onose, Mitsuru Hirao, Nagatoshi Ohki, Takashi Nishida, Koichi Seki, *Member, IEEE*, and Katsuhiro Shimohigashi, *Member, IEEE*

Abstract—**This paper describes a quasi-complementary BiCMOS (QC-BiCMOS) circuit scheme for the low-supply-voltage deep-submicrometer regime. A QC-BiCMOS performs twice as fast as a CMOS even at a 2.5-V supply without a p-n-p bipolar transistor. Key circuits for this excellent low-voltage performance are a separation between the base of the pull-up bipolar and the base of a quasi-p-n-p, and the carefully designed base discharging circuit. Based on these circuits, a quasi-p-n-p combination of a pMOS and an n-p-n bipolar shows an equivalent cutoff frequency over 10 GHz. The delay expressions for the QC-BiCMOS are analyzed and compared with the conventional BiCMOS. A fabricated 0.3-μm fully loaded three-NAND gate verifies more than twice the performance leverage over the conventional BiCMOS and the CMOS at sub-3-V supply.**

I. Introduction

BiCMOS technology is being actively employed in a wide spectrum of ULSI's [1]–[3] for its high-speed, low-power, high-density characteristics. However, performance degradation at low supply voltages is recognized as a major limiting factor for the BiCMOS in the deep-submicrometer regime [4], [5], [7], [12]. The widely used BiCMOS circuits lose the performance leverage over the CMOS at supply voltages below 3 V. To overcome this limitation, new circuits, such as BiNMOS [4] and complementary BiCMOS (C-BiCMOS) [4], [5], [9] have been proposed. However, the BiNMOS has a small performance leverage over the CMOS and the advantages of the C-BiCMOS are considerably offset by the additional process complexity and cost for the p-n-p bipolar fabrication. Therefore, a new circuit that solves the above issues is still strongly desired.

A BiCMOS circuit that discharges output loads through a composite circuit of pMOS and n-p-n (we call this circuit quasi-complementary BiCMOS (QC-BiCMOS) after the analog output circuit name) was proposed in the early stage of BiCMOS development [11] and has been considered by many designers. However, it has not played an important role in actual circuits because of its limited speed leverage over the CMOS. Recently, Raje *et al.* demonstrated that this QC-BiCMOS has a smaller dependence of delay on the supply voltage [6] than that of the conventional BiCMOS circuit. However, it still has a limited performance advantage over CMOS.

The purpose of this paper is to exploit the possibility of QC-BiCMOS with a realistic speed leverage over the CMOS even at below 3-V supply by using new circuit techniques. The performance, design issues, and limitations of this new gate are also examined. First, the basic concept of the QC-BiCMOS is briefly discussed in Section II, and the performance of the QC-BiCMOS is compared with the conventional circuit configurations in Section III. Next, the delay time of the QC-BiCMOS is analytically expressed and compared with the conventional delay expressions in Section IV, and based on these, the design issues are discussed in Section V. Finally, the fabrication results are described in Section VI.

II. QC-BiCMOS: Circuit Concept

The reason for the degraded performance of the conventional BiCMOS circuits at below 4 V is (as discussed in [7] and [10]) briefly reviewed as follows. As shown in Fig. 1(a), when the input changes from low level to high level, the maximum gate–source voltage applied to the nMOS $MN1$ is

$$V_{GS} = V_{CC} - 2V_{BE}$$

where V_{GS} is the gate–source voltage, V_{CC} is the supply voltage, and V_{BE} is the base–emitter voltage when the bipolar transistor is ON (about 1 V, when the bipolar is fully operating). As the supply voltage is reduced, the V_{GS} is reduced significantly by the influence of V_{BE}, resulting in severe drain-current degradation. This is the simple mechanism of the degraded performance of the conventional BiCMOS circuit in the pull-down operation. This conventional circuit is denoted as "Darlington type" in this paper to distinguish it from the other conventional circuits, such as C-BiCMOS and BiNMOS.

Manuscript received March 20, 1991; revised July 1, 1991.

K. Yano is with the Center for Solid State Electronics Research, Arizona State University, Tempe, AZ 85287 on leave from the Central Research Laboratory, Hirachi Ltd., 1-280 Koigakubo, Kokubunji-shi, Tokyo 185, Japan.

M. Hiraki, S. Shukuri, T. Nishida, K. Seki, and K. Shimohigashi are with the Central Research Laboratory, Hitachi Ltd., 1-280 Koigakubo, Kokubunji-shi, Tokyo 185, Japan.

Y. Onose and M. Hirao are with Hitachi Research Laboratory, Hitachi Ltd., Hitachi 319-12, Japan.

N. Ohki is with Hitachi VLSI Engineering Corporation, Ltd., Kodaira, Tokyo 187, Japan.

IEEE Log Number 9103017.

Reprinted from *IEEE J. Solid-State Circuits*, vol. 26, no. 11, pp. 1708–1719, Nov. 1991.

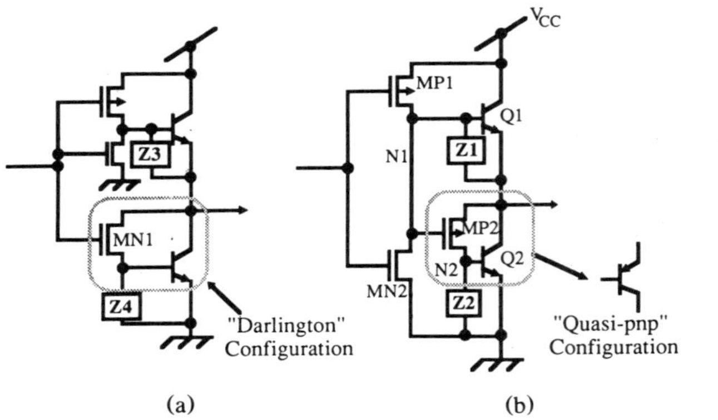

Fig. 1. Schematic circuits of (a) conventional (Darlington-type) BiCMOS inverter and (b) QC-BiCMOS inverter.

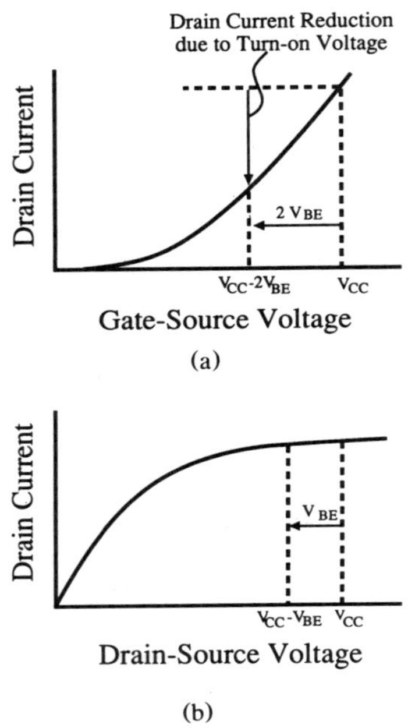

Fig. 2. Current–voltage characteristics of MOSFET's which show the drain-current reduction due to base–emitter turn-on voltage in (a) conventional Darlington-type BiCMOS, and (b) QC-BiCMOS.

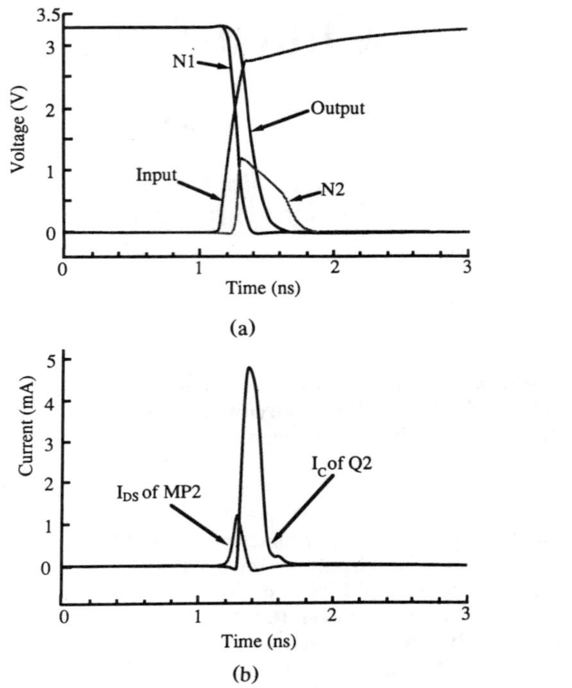

Fig. 3. Simulated (a) voltage waveforms and (b) simulated currents for QC-BiCMOS in the pull-down operation (three-input NAND with 5-fan-out load). QC-BiCMOS-c (Fig. 5) is used.

The main concept behind QC-BiCMOS is the avoidance of this Darlington circuit to eliminate the voltage loss due to V_{BE} [6]. The schematic inverter configuration of the QC-BiCMOS is shown in Fig. 1(b). The QC-BiCMOS features the "quasi-p-n-p" connection, which consists of a pMOS ($MP2$) and n-p-n bipolar transistor ($Q2$). The idea of quasi-p-n-p itself, which consists of p-n-p and n-p-n, is well known in the class-B amplifier in the analog IC's [8]. This original quasi-p-n-p configuration reinforces the poor lateral p-n-p by adding the high-performance vertical n-p-n, which enables class-B high efficiency (in other words, low power). There is a similarity between this original quasi-p-n-p and the quasi-p-n-p in the QC-BiCMOS in the sense that the poor pMOS is reinforced by the powerful n-p-n.

The reason for the smaller delay dependence on the supply voltage is described as follows. The base of the n-p-n bipolar transistor for pull-down ($Q2$) is driven by the pMOS ($MP2$). In this pull-down operation, the drain–source voltage of $MP2$ is reduced by the V_{BE} of $Q2$; however, the gate–source voltage is not affected. As shown in Fig. 2(b), the drain current depends little on the drain–source voltage in the saturation region. Therefore, the influence of the V_{BE} on delay is small. On the other hand, conventional BiCMOS suffers gate–source voltage degradation, which reduces the drain current significantly as shown in Fig. 2(a).

QC-BiCMOS has another attractive feature in that the pull-down waveform is steep until it reaches ground level. The reason for this is explained using Figs. 1(b) and 3. When the output voltage is lower than V_{TP} (threshold voltage of $MP2$), $MP2$ turns off. After this the carriers in the base are lost only through the resistor $Z2$ and intrinsic base current component. By designing the resistance of $Z2$ to an appropriate value, these base carriers are retained until the output is driven to the ground level by $Q2$. (Without the discharging current through the resistance $Z2$, the carriers in the base of $Q2$ are retained for more than a nanosecond in an ordinary transistor with current gain of over 50.) This is demonstrated in the simulated waveforms in Fig. 3. The output dc level is retained on ground level by the current path through the resistor $Z1$.

Although the bipolar transistor ($Q2$) is temporarily saturated just after this pull-down operation, this saturation is relatively shallow. This is because the base-current drive through $MP2$ is already terminated when $Q2$ becomes saturated (Fig. 3). Note that this "shallow" saturation is realized by connecting the $MP2$ source to output, not to V_{CC}. (This heavily saturated logic circuit having pMOS whose source is directly connected to V_{CC} was discussed in [10].) The shallow saturation of QC-BiCMOS is relieved in a short time by the resistor $Z2$ promptly after the pull-down operation. Note that there is a clear distinction between QC-BiCMOS and "saturation logic," such as TTL and IIL, which maintains saturation even after transient.

By contrast, in any conventional BiCMOS (such as Darlington-type BiCMOS, C-BiCMOS) saturation is elaborated to be avoided. Therefore, they lose bipolar drive when the output is lower than the V_{BE} level. Although resistors or small MOSFET's can be employed to lower the output level to lower than V_{BE}, the discharging speed is much slower than that of the bipolar transistors. Because the pull-down waveforms of these conventional circuits are severely degraded in heavily loaded conditions, the maximum frequency is limited by this constraint. Moreover, the maximum gate–source voltage applied to the next-stage pMOS is effectively reduced by the V_{BE}, resulting in large pull-up delay.

When QC-BiCMOS is actually used in the logic circuits, there are many possible configurations. Although one of the simplest configurations (which was discussed in [6] and [11], but is not recognized as a quasi-p-n-p configuration) was examined by many designers, it has not been used in actual circuits because of the limited speed leverage over the CMOS. In this paper, based on the above consideration for the low-voltage operation, we employed some new circuit techniques that may enhance the circuit speed up to a level which has sufficient performance leverage over the CMOS. One is a separation between the base of the pull-up bipolar ($Q1$) and the base of the quasi-p-n-p (the gate of $MP2$). While we recognize the combination of the $MP2$ and $Q2$ as a p-n-p transistor, similar separations have also been used in conventional BiCMOS circuits [4], [5], [9] to improve the speed. Therefore this separation is expected to improve the performance leverage. The other technique is the base discharging circuit ($Z2$), which enables the above-described quasi-full-swing operation. It should be noted that role of $Z2$ is completely different from that of the resistance $Z4$ in the conventional Darlington circuit. ($Z4$ determines the low level of the output. The charge stored in the base is discharged through $MN1$.) Because the referenced circuit in [6] and [11] does not have this discharging circuit, the charge stored in the base of $Q2$ is retained until the next change of input. This causes large short-circuit current during the next transient. The discharging circuit $Z2$ requires careful design, which takes the base charge retention and the relief from the shallow saturation into account. These design issues will be discussed in Section V.

Four configurations of QC-BiCMOS are studied in this paper as shown in Fig. 4. The QC-BiCMOS-a connects simple CMOS logic to a quasi-complementary buffer. This features a small number of transistors and small area

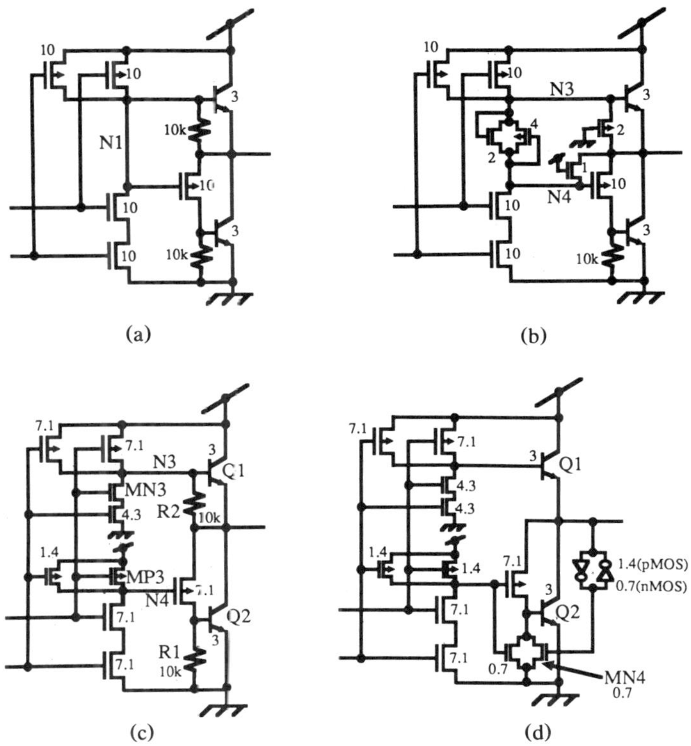

Fig. 4. Typical QC-BiCMOS two-input NAND configurations: (a) QC-BiCMOS-a (direct-connected type), (b) QC-BiCMOS-b (diode-connected type), (c) QC-BiCMOS-c (separate type), and (d) QC-BiCMOS-d (feedbacked MOS discharging type). The numbers are gate widths for MOSFET's and emitter lengths for the bipolar transistors.

(actually it is not useful because of its slow speed, as shown in Section III). The QC-BiCMOS-b features the separation of nodes $N3$ and $N4$ by diode-connected MOSFET's. QC-BiCMOS-c has independent MOS logics for pull-up and pull-down. It can construct a large fan-in logic without suffering short-circuit current through $Q1$ and $Q2$, because of the large degree of freedom in its MOS size design. QC-BiCMOS-d feedbacks the output signal to determine the timing of extracting charge from the $Q2$ through $MN4$. Because of this feedback circuit, the outflow of the base charge is minimized in the pull-down transient.

III. Performance Comparison

In this section the performance of the QC-BiCMOS is compared with the conventional BiCMOS circuits. In this comparison, 0.3-μm BiCMOS parameters shown in Table I are used. These parameters are based on our designed BiCMOS process having a double-polysilicon self-aligned structure. The comparison is done with various threshold voltages of MOSFET's, however, only the result with 0-V threshold MOSFET's is shown here. (The threshold voltage here is defined as the gate voltage which drives 1-nA drain current per 1-μm gate width. This threshold voltage definition gives about 0.2-V lower value than that defined by the extrapolation method [22]. Our actual design goal for the 0.3-μm MOSFET's is 0.3 V.) The reason is as

TABLE I
0.3-μm BiCMOS DEVICE PARAMETERS USED IN CIRCUIT SIMULATIONS

MOSFET	gate oxide thickness		8 nm
	drain current		2.8 mA (nMOS) 2.0 mA (pMOS)
	drain–substrate capacitance		10 fF
Bipolar Transistor	cutoff frequency		24 GHz
	knee current		5 mA
	junction capacitance	C_{TE}	6.5 fF
		C_{TC}	5.0 fF
	base resistance		250 Ω

follows. Above 3 V, this threshold-voltage change from 0.3 to 0 V did not affect the relative circuit performance. And for the comparison below 3-V supply, at which the threshold voltage affects the results, the threshold voltage should be set around 0 V, because the circuits operating below 3-V supply will mainly use 0.2-μm (or less) dimensions. The small leakage current, which still is more than 4 orders of magnitude smaller than the dynamic current, should be justified in most speed-oriented BiCMOS LSI's. For fair comparison, all the threshold voltages of MOSFET's, including those of CMOS, are set to the same value. Note that the speed advantage due to this lower threshold design is more significant for the CMOS than the BiCMOS.

The compared circuits are shown in Fig. 5. Because the delay depends strongly on the gate-width design of the base discharging MOSFET's (e.g., $MN3$ and $MP2$) or resistances (e.g., $R1, R2, R3$), it is important for a fair comparison to carefully set these parameters. In general, the larger these equivalent resistances are (the smaller the gate widths are in the case of MOSFET's), the smaller the delay is. However, too large resistances (or too small width MOSFET's) cause short-circuit current through bipolar transistors (e.g., $Q3$ and $Q4$). Therefore, the resistances are set to the largest value (MOSFET gate widths are set to the smallest value) as long as the short-circuit current is negligible in the fan-out of 5 condition.

The output fan-out (FO) load is set to 5, which is the middle of the optimum loaded condition for the CMOS and for the BiCMOS, unless otherwise specified. This is chosen so that this comparison is fair to both CMOS and BiCMOS. (The optimum fan-out load which minimizes the total delay in a multistage CMOS circuit is 3–4 [13]. On the other hand, for BiCMOS it is 7–10 [14], [21]. This distinction originates from the fact that the delay dependence of BiCMOS is proportional to square root of load capacitance, whereas the dependence of CMOS is proportional to load capacitance.)

First, various QC-BiCMOS variations are compared as shown in Fig. 6 and Table II. The QC-BiCMOS-a, which is 1.4 times slower than the other QC-BiCMOS circuits, has a small leverage over the CMOS. This is because the charging and discharging of node $N1$, which has a larger parasitic capacitance than those of $N3$ or $N4$, is slow. (The parasitic capacitance for $N1$ is 69 fF, whereas the

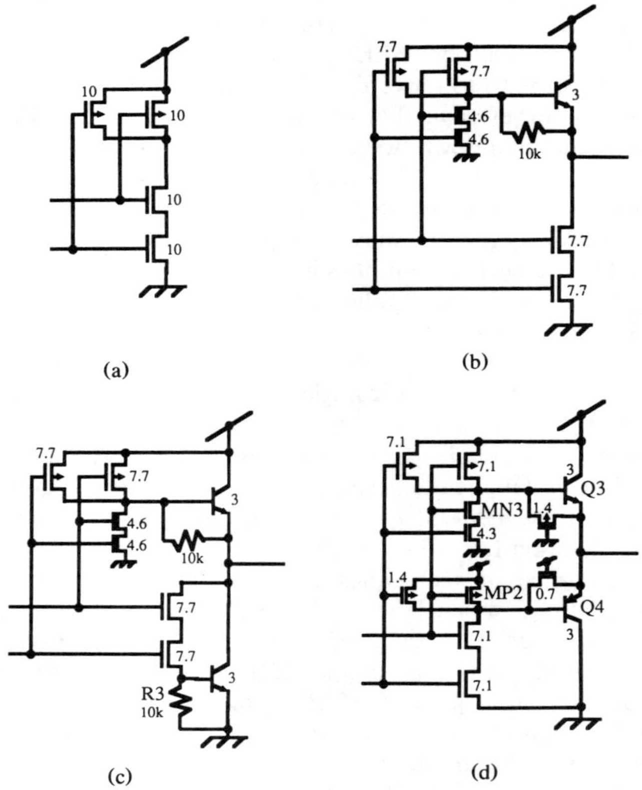

Fig. 5. Compared conventional circuits: (a) CMOS, (b) BiNMOS, (c) Darlington, and (d) C-BiCMOS. The numbers are gate widths for MOSFET's and emitter lengths for the bipolar transistors.

parasitic capacitances for $N3$ and $N4$ of QC-BiCMOS-b are 49 and 30 fF, respectively.) The performance leverage of the conventional BiCMOS over the CMOS is 1.5–2 even at 5 V. Therefore, this 1.4-time slower speed is not acceptable in most actual situations, even if it is advantageous in small areas.

It is clear, then, that a realistic QC-BiCMOS gate should separate the node for pull-up ($N3$) from the node for pull-down ($N4$). In the QC-BiCMOS-b the $N4$ is discharged in a short time without being affected by the parasitic capacitance of the $N3$ in pull-down operation. Similarly, the QC-BiCMOS-c and -d, which have small MOSFET's ($MN3$ and $MP3$), operate quickly without being affected by those large parasitic capacitances.

The QC-BiCMOS-d (feedbacked MOS discharge type) is suitable for operation below 2.5 V. This is because it

TABLE II
COMPARISON AMONG CIRCUITS (THREE-INPUT NAND WITH 5-FAN-OUT LOAD AT 3.3 V)

	Conventional				QC-BiCMOS			
	CMOS	BiNMOS	Darlington	C-BiCMOS	(a)	(b)	(c)	(d)
No. of MOSFET's	6	10	11	14	9	11	15	19
No. of Bipolars	0	1	2	2	2	2	2	2
Gate Area[1]	0.5	0.81	1	1.23	0.98	1.06	1.30	1.39
Delay (ps)	277	198	192	152	191	139	144	138
Power[2] (μW)	146	151	150	151	174	179	151	165

1: Relative area with respect to Darlington type BiCMOS; 2: 100-MHz operation.

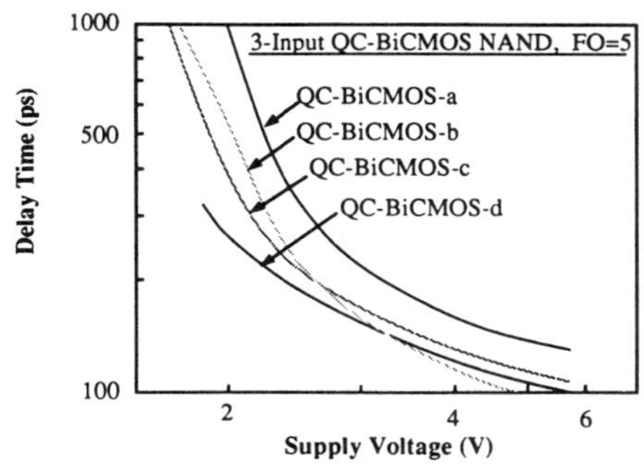

Fig. 6. Delay dependences on supply voltage for various QC-BiCMOS circuits.

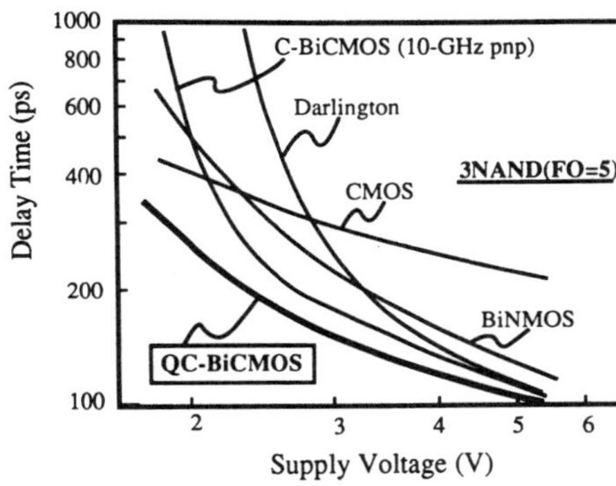

Fig. 7. Delay dependences on supply voltage for various circuits. QC-BiCMOS-d is used.

has a steep pull-down waveform due to the good charge retention capability of bipolar $Q2$. In addition, in the pull-up operation the charge outflow from the base of the bipolar transistor $Q1$ is also small due to the output latch. On the other hand, the QC-BiCMOS circuits having resistance between base and emitter (QC-BiCMOS-b and -c) lose a part of the stored charge in the pull-down transients. Based on these preliminary evaluations, the QC-BiCMOS-d is used in the following comparison with the other conventional circuits unless otherwise noted. However, the main conclusions are valid even if QC-BiCMOS-b or QC-BiCMOS-c is employed as is seen in Fig. 6.

The dependence of the QC-BiCMOS delay time on supply voltage is smaller than any other BiCMOS circuit ever reported as shown in Fig. 7. At 2.5-V supply QC-BiCMOS is twice as fast as CMOS, and 1.5 times faster than that of the CMOS even at 2-V supply. By contrast,

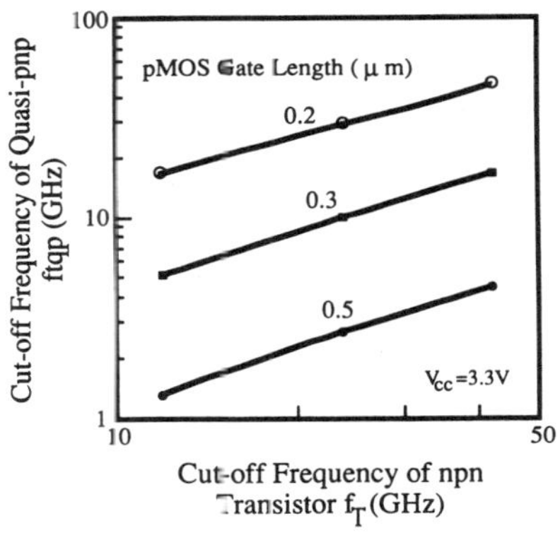

Fig. 8. Dependence of equivalent cutoff frequency of quasi-p-n-p on pMOS gate length and cutoff frequency of n-p-n transistor. QC-BiCMOS-d is used.

the conventional Darlington BiCMOS, BiNMOS, and C-BiCMOS lose performance leverage over CMOS at 2.9, 2.4, and 2.1 V, respectively. Note that an additional 1 V is required for these conventional circuits to be advantageous over CMOS by offsetting the additional process cost and the additional area for bipolar transistors (namely, 3.9, 3.4, and 3.1 V are the minimum feasible supplies for these conventional circuits, respectively).

Next QC-BiCMOS is compared with the C-BiCMOS in more detail. To clarify the performance correspondence between quasi-p-n-p and real p-n-p, the equivalent cutoff frequency of quasi-p-n-p is estimated as follows. The cutoff frequency of the real p-n-p transistor, which provides the same pull-down delay time by replacing the quasi-p-n-p, is defined as the equivalent cutoff frequency of the quasi-p-n-p. Circuit simulation based on a variety of parameters of the pMOS and n-p-n bipolar transistor is compared with the results employing p-n-p transistors as shown in Fig. 8. The cutoff frequency here is referred to as the maximum cutoff frequency with varying collector current. This cutoff frequency was varied by changing the transit time parameter of the bipolar transistor. The p-n-p transistor model in this simulation has very advanced parameters, which are the same parameters as our advanced double-poly-Si n-p-n bipolar transistor (Table I) except the transit time. The result reveals that the quasi-p-n-p generally has a higher cutoff frequency than that of real p-n-p transistors in actual situations in the deep-submicrometer range, even though we assumed very optimistic parameters for the real p-n-p transistor. For example, a combination of a 0.3-μm pMOS and a 20-GHz

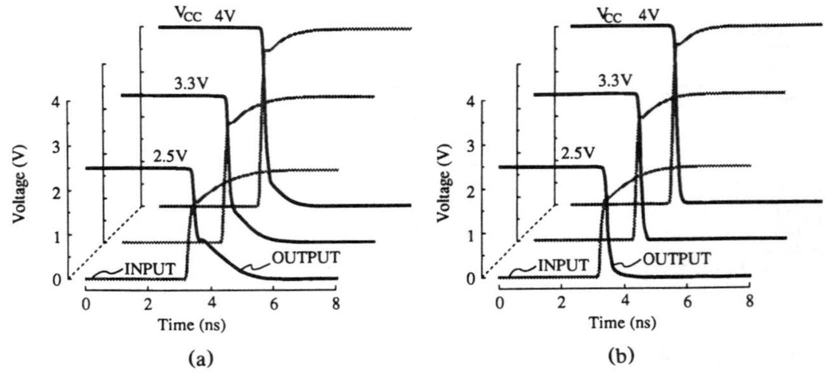

Fig. 9. Simulated input and output waveforms of (a) C-BiCMOS and (b) QC-BiCMOS at various supply voltages. QC-BiCMOS-d is used.

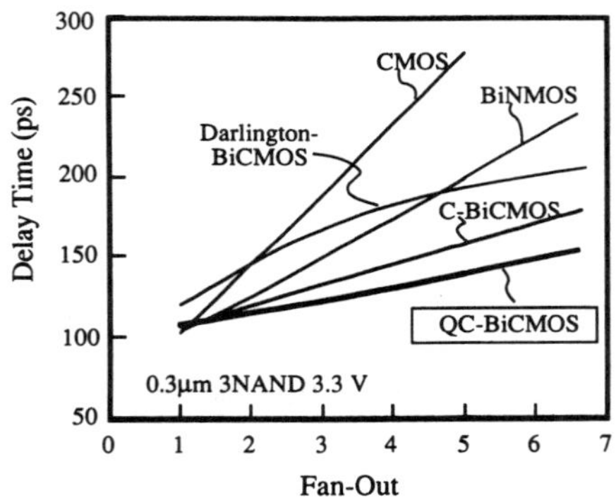

Fig. 10. The dependence of delay time on fan-out load. QC-BiCMOS-d is used.

n-p-n bipolar transistor at 3.3 V shows a cutoff frequency as high as 10 GHz, which is higher than any p-n-p transistor fabricated in the BiCMOS process.

The input/output waveforms for C-BiCMOS and QC-BiCMOS are compared in Fig. 9. As discussed above, the pull-down waveform of QC-BiCMOS is very steep and quasi-full swing. On the other hand, C-BiCMOS has a slow transient in the region near V_{CC} and ground level. Note that the transition from steep (bipolar) discharging to slow (small MOSFET) discharging occurs at a higher level at lower supply voltages. This is because the stored charge in the base of the p-n-p transistor is reduced at low supply voltages, resulting in early termination of bipolar current drive.

The dependence of the QC-BiCMOS delay on load capacitance is also superior to the other conventional circuits as shown in Fig. 10. This is because the output current is amplified by nMOS ($MN2$), pMOS ($MP2$), and n-p-n ($Q2$) in the pull-down operation. In the pull-up operation, the n-p-n ($Q1$) in the QC-BiCMOS drives a larger current than the conventional BiCMOS's because the pMOS ($MP1$) receives quasi-full-swing input.

The gate area of the QC-BiCMOS is larger than that of the C-BiCMOS by the additional area for the pMOS ($MP2$) as shown in Table II. This may limit the use of the QC-BiCMOS in some situations where the available area is strictly limited. However, this increase of area will pose less constraints when it is used in the sea-of-gates gate array because it can be used as the writing area. Power dissipation of the QC-BiCMOS is the same or slightly larger than CMOS and conventional BiCMOS circuits (Table II) as a result of the additional power for the pMOS ($MP2$) drive.

Next the impact of the excellent low-voltage performance of the QC-BiCMOS on the performance of the deep-submicrometer ULSIs is assessed. For this purpose, BiCMOS device parameters ranging from 2 to 0.05 μm are constructed. The current–voltage characteristics for MOSFET's are estimated considering the velocity saturation effect, source–drain parasitic resistances, and mobility degradation in high transverse field conditions. The bipolar transistor model is based on the Gummel–Poon model [15], whose parameters are determined to fit the measured cutoff frequency and base resistance of fabricated bipolar transistors. Horizontal and vertical parameters are proportionally scaled. Supply voltage is chosen to be the maximum tolerable voltage for the hot-carrier degradation. The supply voltage is determined by the empirical formula [16] based on the experimental data.

The dependence of delay time shows characteristic gate length, which minimizes the delay time as shown in Fig. 11. This is because delay increase due to supply-voltage reduction overwhelms the delay reduction due to gate-length reduction at the characteristic length. (This "*J* curve" was discussed by one of the authors in [17] in the context of CMOS scaling limitation.) This minimum gate length is a good index for the scalability of a circuit.

The QC-BiCMOS delay reduction continues to 0.13 μm, which operates at 2-V supply as shown in Fig. 11. The performance leverage over CMOS is also maintained at this gate length. By contrast, the conventional Darlington BiCMOS shows minimum delay time at 0.3 μm. Therefore, the QC-BiCMOS extends the scaling limit of BiCMOS by two generations.

IV. Delay Analysis

Because present circuit simulation capability enables accurate circuit analysis (if the device models are well defined), the analytical approach cannot compete with

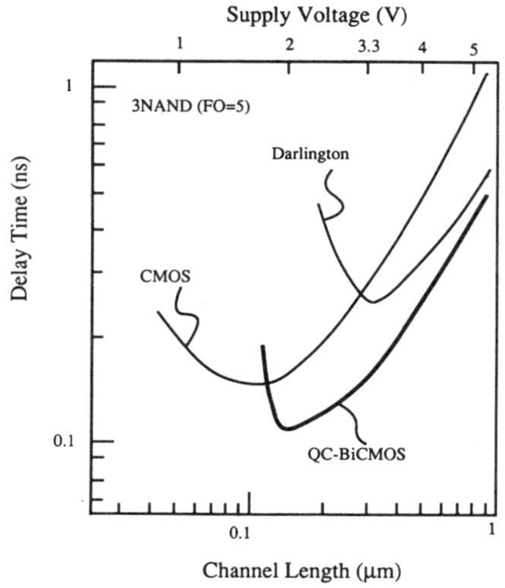

Fig. 11. The dependence of delay time on channel length. QC-BiCMOS-d is used.

the simulation in accuracy in any sense. Therefore, no one uses the analytical approach to obtain an accurate circuit performance prediction. However, even so, the analytical approach gives a designer a perspective and insights which are indispensable when one applies a circuit to different device technology or extends the circuit concept to meet additional requirements. The analytical approach for the conventional BiCMOS circuits has been studied [18]–[20] and is well understood. In this section the delay of QC-BiCMOS is also analytically shown to have a better perspective of QC-BiCMOS operation. The simplified circuit in Fig. 12. is used to derive delay expressions.

A. Pull-Down Delay

As detailed in the Appendix, the pull-down delay time t_{PHL} of the QC-BiCMOS is expressed by

$$t_{PHL} = \left(\frac{3\tau\tau_P C_L V_{CC}}{I_{DN}} \right)^{1/3} \tag{1}$$

where τ denotes transit time of the bipolar transistor ($Q2$), τ_P is the transit time of the pMOS ($MP2$), C_L is the load capacitance, V_{CC} is the supply voltage, and I_{DN} is the drain current of the nMOS ($MN2$). If we compare this equation to that of the conventional BiCMOS circuit [19],

$$t_{PHL} = \left(\frac{\tau C_L V_{CC}}{I_{DN}} \right)^{1/2},$$

it is clear that the QC-BiCMOS delay has a smaller dependence on load capacitance. The QC-BiCMOS delay is proportional to $C_L^{1/3}$, whereas the delay of the conventional BiCMOS is proportional to $C_L^{1/2}$. This is because the current drive is amplified by $MN2$, $MP2$, and $Q2$.

Equation (1) also clarifies that the delay is determined by the product average of the performance of nMOS, pMOS, and n-p-n bipolar. In other words, these three device performances equally contribute to the delay. By contrast, in the conventional BiCMOS gate the performances of the nMOS and n-p-n bipolar transistor determine the pull-down delay. Therefore, QC-BiCMOS delay is less sensitive to the bipolar cutoff frequency.

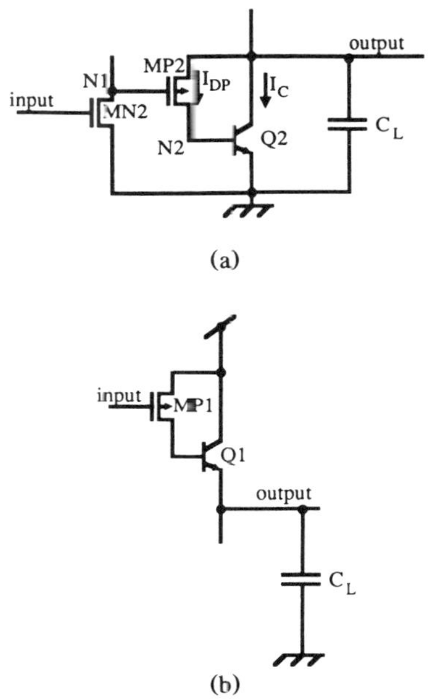

Fig. 12. Simplified QC-BiCMOS circuits for (a) pull-down and (b) pull-up delay analysis.

When the input has a ramped waveform, these delay equations are modified to

$$t_{PHL} = \left(\frac{12\tau\tau_P \tau_N t_r C_L V_{CC}}{0.8 I_{DN}} \right)^{1/4} \tag{2}$$

for the QC-BiCMOS and

$$t_{PHL} = \left(\frac{3\tau\tau_N t_r C_L V_{CC}}{0.8 I_{DN}} \right)^{1/3}$$

for the conventional BiCMOS. Here, t_r is the input rising time. If we compare these equations with those of the steep-input case, the difference is the change of the power from 1/3 to 1/4 for QC-BiCMOS, and from 1/2 to 1/3 for conventional BiCMOS. However, the above conclusions are unchanged.

The dependence of QC-BiCMOS delay on the supply voltage can be understood by using (1) and the relation for the MOSFET drain current in the velocity-saturation limit:

$$I_D = qWC_{ox}v_S(V_{GS} - V_T).$$

The delay equation for QC-BiCMOS explicitly including the supply voltage dependence is

$$t_{PHL} \propto \left(\frac{V_{CC}}{V_{CC} - V_{BE} - V_T} \right)^{1/3}.$$

By contrast, for the conventional BiCMOS gate, delay is

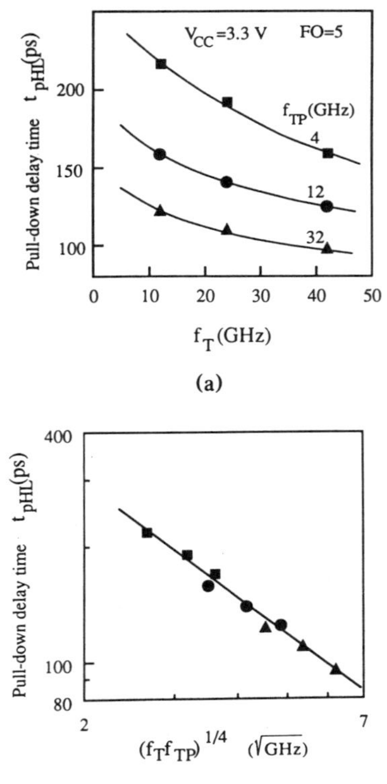

Fig. 13. Pull-down delay time versus (a) f_T and f_{TP}, and (b) $(f_T f_{TP})^{1/4}$, where f_T is the maximum cutoff frequency of $Q2$ (at $V_{CE} = 2$ V), and f_{TP} is the maximum cutoff frequency of $MP2$ (at $V_{DS} = -2$ V).

given by

$$t_{PHL} \propto \left(\frac{V_{CC}}{V_{CC} - 2V_{BE} - V_T} \right)^{1/2}.$$

These equations clearly show that the QC-BiCMOS gate has a smaller supply-voltage dependence than the conventional BiCMOS gates.

The validity of these analyses is verified by using circuit simulation. Pull-down delay is evaluated by changing the parameters of the pMOS and the n-p-n transistor, which constitute the quasi-p-n-p. As shown in Fig. 13(a), the delay depends on both the transit time of the pMOS and n-p-n as expected. Note that these scattered data are arrayed on the line when the delays are plotted with respect to $(f_T f_{TP})^{1/4}$ as shown in Fig. 13(b). This coincides with (2).

The above analyses neglect the effect of parasitic capacitances at nodes $N1$ and $N2$. These capacitances only add the charging/discharging delay to (1) and (2).

B. Pull-Up Delay

The pull-up delay is expressed by the same equation as the conventional BiCMOS, because the circuit is essentially the same:

$$t_{PLH} = \left(\frac{\tau C_L V_{CC}}{I_{DP}} \right)^{1/2}. \quad (3)$$

In this equation, however, the drain current for the pMOS (I_{DP}) is different from that of the conventional BiCMOS, because the pull-down waveform is quasi-full swing in QC-BiCMOS. The drain current expression for pMOS, whose characteristics are not as full in the velocity-saturation region as nMOS, can be expressed by $I_D \alpha (V_{GS} - V_T)^\beta$, where β is 1–2. The drain current for $MP2$ is expressed by

$$I_{DP} \propto (V_{CC} - V_T)^\beta \qquad \text{for QC-BiCMOS}$$

$$I_{DP} \propto (V_{CC} - V_{BE} - V_T)^\beta \qquad \text{for Darlington BiCMOS.}$$

The QC-BiCMOS pull-up delay is smaller than that of the conventional BiCMOS, especially at low supply voltages, although the pull-up circuit itself is the same as the conventional BiCMOS.

C. Low-Voltage Limitation

QC-BiCMOS also has a low-voltage limit, which is analyzed as follows. Equation (1) is based on the assumption that the pMOS ($MP2$) operates in the saturation region. As is evident from Fig. 2(b), the operation point, however, moves to the linear region at low supply voltages due to the V_{BE}. In the linear region, the drain current decreases rapidly with decreasing supply voltage, resulting in degraded performance. The condition for the $MP2$ entering into the linear region is given by

$$V_{CC} = V_{BE} + V_{DS(\text{SAT})}$$

where $V_{DS(\text{SAT})}$ denotes the drain–source voltage for the saturation region. Defining the coefficient γ by $V_{DS(\text{SAT})} = \gamma V_{CC}$ (γ is around 1/2 in the deep-submicrometer MOSFET's, which are used in the scaling analysis shown in Fig. 11), the supply voltage corresponding to this condition is given by

$$V_{CC} = \frac{V_{BE}}{1-\gamma}.$$

Considering that V_{BE} is 0.8–1 V, the transition voltage from saturation to the linear region is about 1.6–2 V. Below this voltage the current drive into the $Q2$ base reduces rapidly, severely degrading the performance. Actually this result coincides with the rapid delay increase below 2 V in Fig. 11.

V. Design Issues

Discharging circuits for the base charge of $Q2$ requires careful design (Fig. 1(b)). The design should meet the following competing conditions. The base charge of $Q2$ should be designed to be retained until the pull-down transient is completed. In this respect, the resistance of $Z2$ is desired to be high. On the other hand, the saturated $Q2$ has to be relieved (the base charge should be discharged through $Z2$) shortly after the pull-down transient. Otherwise, the next pull-up transient causes a large short-circuit current through $Q1$ and $Q2$. In this respect, resistance $Z2$ cannot be designed too large.

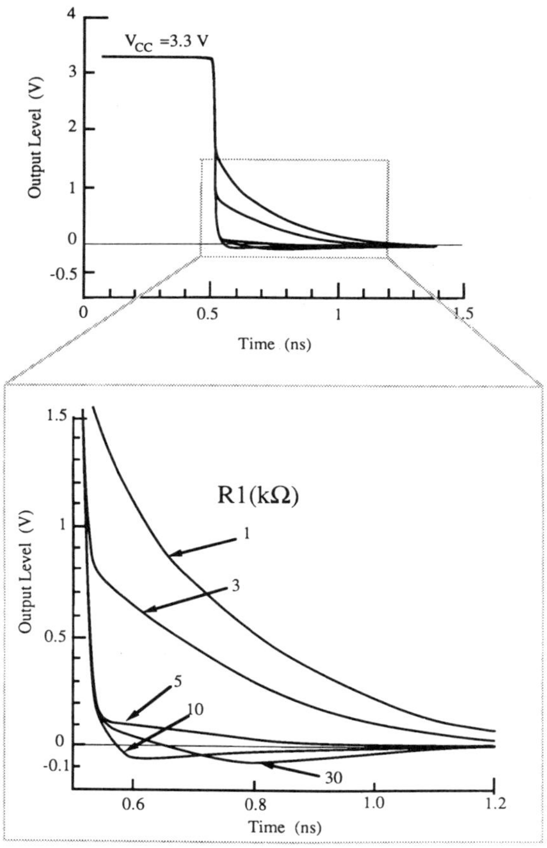

Fig. 14. The influence of resistance $R1$ on the output waveform in the pull-down operation of QC-BiCMOS two-input NAND (QC-BiCMOS-c).

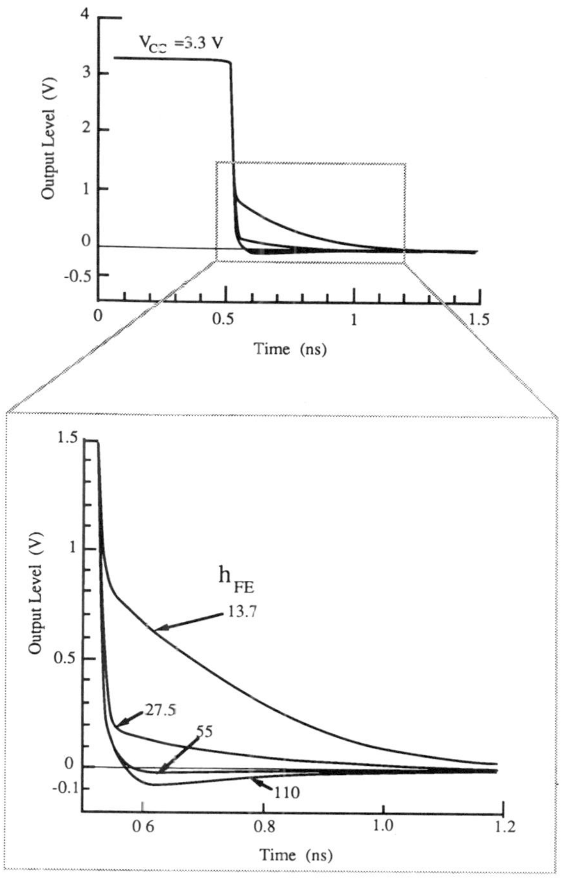

Fig. 15. The influence of the current gain h_{FE} on the output waveform in the pull-down operation of QC-BiCMOS two-input NAND (QC-BiCMOS-c).

The pull-down waveform of the QC-BiCMOS two-input NAND (QC-BiCMOS-c) is simulated by changing the resistance of $R1$ as shown in Fig. 14. When the resistance is smaller than 3 kΩ, a transition from steep transient to slow transient far above ground level is observed. This transition shows that the base charge is discharged through resistance $R1$ before the pull-down is completed. When the resistance is larger than 100 kΩ, this discharging time of the bipolar base becomes long (3 ns), which will restrict the machine cycle. The resistance of 10–30 kΩ is preferred at 3.3-V operation for high-machine-cycle low-power operation.

An undershoot below ground level is observed in some resistance conditions. This is because the electrons stored in the base are relieved to the collector and emitter when the holes in the base are discharged through resistance $R1$. In the equivalent circuit viewpoint, this can be seen as capacitive coupling between the base and collector through diffusion capacitance lowering the output level when the base voltage is discharged. This undershoot is finally diminished by the current through the resister $R2$.

If the feedbacked signal from the output is utilized to determine the active MOS resistance, the above competing requirement can be ideally fulfilled. The QC-BiCMOS-d is advantageous in this respect.

The high current gain of the bipolar transistor is more important in QC-BiCMOS than in the conventional BiCMOS for the good charge retention in the base. The simulated pull-down waveform with different current gain is shown in Fig. 15. A degraded waveform at the current gain below 30 is observed. Because bipolar transistors with current gains above 50 are commonly used in the BiCMOS LSI's, this will not cause severe restrictions to device design. However, too low of a current-gain design has to be avoided in QC-BiCMOS.

VI. Experimental Results

Based on these analyses, loaded-gate chains using the QC-BiCMOS have been fabricated using 0.3-μm double-poly-Si self-aligned BiCMOS technology. The gate length is 0.3 μm, gate-oxide thickness is 9 nm, the threshold voltage is 0.3 V, and the cutoff frequency of the n-p-n bipolar transistor is 20 GHz. The fine patterns, such as gate, contact, and first metal, are defined by electron-beam lithography, and the other patterns are defined by the i-line optical lithography. The interconnection consists of first-level tungsten, second-level aluminum, and third-level aluminum. More details of our device/process technology will be described in another paper. A microphotograph of a 31-stage QC-BiCMOS gate chain is shown in Fig. 16. CMOS and conventional Darlington-type BiCMOS gate

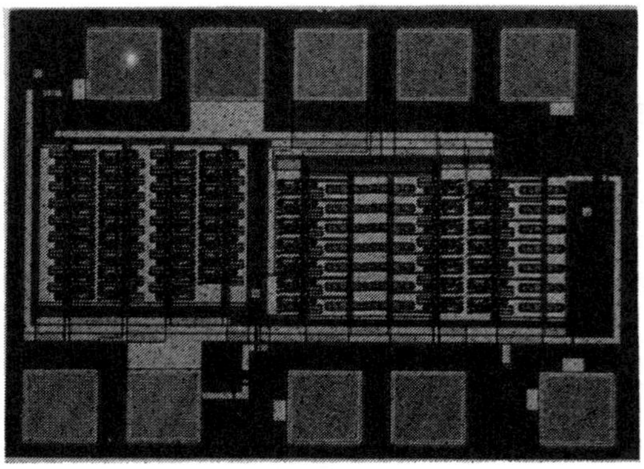

Fig. 16. Microphotograph of the 31-stage three-input NAND chains using QC-BiCMOS.

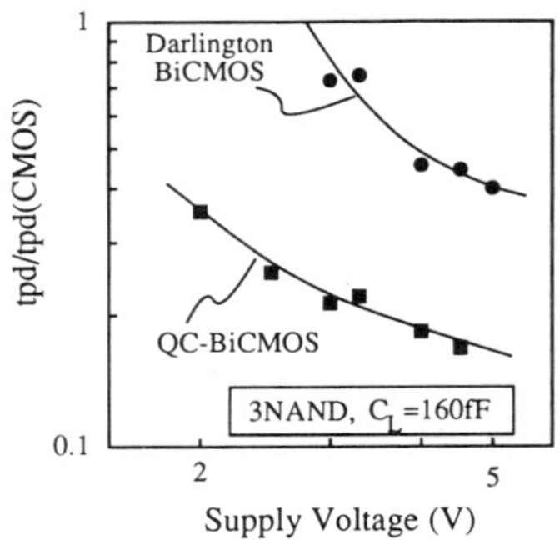

Fig. 17. Measured performance leverage of the conventional and QC-BiCMOS gate over the CMOS. QC-BiCMOS-d is used.

chains are also designed and fabricated on the same wafer. Therefore, they have the same device parameters as those of the QC-BiCMOS. The delays from the input signal (it is actually a regenerated signal by an on-chip inverter circuit) to the output signal are measured on the wafer level using the on-chip source-follower circuits. The measured BiCMOS delay time relative to CMOS is shown in Fig. 17. The QC-BiCMOS-d circuit is used. The fully loaded ($C_L = 160$ fF, FO = 5) three-input NAND gate maintains sufficient performance leverage over the CMOS even below 3-V supply voltage, as shown in Fig. 17. By contrast, the conventional Darlington-type BiCMOS lose its performance advantage over CMOS at 3 V, which coincides with the stimulation result. This verifies the above described excellent low-voltage performance of the QC-BiCMOS over CMOS in the sub-3-V region.

VII. Conclusions

Quasi-complementary BiCMOS circuits for low-voltage (even below 3 V) operation are described. Two key circuit techniques, 1) a separation between the base of the pull-up bipolar and the base of the quasi-p-n-p, and 2) the carefully designed base discharging circuit, enable QC-BiCMOS to have a better low-voltage performance than any other conventional BiCMOS circuits. A novel feature of the full-swing pull-down waveform is realized by the temporarily saturated bipolar operation. The QC-BiCMOS delay expression is derived to clarify the smaller delay dependence on load capacitance ($t_{PHL} \propto C_L^{1/3}$). QC-BiCMOS gate chains are fabricated using a 0.3-μm BiCMOS technology, which verifies that the QC-BiCMOS operates more than twice as fast as CMOS even at 2.5 V. This demonstrates that the QC-BiCMOS breaks the low-voltage operation limit in conventional BiCMOS circuits, and the QC-BiCMOS is a very promising circuit in 0.3-μm (or less) BiCMOS ULSI's.

Appendix

First, the input pulse is assumed to be as steep as can be approximated for a step function. The parasitic capacitances are also neglected for simplicity. The drain current through the nMOS transistor $MN2$ is constant (I_{DN}). The following equations are obtained:

$$\frac{C_L V_{CC}}{2} = \int_0^{t_{PHL}} I_C \, dt$$

$$I_C = \frac{\int_0^t I_{DP} \, dt}{\tau}$$

$$I_{DP} = \frac{\int_0^t I_{DN} \, dt}{\tau_P}$$

where I_C denotes collector current, I_{DP} is the drain current of the pMOS ($MP2$), τ is the transit time of the n-p-n bipolar transistor, and τ_P is the transit time of the $MP2$. Integrating, we get

$$\frac{C_L V_{CC}}{2} = \frac{I_{DN} t_{PHL}^3}{6 \tau \tau_P}.$$

The pull-down delay time is expressed by

$$t_{PHL} = \left(\frac{3 \tau \tau_P C_L V_{CC}}{I_{DN}} \right)^{1/3}. \qquad (4)$$

When the input has a ramped waveform as defined by

$$\frac{V_{CC}}{I_{in}} = \frac{t_r}{0.8\, C_{in}}$$

where I_{in} is the current that drives the input capacitance C_{in} of the nMOS ($MN2$), and t_r is the input rising time, the time-dependent drain current of $MN2$ is expressed by the following equation:

$$I_{DN} = \frac{\int I_{in} \, dt}{\tau_N}.$$

A similar procedure as the above steep-waveform case gives

$$t_{PHL} = \left(\frac{12 \tau \tau_P \tau_N t_r C_L V_{CC}}{0.8 I_{DN}} \right)^{1/4}. \qquad (5)$$

The equation for the conventional bipolar transistor is similarly obtained.

Acknowledgment

The authors wish to thank Dr. T. Masuhara, Dr. H. Sunami, Dr. T. Nagano, Dr. E. Takeda, Dr. H. Higuchi, M. Suzuki, and S. Tachibana for their useful discussions. The authors are greatly indebted to Dr. Y. Kawamoto, Dr. T. Kure, Dr. F. Murai, and the processing staff members for their chip fabrication. The authors are also greatly indebted to S. Morita, A. Kawamata, and T. Kumihashi for their circuit simulations.

References

[1] M. Suzuki *et al.*, "A 3.5-ns, 500-mW, 16-kbit BiCMOS ECL RAM," *IEEE J. Solid-State Circuits*, vol. 24, no. 5, pp. 1233–1237, Oct. 1989.

[2] T. Hotta *et al.*, "A 70MHz 32b microprocessor with 1.0 μm BiCMOS macrocell library," in *ISSCC Dig. Tech. Papers*, 1989, pp. 124–125.

[3] G. Kitsukawa *et al.*, "A 23-ns 1-Mb BiCMOS DRAM," *IEEE J. Solid-State Circuits*, vol. 25, no. 5, pp. 1102–1111, Oct. 1990.

[4] A. Watanabe, T. Nagano, S. Shukuri, and T. Ikeda, "Future BiCMOS technology for scaled supply voltage," in *IEDM Tech. Dig.*, 1989, pp. 429–433.

[5] H. J. Shin *et al.*, "Full-swing complementary BiCMOS logic circuits," in *Proc. Bipolar Circuit Technology Meeting*, 1989, pp. 229-233.

[6] P. Raje, R. Ritts, K. Cham, J. Plummer, and K. Saraswat, "Merged BiCMOS: A device and circuit technique scalable to the sub-micron sub-2V regime," in *ISSCC Dig. Tech. Papers*, 1989, TPM 9.2.

[7] M. Fujishima, K. Asada, and T. Sugano, "Evaluation of delay-time degradation of low-voltage BiCMOS based on a novel analytical delay-time modeling," *IEEE J. Solid-State Circuits*, vol. 26, no. 1, pp. 25–31, Jan. 1991.

[8] P. R. Gray and R. G. Meyer, *Analysis and Design of Analog Integrated Circuits*, 1st ed. New York: Wiley, 1979, pp. 311–313.

[9] H. Shin, "Full-swing complementary BiCMOS logic circuits," in *Symp. VLSI Circuits Tech. Dig.*, 1990, pp. 89–90.

[10] H. J. Shin, "Performance comparison of driver and full-swing techniques for BiCMOS logic circuits," *IEEE J. Solid-State Circuits*, vol. 25, no. 3, pp. 863–865, June 1990.

[11] G. Kitsukawa *et al.*, "Low-power, high-speed BiCMOS memory circuits," in *Extended Abstr. 1984 Int. Conf. Solid State Devices and Mater.* (Kobe, Japan), pp. 233–236.

[12] H. Momose, K. M. Cham, C. I. Drowley, H. R. Grinolds, and H. S. Fu, "0.5 micron BiCMOS technology," in *IEDM Tech. Dig.*, 1987, pp. 838–840.

[13] C. Mead and L. Conway, *Introduction to VLSI Systems*. Reading, MA: Addison Wesley, 1980.

[14] Y. Urakawa, M. Matsui, and K. Ochii, "Optimization of BiCMOS cascade" (in Japanese), presented at the Inst. Electron., Inform. and Commun. Eng. Spring Meeting, C-259, 1988.

[15] H. K. Gummel and H. C. Poon, "An integral charge control model of bipolar transistors," *Bell Syst. Tech. J.*, vol. 49, pp. 827–851, May–June 1970.

[16] E. Takeda, Y. Ohji, and H. Kume, "High field effects in MOSFETs," in *IEDM Tech. Dig.*, 1985, pp. 60–63.

[17] K. Yano, M. Aoki, and T. Masuhara, "PARAMOST—A new parasitic resistance model for deep submicron MOS transistors," in *Extended Abstr. 1986 Int. Conf. Solid State Devices and Mater.* (Tokyo), pp. 85–88.

[18] H. Higuchi *et al.*, "Performance and structure of scaled-down bipolar devices merged with CMOSFETs," in *IEDM Tech. Dig.*, 1984, pp. 694–697.

[19] M. Suzuki, S. Tachibana, and H. Higuchi, "Drivability of BiCMOS logic gate" (in Japanese), in *Inst. Electron., Commun. Eng. Meeting Dig.*, 2-127.

[20] G. P. Rosseel and R. W. Dutton, "Influence of device parameters on the switching speed of BiCMOS buffers," *IEEE J. Solid-State Circuits*, vol. 24, pp. 90–99, 1989.

[21] T. Sugano and T. Iizuka, Eds., *The Design of CMOS ULSIs*. Tokyo: Baihukan, 1989, pp. 35–36.

[22] S. M. Sze, *Physics of Semiconductor Devices*, 2nd ed. New York: Wiley, 1981, p. 324.

Merged BiCMOS Logic to Extend the CMOS/BiCMOS Performance Crossover Below 2.5-V Supply

Rosalyn B. Ritts, *Member, IEEE*, Prasad A. Raje, *Student Member, IEEE*, James D. Plummer, *Fellow, IEEE*, Krishna C. Saraswat, *Fellow, IEEE*, and Kit M. Cham, *Senior Member, IEEE*

Abstract—**This paper will discuss the merged BiCMOS (MBiCMOS) gate, a unique circuit configuration to improve BiCMOS gate performance at low supply voltages. Measured results from both a 2-μm process with triple-diffused n-p-n bipolar transistors and a 1-μm conventional buried-layer/epi process with 15-GHz n-p-n bipolar devices indicate that MBiCMOS provides measured delay advantages over CMOS to below 2.5-V supplies, in a compact four-device cell that does not require any change in the standard BiCMOS processing sequence or any more power than a standard BiCMOS gate. Gates designed for fabrication in a 0.5-μm technology and simulated using measured device parameters show MBiCMOS/CMOS performance crossover voltages of 2.19 V, as compared to a crossover of 2.63 V for standard BiCMOS/CMOS. Use of a modified MBiCMOS gate extends the performance crossover to 1.9 V with the addition of a single implant to implement zero-threshold PMOS devices. With modified MBiCMOS, the useful operating range for BiCMOS technology is extended to the submicrometer sub-2-V regime. A full-swing version of the MBiCMOS gate (FS-MBiCMOS) is introduced. Simulations of 2-μm gates show FS-MBiCMOS/CMOS performance crossover voltages of 2.2 V.**

I. Introduction

A NUMBER OF recent papers have presented and evaluated the idea of merging devices within a BiCMOS design [1]–[5]. Merging has been shown to reduce area and to increase speed by reducing internal capacitances. In previous works, the benefits of merging have been evaluated in standard BiCMOS and BiNMOS circuit configurations. Performance of an inverter with separate PMOS and n-p-n devices has been compared with the performance of an inverter of the same configuration with a single merged PMOS/n-p-n device. In this work, the merged PMOS/n-p-n device is taken as a building block for BiCMOS circuit design. Instead of simply using the concept of merging in a conventional circuit design, a new merged BiCMOS (MBiCMOS) inverter designed specifically to take advantage of the merged PMOS/n-p-n cell is discussed [6].

MBiCMOS maintains a measured delay and power–delay advantage cover CMOS into the 2-V supply range, in a simple four-device gate that does not require any change in the standard BiCMOS processing sequence. In a 2-μm technology, MBiCMOS outperforms CMOS down to 2.6-V supply, while the standard six-transistor BiCMOS gate loses its delay advantage over CMOS at 3.2 V. In a 1-μm technology, the measured MBiCMOS/CMOS crossover occurs at 2.4 V, as compared to a BiCMOS/CMOS crossover of 3.0 V. Gates designed for fabrication in a 0.5-μm technology and simulated using measured device parameters indicate that MBiCMOS can be used to extend the performance crossover voltage to below 2 V in the submicrometer regime. Although the MBiCMOS gate was designed to take advantage of the merged PMOS/n-p-n device in order to maximize the drive per unit area ratio, device merging is not critical to achieve MBiCMOS performance benefits in terms of delay and power. Performance advantages are obtained with or without device merging.

The advantage of MBiCMOS over other recently introduced BiCMOS designs for scaled supplies is that it achieves performance gains at scaled supplies in a compact cell that requires no additional processing, no additional gates, and no more power than a standard BiCMOS design. Other published solutions to the problem of reduced BiCMOS performance at scaled supplies fall into one of three categories. Some, such as ground level shifting [7], positive feedback [8], clamping diodes, and collector–emitter shunts [9], involve adding gates to the basic cell. Others, such as complementary BiCMOS with emitter-follower drivers [2], [10] and complementary BiCMOS with inverter drivers [2], involve the addition of a p-n-p device to the BiCMOS process. Finally, still others avoid the problem by using a bipolar device in the pull-up section of the circuit only (BiNMOS) [5]. All of these techniques are successful in terms of extending the range of usable BiCMOS supply voltages, but unlike MBiCMOS they succeed at the expense of added process complexity (p-n-p devices), lack of performance symmetry (in the

Manuscript received February 28, 1991; revised July 1, 1991.

R. B. Ritts, J. D. Plummer, and K. C. Saraswat are with the Center for Integrated Systems, Stanford University, Stanford, CA 94305-4070.

P. A. Raje was with the Center for Integrated Systems, Stanford University, Stanford, CA 94305. He is now with Hewlett-Packard Laboratories, Palo Alto, CA 94303.

K. M. Cham is with Circuit Technology Research and Development, Hewlett-Packard Company, Palo Alto, CA 94303.

IEEE Log Number 9103014.

Reprinted from *IEEE J. Solid-State Circuits*, vol. 26, no. 11, pp. 1606–1614, Nov. 1991.

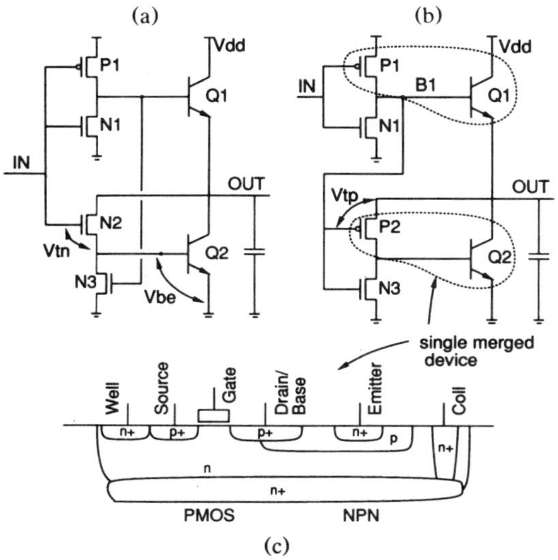

Fig. 1. (a) The conventional BiCMOS gate circuit, with six discrete devices. (b) The MBiCMOS gate circuit, with four discrete devices. (c) The merged PMOS/n-p-n device.

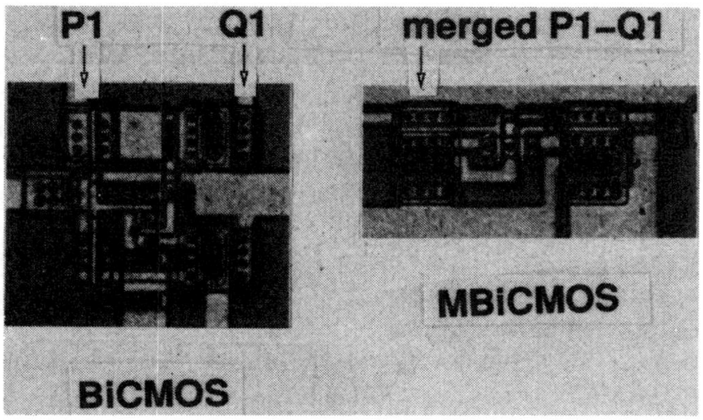

Fig. 2. Microphotograph of BiCMOS and MBiCMOS gates in the 1-μm technology.

case of BiNMOS), or increased power and circuit complexity (due to added devices per gate).

II. The MBiCMOS Gate

A. Gate Design

Fig. 1(a) and (b), respectively, shows a standard BiCMOS inverter circuit [11] and the new MBiCMOS gate. The MBiCMOS gate contains four devices: two merged PMOS/n-p-n devices and two NMOS devices. A device cross section for the merged PMOS/n-p-n device is shown in Fig. 1(c). The PMOS n-well is merged with the n-p-n n-collector, and the PMOS p-drain is merged with the n-p-n p-base. Fig. 2 shows microphotographs of the standard BiCMOS and MBiCMOS gates in a 1-μm process. With equivalent device sizes, the BiCMOS gate occupies 876 μm^2 while MBiCMOS occupies only 582 μm^2, an area reduction of 34%.

For a given process, layout design rules for MOS and bipolar devices also hold for the merged device. In addition, several new layout rules are needed: 1) emitter-n^+-to-base-p^+ spacing, 2) spacing of gate edge to base edge, and 3) for the triple-diffused case, location of n-well and n-collector edges beneath the merged device (for the buried-layer/epi technology, a single n-type implant above the n^+ buried layer is sufficient).

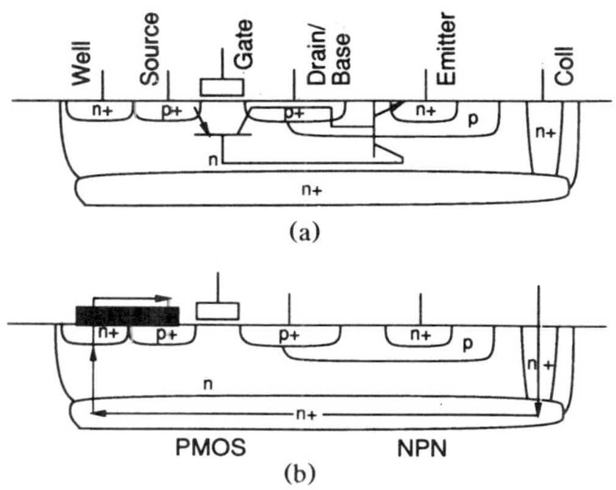

Fig. 3. Design to avoid latch-up in the merged PMOS/n-p-n device. (a) Latch-up path of concern. (b) Current flow path to avoid latch-up.

Due to the lack of isolation between devices, the merged PMOS/n-p-n device is more prone to latch-up than its nonmerged equivalent. The latch-up path of concern contains the intrinsic n-p-n device and the parasitic p-n-p device under the PMOSFET (Fig. 3(a)). We have found that latch-up-free operation can be obtained by using appropriate layout to control the direction of current flow, as reported previously by Momose *et al.* [5] and shown in Fig. 3(b). There is no external contact to source or well. Instead, source current is supplied by the collector contact through the path indicated by arrows. Current flow in this direction prevents latch-up by ensuring that the source/well junction or the emitter–base junction of the parasitic p-n-p, is reverse biased.

B. Static Analysis

Static operation of the MBiCMOS gate is as follows. When the input is low, $P1$ is ON and $N1$ is OFF, so that node $B1$ is at V_{DD} and the output is pulled high through $Q1$. $N3$ is ON, holding the base of $Q2$ at ground and keeping it OFF. This is identical to the standard BiCMOS V_{OH} level. When the input is high, $P1$ is OFF and $N1$ is ON, so that node $B1$ is at ground. $Q1$ and $N3$ are now OFF, while $P2$ is ON, supplying base current to turn on $Q2$. The output low voltage depends on the magnitude of the threshold voltage of $P2$ relative to the base–emitter voltage of $Q2$; V_{OL} is the higher of these two voltages.

Both standard BiCMOS and MBiCMOS have large noise margins similar to CMOS, reduced only by the slightly smaller V_{OH} level and slightly higher V_{OL} level. Since the output levels in both the BiCMOS and MBiCMOS gates do not go rail to rail, the MOS threshold voltages must be evaluated relative to V_{OH} and V_{OL} to ensure that the MOS devices do not provide large leakage currents and thereby degrade circuit performance. It is especially important when the threshold voltage of $P2$ is large and controls V_{OL}, as discussed above.

For optimized performance, the threshold of $P2$ should be low, to increase drive to the base of $Q2$. Modifying $P2$ to be a zero-threshold device increases the current drive available from $P2$ to charge $Q2$ with no change in functionality, at the cost of a single threshold adjusting implant. Note that when the input is low, the base of $Q1$ (also the gate of $P2$) is at V_{dd}, while the output voltage is less than V_{dd} due to the V_{be} of $Q1$. Therefore, $P2$ will remain OFF, even with $V_{tp} = 0$ V, and there will be no degradation in noise sensitivity or increase in leakage current.

C. Transient Analysis

The pull-up portion of the MBiCMOS cell is essentially similar to that of the standard BiCMOS cell (see Fig. 1). The parasitic capacitance at MBiCMOS node $B1$ is slightly larger than the capacitance at the base of $Q1$ in BiCMOS, but the pull-up transient analysis is the same and has been well documented elsewhere [12], [13]. The significant differences occur in the pull-down portion of the circuit. In the BiCMOS gate, a low-to-high step at the input turns NMOSFET $N2$ on, immediately supplying base current to $Q2$. Following a low-to-high input to the MBiCMOS gate, node $B1$ must be pulled low through $N1$ before $P2$ turns on and $N3$ turns off. This does not translate into increased overall gate delay, as will be discussed here and as our experimental results show. Pull-down delay in the MBiCMOS gate is minimized by appropriate sizing of transistor $N1$ to achieve a fast pull-down at $B1$, and by other advantages inherent to the structure of the MBiCMOS gate.

Before comparing the pull-down response of MBiCMOS to BiCMOS, it is first important to note the differences in the pull-up and pull-down responses of the BiCMOS gate itself. Previous works have often presented analytical models for the pull-up case only, stating that the pull-up and pull-down responses in BiCMOS can be modeled with the same set of assumptions. Differences between the pull-up and pull-down responses in the BiCMOS gate are: 1) the gate-to-source voltage of $N2$ during pull-down is degraded by a diode drop at the input and a diode drop between source and ground, and is therefore smaller than the gate-to-source voltage of $P1$ during pull-up; 2) in the pull-up case, $N1$ shuts off quickly when the input goes low, while during pull-down, $N3$ stays on until $B1$ falls, decreasing the total base current to $Q2$; and 3) body effect must be considered for $N2$ during pull-down, unlike $P1$ during pull-up.

The MBiCMOS gate is designed to eliminate the above differences. PMOSFET $P2$ sees a V_{GS} equal to the full supply range. The input is not degraded by a diode drop nor is there a diode drop between the source and ground, as is the case for the standard BiCMOS gate. This increased base drive to $Q2$, illustrated in Fig. 4, translates directly into increased performance, especially at scaled supply voltages. The gate is symmetrical, eliminating body effect for the MOS device supplying base current to $Q2$. Another advantage of MBiCMOS for pull-down delay concerns input capacitance of the cell in relation to the size of NMOSFET $N1$. Input capacitance of the MBiCMOS gate consists of capacitance from only two MOS devices, $P1$ and $N1$. Input capacitance of the standard gate consists of capacitance from three MOS devices, $P1$, $N1$, and $N2$. Thus, it is possible to make $N1$ larger in the MBiCMOS gate than it is in the BiCMOS gate, while maintaining equal input capacitance. This is a key factor for improved switching speeds because it allows for very fast pull-down at node $B1$.

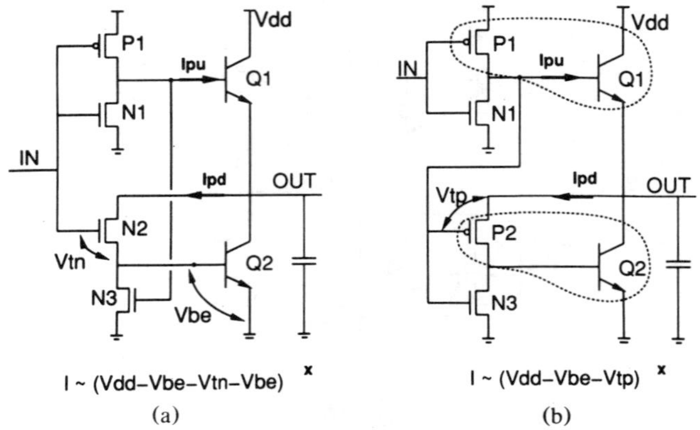

Fig. 4. Current drive during pull-down: (a) BiCMOS, and (b) MBiCMOS.

Another difference between BiCMOS and MBiCMOS operation is in the state of $Q2$ towards the end of the output falling transition. In BiCMOS, the base of $Q2$ is held at the same voltage as the output through $N2$. In MBiCMOS, the output voltage will continue to fall after PMOSFET $P2$ turns off, due to collector current resulting from charge stored in the base of $Q2$. In this case, the base of $Q2$ stays higher than the output voltage, and $Q2$ will saturate. Saturation of $Q2$ is shallow and does not noticeably slow the following output rising transition. During the next output rising transition, saturation is quickly relieved by NMOSFET $N3$. Saturation in $Q2$ is further counterbalanced by the fact that $P2$ begins to conduct in the opposite direction after V_{OUT} falls below the base voltage of $Q2$, helping to pull stored base charge out of $Q2$. Charge removal is especially efficient in the modified MBiCMOS gate in which device $P2$ is a zero-threshold device.

Fig. 5 shows BiCMOS, MBiCMOS, and modified-MBiCMOS performance following both input rising and input falling transitions. Input and output waveforms are shown in Fig. 5(a) and (c), and the pull-up and pull-down current (as labeled in Fig. 4) are shown in Fig. 5(b) and (d). Simulations use models extracted from fabricated 0.5-μm devices. Although a single representative input rising transition is shown in Fig. 5(a), the input levels actually vary slightly for the three gates (as shown in output voltage section of Fig. 5(c)). Similarly, although a single representative input falling waveform is shown in

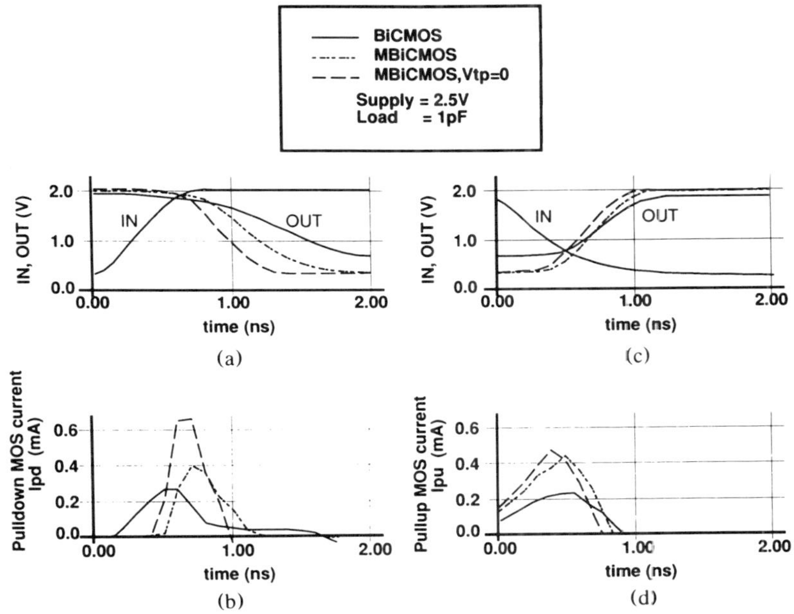

Fig. 5. Transient response to BiCMOS and MBiCMOS gates. (a) Input and output voltages for an output falling transition. (b) Pull-down current during the output falling transition. (c) Input and output voltages for an output rising transition. (d) Pull-up current during the output rising transition.

Fig. 5(c), the input levels used for the actual simulations are the outputs from Fig. 5(a). In the input falling case, there is a wide variation in the slope of the input transitions for the three gates.

Clearly, MBiCMOS pull-down is faster under similar conditions of operation (2.5-V supply and 1-pF load in this case). Increased performance is due to increase MOS drive to $Q2$ during pull-down, as shown in Fig. 5(b). Note in Fig. 5(a) that V_{OL} is lower for MBiCMOS than for BiCMOS. This is due to saturation of $Q2$ in MBiCMOS. Although V_{OL} in a static analysis is equal to the higher of either V_T of $P2$ or V_{BE} of $Q2$, as discussed above, the actual V_{OL} level is equal to the saturated collector–emitter voltage of $Q2$. Pull-up, as shown in Fig. 5(c), is very similar for the three gates. In the MBiCMOS gate, the larger capacitance faced by the current supplying device $P1$ is compensated by the additional current drive, as shown in Fig. 5(d).

III. Experimental Strategy

The MBiCMOS gate was built and tested both at Stanford University and at the Hewlett Packard Research and Development Lab. Both experimental strategy and device processing varied with location. The Stanford process is a 2-μm process with triple-diffused n-p-n bipolar transistors [14], while the HP process is a 1-μm conventional buried-layer/epi process with 15-GHz n-p-n bipolar devices. Fig. 6 describes the experimental strategies used at each facility. A "normalized" comparison is one in which the gates being compared are built to have both the same input capacitance and the same overall silicon area [15].

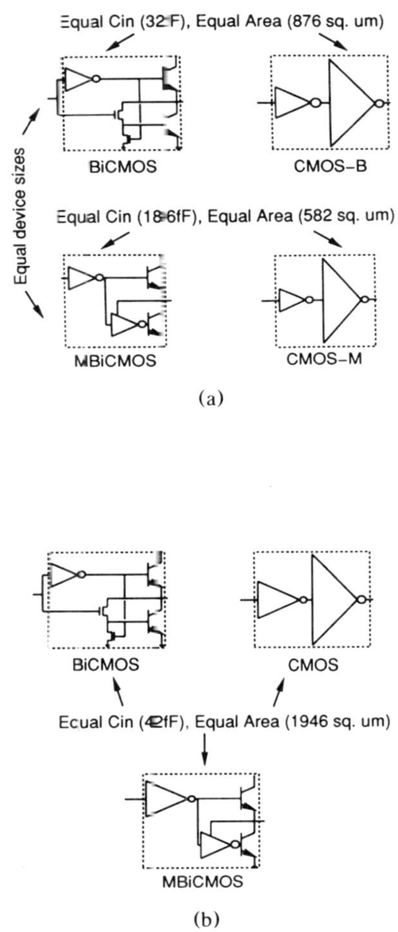

Fig. 6. Experimental strategies. (a) In the 1-μm technology, BiCMOS and MBiCMOS device sizes are equal; MBiCMOS has smaller gate area and input capacitance than BiCMOS. (b) In the 2-μm technology, all gates have equal area and input capacitance.

1um		2um	
MBiCMOS		**MBiCMOS**	
P1	7.8/1	P1	8/2
N1	3/1	N1	8/2
P2	7.8/1	P2	8/2
N3	3/1	N3	6/2
Q1	1x5	Q1	2x8
Q2	1x5	Q2	2x8
BiCMOS		**BiCMOS**	
P1	7.8/1	P1	6/2
N1	3/1	N1	4/2
N2	7.8/1	N2	6/2
N3	3/1	N3	4/2
Q1	1x5	Q1	2x8
Q2	1x5	Q2	2x8
CMOS-B		**CMOS**	
P1	11/1	P1	10/2
N1	7.6/1	N1	6/2
P2	66/1	P2	72/2
N2	45.6/1	N2	38/2
CMOS-M		**BiCMOS-M**	
P1	6.5/1	P1	6/2
N1	4.3/1	N1	4/2
P2	32.5/1	N2	6/2
N2	21.5/1	N3	4/2
		Q1	2x8
		Q2	2x10

Fig. 7. Sizes of the devices in the measured gates.

In the 1-μm technology, BiCMOS and MBiCMOS gates were built with the same device sizes. Due to device merging and the MBiCMOS circuit topology, the MBiCMOS gate in this case had lower input capacitance than the BiCMOS, and occupied 35% less area. Two CMOS gates were built, one normalized to the BiCMOS gate (CMOS-B) and one normalized to the MBiCMOS gate (CMOS-M). In the 2-μm technology, all gates were normalized. In this case, the size of MBiCMOS $N1$ was twice the size of BiCMOS $N1$, to achieve equal input capacitance and to improve circuit performance as discussed above. Device sizes for the fabricated gates are given in Fig. 7.

IV. Measured Results

Fig. 8 shows measured delay versus supply voltage for gates in the 1-μm technology with 1-pF loads. Curves are shown for BiCMOS and its normalized CMOS-B, and for MBiCMOS and its normalized CMOS-M. The delay of the BiCMOS gate is lower than the delay of CMOS-B until the supply is reduced to 3 V, at which point the BiCMOS delay rises sharply above that of CMOS-B. In contrast, MBiCMOS shows a measured delay advantage over its normalized gate, CMOS-M, down to 2.4-V supply. Note that the BiCMOS is slightly faster than MBiCMOS for supply voltages from 5 to 3 V. This is because the size of BiCMOS device $N1$ is the same as that of MBiCMOS device $N1$ in this comparison. MBiCMOS is faster than BiCMOS, even at 5-V supply, in a normalized comparison in which the input capacitances of the two gates are made equal by increasing the size of MBiCMOS device $N1$. As discussed above, a large $N1$ is needed in MBiCMOS to drive the increased internal capacitance at $B1$. This point is demonstrated by the 2-μm data, as shown in Fig. 9.

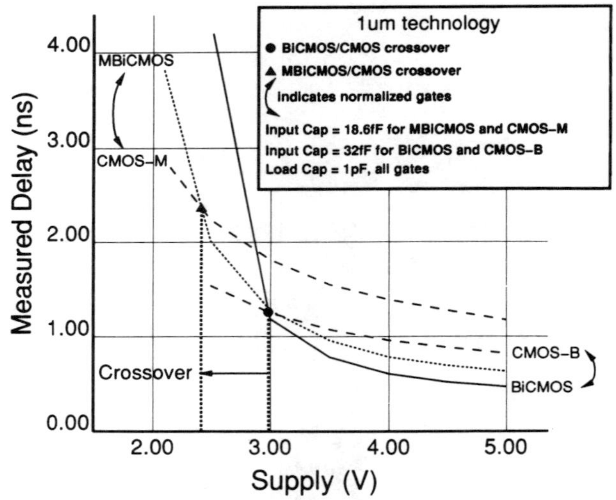

Fig. 8. Measured delay versus supply in the 1-μm technology.

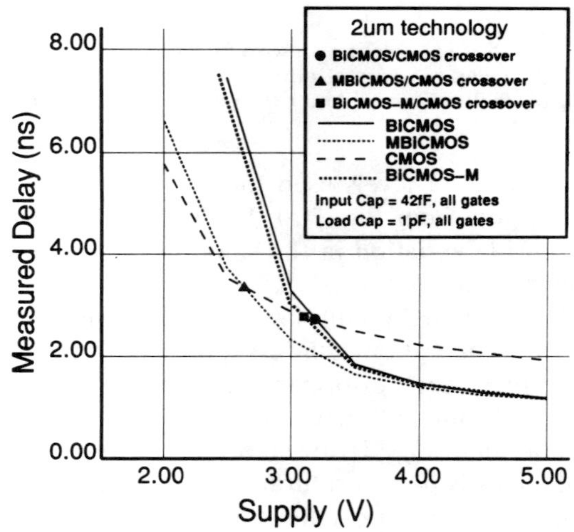

Fig. 9. Measured delay versus supply in the 2-μm technology.

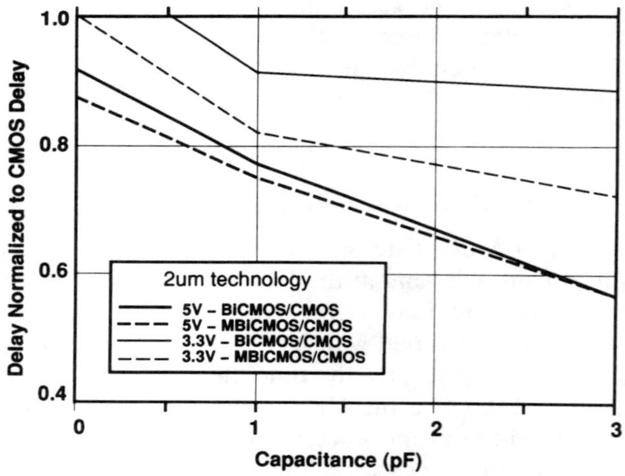

Fig. 10. Measured delay versus capacitance in the 2-μm technology.

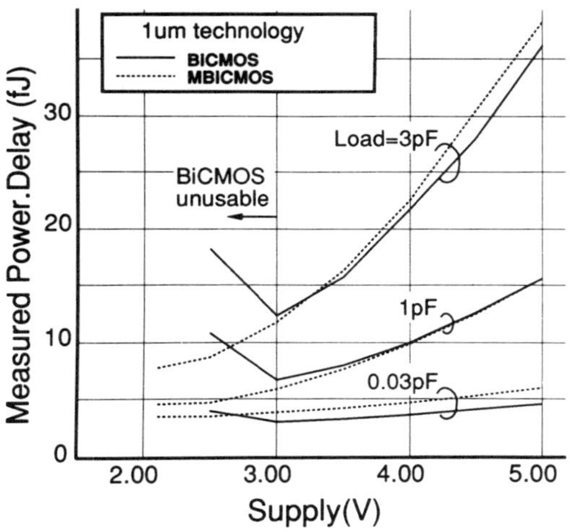

Fig. 11. Measured power–delay product versus supply for 1-μm BiCMOS and MBiCMOS.

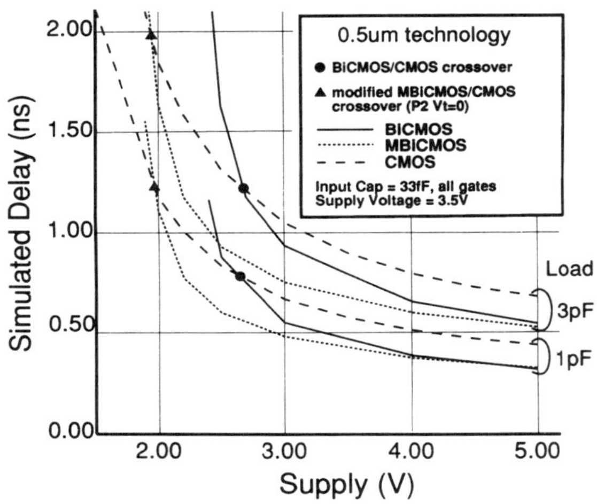

Fig. 12. Simulated delay versus supply for the modified MBiCMOS gate in the 0.5-μm technology.

Fig. 9 shows measured delay versus supply voltage for the 2-μm CMOS, BiCMOS, BiCMOS-M, and MBiCMOS gates with 1-pF loads. BiCMOS-M is a conventional BiCMOS gate (as shown in Fig. 1(a)) with a merged PMOS/n-p-n pull-up device. In this case, MBiCMOS clearly outperforms BiCMOS at all supply voltages. The 2-μm performance-crossover voltage, where the measured BiCMOS delay becomes greater than CMOS, is pushed from 3.2 V for BiCMOS to 3.1 V for BiCMOS-M, and to 2.6 V for MBiCMOS.

Fig. 10 shows measured delay versus load capacitance for the 2-μm BiCMOS and MBiCMOS gates. Data are obtained from ring-oscillator circuits with loads of 0, 1, and 3 pF. Delays are normalized to the measured delays of two-stage CMOS gates with equal input capacitance and equal area in order to emphasize the relative performance advantage. At 5-V supply, both BiCMOS and MBiCMOS outperform CMOS over the range of capacitances tested. At 3.3-V supply, BiCMOS outperforms CMOS only for loads above 0.5 pF, while MBiCMOS continues to outperform CMOS in the unloaded case.

At loads below 0.5 pF, the performances of BiCMOS and MBiCMOS gates relative to that of a single-stage CMOS gate are also of interest. Simulations of this case at 5 V show that BiCMOS outperforms CMOS for loads above 0.8 pF while MBiCMOS outperforms CMOS for loads above 0.5 pF. At 3.3 V, the crossover capacitances for BiCMOS/CMOS and MBiCMOS/CMOS are 1 and 0.7 pF, respectively.

From Fig. 11, the measured power–delay products for 1-μm MBiCMOS and BiCMOS are comparable for high supply voltages. The increase in power–delay product for BiCMOS below 3 V indicates the presence of static power dissipation due to incomplete switching. No such anomaly is observed for MBiCMOS down to 2.1 V.

V. Simulation Results

A. Modified MBiCMOS (with V_t of $P2 = 0$)

The performance of the MBiCMOS gate in a 0.5-μm technology was studied with SPICE simulation. Simulations used device parameters extracted from fabricated 0.5-μm CMOS devices and 18-GHz n-p-n bipolar transistors. The BiCMOS/CMOS and MBiCMOS/CMOS performance crossover voltages in this case were 2.63 and 2.19 V, respectively.

The MBiCMOS/CMOS performance crossover can be extended to 1.9 V by using the modified MBiCMOS gate, in which $P2$ is a zero-threshold device. Fig. 12 shows simulated delay versus supply curves for normalized CMOS, BiCMOS, and modified MBiCMOS gates in the 0.5-μm technology. The supply voltage used for simulations is 3.5 V, the intended supply voltage for this technology. All gates are designed with input capacitance of 33 fF. The BJT sizes in the BiCMOS and modified MBiCMOS gates, and the MOS gate sizes in the second stage of the CMOS buffer, were optimized at each of the loads and at each supply voltage to give the lowest possible delay.

The BiCMOS delay curve crosses over the CMOS delay at 2.63 V, while the MBiCMOS gate shows a lower delay than CMOS down to 1.9-V supply. Since the device sizes for each gate are optimized for each of the loads, the crossover is independent of load. Thus, with the modified MBiCMOS gate, the superiority of BiCMOS for sub-2-V submicrometer technologies is maintained.

Simulations of the modified MBiCMOS gate in the 2-μm technology indicate that the MBiCMOS/CMOS crossover voltage can be reduced to 2.4 V using the zero-threshold technique. This technique can also be used to improve performance at nonscaled voltages. In the 2-μm case, BiCMOS and MBiCMOS delays are comparable at 5-V supply, as shown in Fig. 11. Use of the modified zero-threshold gate at 5-V supply results in simulated delays that are 15% less for MBiCMOS than for BiCMOS.

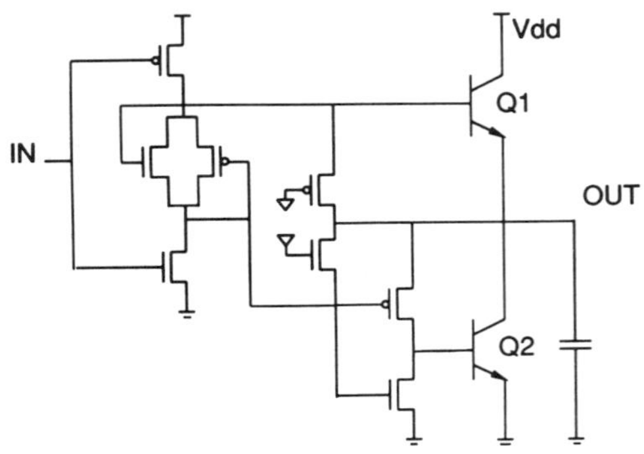

Fig. 13. FS-MBiCMOS gate.

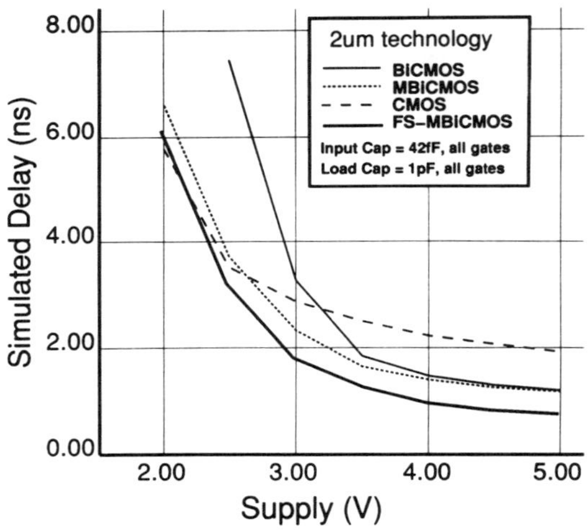

Fig. 14. Simulated delay versus supply for the FS-MBiCMOS gate in the 2-μm technology.

B. Full-Swing MBiCMOS Gate

A full-swing MBiCMOS (FS-MBiCMOS) gate is shown in Fig. 13 This gate is a direct adaption of the full-swing complementary BiCMOS logic (FS-CMBL) gate introduced by Shin *et al.* in [9]. In the FS-MBiCMOS gate, the p-n-p pull-down of the FS-CMBL gate is replaced by a merged PMOS/n-p-n device. Similar performance benefits are achieved without needing to add a p-n-p device to the process. Fig. 14 compares simulated delay versus supply voltage for FS-MBiCMOS, MBiCMOS, BiCMOS, and CMOS gates. Even with the 2-μm models used for these simulations, the FS-MBiCMOS/CMOS crossover voltage is reduced to 2.2 V.

The zero-V_t $P2$ modification is not used in the FS-MBiCMOS gate because of the potential for degraded noise sensitivity and increased leakage current when the output is high. In this case, the output voltage (source of $P2$) and the voltage at the base of $Q1$ (gate of $P2$) are both at V_{dd}. Use of a zero-threshold device is undesirable because of potential current flow in $P2$.

VI. Implementation of Complex Logic

Fig. 15 shows how complex multiple-input logic is implemented in MBiCMOS. In MBiCMOS, unlike BiCMOS, there is no need to repeat the logic in the pull-down path. This leads to a substantial area reduction compared to the same functionality in BiCMOS. A merged NAND gate built in the 2-μm technology achieved a 50% savings in area when compared to a standard NAND gate in the same technology. MBiCMOS can be extended to any desired logic function with significant savings in overall area and no penalty in performance.

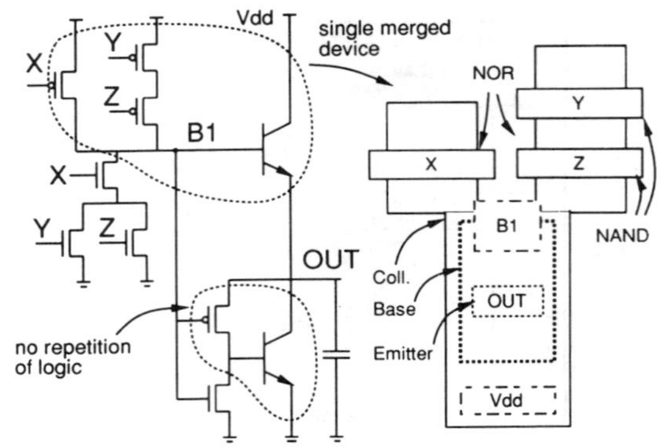

Fig. 15. Compact implementation of NAND/NOR functions in MBiCMOS.

VII. Summary

MBiCMOS, a unique circuit configuration to improve BiCMOS gate performance at low supply voltages, has been discussed in detail. Clear advantages in the delay and the power–delay product of the MBiCMOS device and circuit technique have been demonstrated. These advantages are shown to be process-independent by the variety of experimental strategies used in this work. Although the MBiCMOS gate was designed to take advantage of the merged PMOS/n-p-n device in order to maximize the drive per unit area ratio, device merging is not critical to achieve MBiCMOS performance benefits in terms of delay and power. Performance advantages are obtained with or without device merging.

Fig. 16 shows the BiCMOS/CMOS and MBiCMOS/CMOS crossover voltages as a function of technology. In a normalized comparison (a comparison in which all gates have both equal input capacitance and equal overall area), MBiCMOS outperforms CMOS down to 2.6-V supply in a 2-μm triple-diffused technology and to 2.4-V supply in a 1-μm buried-layer/epi process. The simulated full-swing MBiCMOS/CMOS crossover in the 2-μm technology is 2.2 V. These performance gains at scaled supply voltages are obtained with no additional circuit or process complexity beyond the standard BiCMOS process.

Looking ahead to sub-2-V submicrometer operation, the MBiCMOS concept is extended with the addition of a single threshold-adjusting implant to obtain a zero-threshold PMOS device. This modified MBiCMOS gate achieves a CMOS/MBiCMOS crossover of 1.9 V when simulated in a 0.5-μm technology. These results show that MBiCMOS can successfully be used to extend the useful operating range for BiCMOS technology to submicrome-

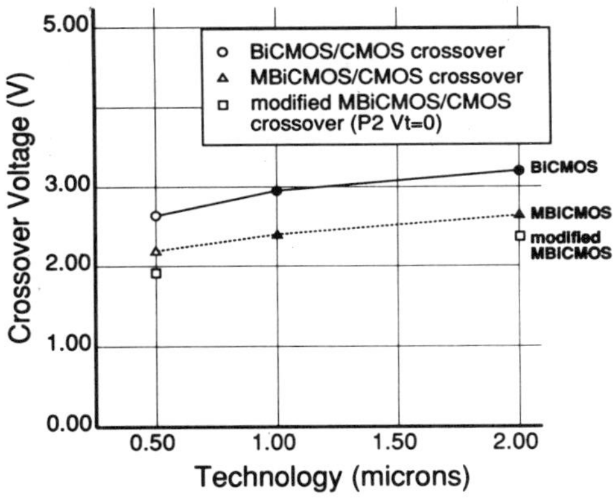

Fig. 16. Supply voltage at which delay becomes greater than CMOS, as a function of technology. Solid symbols indicate measured data. Open symbols indicate simulated data.

ter sub-2-V regimes. The modified MBiCMOS gate is also useful at 5-V supply to increase gate performance.

Acknowledgment

The authors would like to acknowledge Dr. J. Shott for the development of the Stanford BiCMOS process used to fabricate the 2-μm devices and Dr. Y. Nishi of Hewlett Packard for his encouragement and discussions. Thanks are also due to the Stanford and Hewlett Packard lab staff for fabrication of the devices.

References

[1] R. B. Ritts and J. D. Plummer, "Merged BiCMOS with vertical PNP," in *Device Res. Conf. Dig. Tech. Papers*, 1990, p. IIB-8.

[2] S. Ogura *et al.*, "Merged complementary BiCMOS for logic applications," in *Symp. VLSI Technology Tech. Dig.*, 1990, pp. 81–82.

[3] T. Hanibuchi, M. Ueda, K. Higashitani, M. Hatanaka, and K. Mashiko, "A bipolar-PMOS merged basic cell for 0.8 μm BiCMOS sea of gates," in *Proc. IEEE Custom Integrated Circuits Conf.*, 1990, p. 4.2.1.

[4] J. B. Kuo, G. P Rosseel, and R. W. Dutton, "Two-dimensional analysis of a merged BiPMOS device," *IEEE Trans. Computer-Aided Design*, vol. 8, pp. 929–932, Aug. 1989.

[5] H. Momose *et al.*, "Characterization of speed and stability of BiNMOS gates with a bipolar and PMOSFET merged structure," in *IEDM Tech. Dig.*, 1990, pp. 231–232.

[6] P. Raje, R. Ritts, K. Cham, J. Plummer, and K. Saraswat, "MBiCMOS: A device and circuit technique scalable to sub-micron, sub-2V regime," in *ISSCC Dig. Tech. Papers*, 1991.

[7] C. Chen, "Level-shifted and voltage reduced 0.5 μm BiCMOS circuits," in *ISSCC Dig. Tech. Papers*, 1990, pp. 236–237.

[8] Y. Nishio *et al.*, "A BiCMOS logic gate with positive feedback," in *ISSCC Dig. Tech. Papers*, 1989, pp. 116–117.

[9] H.J. Shin *et al.*, "Full-swing complementary BiCMOS logic circuits," in *Bipolar Circuits and Technology Meeting Tech. Dig.*, 1989, pp. 229–233.

[10] A. Watanabe, T. Nagano, S. Shukuri, and T. Ikeda, "Future BiCMOS technology for scaled supply voltage," in *IEDM Tech. Dig.*, 1989, pp. 429–432.

[11] S. C. Lee and D. W. Schucker, "BiCMOS driver circuit," U.S. Patent 4 616 146, Oct. 7, 1986.

[12] E. W. Greeneich and K. L. McLaughlin, "Analysis and characterization of BiCMOS for high-speed digital logic," *IEEE J. Solid-State Circuits*, vol. 23, pp. 558–565, Apr. 1988.

[13] G. P. Rosseel and R. W. Dutton, "Influence of device parameters on the switching speed of BiCMOS buffers," *IEEE J. Solid-State Circuits*, vol. 24, pp. 90–99, Feb. 1989.

[14] J. Shott, C. Knorr, and M. Prisbe, "BiCMOS technology overview," Center for Integrated Systems, Stanford, CA, Stanford BiCMOS Project Tech. Rep., Sept. 1990.

[15] P. Raje, K. Saraswat, and K. Cham, "A new BiCMOS/CMOS gate comparison/design methodology and supply voltage scaling model," in *IEDM Tech. Dig.*, 1989, pp. 433–436.

Paper 1.16

A New Methodology for Design of BiCMOS Gates and Comparison with CMOS

Prasad A. Raje, *Member, IEEE,* Krishna C. Saraswat, *Fellow, IEEE,* and Kit M. Cham, *Senior Member, IEEE*

***Abstract*—A gate comparison methodology is presented to accurately compare the performance of an arbitrary BiCMOS logic gate with a pure CMOS gate. The concept of the sizing plane is introduced, which is a geometrical framework for representation of all possible sizing choices of BiCMOS and multistage CMOS inverters in a given technology. The methodology is translated to geometrical operations in the sizing plane. This serves to highlight the important considerations in making gate comparisons. The methodology is demonstrated by fabrication in a 2-μm BiCMOS technology. The measured delay comparison shows that CMOS suffers larger gate delays than BiCMOS, especially for large loads, but not by as large a factor as would be obtained by previous delay comparison methods. The sizing plane helps in providing insight into the delay trends observed. The sizing plane is also shown to be a powerful tool for BiCMOS gate design where several design constraints can be simultaneously visualized and tradeoffs can be evaluated. The design procedure for some specific design criteria is demonstrated for gates in a 1-μm conventional BiCMOS technology. Finally, a technology comparison methodology is proposed to obtain a ratio of gate delay of BiCMOS over CMOS that is relatively constant independent of sizing dependencies. The methodology is demonstrated for a 0.6-μm BiCMOS technology. It is shown that a size-independent delay ratio of 0.7 is obtained for large fanout loads. For light loads, the CMOS gate is superior to BiCMOS upto a fanout of 3.**

I. Introduction

The BiCMOS gate is a popular and widespread application of BiCMOS technology. Its advantages are low propagation delay even for large loads because of the bipolar totem pole drivers, combined with zero static power dissipation and high input impedance because of the MOS input devices. The BiCMOS inverter has been extensively quoted to demonstrate the performance advantage of BiCMOS technology over CMOS [1]–[7]. In these comparisons the slope of the delay versus load curve is shown to be much lower for BiCMOS than CMOS, purportedly making a case for the superiority of BiCMOS. There are two separate but related deficiencies in such comparisons. First, in order for delay comparisons between specific BiCMOS and CMOS gates to be quantitatively meaningful, the gates must be "normalized" in some sense. Second, delay versus load comparisons of some arbitrarily sized gates cannot be extrapolated to make broad conclusions about the overall performance of BiCMOS technology as compared to CMOS in the context of comparing basic digital logic functions.

This paper presents a gate comparison methodology [8] that specifies for any given BiCMOS gate, the CMOS gate with which it must be compared to make a reasonable and accurate performance comparison (Section II). The concept of the sizing plane (SP) is introduced as a geometrical framework in which the gate comparison methodology is represented (Section III). The sizing plane is also shown to be an elegant platform to represent the constraints and tradeoffs in BiCMOS gate design and this is demonstrated by example for a 1-μm BiCMOS technology (Section IV). To illustrate the comparison methodology, BiCMOS and CMOS gates are fabricated in a 2-μm BiCMOS technology [9]. The measured performance results are presented and interpreted using the sizing plane (Section V). Finally, a technology comparison methodology is proposed that predicts the relative performance of a BiCMOS versus a pure CMOS implementation of any arbitrary block of digital logic (Section VI).

II. Gate Comparison Methodology

Fig. 1 shows delay versus load graph conventionally used for delay comparison between a BiCMOS and a CMOS inverter. For a capacitive load

$$i = C_L \frac{dV}{dt} \Rightarrow \frac{t}{C_L} = \frac{V}{I}.$$

The slope of the delay-load curve (t/C_L) is inversely proportional to the average drive current (I) available from the output driver device to charge or discharge the load. By simply making the MOS devices wider, the slope of the CMOS delay-load curve can be proportionately reduced. Similarly, the slope of the BiCMOS delay-load curve depends on the sizes of the component devices. Clearly, a delay comparison made with no constraints on gate sizing is quantitatively meaningless. In some previous comparisons, the MOS device widths in the BiCMOS and CMOS inverters have been kept equal [2], [3], but this has led to the BiCMOS inverter enjoying a much larger silicon area (schematically represented in Fig.

Manuscript received March 20, 1990; revised April 1, 1991. The review of this paper was arranged by Associate Editor P. Cottrell.

K. C. Saraswat is with the Center for Integrated Systems, Stanford University, Stanford, CA 94305.

P. A. Raje and K. M. Cham are with Circuit Technology R&D, Hewlett-Packard Co., Palo Alto, CA 94304.

IEEE Log Number 9104681.

Reprinted from *IEEE Trans. Electron Devices,* vol. 39, no. 2, pp. 339–347, Feb. 1992.

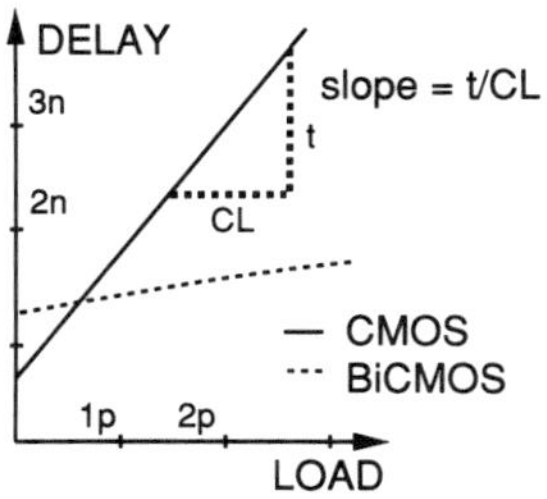

Fig. 1. Typical delay versus load curves for BiCMOS and CMOS inverters.

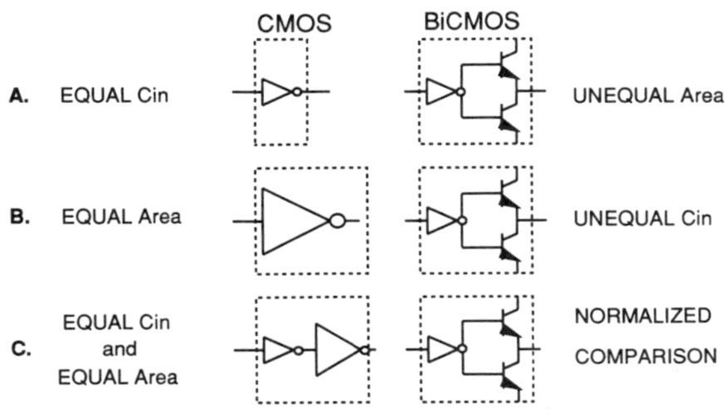

Fig. 2. Schematic representation of previous comparisons of BiCMOS and CMOS gates (Cases *A*, *B*) and the proposed methodology (Case *C*).

2, Case *A*). In some cases, the gate areas have been equalized by increasing the size of the CMOS inverter [1], [5], but this has led to its having a larger input capacitance (Fig. 2, Case *B*).

To accurately compare the performance of a BiCMOS inverter with CMOS, the propagation delay and power dissipation of the gates must be compared for the same load capacitance under the following two constraints: a) The total layout area of the two gates must be equal. This ensures that one gate is not enjoying an ''unfair'' advantage over the other by using up more area. The layout for both gates must be as close-packed as the design rules for the respective technologies allow. b) The input capacitance for both gates must be the same. This ensures that both gates cause the same delay in the previous stage driving this gate. Both these constraints can be met if the CMOS gate used in the comparison is a two-stage buffer instead of a single-stage inverter. The first stage is sized to make the input capacitance equal and the second stage is sized to make the total area equal (Fig. 2, Case *C*).

Gates designed under these constraints are said to be *normalized* to one another. A performance comparison of normalized gates is called a *normalized comparison.* In doing a normalized comparison, a BiCMOS gate is conceptually swapped for a pure CMOS gate in a circuit and the delay and power compared with no perturbation to the area occupied or load presented to the previous stage.

The methodology is not restricted to CMOS gates with only two stages or to comparing simple inverters. For a multistage CMOS gate, the width ratio of successive stages is assumed to be uniform. The number of stages as well as the corresponding width ratio must be determined under the same two constraints. For a fixed area and input capacitance, each equal to that of the BiCMOS gate, the number of stages and by implication the width ratio chosen is that which gives the lowest delay for the load capacitance specified. This requires a simultaneous determination of the number of stages and the width ratio. A full mathematical treatment of this problem is possible but is not detailed in this paper. In most practical cases, the number of stages that is appropriate is limited to two or three and is predetermined separately. When normalizing gates implementing more complex logic functions (NAND, NOR, etc.), the first stage of the CMOS gate is made to implement the same logic function while keeping the input capacitance equal and the subsequent stages are simple tapered inverters, sized so that the total area is made equal. In general the normalized multistage CMOS gate could have its output inverted relative to the BiCMOS gate. This is not a serious impediment because it is mainly the delay of signal propagation through the gate that is of interest. Further, in most logic circuits the input can be made available either inverted or noninverted as desired.

Power is an important consideration in comparing gate performance. Note however that both BiCMOS and CMOS gates dissipate no static power. In both cases, all the power dissipation is dynamic power under the reasonable assumption that there is no through current from supply to ground during switching. This is generally achieved for both BiCMOS and CMOS for input pulses with rise times of the order of the output transient. The dynamic power for BiCMOS can be written as

$$\text{BiCMOS power} = (C_L + C_i + C_p) V_s^2 f. \tag{1}$$

C_L is the output load, C_i accounts for the internal device capacitance, C_p is due to parasitic capacitances, V_s is the output swing, and f is the frequency. For CMOS the dynamic power is

$$\text{CMOS power} = (C_L + RC_{\text{in}} + C_p) V_{dd}^2 f. \tag{2}$$

RC_{in} accounts for the capacitance due to the second stage, if one is present. Now for the same frequency and load, the only differences in (1) and (2) are due to the unequal output swing and the unequal internal capacitances. The internal capacitances tend to be larger for BiCMOS because of the BJT and the additional devices in the gate but the output swing is smaller. To a first order, these two effects generally cancel each other and the powers are comparable. At any rate, the difference in BiCMOS and CMOS power does not have a very strong dependence on gate sizing and therefore power differences are not treated further in this paper.

III. The Sizing Plane

A. Definition

For simplicity, the definition of the sizing plane is developed with reference to a simple inverter (as opposed to a complex logic function). In the BiCMOS inverter (Fig. 3) the widths of MOSFET's $M1$, $M2$, $M3$, $M4$ are assumed to bear fixed ratios to one another. As a typical

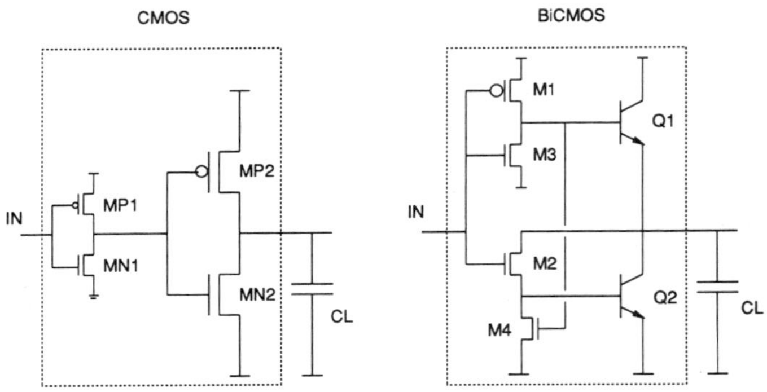

Fig. 3. Detail of normalized CMOS and BiCMOS inverters (two-stage case shown for the CMOS gate).

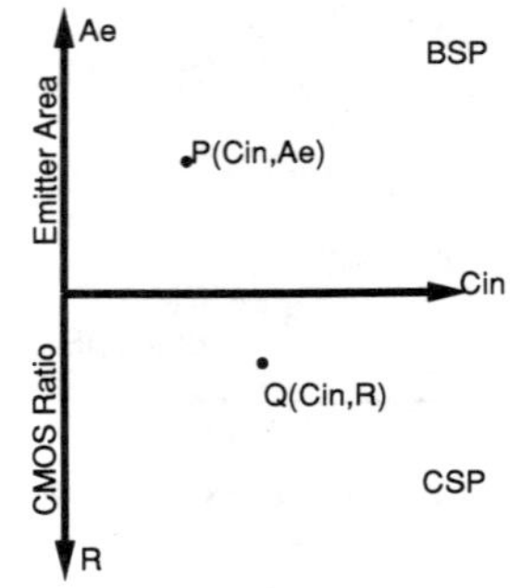

Fig. 4. The combined sizing plane formed from the BiCMOS sizing plane (BSP) and the CMOS sizing plane (CSP) sharing a common C_{in} axis.

example $M1$, $M2$ may have equal widths as also $M3$, $M4$ and the width of $M4$ could be 40% of that of $M2$. This 40% ratio is empirically determined from the requirement of providing a low enough impedance to ground during the turnoff of the BJT while not stealing away an excessive amount of base current in the initial part of the turn on of the BJT. The sizes of BJT's $Q1$ and $Q2$ are assumed to be equal. Also, the MOSFET gate lengths and the BJT emitter widths are assumed to be the minimum possible for the technology. Specifying the input capacitance C_{in} determines the size of the MOSFETs and specifying an emitter area A_e determines the size of the BJT's. The sizes of all devices in the BiCMOS gate are thus fully defined by the two parameters C_{in} and A_e. The term "MOS size" of the BiCMOS gate is used interchangeably with C_{in} and "BJT size" with A_e. The BiCMOS sizing plane (BSP) is defined as the two-dimensional coordinate system with axes C_{in} and A_e (Fig. 4). Each point $P(C_{\text{in}}, A_e)$ in this plane represents a unique BiCMOS gate in the technology. The gate area can be determined for each point from the layout or can be expressed analytically as

$$\text{Area}_B(P) = \mathcal{F}(C_{\text{in}}, A_e) = b_0 + b_1 C_{\text{in}} + b_2 A_e \tag{3}$$

where b_0, b_1, b_2 are constants. The area function $\mathcal{F}$ embodies the layout design rules of the technology to the extent that they affect gate area. The linearity of (3) is based on the reasonable assumption that the gate area increases linearly with MOS size and BJT size for a given technology. For each point P there is a unique value of the propagation delay of the gate (although many points may have the same delay). The delay is specified for a given supply voltage (V_{dd}) and load capacitance (C_L). C_L can either be a fixed capacitance value or expressed as a fanout, in which case $C_L = \text{fanout} \times C_{\text{in}}$. The delay for any point in the BSP can be expressed as

$$t_d(P) = \mathcal{G}(C_{\text{in}}, A_e, C_L, V_{dd}) \tag{4}$$

where the delay function $\mathcal{G}$ is an analytical expression or computed by simulation. The delay function embodies the device characteristics of the BiCMOS technology.

For the multistage CMOS gate (Fig. 3) the widths of the NMOSFET and PMOSFET in each stage are assumed to be in a fixed ratio (usually the inverse ratio of the mobilities). Specifying the input capacitance C_{in} determines the MOSFET widths for the first stage. The CMOS ratio R (also known as the taper of the multistage gate) is defined as the ratio of the widths of MOSFET's in successive stages. Mathematically

$$R = \frac{W_{MNn} + W_{MPn}}{W_{MNn-1} + W_{MPn-1}} \tag{5}$$

where W_{MNn} is the NMOSFET width in the nth stage (W_{WPn} for PMOSFET). Note that the CMOS ratio is assumed to be the same for all stages. Specifying the C_{in} and R then determines the widths of all MOSFET's in the multistage CMOS gate. By analogy with the BSP, the CMOS sizing plane (CSP) is defined as the two-dimensional coordinate system with axes C_{in} and R (Fig. 4). For each point $Q(C_{\text{in}}, R)$ in the CSP the area of the n-stage CMOS gate is expressed analytically as

$$\text{Area}_C(Q) = \mathcal{M}(C_{\text{in}}, R) = c_0 + c_1 C_{\text{in}} + c_2 R C_{\text{in}} + c_3 R^2 C_{\text{in}} \cdots + c_n R^{n-1} C_{\text{in}} \tag{6}$$

where the area function $\mathcal{M}$ is expressed as the sum of the areas of each of the stages. $c_0, c_1, \cdots, c_n$ are constants that depend on the layout design rules of the CMOS technology (typically $c_1 = c_2 \cdots = c_n$). The propagation delay at each point Q is expressed as

$$t_d(Q) = \mathcal{N}(C_{\text{in}}, R, C_L, V_{dd}). \tag{7}$$

The delay function $\mathcal{N}$ embodies the device characteristics for the CMOS technology and can be expressed analytically or computed by circuit simulation.

The combined sizing plane or simply the sizing plane (SP) (Fig. 4) is formed by drawing the BSP and CSP with a common origin and a shared C_{in} axis. The A_e axis of the BSP points upwards and the R axis of the CSP points downwards.

B. The Gate Comparison Methodology in the Sizing Plane

The sizing plane constructed as described above provides a convenient and visually expressive framework to carry out gate comparisons under the constraints mandated by the methodology. Given a BiCMOS gate, the goal is to make a normalized comparison of the gate with a pure CMOS implementation of the same logic function.

The BiCMOS gate is first located in the BSP at the appropriate C_{in} and A_e. Let this point be $P(C_{in}, A_e)$. The CMOS gate normalized to this gate is then located in the CSP at a point $P^*(C_{in}, R^*)$ given by the following rules. The C_{in} values at P and P^* must be equal. Alternatively, P and P^* must lie on the same vertical line. R^* is forced to a value that makes the respective gate areas at P and P^* equal. R^* can be determined in terms of A_e and C_{in} by solving (3) and (6) for a known value of the number of stages n of the CMOS gate. For $n = 2$, R^* can be explicitly solved for as

$$R^* = \frac{b_1 - c_1}{c_2} + \frac{b_2}{c_2}\frac{A_e}{C_{in}} + \frac{b_0 - c_0}{c_2}\frac{1}{C_{in}}. \quad (8)$$

The point P^* constructed in this way is called an "image point" of P. The constraints in the gate comparison methodology thus define a one to one mapping from points in the BSP to the CSP. The ratio of the delay value $\mathcal{G}$ at $P(C_{in}, A_e)$ to the delay value $\mathfrak{N}$ at $P^*(C_{in}, R^*)$ is the desired normalized comparison of the gates.

The usefulness of translating the gate comparison methodology to operations in the sizing plane is enhanced when the above procedure is generalized from single points to a general set of points in the BSP. In its general form, a normalized comparison is the ratio of the delay functions $\mathcal{G}(P)$ and $\mathfrak{N}(P^*)$ with the area functions $\mathfrak{F}(P)$ and $\mathfrak{M}(P^*)$ being equal. In a qualitative sense, a normalized comparison compares gate delay (functions $\mathcal{G}$ and $\mathfrak{N}$) while simultaneously accounting for the layout (functions $\mathfrak{F}$ and $\mathfrak{M}$). There is nothing in the definition of the sizing plane to prevent one from comparing delays along any two arbitrary curves in BSP and CSP. If this is done, the areas of the gates compared will not be equal and this fact must be considered while interpreting the delays. The gate comparison methodology has been described above in the sizing plane starting from a given BiCMOS gate. One could also use a given CMOS gate as a starting point and use an exactly analogous procedure to perform the reverse mapping of a point $Q(C_{in}, R)$ from the CSP to its image point $Q^*(C_{in}, A_e^*)$ in the BSP. This is not detailed further here.

A constant A_e contour in the BSP maps to a hyperbola in the CSP ($R^* \propto k + 1/C_{in}$ from (8)). A constant R contour in the CSP maps to a straight line with a positive slope in the BSP. A vertical segment in either plane maps to a vertical segment with the same abscissa. Using these basic rules rectangular regions can be mapped from one plane to the other.

IV. The Sizing Plane as a BiCMOS Design Aid

The process of designing a BiCMOS gate consists of identifying the gate or set of gates that meet some design criterion. In the context of the BSP, this translates to locating the point, contour, or region that satisfies the design criterion. A design criterion is some combination of one or more constraints on the gate. Examples of these constraints are: limits on gate area, desired values of gate delay, upper bounds on input capacitance, etc. The constraints can be individually represented in the BSP in the form of bounds on the C_{in} and A_e values of the gate or relationships between them. Geometrically, the bounds from *allowed regions* and the relationships between C_{in} and A_e form *contours* in the BSP. The specific constraints that constitute the design criterion can then be simultaneously represented in the BSP to arrive at the gate (or set of gates) that meets the criterion. The allowed regions and contours in the BSP can be constructed using analytical expressions or simulations as convenient. Analytical expressions for gate delay are beyond the scope of this paper and herein the gate delays are obtained by circuit simulation. To make the discussion realistic, the design process is demonstrated using typical design constraints in a conventional 1-μm, 5-V, 9-GHz BiCMOS technology developed at Hewlett-Packard.

A. Constraint Curves

In order to limit the load on the previous stage driving a BiCMOS gate, there is generally an upper limit C_{max} on the input capacitance of the gate. This constraint is represented in the BSP as the vertical line at $C_{in} = C_{max}$ (Fig. 5). C_{max} is 167 fF in this example, corresponding to $W_{M1} = 40$ μm. The "allowed" gate sizes must lie to the left of this line. The total silicon area available to the gate is also usually limited to a maximum value. Given a constant gate area, (3) implies a linear relationship between C_{in} and A_e. The straight line thus drawn in the BSP is called a constant gate area contour (the line marked "const area" in Fig. 5 has a gate area = 1240 μm^2). The line has a negative slope because a decrease in the BJT size can be traded off against an increase in MOS size to maintain the same area. The absolute slope of the line (b_1/b_2) is small because the BJT's generally occupy the major fraction of the area of the BiCMOS gate and a small reduction in the BJT size trades off against a large increase in the MOS size. An upper limit on gate area forces the allowed gates to lie below the constant area contour. To prevent base–collector forward bias in the BJT ($Q1$) the IR drop across the internal collector resistance must be kept below a peak value. It can be shown (Appendix) that to a first order, the constant IR drop curve is a straight line through the origin with a positive slope that is directly proportional to the square of the collector sheet resistance (the line marked "BJT saturation" in Fig. 5). The collector saturation constraint thus forces the BiCMOS gate to lie above this line. As shown in the Appendix, this is not a serious constraint for BiCMOS technologies with a buried layer. For example A_e needs to be larger than 1.6 μm^2 for $C_{in} = 100$ fF ($W_{P1} = 24$ μm).

Points in the BSP that have the same value of gate delay are said to lie on a constant delay contour (the "const delay curves" in Fig. 5). Mathematically, a constant delay contour is defined as the curve $\mathcal{G}(C_{in}, A_e, C_L, V_{dd}) =$

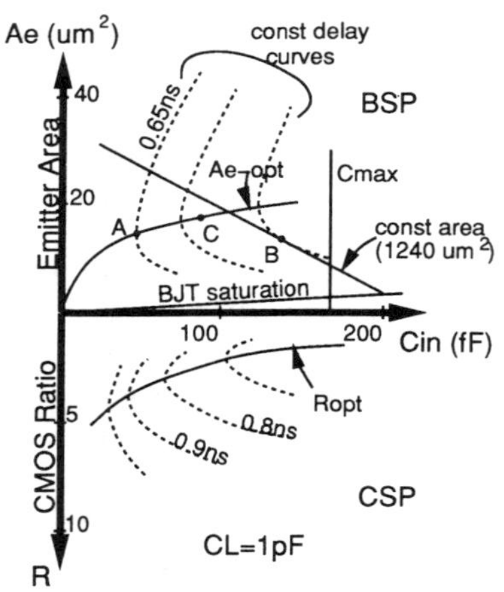

Fig. 5. The combined sizing plane for 1-μm BiCMOS. Various contours, bounds, and design points are shown.

constant, with C_L and V_{dd} specified. In the case of a constant fanout instead of constant load, the fanout and V_{dd} would be specified. Each constant delay contour is "C" shaped with an extremum at a minimum value of C_{in}. The extremum point is called the *turnaround-point*. The locus of turnaround-points is called the "A_e-opt contour." For a range of C_{in} values, the A_e-opt contour is the emitter area for which the gate delay is a minimum. For load expressed as a fanout and hence dependent on C_{in}, the exact shape of the A_e-opt curve is different, but the above definition still applies as do the following qualitative arguments. For a gate with a given C_{in}, increasing the emitter area from small values reduces the delay. This is because the high-level injection effect in the BJT is reduced with increasing A_e and increasing current is available from the BJT to drive the load. With continued increase of A_e, the internal delay of charging up the BJT base node dominates and the total delay increases. The A_e-opt curve thus represents a tradeoff between the high-level injection effect and the internal delay in the BiCMOS gate. The constant delay curves of Fig. 5 are computed using SPICE simulation in the 1-μm BiCMOS technology. Representative values of $C_L = 1$ pF and $V_{dd} = 5$ V are chosen and the example curves shown correspond to inverter delays of 650, 550, and 450 ps from left to right.

B. BiCMOS Design Using the Sizing Plane

To illustrate how the BSP is used to make design tradeoffs, some representative design criteria are chosen and illustrated by example. Consider the design criterion of a target gate delay value, say 650 ps, along with the requirement that the input capacitance be the smallest possible. The design point is then the turnaround-point (leftmost point) on the appropriate constant delay contour. This is the point A in Fig. 5 which has $C_{\text{in}} = 40$ fF, i.e., $W_{M1} = 9.5$ μm and $A_e = 13$ μm^2. If the design criterion consists of a target delay and minimum total gate area, the design point is located where the constant delay contour is tangential to a constant area contour. For example, if the target delay is 450 ps, the constant delay contour for 450 ps is tangential to the area $= 1240$ μm^2 contour at the point B. The desired gate thus has $C_{\text{in}} = 145$ fF, $W_{P1} = 35$ μm, and $A_e = 10$ μm^2. If an input capacitance is specified and the gate delay must be minimized, the design point is on the A_e-opt curve at that C_{in}. For example, if $C_{\text{in}} = 85$ fF, $W_{P1} = 20$ μm, the design point is C on the A_e-opt curve with $A_e = 17$ μm^2. The delay at C is 500 ps and is the minimum achievable delay for the specified C_{in}. If gate delay is to be minimized with no limit on gate area, the design point moves on the A_e-opt curve towards larger and larger C_{in}. However, this must be tempered by the fact that the previous stage driving this gate suffers an increasing delay. To find the design point in this case, the following procedure is applied. The reducing delay of the BiCMOS gate along A_e-opt and the increasing delay of the previous stage are plotted versus C_{in}. The minimum point of the sum of the two delays gives the point of minimum total delay, and the C_{in} at this point specifies the location of the design point on the A_e-opt contour.

All the above arguments are in the context of a specified load (as a fixed value or a fanout number) and a specified supply voltage. For a range of load/supply values, one arrives at a range of design points. It is observed, for example, that the A_e-opt curve moves towards lower A_e values for lower load or for lower supply voltage. The final selection is made by choosing the design point for the load/supply value that is of greatest importance.

Analogous to the A_e-opt curve in the BSP, the R-opt curve is defined as the curve in the CSP along which the multistage CMOS buffer delay as a function of C_{in} is minimum. The equation for the R-opt curve can be simply obtained from textbook analysis of the multistage CMOS gate [10]. An R-opt curve for two-stage gates is shown in Fig. 5 for a 1-pF load for the 1-μm technology. The significance of the R-opt curve in the CSP is that the distance of any point in the CSP from the R-opt curve, measured along a constant C_{in} line is a measure of how much greater than optimum the gate delay is at the point. The further a point is from the R-opt curve, the larger its delay in comparison to the lowest achievable delay by a CMOS gate with the same C_{in}. The R-opt curve in the CSP can also be thought of as the boundary beyond which (i.e., towards larger R values) gates should not be designed. Beyond the R-opt curve, gates have both a larger area and a larger delay for the same C_{in}. The strict rules of normalization expressed in Section II are relaxed to reflect this as follows. If an image point in the CSP lies beyond the R-opt curve (i.e., towards larger R), the image is modified to be the point on R-opt curve at the same C_{in} value. In this situation the CMOS gate is allowed at least as much area as the BiCMOS gate but is not forced to occupy the same area as the BiCMOS gate to the detriment of its delay.

V. Experimental Results

The gate comparison methodology is demonstrated using a BiCMOS technology developed at Stanford as the technology vehicle. The primary features of relevance of this technology are a triple diffused bipolar transistor (no epitaxial layer) and 2-μm drawn minimum feature sizes (L_{eff} = 1.2 μm). Given that the advantage of high-drive capability of BJT's is diluted in this technology because of the large collector resistance in the absence of a buried layer, it is of interest to determine if the BiCMOS driver shows superior performance in a normalized comparison with CMOS.

Three different sizes of BiCMOS gates are chosen as explained below and the corresponding normalized CMOS gates are designed under the constraints described in Section II. The area equalization is done during layout while laying out the gates as aggressively as the design rules allow. Given the area constraint, the only possible value for the number of stages is two for each of the CMOS gates. To obtain a rectangular shape for the CMOS buffer layout the second stage has a folded gate and therefore the ratio R is quantized to be an integer. Ring oscillators are designed with load capacitances of 0, 1, and 3 pF for each of the sizes of BiCMOS and CMOS gates. The CMOS 2 stage gate is noninverting, so in the CMOS ring a small inverter stage is added to enable ringing and a correction is made in the delay measurements to compensate for this inverter. Fig. 6 shows the locations of the BiCMOS gates (1, 2, 3) and the image CMOS gates (1*, 2*, 3*) in the combined sizing plane. The R-opt curve for a 1-pF load is also shown.

Fig. 7 shows the measured delay versus load curves at a supply voltage of 5 V for the gates described above. For reference, the delay load curve for a single-stage CMOS inverter with the same input capacitance as BiCMOS Gate 1 is shown on the same plot. This is the curve that is obtained if the (erroneous) method of Case A of Fig. 2 is used to make the BiCMOS versus CMOS comparison.

The following observations can be made about Fig. 7. The comparison of the single-stage CMOS inverter to the BiCMOS Gate 1 appears to give BiCMOS a substantial performance advantage. If however the normalized CMOS gate (Gate 1*) is compared to Gate 1, the advantage is reduced to approximately 40%. Thus by simply allowing the CMOS gate to enjoy the same area as the BiCMOS gate, the delay versus load comparison appears quite different. In general, the BiCMOS gates have lower delay than the normalized CMOS gates, but the delay ratio (BiCMOS delay to CMOS delay) depends on the gates chosen in the comparison. This highlights the fact that comparison of any one pair of gates cannot be extrapolated to apply to the technology as a whole, even when the comparison is made between normalized gates. The different delay ratios obtained can be explained qualitatively by the locations of the gates and their image points in the combined sizing plane. Gate 1*, which is the image of Gate 1, is located at an R value of 4 (decided by the area constraint) which is close to the R-opt value (3.7) for a 1-pF load for the same input capacitance. The same is true for Gate 2*. Gate 3* however is forced to an R of 2 when the R-opt value at the same C_{in} is 4.5. The delay ratio between Gates 3 and 3* is much higher than that

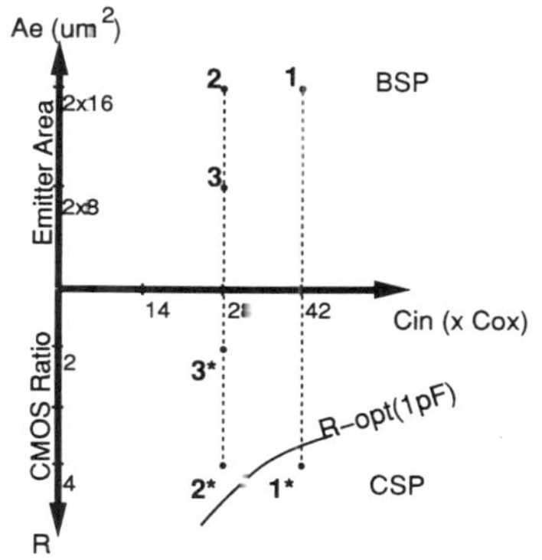

Fig. 6. The sizing plane for a 2-μm BiCMOS technology showing the location of BiCMOS gates (1, 2, 3) and normalized CMOS gates (1*, 2*, 3*) fabricated.

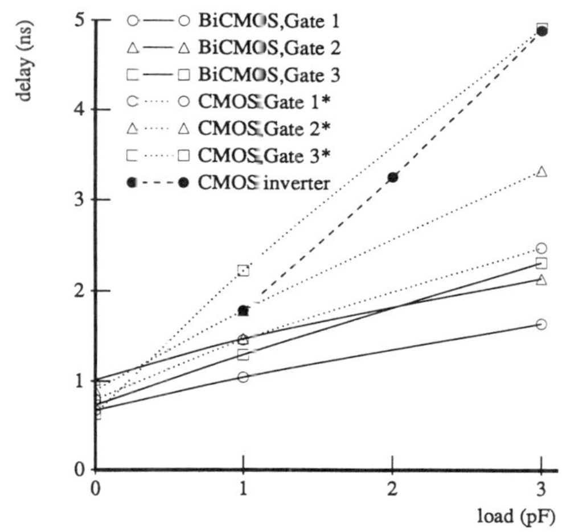

Fig. 7. BiCMOS and (normalized) CMOS delay versus load comparison for a 2-μm BiCMOS technology at V_{dd} = 5 V.

between Gates 1–1* and 2–2*, in particular at 1-pF load. This can be traced to the larger distance of 3* from the R-opt curve for 1-pF load. An examination of the delay-load curves for Gates 2 and 3 provides information about the location of the A_e-opt point at the C_{in} value of the gates. As pointed out earlier (Section IV-A), if the gate delay is plotted versus A_e for a given C_{in}, the delay reduces up to the A_e-opt point and increases thereafter. For C_L = 1 pF, the delay for Gate 2 is larger than that for Gate 3, implying that the A_e-opt point must lie below (i.e., towards smaller A_e) Gate 2. Around C_L = 2 pF, the delays of the two gates are equal. The A_e-opt point must then lie between Gate 2 and Gate 3. Finally, at C_L = 3 pF, Gate 2 has a lower delay than Gate 3 and hence the A_e-opt point must lie above Gate 2 for this load. The above observations imply that the A_e-opt point and by extrapolation the A_e-opt curve as a whole, moves towards larger values of A_e for larger load. This can be predicted by analytical expressions for the gate delay [11].

VI. The Technology Comparison Methodology

Traditionally, the delay versus load curve of some arbitrarily chosen gate sizes has been used to infer performance ratios of BiCMOS over CMOS [1]–[7]. The gate comparison methodology of Section II provides an accurate means to compare specific gates. However, even accurate and normalized comparisons between *specific* gates cannot be extrapolated to make generalized inferences about the delay performance of BiCMOS versus CMOS for digital logic. The following methodology is proposed to arrive at a performance ratio of BiCMOS over CMOS, valid over a wide range of loading conditions and relatively free of specific gate sizing dependencies. This is called the "technology comparison methodology." The procedure for the methodology is illustrated for a 0.6-μm, 4-V, 15-GHz BiCMOS technology.

The logic function is restricted to a simple inverter with the justification that the inverter is the basic logic gate and the performance of the inverter can be extrapolated to the performance of more complex gates. Now the BiCMOS inverter to be chosen for the comparison could be located at any point in the BSP, i.e., both C_{in} and A_e could potentially be arbitrary. In the technology comparison, it is the optimal BiCMOS delay that is of interest. The BiCMOS inverter is then constrained to lie along the A_e-opt curve (see Section IV-A for the definition of the A_e-opt curve). The input capacitance and fanout remain unconstrained at this point. Fig. 8 shows the BiCMOS delay along the A_e-opt curve as a function of input capacitance. The delay is plotted for a range of values of the fanout from 1 to 25. Note that for each fanout there is a separate A_e-opt curve. The remarkable feature of Fig. 8 is that the optimal delay value for a fixed fanout is relatively constant across the range of C_{in}. This feature can be predicted by analytical expressions for the gate delay [12] but the proof is not detailed here. This delay is plotted as a function of fanout in Fig. 9. The small vertical bars at each fanout represent the extent of the C_{in} dependence of the optimal delay. This curve then represents the optimal BiCMOS delay as a function of fanout, independent of sizing dependencies. Note that the delay is independent of sizing provided the BiCMOS gate is optimally sized. Thus strictly, the delay is independent of the input capacitance and the fanout. The emitter area is fixed once the input capacitance is fixed.

For the CMOS case, the gate delay as a function of fanout for a single stage and a two-stage gate can be obtained in a relatively straightforward manner from the well-known textbook [10] behavior of the CMOS gate or from circuit simulations. Fig. 9 shows the CMOS single- and two-stage (optimal) delays as a function of fanout, obtained from SPICE simulation. Note that the one-stage gate has a lower delay than the two-stage gate up to a fanout of 8 rather than 4 as would be expected from the idealized textbook equations. The larger than expected delay for the two-stage case is due to gate overlap capacitance and parasitic source/drain capacitance. The crossover into the three-stage gate occurs at a relatively larger fanout = 30 and is therefore not shown. Hence for fanout < 8, the BiCMOS must be compared with the single-stage CMOS gate and for larger fanout with the two-stage gate. Fig. 10 then shows the ratio of optimal BiCMOS to CMOS gate delay as a function of fanout. The ratio for both one-stage and two-stage CMOS gates is shown. BiCMOS optimal delay is higher than that of the single-stage CMOS gate for fanout <3. For fanout between 3 and 8 the BiCMOS advantage over CMOS (single stage) improves. This delay ratio is a strong function of fanout but does provide a sizing independent ratio of BiCMOS to CMOS single-stage gate comparison. Beyond fanout = 8, the BiCMOS gate is compared to the two-stage CMOS gate. The ratio of BiCMOS optimal delay to two-stage CMOS optimal delay is a weak function of fanout across the whole range of fanouts. The delay ratio is 0.7 and is also

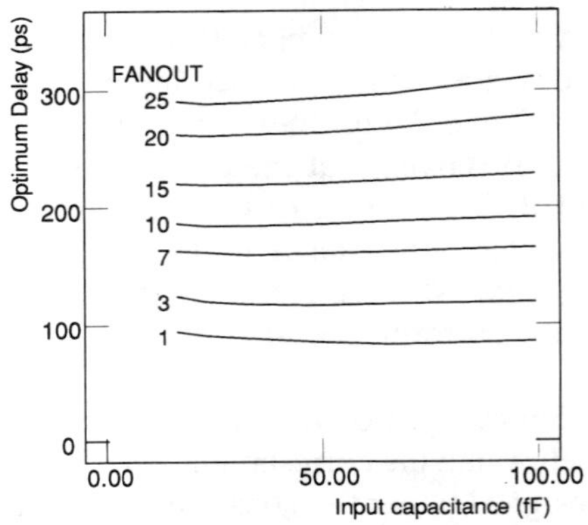

Fig. 8. Optimum BiCMOS delay as a function of input capacitance for a range of fanout. 0.5-μm, 4-V, 15-GHz BiCMOS technology.

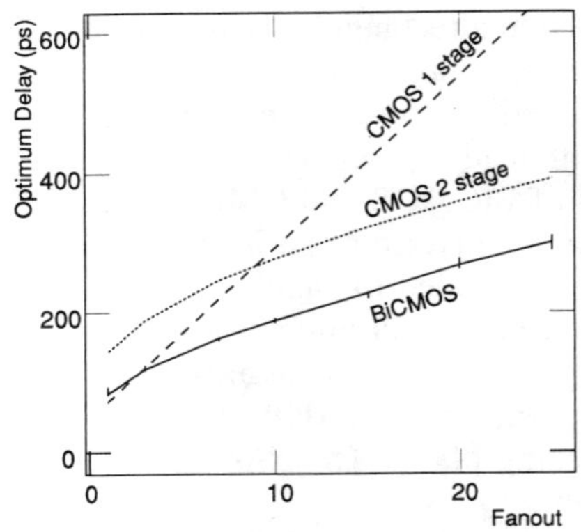

Fig. 9. Optimum 0.5-μm BiCMOS and CMOS single- and two-stage delay as a function of fanout. BiCMOS variation due to C_{in} is in a tight range shown by the vertical bars.

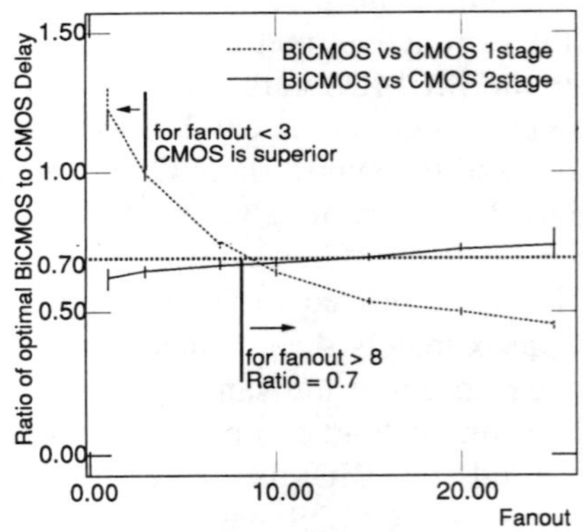

Fig. 10. Ratio of optimum 0.5-μm BiCMOS to CMOS delay. A relatively constant value of 0.7 is obtained for fanout >8.

independent of gate sizing. Therefore except for lightly loaded conditions, 0.7 represents a delay comparison of BiCMOS and CMOS in this technology, insensitive both to gate sizing and fanout. This methodology can be used to evaluate CMOS versus BiCMOS implementations of a digital logic critical path for any technology of interest. Once the optimum ratio as a function of fanout is obtained as in Fig. 10, the delay of each component of the critical path can be evaluated for CMOS versus BiCMOS. In the example 0.6-μm technology, for critical paths with most fanouts smaller than 3, CMOS is clearly superior. If most fanout values are larger than 8, BiCMOS is superior and the single number 0.7 is an accurate numerical value of the superiority. For intermediate fanouts, Fig. 10 can be used to obtain a precise number for the delay ratio.

VII. Conclusion

A new gate comparison methodology has been presented to make quantitatively accurate comparisons between BiCMOS and CMOS gates. The concept of the sizing plane is introduced as an expressive framework to implement the methodology. The sizing plane is also shown to be a useful tool for simultaneous visualization of the constraints and tradeoffs in BiCMOS gate design. Specific design tradeoffs have been demonstrated for a 1-μm technology. The comparison methodology has been demonstrated by fabrication in a 2-μm BiCMOS technology and under a normalized comparison a typical BiCMOS gate shows a 40% improvement over pure CMOS. A technology comparison methodology has been proposed that provides insight into an overall comparison of BiCMOS versus CMOS for digital logic. This methodology has been illustrated for a 0.6-μm BiCMOS technology and the ratio of BiCMOS to CMOS delay for all but very lightly loaded conditions is shown to be 0.7, independent of sizing and loading.

Appendix

The representation of the BJT saturation constraint in the BSP is developed in this Appendix. Let I_B, I_C, R_C be the peak base current, collector current, and collector resistance in the BJT. To avoid forward biasing the base collector junction, the drop across R_C must be less than some value V. V does not need to be as low as the diode drop of 0.7 V because the peak drop across R_C actually occurs before the base voltage has reached its highest value.

$$I_C R_C < V. \tag{9}$$

R_C is related to the emitter area $A_e = W_e \times L_e$ as

$$R_C = R_0 / L_e. \tag{10}$$

I_B is simply the peak saturation current of the MOSFET driving the BJT and directly proportional to the input capacitance C_{in}.

$$I_B = K C_{\text{in}}. \tag{11}$$

The peak value of the collector current depends on I_B in a complex manner [12] but a simplistic expression is used here (J_{KF} is the knee current density).

$$I_C = \sqrt{J_{KF} A_e \beta I_B}. \tag{12}$$

Equation (12) overestimates the magnitude of I_C but does describe it quantitatively [12]. Using (10)–(12) in (9), we get

$$\frac{A_e}{C_{\text{in}}} > R_0^2 \frac{J_{KF} \beta K W_e^2}{V^2}. \tag{13}$$

Equation (13) defines a region in the BSP above a straight line through the origin. Thus it can be seen qualitatively that to avoid collector saturation the emitter area needs to be larger than a certain minimum and the lower bound increases with increasing MOSFET size.

Using more precise collector current expressions for the 1-μm BiCMOS technology with I_B = 1.6 mA for a 10/1 PMOSFET, I_{KF} = 3 mA/μm^2, R_0 = 100 $\Omega \cdot \mu$m, the condition to avoid collector saturation becomes $A_e / C_{\text{in}} >$ 0.016 μm^2/fF. Thus for C_{in} = 100 fF, corresponding to the BiCMOS PMOSFET $M1$ sized at 24 μm, A_e needs to be larger than 1.6 μm^2 to avoid saturation. As seen in Fig. 5 this is much lower than the A_e-opt curve.

References

[1] Y. Nishio, F. Murabayashi S. Kotoku, A. Watanabe, S. Shukuri, and K. Shimohigashi, "A BiCMOS logic gate with positive feedback," in *ISSCC Dig. Tech. Papers*, 1989, pp. 116–117.

[2] M. Kubo, I. Masuda, K. Miyata, and K. Ogiue, "Perspective in BiCMOS VLSI's," *IEEE J. Solid-State Circuits*, vol. 23, no. 1, pp. 5–11, Feb. 1988.

[3] H. Iwai, G. Sasaki, Y. Unno, Y. Niitsu, M. Norishima, Y. Sugimoto, and K. Kanzaki, "0.8 μm Bi-CMOS technology with high f_T ion-implanted emitter bipolar transistor," in *IEDM Tech. Dig.*, 1987, pp. 28–30.

[4] S. Miyaoka, M. Odaka, K. Ogiue, T. Ikeda, M. Suzuki, H. Higuchi, and M. Hirao, "A 7-ns/350-mW 64-kbit ECL-compatible RAM," *IEEE J. Solid-State Circuits*, vol. SC-22, no. 5, pp. 847–849, Oct. 1987.

[5] T. Ikeda, A. Watanabe, Y. Nishio, I. Masuda, N. Tamba, M. Odaka, and K. Ogiue, "High-speed BiCMOS technology with a buried twin well structure," *IEEE Trans. Electron Devices*, vol. ED-34, no. 6, pp. 1304–1310, June 1987.

[6] K. Ogiue, M. Odaka, S. Miyaoka, I. Masuda, T. Ikeda, and K. Tonomura, "13-ns, 500-mW, 64-kbit ECL RAM using Hi-BiCMOS technology," *IEEE J. Solid-State Circuits*, vol. SC-21, no. 5, pp. 681–685, Oct. 1986.

[7] T. Ikeda, T. Nagano, M. Honma, K. Miyata, H. Higuchi, M. Okada, and K. Ogiue, "Advanced Bi-CMOS technology for high speed VLSI," in *IEDM Tech. Dig.*, 1986, pp. 408–411.

[8] P. Raje, K. Saraswat, and K. Cham, "A new BiCMOS/CMOS gate comparison/design methodology and supply voltage scaling model," in *IEDM Tech. Dig.*, 1989, pp. 433–436.

[9] J. Shott and S. Wong, "The Stanford BiCMOS Process," in *Dig. SRC TRC on BiCMOS Tech. and App.*, 1989.

[10] L. Glasser and D. Dobberpuhl, *The Design and Analysis of VLSI Circuits*. Reading, MA: Addison-Wesley, 1985.

[11] P. Raje, K. Cham, and K Saraswat, "BiCMOS gate performance optimization using a unified delay model," in *Proc. Symp. VLSI Tech.*, 1990, pp. 91–92.

[12] P. Raje, "Analysis, design and scaling of BiCMOS for digital logic," Ph.D. dissertation, Stanford University, Stanford, CA, 1991.

Paper 1.17

Optimal Usage of CMOS within a BiCMOS Technology

Larry Wissel and Elliot L. Gould

Abstract—The comparison of CMOS to BiCMOS often seen in the literature shows the delays of single-stage circuits driving a capacitive load, with the BiCMOS circuit exhibiting a bold advantage. This comparison is misleading, as it suggests that the highest possible performance chip design implemented in a BiCMOS technology should use only BiCMOS circuits. When multistage circuits and chip wiring resistance are also considered, CMOS performance is found to be much closer to BiCMOS performance. CMOS circuits are shown to be preferred over BiCMOS circuits for a significant fraction of the chip nets. When nets that can afford a performance decrease are relaxed by using CMOS circuits instead of BiCMOS circuits, the CMOS fraction increases further. High usage of CMOS is desirable for area and yield considerations. Evaluations of the optimal CMOS role in future-generation BiCMOS technologies are expected to show an even larger role for CMOS.

I. Introduction

The traditional CMOS–BiCMOS comparison [1]–[3] shows the delay of a single-stage circuit (e.g., 2-NAND) of each technology driving a capacitive load (Fig. 1), with the slope of each curve representing the output impedance of the circuit. The two curves intersect at the *crossover capacitance* C_s, with BiCMOS faster for loads greater than C_s. Because C_s is small ($\simeq 0.2$ pF), the traditional comparison justifies an all-BiCMOS design point when performance is the primary consideration. Very few nets have $C < C_s$; even for the nets with $C < C_s$, the CMOS advantage over BiCMOS is small.

II. BiCMOS Characteristics

A. BiCMOS Circuit and Waveform

In a BiCMOS circuit such as the 2-NAND of Fig. 2(a), a ONE output level of $V_{DD} - V_{BE}$ is achieved. With the output at this non-CMOS voltage, the p-channel FET's of subsequent (cascaded) circuits do not completely turn off, resulting in a small amount of dc current [1]. On chips with hundreds of thousands of circuits, the total dc current due to non-CMOS voltages is intolerable. Consequently, auxiliary FET's or resistors are added to the basic BiCMOS circuit to complete the voltage transition to a full CMOS level.

Manuscript received August 6, 1991; revised November 4, 1991.
The authors are with IBM General Technology Division, Essex Junction, VT 05452.
IEEE Log Number 9105462.

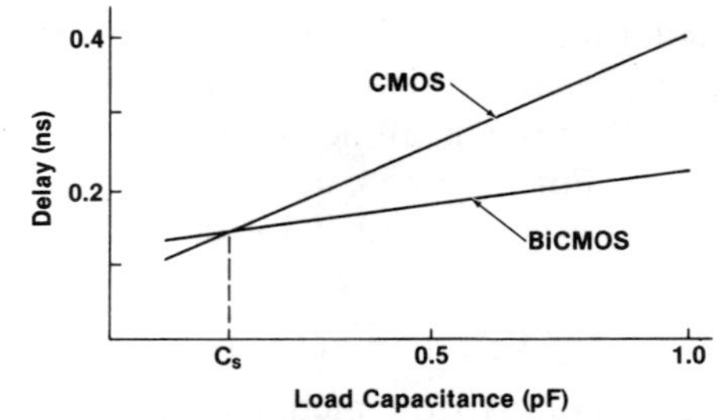

Fig. 1. Single-stage circuit performance comparison.

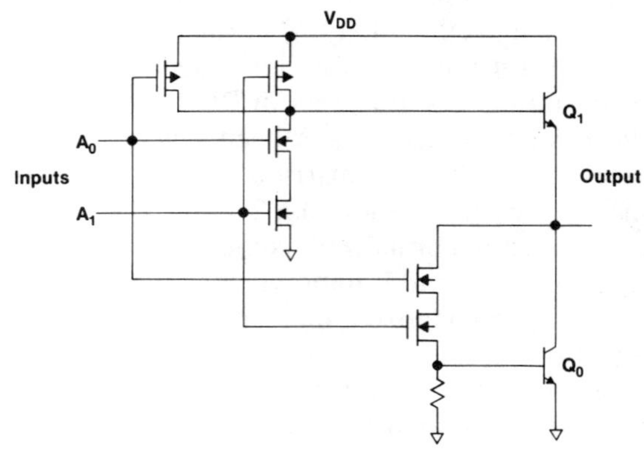

(a)

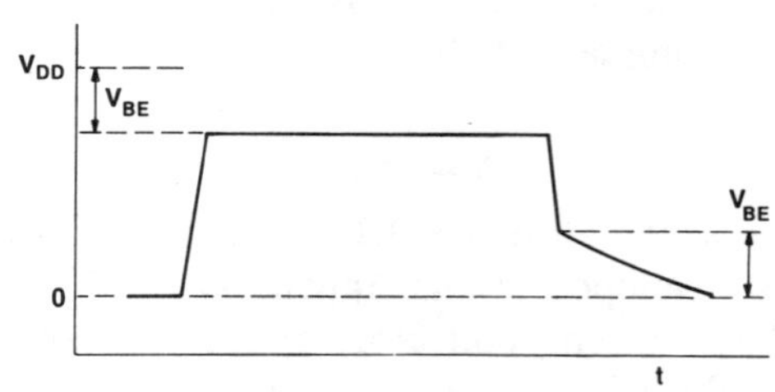

(b)

Fig. 2. Partial-swing BiCMOS. (a) Circuit. (b) Waveform.

A full-swing BiCMOS circuit [3] is shown in Fig. 3(a), with its waveform in Fig. 3(b). Although the output waveform does achieve full CMOS levels, the final transition is slow. The completion of the transition does not even begin until auxiliary inverter I_1 changes state and causes auxiliary FET T_1 to conduct. The falling output waveform of a BiCMOS circuit is complementary to the rising waveform, with a fast partial-swing transition to V_{BE} due to the pull-down n-p-n transistor, followed by a slow finish to ground by the auxiliary devices.

Reprinted from *IEEE J. Solid-State Circuits*, vol. 27, no. 3, pp. 300–306, March 1992.

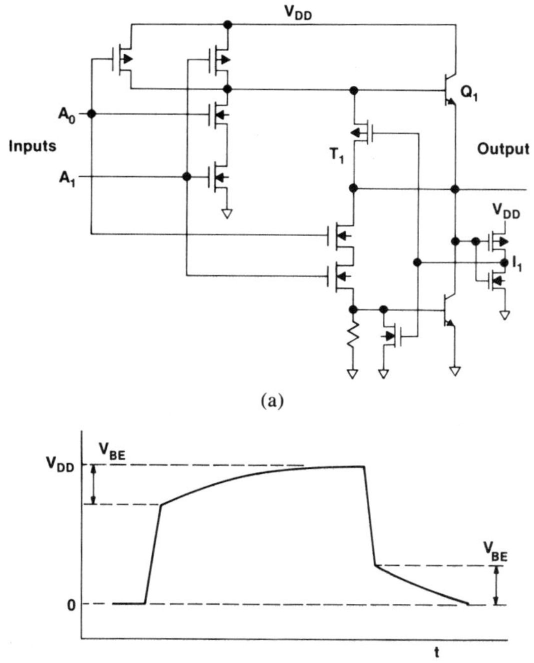

Fig. 3. Full-swing BiCMOS. (a) Circuit. (b) Waveform.

TABLE I
CIRCUIT USAGE STATISTICS

Single Stage		Multistage	
Function	Usage (%)	Function	Usage (%)
2-NAND	12	Latch	14
Inverter	10	2-AND	9
3-NAND	3	Buffer	8
Other	12	Other	32
Total	37%	Total	63%

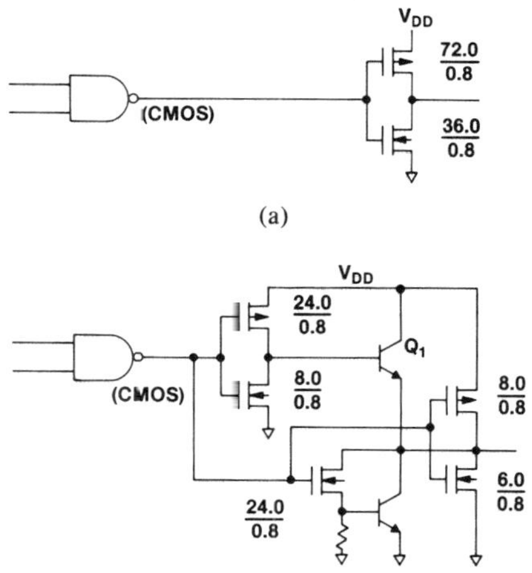

Fig. 4. Multistage circuits. (a) CMOS. (b) BiCMOS.

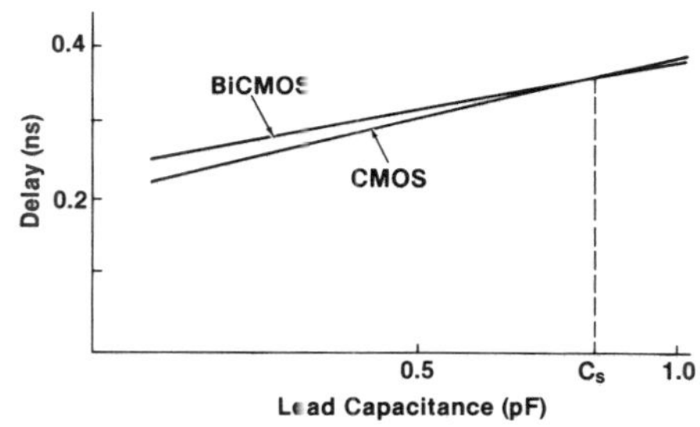

Fig. 5. Multistage circuit performance comparison.

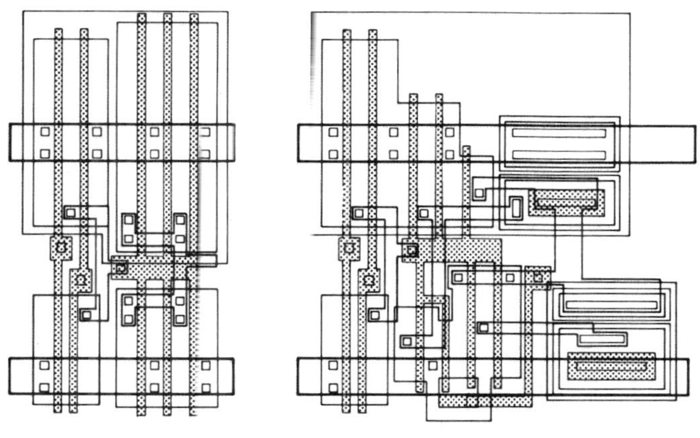

Fig. 6. Multistage circuits (2-AND). (a) CMOS. (b) BiCMOS.

B. Multistage Circuits

Single-stage circuits are not the only elements of a chip; a chip will usually include multi-stage circuits as well. Table I provides statistics of circuit usage from several chips, with the circuits separated into "single-stage" and "multistage" columns. Multistage circuits are found to be dominant.

The 2-AND circuits of Fig. 4 have characteristics that are typical of multistage circuits. In both the CMOS implementation (Fig. 4(a)) and the BiCMOS implementation (Fig. 4(b)), a logic stage (2-NAND) drives an output buffer. FET *width/length* ratios are identified in Fig. 4. Small FET's are required on the 2-NAND to limit input capacitance, but FET's of the output buffers can be made larger. Accordingly, the CMOS–BiCMOS comparison for multistage circuits (Fig. 5) is very different from that for single-stage circuits. The low-output impedance of the multistage CMOS circuit causes the crossover capacitance to be much higher.

Fig. 6 shows the layouts of a CMOS 2-AND and a BiCMOS 2-AND. The CMOS and BiCMOS circuit areas are 700 and 1660 μm^2, respectively, showing a large silicon density advantage for CMOS. Due to wirability limitations at the chip level, the CMOS density advantage may not be fully realized.

C. Effects of Wiring Resistance

The *RC* delay of wiring resistance has not scaled with circuit improvements and has become a significant effect that must be considered as part of a technology evaluation, including a CMOS–BiCMOS performance comparison.

A truncated ramp function with a rise time T_{rise} (Fig. 7) is used to model both the CMOS and BiCMOS waveforms. The CMOS waveform (Fig. 7(a)) has a step height of $V_{SH} = 3.6$ V and a rise time of 0.15 ns; the BiCMOS waveform (Fig. 7(b)) has a step height of $V_{SH} = 2.9$ V and a rise time of 0.08 ns. The fast BiCMOS rise time reflects the low-output impedance of BiCMOS circuits. The use of a reduced-height function for BiCMOS is appropriate because the full-swing finish is too late to have any performance benefits. The critical switching window for subsequent circuits is within the first few tenths of a nanosecond of the transition; the auxiliary devices typically require several nanoseconds to finish the full swing.

Wiring resistance may be modeled as a single resistor between the circuit output and all the load capacitance (wire plus gate). Although this is not a realistic representation of the "trees" and "ladders" actually found on chips, it is a satisfactory model for evaluating the relative characteristics of each circuit type in a resistive environment. A closed-form expression exists for the output voltage of the simple RC net driven by a truncated ramp function [5]. With an input voltage

$$V_I(t) = \frac{V_{SH}}{T_{\text{rise}}}[t \cdot u(t) - (t - T_{\text{rise}}) \cdot u(t - T_{\text{rise}})] \quad (1)$$

applied to the RC network, the output voltage for $t > T_{\text{rise}}$ is

$$V_o(t) = V_{SH}\left[1 - \frac{RC}{T_{\text{rise}}}(e^{T_{\text{rise}}/RC} - 1)e^{-t/RC}\right]. \quad (2)$$

Equation (2) is used to find the time T_{OM} (Fig. 7) at which $V_o(t)$ rises through $0.5V_{DD}$. For a CMOS circuit, $V_O(T_{OM}) = 0.5\ V_{SH}$; for a BiCMOS circuit, $V_O(T_{OM}) = 0.62\ V_{SH}$. Each is parameterized as $V_O(T_{OM}) = h \cdot V_{SH}$ where $h = 0.5$ for CMOS and $h = 0.62$ for BiCMOS (V_{SH} is smaller for BiCMOS, requiring h to be larger to make $h \cdot V_{SH} = 0.5\ V_{DD}$). By subtracting time T_{IM} (Fig. 7) at which $V_I(t)$ rises through $0.5V_{DD}$, the additional RC delay T_{RC} is obtained:

$$T_{RC} = RC \ln\left[\frac{RC}{T_{\text{rise}}}\frac{(e^{T_{\text{rise}}/RC} - 1)}{(1 - h)}\right] - hT_{\text{rise}}. \quad (3)$$

The first term in (3) is T_{OM}; the second term is T_{IM}. Evaluated with $R = 200\ \Omega$, $C = 1.0$ pF, and with the values for h and T_{rise} listed above, the RC delay is 0.143 ns for CMOS, and 0.187 ns for BiCMOS.

The RC delay increase for BiCMOS is larger than that for CMOS by the ratio 0.187/0.143 = 1.31; the greater impact of wiring resistance on the BiCMOS delay is due to the reduced step height of the BiCMOS waveform. In addition to the larger delay increase, the transition time of $V_O(t)$ at the switching threshold $0.5V_{DD}$ is degraded more for BiCMOS than for CMOS (Fig. 7), also impacting the delay of subsequent circuits. The loss of the final V_{BE} of signal amplitude (compared to CMOS) during the critical switching window is a significant limiting factor to the BiCMOS circuit of Fig. 3(a).

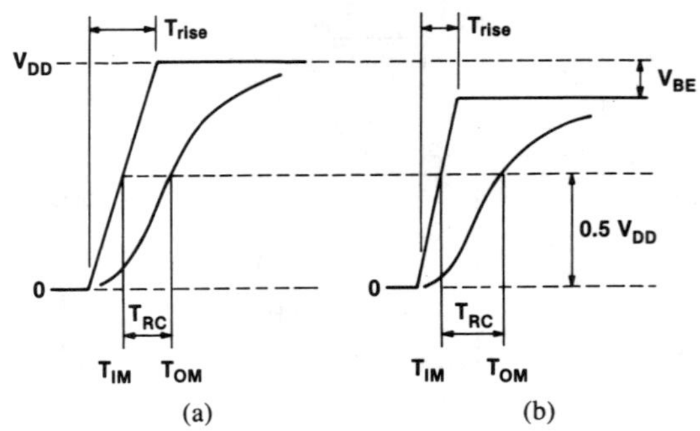

Fig. 7. Effect of resistance on waveforms. (a) CMOS. (b) BiCMOS.

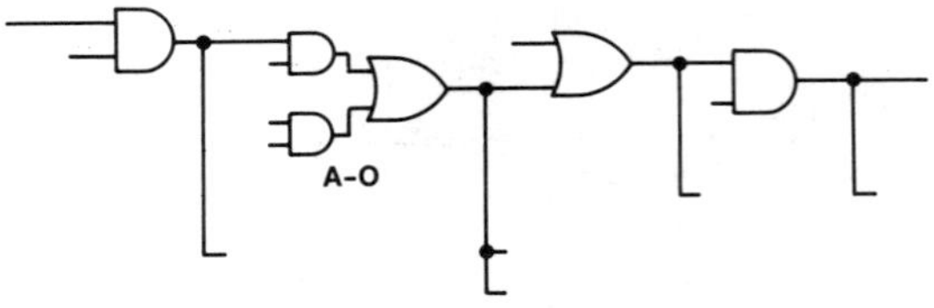

Fig. 8. Logic path.

TABLE II
EFFECT OF RESISTANCE ON PATH DELAY

Circuit Type	R_{Wire} (Ω/mm) 0	50
CMOS	1.35 ns	1.48 ns
BiCMOS	1.23 ns	1.39 ns

A logic path (Fig. 8) was simulated to demonstrate these effects with both CMOS and BiCMOS circuits. A "tree" model of wiring, representative of actual chip wiring, was created. The nets in the path have an average wire length of 3.7 mm, and an average capacitance of 0.8 pF. The delays for the path, with and without wiring resistance, are summarized in Table II.

Resistance changed the BiCMOS delay by 0.16 ns, and the CMOS delay by 0.13 ns. The ratio (0.16/0.13 = 1.23) is consistent with the ratio of delay increments (1.31) found previously with the analytical treatment of the truncated ramp input signal into the simple RC network. However, the impact of wiring resistance on the BiCMOS waveform is expected to be less in the realistic model, which has distributed RC networks, than in the simple model, which has a lumped RC network.

The path delays show that CMOS nearly matches BiCMOS performance. With an average net capacitance of 0.8 pF, the single-stage CMOS–BiCMOS comparison would have predicted a large BiCMOS advantage. However, CMOS is competitive because the circuits are multistage, and the BiCMOS advantage shrinks further due to RC delays of wiring resistance. This comparison shows that CMOS can compete with BiCMOS for loads well beyond what the traditional comparison indicates.

III. QUANTIFICATION OF OPTIMAL CMOS ROLE

With the significant role of CMOS within a CMOS–BiCMOS intermix technology clearly identified, the

CMOS role must be quantified next. The following analysis uses a distribution function for on-chip net capacitances, defines "delay differential" and uses a function to represent it, and defines "delay margin" and uses a distribution function to represent it.

Starting with an all-BiCMOS design, the analysis determines how many nets can be converted from BiCMOS to CMOS ("powered down") without impacting chip performance. This analysis is particularly applicable to a standard-cell design. The goal of the analysis is to evaluate the maximum utilization of CMOS circuits, retaining BiCMOS circuits only where essential. A BiCMOS gate array [1], [2] requires that the n-p-n transistors be uniformly distributed because there is no way of predicting where an n-p-n transistor will be required for unique gate array personalizations. For a standard-cell implementation, however, the n-p-n's can be implemented only where needed.

A. Distribution of Net Capacitances

The distribution of nets versus capacitance ($n_c(C)$ in Fig. 9) on a chip is required, where C is the total capacitance of the net. Various distributions may be proposed [6], but all share the property of being weighted toward low capacitance values as indicated.

A fraction f_1 of the nets are low-capacitance nets ($C \leq C_s$) and can be powered down:

$$f_1 = \frac{\int_{C_{\min}}^{C_S} n_c(C)\, dC}{\int_{C_{\min}}^{C_{\max}} n_c(C)\, dC}. \quad (4)$$

The numerator is the number of low-capacitance nets (capacitance less than C_s) and the denominator is the total number of nets. The shaded area in Fig. 9 represents the low-capacitance nets; the unshaded area represents the high-capacitance nets. High-capacitance nets are driven with BiCMOS circuits due to its performance advantage beyond C_s.

B. Delay Differential

Delay differential is defined as the amount by which CMOS is slower than BiCMOS. CMOS and BiCMOS delays are shown in Fig. 10(a); their difference, which is the delay differential, is shown as the function $\Delta(C)$ in Fig. 10(b). The dashed lines of Fig. 10 show the effect of replacing a single-stage CMOS circuit with a three-stage logical equivalent. Although the function $\Delta(C)$ can be flattened by adding two buffers, the remainder of this evaluation assumes no flattening.

C. Delay Margin

Delay margin τ is defined as allowed delay minus actual delay of a net. None of the nets may have negative delay margin, but many nets may have positive delay margin, indicating that a performance decrease may be tolerated. Nets are distributed versus delay margin as $n_{DM}(\tau)$ in Fig. 11. An aggressive design will reflect a strong weighting toward a zero delay margin. As in Fig. 9, low-capacitance nets are shaded with lines, and high-capacitance nets are unshaded.

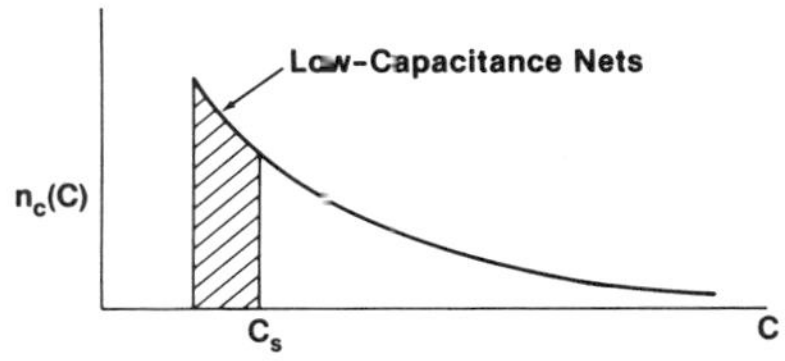

Fig. 9. Distribution of nets versus capacitance.

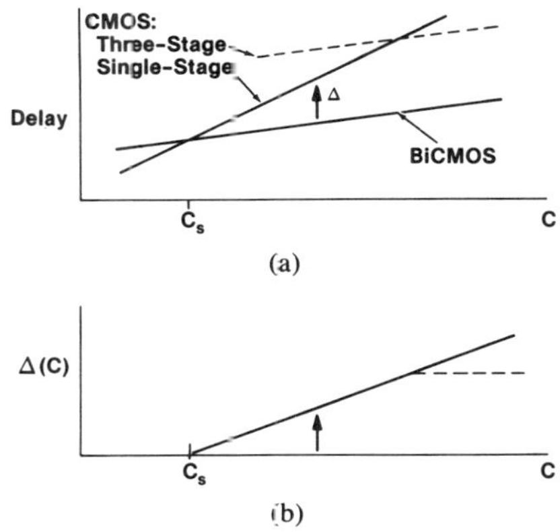

Fig. 10. (a) CMOS and BiCMOS delay. (b) CMOS–BiCMOS delay differential.

Some of the high-capacitance nets (those with $C > C_s$, unshaded in Fig. 9) have positive delay margins. If the delay margin is greater than the performance advantage of BiCMOS over CMOS, the high-capacitance net can be converted to CMOS with no impact to chip performance. Fig. 12 combines Figs. 9, 10, and 11 to show the interrelationship of the distribution of net capacitance, the delay differential, and the distribution of delay margin.

Consider a group dN of high-capacitance nets with a capacitance near C_i (Fig 12(a)): $dN = n_c(C_i)\, dC$. At this load, C_i, the delay differential between CMOS and BiCMOS is $\Delta_i = \Delta(C)$ at the specific point $C = C_i$ in Fig. 12(b) and (c).

A high-capacitance net can be powered down if the delay margin is greater than Δ_i. The probability $P(\Delta_i)$ of the net having enough delay margin is

$$P(\Delta_i) = \frac{\int_{\Delta_i}^{\tau_{\max}} n_{DM}(\tau)\, d\tau}{\int_{0}^{\tau_{\max}} n_{DM}(\tau)\, d\tau}. \quad (5)$$

Fig. 12(c) is the same function as Fig. 11, but rotated 90° counterclockwise to align the delay margin (time) axis to the delay differential (time) axis of Fig. 12(b). In (5), the numerator is the area shaded only with dots; the denominator is the total area not shaded with lines (i.e., either

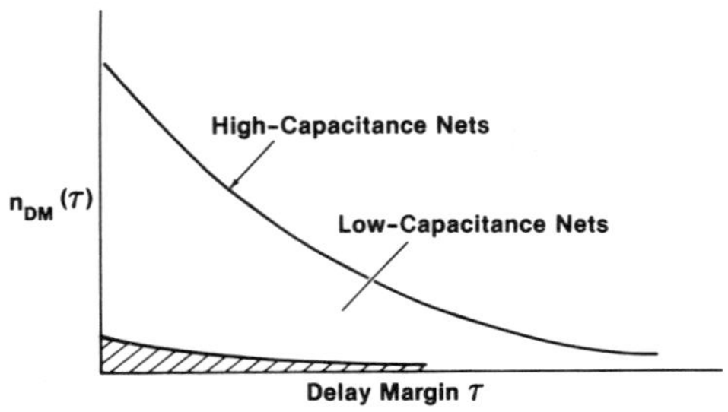

Fig. 11. Distribution of nets versus delay margin.

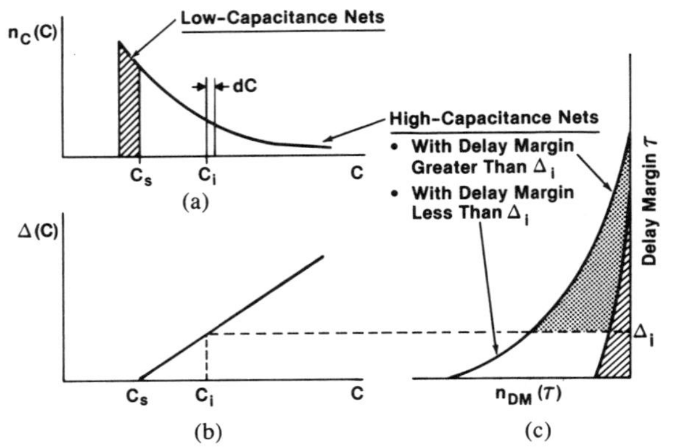

Fig. 12. Interrelationships of (a) net distribution versus capacitance, (b) CMOS-BiCMOS delay differential, and (c) net distribution versus delay margin.

unshaded or shaded with dots only). $P(\Delta)$ is shown in Fig. 13(a). The CMOS-BiCMOS delay differential $\Delta(C)$ can be used to transform the function $P(\Delta)$ into to a function of C, $P_1(C)$ (Fig. 13(b)). The number of nets between $[C, C + dC]$ that can be converted to CMOS is $dN = P_1(C)\, n_c(C)\, dC$. A fraction f_2 of all high-capacitance nets can be converted to CMOS based on a delay margin:

$$f_2 = \frac{\int_{C_S}^{C_{\max}} P_1(C) n_C(C)\, dC}{\int_{C_S}^{C_{\max}} n_C(C)\, dC}. \tag{6}$$

The numerator is the number of high-capacitance nets with enough delay margin for conversion, and the denominator is the total number of high-capacitance nets ($C > C_s$).

The probability of power down based on delay margin (Fig. 13(b)) is multiplied by the net distribution function (Fig. 13(c)) to find additional nets to be driven by CMOS circuits. These additional nets are shaded with dots in Fig. 13(c).

D. Numerical Evaluation

Functional representations of $n_c(C)$, $n_{DM}(\tau)$, and $\Delta(C)$ lead to parameterized expressions for the conversion fractions f_1 and f_2. The distribution $n_c(C)$ is represented as a Pareto–Levy distribution [7]:

$$n_c(C) = n_{c0}\left(\frac{C_{\min}}{C}\right)^2 [u(C - C_{\min}) - u(C - c_{\max})].$$

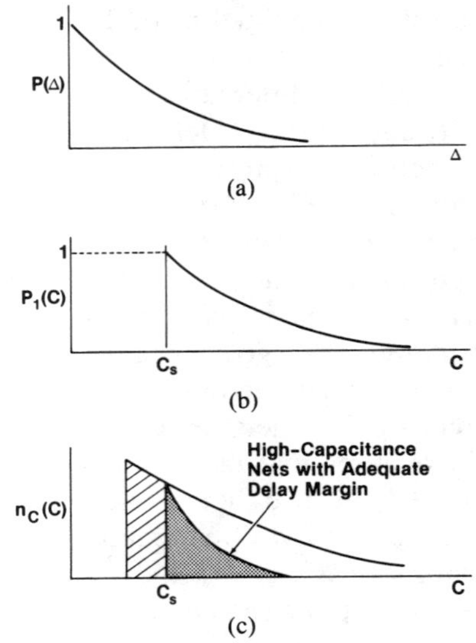

Fig. 13. Conversion of high-capacitance nets from BiCMOS to CMOS. (a) Power-down probability versus delay differential. (b) Power-down probability versus capacitance. (c) Converted nets.

The function $\Delta(C)$ (sloped portion of Fig. 10(b), without dashed-line flattening) is represented as $\Delta = a \cdot (C - C_s)\, u(C - C_s)$. The distribution $n_{DM}(\tau)$ is represented as $n_{DM} = n_{DM0} e^{(-\tau/\tau_1)}$. First, f_1 is obtained from (4) as

$$f_1 = 1 - \frac{C_{\min}}{C_s} \frac{C_{\max} - C_s}{C_{\max} - C_{\min}}. \tag{7}$$

If $C_s < C_{\min}$, there are no low-capacitance nets to be converted based only on the CMOS–BiCMOS comparison (i.e., without taking advantage of delay margin). In a numerical example, if the CMOS–BiCMOS crossover capacitance is $C_S = 0.18$ pF, and the minimum and maximum capacitances are $C_{\min} = 0.15$ pF and $C_{\max} = 5.0$ pF, respectively, (7) gives $f_1 = 0.17$. Using the multistage crossover, $C_S = 0.8$ pF, gives $f_1 = 0.84$. This shows that 17–84% of the nets on a chip have a low enough capacitance so that they can be driven with CMOS instead of BiCMOS without performance loss.

Next, a delay margin is used to further define the fraction of nets that can be driven by CMOS instead of BiCMOS. The functional form of $n_{DM}(\tau)$, with τ_1 being the delay-margin distribution parameter, is used in (5) to obtain an expression for $P(\Delta)$, illustrated in Fig. 13(a):

$$P(\Delta) = e^{-\Delta/\tau_1}. \tag{8}$$

The relationship $\Delta(C) = a \cdot (C - C_s)$ is used to transform $P(\Delta)$ into $P_1(C)$, shown in Fig. 13(c):

$$P_1(C) = e^{-a(C - C_S)/\tau_1} u(C - C_s). \tag{9}$$

With $P_1(C)$, and $n_c(C)$ known, (6) is evaluated for f_2:

$$f_2 = \frac{aC_s}{\tau_1} \frac{C_{\max}}{C_{\max} - C_s} e^{\frac{aC_s}{\tau_1}} \left[A\left(\frac{aC_{\max}}{\tau_1}\right) - A\left(\frac{aC_s}{\tau_1}\right)\right] \tag{10}$$

where

$$A(x) = \int_x^{\infty} e^{-\xi} \left(\frac{1}{\xi}\right)^2 d\xi. \tag{11}$$

As a numerical example, f_2 is evaluated for single-stage circuits with CMOS–BiCMOS delay differential coefficient $a = 0.22$ ns/pF, crossover capacitance $C_s = 0.18$ pF, delay-margin distribution parameter $\tau_1 = 0.05$ ns, and capacitance distribution parameters used previously. These values give $f_2 = 0.45$. For multistage circuits with CMOS–BiCMOS delay differential coefficient $a = 0.03$ ns/pF, crossover capacitance $C_s = 0.80$ pF, and delay-margin distribution and capacitance distribution parameters as before gives $f_2 = 0.63$. Thus, in an aggressive design where more than half of the nets have delay margin of less than 50 ps, CMOS can still be used instead of BiCMOS for 45–63% of the high-capacitance nets.

The total fraction of CMOS nets, f_{CMOS}, is the sum of f_1, the low-capacitance nets, and $f_2(1 - f_1)$, the high-capacitance nets with enough delay margin for conversion: $f_{\text{CMOS}} = f_1 + f_2(1 - f_1)$, which evaluates to $f_{\text{CMOS}} = 0.54$ for single-stage circuits, and $f_{\text{CMOS}} = 0.94$ for multistage circuits.

With the net-capacitance distribution and delay-margin distribution parameters cited, it is possible to implement 54–94% (depending on the fraction of multistage elements) of the circuits in CMOS and still achieve critical performance equivalent to an all-BiCMOS implementation.

IV. Timing Issues in a CMOS–BiCMOS Intermix

The foregoing analysis demonstrates that an intermix of BiCMOS and CMOS circuits can achieve the same performance as an all-BiCMOS chip design. However, a CMOS–BiCMOS intermix chip design requires special tools for timing analysis. The timing tool must be able to account for whether a circuit is driven by a CMOS (full swing) or a BiCMOS (partial swing) input signal. The input has a significant effect on subsequent delay, as demonstrated for a single-stage CMOS book driven by CMOS and BiCMOS inputs shown in Fig. 14. In addition, the timing tool must not only account for the *RC* delay but also be able to apply different *RC* delay and transition time adders for CMOS and BiCMOS circuits.

V. Future Considerations

As power supplies, FET thresholds, and circuit dimensions shrink, V_{BE} (approximately constant) will become a larger percentage of V_{DD}. Thus, partial-swing BiCMOS circuits will suffer an even greater performance penalty compared to CMOS circuits, as indicated in the plot of logic path delay versus V_{DD} with 50-Ω/mm wiring resistance (Fig. 15). New BiCMOS circuits are being developed [8] to be effective at low V_{DD}, but they are increasingly complex (many require the addition of a p-n-p transistor to the process), have reduced noise immunity, and, in most cases, require multiple power-supply lines.

VI. Conclusions

The BiCMOS logic circuit shown in Fig. 3(a) is suited for the niche application of driving high-capacitance nets

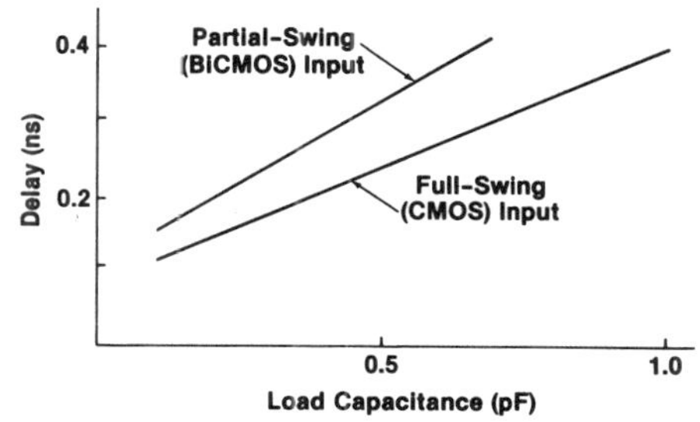

Fig. 14. Circuit delay with different input waveforms.

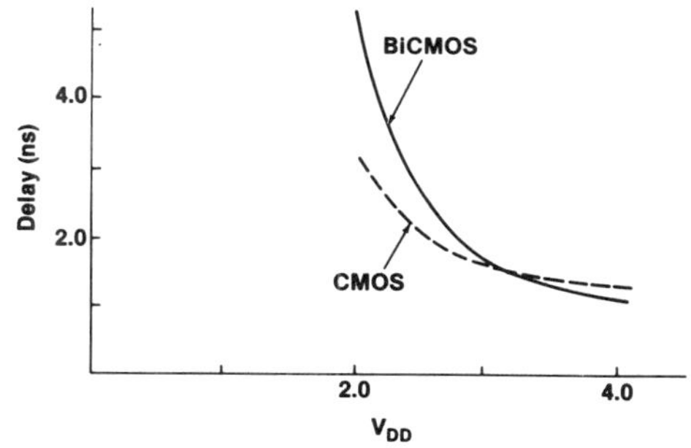

Fig. 15. Delay variation with power supply.

in a high-V_{DD} design, but the circuit is not extendable to future-generation (reduced V_{DD}) technologies. Therefore, new circuit techniques must be developed to gain advantage from BiCMOS in future technologies.

For low-capacitance nets and nets with enough delay margin, CMOS is preferred because of its density, circuit simplicity, lower sensitivity to wiring resistance, lower dI/dt, and clear migration path to the next technology generation. CMOS is expected to remain the dominant circuit type for logic, and designs implemented in a BiCMOS technology will use CMOS extensively.

Within a CMOS–BiCMOS intermix design system, an enhanced timing tool that automatically selects CMOS circuits instead of BiCMOS circuits based on delay margin is required. Such a tool will allow the designer to optimize the CMOS–BiCMOS intermix in the design. The highest possible use of CMOS will reduce chip size, enhance yield, and reduce on-chip switching noise.

References

[1] Yoji Nishio *et al.*, "A BiCMOS logic gate with positive feedback," in *ISSCC Dig. Tech. Papers*, Feb. 1989, pp. 116–117.

[2] J. Gallia *et al.*, "High-performance BiCMOS 100K-gate array," *IEEE J. Solid-State Circuits*, vol. 25, no. 1. pp. 142–149, Feb. 1990.

[3] G. Boudon *et al.*, "A BiCMOS circuit family for a 0.45 μm CMOS/BiCMOS sea of gates," in *1990 IEEE Custom Integrated Circuits Conf. Dig. Tech. Papers*, May 1991, p. 14.5.1.

[4] H. Shin, "Performance comparison of driver configurations and full-swing techniques for BiCMOS logic circuits," *IEEE J. Solid-State Circuits*, vol. 25, no. 3. pp. 863–865, June 1990.

[5] M. Abuelma'atti, "The waveform degradation in VLSI interconnections," *IEEE J. Solid-State Circuits*, vol. 24, no. 4. pp. 1014–1016, Aug. 1990.

[6] P. P. Duchene *et al.*, "Strategies for CMOS/BiCMOS gate usage on sea-of-gates arrays," in *1990 IEEE Custom Integrated Circuits Conf. Dig. Tech. Papers*, May 1991, p. 14.2.1.

[7] W. E. Donath, "Wire length distribution for placements of compute logic," *IBM J. Res. Develop.*, vol. 25, no. 3, pp. 152–155, May 1981.

[8] C.-L. Chen, "Level-shifted and voltage-reduced 0.5 μm BiCMOS circuits," in *ISSCC Dig. Tech. Papers*, Feb. 1990, pp. 236–237.

Paper 1.18

Design and Optimization of Buffer Chains and Logic Circuits in a BiCMOS Environment

Muhammad S. Elrabaa, *Student Member, IEEE*, and Mohamed I. Elmasry, *Fellow, IEEE*

***Abstract*—In this work the design and optimization of BiCMOS buffer chains and multilevel logic circuits are reported. BiCMOS speedup contours are introduced and analytical expressions for the delay are obtained. The speedup contours and the delay expressions were used in the design and optimization of BiCMOS buffer chains. Also, general design guidelines, which can be easily automated, for circuit design in a BiCMOS environment are given. Designing multistage mixed CMOS/BiCMOS buffers, BiCMOS complex logic gates, and multilevel CML gates is also studied and results are reported.**

Nomenclature

C_{in}, C_L	The input and the load capacitances.
L_{eff}, V_{th}, I_{sat}	MOS minimum channel length, threshold voltage, and saturation current.
$C_{\text{S/D}}$, C_{ovl}	MOS source/drain depletion and overlap capacitances.
λ	Minimum feature size.
β_m	Ratio between the PMOS and the NMOS devices in a CMOS buffer stage.
a	Ratio between the length of the S/D diffusions and λ.
f, N	Tapering factor of a CMOS buffer chain and the number of stages.
β, R_B, I_K, τ_F	Bipolar gain, base resistance, knee current, and transit time.
C_{be}, C_{jc}, C_{js}	Bipolar emitter–base junction, collector junction, and substrate capacitances.

I. Introduction

BiCMOS circuit designers are usually faced with the task of selecting the best combination of CMOS/BiCMOS/bipolar circuit structures for the design of critical paths that would render the optimum system performance in terms of speed, power, and area. The answer to this problem is not easy and straightforward in most of the design situations. This is because the BiCMOS technology offers circuit designers an environment that is very rich with different circuit structures to implement buffer chains and logic circuits. This, in turn, complicates both the selection and design processes. This paper attempts to provide a design methodology for BiCMOS circuit design that is comprehensive and yet easy to implement. More specifically it addresses the design and optimization of buffer chains and logic circuits in a BiCMOS environment.

Many researchers compared the performance of CMOS and BiCMOS gates for equal input capacitance [1], [2] and some of them provided design methodologies [2]. However, they either have limited the CMOS buffer to a single stage or equated the area of the CMOS to that of the BiCMOS buffer. However, the design of mixed CMOS/BiCMOS buffer chains has not been explored.

In Section II of this work, a general comparison between optimized CMOS buffer chains and different types of BiCMOS buffers is carried out with out any restrictions. The speedup factor of BiCMOS over CMOS is reported on speedup contours for different BiCMOS technologies. From the results of this comparison, general design guidelines are given. The effects of scaling are studied, and analytical expressions to calculate the crossover capacitance and the speedup contours are provided. Multistage mixed CMOS/BiCMOS buffers are also explored.

In Section III, the performances of complex logic implemented in CMOS and BiCMOS are compared for various conditions of complexity and loading. Other design options for implementing multilevel logic such as CML are considered. Finally, the applicability of the proposed design procedure in Section II to complex logic is evaluated.

II. Buffering Circuits in a BiCMOS Environment

Three types of buffering circuits are available in a BiCMOS technology: a CMOS buffer chain, a single-stage partial-swing BiCMOS buffer (PSBiCMOS) [1], and a single-stage full-swing BiCMOS buffer (FSBiCMOS) [1]. The schematics of the three buffering circuits are shown in Fig. 1. In this work a generic noncomplementary BiCMOS technology was assumed, hence the two types of chosen BiCMOS buffers do not contain p-n-p BJT's. The design problem is stated as follows: given a C_{in} and C_L, design a buffer (or a buffer chain) using CMOS and/or BiCMOS to achieve a given delay and/or area. The design and performance evaluation of the three types are

Manuscript received August 27, 1991; revised December 25, 1991. This work was supported in part by NSERC, ITRC, and MICRONET research grants.

The authors are with the VLSI Research Group, Department of Electrical and Computer Engineering, University of Waterloo, Waterloo, Ont., Canada N2L 3G1.

IEEE Log Number 9106950.

Reprinted from *IEEE J. Solid-State Circuits*, vol. 27, no. 5, pp. 792–801, May 1992.

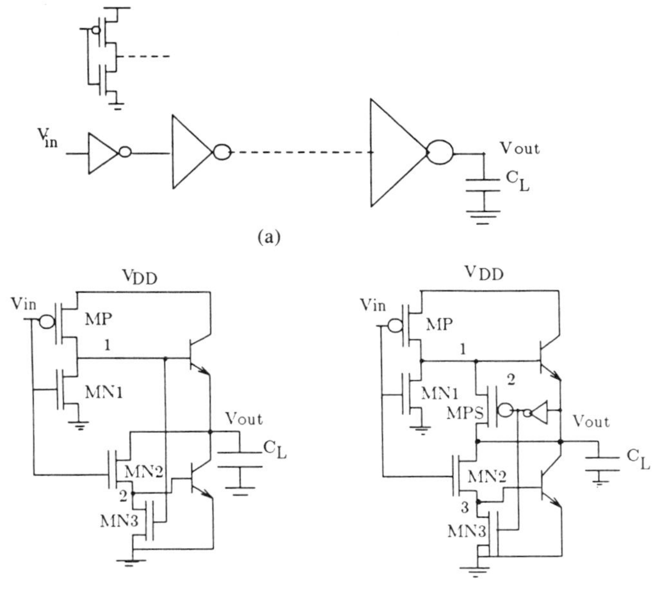

Fig. 1. The three types of buffering circuits considered: (a) the CMOS buffer chain, (b) the partial-swing BiCMOS buffer (PSBiCMOS), and (c) the full-swing BiCMOS buffer (FSBiCMOS).

presented in Sections II-A–E while multistage BiCMOS buffers are considered in Section II-F.

A. The CMOS Buffer Chain

The design procedure used for the CMOS buffer chain is the one given in [6]. The CMOS chain area was calculated as the sum of the channels area and the S/D diffusions area; hence, for N stages

$$A_{\text{CMOS}} = \lambda W_1(\beta_m + 1)(1 + 2a) \sum_{i=0}^{N-1} f^i \tag{1}$$

where W_1 is the width of the first-stage NMOS device.

B. The BiCMOS Buffers

The sizing of the devices in the considered BiCMOS buffers was based on extensive HSPICE simulations. The bipolar transistors sizes were chosen to give a minimum delay for a load capacitance in the midrange and were maintained constant for the different values of C_{in}, i.e., the BiCMOS buffers' area is constant for the same C_{in}. This choice was based on three reasons; to avoid the overhead of sizing the emitter area at each value of C_L considered, the speed of the BiCMOS buffer saturates after a certain emitter area is reached [7], [8], [13] and the optimum emitter area does not change significantly with increasing the input capacitance above a certain value, which is usually very small [2].

C. Simulations Results

The delay of each type of buffering circuits was calculated as the average of the 50% rise and fall times measured from HSPICE [3] simulations. The input capacitance range taken was from 0.05 to 1.0 pF. For each value of the input capacitance, the load capacitance C_L was changed from 0.05 to 5 pF. The simulations were carried out for three generations of BiCMOS technologies and supply voltages. The HSPICE parameters of the reference technology (1-μm L_{eff} and $V_{DD} = 5$ V) are in Table I. Two generations of BiCMOS technology were generated (0.56 μm, $V_{DD} = 3$ V and 0.2 μm, $V_{DD} = 2$ V) by successively scaling the reference technology according to the general BiCMOS scaling rules described in [4] and [5]. The scaling factors for the horizontal and vertical dimensions, base–collector voltage, and the supply voltage (and the threshold voltages), denoted as K_h, K_v, K_m, and K_u, respectively, for the second and third generations are shown in Table II. An identical version of the second-generation technology with unscaled MOS threshold voltage was used for the PSBiCMOS buffer simulations. This is because PSBiCMOS has to be used in a technology with either a MOS threshold voltage $\geq V_{BE(\text{on})}$ or have two types of MOS devices, one with a threshold voltage $\geq V_{BE(\text{on})}$ and the other with a lower threshold value to eliminate the static power dissipation in the CMOS gates driven by the PSBiCMOS. Also, the PSBiCMOS buffer circuit is not considered for the third generation ($V_{DD} =$ 2 V) due to the large deterioration of its performance.

The simulation results are presented as contours of speedup factors of the BiCMOS single buffers over the optimized CMOS buffer chain. Thus, the speedup factor is defined as

$$\text{Speedup factor} = \frac{\text{Delay of Optimized CMOS Buffer Chain}}{\text{Delay of BiCMOS Buffer}}. \tag{2}$$

For the 1-μm, V_{DD} = 5-V BiCMOS technology, the speedup contours for the PSBiCMOS and the FSBiCMOS are shown in Figs. 2 and 3, respectively. Also shown in the same figures are the contours of equal areas of CMOS and BiCMOS and when the CMOS area is ten times that of BiCMOS. Fig. 4 shows the delay of the PSBiCMOS buffer as a function of C_n and C_L. Finally, to examine the effects of scaling, Fig. 5 shows the contour lines of unity speedup factor for both types of BiCMOS buffers for the three generations of BiCMOS technology. Also shown is the area contour where $A_{\text{CMOS}} = A_{\text{PSBiCMOS}} \approx A_{\text{FSBiCMOS}}$. It should be noted that, by definition, the load capacitance C_L at which the speedup factor is unity is referred to as the speed crossover capacitance C_{xov}, while the load capacitance C_L at which the area of the CMOS buffer chain is equal to that of the BiCMOS buffer is referred to as the area crossover capacitance C_{Axov}. The following are concluded from Figs. 2–5.

1) The speed of the BiCMOS buffers saturates after a certain value of C_{in} is reached, and this value increases with increasing C_L (Fig. 4).

2) The speedup factor at a certain value of C_{in} increases with increasing C_L but it begins to decrease as C_L increases further. This is specially obvious from Figs. 2 and 3 at low C_{in}. This is because as the number of stages in the CMOS chain increases (i.e., as C_L increases), its

TABLE I
HSPICE Parameters of the Reference Technology (1 μm, 5 V)

	V_{th}	L_{eff}	$C_{S/D}$	C_{ovl}	Bipolar
NMOS	0.85 V	1.0 μm	4E-4 F/m	2E-10 F/m	C_{jc} = 10.0 fF, C_{js} = 25.0 fF
PMOS	−0.85 V	1.0 μm	5E-4 F/m	2E-10 F/m	C_{be} = 7.5 fF, τ_F = 12 ps
					R_B = 200 Ω, β = 100

TABLE II
The Scaling Factors for the Second and Third Generations of BiCMOS Technology

	K_h	K_v	K_u	K_m	λ (μm)	L_{eff} (μm)
2nd generation	0.667	0.850	0.600	0.600	0.40	0.56
3rd generation	0.375	0.630	0.667	0.667	0.15	0.20

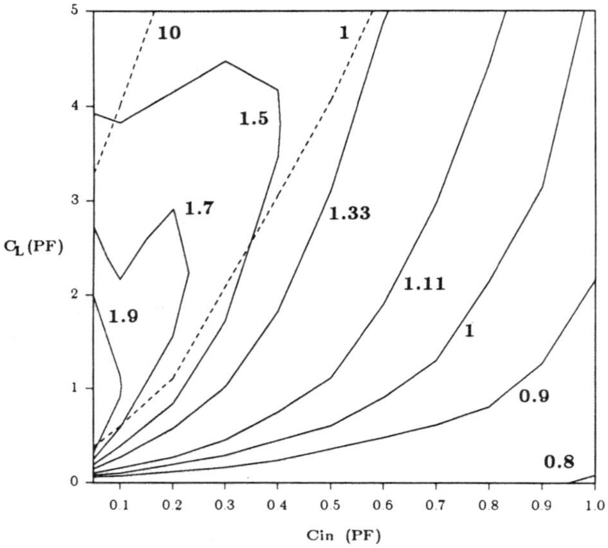

Fig. 2. The speedup factor contours (solid lines) of the PSBiCMOS buffer over the CMOS buffer chain for a 1-μm, 5-V BiCMOS technology plotted on a C_L versus C_{in} plot. The area contours $A_{CMOS} = A_{PSBiCMOS}$ and $A_{CMOS} = 10A_{PSBiCMOS}$ are also shown (dashed lines).

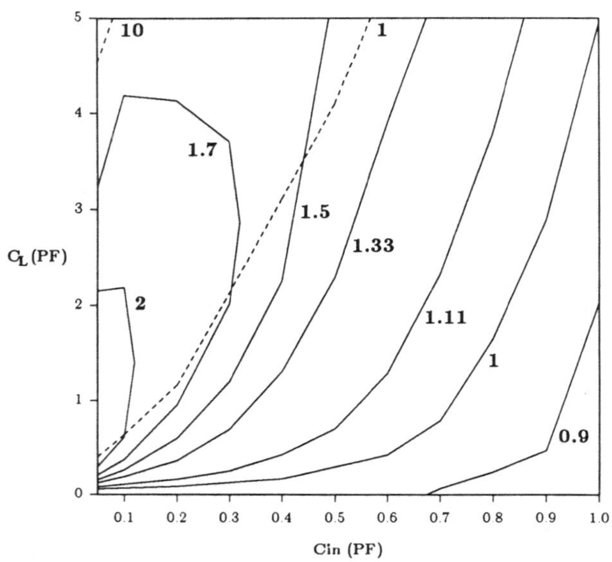

Fig. 3. The speedup factor contours (solid lines) of the FSBiCMOS buffer over the CMOS buffer chain for a 1-μm, 5-V BiCMOS technology plotted on a C_L versus C_{in} plot. The area contours $A_{CMOS} = A_{FSBiCMOS}$ and $A_{CMOS} = 10A_{FSBiCMOS}$ are also shown (dashed lines).

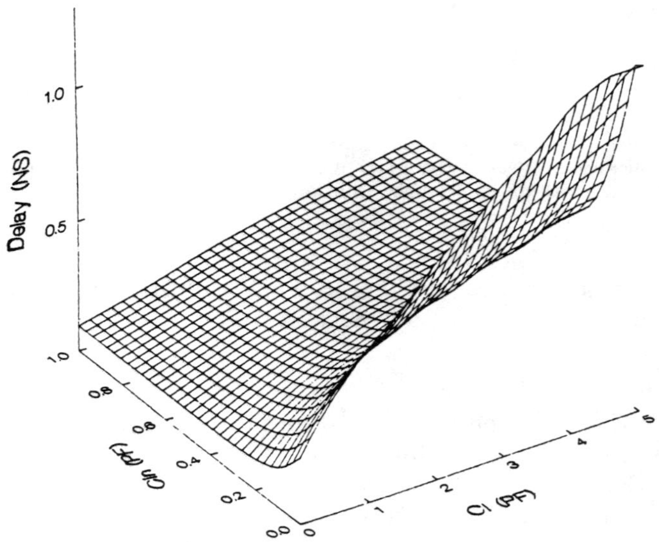

Fig. 4. The delay of the PSBiCMOS buffer versus input capacitance and load capacitance for the 1-μm, 5-V BiCMOS technology.

delay sensitivity to the load capacitance decreases. This is because the delay due to the load is only $1/N$ of the total delay, where N is the number of stages in the CMOS buffer. Fig. 6 shows how while the BiCMOS buffer delay increases linearly with the load capacitance, the CMOS delay increases at a rate that is decreasing with increasing C_L. This is why the BiCMOS buffer speedup starts to decrease at higher C_L. However, the CMOS chain area becomes much larger than that of the BiCMOS MOS buffers.

3) At any value of C_L, the BiCMOS buffers speedup factors decrease as C_{in} increases (Figs. 2 and 3).

4) The FSBiCMOS speedup factor is better than that of PSBiCMOS (Figs. 2 and 3).

5) The crossover capacitance C_{xov} increases rapidly with scaling down of devices and power supplies as shown in Fig. 5. Also, speed degradation with scaling is less in FSBiCMOS than it is in PSBiCMOS. At 2 V the FSBiCMOS is faster than the CMOS chain for small ranges of C_{in} and C_L.

As the above results indicate, it is helpful to the circuit designers to have analytical expressions for the crossover

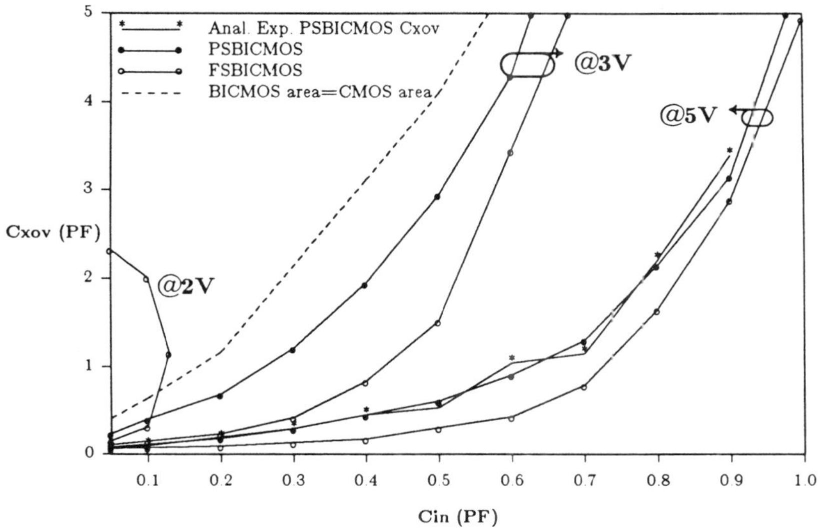

Fig. 5. The crossover capacitance of the PSBiCMOS and the FSBiCMOS buffers versus the input capacitance for the three BiCMOS technologies: 1 μm, 5 V; 0.56 μm, 3 V; and 0.2 μm, 2 V. The area contour $A_{CMOS} = A_{PSBiCMOS}$ is also shown (dashed line). Also shown is the PSBiCMOS crossover capacitance as calculated from the analytical expressions obtained.

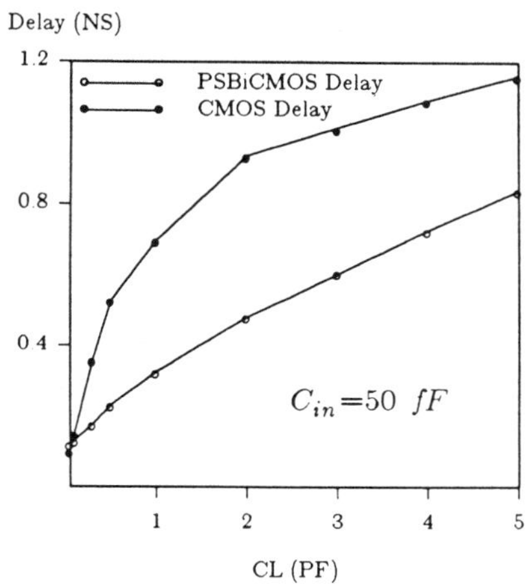

Fig. 6. The delays of the CMOS buffer chain and the PSBiCMOS buffer as a function of the load capacitance C_L. It is clear that as C_L increases, the slope of the CMOS delay decreases, while that of the PSBiCMOS remains constant.

capacitance C_{xov} and the speedup contours. Analytical expressions for the crossover capacitance are obtained next. Similar analysis could be carried out for other speedup contours.

D. Crossover Capacitance Analysis

To find C_{xov} at a certain C_{in}, the delay of the CMOS buffer chain is equated to that of the BiCMOS buffer. Analytical expressions for the delay of a CMOS inverter are given in [6] for long-channel devices and in [9] for short-channel devices. The general form of the delay of stage I is

$$T_{D\text{CMOS}} = \frac{1}{2}\left[\left(\frac{C_L}{K_N}\right)_I G_N + \left(\frac{C_L}{K_P}\right)_{I-1} B_N + \left(\frac{C_L}{K_P}\right)_I G_P + \left(\frac{C_L}{K_N}\right)_{I-1} B_P\right] \quad (3)$$

where $G_{N,P}$ and $B_{N,P}$ are constants and their values depend on whether the devices have long channels [6] or short channels [9]. Also, $K_{N,P}$ is the transconductance constant in the NMOS or PMOS saturation current equation [6], [9]. Note that at each stage both C_L and $K_{N,P}$ are proportional to the devices' widths and input capacitance. Hence, the total delay of a CMOS chain with N stages and a tapering factor f is

$$T_{D\text{chain}} = NC_{in}(g + f)\alpha \quad (4)$$

where g is the ratio of input capacitance to output capacitance of an unloaded CMOS inverter. α is a constant that depends on the technology and the first-stage input capacitance C_{in}:

$$\alpha = \left[\frac{G_N + B_P}{K_N} + \frac{G_P + B_N}{K_P}\right]_{1\text{st stage}}. \quad (5)$$

The delay of the PSBiCMOS buffer has been analyzed extensively [7], [8], [10], [13]. However, due to the complicated nature of the BiCMOS buffer transients, the delay expressions obtained lack precision and simplicity, except for the unified delay model reported in [10] which is simple and shows good correlation with simulations and experiments. But this model still lacks the required accuracy to be used in calculating the crossover capacitance C_{xov}. This is mainly due to the difficulty in choosing the right values of the device parameters such as the gain β, the transit time τ_F, and the different parasitic capacitances. The values of these parameters change during the transients. In this work, the PSBiCMOS delay was fitted by an equation similar to the one in [10]. High-level injection is assumed since this is the case for a wide range of C_{in} and C_L [7]. The general equation for the rise or fall time is given by [10]

$$T_{f(r)} = \frac{\tau_F I_C + A_e C_j V_{BE(\text{on})}}{I_{\text{sat}}} + \frac{C_L V_s}{I_{\text{sat}} + I_C}. \quad (6)$$

I_C is the collector current of the bipolar transistor, which for the high-level injection case equals $\sqrt{\beta I_K I_{sat}}$, A_e is the emitter area, C_j is the junction capacitance at the base, and V_s is the voltage swing. From (6) and noting that I_{sat} is proportional to C_{in}, the general equation to fit the average delay is

$$T_{D_{PSBiCMOS}} = A + BC_L \tag{7}$$

$$A = [A_1(C_{in})^{-1} + A_2(C_{in})^{-1/2}] \tag{8}$$

$$B = \left[\frac{A_3 + A_4(C_{in})^{-1/2}}{C_{in} + A_5(C_{in})^{-1/2} + A_6}\right] \tag{9}$$

where A_1–A_6 are empirical fitting parameters that could be obtained from eight simulations of four different C_{in} and two different C_L values for each C_{in}. These parameters may, in turn, be expressed as functions of explicit circuit parameters in the delay expression such that the circuit could be optimized under different conditions. For example, to study the effect of the emitter area A_e on the circuit performance, the parameters become

$$A_1 = a_1\sqrt{A_e}$$

$$A_2 = a_2 A_e$$

$$A_3 = \text{constant}$$

$$A_4 = a_4\sqrt{A_e}$$

$$A_5 = a_5\sqrt{A_e}$$

$$A_6 = a_6 A_e.$$

Similar treatments could be done such that the delay expression contains other circuit parameters, such as the widths of MOS devices, supply voltage, etc.

The above delay expressions of the CMOS chain and the PSBiCMOS buffer were used to generate the crossover capacitance as a function of C_{in} for the 1-μm, V_{DD} = 5-V BiCMOS technology. The results in Fig. 5 show good agreement between the calculated results and HSPICE simulations. As will be shown later, the ability to generate the contours of equal delay or at different speedup factors is very important in the decision making during BiCMOS buffer design.

E. Design Guidelines

The above results could be used to generate the following design guides. For a given C_{in} and C_L, and the following requirements on delay T_D and area A, the following design guidelines can be used.

1) Minimum T_D and no area constraints: use BiCMOS if $C_L > C_{xov}$ and CMOS otherwise.

2) Minimum A and no delay constraint: use a single-stage CMOS buffer.

3) Minimum T_D and minimum A: use BiCMOS if $C_L > C_{Axov}$ and CMOS if $C_L < C_{xov}$. If $C_{Axov} > C_L > C_{xov}$ then this design requirement cannot be met.

4) Minimum T_D and $A \le A_{max}$: the following procedure can be used:

a) If $A_{BiCMOS} > A_{max}$, then use CMOS. CMOS design under such an area constraint is given by steps 4e)–4i). If $A_{BiCMOS} < A_{max}$, then continue this procedure.

b) Calculate the optimum number of stages and tapering factor (N_o, f_o) of the CMOS chain and then find its area from (1). If $A_{CMOS} \le A_{max}$, then use CMOS if $C_L < C_{xov}$, and BiCMOS if $C_L > C_{xov}$. If $A_{CMOS} > A_{max}$, then continue.

c) Design a CMOS chain under the area constraint using steps 4e)–4i).

d) Calculate the delay of the BiCMOS buffer and that of the CMOS buffer chain from (4) through (9) and use the one which has the minimum delay.

The procedure of CMOS buffer chain design under area constraint A_{max}:

e) Assume that the number of stages $N = N_o - 1$.

f) Calculate the CMOS chain area from (1).

g) If $A_{CMOS} > A_{max}$, then $N = N - 1$ and go to step 4f). Otherwise continue.

h) N_o is set equal to $N + 1$ instead of N if the following condition holds:

$$(N + 1)[g + (C_L/C_{in})^{1/(N+1)}] \le N[g + (C_L/C_{in})^{1/N}]. \tag{10}$$

i) The tapering factor is adjusted using (1) such that $A_{CMOS} = A_{max}$.

Unlike what one might think, this is not a lengthy process, since the optimum number of stages N_o is usually small. For example, for a ratio C_L/C_{in} of 1000, N_o is 5.

5) Minimum A and $T_D \le T_{max}$:

a) If $T_{D_{BiCMOS}} > T_{max}$ use CMOS if $C_L \le C_{xov}$. The design procedure of a minimum area CMOS chain under a delay constraint is given below.

b) Starting with $N = 1$, increment N until a value is reached where the CMOS chain delay $\le T_{max}$. If, at any time, N exceeded N_o or the CMOS chain area exceeded that of the BiCMOS, which is constant, then BiCMOS should be used.

c) The CMOS chain tapering factor is recalculated as

$$f = \frac{T_D}{\alpha N} - g \tag{11}$$

instead of $(C_L/C_{in})^{1/N}$, to give the minimum area at the required delay.

d) Calculate A_{CMOS} using (1) and use the BiCMOS buffer if $A_{BiCMOS} \le A_{CMOS}$ and the CMOS chain otherwise.

F. Multistage BiCMOS Buffer Chains

Unlike CMOS, multistage BiCMOS buffers are seldom used. This is because the advantage of such a practice is not very obvious. However, as the above results indicate, the BiCMOS speed increases with increasing C_{in} until it reaches a certain limit. This means that there could be some situations in which using multilevel BiCMOS buff-

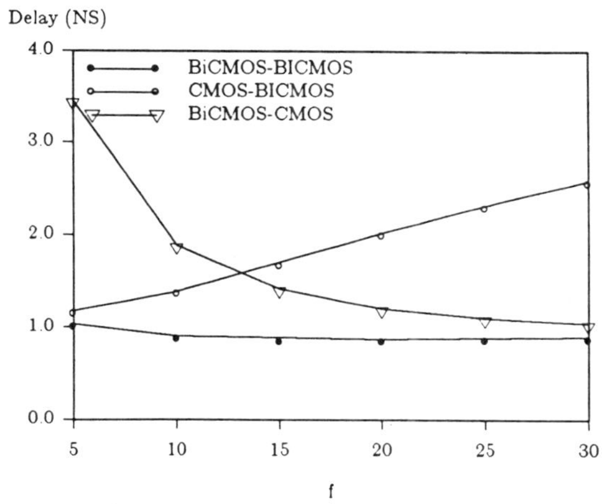

Fig. 7. The delays of the different multistage buffers versus the ratio from the input capacitance of the second stage to that of the first stage. $C_{in} = 20$ fF and $C_L = 5$ pF.

TABLE III
THE PERFORMANCE OF THE DIFFERENT CMOS/BiCMOS BUFFER CHAINS AND THE SINGLE-STAGE BiCMOS BUFFER

	CMOS Chain	Single BiCMOS	BiCMOS–BiCMOS	CMOS–BiCMOS	BiCMOS–CMOS
Delay (ns)	1.465	1.270	0.877	1.176	1.050
Normalized Area*	58.1	3.6	29.3	9.2	33.6

*Normalized to the area of the first stage of the CMOS buffer chain.

ers is advantageous. Such a situation arises when the required C_{in} is very small and C_L is very large. In this case, two BiCMOS buffer stages could be used, one with the required C_{in} and the second with a larger input capacitance, such that the delay is minimum. The value of the input capacitance of the second stage could be calculated using the delay expressions given by (4)–(9) to give the required delay under any constraint. Another option, as seen from the speedup contours, is to use a CMOS stage(s) followed by a BiCMOS stage. This option is useful under a severe area constraint or if the optimum input capacitance of the second stage is $< C_{xov}$ for the given C_{in}. Another option is to use a BiCMOS buffer in the first stage followed by a CMOS chain with high input capacitance. Again, this option could be useful under area constraints or scaling, where BiCMOS is faster than CMOS over a very limited range. Fig. 7 shows a plot of the delay of the BiCMOS chain, the CMOS–BiCMOS chain, and the BiCMOS–CMOS chain versus f, where f is the ratio between the input capacitances of the first and second stages, $C_{in} = 20$ fF, and $C_L = 5$ pF. The number of stages in those buffers was limited to two stages only since there is no need for more stages. This can be depicted from the results above.

The simulations show that the minimum delay of the BiCMOS–BiCMOS chain occur at $f \approx 20$, while the optimum f as calculated from the delay expression was about 25. The delay of the CMOS–BiCMOS chain with $f = 5$ is larger than that of the BiCMOS–BiCMOS with $f = 20$ by about 34%. Its area, however, is about 38% of that of the BiCMOS–BiCMOS chain. The BiCMOS–CMOS chain with $f = 30$ has a delay that is 37% larger than the BiCMOS–BiCMOS minimum delay and an area which is smaller by 23%. Table III shows the optimum delay of each type depicted in Fig. 7 along with that of the CMOS chain and a single BiCMOS buffer. The areas of each type, normalized to the area of the first stage of the CMOS chain, are also shown in Table III. This table shows the trade-offs involved in using the different kinds of buffer chains. At one extreme, the BiCMOS–BiCMOS offers the highest speed. At the other extreme, the CMOS–BiCMOS offers the smallest area at a reasonable speed. Also, as Fig. 7 shows, the BiCMOS–BiCMOS chain speed does not change drastically with f. Therefore, a smaller f, hence a smaller area, could be used without much loss of speed.

III. BiCMOS COMPLEX LOGIC GATES

Two types of complex logic gates are considered. The first type is an AND-OR INVERT gate (Fig. 7(a)) and the second type is a multilevel chain of NAND gates (Fig. 7(b)). The design and simulation results for both types are presented next.

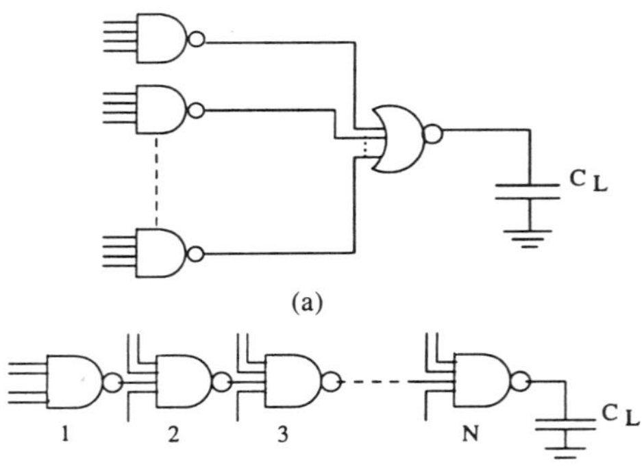

Fig. 8. The two types of complex logic gates considered: (a) a complex logic gate, and (b) a multilevel logic (MLL).

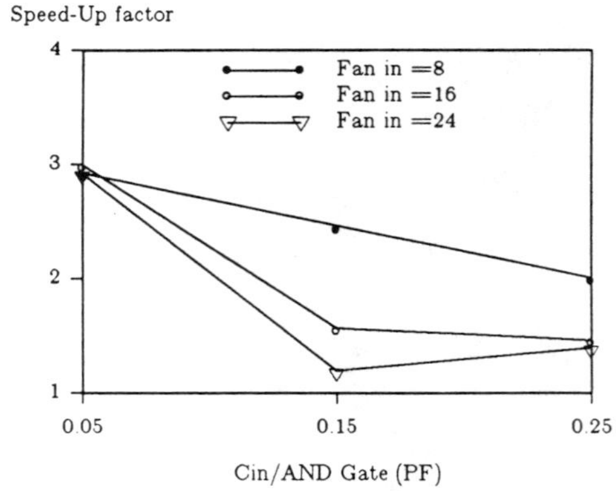

Fig. 9. The speedup factor of the FSBiCMOS complex logic gate over the CMOS counterpart versus the input capacitance of an AND gate for different values of fan-in and $C_L = 0.5$ pF.

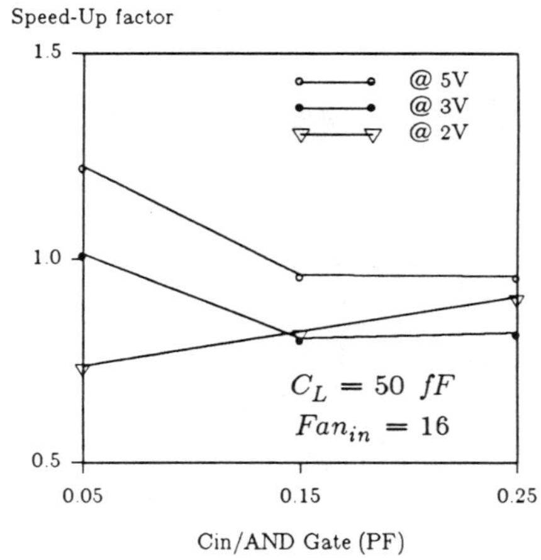

Fig. 10. The speedup factor of the FSBiCMOS complex logic gate versus the input capacitance of an AND gate for the three technology generations. $C_L = 50$ fF and $fan_{in} = 16$.

A. The AND-OR-INVERT Gate

The AND-OR-INVERT is a classical example of a complex logic function that is easy and straightforward to implement in CMOS. For this study, the fan_{in} was changed in multiples of eight, namely 8, 16, and 24, where every four inputs correspond to an AND gate of Fig. 8(a). The AND-OR-INVERT gate considered has been implemented using CMOS and FSBiCMOS, for the three generations of the BiCMOS technology. PSBiCMOS implementation was not considered because the complex gate chosen, due to the unscaled threshold voltage, will either not operate or will operate very poorly as the supply voltage is scaled down. Also the PSBiCMOS will not offer significant savings in area over the FSBiCMOS since the area of both implementations is dominated by the CMOS part.

The speedup factor of the BiCMOS implementation over the CMOS counter part is given in Fig. 9 versus the input capacitance per a single AND gate (C_{in}/AND), for different values of fan_{in}. Fig. 10 shows the speedup factor verus C_{in}/AND for the different BiCMOS technologies at $fan_{in} = 16$. The following can be observed.

1) The speedup factor decreases with increasing C_{in}/AND, for low values of fan_{in} (Fig. 9), which is the same result obtained in Fig. 2.

2) For high fan_{in}, the speedup factor decreases with C_{in} and then increases. This behavior is attributed to the fact that as fan_{in} increases, the sizing of the devices in the BiCMOS gate deviates from the optimum, hence the decrease in the speedup factor with C_{in} is faster. However, as C_{in} increases it becomes possible to size the MOS devices in the BiCMOS gate to give an optimum delay, and hence the increase in the speedup factor. Also another factor that could contribute to the reduction of the speedup factor is that the drivability of the CMOS circuit in the BiCMOS gate is reduced as fan_{in} increases, possibly forcing the BJT's to operate in the very low-current regime with large gain degradation.

3) The speedup factor, in general, decreases with scaling (Fig. 10), as was the case with the buffers. However, it increases with C_{in} for the third generation. This means that, unlike the buffers, for small values of C_{in}, the speedup factor might increase as C_{in} increases or fan_{in} decreases, specially for scaled-down technologies, where the BJT's gain degradation starts at a larger value of V_{BE}.

The above observations suggest that the BiCMOS implementations of complex gates tend to be faster than CMOS at low C_{in} and fan_{in}, and the speedup factor increases with C_L. However, CMOS becomes better with scaling. A similar concept to the speedup contours developed for the buffers in Section II could be applied to complex logic gates by obtaining an equivalent circuit of the complex logic. This is done by substituting the series MOS devices by a single device as in [9] and relating its saturation current to an effective value of C_{in}. By applying the analytical expressions obtained in the previous section the speedup contours could be obtained.

B. Multilevel Logic Gates

The options available in a BiCMOS technology to implement multilevel logic (MLL) are considered next. These options are: CMOS, CMOS with a FSBiCMOS driver, and CML with front/end conversion such that the inputs and outputs are CMOS levels. The input capacitance for all implementations is the same. The fan_{out} for each level of logic is one except for the last level where the fan_{out} and the wire capacitance are represented by a load capacitance C_L.

TABLE IV
THE AVERAGE STATIC POWER OF THE INTERNAL MCSL GATES AND THE END CML/CMOS CONVERSION STAGE FOR THE THREE GENERATIONS OF BiCMOS TECHNOLOGY

BiCMOS Tech.	Int. Gate	Conv. Stage
1.00 μm, 5 V	4.15 mW	14.20 mW
0.56 μm, 3 V	2.40 mW	7.15 mW
0.20 μm, 2 V	1.96 mW	4.74 mW

In the second option, only the last stage of the MLL was implemented in BiCMOS since all other levels have very small output capacitance. In the CML implementation, the basic cell used is the MCSL [11] where the conversion from CMOS to CML and the logic operation are done at the same time, with mixed CMOS and CML levels inputs. In [11] a 16-b adder implemented in MCSL was reported to achieve a speedup factor of about 5 over a similar adder implemented in CMOS. However, there was no mention of the input capacitance of the two implementations and the CML/CMOS conversion stage delay was not included in the total delay. In this work, it will be shown that, for the same input capacitance and including the CML/CMOS conversion stage delay, it is possible to achieve a speedup factor of 4 for six levels of logic at a total power of less than 35 mW. Table IV contains the average static power dissipation per an internal gate and the last conversion stage for the three BiCMOS technologies. The first gate in the MLL implementation using MCSL is shown in Fig. 11(a). An internal MCSL gate is shown in Fig. 11(b). The CML/CMOS conversion circuit at the end of the MLL, Fig. 11(c), is similar to the one in [12] except that a MCSL is used instead of the CML part.

The speedup factors of the CML and the CMOS + FSBiCMOS implementations over the CMOS for different load capacitances, as functions of the number of logic levels, are shown in Fig. 12. Fig. 13 shows the effect of scaling on the speedup factors for low C_L. There are two sets of results for the CML implementation: one where the end CML/CMOS conversion circuit had a fixed size FSBiCMOS buffer as a driver, and the other where the output driver was optimized as in Section II to show the effect of optimizing the output buffer. Also the areas of CMOS + BiCMOS and CML implementations (with optimized output), normalized to that of the CMOS implementation, are shown in Fig. 14.

From the above results, the following can be concluded.

1) For high C_L, the CMOS + BiCMOS implementation offers a good speed-up factor, especially for a low number of logic levels (Fig. 12), at a power level almost equal to that of pure CMOS. CML will offer a larger speedup, which increases as the number of logic levels increases (Fig. 12), at the expense of having a relatively large power dissipation.

2) For low C_L, CML still offers a good speedup factor, which again increases as the number of logic levels increases (Fig. 12).

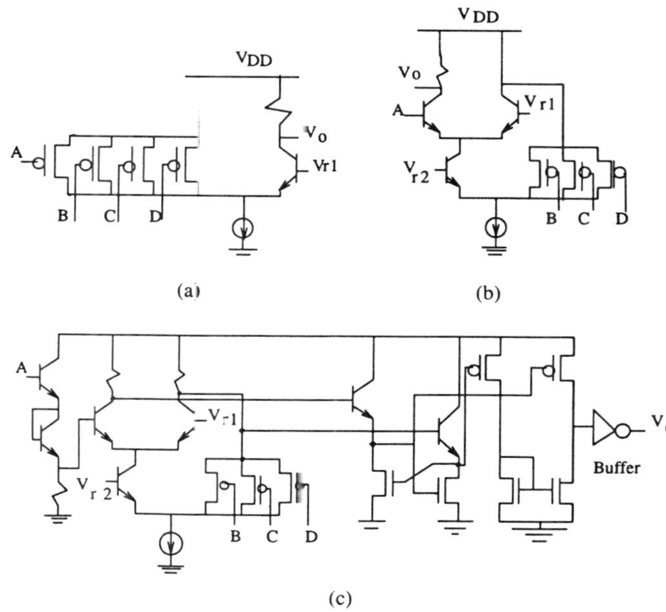

Fig. 11. The different MCSL building blocks used in the CML implementation of the MLL: (a) the first gate in the MLL, (b) an internal gate, and (c) the last gate in the MLL logic with CML/CMOS conversion. The buffer is selected according to the procedure outlined in Section II to give optimum performance.

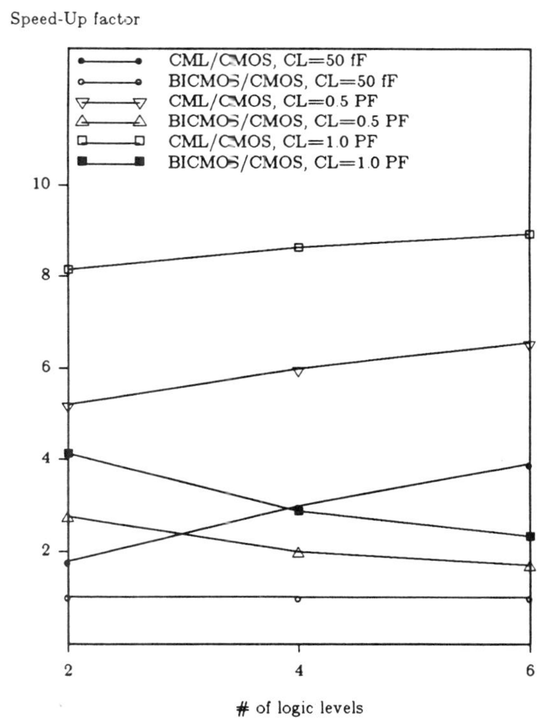

Fig. 12. The speedup factor of the CML and CMOS + BiCMOS implementations of the MLL over the CMOS counterpart versus the number of logic levels and for different C_L, at 5 V.

3) The optimization of the output buffer in the CML/CMOS conversion circuit increases the speedup significantly, especially for the scaled BiCMOS technologies (Fig. 13).

4) The CMOS + BiCMOS implementation does not offer any advantages as the technology scales (Fig. 13).

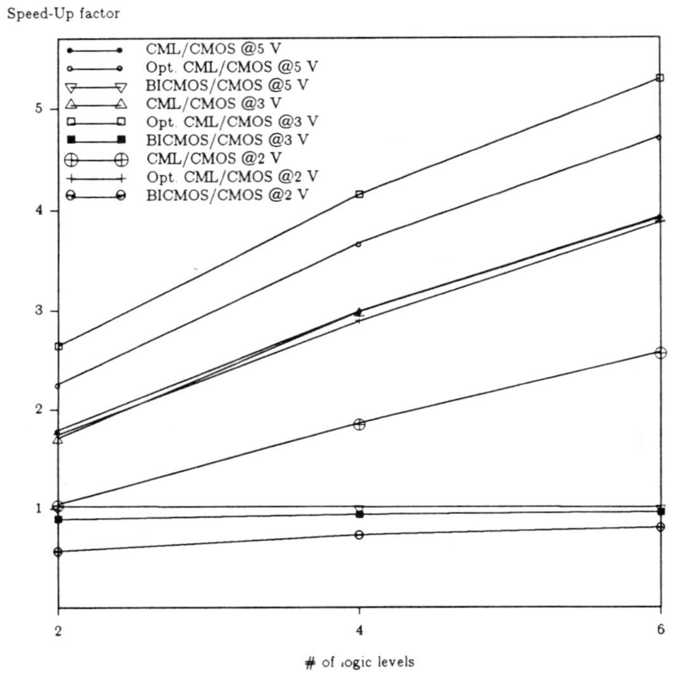

Fig. 13. The speedup factor of the CML and CMOS + BiCMOS implementations of the MLL versus the number of logic levels for the three technologies and at $C_L = 50$ fF.

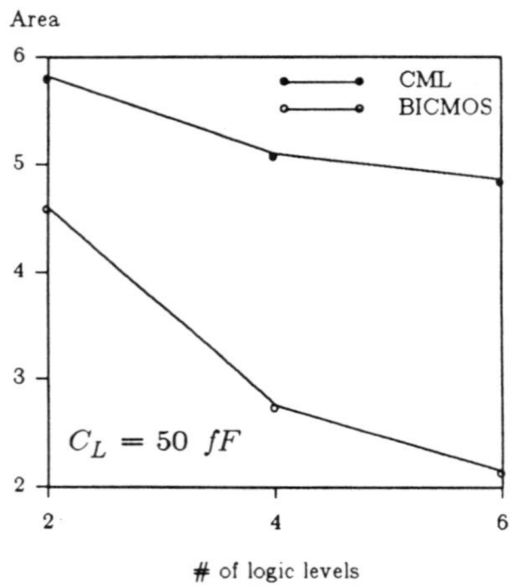

Fig. 14. The areas of the CML and CMOS + BiCMOS implementations of the MLL normalized to that of the CMOS implementation versus the number of logic levels and for $C_L = 50$ fF.

The CML, however, can offer the same speedup with much less power dissipation (Fig. 13 and Table IV).

5) The normalized areas of the CML and CMOS + BiCMOS implementations decrease with increased number of logic levels, and for low values of C_L the CML area is larger than that of the CMOS + BiCMOS (Fig. 14). The area ratio between CML and pure CMOS cannot decrease below a certain limit, and that is the ratio from a CML stage to a CMOS stage. However, this ratio decreases as C_{in} increases since the BJT area does not increase as C_{in} increases.

IV. Conclusion

In this work, several buffering circuits available in a BiCMOS technology have been compared and speedup contours were obtained. These contours show that BiCMOS is faster than CMOS at small values of input capacitance, and the speedup factor increases with the load capacitance until a certain limit is reached after which it starts to decrease. However, at such a point, the area of the CMOS buffer chain is much larger than that of the BiCMOS buffer. General and easy-to-automate design guidelines for buffer chain design in a BiCMOS environment were obtained. The possibility of multistage mixed CMOS/BiCMOS buffers was investigated. It was found that a two-stage BiCMOS buffer can achieve a lower delay than all other buffers available in a BiCMOS technology. The CMOS–BiCMOS buffer chain could achieve a speed closer to that of the BiCMOS–BiCMOS chain at a much smaller area. It was shown that BiCMOS complex logic gates can achieve significant speedup over CMOS for low input capacitances, low fan_{in}, and high fan_{out}. BiCMOS complex gates are more difficult to optimize due to associated limitations on device sizes, which become more severe as the gate complexity increases. Finally, the different options for implementing multilevel logic gates in a BiCMOS technology were compared. It was found that replacing the last stage of a CMOS MLL with a BiCMOS gate offers a good speedup factor for high C_L with no increase in power. However, for scaled-down BiCMOS technologies this is not the case, and to achieve any significant speedup, CML has to be used.

Acknowledgment

The authors thank Dr. S. Rofail for his valuable comments.

References

[1] M. I. Elmasry, A. Bellaouar, and S. H. Embabi, *BiCMOS IC Design*. Norwell, MA: Kluwer Academic, 1991.

[2] P. Raje *et al.*, "A new BiCMOS/CMOS gate comparison/design methodology and supply voltage scaling model," in *IEDM Tech. Dig.*, 1989, pp. 433–436.

[3] *HSPICE User's Manual*, Meta-Software, Inc., Campbell, CA, 1990.

[4] A. Bellaouar, S. Embabi, and M. I. Elmasry, "Scaling of BiCMOS digital circuit structures," in *IEDM Tech. Dig.*, 1989, pp. 437–440.

[5] A. Bellaouar, S. Embabi, and M. I. Elmasry, "Scaling of digital BiCMOS circuits," *IEEE J. Solid-State Circuits*, vol. 25, pp. 932–941, 1990.

[6] N. Hedenstierna and K. Jeppson, "CMOS circuit speed and buffer optimization," *IEEE Trans. Computer-Aided Design*, vol. CAD-6, pp. 270–281, 1987.

[7] G. Rosseel and R Dutton, "Delay analysis for BiCMOS drivers," in *BCTM Tech. Dig.*, 1988, pp. 220–222.

[8] S. H. K. Embabi, A. Bellaouar, and M. I. Elmasry, "Analysis and optimization of BiCMOS digital circuit structures," Univ. Waterloo, Waterloo, Ont., Canada, Int. Rep., 1990.

[9] T. Sakurai and A. Newton, "Delay analysis of series-connected MOSFET circuits," *IEEE J. Solid-State Circuits*, vol. 26, pp. 122–131, 1991.

[10] P. Raje *et al.*, "BiCMOS gate performance optimization using a unified delay model," in *Symp. VLSI Technology Tech. Dig.*, 1990, pp. 91–92.

[11] W. Heimsch *et al.*, "Merged CMOS/bipolar current switch logic (MCSL)," *IEEE J. Solid-State Circuits*, vol. 24, pp. 1307–1311, 1989.

[12] T. S. Yang *et al.*, "A 4-ns 4K × 1-bit two-port BiCMOS SRAM," *IEEE J. Solid-State Circuits*, vol. 23, pp. 1030–1040, 1988.

[13] S. H. K. Embabi, A. Bellaouar, and M. I. Elmasry, "Analysis and optimization of BiCMOS digital circuit structures," *IEEE J. Solid-State Circuits*, vol. 26, pp. 676–679, 1991.

Accurate Delay Models for Digital BiCMOS

Prasad A. Raje, *Member, IEEE,* Krishna C. Saraswat, *Fellow, IEEE,* and Kit M. Cham, *Senior Member, IEEE*

***Abstract*—A detailed transient analysis of the MOSFET–BJT combination prevalent in digital BiCMOS gates is presented. The analysis accounts for high-level injection leading to BJT β roll-off, base pushout leading to BJT f_T roll-off, short-channel behavior of MOS drain current, and parasitic capacitances at the base and output. Based on the transient analysis, a piecewise delay expression is derived that shows excellent agreement with measured gate delay and with SPICE simulated delay. The comparisons are made for a wide range of circuit parameters in the gate, namely, MOSFET/BJT size, load capacitance, and supply voltage for both 1- and 0.6-μm BiCMOS technologies. The model is used to optimally size gates, to predict the nonlinear shape of the delay versus load plot, and to propose circuit and device design guidelines to minimize the delay degradation at reduced supply.**

Nomenclature

β	BJT current gain.
B	Linear proportionality constant in τ dependence on i_c.
C_b	Sum of C_{bc} and C_p.
C_{bc}, C_{be}	Base–collector, base–emitter capacitance of $Q1$.
C_d	Diffusion capacitance of $Q1$.
C_{ib}	Coupling capacitance between input and $B1$.
C_{jp1}, C_{jn3}	Total drain junction capacitance of $M1$, $M3$, etc.
C_L	Total load capacitance at output ($C_{\text{load}} + C_{pe}$).
C_{load}	External load capacitance at output.
C_{ovn}, C_{ovp}	NMOS, PMOS gate–drain overlap capacitance per unit width.
C_{ox}	MOS gate oxide capacitance per unit area.
C_p	Parasitic capacitance at $Q1$ base.
C_{pe}	Parasitic capacitance at $Q1$ emitter (output node).
$E_{\text{crit}P}$	Critical field for velocity saturation in $M1$.
$E_{\text{tra}P}$	Mobility reduction factor for field along the channel of $M1$.
f_T	Unity gain frequency of $Q1$.
i_b, i_c	Transient base, collector current in $Q1$.
I_{cP}	Peak value of $Q1$ transient collector current.
I_k	Knee current of $Q1$.
$I_m[\ ]$	MOS drain current function.
I_{mos1}	Average $M1$ drain current from t_0 to t_1.
I_{mos}	Average $M1$ drain current from t_1 to t_d.
I_{tf}	SPICE parameter controlling τ dependence on current.
L_e	Emitter length of $Q1$.
μ_{0p}	Low-field $M1$ mobility.
μ_p	$M1$ mobility with field dependence included.
R_b, R_c, R_e	BJT base, collector, emitter resistances.
t_0	Time instant when input falls.
t_1	Time when V_{be} reaches V_{diode}.
t_2	Time when tangent to i_c at t_1 reaches I_{cP}.
t_d	Time when output reaches $V_{dd}/2$.
τ	Instantaneous $Q1$ transit time, function of i_c V_{bc}.
τ_f	Low current transit time of $Q1$ without base pushout.
V_b, V_e	$Q1$ base, emitter voltage.
V_{bc}	$Q1$ base–collector voltage.
V_{bcP}	V_{bc} value when $i_c = I_{cP}$.
$\overline{V}_{bc}$	Average V_{bc} from t_1 to t_d.
V_{be}	$Q1$ base–emitter voltage.
V_{dd}	Supply voltage.
V_{diode}	$Q1$ turn-on voltage.
$V_{dp\text{SAT}}$	$M1$ drain saturation voltage.
V_{ih}, V_{il}	Input high, low voltage.
$V_{\text{norm}P}$	Mobility reduction factor for field perpendicular to the channel.
V_{ol}	Initial output low voltage.
V_{sd}, V_{sg}	$M1$ source drain, source gate bias (positive quantities).
V_{tf}	SPICE parameter controlling τ dependence on V_{bc}.
V_{tp}	$M1$ threshold voltage.
X_{tf}	SPICE parameter controlling τ dependence on i_c, V_{bc}.
W_1, W_2, etc.	Gate width of MOSFET $M1$, $M2$, etc.

Manuscript received February 27, 1991; revised November 10, 1991. The review of this paper was arranged by Associate Editor K. Shenai.
P. A. Raje was with the Center for Integrated Systems, Stanford University, Stanford, CA 94305. He is now with Hewlett-Packard Laboratories, Palo Alto, CA 94304.
K. C. Saraswat is with the Center for Integrated Systems, Stanford University, Stanford, CA 94305.
K. M. Cham is with Circuit Technology R&D, Hewlett-Packard, Palo Alto, CA 94303.
IEEE Log Number 9107574.

I. Introduction

NUMEROUS digital logic gates in BiCMOS have been proposed. The BiCMOS gate is the most widely known and variations on this basic gate are BiNMOS,

Reprinted from *IEEE Trans. Electron Devices,* vol. 39, no. 6, pp. 1456–1464, June 1992.

CBiCMOS, MBiCMOS, etc. The gate delay of the BiCMOS gate has been extensively used as a performance measure of BiCMOS technology—to demonstrate the superior load driving capability of BiCMOS [1] and in analyzing the effects of device scaling [2]. Gate delay for an existing technology could be obtained simply by circuit simulation using simulators such as SPICE. However, physical insight into the crucial parameters affecting gate performance is difficult when relying purely on circuit simulation. Further, for an analysis of the effects of scaling, SPICE device models may not be readily available for each of the desired combinations of parameters. For these reasons it is necessary to develop a delay model for BiCMOS gates that accurately accounts for device phenomena during switching.

Previous delay models of the BiCMOS gate have neglected some of the crucial device phenomena that affect the gate delay. In [3] the contribution to the delay from the diffusion capacitance of the BJT is neglected. In [4] the BJT current gain is assumed to be constant thus neglecting high-level injection effects at the base–emitter junction. In [5] the effect of the knee current is greatly simplified by linearizing the current expression. Further, high-level injection effects at the base–collector junction that increase the effective transit time are neglected in [3]–[5]. In [6] the increase of transit time at high currents is included but its dependence on base–collector voltage is not considered. Also, the transient analysis neglects all capacitance at the BJT base node.

The analysis in this work generates a differential equation for the collector current in the BJT from which the output voltage waveform is derived. This single equation is valid in low-level injection and in high-level injection beyond the knee current, accounts for base pushout effects at the base–collector junction, includes the capacitances at the base node, and is applicable with both long- and short-channel MOSFET's in the BiCMOS gate. To obtain the transient waveforms, increasing degrees of complexity of τ are considered, corresponding to increasingly accurate modeling of the base pushout effect. For the most generalized base pushout modeling, a piecewise-linear delay expression for total gate delay is obtained that provides physical insight into BiCMOS gate design. The delay expression lends itself readily to a scaling analysis allowing the variation of any combination of device or technology parameters although the results of the scaling are discussed elsewhere [7], [8].

In Section II, differential equations for i_c and V_e are developed that describe the behavior of the BiCMOS gate during switching. In Section III the differential equations are solved for increasingly complex expressions for τ to give closed-form expressions for i_c and V_e. In Section IV, an expression for the BiCMOS gate delay based on a piecewise approach is developed that fully accounts for the complexity in the transit time. Finally, some applications and insights provided by the delay model are demonstrated in Section V.

II. Gate Transient Equations

A. Equivalent Circuit

The standard BiCMOS gate circuit is shown in Fig. 1(a). For an explanation of the static operation of the gate, see [4]. Fig. 1(b) shows the equivalent circuit of the part of the gate relevant for the output rising transition with the components explained in the Nomenclature. For brevity the model is described here only for the output rising transition of the conventional gate but a similar treatment applies for the output falling transition and for other gate flavors as well. Equations (A1) and (A2) in the Appendix detail the components of the parasitic capacitances C_p and C_{pe}. C_d represents the diffusion capacitance of the BJT due to the forward stored charge which is $=\tau i_c$. The bias dependence of the MOSFET current is emphasized by expressing the current I_m as a *function* of the voltages V_{sg} and V_{sd} (details in (A3)). To a first order, the base resistance R_b does not have a significant impact on the switching transient because it is effectively in series with the much larger MOSFET resistance. For proper operation of the BiCMOS gate, the collector resistance R_c must be low enough that the base–collector junction does not get forward-biased during the transient. For conventional technologies with buried layers this is easily achieved and then the effect of R_c on the transient is negligible. For this reason, R_c is not included in the equivalent circuit. SPICE simulations have confirmed the relative insensitivity of gate delay to R_b and R_c as have results of previous work [6].

B. $t = t_0$ to t_1

Fig. 2 shows schematically the important voltages and currents in the gate for an output rising transient. The input falls from its high to low value at $t = t_0$ and t_1 is the instant at which the base–emitter voltage of the BJT becomes equal to V_{diode}. Initially, the output is at its low value V_{ol}, the input is high, and $B1$ is at 0 V. The falling input edge at $t = t_0$ couples into a "bump" voltage at $B1$ and the base voltage just after the input transition is $V_b(t_0)$ (see (A4)). The output is bumped as a consequence of the input transition to $V_e(t_0)$ (see (A5)).

Just after $t = t_0$, $Q1$ is off, $i_b = i_c = 0$ and C_d is absent from the equivalent circuit of Fig. 1(b). The drain current of $M1$ charges $B1$ until the base–emitter voltage (V_{be}) reaches the forward diode drop at $t = t_1$ as shown in Fig. 2. During this time the output also rises, especially when C_L is small compared to C_{be}. Accounting for this rise is necessary for accuracy of the model at small values of load. Equations (A6) and (A7) give the voltages at the output and $B1$ at $t = t_1$. The MOSFET current (I_{mos1}) charging the base is expressed as the mean of the I_m value between $t = t_0$ and $t = t_1$. The differential equation describing the base voltage from $t = t_0$ to t_1 is

$$I_{\text{mos1}} = \left(C_b + \frac{C_{be}C_L}{C_{be} + C_L}\right)\frac{dV_b}{dt} \tag{1}$$

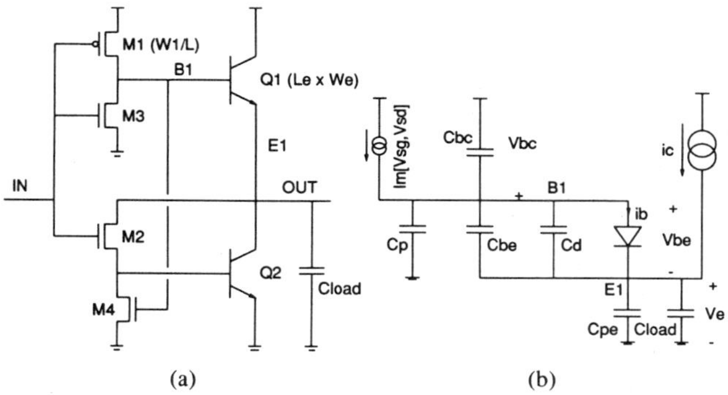

Fig. 1. (a) BiCMOS gate circuit. (b) Equivalent circuit for output rising transient.

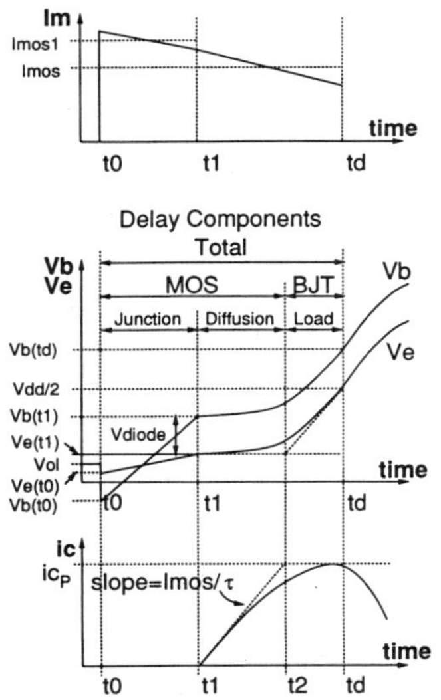

Fig. 2. Schematic output rising transient waveforms for the BiCMOS gate. I_m is the current in $M1$. V_b, V_e are the voltages at the base and emitter. i_c is the BJT collector current.

where

$$I_{\text{mos}1} = 0.5(I_m[V_{dd} - V_{il}, V_{dd} - V_b(t_0)] + I_m[V_{dd} - V_{il}, V_{dd} - V_b(t_1)]). \quad (2)$$

The value of t_1 is therefore

$$t_1 = t_0 + \left(C_b + \frac{C_{be}C_L}{C_{be} + C_L}\right)\frac{V_b(t_1) - V_b(t_0)}{I_{\text{mos}1}}. \quad (3)$$

C. $t = t_1$ to t_d

At $t = t_d$ the output reaches $V_{dd}/2$ (see Fig. 2). During the period from t_1 to t_d, V_{be} is assumed to remain constant at V_{diode} and therefore there is no current through C_{be} in this period. i_c begins to rise and the output rises as a result. Writing Kirchhofff's current law at the base $B1$ (Fig. 1)

$$I_{\text{mos}} = C_b\frac{dV_b}{dt} + \frac{d(\tau i_c)}{dt} + i_b. \quad (4)$$

The physical meaning of (4) is that the total current available from the MOSFET goes towards charging the base parasitic and base–collector capacitances (C_b), towards increasing the stored charge in the BJT base (τi_c) and towards the diode current in the BJT (i_b). It is only this last component that is "helpful" in providing collector current to charge the output. I_{mos} in (4) is the mean value of I_m between $t = t_1$ and $t = t_d$.

$$I_{\text{mos}} = 0.5(I_m[V_{dd} - V_{il}, V_{dd} - V_b(t_1)] + I_m[V_{dd} - V_{il}, V_{dd} - V_b(t_d)]). \quad (5)$$

i_b is written in terms of the collector current such that the expression is valid for low-level as well as high-level injection [9]. This makes the delay analysis elegant by avoiding an artificial separation into low-level injection and high-level injection and also provides accurate modeling of the regime intermediate between low- and high-level injection.

$$i_b = \frac{i_c}{\beta}\left(1 + \frac{i_c}{I_k}\right). \quad (6)$$

Now writing Kirchhoff's current law at the output node gives

$$\frac{d(\tau i_c)}{dt} + i_b + i_c = C_L\frac{dV_e}{dt}. \quad (7)$$

Adding (4) and (7) and using $dV_b/dt = dV_e/dt$, the differential equation for the output voltage V_e is

$$\frac{dV_e}{dt} = \frac{I_{\text{mos}} + i_c}{C_b + C_L}. \quad (8)$$

Using (8) and (6) in (4) gives

$$\frac{d(\tau i_c)}{dt} + i_c\left(\frac{1}{\beta} + \frac{C_b}{C_b + C_L}\right) + \frac{i_c^2}{\beta I_k} = I_{\text{mos}}\frac{C_L}{C_b + C_L}. \quad (9)$$

τ is kept inside the derivative to account for the base push-out effect where τ is not a constant during the transient. The linear term in i_c shows that C_b has a crucial effect on the transient, especially for low loads or large β. The quadratic term accounts for high-level injection in the form of β roll-off beyond the knee current. Equation (9) is thus the central equation describing the BiCMOS gate transient after the BJT is turned on until the time the output reaches half the supply voltage. After i_c is obtained from (9), V_e as a function of time is obtained from (8) and the gate delay is obtained by finding the time t_d at which V_e reaches $V_{dd}/2$.

III. Solving the i_c Differential Equation

The solution to the differential equation for the collector current (9) depends on the assumption made about the nature of the transit time τ. Fig. 3 shows SPICE-simulated i_c and V_e waveforms for a 1-μm BiCMOS technology with three increasingly accurate formulations for τ—constant, linear, and generalized. The transient response of the gate is clearly dependent on the nature of the transit

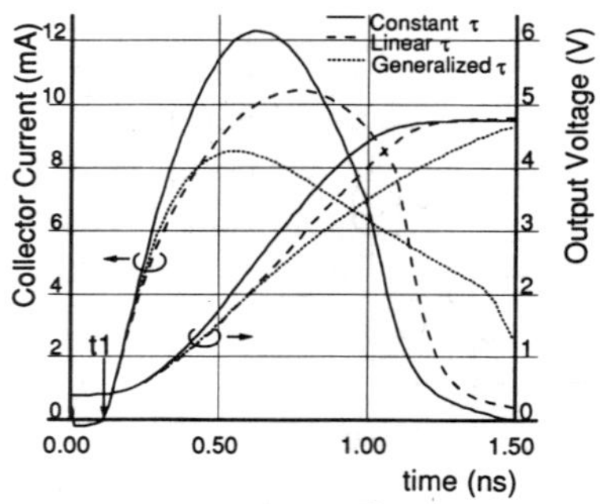

Fig. 3. SPICE simulated i_c and V_e waveforms showing the influence of different τ models.

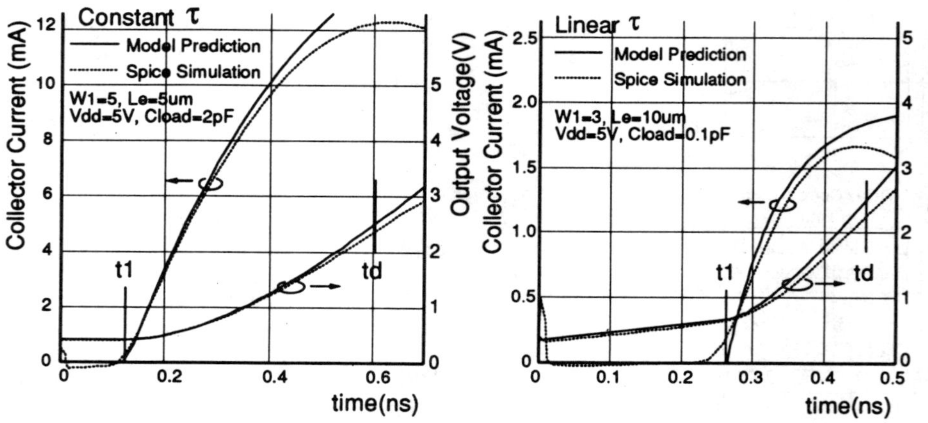

(a) (b)

Fig. 4. Model predicted and SPICE simulated i_c and V_e waveforms for (a) constant τ and (b) linear τ.

time. The meaning of the different τ formulations is detailed in the following subsections and the differential equation for i_c (9) is solved for each of the cases.

A. Constant τ

τ is assumed constant with no dependence on operating conditions. This is the simplest case and the only one considered in [3], [4]. With τ constant, the base pushout effect due to high-level injection at the base–collector junction is neglected. Equation (9) is rearranged as

$$-\tau_f \beta I_k \frac{di_c}{dt} = (i_c - \lambda_1)(i_c - \lambda_2) \quad (10)$$

where

$$\lambda_1, \lambda_2 = 0.5\left(-I_k\left(1 + \frac{\beta C_b}{C_b + C_L}\right) \pm \sqrt{\left(I_k\left(1 + \frac{\beta C_b}{C_b + C_L}\right)\right)^2 + 4 I_{mos} \beta I_k \frac{C_L}{C_b + C_L}}\right). \quad (11)$$

This equation can be solved in a closed form. Using the initial condition $i_c = 0$ at $t = t_1$

$$i_c = \frac{\lambda_1 + \lambda_2}{2} + \frac{\lambda_1 - \lambda_2}{2} \tanh\left(\frac{t + c}{\tau'}\right)$$
$$c = (\tau'/2) \log(-\lambda_1/\lambda_2) - t_1$$
$$\tau' = 2\tau_f \beta I_k/(\lambda_1 - \lambda_2). \quad (12)$$

Using (12) in (8) and the initial condition $V_e = V_e(t_1)$ at $t = t_1$ (A6), the output voltage waveform is

$$V_e = V_e(t_1) + \frac{1}{C_b + C_L}\left[\left(-\frac{I_k}{2}\left(1 + \frac{\beta C_b}{C_b + C_L}\right) + I_{mos}\right) \cdot (t - t_1) + \tau_f \beta I_k \log\left(\frac{\cosh\left(\frac{t + c}{\tau'}\right)}{\cosh\left(\frac{t_1 + c}{\tau'}\right)}\right)\right]. \quad (13)$$

Fig. 4(a) shows the transient i_c and V_e waveforms predicted by (12) and (13) and the SPICE-simulated waveforms for a constant τ. There is excellent agreement between the model and SPICE simulation. The simulated i_c eventually decreases with increasing V_e after $t = t_d$. This is because the rising base voltage leads to a decrease in the current provided by $M1$. In the model prediction, i_c rises asymptotically to a maximum value equal to λ_1. There is no subsequent reduction because I_{mos} in (9), while accounting for the bias dependence of I_m, is a constant. As seen from Fig. 4(a) this is an acceptable approximation because the MOSFET is predominantly in saturation during $t = t_1$ to t_d and I_m varies slowly.

The peak value of i_c for a constant τ is λ_1 (11). Only if $C_L >> \beta C_b$ does the peak value reduce to

$$\lambda_1 = 0.5(-I_k + \sqrt{I_k^2 + 4I_{mos}\beta I_k})$$

which is the expression provided in [6]. For the circuit used in Fig. 4(a) this approximation can be made only if $C_L >> 1.5$ pF. If the approximation is used when $C_L = 2$ pF, the resulting error in the peak current is 25%. This emphasizes the importance of accurately accounting for the capacitance at the base of the BJT.

B. Linear τ

τ is assumed to depend linearly on i_c as $\tau = \tau_f(1 + Bi_c/I_k)$ with the factor B representing the proportionality [9]. Here, the base widening is accounted for by the increased τ at high collector currents. τ is however assumed to have no dependence on the base–collector voltage. A second shortcoming is that the base pushout is tied to the value of the knee current (I_k) while it really depends on the collector doping which has no effect on I_k. Equation (9) is rearranged as

$$-\tau_f\beta I_k(1 + 2Bi_c/I_k)\frac{di_c}{dt} = (i_c - \lambda_1)(i_c - \lambda_2) \quad (14)$$

where λ_1, λ_2 are the same as before (11). The closed-form solution is

$$t = t_1 + \frac{\tau_f\beta I_k}{\lambda_1 - \lambda_2}\left[-\left(1 + \frac{2\lambda_1 B}{I_k}\right)\log\left(1 - \frac{i_c}{\lambda_1}\right) + \left(1 + \frac{2\lambda_2 B}{I_k}\right)\log\left(1 - \frac{i_c}{\lambda_2}\right)\right]. \quad (15)$$

If the factor B is set to 0 in the above equation, it reduces to the solution for the constant τ case (12) as expected. Using (15) the output voltage waveform (8) is

$$V_e = V_e(t_1) + \frac{\tau'/2}{C_b + C_L}\left[\left(1 + \frac{2\lambda_1 B}{I_k}\right)\cdot\left(-i_c - (\lambda_1 + I_{mos})\log\left(1 - \frac{i_c}{\lambda_1}\right)\right) + \left(1 + \frac{2\lambda_2 B}{I_k}\right)\cdot\left(i_c + (\lambda_2 + I_{mos})\log\left(1 - \frac{i_c}{\lambda_2}\right)\right)\right]. \quad (16)$$

Fig. 4(b) shows predicted waveforms from (15) and (16) compared with SPICE-simulated waveforms with τ expressed as a linear function. The circuit parameters for the gate in Fig. 4(b) are intentionally chosen different from those in the constant τ case and demonstrate the rise in V_e during the period from $t = t_0$ to t_1. Note again that the model predictions for i_c and V_e are most accurate until $t = t_d$. Obtaining full accuracy over the full range of the transient would require setting up a much more complicated (second-order, nonlinear, nonhomogeneous) differential equation if the physical effects of high-level injection and base pushout are to be properly modeled. It is considered a good compromise to write the simpler equation that can be solved in a closed form and that also predicts the transient accurately until the output reaches halfway to the supply.

C. Generalized τ

As evidenced by the shape of the f_T versus i_c curve for high i_c and its dependence on the base–collector bias, τ varies with both i_c and V_{bc}. A general expression for τ is written as

$$\tau = \tau_f\mathfrak{F}(i_c, V_{bc}). \quad (17)$$

For example, in SPICE τ is expressed as

$$\tau = \tau_f\left(1 + X_{tf}\exp(V_{bc}/1.44V_{tf})\left(\frac{i_c}{i_c + I_{tf}}\right)^2\right). \quad (18)$$

This expression is convenient to use because the SPICE parameters for a BJT are readily available for any technology and despite its empirical nature, the expression is capable of accurately predicting the shape of the f_T versus i_c curve.

In this case, (9) cannot in general be solved in a closed form. But the peak value I_{cP} reached by the collector current can be obtained by the following approach. Equation (9) is written out (with τ as in (17)) at the instant i_c reaches its peak I_{cP}. At this instant the di_c/dt is zero and dV_{bc}/dt is replaced using (8). The differential equation then reduces to the algebraic equation

$$I_{cP}\tau_f\frac{\partial\mathfrak{F}}{\partial V_{bc}}\frac{I_{mos} + I_{cP}}{C_b + C_L} + I_{cP}\left(\frac{1}{\beta} + \frac{C_b}{C_b + C_L}\right) + \frac{I_{cP}^2}{\beta I_k} = I_{mos}\frac{C_L}{C_b + C_L}. \quad (19)$$

When the SPICE expression (18) is used as the formulation for $\mathfrak{F}$, (19) becomes

$$\frac{\tau_f X_{tf}}{1.44V_{tf}}\exp(V_{bc}/1.44V_{tf})\frac{I_{cP}^3}{(I_{cP} + I_{tf})^2}\frac{I_{cP} + I_{mos}}{C_b + C_L} + I_{cP}\left(\frac{1}{\beta} + \frac{C_b}{C_b + C_L}\right) + \frac{I_{cP}^2}{\beta I_k} = I_{mos}\frac{C_L}{C_b + C_L} \quad (20)$$

where V_{bcP} is the value of V_{bc} when $i_c = I_{cP}$ and is estimated to be the average of V_{bc} between $t = t_1$ and t_d. The above fourth-order polynomial is then solved by Newton–Raphson iteration for I_{cP} using $I_{cP} = \lambda_1$ as the seed value. If X_{tf} is set to zero this is equivalent to treating τ as a

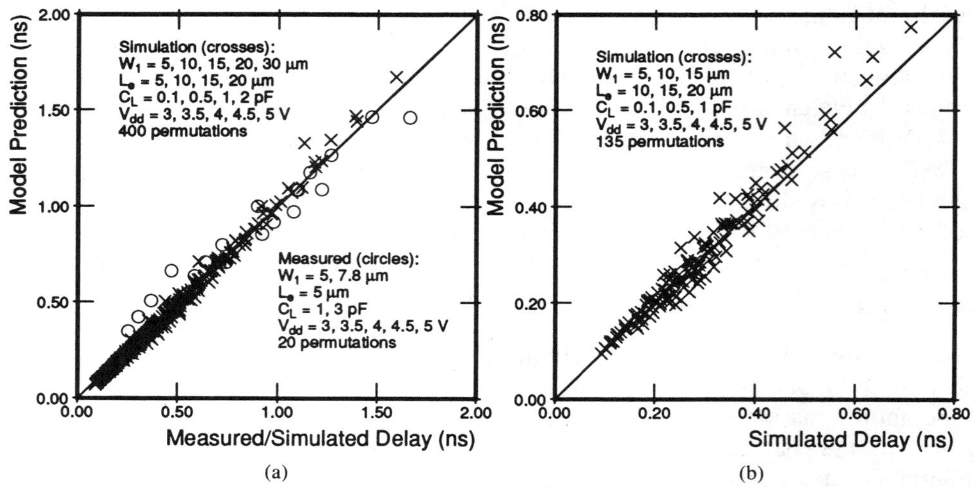

Fig. 5. Correlation plot of delay prediction from the piecewise model (a) with measured data (circles) and SPICE simulation (crosses) for the 1-μm BiCMOS technology and (b) with simulation (crosses) for the 0.6-μm technology.

constant (18). With $X_{tf} = 0$ in (20), the polynomial reduces to a quadratic and the peak collector current is exactly the same as the λ_1 value predicted for the constant τ case earlier. The I_{cP} value from (20) is used in a piecewise delay expression discussed below.

IV. Piecewise Delay Expression

A. Derivation

When the transit time is treated in its generalized form, a piecewise-linear approach is adopted wherein the total delay is modeled as the sum of three nonoverlapping time intervals illustrated in Fig. 2.

1) **Junction delay.** $t = t_0$ to t_1. This is the time required to charge the net capacitance at the base node just after the input falls until the BJT turns on (V_{be} reaches V_{diode}). The junction delay is the same as t_1–t_0 from (3).

2) **Diffusion delay.** $t = t_1$ to t_2. The diffusion delay accounts for the time required to provide the forward stored charge in the BJT. More precisely, this is the time that would be needed for the collector current to reach its peak value I_{cP} at its initial rate of increase. The tangent drawn at $t = t_1$ to the i_c curve reaches I_{cP} at $t = t_2$. I_{cP} is given by (19) and thus accounts for the base pushout by using the generalized form of the τ expression. In the current implementation of the model, I_{cP} is obtained from (20). The diffusion delay is

$$t_2 - t_1 = \frac{\tau_f I_{cP}}{I_{\text{mos}}}. \tag{21}$$

3) **Load delay.** $t = t_2$ to t_d. This is the time taken for the output to reach from its value $V_e(t_1)$ at $t = t_1$ to midway to the supply at a constant rate equal to the rate when the collector current is at its peak value. The implicit assumption here is that the tangent to the V_e waveform at $t = t_d$ intercepts the horizontal line drawn at $V_e(t_1)$ at $t = t_2$. This assumption can be shown to be exactly valid if the i_c waveform is a simple exponential function. In the general case the validity of this assumption is suggested by observation of simulated waveforms. The validity is *verified* by comparison of the predicted total delay with simulation or measurement. Using the rate of rise of V_e from (8), the load delay is

$$t_d - t_2 = \frac{(C_b + C_L)(V_{dd}/2 - V_e(t_1))}{I_{\text{mos}} + I_{cP}}. \tag{22}$$

The expression for the piecewise delay is therefore

$$t_d = (t_1 - t_0) + \frac{\tau_f I_{cP}}{I_{\text{mos}}} + \frac{(C_b + C_L)(V_{dd}/2 - V_e(t_1))}{I_{\text{mos}} + I_{cP}} \tag{23}$$

where $t_1 - t_0$ is obtained from (3), I_{cP} from (20), I_{mos} from (5), and $V_e(t_1)$ from (A6). The apparently artifical separation of diffusion delay and load delay into distinct intervals is justified by observing the last two terms in (4). Around $t = t_1$, i_b is small since i_c is just starting from zero, therefore a large portion the MOSFET current (I_{mos}) goes to supplying diffusion charge in the base $d(\tau i_c)/dt$. When i_c has increased towards its peak value, the rate of change of i_c is small and there is little increase in the diffusion charge. I_{mos} is now largely providing the base current (i_b) that translates into collector current that charges the load. The sum of the junction and diffusion delays is referred to as the "MOS delay" since this delay is the time it takes the MOSFET to provide the junction and diffusion charges to the BJT. Similarly, the load delay is called the "BJT delay" since during this period it is the BJT that is providing the current to charge up the output.

B. Verification

The model is implemented as a set of expressions in the Mathematica environment. The accuracy of the piecewise delay expression is verified by comparing the BiCMOS gate delay predicted by the model with measured delay from loaded ring oscillators and simulated delay from SPICE simulations. Fig. 5(a) is a scatter plot for a 1-μm BiCMOS technology with measured or simulated delay values along the X-axis and model predicted values along the Y-axis. The comparison for measured data is for two gates, each with 1- and 3-pF load and for supply values from 5 down to 3 V. The comparison with simulated data can be carried out for a much larger set of variations in the gate. Each of the circuit quantities MOS width (W_1), BJT emitter length (L_e), load (C_{load}), and supply (V_{dd}) are varied over the values shown in Fig. 5(a) and the gate delay compared for exhaustive permutations of these variables. The model is also verified for the 0.6-μm technology by comparing it with simulation using SPICE parameters measured from individual CMOS and n-p-n devices in the technology. Fig. 5(b) shows the simulated and predicted gate delays for all permutations of W_1, L_e, C_{load}, and V_{dd} indicated. As seen from these figures, there is excellent correlation between model prediction and measured or simulated gate delay data over a wide range of the parameters of the model.

V. Applications of the Model

In this section the delay model, especially the piecewise formulation, is used to gain insight into the effect of various circuit parameters on the gate delay.

A. Optimum Gate Sizing

Unlike conventional CMOS gates where a single variable, the MOSFET width, defines the gate completely, there are at least two variables that define a BiCMOS gate, namely W_1 and L_e (Fig. 1). The assumption is that the sizing of the remaining MOSFET's follows from selection of W_1, and the BJT's are identical. The design space can then be represented in a two-dimensional plane with axes W_1 and L_e, called the BiCMOS Sizing Plane (BSP) [10]. The piecewise delay model offers insights into the choice of a point in the BSP. For a given W_1, increasing the L_e has opposite effects on the components of the total delay. With larger L_e, high-level injection at the base–emitter junction is reduced due to a larger I_k, and base pushout is reduced due to a higher I_{tf}. This increases I_{cP} which *decreases* the BJT delay. The MOS delay *increases* with L_e because the BJT junction capacitance increases with L_e and the diffusion charge is proportional to I_{cP}. There is therefore an optimum $L_e = L_{e\text{-opt}}$ at which the total delay is a minimum. The value of $L_{e\text{-opt}}$ depends on W_1 and is larger for larger W_1. Fig. 6 shows the $L_{e\text{-opt}}$ values as a function of W_1 in the BSP in a 1-μm BiCMOS technology obtained using the piecewise delay model (23). The $L_{e\text{-opt}}$ curve moves towards larger values of L_e for larger load. This is because the BJT delay is larger at larger load. For smaller supply, the $L_{e\text{-opt}}$ curve moves towards smaller L_e because the MOS delay is larger as seen below. Also shown is the delay along the $L_{e\text{-opt}}$ curves. This represents the lowest achievable delay of the BiCMOS gate as a function of W_1, for a given C_{load} and V_{dd}.

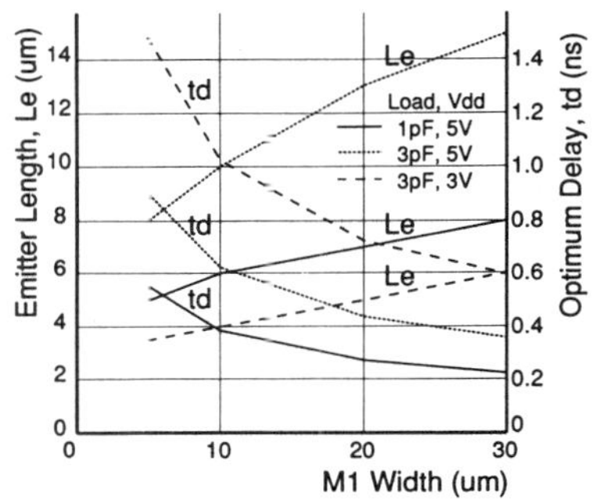

Fig. 6. L_e as a function of W_1 for which delay is a minimum. The minimum delay is also shown. Three sets of data for different load and supply conditions.

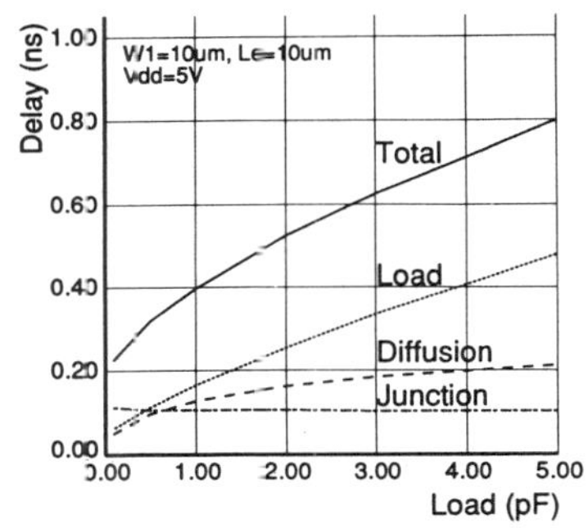

Fig. 7. The components of gate delay as a function of load. Delay versus load is not a straight line.

B. Delay versus Load

Fig. 7 plots the predicted delay versus load curve for a typical gate in 1-μm BiCMOS. The piecewise delay allows one to observe the three components of delay as a function of load. It is seen that the delay versus load curve is not a straight line with constant slope as is commonly assumed [11]. From (23) this suggests that I_{cP} is not constant with changing load. This is confirmed by the shape of the diffusion component of the delay in Fig. 7. The diffusion delay is proportional to I_{cP} and it is seen to rise with increasing load. The exact dependence of I_{cP} on C_{load} is contained in (19). The reason that I_{cP} increases with load is as follows. For larger load, the output rising transient is slower which means that the base voltage rises more slowly. Therefore, a smaller component of I_{mos} is "wasted" in charging the base–collector capacitance (the first term in (4)). Further, the slower reduction in base–collector voltage leads to a slower increase in the transit time and the second term in (4) is also smaller. This allows a larger fraction of I_{mos} towards base–emitter diode current (i_b). The larger i_b leads to a larger i_c (6) and therefore I_{cP} is larger for larger load.

C. Delay versus Supply

It has been observed [1] that BiCMOS gate delay degrades dramatically with supply as compared with CMOS. The quantitative severity of the delay degradation strongly depends on the circuit conditions [12]. Fig. 8(a) shows the delay predicted by the piecewise delay model for gates

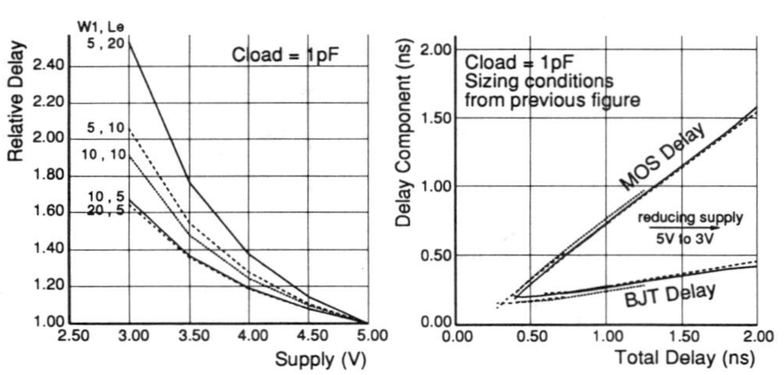

(a) (b)

Fig. 8. (a) Delay relative to 5-V value versus supply for various sizing conditions. (b) MOS and BJT components of delay as supply reduces.

with different W_1 and L_e values. The delay for each gate is plotted normalized to the corresponding value at 5 V. It is clearly seen that the relative delay increase in going from 5 to 3 V supply is minimized for gates with large W_1 and small L_e. The explanation for this is suggested by Fig. 8(b). The piecewise delay expression allows one to plot the MOS and BJT delay components separately versus the total delay for each of the gates of Fig. 8(a). V_{dd} implicitly goes from 5 to 3 V from left to right in Fig. 8(b). As V_{dd} decreases and the total delay increases, the BJT delay rises slowly but the MOS delay becomes the dominant component of the delay for each of the gates. Thus the delay increase of BiCMOS gates at reduced supply is predominantly due to increase in the MOS delay. In the circuit domain, the delay degradation is minimized by designing with wide MOSFET's and small BJT's consistent with other constraints. In the device domain, the MOS delay (23) is minimized by scaling down the MOSFET gate length to increase I_{mos}, reducing the BJT junction capacitance and junction areas to reduce C_b and reducing the transit time by scaling down the base width.

VI. Conclusion

Accurate transient analysis of the BiCMOS gate has been carried out by accounting for the important physical phenomena during gate switching—namely, β roll-off, base pushout, MOSFET short-channel effects, and base and output parasitics. A piecewise delay expression results from simplification of the transient analysis and is verified for a variety of circuit conditions for both 1- and 0.6-μm BiCMOS technologies. The delay model provides insight into the behavior of the gate and a simple, physically intuitive expression for the gate delay. Further, the model predicts the nonlinear shape of the delay versus load curve and a means of selecting optimum device sizes in a BiCMOS gate. The model shows that a large MOS delay is responsible for delay degradation at reduced supply values. Finally, the model provides device and circuit guidelines to reduce the delay degradation.

Appendix

A. Parasitics

The parasitic capacitance at the base $B1$ of $Q1$ comes from the gate–drain overlap and drain junction capacitance of $M1$ and $M3$ and the gate capacitance of $M4$. Thus

$$C_p = C_{\text{ovp}} W_1 + C_{\text{ovn}} W_3 + C_{jp1} + C_{jn3} + 2 C_{\text{ovn}} W_4. \tag{A1}$$

The parasitic capacitance at the emitter $E1$ of $Q1$ is due to the gate–drain overlap and drain junction capacitance of $M2$, the collector–base and collector–substrate capacitance of $Q2$.

$$C_{pe} = C_{\text{ovn}} W_2 + C_{jn2} + C_{bc2} + C_{cs2}. \tag{A2}$$

The junction capacitance terms in the above equations have implicit voltage dependencies. The bias at which the capacitance is evaluated is the mean of the voltage value in the relevant time period.

B. MOSFET Current

The drain current of MOSFET $M1$ accounts for mobility reduction due to vertical and transverse electric fields (parameters $V_{\text{norm}P}$ and $E_{\text{tra}P}$) and saturation voltage reduction due to velocity saturation (parameter $E_{\text{crit}P}$). I_m is expressed as a function of the source–gate (V_{sg}) and source–drain voltage (V_{sd})

$$\begin{aligned} I_m[V_{sg}, V_{sd}] &= \mu_p C_{\text{ox}}(W/L)(V_{sg} - V_{tp} \\ &\quad - V_{sd\text{-eff}}/2) V_{sd\text{-eff}} \\ \mu_p &= \mu_{0p} \frac{1}{1 + (V_{sg} - V_{tp})/V_{\text{norm}P}} \\ &\quad \cdot \frac{1}{1 + V_{dp\text{SAT}}/(E_{\text{tra}P} L)} \\ V_{sd\text{-eff}} &= \min\,[V_{sd}, V_{dp\,\text{SAT}}] \\ V_{dp\,\text{SAT}} &= (V_{sg} - V_{tp}) + E_{\text{crit}P} L \\ &\quad - \sqrt{(V_{sg} - V_{tp})^2 + (E_{\text{crit}P} L)^2}. \end{aligned} \tag{A3}$$

C. V_b and V_e at t_0 and t_1

When there is an abrupt fall in the input, the capacitance ($C_{ib} = C_{\text{ovp}} W_1 + C_{\text{ovn}} W_3 + 0.5 C_{\text{ox}} W_3 L$) between the input and base ($B1$) causes $B1$ to fall. The net capacitance at $B1$ is $C_b (= C_p + C_{bc})$ is parallel with the series combination of C_{be} and C_L. Therefore

$$V_b(t_0) = (V_{il} - V_{ih}) \frac{C_{ib}}{C_b + \dfrac{C_{be} C_L}{C_{be} + C_L}}. \tag{A4}$$

As a result of the "bump" at B, the output gets bumped to

$$V_e(t_0) = V_{ol} + V_b(t_0) \frac{C_{be}}{C_{be} + C_L}. \tag{A5}$$

When the input transition is not abrupt, $V_b(t_0)$ and $V_e(t_0)$ are simply the initial values of 0 and V_{ol}.

The change in V_e from t_0 to t_1 is related to the change in V_{be} during the same period by the ratio of the impedance at the output to that across the base–emitter junction.

At the instant $t = t_1$, the voltage across the base–emitter capacitor is V_{diode}. The voltage at the output is therefore

$$V_e(t_1) = V_e(t_0) + (V_{\text{diode}} - (V_b(t_0) - V_e(t_0))) \frac{C_{be}}{C_L}. \quad \text{(A6)}$$

The base voltage at t_1 is simply a diode drop above the emitter

$$V_b(t_1) = V_e(t_1) + V_{\text{diode}}. \quad \text{(A7)}$$

REFERENCES

[1] H. Momose, K. M. Cham, C. I. Drowley, H. R. Grinolds, and H. S. Fu, "0.5 micron BiCMOS technology," in *IEDM Tech. Dig.*, 1987, pp. 838-840.

[2] A. Bellaouar, S. Embabi, and M. Elmasry, "Scaling of digital Bi-CMOS circuits," *IEEE J. Solid-State Circuits*, vol. 25, no. 4, pp. 932-941, Aug. 1990.

[3] H. J. De Los Santos and B. Hoefflinger, "Optimization and scaling of CMOS-bipolar drivers for VLSI interconnects," *IEEE Trans. Electron Devices*, vol. ED-33, no. 11, pp. 1722-1730, Nov. 1986.

[4] E. W. Greeneich and K. L. McLaughlin, "Analysis and characterization of BiCMOS for high-speed digital logic," *IEEE J. Solid-State Circuits*, vol. 23, no. 2, pp. 558-565, Apr. 1988.

[5] P. Heedley and R. Jaeger, "An analytical model for BiCMOS logic transient response allowing parameter variations," in *Proc. 1989 Custom Integrated Circuits Conf.*, 1989, pp. 13.4.1-13.4.4.

[6] G. Rosseel and R. Dutton "Influence of device parameters on the switching speed of BiCMOS buffers," *IEEE J. Solid-State Circuits*, vol. 24, no. 1, pp. 90-99, Feb. 1989.

[7] P. A. Raje, K. C. Saraswat, and K. M. Cham, "Performance-driven scaling of BiCMOS technology," *IEEE Trans. Electron Devices*, vol. 39, no. 3, pp. 685-694, Mar. 1992.

[8] P. Raje, "Analysis, design and scaling of BiCMOS for digital logic," Ph.D. dissertation, Stanford University, Stanford, CA, 1991.

[9] I. Getreu, "Modeling the bipolar transistor," Tech. Rep., Tektronix Inc., 1979.

[10] P. A. Raje, K. C. Saraswat, and K. M. Cham, "A new methodology for design of BiCMOS gates and comparison with CMOS," *IEEE Trans. Electron Devices*, vol. 39, no. 2, pp. 339-347, Feb. 1992.

[11] T. Ikeda, A. Watanabe, Y. Nishio, I. Masuda, N. Tamba, M. Odaka, and K. Ogiue, "High-speed BiCMOS technology with a buried twin well structure," *IEEE Trans. Electron Devices*, vol. ED-34, no. 6, pp. 1304-1310, June 1987.

[12] P. Raje, K. Cham, and K. Saraswat, "BiCMOS gate performance optimization using a unified delay model," in *Proc. Symp. VLSI Tech.*, 1990, pp. 91-92.

Delay Analysis of BiNMOS Driver Including High Current Transients

Jiann-Shiun Yuan, *Member, IEEE*

***Abstract*—The BiNMOS gate delay analysis including high current transients has been developed. The modeling equations account for high electric field effect in the nMOS transistor and emitter crowding, base pushout, and base conductivity modulation in the bipolar transistor. In examining the switching transient of a BiNMOS driver, base pushout mechanism exhibits the detrimental effect on the gate propagation delay. The present circuit modeling methodology provides a fast turn-around design evaluation of sensitivity of process and device parameters into circuit performance. Computer simulation of a BiNMOS driver using the present analysis is compared with PISCES device simulation in support of physical reasoning.**

I. Introduction

The BiCMOS circuits have gained lots of interest recently because they offer high-speed capability of bipolar circuits and high density and low power dissipation of CMOS circuits. The combinational bipolar and CMOS process technologies allow chip designers the opportunity of mixing ECL and CMOS logic with new evolving digital and linear circuits together to greatly enhance the performance of future integrated circuits.

Steady-state and transient operation for both bipolar and CMOS logics have been well understood over the past years [1]–[3]. The characterization and understanding of the gate logic operation provide proper engineering trade-offs at the process, device, and circuit design levels. Recently, the analysis of BiCMOS digital logic has been explored [4]–[6]. The BiCMOS driver improves the delay time of the common CMOS buffer by much less than a factor of β_0, the bipolar peak current gain, due to high injection effect [5], [6]. In addition to high injection and quasi-saturation resulting from a high collector resistance, the bipolar transistor experiences base pushout, emitter crowding, and sidewall injection at high current transients. These high current effects are not properly accounted for in the typical circuit simulator such as SPICE2. High current effects of a BJT can be analyzed in device simulation, but with tremendous amount of CPU time. Furthermore, constraint of the number of mesh points in numerical simulator prevents the use of a complicate BiCMOS circuit from device simulation environment.

Manuscript received November 8, 1990; revised February 28, 1991. The review of this paper was arranged by Associate Editor M. D. Feuer.
The authors is with the Department of Electrical Engineering, University of Central Florida, Orlando, FL 32816.
IEEE Log Number 9105342.

In this paper, high-current effects such as emitter crowding, base pushout, base conductivity modulation, and sidewall injection are included in the derivation of BiNMOS gate delay. The semi-analytical equations provide physical insight into the relative importance of process, device, and circuit parameters on BiNMOS switching transient. The results obtained describe the optimal gate delay in terms of device doping profiles, such as the epitaxial collector doping density of the advanced bipolar transistors. The present circuit modeling methodology has certain degrees of mixed-mode device and circuit simulators capability and offers a fast turn-around design time.

II. Transient Analysis

The transient behavior of the bipolar devices in the BiNMOS circuit is different from that in standard bipolar circuits [7], [8]. In this section, the gate delay of the BiNMOS inverter is analyzed. The modeling methodology can also be used for BiPMOS logic in BiCMOS circuit.

In order to take into account high electric field effect on the small-geometry MOS device, the analytical model proposed in [9] was adopted. The drain current under triode and saturation regions of the BiNMOS circuit in Fig. 1 can be expressed as

$$I_D = \frac{W_{\text{eff}}\mu_{\text{eff}}C_{\text{ox}}(V_{GS} - V_T - V_D/2)(V_D - V_{BE})}{L_{\text{eff}}[1 + V_D/(E_c L_{\text{eff}})]},$$
$$\text{for } V_D - V_{BE} \le V_{D\text{sat}} \qquad (1)$$

$$I_D = \frac{W_{\text{eff}}\mu_{\text{eff}}C_{\text{ox}}(V_{GS} - V_T - V_{D\text{sat}})E_c}{2},$$
$$\text{for } V_D - V_{BE} > V_{D\text{sat}} \qquad (2)$$

where

$$V_{D\text{sat}} = \frac{E_c L_{\text{eff}}(V_{GS} - V_T)}{(E_c L_{\text{eff}} + V_{GS} - V_T)} \qquad (3)$$

L_{eff} is the effective channel length, W_{eff} is the effective channel width, μ_{eff} is the effective electron mobility, C_{ox} is the oxide capacitance per unit area, V_D is the drain voltage, V_{GS} is the gate–source voltage, V_T is the threshold voltage, and E_c is the critical electric field at which the carrier velocity is saturated ($E_c = 2v_{\text{sat}}/\mu_{\text{eff}}$).

Since the source–body voltage V_{SB} of the nMOSFET affects its threshold voltage, the nonlinear V_T is expressed

Reprinted from *IEEE Trans. Electron Devices*, vol. 39, no. 3, pp. 587–592, March 1992.

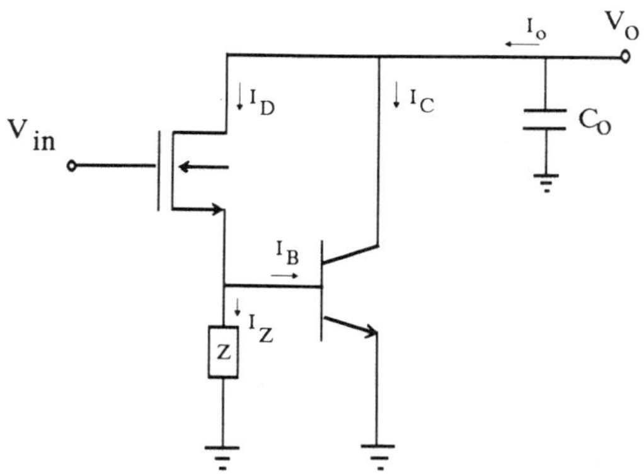

Fig. 1. A BiNMOS driver for transient analysis.

as

$$V_T = V_{T0} + \gamma[(2\phi_f + V_{BE} - V_{B\text{nMOS}})^{1/2} - (2\phi_f)^{1/2}] \quad (4)$$

where V_{T0} is the threshold voltage at $V_{SB} = 0$ V, ϕ_f is the Fermi potential, V_{BE} is the base–emitter applied voltage, $V_{B\text{nMOS}}$ is the body voltage of the nMOS transistor, and γ is the body effect factor.

For the BiNMOS circuit shown in Fig. 1, the drain current supplies the base of the BJT and the parasitic impedance as

$$I_D = I_B + I_Z \quad (5)$$

where I_B is the base current and I_Z is the parasitic impedance current. The impedance current I_Z is equal to V_{BE}/R_P if a resistor R_P is used or $C_P\, dV_{BE}/dt$ if a capacitor C_P is applied.

Simulations and closed-form solutions show that the n-p-n bipolar transistors of the BiCMOS driver have to be operated in high level injection and that the base–emitter voltage is beyond 1.0 V during turn-on transients [5]–[8]. The bipolar transistor operated in this level of injection experiences physical effects of base conductivity modulation, base pushout, and emitter current crowding [10]–[12].

For an n^+-p-n-n^+ advanced bipolar transistor, the collector current I_C is a function of base conductivity modulation, base pushout, and emitter crowding effects and that the base current is influenced by emitter crowding and emitter–base sidewall injection [13], [14]. The sidewall injection or the peripheral base current increases as the emitter width shrinks which results in a reduction of current gain. The base current accounting for the peripheral current and emitter crowding can be modeled as [13]

$$I_B \approx \left(f_{CR} + \frac{P_E X_{JE}}{A_E}\right)\left[\frac{I_S}{\beta_0}\exp\left(\frac{qV_{BE'}}{kT}\right) + C_2 I_S \exp\left(\frac{qV_{BE'}}{kT}\right)\right] \quad (6)$$

where P_E is the emitter perimeter, X_{JE} is the emitter junction depth, A_E is the emitter area ($W_E \times L_E$), $V_{BE'}$ is the base–emitter junction voltage ($V_{BE'} = V_{BE} - I_B R_B - I_E R_E$), C_2 is the base leakage current coefficient, I_S is the saturation current, q is the electron charge, k is Boltzmann's constant, T is the absolute temperature, and f_{CR} is the emitter crowding factor which is defined as the ratio of the emitter current with current crowding to the emitter current without current crowding. Bias-dependent f_{CR} can be calculated based on a partitioned resistance network in [14].

Solving (1)–(6) together numerically gives the base–emitter junction voltage $V_{BE'}$ for determining the collector current. The collector current including high current effects and the collector charging current $C_{JC}\, dV_{BC'}/dT$ is

$$I_C \approx f_{CR} f_{PO} I_S \exp(qV_{BE'}/kT)/q_b - C_{JC}\, d(V_{BE'} - V_o)/dt \quad (7)$$

where

$$q_b = \frac{(\Delta Q_B + Q_{B0})(\delta Q_B + Q_{B0})}{Q_{B0}^2} \quad (8)$$

Q_{B0} is the base charge at $V_{BE'} = 0$ V, ΔQ_B is the incremental base charge resulting from base-width modulation, δQ_B is the excess charge due to base conductivity modulation, C_{JC} is the base–collector junction capacitance, and f_{PO} is the base pushout factor. f_{PO} accounts for base widening and quasi-saturation effects at the high collector current densities and is defined as $W_B/(W_B + \Delta W_B)$, where W_B is the quasi-neutral base width without base widening and ΔW_B is the incremental base width due to base widening [14]–[16]

$$\begin{aligned}\Delta W_B &= 0, && \text{for } J_C \le J_0\\ &= W_C[(J_0 - J_1)/(J_C - J_1)]^{1/2}, && \text{for } J_C > J_0 \text{ and } J_2 > J_C > J_1\\ &= W_C(1 - J_2/J_C), && \text{for } J_C > J_2.\end{aligned} \quad (9)$$

In (9), W_C is the epitaxial collector width, J_C is the collector current density, J_0 is the onset collector current density for base pushout ($J_0 = qv_s[N_c + 2\,\epsilon_s V_{CB}/(qW_C^2)]$), $J_1 = qv_s N_C$, and $J_2 = q\mu_n N_C V_{CB}/W_C$, where v_s is the saturation velocity, ϵ_s is silicon permittivity, N_C is the epitaxial collector doping density, and V_{CB} is the collector–base voltage.

Bias-dependent ΔQ_B due to base-width modulation can be obtained from the voltage-dependent junction capacitances [17], [18]

$$\Delta Q_B = \int C_{JE}\, dV + \int C_{JC}\, dV \quad (10)$$

where C_{JE} is the emitter–base junction capacitance.

The excess base charge δQ_B resulting from conductivity modulation or high injection in the quasi-neutral base is [19]

$$\delta Q_{B0} = Q_{B0}\frac{(2 + C_1 + n_0)C_2 n_0}{(2 + C_1)C_2/C_3 + n_0} \quad (11)$$

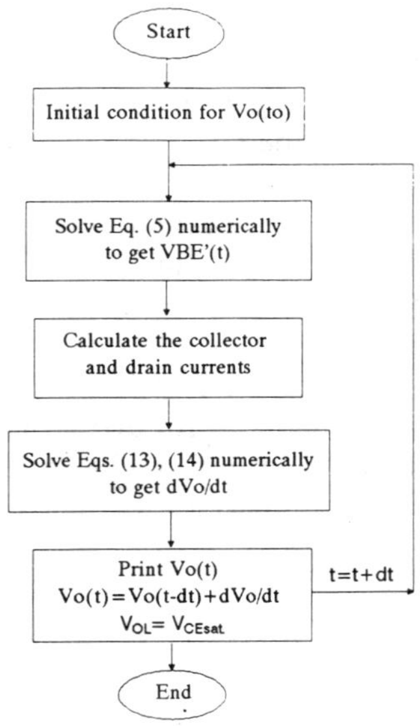

Fig. 2. Flow chart demonstrating the methodology for determining the time-dependent output voltage V_o.

where $C_1 = 2\,[\exp(\eta) - 1]/[\eta \exp(\eta)]$, $C_2 = 1/C_1$, $C_3 = \exp[(\eta - 1)\exp(\eta) + 1]/[\exp(\eta) - 1]^2$, η is the effective base doping gradient, and n_0 is the electron concentration normalized by base doping at the base edge of the emitter–base space-charge region [19].

The discharging current of the interconnect capacitance supports the drain current and the collector current of the BiNMOS circuit as follows:

$$C_o\, dV_o/dt = I_D + I_C \tag{12}$$

where C_o is the output capacitance.

Inserting (1), (2), and (7) into (12) gives the time-dependent differential equations below

$$\frac{dV_o}{dt} \approx C_o \left\{ \left[\frac{W_{\text{eff}}\mu_{\text{eff}} C_{\text{ox}} (V_{GS} - V_T - V_o/2)(V_o - V_{BE})}{L_{\text{eff}}[1 + V_o/(E_c L_{\text{eff}})]} \right] + f_{CR} f_{PO} I_S \exp(qV_{BE'}/kT)/Q_B - C_{JC} \frac{d(V_{BE'} - V_o)}{dt} \right\}, \quad \text{for } V_o - V_{BE} \le V_{D\text{sat}} \tag{13}$$

$$\frac{dV_o}{dt} \approx C_o \left\{ \left[\frac{W_{\text{eff}}\mu_{\text{eff}} C_{\text{ox}} (V_{GS} - V_T - V_{D\text{sat}}) E_c}{2} \right] + f_{CR} f_{PO} I_S \exp(qV_{BE'}/kT)/Q_B - C_{JC} \frac{d(V_{BE'} - V_o)}{dt} \right\}, \quad \text{for } V_o - V_{BE} > V_{D\text{sat}}. \tag{14}$$

The differential equations in (13), (14) can be solved numerically using the modified Euler method [20], [21]. The output voltage is then obtained by using the methodology in Fig. 2. Note that the minimum output voltage V_{OL} is limited by $V_{CE\text{sat}}$ when the BJT is in the saturation region.

III. Results and Discussions

To demonstrate the model utility, the semi-analytical equations developed in Section II are simulated. Applying a rising ramp voltage at the gate of the BiNMOS driver in Fig. 1, the transient response of the NMOS and bipolar transistors in examined. The nMOSFET switches on very rapidly due to its light load, but the n-p-n BJT is not conducting until its base–emitter junction reaches about 0.7 V. Consequently, there is an initial delay and collector–base charging current before V_o begins to fall. During the early part of the fall response, nMOSFET operates in the saturation region. Beyond t_{sat} at which $V_{GS}(t_{\text{sat}}) - V_T = V_o(t_{\text{sat}})$, the nMOSFET operates in the triode region. When the collector current of the BJT is sufficiently large, the BiNMOS driver starts to benefit its high current driving capability.

Since the commonly available SPICE bipolar transistor models are inadequate for simulating high current effects critical to the operation of BiCMOS circuits, the simulated output waveform is compared with PISCES simulation to verify the model utility and accuracy. Fig. 3(a) shows the transient voltage waveforms from the present model (solid dots) and PISCES results (solid line) [8]. Good agreement between their predictions is obtained. The corresponding transient collector current is plotted in Fig. 3(b). Only reasonable agreement of transient current waveform predicted from the model is shown. This is because the capacitances of C_{BE}, C_{GD}, C_{GS}, and two-dimensional collector spreading are neglected in the derivation of the gate delay as well as physical parameters of carrier mobilities; lifetimes, Auger coefficients, and bandgap narrowing constants in PISCES simulation are not tediously adjusted. Nevertheless, the model simulation provides sufficient physical insight and accuracy into analyzing the transient waveforms in BiNMOS circuit. To further demonstrate the utility of the analytical equations, base pushout factor is plotted in Fig. 3(c). In Fig. 3(c) f_{PO} drops sharply to a small value once the BJT turns on, due to a very high forward-biased base–emitter junction and a thick bipolar epilayer.

Examining the input and output waveforms, the pull-down delay time τ can be measured from NMOS transistor turning on ($V_{GS} = V_T$) to the midpoint of the output voltage. Since the design of optimal BiCMOS buffer depends on the intended capacitive load, the pull-down time versus the output capacitance is plotted in Fig. 4. In Fig. 4, τ increases with the output capacitance, as expected.

The high injection in the quasi-neutral base increases the base Gummel number and decreases the collector current. The high injection effect degrades the delay time of BiCMOS driver and was reported previously [4], [5]. To probe the additional high current effects on gate propagation delay, sensitivity of emitter crowding, base pushout, and sidewall injection on the output waveform is shown in Fig. 5. In Fig. 5 the solid line represents the present model, the dashed line represents the present model without sidewall injection effect, the dotted line represents the model without base pushout, and the dashed

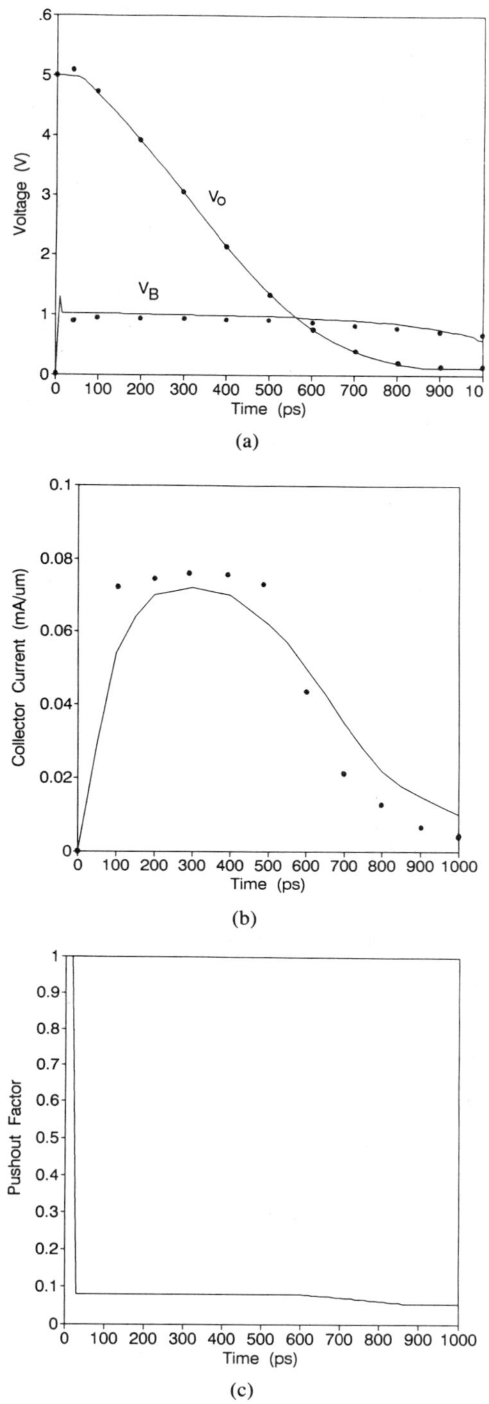

Fig. 3. (a) Voltage waveforms of a BiNMOS driver during pull-down transient. In this plot, the input has a voltage step from 0 to 5 V in 10 ps, $C_o = 0.1$ pF, $N_c = 2 \times 10^{16}$ cm^{-3}, $W_C \approx 1.1$ μm, $l_{eff} = 1$ μm, and $W_E \approx 0.75$ μm. (b) Comparison of transient collector current from the present model and from PISCES simulation. (c) Base pushout factor versus time.

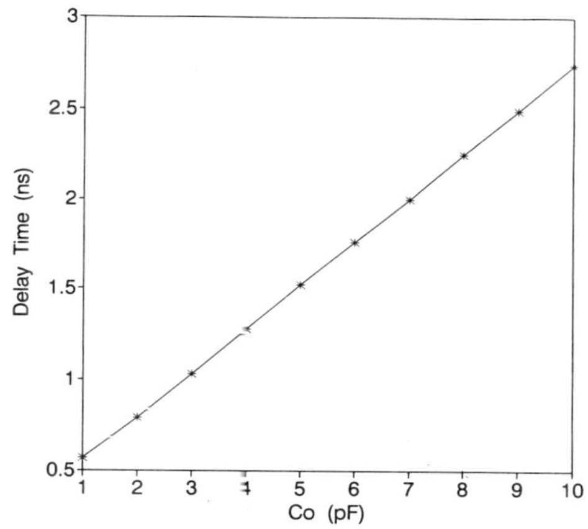

Fig. 4. Plot of pull-down time τ versus the output capacitance C_o. In this plot, $N_C = 2 \times 10^{16}$ cm^{-3}, $W_C = 0.3$ μm, $L_{eff} = 2$ μm, $W_{eff} = 13$ μ, $W_E = 1.5$ μm, $L_E = 4.5$ μm, and the input has a voltage step from 0 to 5 V in 50 ps.

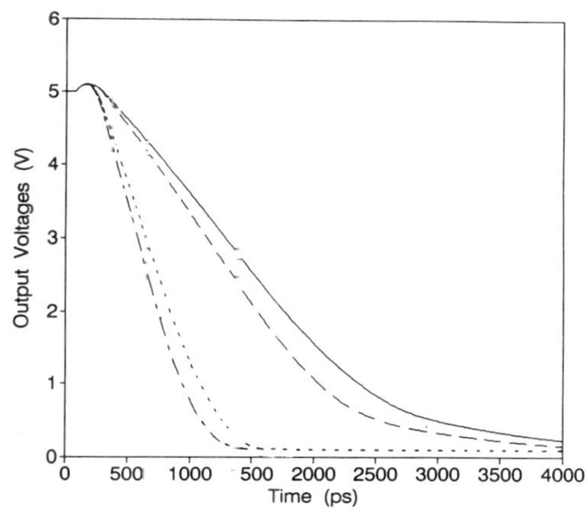

Fig. 5. Sensitivity of emitter crowding, base pushout, and emitter–base sidewall injection on the output voltage. In this plot, $C_o = 5$ pF and N_C, W_C, L_{eff}, W_{eff}, W_E, L_E, and the input waveform are the same as in Fig. 4.

and dotted line represents the model without base pushout and emitter crowding. The delay time predicted by the model without emitter–base sidewall injection is smaller because of an overestimated V_{BE}, due to neglecting of emitter–base sidewall current. The delay time predicted by the model without base pushout (dotted line) is much smaller compared with the present model (solid line). The base pushout reduces BiNMOS driving capability and switching speed due to a decrease of collector current and an increase of base transit time. The comparison between the dotted line and the dashed and dotted line describes the relative significance of emitter crowding in BiNMOS transient response. The emitter crowding results in a nonuniform current density distribution under the emitter, which reduces the collector current and switching speed. The significance of emitter crowding depends on the magnitude of collector current and emitter width.

Since base widening plays a major role in BiNMOS signal propagation, the pull-down delay time versus the epitaxial doping density N_C and epitaxial collector thickness W_C are simulated. For a lightly doped epitaxial collector, τ is degraded due to the base widening effect. When a sufficiently large N_C is applied, base widening is inhibited due to an increase of the onset current for base pushout. This high current effect on the gate delay is shown in Fig. 6, where the pull-down delay has a sharp transition about 10^{17} cm^{-3}. The transition point is determined by the onset current density J_0 and the current densities of J_0 and J_2

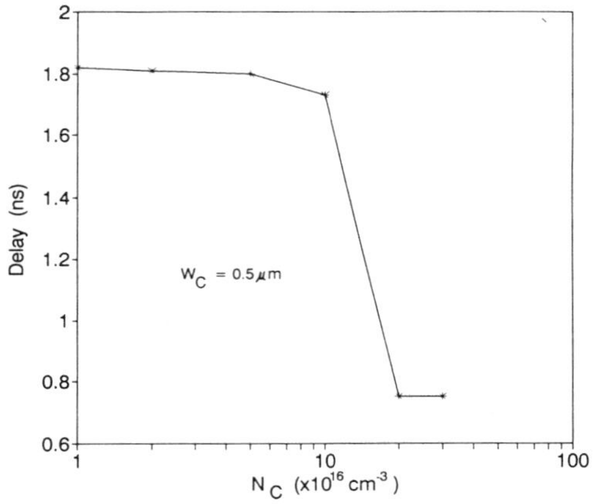

Fig. 6. Plot of the delay time versus the epitaxial doping density N_C at $W_C = 0.5\ \mu m$. For $N_C = 1 \times 10^{17}\ cm^{-3}$ and $V_{CB} = 3.0$ V, $J_0 = 3.67 \times 10^5$ A/cm^2, $J_1 = 3.2 \times 10^5\ A/cm^2$, and $J_2 = 7.3 \times 10^5\ A/cm^2$.

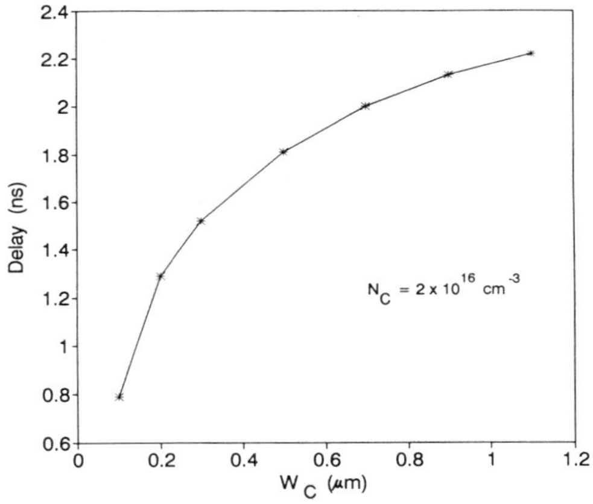

Fig. 7. Plot of the delay time versus the epitaxial collector width W_C at $N_C = 2 \times 10^{16}\ cm^{-3}$.

are a function of time-dependent output voltage as well. Once base pushout is suppressed, the increase of N_C beyond the transition point does not reduce τ any further. The delay time versus the epitaxial layer thickness is also shown in Fig. 7. The delay time increases with W_C due to a lower J_0 and J_2 at a higher W_C. For an optimal collector design, the epitaxial layer doping concentration, epilayer thickness, base–collector junction capacitance, and base–collector junction breakdown voltage should be optimized simultaneously. For typical BiNMOS drivers, the output capacitance is much larger than the base–collector junction capacitance. So, to achieve a lower propagation delay time, W_C should be designed as small as possible and N_C should be designed as large as possible if the breakdown voltage of the base–collector junction is above the supply voltage.

IV. Conclusions

The BiNMOS gate delay analysis including high-current transients has been developed based on a simple circuit modeling methodology. High-current effects such as base widening, emitter crowding, base conductivity modulation, and emitter–base sidewall injection on BiNMOS switching speed are examined. For a low epitaxial collector doping density, the pull-down delay time is significantly increased due to the base widening effect at high currents. The delay time is decreased at a low W_C and a high N_C at which the optimal epitaxial collector current density can be determined. The emitter current crowding increases the gate delay due to nonuniform current density distribution under the emitter which reduces the collector current. The emitter-base sidewall injected current, which is significant in the advanced BJT's with narrow emitter widths, increases the delay time because of a reduction of the effective base-emitter junction voltage and collector current. The transient voltage and current waveforms predicted by the present model are verified against PISCES device simulation in support of physical reasoning. The analytical equations provides a fast turn-around design time in examining BiNMOS driver transient response and optimizing the bipolar doping profiles in BiCMOS circuit. The present modeling methodology, which has certain degrees of mixed-mode circuit simulator capability, evaluates process impact on circuit performance before the actual circuits are built and it can be useful in technological computer-aided circuit designs.

References

[1] D. D. Tang and P. M. Solomon, "Bipolar transistor design for optimized power-delay logic circuits," *IEEE J. Solid-State Circuits*, vol. SC-14, pp. 679–684, August 1979.

[2] D. C. Mayer and W. E. Perkins, "Analysis of the switching of a submicrometer-gate CMOS/SOS inverter," *IEEE Trans. Electron Devices*, vol. ED-28, pp. 886–888, July 1981.

[3] F. S. Lai, "A generalized algorithm for CMOS circuit delay power, and area optimization," *Solid-State Electron.*, vol. 31, pp.1619–1627, Nov. 1988.

[4] H. J. De Los Santos and B. Hoefflinger, "Optimization and scaling of CMOS-bipolar drivers for VLSI interconnects," *IEEE Trans. Electron Devices*, vol. ED-33, pp. 1722–1730, Nov. 1986.

[5] E. W. Greenwich and K. L. McLaughlin, "Analysis and characterization of BiCMOS for high-speed digital logic," *IEEE J. Solid-State Circuits*, vol. 23, pp. 558–565, Apr. 1988.

[6] G. P. Rossel and R. W. Dutton, "Influence of device parameters on the switching speed of BiCMOS buffers," *IEEE J. Solid-State Circuits*, vol. 24, pp. 90–99, Feb. 1989.

[7] J. B. Kuo, G. P. Rossel, and R. W. Dutton, "Two-dimensional analysis of a merged BiPMOS device," *IEEE Trans. Computer-Aided Des.*, vol. 8, pp. 929–932, Aug. 1989.

[8] J. B. Kuo and Y.-W Chen, "Pull-down transient imposed extrinsic base consideration in BiNMOS transistors," *IEEE Trans. Electron Devices*, vol. 37, pp. 1308–1313, May 1990.

[9] C. G. Sodini, P. K. Kuo, and J. L. Moll, "The effect of high fields on MOS device and circuit performance," *IEEE Trans. Electron Devices*, vol. ED-31, pp. 1386–1393, Oct. 1984.

[10] S. M. Sze, *Physics of Semiconductor Devices*, 2nd ed. New York: Wiley, 1981.

[11] J. R. Hauser, "The effects of distributed base potential on emitter-current injection density and effective base resistance for stripe transistor geometries," *IEEE Trans. Electron Devices*, vol. ED-11, pp. 238–234, 1964.

[12] D. D. Tang, "Switch-on transient of shallow-profile bipolar transistors," *IEEE Trans. Electron Devices*, vol. ED-31, pp. 80–82, Jan. 1984.

[13] J. S. Yuan and J. J. Liou, "Circuit modeling and transient emitter crowding and dynamic resistance effects for advanced bipolar transistors," *Solid-State Electron.*, vol. 32, pp. 623–631, Aug. 1989.

[14] J. S. Yuan, J. J. Liou, and W. R. Eisenstadt, "A physics-based current-dependent base resistance model for advanced bipolar transistor," *IEEE Trans. Electron Devices*, vol. 35, pp. 1055–1061, July 1988.

[15] C. T. Kirk, "A theory of transistor cutoff frequency falloff at high current densities," *IRE Trans. Electron Devices*, vol. ED-9, pp. 164–174, Mar. 1962.
[16] D. L. Bowler and F. A. Lindholm, "High current regimes in transistor collector regions," *IEEE Trans. Electron Devices*, vol. ED-20, pp. 257–263, Mar. 1973.
[17] J. J. Liou, F. A. Lindholm, and J. S. Park, "Forward-voltage capacitance and thickness of p-n junction space-charge regions," *IEEE Trans. Electron Devices*, vol. ED-34, pp. 1571–1579, July 1987.
[18] J. J. Liou, "Base-collector junction capacitance of bipolar transistors operating at high current densities," *IEEE Trans. Electron Devices*, vol. ED-34, pp. 2304–2308, Nov. 1987.
[19] H. C. de Graaff and W. J. Kloosterman, "New formulation of the current and charge relations in bipolar transistor modeling for CACD purposes," *IEEE Trans. Electron Devices*, vol. ED-32, pp. 2415–2419, Nov. 1985.
[20] J. S. Yuan and J. J. Liou, "An improved latching pulse design for dynamic sense amplifiers," *IEEE J. Solid-State Circuits*, vol. 25, pp. 1294–1299, Oct. 1990.
[21] Kaiser S. Kunz, *Numerical Analysis*. New York: McGraw-Hill, 1957.

An Analytical Model for the Determination of the Transient Response of CML and ECL Gates

MOUSTAFA Y. GHANNAM, MEMBER, IEEE, ROBERT P. MERTENS, SENIOR MEMBER, IEEE, AND ROGER J. VAN OVERSTRAETEN, FELLOW, IEEE

Abstract—**A rigorous analytical study of the transient response of current-mode logic (CML) and emitter-coupled logic (ECL) gates is presented. The full-range transient response of the gate has been calculated using closed-form analytical expressions. This has been achieved by generalizing and improving the method used by Stork [1] for the determination of the propagation delay. The proposed model is applicable at low-level injection, unity fan-in, and unity fan-out. The delays related to the transit time, to the load, and to the junction capacitances are considered. For ECL gates, the emitter follower delay is also included. Various delays (rise time, propagation delay, etc.) calculated using the proposed model agree perfectly with the results of SPICE computer simulations and with the reported experimental values.**

NOMENCLATURE

α	Common base current gain.
β	Common emitter current gain.
C_c	Transition capacitance of the base–collector junction (F).
C_{bM}	Miller component of the collector junction capacitance in the base circuit (F).
C_{cM}	Miller component of the collector junction capacitance in the collector circuit (F).
C_{cs}	Transition capacitance of the collector–substrate junction (F).
C_e	Transition capacitance of the base–emitter junction (F).
C_π	II capacitance in the emitter follower equivalent circuit (F).
g_m	Transconductance (A/V).
I_b	Base current (A).
I_c	Collector current (A).
I_{co}	Collector reverse saturation current (A).
I_s	Current delivered by the current source (A).
k	Boltzmann's constant (eV/K).
q	Electron charge (C).
q_c	Charge in the collector junction space-charge region (C).
q_e	Charge in the emitter junction space-charge region (C).
q_F	Total mobile charge in the forward active region (C).
R_b	Base resistance (Ω).
R_e	Emitter resistance (Ω).
R_{ef}	Emitter resistance of the emitter follower (Ω).
R_g	Source resistance for the switching transistor (Ω).
R_L	Load resistance (Ω).
R_s	Source resistance for the emitter follower (Ω).
τ_F	Forward transit time (s).
T	Temperature (K).
T_b, T_b'	Time constant of the base circuit of the switching transistor (s).
T_c	Time constant of the collector circuit of the switching transistor (s).
T_d	Propagation delay (s).
t_i	Delay in the collector current response due to the base transit time (s).
$t_{jc,i}$	Delay in the collector current due to junction capacitances (s).
$t(v)$	Delay in the collector voltage response due to the base transit time and to the collector circuit (s).
$t_{jc}(v)$	Delay in the collector voltage response due to junction capacitances (s).
$t_{ef}(v)$	Delay in the response of the output voltage of the emitter follower (s).
$T(v)$	Total delay of the gate (s).
V_c	Voltage at the collector node of the switching transistor (V).
V_{cc}	Collector supply voltage (V).
V_{ee}	Emitter supply voltage (V).
V_{eo}	Voltage at the common emitter node (V).
V_r	Reference voltage (V).
V_s	Voltage swing (V).
V_T	Thermal voltage (V).

Manuscript received April 12, 1989; revised July 13, 1989. The review of this paper was arranged by Associate Editor T. H. Ning.

M. Y. Ghannam is on leave with the Interuniversity Microelectronics Center, Kapeldreef 75, Leuven, Belgium. He is with the Electronics and Communications Department, Cairo University, Guiza, Egypt.

R. P. Mertens and R. Van Overstraeten are with the Interuniversity Microelectronics Center, Kapeldreef 75, Leuven, Belgium.

IEEE Log Number 8930909.

I. INTRODUCTION

SEVERAL earlier reports dealt with the delays in current-mode logic (CML) and emitter-coupled logic (ECL) gates, e.g., [1]–[4]. The motivation behind these studies is the necessity of finding an analytical tool that estimates the gate delay T_d without proceeding via lengthy computer simulations. Also, the analytical expressions give a clear picture to the phenomenon and help in im-

Reprinted from *IEEE Trans. Electron Devices*, vol. 37, no. 1, pp. 191–201, Jan. 1990.

proving and optimizing the device and gate design. Using a sensitivity analysis, a rather complex expression for the gate delay has been proposed [3]. In a simple reasoning, the gate delay T_d of an ECL gate is given by $T_d = 0.7(T_b + T_c)$ [4] where T_b and T_c represent the base and collector time constants, respectively. The definitions of T_b and T_c are [4], however, very approximate. Recently, Stork [1] studied the particular case when T_b and T_c are about equal and presented a closed-form expression for the gate delay based on measurable quantities. In both cases [1], [4], it is assumed that the collector current of the switching transistor increases exponentially toward an assymptotic value equal to the current delivered by the current source connected to the common-emitter node. As it will be shown here, such an assumption is not generally valid.

In this work, we propose a rigorous analytical model to calculate the transient behavior of the collector current and of the output voltage of a CML switch or of an ECL gate. This model is obtained by generalizing and improving the method used by Stork [1] to determine the propagation delay.

II. Analysis

The configuration of a standard CML or ECL gate is displayed in Fig. 1. In the ECL gate an emitter follower output stage is added to the common-mode pair. When the voltage V_{Bl} applied to the base B_l of the left (switching) transistor Q_l is $(V_r - V_s/2)$, where V_r is the reference voltage permanently applied to the base of the right (reference voltage) transistor Q_r, and V_s is the voltage swing, Q_l is off. Hence, the collector current I_{cl} is zero (actually its magnitude is not zero but orders of magnitude smaller than I_{cr}) and the voltage V_c at the collector node of Q_l is at its high state. When B_l is subjected to a positive voltage step raising V_{Bl} to $(V_r + V_s/2)$, the collector current I_{cl} rises to reach finally αI_s, where α is the common base current gain and I_s is the current delivered by the current source connected to the common-emitter node. Consequently, the voltage at the collector node of Q_l drops to the low state. A simple analysis of the time response of the circuit is made possible by decoupling the base circuit from the collector circuit. Each circuit is then studied separately. The decoupling is done by splitting the base–collector (collector) junction capacitance into a collector component C_{cM} and a base component C_{bM} using Miller's rule.

A. Transient Response of The Collector Current: The Base Circuit

In this section, the collector current I_{cl} flowing in the switching transistor Q_l as a response to a positive voltage step V_s applied to B_l is investigated. Computer circuit simulations are carried out and compared to the numerical solution of the time-dependent base circuit differential equation. Finally, an approximate analytical solution of the same equation is elaborated.

Computer simulations of the transient response are performed throughout this work using the SPICE circuit simulator. The choice of SPICE is based on the facts that: 1) the program contains an accurate bipolar model with graded complexity, 2) it has been intensively used previously for bipolar transistor simulations with no major complaints (e.g., [3]), 3) it is simple to use, and finally 4) it is easily accessible.

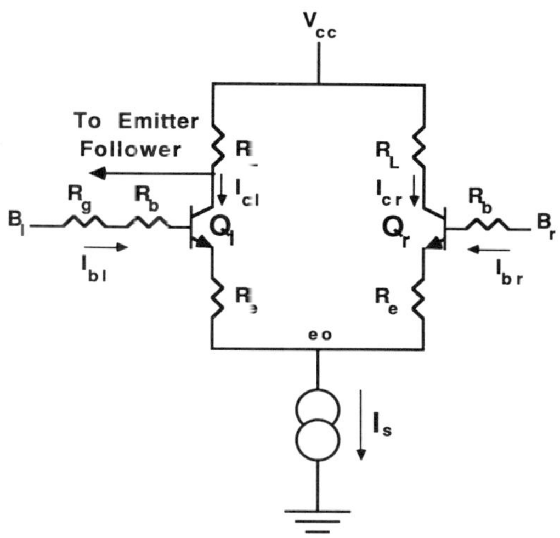

Fig. 1. Configuration of a standard emitter-coupled logic gate or of a current mode logic gate. For the latter, no emitter follower has to be added.

1) Computer Simulations: The collector current $I_{cl}(t)$ determined using SPICE is plotted in Fig. 2. What is to be retained from this example is that I_{cl} increases exponentially with time up to a certain limiting time $t_{\lim}$. At $t_{\lim}$, the magnitude of I_{cl} reaches αI_s. Beyond $t_{\lim}$ I_{cl} becomes clamped at αI_s due to the emitter current boundary condition imposed by the current source. Thus, we can say that the switching transistor Q_l is voltage driven up to $t_{\lim}$ and becomes current driven beyond $t_{\lim}$. The exponential collector current rise can be expressed as

$$I_{cl} = D\alpha I_s[1 - \exp(-t/T_b)] \tag{1}$$

which differs from the conventionally assumed expression [1], [4] by the pre-exponential factor D. As shown in Fig. 2, the voltage at the common-emitter node V_{eo} tends to stabilize shortly after applying the input voltage step V_s at B_l. However, it has to readjust its steady-state value when I_{cl} reaches the forced value αI_s. This explains the step observed in V_{eo} at $t_{\lim}$. In this figure, we assume that $V_{cc} = 5$ V and $V_s = 0.5$ V. The value of V_r follows from the input–output voltage compatibility requirements. Note that if other values are chosen for V_{cc} and for V_r, the result is identical with, however, a vertical parallel shift of the voltage scale.

2) Analytical Formulation: The equivalent circuit of the CML (ECL) gate is shown in Fig. 3. According to the charge-control model, the base and collector currents can be expressed as

$$I_{bl} = \frac{dq_{Fl}}{dt} + \frac{q_{Fl}}{\tau_{BFl}} + \frac{dq_{el}}{dt} + \frac{dq_{cl}}{dt}$$

$$I_{cl} = \frac{q_{Fl}}{\tau_{Fl}} - \frac{dq_{cl}}{dt} \tag{2a}$$

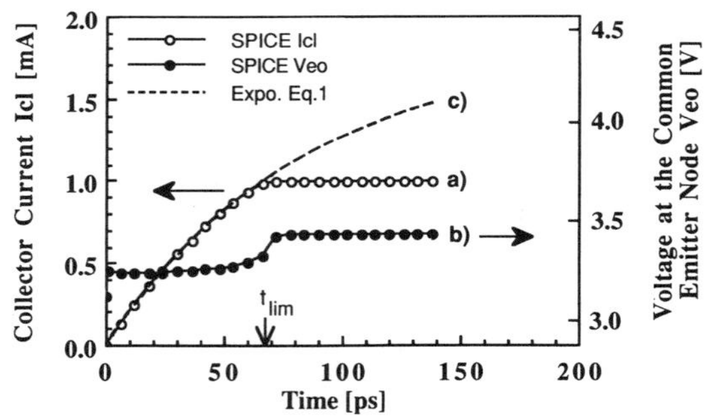

Fig. 2. Time response of: (curve a) the collector current I_{cl} of Q_l and (curve b) the voltage V_{eo} at the common emitter node eo, obtained by means of SPICE simulations. The dashed curve (c) represents a pure exponential rise. Here, τ_F = 15 ps, V_s = 500 mV, I_s = 1 mA, R_g = 500 Ω, R_b = 250 Ω, R_e = 25 Ω, β = 100, V_{cc} = 5 V, V_r = 3.95 V and T = 300 K. These values are used throughout the analysis unless other values are specified.

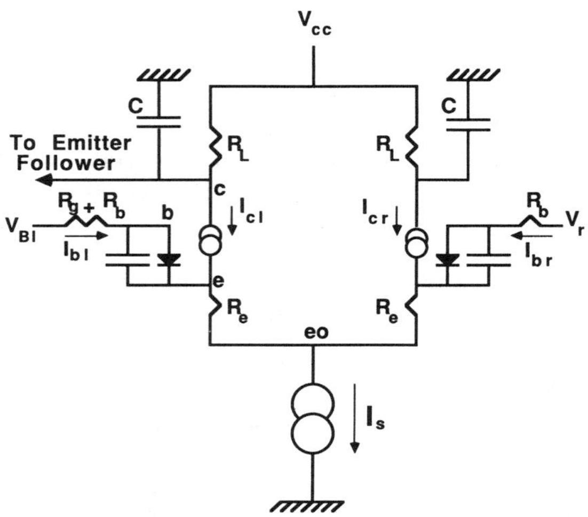

Fig. 3. Equivalent circuit of the ECL (CML) gate.

for the left transistor and

$$I_{br} = \frac{dq_{Fr}}{dt} + \frac{q_{Fr}}{\tau_{BF_r}} + \frac{dq_{er}}{dt} + \frac{dq_{cr}}{dt}$$

$$I_{cr} = \frac{q_{F_r}}{\tau_{F_r}} - \frac{dq_{cr}}{dt} \tag{2b}$$

for the right transistor, where I_b, I_c, q_F, q_e, and q_c stand for the time-dependent base and collector currents, total mobile charge in the forward active region, and the charge in the space-charge region of the base–emitter (emitter) junction and of the base–collector (collector) junction, respectively. The time constant $\tau_F = 1/2\pi f_{T_{\max}}$ represents the total forward transit time with $f_{T_{\max}}$ the maximum cut-off frequency of the transistor. The time constant τ_{BF} is the product of τ_F with the transistor current gain $\beta = \alpha/(1 - \alpha)$ at steady state.

We start the analysis by assuming that $q_F >> q_e, q_c$, or equivalently that the base charging (diffusion) capacitance C_D is much larger than the emitter junction capacitance C_e and than the Miller base capacitance C_{bM}. Consequently, the terms involving q_e and q_c in (2a) and (2b) are omitted. This approximation will be reconsidered later in this paper.

Applying the Kirchhoff voltage law (KVL) to the loop $B_l \rightarrow B_r \rightarrow B_l$ passing through the bases and emitters of Q_l and Q_r, we can write

$$V_{Bl} = I_{bl}(R_g + R_{bl}) + V_{bel} + (I_{bl} + I_{cl})R_{el} - (I_{br} + I_{cr})R_{er} - V_{ber} - I_{br}R_{br} + V_r \tag{3}$$

where R_g is the source resistance, R_b is the base resistance, and R_e is the emitter resistance. As mentioned earlier, a voltage step V_s is applied to B_l raising V_{Bl} to $(V_r + V_s/2)$. Let us assume that Q_l and Q_r are identical, substitute $V_{be(l,r)}$ by $V_T \times \ln\left[I_{c(l,r)}/I_{co(l,r)}\right]$ with I_{co} the collector saturation current and $V_T = kT/q$, and finally apply the Kirchhoff current law (KCL) to the common emitter node eo

$$I_{bl} + I_{cl} + I_{br} + I_{cr} = I_s. \tag{4}$$

Equation (3) can now be rewritten as

$$\tau_F R_t \frac{dI_{cl}}{dt} + \left(\frac{R_t}{\beta} + 2R_e\right) I_{cl} - V_T \ln\left(\frac{I_s\beta}{I_{cl}(\beta + 1)} - 1\right) - I_s\left(R_e + \frac{R_b}{\beta + 1}\right) - V_s/2 = 0 \tag{5}$$

where the total base loop resistance $R_t = R_g + 2R_b + 2R_e$.

Introducing the normalized variable $i = I_{cl}/\alpha I_s$, the first-order nonlinear differential equation (5) becomes

$$\tau_F \frac{di}{dt} = A_1 + B_1 \ln(1/i - 1) - C_1 i \tag{6}$$

where

$$A_1 = \frac{V_s/2I_s + R_e + R_b/(\beta + 1)}{\alpha R_t}$$

$$B_1 = \frac{V_T/I_s}{\alpha R_t}$$

$$C_1 = 1/\beta + 2R_e/R_t.$$

Changing (6) into an integral form, we get

$$\int \frac{di}{A_1 + B_1 \ln(1/i - 1) - C_1 i} = \int \frac{1}{\tau_F} dt. \tag{7}$$

Although (7) has no general closed-form analytical solution, an approximate analytical solution will be elaborated and compared with the result of the numerical integration.

3) Numerical Solution: The results of the numerical integration of (7) are indistinguishable from those of SPICE simulations as shown in Fig. 4. The integration is performed from $i(0) = 0$ (corresponding to $t = 0$) to any i (any t). The maximum value of i equals 1. As mentioned earlier, at $t = 0$ the collector current of the Q_l is not zero, but the fraction $i(0)$ given by

$$i(0) = \exp\left[(-V_s/2 + I_sR_b/(\beta + 1) + I_sR_e)/V_T\right]$$

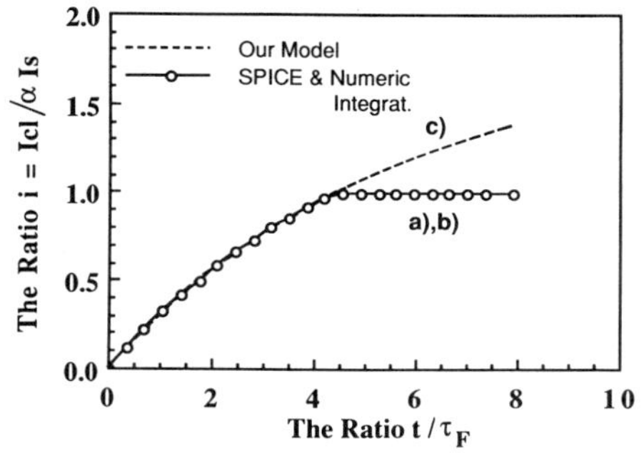

Fig. 4. Comparison between the time response of I_{cl} obtained: (curve *a*) by means of SPICE simulations, (curve *b*) by numerical integration of (7), and (curve *c*) using our analytical model.

is so small such that $i = 0$ at $t = 0$ is an acceptable approximation.

4) Analytical Solution: Using a quite accurate approximation, (7) can be integrated analytically as detailed in the Appendix (Section A). By analogy to (1), the result of the integration can be written in the form

$$i = D'[1 - \exp(-t_i/T_b')] \quad (8)$$

where

$$D' = 0.5 + \frac{V_s/I_s - R_g/(\beta + 1)}{10V_T/I_s + 4R_e + 2(R_g + 2R_b)/(\beta + 1)} \quad (9a)$$

$$T_b' = \frac{\tau_F\beta(R_g + 2R_b + 2R_e)/(\beta + 1)}{5V_T/I_s + 2R_e + (R_g + 2R_b)/(\beta + 1)}. \quad (9b)$$

The source resistance R_g is the output impedance of the preceeding stage and is equal to the load resistance $R_L = V_s/\alpha I_s$ of the switching transistor. By doing this substitution in (9a) and (9b), D' and T_b' become

$$D' = 0.5 + \frac{(V_s/I_s)(1 - 1/\beta)}{10V_T/I_s + 4R_e + 2(V_s/\alpha I_s + 2R_b)/(\beta + 1)} \quad (9c)$$

$$T_b' = \frac{\tau_F\beta(V_s/\alpha I_s + 2R_b + 2R_e)/(\beta + 1)}{5V_T/I_s + 2R_e + (V_s/\alpha I_s + 2R_b)/(\beta + 1)}. \quad (9d)$$

As depicted in Fig. 4, the values obtained for the pre-exponential factor D' and for the time constant T_b' using the expression proposed above are in perfect agreement with their analogs D and T_b determined from SPICE simulations listed in Table I. From now on, we shall use the notations D and T_b for D' and T_b'. When $\alpha = 1$ and $R_e = 0$, we have

$$T_b = \tau_F \frac{V_s}{5V_T}(1 + 2R_bI_s/V_s). \quad (10)$$

This T_b expression differs by a factor $V_s/5V_T$ from the expression proposed by Stork [1] under the same conditions.

TABLE I

VALUES OF D AND T_b RESULTING FROM SPICE SIMULATIONS AND FITTING WITH (1) AND VALUES OF D' AND T_b' OBTAINED BY SUBSTITUTION IN (9c) AND (9d) FOR THREE DIFFERENT VALUES OF V_s

(For generalization, two cases, (a) and (b), with different transistor parameters are considered.)

V_s(V)	0.40	0.50	0.60
D	2.03	2.43	2.74
T_b/τ_F	10.6	11.5	11.9
D'	2.03	2.41	2.79
T'_b/τ_F	10.7	11.5	12.2

$I_s = 1$mA, $R_b = 500\ \Omega$, $\beta = 1000$, $R_e = 0$

(a)

V_s(V)	0.40	0.50	0.60
D	1.54	1.82	2.08
T_b/τ_F	4.90	5.49	6.00
D'	1.55	1.81	2.06
T'_b/τ_F	5.00	5.49	6.00

$I_s = 1$mA, $R_b = 250\ \Omega$, $\beta = 100$, $R_e = 25\ \Omega$

(b)

It is important for the upcoming analysis to define the limiting time $t_{\lim}$ at which i reaches unity and beyond which (1) and (8) cease to be valid. By putting $i = 1$ in (8), we get

$$t_{\lim} = t_1 = T_b \ln \frac{D}{D - 1}. \quad (11)$$

5) Delay Due to the Junction Capacitances $t_{jc,i}$: So far, the emitter junction capacitance as well as the base component of the collector junction capacitance have been neglected. This assumption is valid as long as the sum of these two components is negligible with respect to the diffusion capacitance of the emitter junction. In order to achieve very high cutoff frequencies (higher than 10 GHz), the trend in state-of-the-art silicon bipolar transistors is to have a very thin highly doped base region. Also, the gradients of the impurity profiles are becoming very steep. This trend results in an enhancement of the junction capacitances that causes a relatively significant delay t_{jc}, especially at low current levels.

In order to find an analytical expression for t_{jc}, we assume that the emitter junction capacitance has an average constant value C_e. We continue to assume that the Miller base capacitance C_{bM} is negligible. In this case, (6) has to be modified and becomes

$$\left(\tau_F + \frac{C_eV_T}{\alpha I_s}\frac{1}{i}\right)\frac{di}{dt} = A_1 + B_1 \ln(1/i - 1) - C_1 i. \quad (12)$$

The first term in the left-hand side of (12) has been considered in the previous section and is responsible for the delay t_i caused by the transit time to be calculated from

(8). The second term is responsible for the delay $t_{jc,i}$ caused by the emitter junction capacitance and, as shown in the Appendix (Section B), can be approximately expressed as

1) $i < 0.1$

$$t_{jc,i} = C_e R_t \ln \frac{V_s}{V_s/2 - V_T \ln i + I_s[R_e + R_b/(\beta + 1)]}. \quad (13a)$$

2) $i > 0.1$

$$t_{jc,i} = t_{jc,0.1} + \frac{C_e V_T}{\alpha I_s D} \frac{T_b}{\tau_F} \ln \left[\frac{D - 0.1}{D - i} \left(\frac{i}{0.1} \right) \right]. \quad (13b)$$

Excellent agreement is obtained between the transient response of the collector current evaluated analytically using the expressions proposed here and the results of SPICE simulations as shown in Fig. 5.

Note that so far the Miller base capacitance C_{bM} has been neglected. It will be shown later that when C_{bM} is taken into consideration, t_{jc} can still be represented by (13a) and (13b) with, however, replacing C_e by $C_e + C_c$.

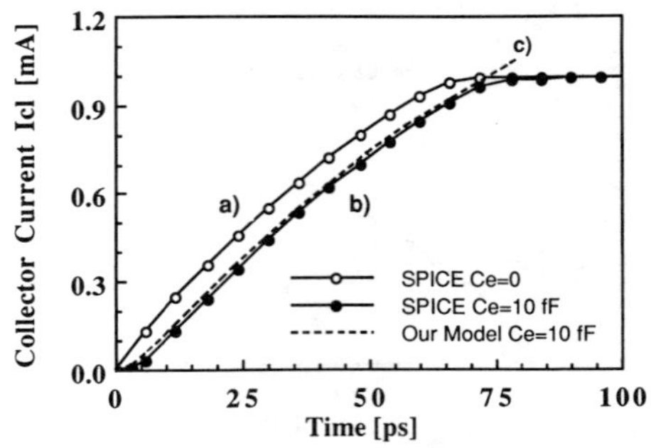

Fig. 5. Time response of the collector current I_{cl}: (curve *a*) obtained by SPICE simulations with $C_e = 0$ and (curve *b*) $C_e = 10$ fF, and (curve *c*) obtained using our analytical model with $C_e = 10$ fF.

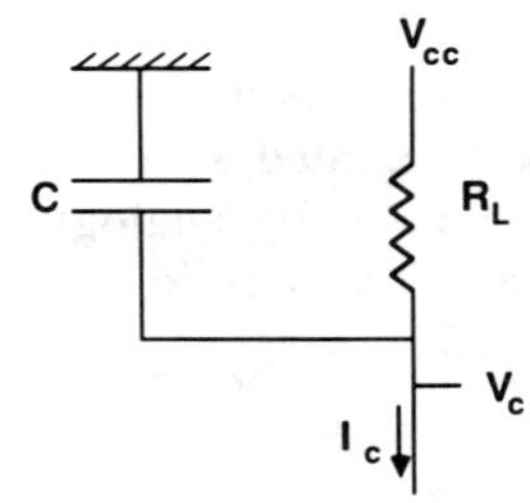

Fig. 6. The *RC* network representing the equivalent collector circuit. The capacitance $C = 3C_c + C_{cs}$ for an ECL gate and $C = 2C_c + C_L + C_{cs}$ for a CML gate.

B. Transient Response of the Voltage at the Collector Node: The Collector Circuit

1) Equivalent Circuit: The collector (load) circuit can be represented by the simple *RC* network shown in Fig. 6. For an ECL gate, the capacitance C is the sum of: 1) the Miller collector component of the collector junction capacitance $C_{cM} = 2C_c$, 2) the collector–substrate capacitance C_{cs}, and finally 3) the input capacitance of the emitter follower C_c. The load resistor capacitance is usually much smaller than C_c and therefore is neglected. The resistance R is equal to the load resistance R_L. Hence, as in [1], the time constant T_c of the collector circuit

$$T_c = RC = R_L(3C_c + C_{cs}). \quad (14)$$

For a CML switch, the emitter follower capacitance C_c is replaced by the load capacitance C_L or by the base input capacitance of the next stage in the case of a multiple-stage CML circuit.

2) Collector Voltage Delays: When the ECL gate is changing its logic state, the voltage at the collector of the switching transistor V_c is equal to $V_{cc} - vV_s$, where v is a positive fraction between 0 and 1 and the voltage swing $V_s = \alpha I_s R_L$. Applying KCL to the collector node, we have

$$\frac{dV_c}{dt} + \frac{V_c}{T_c} = \frac{V_{cc} - I_c(t)R_L}{T_c}. \quad (15a)$$

Taking the Laplace transform of (15a), we get

$$V_c(S) = \frac{V_c(0)}{S + 1/T_c} + \frac{V_{cc}}{ST_c(S + 1/T_c)} - \frac{I_c(S)R_L}{T_c(S + 1/T_c)}. \quad (15b)$$

Since the collector current varies abruptly from the transient exponential behavior to a constant steady-state value at $t_{\lim}$, the collector voltage time response is divided into two distinct regimes with a common boundary condition at $t_{\lim}$. Hence, the Laplace transform $I_c(S)$ of $I_c(t)$ depends on the time spot $t(v)$ at which V_c is to be evaluated.

a) $t(v) < t_{\lim}$: During the first phase, the collector current response is given by

$$I_c(t) = D\alpha I_s(1 - \exp(-t/T_b))$$

$$I_c(S) = D\alpha I_S \left(\frac{1}{S} - \frac{1}{(S + 1/T_b)} \right).$$

The initial condition for the collector voltage at $t = 0$ is

$$V_c(0) = V_{cc}.$$

Substituting $I_c(S)$ and $V_c(0)$ in (15b) we get

$$V_c(S) = \frac{V_{cc}}{S} - D\alpha I_s R_L \left[\frac{1}{S} + \frac{1}{(S + 1/T_c)} \left(\frac{T_c}{T_b - T_c} \right) - \frac{1}{(S + 1/T_b)} \left(\frac{T_b}{T_b - T_c} \right) \right]. \quad (16a)$$

The collector voltage $V_c(t)$ as a function of time is obtained by taking the inverse Laplace transform of (16a) resulting in

$$V_c(t) = V_{cc} - D\alpha I_s R_L \left[1 + \frac{T_c}{T_b - T_c} \exp(-t/T_c) - \frac{T_b}{T_b - T_c} \exp(-t/T_b) \right]. \quad (16b)$$

As mentioned earlier, the collector voltage varies between V_{cc} and $V_{cc} - V_s$. Therefore, we seek a solution for (16b) satisfying

$$V_c(t) = V_{cc} - vV_s \tag{16c}$$

where $t = t(v)$. Rearranging terms and putting $x = T_b/T_c$, the equation to be solved becomes

$$1 + \frac{1}{x-1}\exp\left(-xt(v)/T_b\right) - \frac{x}{x-1}\exp\left(-t(v)/T_b\right) = v/D. \tag{17}$$

We anticipate, as in [1], a possible solution for (17) given by

$$t(v) = f(x, v/D)\sqrt{T_b T_c}. \tag{18}$$

Substituting (18) in (17) we get

$$1 + \frac{1}{x-1}\exp\left(-f\sqrt{x}\right) - \frac{x}{x-1}\exp\left(-f/\sqrt{x}\right) = v/D. \tag{19}$$

For large values of x (greater than a certain value x_{max}), the furthermost pole in (16a) can be neglected. Hence, (19) is reduced to

$$1 - \frac{x}{x-1}\exp\left(-f/\sqrt{x}\right) = v/D \tag{20}$$

resulting in a function f given by

$$f = \sqrt{x}\ln\frac{x}{(x-1)(1-v/D)}. \tag{21}$$

In practice, D is usually larger than 1.4, which means that v/D is always smaller than 0.7. Under this condition, it can be shown that x_{max} is accurately fitted by

$$x_{max} = \frac{1.32D}{v}. \tag{22}$$

For $x > x_{max}$, the error in the value of f determined using (21) is less than 5 percent. When x is smaller than x_{max}, the function $f(x)$ is obtained by solving the transcendental equation (19) at each value of x. The result of this exercise is shown in Fig. 7 for x greater than one, taking v/D as an independent parameter. In this case, the simple linear function

$$f = f' = (0.17x + 2.66)v/D + 0.25 - 0.0092x \tag{23}$$

is a very good fitting the function f as depicted in Fig. 7.

Note that so far we have considered the solution when T_b is greater than $T_c (x > 1)$, which is most probably the situation occuring in state-of-the-art silicon bipolar technology. If T_c is larger than T_b, following the same steps with, however, x equal to T_c/T_b instead of T_b/T_c, we also finally end up with (19). This means that whether $T_b > T_c$ or $T_c > T_b$, the function f always may be expressed by (21) or (23) with, however, the appropriate definition of x.

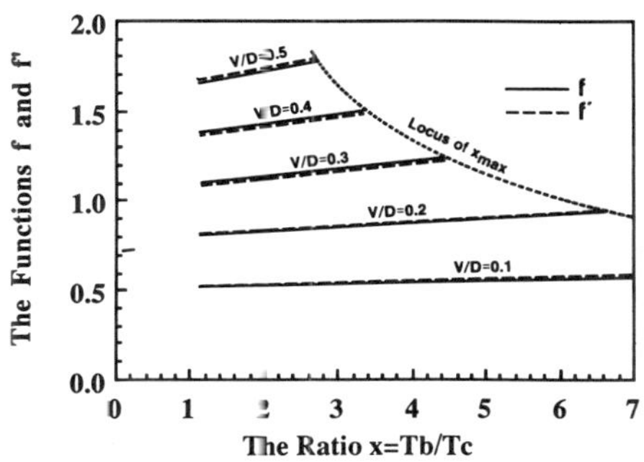

Fig. 7. The function f satisfying (19) versus x (solid line). The ratio v/D is used as an independent parameter. The dashed lines represent f', the linear fitting function given by (23).

From (21) it can be deduced that the role of the function f is twofold. First, it acts as a weight factor emphasizing the role of the largest of the time constants T_b and T_c. Second, the dependence of f on v/D reflects the impact of the transient exponential behavior of the collector current, having a steady-state value $D\alpha I_s$, on the gate delay $t(v)$.

b) $t(v) > t_{lim}$: Expression (18) is valid up to $t(v) = t_{lim}$ only, which puts an upper limit to the value of v submitted to the solution described in the previous section. This upper limit v_{max} is determined by putting

$$t(v) = t_{lim} \tag{24a}$$

$$f\sqrt{T_b T_c} = T_b \ln\frac{D}{(D-1)} \tag{24b}$$

where f is expressed by (21) or (23) for $x > x_{max}$ and $x < x_{max}$, respectively. This results in

1) $x > x_{max}$

$$v_{max} = D\left(1 - \frac{x}{x-1}\left(\frac{D-1}{D}\right)^{\sqrt{T_b/xT_c}}\right). \tag{25a}$$

2) $x < x_{max}$

$$v_{max} = D\,\frac{\sqrt{T_b/T_c}\,\ln\dfrac{D}{D-1} + 0.0092x - 0.25}{0.17x + 2.66}. \tag{25b}$$

During the second phase of the collector current response, the origin of the time variable t is changed from zero to t_{lim} resulting in a new variable $t' = t - t_{lim}$. The collector current is constant during this phase and equal to

$$I_c(t') = \alpha I_s$$

$$I_c(S) = \alpha I_s/S.$$

The initial value of the collector voltage in this second regime is determined by putting $t = t_{lim}$ in (16b) resulting in

$$V_c(t' = 0) = V_{cc} - D\alpha I_s R_L \left[1 + \frac{T_c}{T_b - T_c} \exp\left(-t_{\lim}/T_c\right) - \frac{T_b}{T_b - T_c} \exp\left(-t_{\lim}/T_b\right) \right].$$

In this case, the solution of (15a) is straightforward because the system involves only one pole at $1/T_c$. Once the delay $t'(v)$ is evaluated, the total delay $t(v)$ is obtained by going back to the original variable t resulting in a total delay $t(v) = t'(v) + t_{\lim}$ given by

$$t(v) = T_c \ln \left[\frac{DT_c}{(1 - v)(T_b - T_c)} \cdot \left(\exp\left(t_{\lim}/T_c - t_{\lim}/T_b\right) - 1 \right) \right]. \quad (26)$$

3) Delay Due to the Junction Capacitances $t_{jc}(v)$: We have seen that charging the emitter junction capacitance causes a delay $t_{jc,i}$ in the collector current response. The collector voltage response will be delayed accordingly. During most of this delay, the current is very small and the voltage drop and the load changes very slowly with time. The capacitive current in the load circuit is therefore negligible, and most of the collector current flows in the load resistance. Therefore, we assume that the delay in the collector voltage $t_{jc}(v)$ is approximately equal to the delay in the collector current $t_{jc,i}(i = v)$, given by (13a) and (13b). At higher current levels, the contribution of the delay t_{jc} to the total delay becomes much less significant such that the error introduced when we assume that $t_{jc,i} = t_{jc}(v)$, with $i = v$, is always negligible.

4) Numerical Examples: In this section the transient response of the collector voltage of the switching transistor is calculated using our method and compared to the response obtained using SPICE simulations. The device parameters are chosen such that they represent typical state-of-the-art advanced silicon bipolar technology [5]–[14], and are given in the captions of Figs. 2 and 8. The comparison is carried out for three cases.

a) In SPICE and in our model: $C_{cM} = 2C_c = 16\,fF$, $C_{bM} = 0$ and $C_e = 0$: In this case the collector junction capacitance is not considered as a transistor model parameter in SPICE simulations. Instead, its Miller collector component $C_{cM}(=2C_c)$ is lumped in parallel with the collector substrate capacitance C_{cs} and with the input capacitance of the emitter follower (C_c).

As shown in Fig. 8 (curves A), the transient response of the voltage at the collector node of Q_l determined using our analytical model agrees perfectly with the results of SPICE.

b) In SPICE and in our model: $C_{cM} = 2C_c = 16\,fF$, $C_{bM} = 0$ and $C_e = 10\,fF$: The match between the results of our analytical model and those of SPICE simulations continues to be excellent when the delay caused by C_e is taken into account as shown in Fig. 8 (curves B). As in the A curves, the collector junction capacitance is not considered as a model parameter in SPICE simulations. On the other hand, the emitter junction capacitance $C_e = 10$ fF is considered as a transistor model parameter in SPICE simulations.

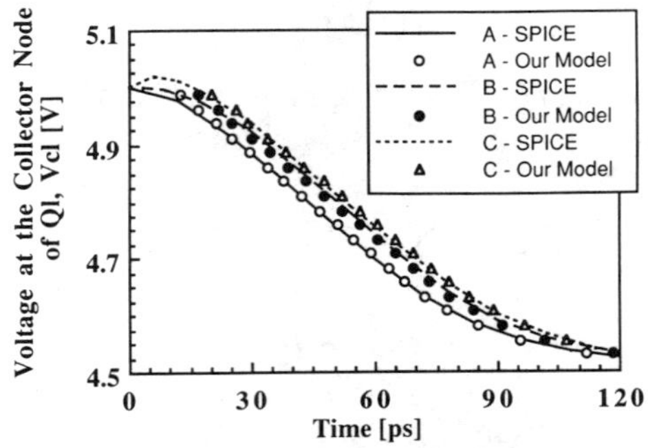

Fig. 8. Comparison between the time response of the collector voltage of the switching transistor Q_l of an ECL gate obtained by SPICE simulations (line plots) and using our analytical model (symbols). In curves A and B: C_c is not a transistor model parameter in SPICE but is replaced by its Miller equivalent $C_{cM} = 2C_c = 16$ fF. In curve C: $C_c = 8$ fF is a transistor model parameter in SPICE, and the base Miller component $C_{bM} = C_c = 8$ fF is added to C_e in our model. $C_e = 0$, 10, and 10 fF in A, B, and C, respectively, and $C_{cs} = 30$ fF in A, B, and C.

c) In SPICE: $C_c = 8\,fF$, $C_e = 10\,fF$; In our model: $C_{cM} = 2C_c = 16\,fF$, $C_{bM} = C_c = 8\,fF$, $C_e = 10\,fF$: In this section, both junction capacitances C_c and C_e are considered in SPICE simulations as transistor model parameters. Thus, in order to make a fair comparison, both Miller components of C_c, i.e., C_{bM} and C_{cM}, are taken into account in the analytical solution. We assume that C_{bM} is equal to C_c and is lumped in parallel with C_e. The agreement with SPICE simulations is remarkable as depicted in Fig. 8 (curves C), which confirms the validity of the C_{bM} assumption.

C. Transient Response of the Output Voltage: The Emitter Follower

An emitter follower (EF) output stage is used in ECL gates to act as a buffer between stages and to insure the input–output voltage compatibility.

1) Equivalent Circuit: The emitter follower configuration and its equivalent circuit during transient are shown schematically in Fig. 9(a) and (b), respectively. The source resistance R_s is the output resistance of the switching transistor, R_L. In an ECL gate, the EF is always "on" with an input voltage V_{bf} swinging between V_{cc} and ($V_{cc} - V_s$). The relation between V_{cc}, V_s, and the reference voltage V_r is imposed by the emitter follower since the output voltage V_0 at the emitter node must equal $V_H = (V_r + V_s/2)$ when V_{bf} equals V_{cc}, and $V_0 = V_L = (V_r - V_s/2)$ when V_{bf} equals ($V_{cc} - V_s$). The slight change in V_{be} accompanying the change of the logic state and the voltage drop in the input circuit are usually negligible with respect to V_s and will be neglected. The values of the emitter resistance R_{ef} of the emitter current and of the emitter bias V_{ee} are chosen such that the output voltage swings between V_H and V_L. Usually, R_{ef} is taken equal to

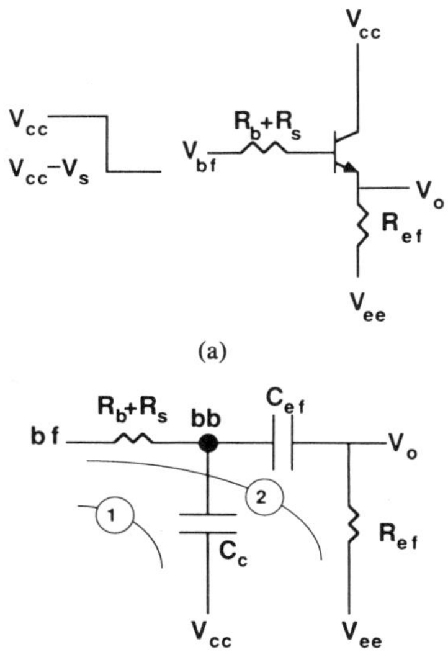

Fig. 9. (a) The emitter follower circuit; (b) equivalent circuit of the emitter follower during switching.

V_s/I_s, and for minimum power dissipation V_{ee} is usually connected to the ground. The emitter follower current is determined accordingly. In special circuitry, such as an ECL gate with totem-pole output, V_{cc} is grounded, which results in negative values for V_r and V_{ee}. In addition, grounded V_{cc} reduces the effect of the supply fluctuations on the logic levels. In the example studied here we started by fixing V_{cc} to 5 V and, consequently, V_{ee} cannot be connected to ground unless a very large emitter follower current is drawn. Since we keep reasonable values for the emitter follower current (1 and 2 mA), the value V_{ee} = 3.2 V follows this choice.

The emitter follower represents an additional circuit that introduces an extra delay to the ECL gate. In Fig. 9(b), the capacitance C_{ef} can be expressed as [15]

$$C_{ef} = \frac{C_\pi}{1 + g_m R_{ef}} \tag{27a}$$

where C_π is the sum of the base charging (diffusion) capacitance C_D and the emitter junction capacitance C_e. The time constant related to the dominant pole of this circuit is given by

$$T_{ef} = C_{ef}(R_{ef} + R_b + R_s). \tag{27b}$$

Since the EF is always "on," we expect that the delay is mainly due to the base charging capacitance and that the delay caused by the junction capacitances C_e and C_c is negligible. With the extremely high cutoff frequencies of today's bipolar transistors, however, the delay caused by the collector junction capacitance C_c might be relatively significant and will be considered in the analysis.

2) Output Voltage Response to an Input Voltage Step: As mentioned above, a rigorous analysis should consider the complete circuit of Fig. 9(b). To simplify the problem, we assume that the output response occurs in two phases.

1) The first phase deals with the circuit consisting of $R_b + R_s$ and C_c and assumes that the impedance consisting of C_{ef} in series with R_{ef} (dominated by R_{ef} during transient) is much larger than the impedance of C_c. In this case, when the input voltage falls suddenly from V_{cc} at $t = 0^-$ to $V_{cc} - V_s$ at $t = 0^+$, the response of the voltage at *bb* is given by

$$V_{bb} = V_H - V_s[1 - \exp(-t/T_{cef})] \tag{28a}$$

where T_{cef} represents the time constant $C_c(R_s + R_b)$. Consequently, the delay $t_{ef1}(v)$ during which V_{bb} falls from V_H to $(V_H - vV_s)$ within this phase is given by

$$t_{ef1}(v) = T_{cef} \ln \frac{1}{(1 - v)}. \tag{28b}$$

This phase continues throughout the transient period, and v is allowed to have values between 0 and 1. Since C_{ef} presents a negligible impedance with respect to R_{ef} during transient, the output voltage V_0 is approximately equal to V_{bb}.

2) In the second phase, it is assumed that the impedance of C_c is much larger than that consisting of C_{ef} in series with R_{ef}. In this case the output voltage V_0 varies almost instantaneously, in response to the voltage step $-V_s$ at *bf*, from V_H to $V_{init} = (V_H - v_{init} V_s)$, where v_{init} is a fraction equal to the potential divider ratio $R_{ef}/(R_{ef} + R_b + R_s)$. This abrupt change in V_0 is due to the fact that the fall time of the input pulse is negligible with respect to T_{ef} given by (27b). Then the output voltage decreases exponentially starting from the new initial value V_{init} heading the final steady-state value V_L. This behavior is described by

$$V_0 = V_L + (V_{init} - V_L) \exp(-t/T_{ef}). \tag{29a}$$

During this phase, the delay $t_{ef2}(v)$ during which V_0 falls from V_H to $(V_H - vV_s)$ is given by

$$t_{ef2}(v) = T_{ef} \ln \frac{(1 - v_{init})}{(1 - v)}. \tag{29b}$$

Note that this phase is limited to values of v greater than v_{init} only.

Finally, the total delay $t_{ef}(v)$ of the output voltage is obtained by superposing the delays $t_{ef1}(v)$ and $t_{ef2}(v)$

$$t_{ef}(v) = t_{ef1}(v) + t_{ef2}(v). \tag{30}$$

The transient response of the emitter follower calculated according to the method described here is plotted in Fig. 10 and compared with the result of SPICE simulations for (curve *a*) a simple case with $C_c = 0$, (curve *b*) for a more the realistic case with $C_c = 8$ fF, and (curve *c*) for a worst case with $C_c = 15$ fF. In cases shown by curves *a* and *b*, the agreement is very reasonable, whereas in case represented by curve *c* the discrepancy starts to be pronounced. This discrepancy is due to the fact that the method of superposition of delays is relatively accurate only when T_{cef} and T_{ef} are widely separated. Note that the discrepancy is particularly significant near the phase-overlap region. On the other hand a very good agreement is observed when either the first phase or the second phase

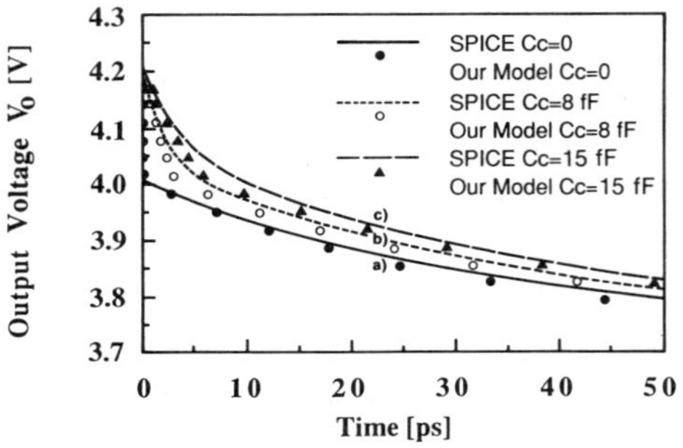

Fig. 10. Transient response of the output voltage V_0 of the emitter follower to a sudden input voltage change from V_{cc} to $V_{cc} - V_s$. The line plots represent SPICE simulations and the symbols represent the results of our model when: (curve *a*) $C_c = 0$; (curve *b*) $C_c = 8$ fF; and (curve c) $C_c = 15$ fF. In these plots, $R_{ef} = R_s = V_s/I_s = 500\ \Omega$, $V_{ee} = 3.2$ V, $C_e = 10$ fF, and the values of the other parameters are identical to those mentioned in the caption of Fig. 2.

dominates the voltage response. Therefore, we estimate that the error in the evaluation of the most interesting delays at $v = 0.1$, 0.5, and 0.9 is relatively small, even in the worst case (curve *c*). Due to the fact that in our model, it is assumed that V_{be} is constant, a slight difference is observed as V_o approaches the final steady-state voltage V_L, as depicted in Fig. 10.

It should be mentioned that, neglecting the collector junction capacitance, we can proceed analytically using KVL and the charge-control model as in Section II-B-1, to obtain an analytical expression for the output voltage transient response of the emitter follower. The resulting expressions for T_{ef} and $V_0(t)$ are identical to (27b) and (29a), respectively. Note that, according to (27a) and (27b), the expression of T_{ef} becomes identical to the expression proposed in [1] when $C_e = 0$, $R_s = R_L = R_{ef} = V_s/I_s$, $g_m R_{ef} >> 1$, and $\alpha = 1$. Finally, the effect of neglecting r_π in the equivalent circuit of Fig. 9(b) is equivalent to neglecting the steady-state base current term. Including this term changes T_{ef} to αT_{ef}.

In reality, the input to the emitter follower in an ECL gate is subjected to a voltage change similar to the response of Fig. 8 rather than to a step voltage change. We studied here, however, the response of the emitter follower to a step input voltage change in order to model the total delay $T(v)$ of the ECL gate as the sum of the transit delay $t(v)$, the junction capacitance delay $t_{jc}(v)$, and the emitter follower delay $t_{ef}(v)$

$$T(v) = t(v) + t_{jc}(v) + t_{ef}(v). \quad (31)$$

III. Application of the Model

In order to estimate accurately the gate delay by applying the model proposed here, low-level injection and unity fan-in and fan-out are prerequisites. The accuracy of the model has been tested by comparing some typical values for the propagation delay measured and reported recently with the values calculated using the model. For example, the delay $T(0.5)$ calculated using the data given in [5] is 87.5 ps, which agrees very well with the reported value 84 ps. In this example, $D = 2.06$, $T_b/T_c = 4.66$, and $t(0.5)$, $t_{jc}(0.5)$, $t_{ef}(0.5)$ represent 56, 27, 17 percent, respectively. In a second example [9], a delay $T(0.5)$ equal to 46.7 ps is calculated, which is in perfect agreement with the measured value of 47.6 ps. In this example, $D = 2.34$, $T_b/T_c = 1.53$, and $t(0.5)$, $t_{jc}(0.5)$, $t_{ef}(0.5)$ represent 62, 26, 12 percent, respectively.

The model proposed here is, *a priori*, general and not restricted to modern technology. Indeed, the propagation delay calculated at three different switching currents differs by less than 10 percent from the measured values in [16].

Finally, the method is applied to cases in which the ECL gate operates in the high-level injection regime. In a first example [6], the calculated delay is 62 ps whereas the measured delay is 80 ps. On the other hand, the propagation delay calculated in another case [7] is 54 ps, which is very close to the reported value of 52 ps. Consequently, one must be very cautious when using the model under high-level injection conditions.

IV. Discussion

Using the expressions proposed here, it becomes rather simple to carry out a sensitivity analysis in which the influence of each parameter involved in the gate delay is investigated [17]. Such an analysis is also useful in validating or disapproving the use of some approximations. For example, to reduce complexity, we have assumed that the emitter junction capacitance has a constant average value C_e independent of the bias condition. The validity of this assumption is investigated by varying C_e and studying carefully the resulting delays. Assuming a step junction and a built-in potential of 0.8 V, the emitter junction capacitance varies between a minimum value $C_{e,\min}$ when the transistor is off to about $1.5C_{e,\min}$ at $i = 0.1$ and finally reaches $2C_{e,\min}$ at full conduction. Usually, $t_{jc,0.1}$ is much larger than the second term constituting $t_{jc,i}$ in (13b). Therefore, the variation of C_e with i for $1 > i > 0.1$ would have a very little influence on the value of $t_{jc,i}$. On the other hand, keeping C_e constant in the region $i < 0.1$ might result in a considerable error in the estimation of $t_{jc,i}$. Two extreme cases are studied. In the first case, C_e is kept constant at $C_{e,\min}$ corresponding to $i = 0$. In the second case, C_e is kept constant at $1.5C_{e,\min}$ corresponding to $i = 0.1$. The analysis reveals that, despite a relatively large difference in t_{jc}, the difference between the total gate delay $T(v)$ calculated in the first case and the total gate delay calculated in the second case is only 9 percent when $v = 0.1$, 5 percent when $v = 0.5$, and 2.5 percent when $v = 0.9$. Since in reality a bias-dependent C_e represents an intermediate situation between these two extreme cases, we consider that a constant C_e is a reasonable assumption.

V. Summary

We propose an analytical model in which the combination of measured (or assumed) quantities (V_s, I_s, τ_F, R_b, R_e, β, R_L, C_c, C_{cs}, and C_e) and four principal figures (D, T_b, $t_{\lim}$, and T_c), easily calculated from these quantities, describes the transient behavior of CML and ECL gates accurately. In this model the total gate delay $T(v)$ is given as the sum of three components: 1) $t(v)$ the delay caused by the transit time and the load circuit of the switching transistor, 2) $t_{jc}(v)$ the delay caused by the junction capacitances of the switching transistor, and 3) $t_{ef}(v)$ the delay caused by the emitter follower (for ECL gates only). Expressions are given for the determination of these three components at any fraction v of the voltage swing. For example, the propagation delay T_d is defined as the total gate delay when the change in the output voltage reaches half the voltage swing as a response to a step input voltage. Therefore, T_d is equal to $T(0.5)$ and is determined at $v = 0.5$.

The junction capacitance delay $t_{jc}(v)$ and the emitter follower delay $t_{ef}(v)$ are calculated by direct substitution in their respective expressions. On the other hand, in order to calculate $t(v)$, two quantities $x_{\max}$ and $v_{\max}$ are defined. The ratio between the base time constant T_b and the collector time constant T_c (or vice-versa to get a ratio greater than one) is compared with $x_{\max}$. Then the fraction v is compared with $v_{\max}$. Depending on the results of these two comparisons an appropriate expression is given for $t(v)$.

Appendix

A. Analytical Integration of (7)

In order to integrate (7) analytically, we proceed through the transformation: $\theta = 2i - 1$, and θ varies between -1 and 1. In terms of θ, (7) becomes

$$\int \frac{d\theta}{A_1' + 2B_1 \ln \dfrac{1-\theta}{1+\theta} - C_1\theta} = \int \frac{1}{\tau_F}\, dt \tag{A1}$$

where $A_1' = 2A_1 - C_1$. Within the range $1 > \theta > -1$ (corresponding to $1 > i > 0$), the denominator in (A1) can be written as $A_1' - (5B_1 + C_1)$, where $\ln[(1-\theta)/(1+\theta)]$ has been replaced by -2.5θ. This approximation introduces a negligible error in the result of the integration. Equation (A1) can now be expressed as

$$\int \frac{d\theta}{A_1' - (5B_1 + C_1)\theta} = \int \frac{1}{\tau_F}\, dt \tag{A2}$$

and can be integrated analytically resulting in

$$i = \left(0.5 + \frac{A_1'}{10B_1 + 2C_1}\right) \cdot \left[1 - \exp\left(-t(5B_1 + C_1)/\tau_F\right)\right]. \tag{A3}$$

Here the boundary condition $\theta = -1$ at $t = 0$ has been used to find the constant of integration.

B. Analytical Determination of the Delay Caused by the Emitter Junction Capacitance

Neglecting q_c, (2a) can be rewritten as

$$I_{bl} = \frac{dq_{Fl}}{dt} + \frac{q_{Fl}}{\tau_{BFl}} + C_e \frac{dV_{bel}}{dt}$$

with

$$\frac{dV_{bel}}{dt} = \frac{V_T}{I_{cl}} \frac{dI_{cl}}{dt}. \tag{A4}$$

Implementing (A4) in the analysis, we finally end up with (12), which can be written in an integral form

$$\int \frac{\tau_F di}{A_1 + B_1 \ln(1/i - 1) - C_1 i} + \frac{C_e V_T}{\alpha I_s} \int \frac{di}{i(A_1 + B \ln(1/i - 1) - C_1 i)} = \int dt. \tag{A5}$$

The first integral in the left-hand side of (A5) is identical to the integral in (A1) and represents the intrinsic delay t_i to be calculated from (8). The second term represents the delay $t_{jc,i}$ caused by the emitter junction capacitance.

In order to obtain an analytical expression for $t_{jc,i}$, we divide the range $1 > i > i(0)$ into two regions:

1) $i < 0.1$: In this region, $\ln(1/i - 1)$ can be approximated by $-\ln i$. We can now write

$$t_{jc,i} = -\frac{C_e V_T}{B_1 \alpha I_s}\left[\int \frac{d(A_1 - B_1 \ln i - C_1 i)}{A_1 - B_1 \ln i - C_1 i} + C_1 \int \frac{di}{A_1 - B_1 \ln i - C_1 i}\right] \tag{A6a}$$

$$t_{jc,i} = \frac{C_e V_T}{B_1 \alpha I_s}\left[\ln \frac{A_1 - B_1 \ln i(0) - C_1 i(0)}{A_1 - B_1 \ln i - C_1 i} - \frac{C\, i}{A_1 - B_1 \ln i - C_1 i}\right]. \tag{A6b}$$

In state-of-the-art silicon bipolar transistors, the terms involving the constant C_1 are usually negligible and can be omitted from the expression.

2) $i > 0.1$: In this region, $\ln(1/i - 1)$ can be approximated by $(2.5 - 5i)$ as in Section II-A-4 (and in Section A above) such that the delay $t_{jc,i}$ can be written as

$$t_{jc,i} = t_{jc,0.1} + \frac{C_e V_T}{\alpha I_s} \int \frac{di}{i[A_1 + 2.5B_1 - (5B_1 + C_1)i]} \tag{A7a}$$

where $t_{jc,0.1}$ has to be evaluated by putting $i = 0.1$ in (A6b). The integral in (A7a) is straightforward and results in

$$t_{jc,i} = t_{jc,0.1} + \frac{C_e V_T}{\alpha I_s (A_1 + 2.5B_1)} \cdot \ln\left[\frac{i}{(A_1 + 2.5B_1) - (5B_1 + C_1)i}\right]. \tag{A7b}$$

On the other hand we know from (A3) and (8) that

$$5B_1 + C_1 = \tau_F/T_b$$

and we can easily prove that

$$\frac{A_1 + 2.5B_1}{5B_1 + C_1} = D.$$

Hence, we can write

$$t_{jc,i} = t_{jc,0.1} + \frac{C_e V_T}{\alpha I_s D} \frac{T_b}{\tau_F} \ln \left[\frac{D - 0.1}{D - i} \left(\frac{i}{0.1} \right) \right]. \quad \text{(A8)}$$

References

[1] J. M. C. Stork, "Bipolar transistor scaling for minimum switching delay and energy dissipation," in *IEDM Tech. Dig.*, 1988, p. 550.

[2] A. Barna, "Analytic approximations for propagation delays in current mode switching circuits including collector–base capacitances," *IEEE J. Solid-State Circuits*, vol. SC-16, p. 597, 1981.

[3] E. F. Chor, A. Brunnschweiler, and P. Ashburn, "A propagation delay expression and its application to the optimization of polysilicon emitter ECL processes," *IEEE J. Solid-State Circuits*, vol. 23, p. 251, 1988.

[4] A. B. Glaser and G. E. Subak-Sharpe, *Integrated Circuit Engineering*. Reading MA: Addison-Wesley, 1979, p. 607.

[5] T. Nakamura, K. Nakazato, T. Miyazaki, and M. Naga, "Integrated 84 ps ECL with I^2L," *ISSCC Tech. Dig.*, 1984, p. 152.

[6] T. Tashiro *et al.*, "An 80 ps ECL circuit with high current density transistor," in *IEDM Tech. Dig.*, 1984, p. 686.

[7] H. Takemura *et al.*, "Submicron epitaxial layer and RTA technology for extremely high speed bipolar transistors," in *IEDM Tech. Dig.*, 1986, p. 424.

[8] K. Kikuchi, S. Kameyama, M. Kajiyama, M. Nishio, and T. Komeda, "A high speed bipolar LSI process using self-aligned double diffusion polysilicon technology," in *IEDM Tech. Dig.*, 1986, p. 420.

[9] K. Washio, T. Nakamura, K. Nakazato, and T. Hayashida, "A 48 ps ECL in a self-aligned bipolar technology," in *ISSCC Tech. Dig.*, 1987, p. 58.

[10] K. Ueno, H. Goto, E. Sugiyama, and H. Tsunoi, "A sub-40 ps ECL circuit at a switching current of 1.28 mA," in *IEDM Tech. Dig.*, 1987, p. 371.

[11] T. C. Chen *et al.*, "A sub-50 ps single poly planar bipolar technology," in *IEDM Tech. Dig.*, 1988, p. 740.

[12] T. Gomi *et al.*, "A sub-30ps Si bipolar LSI technology," in *IEDM Tech. Dig.*, 1988, p. 744.

[13] M. C. Wilson, S. Duncan, and P. C. Hunt, "An ultra high speed trench isolated double polysilicon bipolar process," in *Proc. ESSDERC Conf.*, 1988, p. C4-101.

[14] T. C. Chen *et al.*, "A submicron high performance bipolar technology," in *Symp. VLSI Technol. Dig. Tech. Papers*, 1989, p. 87.

[15] P. R. Gray and R. G. Meyer, *Analog Integrated Circuits*, 2nd ed. New York: Wiley, 1984, pp. 428–440.

[16] N. Hanaoka and A. Anzai, "Perspective of scaled bipolar devices," in *IEDM Tech. Dig.*, 1981, p. 512.

[17] M. Y. Ghannam, R. Mertens, and R. Van Overstraeten, in *Proc. ESSDERC Conf.* (Berlin), 1989, p. 801.

Paper 1.22

Fault Characterization, Testing Considerations, and Design for Testability of BiCMOS Logic Circuits

Aly E. Salama and Mohamed I. Elmasry

***Abstract*—BiCMOS technology is one of the leading candidate technologies for future generations of high-speed VLSI circuits. It is of great importance to study the effect of physical failures on BiCMOS logic circuits and the effect of these failures on the testing techniques currently employed for MOS and bipolar technologies. This paper provides the results of a simulation-based fault characterization study of BiCMOS logic circuits. Based on the fault characterization results, we have studied different techniques for testing BiCMOS logic circuits. Moreover, we present a novel BiCMOS circuit structure that improves the testability of BiCMOS digital circuits.**

I. Introduction

BiCMOS technology significantly enhances speed performance while incurring a small power or area penalty. The main disadvantage of BiCMOS is its greater process complexity, which could result in lower manufacturing yield [1]. This imposes stringent requirements on testing BiCMOS logic circuits [2].

We address the testing problems associated with BiCMOS logic circuits. In Section II, we provide the results of a simulation-based fault characterization study of conventional BiCMOS logic circuits. Based on the fault characterization results, we studied in Section III the effectiveness of stuck-at fault testing, stuck-open fault testing, delay fault testing, and current testing in achieving a high level of defect coverage. Finally, in Section IV, a new circuit is suggested for improving the testability of BiCMOS digital circuits.

II. Fault Characterization of BiCMOS Logic Circuits

A. Defect Modeling

During the fabrication of an IC, physical defects could occur. A defect that causes a change of the logical function of the circuit can be represented by a logical level abstraction known as a *logical fault*. Similarly, a defect that causes a change in a continuous parameter of the circuit can be represented by an abstraction called a *parametric fault* [3].

Since BiCMOS technology combines both CMOS and bipolar devices on the same substrate, BiCMOS devices are subject to most of the defects from which each individual technology suffers. The most prevalent failures in the bipolar technology are [4]: 1) open connections, 2) shorts between connections, and 3) "piped" transistors with excessive emitter-to-collector leakage current. Similarly, defects which can be found in MOS integrated circuits have been well documented [5]. Defects in CMOS circuits include: 1) shorts (e.g., gate oxide shorts), 2) opens (e.g., intragate breaks), and 3) circuit degradation (e.g., threshold voltage variations).

In our study, the defects are modeled and then simulated using the circuit simulator SPICE. In particular, we have simulated the following faults.

1) Short Faults: To model a short between two terminals of a device, we assume a resistive connection between the shorted nodes. The value of the resistance depends on the strength of the defect itself. We confine ourselves here to the case of hard faults where the resistance assumes a very small value $< 10\ \Omega$.

2) Open Faults: To model an open fault of a device terminal, a large resistance $> 10\ \mathrm{M}\Omega$ is inserted between the device terminal and the circuit node to which the terminal would otherwise be connected. For an open gate fault the gate terminal was tied to the corresponding bulk terminal. Recent research [6] suggests that break faults that cause the gate terminal to float in a CMOS logic gate will cause the corresponding transistors to be stuck-on. This is partly covered by the simulated drain–source short.

B. Conventional BiCMOS Logic Circuits

The results of simulating physical defects in a conventional BiCMOS two-input NAND gate, shown in Fig. 1, are given in Table I. Other gate types (inverter, NOR) exhibit similar characteristics. The effects of the simulated faults can be classified into one or more of the following categories: 1) output follows one of the inputs, 2) output stuck at ONE or output stuck at ZERO, 3) output stuck open (at high impedance), 4) input stuck at ONE or input stuck at ZERO, 5) output voltage is logically indeterminate, 6) output voltage has a "soft" logic ONE or a "soft" logic ZERO, 7) an increase in the power supply current (I_{DDQ} fault), and 8) output rise time is increased (output slow to

Manuscript received September 12, 1991; revised February 3, 1992.

A. E. Salama is with the VLSI Research Group, Department of Electrical and Computer Engineering, University of Waterloo, Waterloo, Ont., Canada N2L 3G1, on leave from the Department of Electronics and Electrical Communications Engineering, Cairo University, Giza, Egypt.

M. I. Elmasry is with the VLSI Research Group, Department of Electrical and Computer Engineering, University of Waterloo, Waterloo, Ont., Canada N2L 3G1.

IEEE Log Number 9108182.

Reprinted from *IEEE J. Solid-State Circuits*, vol. 27, no. 6, pp. 944–947, June 1992.

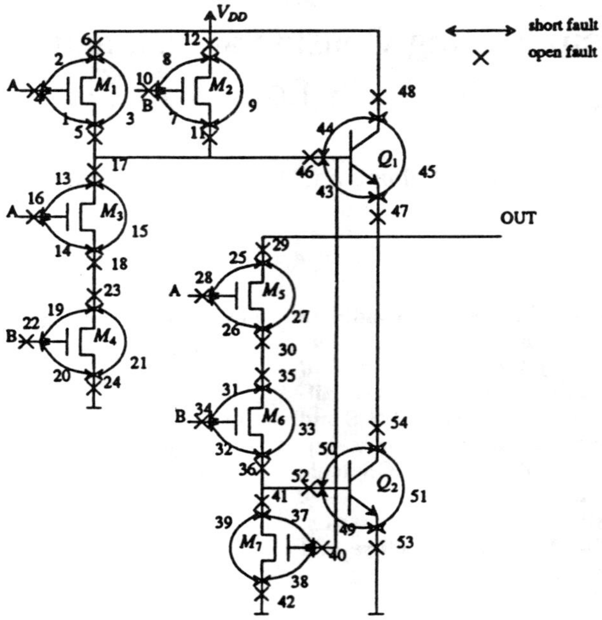

Fig. 1. A two-input BiCMOS NAND gate with fault conditions.

TABLE I
BiCMOS NAND GATE: CHARACTERIZATION OF SIMULATED PHYSICAL FAULTS

Fault Characterization	Simulated Physical Faults
Normal Logic Operation	F_0 (fault-free case), F_{14}, F_{15}, $F_{16} = F_{17} = F_{18}$, F_{19}, F_{21}, $F_{22} = F_{23} = F_{24}$, F_{26}, F_{29}, $F_{39} = F_{49}$, $F_{40} = F_{41} = F_{42}$, F_{43}, F_{48}, F_{54}
O/P stuck at ONE or I/P stuck at ZERO	$F_3 = F_9$, F_{20}, F_{44}, F_{45}
I/P A stuck at ONE	F_2
I/P B stuck at ONE	F_8
O/P stuck at ZERO	F_{37}, F_{50}, F_{51}
O/P stuck open	$F_4 = F_5 = F_6$, $F_{10} = F_{11} = F_{12}$, $F_{28} = F_{29} = F_{30}$, $F_{34} = F_{35} = F_{36}$, F_{38}, $F_{46} = F_{47}$, F_{52}, F_{53}
O/P at intermediate level	F_{27}, F_{31}, F_{32}, F_{33}
O/P follows one of the inputs	$F_1 = F_{13}$, F_7, F_{25}
O/P has a "soft" logic ZERO or a "soft" logic ONE	F_{14}, F_{15}, F_{19}, F_{21}
O/P is slow to rise or is slow to fall	$F_{16} = F_{17} = F_{18}$, $F_{22} = F_{23} = F_{24}$, $F_{39} = F_{49}$, $F_{40} = F_{41} = F_{42}$, F_{43}, F_{48}, F_{54}
An increase in power supply current for certain test vectors	$F_1 = F_{13}$, F_2, $F_3 = F_9$, F_7, F_8, F_{14}, F_{15}, F_{19}, F_{20}, F_{21}, F_{25}, F_{26}, F_{27}, F_{31}, F_{32}, F_{33}, F_{37}, F_{38}, F_{44}, F_{45}, F_{50}, F_{51}

rise) or output fall time is increased (output slow to fall). Fault categories 1–5 are considered logical faults and fault categories 6–8 are considered parametric faults.

C. Other BiCMOS Logic Families

Similar simulations have been carried out for R-type BiCMOS, BiNMOS, and full-swing feedback-type BiCMOS logic gates [11]. The results are similar to those of conventional BiCMOS logic gates [7].

In all simulations the inputs of the gates are assumed driven by a CMOS or a BiCMOS buffer and the output of the gate is connected to a CMOS or a BiCMOS inverter. Moreover, the test vectors have been applied at the normal operating speed of the gate.

D. Comments and Discussion of the Simulated Results

1) BiCMOS circuits have inherited the stuck-open fault model from the CMOS circuits. As such, testing such faults is accomplished by two test vectors.

2) Many physical failures will not affect the logical behavior of the gate. This large number of undetectable failures is a unique feature of BiCMOS circuits due to the redundancy in the basic circuit elements. The degradation in the delay time could generate delay faults.

3) Short faults are usually accompanied by much higher currents than in the CMOS case. A technique based on observing leakage currents, as has been suggested for CMOS circuits, will detect stuck-on faults.

III. Testing BiCMOS Digital Circuits

Test generation depends on the assumed fault model. For MOS and bipolar digital circuits different fault models have been utilized in generating test vectors. We investigate the defect coverage that would result by applying test vectors based on a number of fault models.

A. Stuck-At-Fault Testing

For the two-input NAND gate the minimum stuck-at test set is {11, 01, 10}. According to Table I, only 18.5% of simulated physical faults are modeled as stuck-at faults. Using this test set we detected 46.3% of the modeled physical faults. If we consider the minimum test set in the following order {01, 11, 10}, we detect 59.2% of the faults. This is due to the existence of stuck-open faults.

Thus, the stuck-at fault testing is still viable in detecting an appreciable fraction of the simulated physical faults.

B. Stuck-Open Fault Testing

For the two-input NAND gate a minimum ordered test set for stuck-open testing is given by {11, 01, 11, 10}. Application of this test set to the NAND gate detects about 64.8% of the simulated faults.

C. Delay Fault Testing

Associated with each input and output are two kinds of transition faults, namely, slow to rise (STR) and slow to fall (STF) [7]. The similarity between stuck-open and transition fault testings suggests that a stuck-open test set can be used as a starting point for generating the transition fault test set.

For the two-input NAND gate of Fig. 1, we considered the stuck-open test set BA = {11, 01, 11, 10}. This test set tests for the following transition faults: OUT STR (STF), B STF (STR), and A STF. It is a robust test set. Application of this test set has detected all 14 simulated faults indicated in Table I. Thus, by considering detection of the delay faults, the test set {11, 01, 11, 01} physical fault coverage rises to 90.7%.

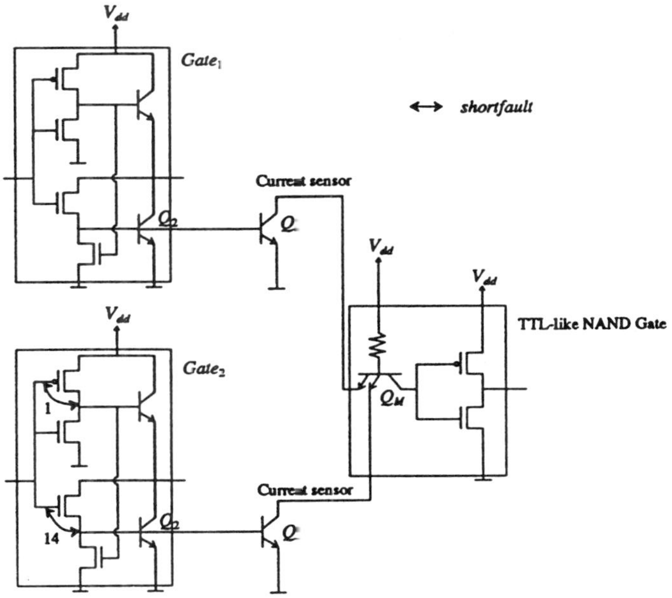

Fig. 2. Current testing circuit.

D. Current Testing

Recently, current testing has emerged as a viable testing technique [9]. As has been indicated in [10], for an independent inverter, NAND, or NOR gate, the minimum stuck-at fault test set is the complete leakage current test set. In Table I, current testing detects 24 simulated faults. Among these faults are five faults that were undetectable using previous testing techniques. Thus, simulated physical fault coverage rises to 100%.

IV. Design for Testability for BiCMOS Circuits

In this section we consider faults that cause abnormal currents. These type of faults could be due to many defects that could be modeled as shorts, bridge faults, transistor stuck-on faults, and even some open faults. We present here a new design-for-testability (DFT) technique for their detection.

A. Current Testing Circuit

The technique presented here is schematically illustrated in Fig. 2. BiCMOS gates are modified with the addition of transistor Q_t whose base and emitter nodes are connected to the base and emitter nodes of the pull-down transistor Q_2. The collector of Q_t is connected to the multiemitter transistor Q_M. The multiemitter transistor is the input of a TTL-like NAND gate. The multiemitter transistor with n emitters will be connected to n different BiCMOS gates, through the added current-mirror like transistor Q_t.

In the case of a fault that causes an abnormal current to pass between V_{DD} and GND and due to the current mirror effect, Q_t will carry current proportional to that which passes through Q_2. This current will discharge the input capacitor of the CMOS inverter and the output will be at logic ONE. Thus, any fault that causes an abnormal current can be detected by checking the output of the added CMOS inverter. A logic ZERO output indicates fault-free operation and a logic ONE output indicates a faulty condition.

V. Conclusions

In this work we have characterized the effect of physical defects on BiCMOS logic circuits. BiCMOS logic is highly susceptible to defects resulting in parametric faults. Our study has shown that to achieve a high level of defect coverage a combined testing strategy should be adopted. Thus, BiCMOS circuits should be tested for stuck-at faults, stuck-open faults, delay faults, and current-induced faults. We also presented a novel design-for-testability technique to improve the testability of BiCMOS circuits.

Acknowledgment

The authors would like to express their appreciation to Dr. S. Embabi of Texas A&M University for fruitful interaction at the early stages of this work.

References

[1] A. R. Alvarez *et al.*, "An overview of BiCMOS technology and applications," in *Proc. IEEE Int. Symp. Circuits Syst.*, 1990, pp. 1967–1970.

[2] M. E. Levitt, K. Roy, and J. Abraham, "Test considerations for BiCMOS logic families," in *Proc. IEEE Custom Integrated Circuits Conf.*, 1990, pp. 17.2.1–4.

[3] F. J. Ferguson, M. Taylor, and T. Larrabee, "Testing for parametric faults in CMOS circuits," in *Proc. IEEE Int. Test. Conf.*, 1990, pp. 436–443.

[4] C. Timoc *et al.*, "Logical models of physical failures," in *Proc. IEEE Int. Test Conf.*, 1983, pp. 546–556.

[5] J. Galiay, Y. Crouzet, and M. Vergniault, "Physical versus logical fault models for MOS LSI circuits: Impact on their testability," *IEEE Trans. Comput.*, vol. C-29, pp. 527–531, 1980.

[6] W. Maly and P. K. Nigh, "Testing oriented analysis of CMOS ICs with opens," in *Proc. Int. Conf. Computer-Aided Design*, 1988, pp. 344–347.

[7] A. E. Salama and M. I. Elmasry, "Testing and design for testability of BiCMOS logic circuits," in *Proc. IEEE VLSI Test Symp.*, 1992.

[8] J. A. Waicukauski, E. Lindbleom, B. K. Rosen, and V. S. Iyengar, "Transition fault simulation," *IEEE Design & Test Comput.*, pp. 32–38, 1987.

[9] P. Nigh and W. Maly, "Test generation for current testing," *IEEE Design & Test Comput.*, pp. 26–38, 1990.

[10] Y. K. Malaiyu and S. Y. H. Su, "A new fault model and testing technique for CMOS devices," in *Proc. IEEE Int. Test Conf.*, 1982, pp. 25–34.

[11] M. I. Elmasry, Ed., *Analysis and Design of BiCMOS Integrated Circuits.* New York: IEEE, 1992.

Paper 1.23

Scaling of Digital BiCMOS Circuits

ABDELLATIF BELLAOUAR, MEMBER, IEEE, SHERIF H. K. EMBABI, STUDENT MEMBER, IEEE, AND MOHAMED I. ELMASRY, FELLOW, IEEE

Abstract —**In this paper we present a generalized first-order scaling theory for BiCMOS digital circuit structures. The effect of horizontal, vertical, and voltage scaling on the speed performance of various BiCMOS circuits is presented. We use the generalized scaling theory for the MOSFET and the constant collector current (CIC) scaling scheme for the BJT. In scaling the bipolar transistor, polysilicon emitter contact and bandgap narrowing are taken into account. A case study for scaling BiCMOS circuits operating at 5- and 3.3-V power supplies shows that scaling improved BiCMOS buffers more significantly than CMOS buffers. Moreover, the low delay-to-load sensitivity of BiCMOS is preserved with scaling.**

NOMENCLATURE

A	Gate area.
A_e	Emitter area.
β_0	Current gain.
C_1	Total parasitic capacitance at the base of bipolar transistor of BiCMOS buffer.
C_{1m}	MOS S/D parasitic capacitance plus base–collector capacitance at the base of bipolar transistor of BiCMOS buffer.
C_2	Total capacitance at the output of CML and CMOS gates.
C_c	Collector depletion capacitance.
C_{cg}	Current-mirror common-gate capacitance.
C_e	Emitter depletion capacitance.
C_j	Junction capacitance per unit area.
C_N	Correction due to bandgap narrowing effect.
C_{ox}	Gate capacitance per unit area.
C_p	Parasitic capacitance at the output of a gate.
D_{nB}	Electron diffusion constant in the base.
D_{pE}	Hole diffusion constant in the emitter.
D_{p1}	Hole diffusion constant in polysilicon emitter constant.
ΔE_g	Bandgap narrowing.
E_c	Critical field for velocity saturation.
ϵ_s	Silicon permittivity.
G	Reduction factor for base resistance.
$I_{D\text{sat}}$	MOS drain saturation current.
I_K	Knee current.
I_o	CML current.
I_{opt}	Optimum current of CML gate.
I_c	Collector current.
K	MOS saturation current factor.
k_β	Scaling factor in the current gain.
k_{cl}	Scaling factor of buffer load.
k_f	Scaling factor of polysilicon emitter contact.
k_h	Horizontal scaling factor.
k_m	Bipolar voltage scaling factor.
k_u	MOS voltage scaling factor.
k_v	Vertical scaling factor.
L_e	Device electrical channel length.
L_{eff}	Effective channel length.
L_{p1}	Hole diffusion length in polysilicon region of emitter contact
N_B	Base doping.
N_E	Emitter doping.
N_{epi}	Collector epitaxial layer doping.
q	Electron charge.
Q_{B0}	Base charge at low voltage.
Q'_{B0}	Base charge per area unit.
r_b	Base resistance.
r_{be}	Extrinsic base resistance.
r_{bi0}	Intrinsic base resistance at low voltage.
r_{bis}	Intrinsic base sheet resistance.
r_c	Collector resistance.
R_i	Input resistance to CML gate.
R_L	Load resistance of CML gate.
t_0	Initial delay to turn on bipolar transistor in CML gate.
t_1	Time to switch the output voltage up to half the voltage swing in CML gate.
t_d	Delay time.
t_{ox}	Gate oxide thickness.
τ_0	Input constant time of CML gate.
τ_f	Base transit time.
τ_i	Time constant in CML gate.
τ_l	Load time constant.
τ_s	Bipolar switching time constant.
V_{BC}	Base–collector voltage.
$V_{BE\,\text{on}}$	Base–emitter voltage to switch on bipolar transistor.
V_{dd}	Power supply.
$V_{D\text{sat}}$	Drain–source saturation voltage.

Manuscript received October 24, 1989; revised February 15, 1990. This work was supported in part by NSERC, NTEL, and ITRC research grants.

The authors are with the VLSI Group, Department of Electrical and Computer Engineering, University of Waterloo, Waterloo, Ont., Canada N2L 3G1.

IEEE Log Number 9036471.

Reprinted from *IEEE J. Solid-State Circuits,* vol. 25, no. 4, pp. 932–941, Aug. 1990.

V_{GS}	Gate–source voltage.
V_{pt}	Base punchthrough voltage.
v_s	Saturation velocity in the silicon.
V_s	Voltage swing.
V_T	Threshold voltage.
W	MOS device width.
W_1	Polysilicon emitter layer thickness.
W_B	Base width.
W_E	Emitter width.
W_{jE}	Emitter junction depth.
x_j	Junction depth.

I. Introduction

THE importance of BiCMOS technology has been widely recognized [1]–[6]. As the technology is scaled down from 1.2–0.8 μm to 0.5–0.25 μm [7]–[10], the effect of this scaling on both device and circuit performance must be studied. Such study should include the different MOS scaling schemes, the scaling of the bipolar transistor, and the trade-off factors related to circuit performance.

In this work we present a generalized first-order scaling theory for BiCMOS digital circuit structures. The effects of horizontal, vertical, and voltage scaling on the speed performance of various BiCMOS circuits are presented. These include different types of BiCMOS pull-up/pull-down buffers, CMOS/ECL and ECL/CMOS converters, ECL gates, and CMOS gates and buffers. The study is based on analytical delay relations for each of these circuits. These relations offer insight into circuit performance. The constant current scaling used for bipolar transistors takes into account vertical, horizontal, and base–collector voltage scaling factors. The scaling approach satisfies punchthrough and Kirk current constraints, which determine the scaling of other related process parameters. Bipolar scaling takes into consideration bandgap narrowing and polysilicon emitter contact. For MOS transistors, a generalized approach is used based on an analytical model for submicrometer devices [11]. An empirical relation for scaling the MOS model is also developed in this work. The impact of scaling the power supply to 3.3 V is considered. It is shown that a certain scaling scheme results in better performance for BiCMOS than CMOS when driving high capacitances, without affecting the improvement factor of CMOS gates.

II. BiCMOS Scaling

MOSFET scaling can follow three theories: the constant electric field (CE) [12], the constant voltage (CV) [13], and the quasi-constant voltage (QCV) [13] scaling theories. Also, other valuable scaling models exist: an empirical scheme [14] and a more generalized theory [15].

The scaling of bipolars has received less attention than that of MOSFET's. The scaling properties and limitations of bipolar circuits have been studied [16]–[19]. Two different schemes have been considered [20]: the constant collector-current density (CJC) scaling and the constant collector-current (CIC) scaling. It was shown that the CIC scaling, where the leakage current resulting from the stress of emitter–base junction has less impact on the current gain, is preferable [20].

TABLE I
Scaling Laws of BiCMOS Device

Parameter	Scaling Factor	Equation
Bipolar		
Horizontal dimensions	k_h	
Vertical dimensions	k_v	
Collector-current density J_c	k_h^{-2}	
Voltage V_{BC}	k_m	
Base doping N_B	$k_v^{-2} k_m$	(A1)
Epi-layer doping N_{epi}	k_h^{-2}	(A2)
Current gain β_0	$k_\beta^{-1} k_m^{-0.75}$	(A3)
Knee current I_K	$k_h^2 k_v^{-2.1} k_m^{0.55}$	(A8)
Intrinsic base resistance r_{bi0}	$k_v^{0.3} k_m^{-0.65}$	(A12)
Collector depletion capacitance C_c	k_h	(A14)
Emitter depletion capacitance C_e	$k_h^2 k_v^{-1} k_m^{0.5}$	(A15)
Substrat depletion capacitance C_s	k_h	[18]
Base transit time τ_f	$k_v^{1.1} k_m^{0.45}$	(A10)
MOS		
Horizontal dimensions	k_h	
Voltages	k_u	
Gate oxide thickness	k_h	
Threshold voltage V_T	k_u	
Oxide capacitance AC_{ox}	k_h	
Junction capacitance AC_j	k_h	[15]
Saturation current I_{Dsat}	$k_h^{-0.4} k_u^{1.3}$	(A16)

The BiCMOS scaling theory must be general, taking into account horizontal, vertical, and voltage scaling factors [21]. For a general scaling of the MOS transistor, the supply voltage is scaled by a factor k_u. For a CIC scaling of the bipolar transistor, the base–collector voltage is scaled down by a factor k_m, where k_m could be equal to or different than k_u. Table I summarizes the basic scaling factors k_h, k_v, k_u, and k_m and their effects on the different device and process parameters that are of special significance for the performance of BiCMOS circuits. Although the depth of the epilayer would affect some bipolar device parameters such as breakdown voltage, collector resistance, and collector–substrate depletion capacitance, only the latter is relevant to the proposed first-order scaling scheme. However, the effect of the epilayer depth on the collector–substrate can be neglected specially for an oxide or trench-isolated bipolar transistor. Therefore, we have not included the effect of scaling the epilayer depth in our study.

The scaling factor of the base doping of the bipolar transistor is to be determined from the punchthrough condition (see (A1) in the Appendix). To avoid the Kirk effect, the epilayer doping has to be scaled by the same factor as the collector-current density. The gain scaling factor depends on k_β^{-1}, which takes into account the bandgap narrowing phenomena and the polysilicon contact technology. k_β^{-1} is proportional to $k_v^{0.5}$, which means that the gain degrades with vertical scaling. However, an analytical study has shown that by controlling the properties of the emitter polysilicon contact the current gain can be increased [22]. To account for the effect of the properties of the emitter polysilicon contact, a new scaling factor k_f is introduced. By decreasing k_f the current gain can be drastically improved as explained in the Appendix. The knee current is not affected by uniform vertical and horizontal scaling. The factor k_v improves the intrinsic base resistance r_{bi0} and the transit time τ_f. Also the factor k_m improves the current gain and the base transit time, but degrades the intrinsic base resistance and knee current. The depletion capacitances show an improvement with horizontal scaling.

For the MOS transistor, it is assumed that the threshold voltage is scaled by the same factor as the supply voltage (k_u) to maintain high transconductance. The gate and junction capacitances are scaled by a factor k_h [15]. If the saturation current for submicrometer devices is to be improved, the effect of horizontal scaling has to dominate over that of the voltage. Details are given in the Appendix.

III. Scaling BiCMOS Circuit Structures

In this section, the effects of scaling on various BiCMOS circuit structures such as BiCMOS pull-up/pull-down, CMOS/ECL and ECL/CMOS converters, and ECL gates are studied. ECL scaling is an important and relevant issue in this study because it is expected that an increasing number of BiCMOS chips will include ECL subcircuits or ECL I/O drivers.

A. BiCMOS Buffer Scaling

Two types of BiCMOS pull-up/pull-down buffers [23] are considered as shown in Fig. 1. The circuit of Fig. 1(a) is an N type which has a logic voltage swing of $V_{dd} - 2V_{BEon}$. Figs. 1(b) and (c) show an R+N type and a feedback (FB) type, respectively. Both produce a logic voltage swing equal to V_{dd}. The delay of each circuit is a function of the logic voltage swing. We will consider two cases. The first case assumes that the voltage swing at both input and output is equal to $V_{dd} - 2V_{BEon}$ and the second case assumes full 0 to V_{dd} swing. The value of the load capacitance determines the mode of operation of the bipolar transistor, i.e., high- or low-level injection.

The delay for the low-level injection case (small load capacitance) is analytically given by [24]

$$t_{dl} = t_{dl1} + t_{dl2} + t_{dl3} + t_{dl4} \tag{1}$$

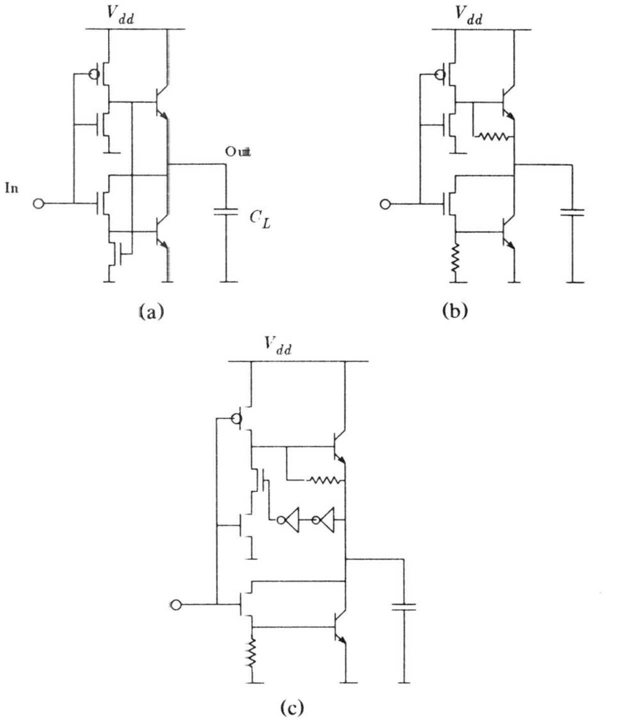

Fig. 1. BiCMOS buffer configurations: (a) N type, (b) R+N type, and (c) FB type.

where

$$t_{dl1} = \frac{C_1 V_{BEon}}{I_{Dsat}}$$

$$t_{dl2} = \frac{\beta_0 \tau_f (C_L + C_p)}{C_L + C_p + \beta_0 C_{1m}}$$

$$t_{dl3} = \frac{V_s}{2 I_{Dsat}} C_{1m}$$

$$t_{dl4} = \frac{V_s}{2\beta_0 I_{Dsat}} (C_L + C_p).$$

The first component, t_{dl1}, is the time required for the bipolar transistor to turn on, t_{dl2} is approximately the time required for the collector current of the bipolar transistor to reach its peak, and the sum of t_{dl3} and t_{dl4} is the time elapsed until the output voltage reaches $V_{dd}/2$.

For the case where $C_L \ll C_{1m}$ and $C_L \ll C_p$, $t_{dl} \approx t_{dl1} + t_{dl3}$. We notice for a 5-V power supply that $t_{dl3} \approx 3.5 t_{dl1}$. On the other hand, if $\beta_0 C_{1m} > C_L > C_p$, $t_{dl2} \approx \tau_f C_L / C_{1m}$ and t_{dl4} can be comparable to t_{dl3}. Therefore all four terms in (1) have to be considered.

At high-level injection (high load capacitance, i.e., $C_L \gg C_{1m}/2(\beta_0 I_K / I_{Dsat})^{1/2}$) and for high MOS current, the delay can be approximated by [24]

$$t_{dh} = t_{dh1} + t_{dh2} + t_{dh3} \tag{2}$$

where

$$t_{dh1} = \frac{C_1 V_{BEon}}{I_{Dsat}}$$

TABLE II
DELAY SCALING FACTORS OF BiCMOS BUFFERS:
LOW-LEVEL INJECTION CASE

Case	Terms of delay	Scaling Factor (Reduced swing case)	Scaling Factor (Full swing case)
$C_L \ll C_{1m}$	$t_{dl1} = C_1 \frac{V_{BE0n}}{I_{Dsat}}$	$k_h^{1.4} k_u^{-2.2}$	$k_h^{1.4} k_u^{-1.3}$
	$t_{dl3} = V_s \frac{C_{1m}}{2I_{Dsat}}$	$k_h^{1.4} k_u^{-0.7}$	$k_h^{1.4} k_u^{-0.3}$
$\beta_0 C_{1m} > C_L > C_p$	$t_{dl1} = C_{1m} \frac{V_{BE0n}}{I_{Dsat}}$	$k_h^{1.4} k_u^{-2.2}$	$k_h^{1.4} k_u^{-1.3}$
	$t_{dl2} \approx \tau_f \frac{C_L}{C_{1m}}$	$k_v^{1.1} k_{cl} k_h^{-1} k_m^{0.45}$	$k_v^{1.1} k_{cl} k_h^{-1} k_m^{0.45}$
	$t_{dl3} = V_s \frac{C_{1m}}{2I_{Dsat}}$	$k_h^{1.4} k_u^{-0.7}$	$k_h^{1.4} k_u^{-0.3}$
	$t_{dl4} = V_s \frac{C_L}{2\beta_0 I_{Dsat}}$	$k_h^{0.4} k_\beta k_u^{-0.7} k_{cl} k_m^{0.75}$	$k_h^{0.4} k_\beta k_u^{-0.3} k_{cl} k_m^{0.75}$

TABLE III
DELAY SCALING FACTORS OF BiCMOS BUFFERS:
HIGH-LEVEL INJECTION CASE

Case	Terms of delay	Scaling Factor (Reduced swing case)	Scaling Factor (Full swing case)
$C_L \gg \frac{C_{1m}}{2}(\beta_0 \frac{I_K}{I_{Dsat}})^{1/2}$	$t_{dh2} \approx \tau_f (\beta_0 \frac{I_K}{I_{Dsat}})^{1/2}$	$k_h^{1.2} k_\beta^{-0.5} k_u^{-1.1} k_m^{0.35}$	$k_h^{1.2} k_\beta^{-0.5} k_u^{-0.65} k_m^{0.35}$
	$t_{dh3} \approx V_s \frac{C_L}{2(\beta_0 I_{Dsat} I_K)^{1/2}}$	$k_h^{-0.8} k_v k_\beta^{0.5} k_u^{0.4} k_{cl} k_m^{0.1}$	$k_h^{-0.8} k_v k_\beta^{0.5} k_u^{0.35} k_{cl} k_m^{0.1}$
$C_L \gg 2\beta_0 \frac{I_K}{V_s} \tau_f$	$t_{dh} \approx V_s \frac{C_L}{2(\beta_0 I_{Dsat} I_K)^{1/2}}$	$k_h^{-0.8} k_v k_\beta^{0.5} k_u^{0.4} k_{cl} k_m^{0.1}$	$k_h^{-0.8} k_v k_\beta^{0.5} k_u^{0.35} k_{cl} k_m^{0.1}$

$$t_{dh2} = \tau_f \left(\frac{\beta_0 I_K}{I_{Dsat}} \right)^{1/2}$$

$$t_{dh3} = \frac{V_s(C_L + C_p + C_{1m})}{2(\beta_0 I_K I_{Dsat})^{1/2}}$$

where t_{dh1} and t_{dh2} are similarily defined as t_{dl1} and t_{dl2} of (1). The delay component t_{dh3} represents the time elapsed between the end of t_{dh2} and the time when the output voltage reaches $V_{dd}/2$. In general, t_{dh1} is negligible compared to t_{dh2} and t_{dh3}. However, for $C_L \gg 2\beta_0 I_K / V_s \tau_f$ (which means that $C_L \gg 1$ pF for typical device parameters), the term t_{dh3} dominates.

Equations (1) and (2) apply for full- and reduced-swing circuits. However, the voltage swing V_s and the saturation current I_{Dsat} differ for both circuits because of the $2V_{BEon}$ losses in the case of the reduced swing circuit. The $2V_{BEon}$ losses have been considered while deriving the scaling laws for both parameters (V_s and I_{Dsat}). As a result, V_s scales as k_u and as $k_u^{1.5}$ for the full- and reduced-swing cases, respectively. The scaling laws of I_{Dsat} are more complicated and their derivation is discussed in detail in the Appendix.

Tables II and III summarize the scaling laws of BiCMOS buffers. We assume as first order that the parasitic capacitances C_1 and C_{1m} scale by a factor k_h, and the load capacitance C_L scales by a factor k_{cl}. The factor k_{cl} takes into account the possible scaling of external loading, for example: buses, driven gates, etc. The tables show that by appropriate scaling of voltage and dimensions, the delay scales down as demonstrated by the test case discussed in Section III-E.

At low-level injection and for very low load capacitance ($C_L \to 0$), the first term of the delay (equation (1)), t_{dl1}, is more degraded by the scaling of the power supply than the term t_{dl3}. Therefore the factor k_u should scale at a slower rate than k_h, to ensure that the delay (when $C_L \to 0$) improves with scaling. For the case where $\beta_0 C_{1m} > C_L > C_p$, the delay-to-load sensitivity is decreased by k_β (term t_{dl4}) and k_m (terms t_{dl2} and t_{dl4}). Hence the factor k_m enhances the performance of BiCMOS buffers at low-level injection provided that the effect of the base resistance on the delay is negligible. To enhance this improvement, the factor k_v should decrease faster than k_h (term t_{dl2}), and the factor k_u should also decrease slower than k_h (term t_{dl4}). We notice that the delays of the buffers with full-swing logic are less de-

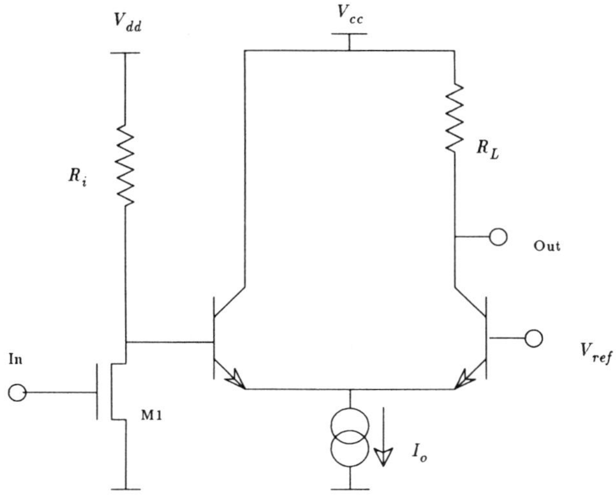

Fig. 2. Simple CMOS/ECL converter.

graded by the power supply scaling than the buffers with reduced swing. For example, for $C_L \ll C_{1m}$ the terms t_{dl1} and t_{dl3} are less degraded by the factors $k_h^{0.9}$ and $k_h^{0.4}$, respectively.

At high-level injection, the current gain scaling factor k_β degrades the term t_{dh_2}, but improves the term t_{dh_3} with the same rate. The effect of k_m is weak compared to the low-level injection case and particularly for high load capacitance. The power supply scaling of the MOS device (k_u) degrades t_{dh_2}, but improves t_{dh_3}. In general, one could conclude that the scaling is more advantageous for low-level injection (particularly for the full-swing case) than for high-level injection. This will be illustrated by the example given in Section III-E.

In this study the effect of saturation of the bipolar due to the voltage drop across the collector resistance was not considered. However, this drop scales by the factor k_h. Hence, its degrading effects could be muted by scaling.

B. CMOS/ECL Converter

A simple CMOS/ECL converter which accepts ECL swings is shown in Fig. 2. Assuming that the voltage swing at the output of the NMOS stage is small (V_s is between 200 and 500 mV), the input resistance R_i can be expressed as [24]

$$R_i = \frac{V_s}{I_{D\text{sat}}}. \tag{3}$$

The delay of the circuit is given by [24]

$$t_d = t_0 + t_1 \tag{4}$$

where t_0 is the turn-on time of the input transistor of the CML pair and t_1 is the time required for the output to reach its midpoint. t_0 is expressed as

$$t_0 = F\tau_0 \tag{5}$$

where F is a function of $V_{BE\,\text{on}}$, V_s, and I_0. As a first-order approximation we can assume that F is not affected by

TABLE IV
CMOS-TO-ECL CONVERTER SCALING FACTORS

Condition/Region		Term	Scaling Factor
$\tau_l \ll \tau_s$	$R_i \ll 2r_b$	t_d	$k_v^{1.4} k_m^{-0.2}$
	$R_i \gg 2r_b$	t_d	$k_h^{0.4} k_u^{-1.3} k_v^{1.1} k_m^{0.45}$
$\tau_l \gg \tau_s$		t_d	k_h
At the minimum ($\tau_l = \tau_s$)	$R_i \ll 2r_b$	t_0	$k_h k_v^{0.3} k_m^{-0.65}$
		t_1	$k_h^{0.5} k_v^{0.7} k_m^{-0.1}$
		I_{opt}	$k_h^{0.5} k_v^{-0.7} k_m^{0.1}$
	$R_i \gg 2r_b$	t_0	$k_h^{1.4} k_u^{-1.3}$
		t_1	$k_h^{0.7} k_v^{0.55} k_u^{-0.65} k_m^{0.2}$
		I_{opt}	$k_h^{0.3} k_v^{-0.55} k_u^{0.65} k_m^{-0.2}$

scaling. The time constant τ_0 is given by

$$\tau_0 = (R_i + r_b)C_1 + R_i C_i \tag{6}$$

where C_i is the parasitic capacitance of the drain of the NMOS in the circuit shown in Fig. 2.

The time t_1 is function of two time constants τ_l and τ_s. The time constant τ_l is given by

$$\tau = R_L C_2 \tag{7}$$

and τ_s is expressed as

$$\tau_s \propto \tau_i \tag{8}$$

where

$$\tau_i = (R_i + 2r_b)\tau_f \frac{I_0}{V_s} + \frac{4U_t}{V_s}\tau_0. \tag{9}$$

If $\tau_l \gg \tau_s$ (for example, in the case of high load capacitance):

$$t_1 \approx \ln(2)\tau_l \approx 0.7\tau_l \tag{10}$$

and if $\tau_l \ll \tau_s$ (high bipolar switching time):

$$t_1 \approx \ln(1.5)\tau_s \approx 0.4\tau_s. \tag{11}$$

The minimum delay occurs when $\tau_l = \tau_s$. This condition determines the optimum current I_{opt}. For this case the time t_1 is given by

$$t_1 \approx 1.2\tau_l \approx 1.2\tau_s. \tag{12}$$

The total delay is dominated by t_1 except for minimum

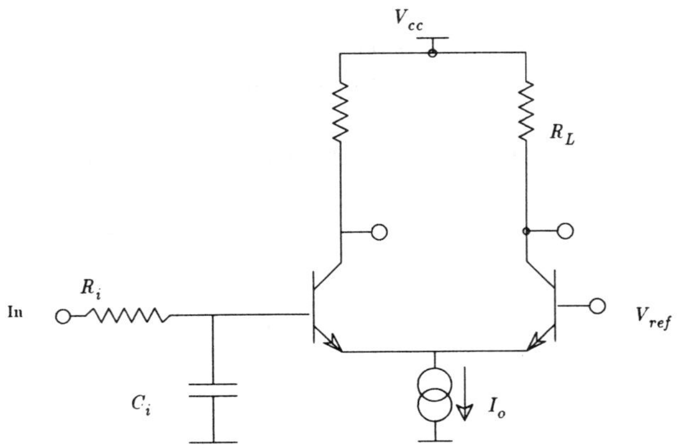

Fig. 3. CML gate.

delay, where the contribution of t_0 is comparable to that of t_1.

Table IV shows the effect of scaling factors on the delay of the CMOS/ECL converter for different regions of operation. In the case where $\tau_l \gg \tau_s$, the delay is given by (10). As C_2 scales by k_h, the delay will scale by the same factor. The first of two terms in (9) dominates (the scaling of $(R_i + 2r_b)\tau_f I_0 / V_s$ depends on whether R_i is greater or less than $2r_b$). The impact of each of these terms on scaling is shown in Table IV for the case where $\tau_l \ll \tau_s$ and $\tau_l = \tau_s$. We notice that the delay improves with scaling the ECL supply only if $R_i \gg 2r_b$, whereas it degrades the delay slightly when $R_i \ll 2r_b$. The scaling of the CMOS power supply has a negative effect on the delay only if $R_i \ll 2r_b$. In short, the delay can be optimized by proper selection of the different scaling factors while taking into consideration R_i and r_b.

C. CML Gate

A simplified CML gate is shown in Fig. 3. The CML gate is similar to the one in Fig. 2, except that the input resistance R_i is the output resistance of the preceding stage. In many cases R_i is equal to the load resistance, which is not affected by the CIC scaling. All of the analysis that we used in Section III-B is valid in this case. Table V lists the scaling laws of the delay in different regions of operation of the CML gate. The scaling effects as discussed related to the CMOS/ECL converter are applicable here except when $R_i \gg r_b$. In this case k_u has no effect on the delay and k_h does not affect the delay when $\tau_l \ll \tau_s$.

D. ECL/CMOS Converter

A schematic of the ECL-to-CMOS translator is shown in Fig. 4. The propagation delay of this circuit is given by [24]

$$t_d = t_d(\text{CML}) + t_d(\text{CMOS}) \tag{13}$$

where

$$t_d(\text{CMOS}) = t_{c1} + t_{c2} \tag{14}$$

TABLE V
CML GATE SCALING FACTORS

Condition/Region		Term	Scaling Factor
$\tau_l \ll \tau_s$	$R_i \ll 2r_b$	t_d	$k_v^{1.4} k_m^{-0.2}$
	$R_i \gg 2r_b$	t_d	$k_v^{1.1} k_m^{0.45}$
$\tau_l \gg \tau_s$	-	t_d	k_h
At the minimum ($\tau_l = \tau_s$)	$R_i \ll 2r_b$	t_0	$k_v^{0.3} k_h k_m^{-0.65}$
		t_1	$k_h^{0.5} k_v^{0.7} k_m^{-0.1}$
		I_{opt}	$k_h^{0.5} k^{-0.7} k_m^{0.1}$
	$R_i \gg 2r_b$	t_0	k_h
		t_1	$k_h^{0.5} k_v^{0.55} k_m^{0.2}$
		I_{opt}	$k_h^{0.5} k_v^{-0.55} k_m^{-0.2}$

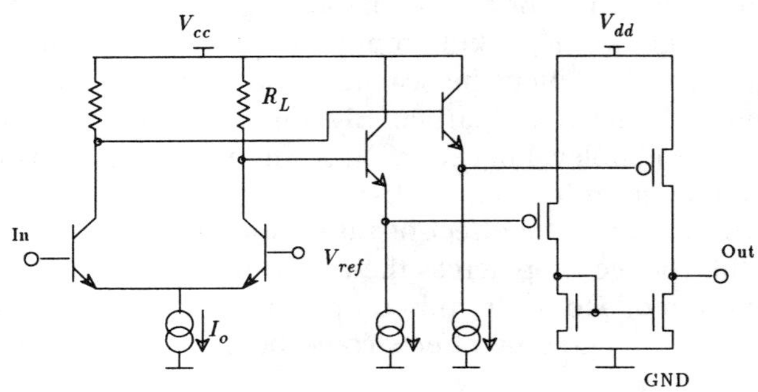

Fig. 4. ECL/CMOS converter.

and

$$t_{c1} = \frac{V_s C_2}{2 I_{D\text{sat}}}$$

$$t_{c2} = 0.7 \frac{C_{cg}(V_{GS} - V_T)}{I_{D\text{sat}}}.$$

In many cases the delays of the emitter followers are neglected. Also the delay of the CMOS shifter t_d(CMOS) dominates over the delay of the ECL stage t_d(ECL), and particularly in the case where the swing of the ECL is small. Assuming minimum gate voltage on the PMOS of the CMOS shifter (corresponding to maximum ECL swing), the gate-to-source voltage is merely a function of the CMOS power supply and hence it will scale by k_u. The capacitances C_{cg} and C_2 scale by a factor k_h. Table VI illustrates the scaling of the delay assuming that the ECL/CMOS converter drives a CMOS gate. The delay of

TABLE VI
ECL-TO-CMOS CONVERTER SCALING FACTORS

Term of delay	Scaling Factor
$t_{c1} = \frac{V_s}{2I_{Dsat}} C_{out}$	$k_h^{1.4} k_u^{-0.3}$
$t_{c2} = 0.8\tau_{cg}$	$k_h^{1.4} k_u^{-0.3}$
$t_d \approx t_d(CMOS) = t_{c1} + t_{c2}$	$k_h^{1.4} k_u^{-0.3}$

this converter scales approximately similar to the delay of the simple CMOS inverter.

E. Case Study

The results of our analyses are illustrated by two examples. In this study, the starting process is a generic 0.8-μm BiCMOS technology (with base width $W_B = 0.1$ μm and $V_{dd} = 5$ V). Two scaling scenarios were considered. In the first, the power supply is unchanged (i.e., $k_m = k_u = 1$). The effective channel length of the MOS device is scaled down to 0.5 μm (i.e., $k_h = 0.625$) and the base width is scaled to 0.05 μm (i.e., $k_v = 0.5$). In the second case, the power supply is scaled to 3.3 V (i.e., $k_m = k_u = 0.66$) allowing for further scaling of L_{eff} to 0.32 μm and W_B to 0.03 μm (i.e., $k_h = 0.4$ and $k_v = 0.3$).

Scaling the base width to 0.05 μm implies that the base and epilayer dopings increase by a factor of 4 and 5, respectively. In the second case, where W_B is shrinked to 0.03 μm, N_B increases by a factor of 7 and N_{epi} by a factor of 12. Assuming a typical profile for the starting process (e.g., $N_B = 1 \times 10^{18}$ cm^{-3} and $N_{\text{epi}} = 1 \times 10^{16}$ cm^{-3}), we find that the scaled profiles satisfy punchthrough and base–collector breakdown limitations.

In both cases, the load capacitance is constant and thus $k_{cl} = 1$. The current gain used in this study is taken to be 100 irrespective of scaling. The assumption that the gain is not affected by scaling is based on the fact that the degradation of the gain due to vertical scaling can be counteracted by improving the properties of the polysilicon emitter contact, which implies that for this case study k_f is equal to 0.75.

Table VII gives the first-order numerical values of scaling the delay at 5- and 3.3-V power supplies. Each of the entries of the table represents the ratio of the delay of the circuit after scaling to the delay before scaling.

Table VII shows that the improvement of the speed of BiCMOS buffers due to scaling is almost independent on the load capacitance, which means that delay-to-load sensitivity of BiCMOS will be preserved under scaling. The table shows that the delay of the BiCMOS buffers with full input voltage swing improves with scaling better than the BiCMOS buffers with reduced voltage swing. However, the output waveform of the circuits shown in Fig. 1 may resemble that of the reduced-swing buffer at high switching frequencies especially when the load capacitance is large (e.g., $C_L > 0.5$ pF). Hence, their delay will scale as that of the reduced-swing circuits. From Table VII we can see that the improvement factor of BiCMOS buffers almost matches that of the CMOS gates. The results of this study imply that BiCMOS buffers driving high capacitance benefit more from scaling than CMOS buffers, both at 5 and 3.3 V. The improvement of the BiCMOS buffer is particularly due to the scaling of the threshold voltage, which minimizes the degradation of the transconductance of the MOSFET's, and the scaling of the current gain of the bipolar transistor. The CML and both the CMOS-to-ECL and ECL-to-CMOS converters show an improvement with scaling especially in the case where the power supply of the CMOS and CML is scaled to 3.3 V.

IV. CONCLUSION

A first-order scaling guide for the BiCMOS process is presented. The process scaling was linked analytically to device parameters and in turn to BiCMOS circuit performance. The effect of scaling on the delay of the BiCMOS circuit and the trade-offs involved were shown (Tables II–VI). The case study presented shows that BiCMOS circuit structures can benefit from scaling provided that the different parameters are scaled according to specific guidelines such as those summarized in Tables I–VI. The case study also shows that under scaling, BiCMOS buffers show almost as much improvement as CMOS gates, whereas CMOS buffers suffer significantly from scaling.

APPENDIX

Bipolar Modeling

The punchthrough voltage between the emitter and the collector is given by [25]

$$V_{pt} = \frac{qN_B W_B^2}{2\epsilon_s}. \tag{A1}$$

This voltage scales by a factor k_m.

The collector-current density increases for current constant scaling, and hence when it is greater than

$$J_c = qv_s N_{\text{epi}} \tag{A2}$$

the base width effectively increases (Kirk effect).

As a first-order approximation, the current gain, for a transparent polysilicon emitter (i.e., $W_{jE} \ll L_p$), can be expressed as [26], [27]

$$\beta_0 \approx \frac{D_{nB} N_E C_N(N_E)}{D_{pE} N_B C_N(N_B)} \frac{1}{W_B} f(\mu_1, W_1) \tag{A3}$$

where $f(\mu_1, W_1)$, which is called the polysilicon term, is given by

$$f(\mu_1, W_1) = \frac{D_{pE} L_{p1}}{D_{p1}} \tanh\left(\frac{W_1}{L_{p1}}\right). \tag{A4}$$

TABLE VII
CASE STUDY: DELAY SCALING FACTORS OF BiCMOS CIRCUITS OF FIGS. 1–4

Scaling scheme	CMOS Buffer †	CMOS Gate ‡	CMOS-to-ECL converter *	CML Gate *	ECL-to-CMOS Converter	BiCMOS Buffer (Reduced swing)	BiCMOS Buffer (Full swing)
V_{DD}=5 V Horizontal scaling to 0.5 μm	0.85	0.50	0.50	0.55	0.50	0.50 †† 0.60 ‡‡	0.50 †† 0.60 ‡‡
V_{DD}=3.3 V Horizontal scaling to 0.3 μm	0.75	0.30	0.40	0.35	0.30	0.45 †† 0.45 ‡‡	0.35 †† 0.40 ‡‡

† CMOS inverter driving large fixed load capacitance

‡ CMOS inverter driving another inverter

* At the minimum delay and $R_s \gg 2r_b$ case

†† corresponds to the first case in Table II ($C_L \ll C_{1m}$)

‡‡ corresponds to the first case in Table III ($C_L \gg \frac{C_{1m}}{2}(\beta_0 \frac{I_K}{I_{Dsat}})^{1/2}$)

The emitter doping, $C_N(N_E)$, and D_{pE} are not subject to scaling. From (A1) we see that the base doping scales as $k_v^{-2}k_m$. The correction factor due to bandgap narrowing in the base is given by [28]

$$C_N(N_B) = e^{-\Delta E_g(N_B)/kT} \tag{A5}$$

where

$$\Delta E_g(N_B) = \alpha \ln\left(\left(\frac{N_B}{N_0}\right)^5 + 1.0\right) \tag{A6}$$

where

$$\alpha = 3.74 \times 10^{-3}\ (\text{eV})$$

$$N_0 = 1 \times 10^{17}\ \text{cm}^{-3}.$$

Assuming that $N_B/N_0 \gg 1$ and substituting from (A6) in (A5), we find that $C_N(N_B)$ scales as $k_v^{1.4}k_m^{-0.7}$.

The electron mobility in the base is related to the base doping by the expression derived by Arora *et al.* [29]. However, this expression is not useful for our scaling study due to its complexity. Instead, we used the following simple relation:

$$\mu_n \approx \mu_{0n}\left(\frac{N_B}{N_{\text{ref}}}\right)^{-0.45} \tag{A7}$$

where N_{ref} is a constant and is equal to 3×10^6 cm^{-3}. Equation (A7) agrees well with Arora *et al.'s* expression for $10^{17} < N_B < 10^{19}$ cm^{-3}, which is the typical range for the base doping concentration. Using (A7) we find that the mobility μ_n scales as $k_v^{0.9}k_m^{-0.45}$.

The polysilicon term $f(\mu_1, W_1)$ is mainly a function of the properties of the polysilicon emitter contact [22], such as Auger recombination lifetime, and mobility of the minority carriers, and the thickness of the polysilicon layer. It is not affected by any of the scale factors k_u, k_m, k_v, k_h, etc. We define a new scaling factor k_f which accounts for any possible variation of $f(\mu_1, W_1)$ due to the change of the properties of the polysilicon. Like the rest of the scaling factors, k_f is assumed to be less than 1. It is defined such that $f(\mu_1, W_1)$ increases when k_f decreases, hence we can say that $f(\mu_1, W_1)$ is proportional to k_f^{-1}.

Accounting for the scaling of N_B, W_B, $C_N(N_B)$, D_{nB}, and $f(\mu_1, W_1)$ we find that the current gain at low-level injection scales as $k_v^{0.5}k_f^{-1}k_m^{-0.75}$. In Table I we take the factor $k_\beta = k_v^{-0.5}k_f$.

Next we consider the forward knee current I_K which is given by

$$I_K = \frac{Q_{B0}}{\tau_f} \tag{A8}$$

where the base charge at lower current is written as

$$Q_{B0} = qA_eN_BW_B \tag{A9}$$

and the forward base transit time is given by

$$\tau_f = \frac{W_B^2}{2\nu D_{nB}}. \qquad (A10)$$

Consider now the base resistance which is approximated by

$$r_b = r_{bi} + r_{be} \qquad (A11)$$

where the contact base resistance is neglected. The intrinsic base resistance r_{bi} is current dependent. At low base–emitter voltage the intrinsic base is expressed as

$$r_{bi0} = \frac{r_{bis}}{G} \qquad (A12)$$

where

$$r_{bis} = \frac{1}{q\mu_p Q'_{B0}}. \qquad (A13)$$

The base resistance reduction factor G is due to the distributed nature of the base current and is dependent on the layout topology and base contact locations. It is not scalable. In many cases, and particularly for submicrometer bipolar devices, r_{bi0} dominates.

The collector and emitter depletion capacitances can be approximated by

$$C_c = A_c\left(q\epsilon_s\frac{N_{\rm epi}}{2V_{bi}}\right)^{1/2} \qquad (A14)$$

$$C_e = A_e\left(q\epsilon_s\frac{N_B}{2V_{bi}}\right)^{1/2} \qquad (A15)$$

MOS Modeling

The important MOS parameters to evaluate BiCMOS performance in our study are capacitances and saturation current. The model for $I_{D\rm sat}$ to be used for scaling submicrometer MOS devices is the engineering model given in [11]:

$$I_{D\rm sat} = Kv_{\rm sat}C_{ox}W(V_{GS} - V_T) \qquad (A16)$$

for $V_{DS} \geqslant V_{D\rm sat}$, and

$$V_{D\rm sat} = (1-K)(V_{GS} - V_T) \qquad (A17)$$

where K is a factor, which tends to go to 1 for deep-submicrometer devices, and is given by

$$K = \frac{1}{1 + \dfrac{E_c L_e}{(V_{GS} - V_T)}}. \qquad (A18)$$

In the above equations, L_e is the device electrical channel length.

Equation (A18) describes the relation between the factor K and the supply voltage and some device parameters. Because of the complexity of this relation, it is difficult to obtain scaling laws for $I_{D\rm sat}$ analytically. Therefore, we had to rely on experimental data [30] and simulation using the BSIM model to develop empirical scaling laws. For $V_{GS} = V_{dd}$, the saturation current was found to scale as $k_h^{-0.4}k_u^{1.3}$. For $V_{GS} = V_{dd} - V_{BE\rm on}$, $I_{D\rm sat}$ is proportional to $k_h^{-0.4}k_u^{2.2}$. These empirical relations are valid for $L_{\rm eff}$ between 0.3 and 1.0 μm, t_{ox} between 50 and 250 Å, x_j less than 0.3 μm, and $V_{GS} - V_T$ greater than $V_{dd}/2$.

References

[1] M. Odaka *et al.*, "A 512kb/5ns BiCMOS RAM with 1kG/150ps logic gate array," in *ISSCC Dig. Tech. Papers*, Feb. 1989, pp. 28–29.

[2] H. Tran *et al.*, "An 8ns BiCMOS 1Mb ECL SRAM with a configurable memory array size," in *ISSCC Dig. Tech. Papers*, Feb. 1989, pp. 36–37.

[3] M. Matsui *et al.*, "An 8ns 1Mb ECL BiCMOS SRAM," in *ISSCC Dig. Tech. Papers*, Feb. 1989, pp. 38–39.

[4] H. Fukuda *et al.*, "A BiCMOS channelless masterslice with on-chip voltage converter," in *ISSCC Dig. Tech. Papers*, Feb. 1989, pp. 176–177.

[5] T. Hotta *et al.*, "A 70MHz-32b microprocessor with 1.0 μm BiCMOS macrocell library," in *ISSCC Dig. Tech. Papers*, Feb. 1989, pp. 124–125.

[6] T. Hotta *et al.*, "1.3 um CMOS/bipolar macrocell library for VLSI computer," *IEEE J. Solid-State Circuits*, vol. 23, no. 2, pp. 500–506, Apr. 1988.

[7] H. Higuchi *et al.*, "Performance and structures of scaled-down bipolar devices merged with CMOSFETs," in *IEDM Tech. Dig.*, Dec. 1984, pp. 694–697.

[8] A. Watanabe *et al.*, "Proposal of high speed bipolar-CMOS merged VLSI technology," *Electron. Commun. Japan*, vol. 71, no. 6, pt. 2, pp. 63–73, 1988.

[9] H. Momose *et al.*, "0.5 micron BiCMOS technology," in *IEDM Tech. Dig.*, Dec. 1987, pp. 838–840.

[10] T. Yamaguchi and T. H. Yuzuriha, "Process integration and device performance of a submicrometer BiCMOS with 16-GHz f_t double poly-bipolar devices," *IEEE Trans. Electron Devices*, vol. 36, no. 5, pp. 890–896, May 1989.

[11] K-Y. Toh, "An engineering model for short-channel MOS devices," *IEEE J. Solid-State Circuits*, vol. 23, no. 4, pp. 950–958, Aug. 1988.

[12] R. H. Dennard *et al.*, "Design of ion implanted MOSFETs with very small physical dimensions," *IEEE J. Solid-State Circuits*, vol. SC-9, pp. 256–266, Oct. 1974.

[13] P. K. Chatterjee *et al.*, "The impact of scaling laws on the choice of n-channel or p-channel for MOS VLSI," *IEEE Electron Device Lett.*, vol. EDL-1, pp. 220–223, Oct. 1980.

[14] J. R. Brews *et al.*, "Generalized guide for MOSFET miniaturization," *IEEE Electron Device Lett.*, vol. EDL-1, pp. 2–4, Jan. 1980.

[15] G. Baccarani *et al.*, "Generalized scaling theory and its application to a 1/4 micrometer MOSFET design," *IEEE Trans. Electron Devices*, vol. ED-31, no. 4, pp. 452–462, Apr. 1984.

[16] P. M. Solomon and D. D. Tang, "Bipolar circuit scaling," in *ISSCC Dig. Tech. Papers*, Feb. 1979, pp. 86–87.

[17] T. H. Ning *et al.*, "Scaling properties of bipolar devices," in *IEDM Tech. Dig.*, Dec. 1980, pp. 61–64.

[18] N. Hanaoka and A. Anzai, "Perspective of scaled bipolar devices," in *IEDM Tech. Dig.*, Dec. 1981, pp. 512–515.

[19] T. H. Ning, "Materials and physics issues in scaling bipolar devices," *J. Vac. Sci. Technol. A*, vol. 5, no. 4, pp. 1394–1395, July/Aug. 1987.

[20] D. D-L. Tang *et al.*, "Junction degradation in bipolar transistors and the reliability imposed constraints to scaling and design," *IEEE Trans. Electron Devices*, vol. 35, no. 12, pp. 2101–2107, Dec. 1988.

[21] A. Bellaouar, S. Embabi, and M. I. Elmasry, "Scaling of digital BiCMOS circuit structures," in *IEDM Tech. Dig.*, Dec. 1989, pp. 437–440.

[22] C. L. Williams *et al.*, "Optimization and temperature dependance of current gain in polysilicon-emitter-contacted bipolar transistors," in *BCTM Tech. Dig.*, Sept. 1988, pp. 50–54.

[23] Y. Nishio *et al.*, "A BiCMOS logic gate with positive feedback," in *ISSCC Dig. Tech. Papers*, Feb. 1989, pp. 116–117.

[24] S. Embabi, A. Bellaouar, and M. I. Elmasry, "Analysis and optimization of BiCMOS digital circuit structures," submitted to *IEEE J. Solid-State Circuits*, 1989.

[25] S. M. Sze, *Physics of Semiconductor Devices*. New York: Wiley, 1981.

[26] T. H. Ning and R. D. Isaac, "Effect of emitter contact on current gain of silicon bipolar devices," *IEEE Trans. Electron Devices*, vol. ED-27, pp. 2051–2055, Nov. 1980.

[27] B. Soerowirdjoeral, "The influence of surface treatment on the electrical characteristics of poly-emitter bipolar transistors," in *IEDM Tech. Dig.*, Dec. 1982, pp. 668–671.
[28] J-C. S. Woo and J. D. Plummer, "Optimization of silicon bipolar transistors for high current gain at low temperatures," *IEEE Trans. Electron Devices*, vol. 35, no. 8, pp. 1311–1321, Aug. 1988.
[29] D. Arora, J. R. Hauser, and D. J. Roulston, "Electron and hole mobilities in silicon as a function of concentration and temperature," *IEEE Trans. Electron Devices*, vol. ED-29, pp. 284–291, Feb. 1982.
[30] M.-C. Jeng *et al.*, "Design guidelines for deep-submicrometer MOSFETs," in *IEDM Tech. Dig.*, Dec. 1988, pp. 386–389.

Paper 1.24

Scaling Rules for Bipolar Transistors in BiCMOS Circuits

G. P. Rosseel and R. W. Dutton

AEL 231A, Stanford University,
Stanford, CA 94305.

Abstract

Scaling rules for bipolar transistors in BiCMOS gates are derived such that the gates maintain their performance advantage over scaled CMOS implementations. These are compared with those for bipolar transistors in ECL gates and are found to be generally similar under realistic scaling assumptions, except for a conflict in the choice of the collector doping concentration.

Introduction

Technological scaling rules must be derived towards circuit applications. Four different classes of BiCMOS circuits can be defined. These are summarized in Table 1. The first class of circuits is called "hybrid circuits" and contains new circuit ideas that use both bipolar and MOS transistors in an innovative way [1-4]. Of these ideas, the BiCMOS gate has become very popular.

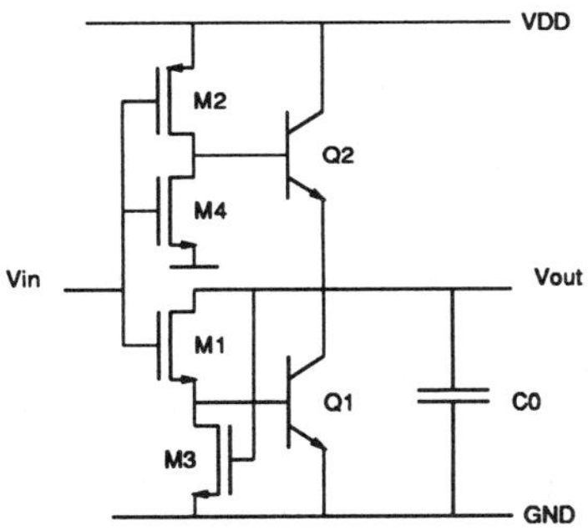

Figure 1: A BiCMOS inverter

This gate is shown in Figure 1 and is about twice as fast as a CMOS gate for driving large load capacitances. It has led to two other classes of BiCMOS circuits. One contains circuits that use only BiCMOS gates such as BiCMOS gate arrays [5]. The other class consists of BiCMOS circuits in which the BiCMOS gate is only used for driving large capacitances [6]. A final class of BiCMOS circuits are circuits which combine CMOS with ECL building blocks. Very fast SRAMs have been built using both ECL and CMOS techniques [7,8]. Other applications are BiCMOS processors with a bipolar data path and CMOS control and memory circuits.

	Class	Examples
1	Hybrid BiCMOS	BiCMOS gate [1], hybrid ECL [3]
2	Only BiCMOS gates	Gate-arrays [4]
3	CMOS + BiCMOS	Carry-look-ahead adder [6]
4	CMOS + ECL	SRAM [7]

Table 1: Classes of BiCMOS circuits.

The bipolar transistors in BiCMOS circuits are used for ECL circuits as well as for BiCMOS gates. We will investigate scaling rules for bipolar transistors for BiCMOS drivers and answer the two following questions : *1) how must the bipolar transistor scale when the CMOS part of a BiCMOS technology scales such that the BiCMOS gate maintains its advantage over the CMOS gate ?* and : *2) are there conflicting requirements between these scaling rules and those for bipolar transistors in ECL circuits ?*

CMOS and Bipolar Scaling Rules

The CMOS technology is assumed to scale following the generalized CMOS scaling rules presented by Baccarani *et al.* [9], which are summarized in Table 2. These rules assume separate voltage (κ) and dimension scaling (λ) and consider both the case of ideal quadratic current-voltage behavior and the case of velocity saturation limited current.

Parameter	Expression	Scaling factor	
Dimensions	W, L, t_{ox}, x_j	$1/\lambda$	
Potentials	V_{dd}, ϕ_j	$1/\kappa$	
Doping Conc.	N_A, N_D	λ^2/κ	
Electric Field	E	λ/κ	
Capacitances	$A.C_{ox}, A.C_j$	$1/\lambda$	
Current	$(W/L)\mu C_{ox}(V_{gs}-V_t)V_{ds}$		λ/κ^2
Current (vel sat)	$k_s W C_{ox}(V_{gs}-V_t)V_s$	$1/\kappa$	
Power	$I_{DD} V_{DD}$	$1/\kappa^2$	λ/κ^3
Gate Delay	$C_g V_{DD}/I_D$	$1/\lambda$	κ/λ^2
Power-delay Pr.	$I_D V_{DD} t_d$	$1/\lambda\kappa^2$	

Table 2: MOS scaling rules. Scaling factors are κ and λ for potentials and dimensions respectively. Left column assumes velocity saturation, the right column a quadratic $I-V$ relation.

Reprinted from *IEDM 1989*, pp. 33.2.1–33.2.4, 1989.

The bipolar transistors for ECL circuits are assumed to scale following the scaling rules derived by Tang, Solomon and Ning [10,11,12] and are summarized in Table 3.

Parameter	Expression	Scaling factor
Dimensions	L_e	$1/\lambda$
Base-width	W_B	$1/\lambda^{0.8}$
Base-doping	N_B	$W_B^{-2} = \lambda^{1.6}$
Current-density	J	λ^2
Coll.-doping	N_C	J
Delay	t_d	$1/\lambda$

Table 3: Bipolar scaling rules. Horizontal scaling factor is λ with λ larger than 1.

Operation of the BiCMOS Gate

It has been shown that the bipolar transistors of a BiCMOS gate can operate under different conditions [13]. When only a small amount of charge must be transferred by the bipolar transistors, the bipolar transistors operate in the low-level injection condition. For large load capacitances, high current effects dominate the delay. These are high-level injection effects and saturation. Both these effects are largely dominated by the collector profile. It is possible then, to identify regions of operation for the bipolar transistor as a function of the load capacitance and the collector doping concentration as shown in Figure 2. For higher load capacitances, base-widening effects will dominate the delay when the collector doping concentration is low, while for higher doping concentrations, saturation effects limit the driving current.

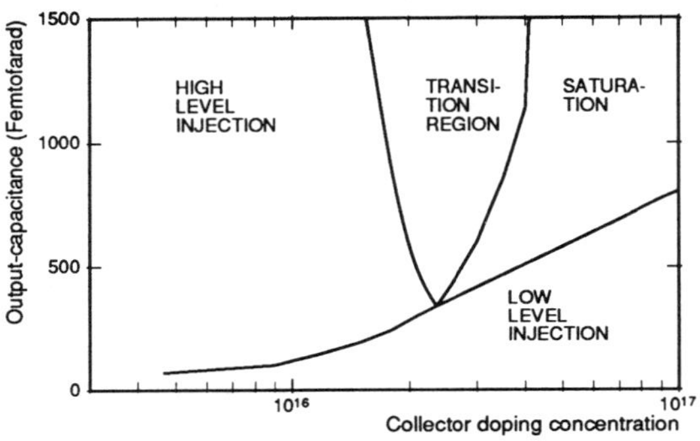

Figure 2: Regions of operation as a function of load capacitance and collector doping concentration. An external collector resistance of 100 Ohms is assumed (for collector plug and buried layer) and the width of the epi is $1\mu m$.

For each of the regions, equations for the delay time were derived and were used in the derivation of the scaling rules. The derivation of these equations and of Figure 2 can be found in [14]. A first-order approximation for the minimum delay of a BiCMOS gate is given by Equation 1 (assuming that all parasitic capacitances are zero). V_s is the switching voltage, C_o is the load capacitance, τ_{Fo} is the base transit time and I is the average current in the MOS transistor providing the base current.

$$t_d = \left(\frac{2 \cdot V_s \cdot C_o \cdot \tau_{Fo}}{I}\right)^{\frac{1}{2}} \quad (1)$$

To first order, a BiCMOS gate is faster than a CMOS gate as long as $2 \cdot \tau_{Fo}$ is smaller than $V_s \cdot C_o / I$.

Derivation of the Scaling Rules

First, the scaling rules for device parameters are identified. Before these rules be defined, two new scaling parameters must be defined which take the V_{be} loss at the input and output into account. Since the output of a BiCMOS gate remains a V_{be} from the ground and supply, the output swing of the BiCMOS gate will scale more than that for a CMOS gate. The scaling factor γ is defined as the ratio between the scaling of the supply voltage and the scaling of the BiCMOS voltage swing and is given by Equation 2.

$$\gamma = \frac{(V_{dd} - 2 \cdot V_{be}) \cdot (V_{dd}/\kappa)}{(V_{dd}/\kappa - 2 \cdot V_{be}) \cdot (V_{dd})} \quad (2)$$

Similarly, the scaling factor α is defined to take the V_{be} loss at the input into account. The current in the MOS transistors of a BiCMOS gate scales more than for a CMOS gate. When ideal MOS behavior is assumed, the MOS transistors operate mostly in the linear region and α is defined as the ratio between the ideal and real MOS scaled transconductances. For velocity saturated operation, the MOS transistors are assumed to operate mainly in the saturation region and α is defined as the ratio between ideal and real scaled MOS currents. The factor α is given by Equation 3 where f is 1 if the pull-up transient is considered and it is 2 for the pull-down.

$$\alpha = \frac{(V_{dd}/\kappa - V_{th})(V_{dd} - V_{th} - f \cdot V_{be})}{(V_{dd} - V_{th})(V_{dd}/\kappa - V_{th} - f \cdot V_{be})} \quad (3)$$

The factors α and γ are always larger than 1.

By applying the scaling rules of Table 2 to the analytical equations for the delay time and by requiring that the performance of the BiCMOS gate scales at least as fast as that for the CMOS gate, general scaling rules for the bipolar transistors can be derived. These are summarized in Table 4. The parameter a in this table models the slope of τ_F as a function of I_c in the high current region assuming a linear dependence. Since α and γ account for non linear scaling effects, it is not possible to achieve fully consistent scaling rules for these two parameters. We have chosen the most conservative rules.

Since BiCMOS gates must be compared with CMOS gates *with the same area*, the bipolar area should scale proportionally to the CMOS area. This means that the emitter area must scale as $1/\lambda^2$. Let's first look at a simple case and assume constant voltage scaling ($\kappa=\alpha=\gamma=1$) and assume that the MOS currents are limited by velocity saturation. The base transit time must then scale as $1/\lambda$, parasitic capacitances also as $1/\lambda$, the resistances may remain constant and high level injection parameters should also remain constant. This results in a bipolar device that is λ^2 times smaller with an f_T that is λ times larger while the maximum f_T occurs at the same current. Since under these assumptions, the f_T of the MOS devices also scales as λ, this proves that the bipolar performance must scale similarly to that of the MOS devices. Similar results are obtained when a quadratic $I-V$ MOS relation is assumed (f_T scales as λ^2).

Parameter	Expression	Scaling factor	
Lateral Dimensions	W_e, L_e	$1/\lambda$	
Emitter Area	A_e	$1/\lambda^2$	
Current gain	β	α/γ	
Base transit time	τ_{Fo}	$\gamma/\lambda\alpha$	$\kappa/\lambda^2\alpha$
Base-emitter cap.	C_{je}	$1/\alpha\lambda\kappa$	$1/\alpha^2\lambda\kappa$
Base-collector cap.	C_{jc}	$1/\alpha\lambda$	
Collector-substr. cap.	C_{js}	$1/\lambda$	
Base resistor	R_b	α/γ	$\alpha\kappa/\lambda$
Collector resistor	R_c	1	κ/λ
High injection current	I_k	$1/\kappa\gamma$	$\lambda/\gamma\kappa^2$
High injection parameter	a	$\kappa\gamma^2/\lambda\alpha$	$\gamma^2\kappa^3/\alpha\lambda^3$
Delay	t_d	$1/\lambda$	κ/λ^2

Table 4 : Bipolar device scaling rules for BiCMOS gate. Horizontal scaling factor is λ and the voltage scaling factor is κ ($\lambda, \kappa, \alpha, \gamma > 1$). Left column assumes velocity saturation, the right column a quadratic $I-V$ relation.

To understand how the bipolar should scale as the voltage scales, it is first assumed that α=1 and γ=1. Again assuming velocity saturation, Table 4 shows that primarily the high injection parameters are affected and their constraints are relaxed under voltage scaling. In the other case (quadratic $I-V$), all constraints are relaxed but those related to high injection parameters even more so.

When α is larger than 1, the MOS current decreases more than desired and this performance degradation must be compensated by a better bipolar device. Table 4 shows that bipolar scaling rules are more severe when $\alpha > 1$. On the other hand, when γ is larger than 1, the voltage swing is reduced more compared to a CMOS gate. As less charge must be transferred through the bipolar device, the bipolar scaling rules are relaxed. For limited voltage scaling the values of the parameters α and γ are about the same; for most parameters, their effects tend to cancel each other. For more substantial voltage scaling, the requirements on the bipolar transistor become very strict if pull-down performance must scale similar to the CMOS performance.

Starting with Table 4, the technological scaling rules for device parameters can now be derived. These are summarized in Table 5. When different technological scaling rules are required by different device parameters, the most strict rule is adopted. The base profile is determined by the transit time. Note that we have assumed an ideal quadratic relationship between τ_{Fo} and W_b. For very small base-widths, the charges in the depletion regions must be included and this results in more strict scaling rules for the base profile. For this case, Tang and Solomon report the relation $\tau_{Fo} \sim W_b^{1.2}$ which is derived using a device simulator [11]. The discussion in the next section is independent of the exact relationship between τ_{Fo} and W_b. The collector profile is mostly determined by the collector resistor and the high injection parameters.

Parameter	Expression	Scaling factor	
Dimensions	L_e, W_e	$1/\lambda$	
Emitter area	A_e	$1/\lambda^2$	
Base-width	W_b	$\gamma^{0.5}/\lambda^{0.5}\alpha^{0.5}$	$\kappa^{0.5}/\lambda\alpha^{0.5}$
Base-doping	N_b	$\lambda\alpha/\gamma$	$\lambda^2\alpha/\kappa$
Coll.-doping	N_c	$\lambda^2/\gamma\kappa$	$\lambda^3/\kappa^2\gamma$
Coll.-width	W_c	$1/\lambda$	$\kappa^{0.5}/\lambda^{1.5}$
Current-density	J	$\lambda^2/\kappa\gamma$	$\lambda^3/\kappa^2\gamma$
Gate-delay	t_d	$1/\lambda$	κ/λ^2

Table 5: Bipolar technology scaling rules. Lateral scaling factor is λ and potential scaling factor is κ with λ and κ larger than 1. Left column assumes velocity saturation, the right column a quadratic $I-V$ relation.

Discussion of the Scaling Rules

The previously derived scaling rules can now be compared to those in Table 3. In the case of constant field scaling (quadratic $I-V$) or velocity saturated MOS behavior, the bipolar scaling rules for BiCMOS circuits generally agree with those for ECL circuits. However when constant voltage scaling is assumed with a quadratic $I-V$ relationship, the requirement for the collector doping concentration for BiCMOS circuits (λ^3) causes an increase in junction capacitances which significantly reduces the performance of ECL circuits using these transistors.

Even for the more realistic cases of velocity saturation or constant field scaling, where the scaling rules for the collector doping profile are similar to these in Table 3, there is a significant difference in the optimal value of the collector doping concentration. This can be best explained by comparing Figures 3 and 4.

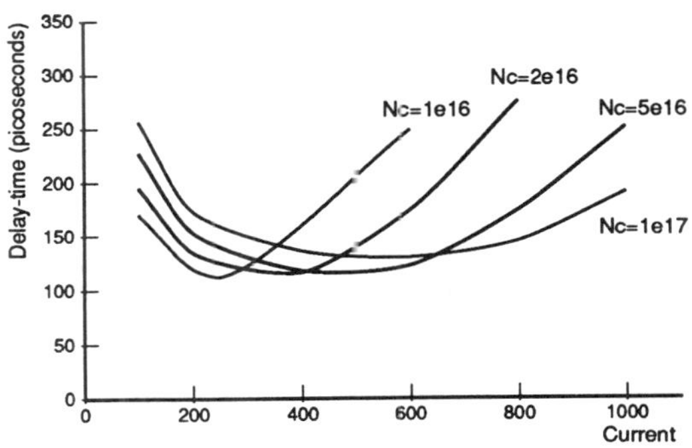

Figure 3: Simulated delay time of ECL gates as a function of collector doping for different bias currents.

Figure 3 shows the simulated delay-time of an ECL gate as a function of collector doping concentration for different bias currents. Figure 4 shows the delay time of BiCMOS gates as a function of collector doping doping concentration, N_c, for two different load capacitors. These simulations were performed with CODECS, a mixed level circuit/device simulator [15] which was used to simulate ringoscillators in which all bipolar devices were described on a numerical level. The delay of an ECL gate decreases as the bias current increases until high level injection effects occur after which the delay increases rapidly. For lower doping concentrations, the minimum delay is achieved for lower currents. Moreover, this minimum is smaller for lower collector doping concentrations. Since power is an important limiting factor for LSI ECL circuits, the current must be reduced and therefore, lower values for N_c are preferred (typically lower than $2e\,16\,cm^3$). For BiCMOS drivers on the other hand, higher collector doping concentrations improve the performance of the gate, especially for larger load capacitances. Figure 4 shows that a collector doping concentration higher than $5e\,16\,cm^3$ is desired. This increased doping concentration is more difficult to scale since breakdown is a limiting factor.

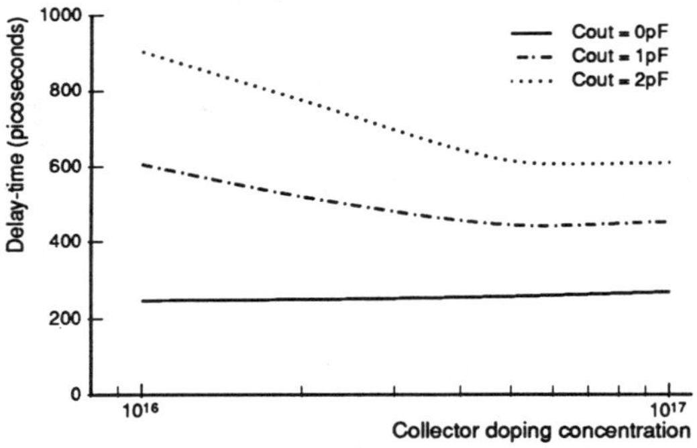

Figure 4: Simulated delay time of BiCMOS drivers as a function of load capacitance for different load capacitances.

Conclusion

Scaling rules for bipolar transistors for BiCMOS drivers were derived such that the BiCMOS gate maintains its advantage over the CMOS gate for driving large load capacitors. These scaling rules were compared with the scaling rules for bipolar transistors for ECL circuits. While the trends are similar, there is a conflict in the requirement for the collector profile. Bipolar transistors for BiCMOS drivers require a high collector doping concentration (typically higher than $5e\,16\,1/cm^3$) while ECL circuits require bipolar transistors with lower value for the collector doping concentration (typically lower than $2e\,16\,1/cm^3$).

Acknowledgement

This work was supported by the Army Research Office (Contract No. DAAL-03-87-K-0077) and SRC (Contract No. 88 DJ 116). The authors thank G. Chin for helpful discussions.

References

[1] S. Lee et al., " Bi-CMOS Circuit for High Performance VLSI," in *1984 Symposium on VLSI Technology Tech. Dig.*, pp. 46-47.

[2] T. Hotta et al., " CMOS/Bipolar Circuits for 60-Mhz Digital Processing," in *IEEE Journal of Solid-State Circuits*, vol. SC-21, pp. 808-815, October 1986.

[3] G. P. Rosseel et al., " A Single-Ended BiCMOS Sense-Circuit for Digital Circuits," in *1989 ISSCC Tech. Dig.*, pp. 114-115.

[4] W. Heimsch et al., " Merged CMOS/Bipolar Current Switch Logic," in *1989 ISSCC Tech. Dig.*, pp. 112-113.

[5] Y. Nishio et al., " A Subnanosecond Low Power Advanced Bipolar-CMOS Gate-Array," in *1984 ICCD*, pp. 428-433.

[6] A. Todesco et al., " A 32-bit BiCMOS adder," to be published.

[7] K. Ogiue et al., " A 13ns/500mW 64Kb ECL RAM," in *1986 ISSCC Tech. Dig.*, pp. 212-213.

[8] T. S. Yang et al., " A 4-ns 4K x 1-bit Two-Port BiCMOS RAM," in *IEEE Journal of Solid-State Circuits*, vol. SC-23, pp. 1030-1040, October 1988.

[9] G. Baccarani et al., " Generalized Scaling Theory and Its Application to a 1/4 Micrometer MOSFET Design," in *IEEE Transactions on Electron Devices*, vol. ED-31, pp 452-461, April 1984.

[10] P. Solomon and D. Tang, " Bipolar Circuit Scaling," in *1979 ISSCC Tech. Dig.*, pp. 86-87.

[11] D. Tang and P. Solomon, " Bipolar Transistor Design for Optimized Power-Delay Logic Circuits," in *IEEE Journal of Solid-State Circuits*, vol. SC-14, pp. 679-684, August 1979.

[12] T. Ning et al., " Scaling Properties of Bipolar Devices," in *1980 IEDM Tech. Dig.*, pp. 550-553.

[13] G. P. Rosseel et al., " Delay Analysis for BiCMOS Drivers," in *1988 BCTM Tech. Dig.*, pp. 220-222.

[14] G. P. Rosseel, "Bipolar Transistors for BiCMOS Circuits," *Ph.D. Thesis*, 1989, Stanford University.

[15] K. Mayaram, "CODECS : A Mixed-level Circuit and Device Simulator," University o f California, Berkeley, Mem. UCB/ERL M88/71.

Paper 1.25

Performance-Driven Scaling of BiCMOS Technology

Prasad A. Raje, *Member, IEEE*, Krishna C. Saraswat, *Fellow, IEEE*, and Kit M. Cham, *Senior Member, IEEE*

***Abstract*—A BiCMOS scaling analysis is carried out to optimize the performance of digital BiCMOS gates while accounting for the conditions that make direct application of conventional BJT and MOS scaling rules invalid. A simple set of technological parameters that directly affect BJT and MOS device characteristics in the context of BiCMOS is identified. Scaling of the technological parameters leads to scaling of the device parameters which in turn appear in a comprehensive, accurate delay model. The scaling of logic gate delay under various possible scenarios is investigated and finally a set of optimal scaling rules is proposed that simultaneously scales all components of gate delay. The optimal rules require a linear reduction in horizontal and vertical dimensions, a linear increase in collector doping, and a square root reduction in supply voltage. When a 1 μm, 5 V, 8 GHz technology is scaled to 0.5 μm, 3.5 V, 18 GHz according to the optimal scaling rules, a 2× improvement in BiCMOS performance is obtained. Despite the supply voltage reduction, BiCMOS maintains a performance advantage of 1.5× over CMOS. Beyond the regime of 0.25 μm, 2.5 V, 39 GHz technology, the performance of the conventional BiCMOS gate does not scale. This is shown to be due to the non scaling of output falling delay because of reduced voltage swing.**

I. Introduction

IN CONVENTIONAL MOSFET scaling [1], the effect of device scaling on circuit performance is measured by the gate delay of the CMOS inverter. In conventional BJT scaling [2], it is the ECL gate that sets the circuit environment in which scaling is carried out. For a BiCMOS application dominated by pure CMOS circuits or pure ECL circuits, device scaling can be carried out simply in the manner already determined for the two separate technologies. However, when the scaling of BiCMOS digital gate performance is of interest, a renewed scaling analysis needs to be performed that accounts for the unique operating conditions in BiCMOS gates. In this work, the conventional BiCMOS gate (Fig. 1(a)) is chosen as the circuit vehicle on which to observe the effects of BiCMOS device scaling. The conventional gate predominates in BiCMOS applications. Furthermore, the basic circuit configuration of a MOSFET driving a BJT is common to other BiCMOS gates as well.

Manuscript received March 20, 1991; revised August 14, 1991. The review of this paper was arranged by Associate Editor K. Shenai.

P. A. Raje was with the Center for Integrated Systems, Stanford University, Stanford, CA 94305. He is now with Hewlett-Packard Corp., Palo Alto, CA 94304.

K. C. Saraswat is with the Center for Integrated Systems, Stanford University, Stanford, CA 94305.

K. M. Cham is with the Circuit Technology R&D, Hewlett-Packard Corp., Palo Alto, CA 94303.

IEEE Log Number 9105352.

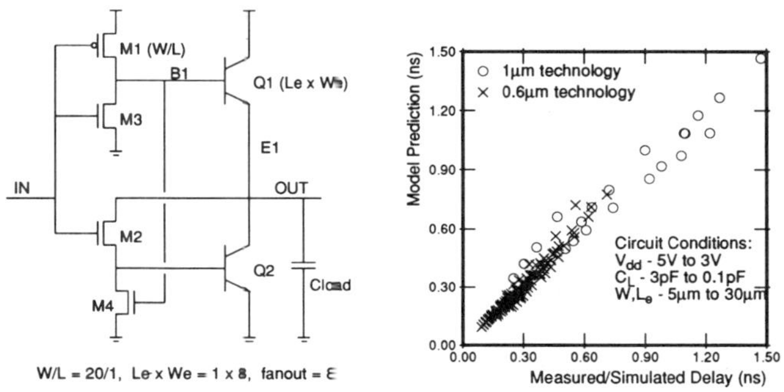

(a) (b)

Fig. 1. (a) BiCMOS gate circuit used for the scaling analysis. Absolute circuit sizes cancel out in the relative delay scaling analysis. (b) Scaling delay expression verification for 1 and 0.6 μm BiCMOS, each with varying circuit conditions.

In BiCMOS scaling, the conventional MOSFET and BJT scaling rules cannot be directly applied. For example, the MOSFET in the BiCMOS gate, unlike in the CMOS gate, does not see a full rail-to-rail input swing. Also, the voltage swing at the drain of the MOSFET is a nonscaling constant equal to the diode drop as opposed to the supply voltage in the CMOS case. The BJT in the BiCMOS gate, unlike in the ECL gate, operates in high-level injection where beta roll-off and base pushout effects dominate. Also, the voltage swing seen by the BJT is much larger than the small swing in the ECL case. Most importantly, MOSFET and BJT scaling cannot be decoupled. For example, the effective capacitance driven by the MOSFET depends on the scaling of BJT junction capacitance and the load capacitance seen by the BJT depends on the scaling of MOSFET oxide capacitance. As another example of coupling, the scaling of MOSFET saturation current determines the base current supplied to the BJT which determines the scaling of BJT collector current.

In [3] the conventional decoupled BJT and MOSFET scaling rules are *assumed* in deriving the effect of scaling on performance. Further, the scaling of epitaxial-layer thickness is neglected and the base pushout effect is neglected in its effect on the gate delay. In [4] the BiCMOS scaling rules are driven by the need to maintain the performance superiority over CMOS but only a first-order delay model is used—parasitic capacitances are neglected as are high-level injection effects.

In this work, various different scaling possibilities are considered that probe the interactions between MOS and BJT scaling. Using an accurate delay model that accounts for several second-order effects in the BiCMOS gate, the

Reprinted from *IEEE Trans. Electron Devices*, vol. 39, no. 3, pp. 685–694, March 1992.

possibilities are examined for their effect on BiCMOS gate delay. A set of scaling rules is proposed that simultaneously scales all components of the gate delay. 0.25 μm, 2.5 V, 39 GHz technology is shown to be a scaling limit of the conventional BiCMOS gate and the primary reason for this limit is identified. The framework for the scaling analysis is presented in Section II. The various scaling scenarios are evaluated and their shortcomings are discussed in Section III. The optimum BJT and MOSFET scaling rules applicable for the BiCMOS gate are presented in Section IV.

II. Scaling Analysis Framework

The scaling analysis begins by identifying basic physical parameters that can be independently scaled. From these, the scaling of device parameters is determined and finally the scaling of the components of the gate delay. This framework is detailed in the following subsections.

A. Technological and Device Parameters

For a first-order scaling analysis, the technological parameters of the BJT and MOSFET that can be scaled are identified. These are horizontal and vertical device dimensions, doping densities, as well as supply and breakdown voltages as listed in Group A and Group B in Tables I and II. The Group A parameters are considered "independent" in the sense that there are no *a priori* constraining relationships between these parameters. They are each assigned separate multiplicative scaling factors. For example, *fwe* is the scaling factor for the emitter width W_e. The Group B parameters are called "dependent" technological parameters and are constrained by the Group A independent parameters as described below.

The scaling factor for base doping is determined from the punchthrough constraint [2]. The punchthrough voltage is $\propto N_B W_B^2$. If the punchthrough voltage is multiplied by *fvbjt* and the base width by *fwb*, the base doping scales as $fwb^{-2}\, fvbjt$. The buried-layer doping is simply scaled along with the collector doping. The scaling of epi thickness is determined by base pushout considerations. The current density when base pushout sets in, J_{push} is given by [5]

$$J_{\text{push}} = v_{\text{sat}}(qN_c + 2\epsilon_{Si}(-V_{bc} + \phi)/t_{\text{epi}}^2). \tag{1}$$

J_{push} is distinguished from the conventionally used expression for the Kirk current density $J_{\text{kirk}} = qv_{\text{sat}}N_c$ which does not include the effect of the epilayer. In order to scale J_{push} proportionally with N_c, the second term in (1) must scale as N_c. Making the assumption that $-V_{bc} + \phi$ scales as V_{bjt}, t_{epi} must scale as $\sqrt{fvbjt/fnc}$. The emitter doping and thickness primarily affect the current gain of the BJT. As a simplification, the properties of the emitter are assumed to scale in a manner that keeps the low current gain constant. This assumption is validated later in Section IV-B. For the MOSFET, the junction depth is made to scale with the gate length. The well/substrate doping density scales in a simple manner to scale the punchthrough voltage with supply and to scale down the drain depletion region with the gate length. The channel implant is scaled in a manner such that the threshold voltage scales with the supply.

The device parameters of the BJT and MOSFET can be expressed in terms of the technological parameters. These expressions result in scaling factors for the device parameters as shown in Group C in Tables I and II. For example, the transit time (τ_f) is proportional to W_B^2/D_{NB}. *ftauf* is therefore given by $fwb^2/fdnb$, which reduces to $fwb^{1.2}/fvbjt^{0.4}$ as shown.

A scaling scenario is defined by a set values for the scaling factors of the independent technological parameters. From these, the scaling factors for the device parameters can be fully determined. Multiplying these with unscaled device parameters from a 1 μm technology, the device parameters for the scaling scenario are found. These scaled parameters are used in the delay expression described below to obtain the scaled value of the delay.

B. Delay Expression

The analytical delay expression for the BiCMOS gate is briefly described here for completeness and details can be found in [6], [7]. The delay of the BiCMOS gate is modeled as a piecewise sum of three components.

$$t_d = \frac{C_{bj}(V_{\text{diode}} + V_{ol})}{I_{\text{mos1}}} + \frac{\tau_f I_{cP}}{I_{\text{mos}}} + \frac{C_L V_s}{I_{\text{mos}} + I_{cP}}. \tag{2}$$

The first term is the *junction* delay which represents the delay in charging the junction plus parasitic capacitance (C_{bj}) at the base of the BJT until the BJT turns on when the base voltage rises to a diode drop (V_{diode}) over the initial emitter voltage (V_{ol}). I_{mos1} is the average MOSFET current during this period. The second term is the *diffusion* delay in providing the forward stored charge ($= \tau_f I_{cP}$) in the BJT by the average MOSFET current (I_{mos}) in this period. The final term is the *load* delay which is the time taken to swing the load and parasitic capacitance (C_L includes both) at the output to half the supply. The emitter current in this period is $I_{cP} + I_{\text{mos}}$. The above expression includes the short-channel effects in the MOSFET, reduced swing at the input and output, bias dependencies of junction capacitances, the effect of parasitic capacitances, and high-level injection effects in the BJT. Details for C_{bj}, I_{mos}, etc., are provided in Appendix I and I_{cP} is discussed below.

I_{cP} is the peak value reached by the transient collector current in the BJT and is a crucial term in the delay expression. If base pushout is neglected, that is, only high-level injection due to the knee current is considered, and if all the MOS current is assumed to provide current into the BJT base, I_{cP} is given by

$$I_{cP} = \sqrt{\beta I_k I_{\text{mos}}} \tag{3}$$

TABLE I

BJT PARAMETERS, SCALING RELATIONSHIPS, FINAL SCALING FACTORS FOR OPTIMUM SCALING, AND PHYSICAL JUSTIFICATIONS

Parameter, Symbol	Scaling Factor (General Case)	Scaling Factor (Optimum)	Equation/Reason
Group A: Independent Technological Parameters			
Emitter width, W_e	fwe	f	independent
Emitter length, L_e	fle	f	independent
Base width, W_B	fwb	f	independent
Collector doping, N_c	fnc	f^{-1}	independent
BJT breakdown voltage, V_{bjt}	$fvbjt$	$f^{0.5}$	independent
Group B: Dependent Technological Parameters			
Base doping, N_B	$fwb^{-2}\, fvbjt$	$f^{-1.5}$	punchthrough $\propto N_B W_B^2$
Buried layer, N_{BL}	fnc	f^{-1}	same as N_c
Epi thickness, t_{epi}	$(fvbjt/fnc)^{0.5}$	$f^{0.75}$	to make $J_{push} \propto N_c$
Group C: Device Parameters			
Base diffusivity, D_{NB}	$fwb^{0.8}\, fvbjt^{-0.4}$	$f^{0.6}$	$D_{NB} \propto N_B^{-0.4}$
Transit time, τ_f	$fwb^{1.2}\, fvbjt^{0.4}$	$f^{1.4}$	$\tau_f \propto W_B^2/D_{NB}$
Knee curr./area, J_k	$fwb^{-2.2}\, fvbjt^{0.6}$	$f^{-1.9}$	$J_k \propto N_B W_B/\tau_f$
Pushout curr./area, J_{push}	fnc	f^{-1}	if t_{epi} is properly scaled
Low current gain, β	1	1	assumed constant
BE junc. cap., C_{be}	$fwb^{-1}\, fvbjt^{0.5}\, fwe\, fle$	$f^{1.25}$	$C_{be} \propto N_B^{0.5} W_e L_e$
BC junc. cap., C_{bc}	$(fnc/fvdd)^{0.5}\, fwe\, fle$	$f^{1.25}$	$C_{bc} \propto (N_c/V_{dd})^{0.5} W_e L_e$

TABLE II

MOS PARAMETERS, SCALING RELATIONSHIPS, FINAL SCALING FACTORS FOR OPTIMUM SCALING, AND PHYSICAL JUSTIFICATIONS

Parameter, Symbol	Scaling Factor (General Case)	Scaling Factor (Optimum)	Equation/Reason
Group A: Independent Technological Parameters			
Gate width, W	fw	f	independent
Gate length, L	fl	f	independent
Oxide thickness, t_{ox}	$ftox$	f	independent
Supply voltage, V_{dd}	$fvdd$	$f^{0.5}$	independent
Group B: Dependent Technological Parameters			
Junction depth, x_j	fl	f	avoid short-channel effects
Doping, N_D	$fl^{-2}\, fvdd$	$f^{-1.5}$	drain/source punchthrough
Group C: Device Parameters			
Mobility, μ_{0p}	$fl^{0.5}\, fvdd^{-0.25}$	$f^{0.4}$	$\mu_{0p} \propto N_D^{-0.25}$
Threshold, V_t	$fvdd$	$f^{0.5}$	appropriate channel implant
Oxide cap, C_{ox}	$ftox^{-1}$	f^{-1}	$C_{ox} = \epsilon_{ox}/t_{ox}$
Junction cap, C_j	fw	f	$C_j \propto (N_D/V_{dd})^{0.5} WL$

where I_k is the knee current $= J_k W_e L_e$. To accurately predict I_{cP}, however, high-level injection effects due to base pushout need to be accounted for. Also, the fact that some of the MOSFET current is diverted to charge the base-collector and parasitic capacitance needs to be accounted for. To account for base pushout, the BJT transit time (τ) is modeled to have a current (i_c) and bias (V_{bc}) dependence, with J_{push} included in the dependence. The expression for τ used in this work is [8]

$$\tau = \tau_f \left(1 + X_{tf} e^{V_{bc}/1.44 V_{tf}} \frac{i_c}{J_{push} W_e L_e}\right) \tag{4}$$

where X_{tf} and V_{tf} are constants. Thus for a larger J_{push}, resulting from larger N_c, the base pushout effect is smaller because of a smaller τ for the same i_c. The above expression for τ is used to obtain a differential equation describing the transient collector current in the BiCMOS gate during switching [6], [7]. The peak value of the collector current I_{cP} is obtained as the root of the following polynomial ($C_b = C_p + C_{bc}$):

$$\frac{\tau_f X_{tf}}{1.44 V_{tf}} e^{V_{bcP}/1.44 V_{tf}} \frac{I_{cP}^2}{J_{\text{push}} W_e L_e} \frac{I_{cP} + I_{\text{mos}}}{C_b + C_L} + I_{cP}\left(\frac{1}{\beta} + \frac{C_b}{C_b + C_L}\right) + \frac{I_{cP}^2}{\beta I_k} = I_{\text{mos}} \frac{C_L}{C_b + C_L}. \quad (5)$$

This value of I_{cP} is used in the delay expression (2) to give the required gate delay with base pushout accounted for.

The delay model has been verified by comparison with BiCMOS gate delay in both 1 and 0.6 μm technologies. The 1 μm technology features 8 GHz conventional BJT's and 5 V MOSFET's. The 0.6 μm technology is a scaled version with 15 GHz BJT's and 4 V MOSFET's. The comparisons are carried out over a wide range of circuit parameters of the model. Supply voltage, load capacitance, MOS, and BJT device widths are varied in multiple permutations and the gate delay predicted by the model is compared with measured or circuit simulated delay. The 1 μm comparisons are with measured gate delays of fabricated gates and the 0.6 μm comparisons are with circuit-simulated delays using measured SPICE parameters extracted from individually fabricated transistors. Fig. 1(b) shows excellent correlation across the range of circuit conditions for both 1 and 0.6 μm technologies.

Using the scaled device parameters in the delay expression provides the value of the scaled delay for the scaling scenario of interest. The scaled delay is observed relative to its basis, unscaled value for the 1 μm case. Since it is the relative delay that is observed, the absolute values of the MOS transistor width or the BJT emitter size do not affect the scaling analysis. The absolute value of the load also cancels out when observing relative delay. For completeness, the values used in the scaling analyses in Section III are mentioned here. The unscaled MOSFET size is $W/L = 20\ \mu\text{m}/1\ \mu\text{m}$. Fanout is chosen = 8, intermediate between light and heavy loads and corresponds to a load = 0.66 pF for the unscaled condition. The BJT emitter size is $W_e \times L_e = 1\ \mu\text{m} \times 8\ \mu\text{m}$, selected to give the lowest gate delay for the given W and fanout in the unscaled case.

C. Simplified Scaling of Delay Terms

By simplifying a few of the delay terms, simplified scaling rules for the delay components can be obtained. These are useful to obtain a first-order qualitative insight into the behavior of scaled gate delay. The simplified rules are shown in Table III and the assumptions made to arrive at them are described below.

To simplify the C_{bj} expression (see (A1)), the load capacitance is assumed to be large and the parasitic capacitance at the base is assumed to be small. C_{bj} reduces to $(C_{bc} + C_{be})$. These two components have different scaling expressions (see Table I) and it cannot be safely assumed that one dominates over the other. Both components are retained and a single scaling expression cannot be written for the general case.

The MOSFET current expression (see (A5)) is simplified by assuming a velocity-saturated MOSFET. Consider the output pull-up case first. In saturation

$$I_m[V_{sg}, V_{sd}] = I_m[V_{sg}] = v_{\text{sat}} W C_{\text{ox}}(V_{sg} - V_{tp}). \quad (6)$$

The MOSFET is assumed to be in saturation until the output reaches $V_{dd}/2$. Therefore

$$I_{\text{mos1}} = I_{\text{mos}} = I_m[V_{dd} - \delta] = v_{\text{sat}} W C_{\text{ox}}(V_{dd} - \delta - V_{tp}). \quad (7)$$

δ is the difference between the input low voltage and ground. V_{dd} and V_{tp} both scale as *fvdd* but δ is a nonscaling constant. The quantity $(V_{dd} - \delta - V_{tp})$ is assigned a scaling factor *fvdd** so that I_{mos} scales as *fw fvdd*/ftox*. The asterisk is equivalent to an exponent greater than 1. Hence if *fvdd* < 1, *fvdd** is smaller than *fvdd*. The exponent is intentionally not given a precise numerical value since this value depends on *fvdd* and also it is only the qualitative nature of I_{mos} scaling that is of interest for the simplified expressions in Table III. For the output pull-down case there is an additional reduction in the MOSFET gate drive due to the diode drop across the pull-down BJT.

$$I_{\text{mos1}} = I_{\text{mos}} = I_m[V_{dd} - \delta] = v_{\text{sat}} W C_{\text{ox}}(V_{dd} - \delta - V_{tn} - V_{\text{diode}}). \quad (8)$$

The scaling of $(V_{dd} - \delta - V_{tn} - V_{\text{diode}})$ equated to *fvdd** with the implicit meaning that the asterisk for the pull-down case represents a larger value than in the pull-up case.

To simplify the expression for the peak collector current, base push-out is neglected and the base current is assumed to be simply I_{mos}. Hence the I_{cP} expression becomes (3). Using the I_k and I_{mos} scaling expressions already determined results in the simplified expression for I_{cP} in Table III. The output swing (from output low to half the supply) is $V_s = V_{dd}/2 - \delta$. This quantity also scales faster than V_{dd} for *fvdd* < 1. Hence in a manner similar to I_{mos}, V_s is assigned the scaling factor *fvdd**. For C_L, if the output is a constant fanout, C_L scales as the MOSFET gate capacitance. If the output is a constant load, the scaling factor is unity.

The simplified expressions for the delay components are then obtained from the equations shown in the righmost column in Table III.

III. Scaling Scenarios

Given the above framework for scaling analysis, various scaling scenarios are investigated to arrive at the optimum set of scaling rules. Note that the actual data plot-

TABLE III
FIRST-ORDER SCALING OF DELAY COMPONENTS WITH SIMPLIFICATIONS

Parameter, Symbol	Scaling Factor (General Case)	Scaling Factor (Optimum)	Equation/Reason
Delay Expression Terms			
Base cap, C_{bj}		$f^{1.25}$	$C_{be} - C_{bc}$
MOS current, I_{mos}	$fw\, fvdd^*/ftox$	$f^{0.5*}$	velocity saturated
Peak collector current, I_{cP}	$(fwb^{-2.2}\, fvbjt^{0.6}\, fwe\, fle \times fw\, fvdd^*\, ftox^{-1})^{0.5}$	$f^{0.3}$	assume $I_{cP} = \sqrt{\beta I_k I_{\text{mos}}}$
Output swing, V_s	$fvdd^*$	$f^{0.5*}$	V_{be} drop reduces V_s
Load cap, C_L	$fw\, fl/ftox$	f	constant fanout
	1	1	constant load
Delay Components			
Junction delay		$f^{1.25}/f^{0.5*}$	$C_{bj}(V_{\text{diode}} + V_{ol})/I_{\text{mos1}}$
Diffusion delay	$(fwb^{0.2}\, fvbjt^{1.4}\, fwe\, fle \times fw^{-1}\, ftox/fvdd^*)^{0.5}$	$f^{1.7}/f^{0.5*}$	$\tau_f I_{cP}/I_{\text{mos}}$
Load delay	$(fwb^{2.2}\, fvbjt^{-0.6}\, fwe^{-1}\, fle^{-1} \times fw\, fl^2\, ftox^{-1}\, fvdd^*)^{0.5}$	$f^{0.7} f^{0.5*}$	$C_L V_s/I_{cP}$ constant fanout
	$(fwb^{2.2}\, fvbjt^{-0.6}\, fwe^{-1}\, fle^{-1} \times fw^{-1}\, ftox^{1}\, fvdd^*)^{0.5}$	$f^{-0.1} f^{0.5*}$	constant load

ted for all of the scenarios below are obtained using the full delay expression. A qualitative, physical explanation for the trends in the data is given with the help of the rough approximations in Table III.

A. Gate Oxide Scaled

Scaling a certain parameter or some restricted set of parameters, say, only the oxide thickness of the MOSFET (t_{ox}) or the base width of the BJT (W_B), is intuitively not optimal. In order to demonstrate this quantitatively, t_{ox} is scaled with all other independent parameters (Group A) in Tables I and II remaining constant. Mathematically, *ftox* varies from 1 down to 0.25 with all other independent scaling factors = 1. The components of total delay (2) and the total delay relative to the unscaled values are plotted as a function of *ftox* in Fig. 2(a). Scaling down t_{ox} leads to an increase in oxide capacitance and thereby a linear increase in I_{mos} (Table III). For constant C_{bj} and ($V_{\text{diode}} + V_{ol}$) this reduces the junction delay component linearly as seen in Fig. 2(a). The peak collector current increases because of the smaller t_{ox} (approximately $I_{cP} \propto ftox^{-0.5}$) but does not increase as fast as I_{mos}($I_{\text{mos}} \propto ftox^{-1}$) and this leads to an overall decrease in the diffusion delay as seen in Fig. 2(a). For a constant fanout, the load capacitance increases as $C_L \propto ftox^{-1}$ and the increase in I_{cP} is slower. The load delay therefore *increases* with reducing t_{ox}. For constant load, the load delay does reduce (not shown). This t_{ox} scaling effectively illustrates that the components of the total delay may be affected in opposite senses when some one physical parameter is scaled, leading to only marginal improvement in the total delay.

B. Base Width Scaled

Next, the BJT base width is scaled with all other parameters remaining constant. *fwb* varies from 1 down to 0.25 with all other independent scaling factors = 1. As a result, the base emitter junction capacitance C_{be} scales as fwb^{-1} (Table I) due to the increased N_B while I_{mos1} remains unchanged. The junction delay increases as shown in Fig. 2(b). Note that the junction delay rises slower than C_{be} does because C_{be} is only one of the components of the total base capacitance C_{bj}. Examining the diffusion delay, τ_f scales as $fwb^{1.2}$ and I_{mos} is unchanged. From Table III, I_{cP} scales as $fwb^{-1.1}$ primarily due to the larger knee current density J_k. Consequently, one might expect the diffusion delay to scale slowly as $fwb^{0.1}$. Fig. 2 shows that the accurately obtained diffusion delay reduces faster than expected suggesting that I_{cP} does not rise as fast as $fwb^{-1.1}$ as predicted by the simplistic (3). This is because with increasing knee current, I_{cP} is limited by the base pushout current which remains constant. The load value and swing are constant, but because of the increasing I_{cP}, the junction delay reduces, albeit scaling slower than $fwb^{1.1}$. Overall, the total delay scales fairly poorly when only the base width is scaled, primarily due to the increased junction delay. Note that the junction delay could be reduced by scaling down the BJT horizontal dimensions.

C. Only MOSFET Scaled

As shown above, it is clearly undesirable to scale only a few select parameters of the MOSFET or BJT. As the next step to overall scaling, the MOSFET is scaled with the BJT remaining unchanged. With reference to the Group A parameters in Tables I and II, $fl = ftox = fmos$ which varies from 1 down to 0.25 and $fw = fwe = fle = fwb = fnc = fvbjt = 1$. The MOS gate width is kept constant since the BJT size is not scaled down. Under constant field scaling the voltage should also be reduced linearly [1]. However the trend in the industry of 1 μm–5 V technology to 0.5 μm–3.5 V to 0.25 μm–2.5 V lies

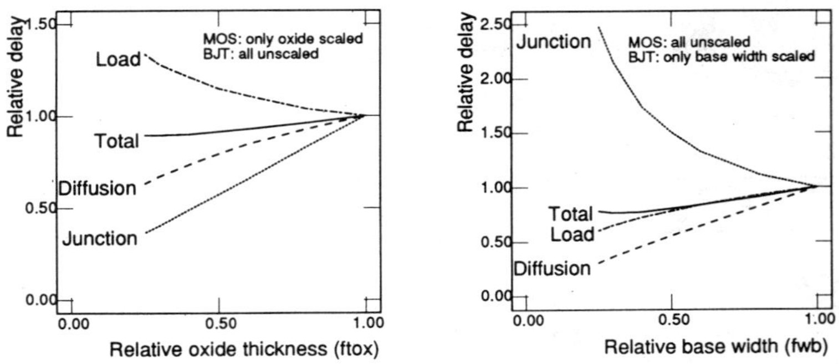

(a) (b)

Fig. 2. Relative delay with (a) only MOS gate oxide scaled or (b) only BJT base width scaled. Total delay scaling is poor because of opposing trends in delay components.

more closely on a square root scaling of voltage with dimensions (quasi-constant voltage scaling [9]). Hence V_{dd} is scaled as $fmos^{0.5}$. The BJT design voltage is kept constant ($fvbjt = 1$) to keep the BJT truly unchanged. The relative delay components are plotted as a function of *fmos* in Fig. 3.

As a first approximation, let $fvdd* = fvdd$. Then from Table III, I_{mos} scales as $fmos^{-0.5}$ and with a constant ($V_{diode} + V_{ol}$) and C_{bj}, the junction delay is expected to scale as $fmos^{0.5}$. The observed junction delay in Fig. 3 is much larger and eventually increases for small *fmos*, and this is due to two reasons. First, the gate bias across the MOSFET really scales as $fvdd*$ and I_{mos} is smaller. Second, the capacitance at the base has a parasitic component due to gate overlap capacitance of the MOSFET's which increases with reduced t_{ox} and this increases C_{bj} for small *fmos*. For the diffusion delay, τ_f is constant while I_{cP} increases slightly due to the larger I_{mos}. Again, if $fvdd*$ were equal to *fvdd*, I_{cP} would scale as $fmos^{-0.25}$ and therefore the diffusion delay would scale as $fmos^{0.25}$. The diffusion delay, however, also scales poorly because of the nonideally scaling I_{mos}. The load delay benefits from the reduced output swing and scales fairly well. Fig. 3 also shows the ideal scaling of pure CMOS delay which is simply *fmos*. The total BiCMOS delay scales poorly compared to the CMOS delay. The conclusion is, the performance advantage that BiCMOS enjoys over CMOS at 1 μm will be rapidly lost if BJT remains unscaled.

IV. Optimal BiCMOS Scaling Rules

From the results above it is clear that the BJT and MOSFET must be scaled down together for optimum delay scaling. The MOSFET horizontal dimensions and oxide thickness are scaled down together by the overall scaling factor f and the supply voltage by $f^{0.5}$. For the BJT, it was shown that scaling down the base width is favorable for the diffusion delay. The problem of increased junction delay when only the base width is scaled is addressed by scaling down the horizontal dimensions as well. Scaling down the BJT design voltage V_{bjt} is favorable in relaxing the base doping increase. This leads to lower τ_f and C_{be}, both favorable outcomes. Since the supply voltage is reduced, the BJT design voltage is also scaled as $f^{0.5}$. Thus all but one of the independent parameters of the BJT and

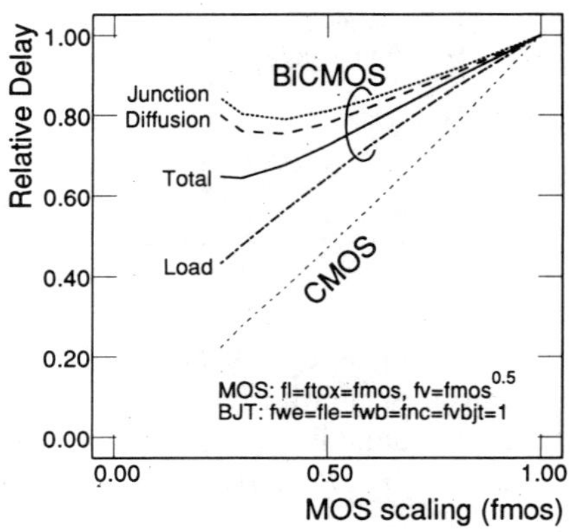

Fig. 3. Relative delay with only MOSFET scaled. BiCMOS delay scales much slower than the pure CMOS delay with the BJT kept unscaled.

MOSFET (Group A in Tables I and II) have been assigned scaling factors. The rule for collector doping (N_c) remains and is treated in detail next.

A. Determining Collector Doping Scaling

The primary effect of collector doping on gate delay is through the base pushout effect. In conventional BJT scaling, N_c is scaled such that the base pushout current density J_{push}, which is $\propto N_c$, scales at the same rate as the current density J_c. For ECL circuits with constant current and linearly reducing dimensions, J_c scales as $1/f^2$ and therefore N_c is scaled as $1/f^2$ as well. Likewise, in the BiCMOS gate, N_c scaling is driven by the need to scale J_{push} and J_c together. However, the collector current behaves quite differently. First, it is a time-varying transient so J_c is defined as the maximum current density $= I_{cP}/(W_e L_e)$. I_{cP} depends on the MOSFET current I_{mos} which itself is scaling and I_{cP} is affected by β rolloff and base pushout (see (5)).

The other factor constraining N_c is avalanche breakdown (BV_{ceo}) in the epi region. BV_{ceo} must scale down no faster than the supply V_{dd}. BV_{ceo} depends on both N_c and t_{epi}. (see (A6) in Appendix II). BV_{ceo} decreases with larger N_c and decreases with smaller t_{epi}. In order for J_{push} to scale with N_c, the epi thickness is already constrained to decrease with larger N_c as dictated by (1). This combined effect reduces BV_{ceo} rapidly with increased N_c.

J_c as a function of overall device scaling is then ex-

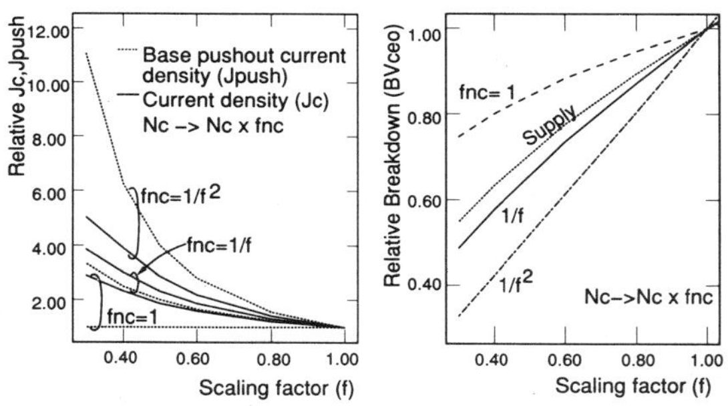

(a) (b)

Fig. 4. BiCMOS collector current density (J_c) and breakdown voltage (BV_{ceo}) for various N_c scaling rules. Linear increase of N_c is the optimum rule.

amined for three different possibilities for N_c: constant ($fnc = 1$), linear increase ($fnc = 1/f$), and quadratic increase ($fnc = 1/f^2$). The scaling of J_c in each case is compared in Fig. 4(a) with the corresponding J_{push} which is simply $= fnc$. The scaling of breakdown voltage is shown in Fig. 4(b) for the same three N_c rules. The scaling of the supply voltage is also shown for comparison.

When N_c is kept constant, the breakdown scales down fairly slowly. However, J_c rises much above J_{push} which remains constant and this scaling rule is, therefore, unacceptable for base pushout reasons. With a quadratic N_c increase, J_c stays much lower than J_{push} but the breakdown scales down much faster than V_{dd} and this scaling rule is unacceptable for breakdown. With a linear increase in N_c, J_c and J_{push} scale approximately together and BV_{ceo} also scales together with V_{dd}. Thus $fnc = 1/f$ is the optimum scaling rule for collector doping from combined base pushout and breakdown considerations. At first glance, this seems physically unintuitive when one considers that high-level injection effects dominate the BiCMOS gate operation. A faster increase in N_c would seem more logical. The simple physical explanation is that due to the scaling down of MOSFET current, the collector current scales down as well. The collector current *density* does increase but only linearly even when emitter area is reducing quadratically.

B. Scaled Device Parameters

With the rule for N_c scaling determined above, all the BJT and MOSFET independent parameters have been assigned scaling rules as listed in the "optimum rule" column of Tables I and II. Table IV shows some representative device parameters for the 1, 0.5, and 0.25 μm technologies. The values for 1 μm technology are actual measured device parameters and those for the scaled technologies are obtained by applying the optimum scaling rules. Note that these are not intended to be precise absolute numbers for the scaled technologies but rather to illustrate the trends in major device parameters required for effective BiCMOS scaling. The 0.5 μm scaled technology has a BJT f_T of 18 GHz and the 0.25 μm technology has f_T = 39 GHz. These values for f_T are determined by scaling each of the components of the emitter collector transit time according to the scaling laws in Tables I and II.

The values in Table IV can be compared with those of BiCMOS technologies published in the literature. Yoshimura *et al.* [10] present a 0.6 μm technology with BJT f_T = 15 GHz. This is precisely the f_T predicted by the scaling model for $f = 0.6$. t_{ox} for the technology is 100 Å, close to the value of 120 Å predicted by the model. Mele *et al.* [11] present a 0.5 μm BiCMOS technology. The BJT f_T = 14 GHz, which at first sight appears small compared to the model predicted value of 18 GHz. However this f_T is for BV_{ceo} = 6.5 V. Data are presented where f_T can reach 17 GHz when BV_{ceo} is reduced to 4 V. If the 0.5 μm BiCMOS circuits are to run at V_{dd} = 3.5 V, this could be an acceptable BV_{ceo} and then there is close agreement in the f_T values. This is confirmed by the data in [12] where a smaller BV_{ceo} = 4.5 V results in an f_T = 18 GHz. The C_{be} and C_{bc} values in [11] are also relatively close to those in Table IV. The assumption of a relatively constant current gain of 100 is also borne out for all three technologies. Konaka *et al.* [13] present a 0.3 μm BiCMOS technology. The f_T = 27 GHz at V_{ce} = 3 V. The model predicted value is 32 GHz. This difference is attributed to the large BV_{ceo} = 5.4 V. For a 0.3 μm technology, the V_{dd} is expected to be lower than 2.7 V so that BV_{ceo} has substantial margin for reduction.

C. Scaled Delay

With these scaling rules, Fig. 5 shows BiCMOS delay as a function of the scaling factor obtained from the full (nonapproximated) delay expression. Both rising and falling delays are shown because significant differences exist between the two. The "optimum" column of Table III shows simple scaling rules for the components of gate delay.

Consider the junction delay. From Table III, if the asterisk is neglected, the junction delay is expected to scale as $f^{0.75}$. The actual scaling in Fig. 5 is much slower. This is directly because of the rapidly reducing I_{mos1}. As a confirmation, it is seen that the junction delay scales much worse in the output pull-down case where the I_{mos1} reduces even faster (Section II-C). Examining the diffusion delay, Table III predicts roughly as $f^{1.2}$ behavior which is what is observed. There is not much difference in the output pull-down and pull-up cases because the difference in I_{mos} is mitigated by the BJT device parameters which also influence the diffusion delay. For the load delay, Table III predicts $f^{1.2}$ scaling. The load delay in Fig. 5 scales down slower. This is because of the faster scaling down of I_{mos} and the faster scaling down of I_{cP} than the simplistic prediction from (3). Here again the output falling delay scales more poorly.

The overall BiCMOS delay scales as shown in Fig. 6 with the CMOS delay scaling also shown for comparison. Absolute delay values are used for more effective comparison between BiCMOS and CMOS, though the actual absolute numbers are not of great significance. The BiCMOS delay does reduce with scaling, even though the

TABLE IV
STARTING FROM MEASURED 1-μm PARAMETERS, THE DESIRED TECHNOLOGICAL AND DEVICE PARAMETERS FOR OPTIMALLY SCALED 0.5 AND 0.25 μm TECHNOLOGIES

Parameter, Symbol	1 μm	0.5 μm	0.25 μm
Technological Parameters			
Emitter width, W_e	1 μm	0.5 μm	0.25 μm
Base width, W_B	0.15 μm	0.075 μm	0.037 μm
Collector doping, N_c	2×10^{16} cm^{-3}	4×10^{16} cm^{-3}	8×10^{16} cm^{-3}
Gate length, L	1 μm	0.5 μm	0.25 μm
Oxide thickness, t_{ox}	200 Å	100 Å	50 Å
Supply voltage, V_{dd}	5 V	3.5 V	2.5 V
Device Parameters			
Transit time, τ_f	17 ps	6.4 ps	2.44 ps
Knee current, J_k	3 mA/μm^2	11.2 mA/μm^2	41 mA/μm^2
BE junc cap, C_{be}	29 fF (A_e = 8 μm^2)	12 fF (A_e = 2 μm^2)	5.15 fF (A_e = 0.5 μm^2)
BC junc cap, C_{bc}	20 fF (A_e = 8 μm^2)	8.6 fF (A_e = 2 μm^2)	3.63 fF (A_e = 0.5 μm^2)
Cutoff frequency, f_T	8 GHz	18 GHz	39 GHz

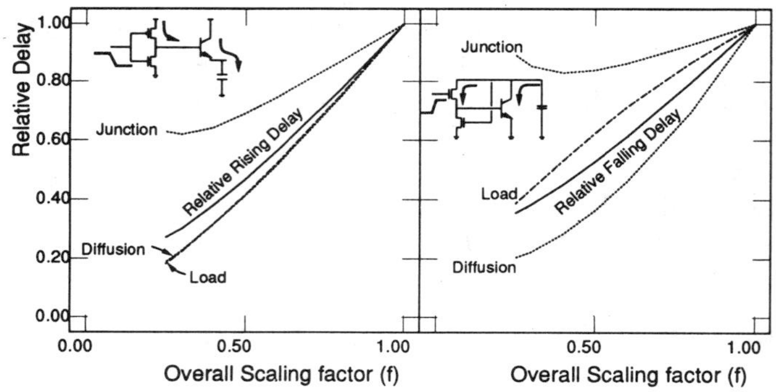

(a) (b)

Fig. 5. Relative values of each of the components of BiCMOS output (a) rising and (b) falling delay.

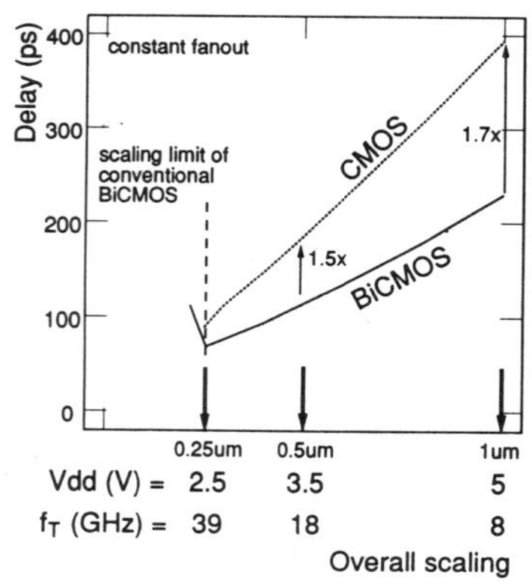

Fig. 6. BiCMOS and CMOS delays under overall scaling. 0.25 μm, 2.5 V, 39 GHz technology is the ultimate scaling limit of the conventional gate.

supply voltage is reduced. This is contrary to the widely held belief that BiCMOS is unusable at 3.5 V supply voltage [14], [15]. The improved behavior is obtained because *device* scaling has also been performed concomitantly with supply voltage scaling. The CMOS delay scales approximately linearly as expected. The performance advantage of BiCMOS over CMOS does reduce slightly with scaling, although CMOS delay is larger than BiCMOS by a factor of 1.5 for the 0.5 μm technology.

The scaling of BiCMOS delay continues until the 0.25 μm technology but the BJTs will have to have an f_T = 39 GHz and the performance advantage over pure CMOS will be only 1.3×. Beyond the 0.25 μm regime there is an *increase* in the BiCMOS delay largely due to the poorly scaling output falling delay. Alternative gate circuits need to be used in the 0.25 μm regime. These circuits must not only have aggressively scaled devices but also a different configuration to overcome loss of gate bias to the MOSFET in the gate [16].

V. Conclusion

A comprehensive scaling framework for BiCMOS scaling is developed that allows the investigation of a variety of alternative scaling scenarios. Scaling only select device parameters or even scaling the MOSFET fully without scaling down the BJT is shown to lead to poor delay scaling. The optimum scaling scenario accounting for the unique operating conditions in the BiCMOS gate is shown to have horizontal and vertical dimensions scaling down linearly, voltage scaling down as the square root, and collector doping scaling up linearly. BiCMOS maintains its performance superiority over CMOS even with reduced supply voltages, showing a 1.5× superiority at 0.5 μm, 3.5 V, 18 GHz BiCMOS technology. There is only a 1.3× advantage at 0.25 μm, 2.5 V technology and advanced BJT's with f_T = 39 GHz are required. With further scaling, the conventional BiCMOS gate suffers an increase in delay. The nonscaling diode drop especially affecting the output falling transition is shown to be the primary cause of the poorly scaling delay. New BiCMOS gate circuits that provide adequate gate bias to the MOSFET can successfully be used in the 0.25 μm regime.

Appendix I
Delay Model Details

The term C_{bj} in (2) is expressed as

$$C_{bj} = C_p + C_{bc} + \frac{C_{be} C_L}{C_{be} + C_L}. \quad \text{(A1)}$$

C_p, the total parasitic capacitance at the base, is expressed as

$$C_p = C_{ovp}W_1 + C_{ovn}W_3 + C_{jp1} + C_{jn3} + C_{ox}W_4L. \tag{A2}$$

$C_{ovp}(C_{ovn})$ are the PMOSFET (NMOSFET) gate overlap capacitances per unit length, $C_{jp1}(C_{jn3})$ are the total source/drain capacitances of $P1$ ($N3$). C_L is sum of load and parasitic capacitance at the output

$$C_L = C_{\text{load}} + C_{ovn}W_2 + C_{jn2} + C_{bc2} + C_{cs2}. \tag{A3}$$

In the above expressions, junction capacitances like C_{bc}, C_{be}, C_{jn2}, etc., have bias dependencies included. Thus C_{bc} is expressed in a manner similar to SPICE [8] with bias being the average reverse bias in the appropriate time interval.

$$C_{bc} = \frac{C_{jc}}{(1 + V_{\text{bias}}/V_{jc})^{m_{jc}}}. \tag{A4}$$

From the above set of equations, it is clear that C_{bj} has contributions from several different sources. In the general case, the scaling behavior of each of the additive terms will be different. For example, from Table I, C_{bc} scales as $(fnc/fvdd)^{0.5}$ *fwe fle* while from Table II, C_{jn3} scales as *fw* and from Table III, C_L scales as *fw fl/ftox*. Therefore, the scaling of C_{bj} cannot be presented as an expression containing simple powers of the independent scaling parameters.

The PMOSFET current is expressed as a function of the source–gate and source–drain bias (V_{sg}, V_{sd}). The NMOSFET expressions are similar.

$$I_m[V_{sg}, V_{sd}] = \mu_p C_{\text{ox}}(W/L)(V_{sg} - V_{tp} - V_{sd\text{-eff}}/2)V_{sd\text{-eff}}$$

$$\mu_p = \mu_{0p}\frac{1}{1 + (V_{sg} - V_{tp})/V_{\text{norm}\,P}} \cdot \frac{1}{1 + V_{dp\,\text{SAT}}/(E_{traP}L)}$$

$$V_{sd\text{-eff}} = \text{Min}\,[V_{sd}, V_{dp\,\text{SAT}}]$$

$$V_{dp\,\text{SAT}} = (V_{sg} - V_{tp}) + E_{\text{crit}\,P}L - \sqrt{(V_{sg} - V_{tp})^2 + (E_{\text{crit}\,P}L)^2}. \tag{A5}$$

$E_{\text{crit}\,P}(E_{\text{crit}\,N})$ is the electric field for velocity saturation in the P(N)MOSFET, $V_{\text{norm}\,P}(V_{\text{norm}\,N})$ and $E_{traP}(E_{traN})$ model the high-field reduction of P(N)MOSFET channel mobility. The term I_{mos1} in (2) is the average $I_m[\ ]$ during the time that the BJT junction capacitance is being charged. I_{mos} in (2) is the average $I_m[\ ]$ from the time the BJT is turned on until the time the output reaches $V_{dd}/2$.

Appendix II
Breakdown Voltage Equations

The equations for the BJT avalanche breakdown in the epilayer (BV_{ceo}) that follow are taken from [17].

$$BV_{ceo} = BV_{cbo}\left(\frac{1}{1+\beta}\right)^{1/n}$$

$$BV_{cbo} = V_{BR}\left(\frac{t_{\text{epi}}}{W_m}\right)\left(2 - \frac{t_{\text{epi}}}{W_m}\right), \quad \text{if } t_{\text{epi}} < W_m$$

$$V_{BR} = 60(N_c/10^{16})^{-0.75}$$

$$W_m = 2 \times 10^4 V_{BR}/E_m$$

$$E_m = \frac{4 \times 10^5}{1 - 0.33\log_{10}(N_c/10^{16})}. \tag{A6}$$

BV_{ceo} can be reduced to a function of N_c and t_{epi}. BV_{ceo} reduces with increasing N_c and with decreasing t_{epi}. t_{epi} in turn reduces with increasing N_c from the scaling constraint (Table I). Hence with BV_{ceo} expressed as a function of N_c alone, BV_{ceo} reduces rapidly with increased N_c.

References

[1] R. Dennard, F. Gaenssler, H. Yu, V. Rideout, E. Bassous, and A. LeBlanc, "Design of ion-implanted MOSFETs with very small physical dimensions," *IEEE J. Solid-State Circuits*, vol. SC-9, no. 5, pp. 256–268, Oct. 1974.

[2] P. Solomon and D. Tang, "Bipolar circuit scaling," in *ISSCC Dig. Tech. Papers*, 1979, pp. 86–87.

[3] A. Bellaouar, S. Embabi, and M. Elmasry, "Scaling of BiCMOS digital circuit structures," in *IEDM Tech. Dig.*, 1989, pp. 437–440.

[4] G. Rosseel and R. Dutton, "Scaling rules for bipolar transistors in BiCMOS circuits," in *IEDM Tech. Dig.*, 1989, pp. 795–798.

[5] R. Whittier and D. Tremere, "Current gain and cutoff frequency falloff at high currents," *IEEE Trans. Electron Devices*, vol. ED-16, no. 1, pp. 39–57, Jan. 1969.

[6] P. Raje, K. Saraswat and K. Cham, "Accurate delay models for digital BiCMOS," accepted for publication in *IEEE Trans. Electron Devices*.

[7] P. Raje, "Analysis, design and scaling of BiCMOS for digital logic," Ph.D. dissertation, Stanford University, Stanford, CA, 1991.

[8] P. Antognetti and G. Massobrio, *Semiconductor Device Modeling with SPICE*. New York: McGraw-Hill, 1987.

[9] P. Chatterjee, W. Hunter, T. Holloway, and Y. Lin, "The impact of scaling laws on the choice of n-channel or p-channel for MOS VLSI," *IEEE Electron Device Lett.*, vol. EDL-1, no. 10, pp. 220–223, Oct. 1980.

[10] T. Yoshimura, S. Yamada, T. Yamauchi, Y. Shimauchi, and K. Inayoshi, "0.6 μm high speed BiCMOS technology with emitter-base self-aligned structure," in *IEDM Tech. Dig.*, 1989, pp. 241–244.

[11] T. Mele, J. Hayden, F. Walczyk, M. Lien, Y. See, D. Denning, S. Cosentino, and A. Perera, "A high performance 0.5 μm BiCMOS triple polysilicon technology for 4 Mb fast SRAMs," in *IEDM Tech. Dig.*, 1990, pp. 481–484

[12] M. El-Diwany, M. Brassington, R. Razouk, P. van Wijnen, and V. Akylas, "Low voltage performance of an advanced CMOS/BiCMOS technology featuring 18 GHz bipolar fT and sub-70 ps CMOS gate delays," in *IEDM Tech. Dig.*, 1990, pp. 489–492.

[13] S. Konaka, T. Kobayashi, T. Matsuda, M. Ugajin, K. Imai, and T. Sakai, "HSST/BiCMOS technology with 26 ps ECL and 45 ps 2 V CMOS inverter," in *IEDM Tech. Dig.*, 1990, pp. 493–496.

[14] H. Momose, K. M. Cham, C. I. Drowley, H. R. Grinolds, and H. S. Fu, "0.5 micron BiCMOS technology," in *IEDM Tech. Dig.*, 1987, pp. 838–840.

[15] A. Watanabe, T. Nagano, S. Shukuri, and T. Ikeda, "Future BiCMOS technology for scaled supply voltage," in *IEDM Tech. Dig.*, 1989, pp. 429–431.

[16] P. Raje, R. Ritts, K. Cham, J. Plummer, and K. Saraswat, "MBiCMOS: A device and circuit technique for sub-micron, sub-2 V regime," in *ISSCC Dig. Tech. Papers*, 1991, pp. 150–151.

[17] S. Sze, *Physics of Semiconductor Devices*, 2nd ed. New York: Wiley-Eastern, 1981.

Part 2
BiCMOS Digital Circuit Applications

BICMOS technology offers digital IC designers a mix of logic families; CMOS, ECL, and BiCMOS, providing the tools to optimize circuit speed, power, and density. As a result, three areas of digital applications have benefited from BiCMOS technology: the area of gate arrays, application-specific ICs (ASICs), and the area of data and signal processors.

This part consists of four sections. Section 2.1 is an introduction and contains an invited overview by Shin. Section 2.2 deals with BiCMOS gate arrays. Section 2.3 deals with BiCMOS application-specific ICs and, finally, Section 2.4 deals with the application of BiCMOS to the design of data and signal processors.

In Section 2.2, the design of state-of-the-art BiCMOS gate arrays is covered in the three papers by Gallia et al., Hara et al., and Yoshino et al. In Section 2.3, two examples are given for the use of BiCMOS in the design of ASICs. The paper by Hara et al. covers the design of BiCMOS standard cell macros, while the paper by Rothermel covers the application of BiCMOS to the design of DPCM coders for HDTV systems.

In Section 2.4, four papers address the topic of applying BiCMOS to data processing, while two papers address its application to signal processing.

Paper 2.1

An Overview of BiCMOS State-of-the-Art Digital Circuits

HYUN J. SHIN
IBM THOMAS J. WATSON RESEARCH CENTER
YORKTOWN HEIGHTS, NY

Abstract—**State-of-the-art BiCMOS digital circuits for logic applications are reviewed in this paper. Applications have diversified into many areas such as gate arrays, ASICs, and data and signal processors. Leading-edge BiCMOS circuits that have been demonstrated include gate arrays with more than 100K gates operating at or above 100 MHz; standard-cell macros for ASICs and processors in a 0.5μm 3.3-V BiCMOS technology; and a 1000-MIPS 32-bit microprocessor and a video signal processor, both having a clock frequency of 250 MHz. BiCMOS progress is being driven by the use of fast high-gain bipolar transistors to improve circuit speed. Most BiCMOS digital circuitry is CMOS-dominant with add-on of specific BiCMOS or bipolar circuitry in appropriate places.**

1. Introduction

Since its introduction in the late 1970s [1], BiCMOS technology has been progressing quite remarkably [2]. Behind this enormous growth is the idea that CMOS circuits, when complemented by bipolar devices, would provide near-ideal, highly integrated, low-power high-speed digital or mixed-analog/digital VLSI circuits that can cover a wide range of applications; BiCMOS would be the one technology to exploit the advantages of CMOS, i.e., high integration-level and low power, with the strengths of bipolar transistors, such as fast switching speed, good drive capability, and large transconductance (g_m), or gain.

Naturally, in the early stage of development, BiCMOS technologies were simple enhancements to CMOS with minimally added process complexity and relatively low-performance npn bipolar transistors [1,3,4]. Therefore, the first applications of BiCMOS were mostly limited to selective use of bipolar devices in CMOS-dominant circuits for improving drive capability of an output buffer, accuracy of a reference generator, or voltage gain of a sense amplifier [5–7]. To go beyond these limited applications in the early BiCMOS technologies, starting from the mid-1980s, substantial efforts have been directed to the development of advanced BiCMOS technologies that offer optimized, small-size, high-performance bipolar transistors [2]. Despite some penalty in increased processing steps and cost, this new breed of sophisticated CMOS-based BiCMOS with uncompromised bipolar devices has led to the proliferation of high-speed digital VLSI and analog applications, including gate arrays, memories, application-specific integrated-circuits (ASICs), and general-purpose microprocessors and signal processors [8–10].

There also have been recent efforts to develop bipolar-based BiCMOS technologies for high-performance bipolar-intensive logic applications to reduce power and increase integration density, especially to increase on-chip memory capacity [9, 11]. Since maintaining high-frequency characteristics (e.g., cut-off frequency f_T) of the bipolar transistors is essential in these applications, the bipolar devices should not be compromised by the addition of the CMOS devices to be used in high-density memory cells. However, the difference between the bipolar-based and the CMOS-based BiCMOS is getting less obvious [2] as new technologies strive to optimize both types of devices.

For BiCMOS technology assessment, a frequently used figure of merit is the relative performance advantage of a BiCMOS logic gate over the gate in pure CMOS; it is typically represented by the gate delay as a function of load capacitance (as in Fig. 3). Until recently (to the 0.8-μm-technology generation), the most widely used BiCMOS gate circuits were the totem-pole type configured with an emitter-follower pull-up and gated-diode pull-down circuitry, as shown in Fig. 1. Figure 2 shows that typical gate delays of these conventional BiCMOS circuits are about 2 times better than those of CMOS in each technology generation marked by the corresponding pattern size, especially for 2.0- to 0.8-μm 5-V technologies [8]. Interestingly, converting CMOS into BiCMOS of equal technology generation for this 2× performance improvement takes only about a year or so, while developing the next CMOS generation for a similar speed enhancement needs approximately three to four years. The CMOS delays improve by

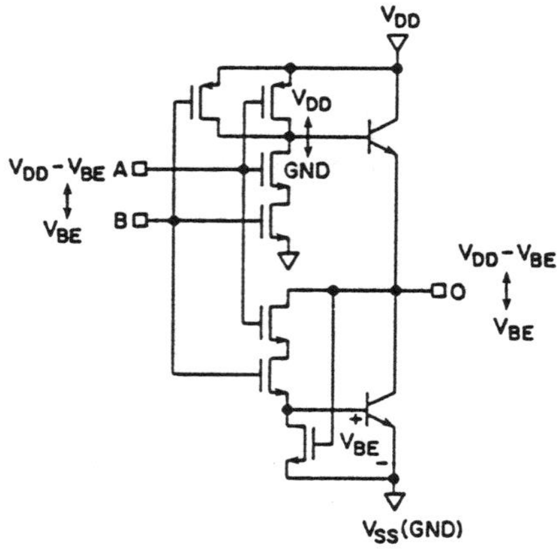

Fig. 1 Conventional BiCMOS logic circuit with emitter-follower pull-up and gated-diode pull-down circuitry ([14]).

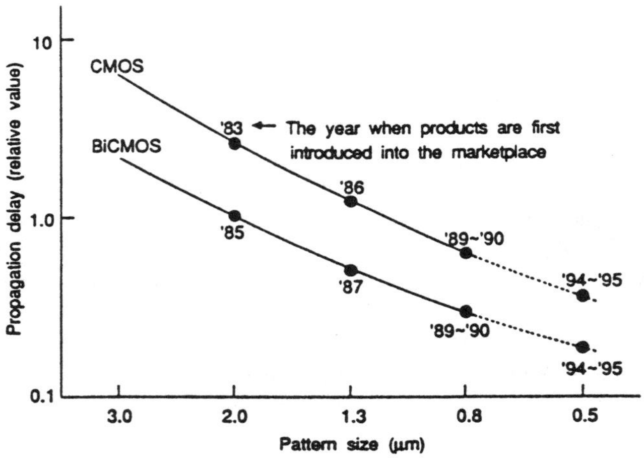

Fig. 2 CMOS/BiCMOS propagation delay versus pattern size ([8]).

a factor of 2 in every generation, roughly following the constant-voltage scaling rule.

Beyond the 0.8-μm into deep sub-micron generations, however, the performance leverage of conventional BiCMOS circuits over CMOS is questionable as shorter channel-length, scaled technologies require correspondingly lower power-supply voltages to ensure reliability of MOS and bipolar devices. Table 1 [12] shows a steady and quite rapid trend in power-supply voltage reduction into the future in accordance with the continued introduction of next-generation technologies: once in approximately every four years. With reduced supply voltages, the conventional BiCMOS gate circuit of Fig. 1 that features gated-diode pull-down circuitry quickly loses its speed advantage because it needs at least 2 $V_{BE} + V_{Tn}$ of supply voltage before turning on the pull-down path [13]. Here, V_{BE} is the base-emitter turn-on voltage (typically 0.8 V) and V_{Tn} is the threshold voltage of the gating nMOSFET. For example, as plotted in Fig. 3, BiCMOS is comparable in delay with CMOS for a 3.6-V 0.5-μm technology, but becomes much worse than CMOS with a 2.5-V 0.25-μm process. In essence, the fast-paced progress in technology has cast a sudden shadow on the future of BiCMOS in digital circuits. The conventional BiCMOS gates will no longer be usable, and viable BiCMOS circuit techniques that still have leverage with this unprecedented reduction of supply voltage are needed. To solve this critical problem, a number of BiCMOS gate circuits that may be extendible into the far sub-micron range with continued speed leverage over CMOS have been proposed [14–18]. These gates primarily utilize pull-down circuits, which are better than the gated diode, as well as full-swing techniques to maximize voltage drive. However, these new circuits require either a more sophisticated complementary bipolar technology [14,18] or use many devices in a large area [16–18]. Some of them are not sufficiently fast or provide only noninverting logic functions [15,18]. So, the question remains: Is it worth investing in BiCMOS rather than pure CMOS, or will the return be marginal with all the increased complexity and cost of processes and decreased integration density?

The answer to this question can be "yes" if BiCMOS or bipolar circuit techniques other than the conventional BiCMOS gates are cleverly applied to specific places in digital circuits. Because each bipolar device requires one V_{BE} for operation (and this value tends to rise as the feature size is reduced), techniques to minimize stacking of bipolar transistors become important. Furthermore, exploiting the main advantages of bipolar devices, such as high speed, high gain, and excellent

TABLE 1 Low-voltage CMOS technology projections ([12]).

	Year				
Technology	1992	1994	1996	1998	2000
Channel length, μm	0.5	0.4	0.3	0.2	0.1
Dynamic RAM Bits, Mb	16	64	64	256	256
Supply voltage, V	3.3/5.0	3.3	2.5/3.3	2.5	2.5
Logic Gates × 1000	200	400	600	1000	2000
Supply voltage, V	3.3	3.3	2.5	2.5	1.5
Processor speed, MHz	50	100	150	200	500

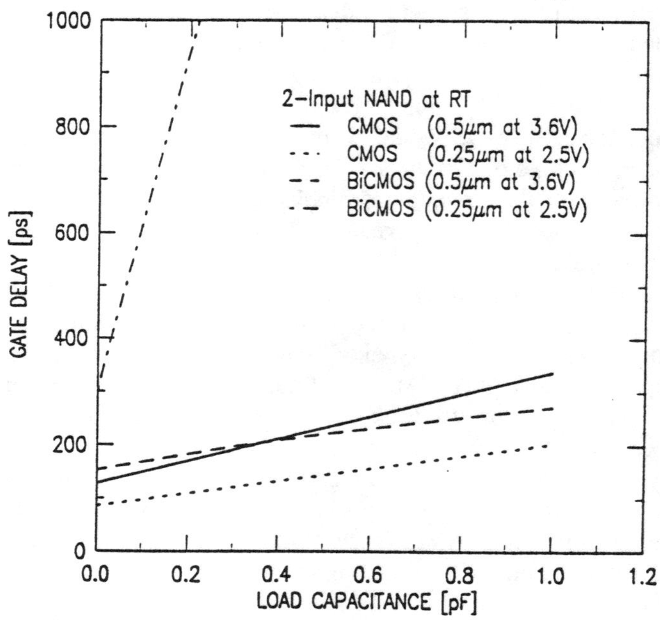

Fig. 3 Extendibility of conventional BiCMOS gate versus CMOS.

drivability, for reducing signal swing and noise and improving sensitivity is beneficial. In fact, these ideas have spurred the development of BiCMOS digital designs where bipolar usage is highly specific and selective. This trend is likely to continue the realization of BiCMOS digital circuits with performance beyond what pure CMOS can offer but with minimal cost penalty.

This paper is an overview of the highlights of the most recent advances in BiCMOS digital circuits. In Section 2, state-of-the-art gate arrays are reviewed. BiCMOS ASICs are looked at in Section 3, and more general purpose data and signal processors are discussed in Section 4.

2. Gate Arrays

Although the main driving force for BiCMOS development has been improved performance of BiCMOS over CMOS, demonstration of gate arrays that employ fast BiCMOS gates for high-speed logic applications has been rather rare. This is because there are many difficult questions to be answered before efficient and cost-effective BiCMOS gate arrays are designed:

1. Which BiCMOS gate circuits offer maximum performance with reasonable processing cost, circuit complexity, and gate density?
2. Are these BiCMOS circuits extendible to future technology generations that will require appropriately lower power-supply voltages, so that no major redesign effort is needed?
3. What are optimum basic cell images for maximum array utilization; versatile macro design such as latches, register files, and memories; and CMOS/BiCMOS gate intermixing?
4. How easily can existing CMOS gate-array macros be converted to BiCMOS macros?

A recent BiCMOS gate array that successfully answers question 1 above is a 100K-gate array implemented in a 0.8-μm 5-V BiCMOS technology [19]. This gate array utilizes the conventional partial- or limited-swing BiCMOS logic circuit shown in Fig. 4(a), and its full-swing versions without feedback control (Fig. 4(b)) or with feedback control (Fig. 4(c)). For 2-input NAND, the BiCMOS gates have intrinsic delays of 200 ps, while CMOS had 100 ps, as shown in Fig. 5. However, the drive factor for the BiCMOS is 17 ps/fan-out, 3.5 times better than the 60 ps/fan-out of the CMOS gate. Even though the performance leverage is big for this 5-V technology, these BiCMOS gate circuits have limited extendibility to future technologies of 0.5-μm channel length and below [13,14], so new circuits need to be developed for these generations.

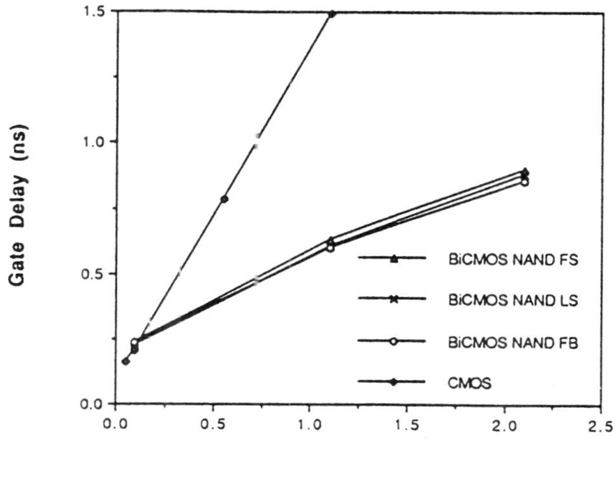

Fig. 5 BiCMOS and CMOS NAND delay versus load capacitance ([19]).

The base cell of this BiCMOS gate array is carefully designed for maximum array utilization and efficient macro layout (Fig. 6). It consists of two independent but identical CMOS sections, each containing two large p-channel MOSFETs, two large and two small n-channel MOSFETs, and a middle bipolar section having two npn transistors as well as four small (two n- and two p-channel) MOSFETs. The bipolar section is equally accessible by each CMOS section. So, in one base cell, two gates can be implemented, one of which may be BiCMOS. Also, due to the small MOSFETs, efficient BiCMOS latch and register-file macros can be designed. The base-cell size is 46.8 × 32 (= 1500) μm^2 and, with this cell, 106K gates can be populated within a 9 × 9-mm^2 array core (Fig. 7). With triple metal layers and local interconnect, this gate array has achieved a remarkably high array utilization of 92% in a test application to a 74K-gate digital filter. The input/output interface is ECL 100K compatible. These design points

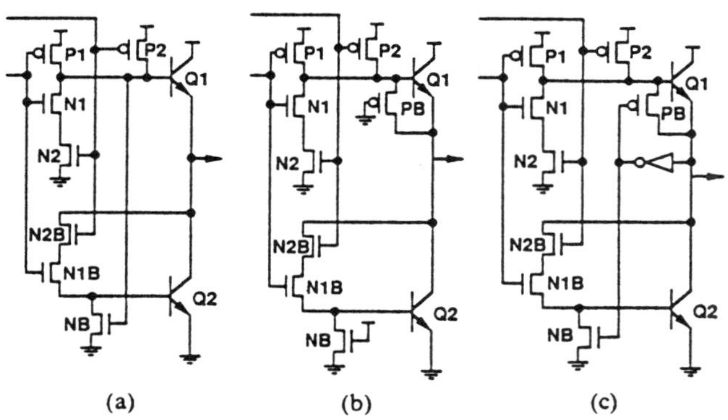

Fig. 4 BiCMOS 2-input NAND circuits: (a) limited swing (LS); (b) full swing (FS); and (c) feedback control (FB) ([19]).

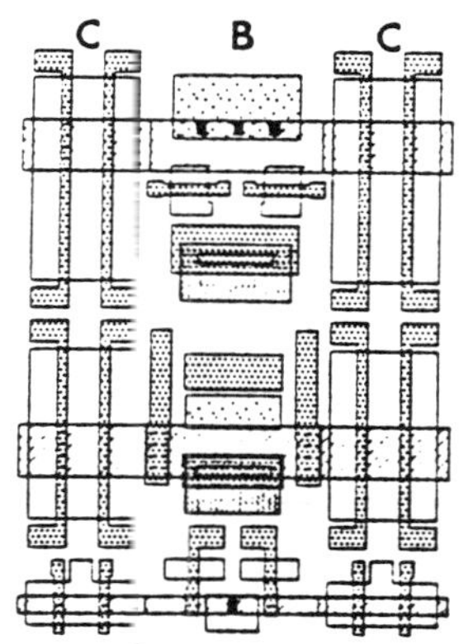

Fig. 6 A BiCMOS base-cell layout ([19]).

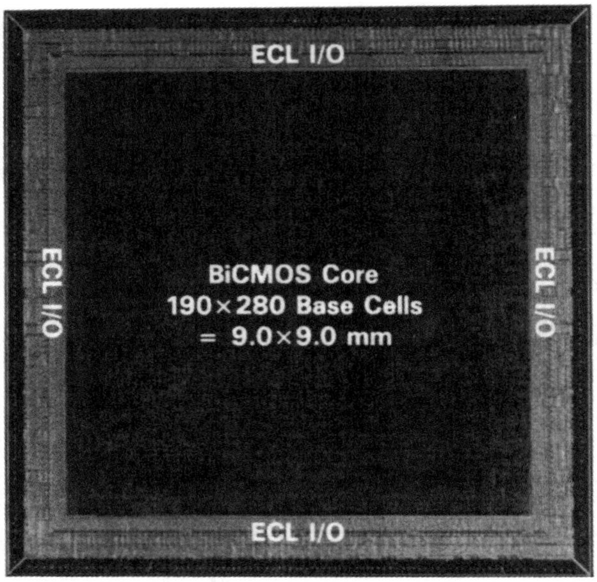

Fig. 7 A 106K-BiCMOS gate array photograph (chip size = 11.4 mm^2) ([19]).

are similarly addressed in the 200K-gate array implemented with a comparable BiCMOS circuit and technology [20].

The 100K-gate array has been applied to a 100-MHz 64-tap FIR filter [21], which dissipates maximum power of 6 W. The filter utilizes 187K MOSFETs and 22K bipolar transistors in an area of 49 mm^2, equivalent to 55K gates or approximately two-thirds of the array core. BiCMOS buffer circuits are selectively used to speed up the critical paths, including the clock and data-bus trees and the carry-save multiply-adder circuit, as shown in Fig. 8. The fast BiCMOS buffer also shows less speed degradation than CMOS at high chip temperature. The performance of this gate-array BiCMOS filter is comparable to a contemporary CMOS filter custom-designed with a module generator, which suggests that the advantage of gate-array design—quick turnaround time—is maintained without sacrificing the performance using BiCMOS technology.

Another state-of-the-art BiCMOS gate array worth noting because it considers questions 1–3 above is a 237K-gate array realized in a more advanced 0.5-μm, 13-GHz f_T, 3.3-V technology [22]. The gate circuit called BiPNMOS (shown in Fig. 9), which is a feedback-controlled full-swing version of the BiNMOS gate, is used. The BiNMOS circuit is one of the earliest BiCMOS circuits [6] and uses a bipolar transistor only for pull-up drive and n-channel MOSFETs for pull-down [15]. As shown in Fig. 10, this gate has a delay of 230 ps (for 2-input NAND with a fan-out of 7 at 3.3 V), 35% faster than pure CMOS, and is suitable for supply voltages down to 2.5 V.

This BiCMOS gate array features an optimized 1400-μm^2 basic cell containing eight small MOSFETs (two p-channel, six n-channel). With the small MOSFETs, memory macros, including a high-speed static RAM, a high-density RAM, a ROM, and a CAM, can be effectively designed, as demonstrated by a 16K-bit 2.7-ns RAM and a CAM using one basic cell for one memory cell, a 4.0-ns high-density RAM with two memory cells per basic cell, and a 3.2-ns ROM integrated on a test chip. A single-ended bit-line sensing scheme with double bit lines and ECL-like BiCMOS sense amplifiers is used to improve the speed.

A gate array that seeks minimal effort for converting existing CMOS gate macros into BiCMOS (specifically in response to question 4 above) has also been described [23]. Although the gate array requires a Schottky diode technology and the gate delay is marginally better than that of CMOS, an existing library of CMOS gates can be used to form BiCMOS macros without modification.

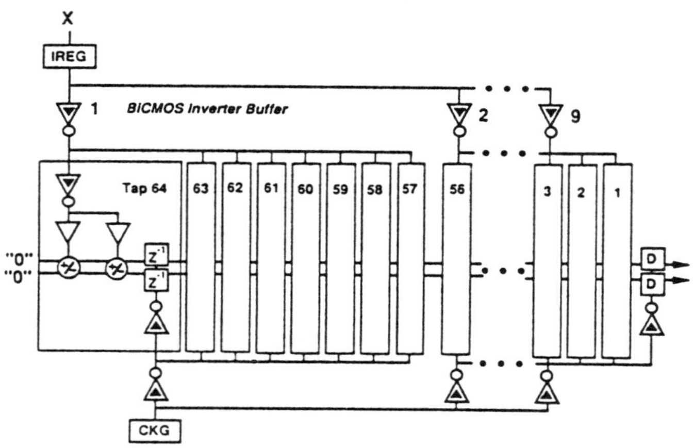

Fig. 8 A 64-tap BiCMOS FIR filter data/clock path ([21]).

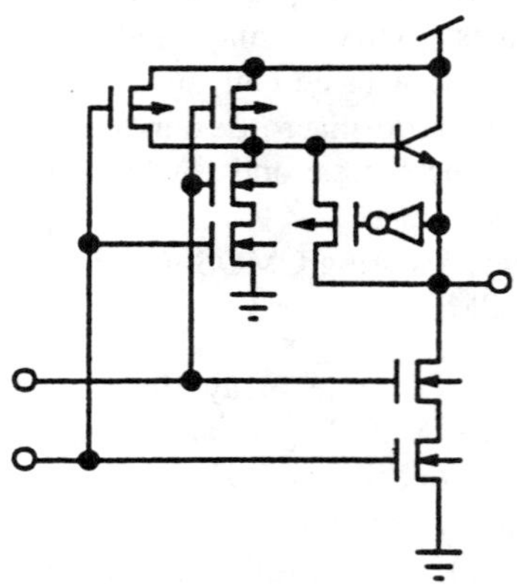

Fig. 9 A BiPNMOS gate ([22]).

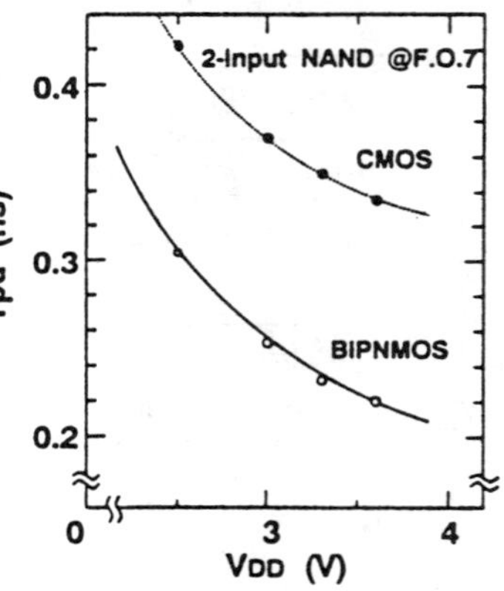

Fig. 10 Supply-voltage dependence of BiPNMOS and CMOS gate delays ([22]).

3. Application-specific Integrated-circuits

For high-speed ASICs, standard-cell macros custom-designed in BiCMOS have been emerging steadily [24, 25]. This is because the use of BiCMOS technology in upgrading performance of functional macros through custom tuning with BiCMOS circuits is believed to be more economical than the use of the next-generation CMOS. Also, unlike basic BiCMOS gate circuits whose extendibility to advanced technologies with lower supply voltages is often questionable, the BiCMOS standard-cell macro circuits are likely to be retained in future technologies and maintain the performance leverage over the CMOS because they are individually tailored to specific needs, such as improving sensing gain and speed, reducing voltage swing, and increasing load drivability. One thing to notice in BiCMOS ASICs is that, for the same reason, the fraction of bipolar devices used relative to the total devices is quite small compared to BiCMOS gate arrays.

A good example is the 54-MHz Differential Pulse Code Modulation (DPCM) coders for HDTV systems implemented in a 2.5-μm BiCMOS technology that is about 30% more expensive than CMOS of the same lithography generation [24]. Even though the minimum feature size is relatively large, the high-speed operation is possible due to architectural modification of the coders as well as some BiCMOS circuit techniques employed. The BiCMOS circuits utilized are a conventional totem-pole-type buffer for driving a significant capacitive load at the input of a Programmable Logic Array (PLA), used for the quantizer, and a single-ended sense amplifier for the PLA. As shown in Fig. 11, the BiCMOS sense circuit is essentially a common-emitter amplifier, which limits voltage swing at the sensing node of the PLA to about 0.8 V and yet prevents the bipolar transistor from saturating. By using this BiCMOS sense amplifier, the delay of the PLA is cut to nearly half that of the CMOS implementation.

A more comprehensive set of BiCMOS standard-cell macros for ASICs that potentially operate at clock rates up to 200 MHz has been demonstrated [25] for a 0.5-μm 3.3-V BiCMOS technology [22]. The macros implemented include a 3.0-ns 72×32-bit 10-port (6-READ, 4-WRITE) register file, a 5.0-ns 32K-byte cache memory, a 2.5-ns table look-aside buffer, and a 3.0-ns 32-bit adder. The chip is shown in Fig. 12. The basic gate in the standard cells used a full-swing version of the BiNMOS circuit that has substantial speed leverage over CMOS at 3.3 V. Interestingly, the register macro does not utilize BiCMOS circuits except for a fast-recovering bit-line load. A CMOS self-aligned threshold inverter is used as a single-ended bit-line sense amplifier. In the cache macro, a circuit technique of intermixing ECL and CMOS is adopted to improve speed, as depicted in Fig. 13. As an extension of this technique to basic BiCMOS standard cells for future 2-V technology generations, a direct-coupled ECL/CMOS circuit has also been explored here.

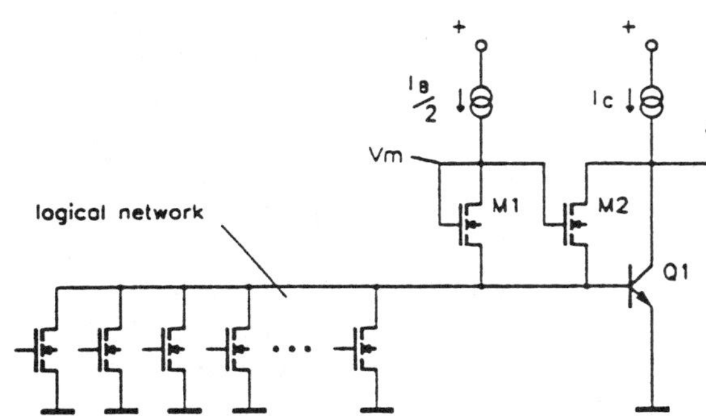

Fig. 11 A BiCMOS unlocked small-swing sense amplifier ([24]).

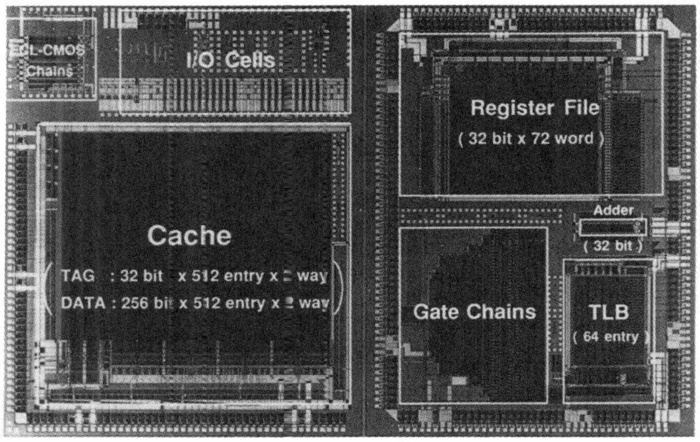

Fig. 12 A 0.5-μm BiCMOS standard-cell chip micrograph ([25]).

4. Data and Signal Processors

Compared with the ASICs discussed in Section 3, general-purpose BiCMOS processors for data and digital signal processing have shown impressive progress in recent years [27–32]. Since the first demonstration of a 70-MHz 32-bit microprocessor in a 1.0-μm BiCMOS technology in 1989 [26], a 1000-MIPS microprocessor operating at 250 MHz and a 250-MHz video signal processor have been reported [29,32]. These BiCMOS processors have been implemented with custom circuit macros that utilize bipolar or BiCMOS circuits only selectively in appropriate places; in fact, the usage of bipolar transistors in the chips ranges merely from 0.4% to 2.7%. This use of a very small proportion of bipolar devices, however, maximizes the performance without degrading the integration level per unit chip area.

In these processors, the most frequently used BiCMOS circuits are buffered logic gates for driving heavy capacitive

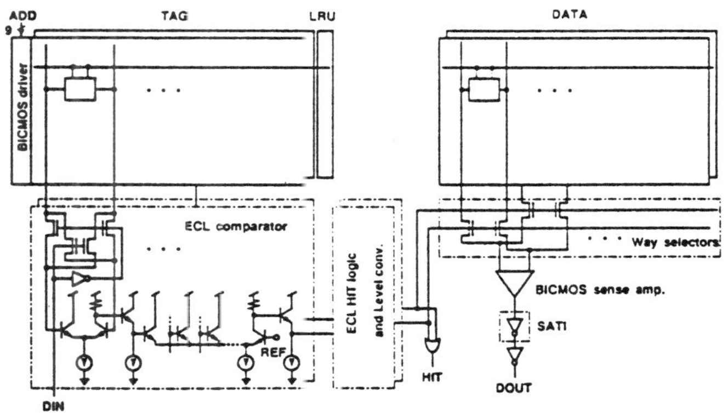

Fig. 13 A Cache macro circuit ([25]).

loads, which differ depending on the technology generation due to the reasons discussed in Section 1. Another BiCMOS circuit utilized is the bipolar-based sense amplifier and bit-line clamp circuit for memory macros that limits the bit-line swing to about 100 mV on READ to minimize noise while maintaining access time and speeding recovery from WRITE. This BiCMOS sense circuit has been adopted for the register file of the 32-bit RISC superscalar microprocessor [27]. With a 0.8-μm 5-V BiCMOS process, the register file can be accessed within 3.0 ns and the microprocessor operates at 40 MHz. In the companion 50-MHz cache controller for the microprocessor [28], BiCMOS techniques are found in the transistor-transitor-logic (TTL) output driver, phase-locked loops (PLLs) for clocks, and cache tag comparator. Basically, they apply the conventional totem-pole BiCMOS gate circuit. Fast BiCMOS gates for a phase/frequency detector in the PLL minimize false glitches and reset periods, resulting in locking up to 90 MHz. The BiCMOS 17-bit tag comparator in Fig. 14 uses a wired-OR emitter-follower for pull-up and an npn pull-down transistor controlled by the NMOS NOR circuitry. This circuit, limiting the highly capacitive NOR output to a swing of only one V_{BE}, makes comparisons in 1.5 ns.

In a 250-MHz 1000-MIPS superscalar microprocessor implemented in a more advanced 0.3-μm 3.3-V BiCMOS technology (Fig. 15) [29], the quasi-complementary BiCMOS (QC-BiCMOS) circuit shown in Fig. 16 [16] is used. The QC-BiCMOS gate uses a quasi-pnp device emulated as a combination of a p-channel MOSFET and an npn transistor for pull-down. (This is also used in the merged BiCMOS gate [17].) Although the circuit is complicated, requiring many devices, it offers more than twice the speed of CMOS at 3.3 V and is extendible to 2.5-V technologies. As illustrated in Fig. 17, this QC-BiCMOS gate is applied to latches and decoders in critical paths of the processor for speed and load drivability. An ECL sense amplifier is used selectively in the cache memory and table look-aside buffer (TLB). Also, QC-BiCMOS-feedbacked massive-input logic [30] is utilized in the arithmetic logic unit (ALU), to achieve addition in 1.2 ns, which is 60% faster than

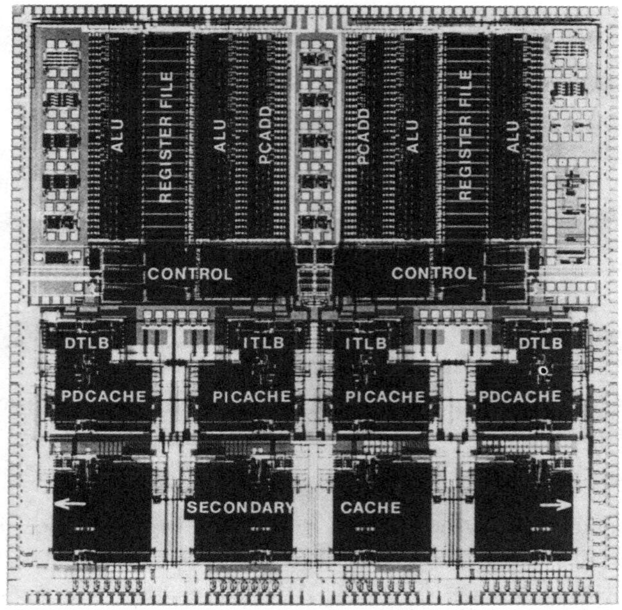

Fig. 15 A 1000-MIPS superscalar BiCMOS microprocessor ([29]).

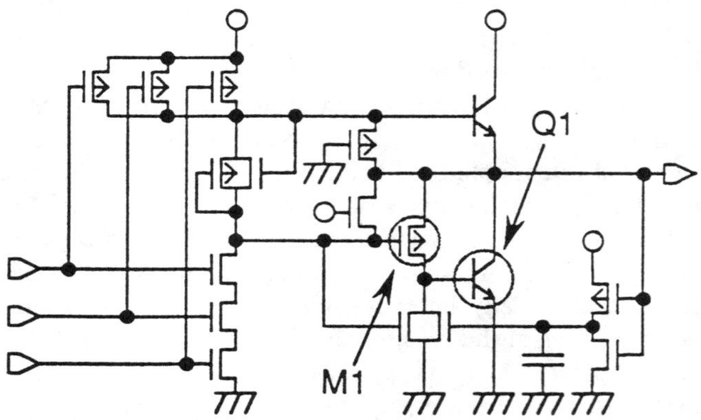

Fig. 16 A quasi-complementary BiCMOS logic circuit ([29]).

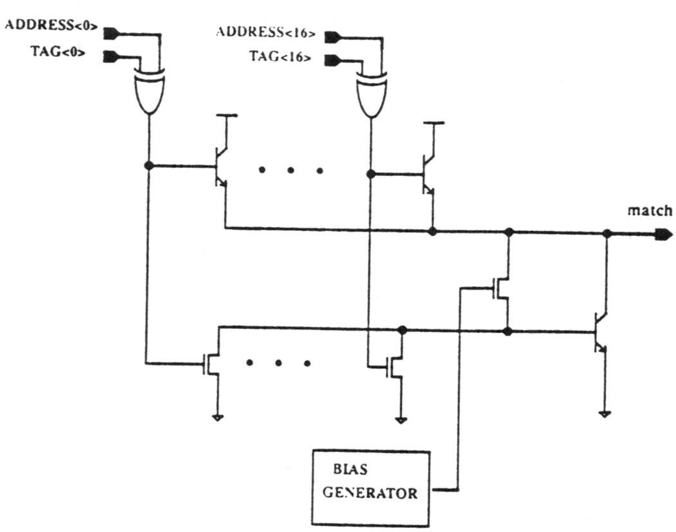

Fig. 14 A 17-bit tag comparator for 50-MHz BiCMOS cache controller ([28]).

CMOS. The QC-BiCMOS-feedbacked massive-input logic gate provides high-speed complex logic with very large fan-in, at the expense of high circuit complexity.

Applying buffered BiCMOS logic gates for driving large capacitive loads is effective in realizing high-performance vector and signal processors. For example, the 64-bit vector-pipelined processor (Fig. 18) implemented in a 0.8-μm 5-V BiCMOS technology [31] operates at 100 MHz to achieve 200-MFLOPS, simply by utilizing BiNMOS-type gates for the Booth decoder in the multiplier/divider, the decoder in the barrel shifters, the word-line decoders in the register file, the clock drivers, etc. With the same technology, a much higher speed (250 MHz) was obtained for the 16-bit video signal processor shown in Fig. 19 [32]. A pipelined ALU with a carry-look-ahead adder, featuring BiNMOS-type logic gates, as well as a pipelined convolver/multiplier with a Booth decoder, utilizing a faster version of the conventional BiCMOS gate, are key contributors for the speed-up. Among the 1.13 million transistors in this video processor chip, only 5,000, or 0.4%, are bipolar devices.

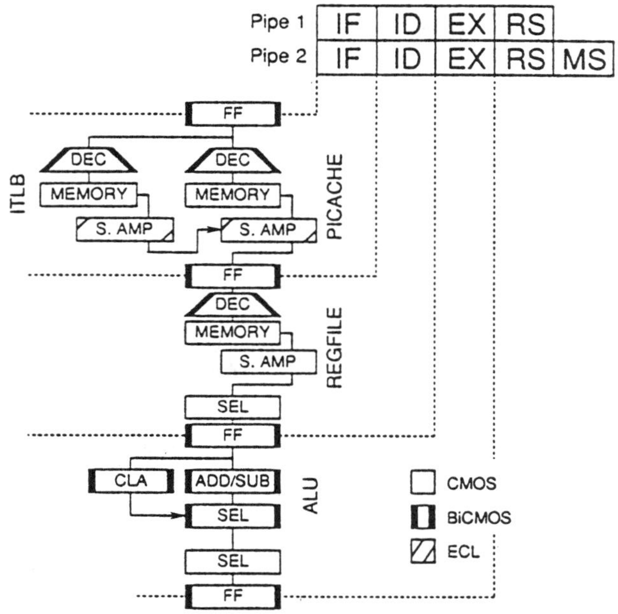

Fig. 17 Pipeline stages and critical paths of 1000-MIPS microprocessor ([29]).

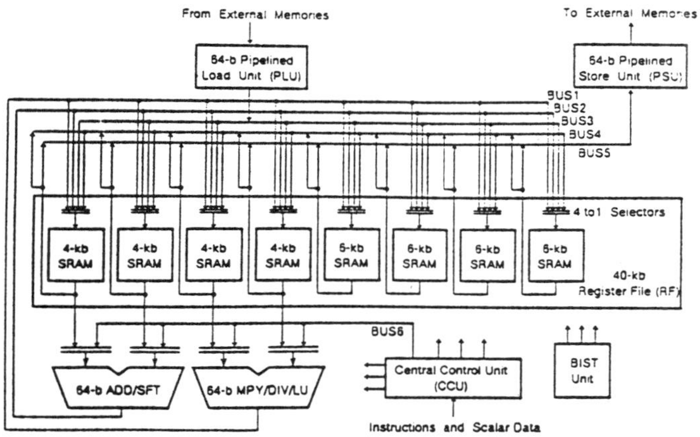

Fig. 18 A block diagram of 200-MFLOPS BiCMOS vector-pipelined processor ([31]).

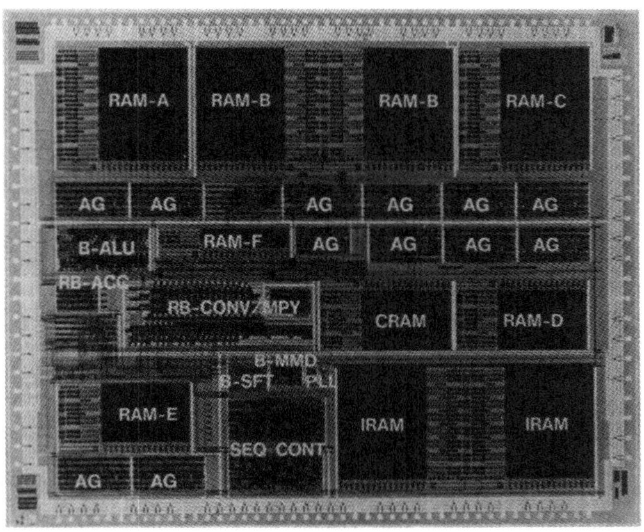

Fig. 19 A photograph of 250-MHz BiCMOS video signal processor ([32]).

5. CONCLUSIONS

There has been explosive progress in recent years in both BiCMOS digital circuits and BiCMOS technologies. Versatile BiCMOS circuit techniques for maximizing performance have been exploited in a variety of applications in high-speed high-density gate arrays, ASICs, and data/signal processors. Innovative BiCMOS logic gates and circuit techniques, such as full-swing techniques for gate circuits and small-swing techniques for sensing, extendible to the next few generations of sub-micron BiCMOS technologies that require supply-voltage scaling, are actively being developed to prolong the life and widen the spectrum of BiCMOS in digital applications.

ACKNOWLEDGMENTS

The author would like to thank Prof. M. I. Elmasry at the University of Waterloo, C.-T. Chuang, and C. J. Anderson for their encouragement. Special thanks are due to L. M. Terman and P. J. Lim for their valuable advice and editorial help.

REFERENCES

[1] G. Zimmer et al., "A fully implanted NMOS, CMOS, bipolar technology for VLSI of analog-digital systems," *IEEE Trans. Electron Devices,* vol. ED-26, pp. 390–396, April 1979.

[2] R. H. Havemann and R. J. Eklund, "Overview of BiCMOS device and process integration," in M. I. Elmasry (ed.), *BiCMOS Integrated Circuit Design,* Part I: Analysis and Design, New York: IEEE Press, 1993.

[3] P. A. Sullivan et al., "High performance bipolar transistors in a CMOS process," *IEEE Trans. Electron Devices,* vol. ED-29, pp. 1679–1680, Oct. 1982.

[4] C. S. Yue et al., "Improved bipolar transistor performance in a VLSI CMOS process," *IEEE Electron Device Lett.,* vol. EDL-4, pp. 294–296, Aug. 1983.

[5] E. L. Hudson and S. L. Smith, "An ECL compatible 4K CMOS RAM," in *ISSCC Dig. Tech. Papers,* Feb. 1982, pp. 248–249.

[6] O. Minato et al., "A Hi-CMOSII 8K × 8b Static Ram," in *ISSCC Dig. Tech. Papers,* Feb. 1982, pp. 256–257.

[7] J. Miyamoto et al., "A 28 ns CMOS SRAM with bipolar sense amplifiers," in *ISSCC Dig. Tech. Papers,* pp. 224–225, Feb. 1984.

[8] M. Kubo et al., "Perspective on BiCMOS VLSI's," *IEEE J. Solid-State Circuits,* vol. 23, pp. 5–11, Feb. 1988.

[9] A. G. Eldin, "An overview of BiCMOS state-of-the-art static and dynamic memory applications," in M. I. Elmasry (ed.), *BiCMOS Integrated Circuit Design,* Part III: BiCMOS Memory Applications, New York: IEEE Press, 1993.

[10] B. H. Leung, "Analog circuit design in BiCMOS technology: An overview," in M. I. Elmasry (ed.), *BiCMOS Integrated Circuit Design,* Part IV: BiCMOS Analog Circuit Applications, New York: IEEE Press, 1993.

[11] Y. Kobayashi et al., "SST-BiCMOS technology with 130-ps CMOS and 50-ps ECL," in *Symp. VLSI Technology Dig. Tech. Papers,* pp. 85–86, June 1990.

[12] B. Prince and R. H. Salters "ICs going on a 3-V diet," *IEEE Spectrum,* vol. 29, pp. 22–25, May 1992.

[13] H. J. Shin, "Performance comparison of driver configurations and full-swing techniques for BiCMOS logic circuits," *IEEE J. Solid-State Circuits,* vol. 25, pp. 863–865, June 1990.

[14] H. J. Shin, "Full-swing BiCMOS logic circuits with complementary emitter-follower driver configuration," *IEEE J. Solid-State Circuits,* vol. 26, pp. 578–584, April 1991.

[15] A. E. Gamal et al., "BiNMOS: A basic cell for BiCMOS sea-of-gates," in *Proc. CICC*, pp. 8.3.1–8.3.4, May 1989.
[16] K. Yano et al., "Quasi-complementary BiCMOS for sub-3-V digital circuits," *IEEE J. Solid-State Circuits*, vol. 26, pp. 1708–1719, Nov. 1991.
[17] R. B. Ritts et al. "Merged BiCMOS logic to extend the CMOS/BiCMOS performance crossover below 2.5-V supply," *IEEE J. Solid-State Circuits*, vol. 26, pp. 1606–1614, Nov. 1991.
[18] M. Hiraki et al., "A 1.5-V full-swing BiCMOS logic circuit," in *ISSCC Dig. Tech. Papers*, pp. 48–49, Feb. 1992.
[19] J. D. Gallia et al., "High-performance BiCMOS 100K-gate array," *IEEE J. Solid-State Circuits*, vol. 25, pp. 142–149, Feb. 1990.
[20] Y. Enomoto et al., "A 200K gate 0.8μm mixed CMOS/BiCMOS sea-of-gates," in *ISSCC Dig. Tech. Papers*, pp. 92–93, Feb. 1990.
[21] T. Yoshino et al., "A 100-MHz 64-tap FIR digital filter in 0.8-μm BiCMOS gate array," *IEEE J. Solid-State Circuits*, vol. 25, pp. 1494–1501, Dec. 1990.
[22] H. Hara et al., "0.5-μm 2M-transistor BiPNMOS channelless gate array," *IEEE J. Solid-State Circuits*, vol. 26, pp. 1615–1620, Nov. 1991.
[23] T. Horiuchi et al., "A new BiCMOS gate array cell with diode connected bipolar driver," in *Proc. IEEE 1990 Bipolar Circuits and Technology Meeting*, pp. 124–127, Sept. 1990.
[24] A. Rothermel et al., "BiCMOS circuits for DPCM coders in HDTV systems," *IEEE J. Solid-State Circuits*, vol. 25, pp. 1470–1475, Dec. 1990.
[25] H. Hara et al., "0.5-μm BiCMOS standard-cell macros including 0.5W 3ns register file and 0.6W 5ns 32KB cache," in *ISSCC Dig. Tech. Papers*, pp. 46–47, Feb. 1992.
[26] T. Hotta et al., "A 70MHz 32b microprocessor with 1.0μm BiCMOS macrocell library," in *ISSCC Dig. Tech. Papers*, pp. 124–125, Feb. 1989.
[27] F. Abu-Nofal et al., A three-million-transistor microprocessor, in *ISSCC Dig. Tech. Papers*, pp. 108–109, Feb. 1992.
[28] B. Joshi et al., "A BiCMOS 50MHz cache controller for a superscalar microprocessor," in *ISSCC Dig. Tech. Papers*, pp. 110–111, Feb. 1992.
[29] O. Nishii et al., "A 1,000MIPS BiCMOS microprocessor with superscalar architecture," in *ISSCC Dig. Tech. Papers*, pp. 114–115, Feb. 1992.
[30] K. Yano et al., "3.3-V BiCMOS circuit techniques for 250-MHz RISC arithmetic modules," *IEEE J. Solid-State Circuits*, vol. SC-27, pp. 373–381, March 1992.
[31] F. Okamoto et al., "A 200-MFLOPS 100-MHz 64-b BiCMOS vector-pipelined processor (VPP) ULSI," *IEEE J. Solid-State Circuits*, vol. 26, pp. 1885–1893, Dec. 1991.
[32] J. Goto et al., "250-MHz BiCMOS super-high-speed video signal processor (S-VSP) ULSI," *IEEE J. Solid-State Circuits*, vol. 26, pp. 1876–1884, Dec. 1991.

Paper 2.2

High-Performance BiCMOS 100K-Gate Array

JAMES D. GALLIA, MEMBER, IEEE, AH-LYAN YEE, MEMBER, IEEE, KWOK KIT CHAU, MEMBER, IEEE, I.-FAY WANG, MEMBER, IEEE, HARVEY DAVIS, SHOBANA SWAMY, VAN M. NGUYEN, MEMBER, IEEE, KAMALESH NATVARLAL RUPAREL, MEMBER, IEEE, KERMIT MOORE, BRIAN CHAE, CARL E. LEMONDS, JR., PAT EYRES, TOSHIAKI YOSHINO, MEMBER, IEEE AND ASHWIN H. SHAH, SENIOR MEMBER, IEEE

***Abstract* — A BiCMOS gate array in 0.8-μm technology has been developed with CMOS intrinsic gate delays of 100 ps plus 60 ps/fan-out and BiCMOS intrinsic delays of 200 ps with a 17-ps/fan-out drive factor. A compact base cell (750 μm^2/gate) has been designed with full bipolar drive capability for the efficient layout of both primitive gates and large-arrayed macros, such as register files and multipliers. A 106K-gate array has been built on a 1.14-cm square chip with ECL I/O capability. The place and route in three levels of metal provide array utilization greater than 90 percent. The gate array was used to implement a 74K-gate filter design with testability features such as JTAG and two-phase scan.**

I. Introduction

THIS PAPER will discuss the circuit technology of a 106K-gate array in submicrometer BiCMOS technology, the largest BiCMOS gate array to date in terms of gate count and density.

BiCMOS is a technology which is very attractive for gate arrays for the reasons of high speed, high density, multiple I/O interfaces, and multiple design methodologies [1]–[4]. In particular, a BiCMOS buffer/driver is very attractive since it maintains the density advantage of CMOS and provides the high drive capability of the bipolar technology. As a result, it has been important to make the high-speed bipolar drive available to all gate sites in the array. This can be achieved for 100K-gate arrays with a 0.8-μm BiCMOS technology that includes high-density interconnect. The multiple-design methodologies in BiCMOS technology allow the designer to choose from and combine the best circuit techniques from ECL, CMOS, or BiCMOS in order to optimize the performance of his system. For high-performance I/O, BiCMOS provides a choice of fast ECL I/O, high-drive TTL, and high-performance CMOS interfaces. BiCMOS gives gate arrays flexibility to all types of embedded modules. The availability of fast high-density memory in BiCMOS [5] also makes the BiCMOS gate array very attractive.

Manuscript received July 28, 1989; revised October 13, 1989.
The authors are with the VLSI Design Laboratory, Texas Instruments Incorporated, Dallas, TX 75265.
IEEE Log Number 8932788.

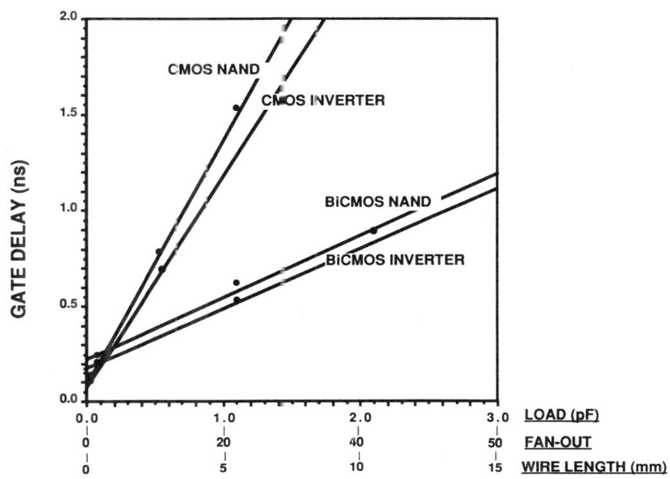

Fig. 1. Delay versus load.

The advantage of BiCMOS for gate arrays is obvious when BiCMOS gate delay is compared to that of CMOS gates, as shown in Fig. 1. This gate-delay comparison is shown for the 0.8-μm CMOS and BiCMOS inverter and NAND gates with approximately the same input capacitance: 0.045 pF for the CMOS gates, and 0.05 pF for the BiCMOS gates. In Fig. 1, the x axis expresses the gate loading in capacitance, equivalent gate loads, or the equivalent wiring length. For example, a 1-pF load is equal to a fan-out of 20 or a routing length of 5 mm. The 0.2-pF/mm wiring capacitance is based on three-dimensional triple-level metal-wiring capacitance simulations and measured data that include three-dimensional fringing.

For a fan-out of two the CMOS and BiCMOS NAND delays are similar: 200 ps for CMOS, and 230 ps for BiCMOS. The extrapolated CMOS intrinsic delay is 100 ps, with a drive factor of 60 ps/fan-out (slope). The BiCMOS NAND has an intrinsic delay of 200 ps; however, the drive factor is 17 ps/fan-out, or 3.5 times stronger than the CMOS gate. When we consider this delay difference in terms of driving long metal lines, such as halfway

Reprinted from *IEEE J. Solid-State Circuits*, vol., 25, no. 1., pp. 142–149, Feb. 1990.

across a 10-mm array, the CMOS inverter delay is 1.3 ns, while the BiCMOS inverter delay is only 0.5 ns.

In CMOS gate arrays it is often necessary to include 2 to 5 times larger drivers in the net list where heavy loads can be predicted. However, this increases the load on the previous stage by the same factor or requires additional stages of delay for buffering. It is difficult to obtain accurate pre-layout capacitance estimates that match the buffer strength to all the post-layout wiring capacitances. Post-layout net-list modifications are often required. The shallow slope of the BiCMOS delay curve means that the delay is less sensitive to the variance in the routing load that we experience in large arrays, and therefore it allows greater tolerance on timing-driven place and route tools.

The CMOS gate delays are faster at less than a fan-out of two because of the smaller intrinsic capacitance within the gate as compared to a BiCMOS gate, which has the internal base–collector and base–emitter capacitance to drive, regardless of the load. An optimum gate array base cell should then be capable of high-performance CMOS delays where loading can be constrained to a fan-out of two equivalent, and also be able to implement a set of interchangeable BiCMOS macros with the drive tolerance for all other loads.

This paper reports on a gate array architecture designed for submicrometer BiCMOS technology. Section II presents simulation and measured data on transistor sizing and alternate NAND gate schematics for BiCMOS gate arrays. Section III describes a BiCMOS base-cell design with a bipolar driver available to each gate site; the efficient layout of both CMOS and BiCMOS gate macros is discussed. Section IV gives the architecture and technology features of the BiCMOS 106K-gate array chip fabricated in 0.8-μm technology and triple-level metal. Results of a 74K-gate design test case are presented in Section V.

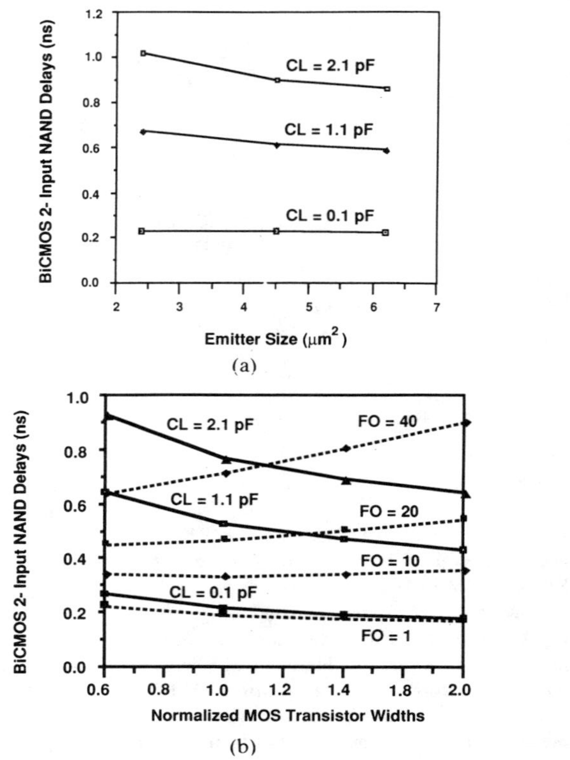

Fig. 2. (a) BiCMOS NAND delay versus emitter size. (b) BiCMOS NAND delay versus relative MOS size.

II. BiCMOS NAND Gate Circuit Design

A. Transistor Sizing

A gate-array base cell with a bipolar drive available for high-capacitance nets must be designed to the density requirements of large arrays. The base-cell design target area was set at 750 μm^2 for an equivalent two-input NAND gate to allow the integration of 100K gates on a 9-mm core or 200K gates on a 12.5-mm core.

The sizes of the base-cell components were optimized for both speed and density. The bipolar transistors are large enough to supply high drive currents, yet small enough to be incorporated into the base cell. Fig. 2(a) shows measured BiCMOS NAND gate-delay data versus emitter area for load capacitances varying from 0.1 to 2.1 pF. For light capacitance loading the delay does not decrease with the greater emitter area, because the increasing parasitic capacitance of the larger n-p-n's slows down the base-node transitions within the gate. The reduction in beta roll-off with larger emitters and larger I_k is a factor which becomes significant as the capacitance loading increases. However, for higher load capacitances, increasing the emitter area greater than 4.5 μm^2 has limited speed advantage. For a 1.1-pF load, the buffer delay changes only 32 ps when the emitter area is increased to 6.2 μm^2.

Fig. 2(b) shows the effect of scaling the MOS gate width for the BiCMOS NAND gate with the n-p-n driver area equal to 4.5 μm^2. The MOS gate width is normalized to a p-channel driver width of 15.6 μm and an n-channel driver width of 13 μm. The simulated BiCMOS NAND delay data is plotted versus two kinds of loading: a) fixed capacitance loads of 0.1, 1.1, and 2.1 pF, and b) a fan-out load, which is a variable capacitance that scales with the higher input capacitance as the gate width is increased. The delay versus constant capacitance load decreases with increasing MOS width, but the delay versus fan-out increases due to the higher input capacitance. Therefore, increasing the MOS widths would have the negative impact of a higher input capacitance loading plus a larger base-cell area.

B. Alternate BiCMOS NAND Circuits

The effect of the BiCMOS NAND circuit type on delay was studied for three different BiCMOS NAND configurations, as shown in the schematics of Fig. 3. The first circuit is the limited swing (LS) type [2]. Here the output levels are $V_{cc} - V_{be}$ (high) and $V_{ee} + V_{be}$ (low). Under switching conditions, the output is limited in swing to a V_{be} drop from the positive and negative power-supply levels. Tran-

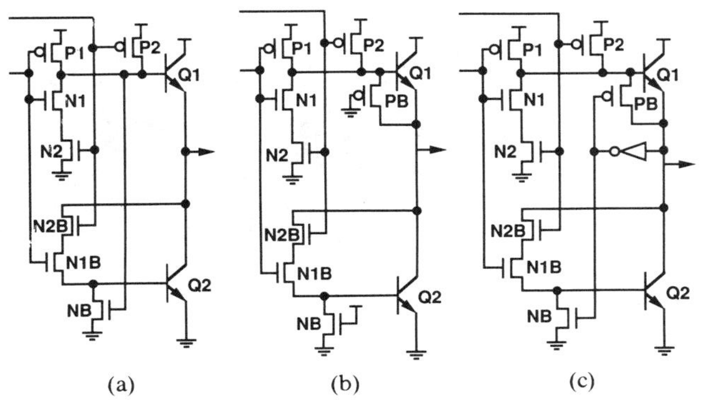

Fig. 3. BiCMOS two-input NAND configurations. (a) Limited swing (LS). (b) Full swing (FS). (c) Feedback inverter (FB).

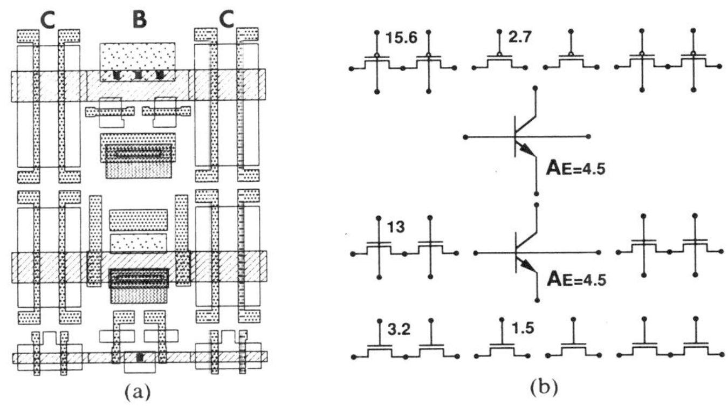

Fig. 5. (a) BiCMOS base-cell layout and (b) schematic.

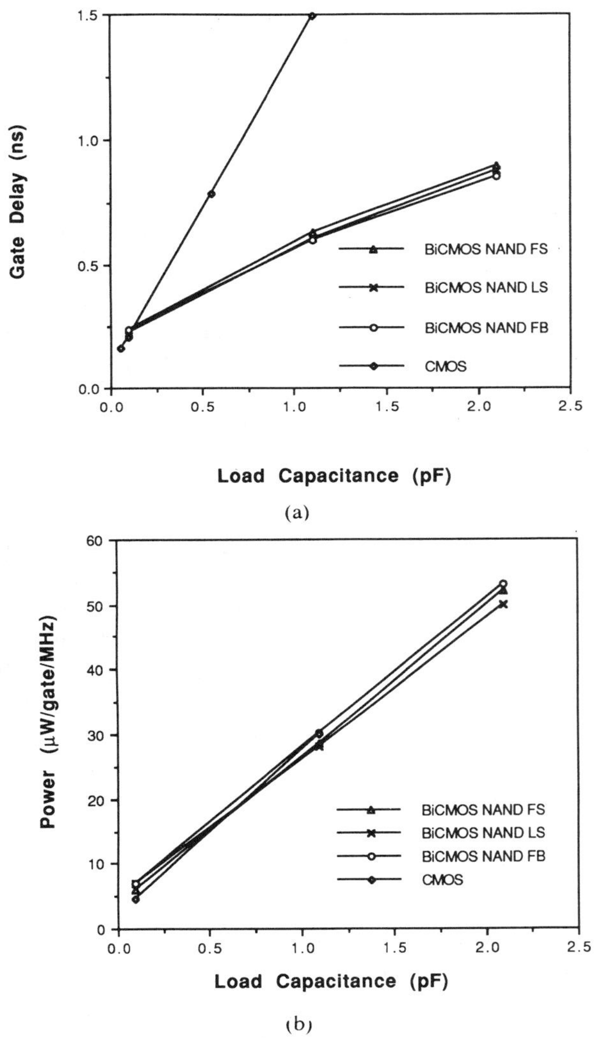

Fig. 4. (a) BiCMOS and CMOS NAND delay versus load capacitance. (b) BiCMOS and CMOS power versus load capacitance.

sistors $P1$, $P2$, $N1B$, and $N2B$ provide the base drive to n-p-n's $Q1$ and $Q2$.

For the full-swing (FS) circuit we have modified the full-swing BiCMOS NAND [3] with transistors PB and NB replacing resistors to switch the output to V_{cc} and V_{ee}. The output low-to-high transition time is enhanced by the use of PB instead of a fixed resistor because of the variable impedance of the p-channel transistor with changing gate bias. PB is OFF as the base of $Q1$ is initially charged high, and gradually turns ON to drive the output to V_{cc}.

In the feedback (FB) circuit, a feedback inverter has been added to control the gates of transistors PB and NB in order to provide full-swing output levels. Because of the switched gates, transistor PB is OFF when the base of n-p-n $Q1$ is charged high, and transistor NB is OFF when the base of $Q2$ is charged high. The effective base drive is increased with PB and NB OFF.

Fig. 4(a) shows the measured gate-delay data of the three BiCMOS NAND types and the CMOS NAND gate with increasing load capacitance. The FB-type BiCMOS NAND was slightly faster than the FS and LS versions because of the higher effective base drive from FS input signals and the switching of transistors PB and NB.

Fig. 4(b) compares the power/gate data of the CMOS and BiCMOS NAND circuits. The CMOS power was measured at 0.1- and 1.1-pF loading. The BiCMOS NAND power was measured at the two capacitance loads and additionally at 2.1 pF. Like CMOS, the BiCMOS NAND's have no static dc current since the bipolar drivers turn ON only during switching. The CMOS NAND power is lower than the BiCMOS NAND's at 0.1-pF loading because of the added capacitance of the bipolar drivers.

The power of the FS and LS BiCMOS NAND's is equal at 1.1-pF loading. At greater loads the LS circuit has slightly lower power. The superior turn OFF of the FS n-p-n drivers helps to lower its power versus the LS version. It should be noted that FS output levels avoid any sensitivity of V_{be} input levels causing CMOS input stage leakage as V_t shifts below 0.8 V with decreasing channel length.

III. BiCMOS Gate Array Macros

A. Base-Cell Layout

Fig. 5 shows the layout of our BiCMOS base-cell and transistor schematic. The cell consists of two independent CMOS sections that share access to a central bipolar

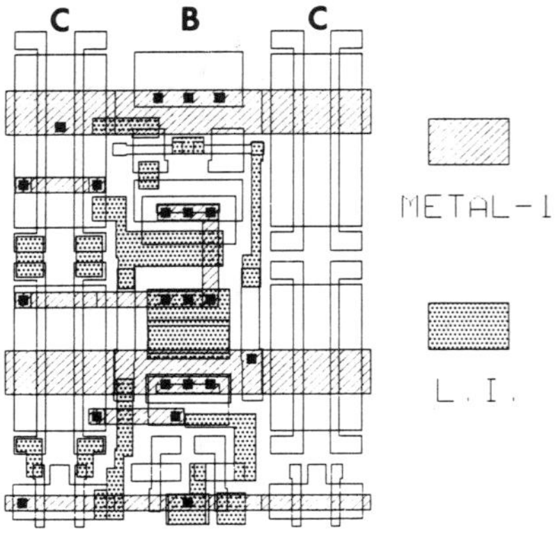

Fig. 6. FS BiCMOS two-input NAND macro layout.

section. Each CMOS section, designated by the C notation, contains two large p-channel transistors and two large and two small n-channel transistors. Each CMOS section is an independent site and can implement a two-input NAND equivalent gate, using the four large MOSFET's. The bipolar area or B section has two n-p-n transistors and four small CMOS devices. The bipolar drivers are readily available to either adjacent C section, depending on the macro selection. The base cell can then implement two gates independently, one of which can be BiCMOS. The base-cell size is $46.8 \times 32 = 1500\ \mu m^2$, or $750\ \mu m^2/\text{gate}$.

Besides component size, the number of routing tracks for the base cell must be sufficient for high-efficiency routing. The base cell has 18 horizontal tracks of metal-1 at 2.6-μm pitch with five power tracks shown. There are ten vertical tracks of metal-2 at 3.2-μm pitch, and 12 tracks of metal-3 at 3.9-μm pitch in line with the metal-1. A unique feature of the cell is that it uses a 2-to-3 ratio of metal-3 tracks to metal-1 tracks. It is designed for a high efficiency with routing of all three metal layers.

B. BiCMOS NAND Macro

The base cell was used to implement the CMOS and all three BiCMOS NAND variations. Fig. 6 shows the FS BiCMOS two-input NAND macro implemented in one C section and one B section of the base cell. In the schematic the transistors $P1$, $P2$, $N1B$, and $N2B$ are the large MOSFET's in the C section that form a CMOS NAND. They provide the base current drive to n-p-n transistors $Q1$ and $Q2$. The remaining small devices serve to discharge the base nodes to turn off the bipolar devices.

For efficient macro layout, all diffusion and poly nodes are silicided to cut the series resistance to contacts. Macro wiring makes use of TiN local interconnect (LI) to connect silicided areas as shown and to minimize the congestion of metal-1 wiring tracks. For example, in the NAND gate layout, LI is used to strap n and p gates together, connect up the small transistors, and provide the base drive to the n-p-n.

With the NAND gate macro layout, seven tracks for metal-1 remain completely available to the three-level router. The base cell is designed to be efficient for both large and small macros with three-level routing. For small macros, an abundance of unoccupied metal-1 tracks eliminates routing congestion around the high density of routing pins that occurs with small macros. For large macros, there are sufficient metal-1 tracks and LI connections to complete the macro wiring with little or no metal-2 usage. This frees the metal-2 and metal-3 planes to be used exclusively for efficient global wiring and power routing.

The placement of all macros can be mirrored horizontally because of the symmetry of the base cell. In Fig. 6, the BiCMOS NAND macro can occupy either C section adjacent to the central bipolar section. The remaining C section is free to be used by an adjacent macro. Because of the base-cell symmetry, each CMOS site has equal access to the bipolar section. Macros then can be composed of one C section, adjacent C and B sections, or any number of C and B sections as long as the CBCCBC pattern is used. This base-cell and macro-layout methodology has permitted a greater than 90-percent gate utilization in the routing results discussed in Section V.

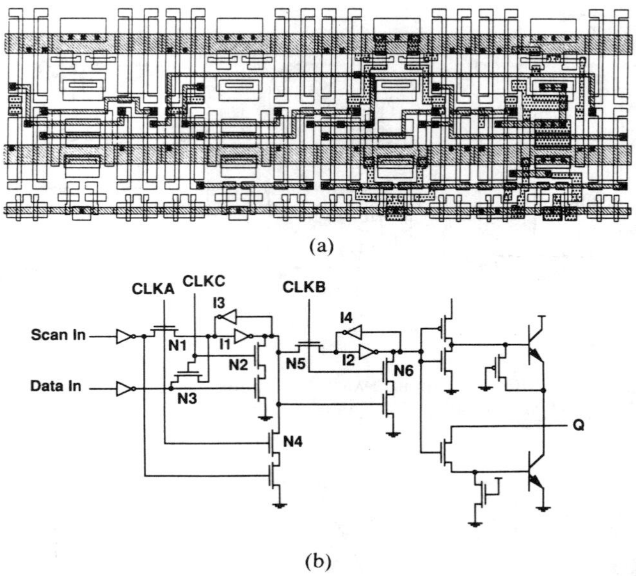

Fig. 7. (a) Two-phase-scan flip-flop macro and (b) schematic.

C. BiCMOS D Flip-Flop

The schematic and layout of a slightly larger macro, a two-phase-scan D flip-flop, are shown in Fig. 7. This circuit is very similar to a CMOS two-phase flip-flop that uses the n-channel transfer gates $N1-N6$ on both sides of the latches to enhance the low-to-high input transition of inverters $I1$ and $I2$ [6]. In this implementation the weak feedback inverters $I3$ and $I4$ are implemented with the small MOSFET's in the base cell, hooked-up with local interconnect. With the more traditional gate-isolation array layout where all transistors are the same size, this basic two-phase flip-flop required the use of five full-size transistors with series connections to realize each weak inverter function.

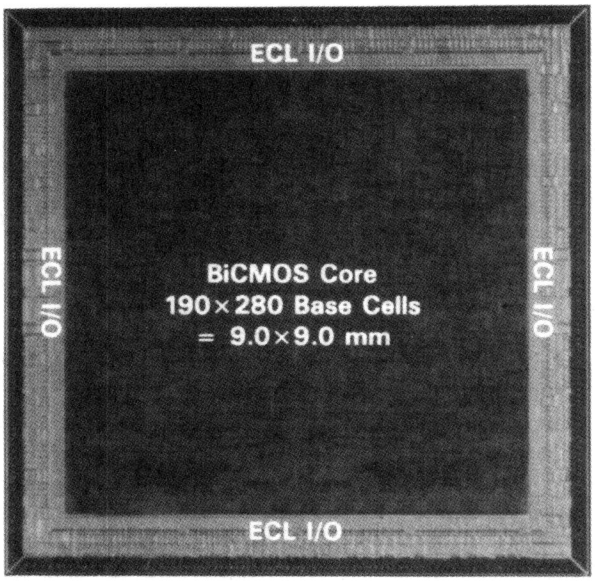

Fig. 8. 106K BiCMOS gate array die photograph. (Chip dimension = $11.4\ mm^2$.)

This flip-flop layout shows that the area of the bipolar section is used for multiple purposes: for the BiCMOS drive of the output, for the metal-2 power-bus channels through the center of the base cell, and for the use of the small MOS transistors for weak inverters.

D. Multiplier and Register File Arrayed Macros

The BiCMOS base cell meets the design requirements for larger arrayed macros for high-density, high-performance logic and memory functions. A 16×16-bit multiplier macro has been designed using ten unique macros. The multiplier uses the modified Booth algorithm for a simulated speed of 18 ns. The circuit uses 3000 gates in a 1.3×1.8-mm array.

A 16×16 register file macro has been created using one base cell for each four-port memory cell with two independent READ and two WRITE ports. The memory cell uses all the MOSFET's in the base cell for differential WRITE, single-ended READ, and zero standby-current operation. For the 16×16 register file, the total gate count is 1.1K, and the typical access time is 3 ns. BiCMOS word-line drivers allow extension of the word width to 64 bits with less than a 1-ns degradation in speed.

IV. 106K-Gate Array Test Chip

A. Array Architecture

A 106K-gate experimental chip with ECL I/O and an array of 190×280 BiCMOS base cells has been fabricated in 0.8-μm BiCMOS technology with three levels of metal, along with TiN local interconnect. The chip, shown in Fig. 8, measures 11.4 mm on a side. The two-gate base-cell size is 46.8×32 μm (1500 μm^2). All three levels of metal are used for signal, power, and clock routing. The metal-1 and metal-3 tracks run parallel, whereas the metal-2 tracks run orthogonal. The periphery contains 200 I/O cells at a 109-μm pitch, each capable of implementing 100K ECL input or output, with translation to the full CMOS levels. Bandgap reference generators and compensated current sources are distributed under the power bus. The power supplies are 0 and −5 V; the ECL and CMOS power buses are kept separate on the chip, for low noise on the ECL circuits. A few of the key characteristics of the array architecture are summarized in Table I.

TABLE I
Gate Array Chip Architecture

Gate Size	46.8μm x 16.0μm	748.8μm2
Basic Cell Size	46.8μm x 32.0μm	1497.6μm2
Chip Size	11.4mm x 11.4mm	130mm2
Base Array	190 rows x 280 columns	53.2K Cells
I/O Pitch		109μm
No. of Tracks:	Metal - 1	18
	Metal - 2	10
	Metal - 3	12

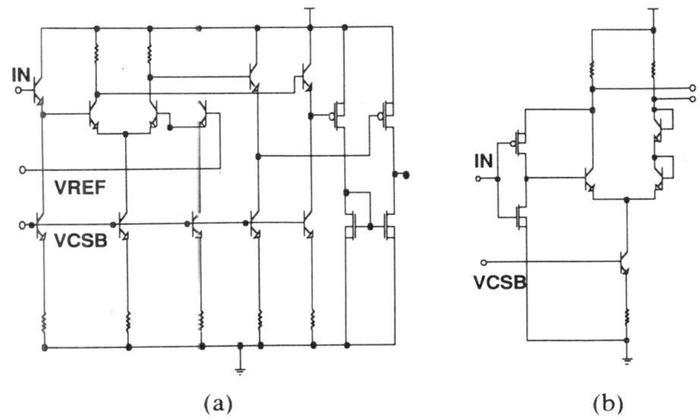

Fig. 9. I/O level translator circuits: (a) ECL to CMOS, and (b) CMOS to ECL.

B. ECL I/O Summary

The 106K-gate test chip features true ECL 100K I/O compatibility. I/O pair delay is 2.5 ns in and out, with level translation. Fig. 9(a) and (b) shows the translator circuit schematics used for the translation between ECL I/O and the CMOS/BiCMOS core-logic levels. For the input cell, an ECL-to-CMOS level translator is followed by a BiCMOS driver to get the input signal to the core. Translation delay is 1.2 ns, followed by a 0.5-ns buffer delay. The ECL-to-CMOS translator features a standard ECL comparator input with emitter followers driving an output stage with p-channel pull-up transistors and an n-channel current mirror pull-down.

Fig. 9(b) shows the circuit implemented to convert the BiCMOS logic swing to that needed to drive the ECL I/O logic [7]. An input V_{ref} signal is not needed, since the input threshold is set by the MOS device-size ratio. For the output cell, to support the JTAG [8] boundary scan with a minimum cost to normal output delay, a multiplexor was integrated with the CMOS-to-ECL translator. The total delay to output is 0.8 ns.

TABLE II
BiCMOS Technology Features

Gate Length	0.8μm
Diffusion Space	1.2μm
Min Emitter Size	1.6μm x 0.8μm
f_T	7GHz
Salicided Gates and Junctions	
TiN Local Interconnect	
CVD with Filled Vias with Stacking Allowed	
Routing Pitches =	2.6μm (METAL-1)
	3.2μm (METAL-2)
	3.9μm (METAL-3)

TABLE III
Filter-Routing Statistics

Design	Single Channel	Double Channel		
Array Utilization	73%	70%	92%	92%
No. of Gates	37K	74K	74K	74K
No. of Components	6K	12K	12K	29K
No. of Nets	8.5K	17K	17K	30K
Avg. Net Length (μm)	712	937	945	697

C. 0.8-μm BiCMOS Technology

The BiCMOS gate array technology is an extension of the double-level metal technology used to implement a very high-speed ECL 256K SRAM [9], which can be imbedded in this array with the third-level metal available for power and signal routing. The key characteristics of the gate array technology are summarized in Table II. Gate lengths are 0.8 μm with titanium-silicide-clad diffusion and poly. TiN local interconnect is available to connect silicide areas over field oxide. The triple-level metal technology features working pitches of 2.6, 3.2, and 3.9 μm for metal-1, -2, and -3, respectively, with CVD tungsten-filled contacts and vias.

V. 74K-Gate Filter-Design Test Case

A two-channel digital filter-design net list comprised of 37K gates per channel was implemented on the 106K BiCMOS gate array along with a macro-array test block. The design was implemented with fully automatic place and route using a library of 38 macros, including BiCMOS and CMOS versions. The hard macros range in size from a single gate to 31 gates. The FS-type BiCMOS driver was used. A soft macro was used to group the two-phase clock generator logic components. Static timing analysis was used to evaluate pre-layout and post-layout performance. The filter test chip was tested with functional pattern sets of 2200 vectors at a maximum clock frequency of 94 MHz.

A. Place and Route Results

Place and route experiments were performed with the single-channel and dual-channel filter design as the available array size and use of large macros were varied. Table III summarizes the routing statistics. The 37K single-channel design was routed with 73-percent utilization and implemented as two independent channels on the test chip. The 74K-gate two-channel design with 12K components (macro placements) and 17K nets was first routed with an array utilization of 70 percent. Utilization is defined as the percentage of available C sites occupied by macros. The 74K-gate design occupied a total of 74 400 C sites and 30 860 B sites. With the tuning of the software parameters, the placement and routing of the 74K-gate design yielded greater than 92-percent array utilization without a significant change in the average or total wire length. A 92-percent utilization was also achieved for the net list, with the large adder and register macros converted into small gate-level macros. The resulting net list consists of 29K components and 30K nets. The average net length decreased due to the addition of the shorter adder nets.

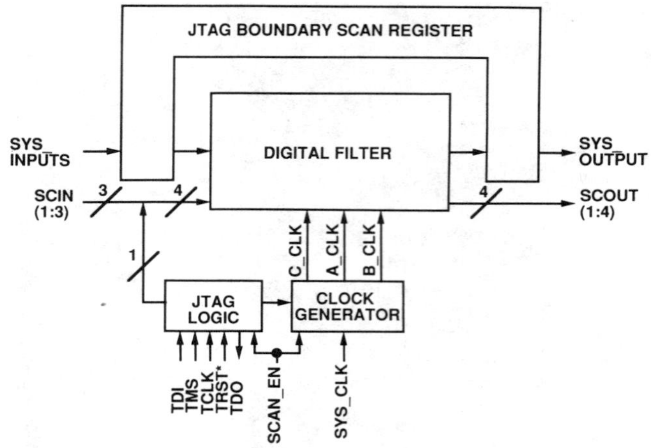

Fig. 10. Filter channel block diagram.

B. Testability

Testability was carefully considered in the design phase of the two-channel filter design with the following additions to the basic filter net list: a) a scan chain for full internal scan, b) JTAG circuitry, and c) a built-in self-test (BIST) function with a 12-bit signature analyzer. The scan and JTAG logic blocks were designed for generic insertion around existing net lists, as depicted in the Fig. 10 block diagram of the filter channel.

The implemented design contains four internal scan chains, three with 512 stages and one with 427 stages. The total overhead of adding the internal scan function was less than 10 percent in area and only one extra pin since all scan-ins and scan-outs make use of the existing functional pins. The two-phase LSSD register shown in Fig. 7 was used in the scan chains.

The JTAG boundary-scan architecture [8] was implemented on this design. Based on the boundary-scan technique, JTAG provides a general framework to check board interconnections (EXTEST mode), to monitor on-line operations (SAMPLE mode), and to test the circuit functionality (INTEST mode). Five extra pins were required for the JTAG protocol: TRST* (TEST RESET), TDI (TEST DATA IN), TMS (TEST MODE SELECT), TDO (TEST DATA OUT), and TCLK (TEST CLOCK). Essentially, the JTAG logic has complete controllability and observability of all the I/O pins of the devices by making use of some special circuits (boundary-scan cell). The internal synchronous 4-bit state machine synchronizes the initiation and termination of the tests and permits the scanning of instructions (data) into the instruction registers. The instruction decoding logic determines the test in question. The JTAG logic was autorouted with the rest of the filter-design net list. The JTAG area overhead is less than 4 percent.

The filter design also includes a built-in self-test structure with a linear-feedback shift register (LFSR) for pattern generation, and a parallel signature analyzer for response compression. The BIST circuits enable a rigorous in-system, high-speed check of the core structure of the design.

VI. Conclusion

In summary, a BiCMOS base cell has been designed for high density and high performance with a 750-μm^2 gate area. Measured performance data showed CMOS intrinsic delays of 100 ps with a drive factor of 60 ps/fan-out, and BiCMOS intrinsic delays of 200 ps with a drive factor of 17 ps/fan-out.

A 106K gate array has been fabricated in 0.8-μm BiCMOS technology using triple-level metal and local interconnect. The array features ECL 100K I/O interface. A 74K filter design was routed and fabricated using a library of CMOS and BiCMOS macros. A routing efficiency of greater than 90-percent utilization was achieved, and a maximum clock frequency of 94 MHz was measured.

Acknowledgment

The authors wish to thank those who helped and supported this effort; in particular, R. Haken, R. Eklund, R. Havemann, C. C. Wei, M. Douglas, B. Gale, T. Bonifield, T. Weaver, C. Huffman, B. Jucha, and D. Bell for the processing; H. Tran for circuit-design consultation; E. Born, G. Stacey, M. Harward, T. Aton, and M. Yao for test and characterization; and M. Craig, D. Forte, and H. Johnson for layout.

References

[1] L. Lin *et al.*, "A 9100 gate ECL/TTL compatible BiCMOS gate array," in *Proc. IEEE 1987 Custom Integrated Circuits Conf.*

[2] P. S. Bennett *et al.*, "High performance BiCMOS gate arrays with embedded memory," in *Proc. IEEE 1987 Custom Integrated Circuits Conf.*

[3] Y. Nishio *et al.*, "A 0.45-ns 7K Hi-BiCMOS gate array with configurable three-port 4.6K SRAM," in *Proc. IEEE 1987 Custom Integrated Circuits Conf.*

[4] A. Wong *et al.*, "A high density BiCMOS direct drive array," in *Proc. IEEE 1988 Custom Integrated Circuits Conf.*

[5] H. Tran *et al.*, "An 8-ns BiCMOS 1-Mb ECL SRAM with a configurable memory array size," in *ISSCC Dig. Tech. Papers*, Feb. 1989, p. 36.

[6] G. Boudon *et al.*, "A 20K CMOS array with 200-ps gate delay," in *ISSCC Dig. Tech. Papers*, Feb. 1988, p. 68.

[7] H. Tran *et al.*, "BiCMOS standard memories," in *BiCMOS Technology and Applications*, T. Alverz, Ed. Boston, MA: Kluwer Academic, 1989, p. 218.

[8] "A test access port and boundary scan architecture," Tech. SubCommittee of the Joint Test Action Group (JTAG), Princeton, NJ, rev. 2.0, Apr. 1988.

[9] H. Tran *et al.*, "A battery back-up 256K ECL RAM," in *ISSCC Dig. Tech. Papers*, Feb. 1988, p. 188.

Paper 2.3

0.5-μm 2M-Transistor BiPNMOS Channelless Gate Array

Hiroyuki Hara, Takayasu Sakurai, *Member*, *IEEE*, Makoto Noda, Tetsu Nagamatsu, Katsuhiro Seta, Hiroshi Momose, *Member*, *IEEE*, Youichirou Niitsu, Hiroyuki Miyakawa, and Yoshinori Watanabe

***Abstract*—A channelless gate array has been realized using 0.5-μm BiCMOS technology integrating more than 2-million transistors on a 14×14.4-mm^2 chip. A small-size PMOS transistor and a small-size inverter are added to the conventional BiNMOS gate to form the BiPNMOS gate. The gate is suitable for 3.3-V supply and achieves 230-ps gate delay for a two-input NAND with full-swing output. Added small-size MOS transistors in the BiPNMOS basic cell can also be used for memory macros effectively. A test chip with four memory macros—a high-speed RAM, a high-density RAM, a ROM, and a CAM macro—was fabricated. The high-speed memory macros utilize bipolar transistors in bipolar middle buffers and in sense amplifiers. The high-speed RAM macro achieves an access time of 2.7 ns at 16-kb capacity. On the other hand, the high-density RAM macro is rather slow but the memory cell occupies only a half of the BiPNMOS basic cell using a proposed single-port memory cell.**

TABLE I
BiCMOS Gate Suitable for 3.3 V

	CBiCMOS	BiNMOS	BiRNMOS	BiPNMOS
Circuit	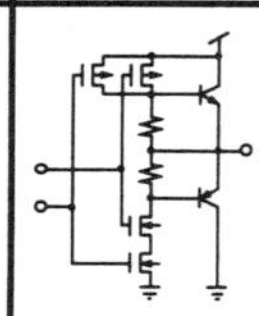	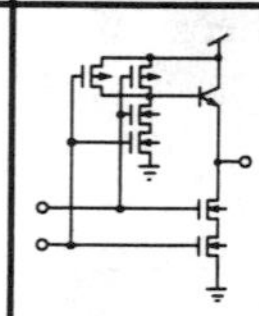	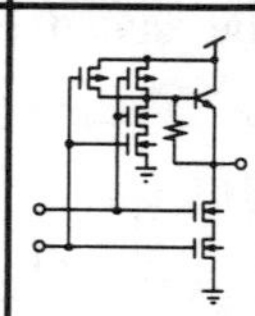	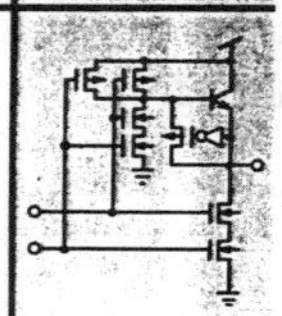
Process	−	++	+	++
Speed	+	+	+	++
Full-swing	+	−	+	++

I. Introduction

In order to ensure high reliability of 0.5-μm gate-length MOSFET's, a power supply voltage of 3.3 V is adopted. Reducing the supply voltage is also necessary for decreasing the power dissipation of highly integrated system LSI's. The low supply voltage, however, makes the conventional totem-pole BiCMOS gate lose its speed advantage over CMOS. In order to maintain the speed advantage over CMOS even at the low supply voltage, a BiNMOS gate [1] and a complementary-BiCMOS gate [2] have been proposed.

Table I shows candidates for the low-supply-voltage BiCMOS gate. In all these gates, improvement is done on the pull-down circuit of an original BiCMOS gate, since the pull-down part is the cause of the low speed in the original BiCMOS gate at low supply voltage. The C-BiCMOS gate requires a p-n-p transistor with a high cutoff frequency, as well as an n-p-n transistor, and hence suffers from an increase in process cost. On the other hand, the BiNMOS gate can be realized with only an n-p-n transistor, but the output does not reach V_{DD}; instead it stays at the $V_{DD}-V_{BE}$ level, where V_{BE} is a base–emitter turn-on voltage. This increases the delay time and leakage current of the next gate, and hence decreases noise margins. A BiRNMOS gate, which has a resistor inserted between the emitter and base of the pull-up bipolar transistor, achieves a full-swing output. However, if the resistance is set small to realize high-speed pull-up, it bypasses the base current and hence degrades the speed. Here, we propose a BiPNMOS gate which is suitable for a low supply voltage. The BiPNMOS gate achieves high speed, low process cost, and full-swing capability. The BiPNMOS gate can maintain a speed advantage over CMOS at 3.3-V supply voltage.

Another problem concerns memory macros in a BiCMOS gate array. High-speed and high-density memory macros are necessary for building a high-performance system on a chip. However, not many kinds of memory macros have been introduced in a BiCMOS gate array, which prevents the user from making highly value-added VLSI's easily. Besides, recent gate arrays tend to use six-transistor CMOS memory cells for a RAM macro [4], [5] but bipolar transistors arranged in BiCMOS basic cells are not utilized effectively. This inefficiency is solved by new circuit technologies, a double bit-line structure [7] with bipolar middle buffer and a BiCMOS sense amplifier with dotted-emitter structure [6], [8]. As for variety of memory macros, several kinds of memory macros are

Manuscript received April 19, 1991; revised July 22, 1991.
H. Hara, T. Sakurai, M. Noda, T. Nagamatsu, K. Seta, H. Momose, Y. Niitsu, and H. Miyakawa are with the Semiconductor Device Engineering Laboratory, Toshiba Corporation, 1, Komukai-Toshiba-cho, Saiwai-ku, Kawasaki 210, Japan.
Y. Watanabe is with Toshiba Microcomputer Engineering Corporation, 1, Komukai-Toshiba-cho, Saiwai-ku, Kawasaki 210, Japan.
IEEE Log Number 9102947.

Reprinted from *IEEE J. Solid-State Circuits*, vol. 26, no. 11, pp. 1615–1620, Nov. 1991.

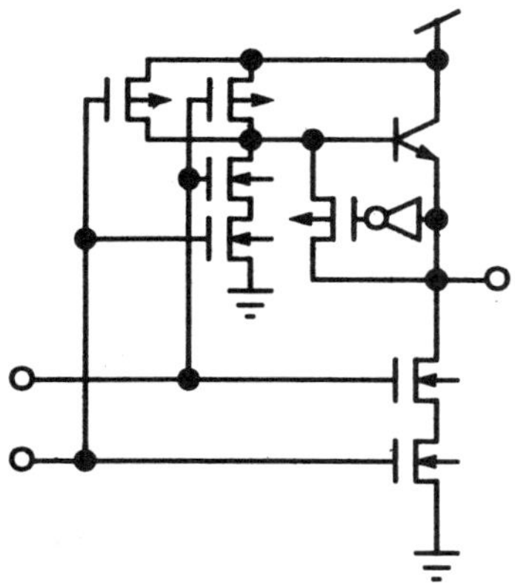

Fig. 1. BiPNMOS gate.

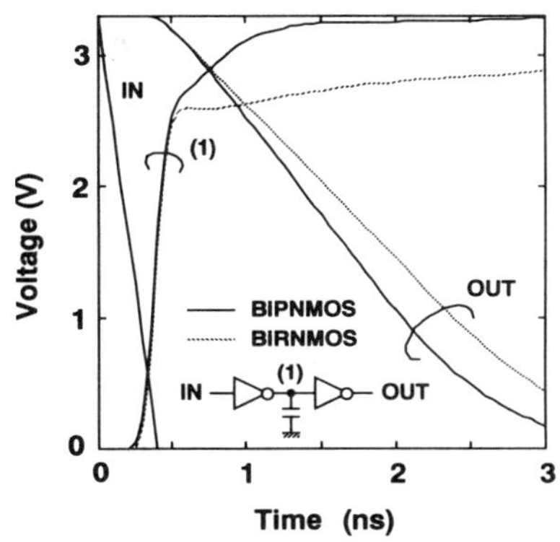

Fig. 2. Output waveforms of BiPNMOS and BiRNMOS gates.

realized including RAM's, a ROM, and a CAM. All these macros use developed BiCMOS circuits.

II. BiPNMOS Gate

Fig. 1 shows a two-input NAND BiPNMOS gate. A small-size PMOS transistor and an inverter are added to the conventional BiNMOS gate. The added PMOS transistor realizes the full swing of the output, while not decreasing the base current of the n-p-n transistor and hence not degrading the speed. When an output terminal is low, the added PMOS is OFF and all current flowing through the PMOS becomes the base current efficiently. When the voltage of the output terminal reaches the threshold voltage of the feedback inverter, the added PMOS turns on and starts pulling up the output to V_{DD}. Even after the bipolar transistor turns off at $V_{DD}-V_{BE}$, the added PMOS keeps ON and realizes a full-swing output. Fig. 2 shows a comparison of the output waveforms of the BiPNMOS gate and the BiRNMOS gate. In this case, the value of the resistor of the BiRNMOS is set to 50 kΩ. As seen from the figure, the full swing of BiRNMOS gate is achieved very slowly. On the other hand, the BiPNMOS gate offers high-speed pull-up to V_{DD}. This reduces the delay and the leakage current of the next gate and increases noise margins.

Fig. 3 shows a layout pattern of a BiPNMOS basic cell. A bipolar transistor and small-size MOS transistors are

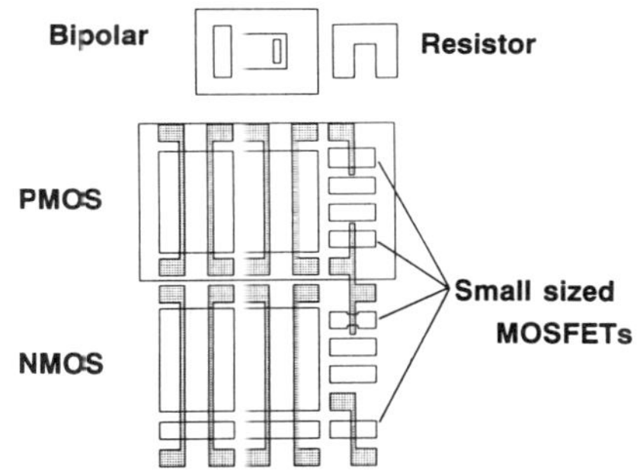

Fig. 3. Basic cell layout of BiPNMOS gate array.

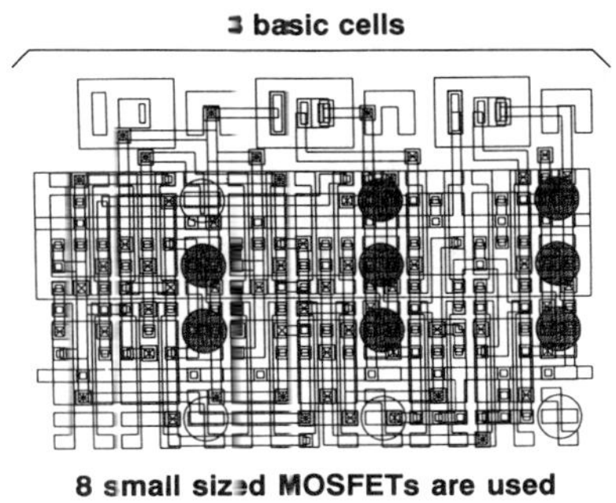

Fig. 4. D-type flip-flop cell layout pattern.

added to the pure CMOS basic cell. These transistors are not only used to construct the basic BiPNMOS gates but also can be used in flip-flops, memory macros, and other cells effectively. Fig. 4 shows an example layout pattern of a D-type flip-flop cell. This D-type flip-flop cell with the BiPNMOS drivers is constructed on three basic cells, and utilizes eight out of 12 small-size MOS transistors.

III. Memory Macros

Recent gate arrays tend to include small MOS transistors in a basic cell to implement a high-density six-transistor CMOS RAM cell. In the present BiPNMOS gate array, small MOS transistors laid out in the BiPNMOS basic cells can be utilized for six-transistor RAM cells. Moreover, the bipolar transistors are utilized for word-line drivers and sense amplifiers efficiently. Four memory macros, namely high-speed RAM, high-density RAM, ROM, and CAM, are designed. A summary of these macros is listed in Table II. One memory cell of the high-speed RAM macro occupies one basic cell. The high-density RAM macro is rather slow but the memory cell occupies half of the basic cell. As for the ROM macro, eight memory cells can be realized in one basic cell and the CAM macro is realized employing one basic cell per bit.

Fig. 5 shows a circuit diagram of the high-speed RAM macro. A RAM cell has a separate write port and a read port and can be used as a two-port RAM cell enabling WRITE and READ operation at the same time. WRITE operation is performed through the DATA line connected to NMOS $N3$, and the dimensions of $N3$ must be selected bigger than $N1$ for stability. High-speed opera-

TABLE II
FEATURES

Chip size	14.0×14.4 mm^2
Internal gate count	237 120 raw gates (CMOS two-input NAND gate)
I/O cell count	1044
Basic cell area	54.4×25.6 μm^2
Supply voltage	3.3 V single
Gate delay	230 ps (typ., BiPNMOS two NAND, fan-out = 7)
Process	1-poly, 3-metal, 0.5-μm BiCMOS process
Bipolar	Poly-Si emitter Emitter size = 0.8×2.8 μm^2
MOS	PMOS = 12.1/0.6 μm NMOS = 12.1/0.5 μm
High-speed RAM	1 memory cell/1 basic cell T_{acc} = 2.7 ns (typ. at 512 word $\times$ 32 b)
High-density RAM	2 memory cells/1 basic cell T_{acc} = 4.0 ns (typ. at 256 word $\times$ 32 b)
ROM	8 memory cells/1 basic cell T_{acc} = 3.2 ns (typ. at 256 word $\times$ 32 b)
CAM	1 memory cell/1 basic cell Data to match = 2.0 ns (typ. at 64 entry $\times$ 32 b)

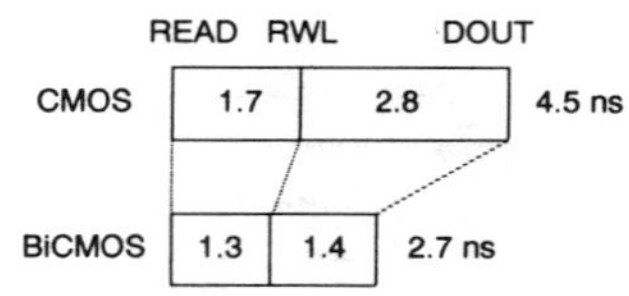

Fig. 6. Simulated access time of BiCMOS RAM macro compared to CMOS RAM.

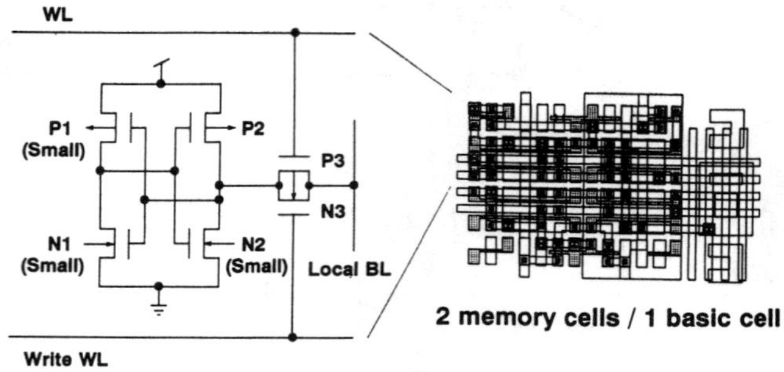

Fig. 7. Circuit diagram and layout pattern of high-density RAM macro.

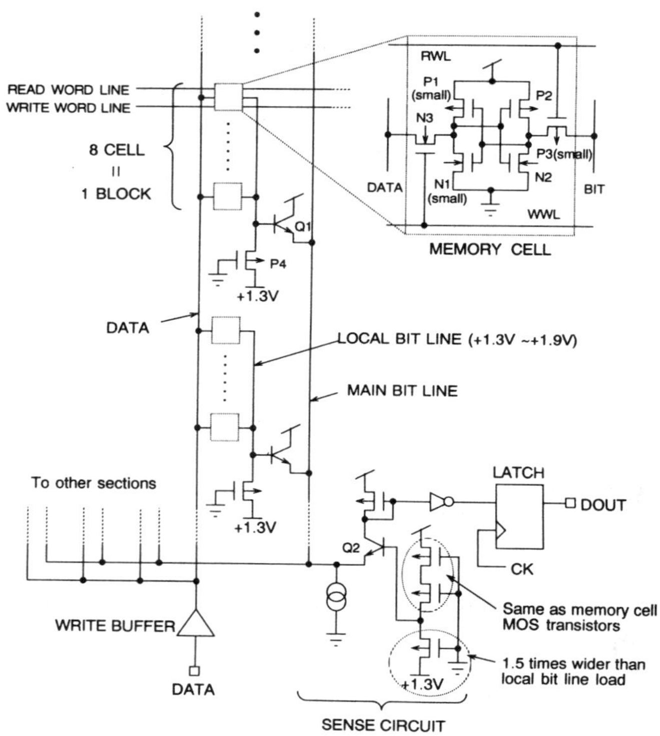

Fig. 5. Circuit diagram of fast RAM macro.

tion in the READ mode is achieved by the following three techniques:

1) double bit lines with bipolar middle buffers,
2) 0.6-V bit-line swing,
3) BiCMOS sense amplifiers with emitter-dotted shared transistors.

Readout operation from a memory cell is carried out through a PMOS ($P3$) connected to the BIT line. A small transistor is used for $P3$ in order to limit the swing of the local bit line. The local bit line is pulled down to the low level of +1.3 V before the readout operation. The local bit-line level is determined by PMOS's $P2$–$P4$ and NMOS $N2$, and only changes to +1.9 V when $P2$ is ON. This 0.6-V swing is transferred to a main bit line through a bipolar middle buffer ($Q1$). The number of memory cells connected to the local bit line is eight (one block). The bipolar middle buffer effectively drives the highly capacitive main bit line. The bipolar middle buffer ($Q1$) is also used as a component of a differential sense amplifier using an emitter dotting technique. The other side of the amplifier is made with the bipolar transistor ($Q2$) whose base is controlled by a self-tracking reference voltage generator. The reference generator uses MOS transistors, which are the same as a memory cell, and 1.5 times wider MOS compared with a local bit-line load to generate a proper reference voltage. The reference generator tracks process, voltage, and temperature variation in a self-tracking way. Fig. 6 shows simulated delay component distributions of the BiCMOS RAM macro and a pure CMOS RAM macro. The delay from READ signal to read word line (RWL) is reduced from 1.7 to 1.3 ns by using BiCMOS drivers. More drastic delay reduction is achieved in the delay from the bit line to the output of the sense amplifier. The current consumption of a sense circuit is 1.3 mA per bit.

In some applications, density is more important than speed. For these applications, a high-density RAM macro, which implements two memory cells in one basic cell, is more suitable. Fig. 7 shows the circuit diagram and the cell layout pattern. The RAM cell realizes high density to adopt a single-port cell and to use the small-size MOS transistors. In READ mode only a PMOS transfer gate turns on to ensure cells stability. In WRITE mode both PMOS and NMOS transfer gates turn on to ensure stable WRITE operation. The sensing scheme is the same as that of the high-speed RAM macro.

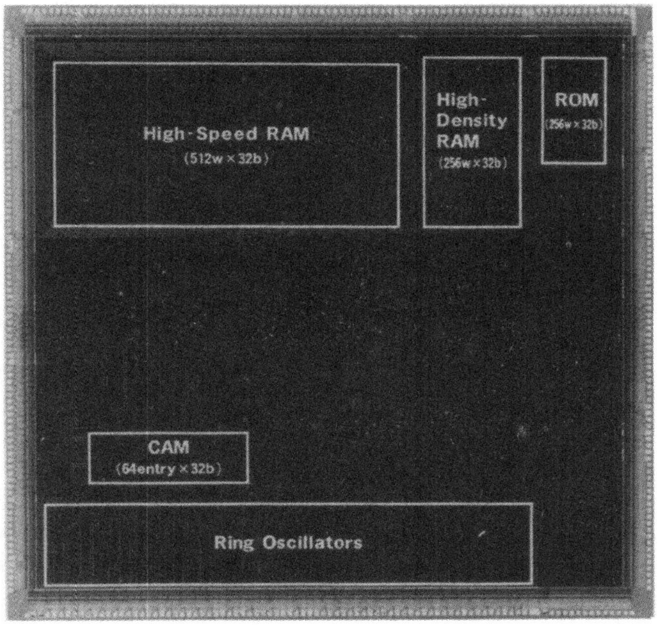

Fig. 8. Microphotograph of evaluation chip.

The ROM macro adopts the double bit line and the same sensing scheme as the RAM's. Four PMOS ROM cells and four NMOS ROM cells can be built using one basic cell. The local bit line for the PMOS ROM cell and that for the NMOS ROM cell are separate, because the bit-line levels are different. Content addressable memory (CAM) is another important macro. Small-size MOS transistors are used for the six-transistor memory cells, so a single CAM cell can be constructed with only one basic cell. The CAM macro employs a double match-line architecture and a BiCMOS pull-up circuit to drive the second match line [13]. The key features of the CAM macro are listed in Table II together with a summary of the present chip.

IV. Experimental Result

Fig. 8 shows a microphotograph of the fabricated $14.0 \times 14.4\text{-mm}^2$ chip. The test chip contains ring oscillators to evaluate the speed for the BiPNMOS gate and the CMOS gate, and memory macros. The test chip is fabricated using a 0.5-μm triple-metal BiCMOS process. Although the gate length for the MOSFET's is 0.5 μm, the basic design rule of 0.8 μm is adopted to achieve high speed and high yield at the same time. A poly-Si emitter bipolar transistor with a cutoff frequency of 13 GHz is adopted, and the rapid thermal annealing (RTA) technique is applied to realize a stable poly-single interface. The features of the present chip are shown in Table II, together with the key process parameters of the 0.5-μm BiCMOS technology. Fig. 9(a) shows a speed dependence of the BiPNMOS gate on fan-out. A BiPNMOS two-input NAND gate realizes a propagation delay time of 230 ps with a fan-out of 7 at room temperature. The BiPNMOS gate shows 35% smaller delay than the pure CMOS gate

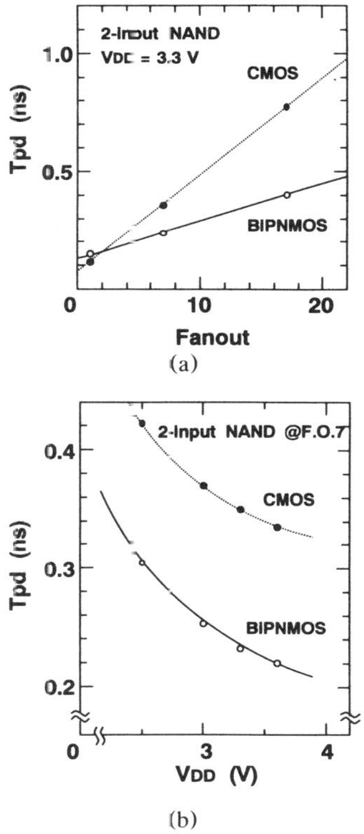

Fig. 9. (a) Propagation delay time versus fan-out. (b) Supply voltage dependence of delay time.

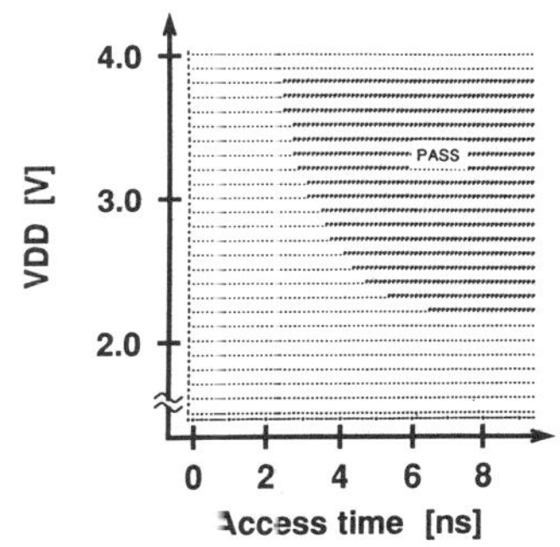

Fig. 10. Schmoo plot of high-speed RAM macro.

even at 3.3 V. Fig. 9(b) shows the speed dependence of the BiPNMOS gate and the pure CMOS gate on V_{DD}. The speed advantage over the CMOS gate is observed down to 2.5 V. Fig. 10 shows a schmoo plot of the high-speed RAM macro and the typical access time of 2.7 ns is seen at 512 word $\times$ 32 b. The speeds of the high-density RAM and ROM macros are 4.0 and 3.2 ns, respectively. These memory macros achieve a high-speed operation using BiCMOS circuit technologies, even in a high-memory-capacity range.

V. Conclusions

A BiPNMOS gate is proposed that assures full-swing and high-speed operation at 3.3 V. The BiPNMOS gate reduces the delay and the leakage current of the next gate and increases noise margins. A two-input NAND gate delay of 230 ps is observed at a fan-out of 7. Small-size MOS transistors, which are added to construct the BiPNMOS gate, are utilized by flip-flop cells and memory macros effectively.

Four BiCMOS memory macros are presented. A double bit-line structure with bipolar middle buffers and a BiCMOS sense amplifier are introduced for high-speed operation. Small-size MOS transistors and a single-port memory cell are used in a high-density memory macro. The typical access time of the high-speed RAM macro is measured to be 2.7 ns at 16-kb capacity.

Acknowledgment

The authors would like to express their appreciation to Dr. T. Iizuka, K. Maeguchi, and H. Nakatsuka for encouragement throughout the work.

References

[1] A. E. Gamal, J. L. Kouloheris, D. How, and M. Morf, "BiNMOS: A basic cell for BiCMOS sea-of-gates," in *Proc. CICC*, May 1989, pp. 8.3.1–8.3.4.

[2] H. J. Shin *et al.*, "Full-swing complementary BiCMOS logic circuits," in *Proc. BCTM*, 1989, pp. 229–232.

[3] Y. Nishio *et al.*, "A BiCMOS logic gate with positive feedback," in *ISSCC Dig. Tech. Papers*, Feb. 1989, pp. 116–117.

[4] H. Veendrick, D. Elshout, D. Harberts, and T. Brand, "An efficient and flexible architecture for high-density gate arrays," in *ISSCC Dig. Tech. Papers*, Feb. 1990, pp. 86–87.

[5] Y. Enomoto, T. Sasaki, S. Tsutsumi, and S. Tone, "A 200k gate 0.8 μm mixed CMOS/BiCMOS sea-of-gates," in *ISSCC Dig. Tech. Papers*, Feb. 1990, pp. 92–93.

[6] L. Tamurat, T. S. Yang, D. Wingard, M. Horowitz, and B. Wooley, "A 4ns BiCMOS translation-lookaside buffer," in *ISSCC Dig. Tech. Papers*, Feb. 1990, pp. 66–67.

[7] T. Sakurai *et al.*, "Double word line and bit line structure for VLSI RAMs," in *Proc. 15th Conf. Solid State Devices & Materials*, 1983, pp. 269–272 (A-7-6).

[8] T. S. Yang, M. A. Horowitz, and B. A. Wooley, "A 4-ns 4K×1-bit two-port BiCMOS SRAM," *IEEE J. Solid-State Circuits*, vol. 23, no. 5, pp. 1030–1040, Oct. 1988.

[9] H. Hara *et al.*, "A 350 ps 50K 0.8 μm BiCMOS gate array with shared bipolar structure," in *Proc. CICC*, May 1989, pp. 8.5.1–8.5.4.

[10] J. Gallia *et al.*, "A 100K gate sub-micron BiCMOS gate array," in *Proc. CICC*, May 1989, pp. 8.6.1–8.6.4.

[11] C. L. Chen, "Level-shifted and voltage-reduced 0.5 μm BiCMOS circuits," in *ISSCC Dig. Tech. Papers*, Feb. 1990, pp. 236–237.

[12] H. Hara *et al.*, "0.5 μm 2M-transistor BiPNMOS channelless gate array," in *ISSCC Dig. Tech. Papers*, Feb. 1991, pp. 148–149.

[13] T. Nagamatsu *et al.*, "A 1.9ns BiCMOS CAM macro with double match line architecture," in *Proc. CICC*, May 1991, pp. 14.3.1–14.3.4.

Paper 2.4

A 100-MHz 64-Tap FIR Digital Filter in 0.8-μm BiCMOS Gate Array

TOSHIAKI YOSHINO, MEMBER, IEEE, RAJEEV JAIN, MEMBER, IEEE, PAUL T. YANG, HARVEY DAVIS, WANDA GASS, MEMBER, IEEE, AND ASHWIN H. SHAH, SENIOR MEMBER, IEEE

***Abstract*—A 64-tap FIR digital filter has been designed using a newly developed filter compiler and fabricated in a 0.8-μm triple-level interconnect BiCMOS gate array technology. The filter has been tested and is fully functional at a 100-MHz clock rate. These results are obtained by combining an optimized architecture and gate array floorplan with submicrometer BiCMOS technology. The filter occupies 49 mm^2, which is approximately two-thirds of the 100K gate array core. The design uses an equivalent of 55K gates (two-input NAND gates). The device input/output are 100K ECL compatible.**

I. Introduction

DIGITAL signal processing techniques are now being applied to high-frequency communication systems, which has resulted in a demand for high-sample-rate FIR digital filters. For IF processing in radio applications, sample rates of 50–100 MHz are needed [1]. Prototypes of these high-speed filters cannot be built with existing programmable devices and custom design can slow down the system development as well as substantially increase the cost.

A possible solution to the above problem is the use of gate array technology. Currently available gate array technologies allow a high degree of complexity and provide very fast gates [2], [3]. The high density of integration can be exploited to build dedicated architectures that achieve high throughput rates. In this paper, a 64-tap FIR filter chip is presented that operates at sample rates in excess of 100 MHz using a 0.8-μm BiCMOS gate array technology.

Even with the availability of the gate array technology, prototype development can be slow if the designer has to manually translate the desired filter specification into an interconnection of gate array cells. The typical user interface to gate array technologies is a place-and-route tool that can automatically generate the layout from a netlist of gate array cells. Therefore, if the translation of the filter specification into the gate array cell netlist can be automated, then the entire design time can be reduced to a matter of hours. To evaluate this approach to rapid prototyping, a FIR filter compiler FIRGEN [4], [5] was used to design the 64-tap filter chip.

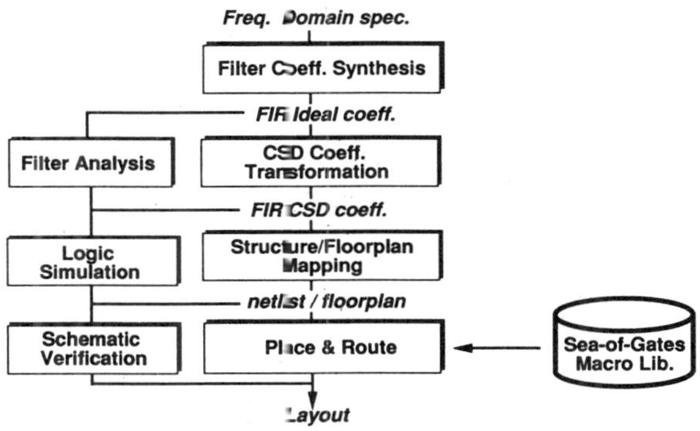

Fig. 1 Compiler flow.

Fig. 1 shows the chip design flow. The input to FIRGEN is a frequency-domain specification of the filter. Using a conventional filter design method [6], [7], a set of ideal filter coefficients is generated. Then the coefficients are quantized using the canonic signed-digit (CSD) [8] code. A filter analysis tool is used to make sure that the CSD implementation meets the original frequency-domain specification. Once the CSD coefficients are finalized, they are input to a structure/floorplan generator to produce the final netlist and floorplan. These are then input to the place-and-route program TANGATE [9].

Section II gives an overview of the filter architecture and the critical BiCMOS circuits used in the chip. Section III discusses the strategy of the physical design including floorplanning. The performance and characteristics of the device are presented in Section IV.

II. Architecture and Circuits

A. Transpose CSD Structure

In the direct form FIR structure the chain of adders is generally the critical path that limits performance. To obtain the higher throughput, extra pipeline registers can

Manuscript received May 10, 1990; revised August 15, 1990. The work done at the University of California, Los Angeles, was supported in part by the State of California MICRO program, DARPA, FBI, and Texas Instruments.

T. Yoshino, H. Davis, and A. H. Shah are with the VLSI Design Laboratory, Texas Instruments Incorporated, Dallas, TX 75230.
R. Jain and P. T. Yang are with the Department of Electrical Engineering, University of California, Los Angeles, CA 90024.
W. Gass is with the Speech and Image Understanding Laboratory, Texas Instruments Incorporated, Dallas, TX 75230.

IEEE Log Number 9039476.

Reprinted from *IEEE J. Solid-State Circuits*, vol. 25, no. 6, pp. 1494–1501, Dec. 1990.

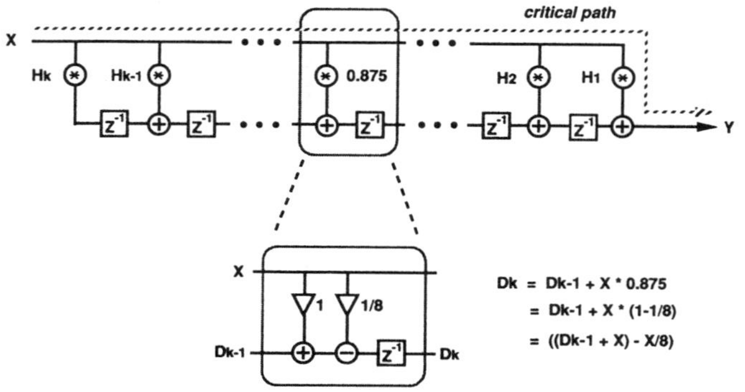

Fig. 2. Transpose form with CSD.

be added to break up the adder chain, however, these introduce not only extra hardware but also extra latency. Fig. 2 shows an FIR filter architecture implemented by the transposed form structure [7]. The transpose form architecture breaks up the chain of adders naturally with the Z^{-1} delay elements. The critical arithmetic delay is reduced to a single multiply–add block without any latency. However, at the layout level, additional delay is introduced due to the long input data bus X which has to be broadcast to several multiplier inputs. For large filters this delay becomes the limiting factor. BiCMOS technology provides efficient means for reducing this delay, as described in Section II-C below.

Since the filter coefficients are fixed, the multiplier in each tap is implemented by dedicated shift–add scheme. The shift operation is accomplished by physically routing (shifting) the data input lines. CSD arithmetic [8] is used to minimize the number of adders.

B. Carry-Save Addition

Fig. 3 illustrates the block diagram of the full chip architecture. The architecture is dedicated to minimize the delay of the data distribution. The number of taps is 64 and the word length of the input and output data is 12 b. The internal word length of the adders is 14 b. The input register IREG and output register OREG isolate the filter circuit from the I/O pads for the input bus X and the output bus Y. Hence, the pad delay is not in the critical path.

To minimize the tap delay caused by adder/subtracters, carry-save addition is used so that carry and sum signals propagate in parallel as shown. The Z^{-1} delay registers have to be doubled to store both of them. Each tap is allowed to use a maximum of four CSD terms (or power-of-2 scalers) to approximate the original coefficient. This results in a maximum of four carry-save adder/subtracter stages in a tap. In this particular design, four CSD terms per coefficient is found to be sufficient.

A vector-merge adder is needed to add the carry and sum signals after the last tap of the filter. This adder can potentially increase the critical path delay and it is isolated by inserting an extra pipeline register D between the final tap and the adder as shown. With this architecture, the total latency is only one clock cycle.

C. BiCMOS Circuits in the Critical Path

The maximum possible sampling rate of the input data X is determined by the propagation delay along the critical path. For the filter architecture in Fig. 3, the critical path can be either the carry-ripple delay in the vector-merge adder, or the data-bus RC delay plus the CSD carry-save multiply–accumulate delay. Due to the large number of taps in this design, there is a heavy loading on the MSB data line so that the critical path is the delay through the MSB data line and the carry-save multiply–accumulate circuits as illustrated in Fig. 2. To speed up the critical path, bipolar buffers are used to drive the data bus and drive the CSD carry-save multiply–adder circuit inputs.

Distributing the input to all 64 taps at the required 100-MHz clock rate requires more than a single BiCMOS driver. A tree structure, shown in Fig. 4, is formed by inserting two levels of BiCMOS inverters in the data line. There are nine first-level buffers. Each one drives six to eight second-level buffers. Each second-level buffer drives the carry-save adders and subtracters in a tap. The clock distribution line to the Z^{-1} registers in each tap is implemented using exactly the same scheme.

The carry-save adder cell used in the filter tap is implemented using CMOS transmission gates, static CMOS inverters, and BiCMOS buffers as shown in Fig. 5. Two BiCMOS inverters are used to drive the carry and sum outputs. One BiCMOS inverter is used to drive the internal pass-transistor tree. To minimize the loading of the second-level data buffer of the MSB bit (Fig. 7), a CMOS inverter is used to receive the X input. The carry-save subtracter cell is implemented similarly with the same transistor count by exchanging the inverter outputs X and $\overline{X}$. The cell is laid out using the gate array base cell and the size is 64.0 μm$\times$140.4 μm (2 by 3 base cells).

The vector-merge adder circuit is designed using a Manchester carry chain scheme to achieve a 14-b carry-ripple addition time of 4 ns. The chain has level boost logic implemented with a bipolar driver after every 4 b in the carry path.

III. Physical Design

All special macrocells are laid out with 64-μm cell width to allow clean pitch matching in taps. Given the macrocell layouts, a netlist describing the architecture of Fig. 3 can be directly input to the gate array place-and-route software. If this is done, the resulting chip layout turns out to be suboptimal in terms of minimizing the interconnection length and potential skew problems. Therefore, in addition to the netlist, floorplanning constraints are generated by the compiler to direct the place-and-route tool.

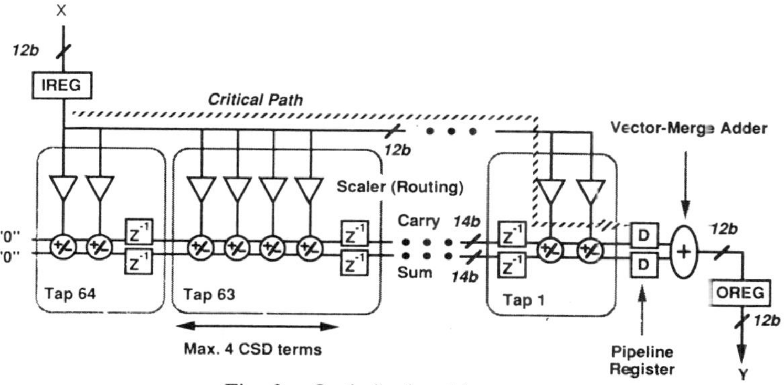

Fig. 3. Optimized architecture.

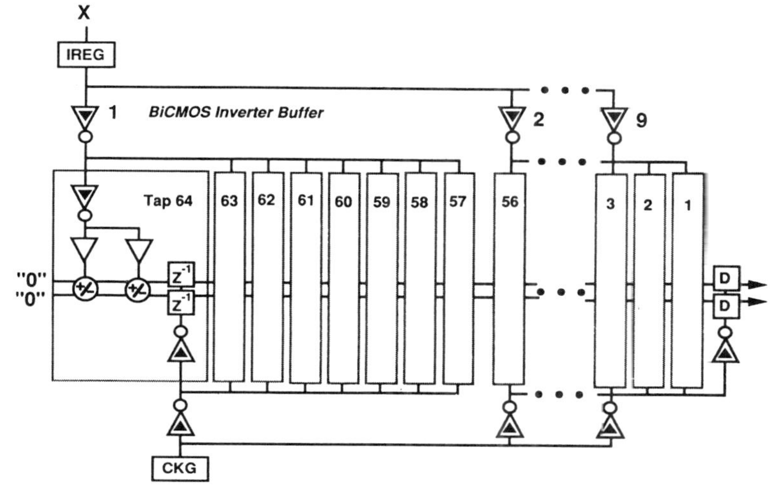

Fig. 4. Data/clock distribution.

A. Global Floorplanning

To form a uniform tree structure and further ensure a high-speed operation in the gate array implementation, the floorplan constraints direct the placement of the cells as well as the placement of the clock and data distribution network. Fig. 6 illustrates the floorplan of the chip, and the clock and data-bus tree structure. To balance the weight of the data and clock trees, the input register IREG and the clock generator CKG are placed on top in the center. The triangles at the top of each column are the first-level drivers inserted in data and clock trees. Their placement is fixed at the top of each column. The filter compiler needs only one control parameter (number of columns) to generate the floorplan discussed above. In this particular design, the number of columns is set to nine.

To minimize the interconnection length and corresponding timing skews, the filter taps are folded in a serpentine fashion and this strategy helps to minimize the skew problems between adjacent taps. The pipeline register, the vector-merge adder, and the output register OREG are placed below the first tap. Since the number of carry-save adder macrocells per tap varies, the number of taps per column of the floorplan is adjusted by the compiler such that all columns have approximately the same height.

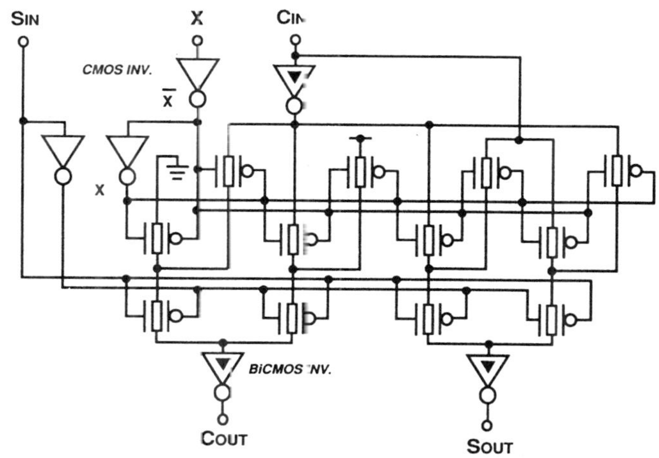

Fig. 5. Carry-save adder cell.

B. Tap Implementation

The floorplanning strategy inside of a tap and wiring of the input data lines are shown in Fig. 7. The tap includes carry-save adders, subtracters, registers and second-level

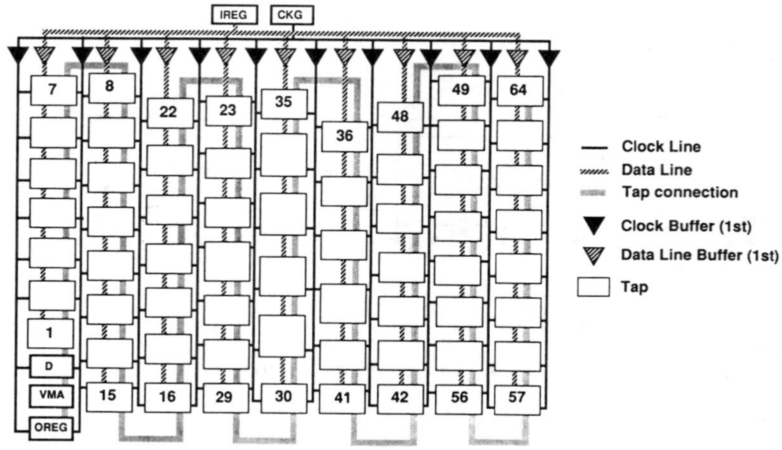

Fig. 6. Filter floorplan.

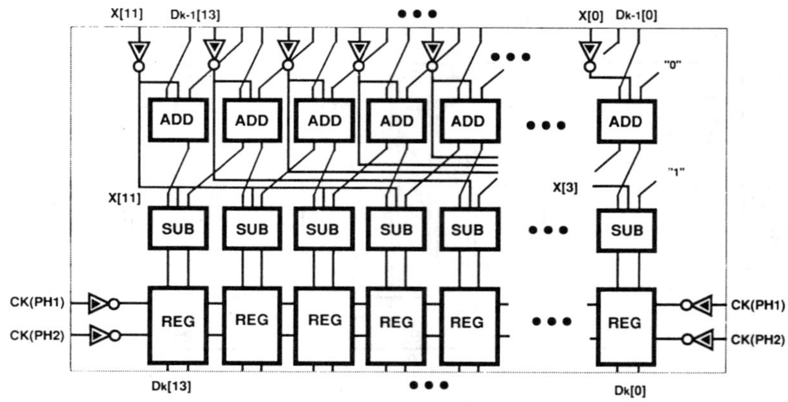

Fig. 7. Tap floorplan.

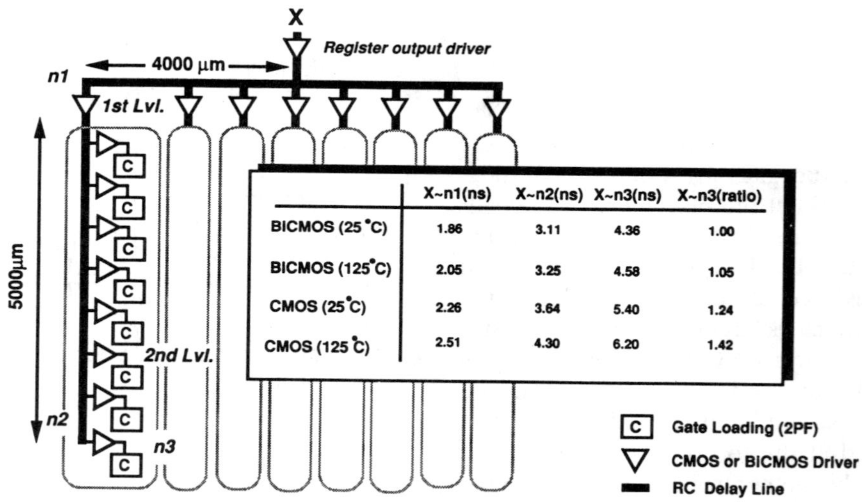

	X~n1(ns)	X~n2(ns)	X~n3(ns)	X~n3(ratio)
BICMOS (25 °C)	1.86	3.11	4.36	1.00
BICMOS (125°C)	2.05	3.25	4.58	1.05
CMOS (25°C)	2.26	3.64	5.40	1.24
CMOS (125 °C)	2.51	4.30	6.20	1.42

Fig. 8. BiCMOS versus CMOS comparison.

data, and clock drivers (Fig. 3). They all have fixed cell locations generated by the compiler. The tap in Fig. 7 implements an operation:

$$D_k(n) = D_{k-1}(n) + 0.875X(n)$$
$$= D_{k-1}(n) + X(n) - 2^{-3}X(n).$$

This is implemented by directly connecting the bit lines X [0:11] and D_{k-1} [0:11] to an adder whose output is connected to a subtracter together with the X data bus shifted by 3 b. This shift realizes the scaling by 1/8. Notice that the shift of the X includes the sign extension. Due to this sign extension, the second-level driver of the MSB bit has the heaviest loading and it varies depending on the coefficient assigned to a tap. The worst-case fan-out in this particular design is equivalent to 40 CMOS in-

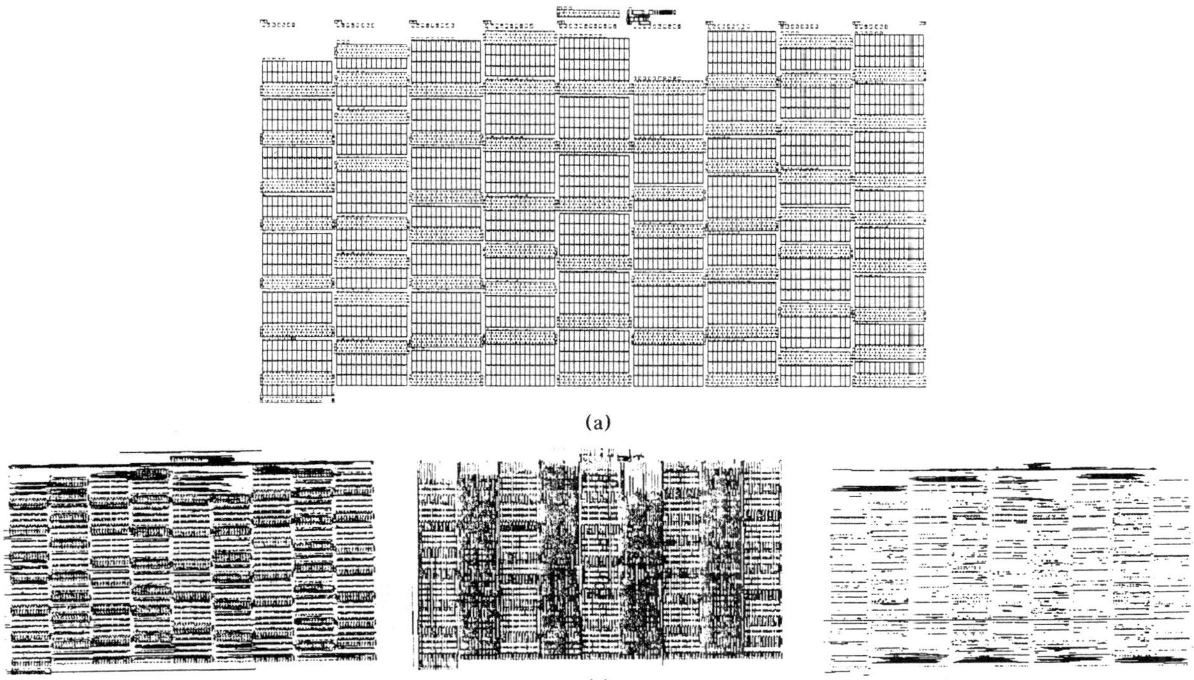

Fig. 9. Filter placement and routing. (a) Placement plot. (b) Metal 1 routing layer plot. (c) Metal 2 routing layer plot. (d) Metal 3 routing layer plot.

verter gates. Note that clock input to the Z^{-1} registers is provided from both sides to equalize the clock distributions to all register cells.

C. BiCMOS Versus CMOS

Based on the actual *RC* delays in the layout, a performance comparison has been made between BiCMOS and CMOS drivers. From the global floorplan (Fig. 6) the data-bus tree structure is modeled in Fig. 8. There are three levels of buffers: root, first, and second. The root level is the output from the input register IREG. The bold line in Fig. 8 shows the actual metal lines that carry the input signal X to the various buffers. The boxes marked C represent the capacitive loading on the second-level buffers due to the adder/subtracter inputs in the tap.

The table in Fig. 8 compares the average propagation delay through each level of buffer at room temperature (25°C) and at 125°C. To have an objective comparison in terms of the cell area, the BiCMOS buffer is implemented using one BiCMOS inverter. The CMOS buffer is implemented by noninverting CMOS buffers consisting of four CMOS inverters (one followed by three parallel inverters). The worst-case (-3σ) model parameters are used in the SPICE simulations for these delays. The BiCMOS technology contributes to the speed improvement by about 24% at room temperature and 42% at 125°C. This relative advantage of BiCMOS is attributable to the smaller degradation of the driving force of the bipolar drivers at high temperature compared to the MOS devices. The comparison shows the speed advantages of BiCMOS gate array technology over CMOS gate array technology for high speed where the device temperature is expected to be high during the operation. Furthermore, since the BiCMOS drivers have lower input gate capacitance, use of BiCMOS drivers provides an advantage in terms of power consumption over CMOS-based drivers at very high frequency.

D. Routing and Placement

Plots of the three different routing layers and the cell placements are shown in Fig. 9. The metal-1 and metal-3 routing tracks are in the horizontal direction and the metal-2 tracks are in the vertical direction. Despite the regular structure of the filter architecture, the place-and-route tool produces an irregular placement in the absence of floorplan constraints [5]. Distribution of the routed wire length is shown in Fig. 10. There are a total of 8000 wires in the chip. Nearly 90% of the wires are shorter than 1000 μm. Without floorplanning the number of wires with large routing length goes up substantially.

With the floorplan constraints generated by the compiler, the place-and-route program generated the final layout in 8 hours on a VAX8800. Since the placement is already defined by the floorplan constraints, 90% of the CPU time was spent on routing. The layout synthesis from the filter specification took only two days of design time.

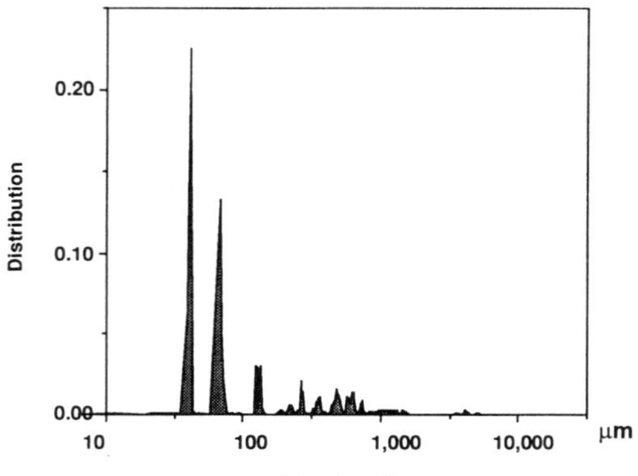

Fig. 10. Routing statistics.

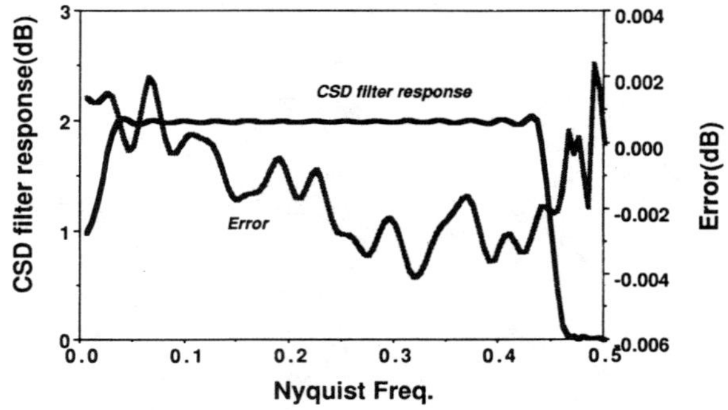

Fig. 11. Filter frequency response.

TABLE I
FILTER SPECIFICATION SUMMARY

Input word length	8 bits
Output word length	12 bits
Clock rate	80 MHz, Maximum

	Band 1	*Band 2*	-	*Band 3*
Lower band edge	0.00MHz	0.262 MHz	36.000 MHz	39.900 MHz
Upeer band edge	0.64MHz	32.000 MHz	36.000 MHz	40.000 MHz
Nominal Gain	1.00 dB	2.00 dB	1.00 dB	0.00 dB
Nominal Ripple	0.10 dB	0.10 dB	-	0.10 dB

TABLE II
FILTER COEFFICIENTS

Tap	*Ideal coeff.*	*CSD coeff.*	*CSD code*
h_1, h_{64}	4.799490e-03	4.760742e-03	(8,10,-13)
h_2, h_{63}	-2.697505e-03	-2.685547e-03	(-8,10,12)
h_3, h_{62}	6.270800e-03	6.225586e-03	(7,-9,11,-13)
h_4, h_{61}	-3.699250e-03	-3.662109e-03	(-8,12)
h_5, h_{60}	5.655850e-03	5.615234e-03	(7,-9,-12)
h_6, h_{59}	-2.702305e-03	-2.685547e-03	(-8,10,12)
h_7, h_{58}	2.224275e-03	2.197266e-03	(9,12)
h_8, h_{57}	4.489130e-04	4.882812e-04	(11)
h_9, h_{56}	-3.669205e-03	-3.662109e-03	(-8,12)
h_{10}, h_{55}	4.704390e-03	4.760742e-03	(8,10,-13)
h_{11}, h_{54}	-1.039155e-02	-1.037598e-02	(-7,-9,-11,-13)
h_{12}, h_{53}	7.896350e-03	7.934570e-03	(7,13)
h_{13}, h_{52}	-1.540555e-02	-1.538086e-02	(-6,12)
h_{14}, h_{51}	7.408850e-03	7.446289e-03	(7,-11,13)
h_{15}, h_{50}	-1.618300e-02	-1.623535e-02	(-6,-11,-13)
h_{16}, h_{49}	1.268780e-03	1.220703e-03	(10,12)
h_{17}, h_{48}	-1.137350e-02	-1.135254e-02	(-6,8,11,-13)
h_{18}, h_{47}	-1.072470e-02	-1.074219e-02	(-6,8,10)
h_{19}, h_{46}	-1.783315e-03	-1.831055e-03	(-9,13)
h_{20}, h_{45}	-2.629355e-02	-2.624512e-02	(-5,8,10,13)
h_{21}, h_{44}	9.258499e-03	9.277344e-03	(7,9,-11)
h_{22}, h_{43}	-4.072820e-02	-4.077148e-02	(-5,-7,-9,12)
h_{23}, h_{42}	1.633385e-02	1.635742e-02	(6,10,-12)
h_{24}, h_{41}	-4.767490e-02	-4.772949e-02	(-4,6,-10,13)
h_{25}, h_{40}	1.290530e-02	1.293945e-02	(6,-8,10,12)
h_{26}, h_{39}	-4.007595e-02	-4.003906e-02	(-5,-7,-10)
h_{27}, h_{38}	-8.246800e-03	-8.300781e-03	(-7,-11)
h_{28}, h_{37}	-9.287300e-03	-9.277344e-03	(-7,-9,11)
h_{29}, h_{36}	-5.929450e-02	-5.932617e-02	(-4,8,-10,12)
h_{30}, h_{35}	6.877450e-02	6.884766e-02	(4,7,-9,11)
h_{31}, h_{34}	-2.083865e-01	-2.089844e-01	(-2,5,7,9)
h_{32}, h_{33}	6.081049e-01	6.083984e-01	(1,3,-6,-10)

IV. DEVICE TEST RESULTS

A. Measured Characteristics

The filter implements a bandpass characteristic for an application in video image signal processing. The filter specifications are summarized in Table I. The Kaiser window filter design method [7] was used to design the filter. The frequency response of the filter implemented by the CSD approach and its error against the ideal coefficients are shown in Fig. 11. Table II lists the ideal filter coefficients and the implemented CSD coefficients. The maximum number of CSD terms in a single tap is four and the average number of terms per tap is three.

The device is fully functional at a clock rate of 100 MHz. The switching waveforms at this frequency are shown in Fig. 12. Fig. 12(a) shows the system clock, Fig. 12(b) shows the input signal provided to the LSB bit of the X, and Fig. 12(c) shows the MSB output of the filter output. The swing level is ECL and all waveforms were measured at pins of a packaged chip.

The device characteristics are summarized in Table III. The device technology is 0.8-μm BiCMOS triple-layer metal sea of gates. With a 100-MHz clock and 5-V power supply, measured maximum power dissipation is 6 W. An area equivalent to 55K gates on the die has been allocated to the filter as shown in the upper half of the die photo in Fig. 13. The filter contains 112.2K NMOS, 74.7K PMOS, and 21.9K n-p-n bipolar active transistors in the array. The utilization of the gate is 66.7%.

B. Comparison With Other Design Methodologies

Table IV shows a comparison of filter chips [10], [11] designed by module generators and the gate array compiler discussed in this paper. The comparison illustrates the uniqueness of the chip performance achieved by this gate array design methodology. The speed performance of this work is comparable to the state-of-art CMOS programmable FIR filter chip [11]. The careful synthesis of the clock and data-bus trees constructed by BiCMOS buffers contributes to this speed performance gain. The power penalty is attributed to the limitation of having to

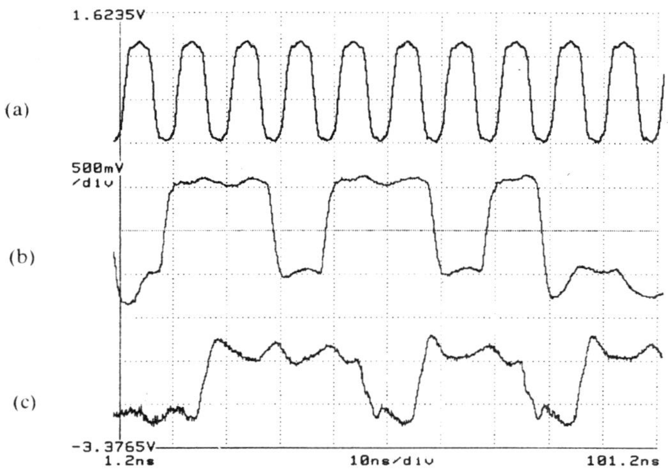

Fig. 12. Switching waveforms.

TABLE III
DEVICE SUMMARY

Filter specification	FIR 64-tap filter, 100 MHz clock rate
Technology	0.8μm BiCMOS triple-layer metal, sea-of-gates
Power supply	Max. 6 watts at 100 MHz
I/O	12-bit input/output, ECL compatible
Core size	8.96 by 5.43 mm^2
Transistor count	112.2K NMOS, 74.7K PMOS, 21.9K NPN

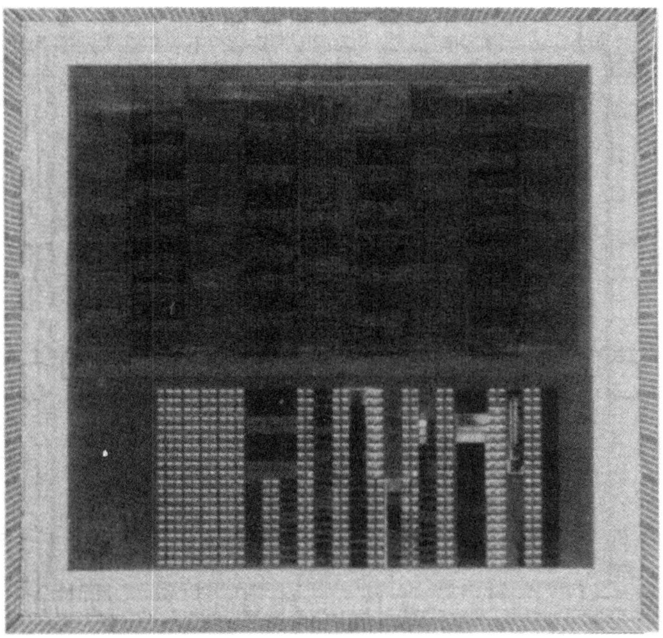

Fig. 13. Chip micrograph.

use fixed transistor sizes. In comparison with the fixed-coefficient FIR filter chip [10] designed to process NTSC composite TV signals, the gate array chip demonstrates the same degree of area efficiency (average tap size), which results from the sea-of-gate structure with the triple layer metal technology. It needs to be pointed out that the gate array approach significantly shortens design cycle time including the development of the cell library.

TABLE IV
FIR FILTER COMPARISON

Design Style	Module Generator [10]	Gate Array Compiler (this work)	Module Generator [11]
Clock Rate	13.5 MHz	100 MHz	100 MHz
Coefficient Type	Fixed	Fixed	Programmable
Coefficient Wordlength	8 bits	14 bits	12 bits
Technology	1.2μm CMOS	0.8μm BiCMOS	0.9μm CMOS
Active Area	80 mm^2 *est.*	49 mm^2	22 mm^2
Number of Taps	15,15,1[illegible],15	64	40
Transistor count	N/A	187K MOS, 22K NPN	240K MOS
Power Dissipation	N/A	6 *W*	3.1 *W*

V. CONCLUSION

A 64-tap 100-MHz sample-rate FIR filter chip has been fabricated successfully in a gate array technology. A combination of compiler techniques and gate array design methodology provides a very fast turnaround time for this design. This facilitates rapid prototyping of high-speed digital signal processing systems that cannot be implemented with programmable devices. The automatic generation of floorplan constraints by the filter compiler results in a relatively compact layout while reducing the place and route time.

CSD carry-save multipliers are used with the transpose structure to reduce the arithmetic delay. Bipolar data-bus drivers are implemented in a tree structure to reduce *RC* delays. The use of 0.8-μm BiCMOS gate array technology provides very fast gates allowing semi-custom techniques to be used for efficient implementation of a relatively complex filter. The 64-tap filter chip achieves a performance comparable to that of a full-custom designed chip while still maintaining compact area and fast turnaround time.

ACKNOWLEDGMENT

The authors thank the staff of Texas Instruments Semiconductor Process Laboratory, in particular T. Holloway and J. Graham for device processing, Prof. H. Samueli, B. Wong, T. Lin, and K. Chau for useful discussions, E. Born and G. Stacy for testing, and C. R. Hewes, R. Haken, and P. Chatterjee for support and encouragement.

REFERENCES

[1] H. Samueli and B. Wong, "A VLSI architecture for a high-speed all-digital quadrature modulator and demodulator for digital radio applications," *IEEE J. Selected Areas Commun.*, Oct. 1990.

[2] J. Gallia *et al.*, "A 100K gate sub-micron BiCMOS gate array," in *Proc. CICC*, May 1989, pp. 8.6.1–8.6.4.

[3] J. Gallia *et al.*, "High-performance BiCMOS 100K-gate array," *IEEE J. Solid-State Circuits*, vol. 25, no. 1, pp. 142–149, Feb. 1990.

[4] P. Yang, T. Yoshino, R. Jain, and W. Gass, "A functional silicon compiler for high speed FIR digital filter," in *Proc. ICASSP*, Apr. 1990, pp. 1329–1332.

[5] R. Jain, P. Yang, and T. Yoshino, "FIRGEN: A computer-aided system for high performance FIR filter integrated circuits," to be published in *IEEE Trans. Acoust. Speech, Signal Processing*, July 1991.

[6] T. W. Parks and C. S. Burrus, *Digital Filter Design*. New York: Wiley, 1987.

[7] A. Oppenheim and R. Schafer, *Discrete-Time Signal Processing*. Englewood Cliffs, NJ: Prentice-Hall, 1989.

[8] H. Samueli, "An improved search algorithm for the optimization of the FIR filter coefficients represented by a canonic signed-digit code," *IEEE Trans. Circuits Syst.*, vol. CAS-34, pp. 1192–1202, Sept. 1987.

[9] *Tangate System User Guide*, Product Release 1.3, Cadence Design Systems Corp., San Jose, CA, May 1989

[10] M. Ishikawa *et al.*, "Automatic layout synthesis for FIR filters using a silicon compiler," in *Proc. IEEE Int. Symp. Circuits Syst.*, 1990, pp. 2588–2591.

[11] M. Hatamian and S. Rao, "A 100 MHz 40-tap programmable FIR filter chip," in *Proc. IEEE Int. Symp. Circuits Syst.*, 1990, pp. 3053–3056.

Paper 2.5

WP 3.1: 0.5µm BiCMOS Standard-Cell Macros Including 0.5W 3ns Register File and 0.6W 5ns 32kB Cache

Hiroyuki Hara, Takayasu Sakurai, Tetsu Nagamatsu, Shin'ichi Kobayashi, Katsuhiro Seta, Hiroshi Momose, Yoichirou Niitsu, Hiroyuki Miyakawa, Tadahiro Kuroda, Kouji Matsuda*, Yoshinori Watanabe*, Fumihiko Sano*, Akihiko Chiba*

Toshiba Corp./*Toshiba Microcomputer Eng. Corp., Kawasaki, Japan

In this paper, BiCMOS standard-cell macros, including a 0.5W 3.0ns register file, a 0.6W 5.0ns 32kB cache, a 0.2W 2.5ns table look-aside buffer (TLB), and a 0.1W 3.0ns adder, are presented based on a 0.5µm BiCMOS technology. These power consumption values are at 100MHz operation. Low power and high speed are crucial for high-performance systems requiring a high level of integration [1]. Several BiCMOS / CMOS circuits achieve high-speed operation with a 3.3V supply. A direct-coupled ECL+CMOS circuit is investigated for use as a BiCMOS standard cell.

A BiNMOS gate is adopted for the basic gates in the library [2]. The BiNMOS configuration is suitable for reduced supply voltage. To restore full output swing, small pMOS devices controlled by the input are placed in parallel with the bipolar pull-ups. Figure 1 shows measured fanout dependence of the propagation delay. The BinMOS 2-input NAND gate has 200ps propagation delay with fanout of 7.

Figure 2 shows a circuit diagram of the register-file macro. The macro enables 6 read-outs and 4 writes at the same time. To realize the 6 read ports with a small memory cell, a single-ended bit line is used. The bit-line swing should be reduced to about 1V for high-speed asynchronous read-out, requiring a sense amplifier. A self-aligned threshold inverter (SATI) sense amplifier is used for the single-ended sense amplifier. The SATI layout can be self-aligned using a narrow pitch and tracks process, voltage, and temperature variation to provide the best operating point for the sense amplifier. The measured worst-case access time is shorter with the self-alignment.

For fast recovery from "0" read, a BiCMOS bit-line load is adopted. Here, an nMOS load is needed to stabilize the bit line swing. For fast write recovery, a circuit is used in which write recovery takes place as soon as the write enable (WE) is turned off. When the macro select (MS) signal is low, the macro consumes no dc power and IDDQ test is possible. This feature is added to all macros without degrading speed. Most of the delay data is obtained by using latches placed at the input and output and by varying the strobe timing to these latches.

A 32kB BiCMOS cache macro employs highly-resistive poly-silicon-load 4T cells for density and speed. Data-bus width is 128b and organization is 2-way set-associative. Figure 3 shows a circuit diagram of the BiCMOS cache macro. Comparators and complicated HIT logic are built with ECL to improve speed. The DATA sense amplifiers are placed after the way selectors to cut power consumption by about a half compared with the conventional architecture [3]. Only 128 sense amplifiers are activated in spite of the 2-way organization. The output of the BiCMOS sense amplifier is connected to the SATI circuit. Address-to-HIT delay is 5ns and address-to-DOUT delay is 8ns. This is the fastest cache reported of this size [4].

The TLB consists of a CAM for directory search and SRAM for physical address fetch. In the CAM, a triple-match-line architecture is used [5]. Because of the NAND structures employed in the design, only the match line of a matched row is activated and the corresponding SRAM word line is enabled. This eliminates complex marginal timing circuits for driving the SRAM word lines in the conventional NOR type match circuit [4]. This TLB makes use of bipolar hit-generation logic to realize a high-speed 64-input OR. The macro shows 2.5ns address-to-HIT delay measured under typical conditions, while simulated delay for the full CMOS implementation with the triple-match-line architecture is 3.3ns and that without the triple-match-line architecture is 4.0ns.

A 32b asynchronous adder is useful as a standard-cell macro, since an appropriate clock is not always available. The salient feature of the adder is the introduction of the chained carry look-ahead (CCLA) circuit, where the outputs of the carry select adders (CSAs) are combined by a Manchester-chain-like circuit to efficiently generate required carries (Figure 4). The MOSFETs in the chain are sized in a geometric series. The macro shows 3.0ns measured addition time, whereas the simulated delay for the full CMOS adder with the CCLA is 3.5ns and that without the CCLA is 4.5ns.

The basic cell library also includes clock drivers, 3V-5V compatible I/Os, and ECL I/Os. The ECL I/O uses a pseudo-ECL configuration shown in Figure 5 for 1ns input delay and 1.4ns output delay, including ECL-CMOS swing conversion. In conventional ECL I/Os an extra delay of 0.5ns is added in shifting signal levels up and down and a more complex process is needed for high-voltage reliability. The SATI circuit reduces the input delay. The characteristics of the macros are summarized in Table 1 and the process/device parameters, in Table 2. Figure 6 shows a micrograph of the chip.

In the future, with the supply voltage to CMOS reduced to 2.0V, conventional BiCMOS / BinMOS buffers will not have sufficient speed advantage over CMOS [6]. In that era, however, ECL signals can drive CMOS gates directly if four voltage rails are used. An ECL buffer chain, and a direct-coupled ECL+CMOS chain in Figure 7 have measured propagation delays of 130ps, 200ps and 400ps, respectively. Since 130ps + 200ps (+70ps) = 400ps, direct connection between ECL and CMOS is possible with small delay overhead.

Acknowledgement

The authors express appreciation to K. Kanzaki and H. Nakatsuka for encouragement.

References

[1] Hotta, T., et al., "A 70MHz 32b Microprocessor with 1.0µm BiCMOS Macrocell Library," ISSCC DIGEST OF TECHNICAL PAPERS, pp. 124-5, Feb. 1989.

[2] Gamal, A. E., et al., "BiNMOS: A Basic Cell for BiCMOS Sea-Of-Gates," CICC, pp. 8.3.1-8.3.4, May 1989.

[3] Suzuki, M., et al, "Sense-Amplifier-Merged Comparator and Selector Scheme for BiCMOS Cache Memories," ESSCIRC, pp. 25-28, Sept. 1991.

[4] Takayanagi, T., et al., "2.6GB/s Bandwidth Cache/TLB Macro for High-Performance RISC Processor," CICC, pp. 10.2.1-10.2.4, May 1991.

[5] Nagamatsu, T., et al., 'A 1.9ns BiCMOS CAM Macro with Double Match Line Architecture", CICC, pp. 14.3.-14-.3.4-, May 1991.

[6] Hara, H., et al., "0.5µm 2M-Transistor BipnMOS Channelless Gate Array", ISSCC DIGEST OF TECHNICAL PAPERS, pp. 148-9, Feb. 1991.

Reprinted from *1992 IEEE ISSCC*, pp. 46–47 and 239, Feb. 1992.

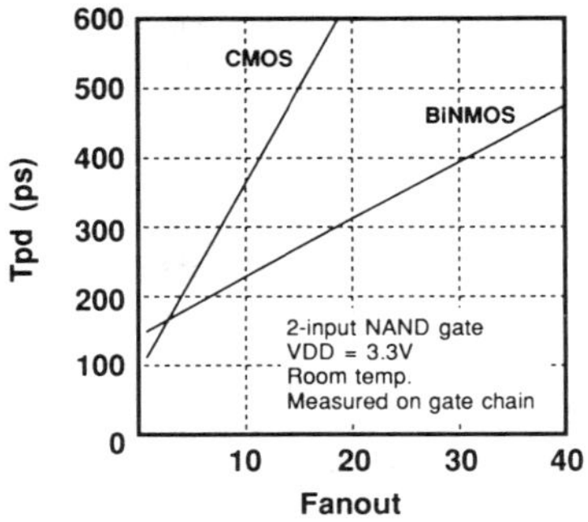

Figure 1: Measured propagation delay vs. fanout.

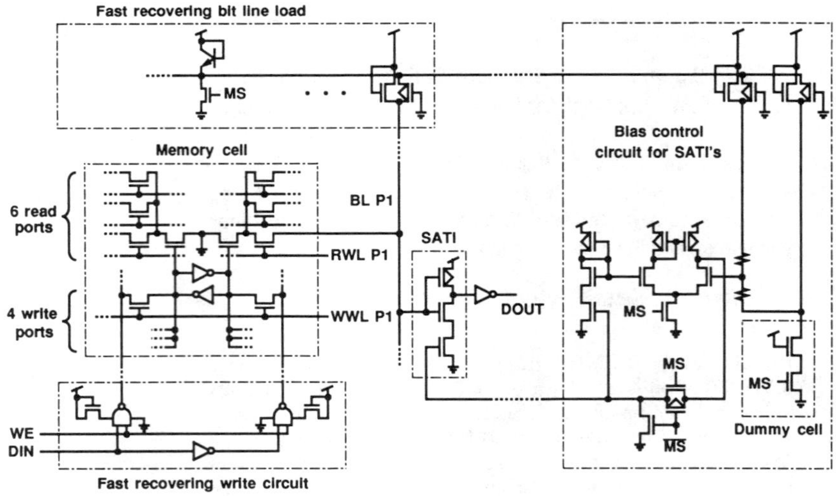

Table 1: See page 215.

Figure 2: Circuit diagram of the register-file macro.

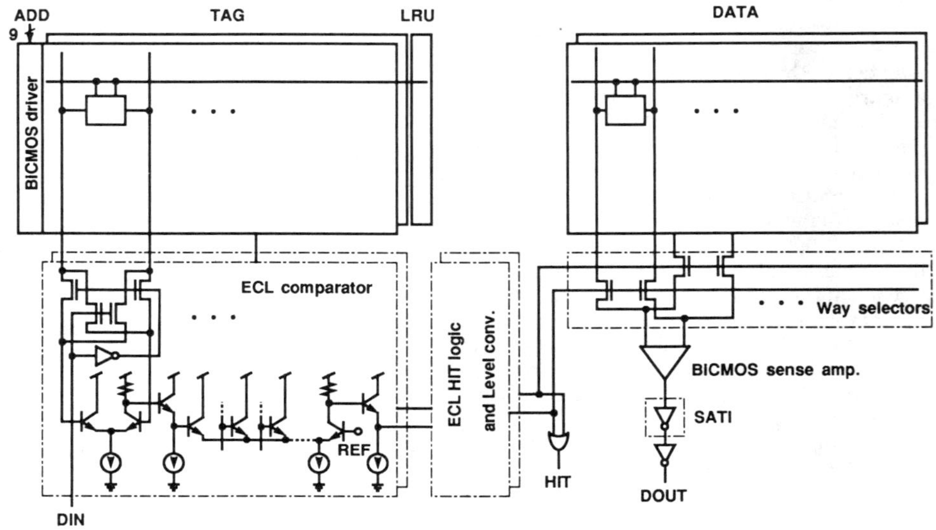

Figure 3: Circuit diagram of the cache macro.

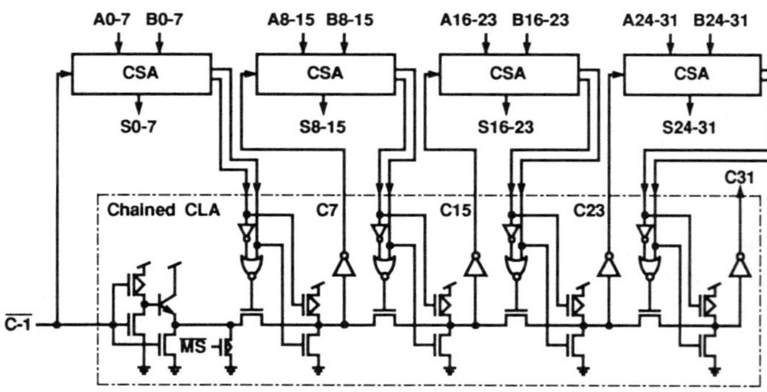

Figure 4: Circuit diagram of the BiCMOS adder.

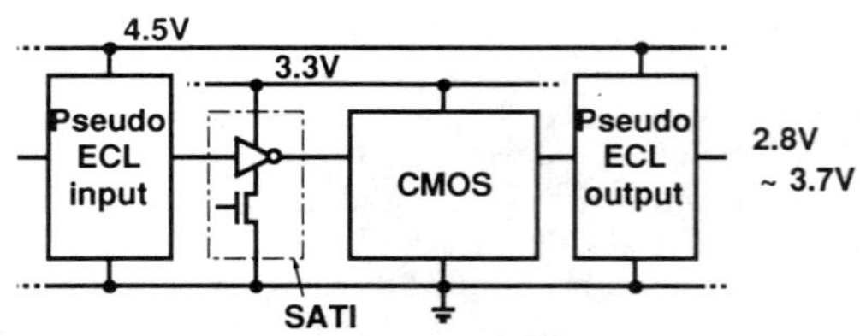

Figure 5: Pseudo-ECL I/O with SATI.

Figure 6: See page 215.

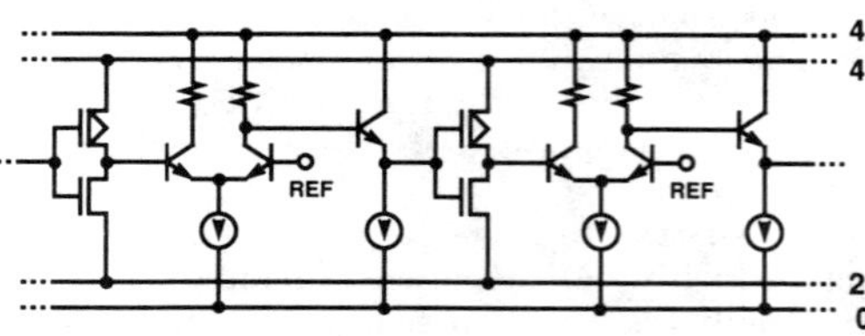

Figure 7: Direct-coupled ECL+CMOS.

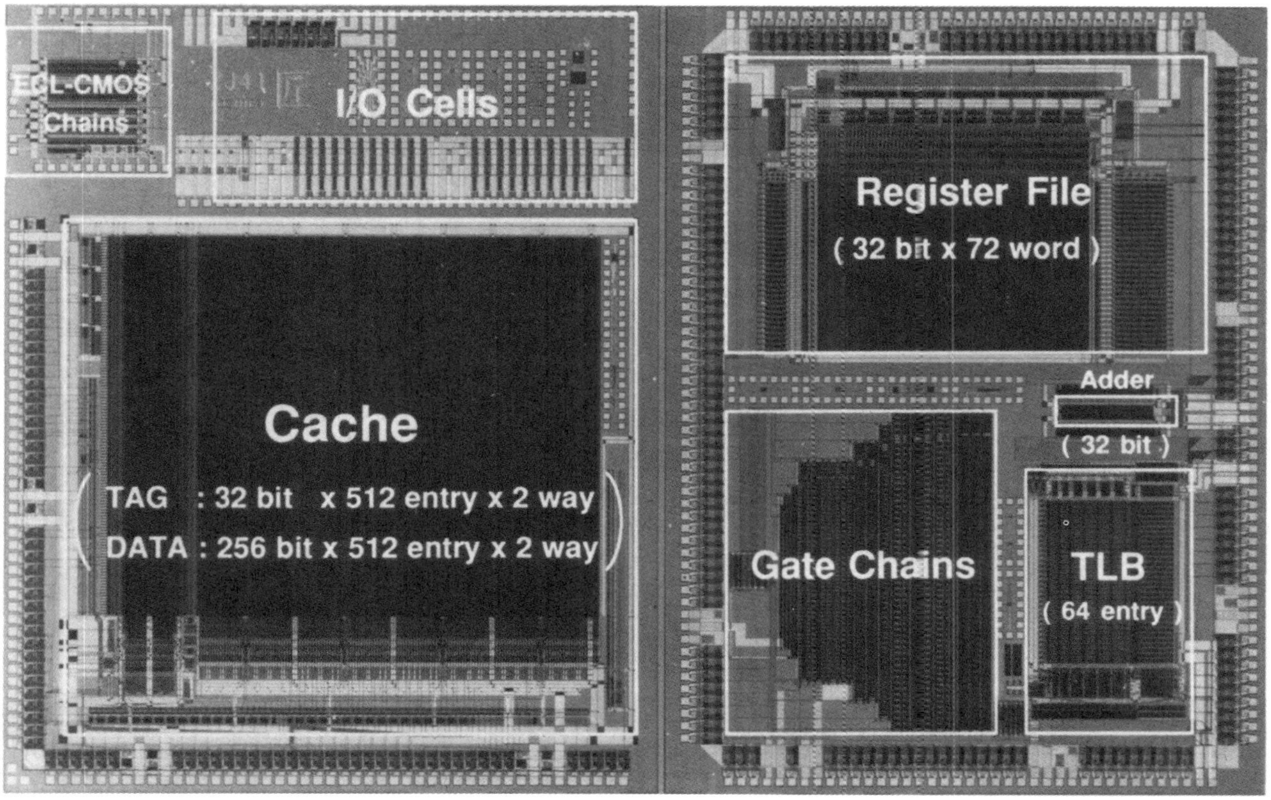

Figure 6: Chip micrograph.

Register file	
Configuration	32bx72w, 6 read, 4 write
Access time (typ.)	ADD-DOUT, 3ns, asynch.
Macro size	3.6x3.8mm^2
Cache	
Configuration	32bx512-entryx2-way TAG 256bx512-entryx2-way DATA
Access time (typ.)	ADD-HIT: 5ns, ADD-DOUT: 8ns
Macro size	6.0x6.5mm^2
TLB	
Configuration	32bx64-entry CAM 31bx64-entry DATA SRAM
Access time (typ.)	DATA-HIT: 3ns, synch.
Macro size	2x3mm^2
Adder	
Configuration	32b asynchronous
Delay time (typ.)	DATA-SUM: 3ns
Macro size	1.2x0.3mm^2

Table 1: Characteristics of BiCMOS macros.

Paper 2.6

BiCMOS Circuits for DPCM Coders in HDTV Systems

ALBRECHT ROTHERMEL, MEMBER, IEEE

Abstract —**Three different strategies have been used to enhance the speed performance of differential pulse-code modultation (DPCM) coders. The first one is the reduction of the number of components located in the time-critical recursive loop. The second one is the use of a BiCMOS technology, and the third one is the development of some new high-speed circuit techniques. Prototype chips prove that a 2.5-μm process is feasible for implementation of DPCM coders operating at 54 MHz. Another test structure shows a BiCMOS sequencer with 64 product terms running at 80 MHz with 125-mW power consumption.**

I. Introduction

Most progress in VLSI-circuit performance is achieved by advances in the production of smaller and faster transistor devices. On the other hand, these highly sophisticated technologies in the range of 1 μm and below are very expensive. Therefore, the production of integrated circuits with processes using optical resolutions of about 2 μm is still economically interesting. Costly processing techniques like wafer stepping can be omitted in this case. Especially in consumer, telecom, and future B-ISDN applications, low-cost IC's will enable a large distribution of the new components.

In this paper some approaches are described to enhance the speed performance of differential pulse-code modulation (DPCM) coders [1], [2]. These coders are very useful for the data reduction of moving picture signals, especially TV, because they remove redundancy from natural pictures without reducing the spatial resolution. The data reduction of digital TV signals becomes very important, if high-definition television (HDTV) signals are to be transmitted.

There are several HDTV production and transmission standards under consideration at this time in Japan, Europe, and the U.S. Therefore, no final definition of a complete DPCM coding system for HDTV is given in this paper. But the ideas presented here will be applicable in many cases, because the described circuits are typical for problems that can be foreseen in HDTV.

Generally, HDTV proposals usually employ a spatial resolution of the TV screen picture that is approximately doubled in the horizontal and the vertical directions compared with the present TV picture. Additionally, the aspect ratio will be changed from 4:3 at present to 16:9 in the future. Owing to this, full-bandwidth HDTV component TV signals require about five times the bandwidth of conventional TV signals [3]. Therefore, a clock frequency of approximately 54 MHz will be needed in HDTV systems.

The principle of a DPCM coder can be seen in Fig. 1. There is a critical timing path in these coders which is defined by a recursive loop in the kernel. The latency time of this loop has to be reduced to 18 ns to make the coder applicable to HDTV systems. The following sections deal with the techniques used to reduce the delay within this recursive loop. In Section II the coder is reconfigured to employ a shorter critical timing path. In Section III the advantages of the use of a BiCMOS technology are discussed. In Section IV some high-speed circuit techniques are described, and Section V presents test results.

II. Reconfiguration of the Circuit

Fig. 1 shows the principle of a two-dimensional DPCM coder. A predictor value is calculated from those two picture elements located above and to the left of the actual picture element. This predictor value is subtracted from the actual sample, and the difference is then quantized, coded, and transmitted to the receiver. The decoder reconstructs the actual sample by adding the predictor value to the received signal.

A significant characteristic of the DPCM coding is the fact that the predictor values in the coder and in the decoder are exactly the same, at least if no transmission errors occur. To achieve this, in the DPCM coder the input signal is reconstructed from the quantized difference signal, and this reconstructed input signal is used for the calculation of the predictor value. This is the reason for the recursive loop in the coder.

It can be shown that due to this coder configuration, the quantization noise is shaped. That means the coder has a high-pass transfer function for the noise added by the quantizer. If this were not the case, the signal-to-noise

Manuscript received April 26, 1990; revised July 16, 1990.
The author is with the Research and Development Laboratories, Department CLI, Thomson Consumer Electronics, D-7730 Villingen, West Germany.
IEEE Log Number 9038920.

Reprinted from *IEEE J. Solid-State Circuits*, vol. 25, no. 6, pp. 1470–1475, Dec. 1990.

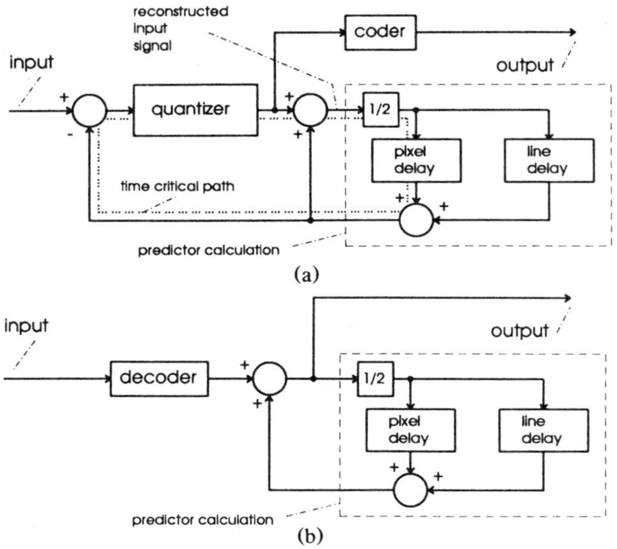

Fig. 1. Principle of a two-dimensional (a) DPCM coder and (b) decoder.

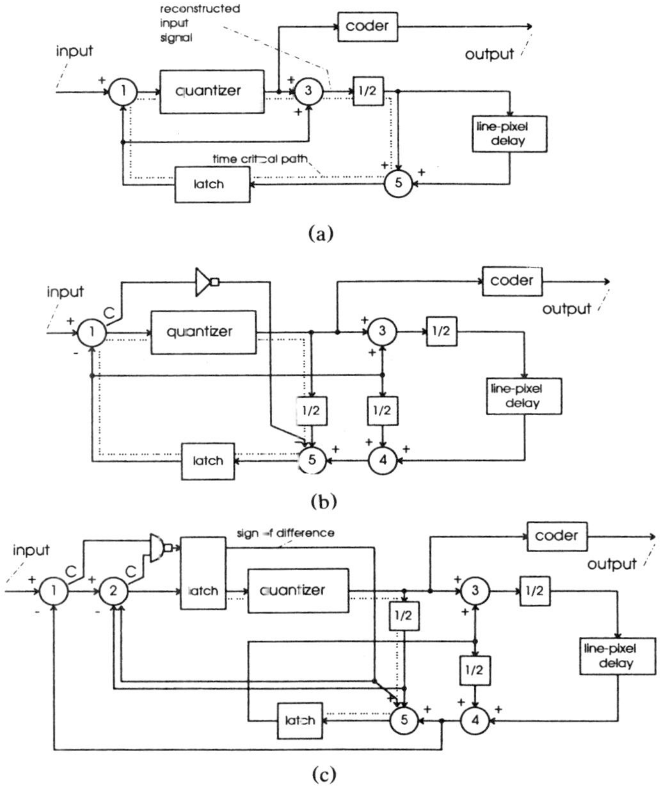

Fig. 2. Reduction of the number of adders in the time-critical path.

ratio of the whole codec would not be high enough for low input signal frequencies.

Some coder designs profit from the fact that the nonlinear characteristic of the quantizer is designed symmetrically. That means that the quantization steps are the same for positive and negative values of the difference signal (which is represented in two's complement). The main advantage is that the size of the quantization table can be nearly halved. Also the sign of the difference signal is not needed in this case for further calculations. Because the sign of the reconstructed input signal is known to be positive all the time, it is possible to add the difference signal in two's complement representation without the sign bit to the predictor value. The correct result is obtained by simply ignoring the sign of the sum signal, because it has to be positive by definition.

It is known that the number of elements in the recursive loop of the coder can be reduced in two steps to only one adder, the quantizer, and a latch (Fig. 2) [4]. Fig. 2(a) shows the basic version of the DPCM coder. In Fig. 2(b) the input signal of the latch is the same as in Fig. 2(a), but the sequence of the additions has been changed by inserting another adder circuit (4). In Fig. 2(c), the first subtractor (1), which calculates the difference between the input signal and the predictor value, is split into two subtractors (1 and 2). The predictor value is not used for this subtraction directly, but it is replaced by the two summands, which are also used to calculate the predictor value in adder #5.

However, in the modified circuits (Fig. 2(b) and (c)), the sign of the difference signal is needed in contrast to the basic structure of Fig. 2(a), because the quantized difference is no longer added exclusively to the predictor value (as it is done in adder #3), but also to other values (in adders #5 and #2). So the sign of the difference signal has to be derived and used as shown in Fig. 2(b) and (c). When the binary difference signal is divided by a factor of 2, the former sign becomes the MSB of the divided signal.

Since the time-critical path of the DPCM coder now includes only one adder, a quantizer, and a latch, we shall consider only this time-critical loop in the following sections.

III. Advantages of a BiCMOS Technology

BiCMOS technologies have been developed by many important IC manufacturers. However, the advantages and disadvantages of these technologies are still under consideration. At this time, BiCMOS is mainly used for some high-speed SRAM's, for high-speed gate arrays, for a number of TTL-compatible interface IC's (74BCD), and for prototyping of mixed analog–digital applications.

The usage of BiCMOS for random logic digital designs seems to be the application with the biggest question marks. Obviously, in pure digital applications a CMOS technology with a significantly reduced feature size will exceed the performance of a BiCMOS technology with a larger feature size. But a significant reduction of the feature size significantly increases production costs as well. This has to be compared with the production costs of a BiCMOS process, which is about 30% more expensive than CMOS at the same feature size.

So it is important to be able to estimate the performance advantage of BiCMOS, compared to the feature size reduction of pure CMOS. This estimation was carried out for digital buffering applications, where BiCMOS circuits have the main advantages [5]. It turns out that the

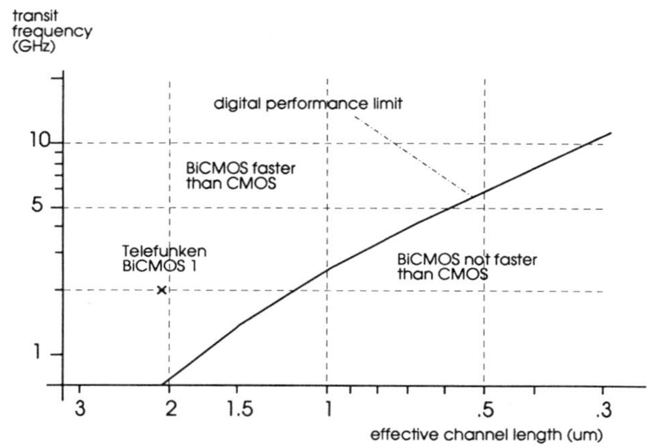

Fig. 3. Performance limit of BiCMOS processes for digital applications.

speed performance of BiCMOS circuits depends mainly on the parameters of the bipolar transistors, namely the transit frequency f_t. The speed of digital CMOS circuits is determined by the effective channel length L_{eff}.

Fig. 3 gives a coarse but useful estimation of this parameter trade-off. It compares the attainable speed using bipolar transistors in BiCMOS buffer circuits on the one hand to the speed of optimized CMOS circuits on the other hand. Both speeds are the same for points on the line denoted "digital performance limit."

The diagram can be interpreted in the following way. If we assume, e.g., a BiCMOS process employing 2-GHz bipolar devices, the maximum possible speed using these transistors in BiCMOS buffers will be comparable to the speed of CMOS circuits with a feature size around 1.2 μm. If the feature size of the BiCMOS process itself is 1.2 μm, obviously the use of the 2-GHz bipolar transistors is not advantageous in digital circuits. This is true even for large load capacitances which have to be driven on the chip, if the CMOS circuits are assumed to be optimized [5].

Another conclusion is that the production costs of the 2-GHz BiCMOS process (with any reasonable feature size, e.g., 2–3 μm) have to be compared to the costs of a 1.2-μm CMOS process, because the maximum attainable speed in both of them will be nearly the same.

Circuit designers can draw the conclusion that advantages in switching speed due to the use of bipolar transistors in a mixed BiCMOS design can be obtained only in those processes that are lying above the decisive line. With processes that are lying below the line in Fig. 3, speed advantages can only be achieved using pure ECL logic with low logic voltage swing, and these advantages decrease rapidly as the distance to the line increases.

Finally, since the process which was used lies clearly above the curve in Fig. 3, some newly developed BiCMOS circuit techniques were employed. We use the BiCMOS 1 process developed by Telefunken Electronic GmbH in Heilbronn, West Germany. The characteristics of this technology are summarized in Table I.

TABLE I
TELEFUNKEN BICMOS 1 TECHNOLOGY

minimum channel length (MOS)	2.5 μm
effective channel length	2.0 μm
gate oxide	40 nm
transit frequency (bipolar)	2 GHz
base resistance	500 OHm
single layer metal/polysilicon	

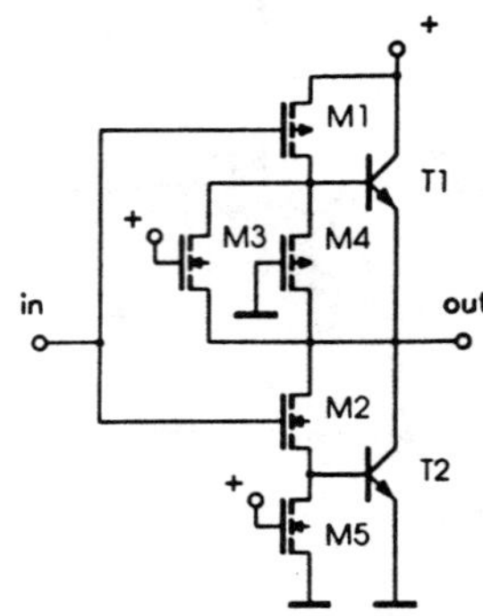

Fig. 4. BiCMOS buffer.

IV. High-Speed Circuit Techniques

BiCMOS circuit techniques have been employed for two purposes, namely buffering and sensing.

A. BiCMOS Buffer

The use of a BiCMOS buffer gives speed advantages only if a significant capacitive load has to be driven [6]. This is the case at the input of the quantizer (approximately 1 pF), which is realized as a PLA circuit. The buffer developed for this purpose is shown in Fig. 4. It is a CMOS-compatible buffer that uses bipolar transistors to amplify the currents of the MOS transistors $M1$ and $M2$.

In all BiCMOS buffer circuits, the base charges of the bipolar transistors have to be removed when they are turned off. This is usually done by switched MOS transistors or by resistors.

The characteristics of this buffer are the MOS transistors $M3$ to $M5$, which are used to discharge the base regions of the bipolar transistors. Because these transistors are not switched, the buffer output reaches the power supply rails, thus guaranteeing high noise margins. The two MOS transistors $M3$ and $M4$ have to be connected in parallel to ensure proper discharging of the base of the bipolar transistor $T1$ during the whole high–low transient at the output node.

If this buffer is designed properly, its output signals profit from the active charge stored in the bipolar transistors during the transients. Due to this charge, the collector current is still flowing for a short time when the base current has already become zero. Thus the output of the buffer approaches the power supply rails very fast.

The utilization of transistors $M3$ to $M5$ instead of resistors has the advantage of lowering the sensitivity to

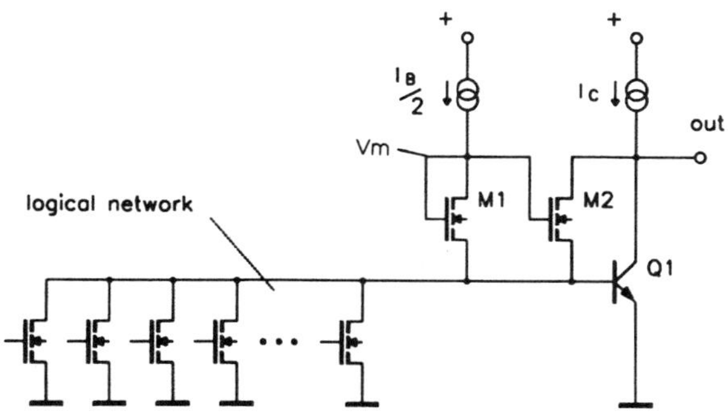

Fig. 5. Sense amplifier, not clocked.

process parameter and power supply voltage variations, because the drain currents of all the MOS transistors increase or decrease simultaneously.

Although the BiCMOS buffer is approximately twice as fast as an optimized cascaded CMOS buffer in this technology, the absolute speed advantage of the BiCMOS buffer is not very large. In our application we gained approximately 1 ns. The advantage of the BiCMOS buffer would be more significant if it were not possible to optimize the buffers for one particular load capacitance (e.g., in gate arrays or in standard cell designs [6]).

B. BiCMOS Sense Amplifier

Larger speed advantages than those gained by BiCMOS buffering can be gained by BiCMOS sense amplifiers. The PLA circuit which realizes the quantizer table consists of an address decoder and a ROM. At the output terminals of the conventional CMOS PLA (or ROM) there is a large capacitance to be charged and only a small charging current from the small MOS transistors of the logic array. Hence, the speed performance is limited.

Therefore, it is advantageous to use a sense amplifier for reading out the PLA with low logic signal voltage swing. This sense amplifier has two purposes: it has to amplify the low logic voltage swing and it has to limit the voltage swing to an appropriate value. Such single-ended sense circuits have already been reported (e.g., [7]). A serious problem in these circuits is the prevention of the bipolar sense transistor from saturation. Usually the sense circuits are clocked for this purpose.

The sense amplifier shown in Fig. 5 is not clocked. This is advantageous in recursive structures because a maximum clock frequency can be achieved with a simple clocking scheme. In this sense circuit saturation of the bipolar transistor is prevented by the current mirror $M1, M2$. When the output of the sense circuit is at a high potential, one transistor in the logical network is conducting current, and the base current of the bipolar transistor is zero. If the transistor in the logical network is now switched off, the whole current ($2 \times I_b/2$, if $M1$ and $M2$ have the same dimensions) flows into the base of the bipolar transistor, thus switching it on.

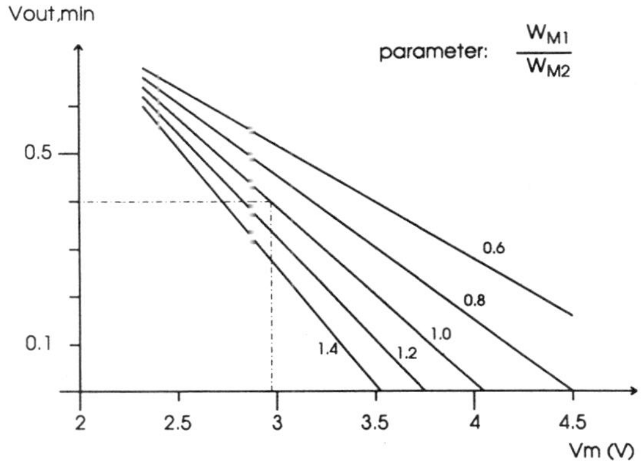

Fig. 6. Estimation of the minimum output voltage of sense amplifier.

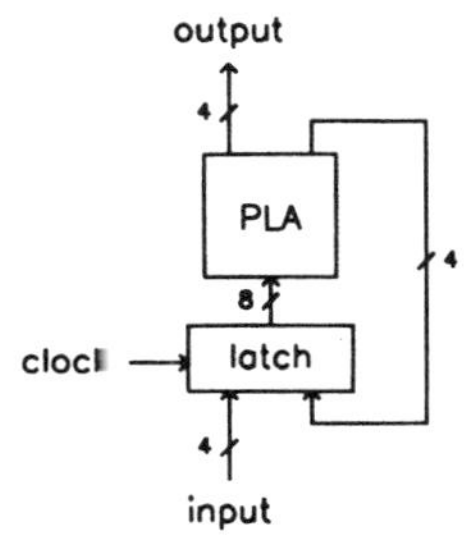

Fig. 7. Sequencer test structure.

Next, the output potential of the sense amplifier goes down. When it arrives at the base potential of the bipolar transistor (approximately 0.7 V), the drain current of $M2$ becomes zero, and the base current of $Q1$ becomes $I_b/2$. When the output goes even lower, the drain and source terminals of $M2$ change their roles, and a part of the drain current of $M1$ is no longer flowing into the base of $Q1$, but via $M2$ to the collector terminal of $Q1$.

The drain current of $M2$ increases very rapidly when the collector–base potential of $Q1$ goes down below 0 V, because in that case the gate–source voltage of $M2$ increases, and the threshold voltage of $M2$ decreases due to the body effect. Therefore we have the surprising result that the bipolar transistor can be properly clamped by an MOS transistor in this case.

With the diagram in Fig. 6, the minimum output voltage can be adjusted by the ratio of $M1$ to $M2$ and the choice of the voltage V_m by the W/L ratio of $M1$. The value of the minimum output voltage is then very insensitive to absolute transistor parameter and temperature variations, because $M1$ and $M2$ are connected in a current-mirror configuration.

V. Results Derived from Test Structures

The first test structure that was realized to evaluate the ideas presented in Sections III and IV is shown in Fig. 7. It is a recursive configuration of a PLA and a latch. With this sequencer structure the sum of the delay times of the PLA and the latch can be estimated simply by increasing the clock frequency. To get a proper comparison of the

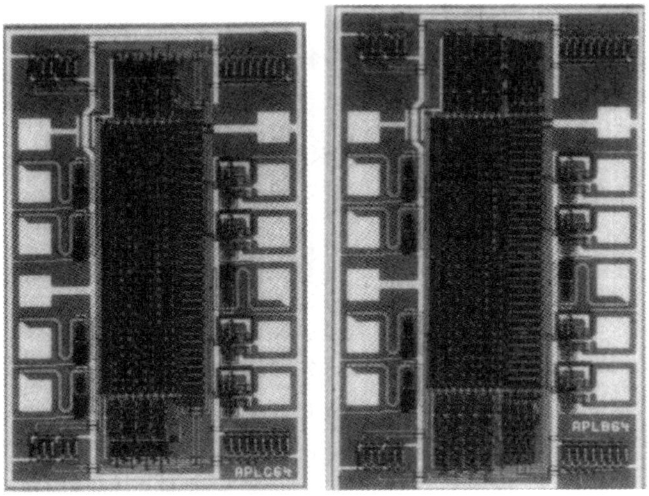

(a) (b)

Fig. 8. Sequencer in (a) CMOS and (b) BiCMOS circuit techniques.

TABLE II
SEQUENCER TEST STRUCTURE RESULTS

organization:	8 inputs, 64 product terms, 8 outputs	
circuit technique:	CMOS	BiCMOS
dynamic latch:	5 ns	5 ns
PLA:	11 ns	6 ns
interconn. and buffers:	2 ns	1 ns
fmax:	55 MHz	80 MHz
operating power (5V):	80 mW	125 mW
active area:	0.60 sqmm	0.72 sqmm

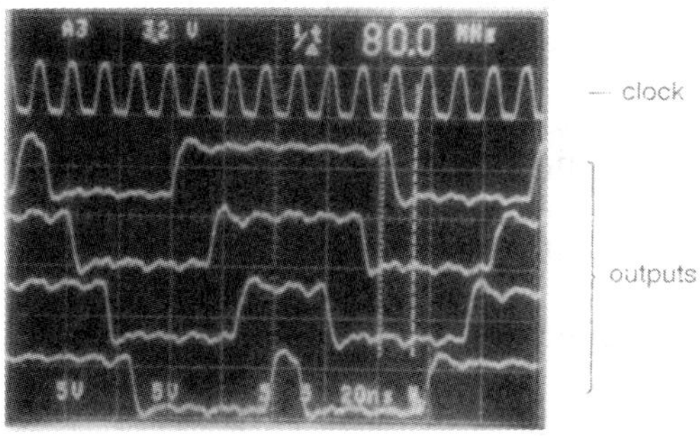

Fig. 9. Sequencer output signals at 80 MHz.

BiCMOS and CMOS circuit techniques, the structure has been realized twice (Fig. 8).

The results for this test structure are summarized in Table II. The partitioning of the delay times between the PLA and the latch has been derived by other tests. It can be seen that the delay of the PLA is almost halved by the use of the BiCMOS sense amplifier.

An example of the output signal at 80-MHz clock frequency is shown in Fig. 9. The output signals represent an arbitrary pattern implemented for easy visual testing.

Finally, the time-critical loop of an assumed HDTV–DPCM coder has been realized. The block diagram is shown in Fig. 10, and a chip micrograph in Fig. 11. The operation of the circuit at 54-MHz clock frequency could

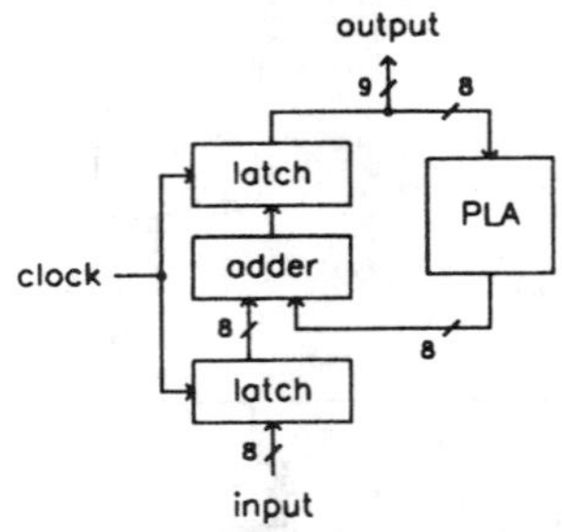

Fig. 10. Block diagram of time-critical loop of DPCM coder.

Fig. 11. DPCM test structure.

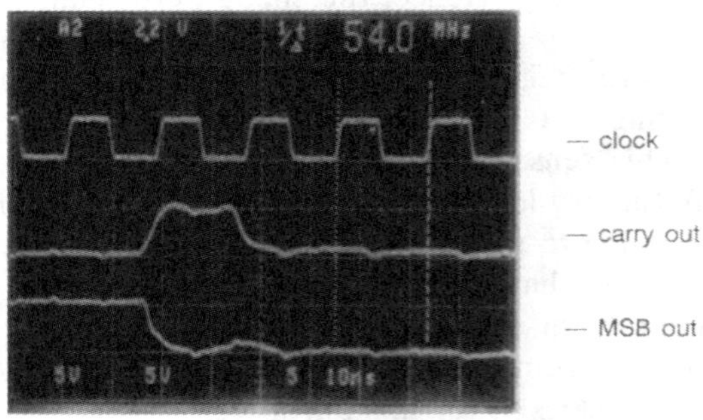

Fig. 12. DPCM test structure output signals.

TABLE III
DPCM TEST STRUCTURE RESULTS

circuit technique:	BiCMOS
latch, PLA, and interconn.:	see Tab. 2
8-bit CMOS adder:	6 ns
fmax:	54 MHz
operating power (5V):	135 mW
active area:	2 sqmm

be verified, and an output signal example is shown in Fig. 12. The carry out and the MSB out bits of the internal adder are shown in this photograph after they have been latched and propagated to the output terminals.

The data of this test structure are summarized in Table III. The 8-b adder was realized as a TGCS configuration [8], which resulted in a very low propagation delay for this pure static CMOS adder concept.

VI. Conclusion

It was shown that DPCM coders for HDTV clock frequencies in the range of 50 MHz can be realized with BiCMOS technologies with a relatively coarse photolithography. As an example, a test structure operated with 54 MHz using a 2.5-μm process. This result was obtained by employing a modified configuration for the coder, some new BiCMOS circuit techniques, and a high-speed CMOS adder structure.

Acknowledgment

The author would like to thank Prof. Zimmer, Prof. Hosticka, Dr. Schardein, Mr. Esser, Mr. Fichtel, and Mr. Lerch from the Fraunhofer Institute of Microelectronic Circuits and Systems, Duisburg, West Germany, for their encouraging support in this work. He also would like to thank Prof. Sieber, Dr. Troester, and Dr. Arndt from Telefunken Electronic GmbH, Heilbronn, West Germany, for their cooperation and help in making this work successful.

References

[1] A. N. Netravali and J. O. Limb, "Picture coding: A review," *Proc. IEEE*, vol. 68, pp. 366–406, Mar. 1980.

[2] H. G. Musmann, P. Pirsch, and H.-J. Grallert, "Advances in picture coding," *Proc. IEEE*, vol. 73, pp. 523–548, Apr. 1985.

[3] H. Gaggioni and D. Le Gall, "Digital video transmission and coding for the broadband ISDN," *IEEE Trans. Consumer Electron.*, vol. 34, no. 1, pp. 16–35, Feb. 1988.

[4] A. Rothermel, "Realization of a DPCM coder for 13.5-MHz sampling rate in CMOS technology," *IEEE J. Solid-State Circuits*, vol. SC-22, no. 6, pp. 1196–1197, Dec. 1987.

[5] A. Rothermel and B. Hosticka, "Speed comparison of digital BiCMOS and CMOS circuits," *Proc. Inst. Elec. Eng.*, vol. 136, pt. G, no. 2, pp. 49–56, Apr. 1989.

[6] A. Rothermel, *Digitale BiCMOS-Schaltungen.*, Munich: R. Oldenbourg, Verlag, 1990.

[7] G. O. Rosseel, M. A. Horowitz, R. W. Dutton, and R. L. Cline, "A single-ended BiCMOS sense circuit for digital circuits," in *ISSCC Dig. Tech. Papers*, Feb. 1989, pp. 114–115, 304.

[8] A. Rothermel, B. J. Hosticka, G. Troester, and J. Arndt, "Realization of transmission-gate conditional-sum (TGCS) adders with low latency time," *IEEE J. Solid-State Circuits*, vol. 24, no. 3, pp. 558–561, June 1989.

Paper 2.7

3.3-V BiCMOS Circuit Techniques for 250-MHz RISC Arithmetic Modules

Kazuo Yano, *Member, IEEE,* Mitsuru Hiraki, *Member, IEEE,* Shohji Shukuri, Makoto Hanawa, *Member, IEEE,* Makoto Suzuki, *Member, IEEE,* Satoru Morita, Atsushi Kawamata, Nagatoshi Ohki, Takashi Nishida, and Koichi Seki, *Member, IEEE*

Abstract—**A quasi-complementary BiCMOS gate for low-voltage supply is applied to a 3.3-V RISC data path. For a parallel RISC processor, the major issues are the construction of arithmetic modules in a small number of transistors and the shortening of the cycle time as well as the delay time. The feed-backed massive-input logic (FML) concept is proposed to meet these requirements. It reduces the number of transistors and the power within the framework of fully static logic 3–4 times. A low-voltage BiCMOS D-flip-flop is also conceived to allow the single-phase clocking scheme, which is favorable for high-frequency operation of RISC's. To demonstrate these circuit techniques, a 32-b ALU is designed and fabricated using 0.3-μm BiCMOS to demonstrate 1.6 times performance leverage over CMOS at 3.3 V.**

I. Introduction

BiCMOS technology has been employed in a wide spectrum of ULSI's because of its fast speed (twice as fast speed as CMOS with almost the same integration level), its capability to interface with the ECL level, and its excellent stability and sensitivity to small input signals in sense amplifiers. Although BiCMOS, to date, has been effectively applied to ECL-interfaced SRAM's [1] and DRAM's [2], its application to fast microprocessors is recent [3], [4]. One strong motivation for this is that the RISC concept is becoming a major factor in fast engineering workstations. Fast device technologies, such as GaAs, bipolar ECL, and BiCMOS, are expected to be applied in microprocessors [5], [6]. BiCMOS is particularly advantageous for parallel-processing architecture, which will play an important role in next-generation processors because of its low-power and high-density characteristics.

However, how much the degraded performance of BiCMOS at low supply voltages [7], [8] affects the value of deep-submicrometer BiCMOS is still unclear. Although new circuits, such as the BiCMOS [9] and C-BiCMOS [9], [10], have been proposed and studied to break conventional limitations, the BiNMOS has a small performance leverage over the CMOS and the advantages of the C-BiCMOS are considerably offset by the additional process complexity and cost for the p-n-p bipolar fabrication. Therefore, a new circuit is still strongly desired.

Another "new" approach to solve this situation is to use quasi-complementary BiCMOS (QC-BiCMOS) [11]–[14]. Although the basic circuit configuration of QC-BiCMOS itself has been examined by many designers since the early stage of BiCMOS development [11], it has not played an important role in actual integrated circuits mainly because of its limited performance leverage over CMOS. However, as we demonstrated in a previous paper [13], [14], QC-BiCMOS gates with two additional circuit techniques (one is a separation between the base of the pull-up bipolar and the base of the quasi-p-n-p, the other is a carefully designed base discharging circuit and full-swing input/output) can have a sufficient performance leverage over CMOS with an excellent low-voltage performance. We demonstrated this by fabricating 0.3-μm three-input NAND gates. However, the feasibility of the QC-BiCMOS in actual larger circuits is unknown. In addition, issues for applying QC-BiCMOS to fast microprocessors are not clarified. Particularly, anticipated high-machine-cycle operation (approaching nominally 250 MHz for 0.3-μm region) will surely pose new issues on circuit design.

This paper discusses the main issues which one inevitably encounters in designing fast low-voltage BiCMOS RISC arithmetic modules. Using this as a vehicle, new circuit techniques to solve these issues are explored. The advantages of the circuit techniques are demonstrated in an experimental 32-b ALU. The issues and low-voltage circuit are discussed in Section II. The design and the fabrication of an experimental ALU are discussed in Section III, followed by the conclusions.

Manuscript received August 8, 1991; revised November 4, 1991.

K. Yano is with the Center for Solid-State Electronics Research, Arizona State University, Tempe, AZ 85287-6206 on leave from the Central Research Laboratory, Hitachi Ltd., Kokubunji, Tokyo 185, Japan.

M. Hiraki, S. Shukuri, M. Hanawa, M. Suzuki, T. Nishida, and K. Seki are with the Central Research Laboratory, Hitachi Ltd., Kokubunji, Tokyo 185, Japan.

S. Morita, A. Kawamata, and N. Ohki are with Hitachi VLSI Engineering Corporation, Kodaira, Tokyo 187, Japan.

IEEE Log Number 9105614.

Reprinted from *IEEE J. Solid-State Circuits,* vol. 27, no. 3, pp. 373–381, March 1992.

II. Low-Voltage Circuit Techniques for 3.3-V RISC

For realizing a high-performance RISC, the conventional precharged circuits, which are widely used in the fast arithmetic modules [15], [16], are unfavorable, because a short cycle time as well as a short gate delay is essential in fast RISC's. In CISC's, precharging can be done while the other circuit is operating because of the longer logic path and the longer machine cycle. However, a RISC, which owes its performance to a shorter machine cycle, cannot afford the additional precharging time. In addition, it is difficult to maintain sufficient noise immunity in a dynamic circuit at a high frequency around 250 MHz, which is our nominal target machine cycle. However, it is also difficult to get enough performance only by the simple static logic in many situations. This makes us conclude that a new circuit which reduces both the delay time and cycle time is expected for high-machine cycle RISC's.

Another important issue for a high-performance RISC is to reduce area and power. Indeed, small area and small power have been pursued in all integrated circuits. However, in the deep-submicrometer RISC's they are directly related with the performance. This is because in a 0.3-μm region (or less), implementing parallel processing elements on a chip is a very attractive (and will probably be a very common) option to achieve high performance. Below eight processors, the number of processors is expected to linearly enhance the total performance of a computer system. In this case the area and power of each processor, which determine the number of processors on a chip, are unprecedentedly important in the deep-submicrometer region. When one sees the BiCMOS gate in this light, its larger area than the CMOS gate may be a considerable disadvantage. Therefore reducing area and power dissipation without degrading the performance is the next challenge for 0.3-μm-level BiCMOS circuit designers.

Based on this consideration, we looked for a way to reduce area and power within the framework of fully static logic. Our approach to solve these issues is to use a circuit named feedbacked massive input logic (FML) shown in Fig. 1. The main features of this circuit are: 1) the separation between the nMOS logic circuit and the pMOS logic circuit by the MOS transistors ($MN1$, $MP1$ in Fig. 1(a)) controlled by the feedback signal from the output; and 2) large input logic is formed in a single stage. The basic operation is explained using the five-input NAND gate as an example (Fig. 1(b)). We start from the case when inputs A, B, C, and D are high level, and input E changes from low to high. Before the transient, $MN6$ and $MN11$ are off. At this moment, nodes $N1$, $N4$ and outputs $N7$, $N5$ are high. When E changes to high, $MN11$ turns on, and therefore node $N4$ is discharged through the path of $MN7$–$MN11$. Because $MP11$ is off at this moment, the voltage of $N4$ goes down to low level without the charge stored in the parasitic capacitance $C7$ (which originates from drain capacitance of $MP6$–$MP10$) being discharged. This is fast because the parasitic capacitance $C7$ is not discharged. The equivalent resistance $Z2$ is designed to be high so that the charge stored in $C7$ would not be discharged at this time. Then $MP13$ turns on, and node $N6$ is charged to turn $Q2$ on. This is followed by the discharging of output $N7$. Simultaneously node $N1$ is discharged through the path of $MN1$–$MN6$ to the ground level. $Q1$ stays off. After the delay of the inverters $I2$ and $I1$, $N5$ becomes low. Then the charge stored in $C7$ is discharged. Simultaneously $MN1$ turns off, however, $N1$ was already discharged, therefore node $N1$ stays low. The operation when input E changes from high to low is similar to this procedure.

In this operation, because the discharging of $N4$ is almost free from the large parasitic capacitances $C7$, the delay is small even if the number of inputs is large. The advantage of this feedback circuit is shown in the simulated result shown in Fig. 2. In this simulation the 0.3-μm BiCMOS parameters shown in Table I are used. These parameters are based on our 0.3-μm process having a double-polysilicon self-aligned structure. The figure shows that the effect of the feedback is larger when the number of inputs is large. In the seven-input logic, the delay with feedback is 1.4 times smaller than that without feedback.

Discharging paths $Z1$ and $Z2$ requires careful design. $Z1$ is necessary to avoid $N1$ being a floating node when A–E are all high and the output is low. If the impedance of $Z1$ is too large, the noise coupled into $N1$ could raise the base of $Q1$. On the other hand, if the impedance of $Z1$ is comparable to the impedance of $MN1$, the effect of the feedback circuit is lost. The impedance of $Z1$ is determined considering this trade-off relation.

Good low-voltage performance down to 2.5 V is possible because of the quasi-complementary-type BiCMOS buffer [12]–[14], which avoids the gate–source voltage loss, which has the been the origin of the degraded low-voltage performance in conventional BiCMOS circuits. In the QC buffer a voltage loss occurs only in the drain–source voltage in the pMOS transistor ($MP2$ in Fig. 1 (a)); the resultant drain-current degradation is small. Large current drive is attained even at low supply voltage. The simulated delay dependences show that the FML circuit is faster than the CMOS circuit down to 2 V (Fig. 3), which has been impossible in conventional BiCMOS circuits (including C-BiCMOS). Note that this excellent low-voltage performance is achieved without a p-n-p transistor.

Because this FML can incorporate very complex logic functions in a single gate, the number of transistors, area, and power are drastically reduced. The simple example shown in Fig. 4 illustrates how the large input logic works to reduce the number of transistors. The six-input AND (or six-input NAND in negative logic) can be constructed with two stages of three-input NAND and two-input NOR. However, constructing the same logic in a single six-input gate reduces the number of MOS transistors to almost half and the number of bipolar transistors to about one-third.

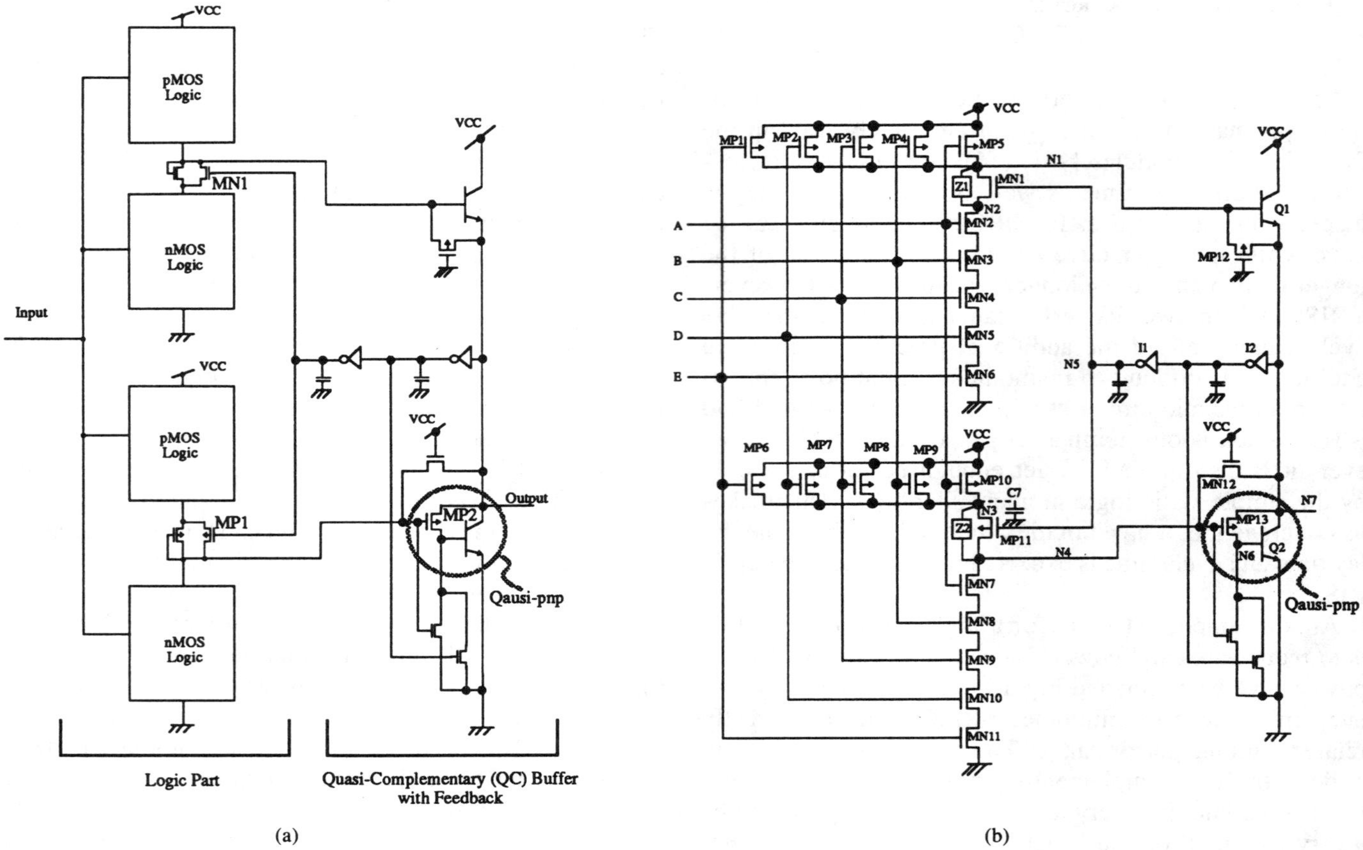

Fig. 1. Circuit diagram of (a) feedbacked massive input logic (FML) and (b) FML five-input NAND logic.

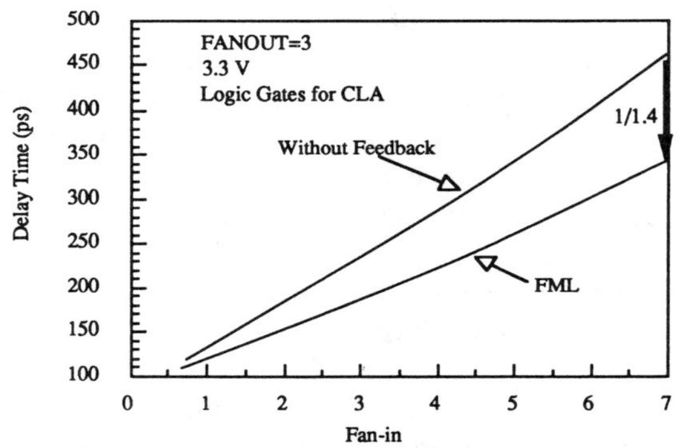

Fig. 2. Simulated dependence of delay time on fan-in.

TABLE I
0.3-μm BiCMOS Device Parameters Used in Circuit Simulations

MOSFET	gate oxide thickness	8 nm
	drain current	2.8 mA (nMOS) 2.0 mA (pMOS)
	drain-substrate capacitance	10 fF
Bipolar Transistor	cutoff frequency	24 GHz
	knee current	5 mA
	junction capacitance	
	C_{TE}	6.5 fF
	C_{TC}	5.0 fF
	base resistance	250 Ω

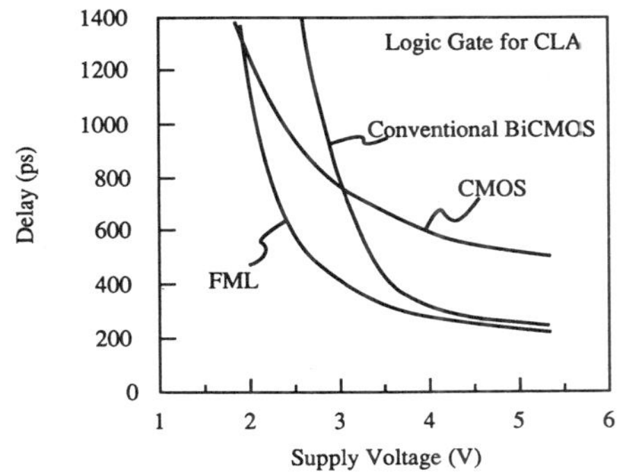

Fig. 3. Simulated dependence of delay time on supply voltage.

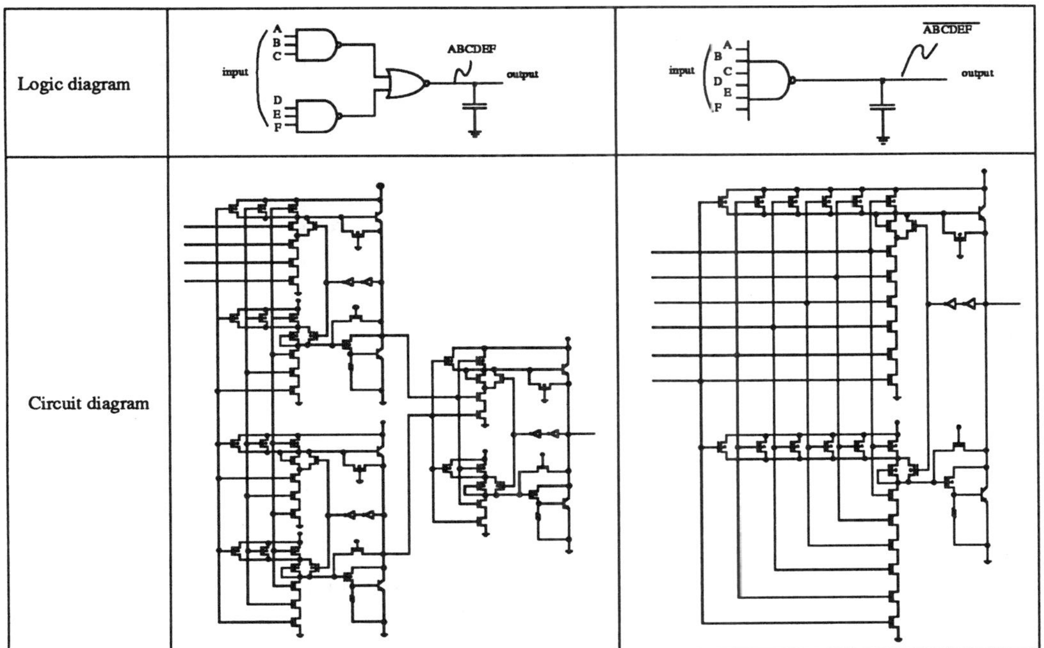

	(a)	(b)
No. MOSFETs	72	37
No. bipolar trs.	6	2

Fig. 4. Two different circuit designs for six-input NAND (AND): (a) two-stage form, and (b) single-stage form.

The effect of reducing the number of transistors by a large fan-in logic is observed in general logics. The effect is evaluated quantitatively using a 16-b carry lookahead unit as a benchmark. The performance of a carry lookahead unit strongly affects the performance of critical paths in ALU's and multipliers. Various 16-b carry lookahead units having different average fan-ins are designed as shown in Fig. 5. The total gate width, transistor count, number of gate stages on the critical path, and total delay are evaluated as shown in Figs. 6 and 7. With increase in average fan-in, the numbers of bipolar transistors and MOSFET's decrease rapidly. For example, if we compare the transistor count in a 7-fan-in design with that of a 3-fan-in design, which is the typical design point for CMOS, the number of bipolar transistors is $1/2.5$ and the number of MOSFET's is $1/2$. The total gate width is also reduced to $1/1.5$. Because the total gate width is a good figure for the power dissipation, the power in the 7-fan-in design is about $1/1.5$ times smaller than that of the three-input logic. Although the feedback circuits in FML increase the area, compared to the simple QC-BiCMOS, reduced area due to the large fan-in design overwhelms the increase. It should be noted that the delay increase of FML is very small even when the average fan-in is large (Fig. 7). By contrast, the delay increases very rapidly in CMOS with an increase in fan-in. This is because with an increase in fan-in, the average load capacitance increases, which severely degrades CMOS speed. The dependence of speed advantage due to the feedback technique in Fig. 7 seems different from that of Fig. 2. This is because the fan-out is fixed in the comparison in Fig. 2, whereas the fan-out is dependent on the average fan-in in Fig. 7. In

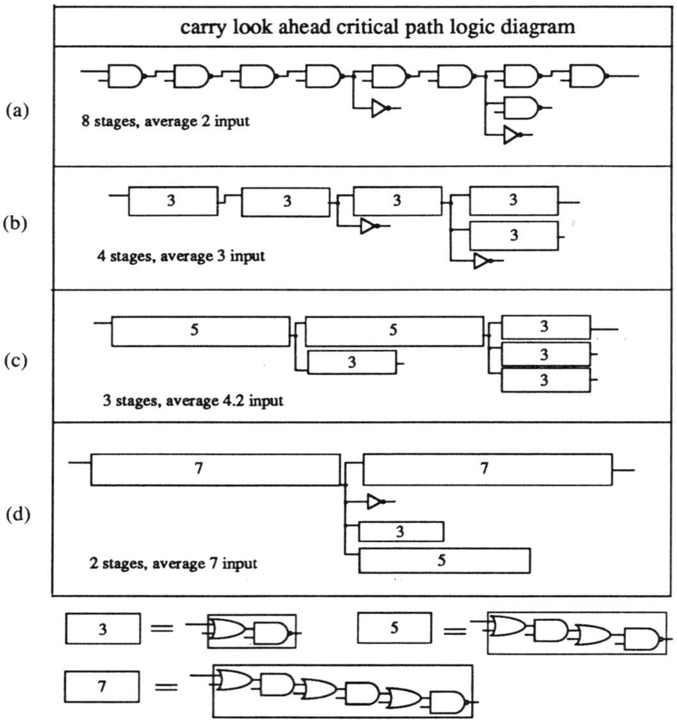

Fig. 5. Critical path logic diagrams for four different design strategies: (a) eight-stage form with average 2 fan-in, (b) four-stage form with average 3 fan-in, (c) three-stage form with average 4.2 fan-in, and (d) two-stage form with average 7 input. The supply voltage is 3.3 V.

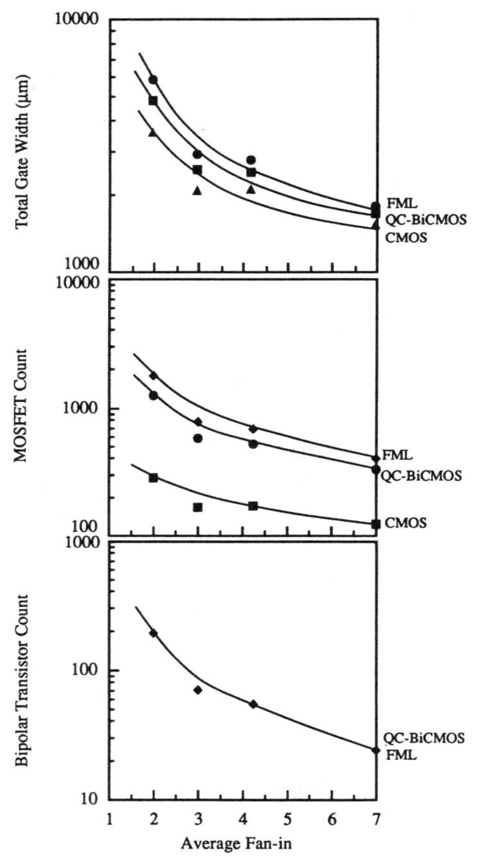

Fig. 6. The dependence of total MOS gate width, MOSFET count, and bipolar transistor count on average fan-in, which corresponds to the various logic configurations in Fig. 5.

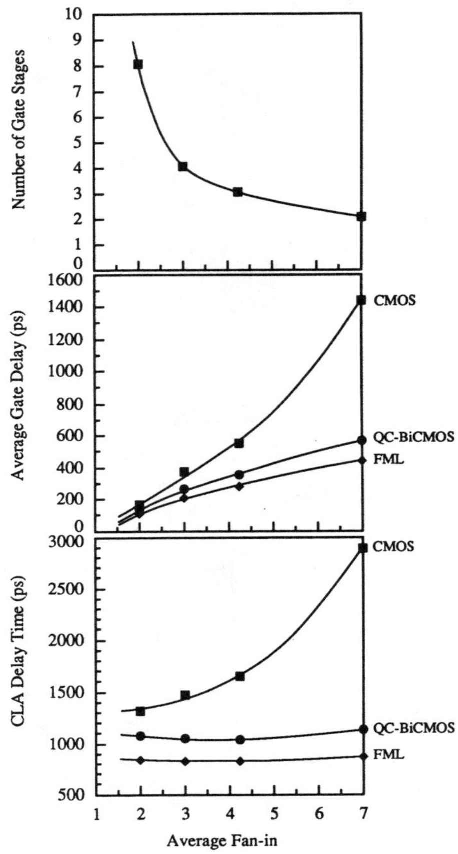

Fig. 7. The dependence of the number of gate stages, average gate delay, and total carry lookahead delay on average fan-in, which corresponds to the various logic configurations in Fig. 5.

general, the advantage of the feedback technique is naturally larger when the fan-out is smaller.

Because the FML circuit operation is static, it is possible to design the cycle time to be the same as the delay time of the critical path. This is beneficial for the high-machine-cycle design of RISC's. The smaller number of transistors due to the FML circuit enables implementing a larger number of processor units in a chip, which again enhances the effective performance of the processor.

Another important issue in high-machine-cycle RISC circuit design is to obtain a fast flip-flop that can be operated within the single-phase clocking environment. In RISC's the pipeline pitch is relatively shorter than those of CISC's. Therefore, the delay of a flip-flop easily contributes to large overhead. This tendency is emphasized when the superpipeline technique or vector execution units [17] are introduced in a microprocessor. When the machine cycle is as short as 4 ns (250 MHz), the conventional two-phase nonoverlapping clocking scheme is difficult because of the additional margin needed to avoid overlapping between different phases. A single-phase clocking scheme is desirable in this respect.

A conventional BiCMOS flip-flop [18], which improves the current drive capability of a CMOS flip-flop by adding a BiCMOS output buffer, is shown in Fig. 8. However, in this conventional circuit, the master latch is pure

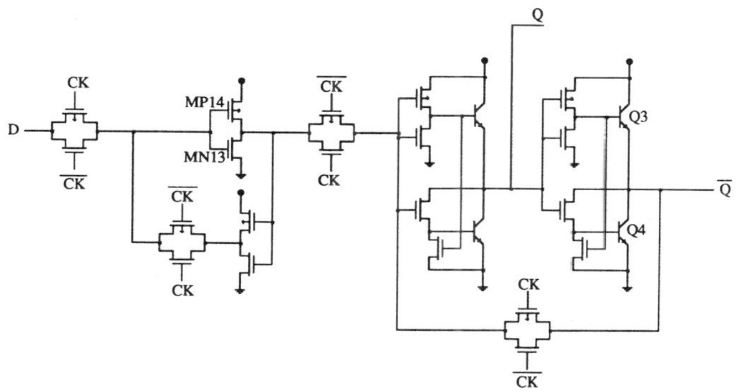

Fig. 8. Circuit diagram of the conventional BiCMOS D-flip-flop [18].

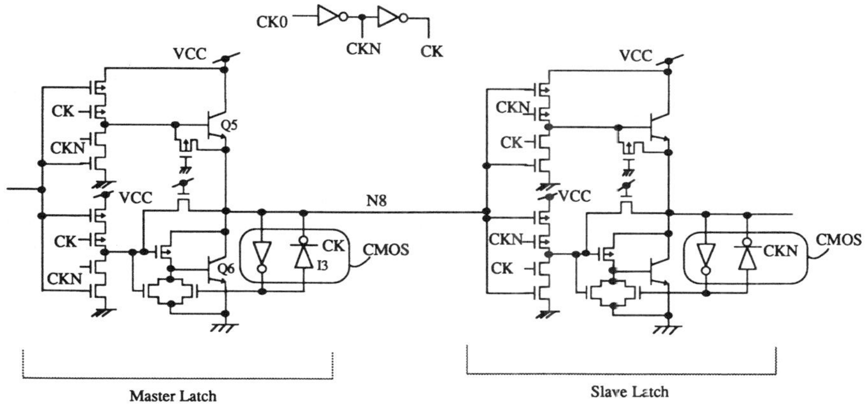

Fig. 9. Circuit diagrams of the new QC-BiCMOS D-flip-flop.

CMOS, which has no performance improvement due to BiCMOS. In addition, when there is a clock skew between CK and $\overline{CK}$, the output driven by the bipolar transistors $Q3$ and $Q4$ is to be inverted by the MOSFET's $MN13$ and $MN14$. When this conflict occurs the operating speed is very low.

Our new D-type flip-flop, which can be used in a single clocking scheme, is shown in Fig. 9. This is based on the concept that the data are retained by the CMOS latch, whereas the critical path in this flip-flop is exclusively constructed from two-stage quasi-complementary BiCMOS. This enables excellent low-voltage operation. Even if there is a clock skew between CK and CKN, the output node is driven quickly by the QC buffer without being affected by the reluctant CMOS latch. The CMOS clocked inverter $I3$ retains the voltage level of $N8$. The superior current drive capability of $Q5$ (or $Q6$) is insensitive to the retaining current drive of $I3$. Input data are transferred quickly without interference from the retaining current. Conventionally the BiCMOS circuit is believed to have a performance advantage over the CMOS only when the load capacitance is large. Although the load capacitance of node $N8$ is small (about 30 fF) in this flip-flop, the performance advantage of BiCMOS over CMOS is relatively large.

The performance of this D-flip-flop is evaluated by circuit simulations as shown in Fig. 10. The total delay (0.31 ns) of the BiCMOS flip-flop required to transmit the data from the input to the output is almost half that of the CMOS even at 2.5 V as shown in Fig. 10. The master latch, which has small load capacitance, has a factor-of-1.4 performance leverage over its CMOS counterpart because of the above-discussed advantage of BiCMOS drivers.

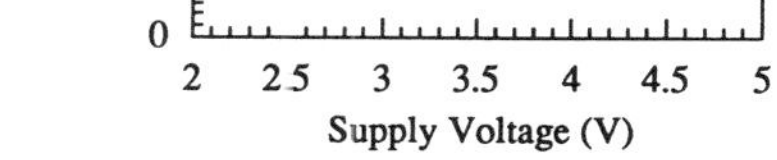

Fig. 10. The dependence of the total delay and write delay of the QC-BiCMOS flip-flop on the supply voltage. "CMOS" represents the circuit, in which the BiCMOS part is replaced by its CMOS counterpart. The load capacitance is 5 fan-out.

III. Design of 3.3-V 32-b ALU

A fully static 32-b ALU employing the above circuits is designed using 0.3-μm BiCMOS rule. The conditional

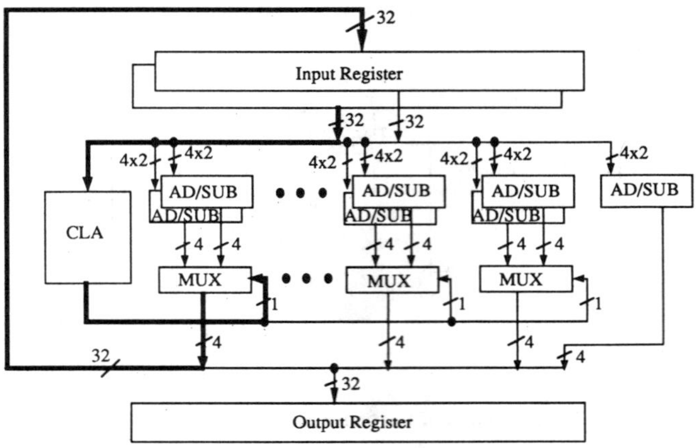

Fig. 11. Block diagram of the designed 32-b ALU. CLA: carry lookahead; AD/SUB: adder/subtractor; and MUX: multiplexer.

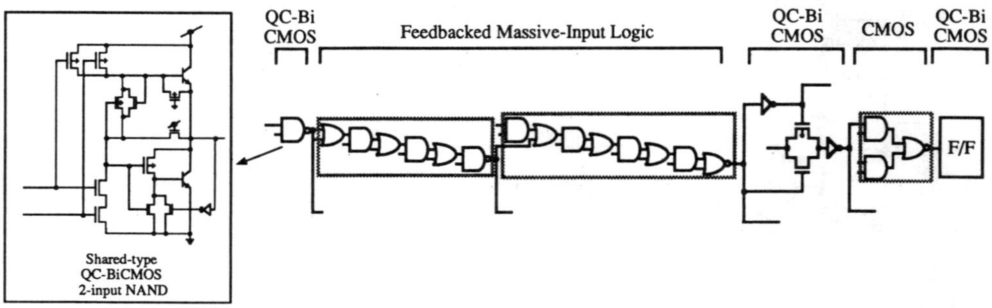

Fig. 12. The critical path of the 32-b ALU.

sum architecture, which selects the two 4-b adder results according to the carry signal generated by the carry lookahead unit, is used as shown in Fig. 11. The critical path consists of only six-state gates, which include QC-BiCMOS gates, FML gates (maximum nine-input gate is used), CMOS multiplexers, and QC-BiCMOS D-type flip-flops as shown in Fig. 12. Because 64 two-input NAND circuits are required in first stage of the carry lookahead unit, the shared QC-BiCMOS circuit discussed in [13] and [14] is used to reduce the area. In the following two-stage large-input carry lookahead logics, the FML technique is fully employed to reduce delay, area, and power. The features of the designed ALU are shown in Table II. Circuit simulation based on our 0.3-μm BiCMOS parameters shows that the adder delay of 1.2 ns is obtained at 3.3-V supply. The power consumption is as small as 39 mW at a frequency of 250 MHz (20% gate activity is assumed). These results are sufficient for 250-MHz RISC operation.

The ALU is fabricated using the self-aligned double-polysilicon 0.3-μm BiCMOS technology. The gate length is 0.3 μm, the gate-oxide thickness is 9 nm, the threshold voltage is 0.3 V, and the cutoff frequency of the n-p-n bipolar transistor is 20 GHz. The fine patterns, such as gate, contact, and first metal, are defined by electron-beam lithography, and the other patterns are defined by the i-line optical lithography. The interconnection consists of first-level tungsten, second-level aluminum, and third-level aluminum. More details of our device/process technology will be described in an another paper. A microphotograph of the fabricated ALU is shown in Fig. 13. The total number of transistors is 13 800, which includes 1000 bipolar transistors. Therefore the percentage of the number of bipolar transistors is only 7%, although almost the entire critical path consists of BiCMOS gates. The delays from the input signal (it is actually a regenerated signal by an on-chip inverter circuit) to the output signal are measured on the wafer level using the on-chip source-follower circuits. Excellent low-voltage performance with a good agreement with simulation is demonstrated as shown in

TABLE II
32-b ALU Design Summary

Architecture	32 b, Conditional Sum
Technology	0.3-μm Double-Poly Si BiCMOS, T_{ox} = 9 nm
Interconnection	Three-Layer Metal (W/Al/Al)
Area	0.6 × 2.8 mm^2
Delay Time	1.2 ns (ADD), 0.31 ns (Register) at 3.3 V
Power	39 mW (250 MHz, at 3.3 V)

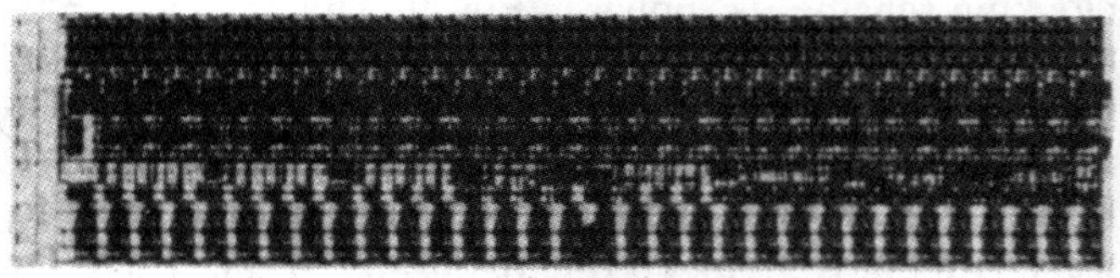

Fig. 13. Microphotograph of the fabricated ALU.

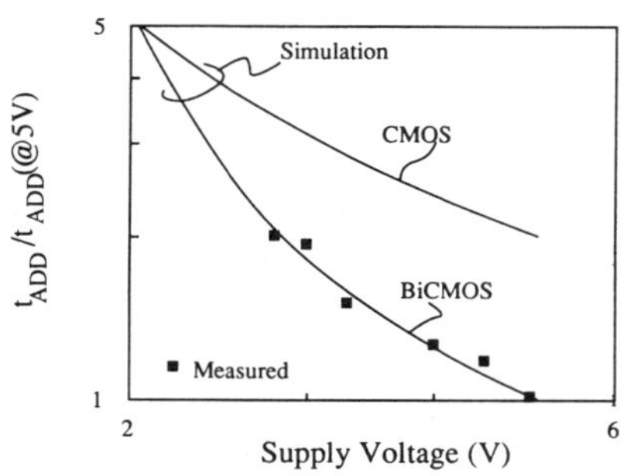

Fig. 14. The dependence of the normalized adder delay time on supply voltage.

Fig. 14. The 1.6-times performance leverage in this actual ALU over CMOS is confirmed, which opens the door to 0.3-μm BiCMOS RISC's and related ASIC's.

IV. Conclusions

Low-voltage high-machine-cycle BiCMOS circuit techniques for deep-submicrometer BiCMOS RISC processors were discussed. The feedbacked massive-input logic (FML) concept reduces the number of transistors and the power within the framework of fully static logic 2–3 times. A quasi-complementary buffer circuit provides unprecedentedly excellent low-supply-voltage performance even below 3 V. Sequential circuits are also able to operate at low voltages with good performance leverage over CMOS due to the novel QC-BiCMOS flip-flop. These demonstrate that the BiCMOS continues to have superior performance over CMOS in the sub-0.3-μm sub-3-V region, which promises 250-MHz-level parallel processing RISC's and ASIC's.

Acknowledgments

The authors wish to thank Dr. K. Shimohigashi, Dr. T. Masuhara, Dr. H. Sunami, Dr. M. Hirao, Dr. E. Takeda, and Dr. T. Nishimukai for their useful discussions and encouragement. The authors wish to express appreciation to T. Hayashi, T. Doi, and O. Nishii for their useful suggestions and comments for VLSI processor design. The authors are greatly indebted to Dr. Y. Kawamoto, Dr. T. Kure, Dr. F. Murai, and the processing staff members for their chip fabrication. The authors are also grateful to Dr. H. Higuchi and S. Tachibana for their valuable discussions and suggestions on BiCMOS circuit design.

References

[1] M. Suzuki *et al.*, "A 3.5-ns, 500-mW, 16-kbit BiCMOS ECL RAM," *IEEE J. Solid-State Circuits*, vol. 24, no. 5, pp. 1233–1237, Oct. 1989.
[2] G. Kitsukawa *et al.*, "A 23-ns 1-Mb BiCMOS DRAM," *IEEE J. Solid-State Circuits*, vol. 25, no. 5, pp. 1102–1111, Oct. 1990.
[3] T. Hotta *et al.*, "A 70MHz 32b microprocessor with 1.0 μm BiCMOS macrocell library," in *ISSCC Dig. Tech. Papers*, 1989, pp. 124–125.
[4] M. Yamashita *et al.*, "200-MHz 16-bit BiCMOS signal processor," in *ISSCC Dig. Tech. Papers*, 1989, 12.8.
[5] E. W. Brown *et al.*, "Implementing SPARC in ECL," *IEEE Micro*, vol. 10, pp. 10–22, Feb. 1990.
[6] D. A. Whitmire, V. Garcia, and S. Evens, "A 32b GaAs RISC microprocessor," in *ISSCC Dig. Tech. Papers*, 1988, p. 34.
[7] H. Momose, K. M. Cham, C. I. Drowley, H. R. Grinolds, and H. S. Fu, "0.5 micron BiCMOS Technology," in *IEDM Tech. Dig.*, 1987, pp. 838–840.
[8] M. Fujishima, K. Asada, and T. Sugano, "Evaluation of delay-time degradation of low-voltage BiCMOS based on a novel analytical delay-time modeling," *IEEE J. Solid-State Circuits*, vol. 26, no. 1, pp. 25–31, Jan. 1991.
[9] A. Watanabe, T. Nagano, S. Shukuri, and T. Ikeda, "Future BiCMOS technology for scaled supply voltage," in *IEDM Tech. Dig.*, 1989, pp. 429–433.
[10] H. J. Shin *et al.*, "Full-swing complementary BiCMOS logic circuits," in *Proc. 1989 Bipolar Circuit Technology Meeting*, 1989, pp. 229–233.
[11] G. Kitsukawa *et al.*, "Low power, high-speed BiCMOS memory circuits," in *Extended Abs. 1984 Int. Conf. Solid State Devices Mater.* (Kobe), 1984, pp. 233–236
[12] P. Raje, R. Ritts, K. Cham, J. Plummer, and K. Saraswat, "Merged BiCMOS: A device and circuit technique scalable to the sub-micron sub-2V regime," in *ISSCC Dig. Tech. Papers*, 1989, TPM 9.2.
[13] K. Yano *et al.*, "Quasi-complementary BiCMOS for sub-3-V digital circuits," in *Tech. Dig. 1991 VLSI Symp. Technology.*
[14] K. Yano *et al.*, Quasi-complementary BiCMOS for sub-3-V digital circuits," *IEEE J. Solid State Circuits*, vol. 26, pp. 1708–1719, Nov. 1991.
[15] K. Makino, "High speed circuit technology for mainframe VLSI," in *Proc. 1987 VLSI Circuit Symp.*, 1987, pp. 93–94.
[16] T. Hayashi *et al.*, "The SDC cell—A novel design methodology for high-speed arithmetic modules using CMOS/BiCMOS precharged circuit," *IEEE J. Solid-State Circuit*, vol. 25, no. 2, p. 403, Apr. 1990.
[17] "INTEL 386TM32 bit architecture," Intel Japan, Apr. 11, 1989.
[18] S. C. Lee, D. W. Schucker, and P. T. Hickman, "BiCMOS technology for high-performance VLSI circuits," *VLSI Design*, pp. 98–100, Aug. 1984.

TA 6.3: A Three-Million-Transistor Microprocessor

Fuad Abu-Nofal, Rick Avra, Kanti Bhabuthmal, Rom Bhamidipaty, Greg Blanck, Andy Charnas, Peter DelVecchio, Joe Grass, Joel Grinberg, Norm Hayes, George Haber, Jim Hunt, Govind Kizhepat, Adam Malamy, Al Marston, Kaushal Mehta, Sunil Nanda, Hoa Van Nguyen, Rajiv Patel, Andy Ray, Jim Reaves, Alan Rogers, Stefan Rusu, Tom Shay, Irwan Sidharta, Terry Tham, Peter Tong, Richard Trauben, Anthony Wong, David Yee, Naeem Maan*, Don Steiss*, Lynn Youngs*

Sun Microsystems, Mountain View, CA / *Texas Instruments

A RISC BiCMOS superscalar microprocessor containing 3.1M transistors executes up to three instructions per clock cycle. Clock frequency is 40MHz with 8mW dissipation. The chip includes a 32b integer pipeline (IU), memory management unit (MMU), 20kB instruction cache (I cache), 16kB data cache (D cache), IEEE-compatible double-precision floating-point unit (FU), and bus interface (BU). The chip supports built-in self-test, internal and JTAG boundary scan, in-circuit emulation for remote symbolic source-code debugging, and hardware to support multiprocessor system architecture. An on-chip phase-locked loop synchronizes external and internal clock edges. Figure 1 shows the chip implemented in a 0.8μm triple-layer-metal salacided BiCMOS process characterized in Table 1. The die is 15.98x15.98mm^2, with 166 active pins and 127 power/ground pins. Table 2 summarizes transistor count and area of major functions. Figure 2 is a block diagram.

Instruction processing consists of an 8-stage pipeline that executes in 4 clock cycles. There are two instruction-fetch stages (F0, F1), three decode stages (D0, D1, D2), two instruction-execute stages (E0, E1), and one result write-back stage (WB). Each stage executes in one clock phase. Instruction issue and execution is strictly in-order with precise exceptions. Figure 3 shows the tasks accomplished by pipeline stages. The I cache is 20kB 5-way set-associative, with 64B line size. It has 4 redundant rows and 16 redundant columns, that are fuse programmable. In each F0 phase, the low-order 12b of virtual address are presented to the cache and data is read out. Simultaneously, the MMU translates the upper 20b of the virtual address to a corresponding 24b physical address. In F1, a tag match operation is done by the I cache, and, on a cache hit, the correct data set is selected from the cache and presented to the IU. To execute up to 3 instructions/cycle, the data path from I cache to IU is 4 words (128b) wide.

The D cache is 16kB 4-way set-associative. It contains 4 redundant rows and 16 redundant columns, that are fuse-programmable. The D cache can be configured to operate in either write-back or write-through modes. 11ns D cache access time allows read data, from load instructions that hit in the D cache, to be used in the following clock cycle (no load/use penalty). The D cache supports single-cycle 64b loads and stores. Both I and D caches are physically-addressed. This makes multiprocessor cache coherency easier to achieve, but the MMU must translate virtual to phyiscal addresses before cache set selection and alignment. The 64-entry fully-associative translation look-aside buffer (TLB) in the MMU supports both data and instruction caches, evaluating twice in each clock cycle using a static bipolar sense-amp with 2.0ns delay from word line high to sense amp output. A 36b match circuit in Figure 4 does a 1.1ns distributed-NOR of match on each bit.

The IU contains a sequential prefetch queue (8 instructions) and a target prefetch queue (4 instructions). Either of the two queues receives up to 4 instructions from the I cache. In the D0 stage, instructions are issued out of the queue in single-, double- or triple-instruction groups. The IU contains a 136 word, 32b, 4-port register file, that does 2 reads and 2 writes, or 4 reads in a single clock phase. This reglster file is time-shared to function as a 6-read, 2-write port memory each clock cycle. To increase reliability for fast accesses, the register file uses standard differential read and write ports. By using metal2 word lines and metal3 bit lines, area is kept to about 1.8% of the chip. To minimize noise coupling while maintaining access speed, a bipolar sense amp/bit-line clamp circuit limits the bit line swing on read to 100mV. This is shown in Figure 5. The npn bitline clamp speeds write recovery, with minimal area and bit line load. Access time from word line high to sense amp output is 3.0ns.

The IU also contains three fully-static 32b ALUs, one 32b adder and one shifter. Each evaluates in one clock phase. The adder evaluates effective addresses for a load or store instruction in the D2 pipe stage. The three ALUs are arranged to allow single-cycle execution of either two independent ALU operations, or one independent plus one dependent ALU operation to execute per clock cycle. Flexible superscalar operations such as one-branch instruction and any two ALU instructions execute in one clock cycle even if the second ALU instruction uses the result of the first as an operand. The evaluation of condition codes is speeded by a dedicated carry chain. The ALUs evaluate in 5ns from data-in to sum-out. The carry-out of the dedicated carry chain is available 4ns after data-in. This allows a compare instruction and a conditional branch in the same cycle. For an untaken branch, the following cycle can execute up to three instructions. For a taken branch, the following cycle executes only the "RISC delay instruction."

The FU contains a 54b static adder and two 32b multipliers that can be used together for double-precision operations. Floating divide and square root and integer multlply and divide are supported by the FU. The execution of floating-point adds can be concurrent with floating-point multiplies. To decouple the floating-point operations from the integer pipeline, the FU contains a 5-port register file with 32 32b entries, and a four-instruction-deep queue. The FU conforms to the IEEE standard 754/1985 spec. Floating-point instructions can be issued to the FU from the IU every cycle. Results are stored into the D cache on a 64b bus. Floating-point multiplies, either single- or double-precision, can be issued on consecutive clock cycles and have three-cycle latency. The FU can accept one floating-point arithmetic instruction together with one floating-point load or store instruction in the same cycle. Figure 6 shows close to peak throughput in a typical floating-point application.

The bus unit contains an 8-entry (64b wide) buffer for all the stores in write-through mode and for write-back buffer. A 32B prefetch buffer detects data misses to consecutive cache blocks and prefetches the next block. The bus unit supports a demuxed pipelined private bus for use with a 1MB external cache and a companion cache controller chip, and a multiplexed, standard, 64b multiprocessor bus. A strap input pin selects the bus configuration at power up. In the first configuration, this chip is part of a 10-chip processor module with eight 128kx9 SRAM chips. These SRAMS are fully-pipelined (addresses, data, and control) resulting in a three-cycle access, but allowing cache block transfers at the clock rate. Byte data parity is supported. In the second configuration, this chip alone implements a complete processor module. Multiprocessor systems are implemented by busing pins together. The companion cache-controller chip in the 10-chip processor module also supports the multiplexed standard multiprocessor bus permitting 10-chip modules and 1-chip modules to have the same electrical and mechanical interface for module-swap upgradability.

Reprinted from *1992 IEEE ISSCC*, pp. 108–109 and 257, Feb. 1992.

Tox	19.5nm
Lgate	0.8μm
Ml	0.8μm width/1.2μm space
M2	0.8μm width/1.2μm space
M3	1.0μm width/1.6μm space
Contact	0.8x0.8μm²
Via1	0.8x0.8μm²
Via2	1.0x1.0μm²
f_T	7.5GHz
h_{fe}	200

Table 1 Process Summary

Block	#MOS	#npn	%area
I$	1200k	2.5k	13
D$	1100k	3.7k	14
FU	323k	3.0k	18
IU	232k	4.6k	15
MMU	118k	2.6k	8
Pads	81k	4.0k	18
EU	50k	0.8k	5
Global wiring	-	-	9
TOTAL	3104k	21k	100

Table 2: Transistor count and area summary.

Figure 1: See page 232.

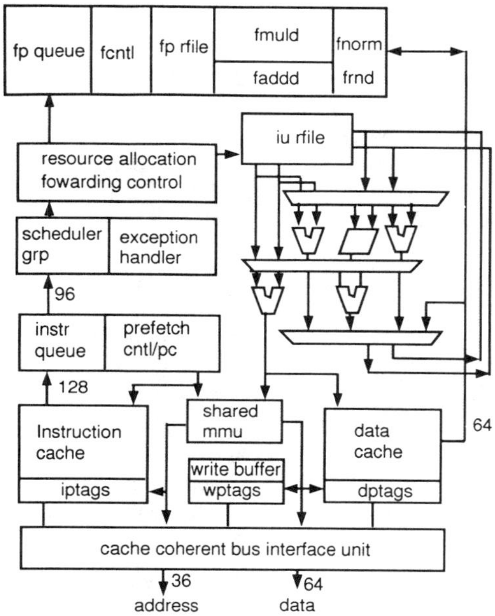

Figure 2: Functional block diagram.

Figure 3: See page 232.

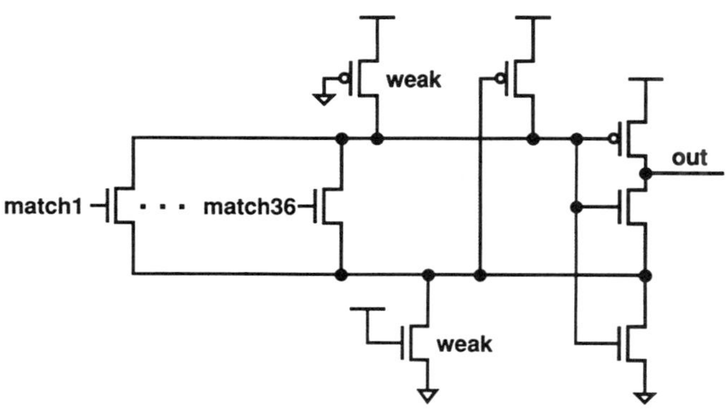

Figure 4: TBL distributed NOR.

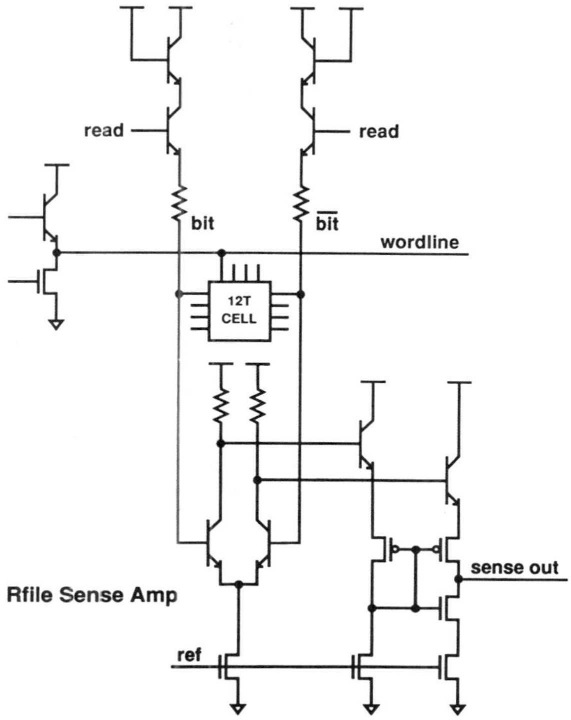

Figure 5: Rfile sense amplifier.

```
ldd    [%l0], %f2
faddd %f2, %f0, %f6
add    %l0,0x8,%l0

---  Break  (max 3 instr.)  ---

ldd    [%l0], %f4
add    %l0,0x4,%l0
fmuld %f4,%f0,%f8

---  Break  (max 3 instr.)  ---

ldd    [%l0+4], %f10
cmp   %l0,0x100
be      loop

---  Break  (max 3 instr. , branch)  ---
```

Figure 6: Super scalar example.

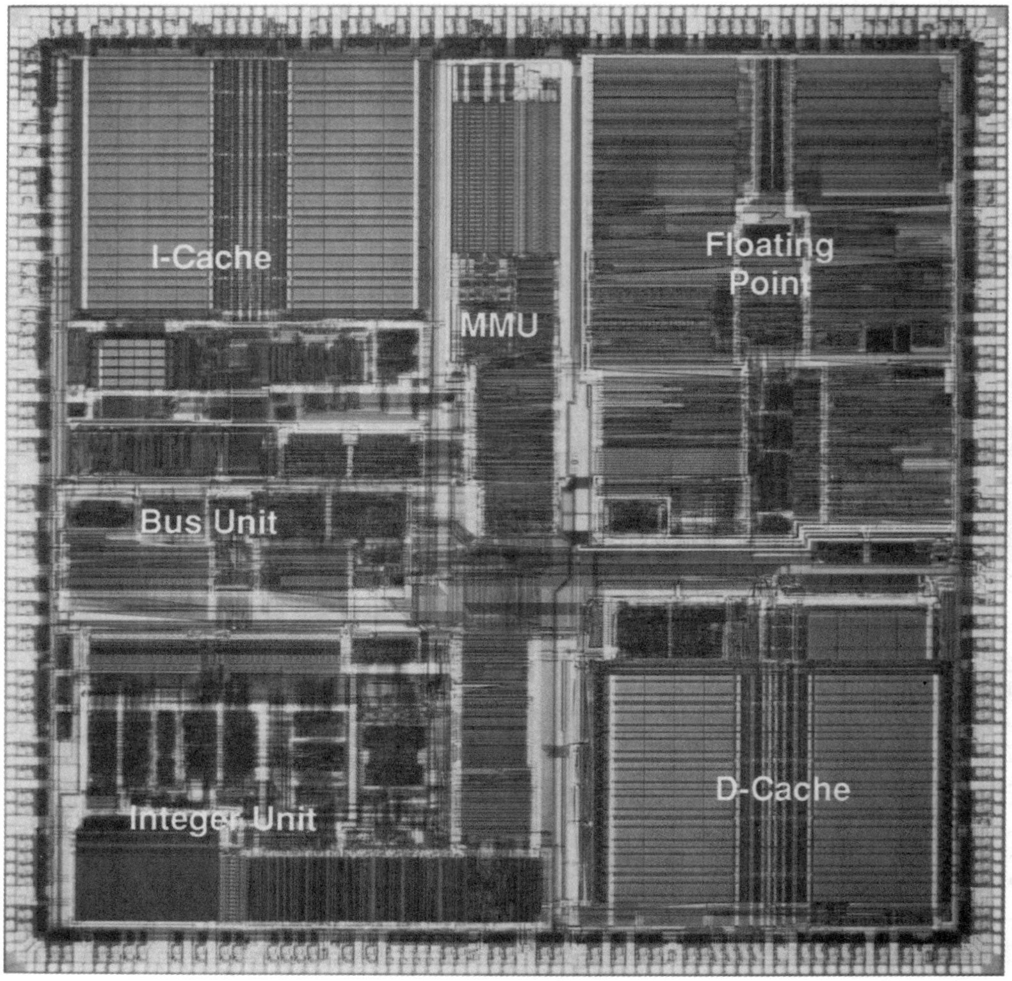

Figure 1: 3M-transistor microprocessor chip.

F0	I cache RAM access
	I cache TLB lookup
F1	I cache match detect
	4-instructions sent to I queue
D0	Issue 1, 2, or 3 instructions
	Select R file indices for LD/ST address registers
D1	Read R file for LD/ST address registers
	Resource allocation for ALU instructions
	Evaluate branch target address
D2	Read R file for ALU operands
	Add LD/ST address registers to form effective address
E0	First stage of ALU
	D cache RAM access
	D cache TBL lookup
	FP instruction dispatch
E1	Second stage of ALU
	D cache match detect
	Load data available
	Resolve exceptions
WB	Writeback into the R file
	Retire store into the store buffer

Figure 3: Pipeline stages.

TA6.4: A BiCMOS 50MHz Cache Controller for a Superscalar Microprocessor

Balakrishna Joshi, R. K. Anand, Curt Berg, Jorge Cruz-Rios, Ashok Krishnamurthi, Nyles Nettleton, Sophie Nguyen, Jim Reaves, John Reed, Alan Rogers, Stefan Rusu, Chuck Tucker, Chung Wang, Ming Wong, David Yee, Jung-Herng Chang

Sun Microsystems Computer Corporation, Mountain View, CA

A BiCMOS 50MHz, 2.2M-transistor cache controller (CC) chip supports up to 2MB of direct-mapped secondary cache for a superscalar microprocessor chip (PU) and interfaces with two multi-processor (MP) buses. One is the MBus, a circuit-switched MP system bus operating at TTL levels [1]. The other is the XBus, a local packet-switched bus operating at either TTL or Gunning-transceiver logic (GTL) levels. In XBus mode, the CC connects to MP buses through bus-watcher chips, up to 4 of which can be connected to the CC to support 4 MP buses. With the XBus interface, the CC can support customized MP buses.

The CC simultaneously handles multiple operations, e.g., one read miss, one write miss, and up to 30 pending writes. It also has special hardware support for block copy and block zero. The CC isolates the processor chip and second-level cache from the rest of the system, allowing faster processor clock operation (e.g. 50MHz) with a slower system clock (e.g. 40MHz). The asynchronous interface between the two clock domains is achieved through multiple FIFOs and synchronizers.

The synchronizer optimizes the C/gm ratio, where C is the capacitance of the metastable node, and gm is the transconductance of the latch inverter. A T_{switch} (synchronizer time constant) of 200ps and a chip mean time between failure (MTBF) of 1 billion years is achieved at 50MHz operation [2]. The chip clock is divided by two, generating master and slave clocks to increase synchronizer settling time. Two synchronizers are used in parallel for each signal. The divided clocks to the two synchronizers in parallel are shifted with respect to each other by a clock cycle to sample incoming signals every cycle. Figures 1a, 1b, and 1c show the synchronized circuit, the synchronizer T_{switch} characteristics, and clocking.

The chip uses dual mode (GTL/TTL) input and output buffers. GTL mode has 1.2V swing to allow fast bus transition [3]. The GTL/TTL input buffer comprises two differential pairs (one for GTL, the other for TTL) capable of large swing outputs ($V_{oh}=V_{CC}-V_{be}$, $V_{ol}=V_{be}$). Their outputs are multiplexed together and drive an emitter follower that delivers the buffered signal. In each mode of operation, the current to the differential pair which is not in use is turned off, avoiding unnecessary power consumption. The GTL/TTL output buffer consists of two drivers (one for GTL, the other for TTL) joined together at their output nodes. In each mode of operation, the buffer not in use is tri-stated. The TTL output driver has conventional configuration, with an npn Darlington pair as the pull-up driver and a nMOS-npn as the pulldown driver. The TTL V_{ol} level is achieved with an additional nMOS pulldown device. Both TTL pullup and pulldown driver are driven by RC networks to control the output slew rate. The GTL output driver is an open-drain buffer capable of sinking 32mA. Its output nMOS device is driven by an RC network to reduce output ringing. In an XBus-based system, a voltage-level sensor is used on the GTL reference voltage supply pin (GTLREF) to differentiate between GTL or TTL mode of operation for system interface transceivers. When GTLREF = 0.8V, the CC is in GTL mode; when GTLREF = 5V, it is in TTL mode.

To minimize clock skew between chips and to reduce impact on system cycle times, the CC is fitted with two phase-locked loops (PLLs), one per clock domain. These PLLs create zero-delay clock buffers for each clock domain within the CC. The PLL oscillator runs at twice the clock speed and is divided by two before distribution around the chip. The PLL is of common construction, with a digital phase and frequency detector (PFD), push-pull charge pump, external capacitor, and five-stage voltage-controlled ring oscillator (VCO). Each of these sections is modified for the BiCMOS process. Precision base resistors moderate the charge-pump and VCO gain, reducing process spread. Also, fast BiCMOS gates in the PFD minimize false output glitches and reset periods. Figure 2 shows characterization results for locking voltage of external capacitor versus the PLL input frequency. The PLL locks up to 90MHz. The two clock domains in the CC can operate either synchronously or asynchronously, making it necessary to control clock skews to within 200ps across clock domains using balanced H-trees. Further, by balancing the RC trees and the topology of clock networks, a maximum clock skew of 250ps is achieved between clock domains.

The 33x8k internal cache tag RAM is synchronous, with 5.0ns column access and 7.8ns row access. The tag RAM is made up of 1.7M transistors. Column redundancy and row redundancy are designed for increased yield in the tag RAM arrays. On column redundancy repair, a pair of faulty adjacent columns are switched with good columns as shown in Figure 3. This allows adjacent column short failures to be repaired by a single repair. The column repair mechanism involves blowing the associated fuses for the bank/block containing the faulty column, and shifting the appropriate number of bits into the column-redundancy-control shift register to "bank switch" all columns to the left of the faulty column, or pair of faulty adjacent columns. On power-up, block shift registers are reset, and redundancy repair data is shifted into block shift registers during CC initialization.

The BiCMOS tag comparator circuit of Figure 4 compares the 17b cache tag to an incoming address in 1.5ns. It uses 17 XOR gates feeding a wired-OR emitter follower. The nMOS NOR structure varies the impedance of the emitter follower pull-downs. The highly-capacitive ORing node (with 17 nMOS pull down transistors) is isolated from the match output and swings less than 1 V_{be}. The bias level at the base of the pull-down npn is set by controlling the impedance of the feedback nMOS with a separate bias generator circuit. This equalizes rise and fall times on match output for fast compare.

The CC supports both JTAG/SCAN and BIST. The JTAG implementation is based on the IEEE P11149.1 boundary scan architecture [4]. The BIST controller uses LFSRs to generate input patterns for an internal scan chain and to compress output vectors into the signature. The scan chain has about 3000 flip-flops. A 17b pattern-generation LSFR and a 31b signature-generation LFSR are used.

The chip is fabricated on a 166mm^2 die and packaged in a 369 pin ceramic PGA dissipating 5.5W. The die has 420 pads with isolated transceiver power supply pads. Power isolation is maintained in the die package. Figure 5 is a die micrograph.

References

[1] "SPARC^{tm}" MBus Interface Specification", Rev. 1.2, Sun Microsystems.

[2] A. Glasser and D. Dobberpuhl, "The Design and Analysis of VLSI Circuits", Addison Wesley, 1985.

[3] B. Gunning et al., "A CMOS Low-Voltage-Swing Transmission-Line Teansceiver", ISSCC DIGEST OF TECHNICAL PAPERS, pp.58-9, Feb. 1991.

[4] "IEEE Standard Test Access Port and Boundary-Scan Architecture, P1149.1", IEEE Computer Society Test Technology Technical Committee, New York, Jan. 1989.

Reprinted from *1992 IEEE ISSCC*, pp. 110–111 and 258, Feb. 1992.

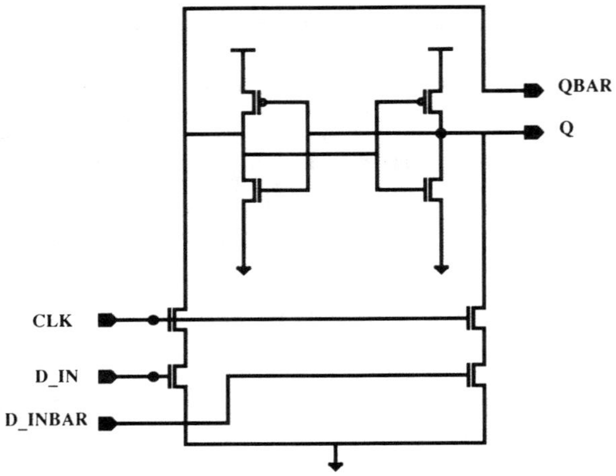

Figure 1a: Synchronizer cell.

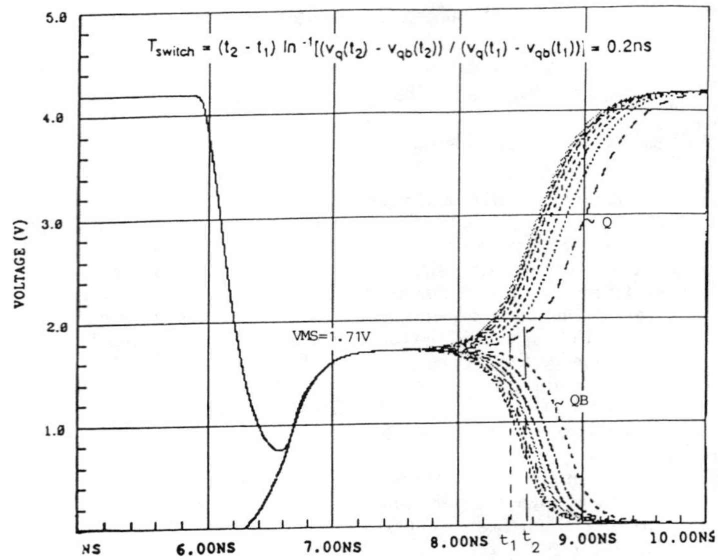

Figure 1b: T_{switch} characterization curves.

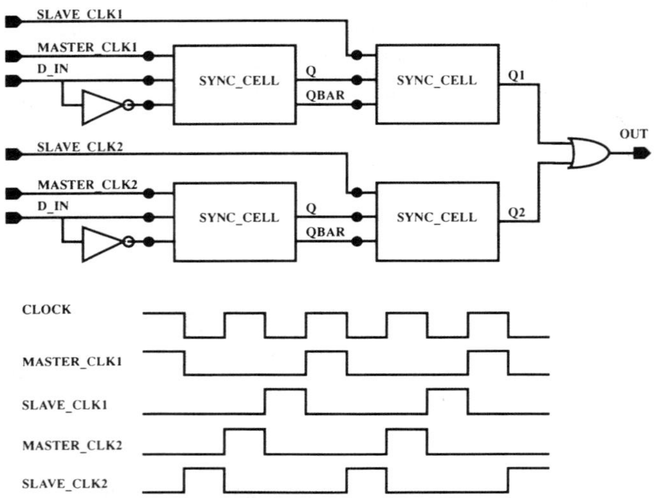

Figure 1c: Parallel synchronizers scheme and clock waveforms.

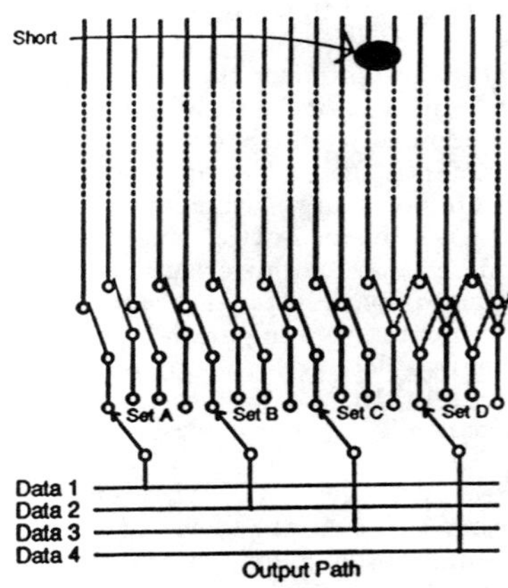

Figure 3: Column redundancy scheme.

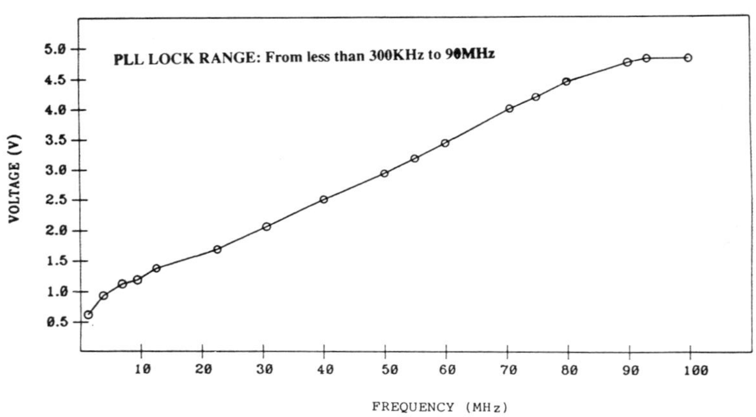

Figure 2: VCO output voltage vs. input frequency.

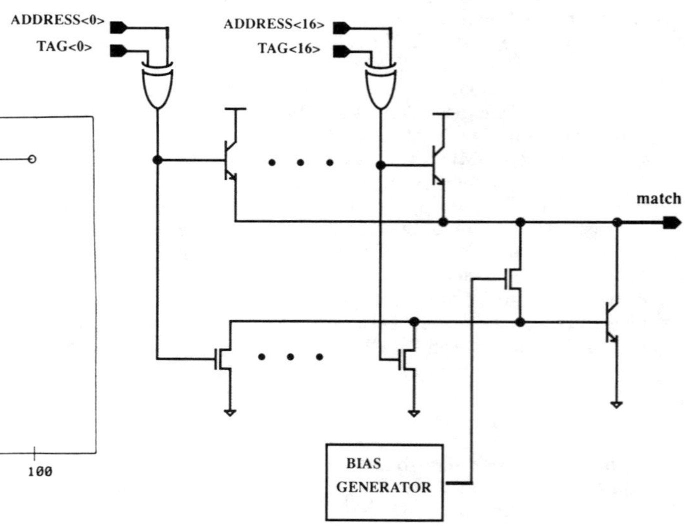

Figure 4: BiCMOS tag comparator circuitry.

Figure 5: See page 235.

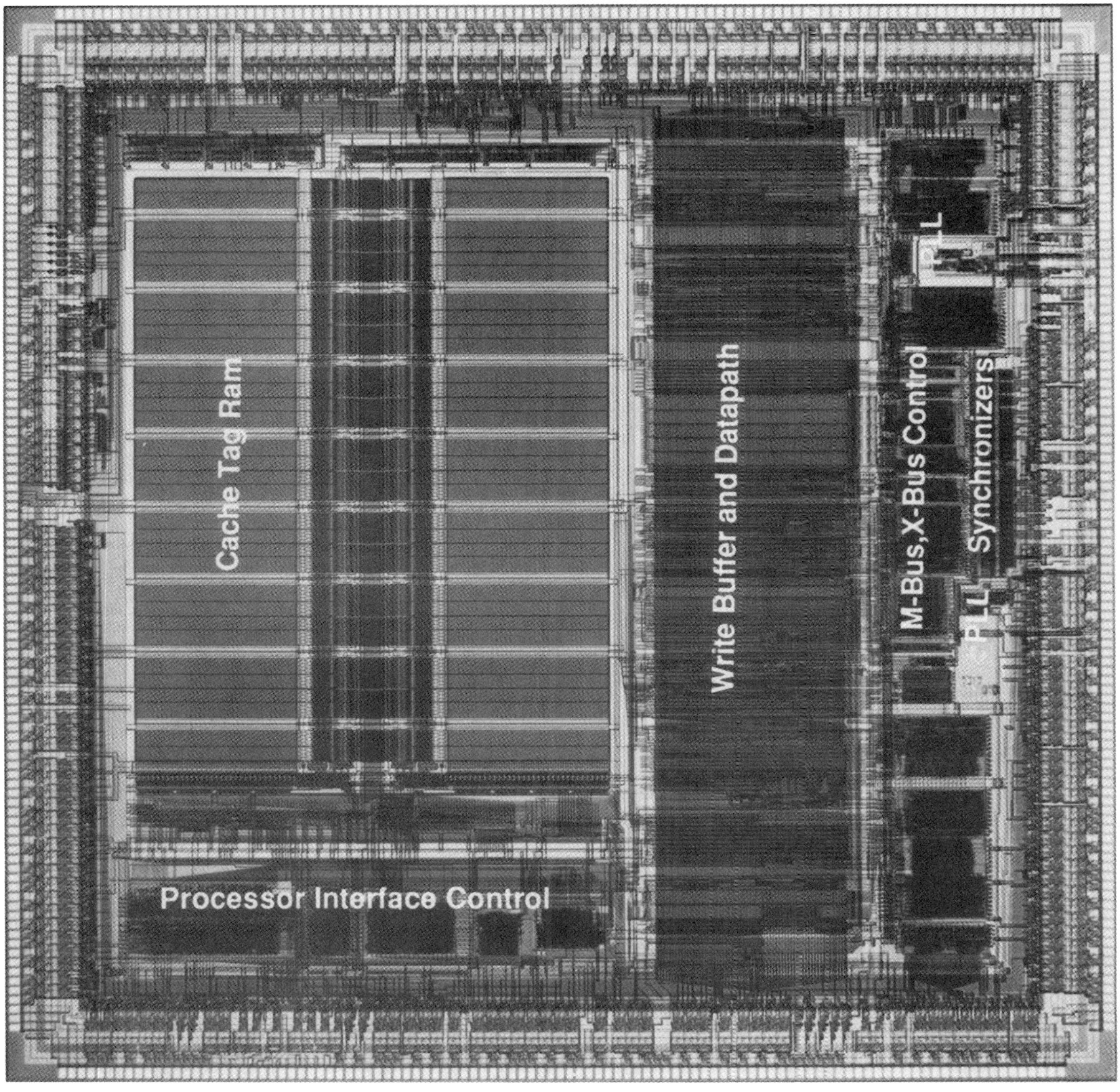

Figure 5: Die micrograph.

TA6.6: A 1,000MIPS BiCMOS Microprocessor with Superscalar Architecture

Osamu Nishii, Makoto Hanawa, Tadahiko Nishimukai, Makoto Suzuki, Kazuo Yano, Mitsuru Hiraki, Shohji Shukuri, Takashi Nishida

Hitachi Central Research Laboratory, Tokyo, Japan

A 1,000MIPS computer, integrated on a single chip and experimentally developed using 0.3μm self-aligned BiCMOS technology, features superscalar processing and pipelined access of interleaved secondary cache [1]. This paper describes circuit techniques of the cache, TLB, register file, and ALU operating at 250MHz.

Figure 1 shows a micrograph of the chip. The block diagram is shown in Figure 2. There are two superscalar processors. Each processor has an 8-port 32wx32b register file, two 32b ALUs, a 30b program count adder, and a control logic composed of 14k transistors. To meet the high rate of processor instruction and data access, four primary caches (1kB per one, 2-way) and four TLBs (64 entry per one, 2-way) are provided. Secondary caches (8kB, 4 banks, 2-way) are also implemented as a part of a shared memory. The 250MHz chip can execute 4 instructions per clock, allowing peak 1,000MIPS performance.

The processor has two 5-stage pipelines. The stages are shown in Figure 3 (pipeline 1 has only 4 stages). In the IF stage, instructions are fetched from the primary cache. In the ID stage, four operands are read from the register file. In the EX stage, ALU operations or memory address calculation are executed. The data cache is accessed in the RS stage. All the results are stored in the register file in the RS stage, except the result of load instruction, stored in the MS stage.

There are three critical paths in the IF, ID, EX stages. Most logic operates with a single-phase clock and edge-trigger flip-flops. Some functions are implemented with BiCMOS. A quasi-complementary (QC) BiCMOS circuit is used [6]. (Figure 4) It features a pMOS (M1) that drives the base of pull-down npn bipolar transistor Q1. For voltage lower than 3.5V, a conventional BiCMOS circuit loses its advantage over CMOS [2]. QC-BiCMOS makes the basic gates more than twice as fast as CMOS with a 3.3V supply [6].

The most critical path is in the IF stage: the read-access path of the primary cache memory and TLB. A 4.6x5.8μm^2 6-transistor full-CMOS memory cell is used. The decoder needs both speed and drivability, so a BiCMOS gate is used. The sense amplifier circuit uses a sense-amplifier-merged comparator and selector (SMCS) scheme that unifies the comparator function and the way selector function into the ECL sense amplifier [4]. The total access time (from virtual address-input to cache data-output) for the SMCS scheme is 2.8ns. The circuit makes it possible to use set associative mapping, giving a higher hit ratio, with a 4ns cycle time.

The ID stage is the read-access path of the register file. The circuit of the register file is shown in Figure 5. The number of ports is 8 (3-write, 5-read) because the superscalar architecture needs 3 ports (1W2R) for pipeline-1 and 5 ports (2W3R) for pipeline-2 [3]. One memory cell may have 1W5R in the same cycle. The BiCMOS gates are used in some places in the decoder, where the fan out number is large. The decoder is faster by 0.2ns than a CMOS decoder.

The register file circuit is optimized for the 5-read case. First, $W_2 = 5W_3$ (as a CMOS NAND, W=width of MOS transistor). The data lines do not use precharging, and read access speed for the H level is limited by the pull-up current. Figure 6 shows the pull-up current. With given W_3, the current for unity ratio, (pMOS width)/(nMOS width), is 90% of that for infinite ratio. Unity ratio is used considering the total delay. The 75x16μm^2 memory cell has 14 transistors. The delay time from read address (A0) to data (N2) is 1.3ns. The waveform of an access path from memory cell inverter loop node to read data is shown in Figure 7, the delay time is 0.85ns.

The EX stage is the ALU operation path. A carry-select adder is used. It has a critical path that one of two 4b conditional sums is selected by a carry generated in a 28b carry lookahead (CLA) circuit. A feedbacked-massive-input logic utilizes the high drivability of the BiCMOS circuit, and can implement the CLA in smaller logical stages [5]. The number of stages is essentially 2. The addition time is 1.2ns.

The specifications of the chip are shown in Table 1. 1.02M transistors are integrated on an 8.1x8.0 mm^2 chip. Metal-3 is used for the word line of the register file, cache memory, and TLB. The reason for use in the register file is that if metal-2 is used for the word line, metal-1 and 3 can not be connected because of the module size. The reason for use in the cache and TLB is that the memory cell size is smaller than that of a cell having data lines on Metal-3. Where these restrictions do not apply, metal-3 with low resistivity is used for Vcc and GND line to minimize the noise and bias of Vcc, GND. Metal-2 and metal-3 are used for global wire, and metal-1 is used for local wire smaller than 0.1mm.

Acknowledgement

The authors thank: T. Masuhara, K. Shimohigashi, and K. Seki for encouragement and discussions; S. Tachibana, A. Kawamata, and S. Morita for chip design; Y. Kawamoto, T. Kure, F. Murai and the process staff members for chip fabrication.

References

[1] Hanawa M., et al., "On-chip Multiple Superscalar Processors with Secondary Cache Memories", ICCD'91, Oct. 1991.

[2] Momose H., et al., "A Supply Voltage Design for Half μm BiCMOS Gates", Symp. VLSI Tech., June, 1989.

[3] Steiss D., et al., "A 65MHz Floating-Point Coprocessor for a RISC Processor", ISSCC DIGEST OF TECHNICAL PAPERS, pp. 94-5, Feb. 1991.

[4] Suzuki M., et al., "Sense-Amplifier-Merged Comparator and Selector Scheme for BiCMOS Cache Memories", European Solid-State Circ. Conf., Sep. 1991.

[5] Yano K., et al., "3.3-V BiCMOS Circuit Techniques for 250-MHz RISC Arithmetic Modules", CICC'91, May 1991.

[6] Yano K., et al., "Quasi-Complementary BiCMOS for Sub-3-V Digital Circuits", Symp. VLSI Circ., Jun. 1991.

Reprinted from *1992 IEEE ISSCC*, pp. 114–115 and p. 260, Feb. 1992.

Figure 1: See page 238.

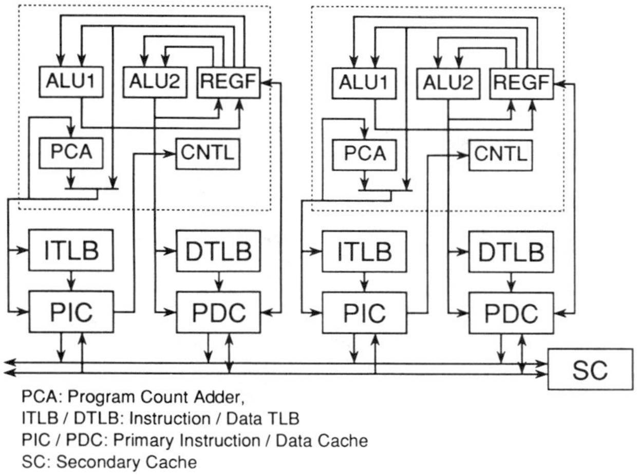

Figure 2: Block diagram of the chip.

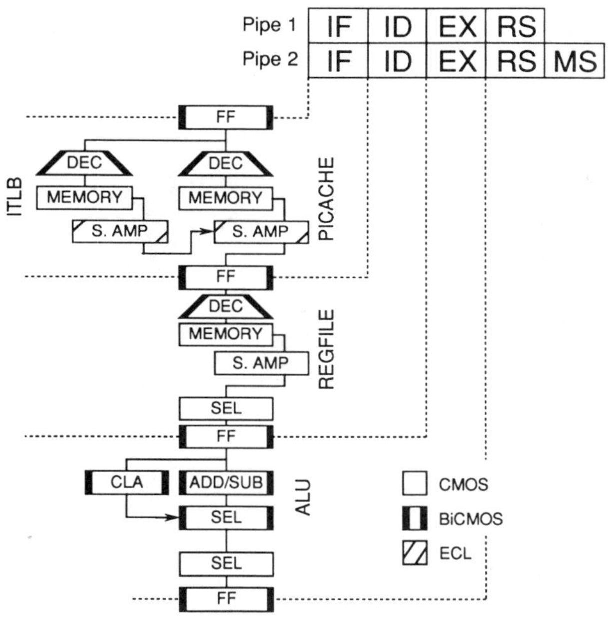

Figure 3: Pipeline stages and processor critical paths.

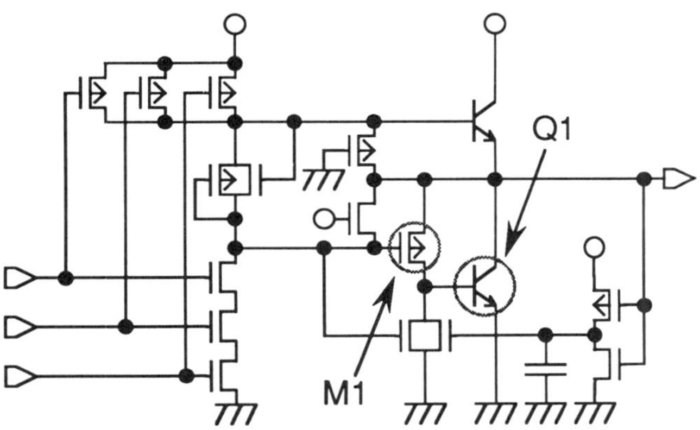

Figure 4: QC-BiCMOS circuit (3-input NAND).

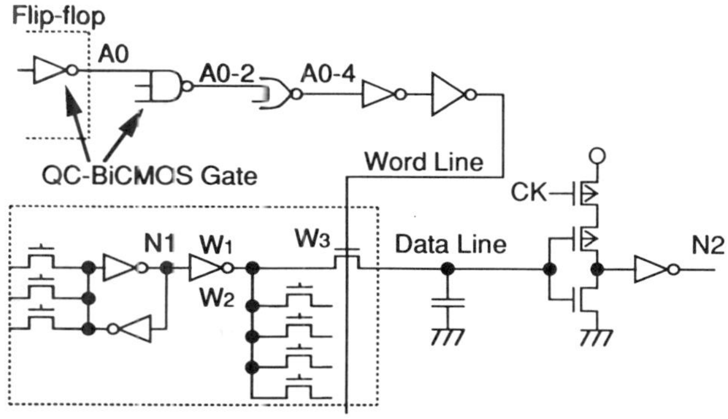

Figure 5: Register file circuit diagram.

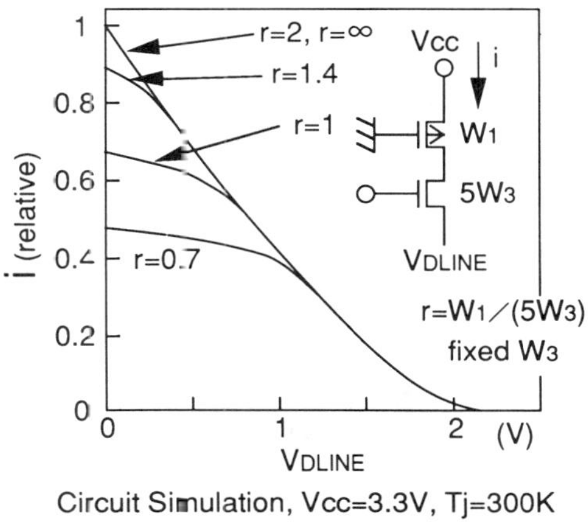

Figure 6: Pull-up current (static).

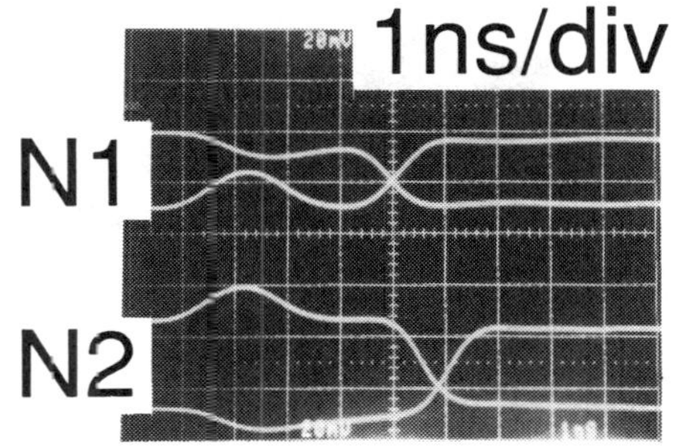

Figure 7: Register file read-access waveform (N1 to N2, Vcc=3.3V).

Process	0.3μm BiCMOS, 3 Metal		
	Line(μm)	Space(μm)	R(mΩ/sq.)
Metal 1 (W)	0.30	0.40	500
Metal 2 (Al)	0.45	0.65	50
Metal 3 (Al)	0.55	0.75	35
Chip size	8.1x8.0mm^2		
No. of transistors	1.02M (20k bipolar)		
Supply voltage	3.3V		
TLB	256-entry (64 entry(2way)x4)		
Primary cache	4kB (1kB(2way)x4, physical)		
Secondary cache	8kB (2kB(2way)x4banks, physical)		

Table 1: Chip specifications.

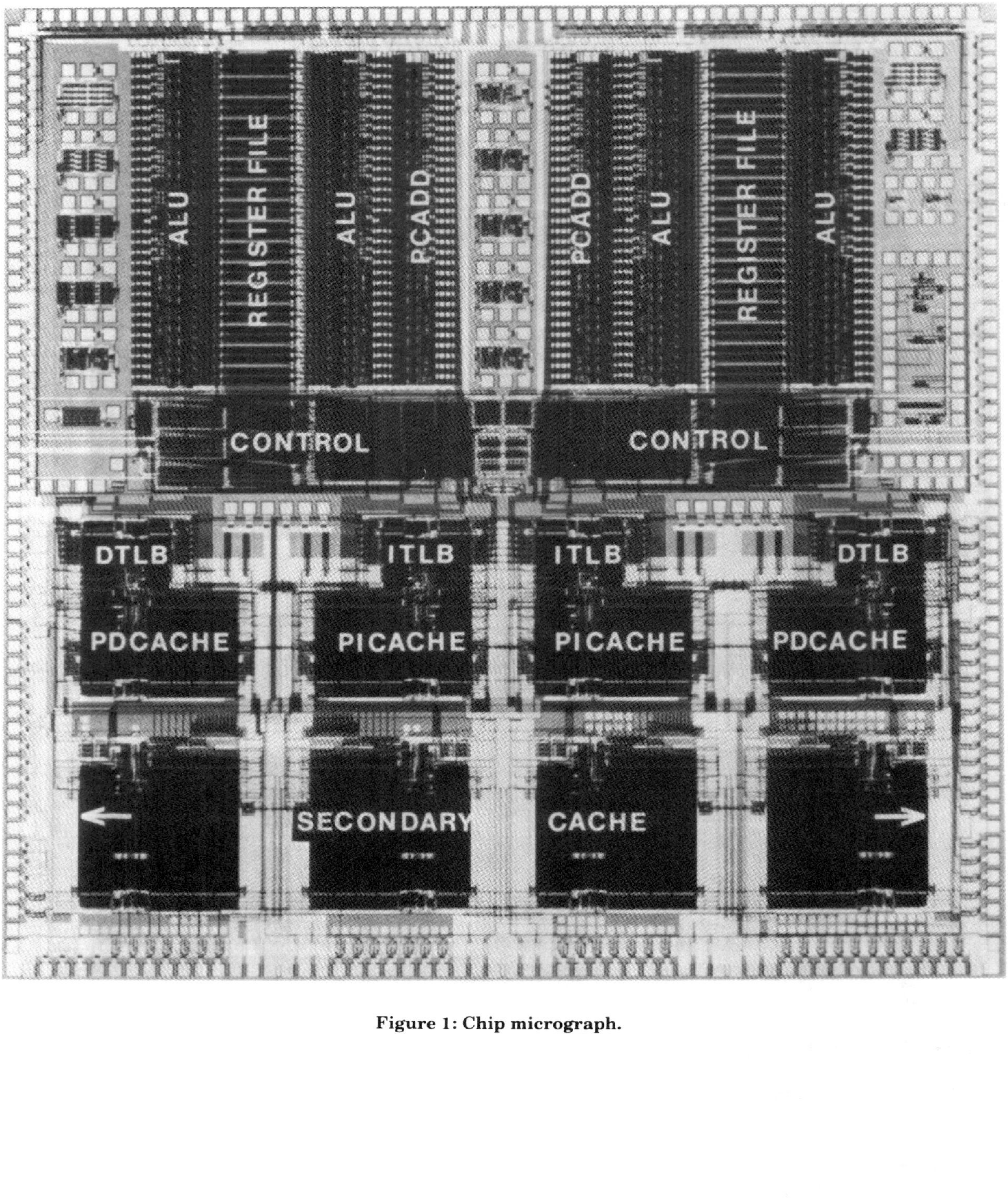

Figure 1: Chip micrograph.

Paper 2.11

A 200-MFLOPS 100-MHz 64-b BiCMOS Vector-Pipelined Processor (VPP) ULSI

Fuyuki Okamoto, Yasuhiko Hagihara, Chie Ohkubo, Naoki Nishi, Hachiro Yamada, and Tadayoshi Enomoto *Member, IEEE*

***Abstract*—The world's first single-chip 64-b vector-pipelined processor (VPP) ULSI has been developed. It executes vector operations indispensable to high-speed scientific computation. The VPP ULSI attains a 200-MFLOPS peak performance at a 100-MHz clock frequency. This extremely high performance is made possible by the integration on the VPP of a 64-b five-stage pipelined adder/shifter, a 64-b five-stage pipelined multiplier/divider/logic operation unit, and a 40-kb register file. Various new high-speed circuit techniques have been also developed for 100-MHz operations. The chip, which was fabricated with a 0.8-μm BiCMOS and triple-layer metallization process technology, has a 17.2-mm×17.3-mm area and contains about 693K transistors. It consumes 13.2 W at a 100-MHz clock frequency with a single 5-V power supply.**

I. Introduction

HIGH-SPEED vector processing for matrix calculation is indispensable to the handling of simultaneous equations, partial differential equations, and other kinds of scientific and engineering computations [1]. Typical vector processors consist of a great number of ECL gate arrays and bipolar memories, which means that these processors are usually very large in size. Further, since bipolar LSI's are high-power devices and produce a great amount of heat, liquid cooling is needed. This, in turn, results in a further increase in size. A 200-MFLOPS vector processor has been reported, for example, which was constructed with a large number of boards each containing 150 ECL gate arrays (170 gates per chip, 5 W per chip) [2]. It required Freon cooling and was approximately 18(W)×28(D)×6(H) feet in size. Such large systems are limited in performance because of the large amount of signal delay produced in the interconnections, not only among boards but also among chips.

In order to overcome these problems and to construct small and fast vector processors at low levels of power consumption, we have developed the world's first single-chip vector-pipelined processor (VPP) ULSI, whose performance, 200 MFLOPS and 500 MOPS at a clock frequency of 100 MHz, is a match for the processing speed of a mini-supercomputer [3]. Four features have played a key role in the VPP development: 1) five pipelined units, including two 5-stage arithmetic units operated in parallel, were integrated in order to achieve a peak performance of 200 MFLOPS and 500 MOPS; 2) a 40-kb register file (RF) with auto index incrementors (AII's) was also integrated, resulting in the elimination of the interconnection delay present among boards and chips; 3) various high-speed circuit techniques were developed to achieve 100-MHz operations; and 4) a 0.8-μm BiCMOS process technology was adopted to reduce power dissipation considerably and to eliminate the need for liquid cooling (only forced air cooling is required), so that system size can be reduced significantly.

In this paper, vector processing features are first introduced, and VPP architecture, pipelined arithmetic units, parallel-pipeline operation, and the register file are then discussed in detail. Various high-speed circuit techniques as well as VPP performance are also described.

II. Vector Processing Features

Vector processing is an operational method in which the same kinds of operations, such as

$$Ci = Ai \text{ op } Bi \qquad (i = 1, 2, 3, \cdots, n) \tag{1}$$

are accomplished on different elements for one or more data vectors, such as $(A1, A2, A3, \cdots, An)$ and $(B1, B2, B3, \cdots, Bn)$. The operations are carried out with the issue of only a single instruction. That is, once a single instruction is issued, the first operation ($C1 = A1$ op $B1$) starts. Subsequently, in the next clock cycle, the second operation ($C2 = A2$ op $B2$) starts. Thus, multiple operations for n data items are carried out in sequence. These operations are easily pipelined and pipelining helps performance.

Vector processing has the following advantages for building high-speed pipelined computers.

1) Ordinary, nonvector computers require more than n instructions just to perform the processing of, for example, (1). This results in a considerable amount of time being consumed by memory access to receive these instructions. Vector processing, on the other hand, requires

Manuscript received May 3, 1991; revised August 26, 1991.
F. Okamoto, Y. Hagihara, C. Ohkubo, H. Yamada, and T. Enomoto are with the System ULSI Research Laboratory, Microelectronics Research Laboratories, NEC Corporation, 1120 Shimokuzawa, Sagamihara-shi, Kanagawa-ken, 229, Japan.
N. Nishi is with the Computer System Research Laboratory, C&C System Research Laboratories, NEC Corporation, 4-1-1 Miyazaki Kawasaki-shi, Kanagawa-ken, 213, Japan.
IEEE Log Number 9103760.

Reprinted from *IEEE J. Solid-State Circuits*, vol. 26, no. 12, pp. 1885–1893, Dec., 1991.

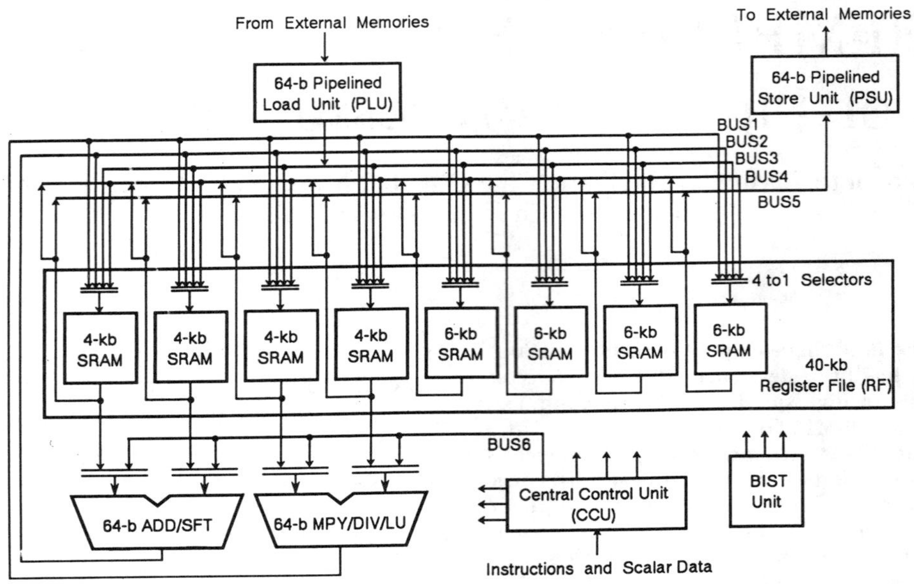

Fig. 1. Block diagram of VPP ULSI.

only one instruction to perform (1), effectively eliminating time-consuming memory bottleneck.

2) Individual calculations for the n vector data in (1) are independent from one another. There are none of the data hazards that generally degrade the speed of pipelines. As a result, the pipeline can be made deeper to produce faster operations.

3) Loop branches, which also degrade speed, are not required. This, too, allows deepening of the pipeline. [4] These advantages are, however, at the expense of making it necessary that instructions be expressed in vector form.

III. VPP Architecture

A block diagram of the VPP is shown in Fig. 1. It consists of five pipelined units, a central control unit (CCU), a built-in self test (BIST) unit, and six 64-b bus lines. The five pipelined units are: 1) an adder/shifter (ADD/SFT), 2) a multiplier/divider/logic operation unit (MPY/DIV/LU), 3) a 40-kb register file (RF), 4) a pipelined load unit (PLU), and 5) a pipelined store unit (PSU). These pipelined units receive and send data every 10-ns clock cycle. The main vector instructions carried out by these pipelined units are listed in Table I. The instructions are 4 bytes: 1 byte for an opcode and 3 bytes mainly for indicating register addresses.

The ADD/SFT and the MPY/DIV/LU are 64-b five-stage pipelined arithmetic units. The ADD/SFT can accomplish 28 instructions, including 64-b floating-point addition, 64- and 32-b integer addition, data shifting, and so on. The MPY/DIV/LU can implement ten instructions, including 64-b floating-point multiplication and division, 32-b integer multiplication, logic operations, and so on. The PLU receives 64-b vector data from the external memories. The PSU sends 64-b vector data to the external memories. Both the PLU and the PSU have two-stage pipelines.

TABLE I
List of Main Instructions of VPP

ADD/SFT	64-b floating-point	addition
		subtraction
		comparison
		accumulation
	32/64-b integer	addition
		subtraction
		comparison
	32-b integer	accumulation
	32/64-b arithmetic shift	
	32/64/128-b logical shift	
	floating-point/integer conversion	
MPY/DIV/LU	64-b floating-point	multiplication
		division
	32-b integer	multiplication
	64-b logical operation	
PLU	load 64-b data from external memories	
PSU	store 64-b data to external memories	
RF	transfer 64-b data between SRAM's	

43 vector instructions are implemented.

The RF consists of four 4-kb (64 word × 64 b) two-port SRAM's and four 6-kb (96 word × 64 b) two-port SRAM's. The 4-kb SRAM's handle communicating 64-b vector data between pipelined arithmetic units. The 6-kb SRAM's are mainly used as temporary registers. The RF also operates in a three-stage pipelined scheme, to transfer 64-b data from one SRAM in the RF to another (the data are read in the first clock cycle, transferred in the second clock cycle, and stored in the third clock cycle).

The CCU receives instructions from an external scalar processor. The CCU decodes and issues the instructions. It also controls parallel pipeline operations, as will be explained later. The BIST unit controls the self-test sequence, in which the pipelined arithmetic units and the RF are tested with random test patterns. Such a BIST scheme makes it easier to detect defects at the time of VPP manufacture.

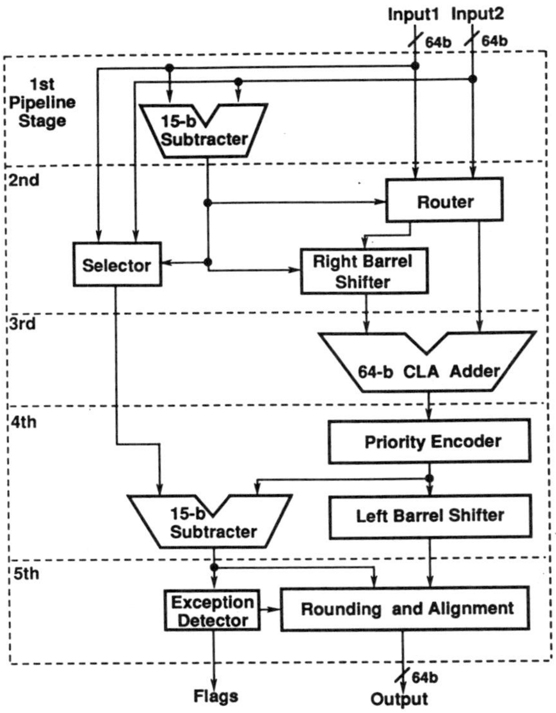

Fig. 2. Block diagram of adder/shifter (ADD/SFT).

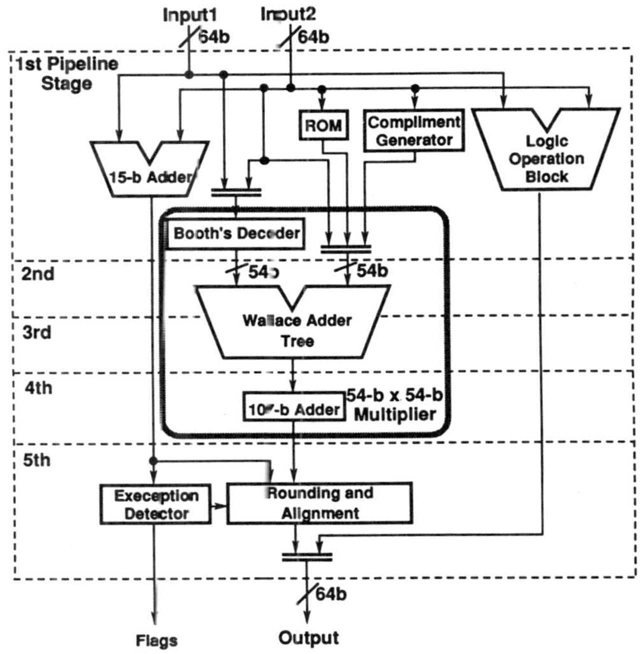

Fig. 3. Block diagram of multiplier/divider/logic operation unit (MPY/DIV/LU).

A VPP is typically operated in the following way. The CCU issues a load instruction given by the external scalar processor. The PLU then sends a memory-request signal to the external memory bank after which the PLU starts to load data vectors from the external memories into the RF. The addresses for the external memories are given by the scalar processor. The CCU then issues an arithmetic instruction, and vector data are continuously supplied from the RF to the ADD/SFT (and/or the MPY/DIV/LU). Calculated results are sent back to the RF every cycle. After arithmetic instructions have been carried out some number of times, the CCU issues a store instruction. After sending the memory-request signal, the PSU stores the data vectors from the RF in the external memories whose addresses have been determined by the scalar processor.

All the pipeline units execute vector instructions in parallel. This results in a peak performance of 200 MFLOPS and 500 MOPS at 100 MHz.

IV. Pipelined Arithmetic Units

A. Adder / Shifter (ADD / SFT)

Fig. 2 is a detailed block diagram of the ADD/SFT, a five-stage pipeline for accomplishing 64-b floating-point operations. The 15-b subtracter in the first stage calculates the difference between the exponents of two input operands. The router in the second stage sends the mantissa of the operand with the smaller exponent to the right barrel shifter, and the other to the 64-b carry lookahead (CLA) adder in the third stage. The right barrel shifter shifts the mantissa to the right by the number of bits given by the 15-b subtracter. The selector selects the larger exponent between two input operands. The 64-b CLA adder carries out absolute addition/subtraction. The fourth stage includes the left barrel shifter, which normalizes the output from the 64-b CLA adder, and the priority encoder, which controls the shifter. This stage also includes the 15-b subtracter, which adjusts the exponent, obtained from the selector, by the number of bits by which the mantissa is shifted. In the fifth stage, rounding and digit alignment are carried out and exceptions, such as overflow, are detected.

The functional blocks mentioned above, which are necessary in order to implement floating-point addition, have been applied to other kinds of instructions, such as integer operations, shift operations, and data-type conversion between floating-point data and integer data. For the left shift instruction, for instance, the operands are shifted to the left by the left barrel shifter in the fourth stage, while simply passing through other blocks. This multifunction scheme reduces the transistor count by 20% over that of a unit in which an adder and a shifter are separately constructed.

B. Multiplier / Divider / Logic Operation Unit (MPY / DIV / LU)

The MPY/DIV/LU shown in Fig. 3 is also a five-stage pipeline. The 15-b adder in the first stage adds the exponents of two inputs. A 54-b×54-b multiplier (from the first stage to the fourth stage) performs mantissa multiplication. The multiplier uses a radix-4 Booth's algorithm. It consists of a modified Wallace adder tree and a 107-b CLA adder. The fifth stage performs rounding and digit alignment and detects exceptions. The 12-b$\times$128-word on-chip ROM and the compliment generator (53-b adder) are provided with the first stage so as to execute division.

The division scheme is based on the Newton–Raphson method, and uses the on-chip ROM which stores approximate reciprocals of divisors in order to improve convergence speed. Iteration of seven vector multiplications are performed to generate quotients. This division scheme permits that an individual quotient to be obtained every cycle.

V. Parallel Pipeline Operation

The central control unit (CCU) achieves a peak performance of 200 MFLOPS and 500 MOPS by controlling the parallel execution of the following vector instructions:

1) an add or shift operation, two READ and one WRITE operation for the RF;
2) a multiply, divide, or logic operation, two READ and one WRITE operation for the RF;
3) a data load from the external memories;
4) a data store into the external memories;
5) a data transfer between SRAM's in the RF.

Since one vector instruction executes n iterative operations, when succeeding instructions are issued before those operations have been completed, succeeding operations will be executed in parallel. While optimum parallelism would be achieved by issuing instruction each cycle, this is not generally possible because of certain "hazards" that prevent an instruction from being issued every cycle.

Fig. 4 shows four timing diagrams: Fig. 4(a) shows the arithmetic instruction timing for the ADD/SFT and the MPY/DIV/LU; Fig. 4(b) shows two instructions independent from one another (these two instructions can be performed in parallel); Fig. 4(c) shows two instructions in which addition results in Ci and multiplication using Ci is carried out (in this case, the multiplication instruction issue must be postponed until the first result of the addition instruction is produced in order to avoid data hazards between two operands); and Fig. 4(d) shows two instructions, addition and shift, both of which use the ADD/SFT (the shift instruction must be initiated after the last operation of the addition instruction). Resource collisions due to simultaneous access by plural instructions, such as in Fig. 4(d), are called "structural hazards." In typical programs, the vector length n is unknown at compile time, that is to say, the instruction executing time is not known until operations are actually under way. For this reason, the CCU has been integrated into the VPP so as to be able to control execution timing during run time.

The CCU consists mainly of an instruction decoder, 13 timers, and a reservation table. The 13 timers are provided for the five pipelined units and eight SRAM's in the RF. Each timer counts the number of clock cycles while its corresponding unit is operating. The reservation table indicates which SRAM's are to be used for writing results

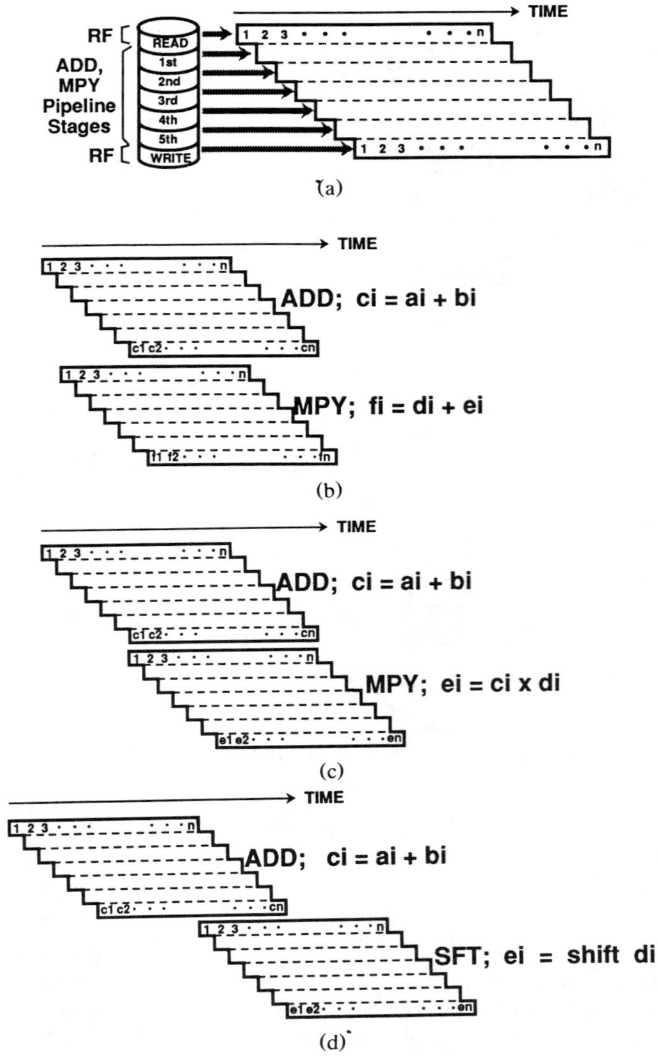

Fig. 4. Timing diagram of vector instructions. (a) Operation timing for ADD/SFT (and/or MPY/DIV/LU). (b) Instructions independent from one another. (c) Multiplication uses addition results. (d) Addition and shift uses the same unit.

of ongoing calculations. Before issuing an instruction, the CCU checks which units are busy (i.e., checking for structural hazards) and determines whether or not the instruction will need to use the results of an ongoing calculation (i.e., checking for data hazards). These decisions are made on the basis of the information regarding the instruction, the timers, and the reservation table. When the CCU detects hazards, the instruction issue is postponed until all necessary conditions are met. In this way, all pipelines operate in parallel without error, and a peak performance of 200 MFLOPS and 500 MOPS at 100 MHz is achieved.

VI. Register-File and Bus-Line Configuration

Without the on-chip register file (RF), the arithmetic units in the VPP would have to communicate directly with slow external memories, and system performance would be degraded. To avoid this, a register file (RF), including four 6-kb (96 word×64 b) two-port SRAM's and four 4-kb (64 word×64 b) two-port SRAM's, and six 64-b bus lines have been integrated into the VPP, as shown in Fig. 1. Further, no instructions that access the external memories have been implemented except load/store instructions (i.e., all the arithmetic instructions operate only on data in the on-chip RF, not in the external memories).

The data from the MPY/DIV/LU, the ADD/SFT, and the PLU are written into the SRAM's in the RF via BUS1, BUS2, and BUS3, respectively. BUS4 is used for data transfer between the SRAM's. These four bus lines are connected to the 4-to-1 selectors at the WRITE ports of the SRAM's. The data produced by the ADD/SFT, for instance, can go into any of the eight SRAM's in the RF via BUS2. Similarly, the data from the external memories can also be loaded into any of the SRAM's via BUS3. BUS5 is used for the store instruction. BUS6 is necessary to transfer scalar data used for such calculation as $Ci = Ai$ op S $(i = 1, 2, \cdots, n)$. As opposed to conventional microprocessors in which all the arithmetic units share a common data bus, the different bus lines are provided to different pipelines in the VPP, so as to make it possible for the pipelines to operate independently.

The RF sends four data sets to the two arithmetic units, and receives two data sets. This means that four READ operations and two WRITE operations are simultaneously performed in one clock cycle. Further, one WRITE and one READ are required for data transfer between the SRAM's, as well as one WRITE for the PLU and one READ for the PSU. Thus, a total of ten 64-b data words must be transferred every clock cycle when all the pipelines are operating in parallel. With conventional methods, all the register addresses would have to be supplied from outside the chip. However, this would not be feasible here because of the extremely large external instruction memories necessary to store the addresses. In addition, such conventional methods would increase the silicon area overhead of wires for address signals.

In answer to this, we have divided the RF into eight 2-port SRAM's and provided two AII's for READ and WRITE operations to each SRAM. Fig. 5 shows a block diagram of the RF with the AII's. Once a start address supplied by the CCU is set in it, the AII starts to calculate the required address to be used every cycle by incrementing the address pointer position by one. Each AII independently calculates the address, so that it is possible to execute eight READ operations and eight WRITE operations every cycle. This gives sufficient bandwidth for the five pipelines operating in parallel, and reduces the *Si* area overhead caused by the address bus.

VII. High-Speed Circuits for 100-MHz Operation

Three new circuit techniques have been developed in order to achieve 100-MHz operation. They are: 1) a predecode scheme for transfer gate selectors, 2) a functional pipeline register, and 3) a two-port SRAM with bypass circuit. Further, BiCMOS logic gates with large capacitive load drivability have been used.

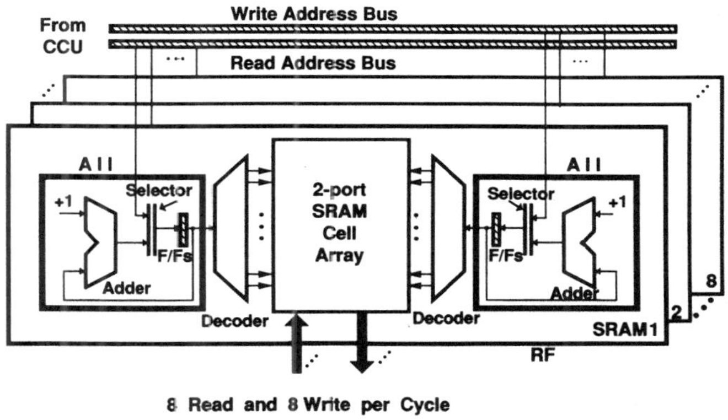

Fig. 5. Block diagram of register file (RF) with auto index incrementors (AII's).

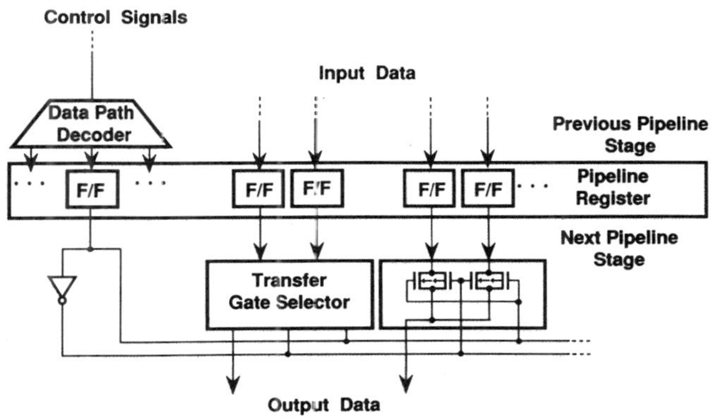

Fig. 6. Predecode scheme for transfer gate selectors.

A. Predecode Scheme for Transfer Gate Selectors

Since the ADD/SFT and the MPY/DIV/LU execute various kinds of instructions, they have many selectors that switch the data path for different instructions. When a next instruction must start without any wasted cycles after a previous instruction ends, the new data path for the new instruction must be formed in the same clock cycle in which the new operands for the new instruction arrives. Therefore, a predecode scheme for the transfer gate selectors has been developed, as shown in Fig. 6. The control signals are decoded one cycle prior to the arrival of the data at the selectors. Consequently, the data paths are formed at the same time that the data reach the selectors. As a result, signal delay time is dependent only on the time required for the signal to pass through the transfer gate. This scheme eliminates the decoding time from the delay of the critical path and takes advantage of the transfer-gate selector, which is 30% faster than the NAND-gate selector.

B. Functional Pipeline Register

Master–slave type flip-flops are conventionally used for the pipeline register as shown in Fig. 7(a). In the VPP, a

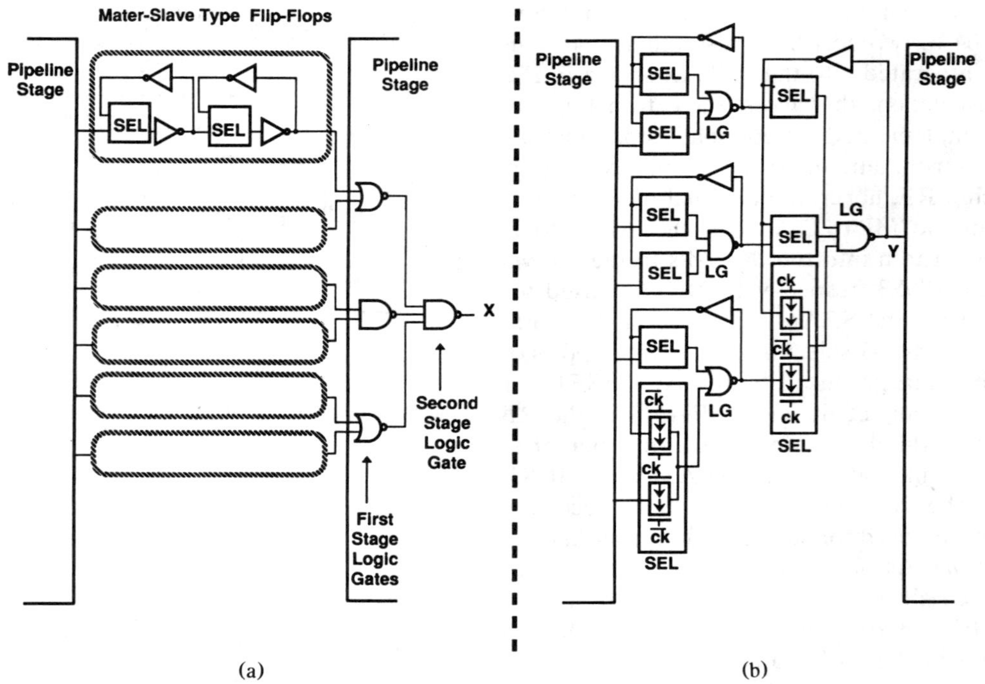

Fig. 7. (a) Conventional pipeline register. (b) Functional pipeline register.

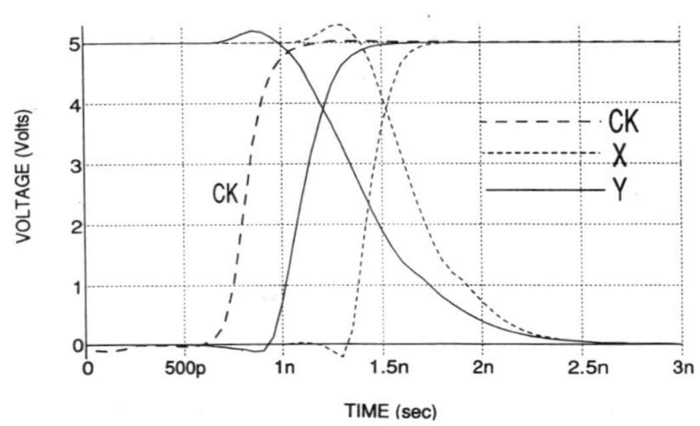

Fig. 8. SPICE plots for the pipeline registers.

new pipeline register has been used to improve pipeline performance, as shown in Fig. 7(b). Its function is equivalent to that for both a conventional pipeline register and the following two-stage logic gates. The new pipeline register consists of a two-stage logic circuit. An individual logic circuit consists of a logic gate (LG), a transfer gate selector (SEL), and an inverter for negative feedback loop. The LG is either a NAND gate or a NOR gate. While the clock pulse (CK) is at a low level, the LG in the first-stage logic circuits carries out logic operation using the input signal from the previous stage. When the CK goes high, a negative feedback loop is formed, and the output from the LG is kept at the same value as it was before. The second stage operates in the same way as the first, but it is controlled by the inverted clock pulse. Thus, the first stage plays both the role of the master part and the first-stage logic gates in Fig. 7(a), and the second stage plays both the role of the slave part and the second-stage logic gate in Fig. 7(b). Advantages of the new pipeline register are: 1) reduction of transistor count over the master–slave flip-flop by a factor as high as 4; and 2) reduction of signal delay in the next pipeline stage by that for two-stage logic, due to inclusion of a two-stage logic function in the pipeline register. Fig. 8 shows the SPICE plots of the waveforms for the clock pulse (CK) and the output signals observed at nodes X and Y in Fig. 7. The new pipeline register is faster than the conventional circuits by 300 ps for rise time and 250 ps for fall time.

C. Two-Port SRAM with Bypass Circuits

An instruction sequence in which the succeeding instruction uses the results of previous instruction, such as $Ci = Ai \times Bi$, $Ei = Ci + Di$ $(i = 1, 2, \cdots, n)$, frequently appears in scientific computation. In order to execute this sequence at high speed, the calculation $Ci + Di$ must start as soon as Ci is obtained. This requires that the data which are ready to be written into the RF must be immediately read out in the same cycle. This operation can be achieved by providing an address correspondence detector and an nMOSFET bypass, as shown in Fig. 9. When the WRITE address coincides with the READ address, the nMOSFET switch closes and the output of the input register is read out through the bypass within 5 ns. Without this circuit, it takes more than 8 ns to read out the WRITE data because input data must be written into the memory cells and then read out. As for the ordinary case where it is not necessary to read out the WRITE data in the same cycle, it takes 7 ns to read out the data which are already stored. This bypass circuit enables linking two pipelines directly, so that chaining is very fast. For in-

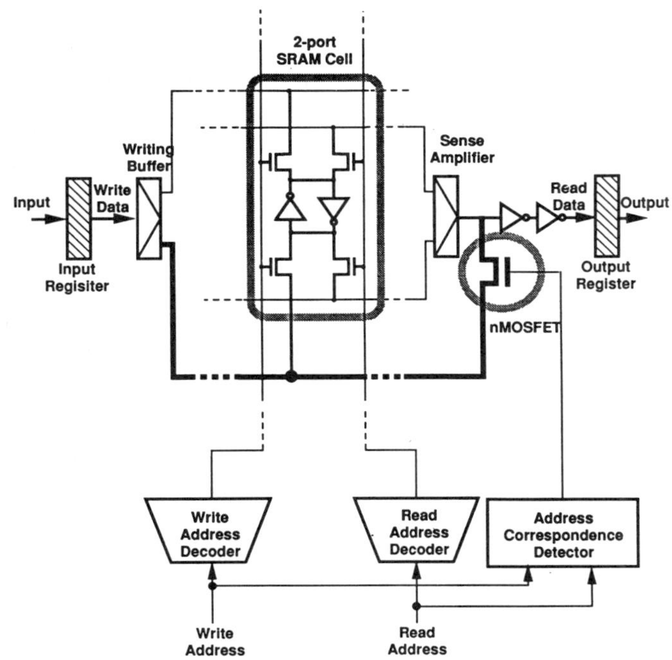

Fig. 9. Two-port SRAM with bypass circuits.

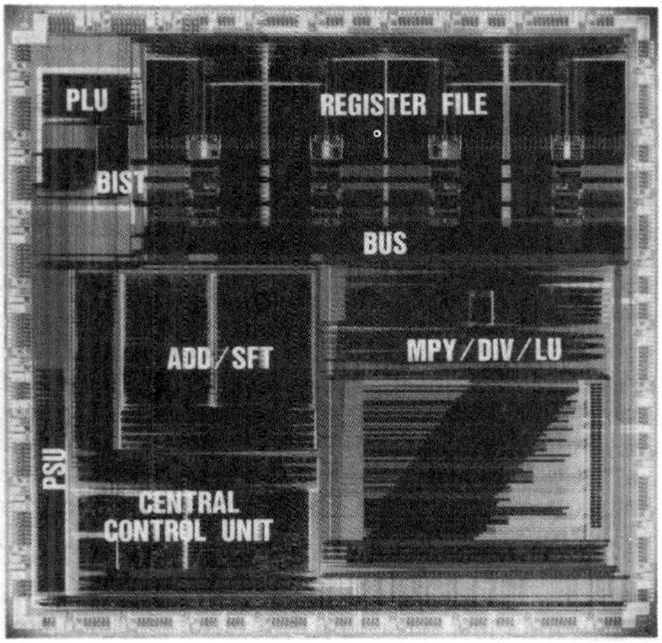

Fig. 10. Photograph of VPP.

TABLE II
MAIN FEATURES OF VPP

Fabrication technology	0.8-μm BiCMOS and triple-level Al	
Chip size	17.2 mm × 17.3 mm	
Number of transistors	total	693K
	(675K MOS + 18K bipolar)	
	ADD/SFT	49K
	MPY/DIV/LU	146K
	Register File	402K
	Central Control Unit	48K
	Others	48K
Data formats	64-b floating point	
	32- and 64-b integer	
	64-b logic	
Number of parallel operations	5 (addition/shifting, multiplication/division/logic operation, load, store, transfer)	
Peak performance rate at 100 MHz	200 MFLOPS, 500 MOPS	
Register file (RF) capacity	40 kb (640 words × 64 b)	
Package	528-pin PGA	
Interface	ECL 10K	
Power supply	5 V	
Power dissipation	13.2 W (at 100 MHz)	

stance, the results of the MPY/DIV/LU can be immediately sent to either of the ADD/SFT inputs.

D. BiCMOS Circuits

The input and output signal interface level for the VPP is ECL 10K level, to enable communicating with the external LSI's at a clock frequency of 100 MHz. Since the ECL I/O interface circuits consist of bipolar transistors, a BiCMOS process technology was adopted. Further, the BiCMOS logic gates were used to drive large capacitive loads (more than about 0.6 pF) quickly. They were used, for instance, in the Booth's decoder, in the decoder which controls the pass transistors in the barrel shifters, in the word decoders in the SRAM's, in the clock drivers, and so on. Each of the BiCMOS gates has a pull-up bipolar transistor and pull-down nMOSFET's.

VIII. CHIP LAYOUT AND PERFORMANCE

Fig. 10 shows a photograph of the VPP ULSI, which was fabricated using a 0.8-μm BiCMOS process with triple Al interconnection technology. The multiplier array of the MPY/DIV/LU, the SRAM's, and the bus lines were laid out manually. Other blocks, such as the ADD/SFT and the CCU, were laid out by the standard cell approach, using an automatic placer and router. The total chip layout was also carried out by the automatic layout tool. The first and the second Al interconnections were used for the signal lines and the power supply lines. The third Al was wired for some of the bus lines and for reinforcement of the power supply so as to reduce resistance. The main features of the VPP are listed in Table II. A 17.2-mm × 17.3-mm chip contains about 675K MOSFET's and 18K bipolar transistors. The peak performance of 200 MFLOPS and 500 MOPS at 100 MHz has been achieved using parallel pipeline operation. The power dissipation at 100 MHz is 13.2 W with a 5-V power supply. Input and output signal interface level conversion is to ECL 10K level. In order to reduce the switching noise, 90 power supply and ground pads were provided in the VPP.

IX. Summary

The world's first single-chip 64-b vector-pipelined processor (VPP) ULSI has been developed using a 0.8-μm BiCMOS fabrication process. The VPP executes vector instructions indispensable to high-speed scientific and engineering computation and achieves a peak performance of 200 MFLOPS and 500 MOPS at a 100-MHz clock cycle. The VPP chip is a 17.2-mm × 17.3-mm chip that contains about 693K transistors and consumes 13.2 W at 100 MHz with a power supply of 5 V. Thus, construction of a 1-GFLOPS vector processor requires only five VPP ULSI's.

Acknowledgment

The authors wish to express their appreciation for the cooperation of all those people involved in this VPP ULSI project.

References

[1] E. W. Kozdrowicki and D. J. Theis, "Second generation of vector supercomputers," *Computer*, vol. 13, pp. 71–83, Nov. 1980.

[2] N. R. Lincoln, "Technology and design tradeoffs in the creation of a modern supercomputer," *IEEE Trans. Comput.*, vol. C-31, no. 5, pp. 349–362, May 1982.

[3] F. Okamoto *et al.*, "A 200 MFLOPS 100 MHz 64b BiCMOS vector-pipelined processor (VPP) ULSI," in *ISSCC Dig. Tech. Papers*, Feb. 1991, FPM15.5, pp. 256–257.

[4] J. L. Hennessy and D. A. Patterson, *Computer Architecture: A Quantitative Approach*. San Mateo, CA: Morgan Kaufmann, 1990, pp. 352.

Paper 2.12

250-MHz BiCMOS Super-High-Speed Video Signal Processor (S-VSP) ULSI

Junichi Goto, Kouichi Ando, Toshiaki Inoue, Masakazu Yamashina, *Member, IEEE*, Hachiro Yamada, and Tadayoshi Enomoto, *Member, IEEE*

***Abstract*—A 250-MHz, 16-b, fixed-point, super-high-speed video signal processor (S-VSP) ULSI has been developed for constructing a video teleconferencing system. Two major technologies have been developed. One is a high-speed large-capacity on-chip memory architecture that achieves both 250-MHz internal signal processing and 13.5-MHz input and output buffering. The other is a circuit technology that achieves 250-MHz operations with a convolver/multiplier, an ALU, an accumulator, and various kinds of SRAM's. A phase-locked loop (PLL) is also integrated to generate a 250-MHz internal clock. The S-VSP ULSI, which was fabricated with 0.8-μm BiCMOS and triple-level-metallization technology, has a 15.5 mm × 13.0 mm area and contains about 1.13 million transistors. It consumes 7 W at 250-MHz internal clock frequency with a single 5-V power supply. A motion picture coding system for the video teleconferencing system, based on CCITT recommendation H.261, can be constructed with three S-VSP ULSI's, and a decoding system can be constructed with two.**

I. Introduction

What is needed for the construction of video signal processing systems, such as video teleconferencing systems, are microprogrammable video signal processor (VSP) LSI's capable of efficient motion picture coding/decoding. The operating clock frequencies for recently reported VSP LSI's have ranged from 13.5 to 40 MHz [1]–[3]. These VSP's have two drawbacks, however. First, their operating speeds are so slow that parallel operation of quite a large number of them [4] would be required for efficient motion picture coding/decoding, which involves complicated processing of huge amounts of data. The second drawback stems from the fact that these VSP LSI's are integrated with memories whose size is insufficient for storing image data. This makes it necessary to add commercially available memories whose cycle times are slower than the VSP LSI's. This, in turn, means that the limits on system speed will be determined by memory speed, not by the VSP speed itself.

Manuscript received May 1, 1991; revised August 26, 1991.

J. Goto, T. Inoue, M. Yamashina, H. Yamada, and T. Enomoto are with the System ULSI Research Laboratory, Microelectronics Research Laboratories, NEC Corporation, 1120 Shimokuzawa, Sagamihara-shi, Kanagawa-ken, 229, Japan.

K. Ando was with the System ULSI Research Laboratory, Microelectronics Research Laboratories, NEC Corporation, 1120 Shimokuzawa, Sagamihara-shi, Kanagawa-ken, 229, Japan. He is now with NEC Yamagata Ltd., Yamagata, Japan.

IEEE Log Number 9103699.

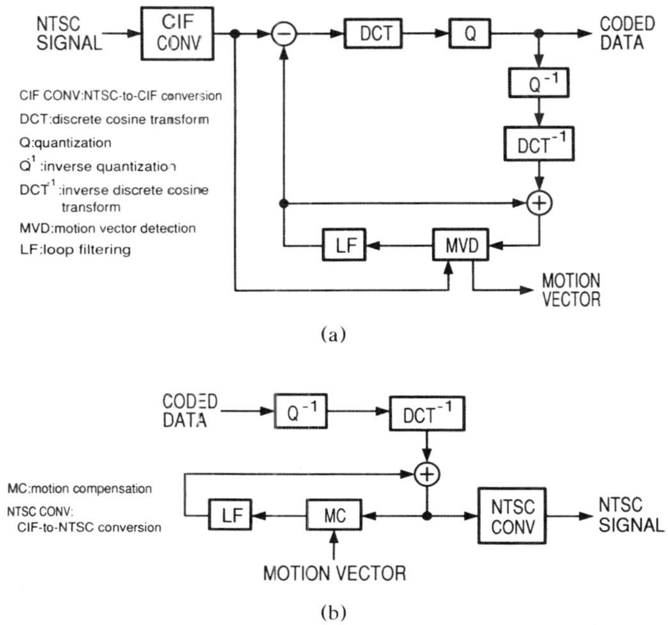

Fig. 1. Block diagram of efficient motion picture coding system. (a) Block diagram for coding system (sender). (b) Block diagram for decoding system (receiver).

In order to overcome these drawbacks, we have developed a 250-MHz 16-b fixed-point BiCMOS super-high-speed video signal processor (S-VSP) ULSI, where several functional units and image data memories are able to operate at up to 250 MHz [5]. It is possible to construct a coding system (Fig. 1(a)) conforming to CCITT H.261 with three S-VSP ULSI's and a decoding system (Fig. 1(b)) with two S-VSP ULSI's. These numbers are much less than one tenth the number of VSP LSI's used in conventional coding and decoding systems [4].

The S-VSP ULSI has been achieved with two major newly developed technologies. The first technology is a VSP architecture with 4-ns cycle time, on-chip two-port memories, each port of which has its own address generator. Those memories are used for both buffer memories and temporary data memories. The S-VSP can exchange image data with off-chip memories at a 13.5-MHz sampling frequency recommended by CCIR 601-1 and can carry out high-speed signal processing at 243 MHz (18 times 13.5 MHz) inside the chip. This means commer-

Reprinted from *IEEE J. Solid-State Circuits*, vol. 26, no. 12, pp. 1876–1884, Dec. 1991.

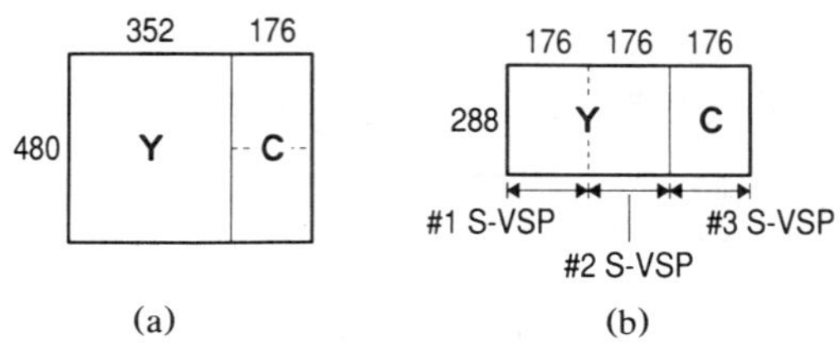

Fig. 2. Number of pixels of luminance signal (Y) and chrominance signal (C) for a single picture frame in (a) NTSC and (b) CIF.

cially available memories can be used as off-chip memories. As a result, signal processing performance has been drastically improved. The second technology is high-speed circuit technology to achieve 250-MHz operations and includes a redundant binary convolver architecture, arithmetic logic unit (ALU) architecture for obtaining an absolute value of the difference of two signals, and negative feedback BiCMOS logic gates.

In this paper we discuss the rationale for a 250-MHz internal operating frequency and describe the construction of the S-VSP ULSI, in addition to introducing the high-speed S-VSP architecture with an on-chip large-capacity data RAM architecture, high-speed circuit technologies, device specifications, and performance data.

II. The Rationale Behind a 250-MHz Maximum Internal Operating Frequency

A single noninterlaced NTSC frame for video teleconferencing systems is converted from two fields of an interlaced format. It consists individually of a 352(horizontal)×480(vertical)-pixel luminance signal (Y) and two 176(H)×240(V)-pixel chrominance signals (C), as shown in Fig. 2(a). In coding to meet CCITT H.261, the single NTSC frame must be converted into the common intermediate format (CIF) frame shown in Fig. 2(b). CIF is a format for video teleconferencing systems, and a single CIF frame consists of a 352(H)×288(V)-pixel luminance signal (Y) and two 176(H)×144(V)-pixel chrominance signals (C). A coding system must, then, carry out both NTSC to CIF conversion and efficient motion picture coding within 33 ms for any single frame.

In order to construct a coding system with three S-VSP ULSI's, we let two S-VSP's handle the 352×288-pixel luminance signal and let one S-VSP handle the two 176×144 chrominance signals, i.e., each S-VSP handles an equal number of pixels (176×288). The number of dynamic steps per pixel, required to carry out all the processes shown in Fig. 1(a), is about 65 (see Table I).[1] However, those dynamic steps include neither overhead nor optional adaptive control. On the basis of previous experience with conventional coding systems, it is estimated that the total number of dynamic steps required to carry out processing that includes both overhead and optional adaptive control would be about 150 steps/pixel. Thus, maximum operating clock cycle time would be 4.38 ns ($=$33 ms/(176×288 pixels×150 steps/pixel)). With the 243-MHz internal operating clock described above (i.e., 18 times the 13.5-MHz CCIR 601-1 sampling frequency), clock cycle time would be 4.12 ns, which meets the 4.38-ns condition. It is for this reason that we have chosen a 250-MHz clock cycle (i.e., 4-ns clock cycle time) as the maximum internal operating frequency of the S-VSP ULSI. In actual efficient motion picture coding operations, the S-VSP operates with a 243-MHz internal clock frequency.

For the decoding system in Fig. 1(b), two S-VSP ULSI's are sufficient to handle a 352×288-pixel luminance signal and two 176×144 chrominance signals. The luminance signal includes 396 16×16-pixel subpictures, the standard size by which the coding and decoding are carried out. Thus, the maximum allowed decoding time for each 16×16-pixel subpicture is 84 μs ($=$33 ms/396). As seen in Table I, the required decoding time for a 16×16-pixel subpicture is about 42 μs, exclusive of overhead and optional adaptive control. If we include these two, we may estimate total required decoding time to be less than the maximum allowable decoding time of 84 μs. Furthermore, a single S-VSP ULSI is sufficient for decoding the chrominance signals because their total number of pixels is half that of the luminance signal.

III. S-VSP ULSI Construction

Fig. 3 shows a block diagram of the S-VSP ULSI, which can accomplish all video signal processings necessary for efficient motion picture coding. The functional units are a 16-b two-stage pipelined arithmetic logic unit (ALU), a 16-b accumulator (ACC), a 16-b × 16-b four-stage pipelined convolver/multiplier (CONV/MPY), a barrel shifter (BSFT), a minimum/maximum value detector (MMD), five two-port data RAM's (2P-RAM-A to 2P-RAM-E; one port is for writing, another for reading), one 1-port data RAM (RAM-F), and a coefficient RAM (CRAM). Each data RAM port has an address generator. The memory cell array organizations, capacities, port counts and main uses of all on-chip memories are summarized in Table II. The S-VSP also has two 16-b input ports for receiving image data from off-chip memories and a 16-b output port for supplying processed results to off-chip memories. Furthermore, a sequence controller (SEQ-CONT) with an instruction RAM (IRAM), a phase-locked loop (PLL), etc. are also integrated. The IRAM stores instructions, which are transferred from a host processor. The SEQ-CONT reads an instruction every clock cycle to generate control signals. The PLL generates a 243-MHz internal clock, which is 18 times higher in frequency than

[1]For CIF-to-NTSC conversion and loop filtering, which are both digital filtering, the number of taps equals the dynamic steps per pixel. As for other processes, the number of required calculations is counted and divided by the number of pixels for which the calculations are carried out. For example, for 8×8 DCT, which is achieved by multiplying three 8×8 matrices, product-sum calculation during convolution is carried out 2×8×8×8 times. Therefore, the number of dynamic steps per pixel is 16 ($=2\times8\times8\times8/8\times8$).

TABLE I

PROCESS TIMES FOR 16×16 PIXELS AND 176×288 PIXELS AND THE NUMBER OF DYNAMIC STEPS PER PIXEL IN CODING AND DECODING FOR THE EFFICIENT MOTION PICTURE CODING AT AN OPERATION CLOCK FREQUENCY OF 243 MHz

Basic processes for coding	Process time for 16×16 pixels (μs)	Process time for 176×288 pixels (ms)	Number of dynamic steps per pixel
NTSC-to-CIF conversion	5.3	1.05	5.00
Motion vector detection	12.6	2.50	11.75
Loop filtering	9.5	1.88	9.00
Interframe difference	1.1	0.22	1.00
8×8 DCT	16.8	3.33	16.00
Quantization	1.1	0.22	1.00
Zigzag scan	2.2	0.44	2.00
Inverse quantization	2.2	0.44	2.00
Inverse 8×8 DCT	16.8	3.33	16.00
Interframe addition	1.1	0.22	1.00
Total	68.7	13.63	64.75
Basic processes for decoding	**Process time for 16×16 pixels (μs)**	**Process time for 176×288 pixels (ms)**	**Number of dynamic steps per pixel**
Zigzag scan	2.2	0.88	2.00
Inverse quantization	2.2	0.88	2.00
Inverse 8×8 DCT	16.8	6.66	16.00
Motion compensation*	0.0	0.00	0.00
Loop filtering	9.5	3.76	5.00
Interframe addition	1.1	0.44	1.00
CIF–NTSC conversion	9.9	16.10	5.00
Total	41.7	28.72	31.00

*In the motion compensation, the 16×16-pixel subpicture, which most closely resembles that in the current frame, is just read out from the preceding frame so that the loop filtering is carried out for it. Because reading out the 16×16-pixel subpicture is involved in the process of the loop filtering with a pipelined method, the process time and the number of dynamic steps are zero.

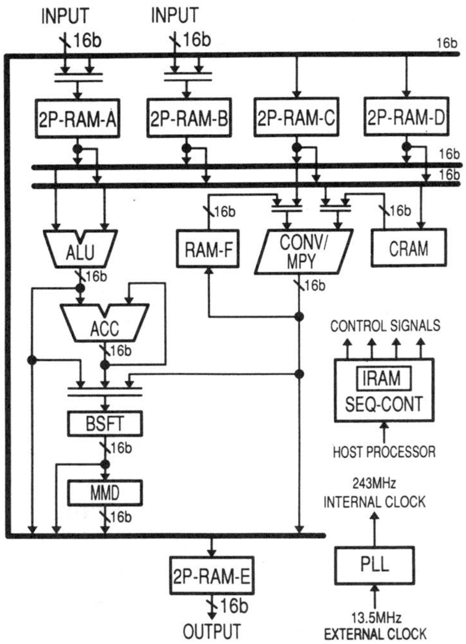

Fig. 3. Block diagram of the super-high-speed video-signal processor (S-VSP) ULSI.

CCIR 601-1 sampling frequency of 13.5 MHz and is synchronous with the internal clock.

The S-VSP has a flexible multistage pipelined architecture. It is controlled by 32-b horizontal instructions, so that it can carry out the variety of processes necessary for efficient motion picture coding and decoding, as shown in Fig. 1(a) and (b), respectively. Discrete cosine transform (DCT), inverse DCT (DCT^{-1}), and loop filtering (LF) are accomplished, for example, by the CONV/MPY. Quantization (Q) is carried out by multiplying the results of the DCT by the reciprocal of the quantization step by the CONV/MPY operating in the multiplier mode. Inverse quantization (Q^{-1}) is also carried out by multiplying the results of Q by the quantization step. Pattern-matching operation, an operation basic to motion vector detection in motion compensation, is achieved by 1) calculating an absolute value of the difference between two sets of data with the ALU, 2) accumulating the absolute values with the ACC, and 3) detecting the minimum accumulated values with the MMD. Motion compensation is a technique effective in reducing the amount of motion picture data which must be transmitted. It involves searching in a 45×45-pixel subpicture in the preceding picture frame for the 16×16-pixel subpicture that most closely resembles the 16×16-pixel subpicture in the current picture frame (a 16×16-pixel subpicture is a standard size with which efficient motion picture coding is carried out). The searching process is conducted in a three-step method [6]. After searching for the subpicture, the direction and distance between the 16×16-pixel subpicture of the preceding picture frame and that of the current one are detected. Direction and distance compose a motion vector. Differences in intensity for every other pixel within the two subpictures, i.e., interframe differences, are then calculated. This process is indicated by ⊖ in Fig. 1(a). The interframe differences are transmitted with the motion vector for each 16×16-pixel subpicture.

TABLE II
MEMORY CELL ARRAY ORGANIZATIONS, MEMORY CAPACITIES, PORT COUNTS, AND USE FOR ON-CHIP MEMORIES

RAM Name	Cell Array Organization	Capacity (kb)	Port Counts (functions)	Main Use
2P-RAM-A	1 kword × 16 b	16	2 (1R + 1W)	Storing subpictures of the current frame
2P-RAM-B	2 kword × 16 b	32	2 (1R + 1W)	Storing subpictures of the preceding frame
2P-RAM-C	1 kword × 16 b	16	2 (1R + 1W)	Storing intermediate results
2P-RAM-D	0.5 kword × 16 b	8	2 (1R + 1W)	Storing intermediate results
2P-RAM-E	0.5 kword × 16 b	8	2 (1R + 1W)	Storing processed results
RAM-F	64 word × 16 b	1	1 (1R/1W)	Storing intermediate results
CRAM	0.5 kword × 16 b	8	1 (1R/1W)	Storing coefficients
IRAM	1 kword × 32 b	32	1 (1R/1W)	Storing instructions

2(1R + 1W): having both READ port and WRITE port.
1(1R/1W): having one port that operates as a READ port on a read-enable signal and as a WRITE port on a write-enable signal.

IV. HIGH-SPEED S-VSP ULSI ARCHITECTURE WITH ON-CHIP MEMORIES

A. Data RAM's

Because the access speeds for commercially available memories are much slower than the S-VSP's 243-MHz operating speed, they are incapable of exchanging data with S-VSP's and cannot be used as off-chip memories for that purpose. A solution would be to provide high-speed memories with sufficiently large capacities (about 2 Mb) to store all the data for both a current picture frame and the preceding one. However, such on-chip memories are as yet out of reach of present VLSI technology. In order to solve the problem, an architecture has been developed with 4-ns cycle time on-chip memories. Seven on-chip memories are for data and one is for instructions. Although their capacities are still small, they are well matched with the 16 × 16-pixel subpicture size with which efficient motion picture coding is carried out. Some of the on-chip memories are used for buffers between off-chip memories and functional units in the S-VSP. That is, those buffers exchange data with off-chip memories at the low rate of CCIR 601-1 sampling frequency and with the functional units at a high rate of up to 250 MHz. Furthermore, a double-buffering method is effective in image data buffering.

Fig. 4 shows the initial phase in image data buffering. 2P-RAM-A stores two 16 × 30-pixel subpictures for a current NTSC picture frame. It is composed of two addressing spaces, which are used alternately in a double-buffering method. At any given time, one space contains a previously transferred 16 × 30-pixel subpicture, and the other is being used to receive a new subpicture of the same size. While this new subpicture is transferred pixel-by-pixel at a 13.5-MHz rate from off-chip memories, which store the current NTSC picture frame, the previously transferred 16 × 30-pixel subpicture is converted to a 16 × 16-pixel CIF subpicture at a 243-MHz rate. The converted subpicture is then temporarily stored in 2P-RAM-C, from which it will be read out for the process involved in efficient motion picture coding. After that, the roles of the two addressing spaces in 2P-RAM-A are interchanged.

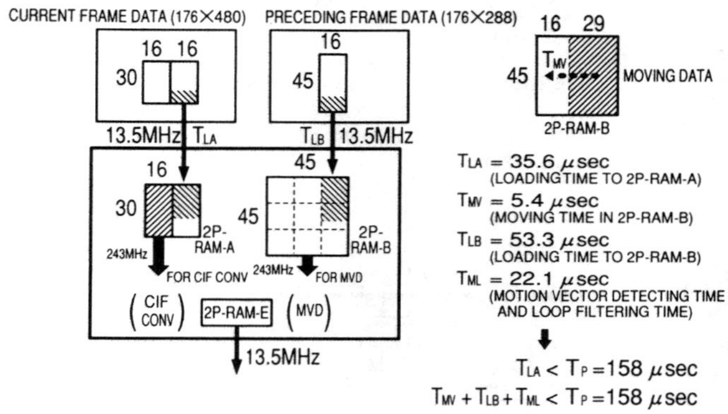

Fig. 4. Image data supply method using 2P-RAM-A and 2P-RAM-B.

For the double-buffering method for 2P-RAM-A, the data loading time must be within the following processing time. Assuming again that 150 dynamic steps per single pixel are required for the coding, the processing time T_P for a 16 × 16-pixel subpicture will be 158 μs (= (16 × 16 pixels) × (150 cycles/pixel)/243 MHz). If a 16 × 30-pixel is loaded into 2P-RAM-A at the 13.5-MHz CCIR 601-1 sampling frequency, its loading time T_{LA} will be 35.6 μs (= 16 × 30 pixels/13.5 MHz), which is well within the limits of processing time T_P.

2P-RAM-B stores a 45 × 45-pixel subpicture of the preceding picture frame for motion vector detection. Because this subpicture must center about the 16 × 16-pixel subpicture after detecting a motion vector, the contents of 2P-RAM-B, (i.e., the 45 × 45-pixel subpicture) must be renewed so that it might center about the next 16 × 16-pixel subpicture. The time required to renew the contents of 2P-RAM-B must be within the processing time. As

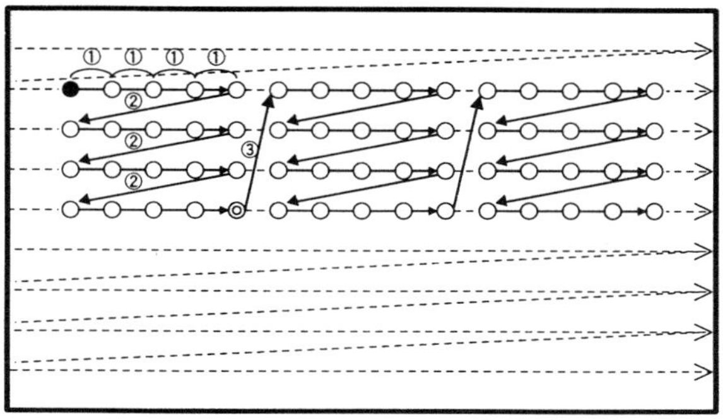

Fig. 5. Accessing pixels in the form of rectangular blocks.

shown in Fig. 4, this is done by moving the 29×45-pixel subpicture data, on the right side of 2P-RAM-B, to the left by 16 pixels and receiving a 16×45-pixel subpicture from an off-chip preceding picture frame memory to the 16×45-pixel empty area. The time required to move a 29×45-pixel subpicture, T_{MV}, is 5.4 μs ($=29\times45$ pixels/243 MHz) and the time required to receive a 16×45-pixel subpicture, T_{LB}, is 53.3 μs ($=16\times45$ pixels/13.5 MHz). The total time, T_{ML}, for both motion vector detection and loop filtering, both of which use data in 2P-RAM-B, is 22.1 μs (see Table I). Therefore, the total time for T_{MV}, T_{LB}, and T_{ML} is 80.8 μs, which is much shorter than the processing time T_P ($=158$ μs) for a 16×16-pixel subpicture. This means that data in 2P-RAM-B can be renewed by the beginning of the next motion vector detection cycle.

2P-RAM-E stores the processed results and supplies them to off-chip memories at a 13.5-MHz rate. Data memories, other than 2P-RAM-A, 2P-RAM-B, and 2P-RAM-E, are used as temporary data memories, such as 2P-RAM-C. Therefore, the S-VSP architecture permits signal processing at a 243-MHz rate inside the chip and allows exchanging data with off-chip memories at a 13.5-MHz rate. Consequently, the data access bottleneck between VSP LSI's and off-chip memories in conventional coding systems has been eliminated. Furthermore, the data transmitting rate between the S-VSP and the off-chip memories is so low that TTL-level I/O buffers can be used.

B. Address Generator

In video signal processing, it is often necessary to generate addresses for every clock cycle so that pixel data aligned in rectangular areas in a memory can be accessed. As shown in Fig. 5, this can be done in a triple-loop operation using the following three displacements: ① a displacement between two pixels on a line, ② a displacement between the end pixel on a line and the beginning of the next line, and ③ a displacement between the end of a rectangular area (two concentric circles in Fig. 5) and the beginning of the next rectangular area. A loop is formed with a loop-control variable to indicate the repeat count, an instruction to increment the loop-control variable, and a conditional branch instruction which exits the loop when the loop-control variable reaches a required value, otherwise repeats the loop. In order to generate address, an adding instruction which adds a displacement to the current address is executed in a loop.

In signal processing with the S-VSP ULSI, one on-chip data memory supplies data to an arithmetic unit and another on-chip memory receives outputs from the arithmetic unit. The loop operation for address generation for the latter memory must wait a given number of clock cycles after the former memory starts. After that, both memories work synchronously with each other. The given number of clock cycles is determined by the number of pipelined stages in the arithmetic unit and the delay time due to buses. However, such loops with different start times cannot be carried out with a single combination of an instruction memory and a program counter. Therefore, we provide an address generator at each on-chip data memory port so that each loop operation for each memory can be carried out independently. This means that each memory acts as if it has its own sequence controller. Each address generator receives seven parameters from the SEQ-CONT. They include the address of the first pixel (black circle in Fig. 5), three displacements, and three loop iteration numbers (the number of pixels in the horizontal direction within the rectangular area, the number in the vertical direction, and the total number of rectangular areas themselves). When a trigger signal is supplied by the SEQ-CONT, the address generator starts to automatically generate addresses every clock cycle on the basis of the seven parameters. In order for two address generators to start at different times, the start trigger signals must be supplied at the different times. Each address generator set a flag when its address generation is completed. This means that signal processing, such as DCT, loop filtering, quantization, etc., has been accomplished. The SEQ-CONT waits until the end of signal processing with a conditional branch instruction that branches while the flag is not set and goes to the next instruction when the flag is set.

V. High-Speed Circuit Technologies

A. Arithmetic-Logic Unit

It is necessary for the ALU in the S-VSP to calculate the absolute value of two input signals because this calculation is basic for motion compensation. It is also required to calculate $X+Y$ and $X-Y$ values. Fig. 6 is a block diagram of one of the conventional methods used for that purpose. It consists of two input registers, a carry look-ahead (CLA) adder (ADD1) with inverters and a selector (SEL1) for its Y input, another CLA adder (ADD2) with inverters for its X input, a pipeline register, and a selector (SEL2). ADD1 calculates either $X+Y$ or $X-Y$ by selecting Y or the inverse of Y with SEL1. ADD2 always calculates $Y-X$. SEL2 selects one of $X+Y$, $X-Y$, and $|X-Y|$ in accord with the instructions. By employing this

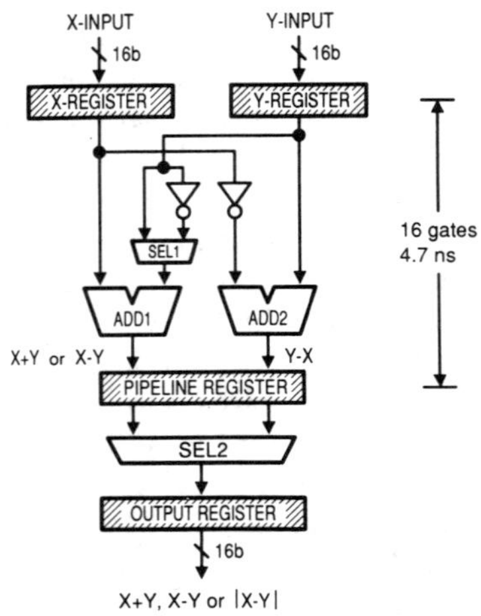

Fig. 6. A conventional method to calculate $X + Y$, $X - Y$, and $|X - Y|$.

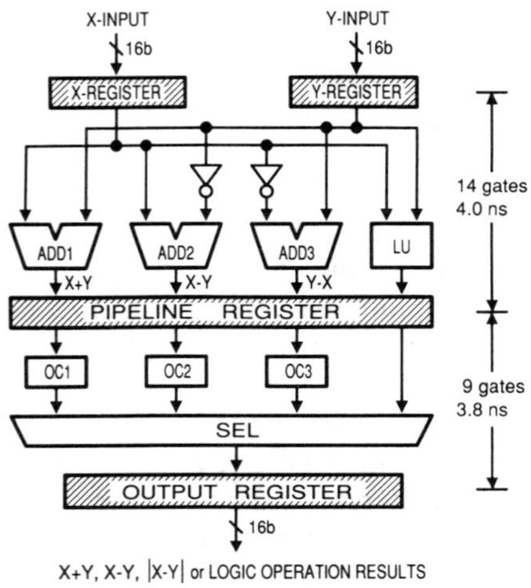

Fig. 7. Block diagram of the 250-MHz two-stage pipelined ALU.

method and using a 0.8-μm BiCMOS process, the critical path gate count in the first pipelined stage is 16 and the propagation delay time is 4.7 ns. Thus, a 250-MHz operation cannot be realized. In the following discussion, delay times are calculated assuming the 0.8-μm BiCMOS process is used.

In order to realize a 250-MHz operation, a 4-ns, 16-b, two-stage pipelined ALU, shown in Fig. 7, has been developed. This two-stage pipelined ALU includes two input registers (X-REGISTER, Y-REGISTER), three CLA adders (ADD1, ADD2, ADD3), a logical operation unit (LU), a pipeline register, three overflow correctors (OC1, OC2, OC3), a selector (SEL), and an output register. A CLA adder has been added so that $X + Y$ and $X - Y$ are respectively calculated by different adders and SEL1 in the Fig. 6 can be taken away. Thus, the critical path gate count in the first pipelined stage has been reduced to 14, which is almost equal to that of the CLA adders. Furthermore, the use of BiCMOS logic gates with pull-down nMOSFET's is effective in helping to improve the carry propagation delay time of the three CLA adders and to provide quick controls for the three overflow correctors and the SEL. As a result, although the hardware size for $X + Y$, $X - Y$, and $|X - Y|$ calculation has become 1.3 times that in Fig. 6, the critical path gate count in the first pipelined stage has been reduced to 14 and the propagation delay time has become 4.0 ns. The propagation delay time of the second pipelined stage is 3.8 ns, which means a 4-ns cycle time. That is, a 250-MHz operating speed can be realized.

B. Accumulator

If accumulation was carried out with a binary adder, accumulation time would be very long, because of continuous carry propagation over several digits for every addition operation during accumulation. In order to eliminate the continuous carry propagation over many digits, an accumulator is constructed using a redundant binary adder, where no continuous carry propagations occur. Thus, a 20-digit addition can be carried out within 1.8 ns. If the accumulator was constructed with a binary adder, the addition time would be as long as 4.5 ns, because continuous carry propagation over 20 digits would be present for the worst case.

It is necessary to convert the redundant binary numbers of accumulated results to binary numbers. Because this conversion is needed to be carried out only once after the completion of accumulation, the converter can be placed in the next pipelined stage of the accumulator, instead of providing it in the accumulating loop, resulting in complete elimination of continuous carry propagation during accumulation. The accumulator and the converter can be considered as a two-stage pipelined binary number accumulator.

C. Convolver / Multiplier

We have previously reported a 16-b, 200-MHz, 5-ns cycle time, two-stage pipelined redundant binary convolver [7]. Its key technology was a four-stage redundant binary adder tree. In the first pipelined stage, the critical path gate count was 16, and the propagation delay time was 5 ns. In the second pipelined stage, these values were nine gates and 4 ns, respectively. In order to achieve 250-MHz convolution and multiplication, we have developed a new 16-b×16-b two-stage pipelined redundant binary convolver/multiplier (CONV/MPY), shown in Fig. 8.

As with our previously reported 200-MHz convolver, this new CONV/MPY includes a redundant binary adder tree so that continuous carry propagation can be completely eliminated from individual product-sum calculations during convolutions. In the new CONV/MPY, however, one of the second-stage adders is used effectively as an accumulator adder. This can eliminate one adder stage, consequently reducing the logic gate count by 4, a drastic reduction. In addition, with this new architecture, a pipeline register can be easily provided almost at the center of the 21 critical path gates (i.e., between the first

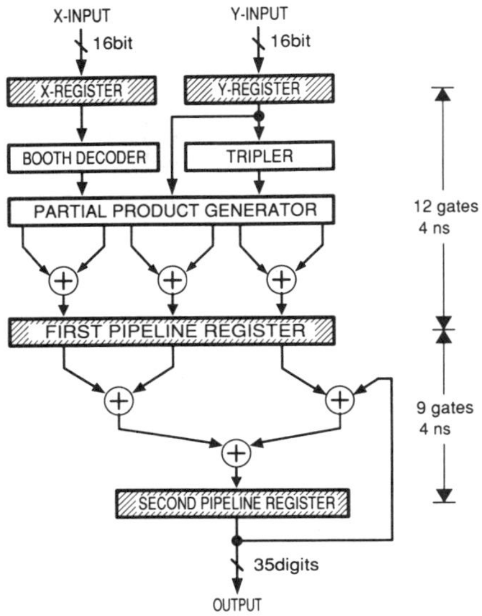

Fig. 8. Block diagram of the newly developed 250-MHz two-stage pipelined redundant binary convolver/multiplier.

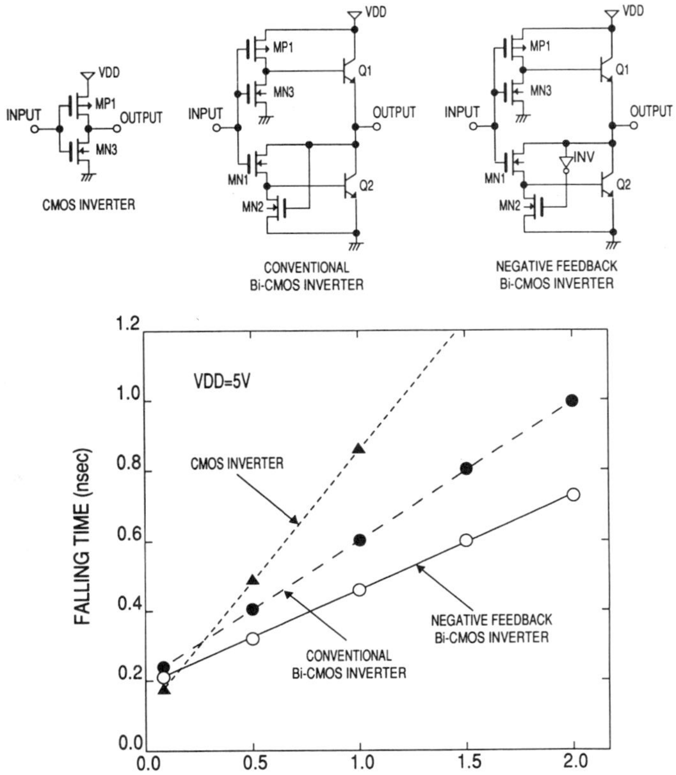

Fig. 9. Calculated falling times as a function of load capacitance for CMOS, conventional BiCMOS, and negative feedback BiCMOS inverters.

and second adder stages) so that pipeline cycle time can be minimized to 4 ns.

In order to improve propagation delay time when driving a large capacitive load, we have developed a BiCMOS logic gate with a negative feedback loop (see Fig. 9). An inverter (INV), provided between the output node and the gate terminal of an nMOSFET ($MN2$), is turned off

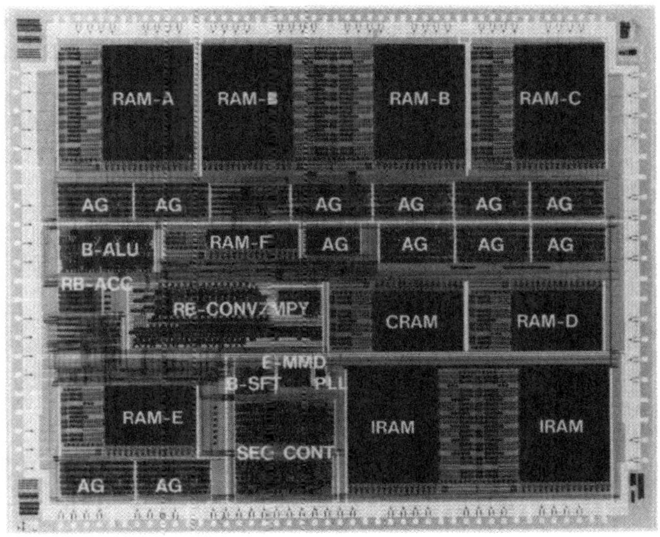

Fig. 10. Photograph of the S-VSP ULSI.

during the output signal falling time, so that current flowing through the $MN2$ is eliminated and the base current for the bipolar transistor ($Q2$) is increased. As a result, the collector current for $Q2$ becomes larger and, consequently, the falling time of the output node becomes faster. Fig. 9 shows calculated falling times for CMOS, conventional BiCMOS, and newly developed negative feedback BiCMOS inverters relative to load capacitance. Falling time of the negative feedback BiCMOS gate is more than 20% faster than a conventional BiCMOS gate.

Because the Booth decoder must drive a large capacitive load, which is the partial product generator of the CONV/MPY, negative feedback BiCMOS NOR gates have been incorporated into the Booth decoder. With CMOS and conventional BiCMOS gates, the propagation delay times of the first pipeline stage are 4.68 and 4.25 ns, respectively, whereas with negative feedback BiCMOS gates the delay time is reduced to 4.0 ns.

The CONV/MPY operates as either a convolver or a multiplier either by connecting or disconnecting the feedback loop from the second pipeline register to the second adder stage. The CONV/MPY produces 35-digit redundant binary results for convolution or multiplication. The redundant binary results are converted to 20-b binary numbers, i.e., rounded off to 20-b binary numbers by a two-stage pipelined converter (not shown). Both the carry generation signals and carry propagation signals for the upper 20 digits and a carry signal from the lower 15 digits are generated in the first pipeline stage. In the second pipeline stage, a 20-b binary number is generated with the signals generated in the first pipeline stage. The delay times of the first and second pipeline stages are 3.65 and 3.64 ns, respectively, with CMOS and conventional BiCMOS gates.

VI. S-VSP's Specification and Performance

Fig. 10 is a photograph of the S-VSP ULSI, which was fabricated with 0.8-μm design rule, BiCMOS, and triple-layer-Al interconnection technologies. The chip size is

TABLE III
FEATURES OF THE S-VSP ULSI

Fabrication technology	0.8-μm BiCMOS and triple-level-Al
Chip size	15.5×13.0 mm^2
No. of transistors	1.13M (MOS: 1.125M, Bipolar: 5K)
Data format	16-b fixed point
Power supply	5 V
Power dissipation	7 W at 250 MHz
Maximum operating frequency	250 MHz (external: 13.9 MHz)
Operating frequency for NTSC	243 MHz (external: 13.5 MHz)
Processing performance at 250 MHz	500 MOPS (convolution) 250 MOPS (multiplication, addition, accumulation)
On-chip memory capacity	121 kb
I/O interface signal level	TTL level
Package	132-pin PGA

TABLE IV
THE MAXIMUM SIGNAL PROCESSING CAPABILITIES FOR INDIVIDUAL FUNCTIONAL UNITS

Operation	Input (b)	Output (b)	No. of Pipeline Stages	Processing Capabilities (MOPS)
Convolution	16×16	35	4	500
Multiplication	16×16	20	4	250
Addition	16	16	2	250
Absolute value accumulation	16	20	1	250

15.5 mm$\times$13.0 mm, in which about 1.13-million transistors, including approximately 5000 bipolar transistors, are integrated. The features of the S-VSP ULSI are summarized in Table III. The S-VSP operates at a clock frequency of as high as 250 MHz with a single 5-V power supply. The total power dissipation is approximately 7 W. The 243-MHz operating clock is generated from a 13.5-MHz external clock being frequency multiplied by the on-chip PLL. The package is a 132-pin PGA. Table IV summarizes processing capabilities for individual functional units.

VII. CONCLUSION

A 250-MHz, 16-b, 1.13-million-transistor, 0.8-μm BiCMOS super-high-speed video signal processor (S-VSP) ULSI has been developed. An efficient motion picture coding and decoding system for video teleconferencing can be constructed with three and two S-VSP ULSI's, respectively. To construct such a fast S-VSP ULSI, two major technologies have been newly developed. The first is an architecture in which integration of 4-ns cycle time memories with their own address generators has been developed. The second is the high-speed circuit technology, used in the ALU, the accumulator, the convolver/multiplier, etc., which can operate at up to 250 MHz. Therefore, the S-VSP ULSI can exchange data at a low transmitting rate of 13.5 MHz with the off-chip memories and carry out signal processing at 243 MHz inside the chip.

ACKNOWLEDGMENT

The authors express their gratitude to H. Takada, O. Fujita, and M. Ishida of the System-ULSI Research Laboratory, Microelectronics Research Laboratories, and to many others who have contributed to this work.

REFERENCES

[1] M. Yamashina *et al.*, "A realtime microprogrammable video signal LSI," *IEEE J. Solid-State Circuits*, vol. SC-22, no. 6, pp. 1117–1123, Dec. 1987.

[2] K. Kikuchi *et al.*, "A single-chip 16-bit 25-ns realtime video/image signal processor," *IEEE J. Solid-State Circuits*, vol. 24, no. 6, pp. 1662–1667, Dec. 1989.

[3] T. Murakami, K. Kamizawa, M. Kameyama, and S. Nakagawa, "A DSP architectural design for low bit-rate motion video codec," *IEEE Trans. Circuits Syst.*, vol. 36, no. 10, pp. 1267–1274, Oct. 1989.

[4] I. Tamitani *et al.*, "A real-time video signal processor suitable for motion picture coding applications," *IEEE Trans. Circuits Syst.*, vol. 36, no. 10, pp. 1259–1266, Oct. 1989.

[5] J. Goto *et al.*, "A 250 MHz 16b 1-million transistor Bi-CMOS super-high speed video signal processor," in *ISSCC Dig. Tech. Papers*, Feb. 1991, pp. 254–255.

[6] T. Koga, K. Iimura, A. Hirano, Y. Iijima, and T. Ishiguro, "Motion-compensated interframe coding for video conferencing," in *Proc. IEEE National Telecommun. Conf.*, vol. 4, Nov. 1981, pp. G5.3.1–G5.3.5.

[7] M. Yamashina *et al.*, "A 200-MHz 16-bit super high-speed signal processor (SSSP) LSI," *IEEE J. Solid-State Circuits*, vol. 24, no. 6, pp. 1668–1674, Dec. 1989.

Part 3
BiCMOS Memory Applications

THE demand for higher density memory chips, operating at higher speeds of operation and at lower supply voltages, is ever increasing. BiCMOS technology offers an option to meet these demands.

This part consists of four sections. Section 3.1 is an invited overview by Eldin on BiCMOS state-of-the-art memory applications. Section 3.2 covers BiCMOS static RAMs. Section 3.3 covers BiCMOS dynamic RAMs and Section 3.4 covers BiCMOS nonvolatile memories.

In Section 3.2, the requirements of a BiCMOS circuit technology for high-speed SRAMs is examined in the paper by Douseka and Ohmori. The following six papers represent the state-of-the-art in BiCMOS SRAMs with density levels up to 4 Mb and performance levels down to 3.8 ns.

Section 3.3 starts with two overview articles: one on trends in DRAM design by Itoh and the other on comparison of CMOS and BiCMOS 1 = Mb DRAM performance by Watanabe et al. The following paper by Watanabe et al examines the development of a BiCMOS technology for high-speed DRAMs. The next three papers deal in some detail with three aspects of BiCMOS DRAM design. The paper by Kobayashi et al. examines the use of bipolar CMOS-merged technology. The first paper by Kawahara et al. deals with reduction techniques for substrate currents and the second paper by Kawahara et al. discusses the use of deep-submicrometer BiCMOS technology for the design of sub-10 ns 4-Mb DRAMs.

Although the application of BiCMOS to the design of nonvolatile memories is not currently widespread, three representative papers are included in Section 3.4. The paper by Smith et al. deals with the design of UV-EPROM, the paper by Kurita et al. examines the design of ROM and the paper by Hu et al. examines the design of EPROMs.

Paper 3.1

An Overview of BiCMOS State-of-the-Art Static and Dynamic Memory Applications

A. G. ELDIN
DEPARTMENT OF ELECTRICAL ENGINEERING
UNIVERSITY OF TOLEDO
TOLEDO, OH

Abstract—**The need for faster, denser, and cheaper dynamic and static random access memory chips has provided the major thrust for the recent development of advanced BiCMOS VLSI technology. The availability of fast and high-density SRAM and DRAM chips has made it possible to implement many functions in a cost-effective manner and has spurred the identification of new system applications. In this paper, an overview of the development, evolution, design strategy, and future challenges and trends of the BiCMOS SRAM and DRAM are presented.**

1. Introduction

Since the introduction of the 1-kb RAM chip in 1970, major development efforts and innovations in processing and design have led to today's 64-Mb RAMS. This 64,000-fold increase in the number of bits per chip is accompanied by a factor of 100 increase in speed. This has dramatically reduced the cost per bit of RAMs and provided the high-performance capability that made the semiconductor memories dominate in the implementation of different storage functions needed by all digital electronic products.

The focus of this paper is on the static and dynamic random access memories (SRAMs and DRAMs). Since the discussion is also applicable to most other memory categories, a brief overview of their basic cell operation is presented. This highlights the similarities and major differences between such categories.

Semiconductor memories now include several functional categories. The nonvolatile category includes ROM, PROM, EPROM, and EEPROM. The volatile category includes SRAM, DRAM, CAM, and serial memory.

The ROM (read only memory) is mask programmable and is therefore programmed only once during fabrication. This can be done using drain-contact, thin-oxide, or threshold-voltage ion implant mask. The PROM (programmable ROM) is programmed only once after fabrication by electrically blowing fuses.

The EPROM (erasable programmable ROM) can be programmed any number of times. It is characterized by fast READ and slow WRITE operations. Programming is done by injecting energetic charge carriers (generated by drain pn junction avalanche breakdown) into the gate oxide where they are trapped on the floating gate. Erasing is done using ultraviolet light, which releases the trapped charges.

The EEPROM (electrically erasable programmable ROM), as the name indicates, is programmed and erased electrically. It differs, however, from the EPROM in the mechanism of programming. In the case of the EEPROM, tunnel current between the floating gate and substrate is used to inject or extract charge carriers into and from the floating gate, depending on the bias of the control gate.

The serial memory consists of a number of cascaded sample and hold circuits. The data shifting is controlled using transmission gates that are driven by a shift clock. The sample and hold circuits are either static or dynamic.

The SRAM consists of a bistable circuit, as shown in Fig. 1, which can be connected to the bit lines through two access transistors. Those are turned on when the word line is selected. Data is read by sensing the differential signal between the bit lines. Writing is realized by forcing the input data on the bit lines.

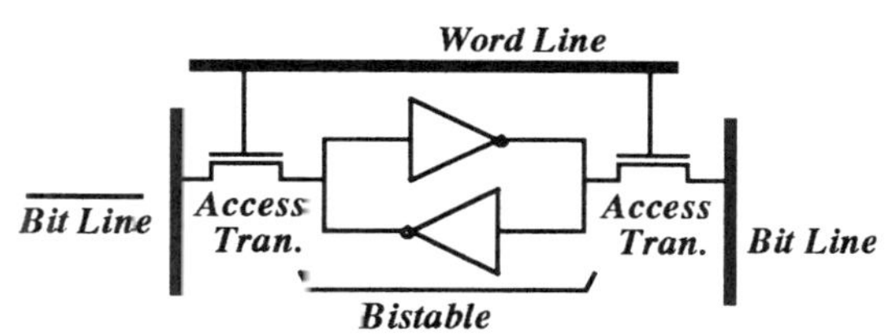

Fig. 1 The SRAM cell.

The author was previously with AT&T Bell Labs., Reading, PA.

The CAM (content addressable memory) uses a writing scheme identical to the SRAM. In the reading operation, data on the input (bit) lines are compared to the data stored in the CAM cell. When all the bits of the input word match all the bits of one of the CAM words, the corresponding match line switches to logic 1. The CAM has fast READ and WRITE operations.

The DRAM cell stores the data temporarily on the storage capacitor C_S as shown in Fig 2. When the word line is selected, the access transistor connects the storage capacitor to the bit line where a READ signal (70–100 mV) is produced by charge sharing between C_S and the bit-line capacitor C_B. Due to leakage currents, the stored data need to be periodically refreshed.

The rationale for BiCMOS technology, its evolution, and future challenges are presented in Section 2. The rapidly growing trend of embedding memories on the chip is discussed in Section 3. In Section 4, the evolution of standard RAM chips is presented. The impact of the revolutionary progress in processing technology and circuit design on the density, power dissipation, and speed of the RAM chips is demonstrated. Different approaches to increase the RAM speed are described to highlight the advantages of BiCMOS and future trends. In Section 5, state-of-the-art BiCMOS RAMs are discussed to illustrate the BiCMOS design strategy and circuit techniques.

2. Evolution of BiCMOS Technology

RAMs have always been the technology driver. They have a large regular structure and require small line pitches and uniform device parameters across the wafer. These characteristics make them ideal for testing and evaluating new processing techniques.

2.1 CMOS RAM Technology

The CMOS technology has become dominant in the fabrication of VLSI chips because it provides high density of integration, low power dissipation, high switching speed and large noise margin. A cross section in a typical DRAM N-well CMOS technology [1], [12] is shown in Fig. 3. The process uses LOCOS isolation, salicided source/drain and gate, and trench capacitor.

2.2 BiCMOS RAM Technology

CMOS and advanced bipolar technologies use many similar processing techniques and steps. Merging the two technologies results in the BiCMOS technology that provides both bipolar and MOS devices on the same chip. BiCMOS combines

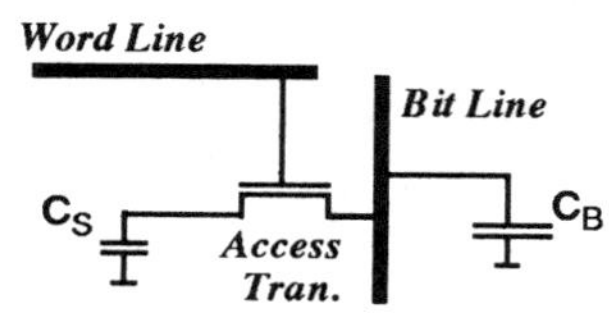

Fig. 2 The DRAM cell.

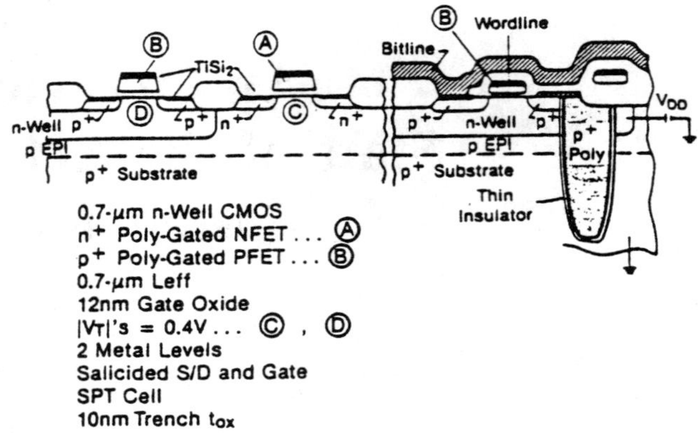

Fig. 3 The CMOS structure [1], [12].

the high-density and low-power dissipation of CMOS with the higher current capability and transconductance of the bipolar transistor. This results in increasing the switching speed by a factor of 1.5–2.0 in the case of driving large capacitive loads.

A typical BiCMOS structure, used in the fabrication of a 4-Mb SRAM [2], is shown in Fig. 4.

A comparison between the two technologies shows that the BiCMOS gate has a larger area than that of the CMOS. BiCMOS also costs more than CMOS because of the increased number of masks and processing steps. A BiCMOS process typically uses three to four additional masks compared to a CMOS process.

2.3 Rationale for BiCMOS

Before the introduction of BiCMOS technology, the situation can be illustrated as shown in Fig. 5.

Bipolar devices are used to implement very high speed-logic functions, whereas CMOS is used to implement very high

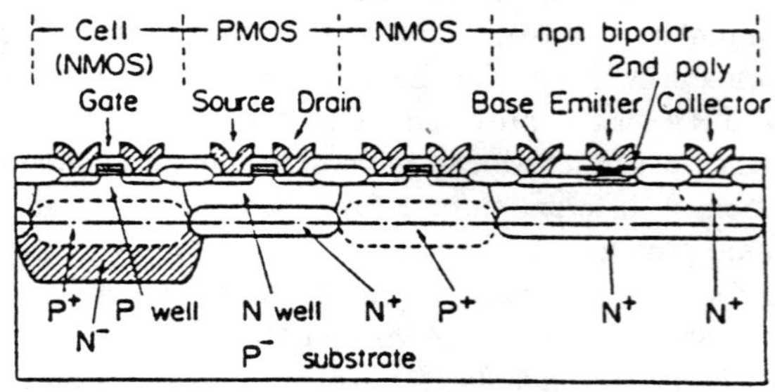

Fig. 4 The BICMOS structure [2].

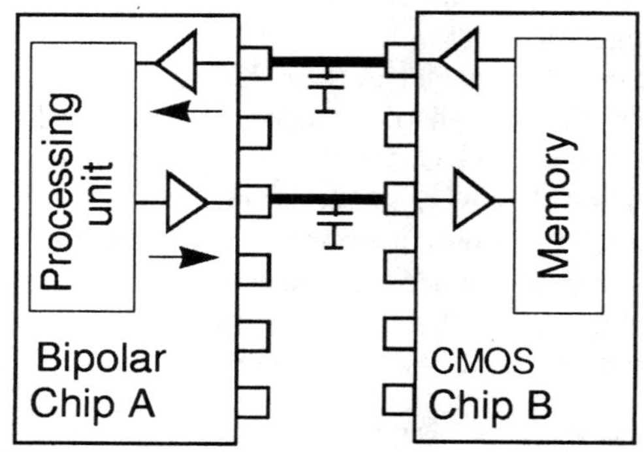

Fig. 5 Additional delay due to off-chip capacitance.

density and low-power functions, such as memories. A request originating at chip A will experience an additional delay due to the interchip communication that is equal to twice the output buffer and the input buffer delays. This accounts typically for an additional 40% delay. The BiCMOS technology solves this problem by integrating both functions on the same chip.

There are three levels of BiCMOS merging. These are *technology* merging, *gate* merging, and *device* merging. Integrating the bipolar and MOS transistors on the same chip (technology merging), results in superior performance in many applications. By eliminating the interchip delays, the system speed is increased.

The mixed use of the MOS and bipolar transistors to form a logical gate is referred to as gate merging. The BiCMOS driver of Fig. 6 illustrates the concept of gate merging. A BiCMOS driver has lower power dissipation compared to a bipolar driver and higher switching speed than the CMOS-equivalent driver. Taking advantage of BiCMOS drivers and gates in the design of the memory block reduces the memory-access time and further increases the system speed.

The third level of merging occurs on the device level (device merging), where the PMOS transistor and the NPN transistor of the BiCMOS driver shown in Fig. 6 share two regions in the structure [16], [17], [27]. The N-well works as the collector of the bipolar transistor and the electrical substrate of the PMOS transistor. The P-base region of the bipolar transistor is merged with the P^+ drain of the PMOS transistor. This merging of the two devices results in a significant reduction in the gate area. Similar merged structures [24]–[26] can be used to implement very high density memory cells and logic functions.

2.4 Future Challenges

BiCMOS, however, faces a challenge in the future as the trend of reducing the supply voltage continues. With scaling down the MOS devices, the supply voltage is reduced. This voltage reduction is driven by the required MOS device reliability.

Lower supply voltage is also necessary to reduce the power dissipation that is desirable for the following reasons. Power dissipation can be the limiting factor for the maximum number of devices per unit area (density of integration). It also can limit the maximum operating speed. Another important reason for reducing the power dissipation is the growing market for battery-operated products.

As the supply voltage is reduced, the conventional BiCMOS gate loses its speed advantage over the CMOS gate. As BiCMOS is scaled to 0.5 μm and towards 0.2 μm by the end of this decade, innovative new BiCMOS circuit structures, compatible with the reduced supply voltage, will be required [6], [23] to maintain BiCMOS speed superiority over CMOS.

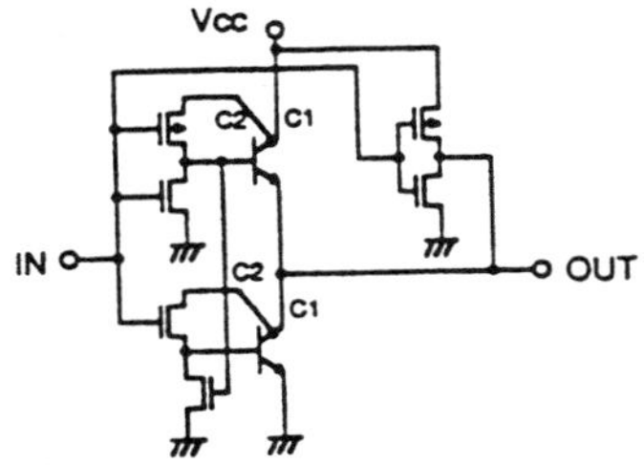

Fig. 6 A BiCMOS driver [14].

3. Embedded Memories

The rapidly growing trend of integrating memories with other processing functions in the design of high-performance application-specific integrated circuits (ASICs) is briefly discussed in this section.

This approach has become practically very efficient due to the dramatic increase in the level of integration (maximum number of devices per chip) and the maturity of sophisticated computer-aided synthesis and verification tools.

Figure 7 illustrates the structure of an ASIC chip where different memory blocks are embedded with logical functions. These functions may be implemented using programmable-logic-array (PLA), standard-cells, or custom-cells design styles.

3.1 Rationale

Embedding memories on the chip increases the system speed [29]. This is achieved by eliminating the input/output delays caused by driving large off-chip loads. Embedded memories have larger memory bandwidth (number of bits accessed per second) as there is no restriction on the number of I/O pins.

Embedded memories increase the ASIC's functionality and level of system integration. They provide lower power dissipation and system cost.

3.2 Automated Synthesis and Verification

To take full advantage of the embedded memory approach, a reliable and efficient memory compiler is needed to automatically generate the customized memory block [21], [22]. The compiler's function is illustrated in Fig. 8. It takes the user's specifications (type, size, organization, speed, aspect ratio,

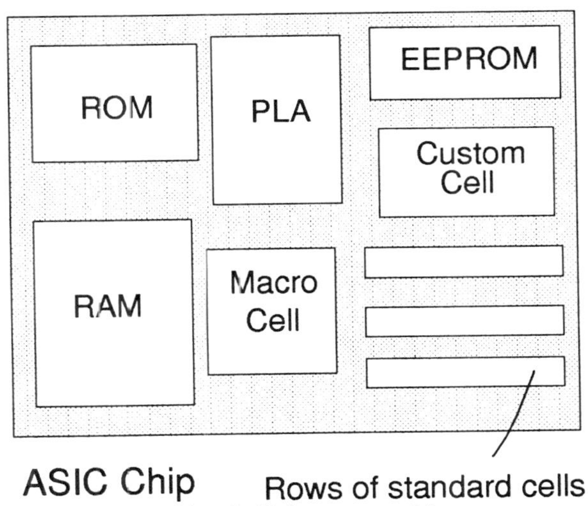

Fig. 7 An ASIC chip with embedded memory modules.

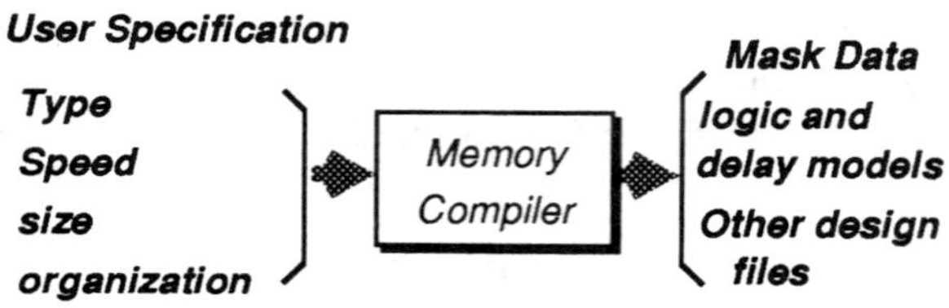

Fig. 8 A memory compiler.

etc.) as an input. The compiler provides as an output the verified design data. This includes the mask data for fabrication, the circuit model for accurate simulation, the logic model, delay parameters, and logic symbol for logic and functional simulation. It also generates the layout boundary and input/output pin and signal tables for automatic placement and routing.

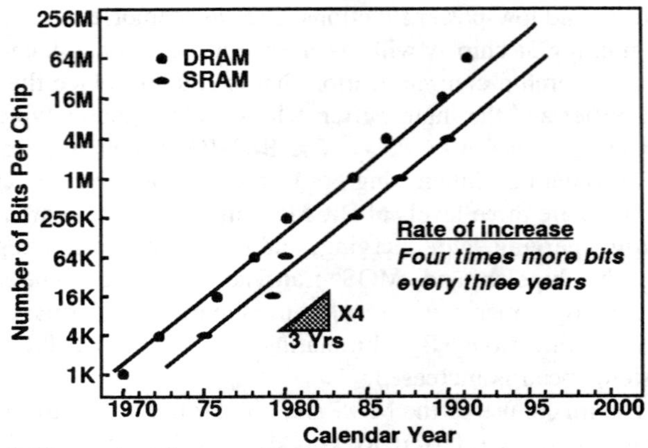

Fig. 9 The number of memory bits per chip.

3.3 Embedded versus Standard Memories

There are major differences between embedded memories and standard memory chips. Embedded memories normally use standard CMOS or BiCMOS processes with limited modifications, while standard memories use processes that are especially tailored to satisfy the requirements of those RAM chips. This mainly impacts the area of the memory cell and makes it larger in the case of embedded memories.

The second difference is in the design philosophy. The embedded memory design should be modular in order to support the automated synthesis of a very large number of different RAM configurations and sizes.

Embedded memories use different testing methods because their address and data terminals are not available at the external pins of the ASIC chip. Therefore, built-in self-testing is usually used. Also, design verification of the generated modules is automated.

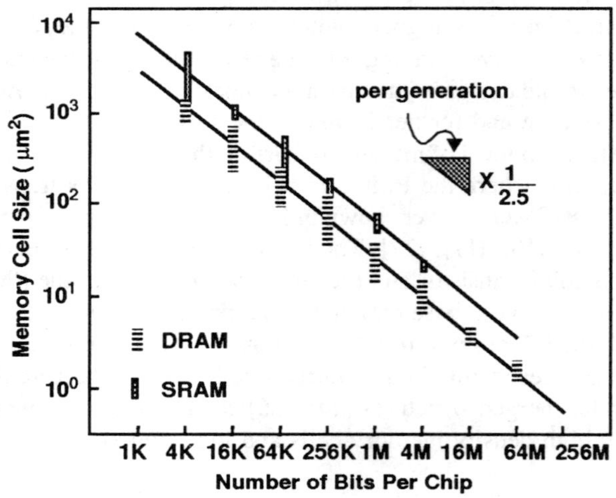

Fig. 10 Memory cell size.

4. Evolution of the Static and Dynamic RAM

The main objectives in the design and fabrication of the standard RAM chips are to achieve:

- largest number of bits per chip,
- highest yield,
- highest speed, and
- lowest power dissipation.

In this section, the progress that has been made during the past two decades to achieve these goals is presented. The processing and design techniques used are discussed.

4.1 The Number of Bits per Chip and the Yield

Figure 9 shows the number of bits per chip versus the calendar year. The number of bits is quadrupled every three years for both the SRAM and DRAM. The rate of increase does not show any sign of slowing down.

In order to increase the number of bits per chip, the size of the memory cell should be reduced and the chip area increased.

Reducing the memory cell size. Figure 10 shows the size of the memory cell versus the number of bits per chip. The size of the memory cell is reduced by a factor of 2.5 per generation[1] of the RAM. In order to achieve this rate of reduction, the following techniques have been used.

- *Smaller feature size.* This requires advances in lithography, processing, and proper scaling of the MOS devices. Figure 11 shows that the feature size has been reduced by a factor of 1.56 per RAM generation.
- *Advanced device isolation techniques.* Isolation techniques include trench isolation and advanced LOCOS isolation.
- *Three-dimensional memory cells.* This includes the use of trench capacitors [1], [12] and transistors, stacked capacitors (DRAM) [3], thin film and polysilicon load transistors (SRAM), and multilayers of polysilicon and metal [4]. Figures 12 and 13 show the three-dimensional DRAM [3] and SRAM [4] cells, respectively.
- *Advanced interconnect and contact technology.* This includes the use of polycides and silicides, vertically

[1] A RAM generation refers typically to a three-year period. It has been the average design and development time needed to quadruple the number of memory bits per chip.

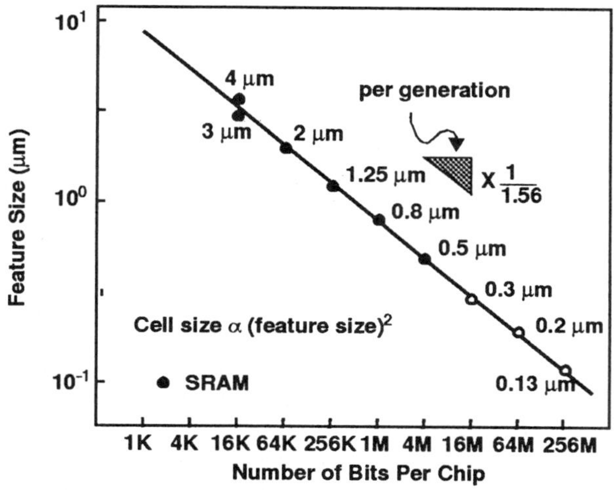

Fig. 11 Process feature size.

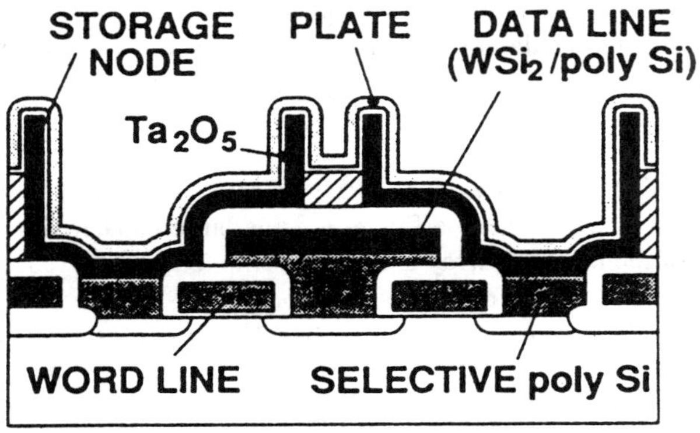

Fig. 12 The DRAM cell [3].

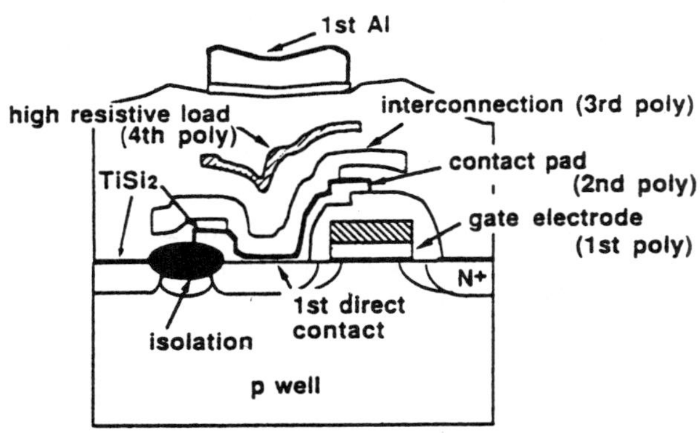

Fig. 13 The SRAM cell [4].

stacked contacts and vias, buried contacts, and high-resistivity polysilicon loads (SRAM).

Increasing the chip area. The other contributing factor to increasing the number of bits per chip is the RAM chip area (die size).

Each process is characterized by a certain defect density. The defect density and the chip area determine the yield. Figure 14 illustrates the dependence of the yield on these two factors and its impact on the cost. The chip cost is determined by the process yield and the number of chips per wafer, which is in turn dependent upon the chip area for a given wafer size.

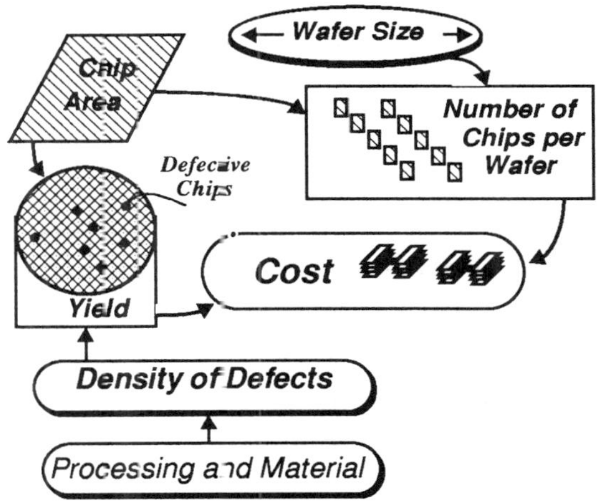

Fig. 14 Dependency of the yield and cost on chip area and defect density.

Therefore, for a given defect density, the yield and cost limit the maximum area of the chip.

Figure 15 shows that the chip size increases by a factor of 1.2 per RAM generation. The 20% increase in the area results from the reduction in the defect density due to improvement in the materials and advancement in the processing techniques. It is noted that there is a kink in the curves. This transition takes place between the 1-Mb and 4-Mb DRAMs and for the corresponding generations of SRAMs. This abrupt increase in the chip area is partly due to the introduction of redundant word and bit lines and error correction code (ECC) circuits.

The redundancy and ECC circuits increase the overall functional yield and reliability. The gain in the yield can be traded off for larger chip area.

The use of redundancy and ECC circuits is summarized as follows.

Once the RAM chip is tested, the addresses of the faulty bit and word lines are stored in fuse bays [5] by blowing the fuses using a laser. When a defective line is accessed, the data on the address bus is compared with the data stored in the fuse bay. If the addresses match, the redundant line is used to replace the selected line in the main memory array.

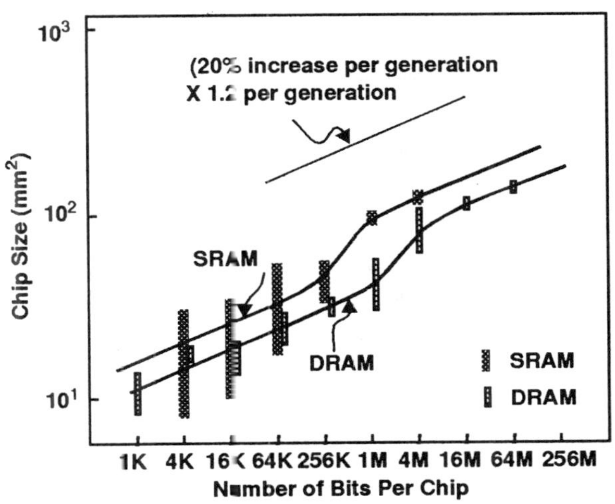

Fig. 15 Memory chips size.

The ECC logic tests the data bits using the check bits. The logic detects which bits of the ECC word are faulty. They are then corrected [5]. Figure 16 shows the impact of using redundancy and ECC logic on the yield.

4.2 RAM speed

Several approaches can be followed, either separately or combined, to increase RAM speed. These can be classified as *technology, device, circuits,* and *architecture* approaches.

Technology approaches.

- *Smaller feature size.* By using a technology that has a smaller feature size, the parasitic capacitances are scaled down. This causes the switching speed of the capacitive nodes to increase.
- *Polycides and silicides.* The technology also provides means for reducing the interconnect delay times by employing polycides and silicides to reduce the resistance of the polysilicon and drain and source regions.
- *Multilayer metallization.* The availability of several layers of metal and polycides further reduces the interconnection length. This reduces the parasitic resistance and capacitance, resulting in shorter delay times.
- *Compact isolation and vertically stacked vias.* Similarly, using advanced compact device isolation techniques, vertically stacked vias and contacts results in a smaller chip area and therefore less parasitic interconnect resistance and capacitance.
- *BiCMOS technology.* BiCMOS technology enables the integration of very high speed functions, implemented, using bipolar transistors, and memory blocks, using CMOS, on the same chip. This eliminates the additional interchip delay times, therefore reducing the effective memory access time.

Device approaches.

- *Scaling down of MOS transistors.* Scaled-down devices provide larger current if the supply voltage is not scaled down. However, scaling down the MOS transistor to 0.5 μm and below requires that the supply voltage be reduced. This, together with the degradation of the carrier mobility with increasing electric field, cause the current available from the scaled-down device to remain almost constant (or slightly reduced). Although the current of the scaled-down device will remain almost unchanged, the transistor input capacitance will be reduced by the scaling factor. Therefore, the scaling down of the MOS transistor below 0.5 μm will cause the rate of increase in the switching speed to be less than that experienced in the previous RAM generations.
- *MOS and bipolar-device merging.* In BiCMOS technology, merging of the MOS and bipolar transistors can result in very high density and fast circuit structures that can be used to implement storage and logical functions [16], [17], [27], [24]–[26].

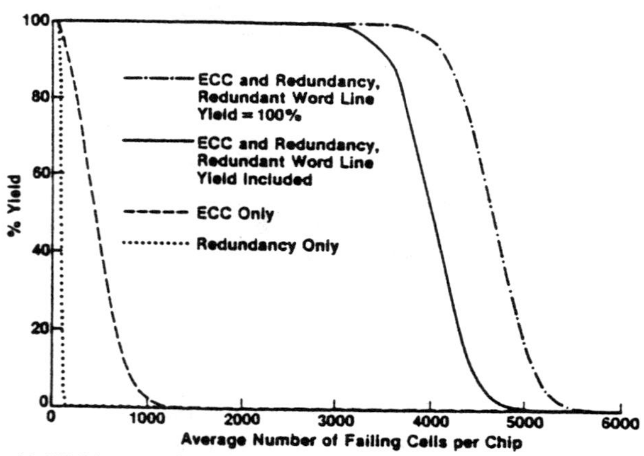

Fig. 16 Yield curves for ECC and word-line redundancy [5].

Circuit approaches. Besides the technology and device approaches, innovative circuit techniques play a key role in achieving very short access time. The following highlights the successfully used circuit techniques.

With the availability of the BiCMOS technology, a much larger flexibility and significant progress has been made to reduce the access time.

- *BiCMOS gates and drivers.* BiCMOS gates and drivers have been used to efficiently drive large capacitive loads of the memory array and off the chip loads, thus reducing the associated delays by a factor of 1.5–2.
- *BiCMOS sense amplifiers.* BiCMOS provides very high gain sense amplifiers that significantly enhance the reading speed.
- *BiCMOS peripheral circuits.* BiCMOS circuits are excellent in accurately generating reference voltage and current sources [11], [13] and have the advantage in generating control signals and clocks [6] while improving the skew by a factor of 2.

Other circuit techniques include the following:

- *Address transition detection (ATD).* The address transition detection (ATD) is used to asynchronously initiate a new cycle in the SRAM if the address is changed. This self-timed operation reduces the access time.
- *Static decoding and direct sensing of DRAMs.* The use of static column decoding for the DRAM enables a fast access mode of operation. The direct-sensing scheme [15], employed in the DRAM design, significantly reduces the access time.
- *Back bias generators.* Back bias generators can be used to increase the reverse voltage across the source/drain to the substrate PN junction, which results in smaller source/drain parasitic capacitances. This enables higher switching speed.
- *Multistage sense amplifiers.* Multistage sense amplifiers and column decoding are used to achieve the minimum access time for large memory arrays.

Architectural approaches. The architectural approaches make use of array partitioning, hierarchical word-line [4], and divided bit-line techniques. This reduces the capacitive loads that need to be driven and provides multistage decoders and drivers to efficiently minimize the decoding and sensing delays.

Figure 17 shows that the access time has been reduced by a factor of 2 per generation of RAM. This rate is expected to slow down due to the reduced carrier mobility and power-supply voltage for MOS devices with channel length less than 0.5 μm. However, the recent progress in the design of BiCMOS circuits that operate at reduced supply voltages [17]–[20] is very promising in maintaining the trend of access-time reduction that has been achieved in the past.

4.3 Reducing the power dissipation

Several techniques are used to reduce the power dissipation.

ATD and switched sense amplifiers. The use of the ATD minimizes the power dissipation in the sense amplifiers by providing a self-timed sense-enable pulse that turns the sense amplifiers ON only during a small part of the READ cycle. The sense amplifiers are enabled when a change in the address occurs and a READ operation is requested. After the data is sensed and latched at the output, the sense amplifiers are disabled to reduce the power dissipation.

Array partitioning. Partitioning the array into blocks and sections enables the activation of only the portion of the memory that contains the destination address. This saves the dynamic power consumed in charging and discharging the bit and word lines in the other blocks of the memory and the associated sense amplifiers.

Lower leakage and MOS subthreshold currents. Improvements in processing result in lower subthreshold current of the MOS transistor. This allows using a large value of the load resistors in the design of the 4-Transistor SRAM cell. This is because, for the static mode of operation, the current supplied by such a resistor is only about 2 orders of magnitude larger than the subthreshold current of the MOS transistor.

Lower subthreshold current and corresponding larger value of the resistors result in significant reduction of the standby power, which is inversely proportional to the value of the resistors.

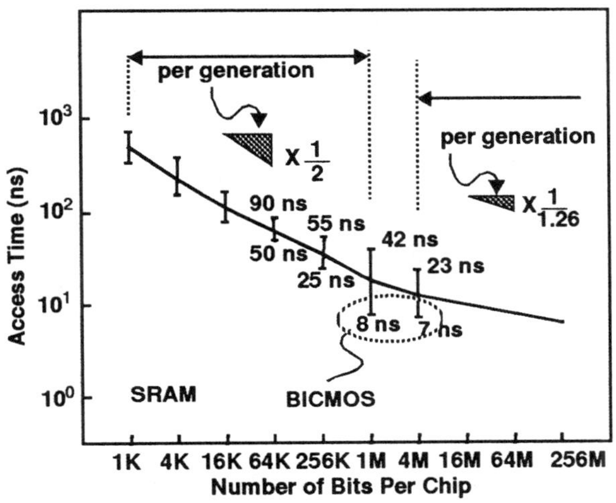

Fig. 17 Memory access time.

Reducing the power-supply voltage. The reduction of the power-supply voltage, a result of scaling down of the MOS transistor, dramatically reduces both the standby and dynamic components of the power dissipation.

Reducing the parasitic capacitance. All the techniques used to reduce the parasitic capacitance, discussed earlier, also reduce the dynamic power dissipation.

5. State-of-the-Art BiCMOS RAMs

In Section 5.1, the BiCMOS RAM design strategy is presented. Then, in Sections 5.2 and 5.3, typical BiCMOS SRAM and DRAM circuits and designs are used as examples to highlight the implementation of such design strategy and the techniques discussed in Section 4.

5.1 Design Strategy

The design strategy of BiCMOS RAMs is summarized in the following points. The justification for the choice between CMOS and BiCMOS circuits to implement different subcircuits of the RAM is provided.

CMOS for high density and low power. In a typical 0.8-μm BiCMOS process, the area of a minimum-size bipolar transistor is 20–40% larger than that of a minimum-geometry MOS device [6]. Therefore, the strategy of the BiCMOS RAM design is to use the CMOS circuits in areas where the high packaging density and low power dissipation are essential (for example, the core section of the memory chip).

BiCMOS for driving large capacitive loads. The relatively large bipolar transistor current per unit area justifies the use of BiCMOS drivers in parts of the chip where large driving currents are required. This is the case in fast switching of large capacitive nodes such as word lines, data lines, (write drivers), and output pads (output buffers).

BiCMOS for high-gain sense amplifiers. BiCMOS is chosen to implement high-gain sense amplifiers in order to speed up the reading operation in the RAM design. The choice of BiCMOS is based on the following reasons.

- *Smaller area.* Comparing the transconductance expressions of both the bipolar and MOS transistors [6] reveals that the area of the MOS transistor must be at least an order of magnitude larger than that of the bipolar transistor to match its transconductance.
- *Higher sensitivity.* In addition, the exponential dependence of the transconductance of the bipolar transistor on the base-emitter voltage leads to a signal-detection sensitivity superior to that of the MOS device [6].
- *Smaller offset voltage.* The bipolar transistors have another advantage resulting from their good matching characteristics. The offset voltage of a typical bipolar differential amplifier is under 1 mV, which compares favorably with an offset voltage of 5–10 mV for the MOS differential amplifiers [6].

BiCMOS for reference sources and I/O interface to TTL and ECL. The use of BiCMOS circuits provides a high flexibility for the input/output interface to TTL, ECL, and CMOS circuits. These voltage levels can be easily generated and converted to each other on the same chip.

BiCMOS circuits are also used for generating reference voltage and current sources accurately [11], [13].

BiCMOS for generation and distribution of clocks. The generation and distribution of the control and other clock signals over the RAM chip is greatly aided by using BiCMOS circuits, while at the same time improving the skew by a factor of 2 [6].

BiCMOS-inherent temperature compensation for deep submicron RAMs. Due to the scaling down of the MOS device [30] and the need for lower power dissipation, the power-supply voltage is reduced. This degrades the noise margin and speed.

The situation is made worse by the increase in the noise voltage due to increased capacitive coupling between the bit lines, word lines, and the devices as a result of the smaller feature size.

With the reduced noise margin, the memory functionality and speed become increasingly sensitive to device-parameter variations due to processing tolerances and operating temperature range.

With increasing temperature, the bipolar transistor current increases, while the MOS transistor current decreases. BiCMOS circuits can be operated over a wider temperature range due to the inherent temperature compensation provided by the mixed use of MOS and bipolar transistors.

The recently reported BiCMOS circuit techniques [17]–[20] are very promising for applications in future deep submicron BiCMOS RAMs.

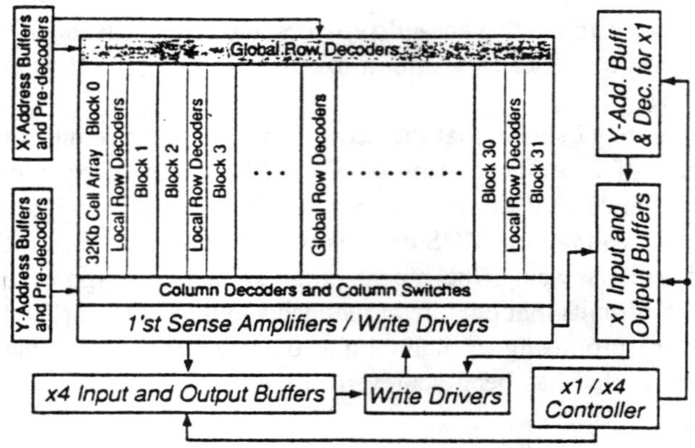

Fig. 18 Chip architecture of a BiCMOS ECL SRAM [10], [28].

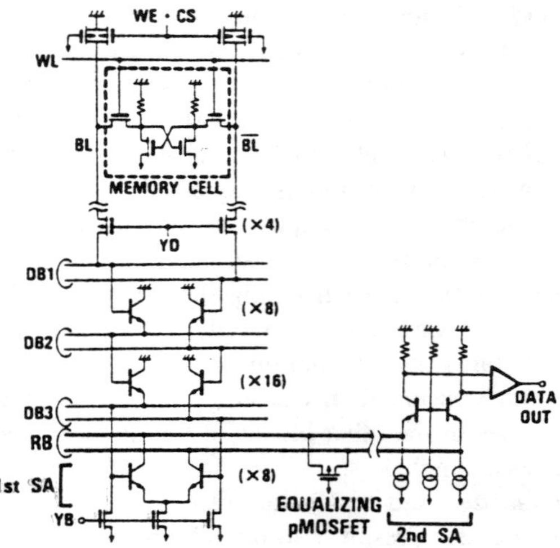

Fig. 19 Bit-line peripheral and sensing circuits [9], [28].

5.2 BiCMOS SRAM

The SRAM design examples presented in this subsection highlight the use of the following design strategy and techniques:

- Array partitioning and hierarchical word-line architecture;
- Redundant rows and columns;
- Hierarchical sensing using multistage BiCMOS amplifiers;
- BiCMOS address transition detection (ATD);
- BiCMOS address buffer with ECL input signal interface;
- On-chip BiCMOS voltage and current sources.

A typical 1-Mb BiCMOS ECL SRAM chip architecture [10], [28] is shown in Fig. 18. A divided word-line architecture is used to obtain both a fast access time and low power consumption. A 1-Mb memory-cell array is divided into 32 blocks. Each block includes 512 rows and 64 columns. The RAM has four redundant columns and two redundant rows in each block. The global word line is connected to four local word-line drivers in each block. Each block is further divided into four subblocks, each consisting of 16-plus-one-spare columns, a pair of the first-stage sense amplifiers, and the WRITE amplifiers.

The bit-line peripheral and sensing circuitry of a 5-ns 1-Mb ECL BiCMOS SRAM [9], [28] is shown in Fig. 19. The differential readout signal from a memory cell is received by the first sense amplifier (SA) through a pair of bit-line selection pMOSFETs and two stages of emitter-follower (EF) wired-OR logic circuits. Then the amplifier readout signal is sent to the second sense amplifier through a pair of READ bus lines (RB). The differential bit-line voltage swing is limited to about 50 mV and the SRAM adopts a four-stage column selection scheme as a hierarchical sensing circuit design in order to increase the speed of sensing.

Another example of a current mirror BiCMOS sense amplifier [7] is shown in Fig. 20. A small bit-line signal is detected by a first-stage current mirror bipolar sense amplifier with MOS-switched current sources. A second-stage CMOS sense amplifier is used and followed by a CMOS data latch.

A BiCMOS ATD circuit is shown in Fig. 21 [7]. All address changes are first converted to current signals. They are connected together in a wired-OR gate configuration, detected by a bipolar transistor, and finally reconverted to a voltage signal by a resistor. The output is amplified by a bipolar differential amplifier with an MOS current source followed by a CMOS current mirror amplifier stage. The voltage is then converted to CMOS levels.

The BiCMOS address buffer (Fig. 22) interfaces with an ECL input signal and provides CMOS voltage levels at the

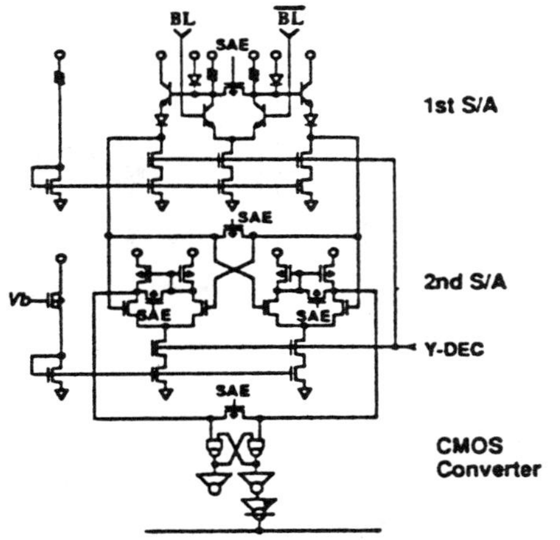

Fig. 20 A BiCMOS sense amplifier [7].

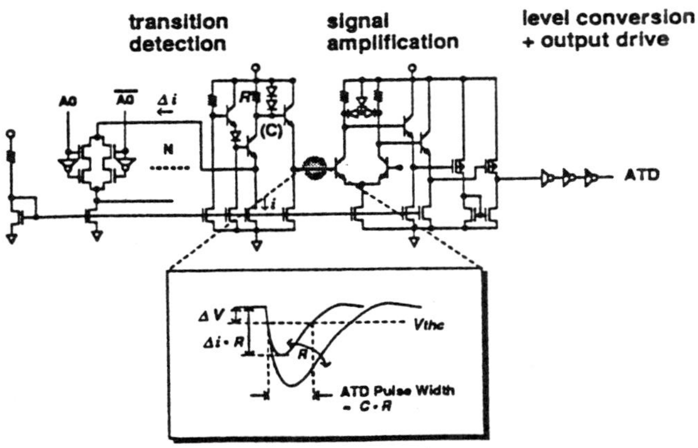

Fig. 21 A BiCMOS address transition detection circuit [7].

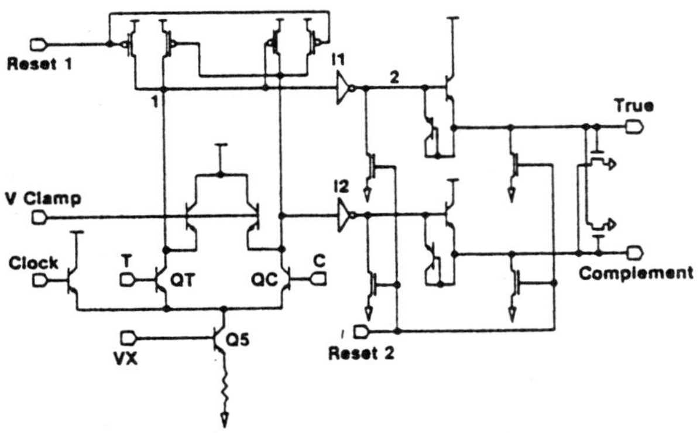

Fig. 22 An ECL-CMOS address buffet [8].

output [8]. Inputs T and C are ECL signals, CLOCK is an ECL clock signal generated on chip, and VX is a bandgap reference voltage of 1.2 V which is generated on the chip. When CLOCK falls below T (assuming T is higher than C), the current set by $Q5$ is switched over to QT, causing node 1 to fall, node 2 to rise, and a true signal of $(V_{dd}-V_{be})$ to be generated.

Figure 23 shows an example of an on-chip intermediate-voltage-level generator of an ECL interface BiCMOS SRAM [11]. The generated voltage level (V_{ss}) is intermediate between the negative ECL supply voltage V_{EE} and 0 V. In this design, (V_{ss}) is used for the MOS memory-cell array.

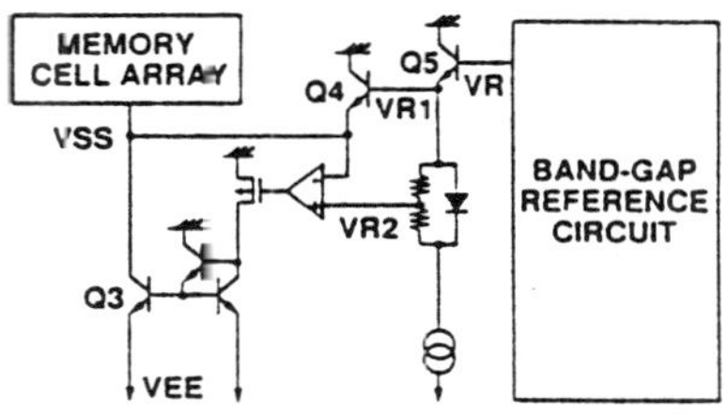

Fig. 23 On-chip voltage generator [11].

5.3 BiCMOS DRAM

In this section, the DRAM design examples demonstrate the use of the following design strategy and techniques:

- Direct-sensing technique using multistage BiCMOS amplifiers;
- Internal power-supply voltage for the MOS DRAM core and ECL interface
- BiCMOS input/output buffers with ECL interface;
- BiCMOS constant current generators;
- Three-dimensional stacked capacitor memory cell;
- Array partitioning and hierarchical word-line decoding.

A BiCMOS-sensing technique called direct-sensing [15] is illustrated in Fig. 24. A highly sensitive bipolar cascode amplifier combined with the CMOS rewrite amplifier are shown. The bipolar amplifier detects the small READ signal on the data line faster than the CMOS rewrite amplifier. In the conventional CMOS sensing scheme, the timing margins required between the word pulse, rewrite amplifier activation pulses, and the Y-decode control pulse connecting the data line to the I/O line are the origin of the relatively long access time associated with the DRAM. The direct-BiCMOS-sensing technique eliminates these timing margins and therefore allows high-speed operation.

Several circuit techniques have been reported to implement a sub-10-ns ECL 4-Mb BiCMOS DRAM using a 0.3-μm BiCMOS process [13]. The design uses an internal power-supply voltage scheme to provide the proper voltage levels for the MOS DRAM core and the BiCMOS ECL interface. It also uses BiCMOS input/output buffers, highly sensitive amplifiers, and voltage limiters.

Figure 25 shows an address buffer circuit with a reset function that is necessary to ensure the precharge operation and a

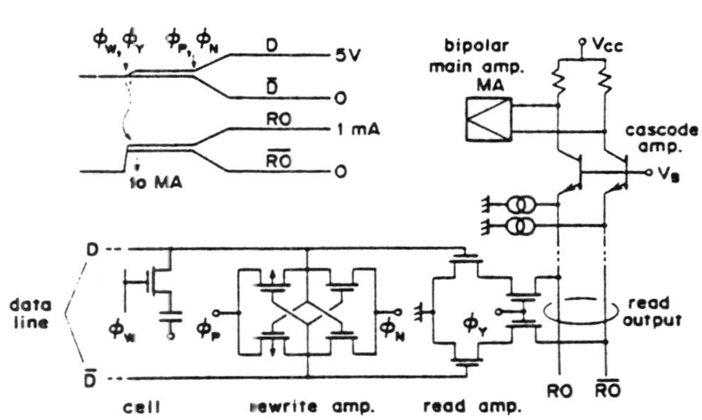

Fig. 24 BiCMOS direct-current sensing [15].

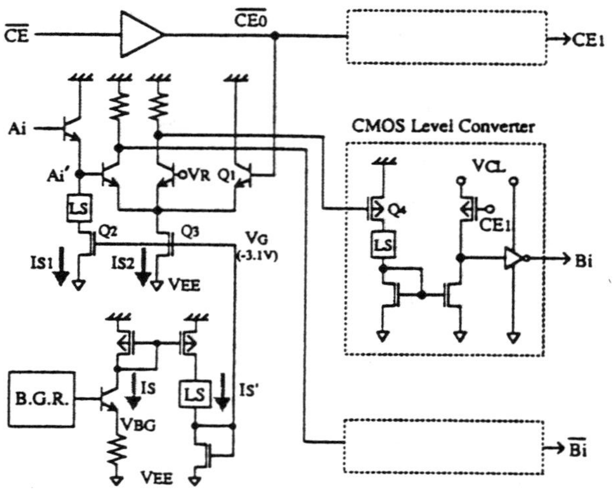

Fig. 25 Address-buffer-associated circuit with reset and level conversion functions [13].

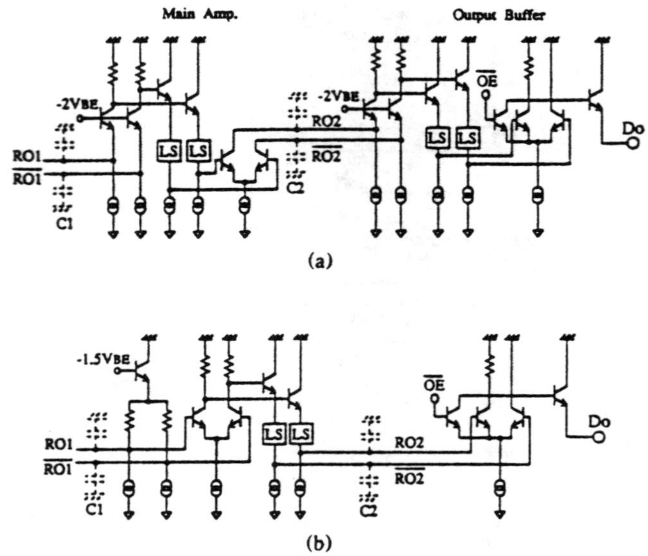

Fig. 27 Main amplifier circuitry [13]. (a) current sensing; (b) voltage sensing.

level converter from ECL level to the internal CMOS level. The figure also shows a BiCMOS (double current mirror) constant-current generator that is less dependent on the MOS device parameter's fluctuations.

Figure 26 shows a 4-Mb sense circuit [13]. It includes an overdrive rewrite amplifier to enhance the speed for the reduced power-supply voltage. It also takes advantage of the direct-sensing technique combined with bipolar cascode connections for both the main amplifier and output buffer.

A double-stage current sensing circuit is shown in Fig. 27(a), whereas a voltage sensing is illustrated in Fig. 27(b). The current sensing is reported to be 10% faster than the voltage sensing [13]. This design achieves a 7.8-ns access time. It uses a crown shaped, stacked-capacitor memory cell. The memory-cell array is divided into blocks and the word line is divided into four sections to reduce the word-line delay.

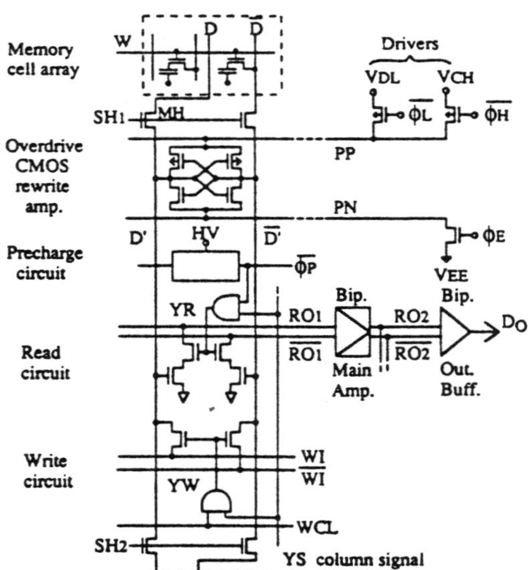

Fig. 26 BiCMOS sense circuitry [13].

6. Conclusions

In the past two decades, the process feature size has been reduced by a factor of 20, the cell size by a factor of 2000, and the access time by a factor of 100. The chip area grew by a factor of 25 and the number of bits by a factor of 64,000. Based on this trend, it is predicted that, by the year 2000, the feature size will be 0.15 μm and the chip area will be 350 mm^2. The DRAM and SRAM cell sizes will be 0.3 μm^2 and 1.2 μm^2, respectively. This will enable the fabrication of a 1-Gb DRAM chip with an estimated access time of 15 ns. SRAM will reach the 256-Mb chip level with an estimated access time of 7.5 ns.

It will be increasingly difficult for the scaled-down MOS devices and the CMOS-only circuits with reduced power-supply voltage to achieve the predicted performance. This is due to the carrier mobility degradation, reduced sense signal swing and noise margins, increased capacitive coupling noise, and the increasing sensitivity of the MOS-device parameters to processing tolerances and temperature changes.

The superior bipolar transconductance and better matching characteristics of the bipolar transistors will provide more sensitive sense amplifiers. BiCMOS circuits are superior in implementing on-the-chip robust constant voltage and current sources, clock generators and drivers, input/output buffers with ECL and TTL interface, and word-line and bit-line drivers. Moreover, BiCMOS circuits have more tolerance to temperature variation.

The bipolar transistor, unlike the MOS device, is a current-driven device. The power-supply voltage reduction will not reduce its current-driving capability.

For these reasons, BiCMOS technology will play an important role in achieving the predicted performance by maintaining the current rate of reduction in the access time for the deep submicron RAM chips.

REFERENCES

[1] W. H. Henkels et al., "A 4 Mb low temperature DRAM," in *ISSCC'91 Tech. Dig.* pp. 30–31, Feb. 1991.

[2] Hiroshi Shimada et al., "10 ns 4 Mb BICMOS TTL SRAM," in *ISSCC'91 Tech. Dig.*, pp. 52–53, Feb. 1991.

[3] Yoshinobu Nakagome et al., "An experimental 1.5 V. 64 Mb DRAM," *IEEE JSSC*, vol. 26, pp. 465–472, April 1991.

[4] Toshihiko Hirose et al., "A 20 ns 4 Mb CMOS SRAM with hierarchical word decoding architecture," *IEEE JSSC*, vol. 25, pp. 1068–1074, Oct. 1990.

[5] H. L. Kalter et al., "A 50 ns 16 Mb DRAM with a 10 ns data rate and on chip ECC," *IEEE JSSC*, vol. 25, pp. 1118–1129, Oct. 1990.

[6] A. Iranmanesh and B. Bastani, "BiCMOS emerges for gate-arrays and memories," *IEEE Circuits and Devices Magazine*, vol. 8, pp. 14–17, March 1992.

[7] Tadahiro Kuroda et al., "Automated bias control (ABC) circuit for high-performance VLSI's," *IEEE JSSC*, vol. 27, no. 4, pp. 649–656, April 1992.

[8] Henry A. Bonges et al., "A 576K 3.5-ns access BiCMOS ECL static RAM with array built-in self-test," *IEEE JSSC*, vol. 27, no. 4, pp. 649–656, April 1992.

[9] Masahide Takada et al., "A 5-ns 1-Mb ECL BiCMOS SRAM," *IEEE JSSC*, vol. 25, no. 5, pp. 1057–1062, Oct. 1990.

[10] Atsushi Ohba et al., "A 7-ns 1Mb BiCMOS ECL SRAM with shift redundancy," *IEEE JSSC*, vol. 26, no. 4, pp. 507–512, April 1991.

[11] Takakuni Douseki et al., "Fast-Access BiCMOS SRAM architecture with a V_{ss} Generator," *IEEE JSSC*, vol. 26, no. 4, pp. 513–517, April 1991.

[12] Walter H. Henkels et al., "A 4-Mb low-temperature DRAM," *IEEE JSSC*, vol. 26, no. 11, pp. 1519–1529, Nov. 1991.

[13] Takayuki Kawahara et al., "A circuit technology for a sub-10-ns ECL 4-Mb BiCMOS DRAM's," *IEEE JSSC*, vol. 26, no. 11, pp. 1530–1537, Nov. 1991.

[14] Goro Kitsukawa et al., "A 23-ns 1-Mb BiCMOS DRAM," *IEEE JSSC*, vol. 25, no. 5, pp. 1102–1111, Oct. 1990.

[15] Kiyoo Itoh, "Trends in megabit DRAM circuit design," *IEEE JSSC*, vol. 25, no. 3, pp. 778–789, June 1990.

[16] T. Hanibuchi et al., "A bipolar-PMOS merged basic cell for 0.8 μm BiCMOS sea of gates," *IEEE JSSC*, vol. 26, no. 3, pp. 427–431, March 1991.

[17] R. B. Ritts et al., "Merged BiCMOS logic to extend the CMOS/BiCMOS performance crossover below 2.5-V supply," *IEEE JSSC*, vol. 26, no. 11, pp. 1606–1614, Nov. 1991.

[18] K. Yano et al., "Quasi-complementary BiCMOS for Sub-3-V digital circuits," *IEEE JSSC*, vol. 26, no. 11, pp. 1708–1719, Nov. 1991.

[19] M. Hiraki et al., "3.3-V BiCMOS circuit techniques for 259-MHz RISC arithmetic modules," *IEEE JSSC*, vol. 27, no. 3, pp. 373–381, March 1992.

[20] C. L. Chen, "2.5-V bipolar/CMOS circuits for 0.25 μm BiCMOS technology," *IEEE JSSC*, vol. 27, no. 4, pp. 485'-491, April 1992.

[21] J. C. Tou et al., "A submicrometer CMOS embedded SRAM compiler," *IEEE JSCC* vol. 27, no. 3, pp. 417–424, March 1992.

[22] H. Shinohara et al., "A flexible multiport RAM compiler for data path," *IEEE JSSC*, vol. 27, no. 3, pp. 300–306, March 1991.

[23] L. Wissel et al., "Optimal usage of CMOS within a BiCMOS technology," *IEEE JSSC*, vol. 26, no. 3, pp. 343–349, March 1992.

[24] A. G. Eldin and M. I. Elmasry, "VLSI Dynamic Memory," United States Patent # 4,791,611. Dec. 13, 1988.

[25] A. G. Eldin and M. I. Elmasry, "New Dynamic Logic and Memory Circuit Structures For BICMOS Technologies," *IEEE J. Solid-State Circuits*, vol. SC-22, pp. 450–453, June 1987.

[26] A. G. Eldin and M. I. Elmasry, "A Novel JCMOS Dynamic RAM Cell For VLSI Memories" *IEEE J. Solid-State Circuits*, vol. SC-15, pp. 715–723, June 1985.

[27] A. Bellaouar and M. I. Elmasry, "Novel Merged BiCMOS Circuit Structures," *IEE Electronic Lett.*, pp. 1555–1556, 1990.

[28] M. I. Elmasry (ed.), "BiCMOS Integrated Circuits Design: with Analog, Digital, and Smart Power Applications," Part III: BiCMOS Memory Applications, IEEE Press, New York: 1994.

[29] M. I. Elmasry (ed.), "BiCMOS Integrated Circuits Design: with Analog, Digital, and Smart Power Applications," Part II: BiCMOS Digital Circuit Applications, paper 2.5. IEEE Press, New York: 1994.

[30] M. I. Elmasry (ed.), "BiCMOS Integrated Circuits Design: with Analog, Digital, and Smart Power Applications," Part I: Analysis and Design, papers 1.25–1.27. IEEE Press, New York: 1994.

Paper 3.2

BiCMOS Circuit Technology for a High-Speed SRAM

TAKAKUNI DOUSEKI AND YASUO OHMORI, MEMBER, IEEE

Abstract —**BiCMOS circuit technology for a high-speed and large-capacity ECL-compatible SRAM is described. To obtain high-speed and low-power operation, a decoder with a pre-main decode configuration having an ECL-interface circuit and a word driver with BiCMOS inverter are proposed. A BiCMOS multiplexer with a single emitter-follower driver is also proposed. An optimization method for memory cell array configuration which minimizes the total delay time and the total power dissipation of SRAM's is presented. Circuit simulation results show that a 64-kbit ECL-compatible SRAM with an access time of less than 7 ns and a power dissipation of less than 1 W is obtainable.**

I. Introduction

HIGH-SPEED and large-capacity ECL-compatible SRAM's play an important role in achieving high-performance computer systems. To realize ECL-compatible SRAM's, BiCMOS technology, which merges bipolar and MOS technology, is required [1]. BiCMOS process technology has been developed to enhance device characteristics. Using a thin epitaxial layer with a buried well, MOSFET's with a 1.0–1.3-μm gate length and bipolar transistors with a high cutoff frequency ($f_T = 2$–6 GHz) were fabricated on the same chip [2], [3].

BiCMOS circuit technology was first applied to logic circuits [4]. With respect to SRAM applications, BiCMOS NAND/NOR gates and BiCMOS inverters [5], [6] have been used in the decoder and driver, and BiCMOS sense amplifiers and emitter-follower circuits [5]–[7] have been used in the multiplexer and the output buffer. These circuits, however, cannot obtain high-speed operation as bipolar circuits because the logic swing and circuit stages are large. The application of bipolar circuits with small logic swings to large-capacity SRAM's has been limited because of their large power dissipation.

This paper describes a BiCMOS circuit technology to overcome the above-mentioned problems. In Section II, a high-speed and low-power decoder and driver are proposed. Section III presents a BiCMOS multiplexer with a single emitter-follower circuit for the data-line driver. The simulated total performance of a 64-kbit ECL-compatible SRAM is presented in Section IV, and a summary is given in Section V.

Manuscript received July 2, 1987; revised September 9, 1987.
The authors are with NTT LSI Laboratories, 3-1 Morinosato Wakamiya, Atsugishi, Kanagawa, 243-01, Japan.
IEEE Log Number 8718227.

II. Decoder and Driver

The delay time of the decoder and driver occupies more than half the access time of SRAM's [5]. To reduce the delay time of the decoder and driver, circuit stages and the logic swing of the decoder must be small with small power dissipation. In this paper, a decoder and driver with a pre-main decode configuration having an ECL-interface circuit and with a BiCMOS inverter are proposed.

Two supply voltages are used in this work. One is a V_{EE} of -5.2 V and the other is a V_{EE1} of -3.0 V. The former voltage is used for the peripheral circuits including the decoder, and the latter voltage is used for the memory cell array and word driver to ensure the reliability of the submicrometer MOSFET's which are used to realize large-capacity SRAM's. "High" and "low" MOS levels are 0 and -3.0 V, respectively. ECL input signals are converted by the decoder and driver to MOS levels.

A. Decoder Circuitry

A pre-main decode configuration with the ECL-interface circuit [8] is shown in Fig. 1. The decoder is composed of a predecoder, a level shifter, and a main decoder. The predecoder detects the ECL level and the main decoder amplifies small signals to the MOS level.

The collector-dotting circuit [9] for the predecoder and the emitter-follower circuit for the level shifter are shown in Fig. 2. The delay time of the predecoder is reduced by detecting the ECL-input level and decoding with a small logic swing. In the level shifter, emitter-follower circuits with bipolar diodes are used to drive the main decoder. The output level is determined by the number of diodes through which the signal passes.

The series-gate circuit for the main decoder is shown in Fig. 3. Two output groups of the predecoder, $D(1)$ to $D(2^n)$ and $C(1)$ to $C(2^m)$, are applied to a two-input NAND gate which consists of an upper gate and a lower gate. Adjusting load resistances, the amplification to boost the output to the MOS level is possible. The delay time of the main decoder is reduced by decoding and amplifying simultaneously in the series-gate circuit. The series-gate circuit has another advantage of low-power operation because current flows only through the selected path.

The total decoder power dissipation is reduced by introducing a block-select configuration. The power dissipation

Reprinted from *IEEE J. Solid-State Circuits*, vol. 23, no. 1, pp. 68–73, Feb. 1988.

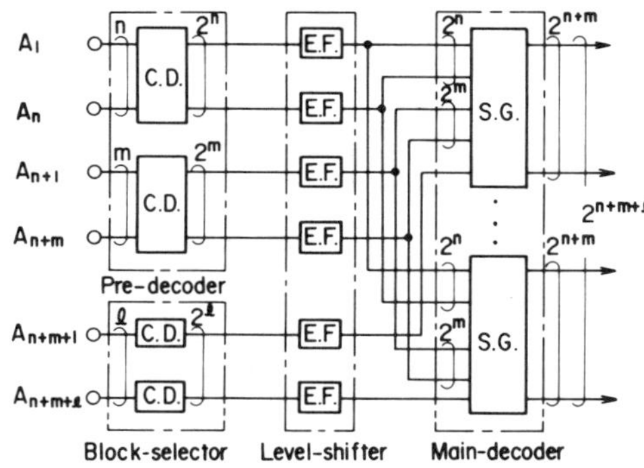

Fig. 1. Schematic diagram of the decoder.

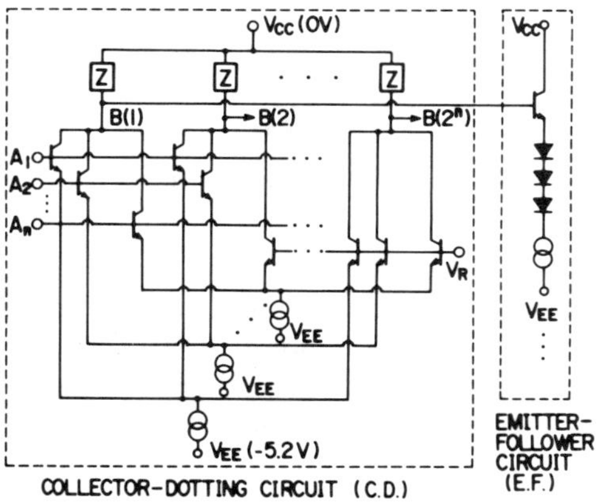

Fig. 2. Predecoder and level shifter.

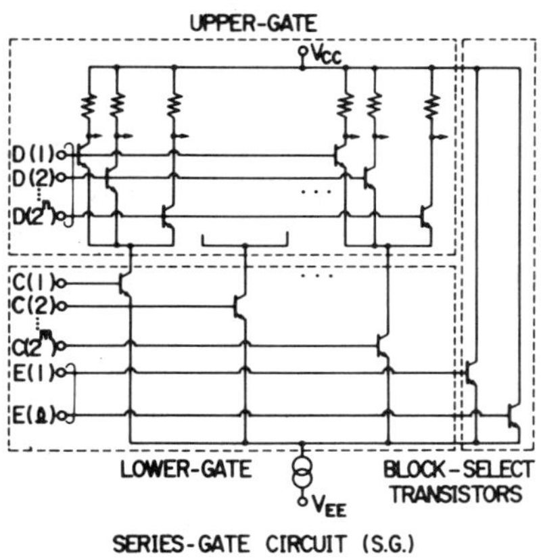

Fig. 3. Main decoder.

of the predecoder is proportional to 2 to the power of input numbers. When the number of the decoder inputs is large, the power dissipation of the predecoder is a dominant factor for that of the decoder. Using the block-select function, the input number of the predecoder can be reduced and the decoder power dissipation decreases.

B. Optimization of Decoder Power Dissipation

Using the block-select function, the power dissipation of the predecoder can be reduced, but that of the main decoder increases in proportion to the number of series-gate circuits. Under the constant delay-time condition, an optimum number of divisions of the main decoder which minimizes the total decoder current is analyzed. When the number of inputs for a predecoder is n and m, and that for a block selector is l, the number of inputs of a decoder k is

$$k = n + m + l. \tag{1}$$

Assuming the collector-dotting current per transistor to be I_0, the current I_{cd} of an n-input collector-dotting current with n current sources is given by

$$I_{cd} = n \cdot I_{so} = n \cdot 2^{n-1} \cdot I_0 \tag{2}$$

where I_{so} is current at one current source.

The total current for the predecoder and the block selector I_1 is given by

$$I_1 = \left(n \cdot 2^{n-1} \cdot w_n + m \cdot 2^{m-1} \cdot w_m + l \cdot w_l\right) I_0 \tag{3}$$

where w_n, w_m, and w_l are weight factors relating to the collector capacitance which is proportional to the input number of collector-dotting circuits. Using the delay time of a two-input collector-dotting circuit as a reference, weight factors are given by $w_n = n/2$, $w_m = m/2$, and $w_l = 1/2$ [10], [11].

The total level-shifter current I_2 is given by

$$I_2 = \left(2^n \cdot v_n + 2^m \cdot v_m + 2^l \cdot v_l\right) A_1 \cdot I_0 \tag{4}$$

where v_n, v_m, and v_l are weight factors relating to the difference of the input capacitance of series-gate circuits. A_1 is a normalizing constant for I_2 with transistor I_0 [10], [11]. Using the delay time of the emitter-follower circuit which drives four inputs of series-gate circuits as a reference, weight factors are given by

$$A_1 = 2/3,$$
$$v_n = \max\left(2^{m+l-2}, 1\right),$$
$$v_m = \max\left(2^{l-2}, 1\right),$$
$$v_l = \max\left(2^{l-3}, 1\right).$$

The main-decoder current I_3 is given by

$$I_3 = 2^l \cdot u_1 \cdot I_0 \tag{5}$$

where u_1 is the weight factor which is determined by the amplification factor of the small-signal level to the MOS level [10], [11]. Using the delay time of a two-input collector-dotting circuit as a reference, the weight factor is given by

$$u_1 = 2.0 \cdot (3.0)/(0.8) \doteqdot 8.0. \tag{6}$$

Using (1)–(6) with an eight-input decoder which corresponds to the decoder of the 64-kbit SRAM, the relationship between the total decoder current and the number of divisions of the main decoder is calculated, as shown in Fig. 4. The total decoder current is normalized with I_0. At the optimum number, the total current is reduced to 60 percent that of a single-unit main decoder.

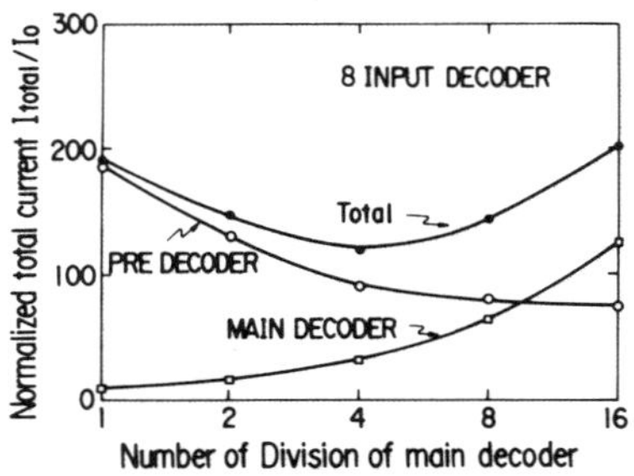

Fig. 4. Decoder current versus the number of divisions of the main decoder.

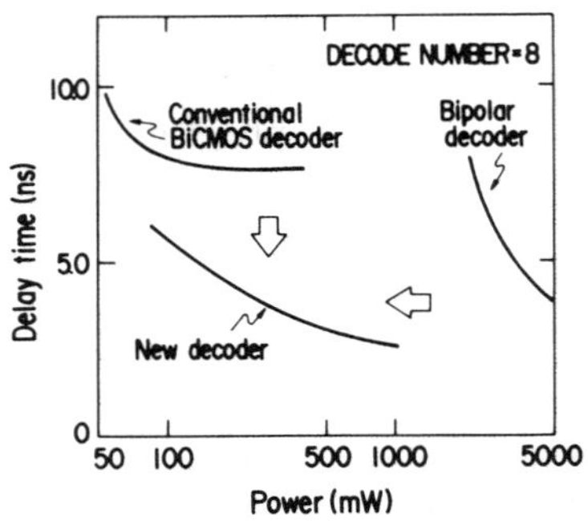

Fig. 5. Delay time versus the power dissipation of an eight-input decoder and a new driver.

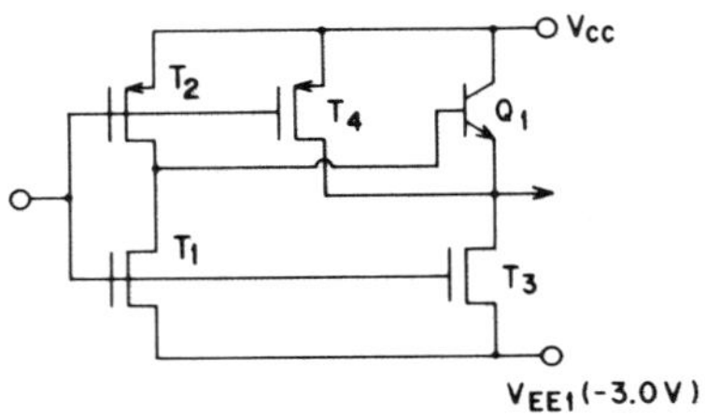

Fig. 6. BiCMOS inverter for the word driver.

C. Decoder and Driver Performance

The delay-power characteristics of an eight-input decoder and driver are simulated using SPICE II, as shown in Fig. 5. Four series-gate circuits were used for a main decoder from the above-mentioned analysis. The BiCMOS inverter shown in Fig. 6 was used for the word driver to drive the MOSFET's memory cell. This inverter can output the "high" MOS level [11]. The new decoder and driver can reduce the delay time to less than half that of a conventional BiCMOS decoder [5], [6] and new driver, and can reduce the power dissipation by more than 80 percent that of a bipolar decoder and driver which is composed of the CMOS inverter and new driver.

III. Multiplexer

Data stored in a memory cell are read out to the output through the multiplexer and the output buffer. To reduce the delay time of these circuits, a small swing operation with small power dissipation is required. A high-speed output buffer was obtained by using a bipolar differential amplifier and an emitter-follower driver [5]. To realize a high-speed and low-power multiplexer, a high-speed data-line driver and low-power column selector are required.

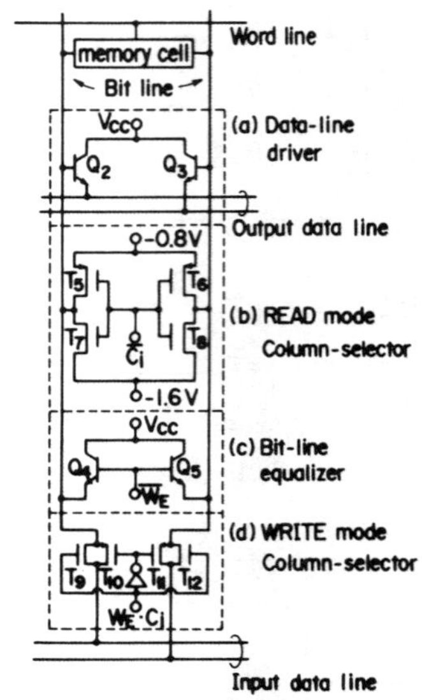

Fig. 7. BiCMOS multiplexer.

A. Circuitry

A BiCMOS multiplexer with a single emitter-follower driver and MOSFET selector is proposed. A schematic diagram of the BiCMOS multiplexer is shown in Fig. 7. This circuit is composed of a data-line driver, a READ-mode column selector, a bit-line equalizer, and a WRITE-mode column selector. The data-line driver is composed of a single emitter-follower circuit and can rapidly drive the large capacitance of a data line. The READ-mode column selector is composed of pull-up transistors with a bit-line selecting function and can operate without increasing the power dissipation of the memory cells. The READ-mode column selector selects a pair of bit lines and selects the data-line driver by pulling them up to -0.8 V and holding the other bit lines at -1.6 V. The bit-line equalizer is composed of an emitter-follower circuit and equalizes the bit lines at -1.6 V when the operation mode changes from the WRITE mode to the READ mode. This results in a small swing operation and reduction of the delay time of the memory cell. The WRITE-mode column selector is composed of transfer-gate circuits. The information in input data lines is sent to the selected pair of bit lines through transfer-gate circuits.

In this multiplexer, the pair of bit lines is selected before the memory cell information appears in it. The delay time of the multiplexer is determined exclusively by the delay time of the data-line driver. The power dissipation of the multiplexer is small because the readout current flows only through emitter-follower circuits on a selected pair of bit lines.

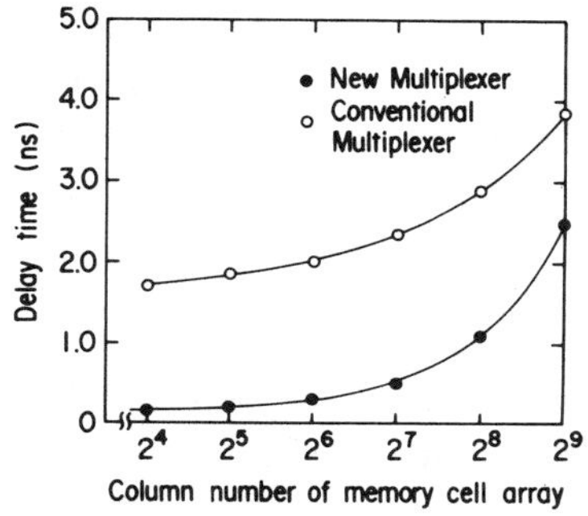

Fig. 8. Delay time versus column number of memory cell array in a multiplexer.

TABLE I
DEVICE PARAMETERS USED FOR THE SIMULATION

BIPOLAR	
• Current Gain (hfe)	100
• Cutoff Frequency (f_T)	2 GHz
CMOS	
• Effective Channel Length	0.5 μm [n] 0.9 μm [p]
• Gate Oxide Thickness	15 nm

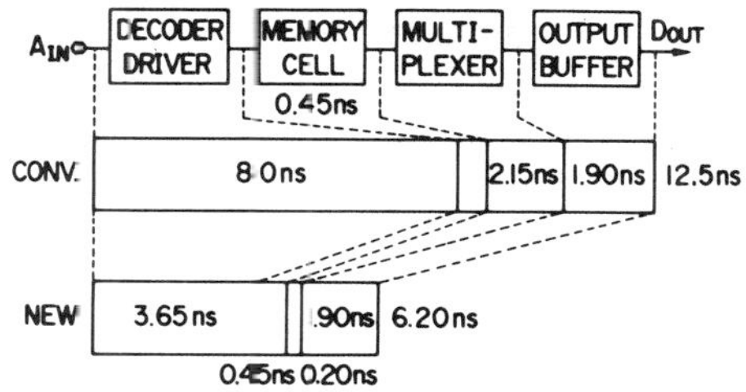

Fig. 9. Simulated access time of a 64-kbit BiCMOS SRAM.

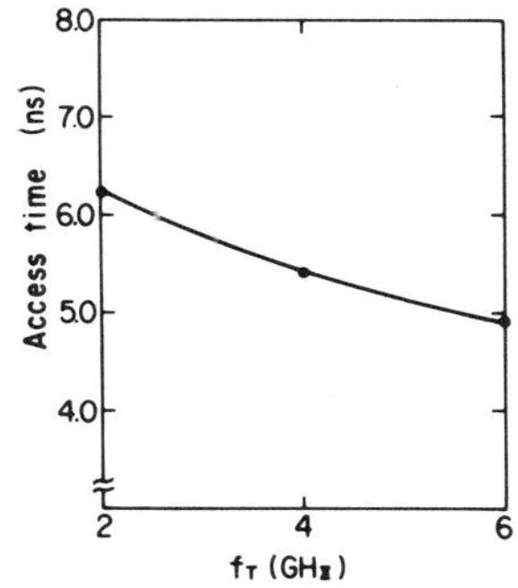

Fig. 10. Access time versus f_T of a 64-kbit BiCMOS SRAM.

B. Performance

The dependence of the delay time in the multiplexer on the capacitance of the data line was simulated with SPICE II. Since the capacitance of the data line is proportional to the column number of the memory cell array, a column number is used instead of the data-line capacitance in Fig. 8. When the column number is 2^5, a delay time of less than 0.2 ns is obtained by using the new BiCMOS multiplexer. The delay time is reduced by more than 80 percent compared with that of the conventional BiCMOS multiplexer [5], when the power dissipation is 52 mW.

IV. SRAM PERFORMANCE

Assuming a submicrometer BiCMOS technology, and using BiCMOS circuits and an optimum cell array configuration (see the Appendix), the performance of 8-kbit× 8-bit SRAM was simulated with SPICE II. The device parameters used for the simulation are summarized in Table I. The same device parameter were used for both conventional circuits [5], [6] and the new circuit. The delay times of each circuit of the SRAM are shown in Fig. 9. In the decoder and the driver, the delay time is less than half that of the conventional BiCMOS decoder and driver, and takes up about 60 percent of the access time. In the multiplexer, the delay time is reduced by more than 85 percent that of the conventional BiCMOS multiplexer and occupies only about 3 percent of the access time. The access time is 6.2 ns, which is half that of the conventional BiCMOS SRAM, and the power dissipation is 995 mW.

The f_T dependence of the access time of this SRAM is shown in Fig. 10. Increasing the f_T from 2 to 6 GHz, it is possible to realize a SRAM with an access time of less than 5 ns. This simulation result was obtained by only varying the value of f_T. More accurate consideration including the reduction effect of parasitic capacitance of the bipolar transistor will show that an access time of sub-5 ns could be realized. This means that the new BiCMOS SRAM has larger potential than the conventional BiCMOS SRAM [5], [6].

V. CONCLUSION

BiCMOS circuit technology for a high-speed and large-capacity ECL-compatible SRAM was proposed. In order to obtain a high-speed and low-power decoder, a pre-main decode configuration with an ECL-interface circuit was proposed. High-speed operation was achieved by combining the ECL level detection and decoding in the predecoder, and combining the decoding and amplification in the main decoder. Moreover, low-power operation was realized by a divided main-decoder configuration with a block-select function.

The BiCMOS multiplexer with a selective single emitter-follower driver was proposed. The single emitter-follower driver and bit-line selective circuits with MOSFET's facilitate a high-speed and low-power operation.

Applying the BiCMOS circuit technology to 64-kbit SRAM's and using a SPICE II circuit simulation, it was shown that an access time of less than 7 ns and a low power dissipation of less than 1 W was obtained.

APPENDIX

The BiCMOS SRAM suffers from an access-time limitation related to the capacitance of the word line, the data line, and the bit line. The BiCMOS SRAM also suffers from a power-dissipation limitation related to the number of row decoders and the memory cells connected to the word line. An optimum cell array structure which minimizes the total delay time and the total power dissipation will be described.

Delay Time

Optimization is carried out on a single block 2^h-bit memory. Assuming the column number of the memory cell array to be 2^i and the number of the output buffer to be 2^j, the row number of the memory cell array is given by 2^{h-i}, where h and j are constant and i is variable. The total delay time of the SRAM is written as

$$t_{pd} = 2^i \cdot C_W/g_m(WD) + 2^{h-i} \cdot C_B/g_m(\text{CELL}) + 2^{i-j} \cdot C_D/g_m(MPX) + t_0 \quad (7)$$

where

C_W	word-line capacitance (/cell),
C_B	bit-line capacitance (/cell),
C_D	data-line capacitance (/cell),
$g_m(WD)$	mutual conductance of word driver (/cell),
$g_m(\text{CELL})$	mutual conductance of memory cell (/cell), and
$g_m(MPX)$	mutual conductance of multiplexer (/cell).

The first term to the right-hand side of (7) is the delay time of the word driver, the second term is that of the memory cell, the third term is that of the multiplexer, and the last term is constant with that of the decoder and the output buffer.

The optimum column number i_{opt} which minimizes the total delay time is derived from the equation $\partial t_{pd}/\partial i = 0$. The optimum column number is given by

$$i_{opt} = (h+j)/2 + (2 \cdot \ln 2)^{-1} \cdot \left(\ln\left(C_B/g_m(\text{CELL})\right) - \ln\left(2^j \cdot \left(C_W/g_m(WD) + C_D/g_m(MPX)\right)\right)\right). \quad (8)$$

In an 8-kbit × 8-bit memory, the parameter values of h and j are 16 and 3. The optimum column number i_{opt} was calculated as 8.

Power Dissipation

Using (1)–(6) for the decoder current, the optimized power dissipation for the k input decoder is approximated as

$$P_{DEC} = \left(A_1 \cdot 2^k + A_2 \cdot k + A_3\right) I_0 \cdot V_{EE} \quad (9)$$

where the first term to the right-hand side of (9) is the power dissipation of the predecoder and the level shifter, the second term and the last term are those of the block selector and the main decoder, and $A_1 - A_3$ are fitting parameters.

The power dissipation of the memory cell array which has a column number of 2^k is given by

$$P_{\text{CELL}} = B_1 \cdot 2^k \cdot I_{\text{CELL}} \cdot V_{EE} \quad (10)$$

where B_1 is a normalized parameter with the voltage V_{EE}, which is supplied to the peripheral circuits of an SRAM, and I_{CELL} is the current of a selected memory cell.

The total power dissipation of the SRAM P_{total} is given by

$$P_{\text{total}} = P_{DEC}(\text{column}) + P_{DEC}(\text{row}) + P_{\text{CELL}} + 2^j \cdot P_0 \quad (11)$$

where the first term to the right-hand side is the power dissipation of the column decoder, the second term is that of the row decoder, the third term is that of the memory cell array, and the last term is constant with that of the input and output buffers. Substituting (9)–(11), the total power dissipation is given by

$$P_{\text{total}} = \left(A_1 \cdot (2^{i-j} + 2^{h-i}) + A_2 \cdot (h-j) + 2A_3 + B_1 \cdot C_1 \cdot 2^i\right) \cdot I_0 \cdot V_{EE} + 2^j \cdot P_0 \quad (12)$$

where C_1 is a normalized parameter with the decoder current I_0.

The optimum column number which minimizes P_{total}, or total power dissipation, is derived from the equation $\partial P_{\text{total}}/\partial i = 0$. The optimum column number i_{opt} is obtained as

$$i_{opt} = (h+j)/2 - (2 \cdot \ln 2)^{-1} \cdot \ln\left(1 + 2^j \cdot (B_1 \cdot C_1/A_1)\right). \quad (13)$$

In this work, a value of 0.25 for the fitting parameter A_1 and a value of 1 for the normalized parameter C_1 are used. The normalized parameter B_1 is given by $B_1 = V_{EE1}/V_{EE} = 3.0/5.2 = 0.58$.

The optimum column number i_{opt} for an 8-kbit × 8-bit memory was calculated as 8.

Acknowledgment

The authors would like to thank N. Ieda and T. Mano for their encouragement. They also thank M. Tanimoto and J. Yamada for helpful discussions.

References

[1] E. L. Hudson and S. L. Smith, "An ECL compatible 4K CMOS RAM," in *ISSCC Dig. Tech. Papers*, Feb. 1982, pp. 248–249.

[2] H. Momose, H. Shibata, S. Saitoh, J. Miyamoto, K. Kanzaki, and S. Kohyama, "1.0-μm n-well CMOS/bipolar technology," *IEEE J. Solid-State Circuits*, vol. SC-20, pp. 137–143, Feb. 1985.

[3] A. Watanabe *et al.*, "High speed BiCMOS VLSI technology with buried twin well structure," in *IEDM Tech. Dig.*, 1985, pp. 423–426.

[4] I. Masuda, Y. Nishio, and T. Ikeda, "High-speed logic circuits combining bipolar and CMOS technology," *Trans. Inst. Electron Commun. Eng. Japan*, vol. J67-C, no. 12, pp. 999–1005, Dec. 1984 (in Japanese).

[5] K. Ogiue, M. Okada, S. Miyaoka, I. Masuda, T. Ikeda, and K. Tonomura, "13-ns, 500-mW, 64-kbit ECL RAM using Hi-BiCMOS technology," *IEEE J. Solid-State Circuits*, vol. SC-21, pp. 681–684, Oct. 1986.

[6] S. Miyaoka, M. Okada, K. Ogiue, T. Ikeda, M. Suzuki, and H. Higuchi, "A 7 ns/350 mW 64K ECL compatible RAM," in *ISSCC Dig. Tech. Papers*, Feb. 1987, pp. 132–133.

[7] J. Miyamoto, S. Saito, H. Momose, H. Shibata, K. Kanzaki, and T. Iizuka, "A high-speed 64K CMOS RAM with bipolar sense amplifiers," *IEEE J. Solid-State Circuits*, vol. SC-19, pp. 557–563, Oct. 1984.

[8] T. Douseki and Y. Ohmori, "BiCMOS circuit technology for a high speed SRAM," in *1987 Symp. VLSI Circuits. Dig. Tech. Papers*, May 1987, pp. 77–78.

[9] W. R. Blood, *MECL System Design Handbook*, 2nd ed. Motorola Inc., Mesa, AZ, 1972, pp. 6–8.

[10] J. A. Narud and C. S. Meyer, "Characterization of integrated logic circuits," *Proc. IEEE*, vol. 52, no. 12, pp. 1551–1564, Dec. 1964.

[11] T. Douseki and Y. Ohmori, "A high speed SRAM using BiCMOS technology," *Trans. Inst. Electron. Inform. Commun. Eng. Japan*, vol. J70-C, no. 6, pp. 783–790, June 1987 (in Japanese).

Paper 3.3

A 5-ns 1-Mb ECL BiCMOS SRAM

MASAHIDE TAKADA, KAZUYUKI NAKAMURA, TOSHIO TAKESHIMA, KOUICHIROU FURUTA, TOHRU YAMAZAKI, KIYOTAKA IMAI, SUSUMU OHI, MEMBER, IEEE, YUMI SEKINE, YUKIO MINATO, AND HISAMITSU KIMOTO

***Abstract*—A 1-Mword × 1-b ECL 10K I/O compatible SRAM with 5-ns typical address access time has been developed using a double-level poly-Si, double-level metal, 0.8-μm BiCMOS technology. To achieve 5-ns address access time, high-speed *X*-address decoding circuits with wired-OR predecoders and ECL-to-CMOS voltage level converters with partial address decoding function, and sensing circuits with small differential signal voltage swing, were developed. The die and memory cell sizes are 16.8 mm × 6.7 mm and 8.5 μm × 5.3 μm, respectively. The active power is 1 W at 100-MHz operation.**

I. Introduction

HIGH-performance computer systems strongly require large-scale, high-speed SRAM's with ECL signal I/O interfaces, which have an advantage in regard to higher speed signal transmission on microprocessor and memory boards than TTL I/O signals. In recent years, BiCMOS SRAM's, composed of ECL interface bipolar circuits and high-density MOS memory cell arrays, have rapidly advanced in memory capacity as well as in access time [1]–[7]. The address access time for such ECL BiCMOS SRAM's mainly consists of the *X*-address decoding time from address input to word-line activation, and the sensing output operation time from memory cell readout to data output. Previously reported 1-Mb ECL BiCMOS SRAM's [5]–[7], fabricated using 0.8-μm BiCMOS technology, have reduced the sensing operation time and attained 6.5–8-ns address access times, mainly by reducing differential signal voltage swing on bit lines and by adopting bipolar sense amplifiers. However, in order to produce a 1-Mb ECL BiCMOS SRAM with a much faster access time of 5 ns, it is particularly necessary to reduce the *X*-address decoding time, as well as the sensing output operation time. This paper describes a 1-Mword × 1-b ECL 10K I/O compatible BiCMOS SRAM with 5-ns typical address access time and 1-W active power at 100 MHz [8]. To achieve 5-ns access time, the following key technologies were developed:

1) *X*-address decoding circuits with wired-OR bipolar predecoders and ECL-to-CMOS voltage level converters with partial address decoding function;

Manuscript received April 18, 1990; revised June 5, 1990.
The authors are with the NEC Corporation, 1120, Shimokuzawa, Sagamihara-shi, Kanagawa-ken 229, Japan.
IEEE Log Number 9037779.

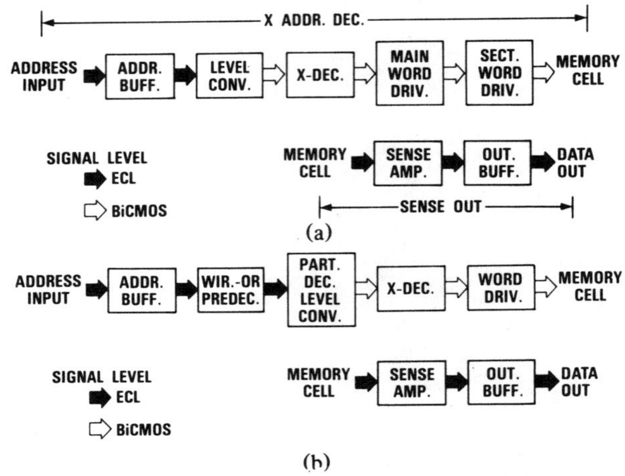

Fig. 1. Read access critical path comparison for conventional and developed ECL BiCMOS SRAM's: (a) conventional SRAM and (b) developed SRAM.

2) sensing circuits with small differential signal voltage swing; and

3) a double-level poly-Si, double-level metal, 0.8-μm BiCMOS process technology and a second-metal-layer wiring for access-time critical path lines.

Section II describes the 1-Mb SRAM chip architecture. In Sections III and IV, new *X*-address decoding circuits and sensing circuits are discussed, respectively. Section V explains fabrication process and device features. Measured chip characteristics are shown in Section VI.

II. Chip Architecture

Fig. 1 shows read access critical path comparison for conventional and developed ECL BiCMOS SRAM's. In a conventional ECL BiCMOS SRAM [6], as shown in Fig. 1(a), the *X*-address decoding time is determined by the delay times from address buffers to level converters, an *X* decoder, and main and section word-line drivers. The sensing output operation time is determined by the delay times from memory cells to sense amplifiers and an output buffer. In the developed 1-Mb ECL BiCMOS SRAM, as shown in Fig. 1(b), wired-OR bipolar predecoders after ECL address buffers, partial decoding level converters, an *X* decoder, and a word-line driver form *X*-address decod-

Reprinted from *IEEE J. Solid-State Circuits*, vol. 25, no. 5, pp. 1057–1062, Oct. 1990.

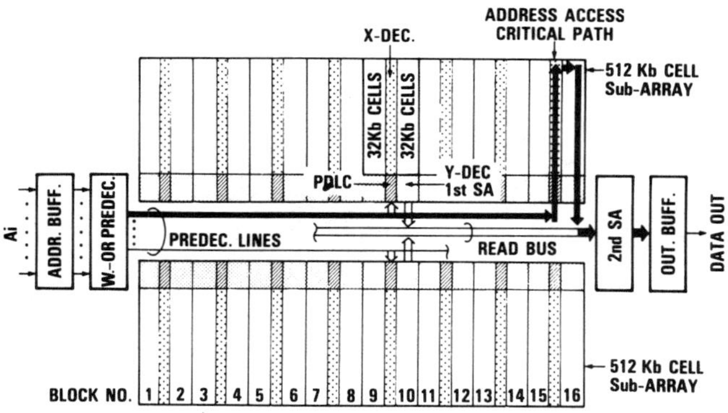

Fig. 2. Block diagram for the main part of developed 1-Mb SRAM.

ing circuits. For the sensing output circuits, a conventional scheme is used. X-address decoding time in the developed SRAM is shortened drastically by using wired-OR predecoders and partial decoding level converters, due to a reduction in the effective decoder gate stage numbers.

A block diagram for the main part of the 1-Mb SRAM is shown in Fig. 2. The memory cell array is divided into two 512-kb cell subarrays. Each subarray consists of 16 32-kb cell arrays with 256 rows and 128 columns. Two 32-kb cell arrays share an X decoder. Address input signals are received by ECL address buffers, then predecoding operations are carried out by wired-OR bipolar predecoders. Predecoded pseudo-ECL signals, with about 1.2-V swing, run along predecoded lines between two 512-kb subarrays. They are received by partial decoding level converters (PDLC's), which are located immediately next to the 32-kb cell array blocks. This wired-OR predecoding scheme achieves high-speed driving of long predecoded signal lines, about 16.5 mm in length [9]. An X decoder is selected by PDLC's.

The readout bit line signal from a memory cell is amplified by a two-stage bipolar differential sense amplifier (SA) and output through an output buffer. The first SA's are placed at local 32-kb cell array blocks. The second SA is placed near the output buffer. The first and second SA's are connected through a pair of read bus lines, about 15 mm in length.

In this block diagram, access-time critical path lines are mainly comprised of predecoded lines, X-decoder input signal lines, bit lines, and read bus lines, which are shown with bold black lines in Fig. 2. In order to achieve high-speed access time, their lines are wired by a second metal layer with low resistivity and low wiring capacitance.

III. X-Address Decoding Circuits

Fig. 3 shows detailed X-address decoding circuits, which consist of an address buffer, wired-OR predecoders, a newly developed PDLC, an X decoder with three-input BiCMOS NAND gates, and a BiCMOS NOR word-line driver. Address input ECL signals are predecoded with only bipolar transistors High (H) and low (L) levels for predecoded signals are -0.9 and -2.1 V, respectively. A PDLC consists of a logical operation portion with three serially connected pMOSFET's, and a level conversion portion with an nMOS current-mirror circuit and load pMOSFET's $Q1$ and $Q2$. When three predecoded pseudo-ECL signals A, B, and C are applied to serially connected pMOSFET's, partial decoding operation $\overline{A+B+C}$ is carried out. On the other hand, level conversion is accomplished by supplying a load current to the current mirror circuit through $Q1$ and $Q2$. Here, $Q1$ is always turned on, while $Q2$ is controlled by the feedback signal from node N. When all input signal levels are at a low level of -2.1 V ($A, B, C =$ L, L, L), voltage at node N is nearly the low supply voltage of -5.2 V, due to mirror current flow. However when either one of the input signal levels is at a high level of -0.9 V, due to negligible mirror current flow, voltage at node N is kept at the high supply voltage of 0 V. A PDLC can save the operational time necessary for one decoding gate stage, because it carries out decoding logic operation and level conversion simultaneously.

Furthermore, the level conversion time itself is increased by controlling the $Q2$ gate by the delayed feedback signal from node N. When node N voltage rises from low voltage V_{OL}, both pMOSFET's $Q1$ and $Q2$ supply the charging current to raise node N voltage. After node N voltage reaches high voltage V_{OH}, $Q2$ is turned off by the high voltage of the delayed feedback signal from node N. Conversely, when node N voltage falls to V_{OL} from V_{OH}, $Q2$ is in the OFF condition. Therefore, the charging current to node N is supplied by $Q1$ alone. As a result, a constant discharging mirror current reduces node N voltage quickly.

The level conversion time as a function of the sum of gate widths $W1$ and $W2$ for pMOSFET's $Q1$ and $Q2$ is shown in Fig. 4, where T_r and T_f are, respectively, rising and falling delay times between a pseudo-ECL signal IN and node N voltage, while input signals levels for the upper two pMOSFET's are set at -2.1 V. As the sum of $W1$ and $W2$ increases, T_r decreases, but T_f increases. However, when the sum of $W1$ and $W2$ is kept constant, T_F decreases, as the ratio of $W2$ to $W1$ increases. In the developed SRAM, the sum of $W1$ and $W2$ is set at 10 μm to keep V_{OL} less than -4.5 V, which is the low supply voltage plus the threshold voltage for the next-stage nMOSFET. The $W2$-to-$W1$ ratio is set at $1/2$ to obtain a T_f value nearly equal to the T_r value. As a result, a level conversion time of about 500 ps for both rising and falling operations was achieved. T_f is, notably, 20% less than that with a 10-μm gate width $W1$ alone. The described high-speed level conversion is successfully realized by setting the feedback delay time at less than 5-ns access time.

In this PDLC, predecoded input signals A and B are used only for 32-kb cell array block selection, while the signal C is for an input address signal for an X decoder in the 32-kb block. In the only case of $A, B, C =$ L, L, L,

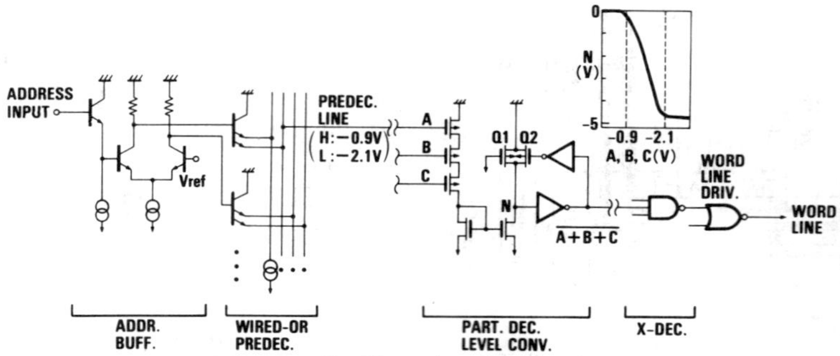

Fig. 3. *X*-address decoding circuits.

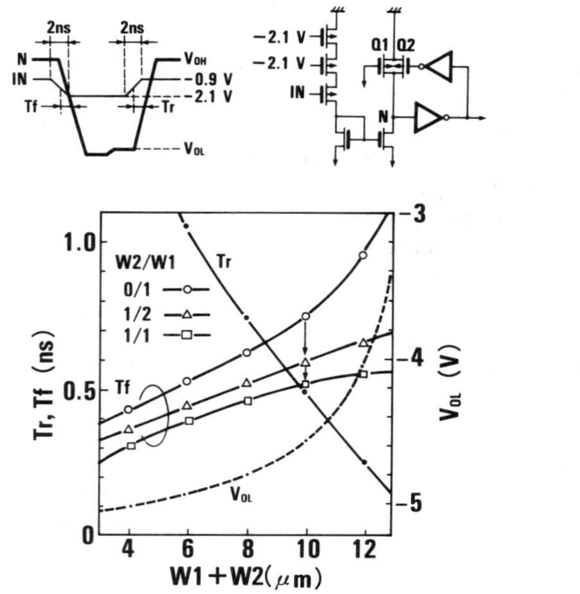

Fig. 4. Level conversion time as a function of the sum of gate widths $W1$ and $W2$ for pMOSFET's $Q1$ and $Q2$.

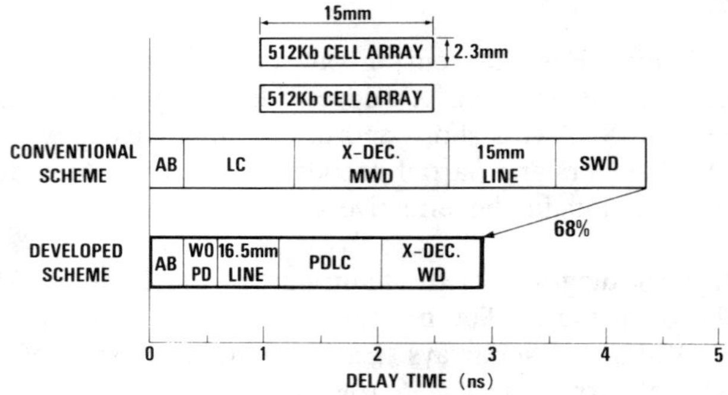

Fig. 5. Comparison of simulated *X*-address decoding delay times for conventional and developed schemes. (*AB*, *MWD*, and *SWD* are address buffers, and main and section word-line drivers, respectively.)

PDLC's are selected and output high-level signals. They are supplied to NAND gates in an X decoder to raise a selected word line through a NOR driver. When signals A, B, and C are skewed, level conversion for selected PDLC with $A, B, C = \text{L, L, L}$ is started by the most delayed input signal. There is no additional delay in the level conversion time due to skewing. For PDLC with $A, B, C = \text{L, L, H}$, if the previous signal level condition is $A, B, C = \text{H, L, L/L, H, L/H, H, L}$ and the only signal C is delayed, the intermediate signal level condition with $A, B, C = \text{L, L, L}$ is generated. As a result, unselected word lines in the same 32-kb block might be activated only during the skewing short time. In the developed SRAM, when the skewing time is more than 700 ps, unselected word lines are activated and pulsed. However, by setting the feedback delay time in PDLC at about 1 ns, negligible overlapping time between above mentioned unselected and selected word-line waveforms is achieved.

Fig. 5 shows a comparison between simulated X-address decoding delay times for conventional and developed schemes in Fig. 1. Here, the memory cell array is assumed to consist of two 512-kb cell arrays, each 15 mm × 2.3 mm in size. The sum of delay times for the wired-OR predecoder (WOPD), 16.5-mm predecoded line, and PDLC is nearly equal to that for the level converter (LC) and 15-mm main word line in the conventional scheme. This means that, in the developed scheme, two decoding gate stages are effectively reduced by the wired-OR predecoder and PDLC. As a result, X-address decoding time for the developed scheme decreases to about 70% of that for the conventional scheme. As the greatest portion of address access time is taken up by X-address decoding, this developed scheme is particularly effective in reducing total access time.

IV. Bit-Line Peripheral and Sensing Circuits

Bit-line peripheral and sensing circuits are shown in Fig. 6. The differential readout signal from a memory cell is received by the first SA through a pair of bit-line selection pMOSFET's and two stages of emitter-follower (EF) wired-OR logic circuits. Then, the amplified readout signal is sent to the second SA through a pair of read bus lines (RB). The sensing speed can be increased by reducing both bit-line capacitance and voltage swing for the differential sense signal. Therefore, the 1-Mb cell array is divided into two 512-kb subarrays, in order to halve the bit-line capacitance. Also, the differential bit-line voltage swing is limited to about 50 mV by optimizing the bit-line load transistor size.

The SRAM adopted a four-stage column selection scheme as a hierarchical sensing circuit design. One out of four pairs of bit lines are connected to a data bus $DB1$

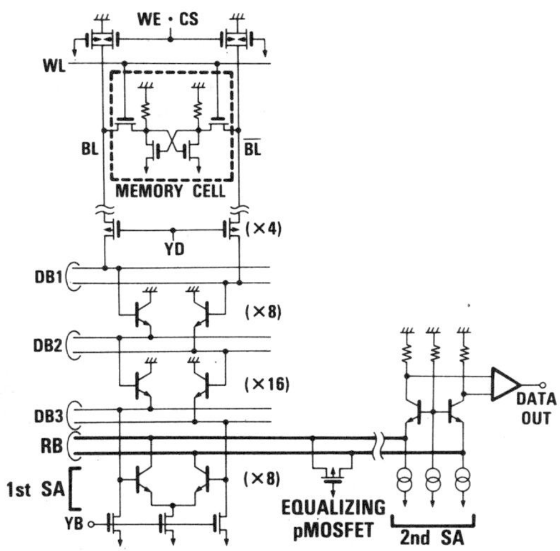

Fig. 6. Bit-line peripheral and sensing circuits.

TABLE I
FABRICATION PROCESS AND DEVICE FEATURES

BIPOLAR	
Emitter Size	0.8 μm × 3 μm
Cut-off Frequency	10GHz
Current Gain	80
CMOS	
Gate Length	0.8 μm (nMOS)
	1.0 μm (pMOS)
|Threshold Voltage|	0.7V
Gate Oxide Thickness	15nm
MEMORY CELL	2 Poly-Si Loads + 4 nMOSFETs
	8.5 μm × 5.3 μm

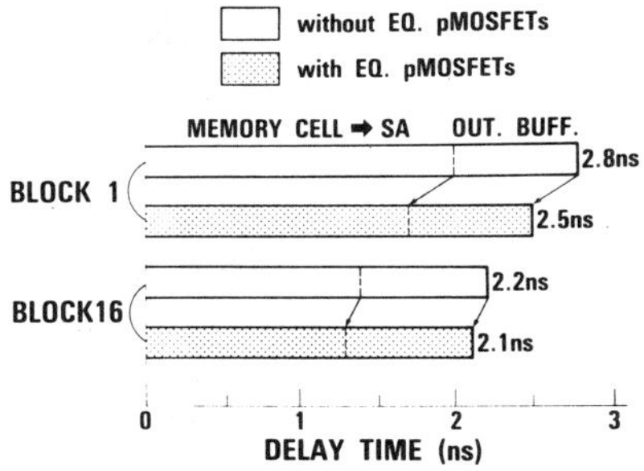

Fig. 7. Simulated delay-time comparison of critical sensing output operations.

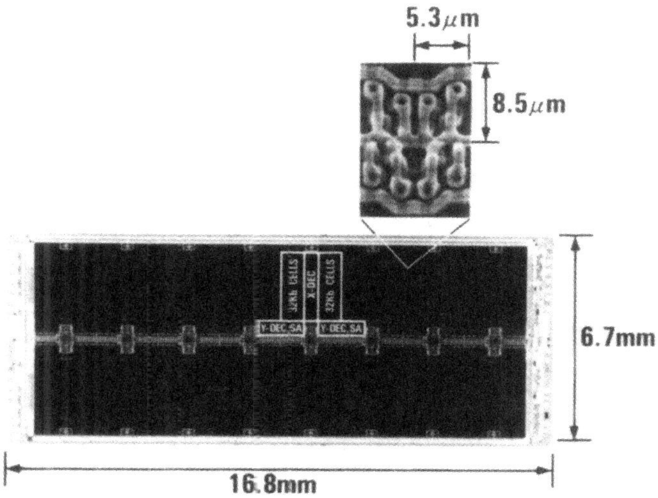

Fig. 8. 1-Mb SRAM microphotograph.

through a pair of pMOSFET's selected by a bit-line select signal *YD*. As a result, a selected differential bit-line signal is sent to the *DB*1 bus, which is charged up at about 0 V. At the same time, unselected *DB*1 buses are biased at −0.8 V. Then, by connecting data buses *DB*2 and *DB*3 to current sources by a column block select signal *YB*, the only *DB*1 differential signal at high voltage is transferred to a *DB*3 bus through one out of eight first EF wired-OR circuits and one out of 16 second EF wired-OR circuits. This circuit structure enables high-speed transfer of a selected differential signal with about 50-mV swing from a pair of bit lines to a *DB*3 bus. Finally, by using the signal *YB*, the *DB*3 differential signal is amplified through one out of eight first SA's and the second SA. This hierarchical sensing scheme also speeds up the sensing time effectively.

In addition, normally-on equalizing pMOSFET's are inserted between a pair of *RB* lines to suppress the voltage swing caused by their parasitic resistance from 250 mV to about 100 mV. This small-signal voltage swing also enables high-speed sensing operations for memory cells located at the far end of *RB* from the second SA. Fig. 7 shows simulated delay-time comparison between critical sensing output operations for memory cells in cell array blocks 1 and 16 in Fig. 2. By inserting equalizing pMOSFET's between about 15-mm *RB* lines, critical sensing output times for the cell array blocks located at both ends of *RB* decrease to 2.1 and 2.5 ns from 2.2 and 2.8 ns, respectively. Due to this reduced delay time of 400 ps for a readout signal on *RB* lines, the read access critical path is formed through cell array block 16, not block 1, as shown in Fig. 2.

V. FABRICATION PROCESS AND DEVICE FEATURES

The SRAM was fabricated by using a double-level poly-Si, double-level metal, 0.8-μm BiCMOS process technology. Table I summarizes the fabrication process and device features. The minimum emitter size for bipolar transistors is 0.8 μm×3 μm. The minimum gate lengths for nMOSFET's and pMOSFET's are 0.8 and 1.0 μm, respectively. The memory cell consists of two high-resistive loads and four nMOSFET's. The cell size is 8.5 μm×5.3 μm. The first and second poly-Si layers are used for MOSFET gates and high-resistive loads, respectively. The first and second metal layers are used for word and bit lines, respectively. Fig. 8 shows a microphotograph of the developed 1-Mb SRAM, 16.8 mm×6.7 mm in size.

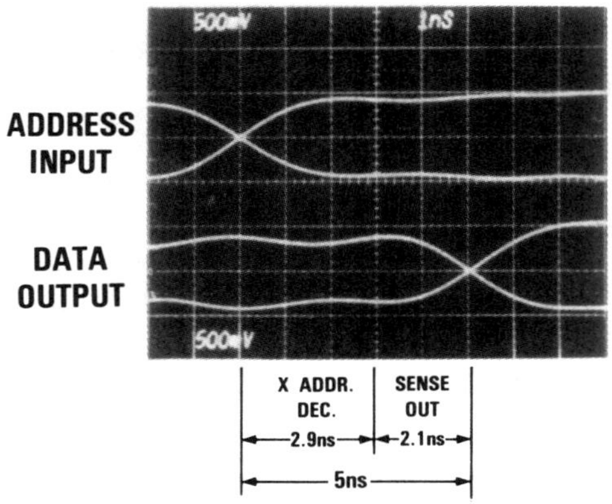

Fig. 9. Address input and data output waveforms for address access.

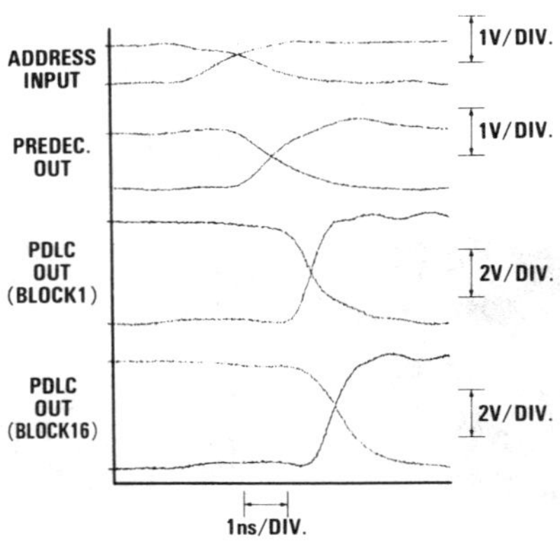

Fig. 10. Internal signal waveforms at a wired-OR predecoder and PDLC's.

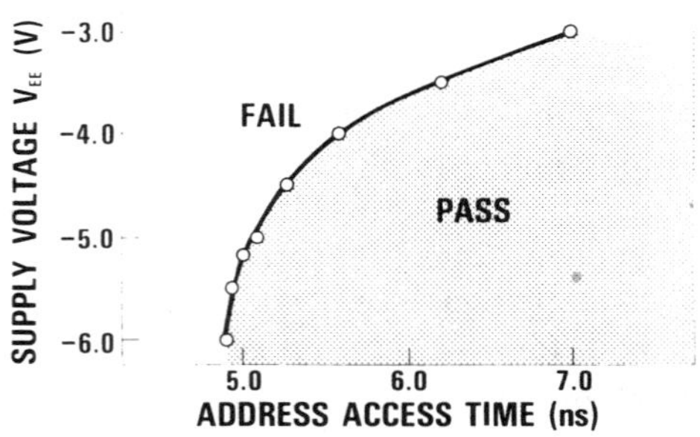

Fig. 11. Address access-time dependency on power supply voltage.

VI. Chip Characteristics

The SRAM is ECL 10K I/O compatible. Fig. 9 shows address input and data output waveforms during a normal operation mode at room temperature with −5.2 V power supply. The measured address access time is 5 ns, which consists of 2.9-ns X-address decoding time and 2.1-ns sensing output operation time. Active power dissipation is typically 1 W at 100-MHz operation. Internal waveforms for output signals from a wired-OR predecoder and PDLC's at cell array blocks 1 and 16 are shown in Fig. 10. These signals were measured using a digital sampling oscilloscope. The delay time from address input to PDLC output at cell array block 16 was observed to be about 2 ns, which agrees well with the simulated delay time. Fig. 11 shows measured address access time dependency on power supply voltage V_{EE}. The SRAM perfectly functioned even at about −3-V V_{EE}, which means that the newly developed circuits in the SRAM have a wide operational margin.

Typical SRAM characteristics are summarized in Table II.

TABLE II
1-Mb SRAM Characteristics

Organization	1Mw × 1b
Technology	0.8μm BiCMOS 2 Poly-Si, 2 Metal
Chip Size	16.8mm × 6.7mm
Cell Size	8.5μm × 5.3μm
I/O Interface	ECL 10K
Address Access Time	5ns
Write Pulse Width	3ns
Power Dissipation	1W @ 100MHz
Soft Error Rate	< 10FIT
Redundancy	2 rows, 2 columns

VII. Conclusion

New high-speed X-address decoding circuits, with wired-OR bipolar predecoders and PDLC's, were developed. In addition, sensing circuits with small differential signal voltage swing, particularly for RB lines, were developed. These two high-speed circuits and a double-level poly-Si, double-level metal, 0.8-μm BiCMOS process technology and a second-metal-layer wiring for access-time critical path lines achieved a 5-ns address access time, 1-W active power dissipation, 1-Mb ECL 10K I/O compatible SRAM.

Acknowledgment

The authors thank T. Enomoto, K. Okada, O. Kudoh, T. Tashiro, H. Mikoshiba, and J. Nokubo for their encouragement and suggestions, and the members of this project for their support.

References

[1] H. V. Tran *et al.*, "An 8-ns 256K ECL SRAM with CMOS memory array and battery backup capability," *IEEE J. Solid-State Circuits*, vol. 23, pp. 1041–1047, Oct. 1988.
[2] R. A. Kertis *et al.*, "A 12-ns ECL I/O 256K×1-bit SRAM using a 1-μm BiCMOS technology," *IEEE J. Solid-State Circuits*, vol. 23, pp. 1048–1053, Oct. 1988.
[3] N. Tamba *et al.*, "An 8-ns 256K BiCMOS RAM," in *ISSCC Dig. Tech. Papers*, 1988, pp. 184–185.
[4] I. Fukushi *et al.*, "A 256-kbit ECL RAM with redundancy," in *ISSCC Dig. Tech. Papers*, 1988, pp. 184–185.
[5] H. Tran *et al.*, "An 8ns BiCMOS 1Mb ECL SRAM with a configurable memory array size," in *ISSCC Dig. Tech. Papers*, 1989, pp. 36–37.
[6] M. Matsui *et al.*, "An 8-ns 1-Mbit ECL BiCMOS SRAM with double-latch ECL-to-CMOS-level converters," *IEEE J. Solid-State Circuits*, vol. 24, pp. 1226–1232, Oct. 1989.
[7] Y. Maki *et al.*, "A 6.5ns 1Mb BiCMOS ECL SRAM," in *ISSCC Dig. Tech. Papers*, 1990, pp. 136–137.
[8] M. Takada *et al.*, "A 5ns 1Mb ECL BiCMOS SRAM," in *ISSCC Dig. Tech. Papers*, 1990, pp. 138–139.
[9] M. Suzuki *et al.*, "A 3.5-ns, 500-mW 16-kbit BiCMOS ECL SRAM," *IEEE J. Solid-State Circuits*, vol. 24, pp. 1233–1237, Oct. 1989.

Paper 3.4

A 3.8-ns 16K BiCMOS SRAM

WOLFGANG HEIMSCH, ASSOCIATE MEMBER, IEEE, ROLAND KREBS, BRUNO PFÄFFEL, AND KLAUS ZIEMANN

Abstract —A 2K×8-bit, ECL 100K compatible BiCMOS SRAM with 3.8-ns (−4.5 V, 60°) address access time is described. The precisely controlled bit-line voltage swing (60 mV), a current sensing method, and optimized ECL decoding circuits permit a reliable and fast readout operation. The SRAM features an on-chip write pulse generator, latches for input and output bits, and a full six-transistor CMOS cell array. Power dissipation is approximately 2 W and the chip size is 3.9×5.9 mm². The SRAM was based on 1.2-μm BiCMOS, using double-metal, triple-polysilicon, and self-aligned bipolar transistors.

I. Introduction

HIGH-SPEED SRAM's, used in cache memories, are one of the most crucial elements determining the performance of mainframe computers. Some years ago this domain was dominated by pure ECL SRAM's. But with increasing memory capacity the disadvantages of these SRAM's have become more and more obvious: high power consumption, large chip area due to the large memory cell, and limited yield. Recently, BiCMOS SRAM's, e.g. [1]–[4], have begun to supersede pure bipolar SRAM's, allowing larger memories without the above-mentioned disadvantages. This paper describes a 16K BiCMOS SRAM which exhibits bipolar performance (address access time of 3.8 ns) combined with the advantages of CMOS technology: a smaller chip area (23 mm²) and reduced power consumption (<2 W). Section II contains a general description of the SRAM. This is followed by detailed information on the read operation (Section III) and write operation (Section IV). In Section V, the BiCMOS technology which was used is briefly introduced and the results are presented in Section VI.

II. General Description

The 16K BiCMOS SRAM (Fig. 1, Table I) is ECL 100K compatible and due to its application as an on-chip memory for bipolar gate arrays, the organization was specified as 2K words×8 bits. The memory cell array consists of eight blocks with 64 rows×32 columns, and the overall size of the memory cell array is 64 rows×256 columns. To obtain a short access time, the word lines are in the first metal layer and the bit lines are in the second metal layer, thereby reducing the bit-line capacitance. The memory cell is a full-CMOS six-transistor cell with a cell area of 432 μm² using 1.2-μm BiCMOS technology. To achieve a short access time, the following measures were taken:

- a large share of ECL circuits are within the time critical path (e.g., the decoders are wholly designed in ECL);
- BiCMOS Totem Pole buffer as a word-line driver;
- small bit-line voltage swing (60 mV);
- bipolar sense amplifiers.

Approximately 54 percent of the access time is needed to convert ECL to CMOS levels (CMOS-level shifter) and for word-line charging. Because of the pure bipolar decoders, sense amplifiers, and output buffers, only 33 percent of the access time is spent in these circuits. The large share of ECL circuits results in a relatively high power dissipation but also in the excellent access time. An important part (16 percent) of the entire power consumption is dissipated in the eight output buffers with latch. If a 1-bit organization was used, power dissipation would be reduced significantly.

An ECL write pulse can be generated by an on-chip write pulse generator (WPG), which generates a write pulse width independent of the pulse width of the trigger signal. An external write pulse signal can be used as an alternative. An ECL multiplexer selects one of the two write pulses. The write bus consists of 16 signal lines and is controlled by 16 pull-down NMOS transistors (200 μm). These are driven by 2×8 complementary CMOS signals which are generated from eight DATA-IN signals.

A system of 14 bias drivers supplies the 16K BiCMOS SRAM with the necessary reference voltages. The SRAM features three operation modes: asynchronous without WPG and clock driven with or without WPG. The clock signal is converted internally to differential logic. Low-output-impedance emitter followers drive the two separate clock nets to limit the load. All SRAM signals except clock (*CLK*) and write select (*WS*) are provided with latches.

Manuscript received July 24, 1989; revised October 16, 1989. This work was carried out within the ESPRIT Project 2430.
The authors are with Siemens AG, Corporate Research and Development, 8000 Munich 83, West Germany.
IEEE Log Number 8933134.

Reprinted from *IEEE J. Solid-State Circuits*, vol. 25, no. 1, pp. 48–54, Feb. 1990.

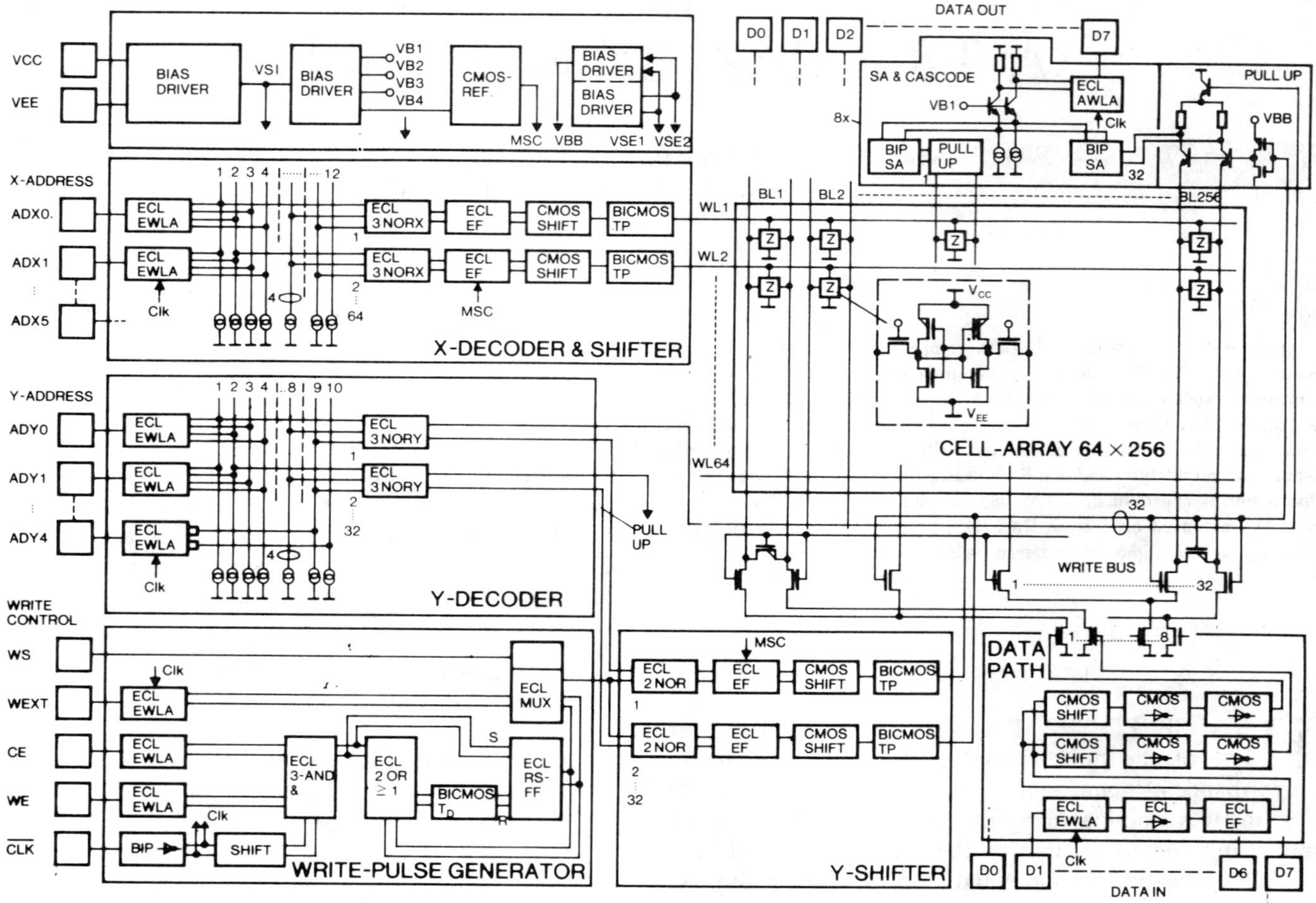

Fig. 1. Overview of the 16K BiCMOS SRAM.

TABLE I
FEATURES OF THE 16K BiCMOS SRAM

Organization	2 k word × 8 bit
Cell array	64 rows × 256 columns
Cell size	432 μm² (6 transistor)
Chip size	3.9 × 5.9 mm²
Address access time	3.8 ns (− 4.5 V, 60°C)
Write pulse width	3.6 ns
Cycle time	8.5 ns
Bit line voltage swing	60 mV
Power dissipation	1.95 W
I/O interface	ECL 100 k

III. Read Operation

A. Word-Line Decoder

Six input bits select one of the 64 word lines (*WL*). The decoding is done in two steps (Fig. 2). In the first step, the input buffers are grouped in two's to provide four decoder lines for each pair of latches. In the second step, an ECL three-input NOR gate permits a 3-out-of-12 code for selecting one of the 64 word lines. For ECL/CMOS level conversion, the ECL voltage swing is amplified in the first instance to 1.8 V by this ECL three-input NOR gate and finally to the full-CMOS voltage swing of 4.5 V by a PMOS current mirror. A special BiCMOS Totem Pole buffer (Section III-B) charges the large word-line capacitance (2 pF).

In addition to the six ECL input buffers, this concept requires 64 BiCMOS Totem Pole buffers, 64 ECL NOR gates, 64 PMOS current mirrors, and 64 ECL emitter followers. Since decoding is mainly performed with ECL, the power dissipation of the decoder amounts to 146.6 mA, which corresponds to 659.7 mW (Table II).

The advantage of this decoder concept is its short delay time, attributable to the small voltage swing of the bipolar transistors. The single delay times (Table III) lead to a word-line decoder delay time of 2.04 ns.

B. Word-Line Driver

We used a special BiCMOS Totem Pole buffer to drive the word line (Fig. 3). During switching, saturation effects may occur in the bipolar transistors. This is caused by the voltage drop at the internal collector resistance. If the output state of the buffer changes, a very large current flows for a short time, causing a significant voltage drop at the internal collector resistance. Under operating conditions this voltage drop may be large enough to make the base–collector diode of the transistor forward biased ($V_{BC} > 300$ mV). To avoid this, the source/drain of the controlling MOS transistor ($M1$, $M3$) is connected to a

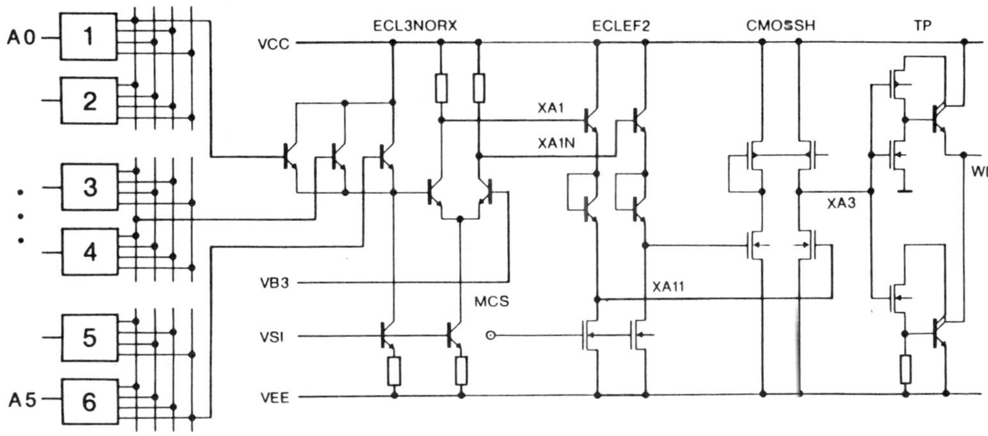

Fig. 2. Word-line decoder.

TABLE II
POWER DISSIPATION OF WORD-LINE DECODER

	current paths	sum
decoder line	12 1.0mA	12.0mA
ECL NOR	64 1.3mA	83.2mA
ECL emitter follower	2x64 0.3mA	38.4mA
PMOS current mirror	1x1.0mA + 63x0.19mA	13.0mA
Total		146.6mA

TABLE III
WORD-LINE DECODER DELAY TIMES

	delay time
ECL NOR gate	176ps
ECL emitter follower	10ps
PMOS current mirror	872ps
BiCMOS Totem Pole buffer	980ps
Total	2038ps

second collector contact of the corresponding bipolar transistor instead of V_{cc}/output (Fig. 3). Therefore source/drain potential is also decreased by the voltage drop at the internal collector resistance.

Another point we wish to mention is the reduced output voltage swing of the Totem Pole buffer. If the output changes from high to low level, the output voltage decreases and so the drain–source voltage of the MOS transistor $M3$ begins to fall. Therefore the base current of transistor $Q2$ is also reduced. Finally the output voltage equals the base–emitter voltage, which is related to the remaining base charge of $Q2$.

Because the low level is approximately one V_{BE} above V_{EE}, it is possible that the access transistors of the unselected memory cells may not be completely turned off. In that case, leakage currents flow and the bit-line potential is not controlled by the selected cell. To suppress this effect,

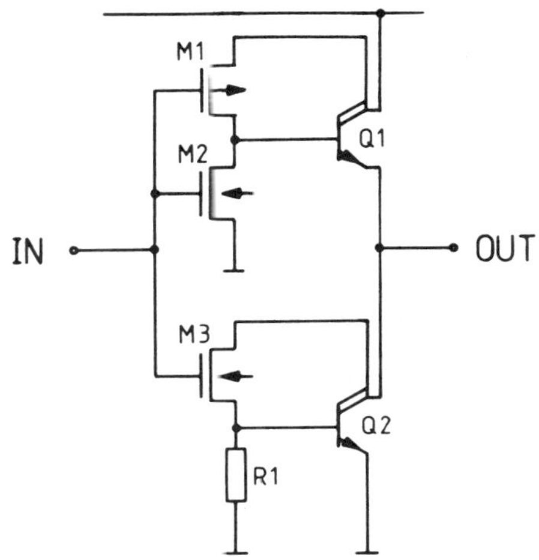

Fig. 3. Word-line driver (BiCMOS Totem Pole buffer).

a modified version of the Totem Pole has to be applied. In this version a resistor ($R1$), which is connected to the base of the bipolar transistor $Q2$ and V_{EE}, discharges the transistor.

C. Bit-Line Decoder

The five input bits of the bit-line decoder select eight out of 256 bit-line pairs. Due to different select modes for the read and write operations, two select signals must be supplied by the bit-line decoder: one select signal with bipolar voltage swing for the read operation and one select signal with CMOS voltage swing for the write operation.

D. Precharge and Sense Circuitry

Another important element determining access time is the precharge and sense circuitry (Fig. 4). The sense amplifier consists of two stages. The first stage, a simple differential pair, senses the voltage difference at the sensing resistors (RS) caused by the current flowing into the selected cell. Every bit-line pair has such a differential pair and 32 of these differential pairs are connected to common collector nodes. These 32 pairs also share one common current source. The second stage of the sense amplifier is a cascode stage amplifying the output voltage of the differential pair to a voltage swing of approximately 200 mV. A cascode stage is used here to reduce the voltage swing of the high capacitively loaded common collector nodes.

This current-sensing method, normally used in bipolar SRAM's, has the advantage of allowing pure ECL column selection. Column selection is performed by the column select transistors (QS). To select, say, the first bit-line pair, the potential of the base $Y1$ must be 200 mV above $Y2 \cdots Y32$. The column select transistors (QS) merely work as emitter followers, so the bases of the corresponding differential pair are also 200 mV higher than the other 31. This causes the current of the common current source to flow through this differential pair, and the voltage difference at the sensing resistors (RS) of the first bit-line pair is sensed.

The bit-line voltage swing is controlled by bipolar pull-up transistors (QL, QR) and an additional PMOS transistor (ME) working as an equilibration transistor. During the write operation, pull-up transistors (QL, QR) and the equilibration transistor (ME) are switched off. Simultaneously two NMOS transistors (ML, MR) are switched on, connecting the selected bit-line pair to the common write bus ($BLLW$, $BLRW$). The write bus itself, which is common to 32 bit-line pairs, is controlled by two large NMOS transistors. These NMOS transistors pull one line of the write bus down to V_{EE}.

Simulated waveforms for the read operation are shown in Fig. 5(a). Here $A0$ is the address input signal, $LEVSH$ is the output signal of the CMOS current mirror converting ECL to CMOS level, WL is the potential of the selected word line, the $DO0$ is the output signal. The corresponding bit-line voltages are shown in Fig. 5(b).

Most of the delay time is spent on ECL/CMOS level conversion, for charging up the word line and sensing the bit-line potential (Fig. 6). With regard to power dissipation, the situation is quite different. Here most of the power is dissipated within the word-line decoder, especially in the NOR gates. A significant amount is also dissipated in the output buffer, which converts the output signal of the sense amplifier into an ECL 100K compatible signal.

IV. Write Operation

As mentioned above, the write operation is performed by connecting the selected bit-line pair to the write bus. This causes one of the bit lines to be pulled down to V_{EE}, depending on the content of data-in. One important feature of this SRAM is the control unit (Fig. 7), which contains a WPG.

An on-chip WPG (Fig. 7) supplies the write control circuit with a write pulse (Fig. 8) independently of external write pulse widths. The possibility of controlling the RAM with an external write pulse is also given by an extra input ($WEXT$). An ECL MUX which selects one of the two

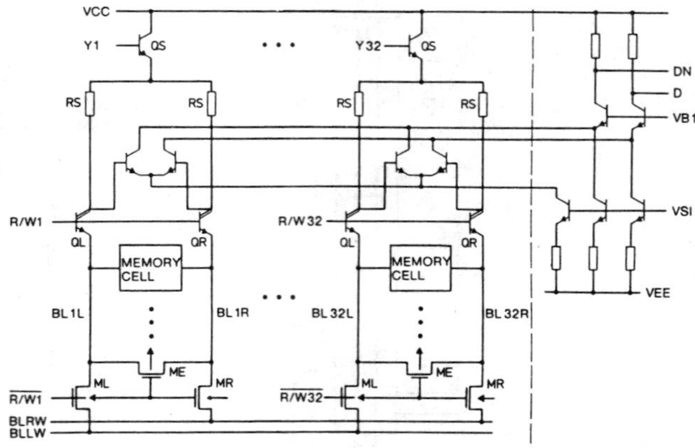

Fig. 4. Precharge and sense circuitry.

write pulses is controlled by the write-select signal WS. This ECL MUX has a powerful output emitter follower for driving 32 NOR gates.

Optimized write pulses meet the following demands:

- The write pulse width is minimal over the entire range of the supply voltage and temperature fluctuations.
- The write pulse width tracks process fluctuations.

To satisfy the last criterion, a delay stage T_d which determines the write pulse width is built up by means of bipolar and MOS elements. This permits tracking of bipolar and MOS parameter fluctuations. The WPG supplies a write pulse approximately twice as long as the minimum possible because of fault tolerances.

The WPG consists of ECL circuits, except for the delay stages T_d. A CMOS WPG proved to be too slow (3-ns delay after write pulse initialization). This was caused by the limited driving capability of a CMOS MUX (due to the large FO) and bad step response of the CMOS circuits (NAND, flip-flop).

The complete WPG works with differential logic instead of reference voltages (due to the symmetrical delays in circuits for rising and falling edges, less complicated wiring, and the absence of noise in bias driver net, which has to provide the reference voltages). It also does not require any emitter followers. The delay stage which determines the write pulse width consists of nine single stages.

V. Technology

The 16K BiCMOS SRAM was based on 1.2-μm BiCMOS. This technology [5] features n-epi, p-well, double metal, triple polysilicon, and a PSA (polysilicon self-aligned) bipolar transistor. The NMOS transistor is an LDD transistor, and the PMOS transistor is provided with an n^+ buried layer to prevent punchthrough. Table IV summarizes the technology data.

VI. Results

Measurements of the samples confirmed the simulations of read access time described in the previous sections. Access time (Fig. 9) was measured to 3.8 ns (−4.5 V,

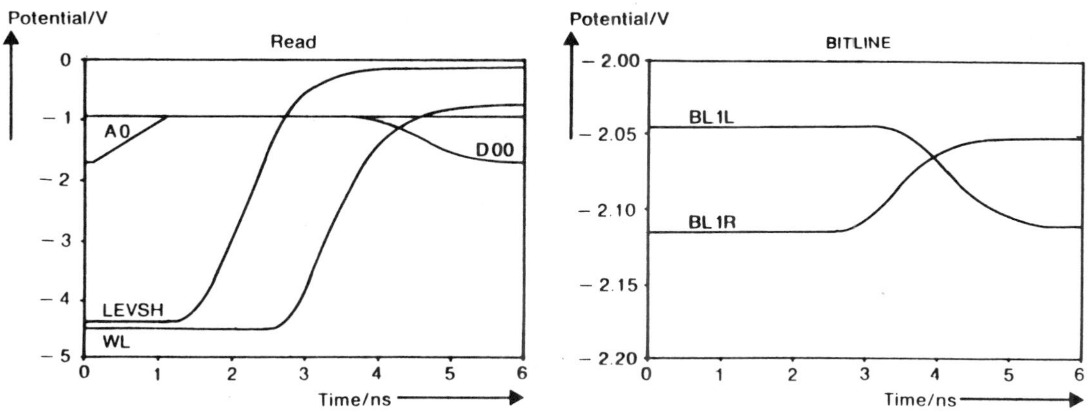

(a) (b)

Fig. 5. (a) Read access waveforms of the time critical path. (b) Bit-line voltages.

TABLE IV
TECHNOLOGY DATA

□ Bipolar		□ CMOS	
Min. emitter size	1.2 μ × 3 μ	Gate length (N-channel)	1.2 μm
hfe	100	Gate length (P-channel)	1.2 μm
Cut-off frequency	12 GHz	Gate oxide thickness	25 nm
Base resistance	210 Ω		
Emitter-base capacitance	13 fF		
Base-collector capacitance	13 fF		
Collector-substrate capacitance	30 fF		

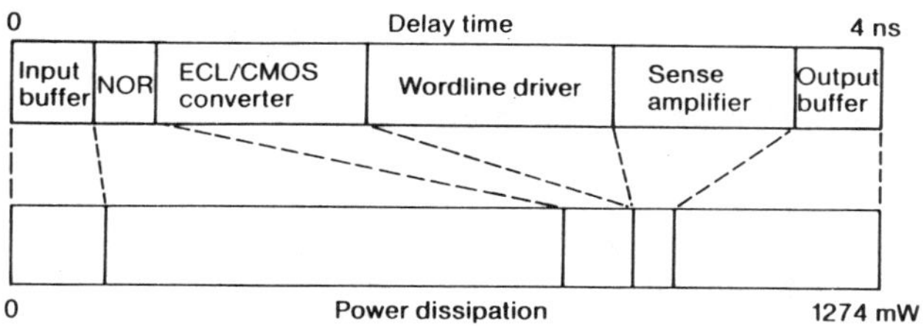

Fig. 6. Distribution of delay time and power dissipation for read operation.

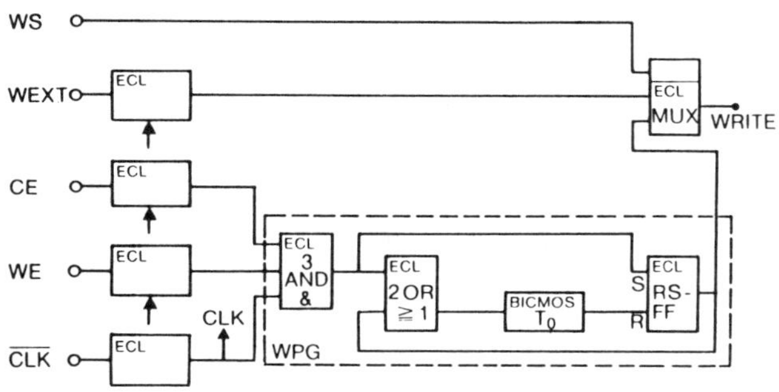

Fig. 7. Write pulse generator.

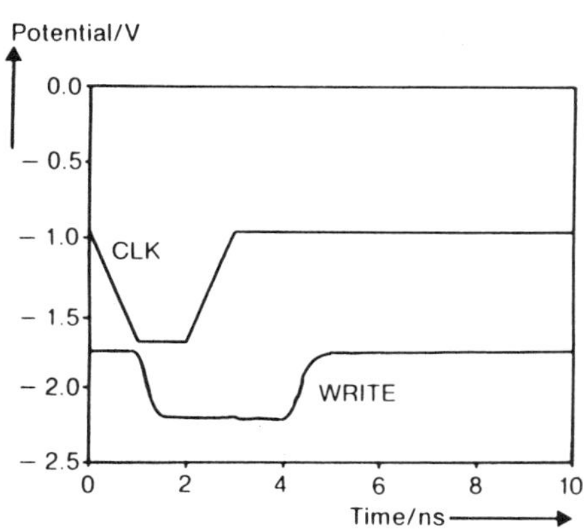

Fig. 8. Write pulse.

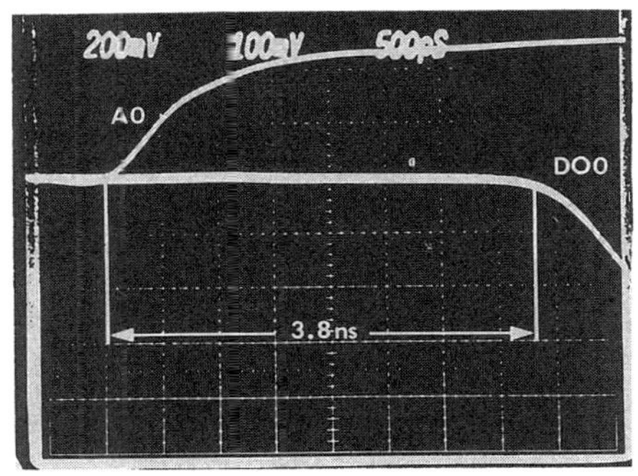

Fig. 9. Measured access time.

60°C), slightly better than simulated. A photo of the layout is shown in Fig. 10.

VII. SUMMARY

A 2K × 8-bit BiCMOS SRAM featuring an access time of 3.8 ns with moderate power dissipation (< 2.0 W) and low area consumption (23 mm^2) is described. The fast access time is mainly achieved by pure ECL decoding and bipolar current-sensing circuits. The bit-line swing is kept down to only 60 mV by a combination of bipolar and MOS transistors. Special Totem Pole buffers, avoiding saturation effects in their bipolar transistors, charge the capacatively loaded word lines of the SRAM. They also guarantee a voltage drop to the lowest possible level on unselected word lines within short time intervals. The use of a six-transistor CMOS cell offers a large gain in terms of lower area consumption and power dissipation.

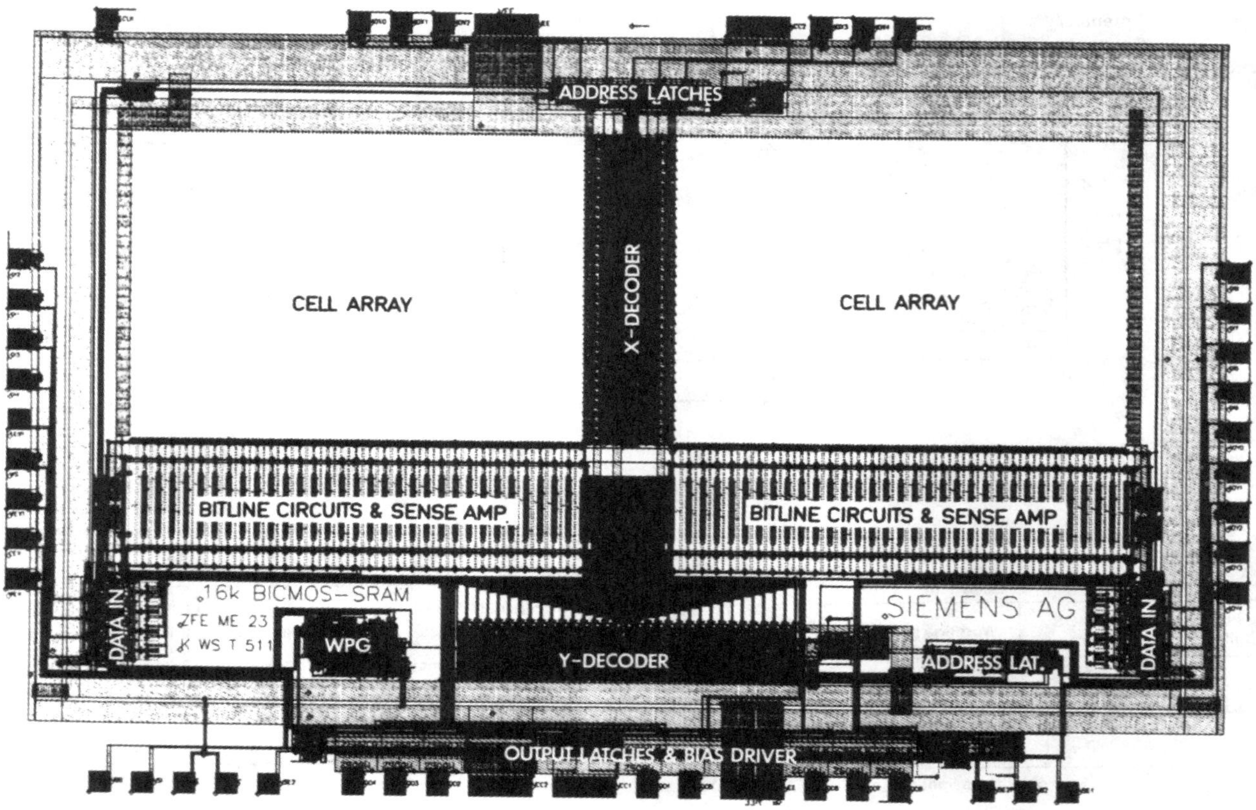

Fig. 10. Photo of layout of the 16K BiCMOS SRAM.

The write operation is optimized by using an on-chip write pulse generator, which provides the SRAM with write pulses which are independent of external conditions.

The SRAM is fully ECL 100K I/O compatible. It has latches for all input and output signals except clock and write select. Thus the SRAM can operate in synchronous or asynchronous mode.

The SRAM is based on a less advanced 1.2-μm BiCMOS technology, featuring two metal layers and three polysilicon layers. Its features are self-aligned bipolar transistors (minimal emitter area $AE = 1.2 \times 3.0\ \mu m^2$) and NMOS and PMOS transistors (both with minimal gate lengths of 1.2 μm).

Acknowledgment

The authors would like to thank H. Klose, B. Hoffmann, and B. Zehner for process characterization, parameter extraction, and many fruitful discussions. They also wish to thank M. Wurm for circuit design and verification.

References

[1] M. Suzuki *et al.*, "A 7ns/350mW 16kb Hi-BiCMOS static RAM," in *Proc. Int. Conf. Solid State Devices and Materials* (Tokyo), 1986, pp. 319–322.

[2] K. Ogiue *et al.*, "A 13-ns, 500-mW, 64-kbit ECL RAM using Hi-BiCMOS technology," *IEEE J. Solid-State Circuits*, vol. SC-21, pp. 681–684, Oct. 1986.

[3] M. Suzuki *et al.*, "A 3.5ns/500mW 16kb BiCMOS ECL RAM," in *ISSCC Dig. Tech. Papers*, Feb. 1989, pp. 32–33.

[4] W. Heimsch, R. Krebs, and K. Ziemann, "A 4ns 16k BiCMOS SRAM," in *Symp. VLSI Circuits, Dig. Tech. Papers* (Kyoto, Japan), May 1989, pp. 67–68.

[5] H. Klose, *et al.*, "Low cost and high performance BiCMOS processes: A comparison," presented at the BCT Meeting, Minneapolis, MN, 1989.

A 4-ns 4K × 1-bit Two-Port BiCMOS SRAM

TSEN-SHAU YANG, STUDENT MEMBER, IEEE, MARK A. HOROWITZ, MEMBER, IEEE, AND BRUCE A. WOOLEY, FELLOW, IEEE

***Abstract*—This paper introduces a two-port BiCMOS static memory cell that combines ECL-level word-line voltage swings and emitter-follower bit-line coupling with a static CMOS latch for data storage. With this cell, referred to as a CMOS storage emitter access (CSEA) cell, it is possible to achieve access times comparable to those of high-speed bipolar SRAM's while preserving the high density and low power of CMOS memory arrays. The memory can be read and written simultaneously and is therefore well-suited to applications such as high-speed caches and video memories. A READ access time of 3.8 ns at a power dissipation of 520 mW has been achieved in an experimental 4K × 1-bit two-port memory integrated in a 1.5-μm 5-GHz BiCMOS technology. The access time in this prototype design is nearly temperature insensitive, increasing to only 4 ns at a case temperature of 100°C.**

I. INTRODUCTION

THE TIME required to access static memory has a significant influence on the throughput achievable in high-performance data processing systems. In such systems, static memories with access times under 5 ns are required if the machine cycle times of emerging high-speed processing units are to be fully exploited. In addition to access time itself, simultaneous multiport access is becoming an increasingly important approach to increasing overall system speed. However, conventional multiport memory designs are typically characterized by a large cell area and relatively slow access.

The highest speed static memories have generally been realized using advanced bipolar technology [1]–[4]. However, the large cell area and standby power dissipation have precluded the scaling of these circuits to increasingly higher levels of integration. The performance of CMOS static memories has been improved dramatically through the continued scaling of technology. However, access times have remained greater than 10 ns, and CMOS memory design continues, therefore, to focus on low-power, high-density applications [5]–[7].

Recently, BiCMOS technology has been used to significantly enhance the speed of CMOS static memory arrays [8], [9]. However, most BiCMOS SRAM's described to date combine conventional CMOS cells with the use of bipolar transistors in the sense amplifiers [10] and for driving large capacitive loads [8]. In these designs, the access time remains limited by factors such as the large voltage swing on the word lines, the limited cell output current, and the number of circuit stages used [11].

This paper introduces a BiCMOS two-port static memory cell, and associated access circuitry, with which it is possible to achieve access times comparable to those of high-speed bipolar SRAM's while retaining the high density and low power of CMOS memory arrays. The cell, referred to as a CMOS storage emitter access (CSEA) cell, combines ECL-level word-line voltage swings and emitter-follower bit-line coupling with a static CMOS latch. Thus, ECL signal swings are maintained throughout the entire READ path. The cell's separate single-ended READ and WRITE ports allow it to be read and written simultaneously, making it especially attractive for multiport memory design. Compared with conventional multiport designs, the CSEA memory offers extremely high speed and small size. The CSEA cell is only one-third larger than a single-port, six-transistor CMOS static memory cell.

To demonstrate the operation of the CSEA cell, a complete 4K × 1-bit two-port SRAM has been designed and integrated in a 1.5-μm BiCMOS technology. This memory operates from a single 5.2-V supply with ECL-compatible I/O. Under nominal operating conditions the measured READ access time for this prototype design is 3.8 ns and the power dissipation is 520 mW. The worst-case access time at a case temperature of 100°C is only 4 ns with a power dissipation of 620 mW. The access time is also nearly independent of variations in power supply voltage and data patterns.

The CSEA cell and the design constraints resulting from the single-ended READ and WRITE operations are considered in Section II. The detailed circuit design of the 4K × 1-bit two-port memory is then presented in Section III. The measured performance of the prototype memory is reported in Section IV, and conclusions are summarized in Section V.

II. CSEA MEMORY CELL

A. Operation

A schematic of the CSEA memory cell and its associated bit-line sensing circuit is shown in Fig. 1. Data are stored

Manuscript received March 21, 1988; revised May 17, 1988. This work was supported in part by a fellowship from IBM and by DARPA under Contract N00014-87-K-0828.

The authors are with the Electrical Engineering Department, Center for Integrated Systems, Stanford University, Stanford, CA 94305.

IEEE Log Number 8822440.

Reprinted from *IEEE J. Solid-State Circuits*, vol. 23, no. 5, pp. 1030–1040. Oct. 1988.

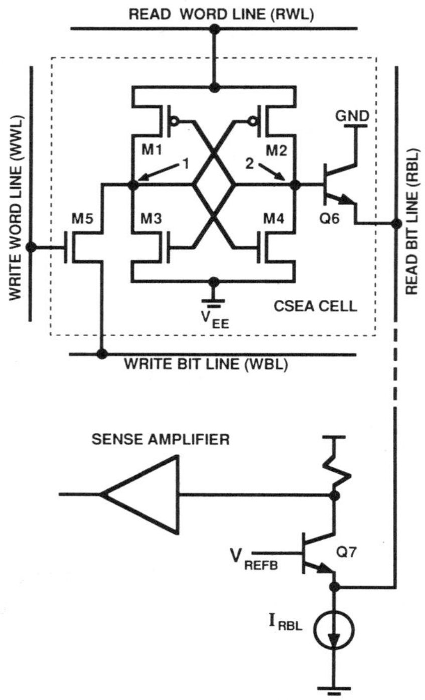

Fig. 1. Schematic of the CSEA memory cell and its associated bit-line sensing circuit.

in an internal CMOS latch formed by the cross-coupled inverters $M1$, $M3$ and $M2$, $M4$. The cell is read through the bipolar emitter follower $Q6$, and is written through the NMOS pass transistor $M5$.

The READ word line (RWL) serves as the positive supply for the internal latch of the CSEA cell, and the cell is read by raising this word line. In this paper, a ONE is defined to be stored in the cell when storage node 2 in Fig. 1 is high; that is, $M2$ and $M3$ are ON, while $M1$ and $M4$ are OFF. When a ONE is stored in the cell, an increase in the potential of RWL is coupled directly to the base of the output emitter follower $Q6$, which forms a differential pair with sense-amplifier input transistor $Q7$. When a ZERO is stored in the cell, the base of $Q6$ is grounded through $M4$ and is unaffected by changes in RWL. Thus, by setting the bit-line reference level V_{REFB} to the midrange of the RWL swing, the bit-line current I_{RBL} will be switched between $Q6$ and $Q7$ depending on the data stored in the cell. A nominal word-line swing of only 550 mV is sufficient to reliably switch the bit-line current.

The pass transistor through which the CSEA cell is written, $M5$, is controlled by the WRITE word line (WWL). Full CMOS logic levels are used on both WWL and the WRITE bit line (WBL). In order to write a ZERO into the cell (node 2 low), $M4$ must be turned on by pulling node 1 high through $M5$ from a high level on WBL. This is more difficult to accomplish than writing a ONE into the cell because $M5$ operates as a source follower and its threshold is increased by the body effect. Therefore, the logic thresholds of the cell inverters must be adjusted to provide equivalent margins for writing either a ONE or ZERO. To provide the drive for a single-ended WRITE, $M5$ is considerably larger with respect to $M3$ and $M4$ than would be the case in a conventional balanced CMOS static cell. As a consequence, care must be taken to avoid disturbing unselected cells in the row where WWL is pulled high. Such WRITE disturbances are avoided by biasing the unselected WBL's to a level close to the logic threshold of the $M2$, $M4$ inverter when RWL is at its low level.

Single-ended sensing of the data stored in the CSEA cell is possible because the full swing on RWL is coupled to the bit line and there is no voltage drop along RWL itself. RWL only supplies the base current for the emitter follower connected to the selected column. This differs from conventional bipolar cells where the bit-line swing is limited to a Schottky-diode forward voltage drop and there is a significant voltage drop along the word line since it must supply both the hold currents and the bit-line current.

The single-ended WRITE operation of the CSEA cell is similar to that in a conventional CMOS six-transistor cell when writing a ONE. In this case, $M5$ operates in the common-source configuration pulling the storage node 1 low. However, to write a ZERO into the CSEA cell, $M5$ operates as a source follower with a correspondingly reduced drive capability. To balance the delays for writing either a ONE or a ZERO, the threshold of the $M2$, $M4$ inverter is set slightly lower than that of the $M1$, $M3$ inverter. Since the READ and WRITE operations are separated in the CSEA memory, the delay associated with the WRITE recovery of the bit lines in a conventional CMOS six-transistor memory is avoided.

The CSEA cell can be selected simultaneously for both reading and writing because the READ and WRITE access are single-ended and access different storage nodes within the cell latch. Even when the WWL for a cell has been selected, the cell can be accurately read at the normal speed (as long as it is not actually being written) because the disturbance at the base of $Q6$ resulting from WWL going high is small. Thus, it is possible to simultaneously read one cell and write another even if they are in the same row.

B. Single-Ended READ

The minimum RWL swing needed to reliably distinguish the data stored in the CSEA cell can be determined by considering the two worst-case conditions depicted in Fig. 2. In Fig. 2(a) a ONE is stored in the cell being read, while all other cells in the same column contain a ZERO. Under these circumstances $Q6$ must switch the bit-line current away from $Q7$. To ensure a particular current sensing ratio $K = I_1/I_2$, where I_1 is the current flowing in $Q6$ and I_2 is the current in $Q7$, the difference between the high level of RWL and the bit-line reference must satisfy the relationship

$$V_{RWL}(\mathrm{HIGH}) - V_{\mathrm{REFB}} > (kT/q)\ln(K) \qquad (1)$$

where k is the Boltzman constant, T is temperature, and q is the electron charge.

As illustrated in Fig. 2(b), the worst case for reading a ZERO occurs when all unselected cells in the column contain a ONE. In this situation the bit-line reference is not compared with the ZERO level stored in the internal latch; instead, the bit-line current is switched between the reference transistor $Q7$ and the emitter-follower output transistors of the unselected cells, all of which have their bases pulled up to the low level of RWL. For this case the current sensing ratio is defined as $K = I_2/I_1$, where I_2 is the current in $Q7$ and I_1 is the total current flowing in the unselected cells of the column. To achieve a specified K, the required difference between the low level of RWL and the bit-line reference is

$$V_{\text{REF}B} - V_{RWL}(\text{LOW}) > (kT/q)[\ln(N-1) + \ln(K)] \quad (2)$$

where N is the number of cells in a column.

It follows from (1) and (2) that the minimum swing on RWL needed to maintain a given current sensing ratio is

$$\begin{aligned}\Delta V_{RWL} &= V_{RWL}(\text{HIGH}) - V_{RWL}(\text{LOW}) \\ &= (kT/q)[\ln(N-1) + 2\ln(K)]. \end{aligned} \quad (3)$$

For $N = 64$ and $K = 100$, the minimum swing is 13.3 kT/q, with $V_{\text{REF}B}$ positioned at 35 percent of the swing below $V_{RWL}(\text{HIGH})$.

The READ access time of the CSEA cell is comparable to that of high-speed bipolar SRAM cells. Assuming a step input on RWL and that charge storage in $Q6$ dominates the capacitance at the storage node, the selection time of the cell can be estimated as

$$T_{\text{SELECT}} = \frac{I_{RBL}\tau_f}{\Delta V_{RWL}/R_{M2}} \quad (4)$$

where τ_f is the transit time of $Q6$ and R_{M2} is the equivalent resistance of $M2$ over the RWL transition. The deselection time can be estimated as

$$T_{\text{DESELECT}} = \frac{\Delta V_{RWL} C_{RBL}}{I_{RBL}} \quad (5)$$

where C_{RBL} is the total parasitic capacitance on the READ bit line. For a design with equal select and deselect times, the cell READ access time is given approximately by

$$T_{\text{READ}} = \left[\tau_f R_{M2} C_{RBL}\right]^{1/2}. \quad (6)$$

In a 1.5-μm 5-GHz BiCMOS technology, the cell access time is approximately 0.5 ns, which is close to that of a bipolar SRAM cell realized using bipolar transistors with a comparable cutoff frequency [12].

C. Single-Ended WRITE

The major design considerations faced with respect to the single-ended WRITE operation are guaranteeing that a selected cell can be overwritten successfully when storing either a ONE or a ZERO and minimizing the disturbance in unselected cells when WWL goes high. In order to change

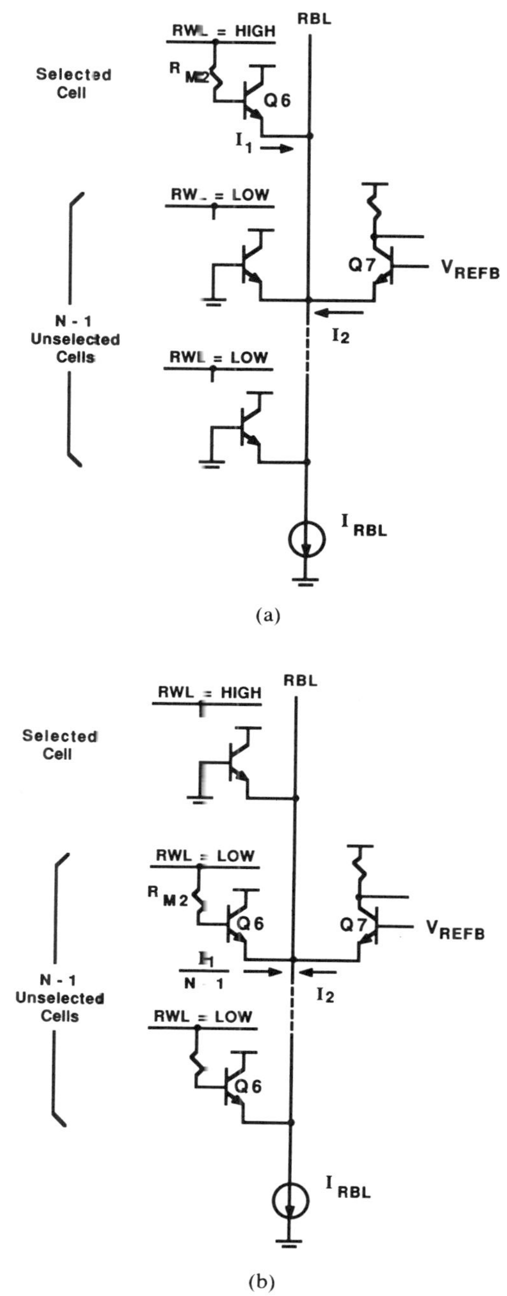

Fig. 2. Worst-case READ considerations for noise margin: (a) reading ONE, and (b) reading ZERO.

the state of the latch, the drive through the pass transistor $M5$ must be strong enough to overcome either the pull-up of $M1$ or the pull-down of $M3$ so that the voltage at the internal node 1 crosses the logic threshold of the $M2$, $M4$ inverter. However, for cells on the same row with WWL high, the unselected WBL's must be biased at a level such that turning on $M5$ will not change the state of the $M2$, $M4$ inverter.

The conditions and equivalent circuits for writing the cell are illustrated in Figs. 3. Writing is most difficult when the RWL is high since the cell is then powered from a higher voltage and the feedback within the latch is stronger. Shown in Fig. 4 are the conditions and equivalent circuits

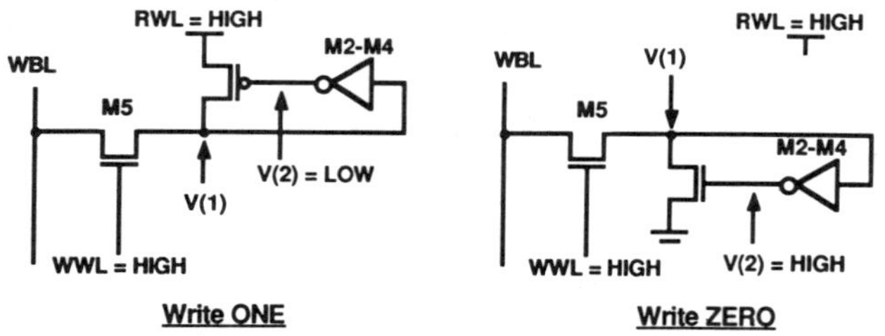

Fig. 3. Conditions and equivalent circuits for single-ended WRITE.

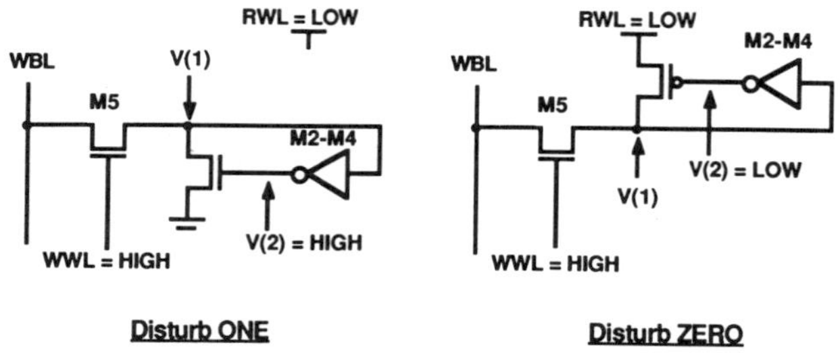

Fig. 4. Conditions and equivalent circuits for WRITE disturbances.

under which WRITE disturbances are evaluated. The disturbance of a cell is more severe when RWL is low because of the reduced margins and lower loop gain of the latch.

Since WWL and WBL are driven at full CMOS levels (GND and V_{EE}), the design parameters governing WRITE and WRITE disturbance that remain to be determined are the level at which WBL is biased for the unselected cells, the voltage at the internal node 1 for the various conditions depicted in Figs. 3 and 4, and the logic threshold of the $M2$, $M4$ inverter. The node 1 voltage $V(1)$ is determined by the relative sizes of $M1$, $M3$, and $M5$, as well as the voltage levels on WWL and WBL. The $M2$, $M4$ inverter threshold depends on both the relative sizes of $M2$ and $M4$, and the level of RWL.

The margins for changing the data stored in a cell, as well as the range within which WBL must be biased for unselected cells, can be determined for a particular design by plotting the transfer curves of $V(1)$ as a function of the WBL voltage level for each of the four conditions depicted in Figs. 3 and 4. These transfer curves for the experimental design reported in this paper are shown in Fig. 5. Curves A and B correspond to writing a ONE and a ZERO, respectively. Curves C and D correspond to the respective disturbance conditions when a ZERO and a ONE are stored in the cell. Also included in Fig. 5, as curves E and F, are the logic thresholds of the $M2$, $M4$ inverter for both a high and low RWL.

In order to change the data stored in a cell from a ZERO (node 2 low) to a ONE, node 1 must be brought below the $M2$, $M4$ threshold. Thus, the WBL voltage must be less than that corresponding to the intersection of curves A and E in Fig. 5; that is, V_{WBL} must lie in region I as identified in the figure. Similarly, to write a ZERO into the cell V_{WBL} must be in region V, above the intersection of curves B and E. Disturbance of an unselected cell severe enough to change its data is avoided by biasing WBL to a voltage in region III, above the intersection of curves C

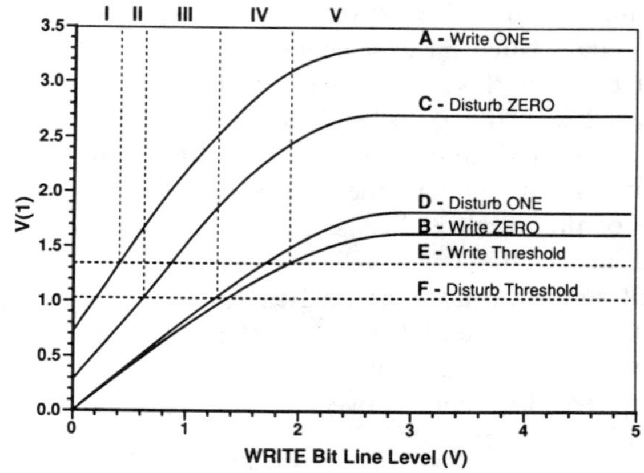

Fig. 5. Transfer curves for single-ended WRITE and WRITE disturbances.

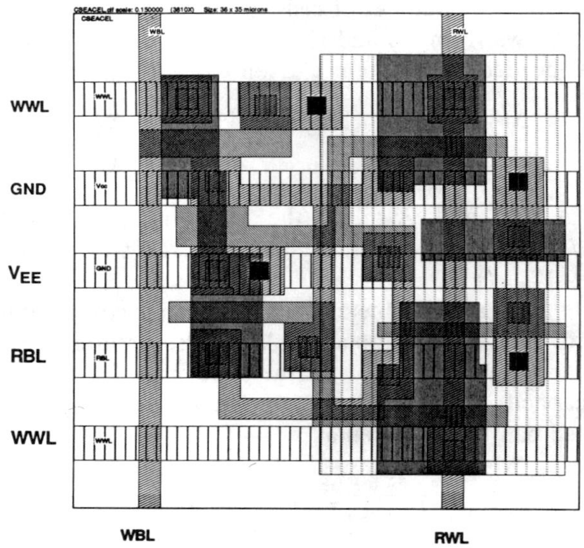

Fig. 6. CSEA cell layout using 1.5-μm design rules and the Manhattan layout style.

and F and below that of curves D and F. For the design to which Fig. 5 applies, writing a ZERO or a ONE requires driving WBL above 2 V or below 0.4 V, respectively, while biasing WBL between 0.7 and 1.2 V for unselected cells ensures that they will not be destructively disturbed. Since the disturb margins determined from Fig. 5 are based on static conditions, they represent a worst-case situation. A WRITE operation that is clocked by a WRITE ENABLE pulse will, in fact, generate much less disturbance in unselected cells than is predicted by the analysis under static conditions.

D. Cell Layout

The size of the CSEA cell is governed largely by the required six interconnections: GND, V_{EE}, RWL, RBL, WWL, and WBL. Shown in Fig. 6 is a layout of the cell using 1.5-μm design rules and a Manhattan layout style. The cell size is 26×25 μm^2. A six-transistor CMOS cell laid out using the same rules and style measures 24×20 μm^2. The separate READ and WRITE ports of the CSEA cell, which prohibit the sharing of signal lines among adjacent rows and columns, account for the increased cell area. Nonetheless, the area of the CSEA cell is comparable to that of single-ended CMOS multiport cell designs [13] and much smaller than that of either high-speed bipolar static memory cells or fully differential multiport CMOS cells.

Since the loading a cell imposes on RWL depends both on the data stored in the cell and on whether or not WWL is high, RWL and RBL are laid out orthogonal to WWL and WBL, respectively. Thus, RWL parallels WBL and RBL parallels WWL. As a result, a selected RWL is loaded by at most a single cell with a high WWL. This arrangement minimizes WRITE interference with the READ operation.

III. Memory Design

A. Chip Architecture

The block diagram of a 4K two-port CSEA memory is shown in Fig. 7. The memory is organized as a 64-row × 64-column array. Separate READ and WRITE address inputs provide direct access to the READ and WRITE ports of the cell array. In addition, separate data input and output paths are used to improve system performance by eliminating the need for multiplexing the data bus, thus reducing the I/O delay in the critical path. The two-port architecture of Fig. 7 allows simultaneous and independent memory access without the overhead encountered in conventional multiport memory designs that is needed to resolve address conflicts. This feature makes the CSEA memory especially attractive in the design of application-specific memories, such as shared memory for multiprocessor communication, cache memory, and video memory.

The READ word lines in the CSEA memory are decoded from the READ X address, $AX0-AX5$, by 64 READ row decoders. There are 32 READ column decoders controlled by $AY1-AY5$. Each column decoder controls two READ bit lines and switches on the bit-line current sources of the selected bit lines. Data in two cells are read using two separate sense amplifiers whose outputs are then multiplexed under the control of the remaining READ column address bit $AY0$. This approach reduces the power dissipation in the column decoders without significantly increasing the READ access time.

Writing of the array is controlled by 64 WRITE row decoders and 64 WRITE column decoders. The WRITE decoders are enabled only while the WRITE ENABLE signal is high in order to minimize power dissipation. All of the external WRITE port signals—WRITE address, WRITE data, and WRITE ENABLE—are converted from external ECL levels to full CMOS internal swings using ECL-to-CMOS translators. Since the delays in these translators balance each other, the memory cycle time depends only on the decoder delay and the time needed to actually write the cell.

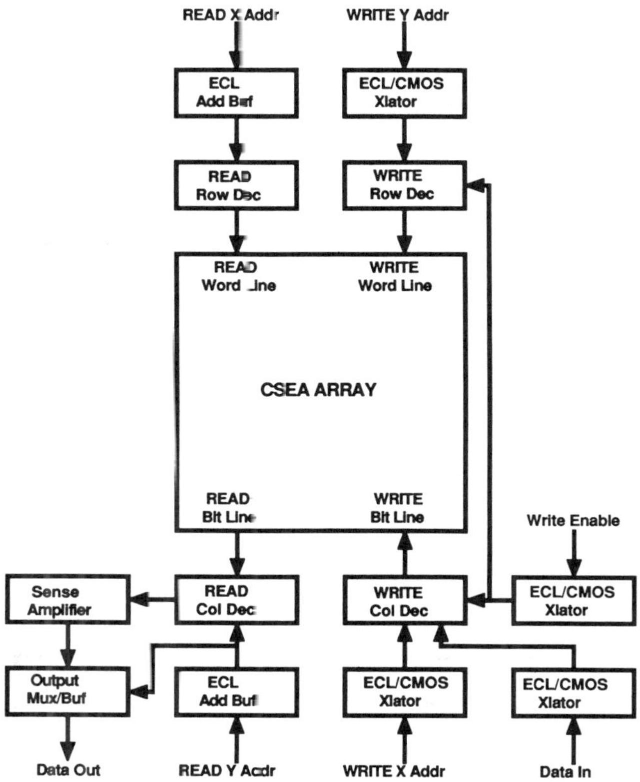

Fig. 7. Block diagram of 4K × 1-bit two-port CSEA memory.

B. READ Path

The READ path consists of input address buffers, row decoders, the memory array, sense amplifiers, and an output multiplexor/buffer. It is comprised entirely of current-switching bipolar circuits operating at small voltage swings, and the signals are all single-ended. The READ path circuits are designed to ensure a voltage swing of 21 kT/q on RWL. Careful generation of the signal levels and references within the READ path is essential to single-ended sensing. The voltage levels are controlled to be proportional to kT/q and independent of variations in supply voltage. Reference levels are generated using either replica techniques or by directly monitoring the levels of the signals that are being compared to the reference. As described below, the latter approach is used to generate the READ bit-line reference voltage.

A circuit schematic for the READ address buffers and row decoders is shown in Fig. 8. Push–pull address input buffers [2] are used to provide fast transitions at the inputs of the row decoders. Row decoding is accomplished using diode decoders. The word lines are driven by Darlington emitter followers tied through resistors to a common pull-down current of 15 mA [4]. The active pull-down current available to discharge a deselected word line is 7 mA, while a static pull-down current of 125 μA is maintained in each of the unselected word lines. Because of voltage-swing attenuation in the diode decoders and Darlington emitter followers, the swing of the internal address lines is designed to be 35 kT/q so as to ensure a swing of 21 kT/q

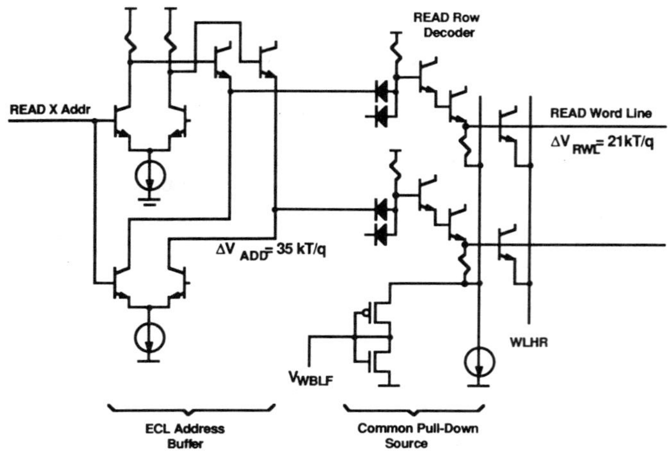

Fig. 8. Circuit schematic of the address input buffer and row decoders in the READ path.

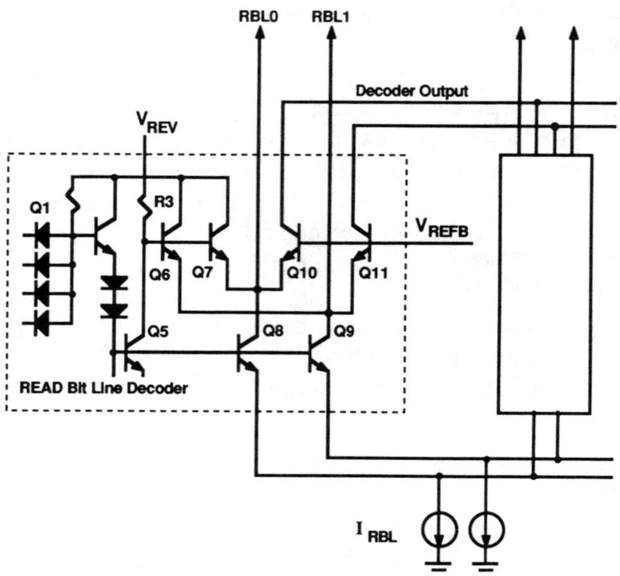

Fig. 10. Circuit schematic of the READ bit-line decoder and column switch.

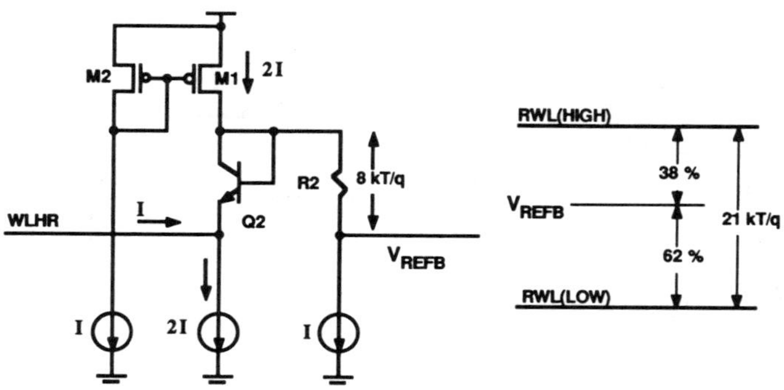

Fig. 9. READ bit-line reference circuit.

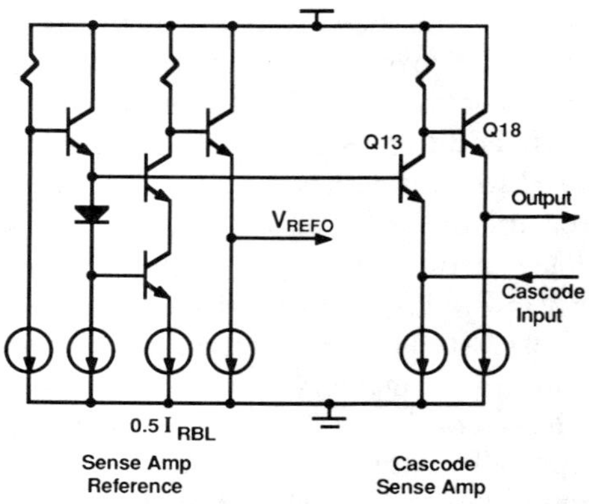

Fig. 11. Single-ended sense amplifier with cascoded inputs.

on RWL. Since the voltage level of the common pull-down node is very close to the low level of the READ word lines, a CMOS inverter with its input and output shorted is connected in parallel with the pull-down current source and used to generate a voltage that approximates the logic thresholds of the memory cell latch inverters. This voltage, V_{WBLF}, is used to bias the unselected WRITE bit lines.

To ensure that the READ bit-line reference $V_{\mathrm{REF}B}$ tracks the high level of RWL in the selected cell, this level is monitored with a wired-OR of emitter followers driven by each of the READ word lines. The wired-OR signal, WLHR, is then fed to the bit-line reference circuit, shown in Fig. 9, where it is used to generate the bit-line reference voltage. Since WLHR is one V_{BE} below the high level of the selected READ word line, it is first shifted up V_{BE} by transistor $Q2$. An accurate replica of the high RWL level is established by forcing the same amount of current in $Q2$ and the wired-OR emitter followers. $V_{\mathrm{REF}B}$ is then generated with a level shift downward of 8 kT/q across resistor $R2$. As indicated in Fig. 9, $V_{\mathrm{REF}B}$ lies 38 percent of the READ word-line swing below the high level of RWL. The generation of the bit-line reference by direct monitoring minimizes reductions in noise margin due to V_{EE} and GND drops, or other common-mode signals.

The schematic of the READ bit-line decoder is shown in Fig. 10. The column decoders are implemented using diode decoders similar to those used for row decoding. Each decoder controls two READ bit lines. When the output of the column decoder is high, the two bit-line current sources are connected to a pair of READ bit lines by turning on $Q8$ and $Q9$. The bit-line reference transistors, $Q10$ and $Q11$ in Fig. 10, form current-switched differential pairs with the output emitter followers of the selected memory cells. To reduce the coupling of the bit-line capacitance to the READ word lines, the unselected READ bit lines are biased so that the base–emitter junctions of the memory cell emitter followers in these columns are reverse biased. This is accomplished by passing a voltage level, V_{REV}, through $R3$, $Q6$, and $Q7$ to the unselected READ bit lines. When a column is selected, $Q5$ is turned on and the voltage drop across $R3$ disables $Q6$ and $Q7$.

The outputs of the bit-line reference transistors $Q10$ and $Q11$ are each connected to the input of a sense amplifier. The schematic of Fig. 11 shows one of the two sense amplifiers along with the shared sense-amp reference network. Since the bit-line reference transistor outputs for one half (32) of the bit-line decoders are connected to each sense-amplifier input, the parasitic capacitance at the input

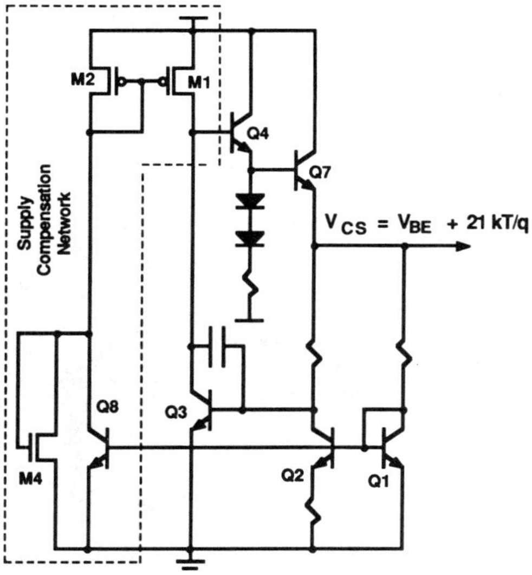

Fig. 12. Circuit schematic of BiCMOS bandgap reference generator with supply compensation.

is quite large. Therefore, a cascode configuration has been used to minimize the voltage swing on the input node. In Fig. 11, $Q13$ is the cascode transistor. A single-ended reference level V_{REFO}, to which the outputs of the sense amplifiers are compared at the inputs of the subsequent output multiplexer/buffer, is provided by the sense-amplifier reference circuit along with biasing for the cascoded sense amplifier itself. V_{REFO} is generated by passing a current of 0.5 I_{RBL}, where I_{RBL} is the READ bit-line current, through a replica of the sense amplifier and the decoder path of the bit-line current. The bit-line current sources themselves are generated within the sense-amplifier circuitry so as to eliminate the concern for power supply drops.

An on-chip bandgap reference generator, shown schematically in Fig. 12, is used to control the voltage swings in the READ path. The output voltage V_{CS} is designed to be $V_{BE} + 21\ kT/q$. An important second-order effect in conventional bandgap circuit designs is the sensitivity of the reference voltage to power supply variations. A change in the supply voltage typically induces a mismatch in current density between the bandgap transistor $Q1$ and the feedback transistor $Q3$. However, in the circuit of Fig. 12 a feedback path consisting of $Q8$, $M2$, and $M1$ is used to ensure close tracking of the current densities in $Q1$ and $Q3$ over the full range of supply voltage and operating temperature. SPICE simulations show that the reference output changes less than 6 mV for supply variations between 4.5 and 5.5 V, and less than 0.014 percent/°C between 25 and 155°C.

C. WRITE Path

The WRITE path of the CSEA memory is similar to that of conventional BiCMOS SRAM's. A schematic of the WRITE path is shown in Fig. 13. ECL-to-CMOS level translators are used to convert the input addresses from external ECL levels to full CMOS internal levels. The WRITE address decoding is accomplished by means of dynamic series decoders that are clocked by the WRITE ENABLE signal. The dynamic decoders occupy a very small area and easily fit into the cell pitch; moreover, they dissipate little power since current flows only in the single selected decoder.

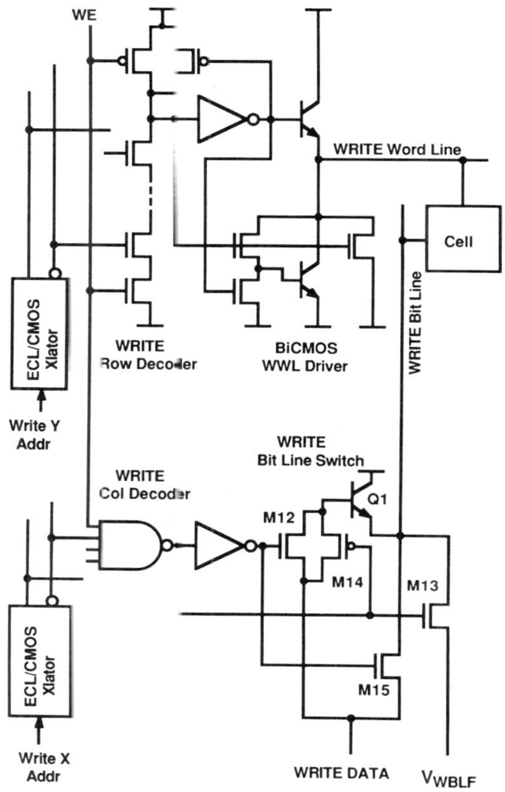

Fig. 13. Schematic of the WRITE path.

The dynamic row decoders are followed by BiCMOS buffers that drive the high capacitance of the word lines. When not selected, the word lines are pulled to V_{EE} by NMOS transistors in parallel with the BiCMOS buffers so as to ensure the complete turn-off of the WRITE access transistors of memory cells in the unselected WRITE rows. Unselected WRITE bit lines are connected to V_{WBLF}, through transistor $M13$ in the schematic of Fig. 13. When a WRITE column is selected, the data are passed onto the WRITE bit line through transistor $M15$. Since the unselected WRITE bit lines are biased in the neighborhood of 1 V, the approximate logic threshold of the memory cell latch inverters when RWL is low, the rise time of the WRITE bit line is considerably longer than its fall time. Therefore, a bipolar transistor $Q1$ controlled by the CMOS switch $M12$ and $M14$ has been used to speed up the rising transition.

A schematic of the ECL-to-CMOS translator is shown in Fig. 14. The propagation delay of this circuit is a critical component of the WRITE access time. The minimum delay is obtained by maximizing the voltage swing at the output of the first differential amplifier $Q4$ and $Q5$. This condition provides the largest possible overdrive to the gates of

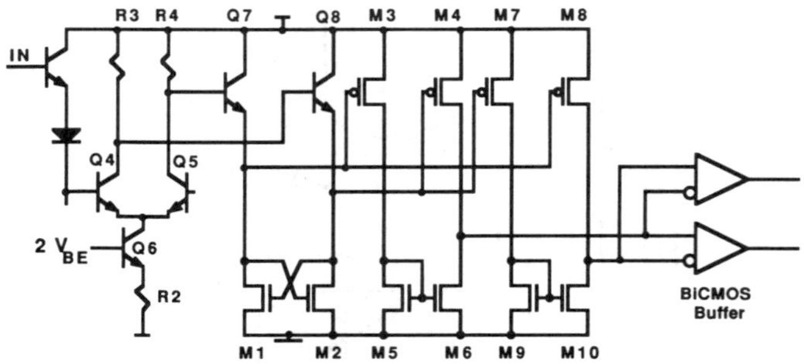

Fig. 14. Schematic of the ECL/CMOS translator.

the following PMOS devices, $M3$, $M4$, $M7$, and $M8$. To prevent saturation of $Q4$ when the external input is high, the current source formed by $Q6$ and $R2$ is biased at $2V_{BE}$. Load resistors $R3$ and $R4$ are then ratioed to $R2$ so as to limit the maximum swing at the collectors of $Q4$ and $Q5$ to $3V_{BE}$. Since the high level of the external inputs is one V_{BE} below GND (the positive supply), the base–collector junction of $Q4$ remains reverse biased under all input conditions over the entire ranges of temperature and supply voltage. The cross-coupled NMOS transistors, $M1$ and $M2$ in Fig. 14, provide an active pull-down at the outputs of the first differential stage, thereby reducing the standby dc current drain in the emitter followers ($Q7$ and $Q8$) of that stage.

IV. Experimental Results

The prototype 4K×1-bit two-port memory design was integrated in a 1.5-μm, 5-GHz, double-metal, double-poly BiCMOS technology. A die photo of this experimental CSEA memory is shown in Fig. 15; the die size is 2.5×3.5 mm^2. In laying out the circuit, particular attention was given to isolating the low-voltage-swing current switching bipolar circuits from power supply noise generated in the CMOS circuits. Separate GND pads, which serve as the positive supply voltages, are used for the bipolar and CMOS circuits, along with three V_{EE} pads. Additional pads for monitoring and, if desired, altering the primary reference levels are included along with internal test pads to aid in the characterization of the memory.

The performance of the experimental memory is summarized in Table I. Shown in Fig. 16 are oscillographs of the measured READ row access time for both worst- and best-case data patterns at room temperature with a supply voltage of −5.2 V. Under these conditions the power dissipation is 520 mW. The worst-case data pattern is that where all the cells in the array other than the one being read store a ONE. In this case, the parasitic capacitances on the READ word lines and bit lines are maximized. The best-case pattern is that where all cells in the array except the one being read store a ZERO. From Fig. 16 it is apparent that no significant data dependency is observed for the READ row access time; for both conditions, the access time is less than 3.8 ns.

Fig. 17 shows the READ column access time at room temperature with the nominal supply voltage. Again, no significant data dependency is observed. The column access time is approximately 2.2 ns.

Shown in Fig. 18 are oscillographs of the READ row and column access responses at a case temperature of 100°C. At this temperature, the measured worst-case READ row and column access times are 4.0 and 2.2 ns, respectively. The power dissipation increases to 620 mW with a supply voltage of −5.2 V because of larger voltage swings. The READ row and columns access times were also measured

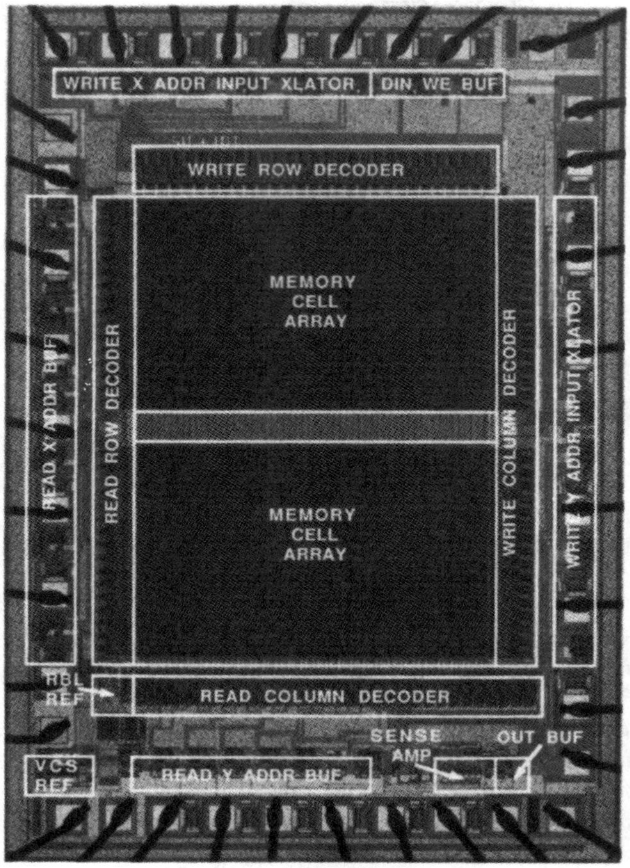

Fig. 15. Microphotograph of the fabricated 4K CSEA memory.

TABLE I
PERFORMANCE SUMMARY OF THE 4K CSEA MEMORY

Technology	1.5μm, 5GHz BiCMOS
Cell Size	26μm x 25μm
Chip Size	2.5mm x 3.5mm
Configuration	4K×1Bit 2-port
I/O Interface	ECL 10K compatible
Power Supply	-5.2 Volts
Power Consumption	520 mW @ 25°C
	620 mW @ 100°C
Read Access Time	3.8 ns @ 25°C
	4.0 ns @ 100°C
Minimum Write Pulse	4.0 ns @ 25°C
Address Setup time	< 1 ns @ 25°C
Address Hold Time	< 1 ns @ 25°C

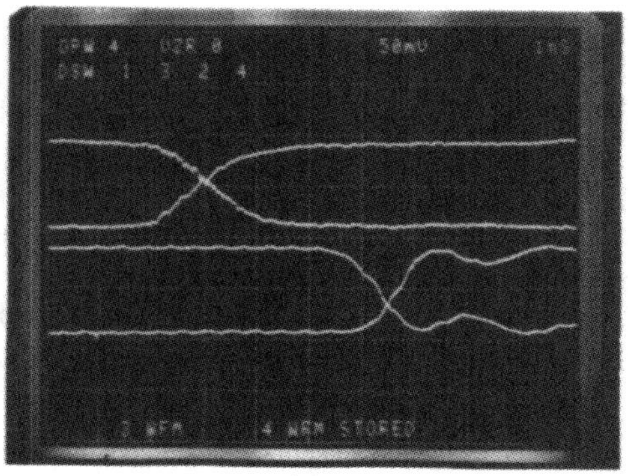

(a)

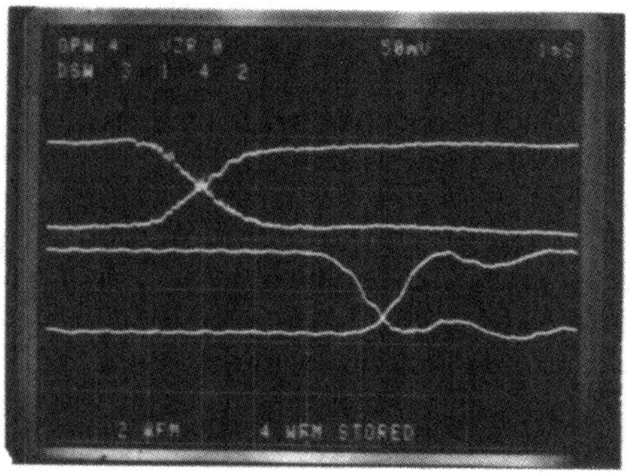

(b)

Fig. 16. Oscillographs of the READ row access time at room temperature: (a) all ONE's, and (b) all ZERO's. Upper trace: address input; lower trace: data output.

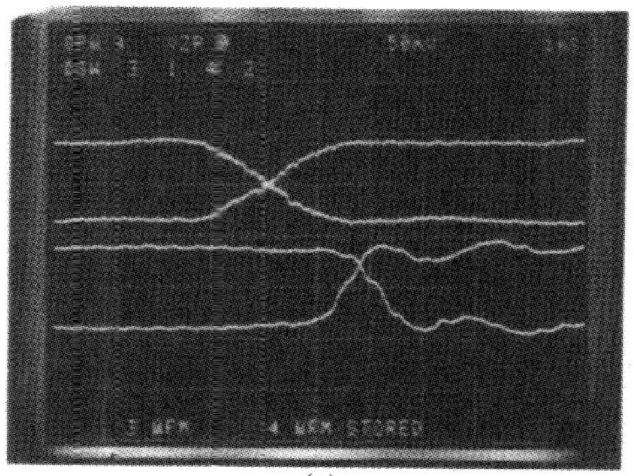

(a)

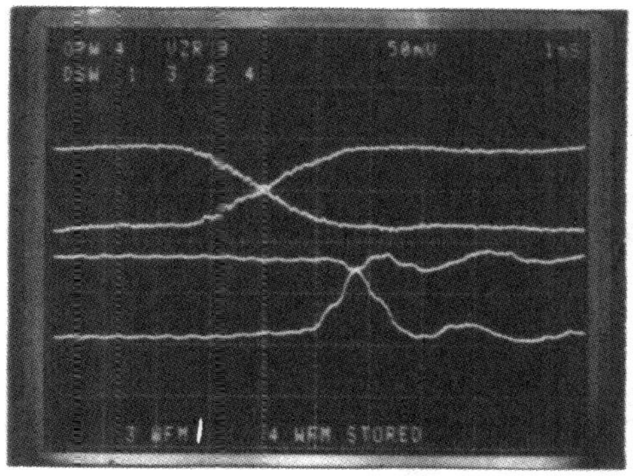

(b)

Fig. 17. Oscillographs of the READ column access time at room temperature: (a) all ONE's, and (b) all ZERO's. Upper trace: address input; lower trace: data output.

with a supply voltage of −4.5 V, and no changes in the access times were observed. The insensitivity of the access times to temperature is a consequence of using temperature-compensated voltage swings and low-temperature-coefficient polysilicon resistors in the READ path.

The independent READ and WRITE capability of the memory is demonstrated in the oscillograph of Fig. 19. In this measurement, one cell is being read while another is simultaneously being written. The cell being read is also chosen so that it is in a WRITE disturbance condition; that is, its WWL has been taken high. Similarly, the cell being written is in a worst-case condition because its RWL is high. This experiment also confirms that the WRITE disturbance does not degrade the READ access time. Because the READ and WRITE ports are separate and independent, there is no arbitration problem between READ and WRITE addresses. When the two ports simultaneously read and write the same cell, the relationship between the READ and WRITE data is purely one of propagation delay; i.e., the READ port reflects the newly written data after an elapsed time approximately equal to the sum of the WRITE ENABLE pulse width and the READ access time.

The measured minimum WRITE ENABLE pulse width is 4 ns. Setup and hold times of the WRITE address signals with respect to the WRITE ENABLE pulse are less than 1 ns. The propagation delays of the ECL/CMOS translators in the WRITE address and WRITE ENABLE paths balance each other during the WRITE operation. Thus, the minimum WRITE ENABLE pulse width is determined by the delay in the dynamic decoder and the time needed to write data into the cell. The minimum WRITE cycle time is the sum of the minimum WRITE ENABLE pulse width and the address setup and hold times. This suggests that the prototype CSEA memory can be cycled at rates above 150 MHz.

Since most of the power in the CSEA memory is dissipated in the peripheral circuits, a 16K two-port SRAM implemented using the same circuit techniques and technology is projected to have a typical READ access time of 4.5 ns with a moderate increase in power dissipation to 750 mW.

V. SUMMARY

This paper has described a new BiCMOS memory cell that combines the density and low power dissipation of CMOS cells with the access times achievable in bipolar SRAMS and independent READ and WRITE access. Key features of the CSEA cell are the use of small, ECL-like, word-line voltage swings and emitter-follower coupling to the bit line in a design wherein a static CMOS latch is used for data storage. The CSEA cell also employs single-ended READ and WRITE access to the cell, thus providing a multiport capability with little increase in cell size.

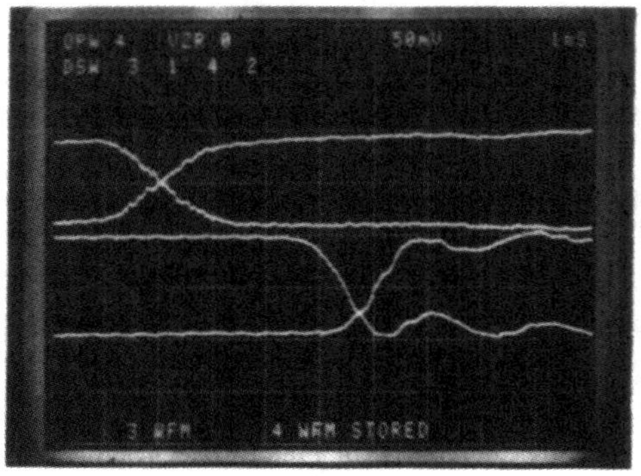

(a)

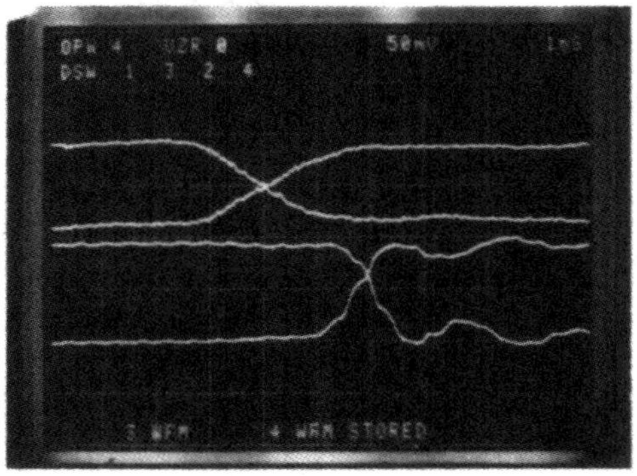

(b)

Fig. 18. READ access time measurement at 100°C case temperature: (a) row access, and (b) column access. Upper trace: address input; lower trace: data output.

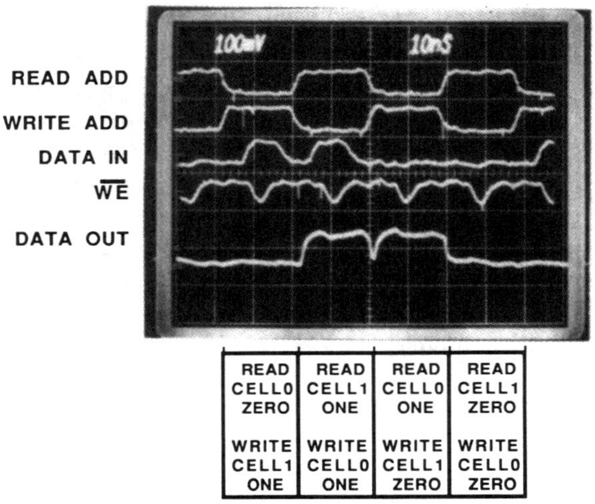

Fig. 19. Memory response to simultaneous READ and WRITE operations.

The CSEA memory cell has been incorporated into the design of an experimental 4K × 1-bit two-port SRAM. This prototype memory was integrated in a 1.5-μm 5-GHz BiCMOS technology, and a READ access time of 4 ns with a power dissipation of 520 mW was achieved. The experimental results confirm that in a BiCMOS technology it is possible to achieve access times comparable to those of bipolar SRAM's in a multiport static memory design that retains CMOS latches for data storage. The design approach adopted in the CSEA memory thus provides significantly enhanced flexibility at the system architecture level without compromising speed, power, or integration density.

Acknowledgment

The authors wish to thank Integrated Device Technology, Inc. for fabricating the circuits. They are especially grateful to F. C. Hsu and C. C. Wu of IDT for numerous helpful discussions. They are also indebted to J. B. Kuo and R. W. Dutton of Stanford, who provided the initial inspiration for this project.

References

[1] K. Kawarada, M. Suzuki, H. Mukai, K. Toyoda, and Y. Kendo, "A fast 7.5 ns access 1-kbit RAM for cache-memory systems," *IEEE J. Solid-State Circuits*, vol. SC-13, no. 5, pp. 656–663, Oct. 1978.

[2] N. Homma *et al.*, "A 3.5-ns, 2-W, 20-mm^2, 16-kbit ECL bipolar RAM," *IEEE J. Solid-State Circuits*, vol. SC-21, no. 5, pp. 675–680, Oct. 1986.

[3] H. Miyanaga, S. Konaka, Y. Kobayashi, Y. Yamamoto, and T. Sakai, "A 0.85-ns 1-kbit ECL RAM," *IEEE J. Solid-State Circuits*, vol. SC-21, no. 4, pp. 501–504, Aug. 1986.

[4] C. T. Chuang, D. D. Tang, G. P. Li, E. Hackbarth, and R. R. Boedeker, "A 1.0-ns 5-kbit ECL RAM," *IEEE J. Solid-State Circuits*, vol. SC-21, no. 5, pp. 670–674, Oct. 1986.

[5] M. Matsui *et al.*, "A 25-ns 1-Mbit CMOS SRAM with loading-free bit lines," *IEEE J. Solid-State Circuits*, vol. SC-22, no. 5, pp. 733–740, Oct. 1987.

[6] N. Okazaki, T. Komatsu, N. Hoshi, K. Tsuboi, and T. Shimada, "A 16 ns 2K × 8 bit full CMOS SRAM," *IEEE J. Solid-State Circuits*, vol. SC-19, no. 5, pp. 552–556, Oct. 1984.

[7] S. T. Flannagan *et al.*, "Two 13-ns 64K CMOS SRAM's with very low active power and improved asynchronous circuit techniques," *IEEE J. Solid-State Circuits*, vol. SC-21, no. 5, pp. 692–703, Oct. 1986.

[8] K. Ogiue *et al.*, "13-ns, 500-mW, 64-kbit ECL RAM using HI-BICMOS technology," *IEEE J. Solid-State Circuits*, vol. SC-21, no. 5, pp. 681–685, Oct. 1986.

[9] S. Miyaoka *et al.*, "A 7-ns/350-mW 64-kbit ECL-compatible RAM," *IEEE J. Solid-State Circuits*, vol. SC-22, no. 5, pp. 847–849, Oct. 1987.

[10] J.-I. Miyamoto *et al.*, "A high-speed 64K CMOS RAM with bipolar sense amplifiers," *IEEE J. Solid-State Circuits*, vol. SC-19, no. 5, pp. 557–563, Oct. 1984.

[11] T. Douseki and Y. Ohmori, "BiCMOS circuit technology for a high speed SRAM," in *Symp. VLSI Circuits Dig. Tech. Papers*, 1987, pp. 77–78.

[12] H. Miyanaga, Y. Yamamoto, Y. Kobayashi, and T. Sakai, "A 1.5 ns 1K bipolar RAM using novel circuit design and SST-2 technology," *IEEE J. Solid-State Circuits*, vol. SC-19, no. 3, pp. 291–298, June 1984.

[13] K. J. O'Connor, "The twin-port memory cell," *IEEE J. Solid-State Circuits*, vol. SC-22, no. 5, pp. 712–720, Oct. 1987.

Paper 3.6

A 7-ns 1-Mb BiCMOS ECL SRAM with Shift Redundancy

Atsushi Ohba, Shigeki Ohbayashi, Toru Shiomi, Satoshi Takano, Kenji Anami, *Member, IEEE*, Hiroki Honda, Yoshiyuki Ishigaki, Masahiro Hatanaka, Shigeo Nagao, and Shimpei Kayano

***Abstract*—A 1-Mb BiCMOS ECL SRAM was fabricated with the 0.8-μm BiCMOS process. An improved ECL buffer having a high-level output of nearly V_{cc} is adopted to eliminate the dc current in the level converter circuit, and the PMOS transistor has a wide operating margin in the level converter.**

The configurable bit organization is realized by using a sense-amplifier switch circuit with no access degradation. A wired-OR demultiplexer for the ×1 output, having the same critical path as the ×4 output circuit, allows for the same access time between the two modes. The ×1 or ×4 mode is electrically selected by the external signal. A simplified programming redundancy technology, named "shift redundancy," is utilized. Address programming is performed by cutting the only one fuse in the "shift redundancy." The RAM operates at the ECL-10K level with an access time of 7 ns, and power dissipation at 50 MHz is 600 mW for the ×1 mode.

I. Introduction

HIGH-DENSITY and high-speed memory devices have greatly contributed to improving the performance of recent computing systems. Much higher density and speed memory devices are in demand for further improvement. BiCMOS devices are advantageous because they have both CMOS's low power and high density and bipolar's high speed.

Recently the density of BiCMOS SRAM's, which use MOS memory cells, has become as large as CMOS SRAM's with fast access time of sub 10 ns [1]–[12]. However, there are some technological problems in the high-speed and high-density memories. One is the increase of test time, another is the decrease of yield, and so on. The configurable bit organization, which was introduced in the 1-Mb CMOS SRAM [14], is one of the most preferable answers to overcome the test-time problems. However, the configurable bit organization may cause access-time degradation, because it needs some additional sensing path switches or increasing of sensing stages. On the other hand, redundancy technologies are widely used to improve the yield. But the address programmings for replacing failed address lines become more complex as the densities increase.

This paper describes a 1-Mb BiCMOS ECL SRAM which has a configurable bit organization with no access degradation and a one-fuse-programming redundancy named "shift redundancy" [11]. In Section II, the circuit technology, which is employed in this ECL SRAM, is explained. The "shift redundancy" technique is described in Section III. The process technology is summarized in Section IV. In Sections V and VI, the chip architecture and characteristics of this ECL SRAM are reported, respectively.

Manuscript received September 7, 1990; revised December 12, 1990.
The authors are with the LSI Research and Development Laboratory, Mitsubishi Electric Corporation, 4-1 Mizuhara, Itami, Hyogo 664, Japan.
IEEE Log Number 9042535.

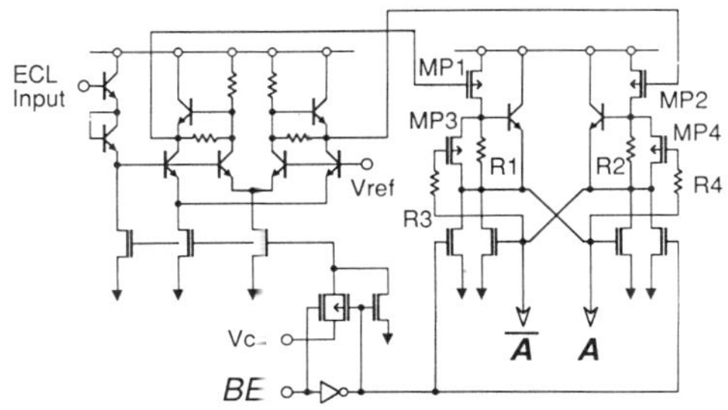

Fig. 1. Circuit diagram of ECL–CMOS converter.

II. Circuit Technology

A. ECL–CMOS Converter

An ECL–CMOS level converter is an important factor of fast access time and low power consumption. Fig. 1 shows the ECL–CMOS level converter circuit used in the present ECL SRAM. The ECL–CMOS level converter consists of the ECL buffer part, which receives the input signal, and the converter part. The ECL buffer part is improved for dc power reduction in the converter part. In this ECL buffer, the load current of the emitter follower with high-level output is cut off by the current switch controlled by the same input signal. Resistors are adopted between the base and emitter electrodes, which pull the high-level output up to V_{cc} after the emitter-follower transistors pull it up to around $-V_{BE}$ and are off. This "high-" level output is suitable for the interfaces to the bipolar buffered level converters, because it eliminates the dc current of the PMOS transistor in the level converter and provides a wide operating margin for the variations of the PMOS-transistor threshold voltage.

The level converter is constructed of two BiNMOS drivers, in which outputs are cross coupled to the gate of each other NMOS transistor [8]. The pull-up bipolar transistor is utilized to obtain enhanced drive capability and low input capacitance. The PMOS transistors $MP3$ and $MP4$ are adopted to shunt the bipolar transistors and to obtain the full-swing CMOS level when the output is high. The polysilicon resistors $R3$ and $R4$, which have the value of several 10 kΩ, are used to make the switching delay of $MP3$ and $MP4$.

Reprinted from *IEEE J. Solid-State Circuits*, vol. 26, no. 4, pp. 507–512, April 1991.

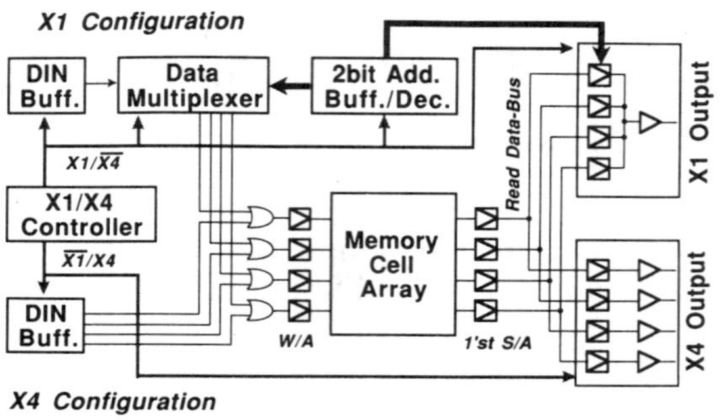

Fig. 2. Block diagram of bit-configuration controller.

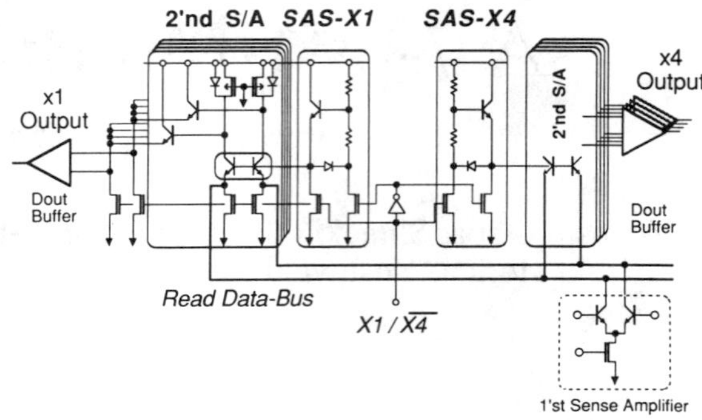

Fig. 3. Circuit diagram of sense-amplifier switch circuit.

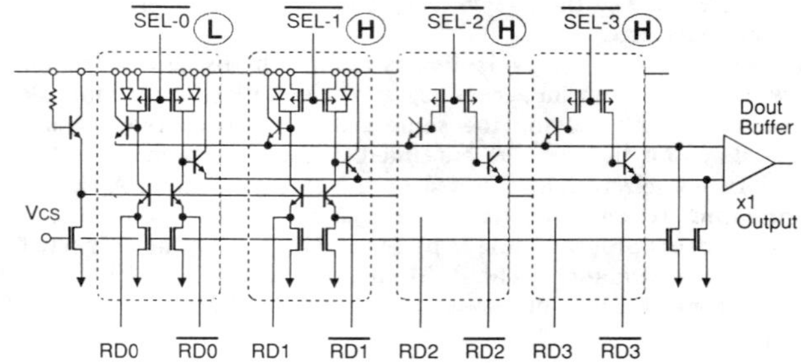

Fig. 4. Circuit diagram of data demultiplexer.

A fast conversion time of 1.3 ns with a capacitive load of 0.3 pF, including the delay time of the ECL buffer, is obtained. The complementary CMOS level outputs are obtained when the buffer-enable signal (BE) is high. On the other hand, when BE is low, the current sources of the ECL circuit are cut off, and both of the emitter-follower outputs rise up to V_{cc} level. Therefore, both outputs of the level converter are at the low level with no power dissipation.

B. X1/X4 Control Circuit

Fig. 2 shows the block diagram of the bit configuration control circuitry. The bit organization is controlled by the external signal and electrically changed to the $\times 1$ or $\times 4$ configuration. The bit-configuration controller, called the $X1/X4$ controller, receives the external control signal and provides the complement signals. The signals are used for cutting off the current in the unused input buffers, for setting the unused level converter output to low, and for switching the sense-amplifier circuits of the $\times 1$ or $\times 4$ mode. Two additional address signals are required for the $\times 1$ mode, and the decoded signals are provided to the data input and output buffers of the $\times 1$ mode. The internal DIN signal pair of the $\times 1$ mode is multiplexed to eight signals by the data multiplexer, and each signal is combined by the OR gate with the internal DIN signals of the $\times 4$ mode. Two second sense amplifiers, one for the $\times 1$ mode and the other for the $\times 4$ mode, are connected to the same read data bus. To obtain the $\times 1$ output, one of four second sense amplifiers is selected by the decoded address signals.

C. Second Sense-Amplifier Switch Circuit

To realize the configurable bit organization, the key technology is switching the sensing paths. In the previous work of the CMOS SRAM, the transmission gates were used as the sensing path switches, and caused the access-time degradation due to the resistance of the transmission gate [14]. The case where ECL series gate is utilized as the switch was also reported [10].

Generally, in the ECL SRAM's, the current output sense amplifier is used for the first sense amplifier, and also in the present ECL SRAM. The second sense amplifier is composed of the current-to-voltage conversion stage such as resistors. The data-bus clamp transistors are adopted to minimize data-bus voltage swing and delay time.

In this ECL SRAM, two second sense amplifiers are connected to the same read data bus, one for the $\times 1$ configuration and the other for the $\times 4$ configuration. Normally, such a connection causes the amplifier outputs to be unstable, because the sensing current is divided into two paths. To avoid this problem, the data-bus clamp transistors are also used as the switches, similar to an ECL circuit. The switch circuitry of the second sense amplifiers is shown in Fig. 3. The sense-amplifier switch control circuits SAS-$X1$ and SAS-$X4$ are adopted to control the clamp transistors. The case of selecting the $\times 1$ mode is demonstrated in Fig. 3. The gray transistors show that they are not in an active state. When the mode control signal $X1/X4$ is set at high level, the output of the SAS-$X1$ is high, and the clamp transistors of the $\times 1$ mode are ON, while the output of the SAS-$X4$ is low, and those of the $\times 4$ mode are OFF. Thus the sensing current flows through only the $\times 1$ mode sense amplifier, and stable operation of the selected second sense amplifier is performed.

D. Data Demultiplexer for X1 Output

Fig. 4 shows the circuit of the second sense amplifiers and the data demultiplexer to obtain the $\times 1$ output. The PMOS transistors are used as loads, and the gate electrodes of those transistors are controlled by the decoded address signals SEL-0 to SEL-3. The outputs of the emitter followers are connected as wired-OR logic. Only one of four transistor pairs is activated, and the complementary outputs are obtained. On the other hand, the other transistors are set in an OFF state, and their outputs are at the low level clamped by the diodes. This circuit realizes the second sense amplifiers and the data demultiplexer in only one stage, thus access degradation is minimized. Minimum access deviation is also realized between the $\times 1$ and $\times 4$ configurations because the sensing paths of the two modes are almost the same.

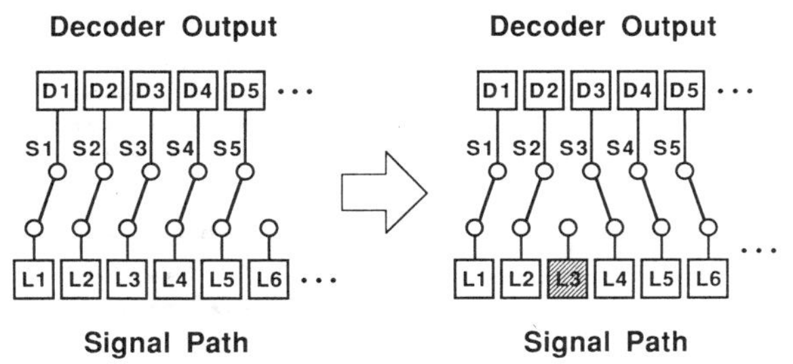

Fig. 5. Concept of shift redundancy.

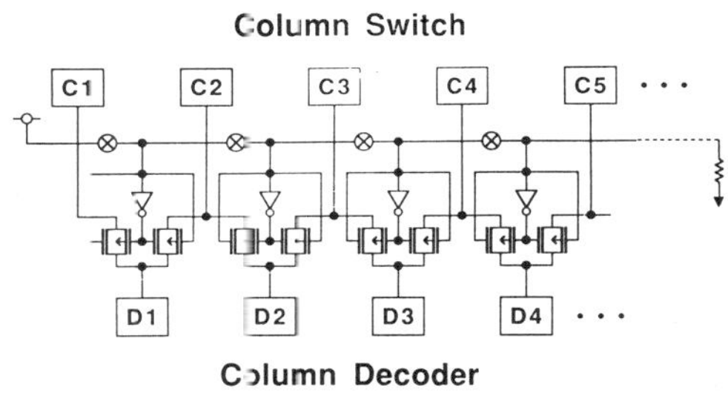

Fig. 7. Detailed circuit of shift redundancy.

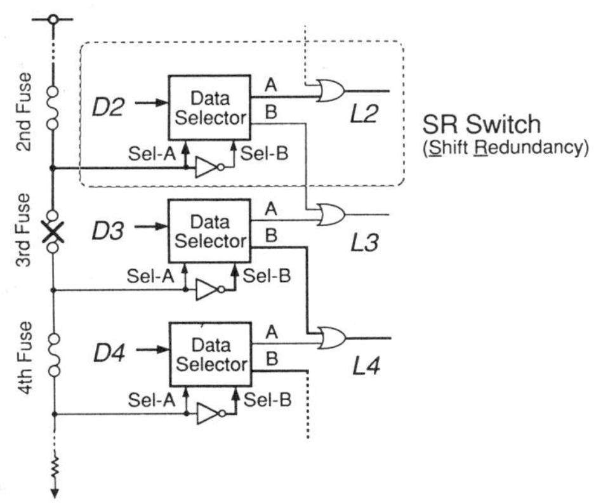

Fig. 6. Block diagram of shift redundancy.

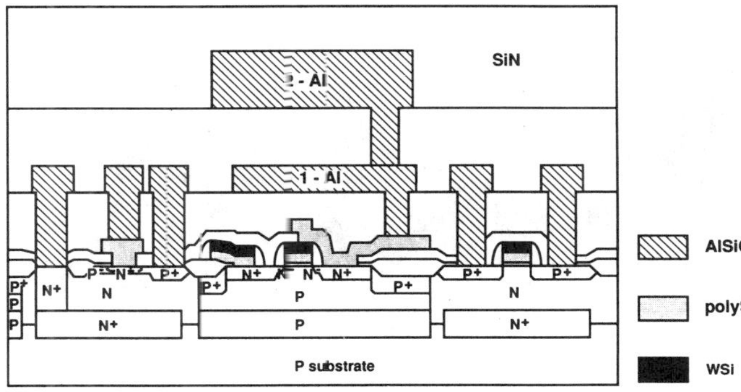

Fig. 8. Cross-sectional view of 0.8-μm BiCMOS process.

III. Redundancy Technology

Fig. 5 shows the concept of the redundancy technology, which is employed in the present ECL SRAM. The outputs of decoders are connected to the signal lines through the switches. When the redundant line is not in use, the output of the decoder $D1$ is connected to $L1$, $D2$ to $L2$, and also the others. Assuming that address line $L3$ contains some failed bits, switch $S3$ turns to $L4$. In the same manner, $S4$ turns to $L5$ and $S5$ to $L6$. In this redundancy technique, if the failed address line is detected, the signal paths shift to the next signal lines one after another. Thus, this scheme is called as "shift redundancy."

Fig. 6 shows the block diagram of "shift redundancy." Switching of the data output path A or B is controlled by select-A and select-B signals. To apply the select-A signals, the fuses are connected in series, and one end of the series fuses is connected to the ground, while the other is connected to V_{ee} through the resistor. If the third line $L3$ contains some failed bits, the third fuse is blown; then the select-A signals of the first and the second lines remain at high level, and those of the third line to the last are set to low. The outputs of the first and second decoders $D1$ and $D2$ come through to the first and second signal lines $L1$ and $L2$, and the outputs of the third to the last decoders come through to the next lines. Then the third line $L3$ is disconnected, and the signal paths are shifted. Therefore, to replace the failed line with the redundant line, only one fuse is blown.

Fig. 7 shows the detailed circuit of the implemented circuit of the "shift redundancy." The transfer gates are used as the data selector. The shift-redundancy circuits are implemented between the fully decoded Y signals and the column switches. As for the row select lines, the shift-redundancy circuits are implemented between the global row decoders and the buffers.

There are three advantages to using this shift redundancy. First, there is no need for a complex address programming circuit to select a redundant line, because the operation is simple and is performed by cutting only one fuse. Second, all the signal paths are identical, thus the access-time degradation is avoided even if a failed address line is replaced. Finally, it is easily implemented at any decoding stage, and the operations are independent of the other groups and/or stages.

IV. Process Technology

The 1-Mb BiCMOS ECL SRAM chip was fabricated with a 0.8-μm BiCMOS technology. It utilizes double-polysilicon and double-metal layers. A cross-sectional view of the RAM is shown in Fig. 8. The minimum emitter size of the bipolar transistor is 1.0 μm$\times$4.0 μm. The cutoff frequency of 8 GHz is achieved by utilizing the polysilicon emitter structure. The gate lengths of PMOS and NMOS transistors are 0.8 and 0.6 μm, respectively. The NMOS transistor has an LDD structure. The thickness of the gate oxide is 15 nm. The 0.6-μm-gate-length NMOS transistors are used only in the memory cells. A memory cell consists of four NMOS transistors and two highly resistive loads of around 3 TΩ. The cell size is 5.4 μm$\times$7.2 μm (38.88 μm^2). The process and device parameters are summarized in Table I.

V. Chip Architecture

Fig. 9 is a micrograph of the 1-Mb BiCMOS ECL SRAM, and Fig. 10 shows the chip architecture. The modified divided-word-line structure [14], [15] is adopted to obtain

TABLE I
PROCESS FEATURES

Process	
0.8-μm BiCMOS	
Double Poly, Double Metal	
Bipolar	
Emitter Size	1.0 μm $\times$ 4.0 μm
Cutoff Frequency	8 GHz
Current Gain	70
CMOS	
Gate Length (p/nMOS)	0.8 μm/0.6 μm
Threshold Voltage	$-$0.7 V/0.7 V
Gate Oxide Thickness	15 nm

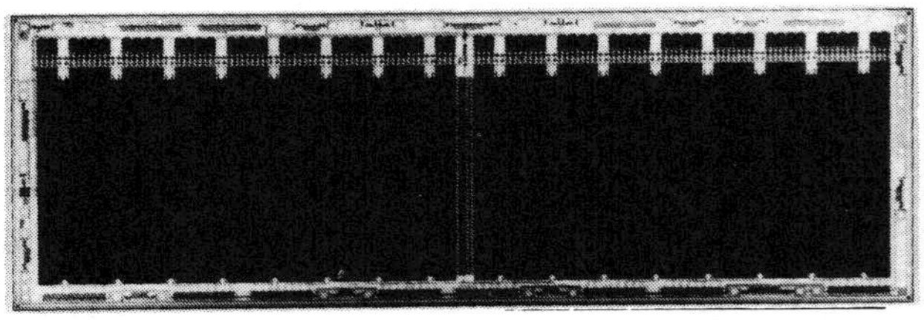

Fig. 9. Chip photomicrograph of the RAM.

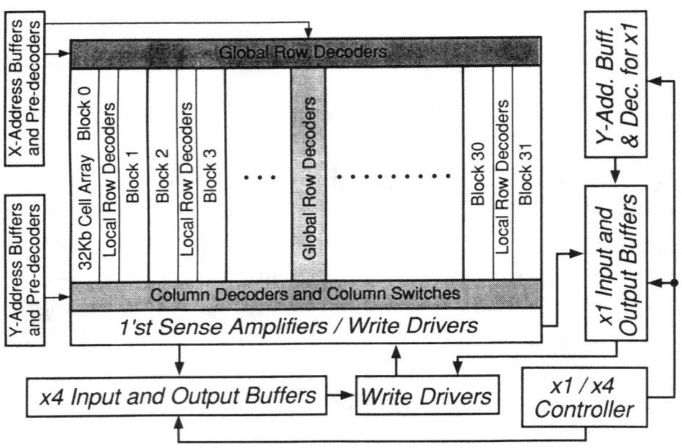

Fig. 10. Chip architecture of the RAM.

TABLE II
SUMMARY OF RAM CHARACTERISTICS

Organization	1M word $\times$ 1 b
	256K word $\times$ 4 b
Address Access Time	7 ns
Power Dissipation	600 mW for $\times$1
	680 mW for $\times$4 (at 50 MHz)
I/O Interface	10K ECL
Redundancy	2 rows and 4 columns
Memory Cell Size	5.4 μm $\times$ 7.2 μm
Chip Size	5.46 mm $\times$ 16.16 mm

both a fast access time and low power consumption. A 1-Mb memory cell array is divided into 32 blocks. Each block includes 512 rows and 64 columns. The RAM has four columns in each block and two rows for the redundancy. The global word decoders are located at the center of the memory cell array to decrease the resistance and the peak current of the signal line to half value. The global word lines are connected to four local word drivers every block.

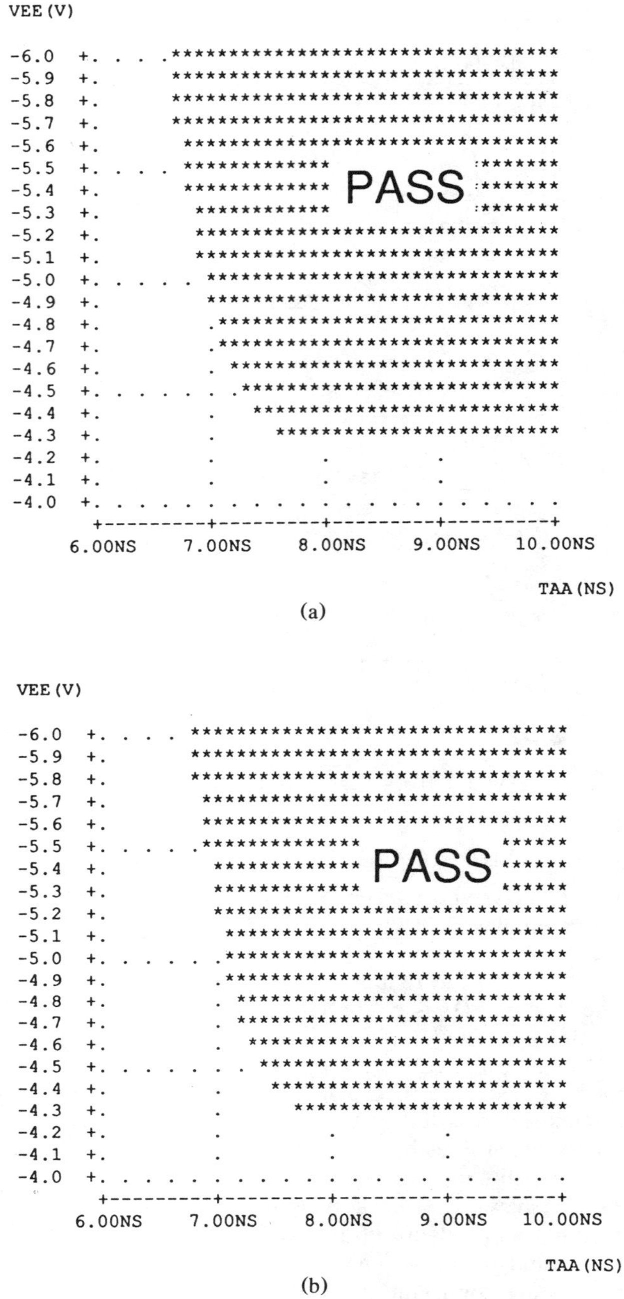

Fig. 11. Access-time schmoo plot: (a) $\times$1 mode, and (b) $\times$4 mode.

The RAM internally has a four bit-wide configuration to realize the configurable bit organization. Each block is further divided into four submatrices. Each submatrix consists of 16 plus one spare columns, a pair of the first sense amplifier, and the write amplifier.

Read and write data buses run along the long side. The data-input buffer and the data-output buffer of the $\times$1 mode are located at the center of the short side, and those of the $\times$4 mode are at the center of the long side. The two bit Y-address buffers and decoders for the $\times$1 mode are located near the data-input and data-output buffers of the $\times$1 mode. The chip size is 5.46×16.16 mm^2 (88.3 mm^2).

VI. Chip Characteristics

The present ECL SRAM is organized as 1M word×1 or 256K word×4. The power dissipations at 50 MHz are 600 mW for the ×1 mode and 680 mW for the ×4 mode. The I/O interface is ECL 10K. The RAM has two rows and four columns for redundancy.

Two schmoo plots of the address access time versus V_{ee} at room temperature are shown in Fig. 11. Fig. 11(a) shows that of the ×1 mode, and Fig. 11(b) shows that of the ×4 mode. Typical address access time of 7 ns was achieved in both modes. The differences between the two modes are within 0.2 ns, thus the configurable bit organization feature can also be used as the ×4 test mode in the ×1 organization SRAM.

VII. Conclusion

A 7-ns 1-Mb BiCMOS ECL SRAM was fabricated with the 0.8-μm BiCMOS technology. The improved ECL buffer, in which the high-level output rises to nearly V_{cc}, is employed to eliminate the dc current in the CMOS level converter part. The cross-coupled type level converter with bipolar buffer is utilized to obtain fast conversion time and enhanced drive capability.

The configurable bit organization is realized with no access degradation by using the second sense-amplifier switch circuit. The wired-OR demultiplexer for the ×1 output realizes the same access time between the ×1 and ×4 modes. The "shift-redundancy" technique simplifies the address programming, the operation of which is to blow only one fuse.

The access time of 7 ns is obtained in both the ×1 and ×4 modes, and the differences between the two modes are within 0.2 ns. The power dissipation at 50 MHz is 600 mW for the ×1 mode, and 680 mW for the ×4 mode.

Acknowledgment

The authors wish to thank Dr. H. Komiya, Dr. T. Nakano, and Dr. N. Tsubouchi for their encouragement.

References

[1] S. Miyaoka *et al.*, "A 7-ns/350-mW 64-kbit ECL-compatible RAM," *IEEE J. Solid-State Circuits*, vol. SC-22, p. 847, Oct. 1987.

[2] H. V. Tran *et al.*, "An 8 ns 256K ECL SRAM with CMOS memory array and battery backup capability," *IEEE J. Solid-State Circuits*, vol. 23, pp. 1041–1047, Oct. 1988.

[3] R. A. Kertis *et al.*, "An 12-ns ECL I/O 256K×1-bit SRAM using a 1-μm BiCMOS technology," *IEEE J. Solid-State Circuits*, vol. 23, pp. 1048–1053, Oct. 1988.

[4] I. Fukushi *et al.*, "A 256K ECL RAM with redundancy," in *ISSCC Dig. Tech. Papers*, Feb. 1988, pp. 184–185.

[5] N. Tamba *et al.*, "An 8-ns 256K BiCMOS RAM," *IEEE J. Solid-State Circuits*, vol. 24, p. 1021, Aug. 1989.

[6] M. Matsui *et al.*, "An 8-ns 1-Mbit BiCMOS SRAM with double-latch ECL–CMOS-level converters," *IEEE J. Solid-State Circuits*, vol. 24, pp. 1226–1232, Oct. 1989.

[7] H. V. Tran *et al.*, "An 8 ns BiCMOS 1 Mb ECL SRAM with a configurable memory array size," in *ISSCC Dig. Tech. Papers*, Feb. 1989, p. 36.

[8] Y. Maki *et al.*, "A 6.5 ns 1 Mb BiCMOS ECL SRAM," in *ISSCC Dig. Tech. Papers*, Feb. 1990, p. 136.

[9] M. Takada *et al.*, "A 5-ns 1-Mb BiCMOS ECL SRAM," *IEEE J. Solid-State Circuits*, vol. 25, pp. 1057–1062, Oct. 1990.

[10] B. Kertis *et al.*, "A 6.8 ns 1 Mb ECL I/O BiCMOS configurable SRAM," in *Symp. VLSI Circuits Dig. Tech. Papers*, June 1990, pp. 39–40.

[11] A. Ohba *et al.*, "A 7 ns 1 Mb BiCMOS ECL SRAM with program-free redundancy," in *Symp. VLSI Circuits Dig. Tech. Papers*, June 1990, pp. 41–42.

[12] K. Fung *et al.*, "An experimental 5 ns BiCMOS SRAM with a high-speed architecture," in *Symp. VLSI Circuits Dig. Tech. Papers*, June 1990, pp. 43–44.

[13] T. Douseki *et al.*, "Fast-access BiCMOS SRAM architecture with a Vss generator," in *Symp. VLSI Circuits Dig. Tech. Papers*, June 1990, pp. 45–46.

[14] Y. Kohno *et al.*, "A 14-ns 1-Mbit CMOS SRAM with variable bit organization," *IEEE J. Solid-State Circuits*, vol. 23, pp. 1060–1066, Oct. 1988.

[15] M. Yoshimoto *et al.*, "A divided word line structure in the static RAM and its application to a 64k full CMOS RAM," *IEEE J. Solid-State Circuits*, vol. SC-18, pp. 479–485, Oct. 1983.

FA13.4: A 6ns 4Mb ECL I/O BiCMOS SRAM with LV-TTL Mask Option

Kazuyuki Nakamura, Takashi Oguri, Takao Atsumo, Masahide Takada, Atsushi Ikemoto, Hisamitsu Suzuki, Tadashi Nishigori, Tohru Yamazaki

NEC Corporation, Sagamihara, Japan

This 4Mb BiCMOS SRAM achieves 6ns and 8ns access times with ECL and TTL I/O interfaces, respectively, using: (1) a BinMOS converter directly connecting the ECL level to the CMOS level, (2) a high-speed BinMOS circuit with low-threshold voltage nMOSFETs, (3) an optimum word decoder, and (4) high-speed bipolar sensing circuits with a 3.3V supply.

The SRAM is applicable to either an ECL100K interface (VCC=0V, VEE=-4.5V) or an LV-TTL interface (VCC=3.3V, VEE=0V), depending upon the mask option used. As shown in Figure 1a, with the ECL100K interface, ECL I/O buffers are supplied at -4.5V VEE, while 4Mb SRAM core circuits are supplied at -3.3V VEE' to insure reliability of 0.55μm MOSFETs. VEE' is generated by an internal voltage converter and separated from VEE by a triple-well BiCMOS process [1]. In this power supply configuration, variations in VEE and VEE' do not adversely affect the speed of conversion from the ECL level to the CMOS level, because the ECL level signal is determined by VCC. With the LV-TTL interface, 3.3V TTL I/O buffers and level converters, which convert bipolar sense amplifier outputs to CMOS level signals, are connected to the SRAM core circuits, as shown in Figure 1b.

Reducing the time for conversion from ECL to CMOS levels is key in achieving high-speed ECL I/O BiCMOS SRAMs. In this SRAM, ECL signals are converted to CMOS signals with a reduced VEE' (-3.3V) swing, instead of VEE swing in conventional BiCMOS SRAMs. Figure 2 shows the ECL to CMOS interface circuits, where ECL address inputs are received by ECL address buffers and predecoded by wired-OR bipolar transistors. The ECL predecoded output signal is connected directly to the CMOS level through a BinMOS inverter (line driver) shared by both the ECL and TTL I/O interfaces. To suppress dc current in the BinMOS line driver, high and low levels of predecoded output voltage are set at VCC-VF and VEE'+Vtn, respectively. Here, VF and Vtn represent, respectively, the bipolar turn-on voltage and the nMOSFET threshold voltage. Figure 3 shows a comparison of delay times for this circuit and conventional schemes [2]. Compared here are the delay times from the ECL address input to the BinMOS line driver output with 1.5pF load capacitance, when internal VEE' changes from -2V to -4.5V. When VEE' is around -3.3V, this ECL to CMOS interface circuit achieves an operation that at its fastest exceeds that of the conventional scheme by 23%.

BinMOS gates with 0.3V Vtn are used in the main word decoder of the 3.3V SRAM core. As the BinMOS gate outputs VEE' as its low level voltage, low Vtn for nMOSFETs may be used without any significant leakage current, enabling higher-speed operation at lower supply voltage. The BinMOS inverter speed at Vtn=0.3V is faster than that at Vtn=0.6V by about 10% at a 3.0V supply voltage. The section word decoder and the word line drivers are composed of CMOS gates, because a small gate layout pitch and a VCC-VEE' swing signal are required. In order to find the combination of gate size and number of stages that will be fastest for a word decoder consisting of BinMOS and CMOS gates, an optimum decoder design method is used. As shown in Figure 4a, each gate delay time tpdi is expressed as intrinsic delay time (tpd0i) + load-capacitance-dependent delay time (tpdfi x Cai / Ki Cgi) + wiring-resistance-dependent delay time (ci x RLi x Cai), as in Equation 1. Total load capacitance Cai is expressed as wiring capacitance (CLi) + next-stage total gate capacitance (fanout number Fi x next-stage gate-size ratio Ki+1 x next-stage unit gate capacitance Cgi+1), as in Equation 2. Total decoder delay time Tpd with n gate stages is given as a sum of individual tpdi, as in Equation 3. The required optimum gate-size ratio Ki is calculated by the steepest descent method, to yield minimum Tpd. Here, the values for tpd0i, tpdfi and ci for CMOS and BinMOS logic gates have been determined by SPICE simulation. Figure 4b shows the relation between decoder delay time and number of gate stages from the BinMOS level converter (line driver) input to the word-line driver output. Using this method, a 5-BinMOS-gate, 2-CMOS-gate word decoder is designed with a minimum delay smaller by 33% than that of a fully-CMOS gate decoder.

Figure 5 shows bit-line peripheral and sensing output circuits. The differential read-out signal from a memory cell is received by the first sense amplifier (SA) through a pair of bit-line-selection pMOSFETs and one stage of wired-OR bipolar transistors. The signal is amplified by cascoded first and second bipolar sense amplifiers through a pair of read data-bus lines. To prevent sensing-speed degradation with 3.3V supply voltage, the number of serially-connected wired-OR stages from the bit-lines to the first sense amplifier was limited to one rather than the two of a previously reported SRAM [3]. In addition, the bit-line voltage in read-out was set at VCC-0.3V using an nMOSFET (MNDL) with 0.3V Vtn. This is because the common emitter node voltage VA (VCC-2VF-0.3V) in the first sense amplifier needs to be set at less than VCC-2VCE-dV [the output voltage swing of the second sense amplifier].

Figure 6 shows a micrograph of the 4Mb SRAM chip. The main word decoders and column decoders are located at the center of the die. The SRAM cell array is divided into 64 64kb array blocks. I/O pads are at the center to mount the die in a lead-on-chip package. This contributes to reduced signal delay and power supply noise. Figure 7 shows a measured ECL100K I/O address access time of 6ns under typical conditions. Active power dissipation at 50MHz is typically 750mW for ECL and 230mW for TTL. The die is 18.87x8.77mm^2, and the memory cell is 5.8x3.2μm^2. Table 1 summarizes the 4Mb SRAM features.

Acknowledgments

The authors thank T. Enomoto, K. Okada, M. Arimura and the other members of this project for support.

References

[1] Douseki T., et al., "Fast-Access BiCMOS SRAM Architecture with a VSS Generator", Symposium on VLSI Circuits Digest of Technical Papers, pp. 45-46, June 1990.

[2] Suzuki M., et al., "A 3.5ns 500mW 16kb BiCMOS ECL SRAM", ISSCC DIGEST OF TECHNICAL PAPERS, pp. 32-33, Feb.1989.

[3] Takada M., et al., "A 5ns 1Mb ECL BiCMOS SRAM", ISSCC DIGEST OF TECHNICAL PAPERS, pp. 138-139, Feb. 1990.

Reprinted from *1992 IEEE ISSCC*, pp. 212–213 and p. 286, Feb. 1992.

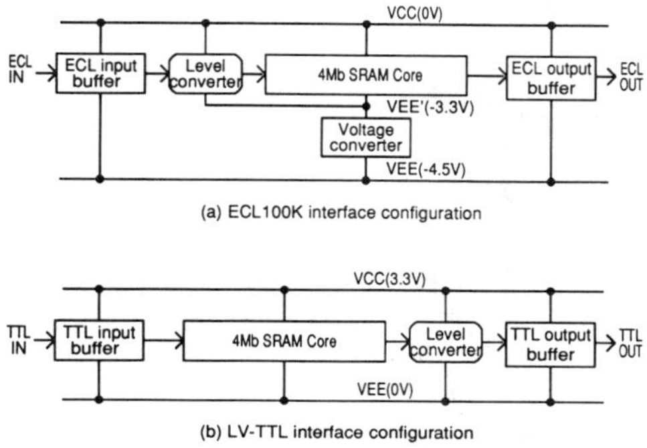

Figure 1: Power supply configuration for ECL100K and LV-TTL interfaces.

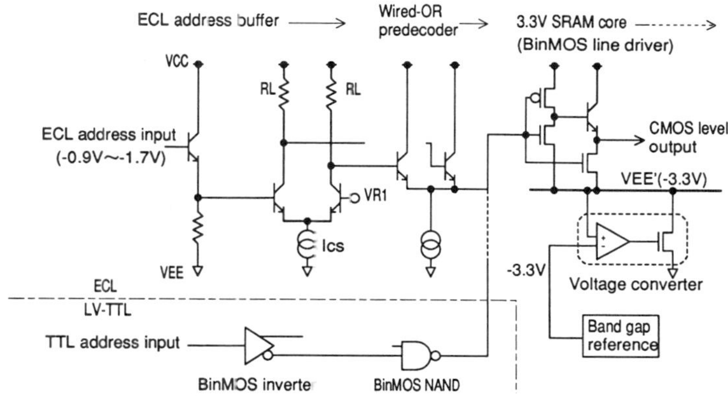

Figure 2: Directly-connected ECL/CMOS signal-interface circuits.

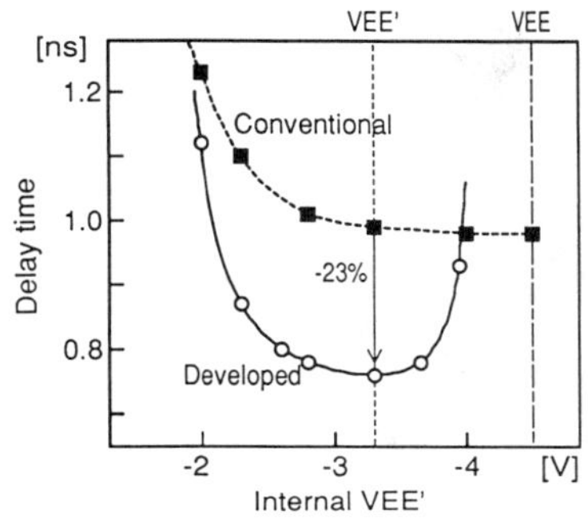

Figure 3: Delay comparison for reported and conventional ECL address input to CMOS signal interface circuits.

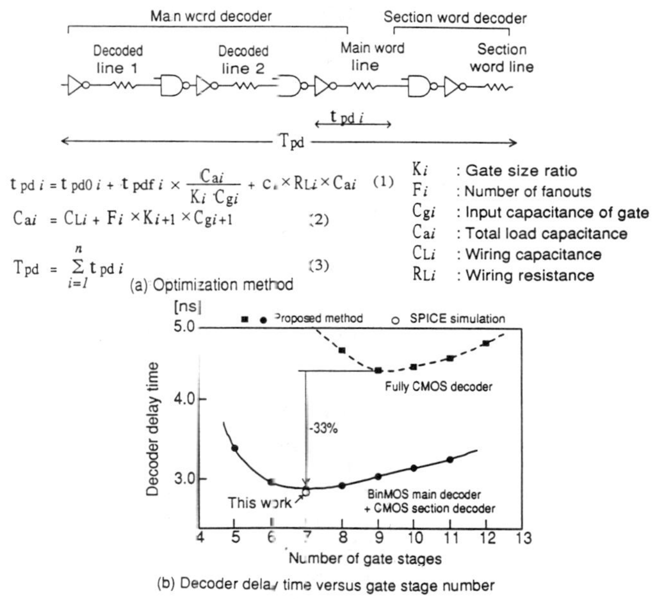

Figure 4: Optimum design of word decoder with BinMOS and CMOS gates.

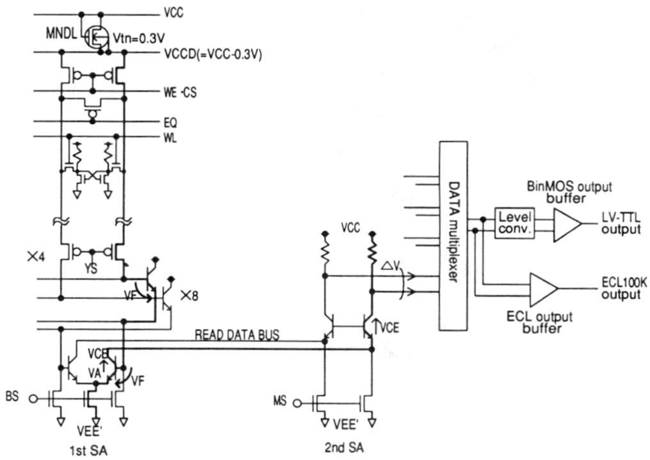

Figure 5: Bit-line peripheral and sensing circuits

Figure 6: See page 302.

Organization	4Mx1b, 1Mx4b
BiCMOS technology	0.55μm 4-poly 2-metal p-substrate triple-well
nMOSFET	L=0.55μm Vtn=0.6V L=0.55μm Vtn=0.3V
pMOSFET	L=0.65μm Vtp=-0.65V
Bipolar	Minimum emitter 0.6x2μm^2 f_T=12GHz, h_{FE}=100
Redundancy	Block, 16kbx4blockx2
Test function	16b parallel test

Table 1: 4Mb SRAM features.

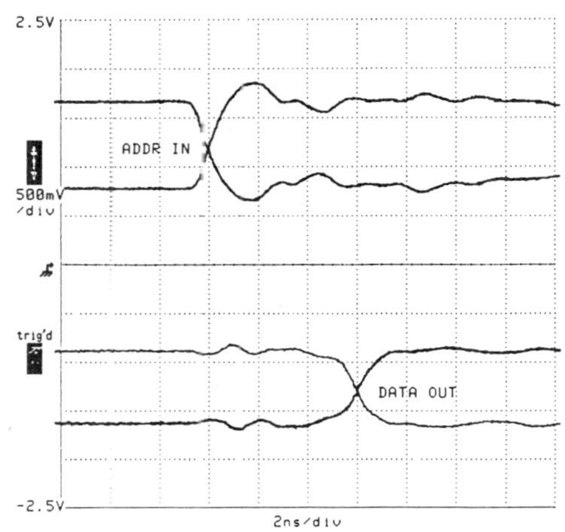

Figure 7: Address access waveforms for ECL100K I/O interface

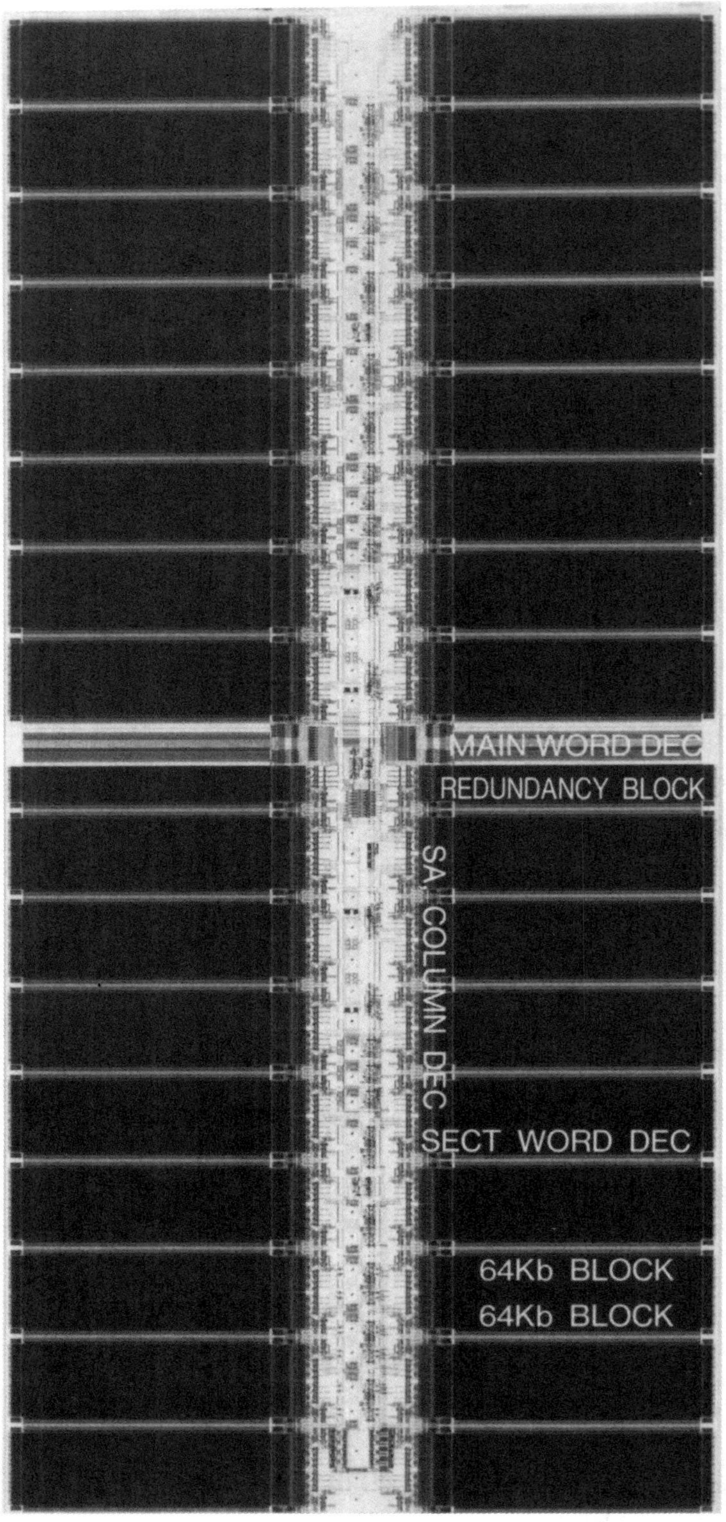

Figure 6: 4Mb SRAM chip micrograph

FA 13.3: A 9ns 4Mb BiCMOS SRAM with 3.3V Operation

Hatsuhiro Kato, Azuma Suzuki, Takahiro Hamano*, Tomohiro Kobayashi, Katsuhiko Sato, Takeo Nakayama, Hiroshi Gojohbori, Takeo Maeda, and Kiyofumi Ochii

Semiconductor Device Eng. Lab., Toshiba Corp.,/*Toshiba Microelectronics Corp., Kawasaki, Japan

This 4Mb TTL SRAM in 0.5μm BiCMOS technology uses scaled-down features of optimized MOS and bipolar transistors and BinMOS circuits to achieve 9ns access and low-power 3.3V operation of a 16b organization.

The decoders and control circuits are implemented by BinMOS gates suitable for low-voltage operation [1]. For CMOS and BiCMOS gates, propagation delay of pull-up, t_{up}, and pull-down, t_{down}, are equal. However, it is difficult to satisfy this condition for BinMOS gates. Figure 1 shows the relation between propagation delay of a BinMOS gate in a ring oscillator and MOS channel width ratio, Wp/Wn, when the gate has output capacitance ten times larger than its input capacitance (fanout=10). There is no cross-over point between t_{up} and t_{down}. The delays cannot be made equal. The criterion for optimization of the BiNMOS gate is to minimize the mean of t_{up} and t_{down}. The minimum is 275ps at the width ratio indicated by an arrow on the abscissa.

There are two bipolar sensing schemes, voltage and current sensing [2,3]. The later is difficult to operate at supply voltages below 3V. Thus the former is used for the first sense amplifier with minimun operating voltage about 2V, as depicted in Figure 2.

The voltage swing of the data line (DL) pair at the collectors of current switch (CS) is increased by parasitic resistance R_{DL} causing signal delay [4]. To reduce the delay, a data transfer stage (DTS) is inserted in the middle of the DL (Figure 2). The DTS is composed of two current switches. Each switch transmits signal on the DL_{left} or DL_{right} pair to the DL' pair. One or the other part of the DTS is activated by the address signal W_{left} or W_{right}. Figure 3 illustrates effect of the DTS. It is possible to reduce effective resistance and capacitance of DL by employing DTS because the DL is separated into two parts, DL_{right} and DL_{left}. The swing of DL pair is 30mV at the cathodes of the twin diodes (TD). The swing is about 60mV at collectors of CS by RI drop along DL. Without the DTS, this swing would be doubled, causing delay. Figure 4 shows the dependence of DL transfer delay on DL length. An arrow on the abscissa indicates 4Mb SRAM DL length. For megabit SRAMs, the DL length exceeds 10mm and the DL delay exceeds 1ns. To reduce this delay is important for fast access. In the case of the 4Mb SRAM, the transfer delay is improved about 50% (except intrinsic delay) by using DTS.

Figure 5 is the SRAM block diagram. The cell array is divided into 32 blocks and each block contains 128 columns and 1024 rows. The block is further separated into upper-byte and lower-byte parts. Each part can be accessed independently by external control signals, /UB and /LB. Row decoders for each byte are located at the middle of the main word line to reduce the RC delay. There is only one transfer gate along the data path from cell to CS, so data transfer RC delay along bit lines is minimized (Figure 2). This configuration is realized with BL multiplexing by activating CS with signal CD2 and by the transfer gates controlled by CD1 and /CD1. One I/O in each block corresponds to 16 BLs. The 16 BLs are reduced to two common BLs by the transfer gate and the last selection is made by activation of the CS for each common BL.

The 0.5μm triple-polysilicon and double-metal BiCMOS process is summarized in Table 1. The first polysilicon layer (polycide) is employed for gate electrodes of MOS transistors, and second polysilicon (polycide) for the emitter electrodes of the bipolar transistors and ground lines in cell array. The third polysilicon is used for cell resistors.

Figure 6 shows typical address access time of 9ns at room temperature and 3.3V supply. Figure 7 is a chip micrograph. The poly-Si-load cell is 3.5x5.7μm². The die measures 8.7x18.8mm². Bit organization is 256kwordx16b and power consumption is 430mW at 50MHz operation, even though the SRAM has 16b organization. Characteristics of the SRAM are summarized in Table 2.

Acknowledgements

The authors thank to T. Matsui, Y. Urakawa, K. Sato, and T. Mitsumori for support and discussions.

References

[1] Watanabe, A., "Future BiCMOS Technology for Scaled Supply Voltage", IEDM Technical Digest, pp.429-432, Dec. 1989.

[2] Kertis, R. A., et al., "A 12ns 256k BiCMOS SRAM", ISSCC DIGEST OF TECHNICAL PAPERS, pp.186-187, Feb. 1988.

[3] Suzuki, M., et al., "A 3.5ns 500mW 16kb BiCMOS ECL SRAM", ISSCC DIGEST OF TECHNICAL PAPERS, pp.32-33, Feb. 1989.

[4] Urakawa, Y., et al., "Data-line wiring delay reduction techniques for high speed BiCMOS SRAM's", Symposium on VLSI Circuits, pp.19-20, May 1991.

Process:	0.5μm BiCMOS, p-sub, n-epi, twin-well 2 polycide, 1 polysilicon, 2 metal layers		
Bipolar:	Emitter size		0.8x4.0μm²
	Maximum cut off frequency		12GHz
	Current gain		100
MOS:	Gate length:	pMOS	0.6μm
		nMOS	0.5μm
	Gate oxide		11nm

Table 1: Process and device parameters

Bit organization	256kwordx16b
I/O interface	TTL
Typical access time	9ns
Power consumption	430mW (50MHz, 3.3V)
Typical operation voltage	3.3V
Minimun operation voltage	2.6V
Cell size	3.5x5.7μm²
Chip size	8.7x18.8mm²
Redundancy	4 rows, 2 columns

Table 2: Characteristics of 4Mb BiCMOS SRAM

Reprinted from *1992 IEEE ISSCC,* pp. 210–211 and 285, Feb. 1992.

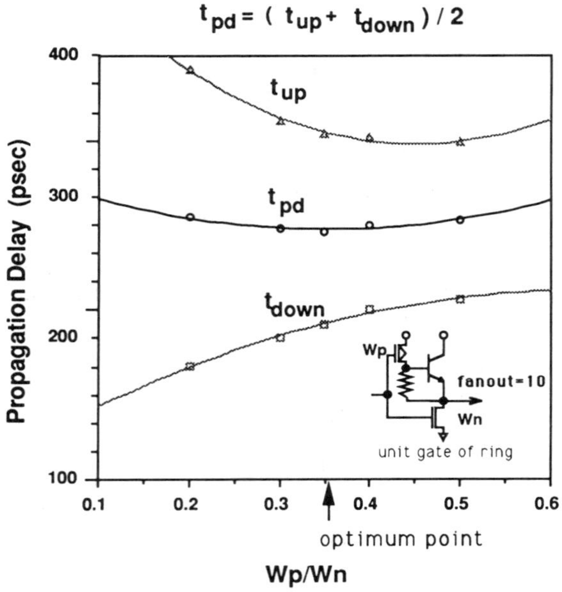

Figure 1: Optimization of BinMOS gate against transistor channel width.

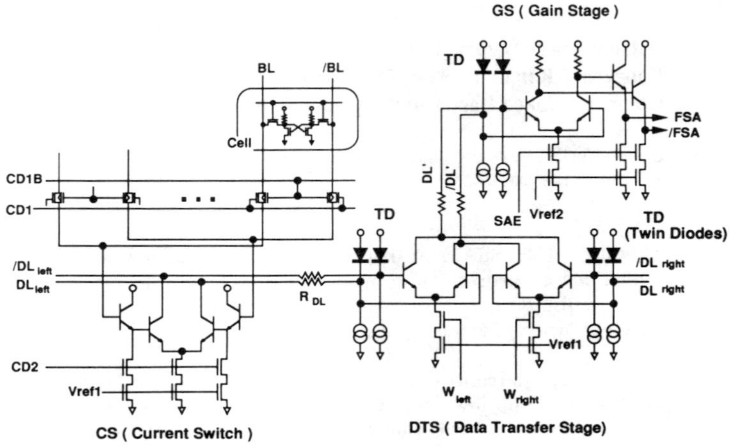

Figure 2: Schematic diagram of sense amplifier.

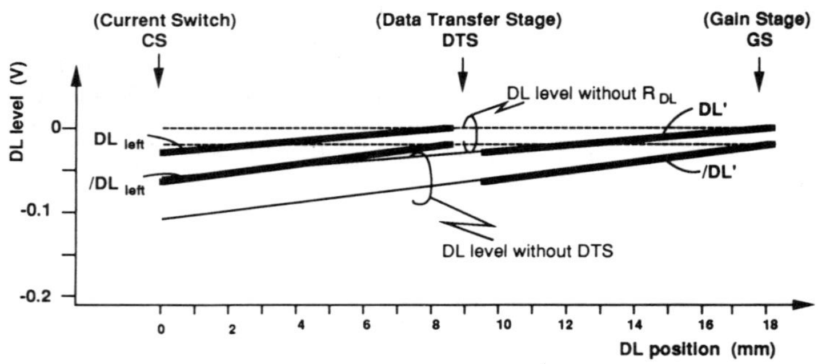

Figure 3: Data transfer stage (DTS) reduces voltage swing of data-line pair (DL).

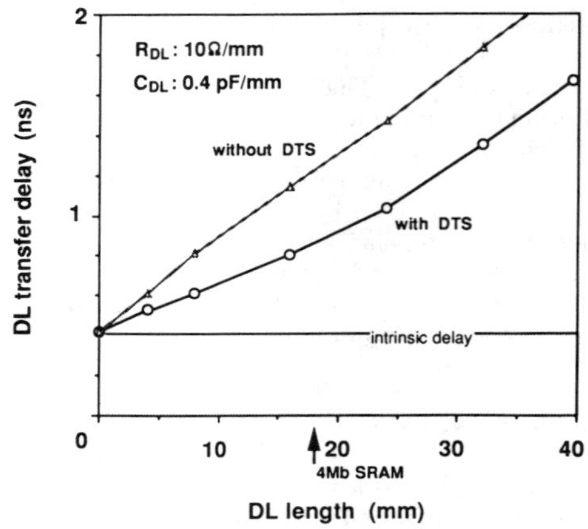

Figure 4: Relation between signal transfer delay and length of DL.

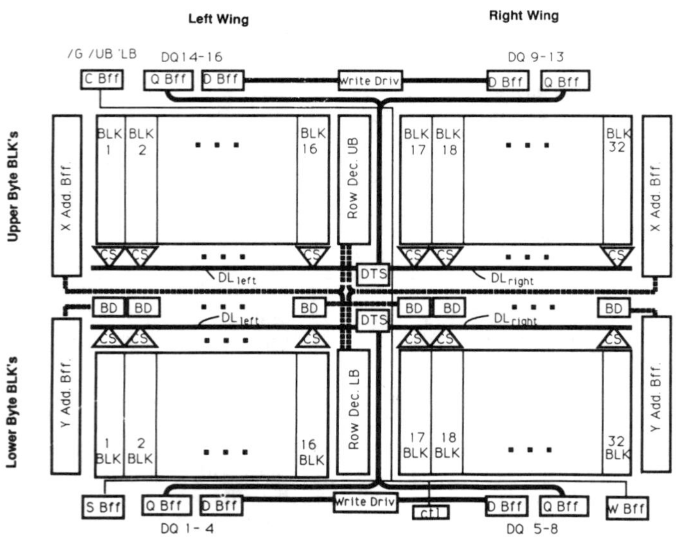

Figure 5: SRAM block diagram.

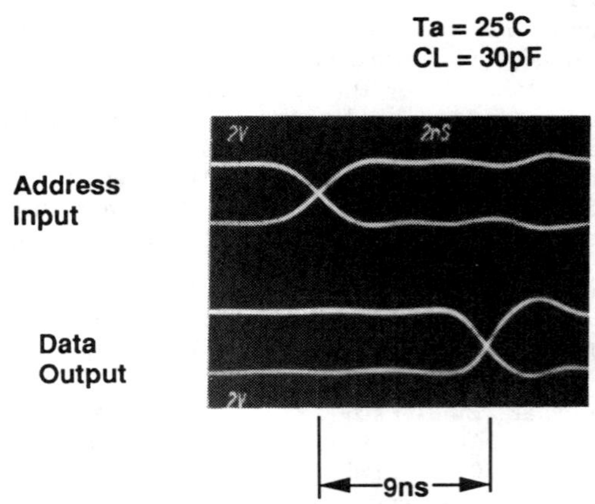

Figure 6: Typical address and output waveforms.

Figure 7: See page 305.

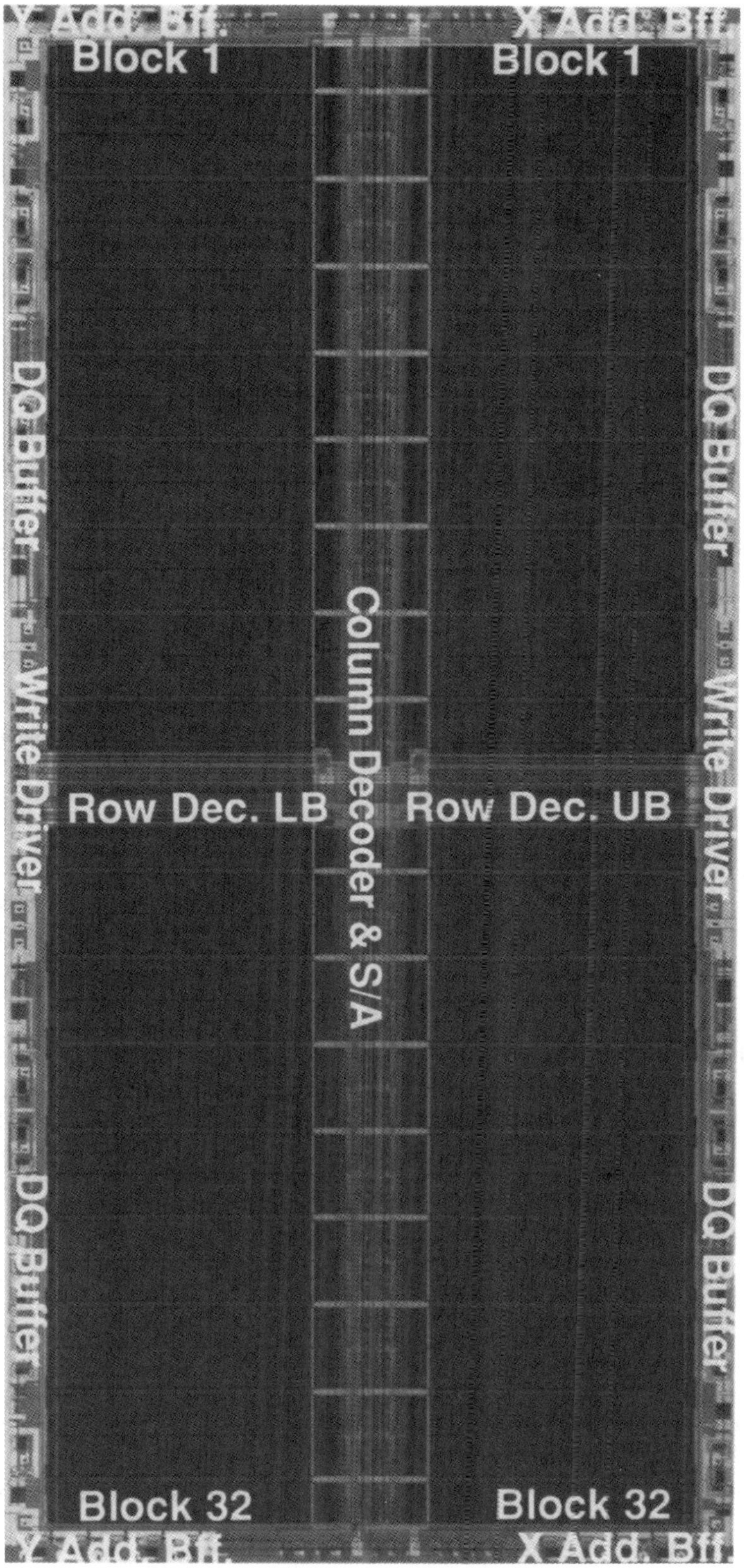

Figure 7: 4Mb TTL BiCMOS SRAM chip micrograph.

Paper 3.9

Trends in Megabit DRAM Circuit Design

KIYOO ITOH, SENIOR MEMBER, IEEE

Abstract—The state of the art in megabit dynamic random access memory (DRAM) circuit and chip design is reviewed in terms of essential design parameters such as signal-to-noise ratio, power dissipation, and speed. The key results obtained are 1) the memory cell signal charge has decreased gradually with an increase in memory cell size, despite the vertically structured cell designs. To offset this decrease, multi-divided data- (bit) line structures, low-power design, and transposition of folded data lines are essential. 2) To reduce power dissipation, an increase in the maximum refresh cycle and multi-divided data lines combined with shared I/O in addition to a reduced operating voltage are efficient. 3) A BiCMOS circuit provides a high-speed access time with low cost due to the high drivability of the driver and the high sensitivity of the amplifier. Based on this discussion, it is predicted that the current DRAM technology might be diversified coexistently to a large-memory-capacity oriented technology and to a high-speed oriented one, posing power-supply standardization as a continuing serious concern.

I. Introduction

THE DENSITY of dynamic random access memories (DRAM's) has quadrupled every three years since their advent about 20 years ago. Developing this higher density has made them cheaper than other types of memories. Low power dissipation and fast access time have been obtained even with an increase in chip size for every successive generation, as shown in Fig. 1. This has also contributed to the DRAM advantage. As a result, 1-Mb DRAM's have now reached maturity in production, and are being followed by the development of 4- and 16-Mb DRAM's [1]–[11]. The resultant DRAM technology has been applied to other successful products such as video RAM's, pseudo SRAM's, ASIC memories, and low-power file memories. Even a 1.5-V DRAM for battery-based applications [12] in addition to a wafer memory [13] has been proposed. Their rapid evolution has been achieved mainly because of the progress in the one-transistor cell (1T cell) [14] and its supporting circuit technologies as well as photolithography. The 1T cell still surpasses gain cells such as three-transistor and four-transistor cells, which were common to 1- and 4-kb generations, with respect to chip size. Smaller chip size and the resulting lower cost due to the 1T cell justify the fabrication process complexity in forming the cell capacitor.

Manuscript received September 8, 1989; revised January 29, 1990.
The author is with the Central Research Laboratory, Hitachi Ltd., Kokubunji, Tokyo 185, Japan.
IEEE Log Number 9035347.

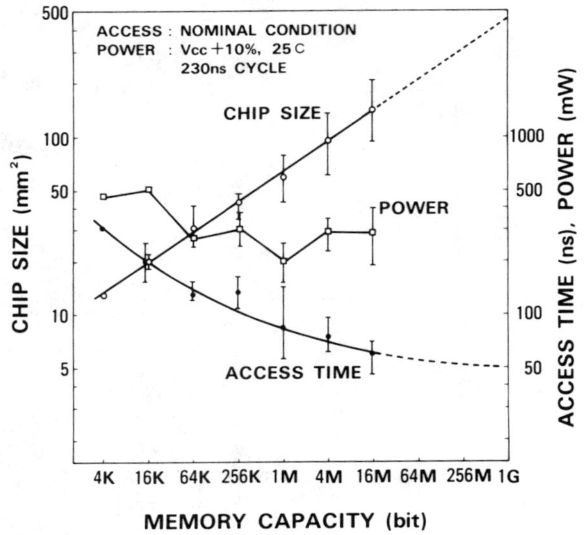

Fig. 1. Trends in standard DRAM development.

Fundamental circuit design parameters of the 1T cell [15]–[17] are summarized as signal-to-noise ratios, power dissipation, and speed. They are summarized in this way because of the following inherent 1T cell characteristics: 1) a small read signal; and 2) charging and discharging the highly capacitive data (bit) lines simultaneously with a high voltage, which invariably introduces many kinds of noises and cause high power dissipation. A high-voltage data-line operation derives from both destructive readout characteristics and the requirement of a refresh operation for 1T cells. Considering DRAM's as an LSI technology driver, speed is also a prime concern. Table I shows key circuit technologies which have contributed to progress in DRAM development. With regard to the signal-to-noise ratio, word voltage bootstrapping [18], [19], a shared amplifier [20], and multi-divided data lines [21] are essential to increase cell signal voltage. Differential sensing [22], a folded data-line arrangement [23], and transposed data lines [4], [24] are representative techniques for reducing noise. For reducing power dissipation, dynamic amplifiers [25], dynamic drivers, CMOS circuits [26], [27], half V_D data-line precharge [28], [29], a shared I/O combined with multi-divided data lines [21], and voltage converters [21], [30], [31] have been playing important roles. Reducing the external power supply V_{CC} from 12 to 5 V at 64-kb generation also contributed to a drastic reduction in power dissipation. To increase speed, multidivision of

Reprinted from *IEEE J. Solid-State Circuits*, vol. 25, no. 3, pp. 778–789, June 1990.

TABLE I
TRENDS IN DRAM CIRCUIT TECHNOLOGIES

TECHNOLOGY / CAPACITY	V_{cc} / MEMORY CELL	SIGNAL TO NOISE RATIO	POWER DISSIPATION	SPEED	OTHERS
1 K	~20V 3T	—	DRAM cell circuits	—	refreshing scheme
4 K	12 3T	differential sense	—	—	address multiplex
16K	12 1T	—	dynamic amplifier dynamic driver	—	—
64K	5 1T	folded data line word boost	5V circuits	—	chip coating V_{BB} generator
256K	5 1T	shared amplifier	—	Al shunted word line	redundancy
1 M	5 1T	multi - divided data line	CMOS circuits half V_D precharge shared I/O voltage converter	two level Al wiring BiCMOS circuits	multibit testing
4 M	5 1T	—	—	—	—
16 M	5 1T	transposed data line	—	—	on-chip tester

V_{cc} denotes the external power supply voltage for the commercially accepted DRAM chips. 1T and 3T are one-transistor cell and three-transistor cell, respectively.

poly-Si word lines shunted by aluminum wiring [32] and two-level aluminum wiring as well as device scaling are important. They contribute to reducing *RC* delay in a chip. Moreover, BiCMOS circuits [33]–[36] are regarded as one of the main contributors. There is no doubt that the refreshing scheme, address multiplex, chip coatings against soft error, a substrate bias (V_{BB}) generator to realize single 5-V supply, and redundancy circuits are also important. Besides, the testing issue is becoming increasingly important with increasing memory capacity. Multibit testing [5], [37], which enables the parallel test for the multiple data read from subarrays to be performed internally, plays an important role in drastically reducing the testing time. ECC circuits [1] and an on-chip microcoded test pattern generator [7] are expected to be the most promising circuits.

Despite its importance, few papers on these circuit trends of the 1T cell have been published so far. This circuit issue [38] is reviewed in this paper, emphasizing megabit DRAM circuits. First, a general description of chip architecture and current trends in this field are presented for clarification of certain points in this paper. Next, the state of the art in megabit DRAM circuits and chip designs is reviewed in terms of essential design parameters described above. Some perspectives for memory capacities greater than 16 Mb are also presented, including the role of BiCMOS technology.

II. GENERAL DESCRIPTION OF CHIP ARCHITECTURE

Fig. 2 shows a logical and conceptual chip architecture with a memory array consisting of m columns by n rows, and k subarrays. Each data line is divided into k sections to obtain a smaller C_D (data-line capacitance). This is because lowering C_D is essential to improving almost all of the important design parameters described later. The internal supply voltage V_D is generated from the external supply voltage V_{CC} by an on-chip voltage limiter (converter). Memory array current and peripheral current are I_A and I_P, respectively. The precharged voltage of data lines is V_P. To reduce power dissipation, V_P normally equals $V_D/2$. This is called the half-V_D precharge. It is well known that for READ operation all the m cells connected to the word line are read simultaneously by applying a word pulse to a selected word line. Next, the resultant m small signals are amplified up to V_D or down to 0 by the amplifiers in accordance with the stored cell information. Then, successive operations of selecting one data line and transferring its data-line voltage to the data output terminal complete a READ operation. Meanwhile, all the amplified data-line voltages on m data lines are fed into the corresponding memory cells along the word line. The refresh operation is simply the READ operation without selecting a data line. Thus, through one refresh operation, m cells on one selected word line are refreshed simultaneously. Therefore, a complete refresh operation for all the memory cells on a chip can be accomplished through the sequential selection of all the n word lines. In this sense, n is called the refresh cycle. The normal READ or WRITE operations are accessed during rest periods in refresh operations, thus avoiding conflict. In order to ensure successful refresh operation, each cell should be refreshed within the maximum refresh time, t_{REFmax}, which is determined by the allowable maximum memory cell leakage current. Thus, the maximum refresh cycle time is t_{REFmax}/n. The ratio β of the maximum refresh cycle time to the minimum normal READ cycle

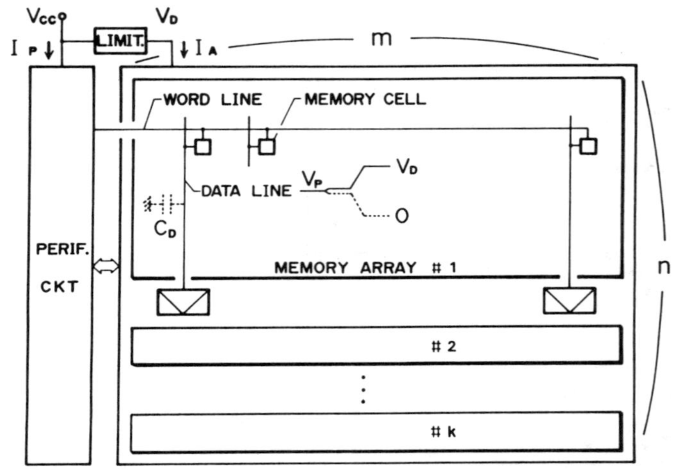

Fig. 2. Logical and conceptual DRAM chip architecture.

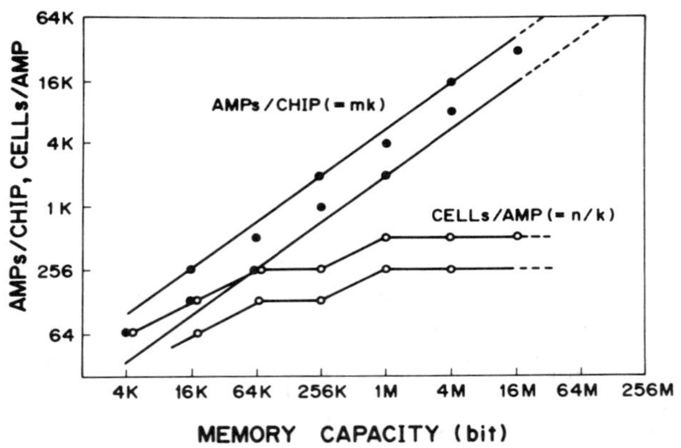

Fig. 3. Trends in the number of amplifiers on a chip and cells physically connected to an amplifier.

time, t_{RCmin}:

$$\beta = (t_{\mathrm{REFmax}}/n)/t_{\mathrm{RCmin}}$$

implies the maximum utility of the chip in terms of the overhead involved by the refresh operation. A larger β is better since it involves less conflict between refresh operations and normal operations. Therefore, longer t_{REFmax} and/or shorter t_{RCmin} are required with increasing n, that is, with increasing memory capacity.

Fig. 3 shows trends in the number of amplifiers on a chip, mk, and the number of memory cells, n/k, physically connected to an amplifier. It is clear that mk increases sharply as the memory capacity increases. Thus, there are expected to be as many as one million amplifiers on a 1-Gb chip. This results from a strict limitation in the number of n/k. In fact, only a gradual increase in n/k can be seen. This is because data lines have to be further subdivided in a larger density memory to offset the larger C_D, compromising with the resulting increase in amplifier and decoder areas.

Even for a given n/k, various multidivisions of data lines have been proposed to obtain a larger memory cell signal, low power dissipation, reduced chip size, and higher speed. Fig. 4 shows the more detailed concepts behind the multi-divided data-line structures [2]–[8], [20], [39], [40] proposed so far. An actual data-line structure consists of a pair of data lines for achieving noise cancellation, which differs from the conceptual one in Fig. 2. One pair of data lines, D and $\overline{D}$, is multi-divided by repeating the basic unit as shown in Fig. 4. One of the resulting subdata lines is activated, while leaving the remaining subdata lines inactive. Fig. 4(a) is the conventional scheme used in the 64- and 256-kb generations. A Y switch, YS, is controlled by a Y decoder for connecting the corresponding subdata lines to the corresponding I/O lines. Fig. 4(b) is the well-known shared amplifier scheme widely used in the 256-kb to 4-Mb generations. This allows two subdata lines to share one amplifier. A small read signal from the subdata line labeled #1-1 is fed into the amplifier input by turning on switch SW_1 and then amplified with switch SW_2 turned off. Then, an amplified signal is transferred to subdata line #1-2 and to the I/O line by successively turning on switches SW_2 and YS. For a given number of amplifiers and switches YS, this scheme offers a doubled signal voltage. This is because the C_D contributing to the cell signal voltage is half that of the conventional one. Note that the C_D contributing to the power dissipation associated with charging and discharging the data lines is maintained the same. This scheme, however, requires a complicated control timing sequence, which results in a slower speed. For a given C_D contributing to cell signal voltage, this scheme provides double the power dissipation with half the number of amplifiers and switches YS. This results in smaller chip size with larger power dissipation. Such is the case for the actual designs shown in Fig. 14. To further improve the performance, two types of shared I/O schemes combined with the shared amplifier have been proposed and used partly for 1- to 16-Mb generations. Fig. 4(c) features the shared I/O lines parallel to the subdata lines, while Fig. 4(d) shows the one across to the subdata lines. Both schemes inevitably introduce two-level metal wirings to avoid an increase in chip size due to the additional wirings of I/O lines and Y control lines, YC's, on the memory cell array. However, the recent progress in development of materials and fabrication processes favors these schemes. Fig. 4(c) allows each selected subdata line to communicate with the common I/O line. Only one YS placed at the end of the common I/O line can control the connection between it and the secondary I/O line in YS. This minimizes the data-line capacitance for both cell signal and power dissipation while enabling a smaller chip size, even for further subdivision, than with the shared amplifier scheme. This scheme, however, still has a disadvantage of additional parasitic capacitance, C_D', on the I/O line. All the capacitances on the I/O lines are charged and discharged simultaneously, which causes a substantial increase in power dissipation. The circuit shown in Fig. 4(d) overcomes this problem. A selected YC can control the connection between each subdata line and each I/O line. Hence, simultaneous connections at the divisions are accomplished. Only one activation of YC and the smaller

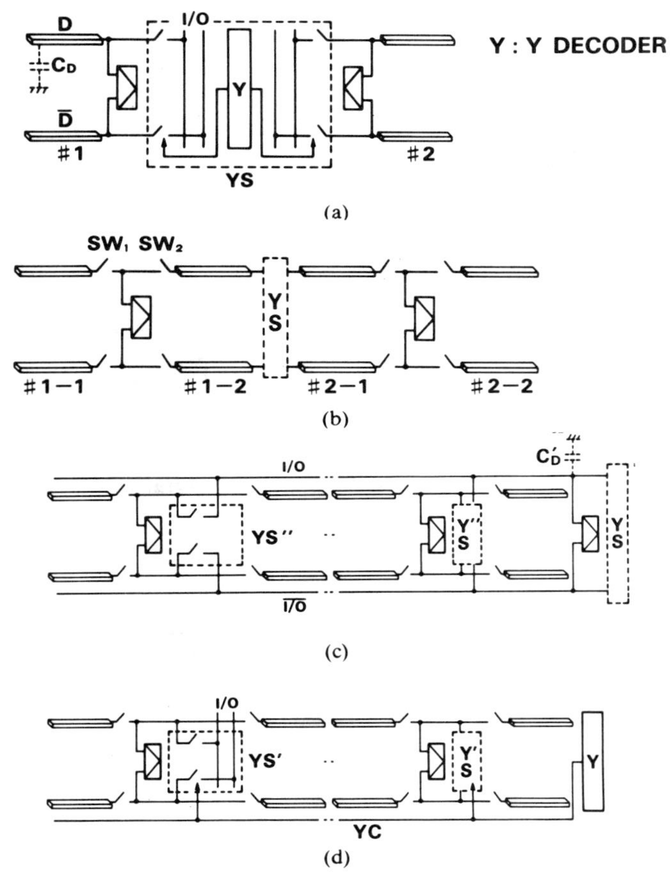

Fig. 4. Concepts of multi-divided data line structure: (a) conventional, (b) shared amplifier, (c) shared I/O lines parallel to subdata lines, and (d) shared I/O lines across to subdata lines.

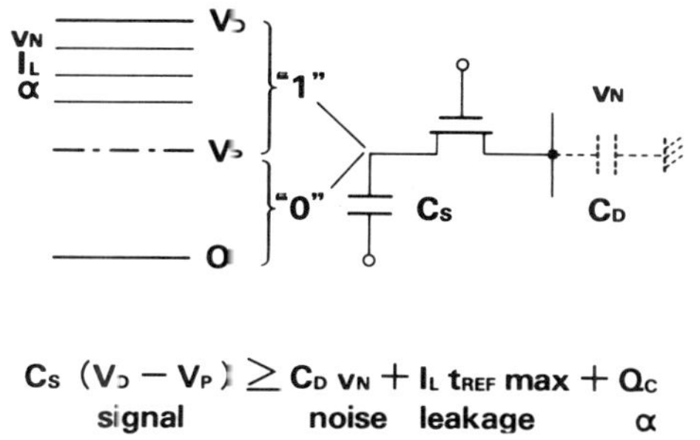

Fig. 5. Memory cell margin. v_N: noise on the data line. I_L: leakage current at the memory cell node. $t_{REF\,max}$: maximum refresh time. Q_c: critical charge. V_P: data-line precharge voltage. $V_P = V_D/2$ in normal case.

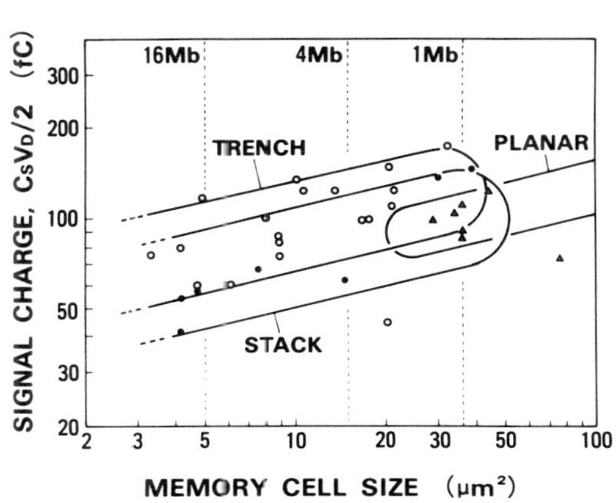

Fig. 6. Signal charge versus memory cell size. Data are from ISSCC.

number of I/O lines are responsible for the lower power dissipation. This scheme is seemingly the best way to reduce power dissipation as well as to reduce chip size. Note that the resulting small C_D through the multidivision contributes to high speed because of the shortened *RC* delay on the data lines.

III. Signal-to-Noise Ratio

Intensive studies and careful design are required to achieve a reliable sensing operation, since DRAM is regarded as a memory having a low S/N ratio as described earlier. The cell margin [16] can be roughly expressed as shown in Fig. 5, using the sum of the various effective noise charges: data-line noise charge, $C_D v_N$, caused by capacitive couplings on the data lines from other conductors, the electrical imbalance between a pair of data lines, and amplifier offset voltage; leakage charge, $I_L t_{\mathrm{REFmax}}$, and critical charge Q_c due to α particle hitting. Maximizing the signal charge, $C_S(V_D - V_P)$, while minimizing the effective noise charges is essential to achieve a stable memory operation. The following are developments to meet this requirement.

1) Signal Charge: A tremendous effort is being made to develop advanced memory cells to obtain a signal charge Q_{sig} as large as possible. Among the new cells, vertically structured cells offering high C_S, such as stacked capacitor cells [41] and trench cells [42], are becoming important for megabit DRAM's. In spite of such cells, however, Q_{sig} decreases gradually with an increase in density, as shown in Fig. 6. This comes from the smaller C_S and the smaller V_D caused by lowering the power dissipation at each generation, as shown in Fig. 7. That is why new techniques that allow lowering the effective noise charges described above are required.

2) Critical Charge: The α-particle-induced soft errors [43] are becoming a major problem. Fortunately, however, it has been found that for the cell mode the critical charge Q_c decreases by reducing the diagonal length of the depletion region in the memory cell, as shown in Fig. 8 [44]. This suggests that a smaller Q_c is obtained by increasing cell density. Thus, reducing the volume of the depletion region is efficient in getting even smaller Q_c. Suppression of soft errors in the data- (bit) line mode is another concern, especially for higher speed operation. This is because the error rate is inversely proportional to the cycle time. A p^+ buried layer under the cell array using a BiCMOS structure [33] has been reported to be one of the most useful ways to suppress the soft errors.

3) Leakage Charge Suppressing the increase in junction temperature T_j, as well as improving the quality of the fabrication process, are essential. This is because a longer t_{REFmax} is required with an increase in memory

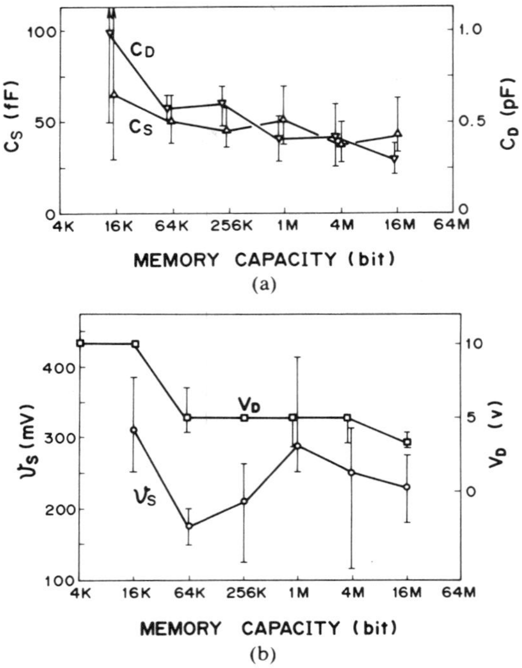

Fig. 7. Trends in memory cell parameters: (a) trends in C_S and C_D, and (b) trends in v_S and V_D. C_D: data-line capacitance contributing to memory cell signal voltage. v_S: cell signal voltage on the data line; $v_S = V_D/2 \cdot C_S/C_D$.

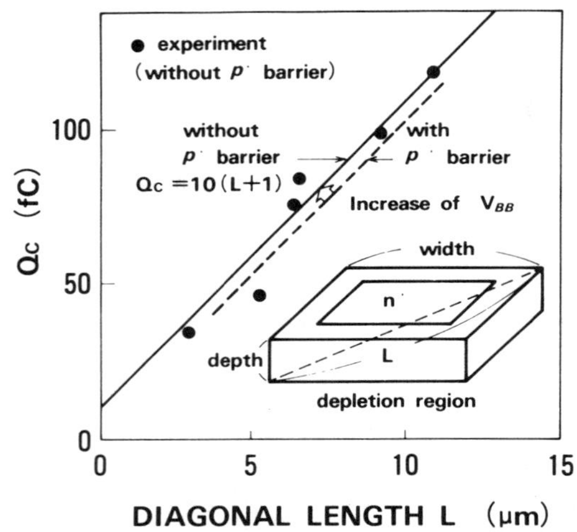

Fig. 8. Critical charge Q_c versus diagonal length L of depletion region in memory cell.

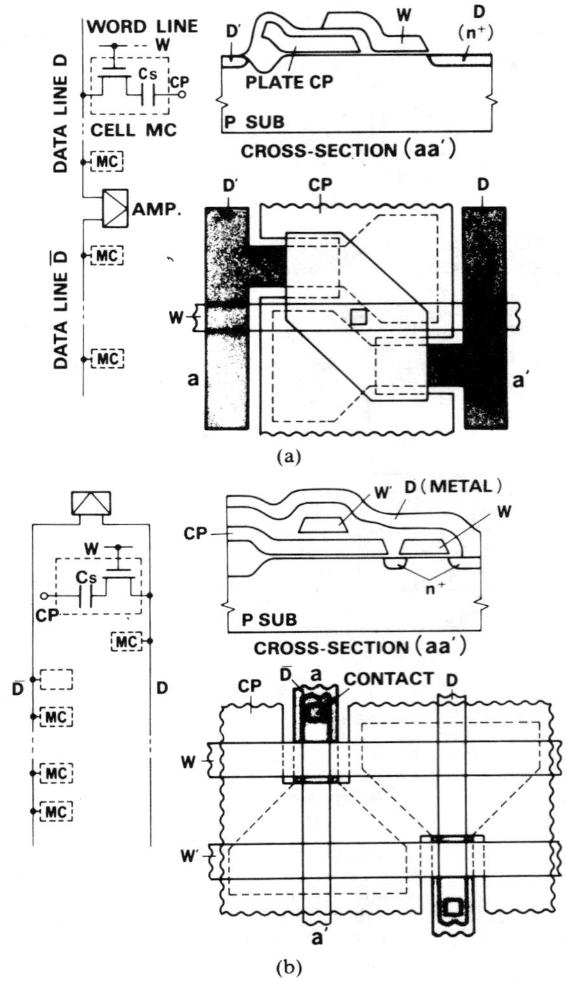

Fig. 9. Typical data-line arrangement: (a) open data line, and (b) folded data line.

capacity, as described later. Thus, low power dissipation and a low thermal resistance package are important to reduce T_j.

4) Data-Line Noise Charge: C_D has been suppressed effectively through the multidivision of data lines, as shown in Fig. 7. In addition, many attempts have been made to reduce the differential noise v_N on the pair of data lines. Obviously, the noise [15], [23] increases in the number of its sources and in its magnitude with an increase in memory capacity. Furthermore, higher speed requirements at every successive generation have increased noise. Hence, the S/N ratio decreases with increases in memory capacity, considering that the actual cell signal v_s is almost constant independent of the memory capacity, as shown in Fig. 7. The S/N ratio of the given memory cell can be degraded by two kinds of internal noise. The first is generated in the memory array. It is higher in larger logical and physical array sizes and is further increased by the scaling down of 1T cells. The other is coupled from peripheral circuits to the memory array. Large spike currents due to peripheral circuit operations, which increase with chip size, are the noise sources during the memory cell operation. These noises can be efficiently suppressed to an acceptable level by employing a folded data-line arrangement.

Figs. 9 and 10 show schematics of the memory cells and array organizations for an open data-line and a folded data-line arrangement [23]. One of the noises in the memory array causes a serious problem. This noise, δD, is related to the data-line–word-line coupling capacitance C_{DW}. A large data-line voltage swing during amplification couples the noise δW on all of the nonselected word lines. Then δW causes noise δD on the given data line D_0. Noise δD depends strongly on the number of memory cells connected to the selected word line and their stored data patterns. It is roughly proportional to $C_{DW}^2\,\delta C/C_D$

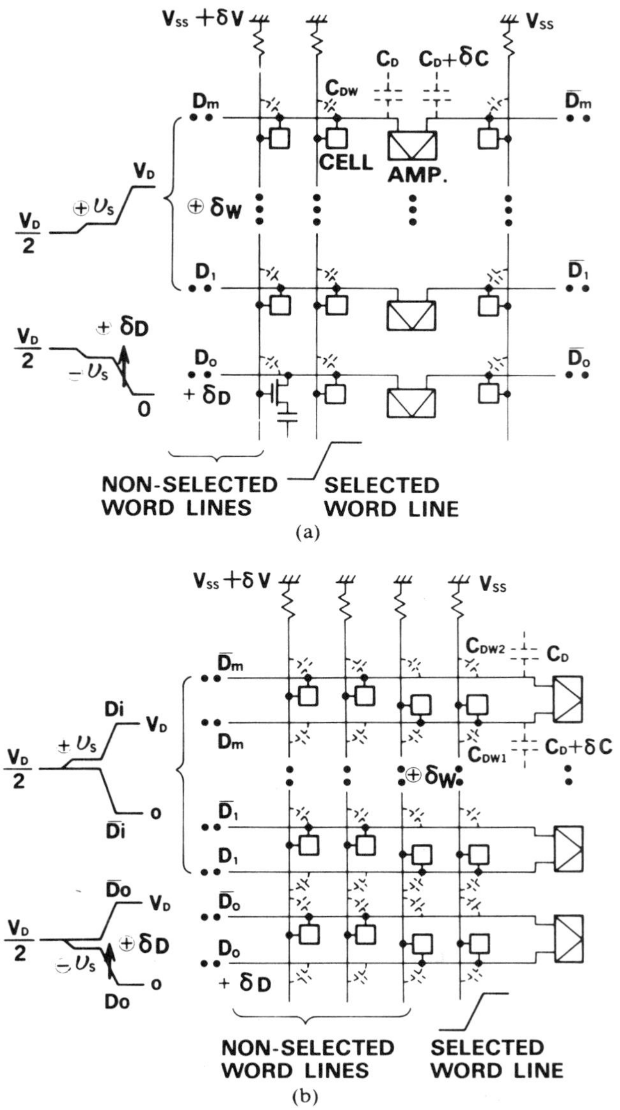

Fig. 10. Schematic memory array and typical noise generations: (a) open data line, and (b) folded data line.

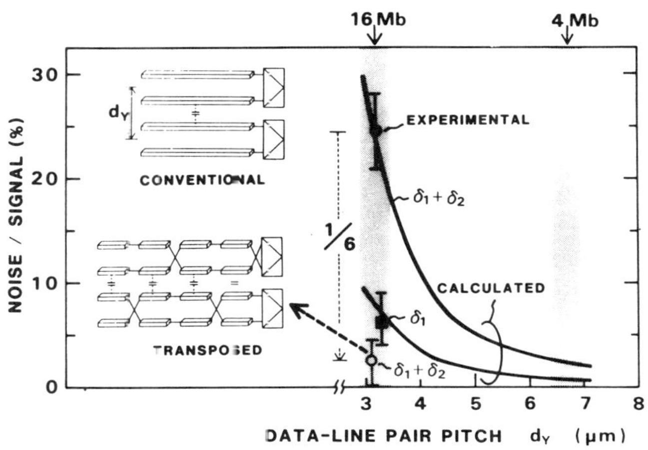

Fig. 11. Interference noise and its suppression with transposed data-line structure.

for the open data line, and to $(C_{DW1} - C_{DW2})^2 \delta C / C_D$ for the folded data line, assuming symmetric data-line waveforms for charging and discharging. Here, δC is the capacitive imbalance of a pair of data lines. Note that the folded data line, due to the closely parallel arrangement of its pair of data lines, produces a smaller δC. It also realizes supreme noise cancellation, which is obvious from the expression of $C_{DW1} - C_{DW2}$. Thus, the folded data line can achieve smaller noise.

For the noise due to the peripheral circuit operations, the resulting ground voltage bounce δV, which is directly coupled to the nonselected word lines, is a prime concern. For the open data line, the δV's belonging to one data-line group $(D_0, D_1, \cdots)$ can differ from those belonging to the other data-line group $(\overline{D}_0, \overline{D}_1, \cdots)$, because of a relatively high ground impedance. The difference in δV between the two groups generates a differential noise in each pair of data lines through C_{DW}. This type of noise may be unacceptably large, even though careful attention is paid to layout in order to lower the impedance. A folded data line, however, can cancel the noises δV in two neighboring cells

The data-line interference noise [4], [24], which has been reported as a serious limiting factor for 16-Mb and beyond, can be classified as being the one generated in the memory array. The noise that becomes prominent with a decrease in data-line pitch is shown in Fig. 11. It has two components: one, δ_1, is at the signal readout from a memory cell, and the other, δ_2, is at the signal amplification. Note that, even for 16-Mb generation having a data-line pitch of around 3 μm, one fourth of the cell signal is lost due to the noise. To overcome this problem, two approaches have been proposed. One is a data-line transposition which minimizes both noise components, as shown in the figure. The other is a data-line shielded cell structure [12] applied to a stacked capacitor cell. One-fourth noise reduction has been reported with this structure. Based on these discussions, it is reasonable to say that the folded data-line arrangement continues to be viable through the megabit DRAM era.

IV. Power Dissipation

A reduction in power dissipation [16], [17] is essential for megabit DRAM designs with respect to the necessity for a longer t_{REFmax} and containment of internal noises. However, there is a definite trend towards increased power dissipation with an increase in memory capacity for a given external supply voltage V_{CC} and for a given cycle time of 230 ns, as shown in Fig. 12. This is due to increases in both peripheral current I_P and memory array current I_A, as shown in Fig. 2. The increase in I_P is caused by an increase in load capacitance in peripheral circuits. This increase in load capacitance is due an increase in the number of decoders and an increase in chip size of about 1.5 times for every successive generation DRAM. The increase in I_A is due to the increase in total data-line capacitance mC_D, charged and discharged si-

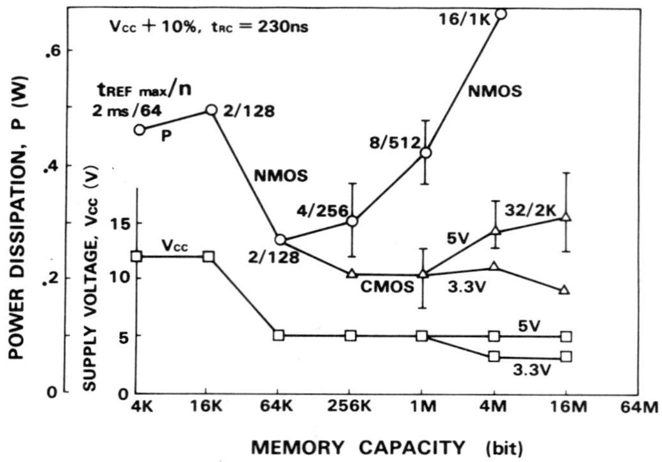

Fig. 12. Trends in power dissipation.

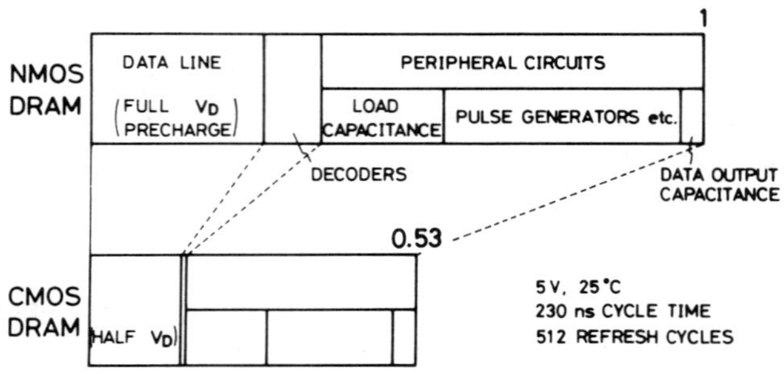

Fig. 13. Power dissipation comparison of 1-Mb NMOS and CMOS DRAM's.

multaneously. To offset the increase in power dissipation, power reduction techniques have been employed in each generation commercial DRAM as explained previously. In particular, the use of CMOS instead of the traditional NMOS and the use of a half-V_D precharge on the data line are extremely useful. Fig. 13 shows a power dissipation comparison of 1-Mb NMOS and CMOS DRAM's [17]. Power dissipation is greatly reduced due to the reduced I_P, which results from CMOS peripheral circuits and CMOS decoders. Furthermore, a half-V_D precharging technique for lowering I_A, which is easily realized by using CMOS amplifiers, remarkably reduces the data-line power dissipation.

Reducing mC_D as much as possible is also vital. One way to meet this requirement is to use the multidivision of data lines, which provides a smaller C_D, as discussed previously. The other way is to double the refresh cycle with the doubled t_{REFmax} at each generation as follows. I_A is expressed using β as

$$I_A = mC_D V_P / t_{\mathrm{RCmin}} = mnC_D V_P(\beta / t_{\mathrm{REFmax}}).$$

To quadruple the original memory capacity mn, there are three alternative memory array organizations, as shown in Table II. They are compared in terms of the refresh cycle, with total data-line capacitance contributing to power dissipation and maximum refresh time realizing a given β. Case (a) affords the same I_A (the same capacitance) as

TABLE II
POSSIBLE MEMORY ARRAY ORGANIZATIONS QUADRUPLING THE MEMORY CAPACITY mn. FIXED VALUES OF C_D AND $t_{\mathrm{RC\,min}}$ ARE ASSUMED

	a	b	c
memory array organization	m; 4n; w; D	2m; 2n	4m; n
refresh cycle	4n	2n	n
capacitance for power dissipation	mC_D	$2mC_D$	$4mC_D$
maximum refresh time realizing a given β	$4t_{\mathrm{REF\,max}}$	$2t_{\mathrm{REF\,max}}$	$t_{\mathrm{REF\,max}}$

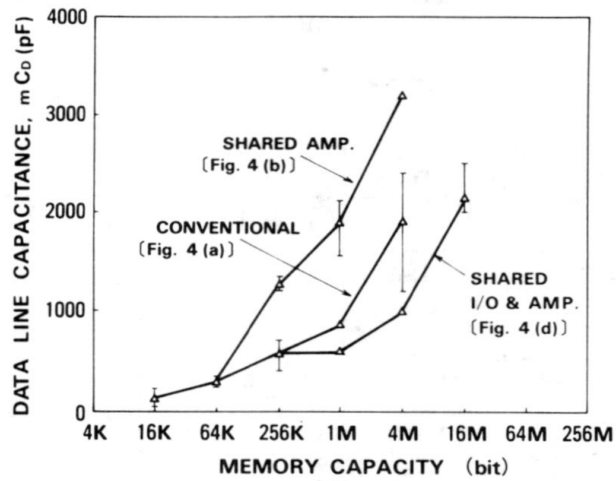

Fig. 14. Trends in total data-line capacitance contributing to power dissipation for the various multi-divided data-line structures.

that of the original memory capacity at the expense of quadrupling the refresh time. This requires, however, a too difficult fabrication process to control the cell leakage current. Case (c), in turn, results in quadrupling I_A even though the requirement for the refresh time is preserved. Thus, case (b) is the best solution in terms of I_A and refresh time. That is why m, i.e. n, has been doubled at each successive generation from 256 kb to 16 Mb, entailing a doubled t_{REFmax} to maintain the constant β, as shown in Fig. 12. For the 16 kb, however, a doubled n was realized at the expense of a halved β. Furthermore, for 64 kb lowering power dissipation with V_{CC} decreasing from 12 to 5 V was efficient enough to maintain the same refresh cycle.

Fig. 14 shows trends in the resultant mC_D for the various multi-divided data-line structures shown in Fig. 4. The shared I/O combined with the shared amplifier scheme is obviously the best. The larger mC_D of the shared amplifier scheme as compared with the conventional one results from the design based on a given signal voltage, described previously. The sharp increase in mC_D with memory capacity for a given scheme is caused by doubling m at each successive generation, although a gradual decrease in C_D is established by further subdivision of data lines, as shown in Fig. 7. In any event, the capacitance becomes more significant as the memory capacity increases. That is why lowering the data-line voltage through an on-chip voltage converter scheme is also important in reducing power dissipation, while preserving the stored voltage as high as possible. Minimizing the data-line voltage using a three-level word pulse combined with plate pushing [12] is one of the most interesting circuits in this respect. For a 1.5-V 16-Mb DRAM, an expected total chip current of 4 mA at a 1-μs cycle time has been reported.

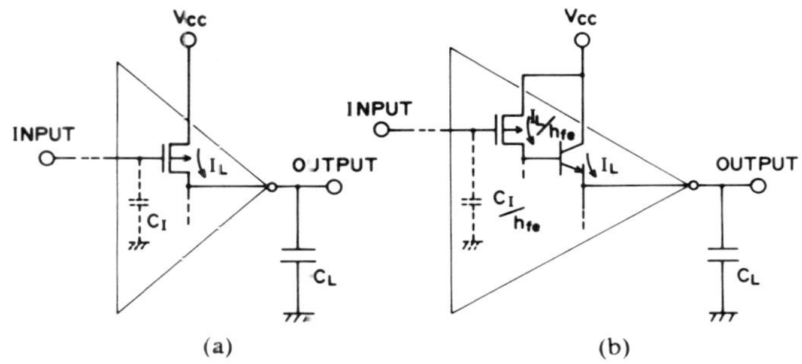

Fig. 15. Schematics of driver circuits: (a) CMOS, and (b) BiCMOS.

V. Speed

It is becoming more difficult to achieve higher speed as memory capacity increases, because an increase in chip size produces an increase in wiring capacitance and wiring resistance [45]. Despite this difficulty, a high speed has been realized, as shown in Fig. 1, using the techniques which reduce *RC* delay time as discussed earlier. A liquid-nitrogen temperature operation provides doubled speed [47] at the expense of additional cooling cost. This enhanced speed is due to lowering the wiring resistance, as well as improving conductance under extremely low temperature. Recently, there has been a tendency toward high speed with relatively small memory capacity. It has been reported that BiCMOS technology [33]–[36] and the PMOS 1T cell [26], [46], [48] provide 512 kb to 1 Mb with 20–30-ns access times.

The high speed of the BiCMOS DRAM results mainly from the high drivability of the driver and the highly sensitive amplifier. The well-known BiCMOS driver provides a larger load current by h_{fe} (h_{fe}: effective current gain is about 7) compared with that of a CMOS driver for a given input capacitance. This implies a smaller input capacitance by h_{fe} for a given load current, as shown in Fig. 15. Thus, the BiCMOS driver has a smaller speed–power product, offering design flexibility. Table III compares CMOS and BiCMOS driver performance. Two key results are obtained. One is that a BiCMOS driver achieves a 23% lower delay time and 28% lower power dissipation, compared with a CMOS driver for the same input capacitance and output rise time. This is due to the inherently small input gate capacitance of the BiCMOS driver and the small number of driver stages required to construct the BiCMOS driver. The other result is that a 1-Mb BiCMOS DRAM, incorporating the BiCMOS driver and a highly sensitive BiCMOS amplifier, provides higher performance in terms of a 36% faster access time and 24% lower power dissipation after fabrication process deviations and temperature changes, as shown in Fig. 16. The resistance to process deviations and temperature changes that the BiCMOS has is responsible for its excellent performance.

BiCMOS technology further enhances access time by using a new BiCMOS sensing technique called direct sensing [36]. A highly sensitive bipolar cascode amplifier, combined with the CMOS read amplifier as shown in Fig. 17, is essential to realize direct sensing. It features sensing of a small read signal on the data line before amplification by the well-known CMOS rewrite amplifier. In conventional DRAM's, the timing margins required between the word pulse, rewrite amplifier activation pulses, and the *Y* control pulse, connecting the data line to the I/O line, are the origin of slow access time. The direct sensing technique eliminates these timing margins, permitting an essentially high-speed operation. Fig. 18 shows relationships between the resulting access time and chip size for a 1.3-μm 1-Mb BiCMOS DRAM chip [36]. Other 1-Mb chips presented at previous conferences are presented for comparison. There seems to be a severe competition in access time between the BiCMOS DRAM [36] and the high speed-oriented CMOS DRAM [48]. However, there is an essential difference in access time, as described above, under the same boundary design conditions. This is because both DRAM's use different types of memory cells (PMOS or NMOS), different memory cell sizes, different array architectures, different aspect ratios of chip length to chip width, etc. In any event, even a 1.3-μm DRAM technology affords a smaller chip size with a competitive access time over 0.8-μm CMOS SRAM. A 0.8-μm DRAM technology could offer a reduced chip size with further access time improvement. Furthermore, BiCMOS technology is preferable to design an ECL interface DRAM. Circuits based on a well-fixed V_{BE} and a bipolar bandgap reference circuit [35] for an on-chip voltage converter also provide design flexibility. Therefore, BiCMOS DRAM technology will be attractive in terms of high-speed access time, low-cost potential, and design flexibility, although the DRAM drawbacks of slower cycle time and higher power dissipation remain unsolved.

VI. Perspectives

Continuous progress on the road toward a gigabit DRAM will surely require innovative circuit technologies, as well as excellent memory cells to solve the previously discussed issues. Future development efforts should also be focused on the pursuit of innovative testing and redundancy technologies. These future technologies must be developed considering the following two issues: possible

TABLE III
COMPARISON OF CMOS AND BiCMOS DRIVERS

DRIVER	NUMBER OF STAGES m	TOTAL GATE CAPACITANCE C_G	DELAY TIME (2V to 2V)	CURRENT CONSUMPTION
CMOS	m = 4, t_r 0.73ns, C_I 0.058pF, C_L 4pF	1.62 pF (1)	1.76 ns (1)	1; I_L, I_D, I_G, I_P
BiCMOS 1	m = 2, 0.73ns, 0.058pF, 4pF	0.20 pF (0.12)	1.35 ns (0.77)	0.72; I_L

I_L: CHARGING & DISCHARGING CURRENT FOR C_L,
I_D: CURRENT CONSUMED IN A DRIVER,
I_G: CHARGING & DISCHARGING CURRENT FOR C_G,
I_P: PENETRATING CURRENT, V_{CC}=4.5V, V_{CL}=4.0V, V_{BB}=-3.0V.

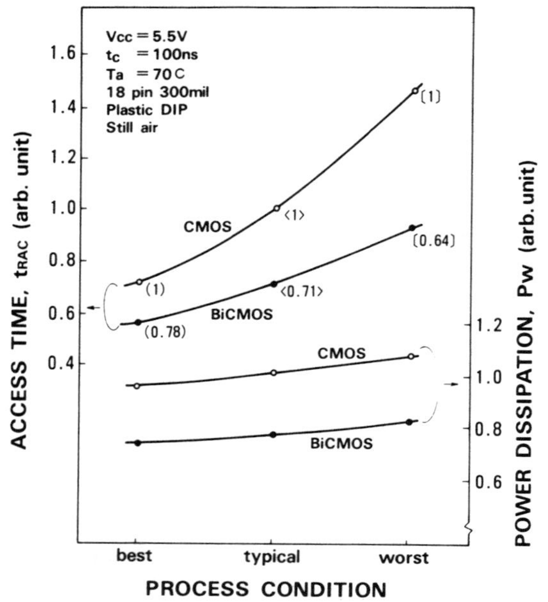

Fig. 16. Dependence of simulated DRAM performance on process deviation.

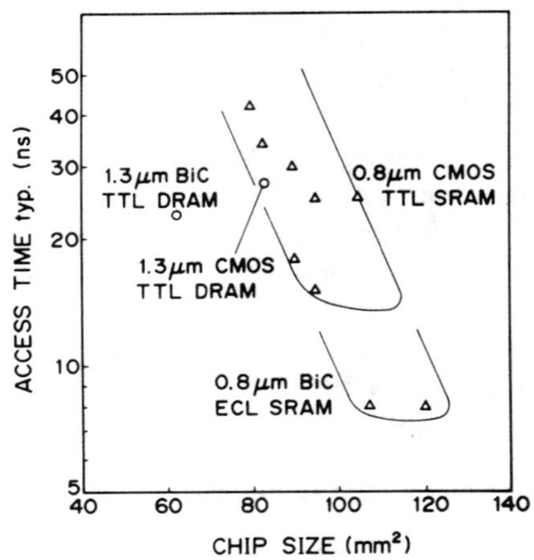

Fig. 18. Access time versus chip size for high-speed 1-Mb RAM's.

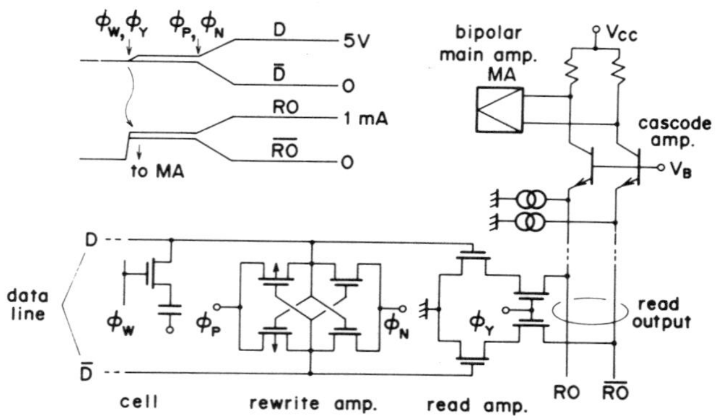

Fig. 17. Concept behind direct sensing scheme using BiCMOS technology.

DRAM technology diversification and power-supply standardization.

Based on the previous discussion, the future of DRAM technology can be envisioned as follows. Current DRAM technology might be diversified so that lower cost and lower power DRAM's with larger memory capacity coexist with higher speed DRAM's with smaller memory capacity. Despite gradual improvement in access time in every successive generation, the 1.5 times increase in chip size makes it difficult to further improve the access time due to the increased *RC* wiring delay described above. For instance, the ability to obtain a high speed of 20 to 30 ns for a 1-Gb chip size as large as 500 mm^2 is questionable. These numbers are extrapolations of the trend so far, as shown in Fig. 1. With such a large chip size, even high-speed transistors no longer contribute to improving speed as in the past, since huge delay times offset the transistor's high speed. Thus, access time will eventually reach an upper limit, as shown by the dotted line in Fig. 1. Such a large memory capacity approach should rather be concentrated on the lowest cost and lowest power dissipation design, utilizing fully the advantages of scaled-down CMOS devices. Speed is inevitably a secondary concern because of the large chip size. Thus,

TABLE IV
Power-Supply Standardization Toward Gigabit CMOS DRAM

		< 4 Mb	16 Mb	64 Mb	256 Mb	1 Gb
DEVICES (Lg/tox)		> 1/20 (μm nm)	0.6/15	0.4/10	0.3/7	0.2/5
MAXIMUM OPERATING VOLTAGE FOR DEVICES		> 5.0V	3.3V	2.3V	1.5V	1.0V
POWER SUPPLY Vcc	I TRANSITION AT EVERY GENERATION WITH MINIMIZING POWER DISSIPATION	—	3.3V	2.3V	1.5V	1.0V
	P(I)/P(II)	—	0.66	0.7	0.45	0.3/0.66
	II MAINTENANCE OF THE SAME Vcc WITH MAXIMIZING USER'S BENEFIT	5.0V	5.0V (CONVERT. TO 3.3V)	3.3V (CONVERT. TO 2.3V)	3.3V (CONVERT. TO 1.5V)	3.3V OR 1.5V (CONVERT. TO 1.0V)

NON—CUSTOM

the power-supply voltage should be as low as possible. TTL and CMOS interfaces are preferable. A high-speed design that still has the low-cost advantage of DRAM's could exist. In this case, a smaller chip with a smaller memory capacity is essential to reduce the *RC* wiring delay. Note that a smaller memory capacity chip should still incorporate the up-to-date technology developed for large memory capacity. Thus, BiCMOS technology enhances the speed in a smaller chip and exclusively enables ECL interface design. In this case, a high power-supply voltage is inevitably required to utilize the high-speed advantage of bipolar current switches. This is justified by the secondary concern of power dissipation even in small capacity memories.

Power-supply standardization continues to be one of the most important, serious, and urgent concerns. This is because the power supply is always closely related to almost all of the design parameters. Furthermore, too much recent rapid scaling down of CMOS devices and emerging technologies, such as BiCMOS, are making standardization difficult. Recognition of a standard's importance and making a long-term commitment to an established power-supply level should be achieved through serious discussion between LSI makers and users. This issue will now be discussed, citing the low-cost CMOS example described above. Table IV shows two possible approaches toward a gigabit CMOS DRAM. As is well recognized, there is no longer any tolerance between the external supply voltage V_{CC} of 5 V and the transistor breakdown voltage, even in the existing 4-Mb DRAM's. Thus, maximum operating voltage, determined by breakdown voltage, should be reduced, in accordance with the depth of device scaling down. Approach I in Table IV is a method to lower the external power supply in every successive generation to the same value as the maximum operation voltage. In this case, a power-supply transition will occur every three years. Such a frequent transition, however, might not be realistic from the user's point of view, as far as noncustom DRAM's are concerned. Users probably might accept maintenance of the same external power-supply level for at least two to three generations.

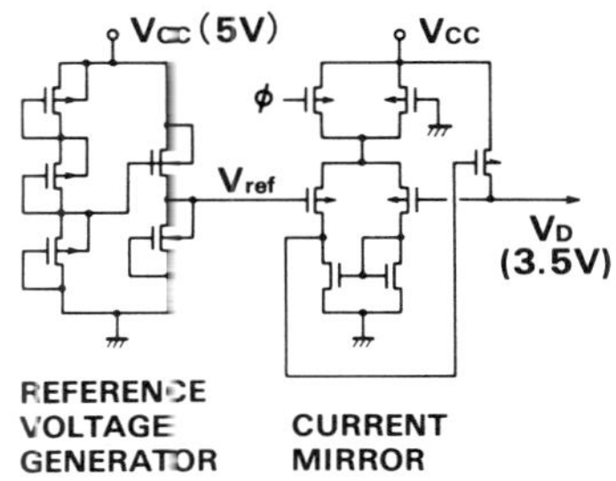

Fig. 19. On-chip voltage converter [30].

This is approach II. An on-chip voltage converter can meet this requirement. It facilitates adjustment of the converted voltage in accordance with the breakdown voltage lowering, while keeping the external power supply the same as shown in Table IV. However, this approach has the drawback of higher power dissipation. This is shown by the power dissipation ratio, $P(\mathrm{I})/P(\mathrm{II})$. In this sense, the long-touted supply voltage of 3.3 V as a standard is seemingly high in terms of power dissipation. The transition to 2.5 V is preferable, as recently proposed by Dennard [49]. Besides settling the power standard to be as low as possible, the challenge of creating low-power circuits suitable for a lower operating voltage must be met to support approach II. Of course, realization of an on-chip voltage converter is also essential.

The keys to achieving an on-chip voltage converter [21], [30], [31] are as follows: suppressing its output voltage bounce, which results from pulsive operation under heavy loads, such as data lines; and dc stabilization against changes in the external power supply, fabrication process, and temperature. Fig. 19 shows an actual voltage converter [30]. To ensure fast response of such a feedback-type regulator during data-line charging, the current mirror has two load transistors. The small conductance transistor is always turned on. The other transistor with a larger conductance is turned on by applying a clock ϕ, which improves the circuit response. This permits minimizing internal voltage V_D bounce due to a large momentary current flow when data lines are charged. The transistor is off during the rest of the cycle to save power. To overcome the V_{ref} variation problem, caused by the PMOS diode threshold variation on this circuit, a bipolar bandgap reference generator has been proposed. Possible remaining problems in this circuit are loop stability and frequency response issues under pulsively changing load capacitance.

VII. Conclusion

It was shown how the S/N ratio, power dissipation, and speed are closely related to continued DRAM progress. Based on the DRAM trends, it was predicted that the current DRAM technology might be diversified in the future so that a large-memory-capacity oriented technology will coexist with a high-speed oriented one. Power-supply standardization poses a continuing serious concern.

References

[1] T. Mano *et al.*, "Circuit technologies for 16 Mb DRAM's," in *ISSCC Dig. Tech. Papers*, Feb. 1987, p. 22.

[2] M. Inoue *et al*, "A 16 Mb DRAM with an open bit-line architecture," in *ISSCC Dig. Tech. Papers*, Feb. 1988, p. 246.

[3] S. Watanabe *et al.*, "An experimental 16 Mb CMOS DRAM chip with a 100 MHz serial read/write mode," in *ISSCC Dig. Tech. Papers*, 1988, p. 248.

[4] M. Aoki *et al.*, "An experimental 16 Mb DRAM with transposed data-line structure," in *ISSCC Dig. Tech. Papers*, Feb. 1988, p. 250.

[5] K. Arimoto *et al.*, "A 60 ns 3.3 V 16 Mb DRAM," in *ISSCC Dig. Tech. Papers*, Feb. 1989, p. 244.

[6] S. Fujii *et al.*, "A 45 ns 16 Mb DRAM with triple-well structure," in *ISSCC Dig. Tech. Papers*, Feb. 1989, p. 248.

[7] T. Takeshima *et al.*, "A 55 ns 16 Mb DRAM," in *ISSCC Dig. Tech. Papers*, Feb. 1989, p. 246.

[8] D. Chin *et al.*, "An experimental 16 Mb DRAM with reduced peak-current noise," in *Symp. VLSI Circuits Dig. Tech. Papers*, May 1989, p. 113.

[9] A. H. Shah *et al.*, "A 4 Mb DRAM with cross-point trench transistor cell," in *ISSCC Dig. Tech. Papers*, Feb. 1986, p. 268.

[10] R. M. Parent *et al.*, "A 4 Mb DRAM with double-buffer static-column architecture," in *ISSCC Dig. Tech. Papers*, Feb. 1987, p. 14.

[11] J. Harter *et al.*, "A 60 ns hot electron resistance 4 M DRAM with trench cell," in *ISSCC Dig. Tech. Papers*, Feb. 1988, p. 244.

[12] M. Aoki *et al.*, "A 1.5 V DRAM for battery-based applications," in *ISSCC Dig. Tech. Papers*, Feb. 1989, p. 238.

[13] N. MacDonald *et al.*, "200 Mb wafer memory," in *ISSCC Dig. Tech. Papers*, Feb. 1989, p. 240.

[14] R. H. Dennard, "Field-effect transistor memory," U.S. Patent 3 387 286, June 4, 1968.

[15] K. Itoh and H. Sunami, "High-density one-device dynamic MOS memory cells," *Proc. Inst. Elec. Eng.*, vol. 130, pt. I, no. 3, pp. 127–135, June 1983.

[16] K. Itoh and K. Kimura, "Power supply constraints in megabit DRAMs of the future," in *ESSCIRC 11th European Solid-State Circuits Conf. Dig. Tech. Papers*, Sept. 1985, p. 130.

[17] K. Kimura *et al.*, "Power reduction techniques in megabit DRAM's," *IEEE J. Solid-State Circuits*, vol. SC-21, no. 3, pp. 381–389, June 1986.

[18] J. M. Lee *et al.*, "A 80 ns 5 V-only dynamic RAM," in *ISSCC Dig. Tech. Papers*, Feb. 1979, p. 142.

[19] S. S. Eaton, "A 5 V-only 2 K×8 dynamic RAM," in *ISSCC Dig. Tech. Papers*, Feb. 1979, p. 144.

[20] S. S. Eaton *et al.*, "A 100 ns 64 K dynamic RAM using redundancy techniques," in *ISSCC Dig. Tech. Papers*, Feb. 1981, p. 84.

[21] K. Itoh *et al.*, "An experimental 1 Mb DRAM with on-chip voltage limiter," in *ISSCC Dig. Tech. Papers*, Feb. 1984, p. 282.

[22] K. U. Stein *et al.*, "Storage array and sense/refresh circuit for single-transistor memory cells," in *ISSCC Dig. Tech. Papers*, Feb. 1972, p. 56.

[23] H. Masuda *et al.*, "A 5 V-only 64 K dynamic RAM based on high S/N design," *IEEE J. Solid-State Circuits*, vol. SC-15, pp. 846–854, 1980.

[24] Y. Nakagome *et al.*, "The impact of data-line interference noise on DRAM scaling," *IEEE J. Solid-State Circuits*, vol. 23, no. 5, p. 1120, Oct. 1988.

[25] P. R. Schroeder and R. J. Proebsting, "A 16 K×1 bit dynamic RAM," in *ISSCC Dig. Tech. Papers*, Feb. 1977, p. 12.

[26] K. Shimohigashi *et al.*, "An n-well CMOS dynamic RAM," in *IEDM Tech. Dig.*, Dec. 1980, p. 835.

[27] R. Chwang *et al.*, A 70 ns high density CMOS DRAM," in *ISSCC Dig. Tech. Papers*, Feb. 1983, p. 56.

[28] N. C. Lu and H. Chao, "Half-V_{DD} bit-line sensing scheme in CMOS DRAM's," *IEEE J. Solid-State Circuits*, vol. SC-19, no. 4, p. 451, Aug. 1984.

[29] H. Kawamoto *et al.*, "A 288 K CMOS pseudostatic RAM," *IEEE J. Solid-State Circuits*, vol. SC-19, no. 5, p. 619, Oct. 1984.

[30] T. Furuyama *et al.*, "An experimental 4-Mbit CMOS DRAM," *IEEE J. Solid-State Circuits*, vol. SC-21, no. 5, p. 605, Oct. 1986.

[31] M. Horiguchi *et al.*, "Dual-operating-voltage scheme for a single-5 V 16-Mbit DRAM," *IEEE J. Solid-State Circuits*, vol. 23, no. 5, p. 1128, Oct. 1987.

[32] T. Fujii *et al.*, "A 90 ns 256 K×1 b DRAM with double level A1 technology," in *ISSCC Dig. Tech. Papers*, Feb. 1983, p. 226.

[33] G. Kitsukawa *et al.*, "An experimental 1-Mbit BiCMOS DRAM," *IEEE J. Solid-State Circuits*, vol. SC-22, no. 5, p. 657, Oct. 1987.

[34] T. Watanabe *et al.*, "Comparison of CMOS and BiCMOS 1-Mbit DRAM performance," *IEEE J. Solid-State Circuits*, vol. 24, no. 3, pp. 771–778, June 1989.

[35] G. Kitsukawa *et al.*, "A 1-Mbit BiCMOS DRAM using temperature-compensation circuit techniques," *IEEE J. Solid-State Circuits*, pp. 597–602 June 1989.

[36] K. Yanagisawa *et al.*, "A 23 ns 1 Mbit BiCMOS DRAM," in *ESSCIRC Dig. Tech. Papers* (Vienna), Sept. 1989, pp. 184–187.

[37] M. Kumanoya *et al.*, "A 90 ns 1 Mb DRAM with multi-bit test mode," in *ISSCC Dig. Tech. Papers*, Feb. 1985, p. 240.

[38] K. Itoh, "Trends in megabit DRAM circuit design," in *Int. Symp. VLSI Technology, Systems and Applications, Proc. Tech. Papers* (Taiwan), May 1989, pp. 21–27.

[39] R. Taylor and M. Johnson, "A 1-Mbit CMOS dynamic RAM with a divided bitline matrix architecture," *IEEE J. Solid-State Circuits*, vol. SC-20, no. 5, p. 894, Oct. 1985.

[40] H. McAdams *et al.*, "A 1-Mbit CMOS dynamic RAM with design-for test functions," *IEEE J. Solid-State Circuits*, vol. SC-21, no. 5, p. 635, Oct. 1986.

[41] M. Koyanagi *et al.*, "Novel high density, stacked capacitor MOS RAM," in *Proc. 10th Conf. Solid-State Devices* (Tokyo), suppl. 18-1, 1979, pp. 35–42.

[42] H. Sunami *et al.*, "A corrugated capacitor cell (CCC)," *IEEE Trans. Electron Devices*, vol. ED-31, no. 6, pp. 746–753, June 1984.

[43] T. C. May and M. H. Woods, "Alpha-induced soft errors in dynamic memories," *IEEE Trans. Electron Devices*, vol. ED-26, pp. 2–9, 1979.

[44] K. Takeuchi *et al.*, "Experimental characterization of α-induced charge collection mechanism for megabit DRAM cells," in *Symp. VLSI Technology Dig. Tech. Papers*, May 1987, p. 99.

[45] J. D. Meindl *et al.*, "Circuit scaling limits for ultra-large scale integration," in *ISSCC Dig. Tech. Papers*, Feb. 1981, p. 36.

[46] N. C. C. Lu *et al.*, "A 20 ns 512Kb DRAM with 83 MHz page operation," in *ISSCC Dig. Tech. Papers*, Feb. 1988, p. 240.

[47] W. H. Henkels *et al.*, "An experimental low temperature DRAM," in *Symp. VLSI Circuits Dig Tech. Papers*, May 1989, pp. 25–26.

[48] S. Dhong *et al.*, "An experimental 27 ns 1Mbit CMOS high-speed DRAM," in *Symp. VLSI Circuits Dig. Tech. Papers*, May 1898, pp. 107–108.

[49] R. H. Dennard, "Power-supply considerations for future scaled CMOS systems," in *Int. Symp. VLSI Technology, Systems and Application Proc. Tech. Papers* (Taiwan), 1989, pp. 188–192.

Paper 3.10

Comparison of CMOS and BiCMOS 1-Mbit DRAM Performance

TAKAO WATANABE, GORO KITSUKAWA, YOSHIKI KAWAJIRI, KIYOO ITOH, RYOICHI HORI, YOSHIAKI OUCHI, TAKAYUKI KAWAHARA, AND TETSURO MATSUMOTO

Abstract —The advantage of BiCMOS technology over CMOS technology in terms of the access time and the power dissipation is demonstrated for a 1.3-μm 1-Mbit DRAM with a TTL interface. Two key results are obtained. One is that a BiCMOS driver achieves a 23 percent lower delay time and 28 percent lower power dissipation compared with a CMOS driver. This is due to the inherently small input gate capacitance of the BiCMOS inverter and the small number of inverter stages required to make the BiCMOS driver. The other result is that a 1-Mbit BiCMOS DRAM incorporating the BiCMOS driver provides higher performance in terms of a 36 percent faster access time and 24 percent lower power dissipation after fabrication process deviations and temperature changes. The resistance to the process deviations and temperature changes that the BiCMOS has is responsible for such excellent performance.

I. Introduction

The bit density of DRAM's has increased by a factor of 4 every three years. In line with this, 16-Mbit CMOS DRAM's have been investigated [1], [2]. In addition to this significant increase in bit density, a reduction in the access time is desired for higher performance microprocessor applications. To meet this requirement, a 32-ns 1-Mbit BiCMOS DRAM has already been reported [3]. Other high-speed DRAM's, such as a 10-ns 1-kbit BiCMOS DRAM [4] and a 20-ns 512-kbit CMOS DRAM [5], have also been proposed.

BiCMOS has been shown to be a promising technology for improving the speed of both SRAM's and gate arrays [6]–[8]. A simple fabrication process for BiCMOS DRAM has been proposed [9]. Therefore, introducing BiCMOS technology to conventional DRAM's should be a promising approach for improving access time. In particular, BiCMOS technology is very convenient for achieving ECL interface DRAM's, which make higher performance systems possible. Even for TTL interface DRAM's, such technology seems to have great potential. However, a quantitative comparison of CMOS and BiCMOS DRAM circuit performance had not been done until this present work, even though such a comparison is especially important for evaluating future DRAM technology [10].

This paper compares simulated CMOS and BiCMOS circuit performance of high-speed 1-Mbit DRAM's with TTL interfaces. The standard nMOS one-transistor memory cell, high-speed array configuration [3], voltage limiter scheme [11], and 1.3-μm fabrication process [9] are assumed for both DRAM's.

In the following sections, CMOS and BiCMOS driver circuit performance is compared with consideration of the design of the driver in a memory chip. Next, CMOS and BiCMOS 1-Mbit performance is compared through analysis of access time and power dissipation. Finally, the dependence of the DRAM performance on fabrication process deviation and temperature change is compared, verifying the advantage of BiCMOS technology.

II. General Description of Driver Design in a Memory Chip

In this section, the importance of designing the driver based on the input gate capacitance is discussed. Such designing is important because the input gate capacitance is closely related to power dissipation and speed.

Fig. 1 shows a schematic of the driver chain in a memory chip. DR_0 is the address driver. DR_1 to DR_{n-1} are the clock drivers for generating various internal clocks. DR_n is the driver for the array or decoder. C_{I0} to C_{In-1} and C_{In} are the input gate capacitances of the DR_0 to DR_{n-1} and DR_n drivers, respectively. C_{W0} to C_{Wn-1} are the wiring capacitances. C_A is the load capacitance of DR_n. R_{W0} to R_{Wn-1} are the wiring resistances.

Achieving the minimum input gate capacitance for each driver is important to designing lower power and higher speed DRAM's. Otherwise relatively large wiring resistances cause large delay times throughout the chip. To satisfy this requirement, a driver consisting of a string of inverters is necessary as shown for DR_{n-1} as an example in Fig. 1. This driver features minimum input gate capaci-

Manuscript received August 8, 1988; revised January 2, 1989.
T. Watanabe, G. Kitsukawa, Y. Kawajiri, K. Itoh, and T. Kawahara are with the Central Research Laboratory, Hitachi Ltd., Kokubunji, Tokyo 185, Japan.
R. Hori is with Musashi Works, Hitachi Ltd., Kodaira, Tokyo 187, Japan.
Y. Ouchi is with Hitachi VLSI Engineering, Kodaira, Tokyo 187, Japan.
T. Matsumoto is with the Device Development Center, Hitachi Ltd., Oume, Tokyo 198, Japan.
IEEE Log Number 8927706.

Reprinted from *IEEE J. Solid-State Circuits*, vol. 24, no. 3, pp. 771–778, June 1989.

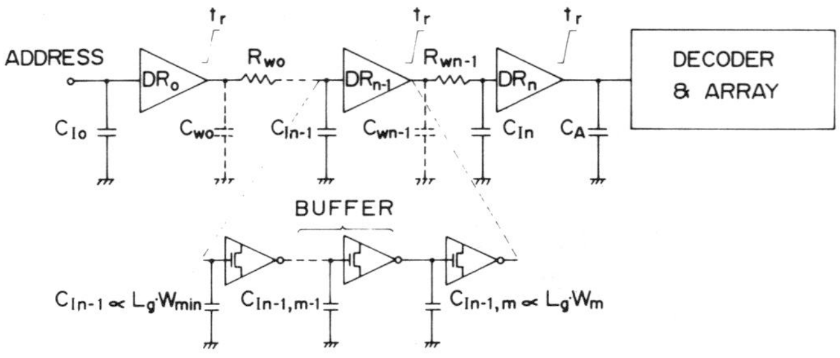

Fig. 1. Concept behind driver design for a memory chip.

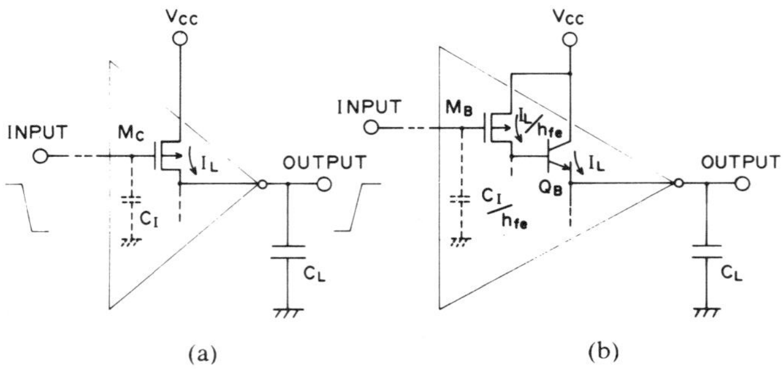

Fig. 2. Schematic of inverter circuits: (a) CMOS, and (b) BiCMOS.

tance for the first-stage inverter, which has the minimum gate width $W_{\min}$ for a given gate length L_g. The driver permits a given output rise time t_r at the final stage independent of the load capacitance. Thus, the m intermediate stage inverters (m is the number of inverters) act as buffers, gradually adjusting the input capacitance mismatching occurring between the first stage and the final stage [12].

Note that the wiring resistance and capacitance for these inverters are negligible because the inverters are close to each other on the chip. Hence, the delay time t_{pd} and power dissipation P_G consumed by charging and discharging of gate capacitance for the driver DR_{n-1} are expressed as

$$t_{pd} = m \cdot t_{pd0} \tag{1}$$

$$P_G \propto \sum_m C_{In-1,m} = C_G \tag{2}$$

where t_{pd0} is the delay time of each inverter. The same equations as (1) and (2) are derived for other drivers. Therefore, lowering the m and total gate capacitance C_G in each driver is important in order to realize high speed and low power.

III. Comparison of CMOS and BiCMOS Driver Performances

A. Input Gate Capacitance in Each Inverter

In this section, the input gate capacitance is investigated for CMOS and BiCMOS inverters, as shown in Fig. 2, since it is the main determinant of the m and C_G for the driver. For simplicity, only pMOS transistors M_C and M_B and bipolar transistor Q_B for charging load capacitance C_L are shown in the figure. Under a given output current, I_L, the bipolar transistor Q_B, having an effective current gain of h_{fe}, enables a reduction in the M_B gate width to $1/h_{fe}$ that of M_C in the CMOS inverter. This is because the M_B drain current is amplified by Q_B. Therefore, the BiCMOS inverter has an inherently smaller input gate capacitance of C_I/h_{fe}, where C_I is the input gate capacitance of the CMOS inverter. Effective current gain h_{fe} is mainly determined by the cutoff frequency f_T of Q_B and the rise time of the load capacitance [13]. The h_{fe} can be expected to be seven in an actual DRAM design, as shown in the next section. This implies a light load capacitance for the preceding inverter. As a result, a smaller m and C_G are attained by using BiCMOS inverters, providing both lower power dissipation and higher speed.

TABLE I
Typical Drivers and Their Performance Formulations

DRIVER	ORGANIZATION	NUMBER OF STAGES m	TOTAL GATE CAPACITANCE C_G
CMOS	CMOS (C_I, m, C_L)	$\log_{\eta_C}\frac{C_I}{\xi_C C_L}+1$	$C_L \xi_C S_C(m)$
BiCMOS 1	BiCMOS	$\log_{\eta_C}\frac{C_I}{\xi_B C_L}+1$	$C_L \xi_B S_C(m)$
BiCMOS 2		$\log_{\eta_C}\frac{\eta_C C_I}{\xi_B \eta_B C_L}+1$	$C_L \xi_B\{1+\eta_B S_C(m-1)\}$
BiCMOS 3		$\log_{\eta_B}\frac{C_I}{\xi_B C_L}+1$	$C_L \xi_B S_B(m)$

$S_C(m)=\frac{1-\eta_C^m}{1-\eta_C}$, $S_B(m)=\frac{1-\eta_B^m}{1-\eta_B}$

B. Comparison of Total Gate Capacitance and Number of Stages in a Driver

Drivers can utilize both CMOS and BiCMOS inverters. In this section, the performances of typical drivers are compared on the basis of the number of stages and total gate capacitance.

Table I shows four types of drivers and their performances. The CMOS driver consists of CMOS inverters only. The BiCMOS 1 driver has a BiCMOS inverter at its final stage. The BiCMOS 2 has BiCMOS inverters at its two final stages. The BiCMOS 3 consists of BiCMOS inverters only. C_I is the input gate capacitance of the first stage. η_C is the ratio of the input gate capacitance of a CMOS stage to that of the succeeding stage. η_B is the ratio of the input gate capacitance of a BiCMOS stage to that of the succeeding stage. ξ_C is the ratio of the gate capacitance of the final CMOS stage to the load capacitance. ξ_B is the ratio of the gate capacitance of the final BiCMOS stage to the load capacitance C_L. In the table, the number of stages m and the total gate capacitance C_G are calculated as a function of the load capacitance C_L and the parameters

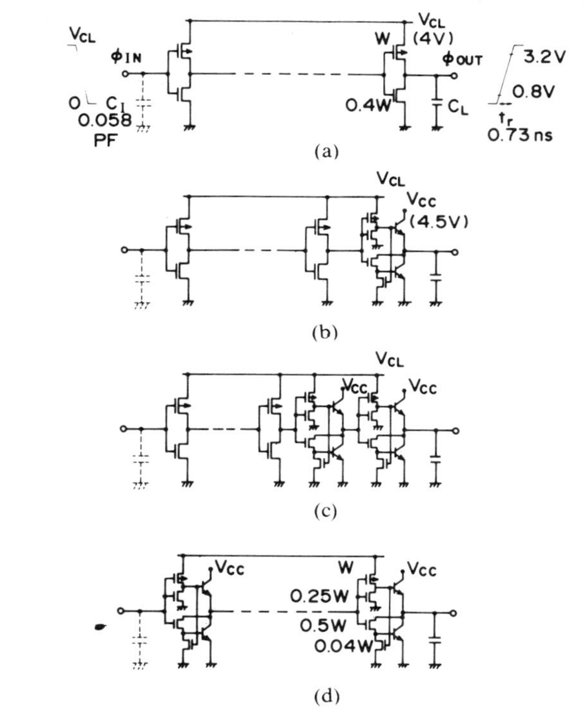

Fig. 3. Circuit diagram of drivers: (a) CMOS, (b) BiCMOS 1, (c) BiCMOS 2, and (d) BiCMOS 3.

TABLE II
1.3-μm Device Parameters

TRANSISTORS	PARAMETERS	VALUES
BIPOLAR	CUTOFF FREQUENCY f_T	5.2 GHz
	DC CURRENT GAIN h_{FE}	49*
	KNEE CURRENT I_K	9.3 mA*
MOS	GATE OXIDE THICKNESS t_{ox}	20 nm
	GATE LENGTH L_g	1.2 μm
nMOS	THRESHOLD VOLTAGE V_T	0.68 V
	SUBSTRATE BIAS COEFFICIENT K_B	0.085 $V^{1/2}$
pMOS	THRESHOLD VOLTAGE V_T	-0.58 V
	SUBSTRATE BIAS COEFFICIENT K_B	-0.19 $V^{1/2}$

* $A_E = 3 \times 8\ \mu m^2$

described above. The derivation of these formulas is explained in Appendix A.

Once the four above-mentioned parameters (η_C, η_B, ξ_C, and ξ_B) are determined, m and C_G are readily calculated by use of the formulas in Table I. The four parameters are derived through computer circuit simulation since the actual transistor characteristics are too complicated for analytical formulation.

The computing procedure is explained using the circuits in Fig. 3 in accordance with those in Table I. First, η_C and η_B are obtained from the minimized delay time condition [12] for the successive stages. This condition is achieved by changing the ratio of the corresponding gate widths for the two successive stages, while maintaining the gate width ratio of the pMOS and nMOS in a stage. As for ξ_C and ξ_B, they are easily determined so as to realize a given rise time for a load capacitance. Note that η_C, η_B, ξ_C, and ξ_B are independent of C_L. Furthermore, ξ_C, and ξ_B are dependent on t_r, while η_C and η_B are independent of t_r.

The four parameters of the drivers in Fig. 3 are calculated through the procedure. The values in the figure represent the design environment for an actual DRAM. They are $C_I = 0.058$ pF, $t_r = 0.73$ ns, $V_{CC} = 4.5$ V, $V_{CL} =$ 4.0 V (voltage limiter approach [3]), $V_{BB} = -3$ V, and the nMOS-to-pMOS gate width ratio for CMOS is 0.4. The figure also shows the gate width ratios for BiCMOS 3, which apply to BiCMOS 1 and 2 as well. These gate width ratios make it possible to offer equal rise and fall times for each stage with logical threshold voltage of $1/2V_{CL}$. Other key environment factors are the device parameters shown in Table II. In addition, the emitter size is set so that a constant emitter current density is realized. Thus, the

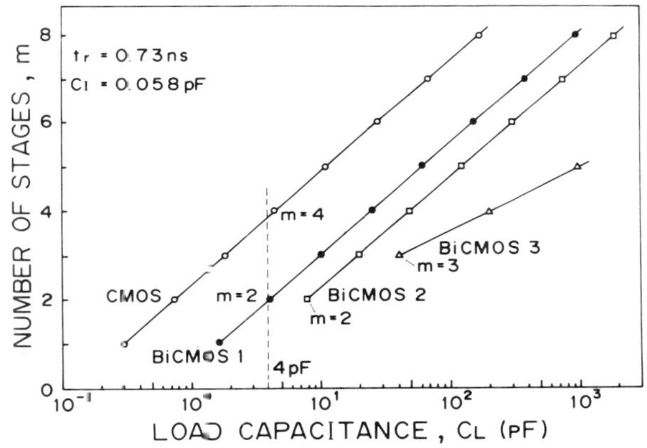

Fig. 4. Number of driver stages versus load capacitance.

parameters in Table I are found to be $\eta_C = 0.40$, $\eta_B = 0.20$, $\xi_C = 0.195$, and $\xi_B = 0.035$ through calculations.

Fig. 4 is the relationship between m and C_L, which is obtained by using the above parameters and the formulas in Table I. Note that $m = 1$ in BiCMOS 2 and $m = 1$ and 2 in BiCMOS 3 are not shown because they are included in the cases of $m = 1$ in BiCMOS 1 and $m = 1$ in BiCMOS 1 and $m = 2$ in BiCMOS 2, respectively. The m increases with an increasing C_L. This is because the resultant larger input gate capacitance at the final stage increases m to maintain a constant C_I. The m in BiCMOS 3 is less sensitive to C_L because η_B is less than η_C. At $C_L = 4$ pF, which is the average load capacitance in the 1-Mbit DRAM [3], $m = 2$ is satisfactory for BiCMOS 1 while $m = 4$ for CMOS. This means that a higher speed can be expected in BiCMOS drivers by eliminating two stages.

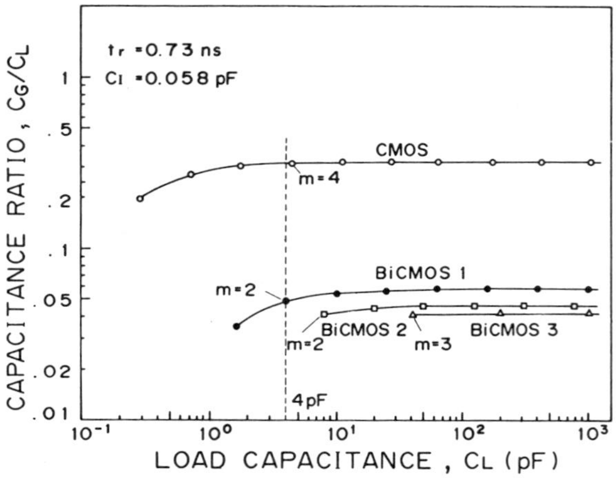

Fig. 5. Capacitance ratio versus load capacitance.

TABLE III
COMPARISON OF CMOS AND BiCMOS 1 DRIVERS

DRIVER	NUMBER OF STAGES m	TOTAL GATE CAPACITANCE C_G	DELAY TIME (2V to 2V)	CURRENT CONSUMPTION
CMOS	m=4, t_r 0.73ns, C_I 0.058pF, C_L 4pF	1.62 pF (1)	1.76 ns (1)	I_L; 1: I_D = I_G + I_P
BiCMOS 1	m=2, 0.73ns, 4pF, 0.058pF	0.20 pF (0.12)	1.35 ns (0.77)	I_L; 0.72

I_L: CHARGING & DISCHARGING CURRENT FOR C_L,
I_D: CURRENT CONSUMED IN A DRIVER,
I_G: CHARGING & DISCHARGING CURRENT FOR C_G,
I_P: PENETRATING CURRENT, V_{CC}=4.5V, V_{CL}=4.0V, V_{BB}=-3.0V.

Fig. 5 shows the relationship between C_G, normalized by C_L, and C_L. C_G/C_L is 0.32 for the CMOS at $C_L = 4$ pF. In contrast, it is 0.05 for BiCMOS 1, resulting in a significant power reduction since a driver with a larger C_G/C_L consumes more power. Note that the differences in the C_G/C_L for various cases of BiCMOS driver is less than 0.02, which implies no substantial difference in the power reduction. However, C_L imposes a restriction on BiCMOS 2 and BiCMOS 3. That is, they are only useful for a C_L much larger than 4 pF. Thus, BiCMOS 1 is the most promising for DRAM design.

The m and C_G/C_L are dependent on t_r, because ξ_C and ξ_B are dependent on t_r. The sensitivity is discussed in Appendix B.

Table III shows a summary of a performance comparison of the CMOS and BiCMOS 1. In addition to the m and C_G obtained above, the delay time and current consumption required for the driver to achieve a rise time of 0.73 ns at $C_L = 4$ pF are shown. I_G and I_L are the charging and discharging currents for C_G and C_L. I_P is the penetrating current, which flows from supply voltage to ground during the input signal transition. Regarding the delay time, BiCMOS 1 has the advantage of speed that is 23 percent (0.4 ns) higher. This is responsible for the smaller number of stages in BiCMOS 1. As for the power dissipation, BiCMOS 1 cuts CMOS power by 28 percent. Note that the current $I_D = I_G + I_P$ in BiCMOS 1, which is consumed in the driver itself, is only one-third the current of CMOS. The smaller I_G in BiCMOS 1 results from the smaller C_G described before. The penetrating current I_P is also smaller in BiCMOS 1, because in the CMOS the I_P at the preceding stages having a large input gate capacitance is not negligible. The I_L in the BiCMOS is also smaller than that of the CMOS, because the output amplitude of the BiCMOS inverter is lower than that of the CMOS inverter by V_{BE} drop. Hence, the BiCMOS driver permits both short delay time and power reduction.

IV. Comparison of CMOS and BiCMOS 1-Mbit DRAM Performance

In this section, a comparison of CMOS and BiCMOS 1-Mbit DRAM performance is described on the basis of the results obtained in the previous section. The same conditions and the same techniques are assumed for both: nMOS one-transistor memory cell, high-speed array configuration [3], TTL interface, address multiplexing, 512 refresh cycles, the same Al wiring width, and voltage limiter approach.

The 1-Mbit BiCMOS DRAM, being a slightly modified version of the previously reported DRAM [3], was actually fabricated. The 1-Mbit CMOS DRAM performance was estimated through a virtual design and circuit simulation. That is, BiCMOS drivers used in the peripheral circuitry were replaced by CMOS drivers having the same C_I and t_r without any changes in the R_W and C_W in Fig. 1, which makes wiring delay the same for both DRAM's. Based on the results in the previous section, this replacement is realized by adding two CMOS inverter stages for each driver. The replacement is possible in spite of the additional stages because the area of the drivers is determined mainly by the Al wiring width for the power supplies, especially for high-speed design.

Below, the array configuration and the inherent peripheral circuitries for each technology are described in detail. Then, the performance of the CMOS and BiCMOS DRAM's is compared.

A. Array Configuration

Fig. 6 shows the chip architecture assumed for both DRAM's. To enhance speed and reduce power dissipation, a divided data line, shared sense amplifier, and I/O line configuration were used [10]. The memory array was divided into four 256-kbit blocks. Each block was divided into two 128-kbit subarrays, which share one sense amplifier per column and one I/O line. Thus, the data line was effectively divided into eight subdata lines. The data line and memory cell capacitances were 333 and 40 fF, respectively. Circuit operation in both the peripheral circuit and memory array was controlled by regulating the voltages provided by the voltage limiter [11]: V_{CL} was 4 V and V_{CH} about 7 V. V_{CL} was supplied to the drivers and other

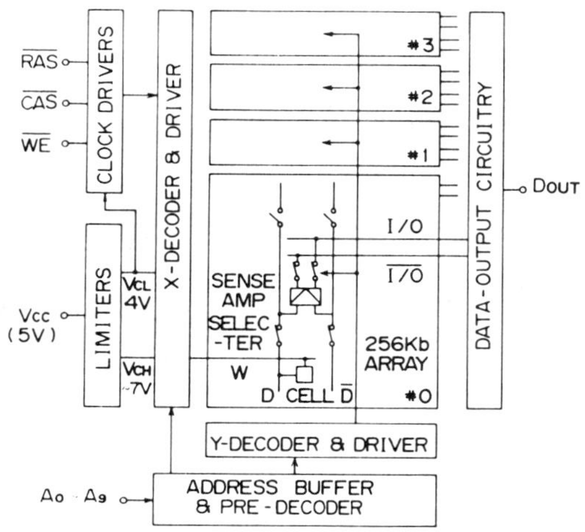

Fig. 6. Chip architecture of DRAM.

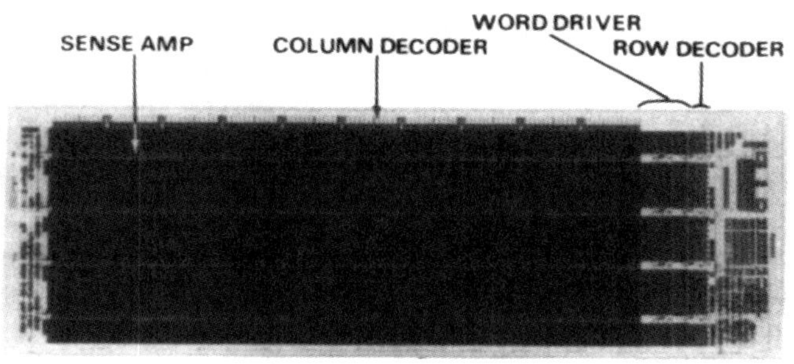

Fig. 7. Photomicrograph of 1-Mbit BiCMOS DRAM.

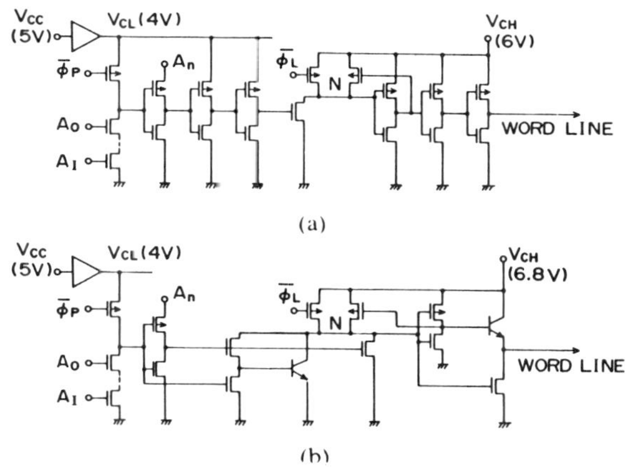

Fig. 8. *X*-decoder and driver circuitries: (a) CMOS and (b) BiCMOS.

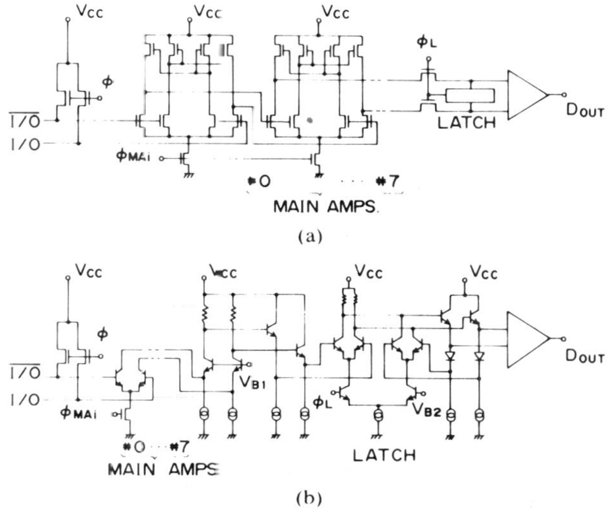

Fig. 9. Data-output circuitries: (a) CMOS and (b) BiCMOS.

circuits for the power reduction. V_{CH} enabled *X*-drivers to provide a word-line voltage above 6 V.

Fig. 7 is a photomicrograph of the 1-Mbit BiCMOS DRAM on which the comparison is based. The chip measures 5.0×14.9 mm^2. Word lines and data lines run along the long and short side of the chip, respectively. Special attention was paid to reducing the wiring resistance in the memory arrays as well as in the peripheral circuitry. Polycide word lines were shunted with the second layer of aluminum. The data lines were formed using the first Al layer.

B. X-Decoder and Driver

Fig. 8 shows circuitries for *X*-decoders and drivers featuring static operation. Both the CMOS and BiCMOS circuits were designed to have the same input gate capacitance for the decoder and the same rise time as that of a 6-V word pulse. As soon as the decoder input signal, A_n, rises to the high level of 4 V, node *N* discharges and the word line is driven quickly to the high level of 6 V. V_{CH} is around 6 V for the CMOS and around 6.8 V for the BiCMOS, which compensates for the diode voltage drop of the BiCMOS. The clocks $\overline{\Phi}_P$ and $\overline{\Phi}_L$ are for precharging the decoder and for discharging the selected word line, respectively. The pull-down of the word line is achieved by using a nMOS for both the CMOS and the BiCMOS, since the word line should be pulled down to ground. Note that the speed of the pull-down does not affect the access time of DRAM's.

C. Data-Output Circuitry

Fig. 9 shows the data-output circuitries that were chosen [14]. The main amplifiers feature a differential amplifier with eight blocks in parallel, one of which belongs to each 128-kbit subarray and is activated selectively by Φ_{MAi}. Each block in the CMOS has two stages of differential amplifiers to amplify the signal for succeeding CMOS circuits.

The output signal of the main amplifier is transferred to the latch circuit. The main amplifier is cut off by Φ_{MAi} after completion of the latch operation to reduce the power dissipation.

D. Comparison of DRAM Performance

Fig. 10 shows an access time comparison of the CMOS and BiCMOS 1-Mbit DRAM's, both with 30-pF data output capacitances. The device parameters in Table II were used in the circuit simulation. As expected, the

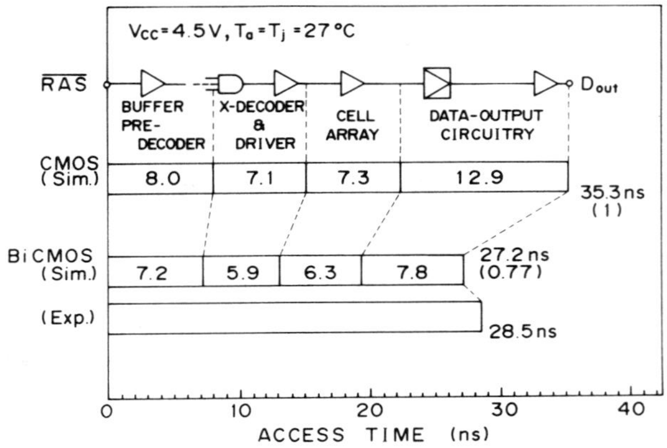

Fig. 10. Access time comparison for 1-Mbit virtually designed CMOS and BiCMOS DRAM's.

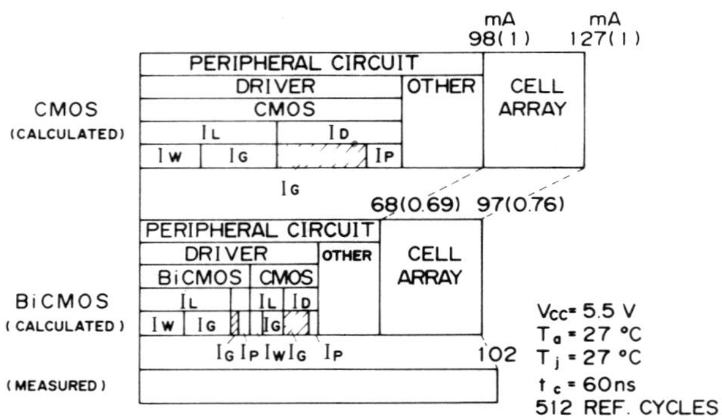

Fig. 11. Comparison of power dissipation of 1-Mbit virtually designed CMOS and BiCMOS DRAM's.

BiCMOS DRAM has the advantage of high speed. The simulated access times of the CMOS and BiCMOS are 35.3 and 27.2 ns, respectively. The BiCMOS reduces the access time by 23 percent. The details are as follows.

The delay times for $\overline{RAS}$ to a word line are 15.1 ns in the CMOS and 13.1 ns in the BiCMOS. This reduction is due to the high speed in the BiCMOS drivers. The delay time in the memory cell array is also smaller in the BiCMOS since the BiCMOS provides a smaller timing margin between the word pulse and sense-amplifier activating clock. This timing advantage results from the immunity from temperature changes and fabrication process deviations, which are described in the next section. The delay time from the main amplifier to the data output is 5.1 ns less in the BiCMOS due to the adoption of high-speed bipolar circuitry, as shown in Fig. 9. The measured access time of the 1-Mbit BiCMOS DRAM is also shown in Fig. 10, in good agreement with the result of the circuit simulation. The improvement over the previous BiCMOS chip performance [3] is mainly due to adoption of the X-driver shown in Fig. 8.

Fig. 11 shows a comparison of current consumption for 1-Mbit DRAM's. I_L and I_D are the current for the charging and discharging of a load capacitance and the current consumed in the drivers, respectively. I_W, I_G, and I_P are the currents for the charging and discharging of the wiring capacitance, the gate capacitance, and the penetrating current, respectively. A V_{CC} of 5.5 V, a cycle time of 60 ns, and an ambient temperature T_a of 27°C were selected. The junction temperature was assumed to be 27°C as well. The BiCMOS DRAM has a 24 percent lower power dissipation. This is mainly due to the lower gate capacitance of the BiCMOS drivers (hatched area) in the peripheral circuitry, as was expected. The current unrelated to the drivers (OTHER) is larger in the CMOS because the seven unselected main amplifiers in the CMOS DRAM require a large standby current to realize high-speed static column operation.

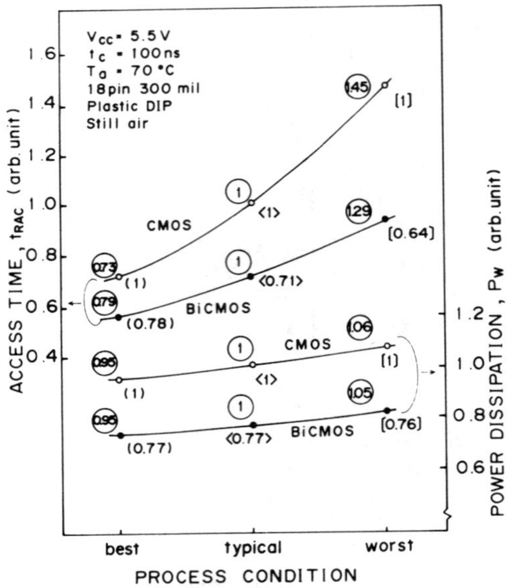

Fig. 12. Dependence of simulated DRAM performance on process deviation.

V. Further Consideration

In this section, the dependence of DRAM performance on fabrication process deviations and temperature change is described. BiCMOS circuitry has been reported to be less sensitive to such factors than CMOS circuitry [11], [15]. Therefore, the BiCMOS DRAM should have an advantage, especially in the worst-case conditions.

Fig. 12 shows the dependence of access time and power dissipation on fabrication process deviation for CMOS and BiCMOS 1-Mbit DRAM's. An internal voltage of 4 V is provided by the on-chip voltage limiter in spite of an external voltage of 5.5 V, as described before. Deviations of 20 percent are assumed to exist for the device parameters shown in Table II. The temperature dependence of the access time is taken into consideration as follows:

The junction temperature T_j and the access time t_{RAC} are derived:

$$T_j = T_a + \Theta_{ja} P_w \tag{3}$$

$$t_{RAC} = t_{RAC0} k_T (T_j - 27) \tag{4}$$

where T_a is the ambient temperature and Θ_{ja} is the heat resistance of the package. Power dissipation P_w is calcu-

TABLE IV
PERFORMANCES OF 1-MBIT VIRTUALLY DESIGNED CMOS AND BiCMOS DRAM'S UNDER THE WORST PROCESS CONDITIONS

	CMOS	BiCMOS
ACCESS TIME t_{RAC}	1	0.64
POWER DISSIPATION P_W	1	0.76
CHIP SIZE	1	0.97

PROCESS WORST, Vcc=5.5V, tc=100ns, Ta=70 °C, 18pin 300mil plastic DIP.

lated at a cycle time t_c of 100 ns. t_{RAC0} is the access time at $T_j = 27°C$. k_T is the temperature coefficient of the access time, which is 0.26 and 0.18 percent/°C for the CMOS DRAM and the BiCMOS DRAM, respectively. These values were obtained experimentally by using a 1-Mbit BiCMOS DRAM [3] and a conventional medium-speed 1-Mbit CMOS DRAM, both fabricated simultaneously on the same wafer.

Note that the access time of the CMOS DRAM varies almost 100 percent when going from the best to the worst conditions, while that of the BiCMOS DRAM varies only 60 percent. Therefore, the advantage of the BiCMOS DRAM regarding the access time is significant for the worst-case conditions. This smaller variation in the access time of the BiCMOS DRAM can ensure a higher yield in mass production. As for the power dissipation, it only depends a little on the fabrication process. It is around 5 percent larger for the worst-case condition than for the typical condition for both the CMOS and BiCMOS DRAM's. This change in power dissipation is mainly caused by the increase in the gate capacitance resulting from the deviation in the gate length.

Table IV summarizes the comparison of the 1-Mbit DRAM's for the worst-case conditions of the fabrication process deviations. The larger chip size of the CMOS DRAM is due to the large area occupied by the *X*-drivers. This table shows the advantage of the BiCMOS DRAM over the CMOS DRAM: 36, 24, and 3 percent for access time, power dissipation, and chip size, respectively.

VI. CONCLUSION

The performances of CMOS and BiCMOS 1-Mbit DRAM were compared, followed by a comparison of CMOS and BiCMOS drivers, in terms of speed and power dissipation. The results obtained are summarized as follows:

1) A driver having a 23 percent lower delay time and 28 percent lower power dissipation is realized at an average load capacitance of 4 pF in a 1-Mbit DRAM chip by using BiCMOS.

2) A 1-Mbit DRAM with a TTL interface having a 36 percent faster access time and 24 percent lower power dissipation is obtained under worst-case conditions for fabrication process deviations and temperature changes.

APPENDIX A
DERIVATION OF EQUATIONS IN TABLE I

The formulas for the CMOS driver in Table I are obtained as follows. The input gate capacitance of the final stage is $\xi_C \cdot C_L$ and that of the mth preceding stage to the last stage is $\eta_C^{m-1} \cdot \xi_C \cdot C_L$. Therefore, C_I and the total gate capacitance C_G can be expressed as

$$C_I = \eta_C^{m-1} \cdot \xi_C \cdot C_L \tag{A1}$$

$$C_G = (\xi_C + \eta_C \cdot \xi_C + \cdots + \eta_C^{m-1} \cdot \xi_C) \cdot C_L. \tag{A2}$$

Hence, m and C_G are given by

$$m = \log \eta_C (C_I / (\xi_C \cdot C_L)) + 1 \tag{A3}$$

$$C_G = C_L \cdot \xi_C \cdot S_C(m) \tag{A4}$$

where $S_c(m)$ is defined as

$$S_C(m) = (1 - \eta_C^m)/(1 - \eta_C). \tag{A5}$$

The formulas for the other drivers (BiCMOS 1, 2, and 3) in Table I can be derived in the same manner.

APPENDIX B
SENSITIVITY OF m AND C_G/C_L TO DESIGN PARAMETERS

The values of m and C_G/C_L, which are important for the comparison of drivers, are dependent on some design parameters (see Table I). Only the dependence on C_L and t_r is examined in the following paragraph. The dependence on t_r is a result of the dependence of ξ_C and ξ_B on t_r. Note that C_i, the minimum input gate capacitance of a driver, is almost uniquely determined by the t_{ox} and the layout rule. η_C and η_B are also uniquely determined by minimizing the delay time for the internal inverter chain of a driver [12].

The dependence of m and C_G/C_L on C_L is shown in Figs. 4 and 5, respectively. Fig. 4 shows that the sensitivity of m on C_L is the same for the CMOS and the BiCMOS 1. This means that the difference in the number of inverters in a driver between the CMOS and the BiCMOS 1 is independent of C_L. Fig. 5 shows that the dependence of C_G/C_L on C_L is low, especially for C_L larger than 4 pF. Even for C_G/C_L less than 4 pF, C_G/C_L for BiCMOS 1 is much less than that of CMOS. It means that the advantage of the BiCMOS 1 in terms of power dissipation is realized for a wide range of C_L.

In order to clarify the dependence of m and C_G/C_L on t_r, the following relations are derived:

$$\xi_C = \xi_{C0} \cdot (t_{r0}/t_r) \quad \text{(B1)}$$

$$\xi_B = \xi_C / h_{fe}$$

$$= \xi_{B0} \cdot (t_{r0}/t_r)^2 \quad \text{(B2)}$$

where ξ_{C0} and ξ_{B0} are the ξ_C and the ξ_B at $t_r = t_{r0}$, respectively. Equation (B1) is easy to understand, since t_r is inversely proportional to the gate width in the CMOS and ξ_C is proportional to the gate width. Equation (B2) is derived from the approximation that h_{fe} is proportional to t_r [13]. From the above two equations and the formulas shown in Table I, the sensitivity of m and C_G/C_L can be derived. For example, for a variation in the value of t_r of 0.1 ns centered at $t_r = 0.73$ ns, the resultant variation in m is 0.15 and 0.3 for the CMOS and BiCMOS 1, respectively, while the resultant variation in C_G/C_L is 0.05 and 0.02. The variation in m can be considered negligible, since m, the number of inverters in a driver, should be an integer in an actual design. Even though the variation in C_G/C_L for BiCMOS 1 is relatively large compared with its value calculated at $t_r = 0.73$ ns, the C_G/C_L value for the BiCMOS 1 (0.07 at $t_r = 0.63$ ns) is still much smaller than that for the CMOS (0.37 at $t_r = 0.63$ ns). Therefore, these results support the assertion that BiCMOS has the advantage in terms of delay time and power dissipation.

Acknowledgment

The authors would like to express their thanks to H. Kawamoto, A. Endo, and T. Masuhara for their constant encouragement and to K. Miyata, N. Momma, T. Ikeda, Y. Kobayashi, M. Ohayashi, H. Miyazawa, and K. Asayama for their contributions in the process development. They would also like to thank H. Tsukada, K. Mihashi, and S. Kato for their technical support.

References

[1] T. Mano *et al.*, "Circuit technologies for 16Mb DRAMs," in *ISSCC Dig. Tech. Papers*, Feb. 1987, pp. 22–23.

[2] M. Aoki *et al.*, "An experimental 16Mb DRAM with transposed data-line structure," in *ISSCC Dig. Tech. Papers*, Feb. 1988, pp 250–251.

[3] G. Kitsukawa *et al.*, "An experimental 1-Mbit BiCMOS DRAM," *IEEE J. Solid-State Circuits*, vol. SC-22, pp. 657–662, Oct. 1987.

[4] S. Watanabe *et al.*, "BiCMOS circuit technology for high speed DRAMs," in *Symp. VLSI Circuits Dig. Tech. Papers*, May 1987, pp. 79–80.

[5] N. Lu *et al.*, "A 20 ns 512Kb DRAM with 83MHz page operation," in *ISSCC Dig. Tech. Papers*, Feb. 1988, pp. 240–241.

[6] J. Miyamoto *et al.*, "A high-speed 64K CMOS RAM with bipolar sense amplifiers," *IEEE J. Solid-State Circuits*, vol. SC-19, pp. 557–563, 1984.

[7] K. Ogiue *et al.*, "A 13ns/500mW 64Kb ECL RAM," in *ISSCC Dig. Tech. Papers*, Feb. 1987, pp. 212–213.

[8] Y. Nishio *et al.*, "A subnanosecond low power advanced bipolar-CMOS gate array," in *Proc. ICCD*, Oct. 1984, pp. 428–433.

[9] Y. Kobayashi *et al.*, "Bipolar CMOS merged structure for high speed M bit DRAM," in *IEDM Tech. Dig.*, Dec. 1986, pp. 802–804.

[10] K. Kimura *et al.*, "Power reduction techniques in megabit DRAM's," *IEEE J. Solid-State Circuits*, vol. SC-21, pp. 381–389, June 1986.

[11] G. Kitsukawa *et al.*, "A 1-Mbit BiCMOS DRAM using temperature compensation circuit techniques," in *Proc. ESSCIRC'88*, Sept. 1988.

[12] C. Mead and L. Conway, *Introduction to VLSI Systems*. Reading, MA: Addison-Wesley, 1980, ch. 1.

[13] H. Higuchi *et al.*, "Performance and structures of scaled-down bipolar devices merged with CMOS FET's," in *IEDM Tech. Dig.*, Dec. 1984, pp. 694–697.

[14] O. Minato *et al.*, "A 20ns 64K CMOS SRAM," in *ISSCC Dig. Tech. Papers*, Feb. 1984, pp. 222–223.

[15] Y. Nishio *et al.*, *Nikkei Electronics*, no. 374, pp. 187–206, Aug. 1988.

Paper 3.11

BiCMOS Circuit Technology for High-Speed DRAM's

Shigeyoshi Watanabe, Koji Sakui, Tsuneaki Fuse, Takahiko Hara, Seiichi Aritome, and Katsuhiko Hieda, *Member, IEEE*

***Abstract*— A new BiCMOS circuit technology featured by a novel bit-line sense amplifier has been developed. The bit-line sense amplifier is composed of a BiCMOS differential amplifier, the impedance-converting means featured by the CMOS current mirror circuit or the clocked CMOS inverter between the bit line and the base node of the BiCMOS differential amplifier, and a conventional CMOS flip-flop. This technology can reduce the access time to half that of a conventional CMOS DRAM access time. Applied to a 1-kb DRAM test chip, a new BiCMOS circuit technology was successfully verified. Furthermore, the sensitivity and area penalty of the new BiCMOS bit-line sense amplifier and future applications to megabit DRAM's are discussed.**

I. Introduction

IN the development of high-density DRAM's such as 1 Mb and larger, it will be increasingly important to achieve high speeds competitive with SRAM's, and in addition to achieve a low cost per bit [1], [2]. In this paper, a new BiCMOS circuit technology for high-speed DRAM's is proposed.

Previously, BiCMOS technology has been mainly proposed for achieving high-speed and low-power SRAM's [3], [4]. BiCMOS differential amplifiers for I/O line sensing and BiCMOS inverters for peripheral circuits remarkably reduce the SRAM access time. The access time is about 1/2 of that of a conventional CMOS SRAM, as shown in Fig. 1(a). Since a SRAM cell has current drive capability without losing the storage data, it can be directly coupled to a BiCMOS differential amplifier. The delay time from a word line (WL) to an input–output line (I/O) is drastically reduced by introducing a BiCMOS differential amplifier as shown by the dotted portion of Fig. 1(a). Furthermore, by employing the BiCMOS driver to the address buffers and row decoders featured by the large load capacitance, the delay time from input to WL can be easily reduced [5].

In the case of DRAM's, the same as in the BiCMOS SRAM case, high-speed BiCMOS circuits are easily introduced for the peripheral circuitry such as address buffer, row decoder, and I/O line sense amplifier [6]–[9]. The access time of the RAM is reduced by 30% with BiCMOS peripheral circuits, as shown in Fig. 1(b). However, a BiCMOS differential amplifier cannot be easily introduced into a bit-line sensing circuit. If a SRAM-type BiCMOS sense amplifier is used for bit-line sensing as shown in Fig. 2, two fatal problems occur. First, memory cell data disappear through the solid line path, because a DRAM cell cannot restore the data by itself. Second, there is a large power dissipation during bit-line sensing. If all the BiCMOS differential amplifiers are activated simultaneously, a very large current flows during bit-line sensing. For example, in the case of 1-Mb DRAM, more than 1K differential amplifiers are activated, causing the peak current to exceed 1 A. To achieve faster access time, it is very efficient to reduce the delay time from WL to I/O preventing these two fatal problems.

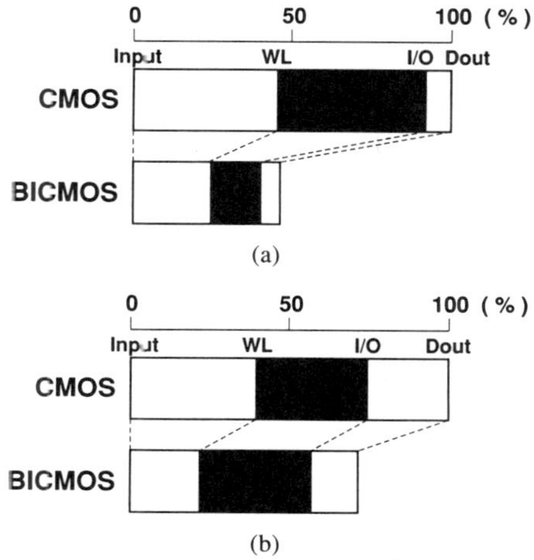

Fig. 1. (a) The access time reduction in SRAM by introducing BiCMOS differential amplifier for I/O line sensing and BiCMOS inverters for peripheral circuits. (b) The access time reduction in DRAM by introducing BiCMOS circuits for the peripheral circuits.

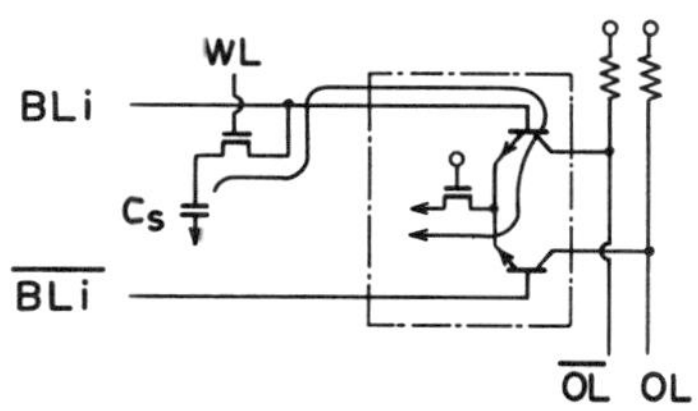

Fig. 2. The SRAM-type BiCMOS bit-line sense amplifier circuit.

To show the effectiveness of BiCMOS technologies, this paper is organized as follows. In Section II the design concept of the new BiCMOS bit-line sense amplifier is described. In Section III the design concept of the new BiCMOS DRAM is described. In Section IV simulation results of new BiCMOS DRAM circuit are described. In Section V the application for a 1-kb BiCMOS DRAM test chip is presented. Finally, Section VI describes future applications to megabit DRAM's.

Manuscript received May 2, 1991; revised July 27, 1992.
The authors are with the ULSI Research Center, Toshiba Corporation, 1, Komukai-Toshiba-cho, Saiwai-ku, Kawasaki, 210, Japan.
IEEE Log Number 9204542.

Reprinted from *IEEE J. Solid-State Circuits,* vol. 28, no. 1, pp. 4–9, Jan. 1993.

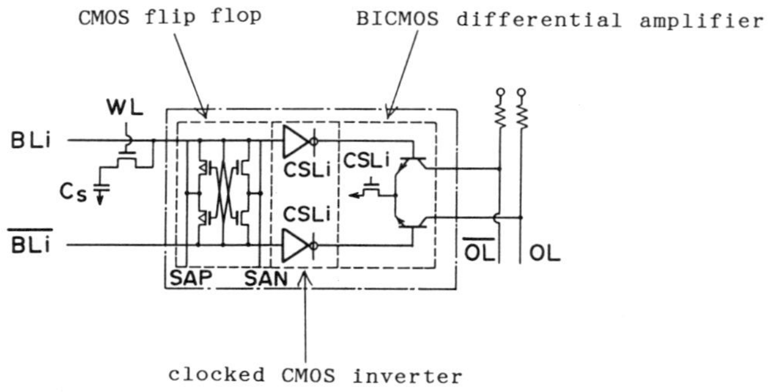

Fig. 3. The newly introduced BiCMOS bit-line sense amplifier circuit.

II. Design Concept of the New BiCMOS Bit-Line Sense Amplifier

Fig. 3 shows a circuit diagram of a newly introduced BiCMOS bit-line sense amplifier. It consists of a BiCMOS differential amplifier and the impedance-converting means between the bit line and the base node of the bipolar transistor and conventional CMOS flip-flop. For the impedance-converting means, the CMOS current mirror circuit or the clocked CMOS inverter are adopted.

Small change in the bit-line voltage, which is generated by the memory cell data, is transferred to the input node of the BiCMOS differential amplifier through the impedance-converting means. The BiCMOS differential amplifier has a high drivability and sensitivity for small signals compared with the conventional CMOS sense amplifier (CMOS flip-flop). Therefore, a high-speed sensing can be performed without losing the memory cell data.

In the case of a conventional CMOS sense amplifier, total noise within the sense amplifier and memory cell array V_{CMOS} can be formulated in the following equation by the sensing speed dv/dt which can be controlled by the driver transistor for the common source line of the sense amplifier, unbalance of bit-line pair capacitances $\Delta C_b/C_b$, unbalance of β of flip-flop pair MOSFET's $\Delta\beta/\beta$, threshold voltage difference between flip-flop pair MOSFET's ΔV_t, and noise voltage caused by the alpha-particle-induced soft error $V_{se\text{-CMOS}}$ [10]:

$$V_{\mathrm{CMOS}} = A \cdot \sqrt{dv/dt}(\Delta C_b/C_b + \Delta\beta/\beta) + \Delta V_t + V_{se\text{-CMOS}}. \tag{1}$$

For a 1-Mb CMOS DRAM the three terms of (1) are 40, 30, and 40 mV, respectively [11], [12]. Therefore, the conventional CMOS sense amplifier cannot be activated, before the voltage difference between bit line and bit-line bar becomes larger than 110 mV. Furthermore, the sensing speed is limited to a fixed value so as not to exceed the first term of (1) beyond 40 mV.

On the other hand, in the case of the newly developed BiCMOS differential amplifier, noise within the sense amplifier and memory cell array, V_{BiCMOS}, can be formulated by using V_{be} difference between pair bipolar transistor ΔV_{be} and noise voltage caused by the alpha-particle-induced soft error $V_{se\text{-BiCMOS}}$ and the offset and noise for the impedance-converting means $\Delta V_{\mathrm{offset}}$.

$$V_{\mathrm{BiCMOS}} = \Delta V_{be} + V_{se-\mathrm{BiCMOS}} + \Delta V_{\mathrm{offset}}. \tag{2}$$

For 1-Mb BiCMOS DRAM, ΔV_{be} is 10 mV. $V_{se-\mathrm{BiCMOS}}$ is 30 mV. The reduction of soft error rate is achieved by the p^+ layer applied to the memory cell array and sense amplifier [6], [7]. The reduction rate is estimated with the combination of the improvement of the measured soft error rate of an experimental 1-Mb BiCMOS DRAM [6] and the potential barrier effect on the total collected charge calculated by using the 3-D device simulator [13]. $\Delta V_{\mathrm{offset}}$ is 30 mV, which is the same value as the ΔV_t. Therefore, the BiCMOS differential amplifier can be activated when the voltage difference between bit line and bit-line bar is as small as 70 mV, which is 2/3 of that of the conventional one. Furthermore, the sensing speed is not limited. This is because in the impedance-converting means such as the CMOS current mirror circuit case, the input stage is separated from the output stage, whereas the input stage is coupled to the output stage for the conventional case. Thus, higher sensing speed results in the false feedback between the input and output stage for the conventional case. Therefore, the advantage of the bipolar transistor that the current drivability is about ten times larger than that of the MOS transistor can be fully used for achieving high-speed sensing. As a result, new BiCMOS differential amplifier can reduce the delay time from WL to I/O line drastically compared with the conventional CMOS case.

A conventional CMOS flip-flop as shown in Fig. 3 restores the memory cell data concurrently. When the signal SAN goes low, the NMOS flip-flop is activated. When SAP goes high, the PMOS flip-flop is activated.

III. Design Concepts of the New BiCMOS DRAM

Fig. 4 shows a circuit diagram of this BiCMOS DRAM. Two bit-line pairs are described in this diagram for simplicity. The BiCMOS inverter is applied to the row decoder circuit. The new BiCMOS sense amplifier is introduced for each bit-line pair. For the impedance-converting means the clocked CMOS inverter is adopted. In addition, a conventional I/O line is separated into an input line (IL) and an output line (OL). The output node of each BiCMOS sense amplifier is connected to this output line (OL). The low level of OL is limited to V_{cc}-V_{be} to prevent bipolar transistors from operating in the saturation region. Since OL is separated from a bit line, a conventional BiCMOS circuit can be connected to the OL without losing the memory cell data. The node of the OL and $\overline{\mathrm{OL}}$ pair is amplified by the BiCMOS differential amplifier and BiCMOS inverters and finally transferred to the D_{out} terminal. By using these circuits, the delay time from WL to D_{out} is reduced.

Furthermore, in order to prevent an increase in the active current, each BiCMOS bit-line sense amplifier is selectively activated by the column select signal CSL. During a write operation, the data are written into the bit-line pair through two transfer gate transistors $T1$ and $T2$.

IV. Simulation Results

Simulated waveforms of main nodes during a read operation are shown in Fig. 5. For the impedance-converting means the clocked CMOS inverter is adopted. A selected word line is activated first, and then the memory cell data are

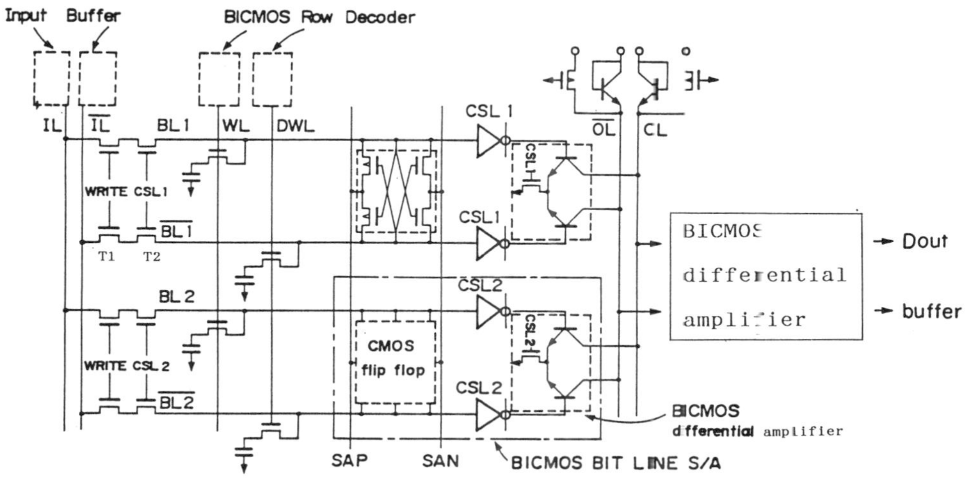

Fig. 4. The circuit diagram of newly developed BiCMOS DRAM.

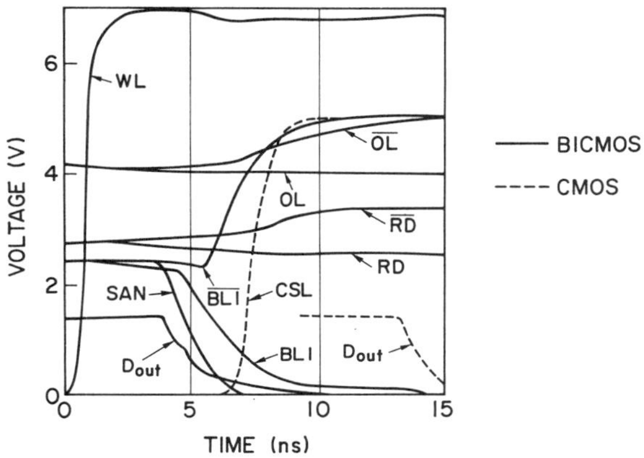

Fig. 5. Simulated waveforms of CMOS and BiCMOS DRAM's at read operation.

transferred to a bit line. A large capacitance of output line (5 pF) is driven within a few nanoseconds. Then this voltage difference between OL and $\overline{\text{OL}}$ is amplified by the next BiCMOS differential amplifier, and finally transferred to the D_{out} terminal. At the same time, when the sense amplifier activating signal SAN is clocked, the conventional CMOS flip-flop starts to restore the memory cell data. As shown in Fig. 5, the D_{out} signal appears before the bit-line restore begins.

In the case of conventional CMOS DRAM, CSL is activated after bit-line restore begins so as not to destroy the memory cell data. As a result, the D_{out} signal of CMOS DRAM is observed 9 ns later than that of BiCMOS DRAM. From this simulation results, it is clear that the new BiCMOS circuit drastically reduce the delay time from WL to D_{out}.

V. Implementation for a 1-kb BiCMOS DRAM Test Chip

The new BiCMOS circuit technology was applied to a 1-kb DRAM test chip to examine a feasibility of this technology. For the impedance-converting means the clocked CMOS inverter is employed. Fig. 6 is a photomicrograph of the test chip. In addition to the BiCMOS circuitry, CMOS DRAM circuitry is also implemented as a reference. In order to avoid the difficulties of the fabrication process technology, a conventional 2-μm BiCMOS process was used. Only two mask steps are added to the conventional CMOS process to implement the BiCMOS structures. The device parameters are given in Table I.

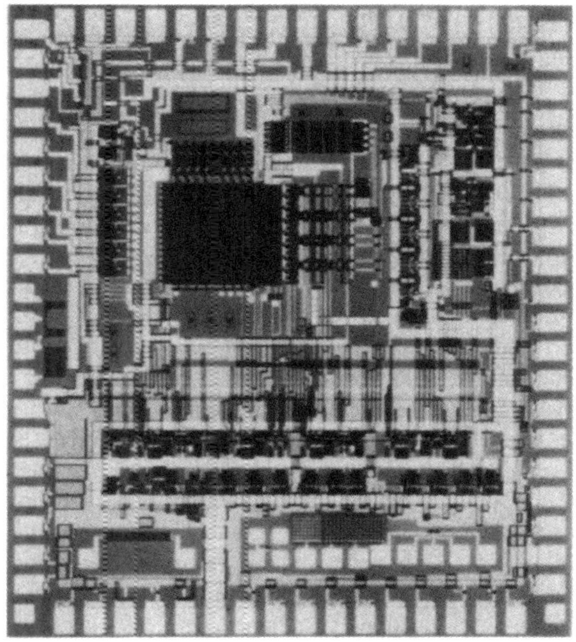

Fig. 6. Photomicrograph of the 1-kb BiCMOS DRAM test chip.

The successful write/read operation is observed and shown in Fig. 7. In addition, timings of internal main nodes were precisely measured by an electron-beam tester. The measured waveforms are shown in Fig. 8. D_{out} is activated before the bit-line restore begins. These results show that the newly introduced BiCMOS sense amplifier operates correctly. The measured delay time of 6 ns from WL to D_{out} shows a good agreement with the simulated results of Fig. 5. The RAM provides 10-ns $\overline{RAS}$ access time under typical operating condition with 5-V supply voltage. As a reference, a CMOS DRAM circuit is also implemented on the test chip. The block

TABLE I
DEVICE PARAMETERS OF 1-kb BiCMOS DRAM TEST CHIP

MEMORY CELL	PLANAR CELL	$C_B/C_S = 10$
BIPOLAR	EMITTER SIZE	2.0 x 5.0 μm
	CUT OFF FREQUENCY	4 GHz
CMOS	GATE LENGTH	2.0 μm
	THRESHOLD VOLTAGE	
	: NMOS	0.8 V
	: PMOS	-0.8 V

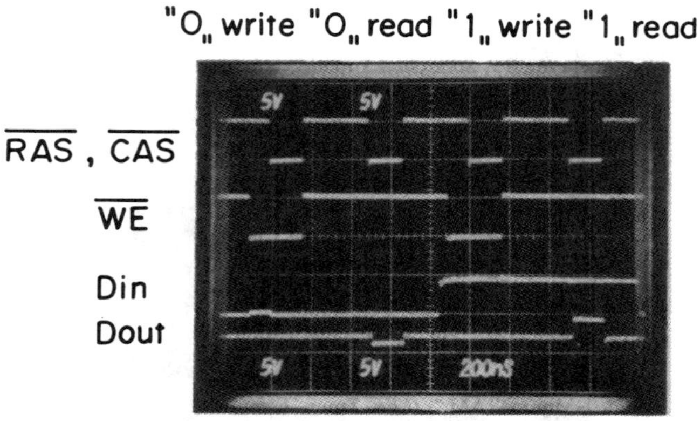

Fig. 7. Input/output waveforms of write/read operation.

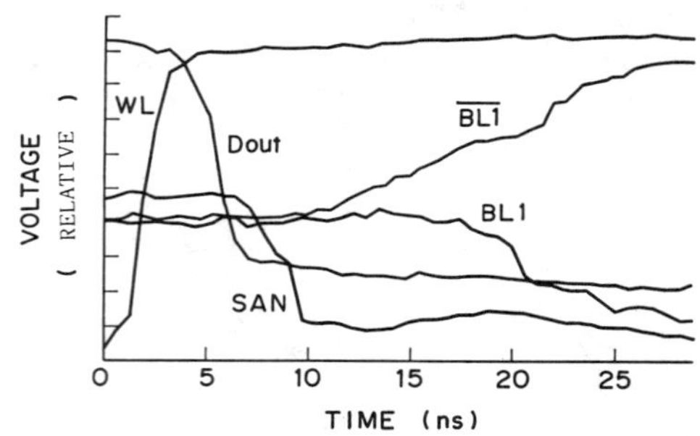

Fig. 8. The measured waveforms by an electron-beam tester.

diagram and timing chart of this CMOS DRAM circuit are shown in Fig. 9. A BL and I/O line are directly connected, when the CSL is activated. A conventional CMOS buffer circuit is used for the data transfer from I/O line to D_{out} pin. The access time of this CMOS DRAM is about 18 ns. In the CMOS DRAM case, if the BL and I/O line are separated by a clocked CMOS inverter, the access time will be reduced to 16 ns, which is still 6 ns slower than the BiCMOS one.

Internal main nodes of CMOS DRAM shown in Fig. 9 are also measured by an electron-beam tester. Measured delay times of BiCMOS DRAM and CMOS DRAM are compared in Fig. 10. As shown in Fig. 10(a), the new BiCMOS bit-line sense amplifier, which is composed of a clocked CMOS inverter, a BiCMOS differential amplifier, and a conventional flip-flop, helps reduce the access time by 4 ns. A BiCMOS differential amplifier, which is connected to the output stage of the BiCMOS differential amplifier for the bit-line sensing,

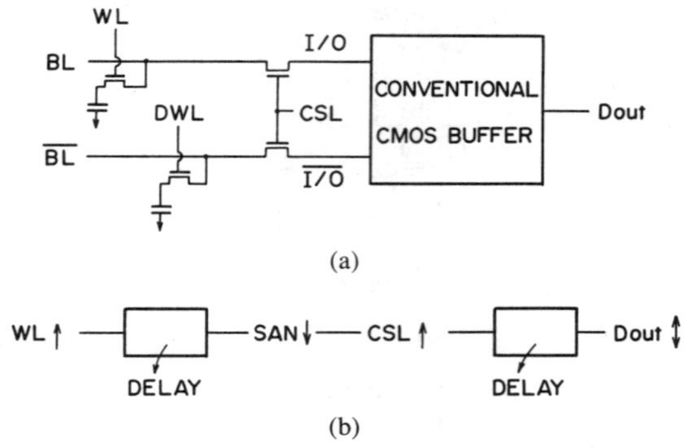

Fig. 9. (a) The block diagram and (b) the timing chart of CMOS DRAM which is implemented in the test chip.

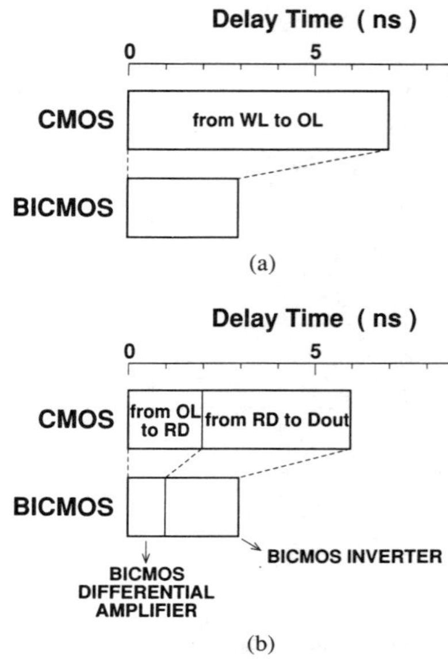

Fig. 10. The measured delay times of BiCMOS and CMOS DRAM's.

	BICMOS	CMOS
RAS ACCESS TIME	0.55	1
CHIP SIZE	1.1	1
ACTIVE CURRENT	1.1	1

- 1.2 μm PROCESS
- $C_B/C_S = 15$
- 1M words x 1b

Fig. 11. Estimated performance of 1-Mb DRAM.

and a BiCMOS inverter help reduce the access time by 3 ns. In total, the access time of BiCMOS DRAM becomes about 1/2 that of conventional CMOS DRAM.

VI. FUTURE APPLICATION TO MEGABIT DRAM's

In order to study the feasibility of the new BiCMOS circuit technology to higher density DRAM's, an estimation of the performance of a 1-Mb DRAM is performed, assuming this BiCMOS circuit technology is applied. It is also assumed that an advanced 1.2-μm process technology is used, C_B/C_S is 15, and the bit organization is 1M words $\times$ 1 b. Estimation results are shown in Fig. 11. The performance of 1-Mb BiCMOS DRAM is compared with that of conventional CMOS DRAM.

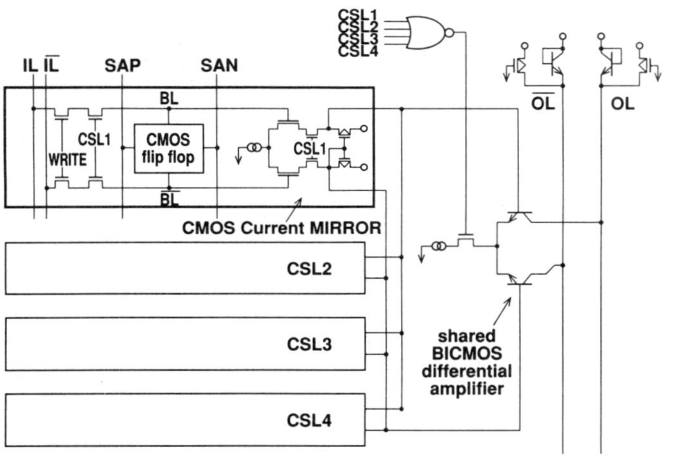

Fig. 12. Optimized BiCMOS bit-line sense amplifier for megabit DRAM's.

For 1-Mb BiCMOS DRAM's, the variation in device parameters and the precharge level of the bit-line pair become serious problems. Therefore, the CMOS current mirror circuit should be employed instead of the clocked CMOS inverter for improving the sensitivity of the new BiCMOS bit-line sense amplifier.

One of the key issue for realizing megabit BiCMOS DRAM is how to achieve the fast access time without increasing the chip size. Two candidates are considered for this purpose. First is the newly introduced BiCMOS bit-line sense amplifier, and second is direct connection with the CMOS current mirror circuit and the BiCMOS differential amplifier for I/O sensing while omitting the BiCMOS differential amplifier for bit-line sensing. The second candidate cannot be employed because of the slow bit-line sensing operation caused by the charging of the large capacitance of OL [14], [15]. Therefore, the newly introduced BiCMOS bit-line sense amplifier featured by the fast access time should be optimized for reducing the chip area penalty. The design rule of the bipolar transistor is two or three generations behind that of the MOS transistor. Thus, the BiCMOS differential amplifier cannot be easily introduced for each bit line without large area penalty. In order to overcome this problem, one BiCMOS differential amplifier for bit-line sensing should be shared with the four columns as shown in Fig. 12. By using this scheme the chip area penalty is limited to 10% compared with the CMOS DRAM.

The active current increases only 10%, because the BiCMOS bit-line sense amplifier is selectively activated by the column select signal.

The write cycle time and read cycle time will be reduced compared with the CMOS DRAM case, because the BiCMOS peripheral circuit is useful for reducing the delay time from $\overline{RAS}$ to WL. By using the BiCMOS buffer with large current drivability, the large load capacitance of address and WL wiring can be successfully driven in a short time. From this estimation, this new BiCMOS circuit technology is verified to be appropriate for use in the megabit DRAM's.

VII. Conclusion

A new BiCMOS circuit technology featuring a novel BiCMOS bit-line sense amplifier has been developed. With the small bit-line noise of 70 mV and large current drivability of the bipolar transistor, the access time can be reduced to half that of a conventional DRAM. By introducing the CMOS current mirror circuit and the shared BiCMOS differential amplifier for bit-line sensing, the fast access time will be successfully achieved with only 10% area penalty for megabit DRAM's.

Acknowledgment

The authors would like to thank F. Masuoka, A. Hojo, K. Ohuchi, F. Horiguchi, Y. Watanabe, and H. Iizuka for their helpful discussions.

References

[1] N. Lu *et al.*, "A 20-ns 128-kbit × 4 high-speed DRAM with 330-Mbit/s data rate," *IEEE J. Solid-State Circuits*, vol. 23, no. 5, pp. 1140–1149, Oct. 1988.
[2] N. Lu *et al.*, "A 22-ns 1-Mbit CMOS high-speed DRAM with address multiplexing," *IEEE J. Solid-State Circuits*, vol. 24, no. 5, pp. 1198–1205, Oct. 1989.
[3] M. Odaka *et al.*, "A 12 ns/350 mW 16Kb ECL compatible RAM," in *Proc. 11th European Solid-State Circuits Conf.*, 1985, p. 166.
[4] J. Miyamoto *et al.*, "A 1.0 μm n-well CMOS/bipolar technology for VLSI circuits," in *ISSCC Dig. Tech. Papers*, 1985, p. 244.
[5] M. Kubo *et al.*, "Prospective on BiCMOS VLSI's," *IEEE J. Solid-State Circuits*, vol. 23, pp. 5–11, Feb. 1988.
[6] G. Kitsukawa *et al.*, "An experimental 1-Mbit BiCMOS DRAM," *IEEE J. Solid-State Circuits*, vol. SC-22, pp. 657–662, Oct. 1987.
[7] Y. Kobayashi *et al.*, "Bipolar CMOS-merged technology for a high-speed 1-Mbit DRAM," *IEEE Trans. Electron Devices*, vol. 36, pp. 706–711, Apr. 1989.
[8] T. Watanabe *et al.*, "Comparison of CMOS and BiCMOS 1-Mbit DRAM performance," *IEEE J. Solid-State Circuits*, vol. 24, pp. 771–778, June 1989.
[9] T. Kawahara *et al.*, "Substrate current reduction techniques for BiCMOS DRAM," *IEEE J. Solid-State Circuits*, vol. 24, pp. 1381–1389, Oct. 1989.
[10] H. Masuda *et al.*, "A 5V-Only 64K dynamic RAM based on high S/N design," *IEEE J. Solid-State Circuits*, vol. SC-15, pp. 846–853, Oct. 1980.
[11] T. Furuyama *et al.*, "A new sense amplifier technique for VLSI dynamic RAM's," in *IEDM Tech. Dig.*, Dec. 1981, pp. 44–47.
[12] G. Sai-Halasz *et al.*, "Alpha-particle-induced soft error rate modeling," in *ISSCC Dig. Tech. Papers*, Feb. 1982, pp. 20–21.
[13] K. Takeuchi *et al.*, "Alpha-particle-induced charge transfer between n regions in high-density trench DRAM with isolated p-well structures," *IEEE Trans. Electron Devices*, vol. 37, pp. 1893–1901, Aug. 1990.
[14] S. Mori *et al.*, "A 45-ns 64-Mb DRAM with a merged match-line test architecture," *IEEE J. Solid-State Circuits*, vol. 26, pp. 1486–1492, Nov. 1991.
[15] Y. Nakagome *et al.*, "A 1.5 V circuit technology for 64 Mb DRAM's," in *VLSI Circuits Dig. Tech. Papers*, June 1990, p. 17.

Paper 3.12

Bipolar CMOS-merged Technology for a High-Speed 1-Mbit DRAM

YUTAKA KOBAYASHI, KYOICHIRO ASAYAMA, MASAYUKI OOHAYASHI, RYOICHI HORI, GORO KITSUKAWA, AND KIYOO ITOH

***Abstract*—This paper proposes a new high-performance bipolar CMOS-merged technology for a 1-Mbit DRAM. A memory cell having a twenty-times-higher soft-error immunity in its bit-line mode and a bipolar transistor having a high-drive ability (f_T = 5.2 GHz at I_c = 1.2 mA) can be realized. The fabrication process is fully compatible with the conventional CMOS DRAM and involves only three additional masking steps. An experimental 1-Mbit BiCMOS DRAM is fabricated using this technology, with a typical access time of 32 ns.**

I. Introduction

SINCE its introduction, the density of dynamic random-access memory (DRAM) circuits has increased fourfold every three years. High speed and high density have been realized primarily through the development of device minimization. However, these cannot be realized through device minimization alone [1] since current density in the device does not increase in proportion to the scaling rule due to an increase of hot-carrier effects and a decrease in breakdown voltage.

Bipolar CMOS is a promising technology for realizing high-speed, high-density DRAM's in which memory arrays are constructed with one transistor and one capacitor-type cell, and peripheral circuits are constructed with BiCMOS circuits. The bipolar transistors can drive high load capacitances such as a word line and can compensate for the lack of drive ability in CMOS circuits.

Several problems must be solved regarding the memory cell and bipolar transistor in the applications of this technology in order to realize high-speed, high-density DRAM's. In the memory cell, α-ray soft error immunity, especially for the bit-line mode, must be high because this error rate increases with decreasing cycle time. In the bipolar transistor, cutoff frequency (f_T) must be high in the high-current region because the load capacitance of such as that of the word line is higher than that of the gate array or SRAM application [2].

This paper proposes a new BiCMOS technology which can lead to a memory cell having a twenty-times-higher soft-error immunity in its bit-line mode as compared to a conventional CMOS DRAM structure and a bipolar transistor having a high-drive ability. The fabrication process is fully compatible with the conventional CMOS DRAM and involves only three-additional masking steps.

Manuscript received November 30, 1987; revised November 10, 1988.

Y. Kobayashi, K. Asayama, and M. Oohayashi are with the Hitachi Research Laboratory, Hitachi Ltd, Hitachi, Ibaraki 319-12, Japan.

R. Hori is with the Musashi Works, Hitachi Ltd, Kodaira, Tokyo 187, Japan.

G. Kitsukawa and K. Itoh are with the Central Research Laboratory, Hitachi Ltd, Kokubunji, Tokyo 185, Japan.

IEEE Log Number 8825925.

II. Structure and Fabrication Process

A cross-sectional view of the new BiCMOS DRAM structure is shown in Figure 1. A discussion of each element in this structure is given in the next section, whereas just a general outline is given here.

The fundamental structure is a twin-well formed in a 1.5-μm epitaxial layer. The memory cell is a planar cell-type and it is formed just above the p^+-buried layer in order to reduce the bit-line mode soft-error rate. To reduce a α-ray soft-error rate, a memory cell with a n^+-buried layer and a p^+-buried layer was already presented by Wordeman *et al.* [3]. However, this structure cannot reduce the bit-line mode α-ray soft-error rate because the depletion region of the bit-line mode does not touch the p^+-buried layer. In the BiCMOS DRAM, the reduction of bit-line mode α-ray soft-error rate is very important. The reduction of the soft error is discussed in Section IV. On the other hand, the p^+-buried layer is not formed under the peripheral NMOSFET, so as to ensure a low K_b. The influence of the p^+-buried layer on K_b is discussed in the next section. Bipolar transistor emitters are formed concurrently with the source and drain of the NMOS in order to keep the number of additional steps at a minimum. An outline of the fabrication sequence is shown in Figure 2. The structure is produced with only three more masking steps than the advanced 1.2-μm CMOS process.

III. Device Technology

A schematic cross-sectional view of two emitter structures in a bipolar transistor and the additional emitter fabrication-process steps necessary are shown in Fig. 3.

In general, a high-performance bipolar transistor is constructed with a poly-crystalline silicon-emitter structure [4]. However, this structure is not suitable for a BiCMOS DRAM structure because several steps, such as poly-crystalline silicon-film deposition, doping, photolithography, and etching, must be added to the conventional CMOS DRAM process in order to realize a bipolar transistor.

A simple emitter (SD emitter) structure, as shown in

Reprinted from *IEEE Trans. Electron Devices*, vol. 36, no. 4, pp. 706–711, April 1989.

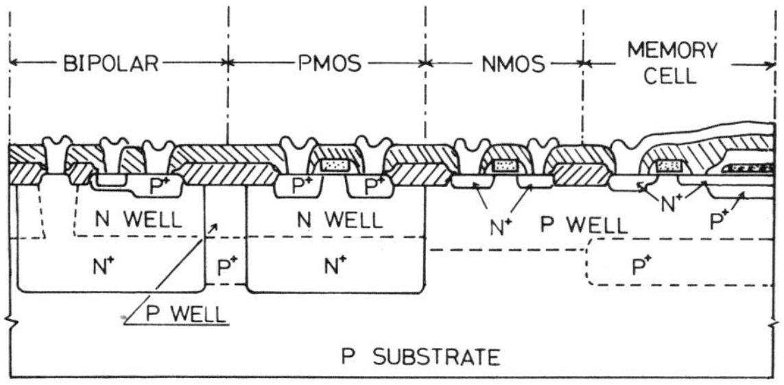

Fig. 1. Cross-sectional view of the BiCMOS DRAM.

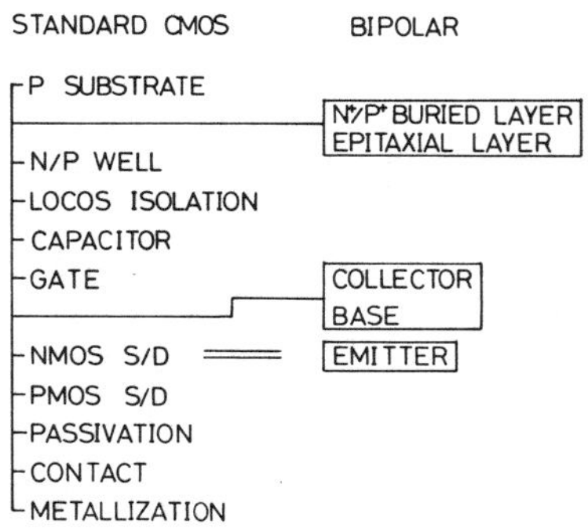

Fig. 2. Outline of the fabrication process.

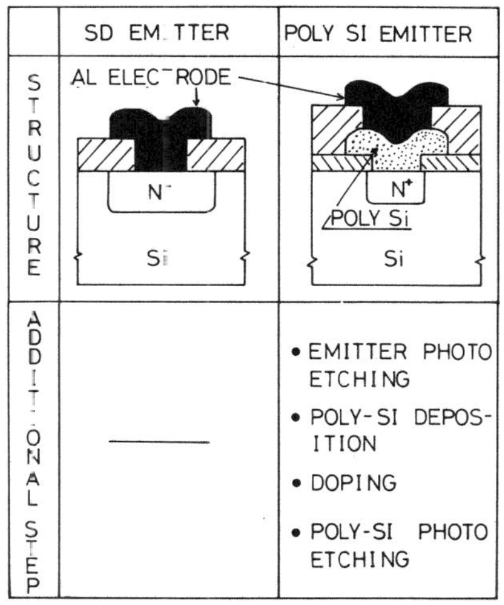

Fig. 3. Choice of emitter structure.

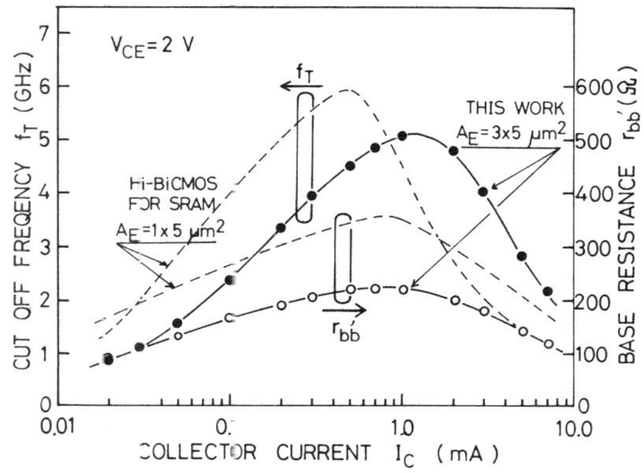

Fig. 4. f_T–I_C and r'_{bb}–I_C characteristics of the bipolar transistor.

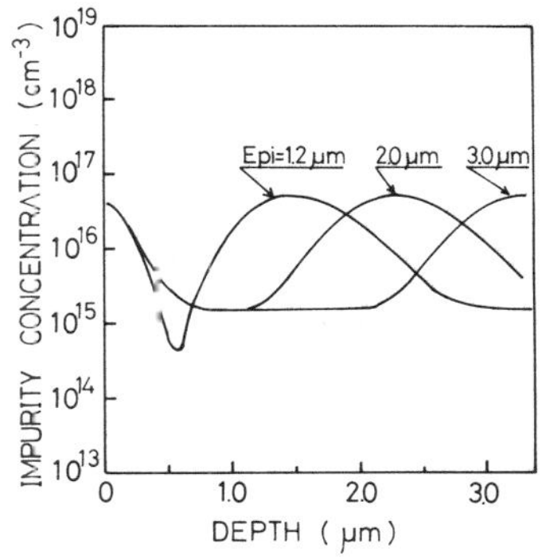

Fig. 5. Epitaxial-thickness dependency of the impurity profile in the p well.

Fig. 3, is proposed for use in the BiCMOS DRAM structure. Since it can be formed concurrently with the source and drain of an NMOSFET, this structure requires no additional process steps. However, a high-performance bipolar transistor has not yet been realized using this structure. To obtain a high cutoff frequency in the high-current region, the planar layout and the doping profile of the bipolar transistor are optimized using three-dimensional two-carrier simulations [5].

Cutoff frequency and base-resistance characteristics of the two-type bipolar transistors are shown in Fig. 4. The maximum cutoff frequency of the SD-emitter bipolar transistor is 5.2 GHz, which is slightly lower than that of the poly-crystalline silicon-emitter transistor. On the other hand, in the high-current region, its cutoff frequency is higher. Moreover, the base resistance of the SD-emitter bipolar transistor is lower than that of the poly-crystalline silicon transistor over the entire collector-current region. This result indicates that a high-performance bipolar transistor can be produced which is compatible with the conventional CMOS DRAM.

Threshold sensitivity, K_b, must be low for DRAM application because back bias is usually supplied and a high K_b prevents high-speed operation. In general, K_b is given by [6] $K_b = \sqrt{2\, \epsilon_{Si}\, q\, N_A}/C_{ox}$ where ϵ_{Si} is silicon permittivity, q is the magnitude of electronic charge, N_A is the acceptor-impurity density in the p well, and C_{ox} is gate capacitance. The equation means that K_b is affected by the impurity profile of the p-well.

Fig. 5 shows the simulation results of the impurity profile when epitaxial thickness is changed from 1.2 to 3.0 μm. The thicker the epitaxial layer, the deeper the p^+-buried layer.

Fig. 6 shows the simulation results of K_b, when the epitaxial thickness is changed from 1.2 to 3.0 μm, based on the profiles shown in Fig. 5. K_b is defined by the gradient of threshold voltage to back-bias voltage and is measured at a -3 V back-bias voltage. It appears that the changes in K_b are caused by changes in epitaxial thickness; it is decreased as the epitaxial-layer thickness increases, and is high as long as the depletion layer touches the p^+-buried layer. This result verifies that the p^+-buried layer causes an increase in K_b.

Fig. 7 shows the threshold sensitivities of two NMOS-

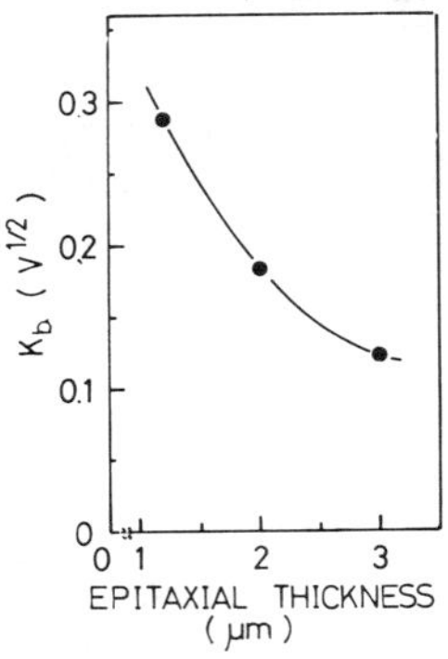

Fig. 6. Epitaxial-thickness dependency of K_b.

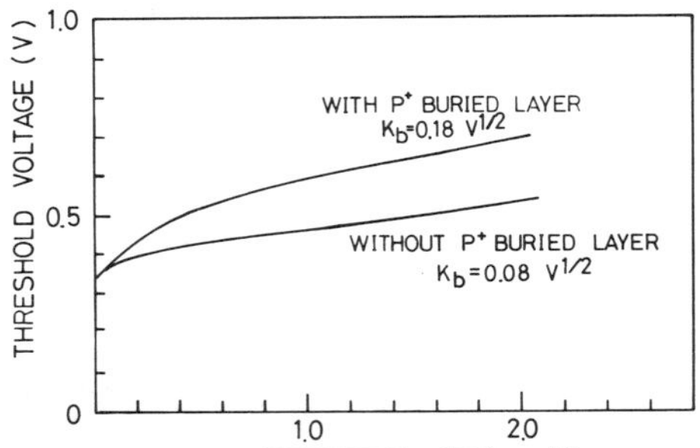

$\sqrt{|V_{bb}| + 2\Phi} - \sqrt{2\Phi}$ ($V^{1/2}$)

Fig. 7. The increase of K_b by the p^+-buried layer.

FET's, one fabricated on the p-well with a p^+-buried layer, and another fabricated on the p-well without the p^+-buried layer. With the p^+-buried layer, K_b is 0.18 $V^{-1/2}$ at a -3 V back-bias voltage. But without it, K_b is 0.08 $V^{-1/2}$ at the same back-bias voltage, which is sufficiently low. The simulation and experimental results indicate that the p^+-buried layer should not be formed under the peripheral NMOSFET in order to design a high-speed DRAM.

The characteristics of each device are summarized in Table I. In the bipolar transistor, h_{FE} is 60, the breakdown voltage between the emitter and collector is 10 V, and knee current is 5.8 mA. In the NMOSFET, the drain current is 4.5 mA at a drain voltage of 5 V, the drain breakdown voltage is 13 V, and the threshold-voltage sensitivity for back bias is 0.08 $V^{-1/2}$ when the physical gate length is 1.2 μm. In the PMOSFET, the drain current is 2.6 mA and the breakdown voltage between the source and drain is 15 V when the physical gate length is 1.5 μm. These values are sufficient for this application.

In order to characterize the drive ability of the BiCMOS, a two-input NAND BiCMOS gate-ring oscillator is fabricated. Fig. 8 shows a photograph and a circuit schematic of the BiCMOS gate used for the two-input gate-ring oscillator. The emitter size is 3 μm square. The external load capacitance (C_L) dependency on propagation delay time is shown in Fig. 9. The external load capacitance was made with a capacitance between an aluminum wire and substrate. The solid line shows the BiCMOS gate, and the dotted line shows the CMOS gate designed to occupy the same gate area as that of the BiCMOS gate.

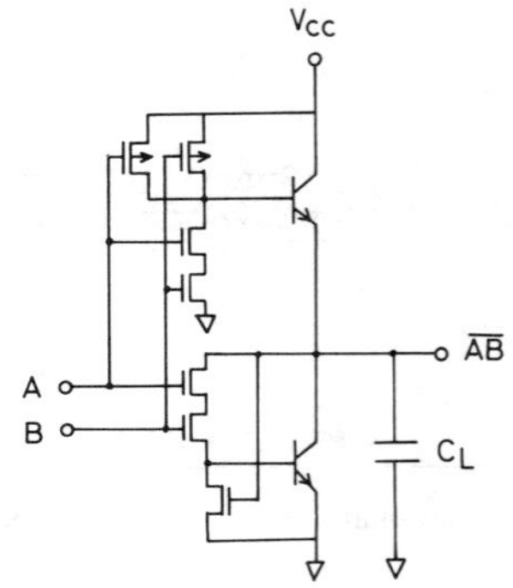

(a)

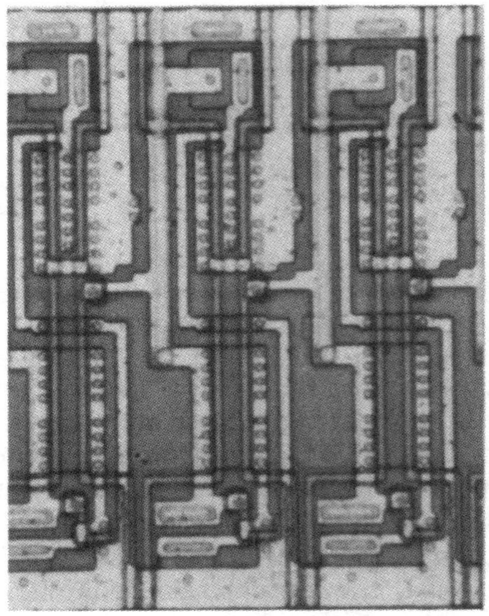

(b)

Fig. 8. (a) Schematics, and (b) the optical micrograph of a two-input NAND BiCMOS gate.

TABLE I
DEVICE CHARACTERISTICS

	Character-istics	Unit	Value
	Emitter	μm²	3 x 5
B	h_{FE}		60
	BV_{CEO}	V	10
I	BV_{CBO}	V	22
	f_T	GHz	5.2
P	I_k**	mA	5.8
	T_{ox}	Å	200
N	W/L	μm/μm	15/1.2
M	V_T	V	0.5
O	I_D***	mA	4.5
S	BV_{DS}	V	13
	K_b	V	0.08
P	V_T	V	-0.5
M O	I_D***	mA	-2.6
S	BV_{DS}	V	15

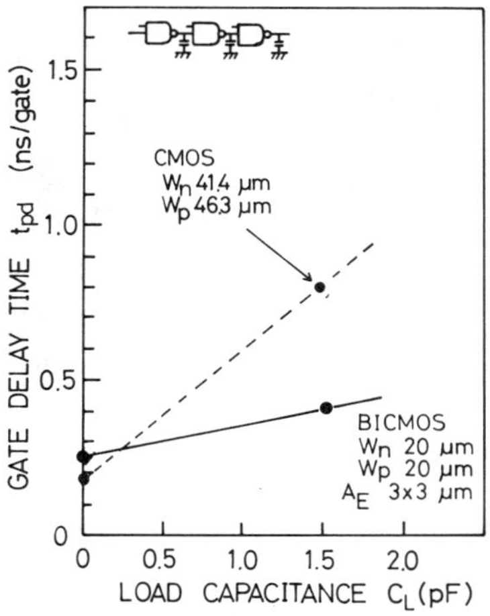

Fig. 9. Gate-delay time characteristics.

Propagation-delay time is half or one-third that of the CMOS gate, with the external load capacitance of 1.5 pF. The result means that the BiCMOS gate constructed with the simple-emitter bipolar transistor has sufficient high-drive ability.

IV. Soft-Error Immunity

The α-ray soft-error rate of the bit-line mode in DRAM's increases with decreasing cycle time. As a high-speed operation is obtained through BiCMOS technology, the α-ray soft-error rate of the bit-line mode must be reduced. A memory cell with a p^+-buried layer is proposed for reducing the soft-error rate.

The α-ray soft error is caused by collecting the charge induced by the α ray. The charge-collecting process has two components, a funneling and a diffusion processs [7]. Cross-sectional views of the conventional bit line and the new one with the p^+-buried layer are shown in Fig. 10. Due to the presence of the p^+-buried layer, the depletion layer stops at the p^+-buried layer. Without it, the depletion layer extends into the deep region. The amount of the collected charge induced by the funneling effect depends on the depletion depth; thus the funneling component of the soft-error rate can be reduced using this p^+-buried layer. The Fermi level of the p^+-buried layer is shallower than that of the p-substrate. The p^+-buried layer acts as a potential barrier for electrons and can also reduce the diffusion component of the soft error.

Fig. 11 shows the induced current and the total amount of the collected electron charge at the bit line, which is obtained by three-dimensional two-carrier simulations. The simulation conditions are as follows: An α-particle hits the bit-line junction through a 3-μm passivation layer; the incident angle is perpendicular to the silicon surface; and the energy of the α-particle is 3 MeV.

In this figure, when the p^+-buried layer is present, the time period during which the current is higher than 0.1 mA is 3.4×10^{-11} s. In its absence, the high-current period is 8×10^{-11} s. Funneling induces a high current in a short time period. This result means that the current in-

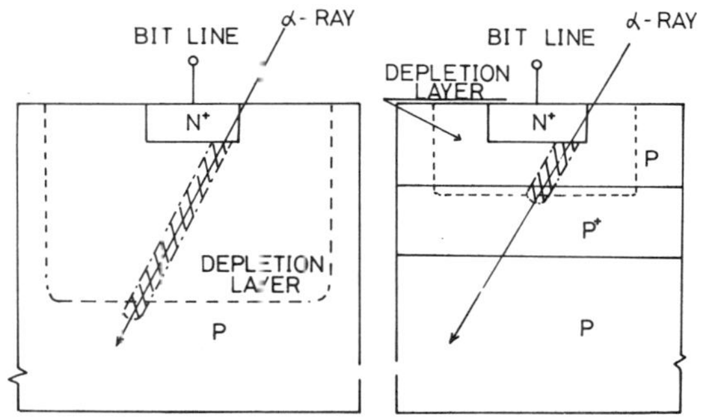

Fig. 10. Cross-sectional view of bit lines.

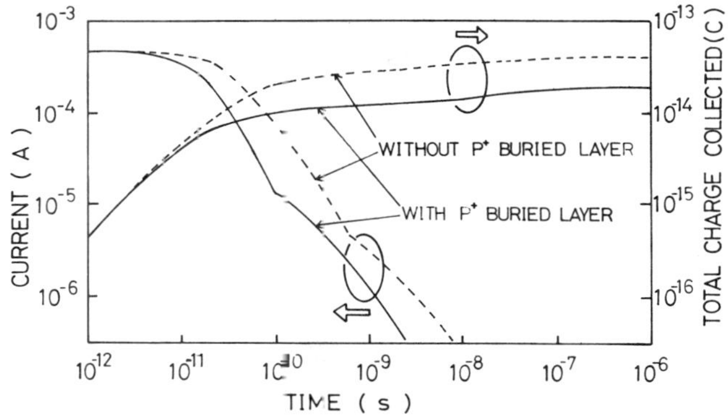

Fig. 11. Simulated results of the current and total collected charge induced by an α ray.

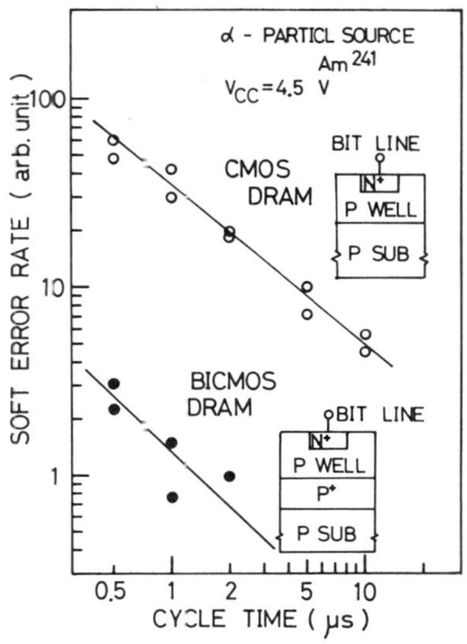

Fig. 12. Bit-line mode soft-error rate of BiCMOS/DRAM.

duced by funneling can be reduced by the p^+-buried layer. Fig. 11 also indicates that the total collected charge, including funneling and diffusion, can be reduced by the p^+-buried layer.

In order to confirm the p^+-buried layer effect for the α-ray soft-error rate, two 1-Mbit DRAM's with and without the p^+-buried layer are fabricated. Fig. 12 shows the soft-error rate obtained with them. The rate of this BiCMOS structure is less than one-twentieth of the CMOS soft-error rate. It is further concluded that this structure

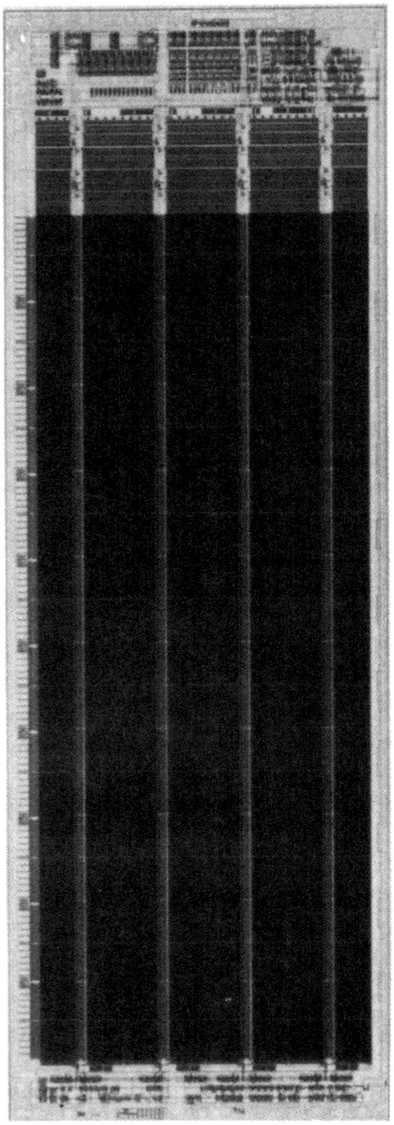

Fig. 13. Optical micrograph of the 1 Mbit BiCMOS DRAM.

maintains almost the same soft error rate even at ten times the speed.

V. An Experimental 1-MBit BiCMOS DRAM

To verify high performance using BiCMOS technology in an actual DRAM, an exnerimental 1-Mbit BiCMOS DRAM is fabricated. Bipolar transistors were used for word-line drivers, address drivers, and sense amplifiers. A micrograph of the 1-Mbit BiCMOS DRAM is shown in Fig. 13. The chip size is 5.0 × 14.9 mm, which is about 16 percent larger than a CMOS chip using the same memory cell and design rule. The increase in chip size is mainly due to widening the Al wires to achieve low wiring resistance and high speed.

Operating waveforms of this chip are shown in Fig. 14. A typical access time (t_{RAC}) from the clock signal, RAS of 32 ns, and an address access time (t_{AA}) from the column address, ADDRESS of 14 ns, are realized with a cycle time of 60 ns.

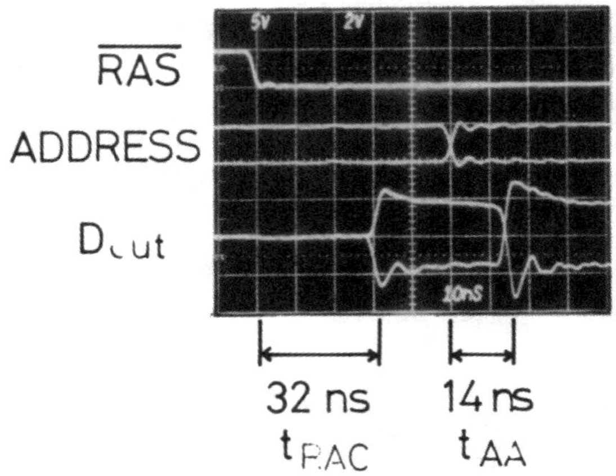

Fig. 14. Operating waveforms of the 1 Mbit BiCMOS DRAM.

VI. Conclusion

A high-speed, high-density BiCMOS DRAM technology was proposed and a 1-Mbit BiCMOS DRAM was fabricated. A high-performance bipolar transistor was fabricated through the BiCMOS DRAM process, which was shown to be compatible with a conventional CMOS DRAM. The α-ray soft-error rate of the bit-line mode was reduced to one-twentieth of the conventional CMOS DRAM using a p^+-buried layer. Typical access time of the 1-Mbit BiCMOS DRAM was 32 ns, which was twice as fast as the conventional CMOS DRAM.

Acknowledgment

The authors would like to express their sincere thanks to H. Kawamoto, T. Matsumoto, T. Ikeda, S. Shimizu, H. Miyazawa, Dr. S. Asai, Dr. M. Momma, and K. Miyata for their guidance and encouragement. They also would like to thank Y. Kawajiri, T. Watanabe, and T. Kawahara for their work in the circuit design of the 1-Mbit BiCMOS DRAM.

References

[1] H. Higuchi, G. Kitsukawa, T. Ikeda, Y. Nishio, N. Sasaki and K. Ogiue, "Performance and structures of scale-down bipolar devices merged with CMOSFETs," in *IEDM Tech. Dig.*, Dec. 1984, pp. 694–697.

[2] A. Watanabe, T. Ikeda, T. Nagano, N. Momma, Y. Nishio, N. Tamba and K. Ogiue, "High speed BiCMOS technology with buried twin well structure," in *IEDM Tech. Dig.*, Dec. 1985, pp. 423–426.

[3] M. R. Wordeman, R. H. Dennard, and G. A. Sai-Halasz, "A buried n-grid for protection against radiation induced charge collection in electric circuits," in *IEDM Tech. Dig.*, Dec. 1981, pp. 41–43.

[4] J. Graul, A. Glasl and H. Murrmann, "High performance transistors with arsenic implanted polysily emitters," *IEEE J. Solid-State Circuits*, vol. SC-11, pp. 491–495, 1976.

[5] H. Masuda, T. Toyabe, H. Shukuri, K. Ohshima, and K. Itoh, "A full three-dimensional simulation on α-particle induced DRAM soft-errors," in *IEDM Tech. Dig.*, Dec. 1985, pp. 496–499.

[6] S. M. Sze, *Physics of Semiconductor Devices.* New York: Wiley, 1969.

[7] C. M. Hsieh, P. C. Murley, and R. R. O'Brien, "A field-funneling effect on collection of α-particle-generated carriers in silicon devices," *IEEE Electron Device Lett.*, vol. EDL-2, pp. 103–105, 1981.

Paper 3.13

Substrate Current Reduction Techniques for BiCMOS DRAM

TAKAYUKI KAWAHARA, GORO KITSUKAWA, HISAYUKI HIGUCHI, YOSHIKI KAWAJIRI, TAKAO WATANABE, KIYOO ITOH, RYOICHI HORI, YUTAKA KOBAYASHI, AND TETSURO MATSUMOTO

***Abstract*—Substrate current (I_{BB}) reduction techniques useful to BiCMOS DRAM's are proposed through a comparison of the I_{BB} characteristics for a BiCMOS driver and a CMOS driver and through a clarification of the I_{BB} generation mechanism. Four key results are obtained. One is that the I_{BB} of conventional BiCMOS drivers is unacceptably high, i.e., three times larger than that of CMOS drivers. Another is that the excessive I_{BB} of BiCMOS drivers is generated by the saturation of bipolar transistors, which results from a collector voltage decrease caused by large amounts of collector current flowing through the collector saturation resistance r_{cs}. The third result concerns proposals for I_{BB} reduction techniques. One features two collector outputs terminals for avoiding the r_{cs} effect and the other features application of a low supply voltage for the base. The last result is that BiCMOS drivers using the I_{BB} reduction techniques make it possible to achieve 1-Mbit BiCMOS DRAM's with I_{BB} levels comparable to those of conventional CMOS DRAM's.**

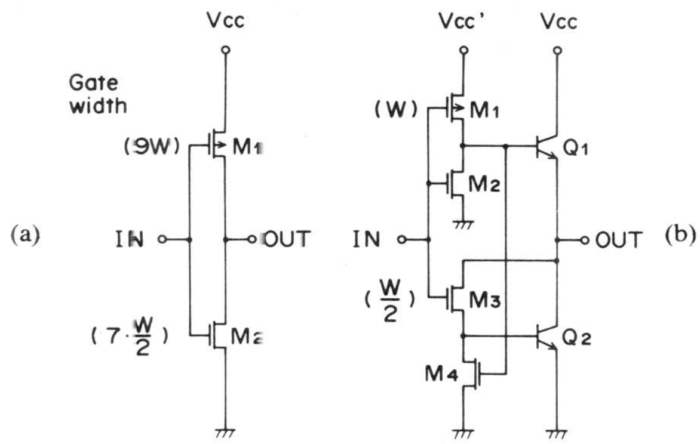

Fig. 1. Conventional (a) CMOS and (b) BiCMOS drivers.

I. Introduction

UP TO NOW, the bit density of DRAM's has been improved by a factor of 4 every three years. In line with this, 16-Mbit DRAM's have been investigated intensively [1]–[4]. However, in addition to these remarkable improvements in bit density, improvements in typical access times are needed for higher performance microprocessor applications. To meet this requirement, a 1-Mbit BiCMOS DRAM [5] has been produced. Careful consideration of the substrate current characteristics inherent to the BiCMOS drivers is key to realizing stable DRAM operation. This is because stabilization of substrate voltage supplied by the on-chip substrate-bias generator is essential for reduction of data-line capacitance. Stable substrate voltage also prevents data retention time degradation by minority-carrier injection and overcomes the undershoot problem of TTL interface circuits. If BiCMOS circuits, especially bipolar transistors, cause an unacceptable amount of substrate current I_{BB}, a stable substrate voltage cannot be provided, causing unstable memory operation. Recently, the I_{BB} increase in BiCMOS gate arrays [6] due to saturation of bipolar transistors has been shown to greatly increase the possibility of latch-up. An I_{BB} model taking bipolar transistor saturation into account has been proposed [7]. However, detailed characteristics of I_{BB} in BiCMOS drivers and the suppression of I_{BB} have not been investigated. This paper addresses those issues.

First, the I_{BB} characteristics of a BiCMOS driver and a CMOS driver are compared, showing unacceptably large amounts of I_{BB} for the BiCMOS driver. Next, the I_{BB} generation mechanism is clarified. Then, I_{BB} reduction techniques are proposed. Finally, these techniques are confirmed to be useful through experiments using a 1-Mbit BiCMOS DRAM and a 1-Mbit CMOS DRAM.

II. I_{BB} Characteristics of Conventional BiCMOS Driver

A. Comparison of I_{BB} in Conventional BiCMOS and CMOS Drivers

Fig. 1 shows schematics of the conventional CMOS and BiCMOS drivers. The driving capability of each of these drivers is equal. In this case, the BiCMOS driver permits the MOS gate widths to be h_{fe} times smaller than those of the CMOS driver because of the effective current gain h_{fe}

Manuscript received August 2, 1988; revised April 14, 1989.
T. Kawahara, G. Kitsukawa, H. Higuchi, Y. Kawajiri, T. Watanabe, and K. Itoh are with the Central Research Laboratory, Hitachi Ltd., Kokubunji, Tokyo 185, Japan.
R. Hori is with Musashi Works, Hitachi Ltd., Kodaira, Tokyo 187, Japan.
Y. Kobayashi is with Hitachi Research Laboratory, Hitachi, Ibaragi 319-12, Japan.
T. Matsumoto is with the Device Development Center, Hitachi Ltd., Oume, Tokyo 198, Japan.
IEEE Log Number 8929848.

Reprinted from *IEEE J. Solid-State Circuits,* vol., 24, no. 5, pp. 1381–1389, Oct. 1989.

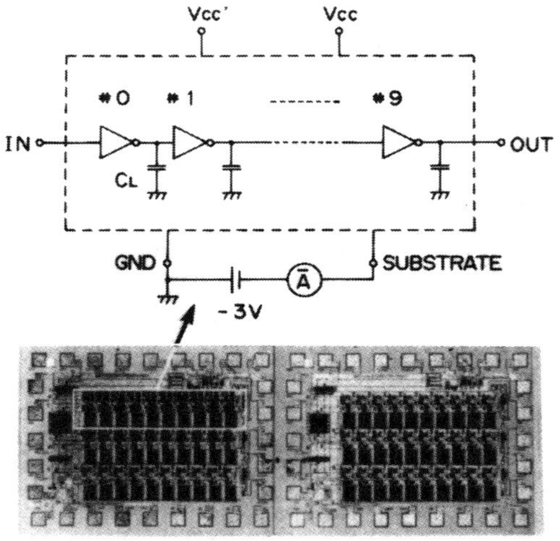

Fig. 2. Configuration of experimental chip.

TABLE I
1.3-μm DEVICE PARAMETERS

Tr's	Parameters		Values
Bipolar	Emitter width		3 μm
	Cutoff frequency,	fT	5.2 GHz
	DC current gain,	hFE	49
	Collector resistance,	rcs	46 Ω
MOS	Gate oxide thickness, Tox		20 nm
	Gate Length,	Lg	1.2 μm
	Threshold voltage	VT (n)	0.68 V
		VT (p)	-0.58 V

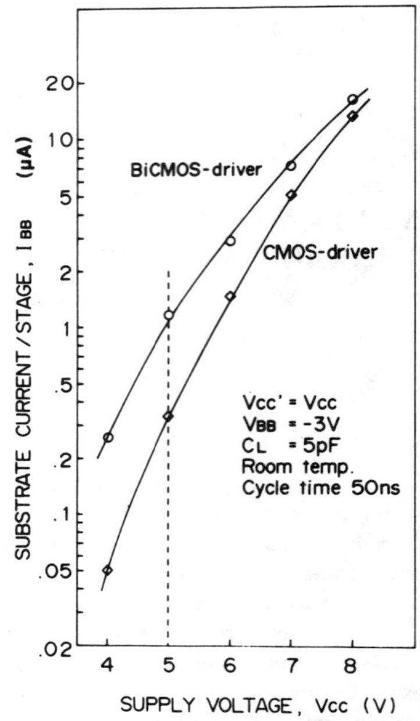

Fig. 3. Comparison of I_{BB} in BiCMOS and CMOS drivers.

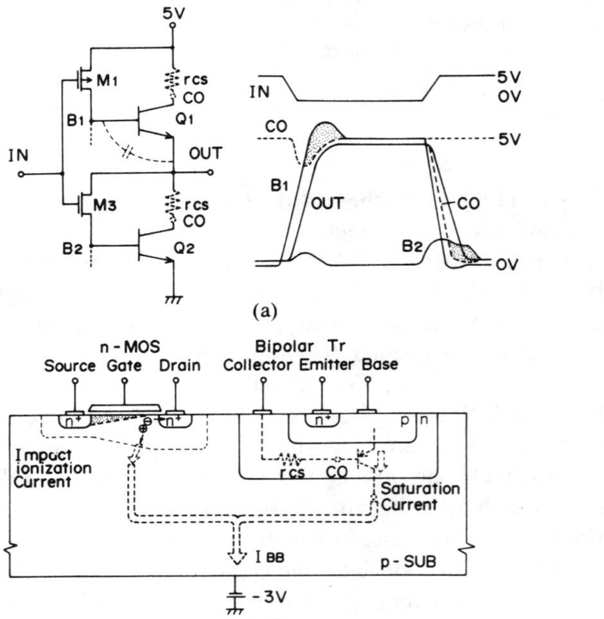

Fig. 4. I_{BB} generation mechanism in BiCMOS driver: (a) waveforms at various nodes, and (b) cross section of the transistors.

of the bipolar transistor. In the case of 1.3-μm BiCMOS technology, in order to obtain both rise and fall times (20–80 percent) of 0.7 ns, the gate-width ratios of the CMOS driver and the BiCMOS driver are designed to be 9 (pMOS) and 7 (nMOS), as shown in Fig. 1. Fig. 2 shows the configuration of an experimental chip for evaluating the I_{BB} of various kinds of drivers which will be described later. Each group of drivers consists of ten stages in series to improve the I_{BB} measurement accuracy. Table I summarizes the device characteristics for this chip, which are slightly modified from the values previously reported [8].

Fig. 3 shows I_{BB} versus supply voltage V_{cc} for the conventional CMOS and BiCMOS drivers shown in Fig. 1. Note that the I_{BB} of the BiCMOS driver is about three times larger than that of the CMOS driver. This high I_{BB} makes it impossible to realize BiCMOS DRAM's with faster cycle times due to the problem of designing an adequate on-chip substrate-bias generator. Therefore, the mechanism for generation of this excessive I_{BB} will be clarified first. Following that, proposals for I_{BB} reduction will be presented.

B. I_{BB} Generation Mechanism in BiCMOS Driver

Fig. 4 shows the schematic waveforms and cross section for the BiCMOS driver. In addition to the well-known I_{BB} generation due to the MOS impact ionization, especially for nMOS [9], there is additional I_{BB} generation due to bipolar transistor saturation.

While Q_1 charges up the output, the large collector current develops a voltage drop across the collector resistance r_{cs}. Thus, the base–collector ($C0$) junction is subject

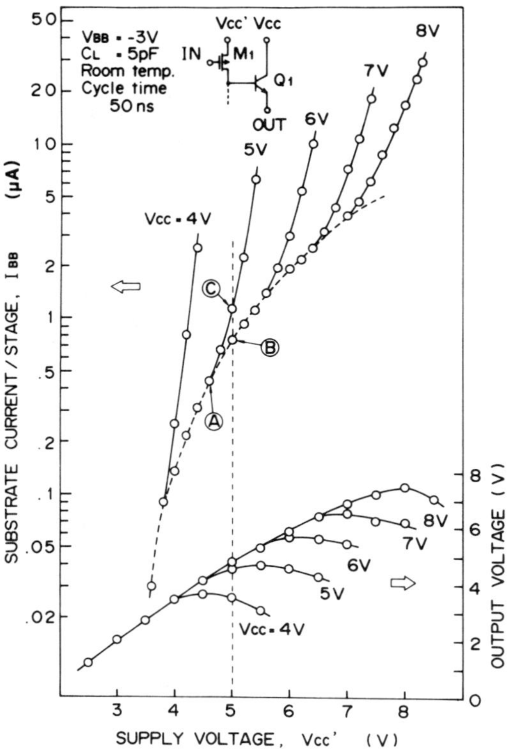

Fig. 5. I_{BB} and output voltage for $V_{cc'}$.

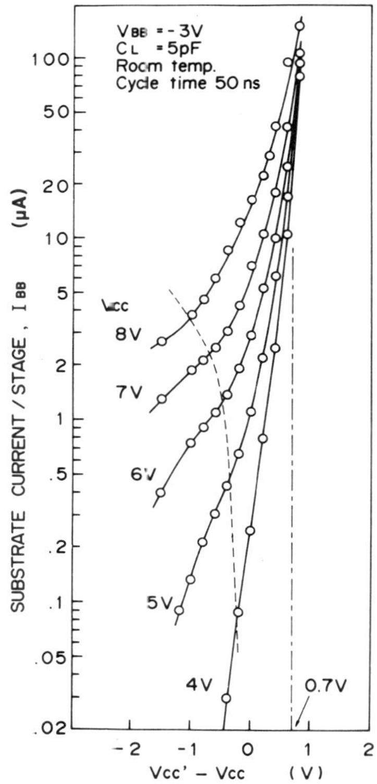

Fig. 6. I_{BB} versus $V_{cc'} - V_{cc}$.

to being forward-biased; i.e., the bipolar goes into saturation. Furthermore, a boost in the base voltage beyond V_{cc}, due to a voltage transient at the output which is coupled through the emitter–base capacitance, enhances the saturation. Shadowed areas in the figure show the forward-biased periods of Q_1 and Q_2. On the condition that this forward bias exceeds 0.7 V, the parasitic p-n-p transistor (composed of the base (p), the collector (n), and the substrate (p)) turns on, resulting in I_{BB} generation. The situation is the same for the pull-down bipolar transistor Q_2. This mechanism is demonstrated experimentally as follows.

Fig. 5 shows the I_{BB} and the output voltage of the BiCMOS driver as a function of $V_{cc'}$ with V_{cc} as a parameter. Note that V_{cc} and $V_{cc'}$ are separated as shown in Fig. 1. At first, I_{BB} increases gradually along the broken line with increasing $V_{cc'}$ for a given V_{cc}. However, I_{BB} starts increasing rapidly beyond a certain $V_{cc'}$. For example, at $V_{cc} =$ 5 V, the rapid increase starts at point Ⓐ. This exponential increase in I_{BB} for $V_{cc'}$ at each V_{cc} proves that the parasitic p-n-p transistor for transistor Q_1 turns on. Note that I_{BB} is the collector current of the p-n-p transistor. Therefore, the broken line represents the I_{BB} generated mainly by bipolar transistor Q_2 and nMOS M_3, since the I_{BB} generated by pMOS M_1 and nMOS M_2 and M_4 is negligible. This is because of the low ionization rate in pMOS and the small discharge drain current of nMOS M_2 and M_4. Thus, the difference between Ⓒ and Ⓑ at $V_{cc} = V_{cc'} = 5$ V is the I_{BB} due to the saturation of bipolar transistor Q_1.

On the other hand, the output voltage increases with $V_{cc'}$ for a given V_{cc}. However, it, in turn, decreases gradually when $V_{cc'}$ is much higher than V_{cc}. This is probably because the rise in the output level by the base stored charge itself, even after stopping the base currently supply from pMOS M_1 [10], does not occur because of base voltage clamping due to bipolar transitor saturation.

Fig. 6 shows I_{BB} versus $V_{cc'} - V_{cc}$, which is a rearrangement of the data in Fig. 5. The broken line shows the condition where I_{BB} starts increasing rapidly. It would appear at the point where $V_{cc'} - V_{cc} > 0.7$ V that the base–collector p-n junction turns on, causing a large I_{BB} flow. However, the experimental data show the rapid increase of I_{BB} even at $V_{cc'} - V_{cc} < 0$. The higher V_{cc} becomes, the larger this discrepancy becomes. Hence, it is expected that the intrinsic base and collector voltages are different from the external voltages. In particular, a large amount of current flow through r_{cs} lowers the voltage at the intrinsic collector node ($C0$ in Fig. 4), named $V_{cc\,\mathrm{eff}}$. The condition of $V_{cc'} - V_{cc\,\mathrm{eff}}$ larger than 0.7 V can occur, even when $V_{cc'} - V_{cc} < 0$. Thus the saturation is more likely to occur at high V_{cc} levels, when larger current flow is produced.

The boosting effect of base voltage beyond $V_{cc'}$ also contributes to the saturation of bipolar transistor Q_1. Fig. 7 shows the waveforms measured with an electron-beam tester [11] and simulated with circuit simulation. Note that the test condition is $V_{cc} = V_{cc'} = 5$ V. It can be seen that the Q_1 base voltage is boosted to 5.6 V during charging up of the load. In addition, the Q_2 collector voltage becomes lower than that of the base voltage during discharging of the load. These effects make the saturation deeper. Fig. 8 shows the classification of I_{BB} generated through the

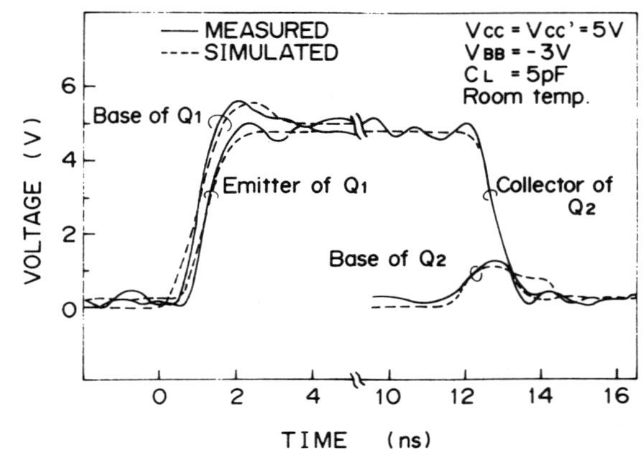

Fig. 7. Waveforms measured by electron-beam tester.

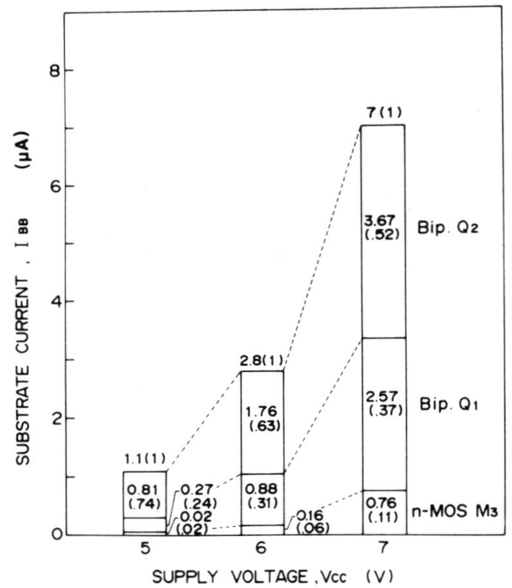

Fig. 8. Classification of I_{BB} for conventional BiCMOS driver.

above-described mechanism, using experimental data which will be described later. It is obvious that the I_{BB} generated from bipolar transistors Q_1 and Q_2 dominates. I_{BB} from these bipolar transistors comprises 98 percent of total I_{BB} with only 2 percent comprised of I_{BB} from MOS at $V_{cc} =$ 5 V.

III. Proposal of I_{BB} Reduction Techniques for BiCMOS Driver

I_{BB} reduction techniques are proposed in this section.

Fig. 9 shows the two proposed circuits (Fig. 9(b) and (c)) and the conventional circuit (Fig. 9(a)) from Fig. 1. The proposed circuits aim to inhibit the bipolar transistor saturation. Fig. 9(b) features the saturation inhibited connection (SIC) to avoid the r_{cs} effect. Since exactly the same voltage is applied at both base and collector, the bipolar transistor Q_1 never goes into saturation, resulting in I_{BB} reduction. Fig. 9(c) shows an application of the lower supply voltage V_{CL}, which is generated using the on-chip voltage limiter technique. A V_{CL} lower than V_{cc} makes it

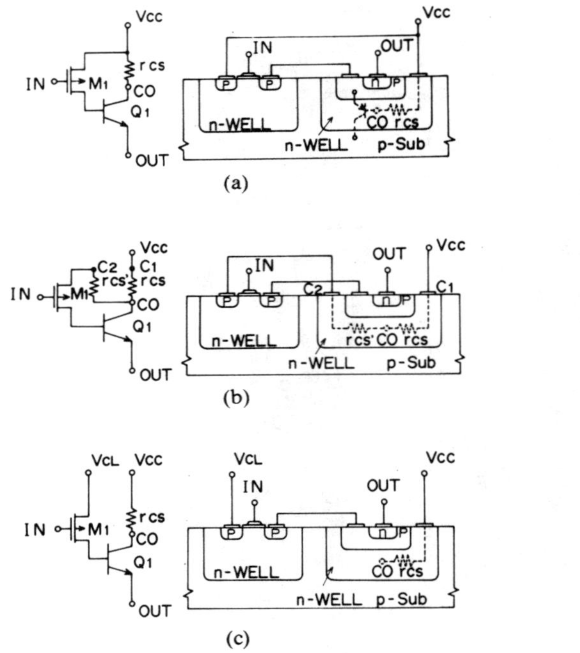

Fig. 9. I_{BB} reduction circuits and cross sections: (a) conventional, (b) proposed, and (c) proposed.

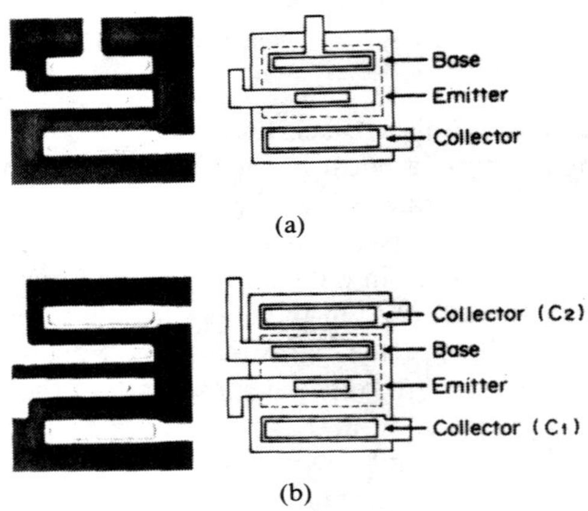

Fig. 10. Microphotograph of bipolar connections: (a) conventional connection and (b) SIC.

difficult for a bipolar transistor to go into saturation. Note that the separated n-well structure for pMOS and bipolar devices, as in previous works [12], [13], prevents unexpected problems. Otherwise the large voltage bounce in the n-p-n collector by a large current flow provides detrimental effects such as the latch-up problem. Microphotographs of the bipolar transistors for the SIC technique and the conventional circuits are shown in Fig. 10.

To evaluate the proposed techniques in terms of I_{BB} reduction, the various configurations of drivers are prepared as shown in Fig. 11. The type A driver incorporates the V_{CL} technique for the pull-up bipolar transistor Q_1. This enables the evaluation of I_{BB} reduction for Q_1 through this technique. The type B driver applies the SIC technique to the pull-down device Q_2 to evaluate I_{BB} reduction for Q_2 through this technique. The type C driver features the SIC technique applied to both Q_1 and Q_2. The

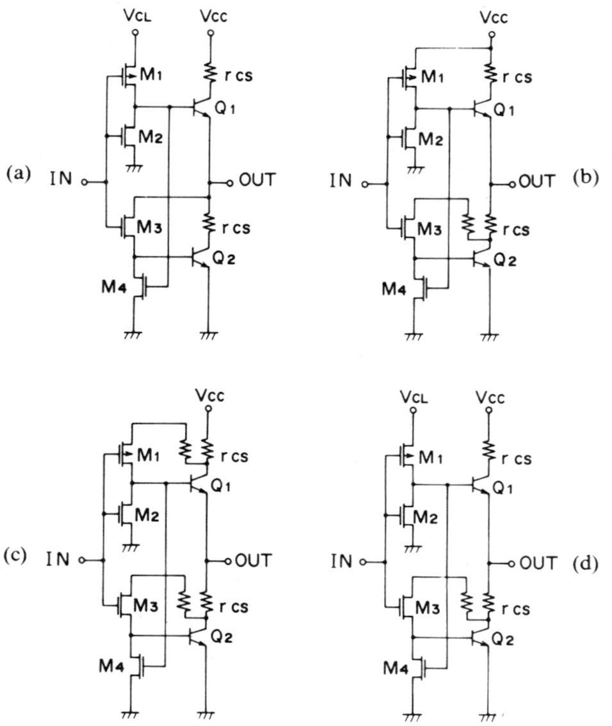

Fig. 11. Configuration of BiCMOS driver for I_{BB} reduction: (a) type A, (b) type B, (c) type C, and (d) type D.

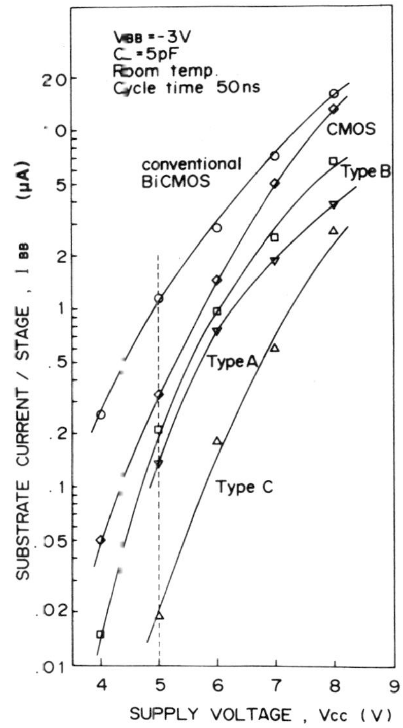

Fig. 12. I_{BB} versus V_{cc} for various drivers. In type A, $V_{CL} = V_{cc} - 1$ V.

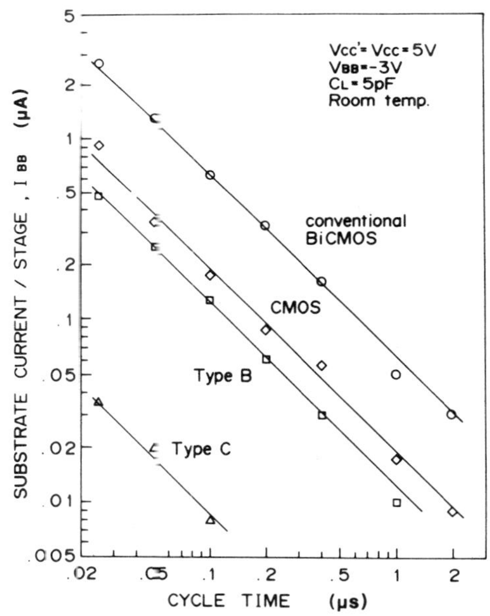

Fig. 13. I_{BB} versus cycle time.

I_{BB} reduction effect by SIC technique can be clarified by comparison with the type B driver. The type D driver takes full advantage of the I_{BB} reduction techniques, since both techniques are applied to Q_1 and Q_2. In spite of a high-level loss in the output swing, the V_{CL} technique still achieves a 23-percent delay time and 28-percent lower power dissipation compared with a CMOS driver with a supply voltage of 4 V, as shown in [14]. On the other hand, for the SIC technique, a transient lowering of the Q_1 base voltage due to a series resistance $r_{cs} + r_{cs'}$ of 92 Ω causes a slight reduction of speed. In fact, the type C driver provides only a 5-percent loss in speed compared to the conventional BiCMOS driver. Each driver has the same driving capability as shown in Fig. 1. The implementation is as shown in Fig. 2.

Fig. 12 shows the measured I_{BB} as a function of V_{cc} for the type A, type B, and type C drivers, compared with that of the conventional BiCMOS and CMOS drivers. The experimental procedure is the same as in Section II-A. At $V_{cc} = 5$ V and $V_{CL} = 4$ V, the type A driver generates an I_{BB} approximately 1/8 that of the conventional BiCMOS driver and about 1/2 that of the CMOS driver. At $V_{cc} = 5$ V, the type B driver also reduces I_{BB}, to about 1/5 that of the conventional BiCMOS driver and about 2/3 that of the CMOS driver. The type C driver reduces I_{BB} drastically, showing I_{BB} levels as low as about 1/50 of that of the conventional BiCMOS driver. At $V_{cc} = 5$ V, the ratio of the type C driver I_{BB} to the CMOS driver I_{BB} is about 1/15. Based on the large reductions in I_{BB} seen with the SIC technique, it was originally expected that the I_{BB} of the type C driver would be 1/7 that of the CMOS driver. This fraction corresponds to the gate-width ratio for the I_{BB} contributing MOS's (M_3 in the type C driver and M_2 in the CMOS driver in Fig. 1). The discrepancy between the expected reduction and the actual reduction results from the fact that the output voltage lowering due to the V_{BE} drop makes the I_{BB} of nMOS M_3 lower. In fact, considering the resultant output amplitude of 4.6 V (shown in Fig. 7) which equals the M_3 drain voltage, an I_{BB} ratio of 1/7 is obtained at the same drain voltage of 4.6 V for the above transistors, M_3 and M_2. This shows that the

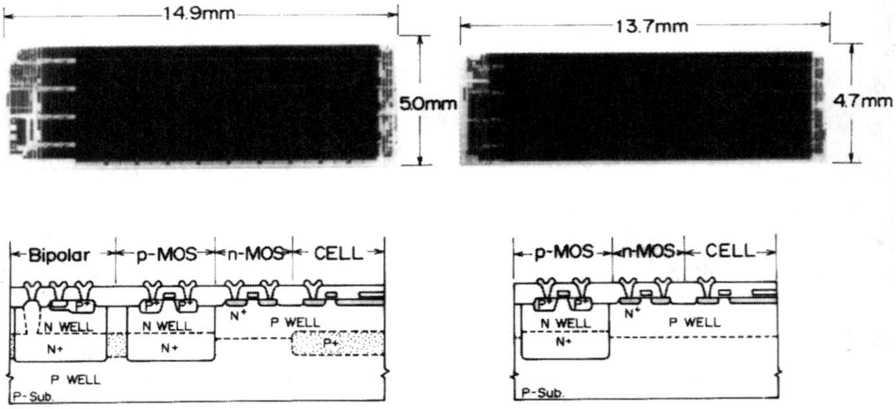

Fig. 14. 1.3-μm, 1-Mbit chip microphotographs and device structures.

bipolar transistor did not contribute to I_{BB}. Hence, M_3 is only one determinant of I_{BB}. The I_{BB} of the type D driver can also be expected to be generated by M_3 in spite of the shortage of data. This will be proven through the 1-Mbit BiCMOS DRAM experiment described later.

Fig. 13 shows the cycle-time dependence of I_{BB} for type B and type C drivers compared with the conventional BiCMOS and CMOS drivers. This is evidence that the I_{BB} of each driver is generated only during transient operation.

IV. Application to 1-Mbit DRAM

In this section, the I_{BB} reduction techniques proposed in the previous section are shown to be useful through a comparison of the 1-Mbit BiCMOS DRAM with a 1-Mbit CMOS DRAM.

Fig. 14 shows chip microphotographs and device structures [5] of BiCMOS and conventional CMOS 1-Mbit DRAM's. Identical memory cells of 3.4×10.6 μm^2 and identical design rules of 1.3 μm are used. The BiCMOS chip size measures 5.0×14.9 mm^2, which is about 16 percent larger than the CMOS chip. The increase in chip size is mainly due to widening the Al wires to low wiring resistance which offers high speed. Bipolar transistor emitters are formed concurrently with the source and drain layers of nMOS. As a result, the structure is produced with only three additional masking steps more than the CMOS process [5], [8]. The characteristics of the bipolar transistor and the MOS of the peripheral circuitry are the same as shown in Table I.

Fig. 15 shows the chip architecture of the BiCMOS DRAM. The memory array is constructed only with CMOS and is divided into four 256-kbit subarrays. The proposed I_{BB}-reduction BiCMOS drivers (type C, D), as well as the CMOS driver, are used for the peripheral circuitry to ensure high speed and low I_{BB} operations. The on-chip voltage limiter is also implemented for I_{BB} reduction, as well as for powerful reduction with a V_{CL} of 4 V. Note that two kinds of power supplies, V_{CL} and V_{CC}, are used for the CMOS drivers. The V_{CH} is a constant voltage of 7 V driving the word line. The chip architecture of CMOS

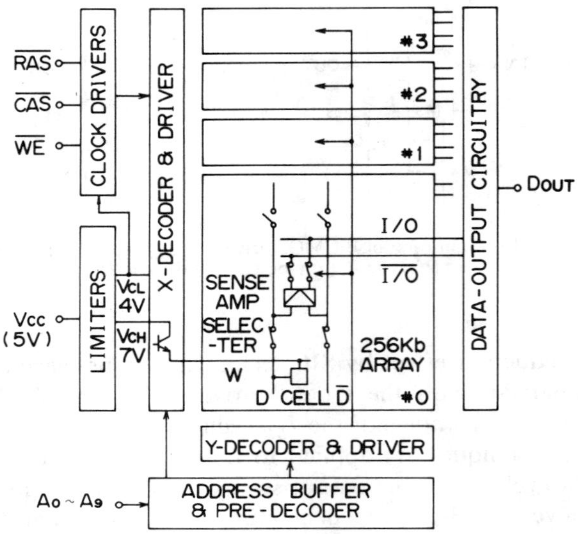

Fig. 15. Chip architecture of BiCMOS DRAM.

DRAM is the same except for the on-chip voltage limiter. A TTL interface 1-Mbit BiCMOS DRAM having a 36-percent faster access time and 27-percent lower dissipation compared to a CMOS DRAM is obtained, as shown in [14].

Fig. 16 shows cycle time versus I_{BB} generated from the peripheral circuitry. Total I_{BB} of the whole chip is also shown. Measurements are performed under an external supply substrate bias voltage V_{BB} of -3 V, while stopping the operation of the on-chip substrate-bias generator. I_{BB} from the peripheral circuitry is measured with memory array associated clocks turned off. The I_{BB} of the BiCMOS DRAM is 0.4 μA and that of CMOS DRAM is 1.1 μA at a cycle time of 400 ns with an I_{BB} ratio of 0.36. This improvement in I_{BB} is explained by the data shown in Fig. 12 and the total nMOS gate width in the peripheral circuitry as follows.

The transistors contributing to I_{BB} are M_2's in CMOS drivers and M_3's in type C and D drivers, since no contribution is made by the bipolar transistor anymore. The amount of their contributions, however, depends on

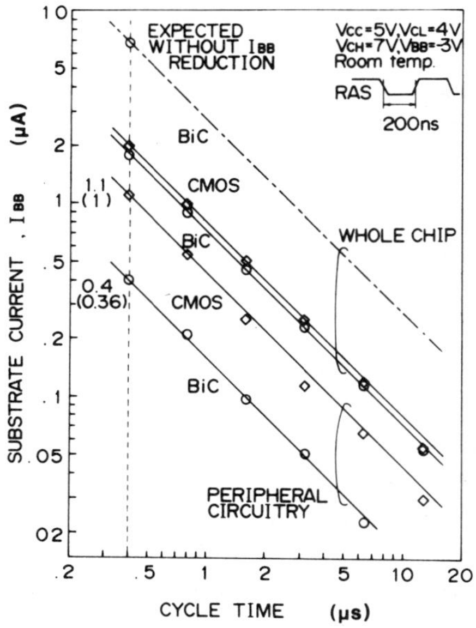

Fig. 16. I_{BB} versus cycle time for 1-Mbit DRAM. Other TTL inputs are fixed at high levels.

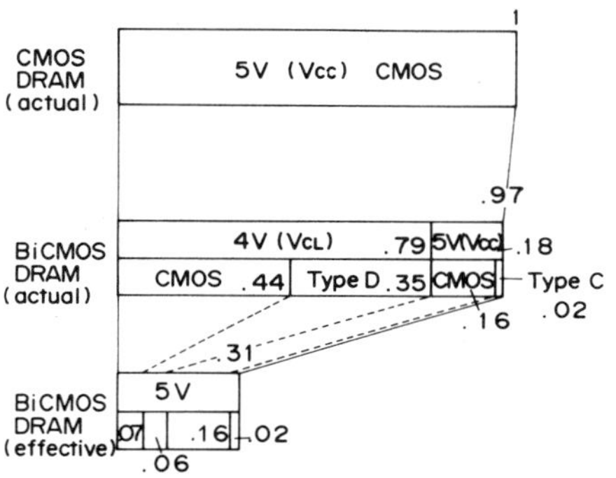

Fig. 17. The gate width of nMOS contributing to I_{BB} in the peripheral circuitry of a 1-Mbit DRAM.

their gate widths and their operating voltage: V_{CL} for type D and some CMOS drivers and V_{cc} for type C and other CMOS drivers. Fig. 17 shows the total nMOS gate-width comparison. The BiCMOS DRAM's gate width is 0.97, normalizing to that of the CMOS DRAM. Note that BiCMOS DRAM operates at two kinds of voltages. Therefore, the comparison must be made at a given voltage of 5 V. Under this condition, the effective (virtual) total gate width of BiCMOS DRAM at 5 V is 0.31, as shown in the figure. This is derived from the experimental result in Fig. 12: the gate width of nMOS operating at 4 V (V_{CL}) in type D and some CMOS drivers is equivalent to one-sixth of that of 5 V (V_{cc}) in type C and other CMOS drivers, for an equal amount of I_{BB}. Thus the I_{BB} ratio of 0.31 based on the gate-width comparison is in good agreement with the measured I_{BB} ratio of 0.36. This implies that the I_{BB} of type D drivers as well as type C drivers consists of only that from nMOS, as expected.

In spite of a remarkable difference in the peripheral I_{BB}, only a slight difference in the whole chip I_{BB} is obtained. Moreover, there is the large amount of whole I_{BB} for both

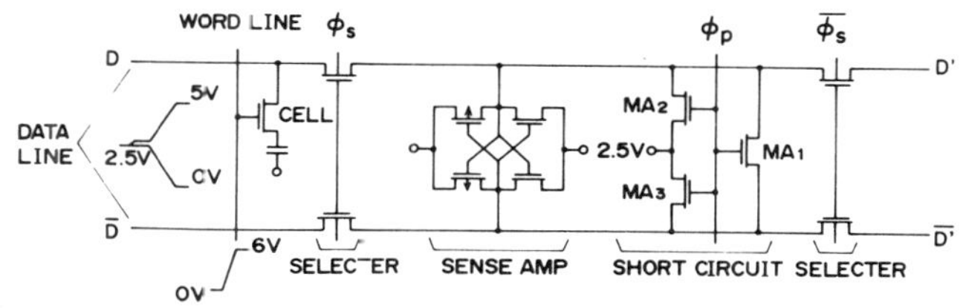

Fig. 18. Array-associated circuits.

BiCMOS DRAM and CMOS DRAM, implying that the array-associated circuitry I_{BB} is dominant. The details will be discussed later. Without the I_{BB} reduction techniques, however, an unacceptably large whole I_{BB} is expected, as shown with a dashed line in Fig. 12. The expected whole I_{BB} is 3.5 times larger than that with I_{BB} reduction techniques since the conventional BiCMOS driver makes the peripheral I_{BB} 50 times larger, as shown in Fig. 12.

NMOS MA_1's in the short circuit which is shown in Fig. 18 are responsible for the array-associated circuit I_{BB}. This is because a large number of MA_1 devices, as many as 2048, start operating simultaneously under such a high drain–source voltage as 5 V. Other circuits do not contribute to the I_{BB} generation, since they operate with a sufficiently small drain–source voltage of 2.5 V. The slight difference in the whole chip I_{BB} between two chips results from the difference in the gate width of nMOS MA_1 between two chips. The gate width in BiCMOS DRAM is twice as large as that in CMOS DRAM to obtain high speed. Hence, this offsets the large difference in peripheral circuitry I_{BB}.

It is concluded that a BiCMOS DRAM having sufficiently small I_B, which is comparable to that of CMOS DRAM, has been achieved through I_{BB} reduction techniques. Hence, a conventional on-chip substrate-bias generator for CMOS DRAM is applicable to BiCMOS DRAM. The on-chip substrate-bias generator can supply a sufficiently large current to the substrate even at a high-speed memory cycle realized exclusively by BiCMOS technology. This is achieved by adjusting pumping frequency in the on-chip substrate-bias generator in accordance with the memory cycle time [15].

V. Conclusions

The results are summarized as follows.

1) The I_{BB} of conventional BiCMOS drivers is unacceptably high: three times larger than that of CMOS drivers.

2) The excessive I_{BB} of BiCMOS drivers is generated by the saturation of bipolar transistors, which results from the collector voltage drop developed by large amounts of collector current flow through the r_{cs}.

3) SIC and V_{CL} techniques to reduce I_{BB} have been proposed and proven to be useful. They inhibit the saturation of bipolar transistors, reducing the I_{BB} of BiCMOS

drivers down to 1/50 of that of conventional BiCMOS drivers.

4) BiCMOS drivers using I_{BB} reduction techniques make it possible to achieve 1-Mbit BiCMOS DRAM's with I_{BB} levels comparable to those of conventional CMOS DRAM's.

Acknowledgment

The authors would like to thank H. Kawamoto, A. Endo, and T. Masuhara for their continuous encouragement, and K. Miyata, M. Momma, T. Ikeda, M. Ohayashi, H. Miyazawa, and K. Asayama for their contributions to process development. They would also like to thank H. Todokoro for his help in using the electron-beam tester, and S. Kato, Y. Ouchi, H. Tsukada, and K. Mihashi for their technical support.

References

[1] M. Aoki *et al.*, "An experimental 16Mbit DRAM with transposed data-line structure," in *ISSCC Dig. Tech. Papers*, Feb. 1988, pp. 250–251.

[2] S. Watanabe *et al.*, "An experimental 16Mbit CMOS DRAM chip with a 100 MHz serial read/write mode," in *ISSCC Dig. Tech. Papers*, Feb. 1988, pp. 248–249.

[3] M. Inoue *et al.*, "A 16 Mb DRAM with an open bit-line architecture," in *ISSCC Dig. Tech. Papers*, Feb. 1988, pp. 246–247.

[4] T. Mano *et al.*, "Circuit technologies for 16Mb DRAMs," in *ISSCC Dig. Tech. Papers*, Feb. 1987, pp. 22–23.

[5] G. Kitsukawa *et al.*, "An experimental 1-Mbit BiCMOS DRAM," *IEEE J. Solid-State Circuits*, vol. SC-22, pp. 657–662, Oct. 1987.

[6] A. R. Alvarez, *et al.*, "Second generation BI-CMOS gate array technology," in *Proc. IEEE 1987 Bipolar Circuit & Technology Meeting*, pp. 113–117.

[7] S. Inohira *et al.*, "Merged bipolar transistor model including the substrate current," I.E.C.E. Japan, Tech. Rep., SDM87-95, pp. 57–62, Oct. 1987.

[8] Y. Kobayashi *et al.*, "Bipolar CMOS merged structure for high speed M bit DRAM," in *IEDM Tech. Dig.*, Dec. 1986, pp. 802–804.

[9] E. Takeda *et al.*, "Submicrometer MOSFET structure for minimizing hot-carrier generation," *IEEE Trans. Electron Devices*, vol. ED-29, pp. 611–618, Apr. 1982.

[10] A. Watanabe *et al.*, "Proposal of high speed bipolar–CMOS merged VLSI technology and its future prospect," *J. I.E.I.C.E. Japan*, vol. J70-C, pp. 1115–1123, Aug. 1987.

[11] H. Todokoro *et al.*, "Electron beam tester with 10 ps time resolution," in *Proc. Int. Test Conf.*, Sept. 1986, pp. 600–606.

[12] M. Odaka *et al.*, "A 12ns/350mW 16Kb ECL compatible RAM," in *ESSCIRC Abstr.*, Sept. 1985, pp. 166–176.

[13] K. Ogiue *et al.*, "A 13ns/500mW 64Kb ECL RAM," in *ISSCC Dig. Tech. Papers*, Feb. 1986, pp. 212–213.

[14] T. Watanabe *et al.*, "Comparison of CMOS and BiCMOS 1-Mbit DRAM performance," *IEEE J. Solid-State Circuits*, vol. 24, pp. 771–778, June 1989.

[15] K. Sato *et al.*, "A 20ns static column 1Mb DRAM in CMOS technology," in *ISSCC Dig. Tech. Papers*, Feb. 1985, pp. 254–255.

Paper 3.14

Deep-Submicrometer BiCMOS Circuit Technology for Sub-10-ns ECL 4-Mb DRAM's

Takayuki Kawahara, *Member, IEEE,* Yoshiki Kawajiri, Goro Kitsukawa, Kazuhiko Sagara, *Member, IEEE,* Yoshifumi Kawamoto, Takesada Akiba, Shisei Kato, Yasushi Kawase, and Kiyoo Itoh, *Senior Member, IEEE*

***Abstract*—A 0.3-μm sub-10-ns ECL 4-Mb BiCMOS DRAM design is described. The results obtained are: 1) a V_{CC} connection limiter with a BiCMOS output circuit is chosen due to ease of design, excellent device reliability, and layout area; 2) a mostly CMOS periphery with a specific bipolar use provides better performances at high speed and low power; 3) the direct sensing scheme of a single-stage MOS preamplifier combined with a bipolar main amplifier offers high speed; and 4) the strict control of MOS transistor parameters has been proven to be more important in obtaining high speed-DRAM's, based on the 4-Mb design.**

I. Introduction

The density of DRAM's has quadrupled every three years since their introduction almost 20 years ago, currently reaching a level of 64-Mb DRAM's (for example, [1]). In addition to such a remarkable increase in bit density, rapid progress has also been made in improving speed. As a result, TTL 4-Mb DRAM's with access times as fast as 17 ns [2], [3] have been reported. However, a recently developed ECL interface 4-Mb BiCMOS DRAM [4] with a sub-10-ns access time is a more outstanding achievement. This ECL DRAM exemplifies the DRAM's high-speed potential, by keeping abreast of the high-speed developments in SRAM's while simultaneously maintaining the DRAM's high-density advantages. Moreover, it enables the use of new circuits obtained through deep-submicrometer BiCMOS technology. These circuits are ECL I/O buffers suitable for DRAM operation, and the overdrive rewrite amplifier which enables a high-speed cycle time of 16 ns to be achieved. A major concern here, however, is the design of on-chip voltage limiters which ensure the reliability of BiCMOS devices as small as 0.3 μm even when using an ECL supply voltages as high as 4.5 V. Such an approach based on limiters remains unexplored even though it could provide a possible approach to future BiCMOS LSI's. Besides, the reason why mostly CMOS periphery was chosen in this DRAM has not been clarified. The direct sensing approach essential to high-speed DRAM's is another concern. The essential differences, however, have not been clarified even though various approaches have been proposed so far. In sub-10-ns regions, the dependence of access time on device parameters is also a prime concern in the design of future high-speed devices.

This paper describes circuit technology by emphasizing these four major areas of concern. First, after two alternative limiter configurations are compared, the limiter and its output circuit are discussed in terms of reliability. Second, a mostly CMOS periphery is compared with a mostly bipolar periphery in terms of speed and power. Third, differences in speed between two direct sensings of the CMOS and bipolar amplifier scheme are discussed. Finally, the dependences of access time on design parameters are discussed using an actually designed 4-Mb chip.

II. Voltage Limiter Configuration

Fig. 1 shows two alternative limiter connections for the ECL BiCMOS DRAM: limiters directly connected to V_{CC} terminals (V_{CC} connection) [4] and those connected to V_{EE} terminals as well as V_{CC} terminals (V_{EE} connection). The ECL BiCMOS DRAM itself features the combination of a MOS DRAM core with specifically used bipolar circuits [4]. In this scheme, the following internal voltage V_L's are necessary: a V_{CL} for the CMOS driver, a V_{CH} for the word-line driver, a V_{DL} to obtain a high level on the data lines, and a V_{EL} to obtain a low level on the data lines in the V_{EE} connection.

In the design of the limiter, the number of limiters dedicated to the data line is important even though the total number of limiters is the same for both. This is because the charging and discharging of heavy data-line capacitive loads make it difficult to regulate the limiter output voltage well. In the V_{CC} connection, it is achieved solely with a single well-regulated V_{DL} voltage, while in the V_{EE} connections, two different limited V_{DL} and V_{EL} voltages are needed. In addition, the different functions for the limiter, the current sources for the L'#, and a current sink for the L''# require a complicated design for the V_{EE} connection. Hence, a V_{CC} connection is preferable. Based on the V_{CC} connection limiters, the following internal voltages were chosen: 4.5 V for the bipolar circuits; 2 V for the CMOS drivers to regulate device velocity saturation and power

Manuscript received August 27, 1991; revised November 8, 1991.

T. Kawahara, Y. Kawajiri, G. Kitsukawa, K. Sagara, Y. Kawamoto, and K. Itoh are with the Central Research Laboratory, Hitachi Ltd., 1-280, Higashi-koigakubo, Kokubunji, Tokyo 185, Japan.

T. Akiba, S. Kato, and Y. Kawase are with Hitachi Device Engineering Corporation, Ltd., Mobara, Chiba 297, Japan.

IEEE Log Number 9106281.

Reprinted from *IEEE J. Solid-State Circuits,* vol. 27, no. 4, pp. 589–596, April 1992.

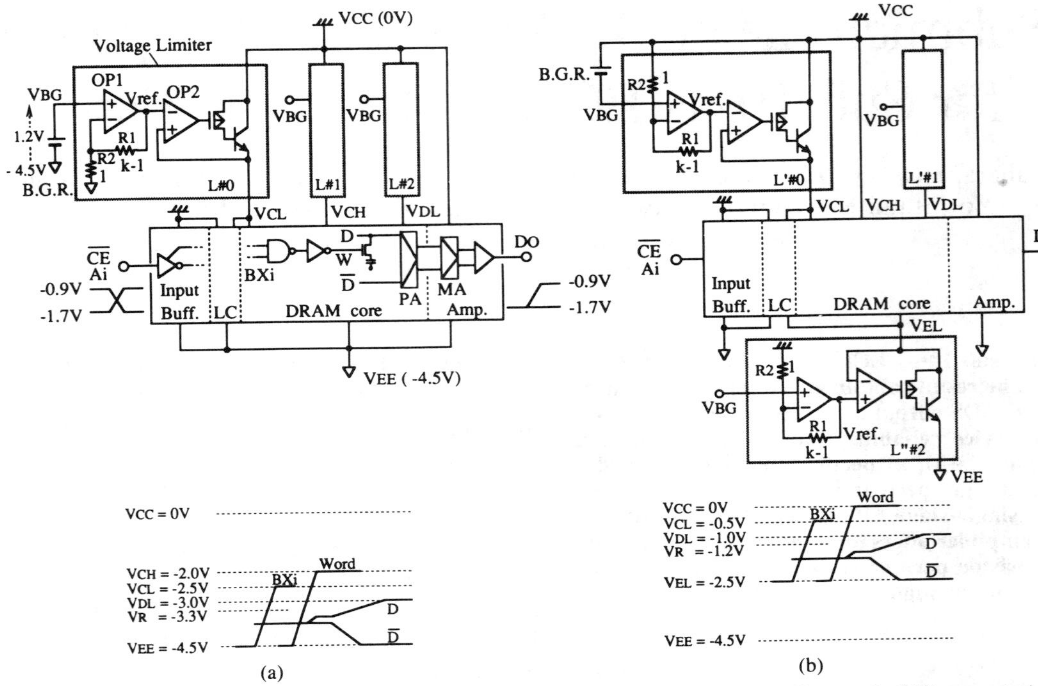

Fig. 1. Comparison of voltage generation scheme for the limiters: (a) limiter connection to V_{CC} terminal (V_{CC} connection); and (b) limiter connection to V_{EE} and V_{CC} terminals (V_{EE} connection). LC: level converter, PA: preamplifier, MA: main amplifier. $k = (V_{CL} - V_{EE})/(V_{BG} - V_{EE})$.

TABLE I
COMPARISONS OF OUTPUT DRIVING SCHEME OF VOLTAGE LIMITER

Features \ Circuit \ Output Trs.			pMOS	Bipolar	Bipolar Darlington	pMOS+Bipolar
Circuit			VR, IL, VL (VCH,VCL,VDL)	VR, IL, VL	VR, IL, VL	VR, IL, VL
Margin for the maximum voltage applicable to pMOS			VB - |VL|	—	—	VB - |VL + VBE|
	VL	VCH	1.0 V			1.8 V
		VCL	0.5 V	—	—	1.3 V
		VDL	0			0.8 V
Margin for bipolar active operation			—	|VL| - VBE	|VL| - 2VBE	|VL| - VBE
	VL	VCH		1.2 V	0.4 V	1.2 V
		VCL	—	1.7 V	1.1 V	1.7 V
		VDL		2.2 V	1.4 V	2.2 V
Output current of op-amp.			0	I_L/h_{FE}	I_L/h_{FE}^2	0
pMOS size (ratio)			1	—	—	$1/h_{FE}$

VB : maximum voltage applicable to pMOS(=3.0V)   VBE : 0.8V

dissipation; 2.5 V for the word-line drivers, being the maximum applicable voltage for the nominal operation to ensure device reliability; and 1.5 V for data lines, which matches power associated with the data line to the charge stored in the cell.

The limiter itself has the following two features. First, the V_{ref} is generated by the up-converter, which compares voltages obtained by the resistor division of the V_{ref} with well-regulated $V_{BG}-V_{EE}$ of 1.2 V. An excellent V_{ref} is obtained due to the well-regulated $V_{BG}-V_{EE}$, the well-controlled resistor ratio derived from the same diffused layers, and the op amp (OP1) with the bipolar-based circuit design, as will be discussed later. The succeeding op amp (OP2) is for isolating well-regulated V_{ref} from a noisy V_L subject to a load operation. Second, the output circuit has been composed of pMOS and bipolar transistors due to the advantages of high reliability and small op-amp layout area as is discussed in the following. Table I shows var-

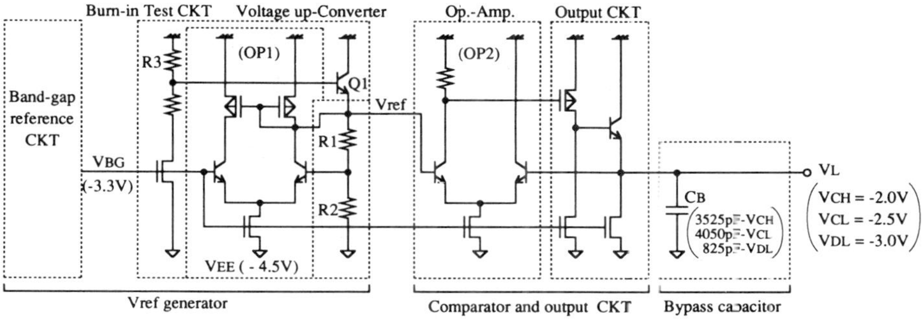

Fig. 2. Voltage limiting circuit.

ious output circuits. A pMOS output, which is commonly used in CMOS DRAM's, causes poor reliability since the drain-to-source voltage of 3 V is almost the same as the transistor breakdown voltage of 3 V. Moreover, a large pMOS transistor is needed to adequately regulate the limiter output voltage to prevent the charging and discharging of heavy data-line capacitive loads. A bipolar output makes the op-amp design difficult, since the base current is still large despite the current reduced by h_{FE}. A Darlington bipolar output, caused by the $2V_{BE}$ drop, creates a narrow margin for active V_{CH} generation. On the other hand, the pMOS and bipolar outputs create a relaxed pMOS drain-to-source voltage due to V_{BE}. Moreover, the bipolar transistor allows the pMOS transistor to transmit a current reduced by h_{FE}. A small pMOS transistor enables the use of a smaller layout area for the limiter, since the bipolar transistor is still relatively small. Thus, an output circuit making use of the pMOS and bipolar transistors was chosen.

Fig. 2 shows the limiter circuit in detail, featuring a stabilized V_{ref} with burn-in test capability. The V_{ref} generator is composed of a bandgap reference circuit [5] and a voltage up-converter. Note that a burn-in test circuit is incorporated, which is an improvement over the racing circuit [5]. The comparator consists of a simple differential amplifier and the above output circuit. The bypass capacitor is for phase compensation during operation [6]. The capacitances are different for the limiters, as shown in the figure. The bypass capacitances are structured by using many 25-pF sets along each V_L line to ensure good phase compensation [6].

Each up-converter generates a different V_{ref} through the $R1$-to-$R2$ ratio depending on the V_{CH}, V_{CL}, and V_{DL}. The up-converter is essentially a bipolar current switch with a MOS current source and MOS load. Its characteristics are largely determined by the bipolar circuit, which is uninfluenced by variations in process. The burn-in test circuit features a MOS current source to prevent any saturation of the bipolar current source [5]. The comparator has been simplified from the previous one [5]. This is because a sufficiently large voltage difference in the V_{CC}-V_L enables the elimination of diodes that prevent saturation. The fact that dc power to the limiter is not a prime design concern in this ECL DRAM has also contributed to the simplification.

III. Choice of Device for Periphery Circuit

A bipolar input buffer is suitable to establish a genuine ECL circuit, although a pseudo ECL CMOS circuit has recently been proposed [7]. However, a CMOS circuit is unable to provide excellent interface specifications due to inherently large MOS device parameter fluctuations and temperature dependence. Despite this, for the remaining circuit from $\overline{CE}$ to word line, there are two alternatives as shown by Fig. 3: a mostly CMOS periphery and a mostly bipolar periphery. In the mostly CMOS periphery, the decoder is constructed using a conventional CMOS logic gate, while in the mostly bipolar periphery it is constructed by means of a bipolar current switch, wired-OR logic, and transistor matrix. The transistor matrix is useful in reducing dc power since the transistor selected from among 512 others requires only one current source of 4 mA. A power switch to reduce the dc current was not installed in the bipolar current switch of the address buffer and predecoder due to the resulting noise generation and speed loss caused by the addition of activation clocks. Note that the location of the level converter differs in both: behind the bipolar input buffer in the CMOS circuit, and behind the bipolar decoder in the bipolar circuit. The word line features static operation [8] from a 2-V input to a large output pulse of 2.5 V.

Fig. 4 shows a comparison of delay time and supply current from $\overline{CE}$ to word line. The delay time is almost the same for both. This implies that the CMOS level converter increasing the 1.6-V pulse to a 2-V pulse is responsible for speed. Moreover, the insufficient slow speed of the mostly bipolar periphery is due to the degraded performance of the bipolar transistor as a result of the low-cost process shown in Table III. The supply current in the mostly bipolar periphery is about seven times larger than that in the mostly CMOS periphery despite the new transistor matrix circuit. The large number of bipolar current switches in the predecoder contributes to the larger current. Due to these advantages, a mostly CMOS periphery was selected for the ECL BiCMOS DRAM.

IV. BiCMOS Direct Sensing Scheme

This section compares the essential differences between the two types of direct sensing schemes in Fig. 5 so far proposed [2], [3]. Sensing is performed directly through

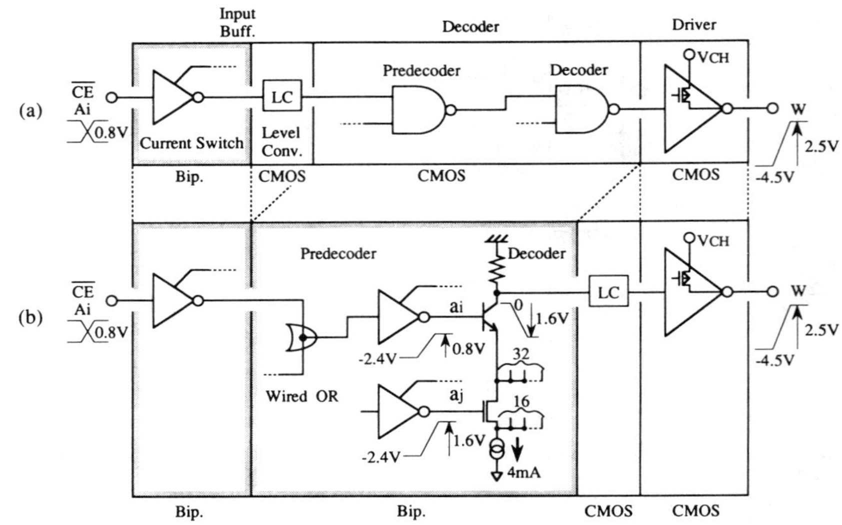

Fig. 3. Device selection for 0.3-μm ECL BiCMOS DRAM: (a) mostly CMOS, and (b) mostly bipolar.

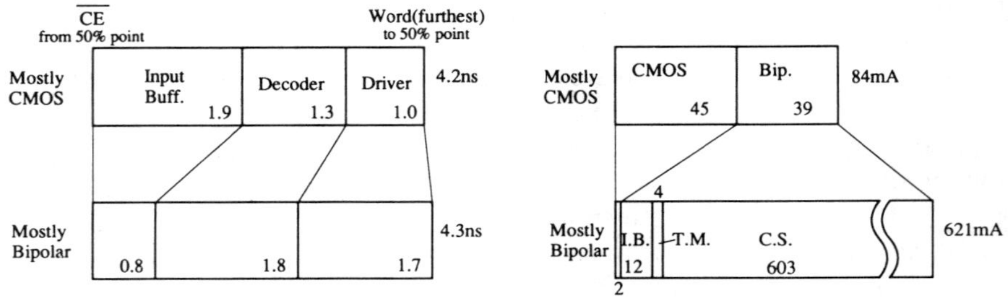

(a) (b)

Fig. 4. Comparisons of (a) delay time and (b) supply current from $\overline{CE}$ to word. I.B.: input buffer, T.M.: transistor matrix, C.S.: current switch. $V_{EE} = -4.5$ V; cycle time = 16 ns.

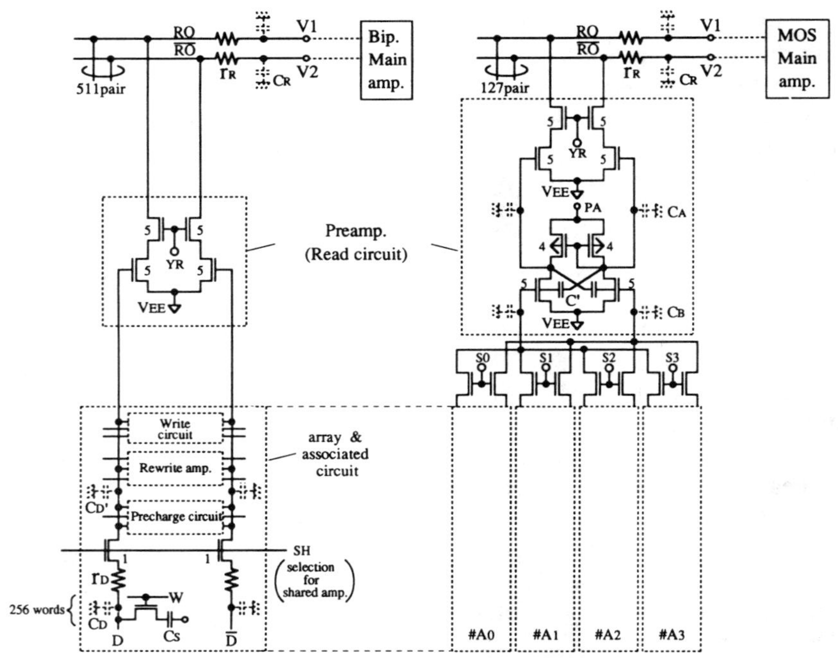

(a) (b)

Fig. 5. Direct sensing schemes: (a) single-stage MOS amplifier, and (b) double-stage MOS amplifier. Numerals show the channel width with given channel length of 0.3 μm.

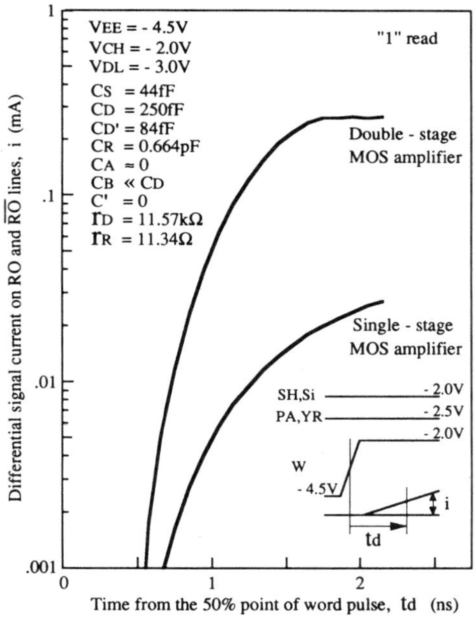

Fig. 6. Differential signal current developed on the *RO* lines. *V*1 and *V*2 in Fig. 5 are clamped at −2.5 V.

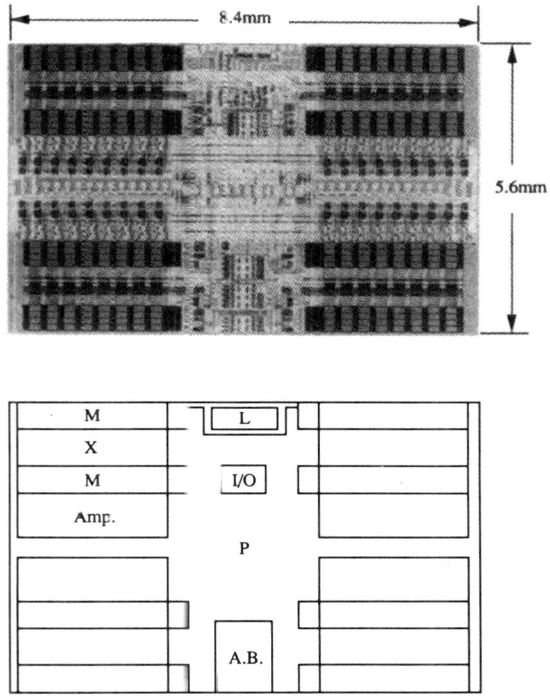

Fig. 7. Microphotograph of 4-Mb BiCMOS DRAM. *M*: Memory array; *X*: *X* decoder and driver; *Y*: *Y* decoder and driver; Amp.: main amplifier, rewrite amplifier driver; *A*.B.: address buffer; I/O: $\overline{CE}$, $\overline{WE}$, $\overline{OE}$ buffer, output buffer; *L*: limiters; *P*: periphery.

the preamplifier (read circuit) without waiting for active restoration, allowing high speed. Fig. 5(a) features a simple single-stage MOS preamplifier assisted by a highly sensitive bipolar main amplifier. For the MOS main amplifier, however, its insensitivity requires boosting through the help of the double-stage MOS preamplifier shown in Fig. 5(b). Note that a double-stage MOS preamplifier should be laid out to correspond to each of the four pairs of data lines due to its larger area. The bipolar main amplifier [4] is assumed to consist of a bipolar current switch, while the MOS main amplifier is assumed to consist of a MOS current switch. Thus, the sensitivity of the bipolar main amplifier is more than ten times that of the MOS main amplifier, taking offset voltage, due to layout mismatch and device fluctuation for pair transistors, into consideration.

Fig. 6 shows differential signal current developed on a pair of common read lines *RO* and $\overline{RO}$ after word pulse is applied. The double-stage MOS preamplifier provides a current about ten times larger than a single-stage MOS preamplifier. Such a large current could be detected with the CMOS main amplifier. Thus, both main amplifiers attain the same speed. In the design of the actual high speed, however, internal timing margins should be taken into consideration. The larger number of critical timings necessary for the double-stage MOS preamplifier adversely affects the speed. Four critical timings (*SH*, *Si*, *PA*, *YR*) are necessary for the double-stage MOS preamplifier while the single-stage MOS preamplifier has two (*SH*, *YR*). The addition of these two extra timing clocks, *PA* and *Si*, causes speed degradation. By the time the signal appears on the data line, array selection clock *Si* and preamplifier activation clock *PA* have to have completed their functions. Otherwise, access time degrades. As a result, two clocks, *Si* and *PA*, must be activated earlier than the word pulse. However, this is almost impossible, due to the various fluctuations in process, power supply voltage, and temperature. Wiring resistances also make this achievement more difficult. The extra timings add another 2 ns to the scheme for the double-stage MOS preamplifier according to the result of simulation. Additional current dissipation of 0.5 mA in the preamplifier is another drawback of the double-stage MOS preamplifier. Thus, a single-stage MOS preamplifier, combined with a bipolar main amplifier, was chosen. It is obvious that the higher speed and stable operation for successive main amplifiers, data output buffers, and ECL interfaces can only be obtained using the predominantly bipolar circuits. That is why BiCMOS technology was chosen.

V. 4-Mb Chip Design

Fig. 7 shows a microphotograph of an ECL 4-Mb BiCMOS DRAM designed with the above circuits. It also incorporates an ECL DRAM address buffer with reset and level-conversion functions to enable ECL interface to be attained and an overdrive rewrite amplifier to achieve fast cycle time [4]. The rewrite amplifier set and read and write circuits are shared between two sets of 256-word lines. The memory cell is a crown-shaped stacked capacitor cell [9]. All memory cells are separated from the periphery by an n-well that reduces noise from the periphery. The increase in process steps over the CMOS is about 10%. The features of the chip are shown in Table II. An access time of 7.8 ns and a current dissipation of 198 mA at a cycle

TABLE II
4-Mb BiCMOS DRAM Design Summary

Organization	4M Words × 1 b
Memory Cell	Data-line shielded STC cell Cell size 0.8 μm × 1.6 μm C_S/C_D = 44/250 fF
Technology	0.3 μm p-sub triple-well CMOS 0.8 μm poly-Si emitter bipolar
Chip Size	8.4 mm × 5.6 mm
Power Supply	−4.5 V
Access Time	t_{ACS} = 7.8 ns typical (simulated)
Cycle Time	t_{RC} = 16 ns typical (simulated)
Supply Current	Active: 198 mA typical (16-ns cycle) (simulated) Standby: 77 mA typical (simulated)
Refresh Cycle	8192/16 ms

TABLE III
Device Parameters

MOS	T_{OX} = 6.5 nm L_{eff} = 0.3(n)/0.4(p) μm V_T = 0.4 V = 0.25 V (Rewrite Amp.)
Bipolar	A_E = 0.8 μm × 3 μm f_T = 9 GHz h_{FE} = 50 C_{TS} = 82.71 fF C_{TC} = 16.8 fF C_{TE} = 8.73 fF r'_{bb} = 73 Ω

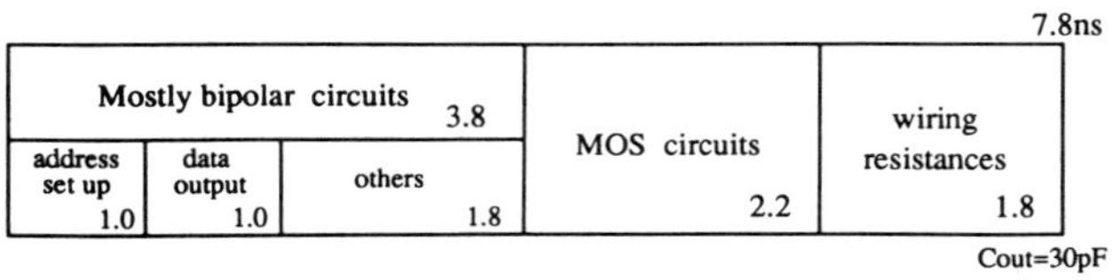

Fig. 8. Components of access time on the actual 4-Mb design: mostly bipolar circuits; ECL interface input/output buffer, highly sensitive amplifiers, voltage limiters, MOS circuits; memory cell array, word drivers, decoders, drivers, level converters.

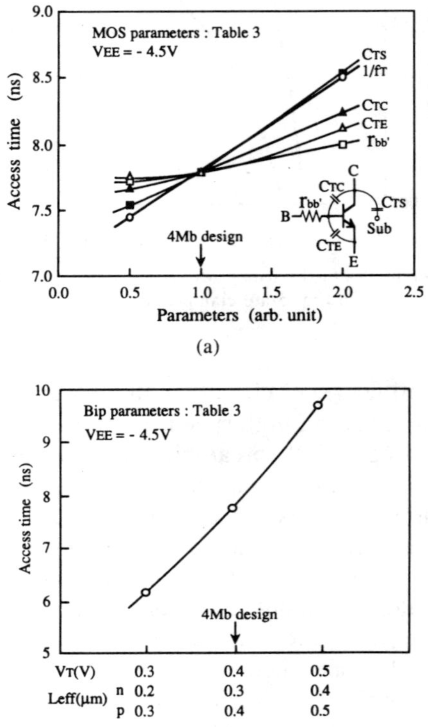

Fig. 9. Access time versus transistor parameters: (a) bipolar transistor parameters, and (b) MOS transistor parameters.

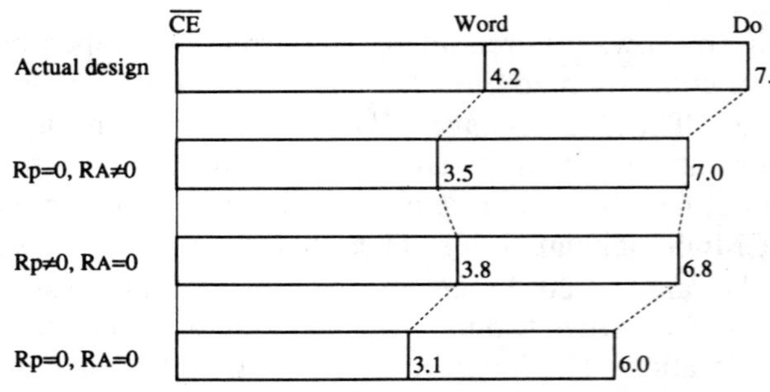

Fig. 10. Dependence of access time on wiring resistance. *RA*: all the wiring resistances of array including word line, data line, and *RO* line; R_p: all the wiring resistances of peripheral circuits.

time of 16 ns were expected. Although the full chip failed to work due to a miscellaneous reason, the results above are based on the transistor parameters, wiring resistances, and parasitic capacitances derived from the actual design of the 4-Mb chip. Moreover, these transistor parameters were obtained through careful evaluation of a small working test chip on the same wafer. Hence, the effectiveness of these technologies was demonstrated on a 0.3-μm ECL 4-Mb BiCMOS chip in spite of its lack of functionality.

VI. Design Parameter Dependence on Access Time

Fig. 8 shows components of access time for the actual design of the 4-Mb chip. It turns out that bipolar circuits dominate despite the chip having the mostly MOS circuits throughout. This is due to the special delay times associated with chip interface specifications. They are an address setup time of 1 ns and a data-output delay to drive the heavy off-chip load capacitance of 30 pF. Wiring resistance including word line contributes as much as 24%, even though special attention was paid to reduce resistance throughout the chip as much as possible. Fig. 9(a) shows access time versus bipolar transistor parameters. These are simulated for variations of a specific parameter against other parameters fixed to the designed values shown in Table III. Access time depends strongly on f_T and C_{TS}. f_T contributes to the speed of the bipolar current switch. Smaller C_{TS} reduces the parasitic capacitance of the *RO*2 line [4], where the collector of many main amplifiers are connected. The dependence of MOS transistor parameters is shown in Fig. 9(b). The variation in corresponding access times is from 6 to 10 ns, which is larger than the variation in the bipolar parameters. Such large variations are expected of a chip using mostly CMOS circuitry. Thus, it is reasonable to say that the MOS transistor parameters should be controlled more strictly than bipolar parameters in order to obtain higher speed.

The dependence on wiring resistance is shown in Fig. 10. If all the resistances of the periphery were zero, the access time would be reduced to 7.0 ns. If array-associated resistances were zero, it could be reduced to 6.8 ns. Hence, the contributions of wiring resistance on periphery and array are almost the same. Note that an access time as fast as 6.0 ns could be obtained if all the resistances in the chip were zero.

VII. Conclusion

A 0.3-μm sub-10-ns ECL 4-Mb BiCMOS DRAM design was described. The results obtained were: 1) the V_{CC} connection limiter with BiCMOS output circuitry was chosen due to its ease of design, excellent device reliability, and layout area; 2) the mostly CMOS periphery with a specific bipolar use provided better performances in both high speed and low power; 3) the direct sensing scheme of the single-stage MOS preamplifier combined with the bipolar main amplifier offered high speed; and 4) the strict control of MOS transistor parameters was proven to be more important in obtaining high speed in DRAM's, based on the design of a 4-Mb design.

Acknowledgment

The authors wish to thank T. Nishida, T. Kaga, and F. Murai for their support during device fabrication. They would also like to thank M. Aoki, K. Shimohigashi, K. Miyazawa, T. Masuhara, H. Sunami, M. Ishihara, H. Kawamoto, and S. Itoh for their helpful suggestions.

References

[1] Y. Nakagome *et al.*, "An experimental 1.5-V 64-Mb DRAM," *IEEE J. Solid-State Circuits*, vol. 26, no. 4, pp. 465–472, 1991.

[2] H. Miwa *et al.*, "A 17 ns 4 Mbit BiCMOS DRAM," in *ISSCC Dig. Tech. Papers*, Feb. 1991, pp. 56–57.

[3] T. Nagai *et al.*, "A 17 ns 4 Mb CMOS DRAM using a direct bit-line sensing technique," in *ISSCC Dig. Tech. Papers*, Feb. 1991, pp. 58–59.

[4] T. Kawahara *et al.*, "A circuit technology for sub-10-ns ECL 4-Mb BiCMOS DRAM's," *IEEE J. Solid-State Circuits*, vol. 26, no. 11, pp. 1530–1537, 1991.

[5] G. Kitsukawa *et al.*, "A 1-Mbit BiCMOS DRAM using temperature-compensation circuit techniques," *IEEE J. Solid-State Circuits*, vol. 24, no. 3, pp. 597–602, 1989.

[6] M. Horiguchi *et al.*, "A tunable CMOS-DRAM voltage limiter with stabilized feedback amplifier," *IEEE J. Solid-State Circuits*, vol. 25, no. 5, pp. 1129–1135, 1989.

[7] K. Ishibashi *et al.*, "A 1.7 V ajustable I/O interface for low voltage fast SRAMs," in *Symp. VLSI Circuits Dig. Tech. Papers*, June 1991, pp. 97–98.

[8] G. Kitsukawa *et al.*, "A 23-ns 1-Mbit BiCMOS DRAM," *IEEE J. Solid-State Circuits*, vol. 25, no. 5, pp. 1102–1111, 1990.

[9] Y. Kawamoto *et al.*, "A 1.28 μm^2 bit-line shielded memory cell technology for 64 Mb DRAMs," in *Symp. VLSI Technol. Dig. Tech. Papers*, June 1990, pp. 13–14.

Paper 3.15

A 3.6ns 1Kb ECL I/O BiCMOS U.V. EPROM

Doug Smith*, John Zeiter*, Terry Bowman, Jeff Rahm, Bob Kertis, Ann Hall, Saul Natan, Len Sanderson, Rob Tromp, Joseph Tsang

National Semiconductor Corp.
1111 39th Ave. S. E., Puyallup, WA 98373

*Motorola Inc.
1300 N. Alma School Rd., Chandler, AZ 85224

1. INTRODUCTION

The combination of high density CMOS and fast bipolar transistors has made it possible to realize high performance ECL I/O electrically programmable read only memories. BiCMOS allows the combination of very high speed ECL circuits with the erasability and testability of U.V. EPROMS. This paper will describe a 1K density EPROM with input and output levels available in ECL 100K or 10K with a masking option. TTL I/O compatible programming and verify modes allow the memory array to be programmed with standard EPROM hardware and software.

2. MEMORY CELL

In deciding upon the memory cell for a fast ECL I/O EPROM, several memory cells were investigated, Figure 1. Traditional bipolar cells, which include metal fuse structures and bipolar collector to emitter shorting structures, have several undesirable characteristics. First, they can be programmed only once, which prohibits full array testing before shipping product. Second, dummy rows and columns are required to check programming and speed of the device. Test guard-band needs to be added to ensure the array will meet the desired speed spec. As a result, customers will observe programming fallout and speed fallout because of incomplete testing. MOS EPROM cells were considered as well but were felt to be too slow because of low cell current. They would also require additional periphery circuitry to translate from ECL to CMOS levels in order to operate the cell. The BiCMOS EPROM memory cell chosen, illustrated in Figure 2, consists of three sections. The first section is a single-poly EPROM element, which is an NMOS transistor, M1, with an MOS capacitor coupled to the floating gate. Programming is done by taking the selected word line (PROW) to a high voltage (12.5v +) and biasing each selected column (PBIT) high enough to generate hot electrons in the programming element M1. Sufficient charge is stored on the floating gate to turn off transistors M1 and M2 during verify and read modes. The second section is a sense circuit which senses the state of the floating gate upon power-up. A small reference current source, transistor M3, acts as the load for transistor M2. If M2 is "on" (the erased state), its drain is at a low voltage and pass-gate M4 is on. If M2 is "off" (the programmed state), the pass gate is turned "off". The third section of the memory cell is the ECL interface portion. One side of the PMOS pass-gate M4 is connected to 1 of 32 ECL word lines, and the other side is tied to the base of an NPN transistor. The emitter is one of 32 on each bit line. ECL access time is minimized by maintaining small signal swings (<800mV) throughout the read path. For accessing a row, the ROW signal swings 800mV and may modulate the base of the cell NPN transistor if the PMOS pass-gate M4 is in the on state.

3. ARCHITECTURE

The BiCMOS EPROM is essentially two devices in one, Figure 3. First, it's an ECL ROM with a high speed ECL read path. The die can be configured to 100K or 10K ECL I/O levels with a metal mask option. Second, it's a U.V. erasable and testable CMOS EPROM. It uses a standard TTL I/O programmer for programming and verification. Figure 4 shows a photomicrograph of the die. The array is organized as 32 rows by 32 columns. Eight columns are dedicated to each of the four outputs. Programming mode is selected by taking the VPP/mode pin to +12.5v (Vcc=6v, Vee=0v). The actual programming pulse is a TTL low level signal on the chip select pin. Holding the VPP/mode pin at a low level (0v) enables the Program Verify mode. This is similar to the VPP/OE- function used in some conventional EPROMs. ECL read mode is activated by holding the VPP/mode pin equal to Vcc, which makes the part pin compatible with existing ECL I/O PROMs. The program/verify/read modes are shown in the truth table of Figure 5.

4. CIRCUIT OPERATION

Connecting two types of input buffers to each input pin is a fairly simple task. Figure 6 shows a TTL input buffer which is controlled by the programmer. The input comes into the source of transistor M1 whose gate is controlled by the TTL input reference signal. When the EPROM is in read mode, ECL input levels control the device. The ECL input levels are above the TTLREF signal, so the TTL input buffers are unmodulated. The ECL input buffer used is shown in Figure 7. It makes use of a bipolar differential pair with emitter followed outputs.

The row decoder used is a bipolar multiple-emitter decoder as shown in Figure 8. This decoder takes its inputs from the ECL input buffer and provides decoded rows which swing 600mV. Thus from the inputs to the decoded rows, all logic is performed with ECL circuitry for increased performance. There are 32 decoded rows with 32 cells per row. When rows are selected, they are in the high state. Traditionally, discharge of a selected row to a deselect state has been performed by a common current source connected through resistors to the individual rows. This is represented by the reference circuit in Figure 8. Given that all row address path delays are

Reprinted with permission from *International Symposium on Circuits and Systems*, pp. 1987–1990, 1990.

not the same and that input address skewing occurs, multiple decoded rows can be selected for a short period of time, Figure 9. With traditional current source discharge of selected rows, the time to discharge selected rows can grow depending on the number of rows selected. This can cause access time push-out compared to individual row path changes. To eliminated this problem, the discharge current source was changed to a voltage source referenced to the top supply. This provides independent equivalent current paths for each decoded row and eliminates the access time push-out. The top supply reference generator is shown in Figure 10. A reference voltage across resistor R2 is mirrored over to the REF signal which is the source of transistor M2. The load for the REF signal is the current through one selected row resistor. As multiple rows are selected, extra current passes through transistor M2 to provide base drive to bipolar transistor Q1. This provides high current gain to handle the discharge of multiple rows, which results in making the REF signal look like a voltage source.

The matrix organization and sensing are shown in Figure 11. A column of 32 cells is selected by one of eight ECL column select signals, COL0 - COL7. Column sensing occurs by steering current through a selected memory cell along the selected column or through the cascode sensing network described by transistors Q1, Q2, Q3 and R1. The output of the cascode sensing network is emitter-followed to drive the input to the ECL output buffer. A MOS current source gated by REF0 determines the column sensing current which is steered to one of the selected columns.

The ECL output buffer, shown in Figure 12, is used to provide 100K temperature compensated output levels. It mainly consists of a differential pair (Q1 and Q2), a temperature compensation network (Q3, Q4 and R1) and the output driving transistor Q5. In order to isolate this output buffer from the I/O pins during TTL program and verify, MOS transistors were added to disable the buffer. The disable function is controlled by the ECLZ signal. It shuts down the differential pair current source, isolates the loads of the differential pair and shorts the base-emitter junction of the output transistor. Separate PMOS switch transistors, M1 and M2, are used to keep the outputs at valid ECL levels during ECL read mode.

Figure 13 shows the programming and verification of the memory cell. Programming of memory cells occurs when Vpp is taken to +12.5v. Cells are programmed when the CSB pin is at a TTL low level. With the data state of "0", programming current will pass through the current source transistor M1, through the column select transistor M2, and through the programmed transistor M3 within the selected memory cell. The programmed cell along the column is selected by the programming row, PROW, which is up at 12.5v. After programming occurs, the CSB pin is taken to a TTL low level by the programmer for verify mode. The programmed transistor within the selected cell is read by a CMOS current sensing circuit which sends TTL information to the TTL output buffer.

In order to maintain constant current in the device over the operating supply range, a bandgap reference generator as shown in Figure 14 is used. The current, I1, which biases the reference generating network needs to be constant over supply variation. To accomplish this, a current mirror, transistors M1 and M2, is used to mirror current from a bipolar current source, Q1 and R1, controlled by the reference signal REF. To ensure successful power-up, a start-up circuit is needed to activate this reference generator since it has two stabel states, activated and deactivated. Figure 15 shows the operating current Iee from 0v to -6v on VEE. At approximately -3.6v, the reference generator regulates the REF signal to less than 3mV/Volt of supply variation. At that point, the operating current is relatively constant.

5. PERFORMANCE

Figure 16 shows the access time of 3.6ns at a supply voltage of -4.5v at 25 C. A shmoo plot of access time vs supply voltage at a temperature of 25 C is shown in Figure 17. The access time at 85 C is approximately 4.3ns. Access time is flat over supply variation due to the constant current sources setting the switching speed of the logic.

6. PROCESS TECHNOLOGY

The die is processed using a twin-well, double-poly, double-metal BiCMOS technology[1] with features down to 1.0u. The main features of the technology are shown in TABLE 1. The cell size is 17.4u x 36.3u. The die size is 79mils x 103mils.

ACKNOWLEDGEMENTS

The authors wish to thank the process development team for their support.

REFERENCES

[1] B. Bastani, et al., "Advanced One Micron BiCMOS Technology for High Speed 256K SRAMs, 1987 Symposium on VLSI Technology Digest of Technical Papers, p. 41-42; May, 1987.

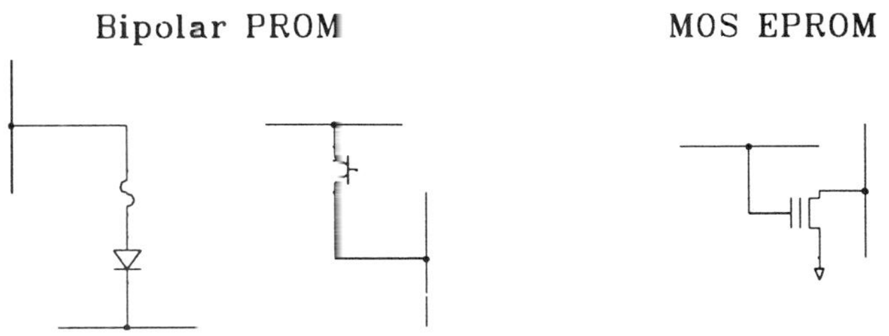

Fig. 1 Bipolar and MOS programmable cells

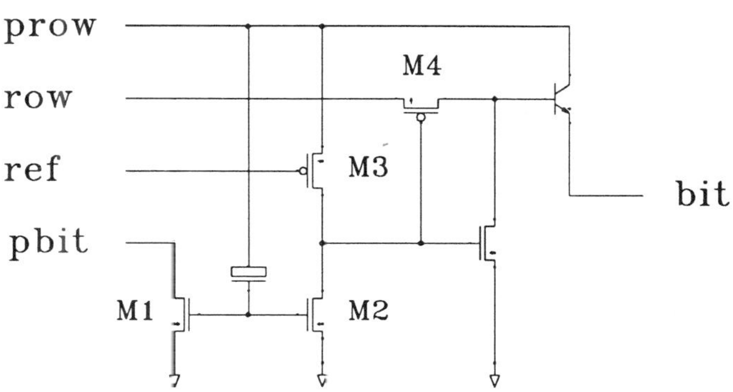

Fig. 2 BiCMOS EPROM cell

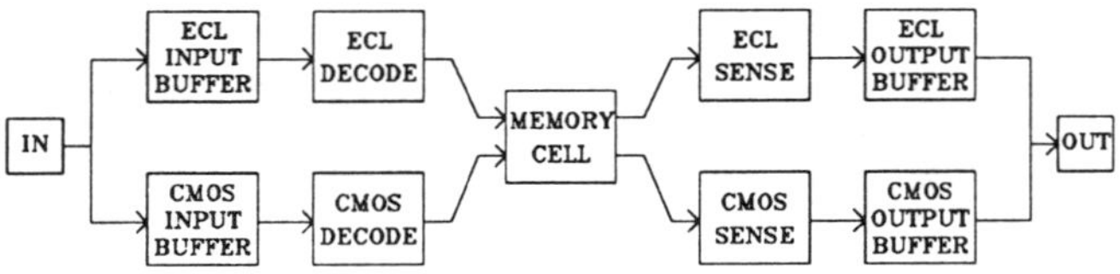

Fig. 3 Device structure

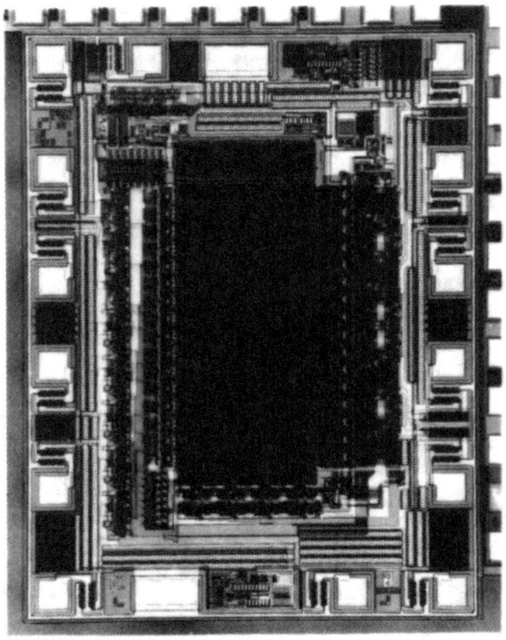

Fig. 4 Photomicrograph of 1K BiCMOS EPROM

VPP pin	CSB pin	Outputs	Mode
Vee+12V	TTL high	high Z	program disable
Vee+12V	TTL low	TTL data in	TTL program
Vee	TTL high	high Z	TTL deselect
Vee	TTL low	TTL	TTL verify
Vcc	ECL high	ECL low	ECL deselect
Vcc	ECL low	ECL	ECL read

Fig. 5 Truth Table

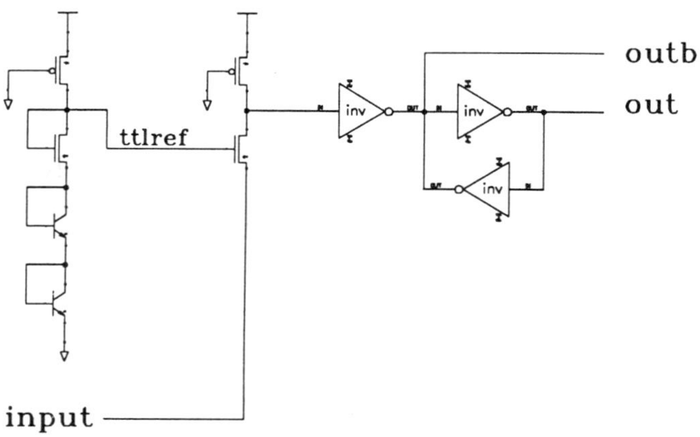

Fig. 6 TTL input buffer

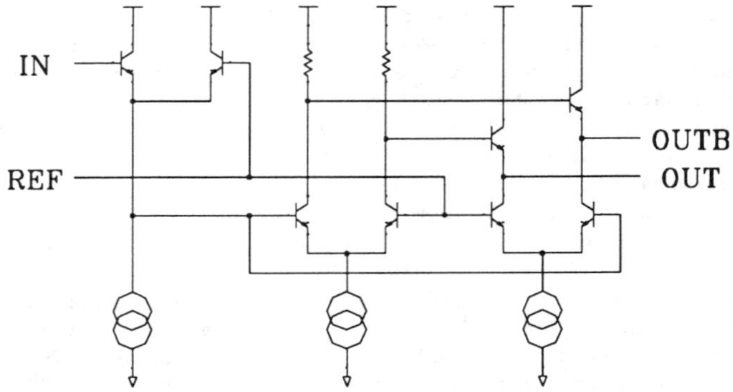

Fig. 7 ECL input buffer

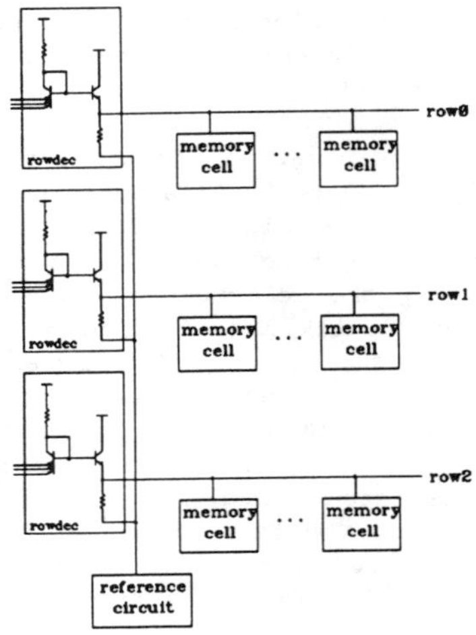

Fig. 8 ECL row decoder

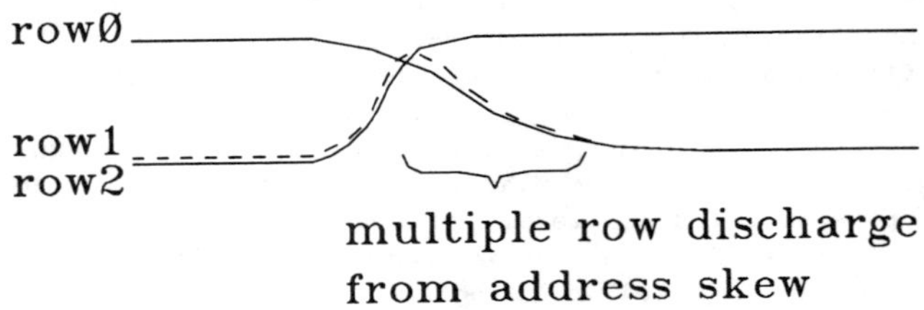

Fig. 9 Multiple row selection

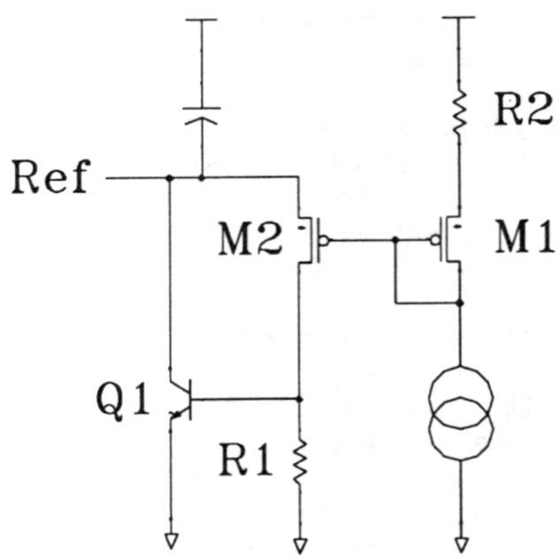

Fig. 10 Top supply reference generator

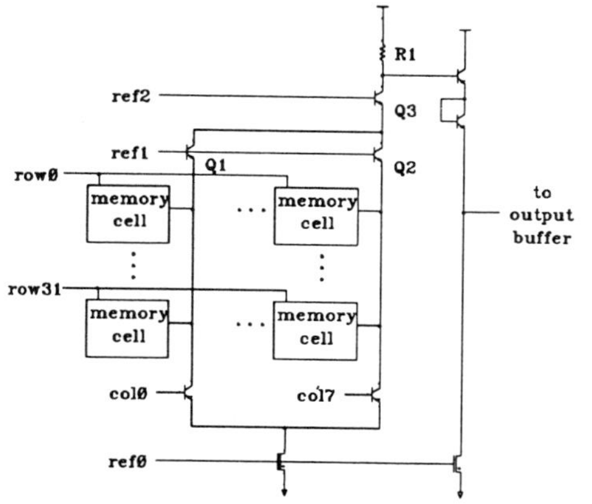

Fig. 11 Matrix organization and sensing

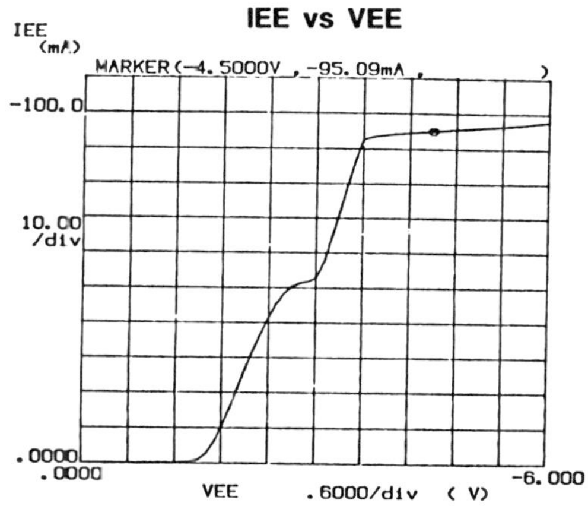

Fig. 15 Operating current vs supply voltage

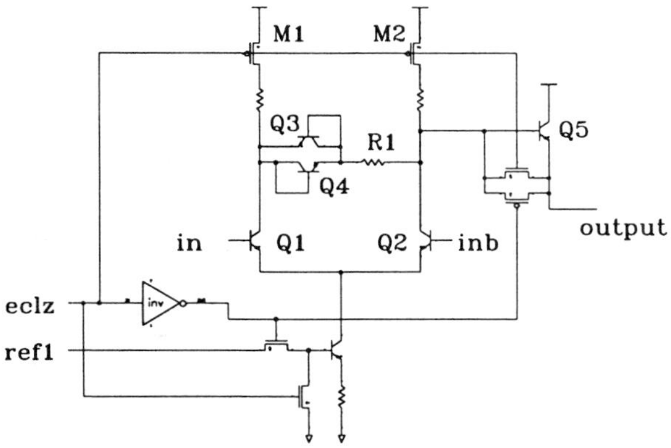

Fig. 12 ECL output buffer

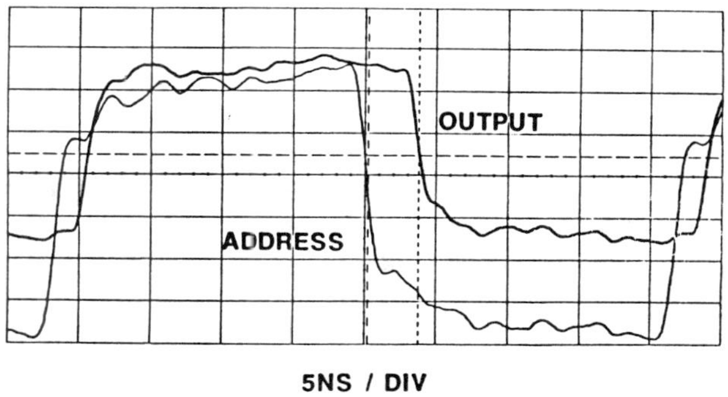

Fig. 16 Waveform of address access time

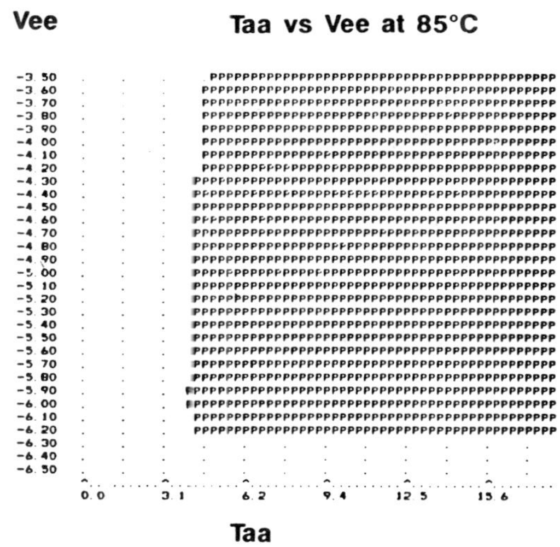

Fig. 17 Access time vs supply voltage, 85 C

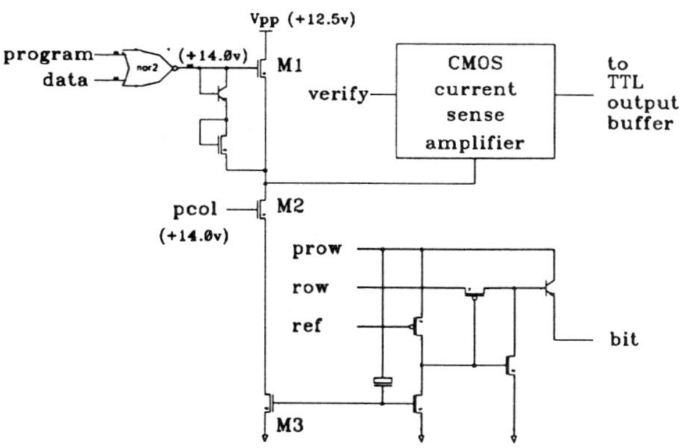

Fig. 13 Cell program and verify

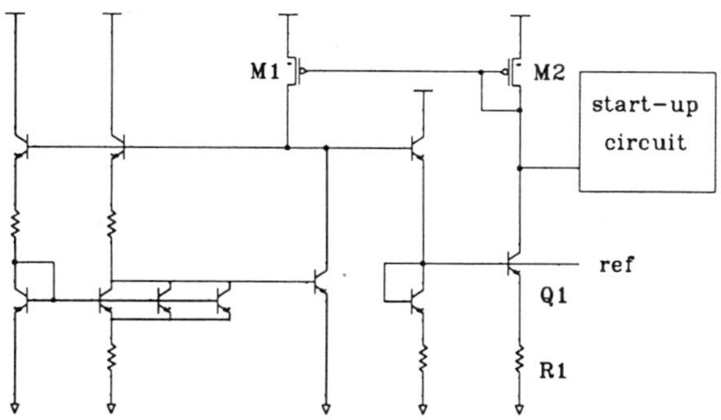

Fig. 14 Bandgap reference generator

Process	**Twin Well**
	Double level polysilicon
	Double level metal
Electrical gate length	**0.8μm**
Gate oxide thickness	**200 angstroms**
Bipolar current gain (h_{fe})	**150 (Ph) / 80 (As)**
Bipolar unity-gain bandwidth (f_t)	**2 GHz (Ph) / 7 GHz (As)**
Contact and via size	**1.1μm x 1.1μm**
Metal-1 pitch	**2.5μm**
Metal-2 pitch	**2.9μm**

Table 1 Process technology

Paper 3.16

Very High-Speed ROM Using Bipolar/CMOS Technology

Koozaburoo Kurita, Takasi Hotta, Members

Hitachi Research Laboratory, Hitachi, Ltd., Hitachi, Japan 319-12

Masahiro Ueno, Member

Hitachi Works, Hitachi, Ltd., Hitachi, Japan 317

and Atsuo Hotta, Member

Device Development Center, Hitachi, Ltd., Oome, Japan 198

SUMMARY

It is a common practice to realize the internal control of a logic LSI by means of a microprogramming method. A high-speed operation of this logic LSI requires a high-speed ROM (micro ROM) which stores this microprogram code. This paper proposes a new circuit configuration derived from a combination of bipolar and CMOS technology for a high-speed micro ROM. In this configuration, the memory cell array is made of MOS transistors while the array periphery consists of a bipolar/CMOS composite circuit in which a high driving capability and a large gm of the bipolar transistor are used so that a high degree of integration and a high speed are attained simultaneously. To confirm the high-speed characteristic of the proposed circuit, a micro ROM of 2 k words × 64 bits has been fabricated with the 2 μm Hi-BiCMOS process mixing the bipolar and CMOS structures. The test results of the fabricated device have a minimum read-out cycle time of 16 ns which is more than twice that of a CMOS circuit having an identical configuration.

1. Introduction

Recently, the requirement for an improved characteristic of a logic LSI has been in wide demand and hence an LSI technology for a high degree of integration and a high speed is desired. As a solution for this requirement, an application of a new LSI technology, Hi-BiCMOS (High Performance Bipolar CMOS) [1], is proposed which makes use of a high degree of integration and low power consumption of the CMOS and a high-speed property of bipolar transistors [2]. The Hi-BiCMOS has been applied to the gate array [3], and memory [4, 5] and its low-power consumption has proven comparable to CMOS and its high speed approaches the ECL. The application of this Hi-BiCMOS to such logic LSI's as processors, requires combining bipolars and CMOS's depending on the internal logic [6].

Logic LSI's such as processors have been decreased in their size but their complexity has increased. At the same time, the internal control of the logic LSI is realized by a microprogram method, in which microprogram instructions are written in an ROM beforehand. The internal control signal is generated by reading out the content of this ROM (micro ROM). Hence, the portion consisting conventionally of random gates is replaced with a regulated micro ROM so that the number of design stages can be reduced substantially.

However, from the viewpoint of speed, the read-out cycle of the micro ROM is a limiting factor for the speed of the entire logic LSI. Hence, it is indispensable to increase the speed of the micro ROM if a high-speed processor or a high-speed logic LSI is to be realized.

This paper presents a new high-speed circuit system based on Hi-BiCMOS technology for realization of a high-speed micro ROM.

Reprinted with permission from *Electronics and Communications in Japan*, Part 2, vol. 71, no. 7, pp. 18–26, 1988.

Table 1. Desired specification of micro ROM

Item	Specification
Process	2 μm Hi-BiCMOS
Operating frequency	40 MHz
Cycle time	25 ms
Word number	2 k words
Word length	64 bits
Total capacity	128 kbits

Table 2. Device characteristics of Hi-BiCMOS

Item		Value	Unit
Bipolar	Emitter size	2×5	μm
	β	100	—
	ft	4	GHz
	R_{B}	100	Ω
MOS	Gate oxide film thickness	35	nm
	L_{E}	2	μm
	V_{th}(n)	0.55	V
	V_{th}(p)	−0.55	V

Further, a test circuit of a micro ROM based on the present approach is evaluated.

2. Configuration of Micro ROM

2.1 Target specifications

The target specifications are shown in Table 1 for discussions of Hi-BiCMOS micro ROM. The capacity of the ROM is 2 k words × 64 bits so that it is sufficient for installation of floating point operations in a microprogramming control in a personal computer. The speed is intended to be more than twice that attained by the CMOS for the identical fabrication rule by means of a 2-μm Hi-BiCMOS process. This process, which is a key for realization of this objective, creates bipolar transistors with β (current gain) of 100 and ft (cut-off frequency) of 4 GHz on a 2-μm rule CMOS. The characteristics of the device by the 2-μm Hi-BiCMOS process are shown in Table 2.

2.2 Memory cell

The planar pattern of the memory cell is shown in Fig. 1. As a result of the need for installing a large capacity, the memory cell is not different from that obtained by a conventional MOSLSI. Because of the need for a high-speed operation, the memory cell is an OR type. To make the structure small, the data are expressed by the existence of the diffusion layer below the polysilicon of the word line [7]. The MOS transistor, the data line by the first Al layer, the contact and the GND line by the diffusion layer are common for both the upper and lower 2 bits. To reduce the signal delay by the resistance of the polysilicon word line, the second Al layer is added. The size of the MOS transistor (*W/L* = channel width/channel length: unit is μm) is 6/2 and the cell size is 8 × 10 μm if the reinforcement of the word line and the GND line is neglected.

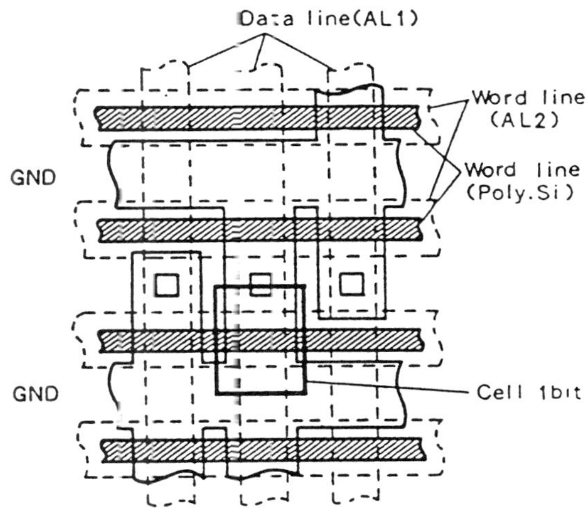

Fig. 1. Memory cell layout.

The configuration of the memory cell array requires consideration of the entire layout of the logic LSI chip on which the micro ROM is installed. In the present study, in consideration of the balance of the lengths in the longitudinal and transverse directions, the configuration is 512 × 256 bits. A 1/3 selector is used which changes the word length from 512 bits to 64 bits.

2.3 Overall configuration

As a large-capacity micro ROM, a dynamic ROM of a precharge configuration is used which is advantageous for reducing the area and the power consumption. Figure 2 shows the block diagram of the micro ROM.

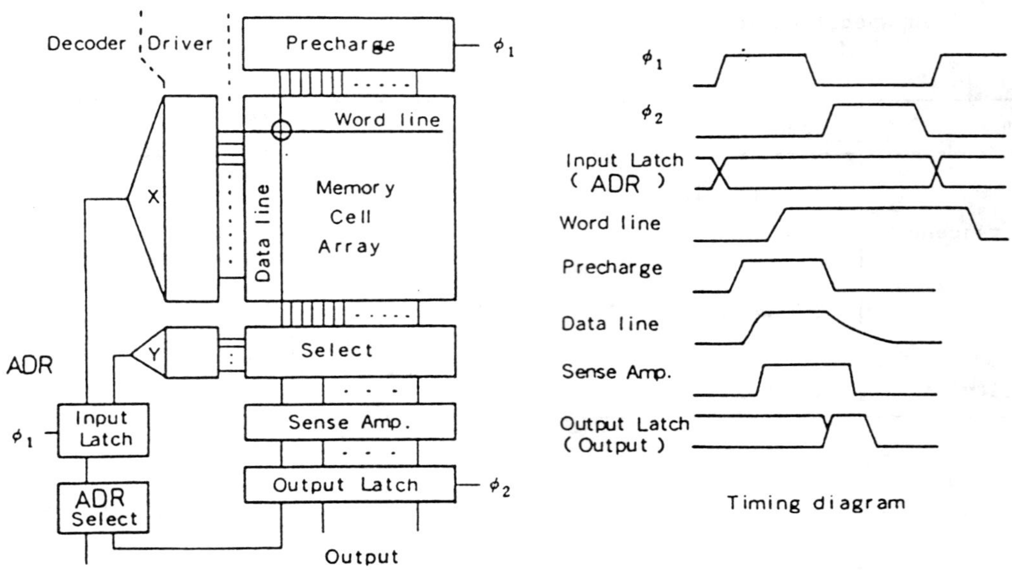

Fig. 2. Block diagram of micro ROM.

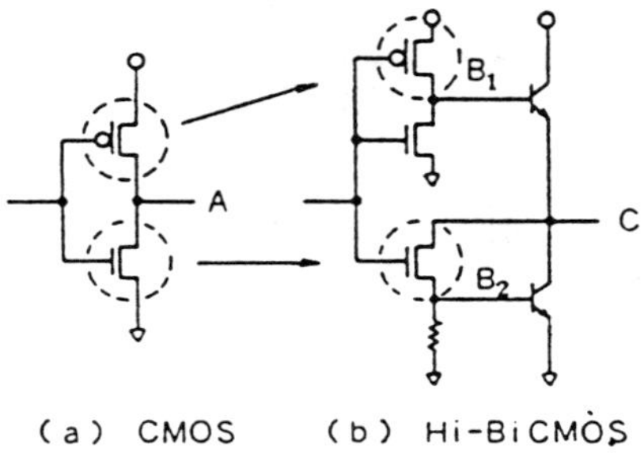

Fig. 3. CMOS and Hi-BiCMOS circuits.

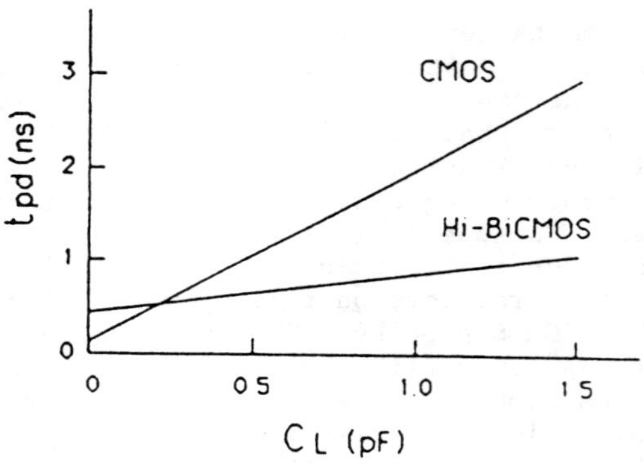

Fig. 4. C_L - t_{pd} characteristics.

When the address is input, it is decoded into the X and Y systems so that the word line and the data line of the memory cell array are selected. The memory cell at the intersection is selected. It drives the data line which is set at a certain level by the precharge. The data are read by the sensor circuit detecting the level of the data line. In the case of cycle operations, the read-out data become the next address and a continuous access occurs.

To increase the speed of the micro ROM, it is necessary to reduce the delay in each internal circuit. It is important especially to reduce the delay when a large parasitic capacitance load appearing in the word line and the data line in a large capacity memory cell array is driven. In the next section, an improvement of the speed by a Hi-BiCMOS circuit is described.

3. High-Speed Operation by a Hi-BiCMOS Circuit

Figure 3 shows the inverter circuits by means of CMOS and Hi-BiCMOS. The Hi-BiCMOS circuit consists of a CMOS circuit with a bipolar output stage. Hence, its current-driving capability is β times that of the CMOS circuit. The delay t_{pd}^{m} of the CMOS circuit driving a load is the sum of the switching delay t_0 of the MOS transistor and the one t_1 required in driving the parasitic capacitance C_m of the MOS transistor at the output A point and the capacitive load C_l:

$$t_{pd}m = t_0 + t_1 \tag{1}$$

t_1 can be expressed as follows if the driving current of the MOS transistor is I_m and the logic threshold value of the circuit is Vlt:

$$t_1 = Vlt \cdot (C_l + C_m)/I_m \tag{2}$$

From Eqs. (1) and (2), $t_{pd}m$ becomes

$$t_{pd}m = t_0 + Vlt \cdot (C_l + C_m)/I_m \tag{3}$$

On the other hand, the delay $t_{pd}b$ of the Hi-BiCMOS circuit driving a load is the sum of the delays t_0, t_2 and t_3. Here, t_0 is the switching delay of the MOS transistor, t_2 is the delay associated with the parasitic capacitance C_m of the MOS transistor at the base (B_1 and B_2) of the bipolar transistor and with the parasitic C_b of the bipolar transistor driven by the MOS transistor, and delay t_3 is caused by the parasitic C_0 at the output point C and the capacitive load C_l driven by the bipolar transistor:

$$t_{pd}b = t_0 + t_2 + t_3 \tag{4}$$

t_2 can be expressed approximately in the following equation which is similar to Eq. (2):

$$t_2 = Vlt \cdot (C_m + C_b)/I_m \tag{5}$$

If the driving current of the bipolar transistor is I_b, then

$$t_3 = Vlt \cdot (C_l + C_o)/I_b \tag{6}$$

Note that the driving current I_b of the bipolar transistor is β times the driving current I_m of the MOS transistor:

$$I_b = \beta \cdot I_m \tag{7}$$

From Eqs. (6) and (7),

$$t_3 = Vlt \cdot (C_l + C_o)/\beta \cdot I_m \tag{8}$$

From Eqs. (4), (5) and (8),

$$t_{pd}b = t_0 + Vlt \cdot (C_m + C_b)/I_m + Vlt \cdot (C_l + C_o)/\beta \cdot I_m \tag{9}$$

Now compare the delays in the CMOS circuit and the Hi-BiCMOS circuit from Eqs. (3) and (9). If the capacitance load C_l is larger than other parasitics, Eqs. (3) and (9) can be approximated as

$$t_{pd}m = t_0 + Vlt \cdot C_l/I_m \tag{10}$$

$$t_{pd}b = t_0 + Vlt \cdot C_l/\beta \cdot I_m \tag{11}$$

The Hi-BiCMOS circuit improves the speed by the effect in which the driving capability is β times that by the bipolar transistor. On the other hand, if the load capacitance C_1 (under the no-load condition) is smaller than other parasitics, Eqs. (3) and (9) can be approximated as

$$t_{pd}m = t_0 + Vlt \cdot C_m/I_m \tag{12}$$

$$t_{pd}b = t_0 - Vlt \cdot (C_m + C_b)/I_m \tag{13}$$

Hence, the delay increases in the Hi-BiCMOS circuit due to the increase of the parasitic capacitance by the bipolar transistor.

Figure 4 shows the results of the circuit simulation for the relationship of the delay to the load of the CMOS or Hi-BiCMOS circuit. For Fig. 4, the size (W/L) of the MOS transistors in both the CMOS and the Hi-BiCMOS circuits is 20/2. The inequality of the delays of both circuits is reversed at the load value of 0.2 pF. This means that the use of a Hi-BiCMOS does not guarantee a higher operating speed. It is necessary to use discriminately the CMOS circuit and the Hi-BiCMOS circuit depending on the load.

4. High-Speed Operating System of Micro ROM's

Here, high-speed operating systems of each part of micro ROM's are described. A Hi-BiCMOS circuit is described which combines bipolar transistors and MOS transistors for realization of a high-speed operation.

4.1 Word line driving system

The parasitic capacitance of the word line is extremely large, 8 pF, because it connects the gates of 512 MOS transistors. Hence, driving by a bipolar transistor is more advantageous. In the Hi-BiCMOS circuit in which a bipolar transistor is used in the output stage, the output amplitude is decreased by Vbe (the base-to-emitter voltage of the bipolar transistor) on the power supply side and the grounded side. This is particularly troublesome in the case of "L" level. A malfunction is a possibility when

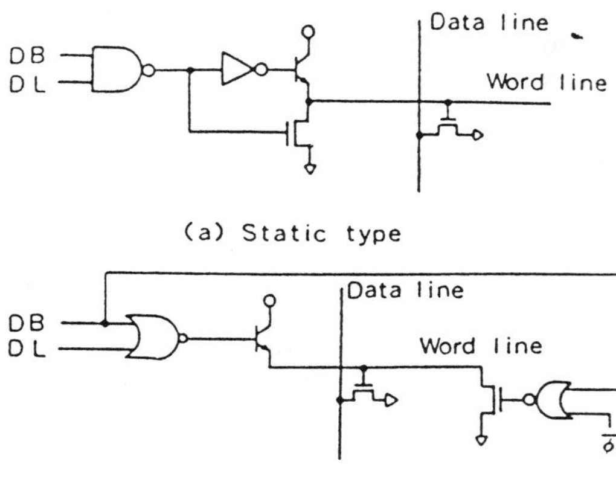

Fig. 5. Word line driver.

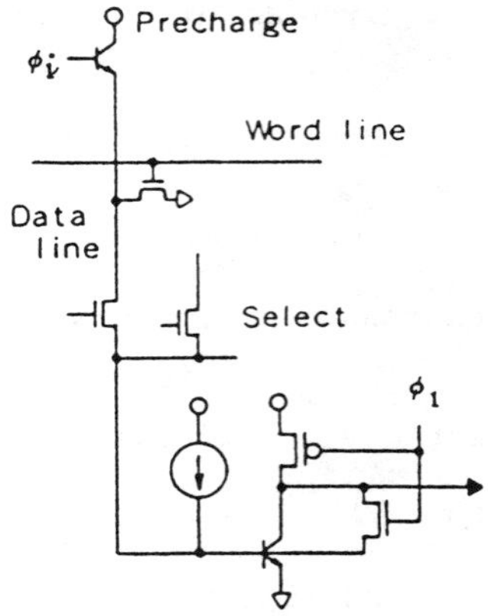

Fig. 7. Data line sense amp.

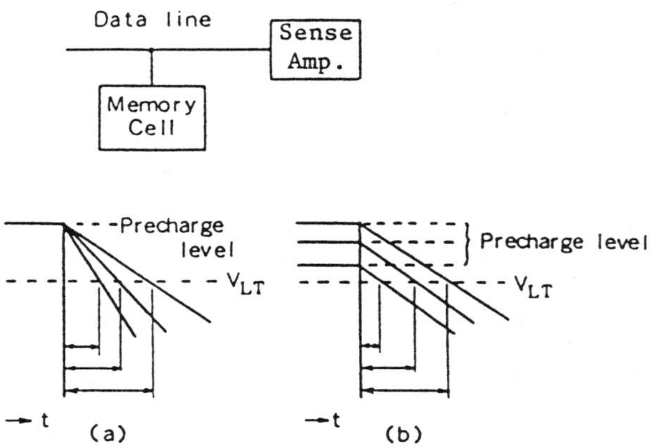

Fig. 6. Concept of high-speed access.

the MOS transistor of the memory cell does not turn off. Hence, the extraction side of the word line driving must be done by an MOS transistor.

The word line driving system is shown in Fig. 5. Figure 5(a) shows the static driving system in which a bipolar transistor and MOS transistors are combined. The delay of the present circuit as calculated by a circuit simulation is 6.8 ns for the rising of the word line, and 7.6 ns for the falling via a decoding of the *X* system from the address. The extraction by the MOS transistor is found to be a bottleneck. This is because this extraction is activated by the selection signal of the word line driver. After the address is switched in the micro ROM operation, the "L" level of the word line does not cause any problem.

A driving method is proposed in which the MOS transistor on the extraction side is synchronized with the clock. Figure 5(b) shows the proposed synchronized driver. The delay of the circuit was obtained by a circuit simulation. The rise from the address is 6.9 ns and the fall is 4.6 ns. Since the MOS transistor on the extraction side is operated by the rising, the speed is high. In this circuit, the extraction by the MOS transistor is operated only on the block selected by the previous address. Therefore there is a smaller probability that the driver selected by the next address drives the bipolar transistor and the MOS transistor executes extraction.

A high-speed operation is now made possible by the Hi-BiCMOS word line driver in which a bipolar transistor with a large driving power and an MOS transistor synchronized with the clock are combined.

4.2 Data line access system

Since only 256 MOS transistors are connected, the parasitic capacitance of the data line is only 4 pF which is smaller than that for the word line. However, since the memory cell to drive the data line is made of MOS transistors with extremely small driving capability, high-speed operations are difficult. In the case of an MOS transistor of the size of 6/2, the on-resistance of the MOS transistor is about 4 kΩ. Hence, the discharge time constant of the data line by the MOS transistor is as much as 16 ns.

By means of a PMOS transistor, the data line is precharged to the power supply voltage and the charge of the data line is discharged by the memory cell (NMOS transistors).

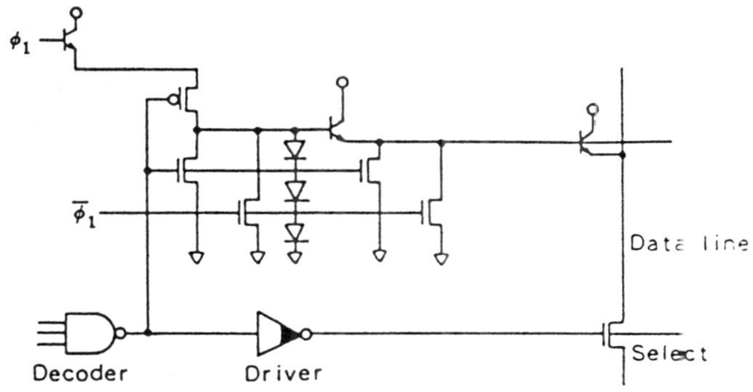

Fig. 8. Precharge circuit.

Subsequently, the CMOS inverter is used for sensing. The delay in the ROM in this CMOS configuration is simulated. The result from the memory cell driving to the sense output is 19.7 ns which cannot attain the objective.

The delay at the data line access is determined by the speed at which the memory cell is driven from the data line level set by the precharge circuit ot the level of the sense circuit. Hence, for a high-speed access, two methods as shown in Fig. 6 can be conceived:

(a) the driving capability of the memory cell is increased;

(b) the amplitude of the data line signal is decreased.

However, in method (a), the channel width of the MOS transistor must be increased in order to increase the driving capability. Hence, the size of the memory cell array becomes too large for high-density integration. Also, since the parasitic capacitance of the MOS transistor is increased, the load of the data line becomes larger. Therefore, this method is not very effective for high-speed operations. On the other hand, in method (b), a high-speed operation is possible without increasing the size of the MOS transistor as long as the potential setting of the data line is accomplished. The present study uses this method.

Figure 7 shows the circuit to realize a low-amplitude operation of the data line. A bipolar transistor with a high gm is used as a sense circuit of the data line so that the amplitude of the data line is suppressed to Vbe. In this circuit, whether or not the memory cell extracts the base current supplied to the emitter grounded bipolar transistor controls the on-off operation of the bipolar transistor. During the precharge, the PMOS transistor used as a load is off so that the straight-through current in the sense circuit is eliminated. On the other hand, by switching on the NMOS transistor, the input and output are short-circuited and the precharge voltage of the data line is clamped by Vbe of the bipolar transistor in the sense circuit so that accurate precharge becomes possible.

When the delay of the present circuit is obtained by simulation, the result is 2.5 ns from the driving of the memory cell to the sense output. It is now possible to obtain a high-speed operation by the data line access method in which a bipolar transistor with a large gm is used in the sense circuit so that the data line is operated with a low amplitude.

4.3 Data line precharge method

There is no bottleneck in terms of the speed for the data line precharge if a bipolar transistor is used since only the 4-pF data line is driven. However, as described in section 3.1, the driving mechanism of the word line is operated by the address signal. Therefore, a dc current flows from the precharge circuit via a memory cell connected to the driven word line, and as a result the power consumption is increased.

Figure 8 is the precharge circuit proposed here to resolve the forementioned problem. The precharge circuit precharges only the data line selected by the *Y* selector. The power consumption in the memory cell array can then be reduced to one-eighth. However, the precharge circuit must operate at a high speed since the precharge circuit is activated by the decoded signal of the *Y*

Fig. 9. Microphotograph of the chip.

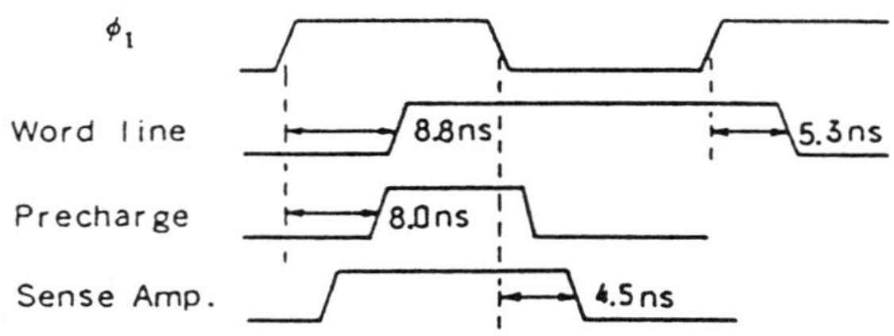

Fig. 10. Measured delay time.

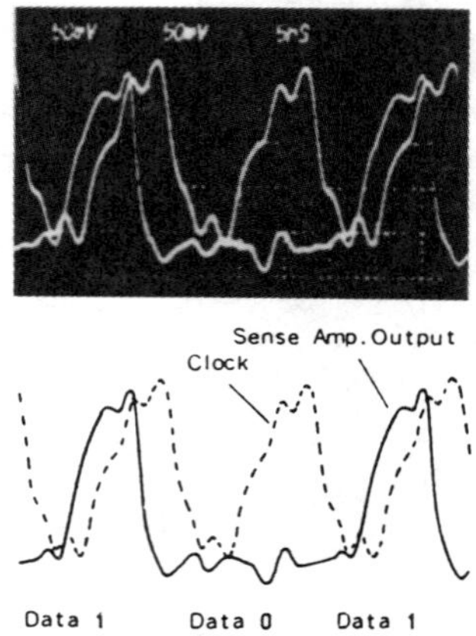

Fig. 11. Sense amp. Output waveform.

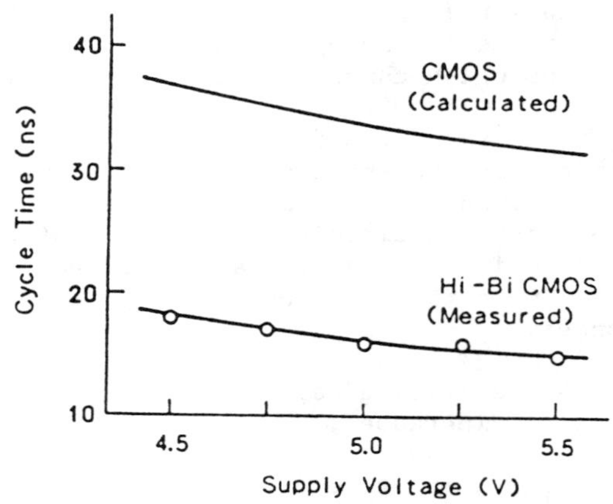

Fig. 12. Micro ROM cycle time.

system. In this circuit, bipolar transistors with a Darlington connection are used for precharge. Then the current supplied by the precharge circuit is reduced to 1/β and hence a high-speed precharge is possible. Also, as described in section 3.2, the potential of the data line must be Vbe. .Hence, the bases of the bipolar transistors with a Darlington connection are clamped by three diodes. The delay of the present circuit obtained by the simulation is 5.9 ns from the address to the data line precharge via a Y decoding. Thus, it is clear that a low power consumption and a high-speed operation are possible by a method in which the bipolar transistors with a Darlington connection are used for selective precharge of the data line.

4.4 High-speed operation of peripheral circuits

In addition to the foregoing, one should select the CMOS and the Hi-BiCMOS depending on the magnitude of the varying load as discussed in section 3 to obtain the optimum driving. In this case, the circuit parameters of the CMOS and Hi-BiCMOS circuits must be considered. The capacitive load by the fan-out number at the circuit output and the wiring length is estimated to be about 0.5 pF.

5. Results of Trial Fabrication

Based on the foregoing discussions, a micro ROM has been fabricated with Hi-BiCMOS technology. Figure 9 shows the microphotograph of the LSI loaded with Hi-BiCMOS micro ROM. The upper portion in Fig. 9 is the 2 k words × 64 bits micro ROM.

5.1 Layout

The memory cell shown in Fig. 1 is used in the memory cell array. GND reinforcement lines are supplied for each 16 bits of data lines. Including the connections to the substrate potential, word line polysilicon, and the second Al layer, the actual size of one bit of the memory cell is 9.2 × 10.6 μm and the size of the memory cell array is

4.74 × 2.72 mm. The entire size of the micro ROM including the peripheral circuit is 5.77 × 3.22 mm.

5.2 Results of evaluations

Figure 10 shows the measured delay time of the internal circuit of the fabricated micro ROM. The high-speed capability of the Hi-BiCMOS described in section 4 is confirmed. Figure 11 shows the output waveforms of the sense circuit at the highest operating speed in the case where the micro ROM is read repeatedly between two addresses. The sense circuit sends out "1" and "0" repeatedly as the data. Figure 12 shows the measured power supply voltage dependence of the cycle time at the maximum operating speed. For comparison, Fig. 12 also shows the simulation results for the micro ROM based on the CMOS with an identical configuration. For the Hi-BiCMOS micro ROM, the cycle time is 16 ns at the power supply voltage of 5 V and is more than twice as fast as than CMOS. The variation of the cycle time is -7 to +13 percent for the power supply voltage variation of ±10 percent. Even with the operating margin, the desired cycle time of 25 ns can be realized.

In the foregoing, we proved by trial fabrication the high-speed operation of the Hi-BiCMOS micro ROM.

6. Conclusions

As an example for realization of a high-speed micro ROM, the Hi-BiCMOS technology is used in this paper since it has both the high degree of integration and low power consumption provided by CMOS and the high-speed capability of bipolar transistors. From comparison of the Hi-BiCMOS circuit and the ROM circuit, it is found that the Hi-BiCMOS is more advantageous for a heavy load while the ROM circuit is more advantageous for a light load. The Hi-BiCMOS word line driver, data line sense circuit, and data line precharge circuit have been developed which make use of high driving capability and large gm values of bipolar transistors.

To confirm the high-speed operation of the developed circuit, a 128-k bit Hi-BiCMOS micro ROM has been fabricated. The cycle time of 16 ns at the highest operating speed is obtained at the power supply voltage of 5 V. The variation of the cycle time is -7 to +13 percent for the power supply voltage variation of ±10 percent. Hence, the high-speed operation of the Hi-BiCMOS circuit has been proven.

Acknowledgement. The authors thank Director Hosozaka and Dr. Ikeda of the Device Development Center, Director Masuda, and Dr. Maejima of Hitachi Laboratory for their guidance. They also thank the staff members who assisted in the design, fabrication and testing of the LSI.

REFERENCES

1. H. Higuchi, G. Kitukawa, T. Ikeda, Y. Nishio, N. Sasaki and K. Ogiue. Performance and structures of scaled down bipolar devices merged with CMOS FETs. IEDM Dig. Tech. Papers, pp. 694-697 (Dec. 1984).
2. Masuda, Nishio and Ikeda. High-speed logic circuits by composite bipolar / CMOS. I.E.C.E. Trans. (C), J67-C, 12, pp. 999-1005 (Dec. 1984).
3. Y. Nishio, I. Masuda, T. Ikeda, M. Iwamura, K. Ogiue and Y. Suzuki. A sub-nanosecond low power advanced bipolar CMOS gate array. Proc. of the ICCD, pp. 428-433 (Oct. 1984).
4. K. Ogiue, M. Okada, I. Masuda, T. Ikeda, T. Yasui, Y. Suzuki and H. Uchida. A 15 ns/250 mW 64 k static RAM. Proc. of the ICCD, pp. 17-20 (Oct. 1985).
5. K. Ogiue, M. Okada, S. Miyaoka, I. Masuda, T. Ikeda, K. Tomomura and T. Ohba. A 13 ns/500 mW 64 kb ECL RAM. ISSCC Dig. Tech. Papers, pp. 212-213 (Feb. 1986).
6. T. Hotta, I. Masuda, H. Maejima and A. Hotta. CMOS/Bipolar circuits for 60 MHz digital processing. ISSCC Dig. Tech. Papers, pp. 190-191 (Feb. 1986).
7. Uchida, Taniguchi, Toyooka, Nakano and Toida. Microcomputer Memory, pp. 35-64. Maruzen (Sept. 1985).

BiFAMOS Technology for High Speed Mega-Bit EPROMs

G.J. Hu, L. C. Tran, P. Keshtbod, J. Segal, K.H. Park, T. Amin, B. Prickett, S. C. Tsao, J. Yen, E. Smith, J. Bornstein, and A. R. Alvarez

Cypress Semiconductor
North First St., San Jose, CA 95134

Introduction

This paper describes an advanced BiCMOS Floating gate Avalanche MOS (BiFAMOS) technology for high speed and high density EPROM applications. It is of great interest to develop chips with access time of 20ns or less to support 33 to 50 MHz systems without SRAM interface. Since channel hot carrier injection is used in EPROM programming, the high current and high voltage programming condition limit the size of the FET that can be used in the cell and hence the low cell current directly impacts speed. A two-transistor cell was introduced[1,2] to overcome this speed limitation but at the expense of area. Here, BiCMOS technology is used for the first time to alleviate this speed/area tradeoff. Fast bipolar sense amplifiers and high current NPN drivers allow very high speed access without large cell current. With a 1-T cell, an access time of 12 ns is obtained on a 1 M-Bit EPROM, demonstrating that this 0.8μm BiFAMOS is capable of sub-20ns 4M EPROMs.

Process and Technology

The fabrication process of this technology starts with a 1.6 μm thin epi layer grown after the buried n and p type layers are implanted. Utilizing a self-aligned twin well process, a twin retro-graded well profile is formed. This structure greatly enhances the latch-up immunity and provides a low resistance collector for the NPN transistors through a deep n+ contact. After field isolation and first gate oxide are formed, phosphorus doped first poly is defined for the floating gate of the FAMOS stack. A thin layer of poly is then deposited after the oxide-nitride-oxide interpoly dielectric and the second gate oxide. At this step, the NPN base and N &P MOS threshold voltage adjustment are implanted. An arsenic-doped poly emitter, also used as the MOS gates, is then deposited immediately after the emitter window is opened. Because the emitter window is larger than the poly emitter electrode, slight silicon gouging takes place during the overetch of the second composite poly. Both NMOS and PMOS are made with LDD structures. The oxide spacer also separates the extrinsic p+ base from the lightly doped intrinsic base area resulting in excellent BVebo and low Ceb. The NPN emitter is isolated by field oxide to form a "walled" transistor. This saves one masking step from the conventional approach. Low emitter resistance is ensured by using a Titanium silicided poly of 1 Ω/square. To further speed up the circuit and improve density, two levels of metal are utilized. Reflowed BPSG below first metal and spin-on-glass below the 2nd metal are used for planarization. Good electromigration reliability is achieved by using Ti/TiW contact metallurgy. The final cross sectional view is shown in Fig. 1 and the key technology parameters are summarized in Table 1.

Device Design and Characterization

Peripheral FET's are fabricated with the 2nd layer polycide. Both N and PMOS are designed for speed and hence relatively heavy LDD doses are used. Circuit limited hot carrier reliability was evaluated and >10 year operating life time has been achieved. Device design tradeoffs between FET, EPROM, and NPN performance have been resolved. The EPROM S/D to gate overlap and the S/D resistance are critical to the device and circuit operation. Because of the high oxide growth rate in the heavily implanted EPROM S/D, the final active doping level is very low unless a high temperature treatment is used. High temperature process is known to degrade the poly emitter device due to the breakdown of interfacial oxide and the difficulty in controlling the implanted base profile. With an optimized 950°C thermal cycle, which delivers the required 50Ω/square FAMOS S/D resistance and satisfactory programming characteristic, the base implant condition has to be changed to compensate for the deeper emitter. This process change results in lower BVebo but higher emitter efficiency (no indication of interface oxide elimination). A blanket high energy punchthrough implant is necessary to control the NMOS short channel effect. Without N-Well profile optimization, this implant adversely affects the NPN and results in low level leakage and smaller current gain. A lower energy implant is not effective in controlling NMOS punchthrough and is detrimental to PMOS. A lower dose or eliminating the implant requires heavy p-well increaseing junction capacitance and body effect, adversely affecting high voltage applications.By carefully balancing the high temperature process, deeper punchthrough implant and moderately doped wells all electrical requirements are satisfied. The FET sub-Vt characteristics and the NPN Gummel plot are depicted in Figs. 2 and 3.The gate swing is 95 mV/decade for both FET's and NPN current gain of >130 are obtained. The measured ft of a NPN vs. the collector current is recorded in Fig. 4.

FAMOS Cell

The FAMOS cell consists of a double poly single transistor without an LDD. Cell channel length is designed to be shorter than the NMOS for higher read current. The short channel effect is particularly severe on the cell due to 1) the floating gate being controlled only through the coupling of the top gate and 2) programming voltage being greater than Vcc. This problem is overcome by applying a bias on the source during programming. This approach is successfully implemented and is scalable to successive generations of technology.Figure 6 illustrates the programed Vt and the punchthrough voltage as a function of Leff. The source biasing scheme works for an Leff design window from 0.35μm to 0.65μm with sufficient margin. Programming speed as a function of channel length is shown in Fig. 5. Snapback on the short device explains the discontinuity in the programming speed. This technology can program to a Vt of 6V within 10 μs. Program disturb characteristics are shown in Fig. 7, where the turn-back of the curve at Vd of 9.5V indicates the on-set of hole injection. This provides sufficient margin for the programming loadline design. Evaluation of read disturb guarantees a minimum 10 year lifetime.

Circuit Performance

CMOS and BiCMOS ring oscillators were studied to demonstrate and calibrate the 1M circuit simulated speed. Table 2 lists the result of both loaded and unloaded circuits measured at 25°C. The typical CMOS and BiCMOS gate delay are 117ps and 257ps respectively. The loaded TTL BiCMOS ring oscillator with equivalent fanout of 11 has a gate delay of 445ps. A 1-T cell with 9.4 μm2 in size has been applied to a 1 M-Bit EPROM. The 64Kx16 architecture gave a die size of 4.8 mm x 6.2 mm. This chip resulted in 12 ns typical access time at 162 mA Icc, which is 25% faster and 2.5x smaller in die size as compared with the published work[2] using a comparable 0.8μm CMOS technology. This result clearly demonstrates that BiCMOS is a viable technology for high performance non-volatile applications to the 4M density and beyond.

Acknowledgement

The effort in the development of the base BiCMOS process by S. Y. Pai and his team was of great help in this work. Also the contribution from G. Gray, M. Cleeves and R. Shrivastava are greatly appreciated.

References

1. D. Hoff et al.,"A 23-ns 256K EPROM with double-layer metal and address transition detection," ISSCC, pp130, 1989
2. N. Matsukawa et al., "Process technology for a 16ns 1Mb CMOS EPROM," VLSI Tech. Symp. pp127, 1990.

Reprinted with permission from *1992 Symposium on VLSI Technology Digest of Technical Papers,* pp. 42–43.

Table 1 Process and Device Parameters

Technology	0.8μm retrograde twin-well, one poly, one polycide, double metal BiCMOS
Gate Oxide	19.5 nm
Min. Poly L	0.8 μm(n), 1.0 μm(p), 0.7 μm(cell)
Cell Size	9.4 μm2
Vt	0.7V(n), 0.8V(p), 1.6V(cell)
Vpp	12.75V
Vtf	17V(n/p)
BN+	25Ω/square
Min. Emit. W	1μm
ft	8 GHz
BVces	>7V
M1/M2 Pitch	2.6 μm

Table 2 Ring Oscillator Gate Delay

CMOS	BiCMOS	BiCMOS(loaded)
117ps	257ps	445ps

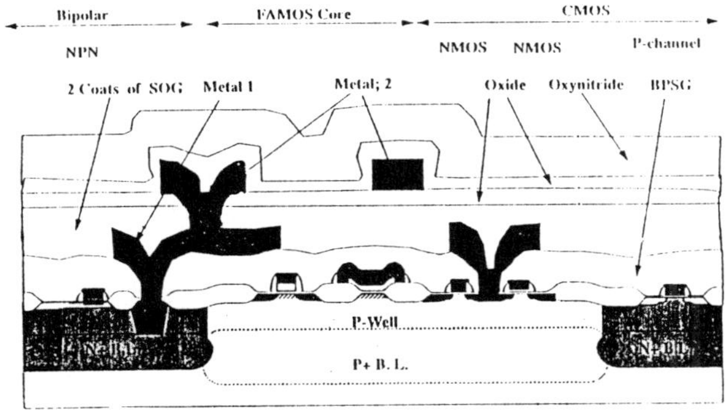

Fig. 1 Cross sectional view of BiFAMOS devices.

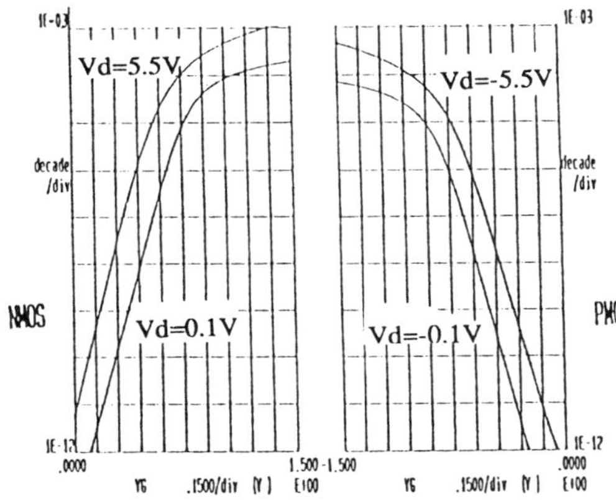

Fig. 2 Sub-Vt Id-Vg characteristics of NMOS and PMOS with channel length of 0.65μm and 0.75μm respectively.

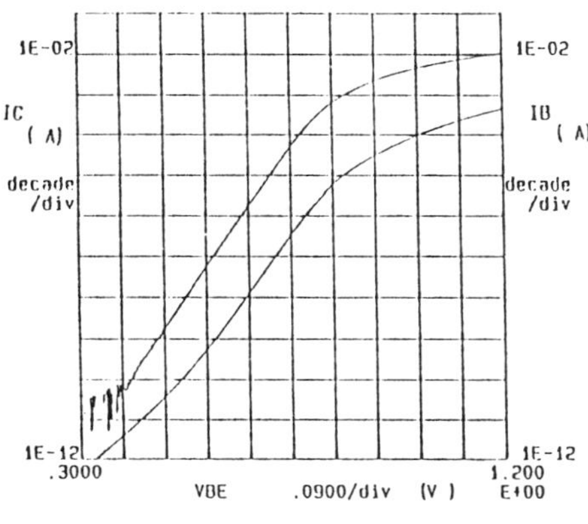

Fig.3 Gummel plot for a 1μm x 6μm NPN. The low level base leakage is due to punchthrough implant.

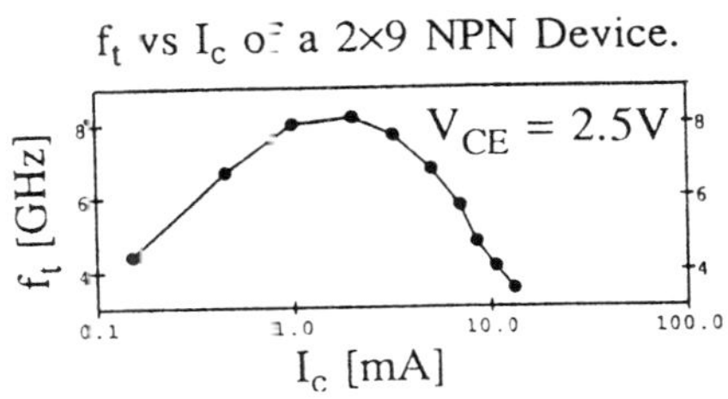

Fig. 4 Measured ft of NPN as a function of collector current

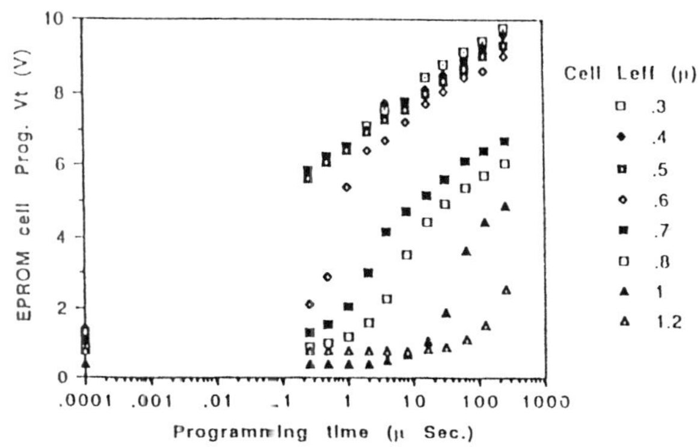

Fig. 5 Programming speed vs. channel length. Programming Vg=13V, Vd=7.5V.

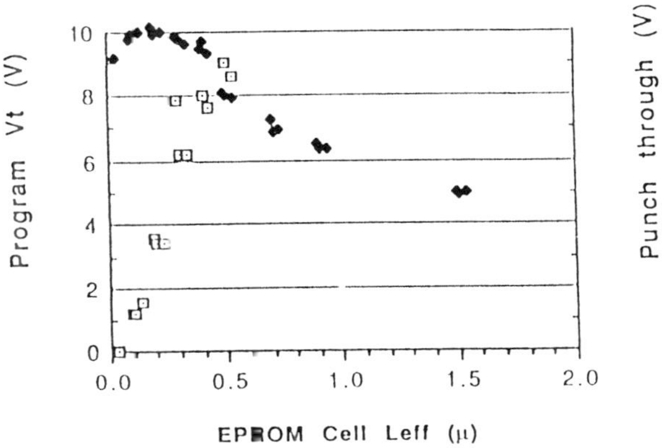

Fig. 6 Program Vt and punchthrough Voltage as a function of cell channel length. Measurements are with 640Ω at source.

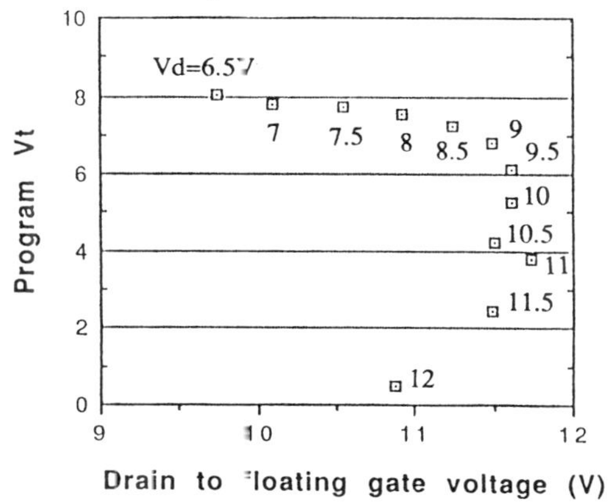

Fig. 7 Program Vt vs. drain to floating gate potential. Absolute drain voltages are marked next to data.

Part 4
BiCMOS Analog Circuit Applications

TODAY, BiCMOS technology has gained a wider acceptance in the design of high-performance digital integrated circuits than in the design of high-performance analog and mixed analog/digital integrated circuits. This is mainly because of the complex design trade-off issues associated with the latter. However, research results have begun to emerge demonstrating that the design issues can be overcome and BiCMOS can be successfully applied to the design of efficient and high-performance analog integrated circuits.

This part consists of three sections. Section 4.1 is an invited overview by Leung. Sections 4.2 and 4.3 discuss BiCMOS analog circuit design and its applications, respectively.

Section 4.2 starts with two general review articles on VLSI analog circuit design, one by Vittoz and the other by Hosticka and Brocherde. The section ends with a paper by Wooley, specifically addressing BiCMOS analog circuit techniques.

Section 4.3 contains ten papers on the application of BiCMOS analog circuits. Sample and Hold circuits are the topic of the three papers by Tsugaru et al., Wakayama et al. and by Mercer. Video circuit application is the topic of the paper by Nayebi and Wooly. Finally, the six papers that follow cover the application of BiCMOS to the design of high-speed comparators, phase-locked loops, operational amplifiers, switched capacitor filters, continuous-time filters and wide-band amplifiers.

Paper 4.1

Analog Circuit Design in BiCMOS Technology: An Overview

B. LEUNG
DEPARTMENT OF ELECTRICAL AND COMPUTER ENGINEERING
UNIVERSITY OF WATERLOO
WATERLOO, ONTARIO, CANADA N2L 3G1

Abstract—**Even though mixed-signal designs in BiCMOS technology have been carried out for some time, implementations of analog-circuit functions in BiCMOS technology are just beginning. This paper will serve to illustrate how selective combinations of bipolar and MOS devices have been used to improve performance in a variety of analog circuits.**

1. Introduction

In view of the performance advantages BiCMOS technology offers in digital circuits such as static memories and ASIC, it is likely to become the mainstream technology. In addition, this technology is fast becoming the norm for large-scale, mixed-signal, integrated-circuit design where dense and complicated logic must coexist with high-accuracy, low-noise, high-speed signal processing. Applications range from audio, medical imaging, video processing, and industrial control, through telecommunications and disk-drive electronics. Finally, the mixing of bipolar and MOS devices toward the realization of specific analog-circuit functions has just begun to emerge.

High-speed, high-precision analog circuits have traditionally been implemented using bipolar transistors, due to their higher transconductance per unit area, lower noise levels, and better matching properties between adjacent devices. MOS transistors, on the other hand, offer simple analog switches with zero dc-offset voltage, high-impedance charge storage nodes, and complementary transistors. This synergy can be put to good use in basic analog circuits such as comparators and operational amplifiers, as well as in systems such as sample/hold amplifiers and multiplexes.

Two additional features are found in analog BiCMOS circuits. The first one is that the circuit sometimes operates at a wider voltage supply range than the single 5-V supply. This may be due to systems requiring bipolar signals centered on ground, thus making a negative supply necessary to avoid level-shifting. Higher supplies may also be required in precision-analog and high-resolution converters in order to improve the signal-to-noise ratio. Circuits may also make use of the available system supply.

A second feature is the availability of other active and passive components. Two common passive elements are high-resistance resistors and highly linear double-poly capacitors. Isolated vertical pnp transistors of comparable speed and voltage performance to the npn transistors are also available in some processes.

This paper illustrates some design techniques available toward the merging of bipolar and MOS transistors. Section 2 reviews operations amplifiers realized in BiCMOS technology. Comparators are discussed in Section 3. Finally, analog subsystems such as analog-to-digital converters will be discussed in Section 4.

2. Operational Amplifiers

BiCMOS technology has been used to create operational amplifier (op-amp) designs with characteristics that were either not possible in pure CMOS or bipolar technologies or are an improvement upon them in a spectrum of application areas. These application areas range from video-speed signal acquisition (e.g., sample-and-hold and track-and-hold) and processing circuits to data-conversion circuits for digital-audio to low-power and low-current circuits for the telephone handset. The known usefulness of BiCMOS technology and its potential for wider use will be illustrated by reference to three published op-amps:

1. Op-amps used in low-voltage (up to 1.5 V) telephone applications [1].
2. A high-speed and high-gain op-amp for switched capacitor (SC) filters [2].
3. A track-and-hold amplifier [3].

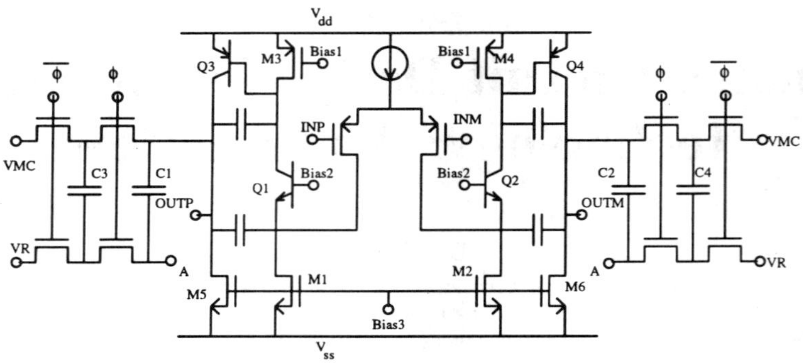

Fig. 1 Differential BiCMOS op-amp for low supply voltage.

The potential use of this technology for transconductor stages and for gain-boosting in cascoded output stages is also discussed. Another example includes a class-AB BiCMOS op-amp [17].

One of the most successful uses of BiCMOS technology has been in the design of very low voltage and low-power circuits for telephone handset applications [1]. These circuits combine a differential switched-capacitor architecture to realize precision filtering, high PSRR, and acceptably low total harmonic distortion (THD) at very low voltages. Figure 1 shows a differential op-amp configuration with dynamic common-mode feedback (CMFB) circuit. It consists of a folded-cascode stage with a PMOS source-coupled differential-pair feeding current into a cascode node. The cascode device is a vertical npn and the active load is a simple PMOS current- source. The second stage consists of a vertical pnp and an npn load. The vertical pnp provides high transconductance and thereby high overall gain for the op-amp. A key feature in the design is the use of a modified Miller compensation scheme with the use of a Miller capacitor C6, returning to the low-impedance cascode node. This capacitance improves the unity-gain frequency of the op-amp by increasing the high-frequency conductance of the op-amp. This becomes necessary because the outputs are loaded by the dynamic CMFB circuitry. Another variant of the op-amp that can drive resistive loads such as the differential-to-single-ended converter has been designed [1]. Here the output-gain stage consists of an npn driver and a PMOS load that is driven by a simple emitter-follower.

Karanicolas et al. [2] report a high-frequency fully differential BiCMOS op-amp for high-frequency switched-capacitor applications. This op-amp features a unity-gain bandwidth of 90 MHz, a phase-margin of 45° and a slew-rate of 150 V/μs with a load of 7 pF using a 2-μm BiCMOS process. This op-amp is essentially a two-stage op-amp. The input stage consists of PMOS input devices and npn emitter-degenerated loads. The emitter degeneration helps to reduce the input-referred noise voltage by reducing the transconductance of the transistor without causing it to saturate. The second gain stage (the output stage) consists of a cascoded npn driver with a cascode PMOS load, resulting in a large gain. This stage is driven by an emitter-follower, which prevents loading of the input stage. Compensation is achieved by using a pole-splitting capacitor in series with a zero-nulling resistor. The high unity-gain bandwidth is due to the high value of the transconductance of the second stage.

Nayebi and Wooley [3] have explored the use of BiCMOS technology for high-speed (video-) sampling systems. Their research concludes that systems using BiCMOS technology can produce video-sampling systems with 15-ns acquisition time and 10-bit resolution. The salient feature in the system consists of a closed-loop track-and-hold amplifier. The basic track-and-hold amplifier architecture consists of an operational transconductance amplifier (OTA) and two sampling switches, followed by an integrating amplifier. The hold capacitor stores the sampled input during the hold mode and also serves as the compensation capacitor of the overall amplifier during the track mode.

The OTA shown in Fig. 2, is a folded-cascode gain stage with vertical npn cascode transistors and was shown to exhibit superior performance in terms of the bandwidth, dc-gain and voltage swing. This design ensures large nondominant poles at the cascode nodes, thus improving the bandwidth/phase-margin of the amplifier. The design also improves its phase-margin by avoiding the use of current-mirrors in the signal path and thus avoiding a ''mirror-pole.''

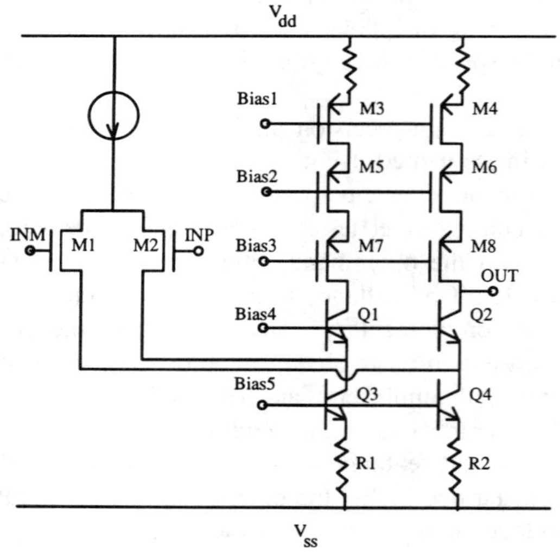

Fig. 2 BiCMOS folded cascode op-amp without current mirror.

The integrating amplifier is a wideband amplifier with a dc-gain of 1000 and is capable of driving a 50-Ω load with a swing of ±1 V. It is made up of a gain stage, a wideband unity-gain buffer, and a wideband class AB output stage. The gain stage is essentially the same as the OTA but with smaller bias currents because of reduced load capacitance.

The application of BiCMOS technology toward enhancing the gain of an op-amp has been explored recently [8]. Traditionally, the output resistance of cascode stages can be increased by applying negative feedback around the cascode device, thus increasing its effective transconductance by a factor equal to the loop gain [12]. This technique was used in [10, 11] to boost the dc-gain of a fully differential folded-cascode CMOS operational amplifier. A dc-gain of 90 dB using a CMOS process was obtained. It is expected that an even higher dc-gain along with higher bandwidth can be obtained by applying the same boosting technique to a BiCMOS folded-cascode amplifier [8]. Figure 3 shows the schematic of the gain-enhanced op-amp. Due to the low input resistance of the bipolar cascode device, an NMOS transistor was used to drive the npn cascode device in a Darlington configuration. This configuration, when compared to the conventional BiCMOS folded-cascode amplifier, also has better frequency response. Spice simulations, using parameters from a 0.8-μm BiCMOS process, show a dc-gain of 120 dB and a unity-gain bandwidth of 250 MHz.

Next, the use of the BiCMOS technology for transconductor stages is discussed. Low-distortion transconductance amplifiers are vital to continuous-time filters with wide dynamic range. With the development of BiCMOS technology, some limitations inherent to CMOS transconductors may be overcome by using the available bipolar transistors on the same chip [5].

Some of the interesting features of bipolar devices that may be useful in such circuits are the high β, the high Early-voltage, the high g_m, the low noise level, and the absence of body-effect. Some of these characteristics have been used in [4] to realize a very linear transconductor using bipolar transistors as cascode devices. This also improves the bandwidth due to the higher nondominant pole associated with the cascode device. The performance results are summarized as follows

- THD less than 0.15% with a 3 V_{pp} differential input signal from a dual-rail 2.5-V supply.

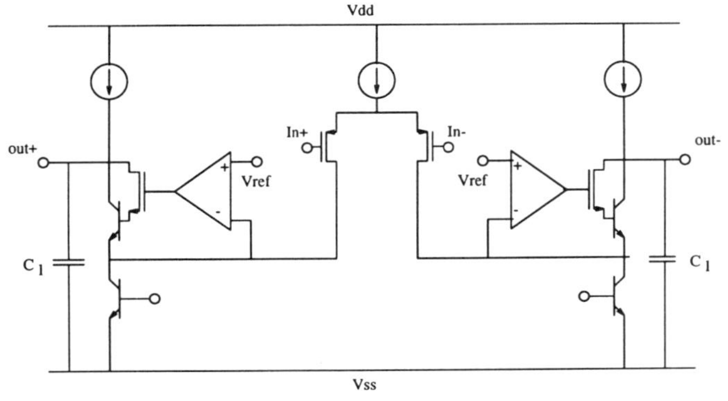

Fig. 3 Gain-boosting stage.

- Very high second pole location (i.e., small excess phase).
- A gain of more than 50 dB.

Negative feedback can also be applied to a transconductor based on the triode region of operation of an MOS device to improve its performance [4]. One major cause of nonlinearity in this type of transconductor is due to the variation of the drain voltage. This voltage variation can be minimized by applying negative feedback around the cascode device [8], hence reducing the impedance of the cascode node. Figure 4 shows a simplified schematic of the BiCMOS transconductor circuit. The impedance looking into the emitter of the cascode device (Q1) in Fig. 4 is reduced by a factor of A + 1 by using negative feedback.

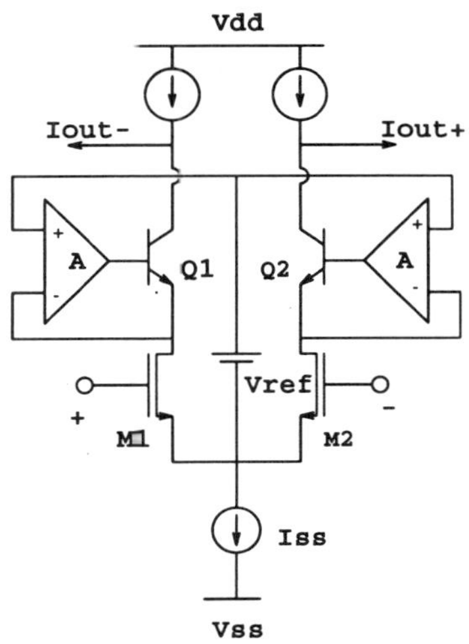

Fig. 4 BiCMOS transconductor.

Spice simulations [8], using parameters from a 0.8-μm process, show a total harmonic distortion of 0.022% (−73 dB) for a 2 V_{pp} differential input signal, and less than 0.08% (−62 dB) for inputs up to 4 V_{pp} from a single 5-V supply. A factor of 2 improvement of the linearity was observed over the case where no feedback was applied [4].

3. Comparators

Comparator design plays a crucial role in the overall system performance of flash analog-to-digital converters (ADCs). Through offset cancellation and self-calibration, MOS comparators are capable of achieving accuracies in excess of 14 bits at frequencies in the audio range. However, the highest-speed comparators have always been implemented with silicon bipolar technologies. The speed and resolution of CMOS comparators is limited by the inherent MOS characteristics of low transconductance and relatively large device and threshold mismatches, making offset cancellation essential for applications requiring more than 8-bit precision. For low-speed applications, a comparator can be obtained using configurations similar to MOS uncompensated operational amplifiers. In such cases, the precision can be efficiently improved by broadbanding the operational amplifier. For high-speed applications, a cascade of linear and regenerative amplifiers are used, since the comparators need not be linear [6].

Figure 5(a) shows a bipolar comparator that has found widespread applications in high-speed flash bipolar ADCs. It consists of a broadband preamplifier cascaded with a regenerative latch. Both are implemented with a bipolar transistor pair. It was found [7] that the speed–power trade-off is dictated by the current needed for the preamplifier to recover from an overdrive condition rather than by the regeneration current. Consequently, the comparator is implemented with direct regeneration and without any preamplification.

In the circuit shown in Fig. 5(b) [6], BiCMOS technology attributes were exploited in the sense that all MOS devices are used as switches, whereas the bipolar devices constitute the latch. The circuit operates in two modes, sampling and regeneration. In the sampling mode, the latch is disabled by disconnecting its "tail current." The input is then sampled on capacitors C1–C4 to create an imbalance across the bases of the two transistors. In the regenerative mode, the bases are isolated from the input signal and the two transistors are cross coupled through capacitors C3 and C4. The initial voltage difference between the bases of Q1 and Q2 is regeneratively amplified at the collectors. This scheme reduces the input-power dissipation since the latch is powered during only one-half of the cycle. The experimental measurements of such a comparator fabricated in a 0.8-μm BiCMOS technology show a maximum operating frequency of 200 MHz, an input offset voltage of 2.1 mV, and a power dissipation of 1.6 mW.

For high-precision comparators (10–14 bits), self-calibration techniques have been used in CMOS implementations. In a BiCMOS environment, self-calibration techniques can be combined with high-speed low-offset bipolar-circuit techniques to achieve high conversion rates together with high precision. A 12-bit 10-MHz self-calibrated BiCMOS comparator is shown in Fig. 6 [9]. In this circuit, a 200-μV input-referred offset voltage at 10-MHz conversion rate was obtained through the use of a bipolar latch combined with self-calibration. The comparator consists of a bipolar preamplifier (driven by MOS buff-

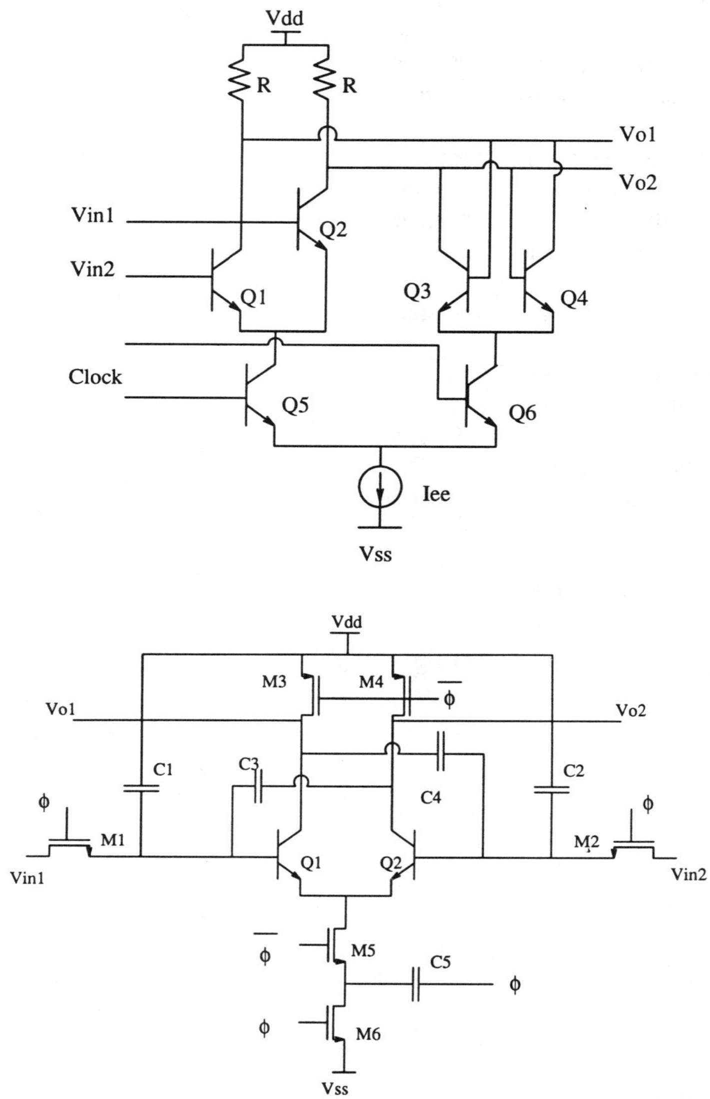

Fig. 5 (a) Bipolar latched comparator; (b) BiCMOS regenerative amplifier [6].

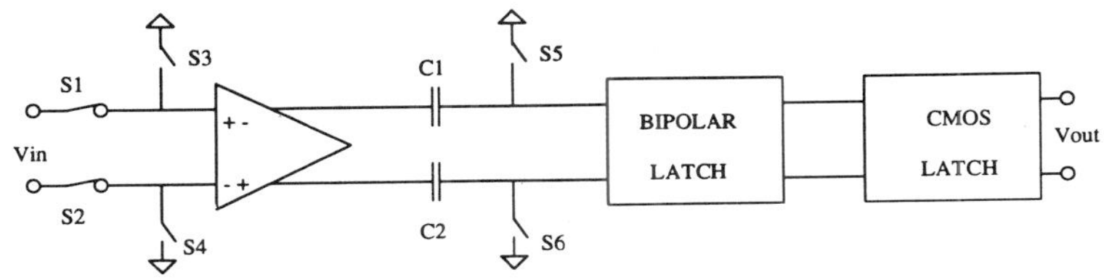

Fig. 6 BiCMOS self-calibrated comparator [9].

ers), offset storage capacitors C1 and C2, a bipolar latch, and an MOS latch. In the calibration mode, the input-offset voltage of the preamplifier is amplified and stored on capacitors C1 and C2. In the comparison mode, the two latches are sequentially strobed to generate CMOS levels at the output. The BiCMOS latch is used because of its low offset voltage compared to an MOS latch. The residual input-referred offset voltage is then dominated by the bipolar-latch offset divided by the preamplifier gain. The bipolar latch is similar to the one previously described, with zero static-power dissipation. This comparator was fabricated in a 2-μm BiCMOS process and experimental results show a 200-μV input-offset voltage with a maximum of 10 MHz comparison rate and a power dissipation of 1.7 mW.

The presence of both bipolar and MOS transistors on the same chip makes it possible to combine the complementary advantages of both devices, namely, the zero-offset MOS switching property and the high speed and good matching properties of bipolar transistors. Furthermore, self-calibration techniques that are well established in CMOS technology can now be applied to bipolar circuits in a very compact fashion without making the design too complicated.

4. Analog Subsystems

Recent interest in applying digital signal processing in such fields as high-definition television, moving target radar, and digital audio has created a motivation for improving the performance of high-speed, high-precision analog-to-digital (A/D) and digital-to-analog (D/A) converters by applying BiCMOS circuit techniques. The converters used in these systems must meet stringent performance requirements with regard to resolution, speed, power dissipation, and area. By using selective combinations of bipolar and MOS devices, significant performance enhancements have been realized in the fields of D/A and A/D conversion.

As mentioned in Section 3, BiCMOS comparators have been used recently in realizing high-speed, high-resolution flash A/D converters. Although these converters provide the performance required at high speeds, excessive power dissipation and area consumption are major drawbacks. In a flash converter, the comparator count increases exponentially with the required resolution. Multistage architectures offer an alternative to flash converters with reduced comparator count.

A recently reported design [13] demonstrates how BiCMOS technology offers new design opportunities for multistage A/D converters. This reference describes a 10-bit two-step parallel BiCMOS ADC with a maximum conversion rate of 30 MHz that was fabricated in a 1.2-μm BiCMOS technology. It features a 6-bit coarse comparison stage and a 5-bit fine stage (1-bit overlap of the coarse and fine comparisons). The key feature of this design is the BiCMOS sample-and-hold (S/H) circuit, which is shown in Fig. 7. The input signal is sampled by a diode bridge, held in a capacitor and buffered by a BiCMOS op-amp. The high input impedance of the MOS-input transistors eliminates droop-rate degradation traditionally caused by the base current of bipolar transistors in sample-and-hold circuits. The S/H circuit achieves a droop rate of 0.2 mV/μs. Bipolar transistors are used in the output stage of the S/H circuit for their current-drive capability. This chip consumes 750 mW from a dual-rail 5-V supply and occupies an area of 25.3 mm^2.

A pipelined A/D converter reduces the hardware complexity further without sacrificing speed by performing several operations concurrently. A 10-bit, 20-MHz BiCMOS pipelined A/D using current-mode signals has recently been reported [14]. The use of current-mode signals allows the use of current-summing techniques, which substantially increases system speed. It also eliminates the need for difference amplifiers. A key component of this design is the wide bandwidth differential BiCMOS track-and-hold amplifier, as shown in Fig. 8. An n-channel MOS transistor (M1) is used as the hold switch. The integrator has MOS inputs to minimize droop and an emitter-follower output stage to drive the hold capacitor. All transconductance stages consist of resistively degenerated differential pairs to provide good linearity. The collector node of Q1 is bidirectionally clamped with respect to the BIAS2 voltage, providing a controlled voltage limit during the hold mode when the feedback loop is opened. This improves the acquisition time.

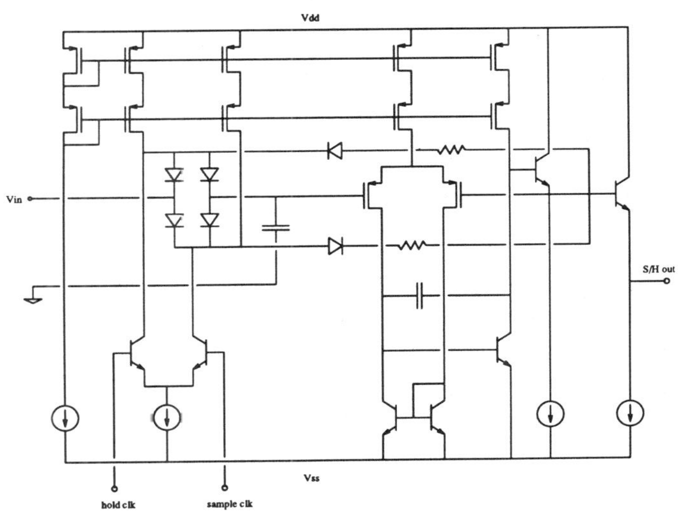

Fig. 7 BiCMOS sample-and-hold circuit.

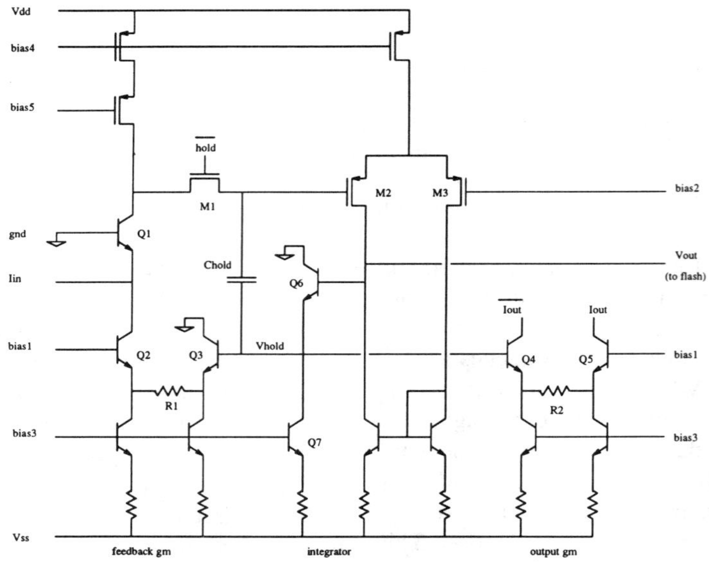

Fig. 8 BiCMOS differential track-and-hold amplifier.

Recently, BiCMOS circuit techniques have also been applied to the design of digital-to-analog converters (DACs) based on current steering and charge redistribution techniques. Current steering D/A converters exploit the high speed of current switching to achieve high-speed conversions. Conversion is achieved by generating a set of currents and selectively switching and summing them to provide an analog output. In a recently reported design [15], a new BiCMOS current cell and switch for current steering DACs was proposed. The BiCMOS self-calibrated current cell offers higher output resistance, smaller minimum voltage, and shows a factor of 2 improvement in simulated current accuracy. The current cell is shown in Fig. 9(a) and its operation is as follows: When the switches are operated as shown in the figure, the difference between the reference current and the collector current of QN1 flows into the gate of MN1 and charges up the gate-source capacitance. MN1 acts as a source follower with source resistance approximately equal to $1/g_m$ of QN2. As the gate voltage of MN1 increases, the collector current of QN1 will increase until the collector current of QN1 equals the reference current, thereby completing the calibration process. Because bipolar transistors have a larger Early voltage than MOS devices, the output resistance of the cell is increased. Additionally, the available output-voltage swing is limited by $V_{ce(sat)}$ and not $V_{ds(sat)}$. Finally, the collector-base capacitance (C_{bc}) of the bipolar transistor is lower than the gate-drain capacitance (C_{gd}) of the MOS transistor, so that the coupling of output-voltage variations back to the storage node is reduced. This lower output capacitance also translates into faster current switching. The current error of this cell can be further reduced with the addition of a main current source in parallel with the new current cell that carries about 90% of the reference current. Since the calibrated cell carries only 10% of the reference current, the g_m of QN1 and QN2 are both reduced by a factor of 10 and the g_m of MN1 is reduced by $\sqrt{10}$. The transconductance of the overall stage is therefore reduced by $\sqrt{10}$. This allows the size of MN1 to be optimized to reduce W/L for the same $V_{ds(sat)}$, resulting in a final reduction of the overall transconductance by a factor of 8, while keeping C_{gs} of MN1 constant. The new current cell was simulated in Spice, along with a main current source that carries 90% of the reference current, using parameters from a 0.8-μm BiCMOS process and indicating an absolute error of less than 0.2%.

In [15], a new BiCMOS current switch was also proposed. This switch exhibits no base current error and achieves close to a factor of 2 improvement in switching speed when compared to a MOS switch. The switch is shown in Fig. 9(b). Transistor M1 is used to return the base current to the switch output, thereby eliminating any DAC nonlinearity due to β mismatches between the bipolar transistors. Resistors R1 and R2 are used to pull down the bases of Q1 and Q2. Since they are returned to the emitters of Q1 and Q2 rather than to ground, the voltage swing at the bases of Q1 and Q2 is reduced. M1 is also used as a level-shifter for turning on Q1. The BiCMOS current switch was simulated using parameters from a 0.8-μm BiCMOS process and achieved a switching delay of 0.275 ns.

BiCMOS technology has also been applied to the design of a charge redistribution DAC, which operates on the principle of charge division. The availability of high-quality MOS capacitors in a BiCMOS process and the high input impedance of MOS transistors makes BiCMOS a good process technology for charge redistribution DACs. In a recently reported design [16], an 18-bit BiCMOS bit stream D/A converter achieving a total harmonic distortion of −102 dB and a dynamic range of 108 dB has been realized. The chip contains two identical conversion channels, each consisting of a switched-capacitor D/A network, digital inputs, and two high-performance BiCMOS operational amplifiers. The input circuits, clock generation circuits, and the switched-capacitor D/A network are all based on MOS devices. The BiCMOS circuit techniques come into play in the design of the high-performance op-amp. The op-amp is a class A Miller-compensated design using a PMOS input stage with large input transistors for low input noise. A Miller-compensated second stage is used to boost the gain,

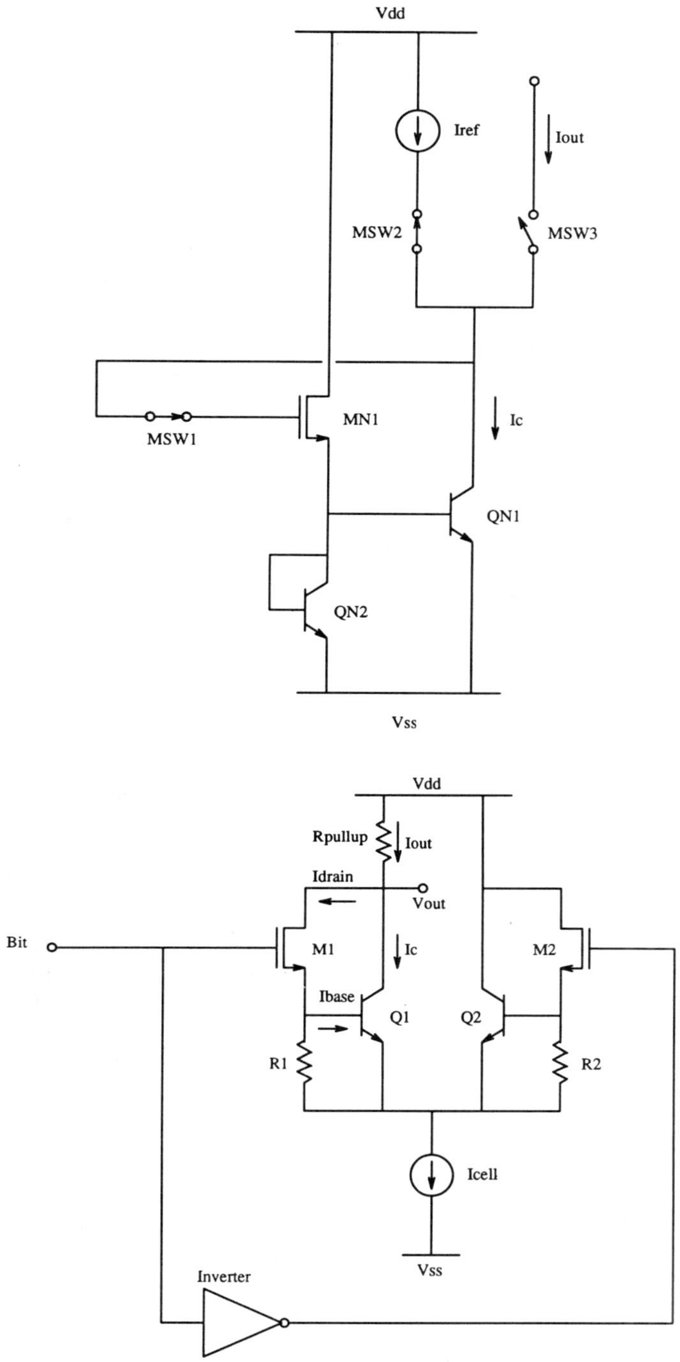

Fig. 9 (a) BiCMOS self-calibrated current cell; (b) BiCMOS-based current switch.

followed by a bipolar output buffer. The chip occupies 7.3 mm^2 and is fabricated in a 2.6-μm BiCMOS process.

5. Conclusions

The unique properties of bipolar and MOS devices bring about a synergy in the realization of analog-circuit functions that rivals similar designs as demonstrated in conventional CMOS technology. As BiCMOS becomes the mainstream technology and designers grow more accustomed to innovate in this environment, new cell designs will emerge in which bipolar and MOS transistors are inextricably merged, resulting in significant improvements.

Acknowledgments

The author would like to thank Seddik Hadri, Richard Romanczyk, and Subhajit Sen for their contributions.

References

[1] R. Castello and L. Tomasini, "1.5-V high-performance SC filters in BiCMOS technology," *IEEE J. Solid-State Circuits,* vol. 26, no. 7, pp. 930–936, July 1991.

[2] A. N. Karanicolas, K. E. O., J. Y. A., Wang, H. Lee, and R. L. Reif, "A high-frequency fully differential BiCMOS operational amplifier," *IEEE J. Solid-State Circuits,* vol. 26, no. 3, pp. 203–208, March 1991.

[3] M. Nayebi and B. A. Wooley, "A 10-bit video BiCMOS track-and-hold amplifier," *IEEE J. Solid-State Circuits,* vol. 24, no. 6, pp. 1507–1516, Dec. 1989.

[4] R. Castello et al., "A very linear BiCMOS transconductor for high-frequency filtering applications," in *Proc. of IEEE International Symposium on Circuits and Systems,* pp. 1364–1367, 1990.

[5] S. D. Willingham and K. W. Martin, "BiCMOS components for video-rate continuous-time filters," in *Proc. of IEEE International Symposium on Circuits and Systems* pp. 2691–2694, 1992.

[6] J. T. Wu and B. A. Wooley, "A 100-MHz pipelined CMOS comparator," *IEEE J. Solid-State Circuits,* vol. SC-23, pp. 1379–1385, Dec. 1986.

[7] P. J-W. Lim, "Performance limits of circuits for analog-to-digital conversion." Ph.D. thesis, Stanford University, March 1991.

[8] S. Hadri, and B. Leung, "Impedance boosting technique based on BiCMOS technology," *IEEE J. Solid-State Circuits,* vol. 28, no. 2, pp. 157–161, Feb. 1993.

[9] B. Razavi and B. A. Wooley, "A 12-Bit 10-MHz BiCMOS comparator," in *IEEE 1991 Bipolar Circuits and Technology Meeting Tech. Dig.,* pp. 289–292, 1991.

[10] K. Bult and G. Gleen, 'A fast-settling CMOS op amp for SC circuits with 90-dB DC gain," *IEEE J. Solid-State Circuits,* vol. 25, pp. 1379–1384, Dec. 1990.

[11] M. Yotsuyanagi, T. Etoh and K. Hirata, "A 4th-order bandpass sigma–delta modulator," in *Proc. of IEEE Custom Integrated Circuits Conference,* pp. 16.7.1–16.7.4 1992.

[12] W. Guggenbuhl et al., "A high-swing, high-impedance MOS cascode circuit," *IEEE J. Solid-State Circuits,* vol. 25, no. 1, pp. 289–298, Feb. 1990.

[13] A. Matsuzawa, M. Kagawa et al., "A 10b 30 MHz Two-Step Parallel BiCMOS ADC with Internal S/H," in *IEEE ISSCC Digest of Technical Papers,* pp. 162–163, Feb. 1990.

[14] P. Real, D. H. Robertson, C. W. Mangelsdorf, and T. L. Tewksbury, "A wideband 10b 20 Ms/s pipelined ADC using current mode signals," *IEEE J. Solid-State Circuits,* vol. 26, no. 8, pp. 1103–1108, Aug. 1991.

[15] B. Leung, "BiCMOS current cell and switch for digital to analog converters," *IEEE J. Solid-State Circuits,* vol. 28, no. 1, pp. 68–71, Jan., 1993.

[16] B. M. J. Kup, E. C. Dijkmans, P. J. A. Naus, and J. Sneep, "A bit stream digital to analog converter with 18-b resolution," *IEEE J. Solid-State Circuits,* vol. 26, no. 12, pp. 1757–1762, Dec. 1991.

[17] S. Sen, B. Leung, "A low-power class-AB BiCMOS op-amp using pseudo-PNP transistors, ' in *Proc. IEEE International Symposium on Circuits and Systems,* pp. 1136–1139, 1993.

Paper 4.2

FUTURE OF ANALOG IN THE VLSI ENVIRONMENT

Eric A.Vittoz

Swiss Center for Electronics and Microtechnology (CSEM)
Maladière 71, CH 2007 Neuchâtel
Switzerland

ABSTRACT

Analog circuits will remain **irreplaceable** for the implementation of the interface circuitry between digital processing and the external world. Most of their characteristics can be improved with scaled-down processes. Analog circuits will also **compete** with digital and will remain advantageous for low-precision signal processing. Since little precision is required to carry out many cognitive tasks, analog VLSI will **dominate** in the implementation of many types of neural networks.

INTRODUCTION

Present technologies allow complete systems to be implemented on a single chip. This trend will be reinforced with the advent of scaled-down processes and larger wafer size. Even if most of these systems use digital techniques, they need to be interfaced with the external world by some analog circuitry mainly located at the periphery of the chip. The impact of process scaling on the performance of this irreplacable analog circuitry will be discussed.

Even though scaling is giving an increasing advantage to digital in the competition with analog to implement signal processing circuits, it will be shown that analog will stay the best candidate for low precision processing. Now, many of the new tasks to be carried out by advanced processing systems will be of cognitive nature. They will be increasingly inspired by the brain structure and organization. Since most of these neuromorphic systems are based on massive parallelism and connectivism instead of precise computation, analog will dominate in their implementation on silicon.

INTERFACE CIRCUITS

In addition to AD and DA conversion, the interface circuits needed between digital processing and the external world include amplifiers, prefilters, normalizers, compressors, power drivers and even compatible sensors. They do not occupy a significant percentage of chip area, but their characteristics are often very critical for the overall performance of a system-on-the-chip.

Scaling down the process to increase the density of digital function will have an effect on the performance of these interface circuits.

This effect will be examined by assuming constant field scaling (CEF), which is the most natural scaling law, the most important consequences of which are reported in Table 1 [1].

Vertical dimensions	x	K^{-1}
Voltages	V	K^{-1}
Impurity concentration	N	K
Capacitance per unit area	C'	K
Lateral dimensions	W, L	$\geq K^{-1}$

Table 1: Constant-field scaling by factor K

The following key parameters will be affected by scaling:

Transconductance. Important since it is a measure of the activity of the transistor, which have a direct effect on its minimum noise, its driving capability and its bandwidth for a given capacitive load. If the device is operated in strong inversion its transconductance may be expressed as:

$$g_m \sim (I_D C_{ox} W/L)^{1/2} \qquad (1)$$

where I_D is the drain current, C_{ox} is the oxide capacitance per unit area and W, L are the width and length of the channel. If W/L is increased, g_m stops increasing when the device starts operating in weak inversion. This maximum value of transconductance only depends on the drain current:

$$g_{mmax} \sim I_D \qquad (2)$$

Dynamic range. It can be defined as the ratio of the maximum signal voltage V_S to the noise voltage. If low frequency 1/f noise is negligible, the noise power spectral density is inversely proportional to the transconductance g_m. Thus, for a fixed bandwidth, the dynamic range is given by

$$DR \sim V_S(g_m)^{1/2} \qquad (3)$$

Mismatch. Two contributions must be distinguished.

The contribution of random parameter fluctuations is proportional to the variance of these fluctuations and inversely proportional to the area if spatial correlation can be neglected. If the amplitude of the fluctuations is assumed to be independent of the scaling factor, the RMS value of this contribution is given by

$$\sigma_{ms} \sim (WL)^{-1/2} \qquad (4)$$

The second contribution is that of edge fluctuations which affect the respective values of WL or W/L of the devices to be matched. For a fixed amplitude of edge fluctuations, the RMS relative value of this contribution can be shown to be proportional to

$$[1/(WL^2) + 1/(W^2L)]^{1/2}$$

Now, the amplitude of these edge fluctuations can reasonably be assumed to be reduced proportionally to the scaling factor. Indeed, the line definition must be K times better to allow the line width be reduced by a factor of K. Thus, for a given W/L ratio, the RMS value of the contribution of edge fluctuations to mismatch can be expressed by

$$\sigma_{me} \sim KW^{-3/2} \qquad (5)$$

Low-frequency 1/f noise. For a fixed value of f_{max}/f_{min}, the 1/f noise voltage component is given by:

$$V_{1/f} \sim (C_{ox}{}^m WL)^{-1/2} \qquad (6)$$

where m is a factor ranging from 1 to 2.

Reprinted with permission from *International Symposium on Circuits and Systems*, pp. 1372–1375, 1990.

Charge injection. The amount of charge that is released in a holding capacitor C when the sampling switch is blocked is proportional to the total charge q that forms the channel. The latter is related to the on-conductance g of the switch by

$$q = gL^2/\mu \qquad (7)$$

where μ is the mobility of the charge carriers. Thus, the relative voltage variation across C, which is a measure of the degradation of the stored value V_S by charge injection, is given by:

$$\Delta V/V_S \sim gL^2/(CV_S) \qquad (8)$$

The third column of Table 2 shows how these parameters are affected by applying systematically (also to the lateral dimensions) the CEF scaling law according to Table 1. The area of the analog part will then be reduced by K^2, but its performance will be degraded by the degradation of most of the key parameters.

		Full scaling	Constant area: constant current	Constant area: constant power
Transconductance	g_m	1	$K^{3/2}$	K^2
Max. transcond.	g_{mmax}	K^{-1}	1	K
Dynamic range	DR	K^{-1}	$K^{1/2}$	K
Max. dyn. range	DR_{max}	$K^{-3/2}$	K^{-1}	1
Mismatch, surface	σ_{ms}	K	1	
Mismatch, edge	σ_{me}	$K^{1/2}$	K^{-1}	
1/f noise	$V_{1/f}$	$K^{1-m/2}$	$K^{-m/2}$	
Charge injection	$\Delta V/V_S$	1	const. speed: K^{-1}	

Table 2 : Effect of scaling on analog circuits

Now, the interface circuits do not occupy a large area on the chip, and thus they do not need to be reduced in dimension together with the digital part. If the area is kept constant, then the effect of scaling is shown in the last columns of Table 2.

If constant current can be further accepted, g_m increases with $K^{3/2}$ whereas its maximum value (in weak inversion) is limited by the available current. If constant power is acceptable, the reduction of voltage can be compensated by an increase of current and g_m increases with K^2. These variations of transconductance with K are sketched on Fig 1.

The degradation of dynamic range due to a reduction of signal amplitude V_S can be avoided by keeping constant power. Mismatch due to surface fluctuations stays the same whereas that due to edge fluctuations is drastically reduced. Low-frequency 1/f noise is improved.

Charge injection does not depend on area. Keeping the same amount of relative charge injection $\Delta V/V_S$ would allow g/C and therefore the speed to be increased by the factor K. Keeping the same speed reduces $\Delta V/V_S$ by K.

These results show that instead of being degraded by process scaling, as is often stated by implicitely referring to the full scaling intended for digital circuits, the performance of analog interface circuits will benefit from scaling, provided they are carefully redesigned in an optimum manner.

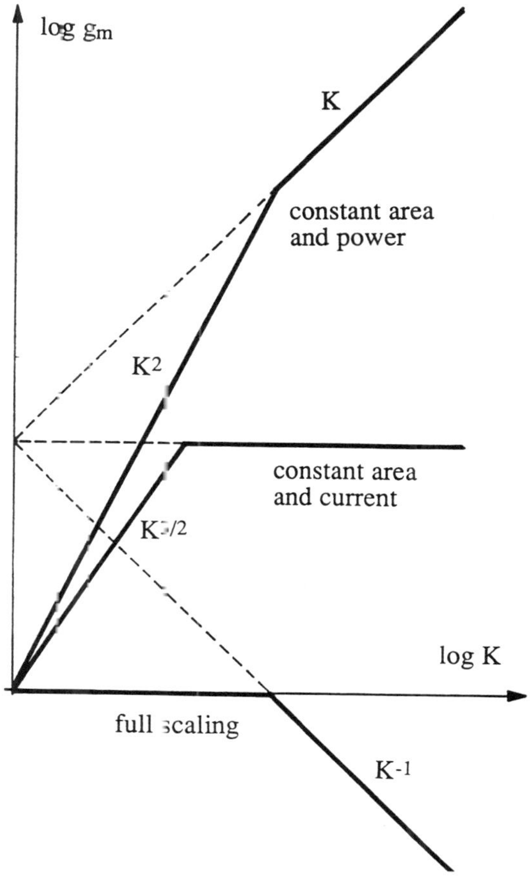

Fig. 1. Effect of CEF scaling on transconductance g_m

SIGNAL PROCESSING

The evolution of scaled-down VLSI processes steadily moves the boundary between digital and analog implementations of signal processing systems in favor of more digital. However, with any given process, analog keeps its advantage for the realization of some tasks. As an example, the comparative merits of analog and digital implementations of filters will be discussed.

One important criterion for comparison is the power consumption per pole This power obviously depends on the required dynamic range defined as the maximum signal-to-noise ratio.

Assuming an analog implementation with one capacitance C per pole and infinite efficiency of the active part of the circuit, the absolute minimum power needed to achieve an RMS value V_S of signal at frequency f is

$$P = 8fCV_S^2 \qquad (9)$$

Now the minimum total noise variance V_N^2 is only limited by C according to

$$V_N{}^2 = kT/C \qquad (10)$$

Thus, the minimum possible energy per cycle and per pole can be expressed as

$$P/f = 8kT(V_S/V_N)^2 \qquad (11)$$

Much more detailed considerations have produced approximately the same limit for low-Q switched capacitor filters, f being the edge frequency of the filter [2]. This limit, which does not depend on the process and is therefore independent of scaling, is reported in Fig.2. It shows that a large increase in power is needed to improve the signal-to-noise ratio in an analog implementation (a factor 10 for 10dB).

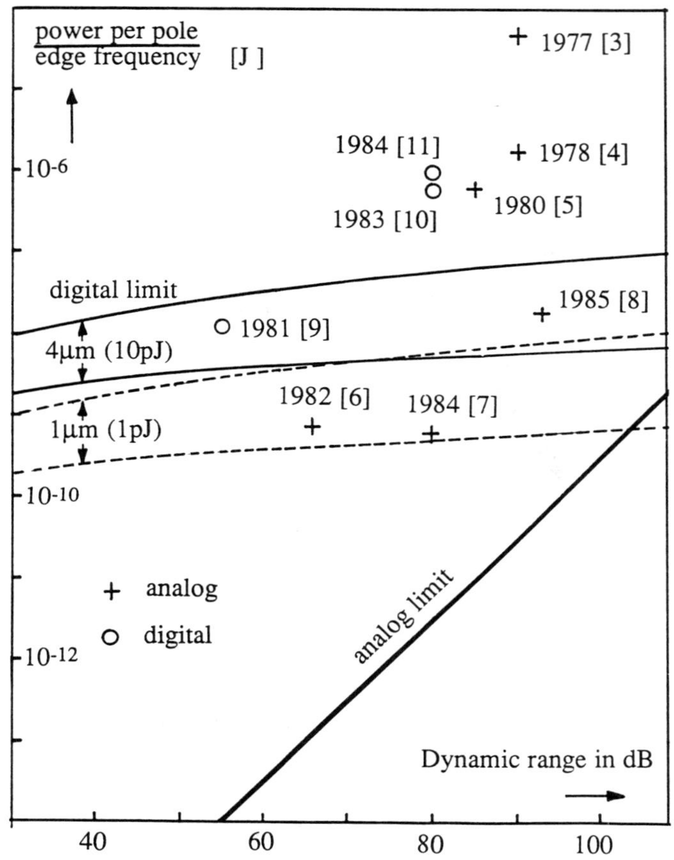

Fig.2. Comparison of analog and digital filters

Data obtained from experimental results on SC filters published since 1977 are reported on the same plot. They show that considerable improvements have been achieved between the very first result which was 6 orders of magnitude above the asymptote and the best one reported until now, which is 4 orders of magnitude better. No further improvements have been obtained during recent years, which were rather devoted to improving other features. No doubt that improved design techniques will allow the limit to be approached closer in the future.

The maximum signal-to-noise ratio of a sinusoidal signal digitized with n bits is given by

$$(V_S/V_N)^2 = 3/2.2^{2n} \quad \text{or} \; (6n + 1.76)\,\text{dB} \qquad (12)$$

The power consumed by a digital filter depends on the number of operations required per cycle, which is essentially a function of the way the necessary multiplications are carried out, and on the energy required for each operation, which can be reduced by scaling.

The implementation of each pole is assumed to require at least an n-bit shift register, an n-bit full-adder and two coefficient multipliers. If the shift of 1 bit is defined as the elementary operation (switching of about 14 transistors in CMOS), the full adder requires about 3 operations per bit. The most simple multiplication is the wired shift (powers of 2) which requires no energy, whereas 2 complete multiplications require N^2 full adders. Assuming a clock frequency 10 times the signal frequency, the total number of operations per cycle then ranges from 40n to 40n + $30n^2$. These numbers do not depend on the amount of parallelism that is used in the architecture. The energy per operation is evaluated at 10pJ for a conservative 4µm 5 volt process and at 1pJ for a 1µm 3 Volt process. The corresponding ranges of minimum energy per cycle and per pole are also reported on Fig. 2.

These limits are very slightly dependent on dynamic range. Data from conservative processes extracted from publications show that they have been already attained by most of the early digital filter implementations. The only way to reduce the power is to use scaled-down processes.

Comparison of the limits for analog and digital clearly shows that analog filters may consume much less power than their digital counterparts if a small dynamic range is acceptable. But analog becomes very power inefficient when a large dynamic range is needed. Comparison of chip area leads essentially to the same qualitative results [2], which may be thought to be valid for most signal processing functions, since they are based on very fundamental considerations.

Now noise and precision are closely related in digital circuits. The precision of analog circuits is usually related to the precision of device matching (.1 to 10%), but it is ultimately limited by noise. It may thus be concluded that the decision breaking point between analog and digital signal processing does not depend on the required speed or on the allowed power consumption, but on the required precision. Analog remains potentially advantageous over digital for low-precision processing.

This potential will be exploited further by developing new circuit techniques. The insensitivity to process and the reliability will be improved by a more systematic use of "healthy" circuits. Design, simulation, layout and test of analog circuits and systems will be facilitated by further improvement in advanced analog CAD tools, in which the design knowlwedge is progressively accumulated [12].

NEUROMORPHIC SYSTEMS

The idea of taking inspiration from the brain's structure and organization to build advanced systems has attracted a lot of interest. These artificial neural networks should be able to carry out tasks that are not easily or not at all tackled by traditional computer approaches, in particular tasks of a cognitive nature.

Now, precision is not required to solve a large variety of these tasks and therefore analog solutions will be more efficient, according to the conclusion derived above. It may thus be predicted that analog VLSI will dominate in the implementation of many types of artificial neural networks.

Transistors in digital circuits are only operated as switches. In analog circuits, just a very small number of transistors may already provide a wide variety of operations such as current sourcing and copying, control of a conductance by a voltage or by a current, amplification, multplication, normalisation, nonlinear function generation (quadratic, exponential, hyperbolic tangent), sampling of signals, storage of information, light sensing. Furthemore the addition of signals can be carried out by simply adding currents and the distribution of information only requires a potential on a single wire. Processing can be carried out in a continuous manner which eliminates the distortions due to undersampling.

Analog circuits will be used much beyond their linear capabilities since nonlinear functions are omnipresent in neural systems. They will allow the various possibilities offered by transistors and other elementary devices to be exploited in an

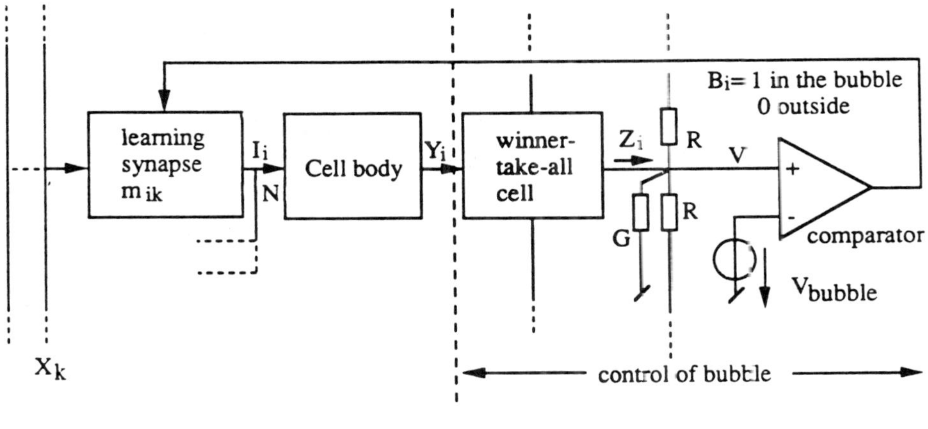

Fig.3. Cell of an analog Kohonen map with limited number of physical lateral interconnections.

opportunistic manner, in order to build cells that provide sophisticated functions on a very small chip area with a very low power consumption. This will allow very large neural networks to be implemented on single chips. These networks will be able to carry out wide-scale collective computation by means of true physical parallelism.

Several analog neural systems have already been successfully implemented on silicon [13][14][15], many of them still exploratory. The important problem of storing analog values is being attacked in various ways and will no doubt find solutions sufficiently dense to build large learning systems.

Adequate architectures will allow the required functional connectivity to be implemented with a limited number of separate physical interconnections between cells [16]. Fig. 3 shows as an example the block diagram of one cell of an analog Kohonen map. The interconnections required to achieve the collective behaviour of the network are limited to the single common wire on which the activities of all the cells are compared by a winner-take-all and a two-dimensional resistance-conductance network. This network controls the radius of the group of cells ("bubble") which must learn the input vector [17].

REFERENCES

[1] S.Wong and C.Salama, "Impact of scaling on MOS analog performance", IEEE Journ. Solid-State Circuits, Vol.SC-18, p.106, Febr.1983.

[2] P.Gray and R.Castello, "Performance limitations in SC filters", in *Design of MOS VLSI Circuits for Telecommunications* edited by Y.Tsividis and P.Antognetti, Prentice-Hall, Englewood Cliffs, 1985.

[3] B.Hosticka *et al*, "MOS sampled recursive filters using switched capacitor integrators",IEEE Journ. Solid-State Circuits, Vol.SC-12, pp.600-608, Dec. 1977.

[4] D.Alstott *et al*, "MOS switched capacitor ladder filters", IEEE Journ. Solid-State Circuits, Vol.SC-13, pp.806-814, Dec.1978.

[5] I.A.Young, "A low-power NMOS transmit/receive IC filter for PCM telephony", IEEE Journ. Solid-State Circuits, Vol.SC-13, pp.997-1005, Dec.1980.

[6] F.Krummenacher, "Micropower switched capacitor biquadratic cell", IEEE Journ. Solid-State Circuits, Vol.SC-17 pp.507-512, June 1982.

[7] Masuda *et al*. "CMOS sampled differential push-pull cascode operational amplifier", Proc. ISCAS 1984, May 1984, p.1211.

[8] R.Castello and P.Gray, "A 350 µW fifth-order low-pass SC filter", Tech.Dig.ISSCC '85, p.276, 1985.

[9] H.J.M.Veendrick, "An NMOS dual-mode digital low-pass filter for color TV", IEEE Journ. Solid-State Circuits, Vol.SC-16,pp.179-182, June 1981.

[10] J.K.J.Van Ginderdeuren, "Compact NMOS building blocksand a methodology for dedicated digital filter applications", IEEE Journ. Solid-State Circuits, Vol.SC18,pp.306-316, June 1983.

[11] P.A.Reutz *et al*, "Computer generation of digital filter banks", Tech. Dig.ISSCC 84, p.20, 1984.

[12] M.Degrauwe *et al*, 'IDAC: an interactive design tool for analog CMOS circuits", IEEE Journ. Solid-State Circuits, Vol.SC-22, pp.1106-1116, Dec. 1987.

[13] H.P.Graf *et al*, "A CMOS associative memory based on neural networks", Tech.Dig. ISSCC '87, p.304, 1987.

[14] C.Mead, *Analog VLSI and neural systems*, Addison-Wesley, Reading, 1989.

[15] C.Mead and M.Ismail, *Analog implementation of neural systems*, Kluwer Academic Publishers, Norwell, 1989.

[16] E.Vittoz, "Analog VLSI implementation of neural networks", Proc. ISCAS 1990.

[17] E.Vittoz *et al*., "Analog VLSI implementation of a Kohonen map", Proc. Journées d'Electronique, EPFL,Lausannne, October 1989.

Paper 4.3
THE ART OF ANALOG CIRCUIT DESIGN IN A DIGITAL VLSI WORLD

B. J. Hosticka and W. Brockherde

Fraunhofer Institute of Microelectronic Circuits and Systems
Finkenstr. 61, D-4100 Duisburg 1
Federal Republic of Germany

Abstract

This contribution surveys issues and challenges of analog design in VLSI. Since today's VLSI is still largely digital, the analog design plays a rather peripheral role. Due to an increasing tendency towards single-chip systems, however, the inclusion of analog circuits in VLSI is becoming more and more important. Nevertheless, the operation of analog and digital circuits on a common substrate harbors many pitfalls.

Introduction

Reaching performance limits in standard technologies, use of advanced submicron, and introduction of new technologies, such as BiCMOS, call not only for a consequent design effort but also require full exploitation of design ingenuity. This is true above all for analog circuits. There are actually six reasons to include the analog design into a VLSI design repertoire: 1. digital chips (i.e. containing e.g. a microprocessor or a DSP) must be interfaced with the real "analog" world via sensors and actuators and, hence, need analog-digital interface electronics [1], 2. analog signal preprocessing can often relax requirements on subsequent digital signal processing by a convenient reduction of information redundancy, 3. low-power chips can benefit from high functionality of analog circuits thus saving chip area and power dissipation [2], 4. design of high-speed digital circuits strongly resembles the analog design and can greatly benefit from analog know-how, 5. high-speed digital transmission and storage need analog pre- and postprocessing [1], and 6. RF reception and transmission also require analog signal processing [3].

Choice of Analog Circuit Techniques

The circuit design techniques are of course to a certain degree dictated by the underlying technology. CMOS technologies, with the MOS device being an excellent analog switch, allow use of analog discrete-time signal processing in form of switched-capacitor (SC) circuits, provided highly linear capacitors with precisely defined ratios are available [4]. On the other hand, the intrinsically high transconductance of the bipolar device allows resistive loading in bipolar circuits and thus enables implementation of RC continuous-time signal processing. In addition, the welldefined exponential characteristic of the bipolar transistor yields easy design of analog multipliers and its 1/f-noise is considerably lower than that of its MOS counterpart. Further, the high transconductance of the bipolar transistors used in the input stages of operational amplifiers contributes to their low offset voltage but it also decreases their slew-rate. Needless to say that the mixed bipolar-CMOS (called BiCMOS) technology is almost ideally suited for building analog circuits as it combines advantages of both types of technologies [5].

Both circuit techniques, discrete-time and continuous-time, however, are plagued by numerous practical problems. Thus the switching process in the SC circuits causes mixing of unwanted undersampled signals, such as noise or interference, into the baseband and can significantly degrade their performance. Careful bandlimiting and shielding is simply a must in these circuits, especially when asynchronous operation cannot be avoided. But even at synchronous operations, the input bandlimiting must be provided to prevent aliasing, and a proper optimization of the noise-speed trade-off is necessary in order to minimize the increase of the aliased noise in the baseband. MOS-designs requiring low 1/f-noise must employ additional noise reduction schemes, such as chopping or correlated-double or multiple sampling [6,7]. While the offset of MOS operational amplifiers tends to be higher than that of bipolar amplifiers, it is usually harmless unless it reduces the dynamic range. Fortunately, the offset can be often easily cancelled in SC circuits.

Reprinted with permission from *International Symposium on Circuits and Systems*, pp. 1347–1350, 1990.

The continuous-time circuits, on the other hand, suffer from the poor correlation between the absolute resistor and capacitor on-chip values: their time constants must be either externally trimmed or adjusted by additional tuning circuits which increases the chip complexity and calls for an additional design effort. Besides, the low slew-rate of bipolar operational amplifiers may increase the distortion at higher frequencies unless appropriate precautions are taken. An additional drawback of bipolar operational amplifiers is that their input base terminals exhibit a finite input resistance and require DC bias currents.

Design Issues

As indicated, even in today's modern monolithic silicon technologies, the control of absolute parameters is still poor, at least measured by requirements laid down by most analog designs. Thus the absolute values of resistors and capacitors can vary by tens of percents. Moreover, some components, e.g. resistors, may display a not negligible temperature and voltage dependence. This makes the design of on-chip references (e.g. for time, voltage, current, temperature, etc.) a rather tricky matter. At the same time, the excellent reproduction of geometry ratios - as far as they are defined by lateral dimensions within a single mask - allows analog designers to rely on relative component values. The ultimate matching accuracy is again determined by the underlying lithography and fabrication technology (and to a certain degree by the layout techniques and precautions). If the desired component accuracy exceeds the maximum achievable precision in a given technology, special design measures must be introduced. There are basically two ways how to deal with this problem: 1. circuit redesign towards structures less sensitive to the component mismatch, and 2. algorithmic approach, i.e. introduction of additional algorithms into the signal processing, either on-chip (e.g. self-calibration or oversampling) or off-chip (e.g. trimming). Both techniques usually increase the circuit complexity and require an additional design effort but the only alternative is to improve the technology which goes beyond the scope of this contribution. The matching techniques prompt the analog designer to a consequent use of ratio techniques whenever feasible and reduce the requirement of on-chip references to a single one per physical quantity, because its results can be easily replicated and multiplied all over the chip.

Voltage and temperature references with decent performance can be only provided when bipolar devices are present (since the parasitic substrate transistors will do, this can be usually solved in CMOS), but a precise time reference can only be delivered externally (e.g. from a quartz oscillator) in form of a precise digital clock. This, of course, explains the wide popularity of the SC circuits in analog signal processing.

Crosstalk

One of the most severe constraints which the analog circuits are subjected to is the presence of the common substrate. All on-chip components are capacitively isolated, either by a silicon dioxide or by pn-junctions, which means that all the components (transistors, resistors, capacitors, interconnection lines, etc.) exhibit stray capacitances. The pn-junction capacitances are even nonlinear and add some temperature-dependent leakage currents. The strays not only introduce interconnection delays in connection with the interconnect resistances, but tend to affect the circuit performance because they may add to the nominal capacitor values unless stray-insensitive circuit techniques are employed. The second effect of strays is that they can cause crosstalk, either directly or via substrate [8]. Typical examples of the former is the clock-feedthrough and capacitive coupling between adjacent interconnection lines. The latter represents an RC coupling and is especially deleterious at high frequencies.

The clock-feedthrough is caused by the injected charge from parasitic gate capacitances of analog MOS switches and is especially harmful when it is signal-dependent because it increases the distortion. A signal-independent clock-feedthrough can be treated as DC offset. Capacitive coupling between neighboring wires can occure at quiet interconnection lines exhibiting high impedance: such lines must be carefully shielded from the noisy ones (e.g. by dummy lines located sideways and connected to a low-impedance source). The RC coupling via substrate can be prevented by careful shielding of sensitive lines by placing low-impedance polysilicon plates or diffusion wells underneath and shielding of other capacitive nodes with guard rings [9]. The best method seems to be the use of dedicated analog terrain, if possible (e.g. use of analog tubs or area separation by guard rings). Anyway, extreme care must be taken when executing the layout, especially during the signal routing. It is essential to maintain a sufficient distance between the sensitive and the noisy circuits and to keep the substrate area in the vicinity of sensitive analog parts at low impedance, e.g. by inserting additional substrate bias contacts. It is important to recall that an undersampled crosstalk interference can also be aliased into the baseband in the SC circuits.

Careful gain management can improve the circuit performance but it is not without problems: if the voltage amplification is too low the signal-to-noise ratio decreases, but a high gain is causing high voltage swings and may degrade the linearity due to limitation set by the power supply voltages.

The final way now an interference can reach an analog circuit is coupling via power supplies, common biasing, and ground. Although we always try to use dedicated power supply voltages for analog and digital circuits, in some circumstances this may not be viable (e.g. if only one battery source is available for the circuit operation). The analog circuits then must be designed to provide a high power-supply rejection which is not always easy, especially at high frequencies. A proper selection of architecture of operational amplifiers used in the design can avoid many problems [10].

The crosstalk problem may be alleviated by employing differential-path techniques instead of the signal-ended path, but they reject only a common-mode interference and their effectiveness is limited by the path mismatch [6]. The differential-path technique also improves the power-supply rejection and doubles the differential voltage swing, i.e. the dynamic range increases.

The use of common biasing networks means chip area and power savings, but it may also create an undesired coupling among the biased circuits [11]. For high performance, a local bias is preferrable.

Although the separation of analog and digital ground is imperative, it is not always easily accomplished in practice. We must try at all cost to maintain its low-impedance character (especially in the circuits employing the single-ended path where it has to conduct return currents) and shield it in order to prevent any crosstalk. Extreme caution is necessary in case of single power supply operations when a mid-point voltage reference instead of ground forms the analog common: their output impedance should be sufficiently low, even at higher frequencies.

Future Challenges in Analog Design

The future development of monolithic technologies is bound to bring more problems to analog designers. The unavoidable development towards submicron silicon will cause futher reduction of the device dimensions and power supply voltages. Thus the future analog designers will battle an increased thermal noise (due to diminishing kT/C), an increased crosstalk (due to shorter distances between lines), a decreased voltage gain (due to shorter MOS channel lengths), and a reduced voltage swing. Nevertheless, maximizing the dynamic range will not be the only problem. Since the chip environment cannot be easily down-scaled, further problems will arise at the chip boundaries. An increased economic pressure will call for locating as much electronics as possible on a single VLSI chip. This means not only microprocessors, digital signal processors, etc., but sensors and actuators as well. It can be easily foreseen that creating the interface between these components will become one of the important tasks of the future analog designer. At the same time, he can contribute to other areas, such as analog signal processing, high speed digital design, etc.

The most important challenge of the future, however, will be his professional education. As the practice has shown, the know-how transfer in analog design is rather difficult. Although modern CAD is capable of providing significant design support, many designs may still present considerable problems: complex systems realizing involved algorithms, analog circuits, low-power low-voltage circuits, high speed digital circuits, etc. Today's CAD greatly aids circuit analysis and optimization, but synthesis and, especially innovative designs, still remain a rather obscure subject. Good designers often rely on intuition which greatly helps constraining parameter space when exploring novel design ideas. Hence, a good designer does not simply try out all imaginable ideas and tests all possible cases but can arrive at the "right" idea within a relatively short time span. This is important as the circuit designer faces a large catalogue of issues, as we have just described. But at the same time, intuition is extremely difficult to teach. Novel CAD approaches, such as expert systems [13], may somewhat alleviate the situation, but improved methods for professional education and training could help acquiring proper circuit skills too.

Acknowledgement

The authors gratefully acknowledge useful discussions with Y. Manoli and U. Schöneberg.

References

[1] R.W. Brodersen and P.R. Gray, "The role of analog circuits in future VLSI technologies", in Dig. Tech. Papers ESSCIRC' 83 (Lausanne, Switzerland), pp. 105-110, Sept. 1983.

[2] B.J. Hosticka, "Performance comparison of analog and digital circuits", Proc. IEEE, vol. 73, no. 1, pp. 25-29, Jan. 1985.

[3] B.-S. Song, "CMOS RF circuits for data communications applications", IEEE J. Solid-State Circuits, vol. SC-21, no. 2, pp. 310-317, Apr. 1986.

[4] D.J. Allstot and W.C. Black, "Technological design considerations for monolithic MOS switched-capacitor filtering systems", Proc. IEEE, vol. 71, no. 8, pp. 967-986, Aug. 1983.

[5] G. Zimmer, W. Esser, J. Fichtel, B. Hosticka, A. Rothermel, and W. Schardein, "BiCMOS: technology and circuit design", Microelectronics Journal, vol. 20, no. 1-2, pp. 59-75, 1989.

[6] K.-C. Hsieh, P.R. Gray, D. Senderowicz, and D.A. Messerschmitt, "A low-noise chopper- stabilized differential switched-capacitor filtering technique", IEEE J. Solid-State Circuits, vol. SC-16, no. 6, pp. 708-715, Dec. 1981.

[7] W. Buttler, B.J. Hosticka, and G. Lutz, "Low noise - low power CMOS readout systems for silicon strip detectors", in Dig. Tech. Papers ESSCIRC'88 (Manchester, England), pp. 171-174, Sept. 1988.

[8] G.H. Warren and C. Jungo, "Noise, crosstalk and distortion in mixed analog/digital integrated circuits", in Proc. IEEE 1988 Custom Integrated Circuits Conf., pp. 12.1.1-12.1.4, May 1988.

[9] L.D. Smith, H.R. Farmer, M. Kunesh, M.A. Masetti, D. Willmott, R. Hedman, R. Richetta, and T.J. Schmerbeck, "A CMOS-based analog standard cell product family", IEEE J. Solid-State Circuits, vol. SC-24, no. 2, pp. 370-379, Apr. 1989.

[10] B.J. Hosticka, W. Bockherde, and M. Wrede, "Effects of the architecture on noise performance of CMOS operational amplifiers", in Proc. 6th European Conf. on Circuit Theory and Design ECCTD'83 (Stuttgart, West Germany), pp. 238-241, Sept. 1983.

[11] C. Laber, C.F. Rahim, S.F. Dreyer, G.T. Vehara, P.T. Kwok, and P.R. Gray, "Design considerations for a high-performance 3-μm CMOS analog standard-cell library", IEEE J. Solid-State Circuits, vol. SC-22, no. 2, Apr. 1987.

[12] C.-K. Wang, R. Castello, and P.R. Gray, "A scalable high-performance switched-capacitor filter", IEEE J. Solid-State Circuits, vol. SC-21, no. 1, pp. 57-64, Febr. 1986.

[13] L.R. Carley and R.A. Rutenbar, "How to automate analog IC designs", IEEE Spectrum, vol. 25, no. 8, pp. 26-30, Aug. 1988.

BiCMOS ANALOG CIRCUIT TECHNIQUES

Bruce A. Wooley

Stanford University
Center for Integrated Systems
Stanford, CA 94305-4070

Introduction

Owing to the performance advantages it offers in applications such as high-speed static memories and semi-custom circuits, BiCMOS appears likely to establish itself a mainstream technology for VLSI. To date this technology has, for the most part, been used to implement digital circuit functions, and the performance benefits relative to CMOS have derived largely from the buffering and current drive capabilities of bipolar emitter followers. Design techniques that more fully exploit the potential of BiCMOS across a broad range of applications have only just begun to emerge. There follows a brief description of some such techniques that have been suggested for enhancing the performance of analog circuits.

The fastest, and in many cases most precise, monolithic analog circuits have traditionally been implemented using silicon bipolar technology. However, the emergence of BiCMOS offers new opportunities to improve the performance of analog circuits by combining the complementary advantages of bipolar and CMOS technologies. Relative to MOS transistors, bipolar devices exhibit superior threshold matching, higher transconductance, and lower noise levels. Because of these characteristics, bipolar circuits are especially well-suited to providing low-noise amplification over large bandwidths. CMOS technology offers the very considerable advantages of simple, zero-offset analog switches, high-impedance charge storage nodes, and complementary transistors. These attributes allow for the extensive use of analog sampling, pipelining, and multiplexing, as well as offset cancellation and self-calibration. The following circuits illustrate how selective combinations of bipolar and CMOS devices can be used to achieve significant performance enhancements in a variety of analog applications.

Operational Amplifiers

The performance of charge-redistribution analog MOS circuits widely used in data conversion and filtering applications is often limited by the operational amplifiers used to implement switched-capacitor integrators and amplifiers. In cases where only a small capacitive load must be driven, a single-stage CMOS folded-cascode amplifier of the type illustrated in Figure 1 is especially attractive in applications demanding the largest possible bandwidth[1]. The stability of this topology is assured, with compensation being provided by the load capacitance itself. The unity-gain bandwidth of the amplifier is generally limited by the nondominant pole associated with the cascode transistors. In a BiCMOS technology this bandwidth can usually be increased simply through the use of bipolar cascode transistors, as shown in Figure 2[2].

Since npn bipolar transistors are used as the cascode devices in the circuit of Figure 2, the input transistors are necessarily PMOS devices. The use of PMOS transistors in the input pair is consistent with providing a virtually infinite input resistance and achieving as low as possible noise levels while using MOS input devices. In particular, the 1/f noise in PMOS transistors is much lower than that found in NMOS devices. The increased transistor size, and associated capacitance, needed to obtain an input-pair transconductance comparable to that obtained with NMOS input transistors is generally not a serious limitation as long as the signal source impedance is low.

In situations where a significant, and perhaps variable, capacitive or resistive load must be driven, an output stage must be added to the basic transconductance amplifier to prevent gain degradation and, for some transconductance configurations, closed-loop instability. This output stage can generally be implemented much more effectively using bipolar, or a combination of bipolar and MOS, transistors than is possible using only MOS devices. In particular, as in digital circuits, bipolar emitter-followers are far superior to MOS source-followers for providing transitions between high and low impedance nodes over large bandwidths and with lower, and less variable, dc voltage drops[3].

In addition to using bipolar transistors to improve the performance of operational amplifiers in applications that exploit the advantages of CMOS technology, MOS devices can be used to enhance the performance of typical precision bipolar amplifier configurations. In particular, input PMOS transistors can be used to increase the input impedance and the

Reprinted with permission from *International Symposium on Circuits and Systems*, pp. 1983–1986, 1990.

input stage bias current, thereby avoiding the input-stage slewing limitations characteristic of bipolar designs. The use of PMOS devices is, of course, likely to degrade both offset and noise performance. However, because of the availability of MOS devices for storing analog signals, the added offset can often be cancelled at the subsystem level. PMOS transistors can also be used as current sources and active loads, replacing lateral pnp transistors. It also seems likely the PMOS devices can be substituted for lateral pnp transistors in the signal path with substantial performance benefits.

Comparators

The performance of data conversion, instrumentation, and digital communications systems is often influenced significantly by the speed and precision with which the function of comparison can be performed. At audio frequencies and below, analog MOS circuits have proven especially effective in achieving accuracies of 14 or more bits through the use of offset cancellation and self-calibration techniques. In such cases, the amplification required in the comparator is typically obtained using circuit configurations similar to those of MOS operational amplifiers. Consequently, in a BiCMOS technology the bandwidth of such high-precision comparators can be increased by the same methods noted above for broadbanding operational amplifiers.

While pipelined cascades of regenerative or linear CMOS amplifiers have been used to achieve moderately high comparison rates at a precision of 8 to 10 bits, their speed and precision remain limited by the inherent MOS limitations of low transconductance and relatively poor device threshold matching. As a consequence, the design and implementation of high-speed comparators remains dominated by silicon bipolar technology. Shown in Figure 3 is a circuit schematic representative of the type of bipolar comparator that has found widespread use in high-speed data conversion applications[4]. In this configuration a current switching latch is typically preceded by at least one stage of preamplification, while emitter followers are used for level shifting and buffering between the stages. The preamplifier both reduces the sensitivity of the circuit to mismatch offsets in the latch and minimizes the coupling of regeneration transients in the latch back into the input source.

One use of BiCMOS technology in circuits such as that in Figure 3 has been to replace the current-switching latch with a clocked CMOS latch coupled to the preamplifier through MOS switches[5]. With this approach the preamplifier is used to establish an imbalance at the differential inputs of the latch while the latch itself is disabled. Then, during regeneration, the latch is decoupled from the preamplifier and enabled. In this way static power dissipation in the latch is avoided; moreover, the latch itself provides a conversion to full CMOS digital levels.

When multiple comparators of the type represented by Figure 3 are used in high-speed flash and subranging A/D converters, the sampling of the input signal occurs within the comparator latches. Consequently, delay mismatch among the comparators can impose a significant limitation on the rate at which the input signal can be accurately sampled and quantized. In a BiCMOS technology, the availability of MOS switches provides an effective means of implementing the sampling function prior to comparison, either directly at the inputs of each comparator or as a single track-and-hold amplifier preceding the overall converter.

The current needed for overdrive recovery in the preamplifier of comparators such as that shown in Figure 3 is generally much larger than that required for fast regeneration within the latch. Thus, power dissipation constraints associated with the preamplifier often constrain the speed with which comparisons can be performed in applications, such as flash analog-to-digital converters, where many comparators must be used.

Shown in Figure 4 is the first stage of a BiCMOS circuit wherein comparisons are performed directly by means of regenerative amplification, without the use of a preamplifier[6]. In this circuit the input voltage is first sampled onto capacitors through MOS switches M1 and M2. The difference between the sampled inputs is then detected by means of a regenerative amplifier that dissipates no static power. The amplification is provided by the bipolar differential pair Q1 and Q2, while all of the MOS transistors shown in the schematic simply operate as switches.

The circuit of Figure 4 operates as follows. In the sample mode the clock ϕ is high, and thus switches M1–M4 are closed while M5 is open. The input voltages are thereby sampled onto capacitors C1–C4, establishing a voltage imbalance at the bases of the bipolar transistors Q1 and Q2. During this time Q1 and Q2 remain inactive, and do not drain current from the sampling capacitors, because their common emitter is open. Also, during this period the collectors of Q1 and Q2 are precharged to the positive supply, and capacitor C5 is precharged by the closing of switch M6.

In the transition to the regeneration mode MOS switches M1–M4 and M6 open, while M5 is closed. The capacitor C5 is thus connected between ground and the emitters of Q1 and Q2 and begins to discharge their common emitter node. In this fashion a fixed, predetermined amount of charge is transferred through the bipolar differential pair. The initial voltage difference stored at the bases of Q1 and Q2 results in a differential discharging of the series combinations of capacitors C1, C3 and C2, C4. As a result of the cross-coupling provided by capacitors C3 and C4, the initial voltage difference between the bases of Q1 and Q2 is regeneratively amplified at the collectors of these transistors. This amplified voltage difference is then latched in a subsequent stage. The size of capacitor C5 relative to capacitances C1–C4 is chosen so that transistors Q1 and Q2 will not saturate during regeneration.

Subsystems

While the examples cited above illustrate the combined use of bipolar and MOS transistors to realize specific circuit functions, BiCMOS technology may also offer significant benefits at the subsystem and system level. For example, a complete tracking servo system for a VCR, comprising a broad spectrum of analog and digital circuit functions, has already been implemented on a single BiCMOS chip[7]. Among the analog functions included in this system are a low-noise amplifier, a gain control amplifier, balanced demodulators, rectifiers, and a subtractor. These precision functions are implemented primarily with bipolar transistors, although CMOS devices are used extensively as switches for analog multiplexing and in BiCMOS gates that provide some of the control signals.

The tracking servo also includes sample-and-hold amplifiers and three switched-capacitor bandpass filters implemented using combinations of bipolar and MOS devices. In the filters high-performance BiCMOS operational amplifiers with MOS input transistors are used in traditional CMOS switched-capacitor configurations. The digital functions in the servo include an oscillator, a clock generator, a four-mode counter, and several dividers, all realized using BiCMOS gates.

A recently reported 10-bit video A/D converter based on a subranging architecture further demonstrates how BiCMOS technology offers new design opportunities at the subsystem level[5]. In a subranging converter, quantization of an input sample is accomplished in two stages without the need for fast, precision analog subtraction. Typically, a single reference resistor string similar to that in a fully parallel (flash) converter is used. The input is first coarsely quantized using a set of "coarse" comparators together with appropriate outputs from the reference network. On the second clock cycle, the most significant bits of the quantization, as determined during the first cycle, control the multiplexing of a particular range of voltages from the reference network onto the reference inputs of the "fine" second stage comparators that are used to determine the least significant bits.

As an alternative to a single resistor string, separate coarse and fine reference networks can be used in a subranging configuration, with the reference voltages for the fine network derived from the coarse resistor string. However, this voltage transformation must be precise and is difficult to accomplish in a fully CMOS design. In the BiCMOS circuit described in reference [5] the voltage transformation is performed using bipolar emitter followers in which a constant current is maintained. The voltage transfer from the coarse to fine networks is thus accomplished with less than 0.5 mV of offset error.

Conclusion

Because of the distinct and unique advantages bipolar and MOS devices bring the realization of analog circuit functions, it seems likely that the application of BiCMOS technology to the implementation of these functions will yield benefits rivaling those already being demonstrated in digital applications. However, analog BiCMOS circuit design is only in its earliest formative stages. It remains for creative designers to unlock the full potential of this emerging technology -- a technology that may well dominate the mainstream of future semiconductor microelectronics.

References

[1] T. C. Choi, et al., "High-Frequency CMOS Switched-Capacitor Filters for Communications Application," *IEEE J. Solid-State Circuits*, vol. SC-18, pp. 652-664, Dec. 1983.

[2] M. Nayebi and B. A. Wooley, "A 10-bit Video BiCMOS Track-and-Hold Amplifier," *IEEE J. Solid-State Circuits*, vol. 24, pp. 1507-1515, Dec. 1989.

[3] B. J. Rodgers and C. R. Thurber, "A Monolithic ±5 1/2-Digit BiMOS A/D Converter," *IEEE J. Solid-State Circuits*, vol. 24, pp. 617-626, June 1989.

[4] T. Wakimoto, et al., "Si Bipolar 2-GHz 6-bit Flash A/D Conversion LSI," *IEEE J. Solid-State Circuits*, vol. 23, pp. 1345-1350, Dec. 1988.

[5] Y. Sugimoto and S. Mizoguchi, "An Experimental BiCMOS Video 10-Bit ADC," *IEEE J. Solid-State Circuits*, vol. 24, pp. 997-999, Aug.1989.

[6] P. J. Lim and B. A. Wooley, "An 8-bit, 200-MHz BiCMOS Comparator," *IEEE J. Solid-State Circuits*, vol. 25, Feb. 1990.

[7] S. Mizoguchi, et al., "Bi-CMOS Tracking Servo LSI for 8-mm VCR," *Proc. IEEE 1988 Custom Integrated Circuits Conf.*, pp. 12.7.1-12.7.4, May 1988.

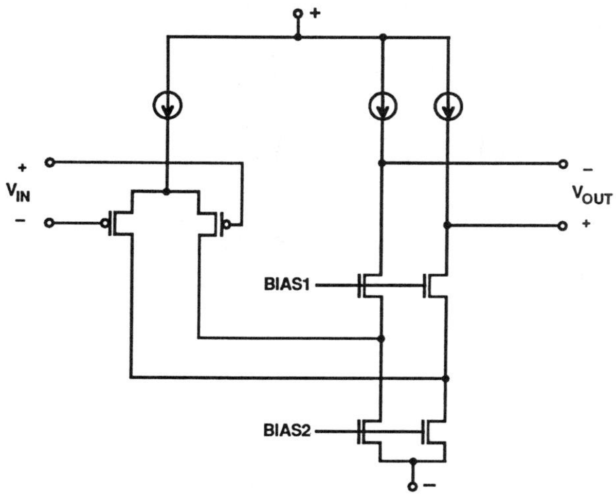

Figure 1. CMOS folded-cascode amplifier.

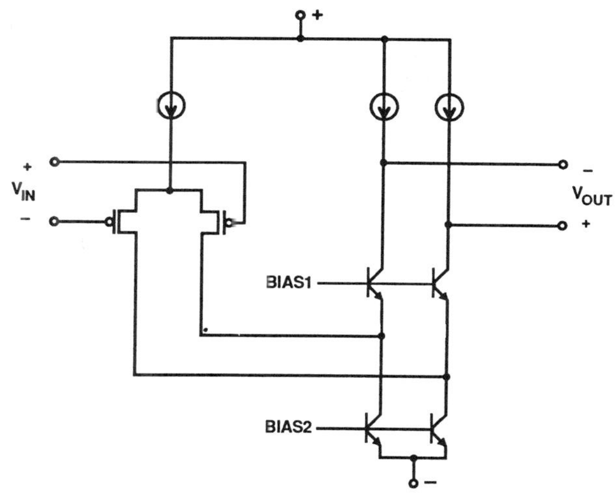

Figure 2. BiCMOS folded-cascode amplifier.

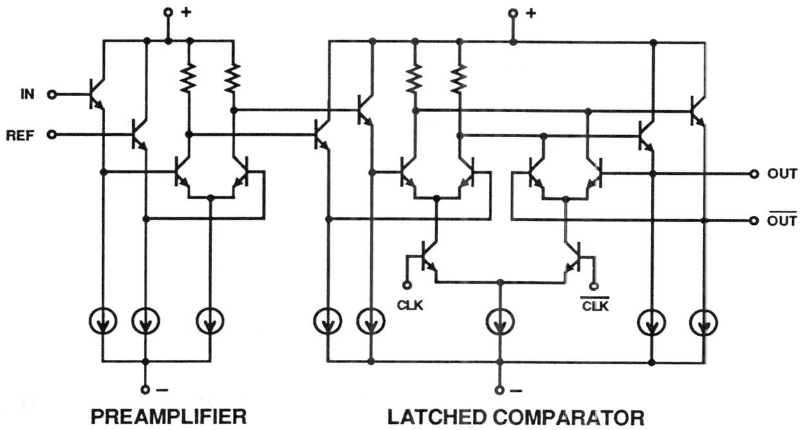

Figure 3. High-speed bipolar comparator.

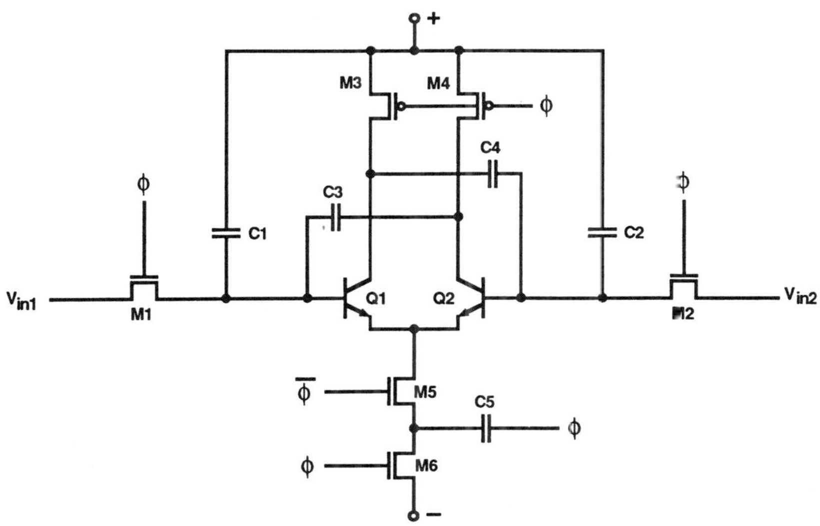

Figure 4. BiCMOS sampled-data regenerative amplifier.

Paper 4.5

A Single-Power-Supply 10-b Video BiCMOS Sample-and-Hold IC

KAZUNORI TSUGARU, YASUHIRO SUGIMOTO, MEMBER, IEEE, MAKOTO NODA, TAKAO ITO, MEMBER, IEEE, AND YOSHITO SUWA

Abstract—A +5-V single-power-supply 10-b video BiCMOS sample-and-hold IC is described. Video speed, low power, and 10-b accuracy sample-and-hold operation has been achieved using a complementary connected buffer format sample switch. A high-speed p-n-p transistor used in the sample switch is formed by a combination of n-p-n and PMOS transistors. The sample-and-hold operation is accomplished by feeding back the hold capacitor voltage to the sample switch inputs, so that the inputs transfer symmetrically for the hold capacitor voltage at any input level. The sample-and-hold IC has been implemented in 1.2-μm BiCMOS technology and evaluated. The following results have been obtained: 185-MHz 3-dB bandwidth at 22-pF hold capacitor, 63-dB signal-to-noise ratio at 8-MHz full-scale input, 20-ns acquisition time at 1-V step input, 15-ns switching settling time, and 0.1% linearity error. Power dissipation is 150 mW.

I. INTRODUCTION

THE DEMAND for a high-resolution video A/D converter has been increasing. A two-step parallel conversion scheme is most effective from the A/D converter power dissipation and cost viewpoints. In this case, however, a high-accuracy and low-power sample-and-hold circuit, operating at video frequency and above, is required as a pre-circuit for the A/D converter. Considering a future one-chip system containing the A/D converter, +5-V single power supply operation is also required for the sample-and-hold circuit. So far, several monolithic high-accuracy video sample-and-hold IC's have been developed [1]–[3]. However, they have used a comparatively large amount of power because a hold capacitor has been driven by a current source in them. Also, they all have needed two power supplies to maintain the 2-$V_{p\text{-}p}$ input full-scale range and to obtain high-accuracy characteristic. Previously, the authors developed a bipolar sample-and-hold IC using a complementary connected buffer as the sample switch to reduce power dissipation [4]. As a result, 10-b accuracy and a 20-MHz maximum conversion rate have been obtained at 400-mW power dissipation. However, the sample-and-hold IC also needed two power supplies and used high-speed p-n-p transistors. A p-n-p transistor, made by a standard process, has poor frequency performance, making video signals difficult to handle. This paper describes a single-power-supply 10-b video sample-and-hold IC, using 1.2-μm BiCMOS technology, that utilizes equivalent high-speed p-n-p transistors formed by a combination of n-p-n and PMOS transistors [5].

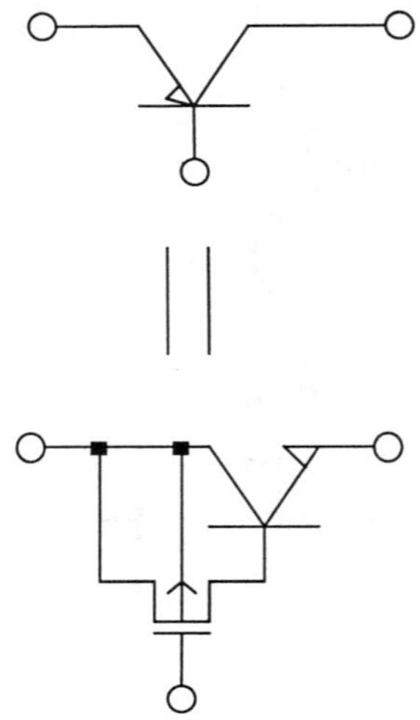

Fig. 1. Basic equivalent high-speed p-n-p transistor configuration.

II. EQUIVALENT HIGH-SPEED P-N-P TRANSISTOR REALIZATION

The authors proposed a sample-and-hold circuit that utilized a complementary connected buffer sample switch formed by an n-p-n and a p-n-p transistor, and obtained 10-b accuracy and a 20-MHz maximum conversion rate [4]. As a large output dynamic range can be obtained according to the circuit configuration, it is advantageous for single-power-supply operation. In BiCMOS technology, which is applicable for a one-chip system realization, however, it is difficult to supply a high-speed p-n-p transistor. So, the authors considered an equivalent high-speed p-n-p transistor, shown in Fig. 1. The equivalent high-

Manuscript received September 4, 1989; revised February 5, 1990.
K. Tsugaru and M. Noda are with the Semiconductor Device Engineering Laboratory, Toshiba Corporation, 1, Komukai-Toshiba-cho, Saiwai-ku, Kawasaki 210, Japan.
Y. Sugimoto is with the Semiconductor System Engineering Center, Toshiba Corporation, 1, Komukai-Toshiba-cho, Saiwai-ku, Kawasaki 210, Japan.
T. Ito and Y. Suwa are with the Microelectronics Center, Toshiba Corporation, 1, Komukai-Toshiba-cho, Saiwai-ku, Kawasaki 210, Japan.
IEEE Log Number 9034823.

Reprinted from *IEEE J. Solid-State Circuits*, vol. 25, no. 3, pp. 653–659, June 1990.

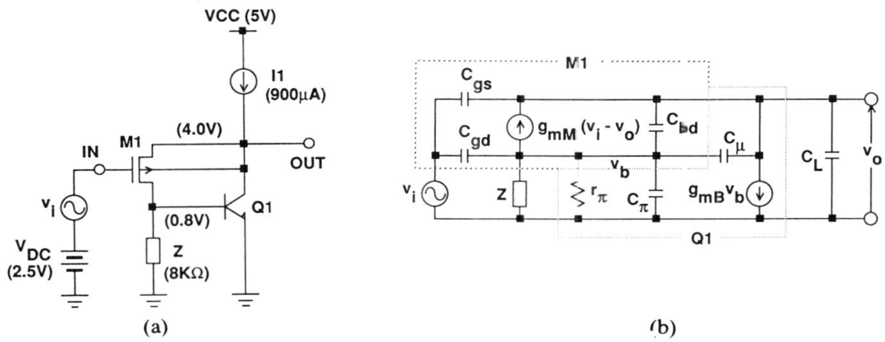

Fig. 2. Emitter-follower circuit, using the equivalent high-speed p-n-p transistor. (a) Circuit configuration. (b) Small-signal equivalent circuits.

TABLE I
n-p-n BIPOLAR TRANSISTOR PARAMETERS IN 1.2-μm BiCMOS PROCESS

Parameter	Symbol	Measured Value
Forward DC Current Gain	h_{fe}	89.2
Transconductance	g_{mB}	28.9 mS
Input Resistance	r_π	2.87 kΩ
Diffusion Capacitance	C_d	736 fF
Base-Emitter Junction Capacitance	C_{be}	66.3 fF
Base-Collector Junction Capacitance	C_μ	11.1 fF
Collector-Substrate Junction Capacitance	C_{cs}	75 fF
Base Transit Time	τ_F	25.5 ps
Base-Emitter Input Capacitance	C_π	802 fF

$C_\pi = C_d + C_{be}$, $C_d = \tau_F \times g_{mB}$

TABLE II
PMOS TRANSISTOR PARAMETERS IN 1.2-μm BiCMOS PROCESS

Parameter	Symbol	Measured Value
Transconductance	g_{mM}	0.60 mS
Drain-Bulk Junction Capacitance	C_{db}	192 fF
Gate-Channel Capacitance	C'_{gs}	93.5 fF
Gate-Source Overlap Capacitance	C_{gsovl}	8 fF
Gate-Drain Overlap Capacitance	C_{gdovl}	8 fF
Bulk-Substrate Junction Capacitance	C_{bss}	224 fF

$C_{gs} = C'_{gs} + C_{gsovl}$

speed p-n-p transistor is formed by the combination of an n-p-n and a PMOS transistor. As the gate–source voltage for the PMOS transistor can be set to about -1 V under actual operating conditions, the circuit shown in Fig. 1 is equivalent to a p-n-p transistor with -1-V base–emitter voltage.

It is very important to analyze the equivalent high-speed p-n-p transistor because the p-n-p transistor performances decide the characteristics for the sample-and-hold circuit, for example, acquisition time, distortion level, etc. As the equivalent p-n-p transistor is actually used in the emitter-follower circuit configuration, the authors have analyzed the circuit shown in Fig. 2(a). DC current flow in the PMOS $M1$ is controlled by an impedance Z to improve frequency response. The simplified small-signal equivalent circuit for Fig. 2(a) is shown in Fig. 2(b). Parameter meanings are as follows:

v_i	input signal,
C_{gs}, C_{gd}, C_{db}	PMOS $M1$ gate–source, gate–drain, and drain–bulk capacitances,
g_{mM}	PMOS $M1$ transconductance,
r_π	n-p-n $Q1$ base–emitter resistance,
C_π, C_μ	n-p-n $Q1$ base–emitter and base–collector capacitances,
g_{mB}	n-p-n $Q1$ transconductance,
C_L	$M1$ bulk–substrate capacitance (C_{bss})+ $Q1$ collector–substrate capacitance (C_{cs}).

In the actual calculation, the 1.2-μm BiCMOS process device parameters have been used, considering IC implementation. The analyzed operating point is also shown in Fig. 2(a). The $M1$ channel length and width are 1.2 and 90 μm, respectively The $Q1$ emitter size is 3×5 μm^2. Device parameter values are shown in Tables I and II.

The small-signal analysis results are discussed next.

A. Output Impedance Z_o

As indicated in Fig. 2(b), the output impedance in the low-frequency region is obtained as

$$Z_o = \frac{v_o}{i_o} = \frac{1}{g_{mM}[1 + g_{mB}(Z//r_\pi)]}. \tag{1}$$

As $Z \gg r_\pi$ and $h_{fe} \gg 1$,

$$Z_o \sim \frac{1}{g_{mM} \cdot h_{fe}}. \tag{2}$$

This shows that the output impedance shown in Fig. 2(a) is reduced by $1/h_{fe}$ of that for the source-follower circuit, using PMOS $M1$. Using the values shown in Tables I and II, $Z_o = 18.7\ \Omega$ is calculated. This value is comparable to that for a p-n-p emitter-follower circuit.

B. Voltage Gain G(s)

From Fig. 2(b),

$$G(s) = \frac{v_o(s)}{v_i(s)} = \frac{a_0 + b_1 \cdot s + b_2 \cdot s^2}{a_0 + a_1 \cdot s + a_2 \cdot s^2} \tag{3}$$

$$a_0 = g_{mM} \cdot g_{mB} \tag{4}$$

$$a_1 = g_{mM} \cdot (C_\pi + C_{gd}) + g_{mB} \cdot (C_\mu + C_{bd} + C_{gs}/h_{fe} + C_L/h_{fe}) \tag{5}$$

$$a_2 = (C_\mu + C_{bd}) \cdot (C_\pi + C_{gd}) + (C_{gs} + C_L) \cdot (C_\pi + C_\mu + C_{bd} + C_{gd}) \tag{6}$$

$$b_1 = g_{mM} \cdot (C_\pi + C_{gd}) + g_{mB} \cdot (C_{gs}/h_{fe} - C_{gd}) \tag{7}$$

$$b_2 = C_{gd} \cdot (C_\mu + C_{bd}) + C_{gs} \cdot (C_\pi + C_\mu + C_{bd} + C_{gd}) \tag{8}$$

are obtained as $Z \gg r_\pi$ and $h_{fe} \gg 1$.

In the low-frequency region, from (3),

$$G(0) = \frac{a_0}{a_0} = 1 \tag{9}$$

is obtained. This indicates the emitter-follower operation.

For higher frequency, the authors have attempted to simplify (3) using the values shown in Tables I and II. Then, below about 700 MHz, (3) is approximated as

$$G(s) \sim \frac{a_0}{a_0 + a_1 \cdot s}. \tag{10}$$

From this equation, the 3-dB bandwidth for the emitter-follower circuit is obtained as

$$f_{-3\,\mathrm{dB}} = \frac{a_0}{2 \cdot \pi \cdot a_1} = \frac{g_{mM} \cdot g_{mB}}{2 \cdot \pi \cdot [g_{mM} \cdot (C_\pi + C_{gd}) + g_{mB} \cdot (C_\mu + C_{bd} + C_{gs}/h_{fe} + C_L/h_{fe})]} \sim \frac{g_{mM} \cdot g_{mB}}{2 \cdot \pi \cdot (g_{mM} \cdot C_\pi + g_{mB} \cdot C_{bd})}. \tag{11}$$

Fig. 3 shows the results calculated using (3) and (10), and the SPICE simulation result. From these results, the 470-MHz 3-dB bandwidth is obtained. This value is approximately consistent with that calculated from (11), which is 458 MHz.

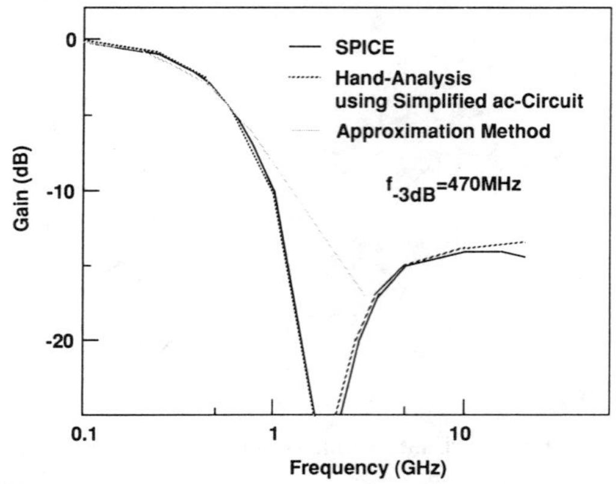

Fig. 3. Emitter-follower circuit frequency responses, using the equivalent high-speed p-n-p transistor.

Equation (11) shows that the 3-dB bandwidth is mainly decided by g_{mM} and C_{bd} for PMOS $M1$ in the 1.2-μm BiCMOS process, because $g_{mB} \cdot C_{bd}$ is dominant in the denominator in (11), and it is necessary to increase g_{mM} and to decrease C_{bd} in order to increase $f_{-3\mathrm{dB}}$. It is considered that the device miniaturization is effective for that. Using 0.8-μm BiCMOS technology, $f_{-3\mathrm{dB}}$ increases to about 1.5 GHz, because g_{mM} increases to 1.5 times and C_{bd} decreases to 0.45 times. A p-n-p transistor, with gigahertz bandwidth, may be realized by the configuration shown in Fig. 1.

III. Sample-and-Hold Circuit Design

The authors have designed a +5-V single-power-supply sample-and-hold circuit using the equivalent p-n-p transistor shown in Fig. 1. A schematic of the BiCMOS sample-and-hold circuit is shown in Fig. 4. The circuit is based on the one in [4] and consists of a complementary connected buffer format sample switch, a hold capacitor C_h, and buffer circuits to feed back the hold capacitor voltage V_{Ch} to the sample switch in the holding mode. The equivalent high-speed p-n-p transistors are used in the input stage and output stage for the sample switch and the feedback buffer circuits. In order to suppress the V_{GS} variation for PMOS transistors $M2$, $M6$, and $M11$ due to the input (V_{IN}) level, cascode-connected PMOS transistors ($M3$, $M7$, and $M12$) are used instead of $M1$

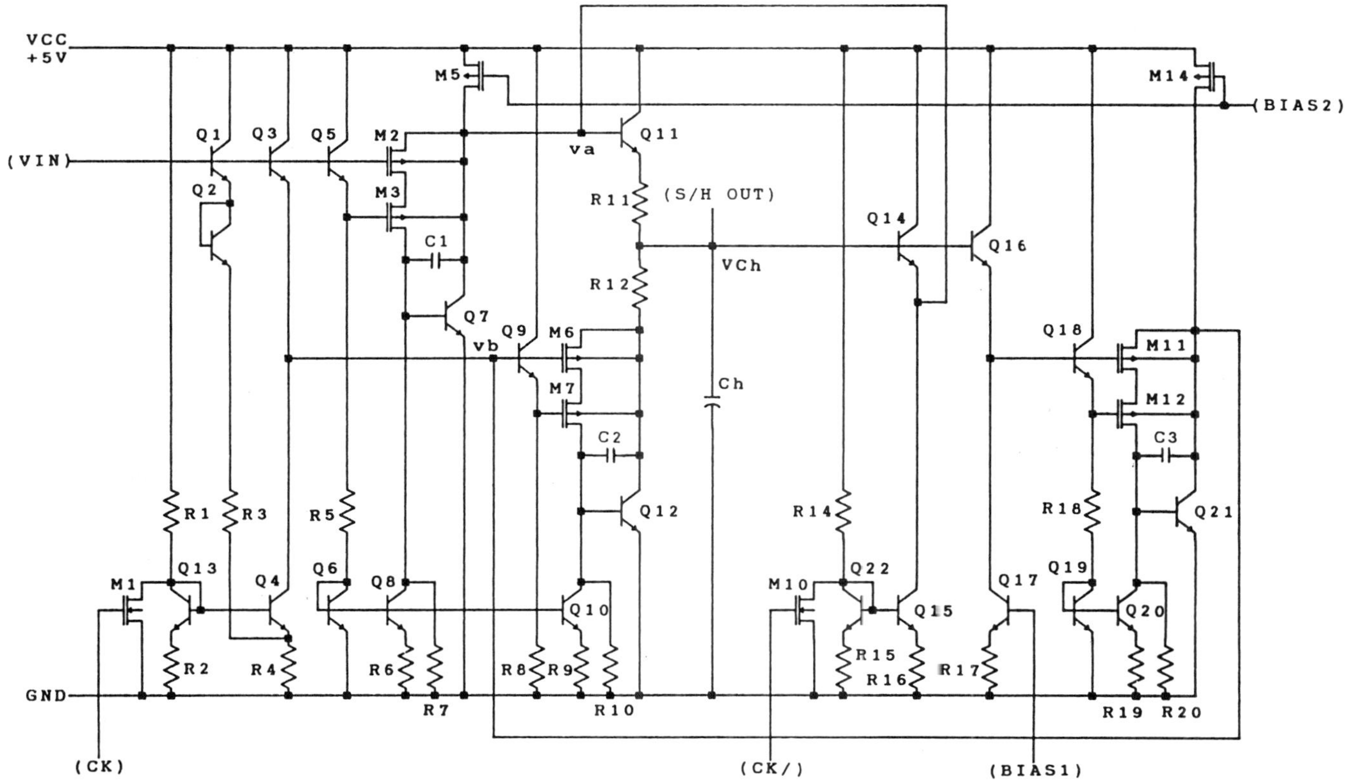

Fig. 4. BiCMOS sample-and-hold schematic.

in Fig. 2, and a parallel-connected resistor and voltage-to-current amplifier are used as impedance Z in Fig. 2. PMOS $M6$ dimensions are $W/L = 120\ \mu m/1.2\ \mu m$ to reduce the output impedance in the sample switch. Idling current for the output stage is 1.4 mA. Capacitors $C1$, $C2$, and $C3$ are used to reduce ringing voltage in the sample-and-hold operation.

The sampling mode and the holding mode are explained using Figs. 5 and 6. Fig. 5 shows the transition waveforms from the holding to the sampling mode and the equivalent circuit in the sampling mode. v_a and v_b are the base voltages for n-p-n $Q11$ and n-p-n $Q9$, respectively. v_a and v_b, which are held symmetrically for V_{Ch} in the holding mode as described later, rise up to $V_{IN} - V_{GSM2}$ due to $M2$ activation, and go down to $V_{IN} - V_{BEQ3}$ due to $Q3$ activation. Then, $Q11$ and the equivalent p-n-p transistor, formed by $M6$ and $Q12$, construct a complementary connected buffer, as shown in Fig. 5(b). Hold capacitor C_h is charged by the emitter-follower circuit for $Q11$ or discharged by the emitter-follower circuit for the equivalent p-n-p transistor. It is predicted that a small acquisition time will be obtained by the broad bandwidth and the small output impedance for the sample switch. According to SPICE simulation, a 15.5-ns acquisition time at a 1-V step, a 244-V/μs rising slew rate, and a 300-V/μs falling slew rate have been obtained.

The transition waveforms from the sampling to the holding mode, and the equivalent circuit in the holding mode are shown in Fig. 6. In the holding mode, v_a and v_b are transferred to $V_{Ch} - V_{BEQ14}$ and $V_{Ch} - V_{BEQ16} - V_{GSM11}$

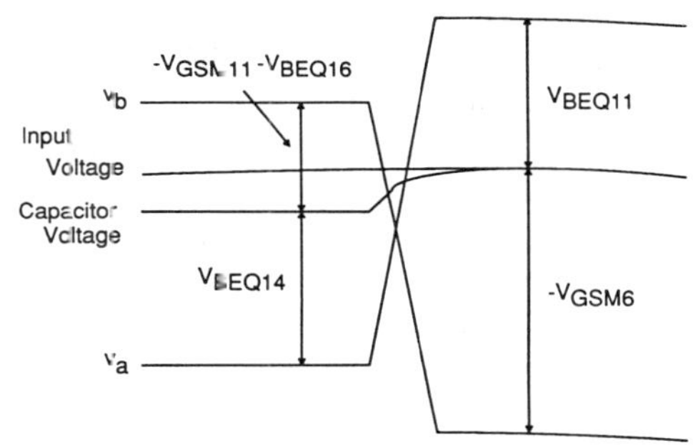

(a)

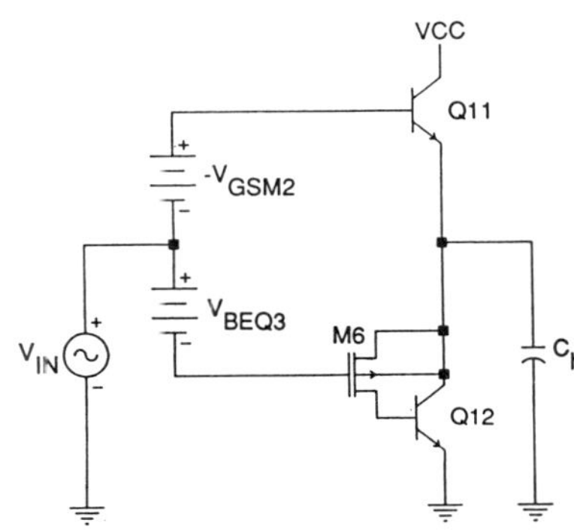

(b)

Fig. 5. (a) Waveform and (b) equivalent circuit in the sampling mode.

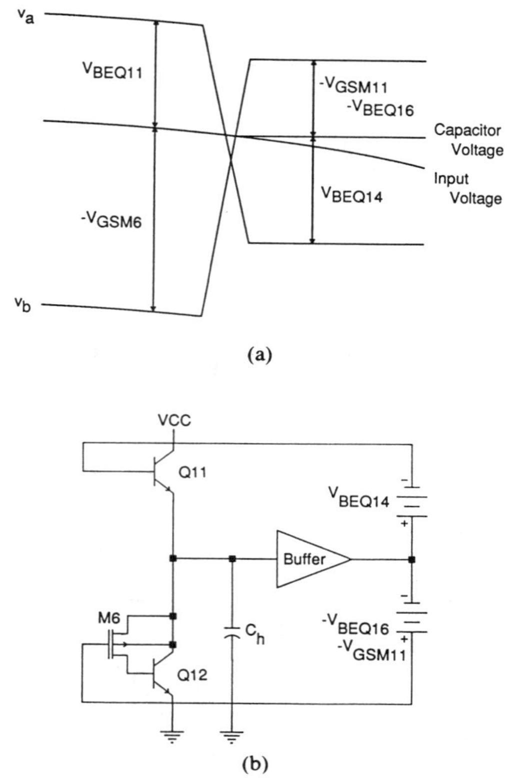

Fig. 6. (a) Waveform and (b) equivalent circuit in the holding mode.

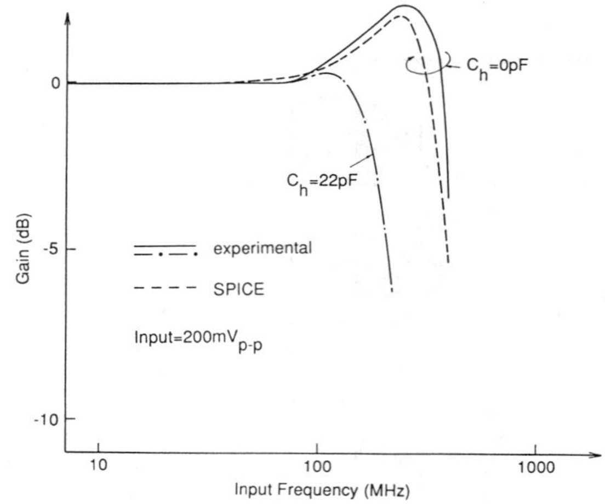

Fig. 7. Frequency responses in the sampling mode.

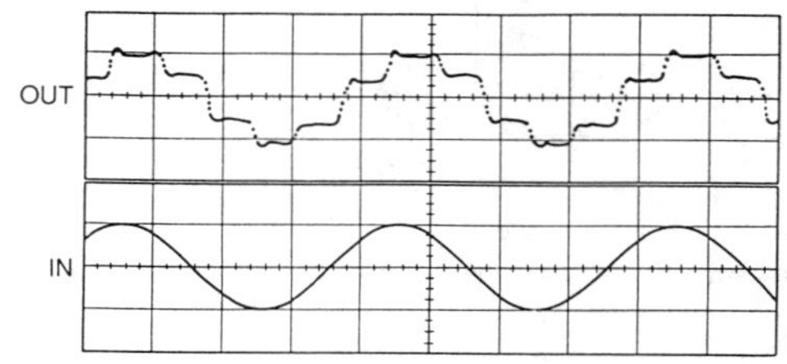

Fig. 8. Typical input and output waveforms for the sample-and-hold.

by activating the feedback buffer circuits, respectively. The $Q11$ base–emiter voltage and the $M6$ gate–source voltage are reverse biased at $-V_{BEQ14}$ and $-V_{BEQ16}-V_{GSM11}$, respectively. As a result, $Q11$, $M6$, and $Q12$ are cut off and V_{Ch} is held. It is seen from the transition waveforms that V_{BEQ11} and V_{GSM6} are always biased, under the same conditions, both in the sampling and in the holding mode. For this reason, the current injected to C_h and the current extracted from C_h are constant, and the distortion due to charge offset is suppressed. By tuning the $Q11$ and $M6$ dimensions, pedestal offset is also decreased. Additionally, as the $Q11$ base and the $M6$ gate are kept in a low-impedance condition by the feedback buffer circuits in the holding mode, the hold-mode feedthrough is suppressed. According to SPICE simulation, a 7-ns switching settling time, a 21-mV pedestal offset, and a -79-dB hold-mode feedthrough at 10-MHz 1-V_{p-p} input have been obtained.

The full-scale input range has been set to 2 to 3 V (1 V_{p-p}). Then, the minimum voltage values for V_{DSM5}, V_{DSM4}, V_{CEQ4}, and V_{CEQ15} are, respectively, about -1, -1.3, 1.2, and 1.5 V. These values are sufficient for current-source operation, considering voltage supply variations.

IV. Experimental Results

The BiCMOS sample-and-hold IC has been fabricated in 1.2-μm BiCMOS technology with double metal layers and evaluated. An n-p-n transistor has an ion-implanted emitter with 3-μm stripe length. The maximum f_T, V_A, and I_K values are 6 GHz, 50 V, and 8 mA, respectively. The channel length and threshold voltage for the MOS transistor are 1.2 μm and 0.8 V, respectively. A poly-Si resistor can be utilized. A double-layer poly-Si capacitor can also be utilized, but it has not been used for this sample-and-hold IC. The IC has been evaluated using an external capacitor for the hold capacitor.

Fig. 7 shows the frequency responses in the sampling mode. These responses have been measured at $C_h = 0$ and 22 pF. The input amplitude is about 200 mV_{p-p} in the 50-Ω termination condition. The SPICE simulation result at $C_h = 0$ pF is also shown for comparison. A 400-MHz 3-dB bandwidth has been obtained at $C_h =$ 0 pF. It is considered that the peak around 250 MHz results from the stray capacitance connected to the emitter for the equivalent p-n-p transistors. The 3-dB bandwidth decreases to 185 MHz at $C_h = 22$ pF.

Fig. 8 shows typical input and output waveforms at $C_h = 22$ pF. The input frequency is 5 MHz and the sampling rate is 30 MHz. It is seen that the 1-V_{p-p} input signal is sampled and held for +5-V single power supply. The rising and falling slew rates are 242 and 276 V/μs, respectively.

Fig. 9 shows the input signal level signal-to-noise ratio (SNR) dependence at $C_h = 22$ pF. Bandwidth is 15 MHz. The full-scale input level is 1 V_{p-p}. The SNR at an

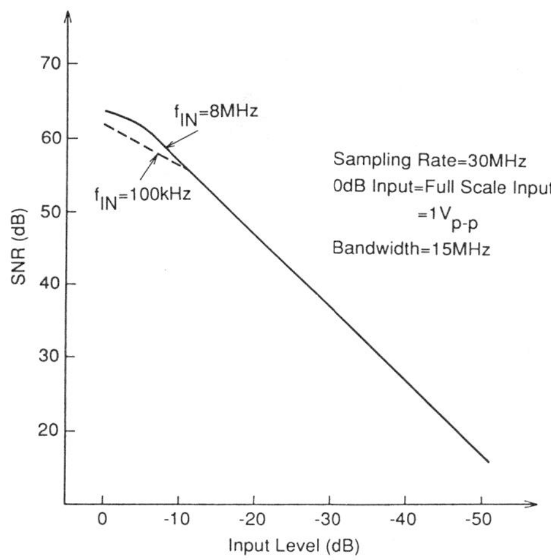

Fig. 9. Input signal level SNR dependences.

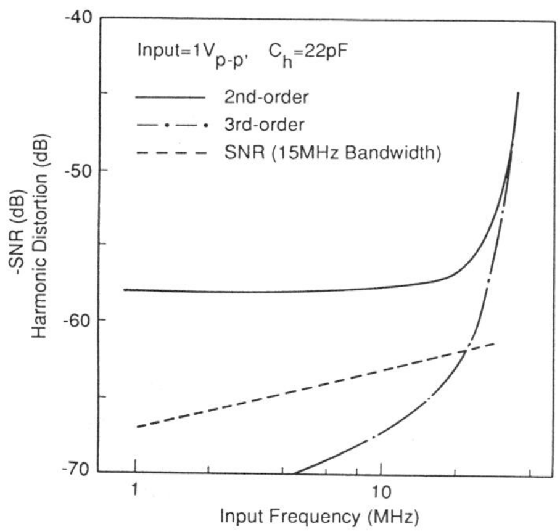

Fig. 10. Harmonic distortion and SNR frequency responses.

8-MHz input frequency is approximately equal to that at a 100-kHz input frequency. The SNR values at 8-MHz full scale and -20-dB inputs are 63.7 and 46.8 dB, respectively. The reason for SNR degradation at a 100-kHz input larger than -10 dB seems to be jitter in the input signal, but has not been clarified yet.

Fig. 10 shows the frequency responses for the harmonic distortion level in the sampling mode and SNR in the sample-and-hold operation. The second-order distortion is dominant and is -58 to -59 dB between 1 and 10 MHz. The SNR at 15-MHz bandwidth is larger than 62 dB below 10 MHz.

The aperture uncertainty evaluation refers to the method in [6]. The results obtained, using a 12-b ADC, are shown in Fig. 11. The sampling rate is 1.4 MHz and the input frequencies are 14.013672 and 28.013672 MHz. The hold capacitor value is 22 pF. From the results, 43-ps jitter and 297-μV_{rms} voltage noise have been calculated for the evaluated sampling system. As the 43-ps jitter contains the input signal jitter and the clock signal jitter, the aperture uncertainty for the sample-and-hold can be expected to be less than 43 ps.

The characteristics for differential gain and differential phase are important for TV signal processing. They are

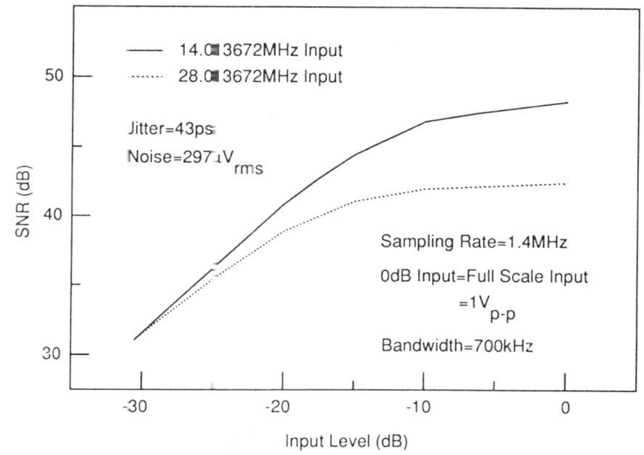

Fig. 11. Input signal level SNR dependences.

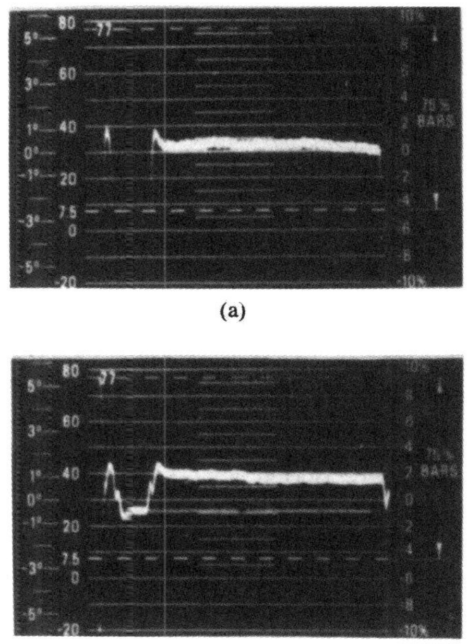

(a)

(b)

Fig. 12. (a) Differential gain (= 0.5%) and (b) differential phase (= 0.3°) at a 20-MHz sampling rate.

TABLE III
MAIN SAMPLE-AND-HOLD CHARACTERISTICS

Input Range	1 V	Acquisition Time	20ns (1V step)
Supply Voltage	+ 5 V	Switching Settling Time	15 ns
Power Dissipation	150 mW	Aperture Uncertainty	< 43 ps
Linearity	0.1 %	Hold Mode Droop Rate	14 mV/μs
SNR	63 dB (15MHz bandwidth)	Hold Mode Feedthrough	- 62 dB (5MHz, 1Vp-p)
Differential Gain	0.5 %	Pedestal Offset	10 mV
Differential Phase	0.3 degree	Die Area	370 x 740 μm^2

shown in Fig. 12(a) and (b). The differential gain is 0.5% and the differential phase is 0.3° at a 20-MHz sampling rate. The hold capacitor value is 22 pF.

The main BiCMOS sample-and-hold IC characteristics are listed in Table III. The acquisition time to 0.1% is 20 ns at 1-V step and the switching settling time to 0.1% is 15 ns. The hold-mode feedthrough is less than -60 dB at 1-V_{p-p} input below 10 MHz. The hold-mode droop rate is 14 mV/μs. The power dissipation is 150 mW. Fig.

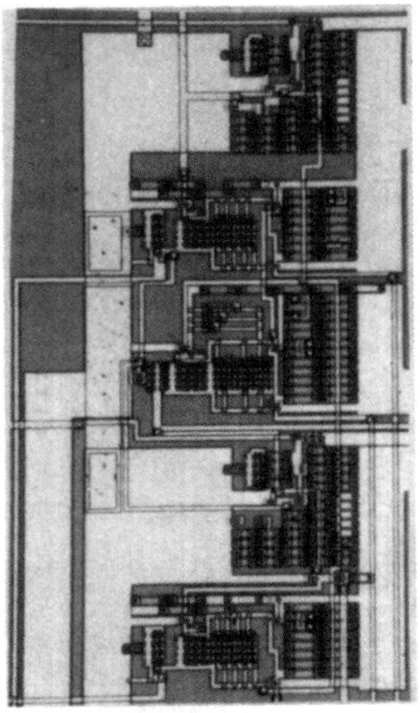

Fig. 13. Sample-and-hold microphotograph.

13 shows a microphotograph of the sample-and-hold circuit shown in Fig. 4. The chip area is 0.27 mm^2.

V. Conclusion

A +5-V single-power-supply BiCMOS sample-and-hold IC has been implemented in 1.2-μm BiCMOS technology and evaluated. It has been clarified that an equivalent p-n-p transistor, formed by the PMOS and n-p-n transistor combination, has a 3-dB bandwidth broader than 400 MHz. The BiCMOS sample-and-hold circuit, designed using the equivalent p-n-p transistors and the hold capacitor voltage feedback buffer circuits, has realized 10-b accuracy and video speed operation with +5-V single power supply. The power dissipation is only 150 mW.

Using this sample-and-hold IC and a two-step parallel format 10-b A/D converter, video signal 10-b A/D conversion is possible with +5-V single power supply and low power consumption. They are applicable to high-quality digital video equipment, such as high-definition TV.

Acknowledgment

The authors would like to thank H. Yamada and T. Sugimoto for their encouragement and advice.

References

[1] K. Tanaka, F. Ishikawa, K. Abe, and K. Koma, "A 40MS/s monolithic S/H IC," in *ISSCC Dig. Tech. Papers*, Feb. 1983, pp. 190–191.

[2] T. Shimizu, M. Hotta, and K. Maio, "A 10b 20MHz two-step parallel ADC with internal S/H," in *ISSCC Dig. Tech. Papers*, Feb. 1988, pp. 224–225.

[3] M. Nayebi and B. A. Wooley, "A 10-bit video Bi-CMOS track-and-hold," in *ISSCC Dig. Tech. Papers*, Feb. 1989, pp. 68–69.

[4] Y. Sugimoto, M. Nakamura, and K. Misio, "High speed sample & hold IC," *IEICE Tech. Rep.*, vol. 86, no. 116, SSD86-43, 1986.

[5] K. Tsugaru, M. Noda, Y. Sugimoto, T. Itoh, and Y. Suwa, "A single power supply 10bit video BI-CMOS sample-and-hold IC," in *Proc. ESSCIRC*, 1989, pp. 84–87.

[6] M. Shinagawa, Y. Akazawa, and T. Wakimoto, "Jitter analysis of high speed sampling systems," in *Symp. VLSI Circuits Dig. Tech. Papers*, 1989, pp. 95–96.

FA12.5 A 1.2μm BiCMOS 100MHz Sample-and-Hold Circuit with a Constant-Impedance Slew-Enhanced Sampling Gate

Myles H. Wakayama, Hiroshi Tanimoto, Takahiro Tasai* and Yoshihiro Yoshida

Toshiba Corporation/*Toshiba Information Systems(Japan) Corporation, Kawasaki, Japan

The diode bridge and related diode-switch-based sampling gate architectures offer high resolution, high-speed sampling and high bandwidth [1,2]. A drastically varying input impedance, low during sample and high during hold, however, typically requires the use of a high-power input buffer that often prescribes clamping circuits to avoid saturation recovery problems for over-range signals. Furthermore, slew performance of the diode bridge is limited by the standing current of the diodes in the sample mode, forcing a tradeoff between slew rate and sampled offset pedestal.

The constant-impedance sampling-gate architecture shown in Figure 1 uses two pairs of diodes to form two switches that perform alternate-cycle sampling. While Output1 is in the sample mode, Output2 is in the hold mode, and vice versa. As with diode-bridge sampling gate-architectures, 460MHz input bandwidth and 150MHz sample rate are achieved.

The sample-mode function of Output1 is implemented through a clock signal such that transistor Q3 is on and Q4 is off. The current I_{BIAS}, switched through Q3, divides evenly between input transistor Q1, switch diodes D1 and D2, and OUT1 of the current-mirror circuit, where the current mirror circuit is biased by the collector of Q1 (Q2 is off). Sampling diodes D1 and D2 are biased by OUT2 of the current mirror. Carefully matching devices D1 and Q1 ensures that Output1 tracks the input voltage. Transistor Q2 and sampling diodes D3 and D4 are reverse biased, and thus, Output2 is in the hold mode. The sampling gate input impedance for this mode is the parallel combination of the impedance looking into the base of Q1 biased at $I_{BIAS} / 3$ and the base of Q2 at zero bias current.

For the hold mode of Output1, Q4 is on and Q3 is off with I_{BIAS} switched through Q4 and divided evenly between input transistor Q2, switch diodes D4 and D3, and OUT2 of the current mirror circuit, where the current mirror circuit is now biased by the collector of Q2 (Q1 is off). Again, by matching devices Q2 and D4, Output2 tracks the input voltage, with Q1, D1 and D2 in reverse bias and Output1 in the hold mode. The sampling gate input impedance in this mode is the same as in the sample mode of Output1 except with Q1 at zero bias current and Q2 biased at $I_{BIAS} / 3$. The input impedance therefore is constant and independent of mode.

In low-slew operation, the current mirror circuit is biased alternately by Q1 or Q2 at essentially constant current. In high-slew operation, however, the currents of Q1 and Q2 vary considerably. This variation in bias current can be used advantageously to enhance the slew rate of the sampling gate by use of the current mirror. For example, if the input signal is slewing upward during the Output1 sample mode, with Q1 on and Q2 off, the current through Q1 will exceed $I_{BIAS}/3$, increasing the output currents of the current mirror circuit, thereby increasing the available slew current to charge C_{HOLD1}. In a similar fashion, the current to discharge C_{HOLD1} is increased for a downward-slewing input signal. Designing the current mirror circuit with an ac gain greater than one further capitalizes on the slew enhancement technique. The measured slew rate of the fabricated sample-and-hold IC was (+430V/μs,-510V/μs) with a C_{HOLD} of 1pF.

Figure 2 is a block diagram of the overall sample-and-hold IC fabricated in a 1.2μm BiCMOS process with 6GHz.-f_T npns. In this implementation only one sampling-gate output is used.

The MUX at the front end improves hold mode feedthrough rejection and, without the use of clamping circuits, prevents corruption of the held signal when the input swings 2V_{be} above the held signal. Integrating the layout of the MUX with the sampling gate reduces offsets due to process and temperature gradients across the die.

The MOS buffer and output buffer circuits are shown in Figure 3. The output buffer is a Darlington configuration with a parallel resistor (R_{LAMP}) to damp ringing with highly-capacitive loads. The linearizing feedback loop (dotted) oscillated on test and was disabled for all measurements, with the Darlington biased externally through a 1kΩ resistor. The MOS buffer is a simple source-follower arrangement with transistor QUB2 added to bootstrap the drain of MUB1, eliminating its nonlinear effect on the gate-source voltage of MUB1 and improving the slew performance of the source follower. For a positive-slewing signal, MUB1 drain current decreases, increasing the current flowing into the emitter of QUB2, biased by the lower current source. This increases the drain current of MUB2 through the current mirror formed by MUB2 and MUB3 to slew the output upward. In a similar fashion, the drain current of MUB2 is reduced for negative-slewing signals allowing MUB1 to pull the output downward more easily. The back gate of MUB1 is driven separately from its source through EUB1 to overcome the large n-well-to-substrate parasitic capacitance. The positive level shift of the MOS buffer compensates for the negative level shift of the Darlington output buffer resulting in zero dc offset through the two stages. This dc offset is zeroed by adjusting the bias current of the MOS buffer either manually or by an offset correction amplifier included on-chip. The offset correction amplifier senses the MOS buffer input and output buffer output and generates a current to bias the MOS buffer. Although the offset correction amplifier requires a high CMR over the full input range, bandwidth requirements are minimal, with its several second time constant set by an external capacitor.

Figure 4 is the output waveform for a 2Vp-p 25MHz input signal sampled at 150MHz clock frequency. At 100MHz clock frequency, the measured sampled distortion for a 20.001MHz 2Vp-p input signal is shown in Figure 5 (beat frequency test) [3]. Other measured data are listed in Table 1. All measurements are into 50Ω to ground through a 1kΩ resistor in series with the output. The IC is assembled in a QFP 44-pin ceramic flat package mounted in a custom test socket with measured 3dB bandwidth greater than 500MHz. Figure 6 is a micrograph of the 3.5mm-square sample-and-hold test chip that dissipates 250mW with +4.5V and -3.5V power supplies.

References

[1] Chambers, M. J., and L. F. Linder, "A Precision Monolithic Sample-and-Hold for Video Analog-to-Digital Converters", ISSCC DIGEST OF TECHNICAL PAPERS, pp. 168-9, Feb. 1991.

[2] Moraveji, F., "A 14b, 150ns Sample-and-Hold Amplifier with Low Hold Step", ISSCC DIGEST OF TECHNICAL PAPERS, pp. 166-7, Feb. 1991.

[3] Poulton, K., et al., "A 2 Gs/s HBT Sample and Hold", GaAs IC Symposium Technical Digest, pp. 199-202, 1988.

Reprinted from *1992 IEEE ISSCC*, pp 200–201 and 282, Feb. 1992.

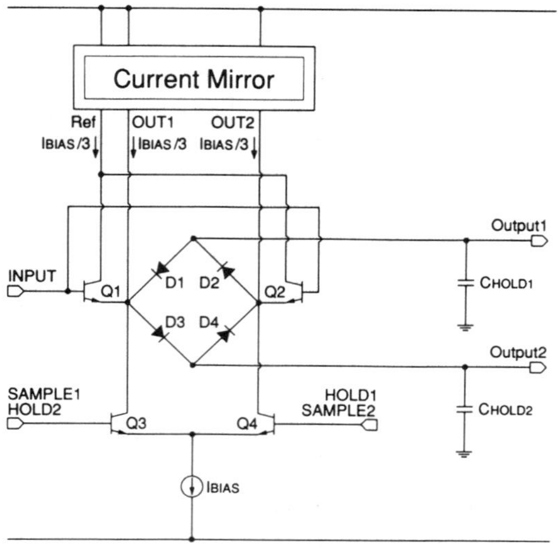

Figure 1: Sampling gate.

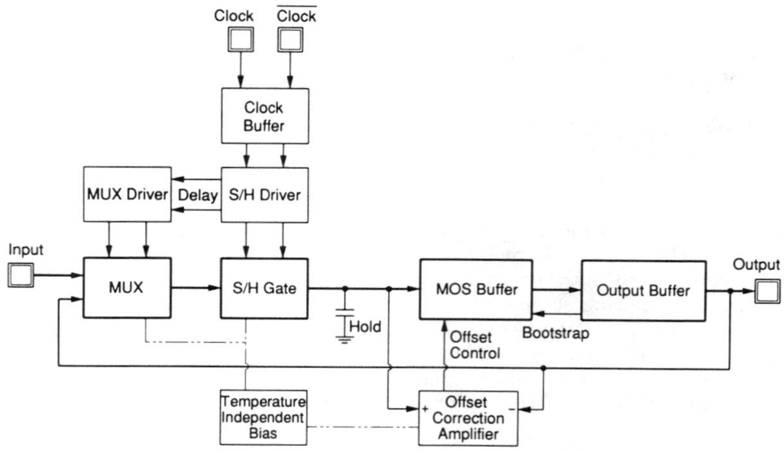

Figure 2: Sample-and-hold block diagram.

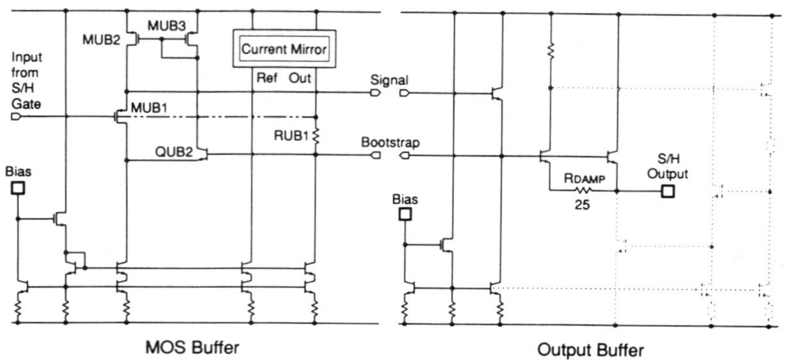

Figure 3: MOS buffer and output buffer circuits.

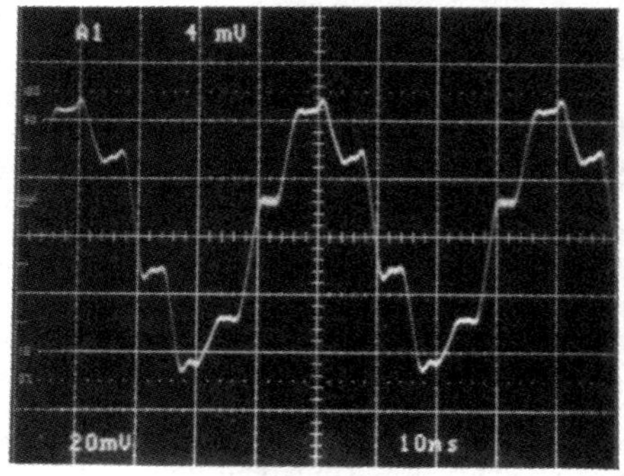

Figure 4: Sampled waveform at 25MHz with 2Vp-p input and 150MHz clock.

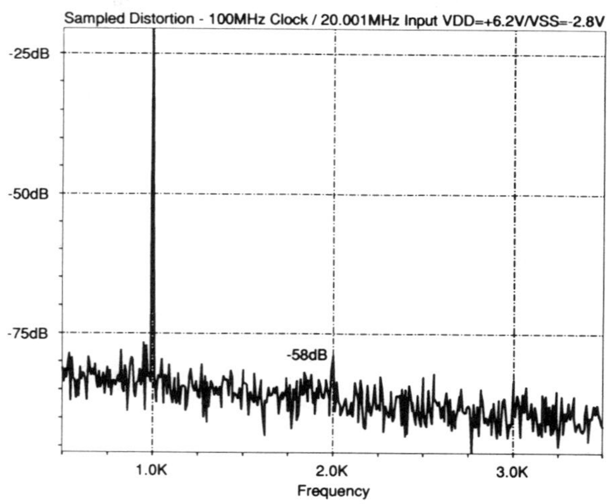

Figure 5: Measured sampled distortion with 20.001MHz 2Vp-p input & 100MHz clock.

C_{HOLD}	1pF
Input range	±1V
Input bias current	<4μA
Input impedance (sampling)	>15MΩ / 2pF
Droop rate	-0.5V/s max
Clock feedthrough to input (100MHz 100mV_{p-p} clock)	1.34mV_{rms} into 50Ω
dc offset (sampling)	+4mV
FS bandwidth (sampling)	50MHz (2V_{p-p})
SS bandwidth (sampling)	460MHz (200mV_{p-p})
Sample pedestal	+4mV
Sample node slew rate	434V/μs, -511V/μs w. 2V step
THD (sample, 20MHz 2V_{p-p})	-54dB (+4.5/-3.5V supplies) -61dB (+6.2/-2.8V supplies)
Sample distortion (100MHz clock, 20.001MHz 2V_{p-p})	-51dB (+4.5/-3.5V supplies) -58dB (+6.2/-2.8V supplies)
Acquisition time, 2V step	0.1% FSR 15ns
Input noise (100MHz input bandwidth)	260μV, 68.7dB SNR
Hold feedthrough rejection	63.4dB (10MHz 2V_{p-p} input)
Power dissipation	250mW (+4.5/-3.5V supplies) 300mW (+6.2/-2.8V supplies)

Table 1: Measured Results

Figure 6: **See page 395.**

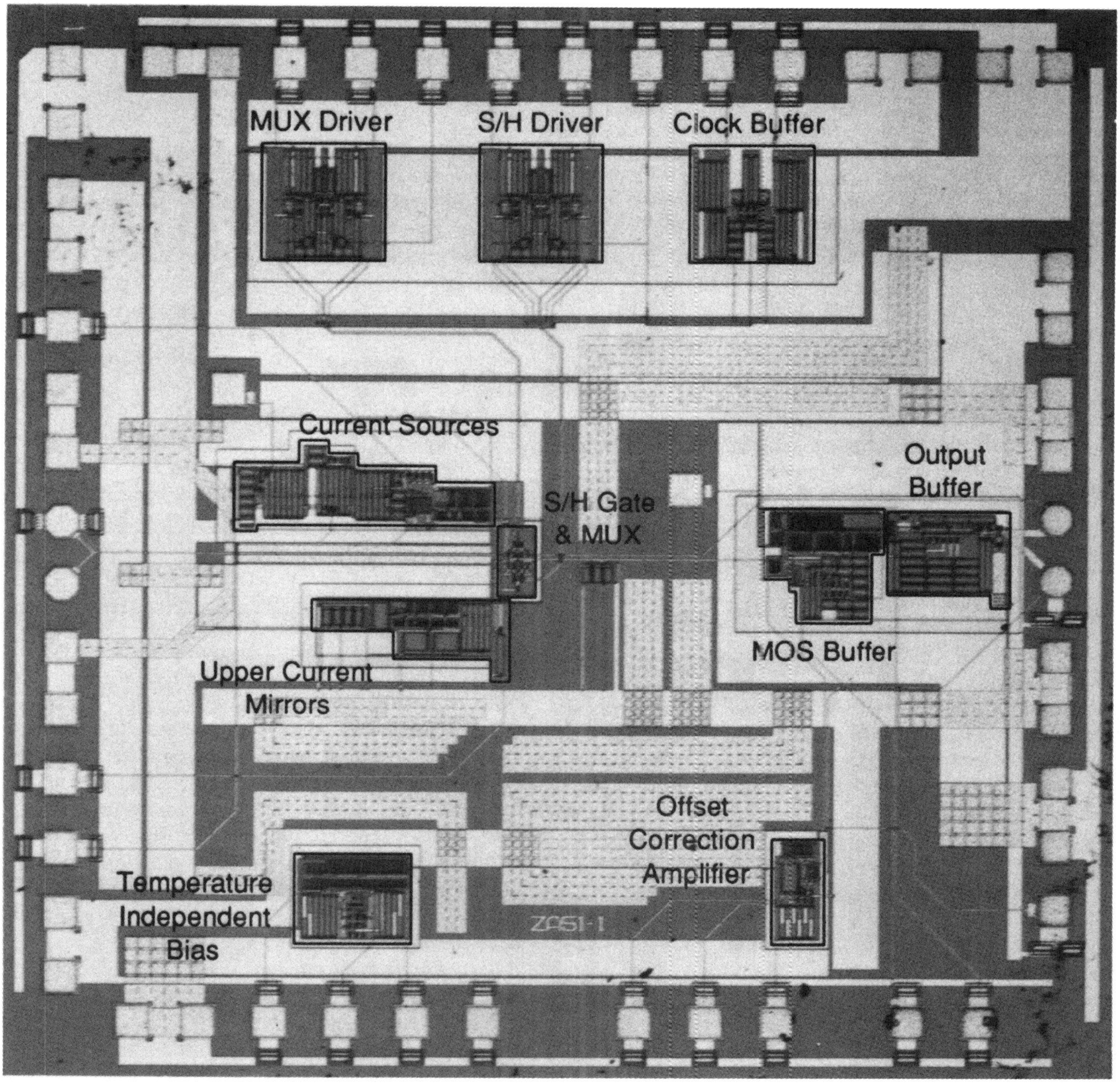

Figure 6: Chip micrograph.

Paper 4.7

A 12-b 750-ns Subranging A/D Converter with Self-Correcting S/H

Douglas A. Mercer, *Member, IEEE*

***Abstract*—A 12-b 750-ns subranging analog-to-digital converter using a multiple-flash architecture is described. An internal sample/hold (S/H) amplifier uses self-correction to achieve ±0.5-mV hold-step error and acquisition time of 250 ns to 0.01% of a 5-V step. The 29.2-mm² device includes a 2.5-V reference and is built on a 2-μm 10-V BiCMOS process.**

I. Introduction

TRADITIONALLY, high-speed high-resolution sampling analog-to-digital converters were constructed using multichip hybrid methods and consumed substantial power. The arrival of high-speed fine-geometry BiCMOS process technology has allowed such devices to be implemented monolithically. This paper presents results of a 12-b BiCMOS sampling analog-to-digital converter IC. The ADC achieves full accuracy with an overall conversion time of 750 ns including sample acquisition. In addition, the design provides a 2.5-V reference output that can be used as the main reference for the ADC.

The multistage subranging circuit performs the analog-to-digital conversion in four steps and includes digital error correction [1], [2]. Without this error correction each of the flash comparators must resolve their respective inputs to less than one 12-b LSB. Requiring large numbers of accurate comparators causes an increase in the power and area needed. The error correction, in addition to relaxing the accuracy requirements for the comparators in each flash, allows the flash converters to be clocked before the sample/hold (S/H) amplifier and subtraction DAC's in the earlier stages have settled fully to 12 b. This can amount to a significant time savings. The signal paths that impact the conversion time were minimized. The slower flash output encoding and correction logic functions were delayed to the end of the conversion cycle and are performed while the S/H amplifier is acquiring the next sample.

A 2-μm feature size BiCMOS process with 3-GHz polysilicon emitter n-p-n transistors was used to implement the design. The resulting die measures 29.2 mm² and nominally dissipates 600 mW.

Manuscript received May 1, 1991; revised July 16, 1991.
The author is with Analog Devices Semiconductor, Wilmington, MA 01887.
IEEE Log Number 9103056

II. Functional Description

The circuit in the block diagram of Fig. 1 shows the overall signal flow and indicates the signal ranges at various points. The on-chip 2.5-V bandgap serves as the master reference for the ADC and optionally provides bipolar offset current through resistor $R2$. The on-chip reference is not hard wired to the analog reference input, allowing the optional use of an external voltage reference. The conversion timing is controlled by an on-chip clock generator. A conversion is started by the single encode digital input. The clock indicates when new valid output data are available and the conversion is complete on the DAV digital output. After the analog input is sampled by the S/H amplifier, it is applied to the input resistors of the converter section. By utilizing four flash converters, the input signal is not recirculated through the same circuitry multiple times as in previous designs [1]. This is a major key to achieving the high conversion rate. As in other multistage subranging converters [2], the analog-to-digital conversion process takes place in a series of steps or flashes. In this implementation there are four independent arrays of comparators. Other multistage configurations [2] use a single summing node for the three DAC outputs. Such a scheme requires that the second and third DAC's be turned off while the second flash is measuring the residue on the summing node. In the configuration described here each of the two 12-b accurate subtraction DAC's has a separate output node. By using a differential input amplifier to obtain the second residue voltage at the output of the residue amplifier, the ADC2–DAC2 combination can get a head start before ADC1–DAC1 is latched.

The objective is to make a very fast guess or approximation of the input with the first 3-b flash (ADC1). The output of each comparator in this flash directly switches one current source of a segmented DAC (DAC1). This provides a fast direct path from the flash to the DAC. The output current of the DAC flowing through the silicon-chrome thin-film input resistor $R1$ creates a voltage. This 12-b accurate voltage is subtracted from the analog input. The remainder, or residue voltage, appears on node OUT1. With such an arrangement, even when the comparators in the flash are not latched, the subtracted voltage will be approximately correct. The DAC output current will be exact when the flash is latched.

Reprinted from *IEEE J. Solid-State Circuits*, vol. 26, no. 12, pp. 1790–1799, Dec. 1991.

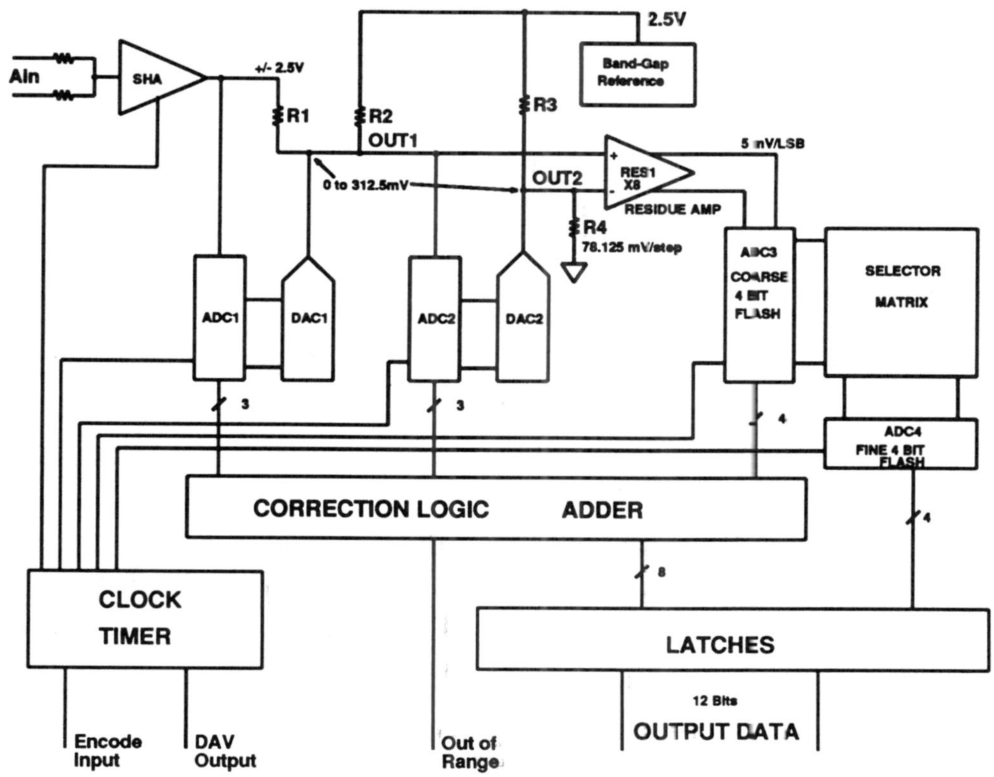

Fig. 1. System block diagram.

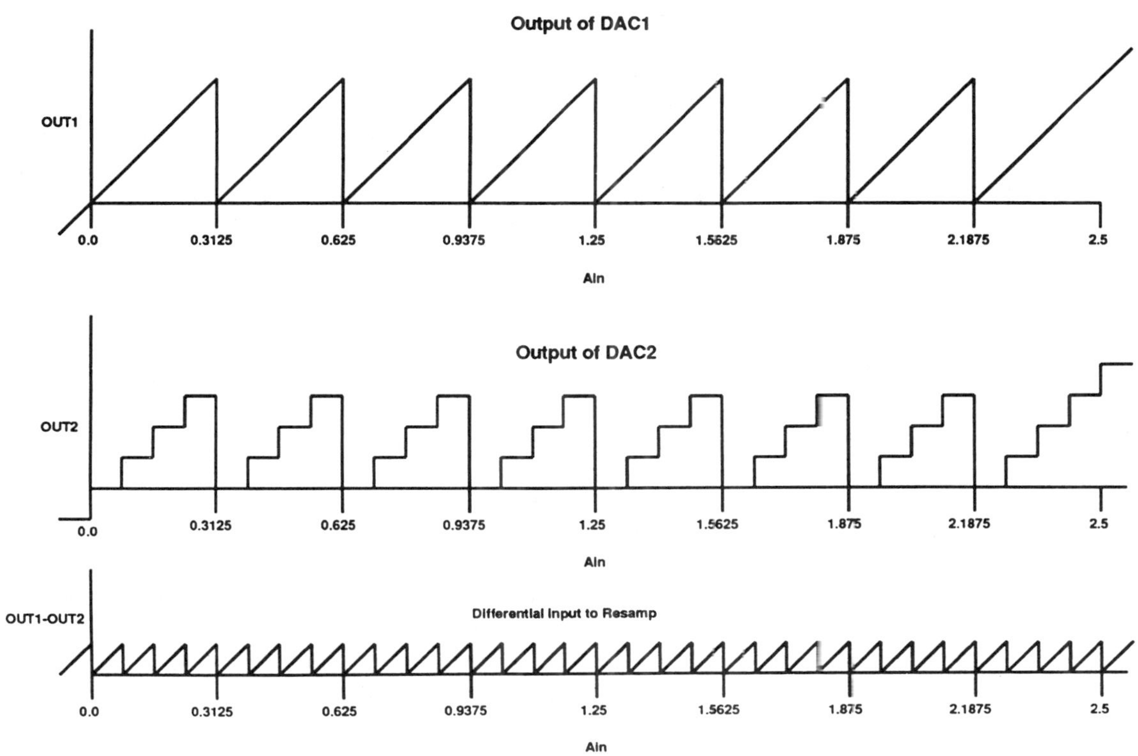

Fig. 2. Internal residue waveforms.

This keeps the voltage at the node OUT1 between 0 and 312.5 mV (one DAC step). This tracking action can be seen in Fig. 2, which plots the residue node OUT1 versus the analog input.

Simultaneously a second flash, ADC2, is tracking a buffered version of the voltage on node OUT1. This second flash also directly switches a segmented current output DAC, DAC2. The output current of DAC2 is converted into a voltage with thin-film resistors $R3$ and $R4$. This voltage, at node OUT2, is a stair step approximation of the voltage at node OUT1. This approximation is within one step of DAC2 or 78.125 mV of node OUT1, as indicated in Fig. 2.

The first two flashes are the same except that ADC2 is not a full 3 b. ADC2 divides the 312.5-mV steps of DAC1 into four smaller steps plus one extra step on each end. This extra range is used by the digital correction logic to correct for possible linearity errors in the results from the first flash converter.

While the DAC's must be 12-b accurate, the first two flash converters need be only accurate to 6 b because of the digital correction in subsequent stages. After the 12-b accurate voltages from DAC1 and DAC2 are subtracted from the input, the resulting signal can be converted at 8-b accuracy. The third flash must be only 8 b linear, but slope or gain error must be minimized. The full scale of the third flash, referred to the input of the residue amplifier, must match the gain or average step size of the two DAC's to less than 1 LSB at the 12-b level to achieve no missing code performance.

An 80-MHz open-loop differential voltage amplifier, RES1, amplifies the difference between the outputs of two DAC's by a factor of 8. A 12-b LSB will now be approximately 5 mV. The untrimmed offset of the bipolar comparators is typically less than 1 mV, which contributes only a small linearity error. The differential output of this residue amplifier drives a two-step flash converter with 8 b of resolution. After the first two stages have quantized the input signal to 5 b, only 7 b of resolution are required for a 12-b overall results. However, extra resolution is again used to detect and correct for errors in the first two flashes.

III. Sample/Hold Design

Fig. 3 is a block diagram of the S/H amplifier. This circuit design includes a self-correcting architecture [3]. By compensating for charge injection and amplifier gain and offset errors, this circuit corrects for internal errors while in the hold mode. Due to the nature of the design, the output in the sample mode is not optimized to accurately track the input. However, in the hold mode, the internal feedback loop is configured to produce an output identical to the input signal at the instant of sampling. The S/H amplifier makes optimal use of the BiCMOS process. The MOS devices are used for switches and low leakage input stages with the bipolar devices used for high-bandwidth buffer and output stages.

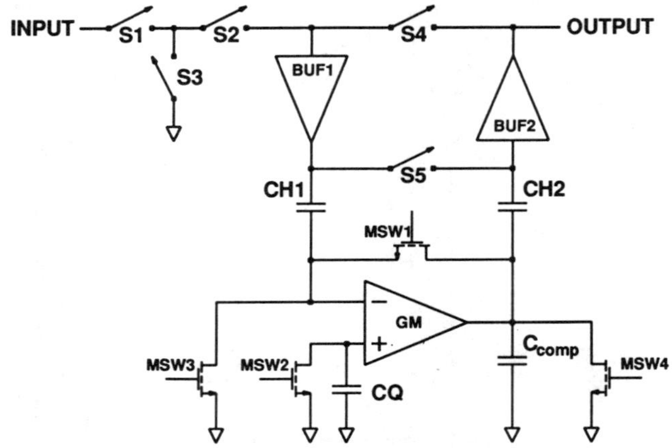

Fig. 3. S/H block diagram.

The configuration in the sample mode is as follows. CMOS switches $S1$, $S2$, and $S5$ are closed; $S3$ and $S4$ are open. This provides a direct path from input to output through the two unity-gain buffers. NMOS switch $MSW1$ is on, closing the feedback around the transconductance amplifier GM. Switch $MSW2$ is also on, grounding the noninverting input of GM. Two additional NMOS switches, $MSW3$ and $MSW4$, are turned on for a short time, approximately 50 ns, at the beginning of the acquisition of a new sample. The large charging current, which is needed to charge capacitors $CH1$ and $CH2$, now flows through these switches to ground rather than from the output of amplifier GM. This shortens the time to slew the 3-pF hold capacitors $CH1$ and $CH2$ and limits the transient voltage excursion at the output of amplifier GM. This can shorten the overall acquisition time by as much as 30% when acquiring large input steps. Clamping these nodes during slew, in effect, band limits the signal seen by amplifier GM. Switches $MSW3$ and $MSW4$ are opened again during the final part of the acquisition phase allowing the amplifier to settle the correct voltage across the hold capacitors. During the final part of acquisition, amplifier GM forces the voltage at the bottom of capacitors $CH1$ and $CH2$ to ground potential. The offset of GM and the first buffer is sampled along with the input signal on $CH1$. The input is also sampled on capacitor $CH2$ through $S5$.

The configuration in the hold mode has switches $S1$, $S2$, and $S5$ open; switches $S3$ and $S4$ are closed. Switch $MSW1$ is turned off first, sampling the input on $CH1$. Switches $S1$, $S2$, and $S3$ comprise a "T" switch, which is used to improve the input isolation when in the hold mode. The feedback path around amplifier GM now includes $CH1$, $CH2$, and the two buffer stages. Capacitor $CH2$ acts as a coupling capacitor for the output of GM to the input of BUF2. Since $CH2$ also has the input sampled on it, the output of amplifier GM only swings the amount needed to correct for the combined offset of the two buffers. The offset and open-loop gain required in amplifier GM for greater than 12-b dc linearity is easily satisfied by a straightforward single-stage design. This would

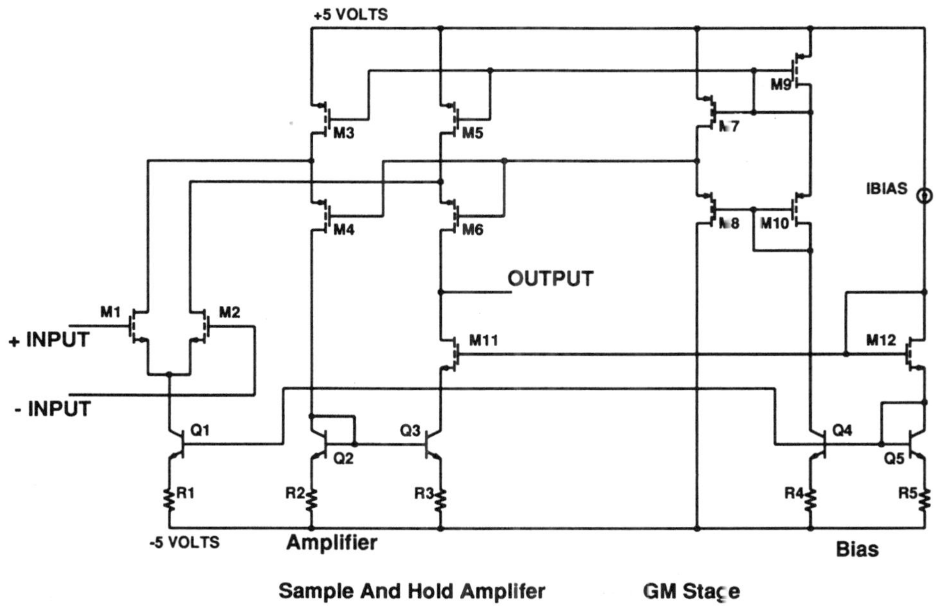

Fig. 4. S/H *GM* amplifier stage.

not be possible with a more conventional S/H design. Moreover, much higher bandwidths are possible with a simple, single gain stage.

As sampling switch *MSW*1 turns off, some error charge is injected onto the hold capacitor *CH*1. Switch *MSW*2 is turned off at the same time and the error charge it injects onto capacitor *CQ* will introduce an equal error voltage at the noninverting input of *GM*. The switches *MSW*1 and *MSW*2, and capacitors *CH*1 and *CQ*, are matched providing a first-order cancellation of the charge injection. The gate drive to the NMOS sampling and compensation switches, *MSW*1 and *MSW*2, respectively, is subregulated from the main power supplies to improve power supply rejection and to minimize the variability of the size of the uncompensated pedestal error.

The loop is compensated by the total load capacitance seen at the output of the *GM* amplifier. While in the sample configuration, this load would normally just be the parallel combination of *CH*1 and *CH*2. However, in the hold mode these two capacitors change functions and *CH*1 becomes an input coupling capacitor and *CH*2 becomes an output coupling capacitor. In this configuration amplifier *GM* is no longer properly compensated. By adding some extra capacitance in the form of C_{comp} we ensure that the loop will remain stable while in the hold mode.

More detail of amplifier *GM* is shown in Fig. 4. Low droop is achieved by using a low leakage current NMOS input stage. The difference current from the input stage is folded in the PMOS cascode section *M*3–*M*6. The gate bias of the cascodes is generated from a reduced headroom current mirror *M*7–*M*10. A simple emitter degenerated n-p-n mirror is used to reflect the PMOS current. The systematic offset caused by the base current error from *Q*1 and *Q*3 is removed by the autozeroing action of

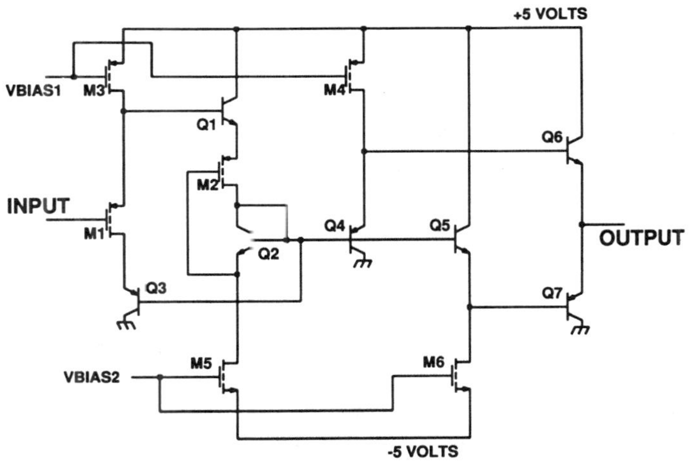

Fig. 5. S/H buffer amplifier.

the design. This simple mirror affords the highest bandwidth with fewer high-frequency poles than a more accurate mirror. To maximize gain the output of *Q*3 is cascoded by NMOS *M*11.

More detail of the unity-gain buffers is shown in Fig. 5. While the input-to-output offset of the buffers is canceled by the feedback loop in the hold mode, large offsets will limit the total swing available from the 5-V power supplies. Some care was taken to minimize the offset of these buffers, thereby improving the track mode performance. The gate-to-source voltage of input FET *M*1 is canceled by a matching FET *M*2. The drain–source voltage of *M*1 is bootstrapped by substrate p-n-p device *Q*3 to be the same as that of *M*2. MOS current sources were used to maximize the possible output swing. This also avoids possible saturation of bipolar current sources at the extremes of the output swing.

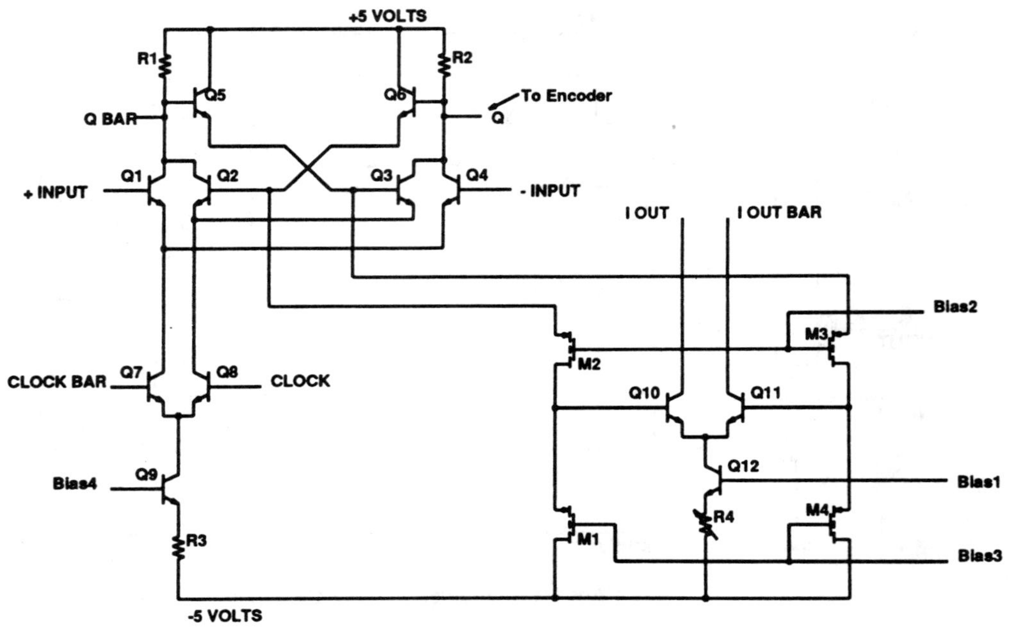

Fig. 6. Flash-DAC stage.

A conventional class-AB output stage supplies the required output drive current. The 3-dB bandwidth of these buffers is in excess of 80 MHz and is well beyond the unity-gain crossover of the overall feedback loop. The excess phase contributed by the buffers is not significant except under heavy capacitive output load conditions. This is never the case in that the output only drives the input resistors of the ADC.

IV. Flash/DAC Design

As mentioned earlier, each comparator in ADC1 and ADC2 directly drives one segment of a current output DAC. Fig. 6 shows the combination of one comparator with one DAC current source and switch. Each of the seven segments in DAC1 is 625 μA. Each of the five segments in DAC2 is 156.25 μA. A conventional cross-coupled latching comparator design was chosen for the flash. Input offset and gain are relatively unimportant because the first two flash codes are digitally corrected. The accuracy requirements are relaxed to the 6-b level. This allows the use of a single-stage design. Bandwidth and latching speed are most important in this application.

The latch is controlled by switching the tail current from n-p-n $Q9$ through either $Q7$ or $Q8$ to the outer n-p-n pair $Q1$, $Q4$ or the inner pair $Q2$, $Q3$. The output of the latch is taken at nodes Q and QB. The emitters of $Q5$ and $Q6$ drive both the positive feedback to the bases of $Q2$ and $Q3$ and the sources of common-gate devices $M2$ and $M3$. These PMOS devices provide the necessary level shift and a fast path from the latch to the DAC switch. Devices $M1$ and $M4$ act as loads and the difference in their gate-to-source voltage appears at the bases of $Q10$ and $Q11$, which form the precision DAC switch. When node Q is higher than QB, $M2$ has more drain current than $M3$. The gate-to-source voltage of $M1$ will be higher than that of $M4$ and thus the base of $Q10$ will be higher than that of $Q11$. $Q10$ will be turned on directing the collector current of $Q12$ to the I_{out} node. Even when the comparator is not latched, there is sufficient gain from the input to the I_{out} node such that only one of the DAC segments will ever be partially on or off. All other segments will be fully switched on or off.

The binary output of the flash is encoded through another path from the Q and QB outputs. Emitter-coupled wired-AND/wired-OR logic encodes the thermometer-code flash output into binary form. An ECL-to-CMOS level translation circuit is used prior to the CMOS correction logic.

V. Residue Amplifier Design

The residue amplifier, which has a fixed gain of 8, is simply a voltage-to-current converter. The output, which is a differential current, is turned back into a differential voltage with a pair of load resistors. The voltage gain is determined by the ratio of two thin-film resistors, the g_m resistor and the output loads. There is no negative feedback around the entire amplifier requiring large forward or open-loop gain for accuracy. This provides much faster settling behavior and recovery from overload than a design that uses a conventional high-gain operational amplifier.

Referring to the schematic of the residue amplifier, Fig. 7, the differential input voltage is applied to the base of n-p-n $Q3$ and the base of n-p-n $Q4$. The circuit is completely symmetric. Looking at the left half, a cascoded PMOS current source consisting of $M5$ and $M6$ supplies a fixed current into the collector of the input n-p-n $Q3$. Any difference between the collector current of $Q3$ and

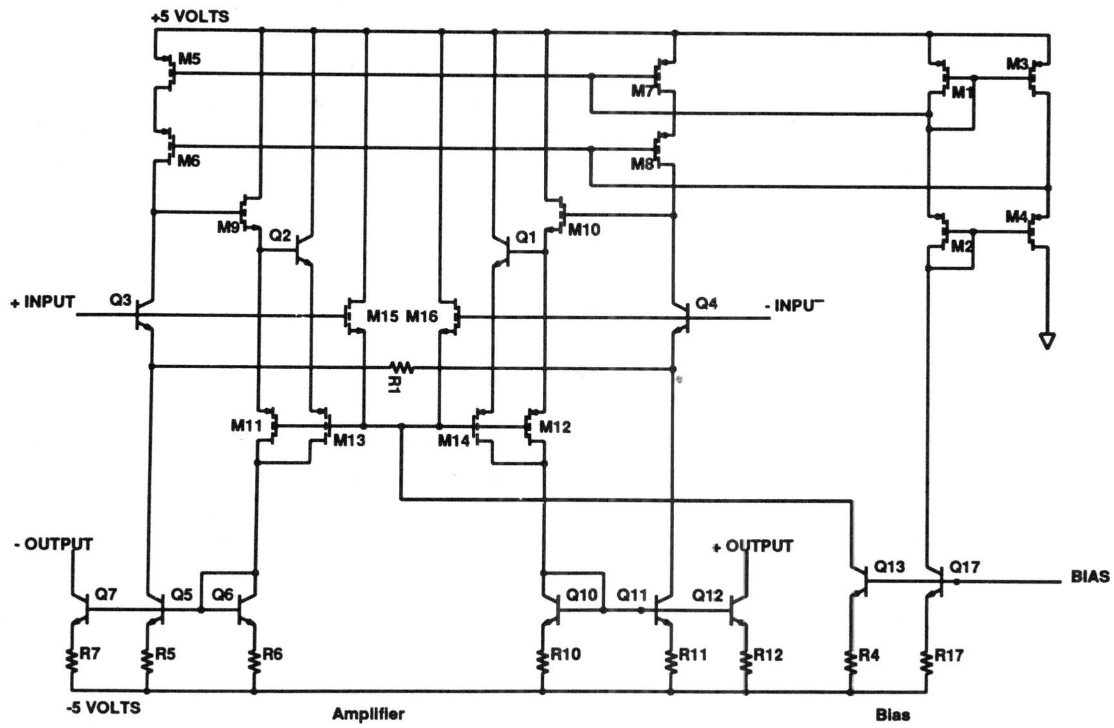

Fig. 7. Gain of eight residue amplifiers.

the current supplied by PMOS $M6$ causes the gate of NMOS $M9$ to move. The combination of $M9$ and $Q2$ pulls up on the source of PMOS $M11$ and $M13$. This sources current into diode-connected n-p-n $Q6$. $Q5$ mirrors this current and sets the current in $Q3$ to that of $M6$. N-P-N $Q7$ makes another copy of this current which, when tied to a load resistor, produces an output voltage. The two identical sides are tied together by g_m resistor $R1$. $R1$ sets the voltage-to-current transconductance. As the differential input voltage changes, the action of the loop forces the current in $Q3$ ($Q4$) to be constant and also the V_{be}. This insures that an accurate version of the differential input voltage is impressed across resistor $R1$. In order to maintain the constant current in $Q3$ ($Q4$), a differential current flows in the two feedback mirrors. This current also appears as a difference in the current in the two output devices $Q7$ and $Q12$. The voltage gain is thus set by the ratio of $R1$ to the output load resistors (not shown in Fig. 7).

NMOS devices $M15$ and $M16$ sense the common-mode voltage of the input and provide the common-gate voltage of PMOS sources $M11$–$M14$. This maintains a relatively constant base–collector voltage on the input n-p-n devices over a range of common-mode input voltages. Again a reduced headroom PMOS current source arrangement, comprised of devices $M1$–$M4$, is used to bias PMOS devices $M6$ and $M8$. The bias current supplied by n-p-n $Q17$ is nominally constant with temperature. This is necessary to insure sufficient standing current for all combinations of inputs and temperature.

VI. Two-Step Flash Design

In this two-step 8-b flash, coarse and fine 4-b flashes are connected together by a selection matrix (see Fig. 1). A differential reference resistor ladder is built from 16 coarse resistors that each have eight fine resistors in parallel. The selection matrix consists of 128 n-p-n differential pairs arranged in 16 rows of eight columns. The switch function is performed by directing eight tail currents to only one of the 16 rows at a time. The thermometer output code of the third flash is converted into a bubble (or one of) code and used to select the matrix row within which the input signal lies.

An interpolation technique is employed to resolve 16 levels with only eight resistor taps [4]. In a bipolar technology the comparator preamps are made with the common differential long tail pair. At least one of these differential pairs is needed for each comparison to be made. By using interpolation the number of resistors and preamps is reduced by a factor of 2.

If we examine two n-p-n differential pairs, shown in Fig. 8, the outputs are on nodes $C1$–$C4$. Note that $R5$ and $R6$ represent the ladder in the flash. Currents $I3$ and $I4$ represent reference currents used to create the necessary tap voltages in the resistor ladder. The code edges occur at the points where the comparator inputs are balanced. Differential pair $Q1$ and $Q2$ are balanced when V_P is -10 mV and $Q3$ and $Q4$ are balanced when V_P is $+10$ mV with respect to V_N. That is to say that the differential voltage at $C1$ and $C2$ is zero when the voltage on the base of $Q1$ is equal to the voltage on base of $Q2$.

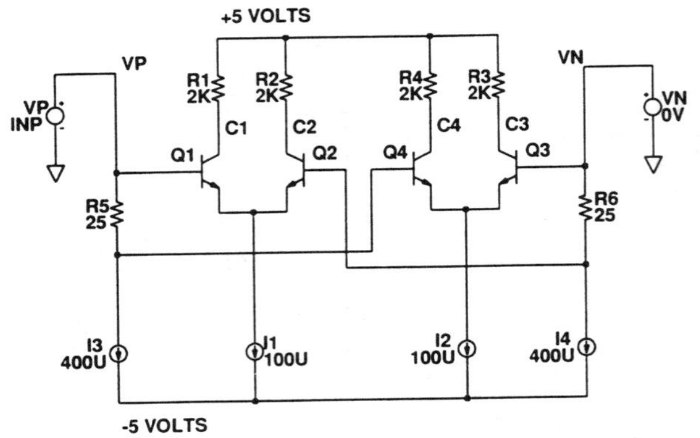

(a)

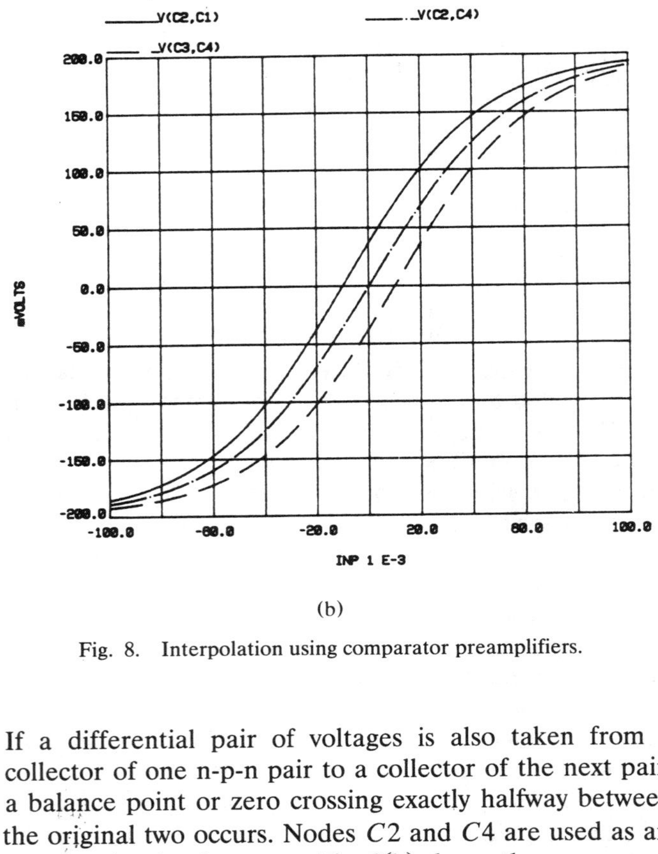

(b)

Fig. 8. Interpolation using comparator preamplifiers.

If a differential pair of voltages is also taken from a collector of one n-p-n pair to a collector of the next pair, a balance point or zero crossing exactly halfway between the original two occurs. Nodes $C2$ and $C4$ are used as an additional pair of outputs. Fig. 8(b) shows the zero crossing of the differential output voltages. The two normal outputs cross zero at +10 and −10 mV and the interpolated signal crosses at 0 mV. This interpolation technique only works well when the actual crossings are closely spaced, by at most no more than a few tens of millivolts. The V_{be} matching requirements are no more stringent than for a single differential pair. The matching of the tail currents ($I1$ and $I2$), however, does become more important.

To complete the fourth stage, 15 single-stage latching comparators are attached to the common output collectors of the matrix. The switch differential pairs of the matrix also perform the preamplifier function in the last flash. The output of this last flash provides the 4 LSB's without going through the error correction adders. The way the resistor ladder is configured, with the fine taps in parallel with the coarse taps, insures that the voltages are monotonic. Errors between the third and fourth flashes will be small and mainly due to offsets in the matrix differential pairs.

VII. Voltage Reference Design

This is a design based on a n-p-n transistor bandgap core [5], [6]. A detailed schematic of the reference is shown in Fig. 9. A curvature correction feature [7], which cancels most of the parabolic curvature inherent in bandgap references, is included. The proportional-to-absolute temperature (PTAT) current in the reference core is used in conjunction with diffused resistor $R3$ to add this correction. Diffused resistors, made using the NMOS source/drain material, exhibit a strong positive temperature coefficient. The small voltage across this resistor has a term that is proportional to temperature squared, and when added in series with the reference core, compensates the second-order curvature of the V_{be} itself.

The amplifier closing the loop around the reference core has some significant design constraints. In the ADC, as the DAC's switch, large current steps are injected into the output of the reference. This requires that the output stage have a very low high-frequency output impedance. There are two possible approaches to achieve low output impedance. One approach is to use a voltage buffer output stage, like a class-A n-p-n emitter follower or class-AB push–pull stage. These typically are biased at a high current level to obtain a very low R_e. This is costly in power. The emitter resistance R_e of a bipolar transistor has a range of frequencies where it becomes inductive. Combined with certain values of load capacitance, the output can ring or even oscillate. An alternative approach is a current source output. Such an output stage has, inherently, a very high output impedance. Feedback lowers the output impedance. This works up to the bandwidth limits of the forward gain of the amplifier. It is in this case, however, possible to use a large capacitor to lower the impedance at high frequencies without introducing any oscillations. The output RC time constant becomes the dominant pole in the system.

An output transconductance amplifier topology was chosen with a PMOS input differential pair consisting of devices $M5$ and $M6$. The amplifier provides feedback to set the operating point of the reference core. The feedback loop's dominant pole is set by the large external capacitor C_{ext} attached from the output REF OUT to ground. This capacitor, on the order of 1 μF, has the effect of filtering the output noise of the reference. It also acts to keep the voltage steady when fast load current changes occur. This configuration has the advantage that

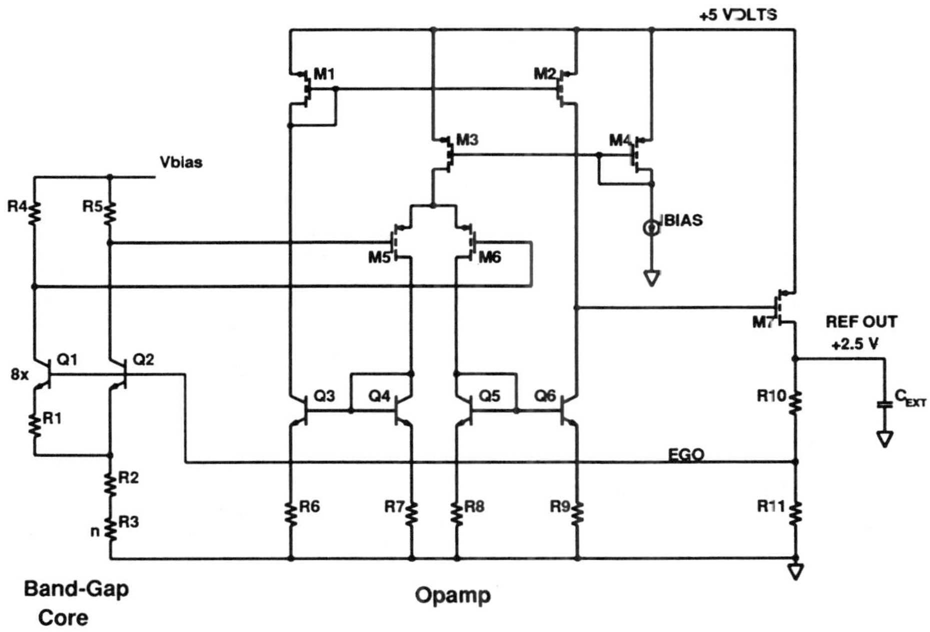

Fig. 9. Curvature corrected bandgap voltage reference.

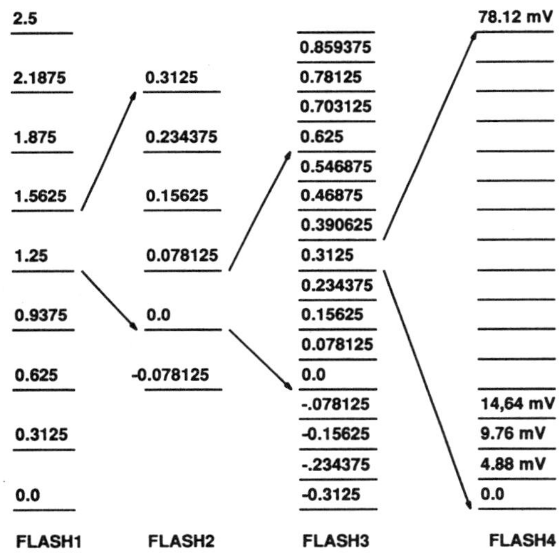

Fig. 10. Overlapping flash input ranges for error correction.

it has inherent short-circuit current limiting. When shorted to a voltage greater than the output, the loop shuts down the PMOS output device $M7$. The amount of current available is just what flows in the feedback resistors ($R10$, $R11$) to ground. When shorted to a voltage less than the output, the current is limited by the size of the PMOS output transistor $M7$. If the output is shorted directly to ground, then the op amp will shut down since I_{bias} is derived from the reference output voltage.

VIII. Correction Logic Design

As stated earlier, subranging ADC's without any error correction require that the flash comparators also be accurate to 12 b. Digital error correction methods, such as used in this design, have been published in the literature [1], [2]. The method used in this design is much the same.

As shown in Fig. 10, by extending the range of the second and third flashes to overlap more than one step in the first and second flashes, we are able to detect when an error is made and thus recover from it. In this technique we are, in a way, producing redundant codes used by the correction logic to obtain the actual digital output. To simplify the correction logic and require only an adder and not both an adder and a subtracter, the binary output from the first two flashes is encoded minus one. This allows the digital value to be left "as is" (add 0) when correcting errors that cause under-range in the later flashes. If no error is detected, 1 is added to the value to get the correct result. If an error is made which causes over-range in the later flashes, 2 is added. Such a digital error correction scheme produces extra output codes beyond the normal range of the 12-b ADC. Internal logic suppresses these codes. In the process, the logic provides an overall out-of-range indication. A logic output signal is set low for analog inputs within range and is set high for analog inputs more negative than 1 LSB below the first code transition and more positive than 1 LSB above the last transition.

IX. Measured Results

The design was fabricated in a 2-μm BiCMOS process featuring 3-GHz f_t n-p-n bipolar devices, N-well CMOS, and laser-trimmable thin-film resistors. Fig. 11 shows a die photo of the converter. The chip measures 5.86 mm × 4.98 mm. Table I lists typical performance data. Important performance achievements are the hold-mode

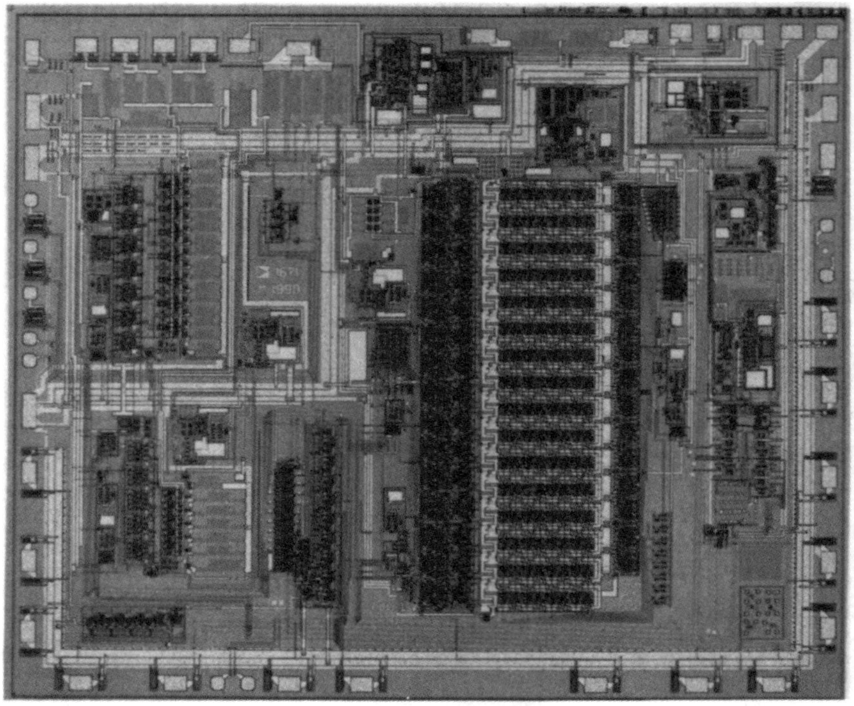

Fig. 11. Die photo.

TABLE I
PERFORMANCE

Max Conversion Rate	1.25 MHz
Large-Signal BW	8.5 MHz
Small-Signal BW	20 MHz
Reference Drift	15 ppm/°C
Unipolar input range	0 to 2.5/0 to 5
Bipolar input range	−2.5 to 2.5/−5 to 5
Power supplies	+/− 5 V
Power Dissipation	600 mW
S/H Amplifier	
Droop Rate	20 nV/μs
Aperture Delay	15 ns
Aperture Jitter	40 ps
Pedestal	0.5 mV
Feedthrough	−80 dB at 500 kHz
	−74 dB at 1 MHz
	−68 dB at 2 MHz

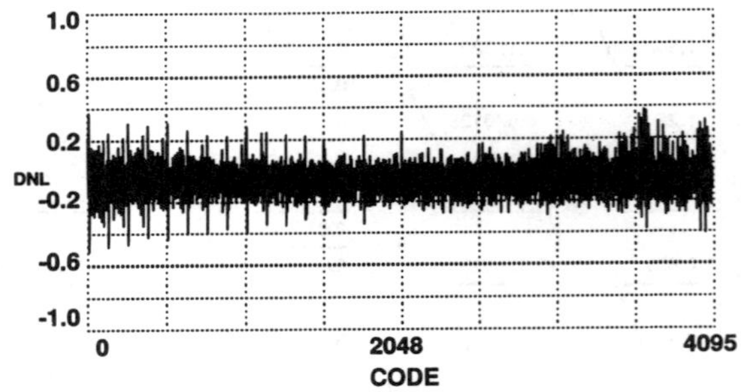

Fig. 12. Differential nonlinearity.

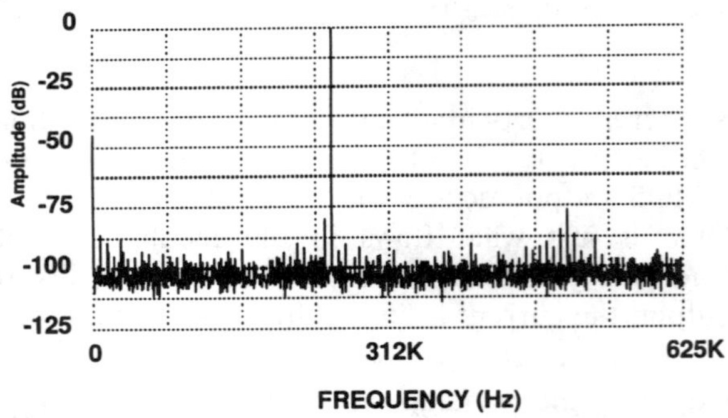

Fig. 13. 2048-point FFT 12-b ADC at 1.25 Msample/s, $f_{in} = 250$ kHz.

pedestal and aperture jitter in the S/H amplifier section. Fig. 12 is a differential nonlinearity (DNL) plot of the ADC. The worst-case DNL is less than ± 0.5 LSB. A 2048-point FFT showing the spectral content of the complete sampling ADC for a 250-kHz full-scale input and a 1.25-Msample/s conversion rate is shown in Fig. 13. It can be seen that the nonharmonic content is at or near the quantization noise level of a 12-b ADC. The harmonic content is dominated by the second and third harmonic. Factors such as hold-mode feedthrough, ADC integral nonlinearity, and amplifier slew rate in the S/H amplifier contribute to this distortion.

ACKNOWLEDGMENT

The author wishes to thank the Converter Design Group and specifically the important contributions of G. Miller for his original concept for the S/H amplifier, C. O'Connor for the first silicon implementation of the S/H amplifier, M. Timko for clock and correction logic design, P. Mason and P. Sheppard for layout design. D. Kelly for his help with characterization, and B. Scharf for process development.

References

[1] J. Fernandes, S. R. Lewis, A. M. Mallinson, and G. A. Miller, "A 14-bit 10-μs subranging A/D converter with S/H," *IEEE J. Solid-State Circuits*, vol. 23, pp. 1309–1315, Dec. 1988.

[2] M. Kolluri, "A 12-bit 500-ns subranging ADC," *IEEE J. Solid-State Circuits*, vol. 24, pp. 1498–1506, Dec. 1989.

[3] D. Mercer and P. Real, "A 14b linear, 250ns sample-and-hold subsystem with self-correction," in *ISSCC Dig. Tech. Papers*, vol. 34, Feb. 1991, pp. 164–165.

[4] C. Mangelsdorf, "Parallel analog-to-digital converter," U.S. Patent 492227, May 8, 1990.

[5] A. P. Brokaw, "A simple three-terminal IC bandgap reference," *IEEE J. Solid-State Circuits*, vol. SC-9, Dec. 1974.

[6] A. P. Brokaw, "Solid-state regulated voltage supply," U.S. Patent 30586, Apr. 21, 1981.

[7] A. P. Brokaw, "Band-gap voltage reference with curvature correction," U.S. Patent 4250445, Feb. 10, 1981.

Paper 4.8

A 10-bit Video BiCMOS Track-and-Hold Amplifier

MEHRDAD NAYEBI AND BRUCE A. WOOLEY, FELLOW, IEEE

Abstract —**A 1.2-μm BiCMOS technology has been used to implement a monolithic video track-and-hold amplifier that settles to an accuracy of 10 bits in 15 ns. This level of performance is competitive with hybrid track-and-hold circuits and surpasses previously reported monolithic implementations by nearly two orders of magnitude. The amplifier's design is based on a closed-loop topology incorporating two BiCMOS folded-cascode gain stages, an NMOS sampling switch, and a BiCMOS switch driver with 1-ns transitions between ±4 V. The circuit operates from ±5-V power supplies and is capable of driving a 50-Ω load with ±1-V swings. For a fully differential implementation, the power dissipation is 1.2 W. The amplifier can be integrated either as a stand-alone track-and-hold circuit or as the front end of an analog-to-digital conversion system for video and high-speed instrumentation applications.**

I. Introduction

THE ABILITY to track and then sample an analog waveform is often an essential function in the signal acquisition interfaces for digital video systems and high-speed instrumentation. In such interfaces, analog signals must often be sampled and stored accurately prior to quantization by either sequential or parallel means [1]. In sequential A/D converters the quantization is accomplished in two or more clock cycles using a pipelined architecture. Thus, the analog input must be sampled and then held for several clock cycles to within the accuracy specified for the overall system. In fully parallel converters, the use of a track-and-hold amplifier significantly reduces the aperture and sampling uncertainties of the conversion, while at the same time easing the performance demands on the subsequent quantizer.

Track-and-hold amplifiers are also used to provide deglitching following digital-to-analog conversion. In video D/A converters, significant glitch energy may appear at the analog output, usually as a result of mismatch in the turn-on and turn-off of switches at major bit transitions. In applications such as CRT display drivers and signal generation in automatic test systems, it is often essential that D/A converter output transitions be monotonic and as smooth as possible.

Manuscript received May 2, 1989; revised July 5, 1989. This work was supported in part by the U.S. Army Research Office under Contract DAAL03-87-K-0111.

M. Nayebi is with the Center for Integrated Systems, Stanford University, Stanford, CA 94305, and with the Vanguard Semiconductor Division, California Micro Devices, Inc., Milpitas, CA 95035.

B. A. Wooley is with the Center for Integrated Systems, Stanford University, Stanford, CA 94305.

IEEE Log Number 8930820.

High-performance track-and-hold circuits have traditionally been implemented in hybrid technologies that provide components such as open transmission lines for nearly 100-percent sampling efficiency, step recovery diodes for fast switching transitions, and low parasitic Schottky-diode bridges for use as sampling switches [2]. This paper presents a monolithic alternative to hybrid realizations. A VLSI BiCMOS technology has been used to implement a monolithic track-and-hold amplifier that settles to within an accuracy of 10 bits in 15 ns, thus surpassing the performance of previously reported monolithic circuits by nearly two orders of magnitude. The amplifier is based on a closed-loop topology comprising an NMOS sampling switch, a BiCMOS amplifier with a unity-gain bandwidth of 250 MHz, and a BiCMOS switch driver that provides 1-ns transitions between ±4 V. The circuit operates from ±5-V power supplies and is capable of driving a 50-Ω load with a voltage swing of ±1 V. The power dissipation is 630 mW for a single-ended amplifier and 1.2 W for a fully differential configuration.

The availability of both bipolar and MOS devices was essential for the design of the track-and-hold circuit reported in this paper. Bipolar transistors were needed to implement the broad-band amplifiers and high-speed digital drivers, while MOS transistors were essential for realizing the analog switches and the high input impedances needed to achieve a low droop rate.

II. Architecture

A. Basic Configurations

Three basic topologies for implementing a track-and-hold circuit are illustrated in Fig. 1 [3]–[6]. Fig. 1(a) is an open-loop configuration, while Figs. 1(b) and (c) illustrate closed-loop architectures. In the open-loop design of Fig. 1(a), the unity-gain input buffer $A1$ isolates the input from the hold capacitance C_h, while the buffer $A2$ isolates C_h from the output load impedance and provides the output current drive required for the overall amplifier. The principal advantage of this architecture is its speed. The open-loop configuration is intrinsically faster than closed-loop topologies because no feedback is employed and because the acquisition time at the hold capacitor is limited only by the input buffer and the sampling switch bandwidth. How-

Reprinted from *IEEE J. Solid-State Circuits*, vol. 24, no. 6, pp. 1507–1516, Dec. 1989.

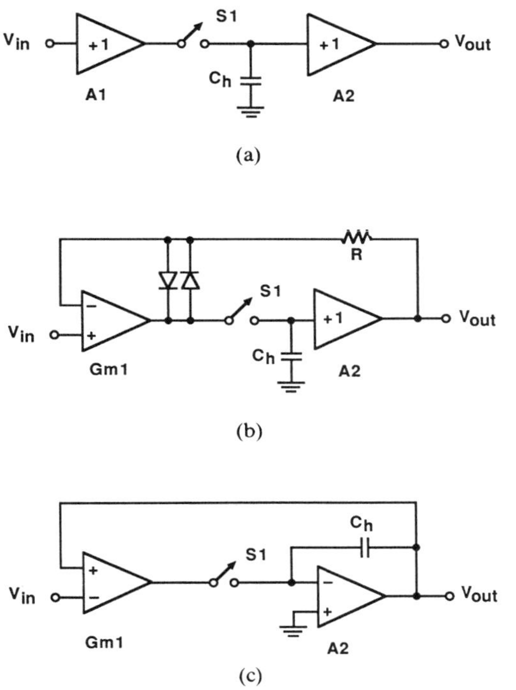

Fig. 1. Track-and-hold amplifier topologies. (a) Open loop. (b), (c) Closed loop.

ever, the accuracy of the open-loop configuration is substantially lower than can be achieved with closed-loop alternatives for at least two reasons. First, the offsets and gain errors of the two unity-gain buffers and the sampling switch all appear unattenuated when referred to the input. Second, both the input and output of the sampling switch operate at the level of the input signal; as a consequence, switch charge injection during turn-off and the switch on resistance will be signal dependent, thereby introducing both harmonic distortion and aperture delay uncertainty.

In the configuration of Fig. 1(b), the closed feedback loop and the gain of the input amplifier serve to significantly reduce the input-referred offset and the gain error. However, as in the open-loop system, the sampling switch input and output follow the input voltage, resulting in signal-dependent charge injection. Moreover, the overall bandwidth of the system is reduced.

The closed-loop architecture of Fig. 1(c) consists of a transconductance input stage *Gm*1, the sampling switch, and an integrating amplifier *A*2, with low output impedance, all enclosed within the feedback loop. In this configuration, the input-referred offset is substantially lower than that of open-loop systems because all offsets other than that of the first stage are reduced by the gain of the transconductance amplifier. Also, the large feedback loop gain results in significantly lower closed-loop gain error and nonlinearity than can be achieved in an open-loop system. Owing to the use of an integrating hold capacitor in the configuration of Fig. 1(c), the input and output of the sampling switch are always driven to virtual ground when the switch is closed. The hold capacitance C_h serves both to store the sampled input during the hold mode and

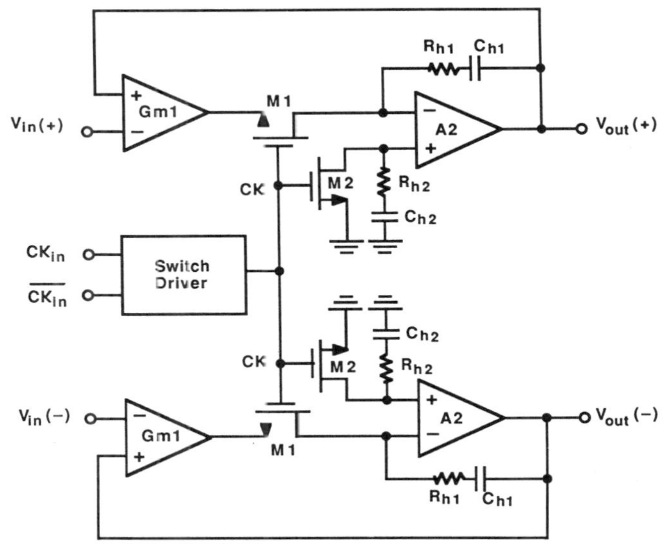

Fig. 2. Fully differential BiCMOS track-and-hold circuit.

to provide frequency compensation of the overall amplifier in the track mode. The speed achievable with this topology remains lower than that of an open-loop configuration.

Because of its superior accuracy, the topology of Fig. 1(c) was adopted for the track-and-hold amplifier described herein. BiCMOS circuits and technology were then used to mitigate the bandwidth limitations of this configuration.

B. Fully Differential Track-and-Hold

Fig. 2 shows a fully differential system comprising two identical closed-loop track-and-hold amplifiers together with a common switch driver. Within each feedback loop only two gain stages, the input transconductance amplifier and the input stage of the integrating amplifier, are used so as to preserve a high bandwidth with adequate phase margin. Each gain stage has a dc gain of approximately 60 dB to ensure a system linearity of at least 10 bits at full-scale voltage swings over variations in temperature and processing. The use of only two gain stages ensures that adequate pole splitting is achieved by the hold capacitor C_{h1}.

Transistors *M*2 and capacitors C_{h2} in Fig. 2 are used to balance the charge injection that occurs during the turn-off of the sampling switches *M*1. The residual error that results from the mismatch in the impedances seen at the inputs of *M*1 and *M*2 is then further reduced through the use of the fully differential overall configuration.

The value of the hold capacitance C_{h1} actually used in the track-and-hold circuit depicted in Fig. 2 is 1.3 pF. This is large enough to provide adequate pole splitting. The use of PMOS transistors at the input of the integrating amplifier *A*2 ensures an adequately low droop rate. The resistor R_{h1} used in series with the hold capacitor provides the compensation needed to ensure the stability of the overall feedback loop.

C. Accuracy and Speed

A number of factors significantly influence the accuracy with which signals can be sampled and held by the track-and-hold system illustrated in Fig. 2. Pedestal error at the output of the sampling switch is induced during the transition from the track mode to the hold mode by switch charge injection that results from both gate overlap capacitance and trapped channel charge. The pedestal error is defined as the ratio of the induced voltage error to the full-scale input voltage range. The finite open-loop gain of the track-and-hold circuit translates directly into a gain error in the overall sampling process. Further, high open-loop gain is relied upon to reduce the closed-loop nonlinearity. Finite common-mode rejection also contributes directly to a gain error. Roll-off in the open-loop gain with frequency limits the accuracy with which signals can be sampled as the signal frequency is increased. Parasitic coupling capacitances in the sampling switch contribute to a feedthrough error while the switch is off. Crosstalk of digital switching transients and power supply noise coupled into the back gate of the sampling switch are additional potentially troublesome sources of error. Finally, care must be taken to prevent electromagnetic coupling from the high current output buffers into the analog signal path.

A closed-loop architecture was chosen in order to meet the accuracy objectives. As a consequence, bandwidth became the highest priority in the design of the amplifier circuits themselves. The two strategies adopted to achieve the highest possible bandwidth were to avoid the use of MOS transistors, except as input devices, and current mirrors in the signal path. Also, care was taken to limit the number of high-frequency nondominant poles.

III. Signal-Path Circuit Design

A. Transconductance Amplifier (Gm1)

To achieve the required dc gain together with the highest possible unity-gain bandwidth, a single-stage BiCMOS folded-cascode configuration was used to implement the transconductance amplifiers *Gm*1 in Fig. 2. Fig. 3 shows a conventional CMOS folded-cascode amplifier wherein a current mirror is used to provide differential-to-single-ended conversion. The unity-gain bandwidth of this amplifier is limited by the nondominant poles associated with the cascode transistors *M*5 and *M*6 and the current mirror *M*7 and *M*8.

Fig. 4 shows a BiCMOS alternative to the CMOS amplifier of Fig. 3. In this circuit, bipolar transistors have been used to significantly broad-band the nondominant pole associated with the cascode devices. Furthermore, the current mirror has been eliminated and a single-ended output is obtained by using only one of the complementary outputs, despite the consequent reduction in gain by a factor of 2.

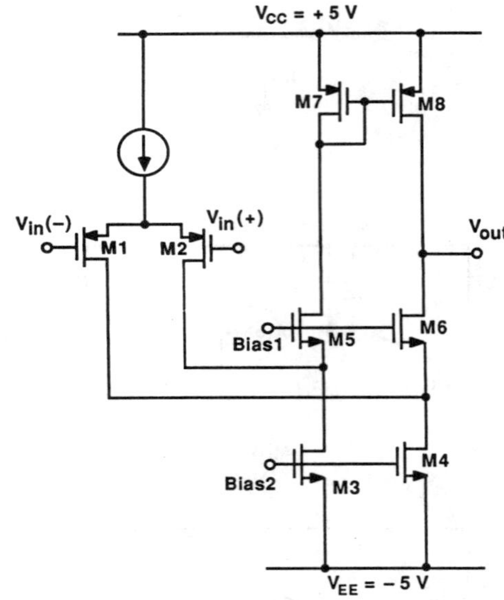

Fig. 3. CMOS folded-cascode amplifier.

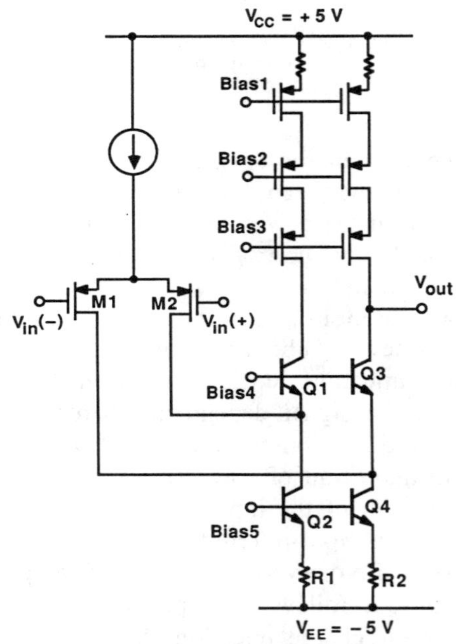

Fig. 4. BiCMOS folded-cascode amplifier without a current mirror.

Elimination of the current mirror in the folded-cascode circuit results in some degradation of the common-mode rejection. Therefore, a separate common-mode feedback circuit outside the critical signal path is used to cancel common-mode signals in a fashion similar to the current mirror but without introducing an additional nondominant pole in the signal response. This is illustrated in the more complete transconductance amplifier schematic given in Fig. 5. In this circuit, the common-mode drain current of input transistors *M*1 and *M*2 is replicated as the combined drain currents of *M*3 and *M*4. The network formed by transistors *M*5, *Q*5, and *Q*6 together with resistors *R*3–*R*5 then alters the bias currents in the collectors of *Q*2 and $Q4$ so as to compensate for change in the $M1$ and $M2$ drain currents.

Since n-p-n bipolar transistors are used as cascode devices in the folded-cascode configuration, the input transistors are necessarily PMOS devices. In the closed-loop unity-gain configuration, where the amplifier is driven from a low impedance source, this will not degrade the bandwidth. Nonetheless, to achieve the maximum overall bandwidth for a given input g_m, minimum channel length PMOS input transistors were used in the circuit of Fig. 5. In addition, the n-well for PMOS transistors, $M1$ and $M2$ in Fig. 4, was tied to their common source to eliminate the backgate dependence of their threshold voltages. The size of $M1$ and $M2$ is 480 μm/1.2 μm.

Frequency compensation of the transconductance amplifier is provided by the hold capacitance C_{h1} in series with resistor R_{h1} in the circuit of Fig. 2. The purpose of R_{h1} is to introduce a zero in the vicinity of the nondominant pole associated with the bipolar cascode transistor, $Q3$ in Fig. 4. The bandwidth of the transconductance amplifier is then limited by the nondominant pole associated with the input PMOS transistor. However, this pole does not significantly constrain the bandwidth of the overall track-and-hold amplifier. The overall bandwidth is instead limited by poles associated with the integrating amplifier $A2$.

Triple cascode current sources implemented with minimum channel length PMOS transistors were used to achieve an output resistance on the order of 1 MΩ. The size of these devices is 120 μm/1.2 μm. The bias currents in $M1$, $M2$, $Q1$, and $Q3$ are all 1.5 mA.

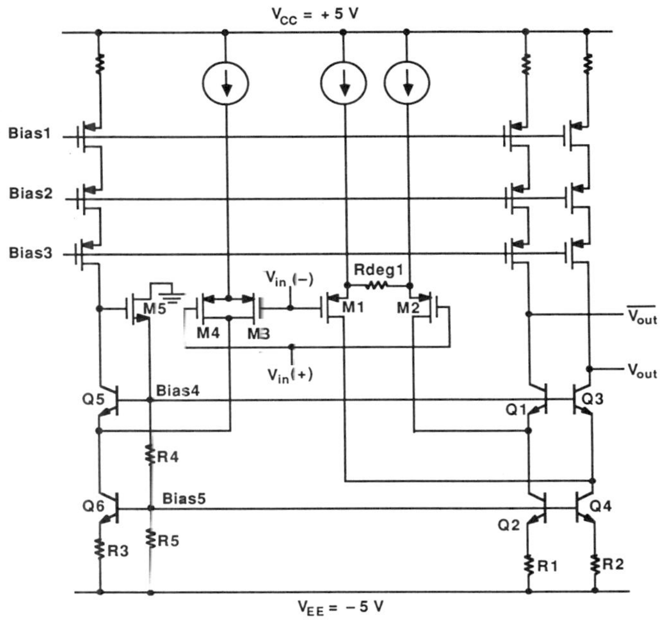

Fig. 5. Transconductance amplifier $Gm1$.

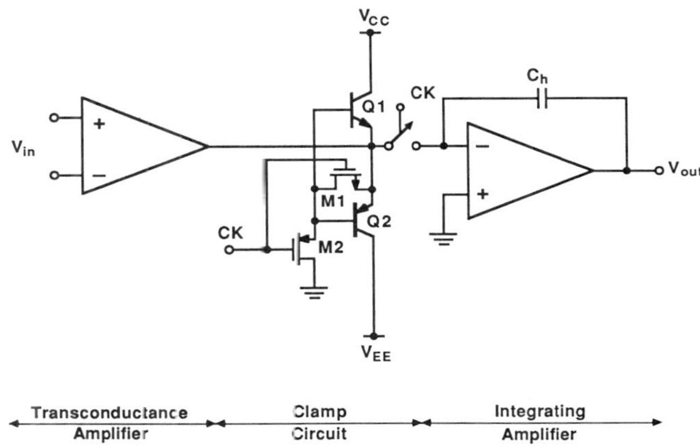

Fig. 6. Voltage clamp at transconductance amplifier output.

B. Clamp Circuit

While the input to the sampling switch is driven to virtual ground in the track mode, during hold large voltage swings could appear at this node. Such swings might result in both a substantial acquisition delay when switching to the track mode and significant switch feedthrough during hold. To avoid these difficulties, the circuit shown in Fig. 6 is used to clamp the voltage swing at the output of the transconductance amplifier. Transistor $Q2$ is implemented as a lateral p-n-p transistor using PMOS source/drain diffusions. This circuit is activated only when the sampling switch is open and the input to the switch departs from ground by more than 0.8 V.

C. Integrating Amplifier A2

A simplified schematic of the integrating amplifier, $A2$ in Fig. 2, is given in Fig. 7. This circuit is a broad-band operational amplifier with a dc gain of approximately 60 dB and the capability of driving a 50-Ω load with a maximum voltage swing of ±1 V [7]. It combines a single gain stage with a tapered intermediate unity-gain buffer and a class AB output stage. It also includes a feedforward path with relatively low dc gain that bypasses the intermediate stage at high frequencies, thereby improving the amplifier's phase margin.

The input stage of the integrating amplifier is identical to the BiCMOS folded-cascode circuit used to implement the transconductance amplifier $Gm1$ except for the device sizes and current levels. The device sizes and currents are half those in the transconductance amplifier. The output of the folded-cascode gain stage in Fig. 7 drives a minimum-geometry bipolar emitter follower. The total parasitic capacitance at the gain stage output, node 1 in Fig. 7, is only 0.12 pF. The compensation capacitor C_C, together with the total parasitic capacitance to ground at node 1, C_L, determines the bandwidth of the input stage.

A resistance, R_C in Fig. 7, is included in series with the compensation capacitor in order to compensate for the phase-margin degradation that results from the nondominant pole associated with the cascode transistor $Q3$. Fig. 8 shows a simple model of the gain stage wherein the poles associated with the input and cascode transistors have

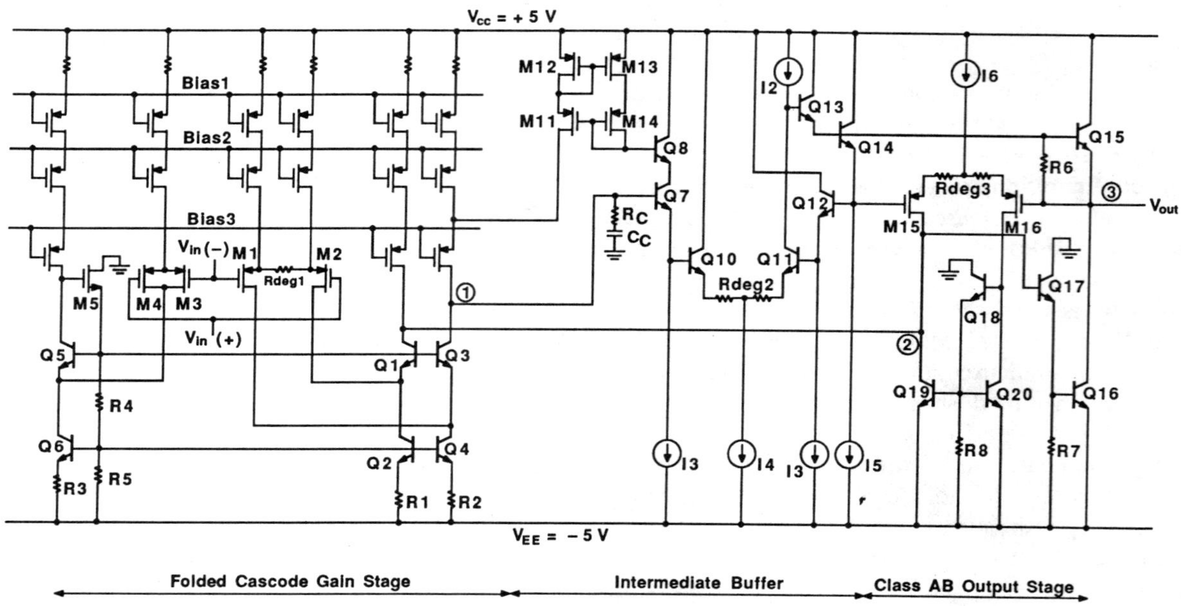

Fig. 7. Integrating amplifier $A2$.

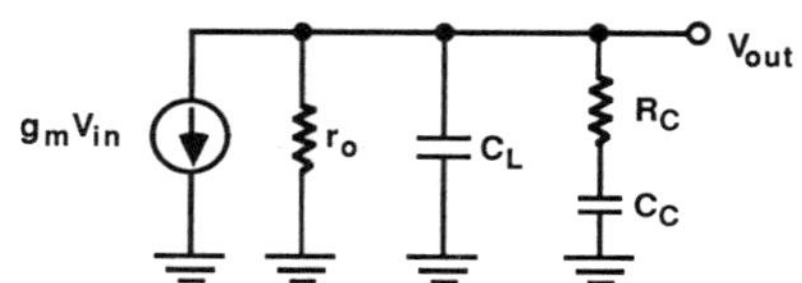

Fig. 8. Model of integrating amplifier compensation network.

been neglected. Resistor r_o is the output impedance seen at node 1 in Fig. 7, while C_L is the parasitic capacitance to ground at that node. The transfer function of this simple model is

$$\frac{V_{\text{out}}}{V_{\text{in}}} = \frac{G_m r_o (1 + s/z_{\text{Lead}})}{(1 + s/p_1)(1 + s/p_{\text{Lead}})} \tag{1}$$

where

$$p_1 = -\frac{1}{r_o (C_L + C_C)} \tag{2}$$

$$z_{\text{Lead}} = -\frac{1}{R_C C_C} \tag{3}$$

$$p_{\text{Lead}} = -\frac{1}{R_C \cdot \dfrac{C_L C_C}{(C_L + C_C)}} \tag{4}$$

The pole p_1 is the pole associated with the capacitive load on the amplifier $A2$. The resistor R_C introduces the pole–zero pair p_{Lead} and z_{Lead}. To obtain a significant improvement in phase margin, the zero z_{Lead} must lie well within the pole p_{Lead}. That is, the ratio

$$\frac{p_{\text{Lead}}}{z_{\text{Lead}}} = 1 + \frac{C_C}{C_L} \tag{5}$$

should be at least a factor of 4. However, C_C should be no larger than absolutely necessary so as to preserve the bandwidth of $A2$. For this design, C_C is approximately 0.4 pF. Ideally, z_{Lead} should be placed so as to cancel the cascode pole. To approximately achieve this, the value of R_C is 300 Ω.

D. Output Buffer

The output buffer of the integrating amplifier consists of a tapered intermediate stage and a class AB output stage. The intermediate stage is simply a single differential pair with unity-gain feedback and tapered input and output emitter-follower buffers. Individual stages in the output buffer, $Q7$, $Q10$, and $Q13$–$Q15$ in Fig. 7, are tapered so as to drive a 50-Ω output load. Bipolar transistors are used throughout the signal path to minimize the phase-margin degradation. The unity-gain differential pair, rather than simply a cascade of emitter followers, is used in the intermediate stage to compensate for the dc level shifts that accompany the followers. A base current cancellation circuit, $Q8$ and $M11$–$M14$, is used to compensate for the load of the $Q7$ base current on the input stage, thereby preventing a large systematic offset at the amplifier input.

In the class AB output stage, local feedback provided by $M15$, $M16$, and $Q16$–$Q20$ is used to maintain a constant transconductance for output transistor $Q15$ under varying load conditions. The output voltage follows the voltage at the emitter of $Q14$ under all load conditions. The output stage also provides a high-input impedance node, node 2 in Fig. 7, whereby a signal can be fed to the output through emitter follower $Q17$. The inverting output of the gain stage is fed forward directly to this node to improve the amplifier phase margin. Fig. 9 provides a simple illustration of this approach. At low frequencies the gain of the feedforward path is much less than that of the main signal path. However, the gain of the main path will fall faster

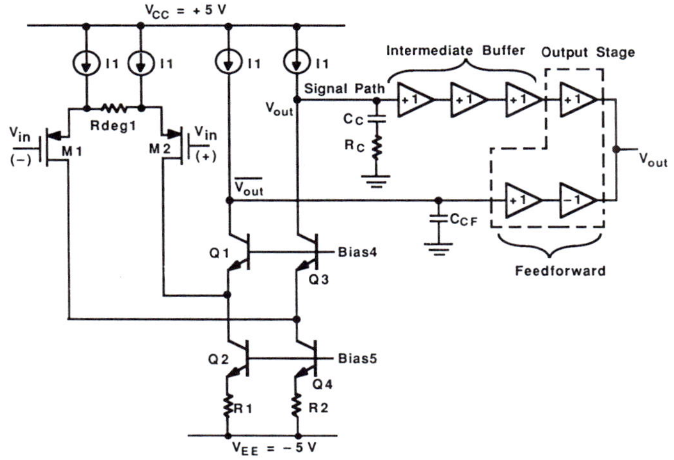

Fig. 9. Simplified model of feedforward in amplifier $A2$.

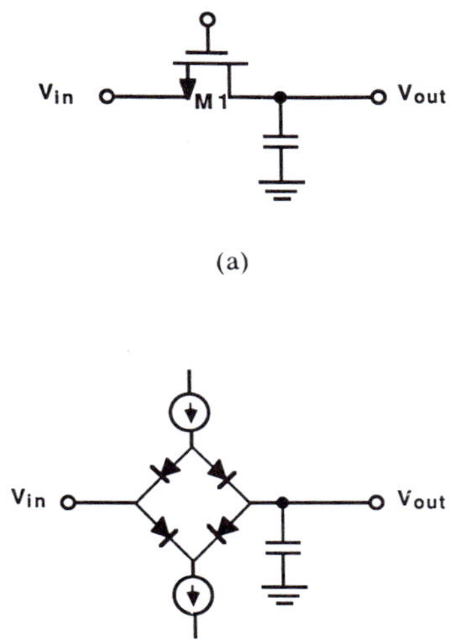

(a)

(b)

Fig. 10. Sampling switches: (a) NMOS transistor and (b) diode bridge.

with increasing frequency than that of the feedforward path owing to the greater number of stages. Consequently, the feedforward path will sustain the gain at higher frequencies and the equivalent number of stages in the output buffer is reduced from four to two, with a corresponding improvement in phase margin.

IV. Sampling Switch

The sampling switch can be implemented, as depicted in Fig. 10, by means of either an MOS transistor or a junction diode bridge. These two alternatives were compared on the basis of three parameters: the switch bandwidth B, the pedestal error P, and the switch feedthrough F [8], [9]. Based on the simple switch model shown in Fig. 11, the bandwidth is given by

$$B = \frac{1}{2\pi R_{on} \cdot C_h} \tag{6}$$

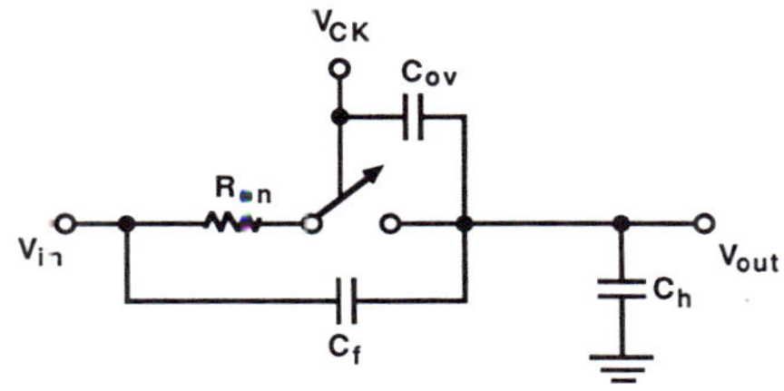

Fig. 11 Sampling switch model.

where R_{on} is the on resistance of the sampling switch and C_h is the hold capacitance. The pedestal error resulting from charge injection during turn-off of the switch is defined as

$$P = \frac{\delta V_{out}}{V_{FS}} = \left(\frac{C_{ov}}{C_h}\right) \cdot \left(\frac{V_{ck}}{V_{FS}}\right) \tag{7}$$

where δV_{out} is the absolute value of the error injected at the switch output, V_{FS} is the full-scale signal range, V_{ck} is the clock voltage swing during switch turn-off, and C_{ov} is the coupling capacitance between the control terminal (gate) of the switch and the switch output. The signal feedthrough when the switch is off is given by

$$F = \frac{C_f}{C_h} \tag{8}$$

where C_f is the coupling capacitance between the input and output of the sampling switch when the switch is off. Based on these three parameters it was found that a 60-μm/1.2-μm NMOS switch provides a better trade-off between on-resistance and overlap capacitance than a p-n junction diode bridge. R_{on} for an NMOS switch this size is 50 Ω at a gate-to-source voltage of 4 V, and the overlap capacitance is only 11 fF. The channel transit time of 30 ps for such a switch (in comparison with a gate turn-off transition time of 1 ns) minimizes the trapping of channel charge when the switch is turned off. Compared with the NMOS switch, the diode bridge introduces significant additional sources of error such as V_{BE} mismatch, pull-up/pull-down current mismatch, and timing mismatch in the switching of the pull-up and pull-down currents.

The sampling switch driver translates a complementary ECL clock signal into a single ±4-V signal capable of driving the switches with 1-ns transition times. A simplified schematic of the switch driver is shown in Fig. 12. This circuit consists of an input buffer, a voltage translator, and two tapered BiCMOS drivers. The voltage translator translates the low voltage swing clock input into a ±4-V single-ended signal with a low current drive capability. The tapered BiCMOS stages then provide an output signal capable of driving the sampling switches. Fig. 13 shows the output of the switch driver as measured with a high-impedance FET probe with 20X attenuation. The pulse width is 5 ns with 1-ns transition times at a clock rate of 58 MHz.

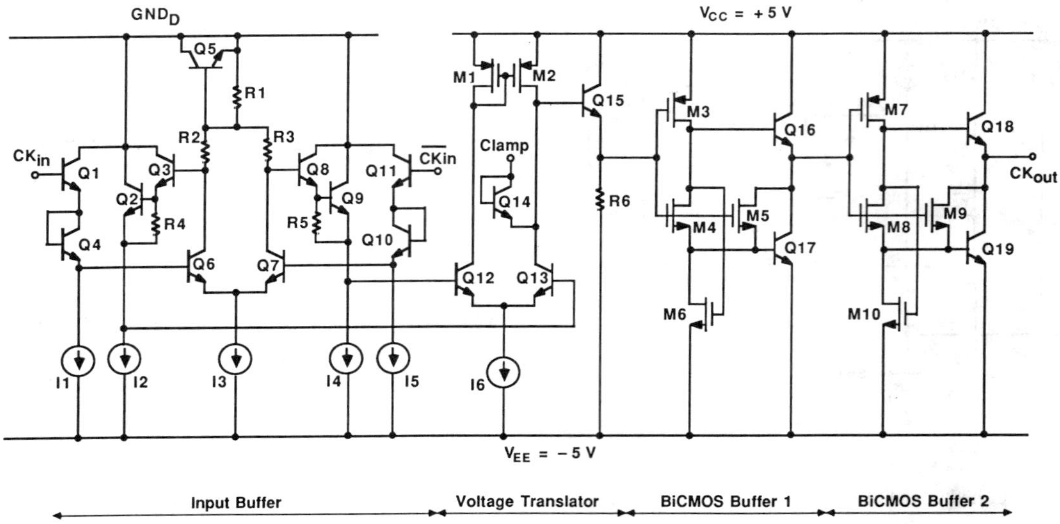

Fig. 12. Sampling switch driver.

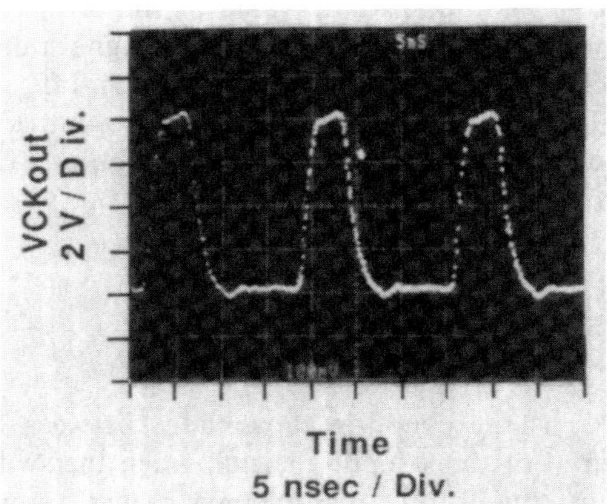

Fig. 13. Switch driver output, attenuated by 20×.

TABLE I

Bipolar Parameters

Minimum Emitter Size: 1.1 μm × 3 μm

Parameter	Value
ß	120 V
BV_{CBO}	14 V
BV_{CEO}	10 V
V_A	20 V
f_T	7 GHz
C_{js} (area)	0.12 fF/μ^2
C_{js} (peri)	1.03 fF/μ^2

CMOS Parameters

Minimum Channel Length = 1.2 μm

Parameter	n-channel	p-channel
V_{TO}	+0.8 V	−0.95 V
μ_O	630 cm^2/V-sec	160 cm^2/V-sec
t_{OX}	200 Å	200 Å
L_D	0.1 μm	0.15 μm

V. Layout

Careful layout is crucial to achieving the accuracy and speed objectives set forth for the track-and-hold amplifier. In this circuit, separate analog and digital grounds are used along with separate supply voltage inputs for the analog and digital portions of the circuit. In addition, separate pads are used to supply current to the output stage of the integrating amplifiers so as to avoid resistive coupling into the gain stage. On-chip bypass capacitors are included with every power supply pad, and all analog ground pads are connected to a 750-μm-wide ground bus implemented with first-level metal. Also, a ground pad is included adjacent to each of the critical signal pads, such as those for the analog input and output, so as to reduce the ground loop inductance. Ground biased n-well diffusions have been used under all critical interconnections and passive components to shield them against noise coupling from the substrate. Examples of elements shielded in this fashion are the input and output signal pads, the clock input pads, and critical capacitors. A ground biased n-well diffusion also surrounds each of the transconductance amplifiers, the integrating amplifiers, the sampling switches, and the switch driver.

VI. Experimental Results

The track-and-hold amplifier was integrated in a 1.2-μm BiCMOS technology with two layers of polysilicon and two layers of metal [10]. Typical device characteristics for this technology are summarized in Table I. The minimum emitter size is 1.1×3 μm^2 and the minimum drawn channel length is 1.2 μm. A die photo is shown in Fig. 14. The chip size is 1.5×2.5 mm^2. The amplifier operates from ±5-V supplies, and the power dissipation of the fully differential configuration is 1.2 W.

The experimentally measured dc and ac characteristics of the track-and-hold circuit are summarized in Table II. The typical dc gain measured for each of the two folded-

Fig. 14. Die photo of track-and-hold amplifier.

TABLE II

DC Measurements

Parameter	Value
DC gain / Stage	58 dB
Offset Voltage	10 mV
CMRR (Vos = 10 mV)	75 dB
(Vos = 0 mV)	110 dB
PSRR	65 dB
Pedestal	0.5 mV
Droop Rate	20 μV / msec
Output Current	± 30 mA
Voltage Swing	± 1 V

AC Measurements

Parameter	Value
Sample Mode Settling	15 nsec
Hold Mode Settling	12 nsec
Overall Bandwidth ($R_L = 50\Omega$)	250 MHz
Min. Aperture Window	5 nsec
Aperture Uncertainty	20 psec
Input Noise	18 nV / $\sqrt{Hz}$

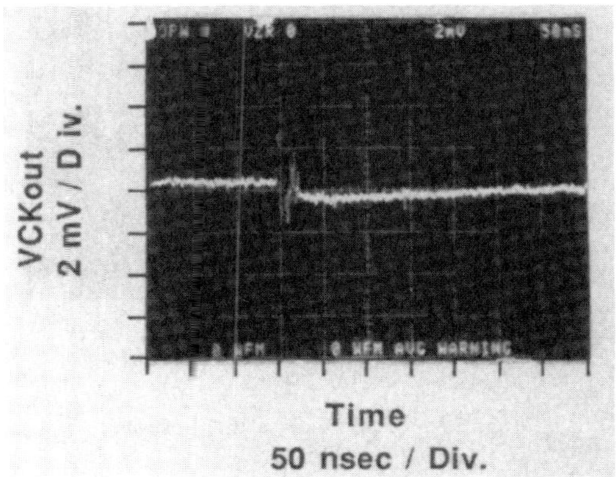

Fig. 15. Pedestal error of fully differential amplifier.

cascode gain stages is 58 dB. The offset of the overall amplifier is typically 10 mV, and the CMRR with a 10-mV offset is 75 dB. Fig. 15 shows the pedestal error in the fully differential configuration that results from charge injection in the sampling switch. In this figure the differential output of the track-and-hold amplifier is displayed when the sampling switch opens; the error induced is approximately 0.5 mV.

The response of the experimental circuit in the track mode to a 1-V input step is shown in Fig. 16. The slew rate is 0.5 V/ns, the overshoot is 20 percent, and the output signal settles to within 1 mV of its final value in 15 ns. Fig. 17 shows the hold mode settling response of the amplifier. The transient resulting from the turn-off of the sampling switch settles to within 1 mV of its asymptotic level in less than 12 ns.

As indicated in Table II, the 3σ value of the aperture uncertainty, as measured indirectly, is 20 ps. This uncertainty is primarily a consequence of noise from the digital power supplies and switch driver components perturbing the switch driver transition edges. In the linear mode of operation, the switch driver can be viewed as an amplifier consisting of four gain stages. The distinction from a linear amplifier is that the output of each gain stage reaches the slew limit, and therefore the overall variation in delay is the root-mean-square sum of the delay variations in all four stages:

$$\sigma\tau_{DEL} = \left(\sigma\tau_{IB}^2 + \sigma\tau_{TR}^2 + \sigma\tau_{OB1}^2 + \sigma\tau_{OB2}^2\right)^{1/2} \tag{9}$$

where, referring to the sampling switch driver schematic in Fig. 12, $\sigma\tau_{IB}$ is the input buffer delay uncertainty, $\sigma\tau_{TR}$ is the voltage translator delay uncertainty, and $\sigma\tau_{OB1}$ and $\sigma\tau_{OB2}$ are the delay uncertainties of first and second BiCMOS buffer stages, respectively. To determine the

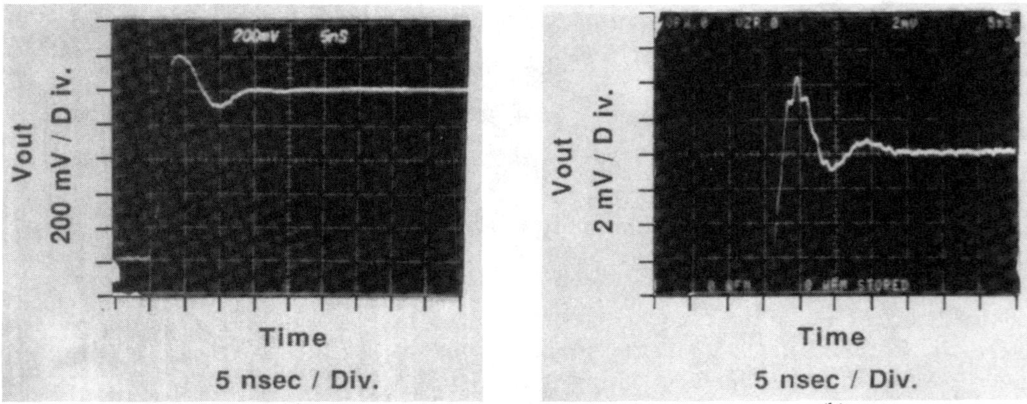

(a) (b)

Fig. 16. Response in track mode to a 1-V input step.

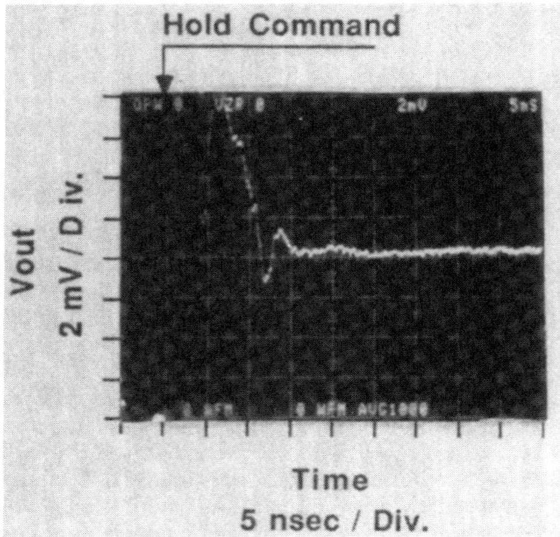

Fig. 17. Hold mode settling response.

overall delay uncertainty experimentally, the small-signal gain and bandwidth were first measured by operating the switch driver in the linear region where there is no clipping of the output. In this fashion a dc gain of 52 dB and −3-dB frequency of 200 MHz were measured for the switch driver circuit. Next, the output noise power density was measured to be a maximum of −70 dBm/Hz, which is equivalent to an input noise power density of −122 dBm/Hz or an input-referred voltage noise density of 173 nV/$\sqrt{\text{Hz}}$. Using a 300-MHz noise bandwidth, this translates into an input-referred voltage noise of $\sigma V_{\text{input}} = 3$ mV for the entire switch driver circuit. It was then verified that for this design the slew rate of each of the four stages in the switch driver was 5 V/ns. As a consequence, every millivolt of the measured input-referred voltage noise corresponds to a 0.2-ps uncertainty in all gain stages except the first stage. Assumng a minimum 0.5-V/ns transition time for the clock input, every millivolt of input-referred voltage noise corresponds to a 2-ps delay uncertainty in the first stage. Based on this approach, the estimated 3σ value of the delay-time uncertainty amounts to less than 20 ps.

To measure the input-referred voltage noise of the overall track-and-hold amplifier during the track mode, a low-noise RF amplifier with 22-dB gain and −3-dB frequency of 1 GHz was used following the unity-gain track-and-hold amplifier. The overall output noise power density was measured to be −120 dBm/Hz. The output noise power density due to the RF amplifier itself was verified to be well below the overall output power; it was measured as −134 dBm/Hz. Therefore, the input noise power density of the track-and-hold is approximately −142 dBm/Hz, which is equivalent to an input-referred voltage noise of 18 nV/$\sqrt{\text{Hz}}$.

VII. Summary

A video BiCMOS track-and-hold amplifier has been designed and integrated in a 1.2-μm digital VLSI BiCMOS technology. Among several architectures studied, a closed-loop configuration was found to provide the best compromise between speed and accuracy. Experimental measurements confirm that the amplifier settles to an accuracy of 10 bits in 15 ns. To achieve this speed in a closed-loop topology, considerable effort was devoted to broad-banding the circuits themselves. Only two gain stages were used within the feedback loop, and each of these was implemented as a BiCMOS folded-cascode transconductance amplifier. In order to drive 50-Ω load impedances with tolerable phase-margin degradation, an output buffer comprising a closed-loop intermediate unity-gain stage and a class AB output stage was used. A fully differential configuration was used for the overall circuit in order to balance the charge injection from the NMOS sampling switch during turn-off. A common switch driver was used to drive both sampling switches with a ±4-V swing and 1-ns transition times.

The experimental results summarized above demonstrate that through the use of a high-performance, low-voltage BiCMOS technology, it is possible to monolithically realize analog circuit functions at levels of performance heretofore obtainable only in hybrid implementations. This performance can thus be exploited not only in stand-alone applications, but also as part of combined digital and analog VLSI systems. Insofar as the track-and-hold function itself is concerned, by using more aggressive design rules settling times below 10 ns appear readily achievable.

Acknowledgment

The authors wish to thank Fairchild Semiconductor Corporation and National Semiconductor Corporation for their support of this work and for the fabrication of the track-and-hold amplifier. Special thanks are due to A. Ley for ensuring continuation of the project under varying business conditions, N. Bhandari for his guidance during project definition, T. Bowman for his support during the design rule checking and layout verification and for orchestrating all the test chips, B. Bastani and C. Lage for fabrication and technology support, L. Bouknight for his device modeling support, and T. Lau and C. Yee for their technical support during testing and evaluation.

References

[1] D. H. Sheingold, Ed., *Analog-Digital Conversion Handbook*. Englewood Cliffs, NJ: Prentice-Hall, 1986.

[2] K. Rush and D. J. Oldfield, "A data acquisition system for a 1-GHz digitizing oscilloscope," *Hewlett-Packard J.*, pp. 4–11, Apr. 1986.

[3] R. J. van de Plassche and H. J. Schouwenaar, "A monolithic high-speed sample-hold amplifier for digital audio," *IEEE J. Solid-State Circuits*, vol. SC-18, pp. 716–722, Dec. 1983.

[4] G. Erdi and P. R. Henneuse, "A precision FET-less sample-and-hold with high charge-to-droop current ratio," *IEEE J. Solid-State Circuits*, vol. SC-13, pp. 864–873, Dec. 1978.

[5] J. Merkelo and R. D. Hall, "Broad-band thin-film signal sampler," *IEEE J. Solid-State Circuits*, vol. SC-7, pp. 50–54, Feb. 1972.

[6] K. R. Stafford, P. R. Gray, and R. A. Blanchard, "A complete monolithic sample/hold amplifier," *IEEE J. Solid-State Circuits*, vol. SC-9, pp. 381–387, Dec. 1974.

[7] M. Nayebi and B. A. Wooley, "A 0.5 GHz BiCMOS operational amplifier," in *1988 Symp. VLSI Circuits Dig. Tech. Papers*, Aug. 1988, pp. 5–6.

[8] W. B. Wilson, H. Z. Massoud, E. J. Swanson, R. T. George, Jr., and R. B. Fair, "Measurement and modeling of charge feedthrough in n-channel MOS analog switches," *IEEE J. Solid-State Circuits*, vol. SC-20, pp. 1206–1213, Dec. 1985.

[9] J. J. Corcoran *et al.*, "High speed sample and hold circuits for waveform acquisition," presented at the IEEE Solid-State Circuits and Technology Committee Workshop, May 1987.

[10] B. Bastani *et al.*,"Advanced one micron BiCMOS technology for high speed 256K SRAMs," in *1987 Symp. VLSI Technology Dig. Tech. Papers*, May 1987, pp. 41–42.

Paper 4.9

An 8-bit 200-MHz BiCMOS Comparator

PETER J. LIM, STUDENT MEMBER, IEEE, AND BRUCE A. WOOLEY, FELLOW, IEEE

Abstract —**This paper introduces the design of a fully differential BiCMOS comparator suitable for application in data conversion, instrumentation, and communication systems operating at video frequencies and above. By exploiting the advantages presented by the integration of both bipolar and CMOS devices within the same technology, the comparator dissipates less power than conventional bipolar designs without sacrificing operating speed. The comparator includes an input stage that combines MOS sampling with a bipolar regenerative amplifier. This stage dissipates no static power and, because the amplification is provided by a bipolar differential pair, no offset cancellation is needed to achieve 8-bit precision. Furthermore, the need for preamplification, and the attendant power–delay penalty associated with preamplifier overdrive recovery, is avoided.**

An experimental version of the comparator, consisting of the BiCMOS regenerative input stage followed by a current-switched latch, has been integrated in a 0.8-μm BiCMOS technology with an area of 140×75 μm. This circuit performs comparisons to a precision of 8 bits at rates up to 200 MHz. The entire circuit dissipates only 1.6 mW at the maximum clock rate while operating from a single 5-V supply. The full-scale input range is 1.5 V. Recovery from an input overdrive of 1 V to a resolution of 8 bits has been experimentally verified for comparison rates up to 150 MHz.

I. Introduction

SILICON BIPOLAR technology has traditionally dominated the field of monolithic high-speed data conversion circuits. However, the emergence of BiCMOS as a viable mainstream VLSI technology offers new opportunities for improving the performance of such circuits by combining the complementary advantages of each technology. Relative to MOS devices, bipolar transistors exhibit superior threshold matching, higher transconductance, and lower noise levels. These attributes make bipolar circuits especially well-suited to providing low-noise amplification over large bandwidths. Alternatively, CMOS technology provides the substantial advantages of simple zero-offset, low-leakage analog switches, high-impedance charge storage nodes, and complementary devices. These CMOS characteristics allow for the extensive use of analog sampling and pipelining. Through the selective use of these attributes, BiCMOS technology offers the potential of significant performance improvements in a wide variety of circuit applications, including analog-to-digital (A/D) conversion.

Manuscript received July 25, 1989; revised October 6, 1989. This work was supported by the U.S. Army Research Office under Contract DAAL03–87–K–0111.

The authors are with the Center for Integrated Systems, Stanford University, Stanford, CA 94305.

IEEE Log Number 8932700.

The function of comparison is a crucial, and often limiting, component in the design of high-speed data conversion systems. This paper describes the design of a low-power, high-speed BiCMOS comparator appropriate for use in analog-to-digital converters and a variety of instrumentation and communication systems operating at video frequencies and above. In particular, the circuit is suitable for use in the design of fully parallel A/D converters with a resolution of 8 bits and conversion rates above 100 MHz. The comparator includes a novel input stage that combines MOS input sampling with a bipolar regenerative amplifier that dissipates no static power. There is no need for preamplification or offset cancellation, and the delay penalty associated with preamplifier overdrive recovery is thereby avoided. The comparator is fully differential and has been integrated in a 0.8-μm digital BiCMOS technology [1]. This experimental prototype performs comparisons accurate to 8 bits at sampling rates as high as 200 MHz, while dissipating only 1.6 mW from a single 5-V supply.

In Section II, several aspects of CMOS and bipolar comparator designs are examined. The BiCMOS comparator input stage is then introduced in Section III, and the design of the entire comparator is described in Section IV. Experimental results are presented in Section V.

II. High-Speed Comparison

Comparison is fundamentally an amplification function whereby digital outputs are generated in response to small analog input differences. This amplification need not be linear nor continuous in time. A study of various generic approaches to comparator design [1] led to the conclusion that, in order to achieve the minimum power–delay product, the amplification required in a comparator is best obtained by means of regeneration. In CMOS comparators analog pipelining can also be used to achieve substantial increases in throughput for applications where the pipeline latency can be tolerated. The use of pipelined cascades of regenerative or linear CMOS amplifiers to achieve moderately high-speed comparison rates has already been demonstrated [2]–[6]. Nevertheless, in comparison with bipolar alternatives, the speed and resolution of CMOS comparators remain limited by the inherent MOS characteristics of low transconductance and relatively large device mismatches. For applications requiring a precision of

Reprinted from *IEEE J. Solid-State Circuits,* vol. 25, no. 1, pp. 192–199, Feb. 1990.

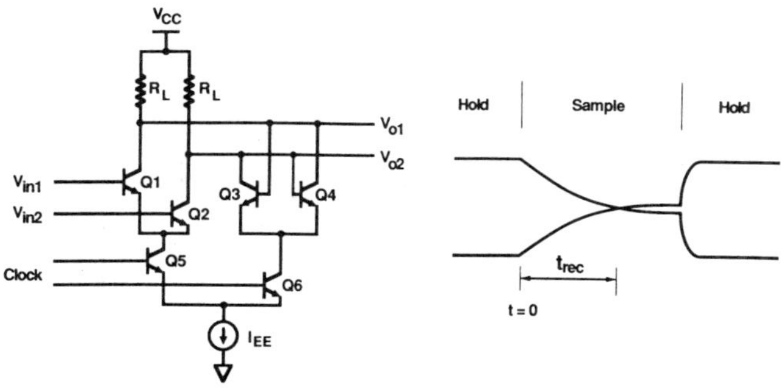

Fig. 1. Typical current-switching bipolar comparator.

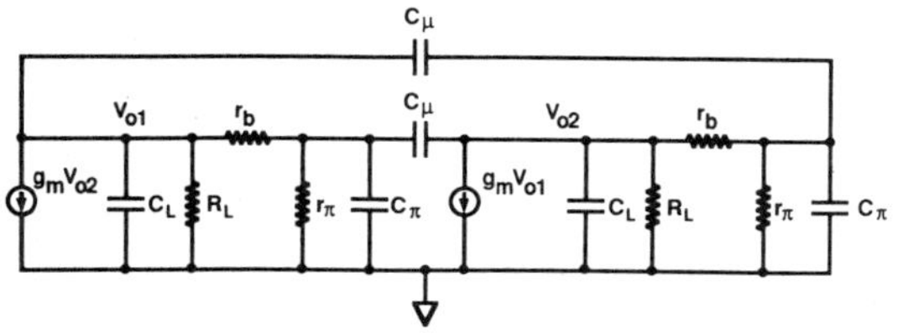

(a)

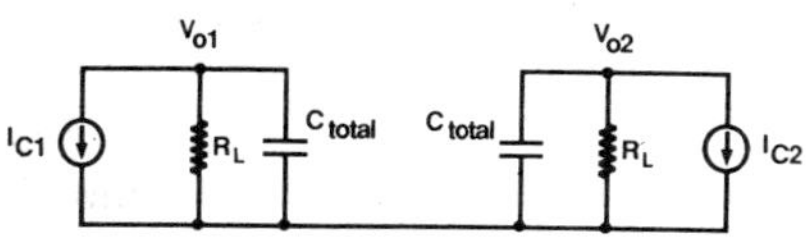

(b)

Fig. 2. Linearized model of bipolar comparator during (a) regeneration, and (b) sampling.

8 bits or more, the mismatch in threshold voltages and device geometries between adjacent MOS transistors makes offset cancellation essential. Furthermore, the relatively low transconductance of MOS devices results in lower speed and higher power dissipation than are achievable in a bipolar technology. Consequently, the highest speed comparators implemented to date have been based on bipolar transistor technology.

Shown in Fig. 1 are a circuit schematic and output waveforms representative of a typical current-switching bipolar comparator. The basic configuration combines a broad-band preamplifier with a regenerative latch, each being implemented with a differential bipolar transistor pair. This type of comparator has found widespread application in high-speed, fully parallel bipolar A/D converters [7]–[10]. In most such converters sampling occurs within the comparator latches, rather than prior to preamplification. In the circuit of Fig. 1, the steering of a single-source current I_{EE} between the two differential pairs $Q1, Q2$ and $Q3, Q4$ defines the sample and latch modes. In the sample mode, the output voltage in response to small input differences is $A_L V_{in}$, where A_L is the gain of the amplifier and V_{in} is the differential input $V_{in1} - V_{in2}$. In the hold mode, this voltage is regeneratively amplified and latched.

To determine the trade-off between power and speed in the comparator of Fig. 1, this circuit can be analyzed on the basis of simplified linearized models [10]. Such a model of the comparator during regeneration is shown in Fig. 2(a). The s-domain characteristic equation for this circuit

$$s^2 r_b(C_L C_\mu + C_\pi C_\mu + C_L C_\pi) + s\left[C_\pi\left(1 + \frac{r_b}{R_L} - g_m r_b\right) + C_\mu\left(4 + \frac{r_b}{R_L} - g_m r_b\right) + C_L\right] - \left(g_m - \frac{1}{R_L} - \frac{1}{r_\pi}\right) = 0. \quad (1)$$

For low power applications, $r_b \ll r_\pi$ is assumed. An approximate solution for the regeneration response is then

$$V_{o2} - V_{o1} = A_L V_{in} \exp(t/\tau_{reg}) \quad (2)$$

with the regeneration time constant

$$\tau_{reg} = \frac{C_\pi(r_b + R_L) + C_\mu[4R_L + r_b(g_m R_L + 1)] + C_L R_L}{g_m R_L - 1}. \quad (3)$$

When $R_L \gg r_b$ the regeneration time constant reduces to

$$\tau_{reg} = \frac{C_{total}}{g_m}\left(\frac{g_m R_L}{g_m R_L - 1}\right) \quad (4)$$

where $C_{total} = C_\pi + C_L + 4C_\mu$. For typical current values the regeneration time constant is very short and is not the dominant factor governing the speed versus power trade-off in a low-power bipolar comparator design. Instead, as demonstrated below, the power dissipation needed to achieve a given conversion rate is determined largely by the current needed in the preamplifying stage to recover from an overdrive condition. This recovery time is defined in Fig. 1 as t_{rec}. The worst-case recovery time is the time required in the sample mode for an input voltage difference of 1/2 LSB to change the output of the latching comparator from one logic state to the midpoint of the logic states.

Fig. 2(b) shows a simplified model of the bipolar comparator in the sample mode. For purposes of this analysis, the output load capacitance is represented approximately by C_{total} as defined above. If the ohmic emitter resistance r_e is much smaller than $1/g_m$, then the collector currents in the sample mode are

$$I_{C1} = \frac{\alpha_F I_{EE}}{1 + \exp(-qV_{in}/kT)} \quad (5)$$

$$I_{C2} = \frac{\alpha_F I_{EE}}{1 + \exp(qV_{in}/kT)}. \quad (6)$$

For $t \geqslant 0$ the output voltages are then given by

$$V_{o1} = V_{CC} - \alpha_F R_L I_{EE} - (I_{C1} R_L - \alpha_F R_L I_{EE}) \cdot \left[1 - \exp\left(-\frac{t}{R_L C_{total}}\right)\right] \quad (7)$$

$$V_{o2} = V_{CC} - I_{C2} R_L \left[1 - \exp\left(-\frac{t}{R_L C_{total}}\right)\right]. \quad (8)$$

The output crossing $V_{o1} = V_{o2}$ occurs when

$$-\exp\left(-\frac{t_{rec}}{R_L C_{total}}\right) = \tanh(-qV_{in}/2kT)\left[1 - \exp\left(-\frac{t_{rec}}{R_L C_{total}}\right)\right] \quad (9)$$

i.e.,

$$t_{rec} = R_L C_{total} \times \ln\left(1 + \frac{1}{\tanh(qV_{in}/2kT)}\right). \quad (10)$$

At room temperature and for $V_{in} = \frac{1}{2}\text{LSB} = 3$ mV,

$$t_{rec} = 2.9 R_L C_{total}. \quad (11)$$

This recovery time is now compared to the time required for regeneration. The output swing can be any reasonable value, and for this analysis is chosen as $I_{EE} R_L = 0.8$ V. For this choice

$$t_{rec} = \frac{2.32 C_{total}}{I_{EE}}. \quad (12)$$

Since the current through each device in the linearized model of Fig. 2(a) is $I_{EE}/2$, it follows from (4) that the regeneration time constant is

$$\tau_{reg} = \frac{2kTC_{total}}{qI_{EE}} \quad (13)$$

which at room temperature becomes

$$\tau_{reg} = \frac{0.057 C_{total}}{I_{EE}}. \quad (14)$$

Equations (12) and (14) illustrate that for the same tail current, and at room temperature, the recovery time is more than 40 times the estimated regeneration time constant. Thus, the current required for overdrive recovery is much larger than that needed for sufficiently fast regenerative amplification. This result has been the motivation for a study of performing comparisons using regenerative amplification without input preamplification and has led to the design of the BiCMOS regenerative amplifier that combines MOS input sampling with high-speed bipolar regenerative amplification.

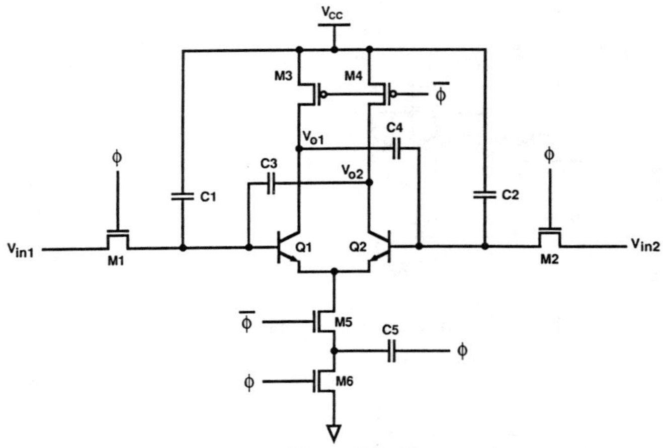

Fig. 3. BiCMOS regenerative amplifier.

III. BiCMOS Regenerative Amplifier

In order to fully exploit the advantages offered by a BiCMOS technology, a novel regenerative amplifier that forms the input stage for a BiCMOS comparator is introduced. A simplified schematic of this amplifier is shown in Fig. 3. The basic amplifier consists of a differential bipolar transistor pair $Q1$ and $Q2$ with positive feedback provided by cross-coupling capacitors $C3$ and $C4$. The current required for the operation of the bipolar transistors is not provided by a constant tail current but rather by the charging and discharging of capacitor $C5$. During each clock cycle a fixed amount of charge is steered through the differential transistor pair. All MOS transistors shown in this schematic operate as switches. To describe the operation of the BiCMOS amplifier, the circuit configurations established by the switches for each of the two modes of operation are shown in Fig. 4.

During the sample mode, the input voltages are sampled onto capacitors $C1$–$C4$. The collectors of the bipolar transistors $Q1$ and $Q2$ are pulled to the positive supply, and capacitance $C5$ is charged to the full supply voltage with the polarity indicated in Fig. 2(a). Because the common emitter of bipolar transistors $Q1$ and $Q2$ is open, the differential pair is inactive during this mode. In the transition to the regeneration mode, the input voltages are disconnected from the sampling capacitors and the collectors of $Q1$ and $Q2$ are disconnected from the positive supply. Capacitance $C5$ is then connected to the common emitter node, and the terminal of $C5$ previously precharged to the positive supply is switched to ground. A predetermined amount of charge stored on capacitor $C5$ is thus transferred through the bipolar differential pair. The initial voltage difference that has been stored at the bases of the bipolar transistors in the sample mode results in a differential discharging of the series combinations of capacitors $C1,C3$ and $C2,C4$. Owing to the cross coupling provided by capacitors $C3$ and $C4$, the initial voltage difference between the bases of $Q1$ and $Q2$ is regeneratively amplified at the collectors of these transistors. The collector nodes serve as the outputs of the amplifier.

Fig. 5 shows typical simulated voltage waveforms at the collectors of $Q1$ and $Q2$ in the circuit of Fig. 4. In this simulation, the differential input voltage changes from $+1$ V to -1 mV between consecutive 100-MHz clock cycles. During regeneration in the first clock cycle, the 1 V previously sampled onto the capacitors at the bases of the bipolar transistors causes all the charge transferred onto $C5$ to flow through $Q1$. Thus, the $Q1$ collector voltage falls, and the voltage at the collector of $Q2$ remains high. The collector remaining high actually rises above the 5-V supply due to the channel-charge injection and capacitive feedthrough from the fast turn-off of the PMOS pull-up device $M4$.

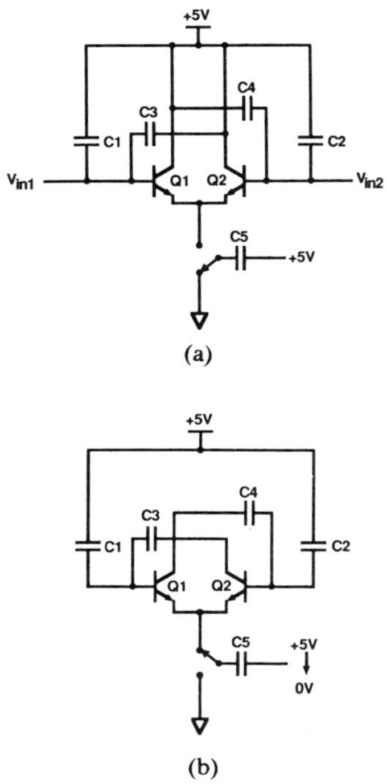

Fig. 4. Modes of operation. (a) Sampling mode. (b) Regeneration mode.

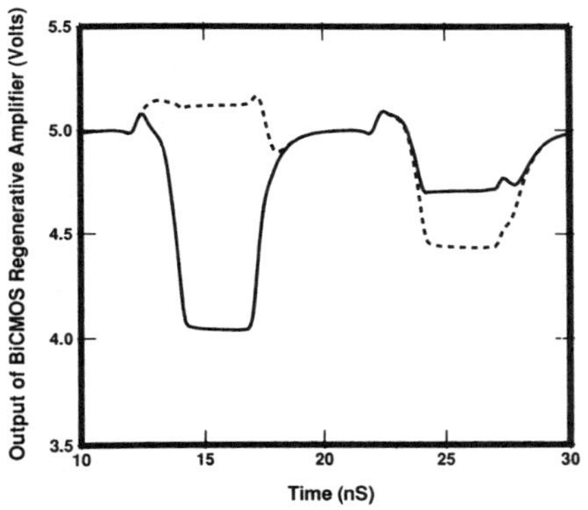

Fig. 5. Simulated output waveforms for the BiCMOS regenerative amplifier. Clock frequency of 100 MHz; differential inputs of $+1$ V and -1 mV.

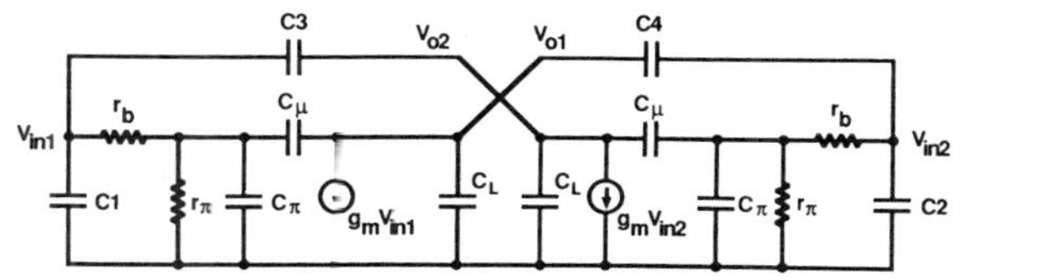

Fig. 6. Linearized model of the BiCMOS amplifier during regeneration.

During the second clock cycle, the collectors of $Q1$ and $Q2$ are pulled to 5 V as the -1 mV is sampled. In the transition to the regeneration mode, both collectors fall, as the voltage imbalance is regeneratively amplified, until capacitance $C5$ is charged to the point where both $Q1$ and $Q2$ are turned off.

Care must be taken to ratio the capacitances in the circuit of Fig. 4 so that the output levels do not fall too low. Otherwise, the output voltage might fall out of the input voltage range required by the following latch stage. A conflicting concern is that the collectors not fall far enough, thereby not allowing sufficient amplification of the differential input. Saturation of the bipolar transistors is easily avoided since the total charge that can be stored in the forward-biased base–collector junction is at most the charge stored on $C5$ during the previous sample mode.

A linearized model of the BiCMOS amplifier during regeneration is shown in Fig. 6, wherein it is assumed that $C1 = C2$ and $C3 = C4$. r_b is assumed to be negligible in the following calculations. The characteristic equation in the s-domain for this circuit is

$$s\left[(C1 + C3 + C_\pi + C_\mu)(C_L + C3 + C_\mu) - (C3 - C_\mu)^2\right]$$

$$- g_m(C3 - C_\mu) + \frac{C_L + C3 + C_\mu}{r_\pi} = 0. \tag{15}$$

The response of the circuit thus takes the form

$$V_{o2} - V_{o1} = KV_{in}\left[\exp(t/\tau_{reg}) - 1\right] \tag{16}$$

where K is a constant. The regeneration time constant τ_{reg} is given as

$$\tau_{reg} = \frac{(C1 + C3 + C_\pi + C_\mu)(C_L + C3 + C_\mu) - (C3 - C_\mu)^2}{g_m(C3 + C_\mu) - (C_L + C3 + C_\mu)/r_\pi}. \tag{17}$$

For most bipolar devices, the current gain β is greater than 50, and thus the second term in the denominator can be ignored. Furthermore, if C_μ is neglected, the regeneration time constant simplifies to

$$\tau_{reg} \approx \frac{C1 + C_\pi + C_L + C_L(C1 + C_\pi)/C3}{g_m} \tag{18}$$

where C_L includes the collector-to-substrate capacitance as well as the drain-to-substrate capacitance of the PMOS pull-up switch in the circuit of Fig. 3. Compared to the regeneration time constant for a typical bipolar comparator design as given in (4), the result in (18) shows that there is only a slight increase in the regeneration time constant. This increase is dependent on the size of the sampling capacitors and for this design is less than a factor of 2.

When used as the input stage in the design of a BiCMOS comparator, the regenerative amplifier of Fig. 7 avoids the need for preamplification. This reduces the required power, since overdrive recovery constraints in the preamplifier are eliminated. A further reduction in power dissipation in the input stage is expected because power is used only during the regeneration mode, i.e., during half the cycle.

In applications such as flash ADC's, strobe kickback into the input source can be of significant concern. For a bipolar comparator circuit utilizing the current-switch latch configuration of Fig. 1, the main source of kickback error is the transient associated with the rapid switching of the regenerative stages. Consequently, bipolar comparator designs often employ an additional preamplifying stage preceding the circuit in Fig. 1 to reduce the coupling to the input. In the BiCMOS design of Fig. 3 the regeneration transient is isolated from the input by the shutting off of the input MOS switches. However, the rapid turn-off of the MOS devices itself introduces error through capacitive coupling of the clock and the injection of channel charge into the input source. This error can be cancelled to first order by means of a variety of techniques, including the use of dummy switches [11].

The increased input capacitance of the proposed comparator design relative to bipolar alternatives has potential disadvantages in applications such as A/D converters. However, this is a less serious concern than in fully CMOS designs because of the availability of bipolar and MOS transistors for implementing either a sample-and-hold circuit or an input buffer amplifier.

IV. Comparator Design

A full schematic of the BiCMOS comparator is shown in Fig. 7. In this circuit, the outputs of the BiCMOS regenerative amplifier input stage are buffered via two source followers and then fed to the inputs of a low-power, current-switched latch. The latch is clocked by means of an MOS transistor differential pair $M9$ and $M10$ so as to allow operation from CMOS-level clocks. The output of the latch is buffered by two emitter followers that provide ECL output levels. All capacitors in the circuit are gate-oxide capacitances and occupy little die area in a digital process; they are formed by connecting the source and drain of an MOS transistor together and applying a suitable voltage between the gate and diffusion nodes so as to form an inversion layer under the gate. MOS transistor $M7$ is placed between $M5$ and $M6$ to prevent a direct path between the common emitter node of bipolar transistors $Q1$ and $Q2$ and ground; such a path would result in extraneous power dissipation and the possible saturation of the bipolar devices.

The major source of input offset error in this comparator arises from differential errors in sampling voltages onto capacitors using MOS switches. These differential errors are the result of mismatches between sampling capacitance values, gate-diffusion overlap capacitances, and the injected channel charge of the MOS switches. Among the mismatch errors, preliminary calculations indicate that the most significant error is that due to differential channel-charge injection effects from the fast turn-off of the MOS sampling devices [12]–[14]. The total injected charge in an MOS switch is a function of a number of parameters—device active area, channel length, gate-voltage slew rate, carrier mobility, and input and output impedances. However, in the absence of a detailed experimental characterization of the mismatch between adjacent pairs of MOS transistors, the total mismatch has been lumped into two variables and is modeled as mismatches in the channel length and channel width. A pessimistic three-standard-deviation value of 0.06 μm for both variables was chosen [15]. Since these mismatches in geometry model the mismatch of many parameters, not just the lengths and widths, no correlation in these variables is assumed. Consequently, for a given allowable differential input-voltage error due to sampling the input voltage onto capacitors, the size of the MOS sampling switches is constrained. This, in turn, limits the speed at which voltage sampling to a given precision can be performed. For the circuit of Fig. 7, the error due to the turn-off of $M1$ and $M2$ cannot be reduced, and thus $M1$ and $M2$ must be sized in relation to the total sampling capacitance $C1$ through $C4$.

In the case of the PMOS pull-up devices $M3$ and $M4$ in Fig. 7, where a common voltage (+5 V) is being "sampled," the differential channel-charge injection error can be reduced through the use of MOS transistor $M8$ and resistor $R1$. $R1$ delays the turn-off clock transition to the gate of $M8$ so that it turns off approximately 200 ps after the turn-off of transistors $M3$ and $M4$. Thus, $M8$ serves to equalize some of the differential injection error resulting from the turning off of $M3$ and $M4$. MOS transistor $M8$ is implemented as two equal geometry transistors connected in parallel in a manner that minimizes process-induced mismatches in coupling capacitances to the collector nodes of bipolar transistors $Q1$ and $Q2$.

For circuit simulations, a single lumped model for MOS transistors was used, since the transit time of minimum channel-length MOS switches is much less than the expected clock transition time. Some care was taken to ensure that the charge was being conserved in these simulations.

V. Experimental Results

The comparator was fabricated in a 0.8-μm digital BiCMOS technology in an area of 140 by 75 μm. A chip photograph showing the comparator cell along with input

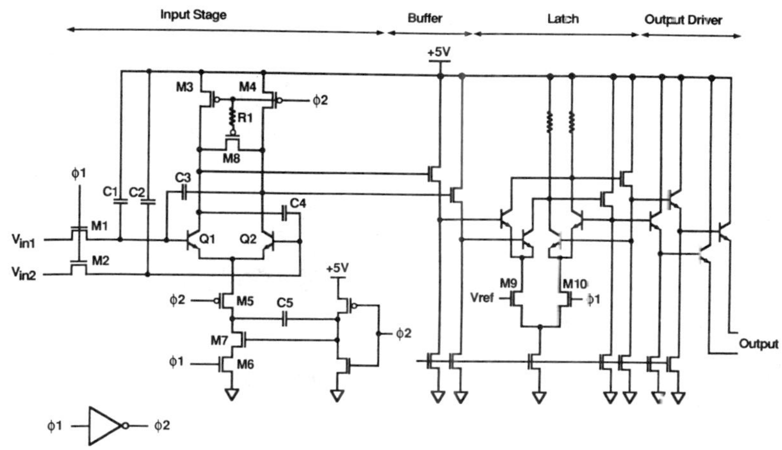

Fig. 7. Schematic of the BiCMOS comparator.

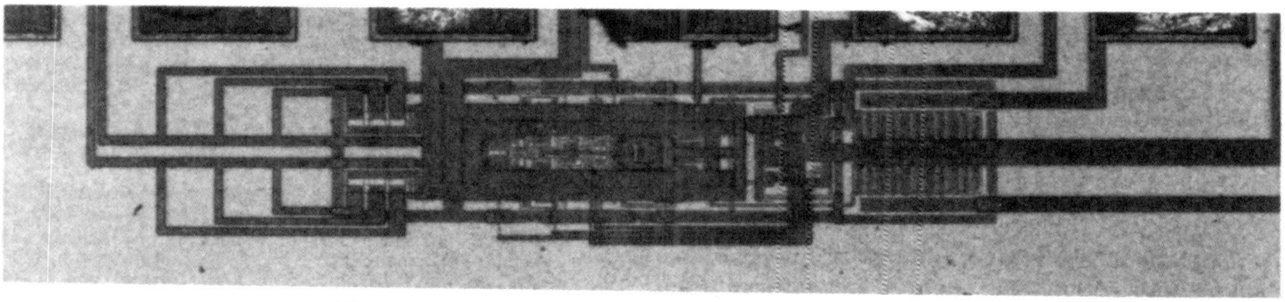

Fig. 8. Chip photograph of the experimental BiCMOS comparator.

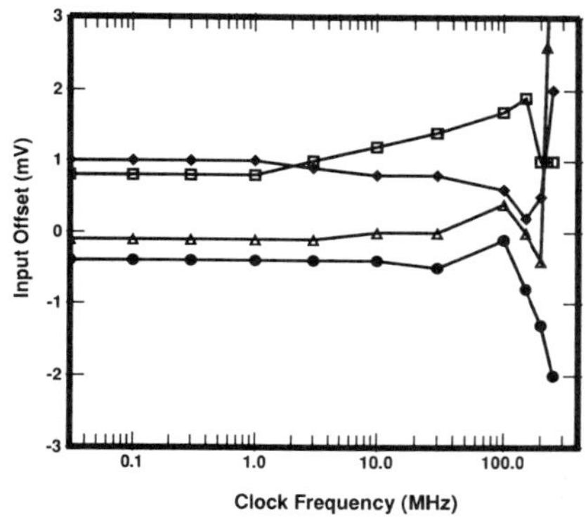

Fig. 9. Measured input offset voltage as a function of clock frequency.

TABLE I

Technology	0.8 μm BiCMOS
Comparator Cell Size	75μm × 140μm
Max. Operating Freq.	200 MHz
Input Capacitance	150 fF
Input Offset Voltage	2.1 mV
Power Supply	+5 Volts
Power at 200 MHz	
Input stage	0.7 mW
Buffers	0.3 mW
Latch	0.6 mW
Total Power	1.6 mW

termination resistors, clock drivers, and output-driving emitter followers is shown in Fig. 8.

The input offset voltage of the comparator was measured using dc input voltages. Fig. 9 shows plots of the measured offset voltage for four typical chips as a function of the sampling rate. The input common-mode voltage was 1 V. The measured comparator offset voltages are constant over a broad frequency range and are typically less than 1 mV. The offset voltages begin to increase at clock frequencies above 100 MHz, but remain within ± 3 mV ($\pm 1/2$ LSB) at a sampling rate of 200 MHz.

To test the input overdrive recovery of the comparator, one input was held at a constant voltage while the differential voltage was stepped from a large difference to within the minimum-required resolvable difference. For this circuit the differential input voltage was changed from $+1$ V to ± 3 mV ($\pm 1/2$ LSB) on consecutive clock cycles, while the output waveform was monitored so as to verify that the input difference was being correctly resolved following the input overdrive. The input overdrive recovery has been experimentally verified at a clock rate of 150 MHz. Verification at higher clock rates was not possible due to test instrumentation limitations. Fig. 10(a) and (b) shows oscilloscope photographs of typical input and output waveforms for the overdrive recovery test at 150 MHz. The waveforms show that the output correctly indicates the

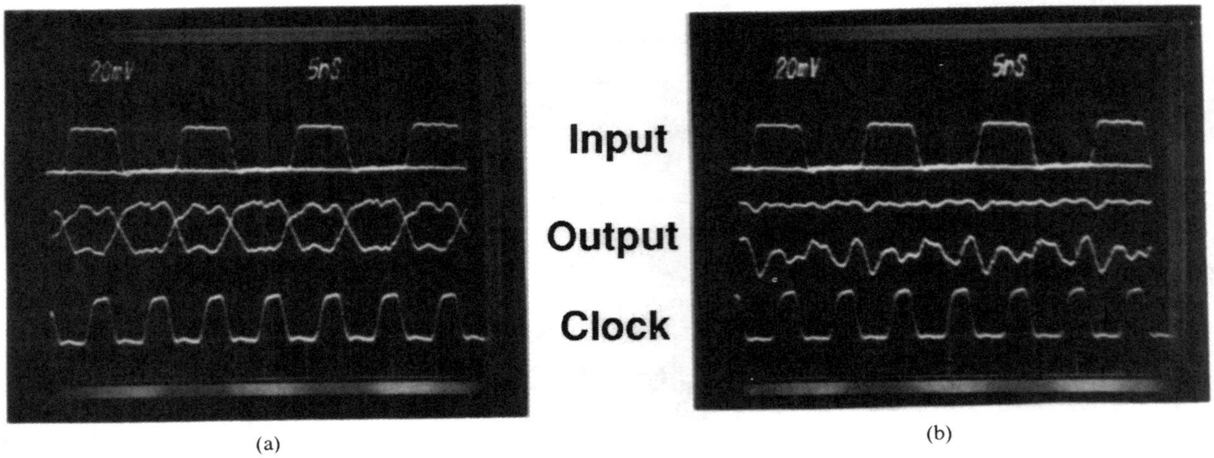

Fig. 10. Oscilloscope photograph of typical waveforms for the input overdrive recovery test. The differential input changes from (a) +1 V to −3 mV, and (b) +1 V to +3 mV.

polarity of the input voltage waveform following the input overdrive.

The performance of the BiCMOS comparator is summarized in Table I.

VI. Conclusion

The integration of both MOS and bipolar devices within the same technology presents opportunities for significant improvements in performance over CMOS and bipolar comparator designs. CMOS designs traditionally operate at lower comparison rates because of the relatively low transconductance of MOS transistors. Bipolar comparator designs are capable of high-speed operation, yet power dissipation is often determined by requirements for the preamplifying input stage. A novel BiCMOS input stage has been introduced which combines MOS input sampling with bipolar regenerative amplification. This circuit dissipates no static power and offers substantial power savings over bipolar comparator designs without a reduction in the comparison rate.

Acknowledgment

The authors wish to thank members of the Semiconductor Process Laboratory and the VLSI Design Laboratory at Texas Instruments, Inc. for the fabrication of the experimental circuits; they are particularly indebted to M. Harward and T. Holloway. They also gratefully acknowledge numerous helpful discussions with Prof. M. Horowitz, Dr. J.-T. Wu, and Dr. T.-S. Yang regarding the design, layout, and testing of the circuit.

References

[1] R. H. Haveman *et al.*, "An 0.8-μm 256K BiCMOS SRAM technology," in *IEDM Tech. Dig.*, Dec. 1987, pp. 841–843.

[2] J. T. Wu and B. A. Wooley, "A 100-MHz pipelined CMOS comparator," *IEEE J. Solid-State Circuits*, vol. 23, pp. 1379–1385, Dec. 1988.

[3] T. Kumamoto *et al.*, "An 8-bit high-speed CMOS A/D converter," *IEEE J. Solid-State Circuits*, vol. SC-21, pp. 976–982, Dec. 1986.

[4] T. Matsuura, T. Tsukada, and S. Ohba, "An 8-bit 20-MHz CMOS half-flash A/D converter," in *ISSCC Dig. Tech. Papers*, Feb. 1988, pp. 220–221.

[5] B. J. McCarroll, C. G. Sodini, and H. Lee, "A high-speed CMOS comparator for use in an ADC," *IEEE J. Solid-State Circuits*, vol. 23, pp. 159–165, Feb. 1988.

[6] A.Yukawa, "A CMOS 8-bit high-speed A/D converter IC," *IEEE J. Solid-State Circuits*, vol. SC-20, pp. 775–779, June 1985.

[7] M. Inoue, H. Sadamatsu, A. Matsuzawa, A. Kanada, and T. Takamoto, "A monolithic 8-bit A/D converter with 120 MHz conversion rate," *IEEE J. Solid-State Circuits*, vol. SC-19, pp. 837–841, Dec. 1984.

[8] Y. Yoshi, K. Asano, M. Nakamura, and C. Yamada, "An 8-bit 100-Ms/s flash ADC," *IEEE J. Solid-State Circuits*, vol. SC-19, pp. 842–846, Dec. 1984.

[9] B. Peetz, B. D. Hamilton, and J. Kang, "An 8-bit 250 megasample per second analog-to-digital converter: Operation without a sample and hold," *IEEE J. Solid-State Circuits*, vol. SC-21, pp. 997–1002, Dec. 1986.

[10] B. Zojer, R. Petschacher, and W. A. Luschnig, "A 6-bit/200-MHz full nyquist A/D converter," *IEEE J. Solid-State Circuits*, vol. SC-20, pp. 780–786, June 1985.

[11] R. E. Suarez, P. R. Gray, and D. A. Hodges, "All-MOS charge redistribution analog-to-digital conversion techniques—Part II," *IEEE J. Solid-State Circuits*, vol. SC-10, pp. 379–385, Dec. 1975.

[12] B. J. Sheu and C. Hu, "Switched-induced error voltage on a switched capacitor," *IEEE J. Solid-State Circuits*, vol. SC-19, pp. 519–525, Aug. 1984.

[13] W. B. Wilson, H. Z. Massoud, E. J. Swanson, R. T. George, and R. B. Fair, "Measurement and modeling of charge feedthrough in n-channel MOS analog switches," *IEEE J. Solid-State Circuits*, vol. SC-20, pp. 1206–1213, Dec. 1985.

[14] G. Wegmann, E. A. Vittoz, and F. Rahali, "Charge injection in analog MOS switches," *IEEE J. Solid-State Circuits*, vol. SC-22, pp. 1091–1097, Dec. 1987.

[15] K. R. Lakshimikumar, R. A. Hadaway, and M. A. Copeland, "Characterization and modeling of mismatch in MOS transistors for precision analog design," *IEEE J. Solid-State Circuits*, vol. SC-21, pp. 1057–1066, Dec. 1986.

Paper 4.10

A BiCMOS PLL-Based Data Separator Circuit with High Stability and Accuracy

Shyoichi Miyazawa, *Member, IEEE*, Ryutaro Horita, *Member, IEEE*, Kenichi Hase, *Member, IEEE*, Kazuo Kato, *Associate Member, IEEE*, and Shinichi Kojima

Abstract —**This article describes a data separator which can work in Winchester disk drives at a read/write speed of up to 30 Mb/s. To realize high stability and accuracy in reproducing data in high-speed transfers, not only a digital synchronization field detector is used, but also an analog dual-mode phase-locked loop (PLL) which has a new phase detector. The new phase detector has constant gain in the data field, independent of pattern. In addition to the above circuit, the IC incorporates a run-length-limited (RLL) 2–7 code encoder/decoder and a write compensator. By using the 2-μm BiCMOS process, the total power consumption is kept as low as 400 mW even at the high transfer rate of 30 Mb/s.**

I. Introduction

A DATA separator has been developed to meet the requirements of the Winchester disk drive industry which has increased capacity and decreased size.

As capacities increase, read/write circuit performance becomes critical. To meet this requirement, a monolithic read/write IC has been developed [1].

The write circuit consists of an encoder using the run-length-limited 2–7 (RLL 2–7) code, which is a form of data compression code, and a write compensation circuit.

The read circuit consists of a data separator incorporating an analog phase-locked loop (PLL), a sync field detector, and a decoder of the RLL 2–7 code. The output signals from the magnetic disk contain large jitter because of the S/N characteristics of the disk, head, and amplifier, and bit-to-bit interference due to peak shift. The data separator therefore is required to separate data and clock timing from such signals.

In a conventional PLL circuit, we could not design the filter of the PLL optimally to cancel the effects of jitter. The reason is that, because data are random in the data field and the output signal includes frequency deviations, the gain of the PLL other than the filter varies due to frequency deviation. Furthermore, in the sync field, the PLL must lock quickly without erroneous locking to the output signal of the fixed pattern.

Thus the demand is for the development of a data separator that locks quickly in the sync field and makes the gain of the circuit, except the filter, fixed in the data field where frequency deviation occurs.

Additionally, as data records are written over old ones on the magnetic disk, write splicing occurs as the speed of rotation constantly fluctuates. If the data separator starts to lock in the write splice area, the wrong signal coming from the write splice may cause the PLL to run out of control.

Therefore, the data separator must lock in the sync field and, for this reason, a sync field detector is required. The conventional sync field detector was an analog circuit comprising passive devices like resistors and capacitors [4]. In such systems, individual tolerances and the temperature coefficients of components affect the accuracy of detection. So, demand arose for a digital sync field detector which would be temperature stable and accurate.

This article describes a data separator which will lock quickly, precisely, and reliably, combined with a digital sync field detector with high accuracy.

II. Study and Design of the PLL

A. Conventional PLL Used in Magnetic Disk Units and Its Problems

The conventional PLL used to read data from magnetic disks consists of a phase detector (PD), loop filter (LF), and VCO. The input and output are digital signals. In order to realize high-speed locking in the sync field and stability against jitter in the data field, natural angular frequency ω_n and damping factor ζ are optimized by switching the output current of the phase detector and the constant of the loop filter. By making use of the fact that signals with a fixed frequency are recorded in the sync field, the phase detector compares the frequency and phase in the sync field to attain a wide capture range and prevent erroneous locking. In the data field, it detects the phase difference of the VCO clock and the input data whose frequency ranges from 1/3 to 1/8 of the VCO clock.

However, the phase detector detects a phase difference only when there is input data, and produces a pulse current of a fixed intensity I_p for a duration that is proportional to the phase difference. So, the open-loop transfer function $G(s)$, natural angular frequency ω_n, and damping factor depend on the ratio N of frequency of the VCO clock to that of the input data as follows:

$$G(s)=\frac{I_p\cdot K_o\{1+S(C1+C2)R\}}{2\pi N\cdot S^2C1(1+SC2R)} \tag{1}$$

$$\omega_n=\sqrt{\frac{I_p}{2\pi}\,\frac{K_o}{C1\cdot N}} \tag{2}$$

$$\zeta=\frac{(C1+C2)R}{2}\sqrt{\frac{I_p\cdot K_o}{2\pi C1\cdot N}}\qquad (3\leqslant N\leqslant 8). \tag{3}$$

Manuscript received March 8, 1990; revised October 10, 1990.

S. Miyazawa, R. Horita, and K. Hase are with the Microelectronics Products Development Laboratory, Hitachi, Ltd., 292 Yoshida-cho, Totsuka-ku, Yokohama 244, Japan.

K. Kato is with the Hitachi Research Laboratory, Hitachi, Ltd., Ibaraki, Japan.

S. Kojima is with Takasaki Works, Hitachi, Ltd., Gunma, Japan.

IEEE Log Number 9041213.

Reprinted from *IEEE J. Solid-State Circuits*, vol. 26, no. 2, pp. 116–121, Feb. 1991.

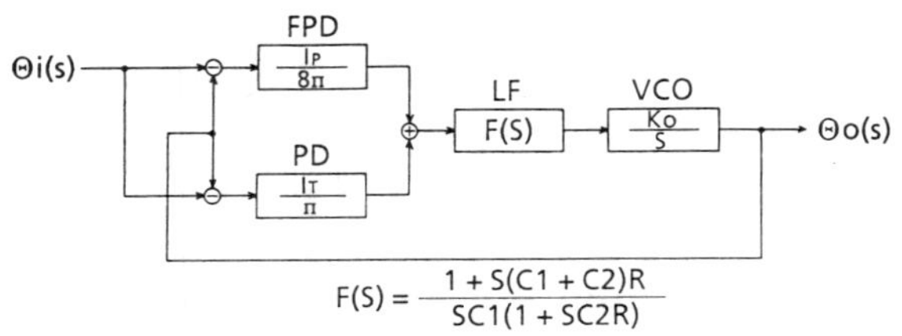

Fig. 1. PLL block diagram.

For this reason, the open-loop frequency response varies in the data field where the frequency ranges from 1/3 to 1/8 of the VCO clock frequency, and optimization is difficult. The system producing pulse current with a fixed intensity I_p will produce a current pulse every time it compares the phase, even after the completion of phase synchronization. So, the output voltage waveform of the loop filter that controls the VCO varies due to the current pulse, and the VCO clock includes jitter.

The VCO clock is used to produce the detection window the decoder uses to read data, and jitter reduces the decoding margin and degrades the data read margin of the device.

B. Solution to Problems and Circuit Design

1) The PLL shown in Fig. 1 is proposed to solve the above problems. A frequency phase detector (FPD) is used in the sync field where data are written with the same frequency, and the output is a conventional pulse current with a fixed intensity I_p. This realizes a wider capture range, preventing erroneous locking and faster phase locking. In the data field where the frequency ranges from 1/3 to 1/8 of the VCO clock frequency, a phase detector is used. The output is a dc current which is proportional to the phase difference of the input data and VCO clock and lasts until the next time–phase difference is detected. It realizes a frequency response that does not depend on the frequency ratio N of the VCO clock to the input data, and reduces the degradation of the data read margin caused by VCO jitter during synchronization.

2) A frequency phase detector compares the VCO clock divided by four with the rising edge of the read data (RAWRD) and, if the VCO clock is earlier, allows current with a fixed intensity I_p to flow from the loop filter during the time interval just as long as the phase difference.

If the phase difference is zero, no current flows to or from the loop filter. If the VCO clock is later, it allows current with a fixed intensity I_p to flow to the loop filter during the time interval just as long as the phase difference. This circuitry presents a wide range of linear comparison and a wide capture range (actually limited by the dynamic range of VCO output and parameters of the circuit), and is suitable for the synchronization when phase synchronization is required to be complete in a few microseconds. Fig. 2 shows the resulting measurement of phase step response at 30 Mb/s (acquisition time: less than 2 μs).

3) Fig. 3 shows the schematic of a PLL using a phase detector and Fig. 4 shows its timing diagram. The signal T_c is generated between the rising edge of the read data and the falling edge of the next VCO clock. The signal T_s is generated between the falling edge of the VCO clock and the rising edge of the next VCO clock. The signal T_D is generated between the rising edge of the VCO clock and the

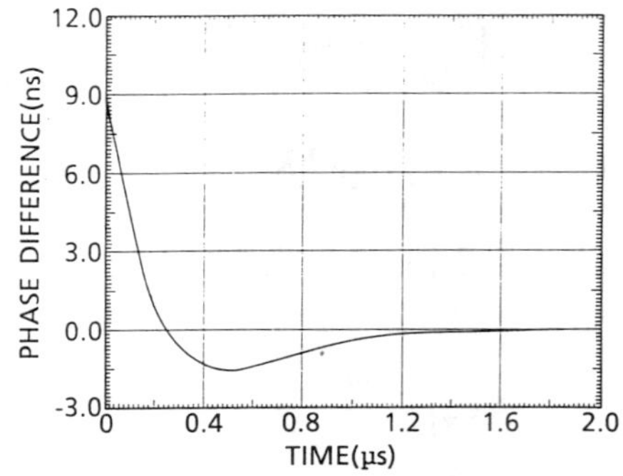

Fig. 2. Measured phase step response using frequency phase detector.

falling edge of the next VCO clock. While the signal T_c is coming out, SW1 is closed and current I_c flows from the capacitor C_c. Then the signal T_s causes SW2 to close and the output voltage V_{op} of the operational amplifier is sampled and held by the capacitor C_s.

A differential amplifier converts the voltage V_{op} to current I_D. While the signal T_D is coming out, SW3 is closed and current I_D flows to the capacitor C_c. We analyze this process quantitatively as follows. Suppose we sample the phase difference n times. Then $I_{D(n)}$ and $V_{op(n)}$ at the nth sampling are

$$I_{D(n)} = g_m \cdot V_{op(n)} \tag{4}$$

$$V_{op(n)} = V_{op(n-1)} + \frac{T_c \cdot I_c - T_D \cdot I_{d(n-1)}}{C_c} \tag{5}$$

where g_m is the gain of the differential amplifier. These equations may be rewritten as follows:

$$I_{D(n)} = \frac{T_c \cdot I_c}{C_c} g_m + \left(1 - \frac{T_d \cdot g_m}{C_c}\right) \cdot I_{D(n-1)}. \tag{6}$$

This recursive equation can be solved as follows:

$$I_{D(n)} = \frac{T_c}{T_D} I_c \cdot \left\{1 - \left(1 - \frac{T_D \cdot g_m}{C_c}\right)^{n-1}\right\} + \left(1 - \frac{T_D \cdot g_m}{C_c}\right)^{n-1} \cdot I_{D(1)}. \tag{7}$$

This indicates that if the condition

$$\frac{T_D \cdot g_m}{C_c} \geqslant 2 \tag{8}$$

holds, the expression of current I_D will diverge; otherwise it will converge. If the condition

$$\frac{T_D \cdot g_m}{C_c} = 1 \tag{9}$$

holds, $I_{D(n)}$ becomes as follows:

$$I_D = I_{D(n)} = \frac{T_c}{T_D} I_c. \tag{10}$$

So, if the circuit is designed so that the output current I_T meets the following:

$$I_T = I_D - I_c \tag{11}$$

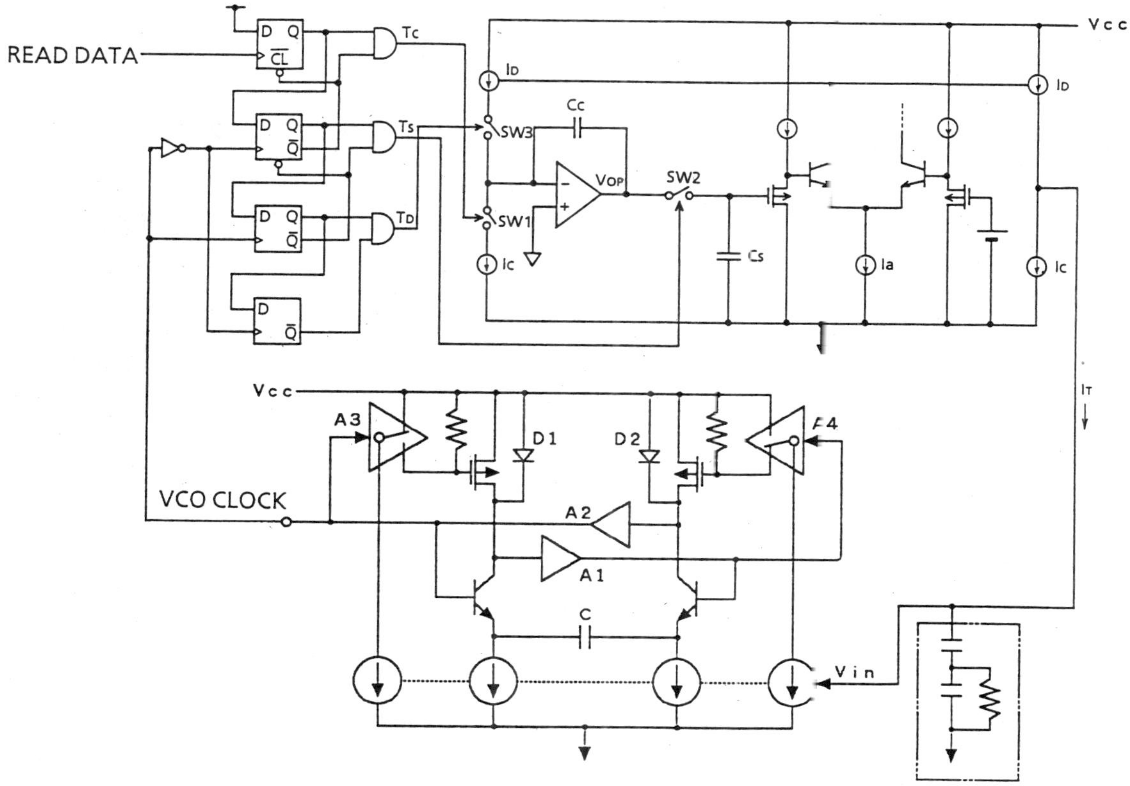

Fig. 3. Schematic of PLL using phase detector.

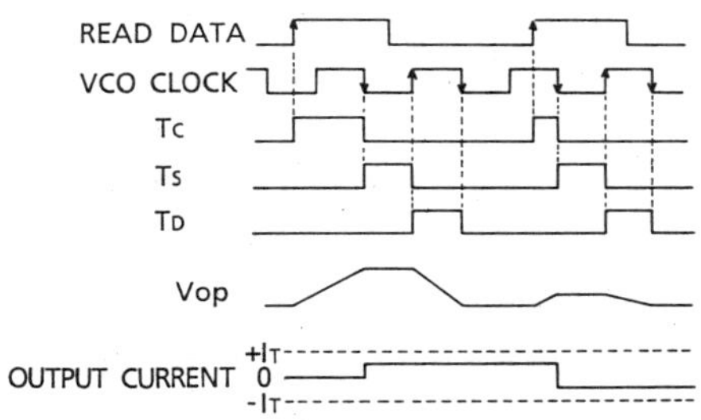

Fig. 4. Timing diagram for phase detector.

the output current I_T is

$$I_T = \frac{T_c - T_D}{T_D} I_c. \tag{12}$$

As the phase difference $\Delta\phi$ is given as

$$\Delta\phi = T_c - T_D \tag{13}$$

the output current I_T is dc current which is proportional to the phase difference $\Delta\phi$.

Next we examine (9) used as a condition for the quantitative analysis of the phase detector. T_D corresponds to the half period of the VCO clock and remains fixed if the VCO clock frequency is fixed and its duty is controlled. Deviation of the capacitance C_c can be reduced, even considering the mask precision, the thickness of spin-on glass (SOG)+ phosphosilicate glass (PSG), and its dielectric constant. The gain g_m of the differential amplifier is given as follows:

$$g_m = \frac{I_a}{V_T} = \frac{qI_a}{kT} \tag{14}$$

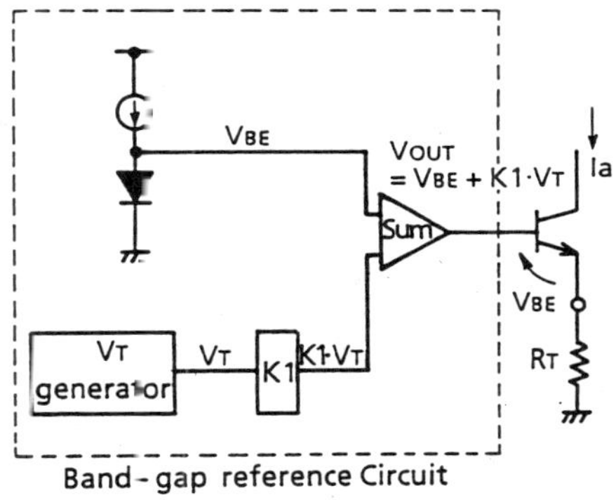

Fig. 5. Reference current source.

where

- q electric charge of an electron,
- k Boltzman constant,
- T absolute temperature (Kelvin).

This has a temperature dependence of about −3300 ppm/°C if I_a is fixed. To cancel this temperature dependence, a bandgap reference circuit as shown in Fig. 5 is used. Its output voltage is applied to resistor R_T through an emitter follower. Then the current I_a of the differential amplifier is

$$I_a = \frac{V_{\text{out}} - V_{BE}}{R_T} = \frac{K_1 \cdot V_T}{R_T} \tag{15}$$

where K_1 is a constant. So, the gain g_m of the differential

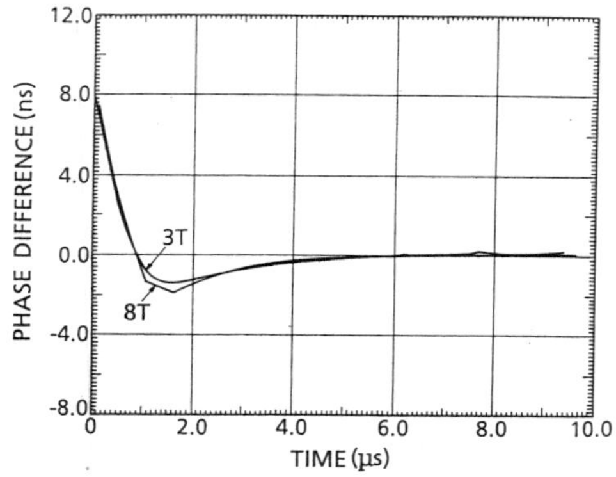

Fig. 6. Measured phase step response using phase detector.

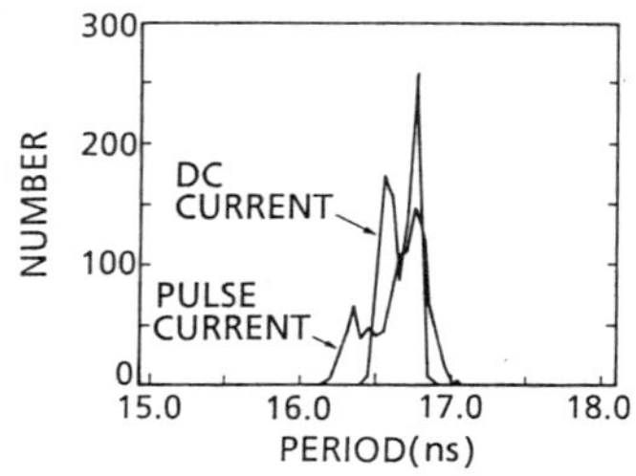

Fig. 7. Measured VCO period.

amplifier is

$$g_m = \frac{1}{V_T} \cdot \frac{K_1 \cdot V_T}{R_T} = \frac{K_1}{R_T}. \tag{16}$$

K_1, which is given as a ratio of two resistors, does not vary with temperature. Therefore it is possible to make the gain g_m independent of temperature by connecting the resistor R_T externally. It is also possible to make (9) independent of temperature and to cause the phase detector to produce a dc current that is proportional to the phase difference $\Delta\phi$.

When we use the circuitry described above, the open-loop transfer function $G(s)$, characteristic frequency ω_n, and damping factor ζ of the system are as follows in the data field:

$$G(s) = \frac{S(C1+C2)R \cdot I_T \cdot K_o + I_T \cdot K_o}{S^2 C1 \cdot \pi(1+SC2R)} \tag{17}$$

$$\omega_n = \sqrt{\frac{I_T \cdot K_o}{\pi \cdot C1}} \tag{18}$$

$$\zeta = \frac{(C1+C2)R}{2} \cdot \sqrt{\frac{I_T \cdot K_o}{\pi \cdot C1}}. \tag{19}$$

They do not depend on the ratio N of frequency of the VCO clock to that of the read data. So, you need not consider the effect of N and can set the system in an optimal condition. Fig. 6 shows two curves of phase step response measured with an experimental circuit when $N=3$ and 8. These curves are almost equal, though differences of the sampling intervals exist. Fig. 7 compares VCO jitter at the time of phase synchronization with the conventional current pulse system and the dc current system. VCO jitter is lower by 500 ps with the dc current system, and the decode margin can be increased by the same amount.

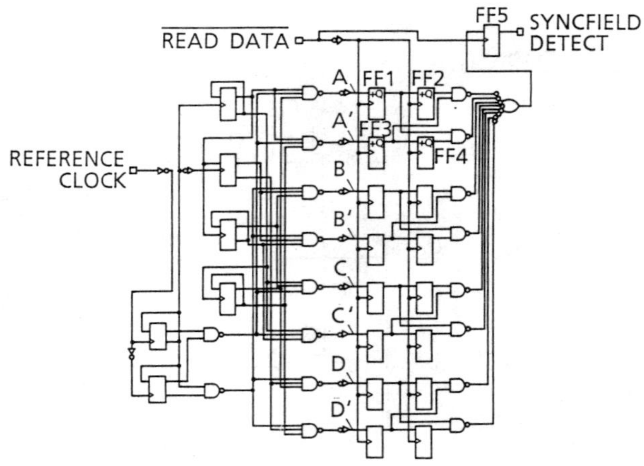

Fig. 8. Circuit diagram of sync field detector.

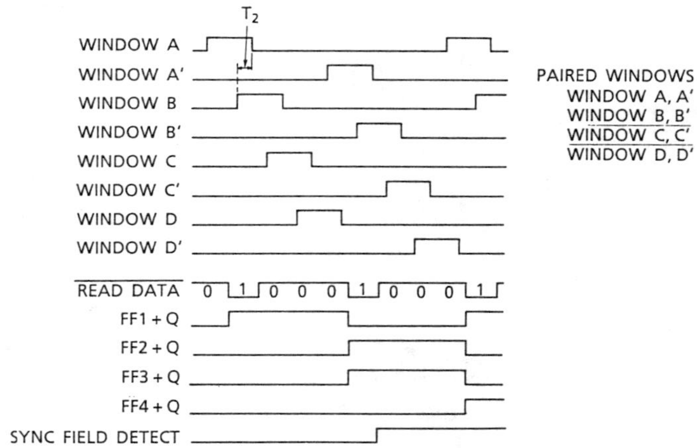

Fig. 9. Timing diagram of sync field detector.

III. Digital Sync Field Detector

The sync field detector was conventionally an analog circuit composed of a monostable multivibrator, and it has the following problems:

1) in the analog circuit, individual differences of components are great (IC's, C, R), so adjustment is needed for accurate operation;
2) error due to temperature drift cannot be ignored (measured as 600 ~ 1000 ppm) and the design must allow a sufficient margin;
3) if the pattern written in the sync field is not the highest frequency of the data code, the conventional system will fail since it will detect all frequency patterns that are higher than that frequency as sync fields.

The digital sync field detector has been developed to solve these problems.

Fig. 8 is the circuit diagram of the sync field detector and Fig. 9 is a timing diagram. Eight-phase window pulses generated by the reference clock are classified into four pairs so that the difference in phase of each pair is $4T_0$ (T_0: the period of reference clock). To detect the pattern of sync fields, the interval of 1's is detected when two neighboring 1's in the input data exist in a paired window. Fig. 9 shows

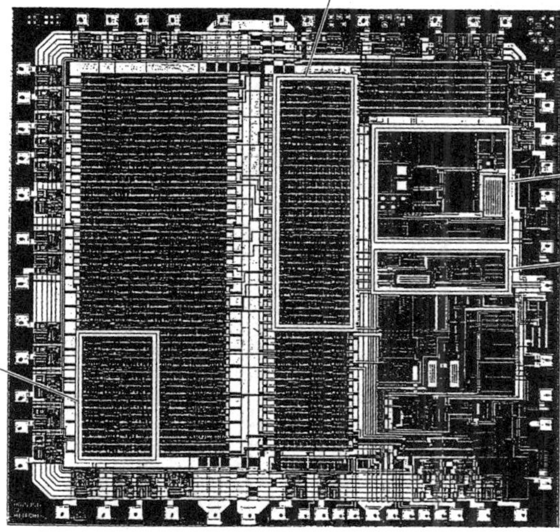

Fig. 10. Photograph of the data recovery circuit.

the eight-phase window pulses paired into four, and the timing of the case where the read data is a $4T$ pattern (repeat of "1000"). The figure shows that the sync field detect signal works when a $4T$ pattern is detected.

The window pulses generated from the reference clock are asynchronous with the read data and they have different frequencies and phases. The data read from a disk includes much jitter. To secure operation whatever the phase relation, arrangement is made so that windows are continuous and neighboring windows overlap. Fig. 9 shows windows A and B overlap. Permissible jitter T_1 is within $\pm 50\%$ of the data bit length T as it is determined from the capability of the following data separator. Overlap T_2 of windows and permissible jitter T_1 are as follows:

$$T_2 = |T_1|. \tag{20}$$

On the other hand,

$$T_1 = \pm 0.5T. \tag{21}$$

So, overlap $T2$ is as follows:

$$T_2 = 0.5T. \tag{22}$$

At this time, permissible frequency deviation $\Delta f\theta$ is found to be 12.5% from

$$\Delta f\theta = \frac{T_2}{4T} \cdot \frac{1}{T} \tag{23}$$

Though this circuit is designed to detect the $4T$ pattern, it may detect a $3T$ or $5T$ pattern incorrectly due to the overlap T_2. But this happens only when the phase relation is such that data enter in the specific 50% area of windows for the time of 6 or 8 bytes, and this does not actually happen because of the frequency deviation and jitter actually involved in read data.

Since this is a digital circuit, adjustment is not necessary and a very accurate sync field detector can be realized with the effects of temperature drift eliminated.

IV. Operation of PLL

Operation of the PLL is controlled through the READ GATE (RG) and PHASE SYNC (PS) terminals. The PLL follows the reference clock (REFCLK) when RG is negated.

TABLE I
Device Characteristics

General	Technology	2-μm BiCMOS
	Supply Voltage	5 V, $\pm 5\%$
	Power Consumption	400 mW (typical)
	Chip Size	4.7×4.9 mm^2
	Data Transfer Rate	30 Mbs
PLL	Acquisition Time	2 μs
	Jitter Output	500 ps
	Capture range	$\pm 10\%$
	Lock Range	$\pm 10\%$
Sync Field Detect	Tolerance to Input Jitter	± 0.5T

When PS is asserted according to the sync field detect (SYNCDETECT) output of the sync field detector or an external signal while RG is asserted, the PLL starts comparing the frequency and phase of the read data (RAWRD) and locking with the charge pump in a high-gain state. After a fixed time of 6 or 8 bytes when PS is disasserted, the PLL switches from the frequency phase comparison mode to the phase comparison mode and operates for T/I conversion at normal gain. When RG is disasserted again, the PLL returns to follow REFCLK. The NRZRD terminal remains at Hi-Z when RG is disasserted, produces L after RG is asserted as long as the high-gain state continues and, in the normal-gain state, produces read data decoded by the conversion rules of the RLL 2–7 code.

V. Results

This IC using the 2-μm BiCMOS process incorporates the functions of a PLL, sync field detector, RLL 2–7 encoder/decoder, and is realized as a chip of 4.9×4.7 mm^2. Fig. 10 shows a photo of the IC.

The PLL in this IC takes 2 μs until it locks by using a frequency phase detector. This is twice as fast as conventional data synchronizer/2-7RLL ENDEC; jitter is also halved in the stable state. The sync field detector is adjustment-free because it is a digital circuit. The power consumption of the IC is reduced to 400 mW thanks to the 2-μm BiCMOS process.

Table I summarizes the performance of the IC.

VI. Conclusion

A high-speed read/write IC suitable for magnetic disk drives has been developed using a dual-mode analog PLL and digital sync field detector.

The dual-mode analog PLL has a wide decode margin, locks up quickly, and operates stably without being affected by the frequency deviation of data. The digital sync field detector is adjustment-free and detects sync fields very accurately.

The results of this research ensure the feasibility of the development of very large mixed analog–digital LSI incorporating the hard disk control function, analog signal amplifiers, and a read/write circuit using the BiCMOS process.

Acknowledgment

The authors wish to thank Dr. S. Kawamura and N. Tanaka for their encouragement, and H. Sato and T. Sase for their technical suggestions.

References

[1] J. T. Kellis and S. Mehrozra, "A 15-Mb/s data separator and write compensation circuit for Winchester disk drives," in *ISSCC Dig. Tech. Papers*, Feb. 1986, pp. 232–233.

[2] K. Kato *et al.*, "A low power dissipation PLL IC operating at 128 MHz clock," in *IEEE 1987 CICC Conf. Rec.*, May 1987, pp. 651–654.

[3] R. W. Wood, "Magnetic recording systems," *Proc. IEEE*, vol. 74, no. 11, pp. 1557–1569, Nov. 1986.

[4] "Data synchronizer/2, 7 RLL ENDEC with write precompensation SSI 32D535," product manual, Silicon Systems.

Paper 4.11

A High-Frequency Fully Differential BiCMOS Operational Amplifier

Andrew N. Karanicolas, *Student Member, IEEE,* Kenneth K. O, John Y. A. Wang, Hae-Seung Lee, *Member, IEEE,* and Rafael L. Reif, *Senior Member, IEEE*

***Abstract*—This paper presents a high-frequency fully differential BiCMOS operational amplifier designed for use in switched-capacitor circuits. The op amp is integrated in a 3.0-GHz, 2-μm BiCMOS process with an active die area of 1.0 mm × 1.2 mm. This BiCMOS op amp offers an infinite input resistance, a dc gain of 100 dB, a unity-gain frequency of 90 MHz with 45° phase margin, and a slew rate of 150 V/μs. The differential output range is 12 V. The circuit is operated from a ±5-V power supply and dissipates 125 mW. The op amp is unity-gain stable with 7 pF of capacitive loading at each output. The op amp is a two-stage, pole-split frequency compensated design that employs a PMOS input stage for infinite input resistance and an n-p-n bipolar second stage for high gain and high bandwidth. The frequency compensation network serves both the differential- and common-mode amplifiers so the differential- and common-mode amplifier dynamics are similar. A dynamic switched-capacitor common-mode feedback scheme is used to set the output common-mode level of the first and second stages.**

I. Introduction

THE increasing interest in high-performance integrated analog and digital mixed processing systems drives the need for high-performance switched-capacitor analog circuits such as switched-capacitor filters [1] and analog-to-digital converters [2]. Such analog circuits often require high-performance operational amplifiers with a high-frequency capability and large dc gain [3]. The key advantage of implementing mixed signal systems in BiCMOS process technology is that for analog circuits, bipolar devices offer a transconductance that is generally higher compared to MOS devices, leading to, for a given area, higher speed and higher gain. The CMOS devices offer high-density logic in addition to capability for switched-capacitor analog circuit architectures. A fully differential pipeline A/D architecture is desirable for implementation in a mixed signal environment. This paper describes a fully differential BiCMOS operational amplifier that is designed for use in switched-capacitor analog circuits.

This paper is organized as follows. A brief description of the BiCMOS process will be presented in Section II. This section will give an overview of the process along with a summary of key BiCMOS device parameters. A more complete description of the process is given in [4]–[6]. The op-amp circuit will be presented in Section III. This section will describe the design of the first and second stages, the common-mode feedback network, and the biasing circuit. In Section IV, measurements of the op-amp unity-gain frequency, gain–bandwidth product, and step response will be shown.

Manuscript received August 3, 1990; revised November 2, 1990. This work was supported by the Semiconductor Research Corporation under Contract 89-SP-080.

A. N. Karanicolas, J. Y. A. Wang, H.-S. Lee, and R. L. Reif are with the Department of Electrical Engineering and Computer Science, Massachusetts Institute of Technology, Cambridge, MA 02139.

K. K. O was with the Massachusetts Institute of Technology, Cambridge, MA 02139. He is now with Analog Devices, Inc., Wilmington, MA 01887.

IEEE Log Number 9041556.

II. BiCMOS Process Description

A. Process Overview

The focus of the BiCMOS process is on building high-precision and high-speed analog/digital mixed systems for 10-V operation [5]. The process design reflects an emphasis on high-performance analog circuit capability in addition to high-density CMOS digital circuit capability. The devices available in the process are: NMOS, PMOS, vertical n-p-n, vertical p-n-p, and poly-n^+ capacitor. A cross section of the devices in this process is shown in Fig. 1.

The process development approach is to integrate the bipolar devices into an existing twin-well CMOS process. The vertical n-p-n devices are fabricated in a selective epitaxial layer deposited after the LOCOS field oxidation step of the CMOS process [4]. This epitaxial layer is deposited on n^+ buried islands, located in p-wells. The nonoptimized vertical p-n-p bipolar structures are formed by depositing the selective epitaxial layer on CMOS p-wells, without the n^+ buried islands, thus using the wells as the collectors. Since there is no buried layer for the p-n-p, its collector resistance is much higher than its n-p-n counterpart. The interconnect system for this BiCMOS process uses a single level of metal and polysilicon. The op amp is fabricated in twin-well BiCMOS built on p/p^+ epitaxial wafers permitting independent connection of the PMOS n-well substrate.

B. Key BiCMOS Device Parameter Summary

Table I summarizes the key CMOS device parameters in the BiCMOS process [5]. The CMOS devices feature a 2-μm drawn channel length and a gate-oxide thickness of 23 nm. For the NMOS and PMOS devices, the respective values for the zero-bias threshold voltage are 0.7 and −0.75 V. The subthreshold slopes are 90.7 and 86.5 mV/decade while the back-body coefficients are 0.66 and 0.26 $V^{1/2}$ for the NMOS and PMOS devices, respectively. The drain–source breakdown voltages of the NMOS and PMOS devices were 7.0 and −20 V, respectively. The threshold voltages of NMOS and PMOS field transistors are 12.9 and −16.7 V, respectively.

Reprinted from *IEEE J. Solid-State Circuits,* vol. 26, no. 3, pp. 203–208, March 1991.

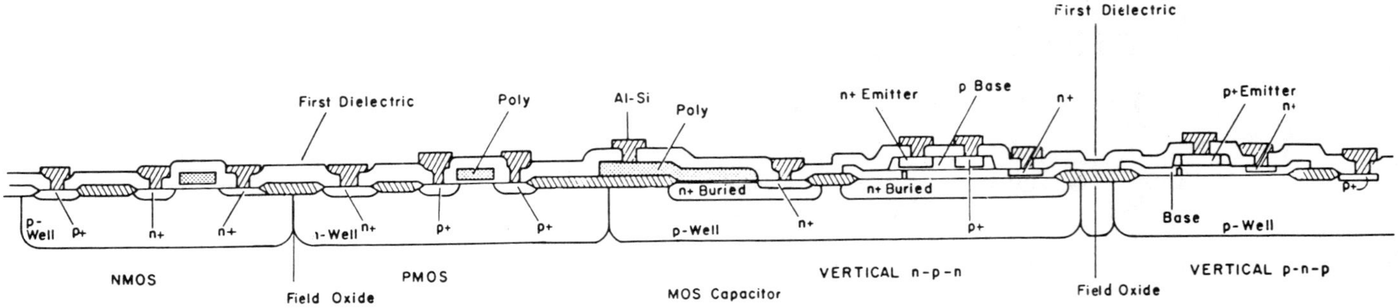

Fig. 1. Cross section of devices available in the BiCMOS process.

TABLE I
SUMMARY OF KEY CMOS DEVICE PARAMETERS IN THE BiCMOS PROCESS

Parameter	NMOS	PMOS
$L_{(DRAWN)}$	2 μm	2 μm
t_{OX}	23 nm	23 nm
V_{T0}	0.70 V	−0.75 V
Substrate Slope	90.7 mV/decade	86.5 mV/decade
Back-Body Coefficient	0.66 $V^{1/2}$	0.26 $V^{1/2}$
V_{DS} Breakdown Voltage	7.0 V	−20 V
V_T (Field)	12.9 V	−16.7 V

TABLE II
SUMMARY OF KEY BIPOLAR DEVICE PARAMETERS IN THE BiCMOS PROCESS

Parameter	n-p-n	p-n-p
Emitter Area (Min)	$(4.5\ \mu m)^2$	$(9.5\ \mu m)^2$
h_{FE} (Max)	70	70
Early Voltage	44 V	125 V
R_C	165 Ω	7.5 kΩ
R_E	11.5 Ω	11.2 Ω
R_B	250 Ω (est.)	n.a.
BV_{CBO}	33 V	> 30 V
LV_{CEO}	9.2 V	> 30 V
BV_{EBO}	4.0 V	33 V
C_{jeo}	77 fF	33 fF
C_{jco}	55.8 fF	51 fF
C_{jso}	320 fF	288 fF
f_T (Max)	3.0 GHz	200 MHz

Table II shows a summary of the key bipolar device parameters in the BiCMOS process [5]. Although the p-n-p device is not optimized, the f_T and R_C for this device can be improved by optimizing the epitaxial thickness and the p-well collector doping. The minimum-size n-p-n bipolar transistor has an emitter area of 4.5 μm×4.5 μm, a maximum h_{FE} of 70, and an Early voltage of 44 V. The series collector resistance R_C is 165 Ω and the series emitter resistance R_E is 11.5 Ω. The series base resistance R_B is estimated as 250 Ω based on op-amp performance. The collector–base breakdown voltage BV_{CBO} and the sustaining collector–emitter breakdown voltage LV_{CEO} are 33 and 9.2 V, respectively. The emitter–base breakdown voltage BV_{EBO} is 4 V. The maximum f_T is 3.0 GHz for a 4.5-μm×18-μm device with $I_C = 5$ mA and $V_{BC} = -4$ V.

III. Op-Amp Circuit Description

The BiCMOS op-amp performance objectives are derived from the requirements of high-performance switched-capacitor analog circuits. The key design goals are a high unity-gain frequency, high dc gain, and an infinite input resistance. The op amp is a fully differential, two-stage design. The fully differential architecture enables good high-frequency power supply rejection ratio in addition to increased dynamic range.

In Figs. 2, 3, and 4, the substrates for NMOS and PMOS devices are implicitly connected to V_{SS} and V_{DD}, respectively, unless otherwise indicated. The p-well substrates for all n-p-n devices and poly-n$^+$ capacitors are connected to V_{SS}.

A. First- and Second-Stage Design

Fig. 2 shows the basic op-amp circuit with the main analog signal path emphasized in bold. The op amp employs a PMOS differential pair in the first stage composed of devices M_1 and M_1'. The PMOS devices provide infinite input resistance as well as lower threshold voltage hysteresis [7] and lower $1/f$ noise power spectrum density (PSD) compared to NMOS devices. The source and substrate of the first-stage PMOS differential pair devices are connected together. This prevents the body effect from coupling source–substrate noise voltage to drain noise current, and thus, to an equivalent gate–source noise voltage.

Observing the right-half portion of Fig. 2, the input stage is actively loaded with transistor Q_2. The emitter resistance R_N is used to reduce the input-referred thermal noise due to Q_2 by reducing the effective transconductance of Q_2. However, the value of R_N cannot be increased arbitrarily as an excessively large value would lead to saturation of Q_2 for the bias currents of interest. As a result, R_N is chosen to reduce the effective transconductance of Q_2 while satisfying bias constraints.

The second stage is a bipolar design consisting of common-emitter amplifier Q_5 and cascode device Q_6. It can be shown that for a frequency-compensated two-stage amplifier, the nondominant pole frequency is $p_2 \approx -G_{m2}/C_L$ [8], where G_{m2} and C_L are the transconductance and load capacitance, respectively, of the second stage. The high transconductance of bipolar devices is used in the second stage to permit a high bandwidth by pushing the nondominant pole out to a high frequency. In addition, this high transconductance permits a high dc gain. The second stage employs a cascode structure to provide a large output resistance for high dc gain. The input impedance to the second state is relatively low compared to the output impedance of the first stage. As a result, transistor Q_3 is used as an emitter follower interposed between the first and second stages to minimize loading effects. The op amp uses a pole-splitting frequency compensation capacitor C_c and nulling resistor

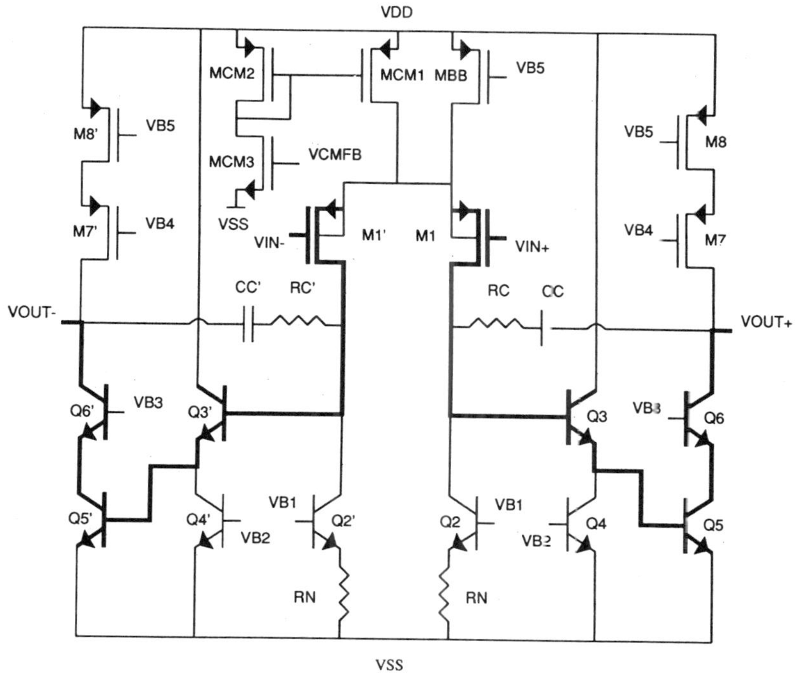

Fig. 2. Main signal path of the BiCMOS op-amp circuit.

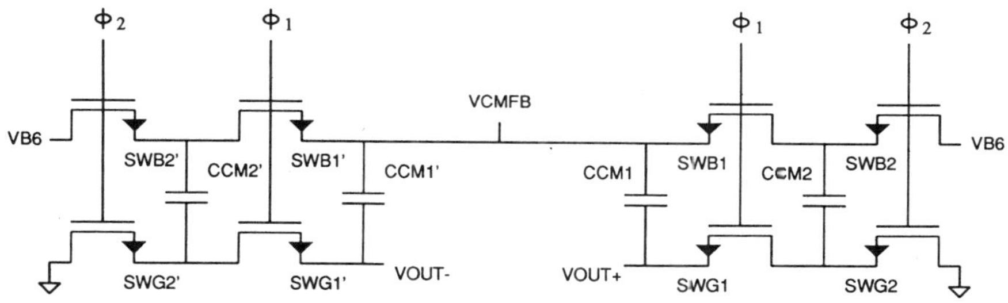

Fig. 3. Switched-capacitor common-mode feedback circuit.

R_c. A low-impedance output buffer is not needed for this op amp for the intended switched-capacitor applications.

The value of R_B for devices Q_5 and Q_5' in the second stage is important because of the pole contributed by R_B and C_π, respectively, for Q_5 and Q_5'. An excessive value of R_B for these devices can lead to peaking in the op-amp frequency response magnitude in two-stage op-amp designs because of a failure in pole splitting [9]. This peaking, of sufficient magnitude, can compromise the stability of the op amp. In the BiCMOS op-amp circuit, the bipolar devices in the second stage use a p^+ base surround around the emitter in order to minimize R_B and thus minimize the peaking.

B. Common-Mode Feedback Scheme

The fully differential op amp requires a common-mode feedback loop to set the output common-mode level. This op amp uses a dynamic common-mode feedback scheme for this purpose [1], [10]. As the op amp is a two-stage design, the common-mode feedback circuit controls the first-stage bias current to set the common-mode levels of both the first- and second-stage outputs [11].

The input of the common-mode amplifier is shown in Fig. 2 as node V_{CMFB} at the gate of transistor M_{CM3}. An increase in this voltage causes each of the op-amp output voltages to decrease. If a disturbance causes the output common-mode level to increase, the common-mode sense circuit will correspondingly increase voltage V_{CMFB}. This causes the op-amp output common-mode level to respond opposite to the disturbance, thus stabilizing the output common-mode level. It can be seen from Fig. 2 that the common-mode signal path is composed of the first and second stages of the differential amplifier in addition to devices M_{CM1}, M_{CM2}, and M_{CM3}. The frequency compensation network serves both the differential-mode and common-mode amplifiers. As a result, the common-mode and differential-mode amplifier dynamics are comparable. The unity-gain frequency of the common-mode amplifier is made slightly lower than that of the differential-mode amplifier. This is to ensure proper phase margin for the common-mode amplifier as the common-mode loop transmission has more high-frequency poles than the differential-mode loop transmission.

The dynamic common-mode output level sense circuit is shown in Fig. 3. Capacitors C_{CM1} and $C_{CM1'}$ are used to derive voltage V_{CMFB} from $V_{\text{OUT}+}$ and $V_{\text{OUT}-}$. Voltage V_{CMFB} represents the output common-mode level. By using capacitors to sense the output common-mode level, the differential- and common-mode dc gains of the op amp are not affected. In addition, the use of a capacitor sense network does not degrade the differential output range. In order to

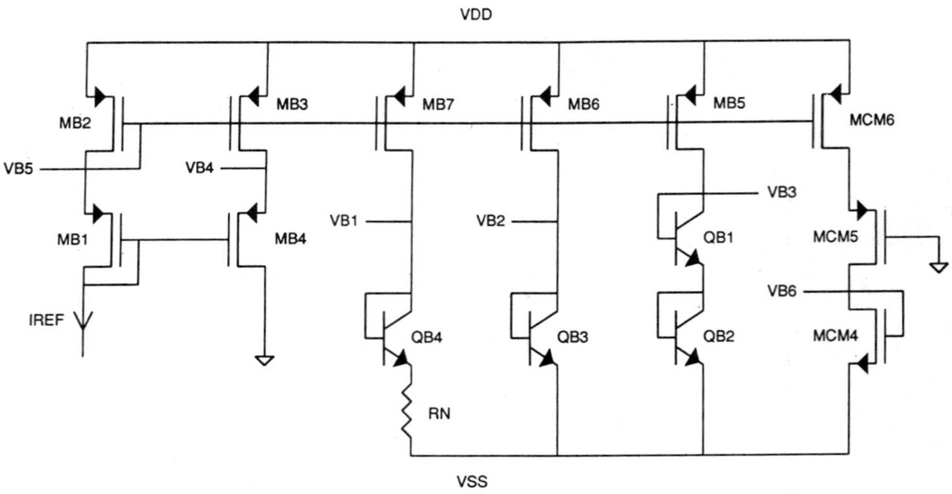

Fig. 4. Bias circuit.

provide a frequency-independent common-mode sense function with the capacitor network, the input to the common-mode amplifier must also be capacitive. As a result, a MOS device is used for this input, namely M_{CM3}. The capacitive sense network is periodically refreshed by capacitors C_{CM2} and $C_{CM2'}$ using a MOS switch network and nonoverlapping clock phases ϕ_1 and ϕ_2. The MOS switch network is composed of NMOS devices $SW_{B1}, SW_{G1}, SW_{B2}, SW_{G2}$ and $SW'_{B1}, SW'_{G1}, SW'_{B2}, SW'_{G2}$. The periodic refresh is necessary for two reasons. First, since V_{CMFB} is a capacitive node, the initial charge at this node is not determined. The refresh provides a means to establish the charge at this node such that the common-mode output voltage is at the desired level. Second, the refresh compensates for leakage currents at node V_{CMFB}. Between refresh cycles, the leakage currents are integrated by the common-mode amplifier and capacitors C_{CM1} and $C_{CM1'}$. Without refresh at sufficient rate, the deviation in the output common-mode level, with respect to a desired level, becomes excessively large between refresh cycles. Thus, the refresh rate must be high enough so that this deviation is at an acceptable level. Bias voltage V_{B6} adjusts the set point for the common-mode feedback loop. This set point is determined such that the output common-mode level is at ground potential, as split power supplies are used.

C. Bias Circuit

Fig. 4 shows the bias circuit for the op amp. Current sources with an "improved cascode" [8] structure are used to provide larger output compliance along with higher effective output resistance for the PMOS active loads in the second stage. As shown in Fig. 2, devices M_7, M_8 and M'_7, M'_8 compose these active loads. The bias reference portion for these devices, and the remainder of the bias circuit, is composed of devices M_{B1}, M_{B2}, M_{B3}, and M_{B4}. Emitter resistance R_N is included between the emitter of Q_{B4} and V_{SS} in order to maintain a 1:1 current mirror ratio between Q_{B4} and Q_2, Q'_2. Devices Q_{B3} and Q_4, Q'_4 compose a current mirror used to bias the emitter followers Q_3, Q'_3. Diode-connected devices Q_{B1} and Q_{B2} are used to derive a bias voltage for cascode devices Q_6, Q'_6. A single external current source is used to bias the op amp.

TABLE III
SUMMARY OF BiCMOS OP-AMP MEASUREMENTS

Parameter	Measured Value
dc gain	100 dB
unity-gain frequency	90 MHz
phase margin	45°
slew rate	150 V/μs
capacitive loading per output	7 pF
differential output range	12 V
power supply	$\pm$5 V
power dissipation	125 mW
active die area	1.0 mm $\times$ 1.2 mm

IV. OP-AMP MEASUREMENTS

Experimental results are summarized in Table III. The measured parameters are as follows: dc gain of 100 dB, unity-gain frequency of 90 MHz, phase margin of 45°, slew rate of 150 V/μs, capacitive loading of 7 pF per output, differential output range of 12 V, $\pm$5-V power supply, and 125-mW power dissipation. A 100-μA external current source was used to bias the op amp. The op amp is unity-gain stable with the above capacitive loading. The dynamic switched-capacitor common-mode feedback was engaged for all op-amp measurements. The associated clock period was 6 μs for ϕ_1 and ϕ_2, respectively. For linear operation, each op-amp output can swing to within 2 V of a $\pm$5-V power supply for zero output common-mode level. Thus, each output can swing 6 V so the differential output range is 12 V. For $\pm$3.5-V power supply, the differential output range is 6 V. Although the differential input-referred thermal noise PSD was not measured, the simulated value was 1.21×10^{-16} V^2/Hz or 11 nV/$\sqrt{\text{Hz}}$. A photograph of the op-amp die is shown in Fig. 5. The active die area is 1.0 mm $\times$ 1.2 mm.

In Fig. 6, the results of the unity-gain frequency measurement are shown. For this measurement, the op amp is operated in an open-loop configuration. The top waveform is the differential input at 500 mV/div and the bottom waveform is the differential output at 500 mV/div. The time scale is 5 ns/div. With an input frequency of 90 MHz, it is seen that the input and output magnitudes are equal. The output waveform is 135° phase shifted with respect to the input

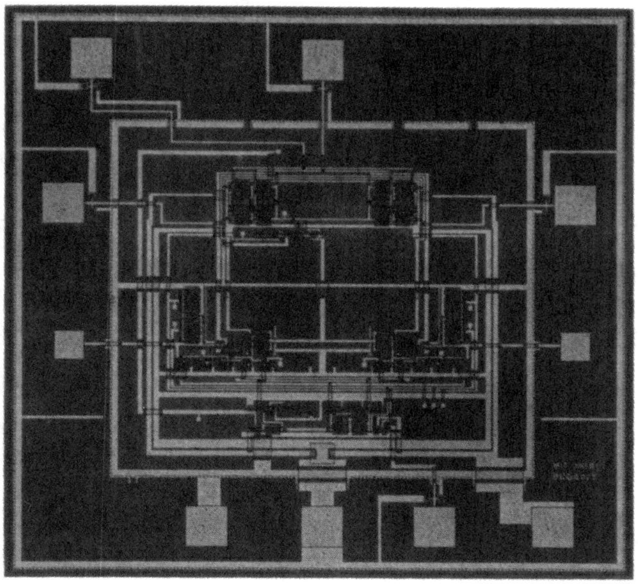

Fig. 5. Die photograph of the BiCMOS op amp.

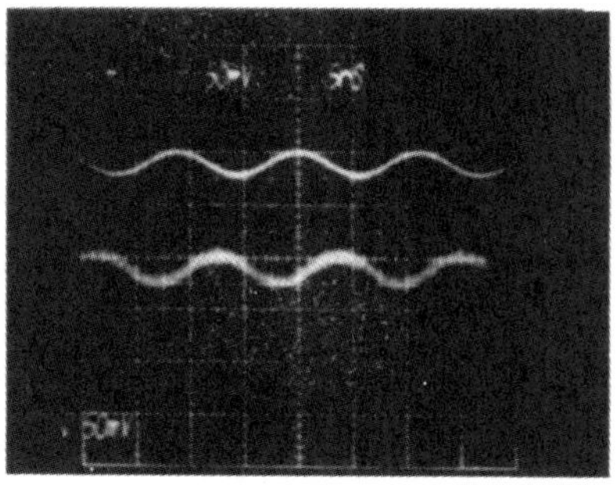

Fig. 6. Unity-gain frequency measurement. Top: differential input, 500 mV/div. Bottom: differential output, 500 mV/div. Time scale: 5 ns/div. Unity-gain frequency is 90 MHz with 45° phase margin.

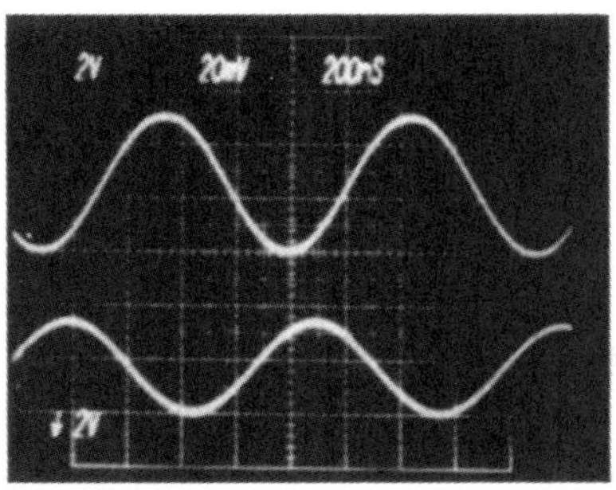

Fig. 7. Gain–bandwidth product measurement. Top: differential input, 20 mV/div. Bottom: differential output, 2 V/div. Time scale: 200 ns/div. Gain–bandwidth product is 110 MHz.

waveform. Thus, the unity-gain frequency is 90 MHz with a 45° phase margin.

In Fig. 7, the op-amp is configured for a closed-loop gain of −100. The top waveform is the differential input at 20 mV/div and the bottom waveform is the differential output at 2 V/div. The time scale is 200 ns/div. With an input frequency of 1.1 MHz, the output magnitude is reduced 3 dB compared to a lower frequency case, for constant input magnitude. In addition, the output waveform is about 45° phase shifted with respect to the input waveform, compared to a lower frequency case. Thus, the −3-dB frequency of the closed loop is 1.1 MHz. This indicates a gain–bandwidth product of 110 MHz for the op amp.

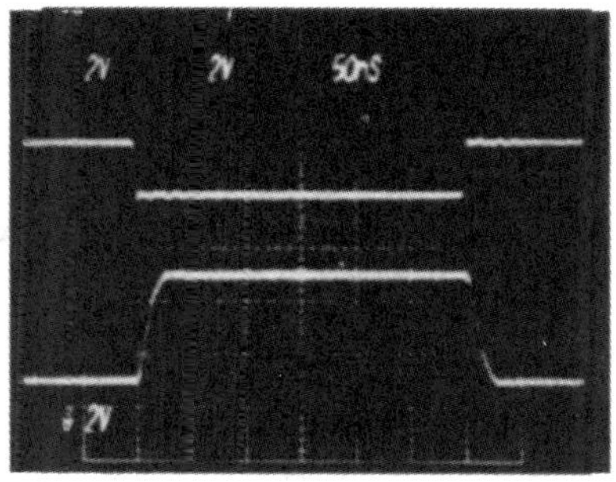

Fig. 8. Slew-rate measurement. Top: differential input, 2 V/div. Bottom: differential output, 2 V/div. Time scale: 50 ns/div. Slew rate is 150 V/μs for positive and negative output changes.

In Fig. 8, the op amp is configured for a closed-loop gain of −2. The top waveform is the differential input at 2 V/div and the bottom waveform is the differential output at 2 V/div. The time scale is 50 ns/div. Upon application of a step at the input, it is seen that the output exhibits slew-rate limiting. The step response shows a slew-rate limit of 150 V/μs for positive and negative output changes. In this closed-loop gain of −2 configuration, the loop transmission is about 1/5 to 1/6 of the op-amp loop transmission because of attenuation from the feedback network, op-amp input capacitance, and parasitic capacitance from the input pins on the chip carrier. As a result of the reduced loop transmission, the closed loop exhibits nearly a single-pole response after the slew-rate limit portion is complete for this configuration.

The op amp was configured in a closed-loop unity-gain connection in order to measure the differential-mode power supply rejection ratio (PSRR). A PSRR of greater than 45 dB was measured for frequencies up to 1 MHz for the positive and negative power supply rails.

Op-amp circuits were operated from ±3.5-V up to ±5.5-V power supply, using the same common-mode feedback clock period and bias current, without significant change in the performance parameters above. At ±5-V power supply operation, it is expected that NMOS device M_{CM3} will be operating near the NMOS drain–source voltage breakdown limit of 7 V. The drain–source voltage of M_{CM3} could be reduced by placing an NMOS, or n-p-n, cascode device in series with the drain of M_{CM3}.

V. Conclusion

A fully differential BiCMOS op amp for use in high-performance switched-capacitor analog circuits has been presented. The op amp is integrated in a 3.0-GHz, 2-μm BiCMOS process. A dc gain of 100 dB, unity-gain frequency of 90 MHz, and slew rate of 150 V/μs have been demonstrated. A dynamic common-mode feedback scheme has been used to set the output common-mode level of the first and second stages.

Acknowledgment

The authors thank the staff of the M.I.T. Integrated Circuits Laboratory for fabrication of the BiCMOS op amp and

Prof. C. G. Sodini at the Department of Electrical Engineering and Computer Science at M.I.T. for useful suggestions.

References

[1] R. Castello and P. R. Gray, "A high-performance micropower switched-capacitor filter," *IEEE J. Solid-State Circuits*, vol. SC-20, no. 6, pp. 1122–1132, Dec. 1985.

[2] B.-S. Song, M. F. Tompset, and K. R. Lakshmikumar, "A 12-bit 1-M sample/s capacitor error-averaging pipelined A/D converter," *IEEE J. Solid-State Circuits*, vol. 23, no. 6, pp. 1324–1333, Dec. 1988.

[3] P. W. Li, M. J. Chin, P. R. Gray, and R. Castello, "A ratio independent algorithmic analog-to-digital conversion technique," *IEEE J. Solid-State Circuits*, vol. SC-19, no. 6, pp. 1138–1143, Dec. 1984.

[4] K. K. O, H.-S. Lee, R. Reif, and W. Frank, "A 2-μm BiCMOS process utilizing selective epitaxy," *IEEE Electron Device Lett.*, vol. 9, no. 11, pp. 567–569, Nov. 1988.

[5] K. K. O, H.-S. Lee, and R. Reif, "A BiCMOS process utilizing selective epitaxy for analog/digital applications," *IEEE Trans. Electron Devices*, vol. 36, no. 7, pp. 1362–1369, July 1989.

[6] K. K. O, "BiCMOS technologies for analog/digital applications utilizing selective silicon epitaxial growth," Ph.D. dissertation, Dept. EECS, Mass. Inst. of Technology, Cambridge, Sept. 1989.

[7] T. L. Tewksbury, H.-S. Lee, and G. A. Miller, "The effects of oxide traps on the large-signal transient response of analog MOS circuits," *IEEE J. Solid-State Circuits*, vol. 24, no. 2, pp. 542–544, Apr. 1989.

[8] P. R. Gray and R. G. Meyer, *Analysis and Design of Analog Integrated Circuits*, 2nd ed. New York: Wiley, 1984.

[9] J. E. Solomon, "The monolithic op amp: A tutorial study," *IEEE J. Solid-State Circuits*, vol. SC-9, no. 6, pp. 314–332, Dec. 1974.

[10] D. Senderowicz, S. F. Dreyer, J. H. Huggins, C. F. Rahim, and C. A. Laber, "A family of differential NMOS analog circuits for a PCM codec filter chip," *IEEE J. Solid-State Circuits*, vol. SC-17, no. 6, pp. 1014–1023, Dec. 1982.

[11] S. D. E. Pomeroy, "High resolution, high speed analog-to-digital conversion techniques," Ph.D. dissertation, Dept. EECS, Mass. Inst. of Technology, Cambridge, Aug. 1989.

Paper 4.12

1.5-V High-Performance SC Filters in BiCMOS Technology

Rinaldo Castello, *Member*, *IEEE*, and Luciano Tomasini

***Abstract*—Line-fed telecommunication circuits like battery-operated circuits must work from very low supplies. Contrary to most battery-operated systems in telecommunication applications, high performance is also required. This paper describes two low-voltage switched-capacitor (SC) filters: one can operate from a minimum supply of 1.5 V and the other from a minimum supply of 2 V (for typical parameter values). Both filters use a fully differential architecture and are fabricated on a standard BiCMOS technology. The lowest supply filter, operated from a 2-V supply, has an SNR of 92 dB and a THD of −70 dB for a 2.4-V_{pp} differential signal. Power consumption and area per pole are 60 μW and 0.18 mm^2, respectively, with a clock frequency of 447 kHz.**

I. Introduction

In the past few years several SC filters operating from very low supplies have been reported in the area of battery-operated systems [1], [3], [6], [7]. In most of these systems the key requirements are low power consumption and small silicon area. Other performance specifications like dynamic range, linearity, PSRR, etc. are often less critical.

In line-fed telecommunication circuits, as for battery operation, the value of the supply voltage can be as low as 2 V. In this application, key objectives are not only power and area minimization but also large dynamic range, good linearity, and good PSRR. Another desirable objective is the utilization of a standard technology, since keeping the cost low is key in the large-volume telecom market. For these reasons few SC filters intended for line-fed high-performance telecommunication systems have been reported to date [6]. Furthermore, these filters require a supply voltage of 3 V or more, which may not always be available.

In this paper two SC filters that achieve a signal-to-noise ratio of about 90 dB are reported. One can operate with a 1.5-V supply and the other with a 2.5-V supply. Both circuits are realized in a standard BiCMOS technology featuring two metal levels of interconnect and a single layer of poly.

Manuscript received November 29, 1990; revised March 6, 1991.
R. Castello is with the Dipartimento di Electtronica, Universitá di Pavia, 27100 Pavia, Italy.
L. Tomasini is with SGS-Thomson Microelectronics, 20041 Agrate Brianza, Italy.
IEEE Log Number 9100039.

The adopted BiCMOS technology, although very streamlined, is more costly than a standard CMOS one. The use of BiCMOS is, however, motivated at the system level. This is due to the need for realizing some critical functions, other than SC filtering, like low noise amplification, large current driving, and others.

The two filters do not use any supply voltage multiplication so as to avoid extra external components. The 1.5-V solution needs a clock voltage multiplier to ensure proper operation under all conditions. This, however, can be realized on chip without any extra external components [3]. Some other measured results are a power per pole of 60 μW, an area per pole of 0.18 mm^2, a THD of −70 dB for a 1.5-V_{pp} signal, and a PSRR of more than 50 dB over the full telephone band. These results are comparable with what is achievable by 5–10-V circuits.

The structure of the paper will be as follows: in Section II, the characteristics of the BiCMOS technology used are described and the motivation for choosing such a technology in terms of the overall system requirements is discussed. In Section III, the system requirements and intended applications for the SC filters are explained. In Section IV, the design of the various operational amplifiers used and their specifications are reported. Finally, in Section V, an extensive set of experimental results is given.

II. High-Performance BiCMOS Technology

The incorporation of both CMOS and bipolar structures on the same chip offers the possibility of exploiting the specific advantages of these technologies. These are: high speed, high driving capability, and high precision for bipolar, and high integration density and low power consumption for CMOS. For mixed technologies to be successful, however, their advantages must surpass the unavoidable drawbacks like technology complexity, extra costs, additional testing problems, and the difficulties of simulation.

As more complex systems are integrated on a chip, the requirements posed to the technology become more stringent. In this situation, the advantages of BiCMOS technology in terms of driving capability, high density, and low power consumption become more and more important. On the other hand, the main drawback of BiCMOS, i.e., process complexity, is becoming less significant. This is due primarily to the increasing complexity of CMOS that

Reprinted from *IEEE J. Solid-State Circuits*, vol. 26, no. 7, pp. 930–936, July 1991.

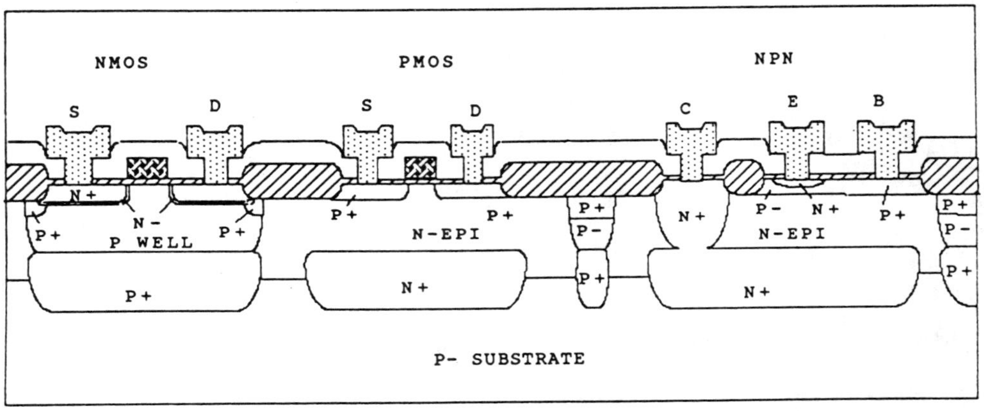

(a)

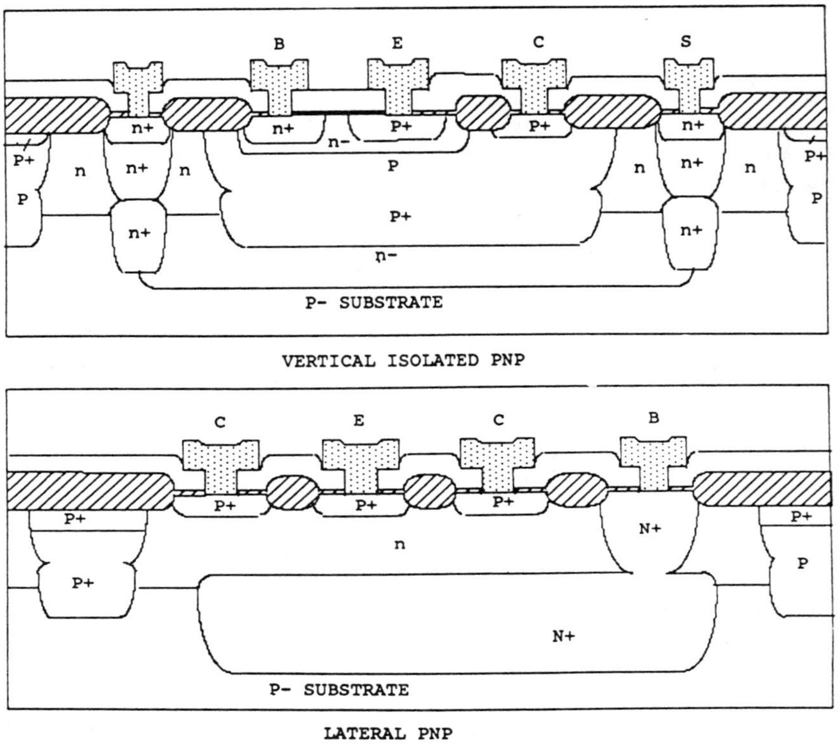

(b)

Fig. 1. (a) Cross section of CMOS and n-p-n. (b) Cross section of vertical and lateral p-n-p.

now demands on the average 14 masking steps. As a consequence a sophisticated BiCMOS technology can be obtained with only four extra masks. The SC filters reported in this paper are part of a complex mixed analog–digital system, i.e., a programmable telephone chip. A telephone chip is a typical example where the extra cost of BiCMOS is outweighed by its advantages. In fact, bipolar devices are the most suitable to accommodate the large dc current (up to 110 mA) and to achieve the high current driving (several milliamperes) required. Bipolar devices are also superior in terms of noise and they are more suitable for very low supply operation. MOS devices, on the other hand, can realize particular functions like precise filtering, programmability, and generally digital functions in a more economical fashion both in terms of chip area, power consumption, and reproducibility.

A cross-sectional view of our BiCMOS technology, called HF2CMOS, is shown in Fig. 1. The n-channel and p-channel CMOS devices and the n-p-n are shown in Fig. 1(a) while the lateral and vertical isolated p-n-p are shown in Fig. 1(b). The HF2CMOS process is a standard 14-mask technology that features two layers of metal interconnect plus one layer of poly, which is also used to realize poly-to substrate capacitors. Four additional masks are available if required: one for poly-to-poly capacitors, one for precision resistors, one for vertically isolated p-n-p's with $f_T = 3$ GHz, and one for EEPROM's.

The main features of the process are summarized in Table I. They are the availability of 2-μm CMOS devices

TABLE I
HF2CMOS TYPICAL CHARACTERISTICS

Device	Characteristics
n-p-n	$f_T = 7$ GHz, $BV_{ceo} = 15$ V
Vertical p-n-p	$f_T = 2$ GHz, $BV_{ceo} = 15$ V
Lateral p-n-p	$f_T = 200$ MHz, $BV_{ceo} = 15$ V
CMOS	$L_{min} = 2$ μm, $V_{th} = 0.8$ V
Rimplanted	Precision $\pm 20\%$
Double Poly	
Double Metal	
EEPROM	

with ± 800-mV thresholds combined with a high-performance bipolar n-p-n ($f_T = 7$ GHz). Only lateral p-n-p's are available without extra steps but with reduced frequency capability. In our application, to keep the technology as simple as possible, only one poly layer is utilized (the double poly is the most costly of the extra features).

III. SC Filter Requirements

The two SC filters reported are intended for different applications, which implies quite different requirements. The first circuit will be used to filter out the high-order harmonics of the signal at the output of a dual-tone multifrequency generator (DTMF) [2].

A biquadratic cell with a bandwidth of 3 kHz and a Q factor of 0.8 is sufficient to guarantee system requirements. In this case the signal amplitude is less than 100 mV_{pp} and no critical noise specifications are required. The most demanding specification is the spurious out-of-band level, which should be below -70 dBm up to 1 MHz. As a consequence, a high clock frequency of 447 kHz is used. This allows a large reduction of the residual output clock using a second-order continuous-time smoothing filter with a nominal cutoff frequency of 25 kHz located after the SC filter.

To further reduce clock feedthrough, a fully differential structure with a continuous-time common-mode feedback (CMFB) circuit is implemented. Finally, since the main supply voltage for this application is guaranteed to be more than 2.8 V, no clock amplitude multiplication [3] is utilized. This further reduces the residual clock energy at the output but increases the on resistance of the MOS switches used in the filter. To improve this situation an output common-mode voltage close to the negative rail (+1 V) is used. This is a viable solution due to the differential structure used and to the small signal level to be processed. A simple CMFB circuit also results.

The second SC filter is used in different portions of a telephone chip to shape the frequency content of the signal. One possible application is the active synthesis of a precise impedance that can be used to perform either line termination or side-tone cancellation [2]. The desired impedance is obtained by placing the SC filter in a feedback loop which includes a precision external resistor. The impedance value as a function of the frequency can be programmed externally or automatically if the system is driven by a microprocessor.

In this application dynamic range maximization is the key requirement since a 85 dB or more signal-to-noise ratio is required by the telephone system. As a consequence, a fully differential topology to double the swing and large unit capacitors (1 pF) to reduce kT/C noise contribution are used. A careful op-amp design to minimize extra noise contributions is also employed.

To be able to satisfy most administration requirements, full spec operation down to 1.5 V for typical conditions (1.8-V worst case) is targeted. Therefore, clock amplitude multiplication becomes mandatory. The feasibility of on-chip clock amplitude multiplication has been demonstrated before [3]. A scheme of the clock amplitude multiplier similar to that reported by Krummenacher *et al.* [3] was designed and extensively simulated. The circuit achieves the required performance; however, due to layout resource limitations, the circuit was not implemented in the first version of the filter. In the actual realization a clock signal higher than the supply was provided externally.

To achieve maximum swing and good linearity for a 1.5-V supply, a dynamic CMFB circuit is used. A high clock rate of 447 kHz is chosen in order to be able to perform antialiasing with a simple passive filter and to guarantee stability under all conditions when the filter is placed in a feedback loop.

A conceptual schematic of the entire filter chain for both the DTMF and the impedance synthesis application is shown in Fig. 2. In both cases, single-ended-to-differential conversion is performed on chip by sampling the signal with opposite signs and during opposite phases on two identical capacitors. This operation is equivalent (to the first order) to a doubling of the sampling frequency, which simplifies antialiasing. After the SC filter, differential-to-single-ended conversion and output smoothing are performed as shown in the figure. The smoothing filter is made up of a classical Rouch filter that implements a second-order low-pass transfer function with a cutoff frequency of around 25 kHz. The nominal values of the capacitors and resistors used in the filter are: $C_1 = 32$ pF, $C_2 = C_3 = 9$ pF, $R_1 = R_3 = 440$ kΩ, $R_2 = R_4 = 490$ kΩ, and $R_5 = R_6 = 110$ kΩ.

IV. Amplifier Design

Several different operational amplifiers are used in the two filtering chains depending on the requirements such as the current driving and the output swing. As an example, in the impedance synthesis filter, maximum output swing is required to achieve the maximum signal-to-noise ratio while in the DTMF filter an output swing of only a few hundred millivolts is required.

The basic schematic of the core amplifier used in both filters when a low current driving capability is required is shown in Fig. 3. In this figure the dynamic CMFB circuit used in the filter that must operate down to 1.5-V supply is also shown.

The two-stage design uses a modified Miller compensation scheme. The input stage is folded toward the positive

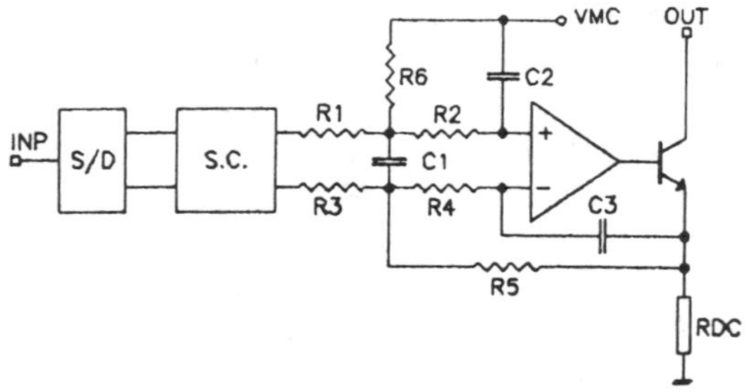

Fig. 2. Filter chain.

supply via the bipolar cascode device $Q1$ ($Q2$). The active load is realized with the simple current source $M3$ ($M4$). The second stage provides most of the amplification and it is made up of the vertical p-n-p $Q3$ ($Q4$) loaded by the simple current source $M5$ ($M6$). The minimum operating voltage of the amplifier is limited by the input stage and is given by one p-type threshold voltage plus two overdrives, i.e., 1.5 V in the typical case. At this low supply and for worst-case conditions, the amplifier still operators properly but with a severely degraded performance.

Proper operation with minimum supply can be achieved by referring the input common-mode voltage to the negative supply and making it track a p-type threshold voltage plus two overdrives. This can be done because in an SC filter it is possible to have different input and output common-mode voltages. Moreover, in a differential configuration this choice does not degrade the PSRR.

The use of an n-p-n cascode at the drain of the input devices provides a low-impedance (nominally 12 kΩ) wide-band node where part of the Miller compensation capacitors can be returned. This modified Miller compensation scheme increases the high-frequency conductance seen at the output and therefore moves the second pole up in frequency. This is quite useful in our case where, due to the low noise required and the use of a dynamic CMFB circuit, a capacitive load of several picofarads appears at the output. In this situation, to further increase the achievable bandwidth with a small power consumption, a two-stage architecture with a bipolar device at the output is used. This is due to the higher g_m/I ratio of a bipolar device compared with that of an MOS.

Combining these two techniques, and using C_5 (C_7) = 2.2 pF and C_6 (C_8) = 1 pF, a bandwidth of 2.5 MHz with a total supply current of 30 μA is obtained for a load capacitance of 7 pF. Furthermore, the use of a two-stage structure gives a nominal output swing equal to 3.2 V_{pp} when a 2-V supply is used.

For the impedance synthesis filter where the maximum output swing is required, the dynamic CMFB shown in Fig. 3 is used. A dynamic CMFB circuit is chosen for its better linearity and larger output swing. It is realized with two fixed capacitors ($C1$ and $C2$) of 1 pF and two switched ones ($C3$ and $C4$) of 0.5 pF. To have the necessary negative feedback in the common-mode path, a wide-band signal inversion is required, which is realized by $M7$ and $M8$.

When a lower output swing is required like in the DTMF filter, the same core amplifier with a continuous-time CMFB circuit is used as shown in Fig. 4. Because of the small output swing, an emitter degenerated differential pair can be used as the CMFB circuit. A degeneration resistance R of 25 kΩ and a bias current I of 4 μA are used in each differential pair. Notice that the nonlinearities are, to the first order, cancelled because there are two differential pairs operating in opposite phases and the two outputs are added. Since only one current output is available from the CMFB circuit, the feedback point is moved to the input stage. This slightly increases the excess of phase in the CMFB loop.

When a resistive load appears at the output of the amplifier, such as when driving the differential-to-single-ended converter, the scheme of Fig. 5 is used. In the figure the case of a dynamic CMFB circuit is shown. The same input configuration and compensation structure as the core amplifier is used for the reasons explained previously. On the other hand, the output stage is made up of an actively loaded n-p-n biased at a current level of 10 μA and of an emitter follower running at 2 μA so as to achieve higher gain and higher current driving.

One more type of amplifier is used in the differential-to-single-ended converter to drive the line. This circuit is shown in Fig. 6. Differential-to-single-ended conversions are obtained with a bipolar mirror in the input stage. This is done to allow working with a 1.5-V supply because of the unacceptably large PMOS device voltage threshold in the worst case. To reduce the output noise, resistor degeneration is used in series with the bipolar mirror. p-n-p's are used at the input to satisfy the required input common-mode range. Notice that the base current noise of the p-n-p's can be a dominant noise contribution if excessively large resistors are used in the Rouch filter.

The third stage is used to buffer the large current load from the amplifying stage. To insure stability a nested Miller [4] compensation is implemented. The noninverting stage $M3$, $M5$ is inserted between the first and the second stage to have a negative feedback on the compensation loop. The circuit drives the base of a power n-p-n transistor that can carry up to 100 mA. A bandwidth of 2 MHz for a load of 10 pF with a power consumption of 75 μW was simulated.

V. Experimental Results

A microphotograph of the lower voltage filter chain including the smoothing filter is shown in Fig. 7. The area of the biquadratic SC cell is 0.36 mm^2. All experimental results given below refer to this structure with a 2-V supply unless otherwise stated.

The measured filter response is shown in Fig. 8 for a clock of 447 kHz. Measurements and simulations agree within 0.2 dB of error in the passband and 0.5 dB in the stopband. The PSR for the negative supply is shown in Fig. 9 from 100 Hz to 500 kHz. The PSR stays below −50 dB up to 10 kHz. Notice that the PSR was measured at

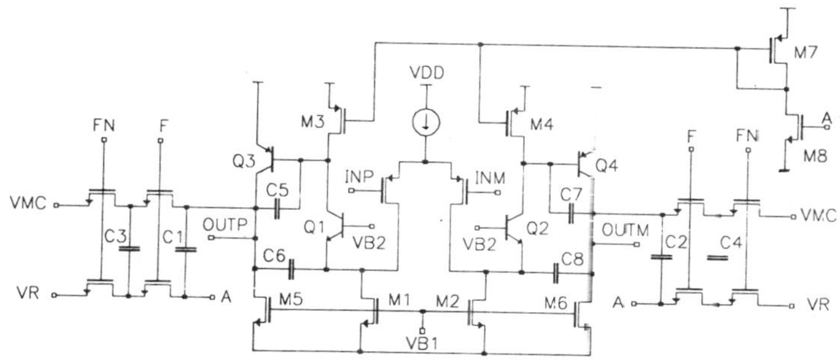

Fig. 3. Core amplifier with dynamic CMFB circuit.

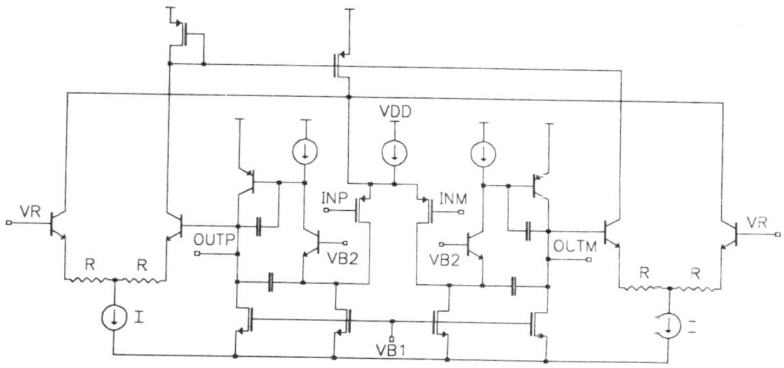

Fig. 4. Core amplifier with continuous-time CMFB circuit.

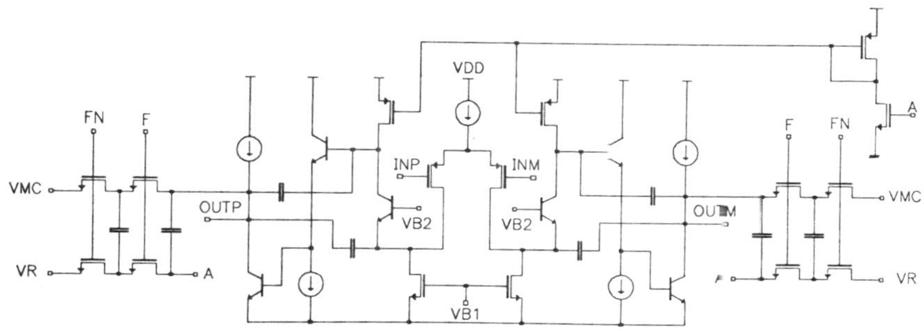

Fig. 5. Buffer amplifier with dynamic CMFB circuit.

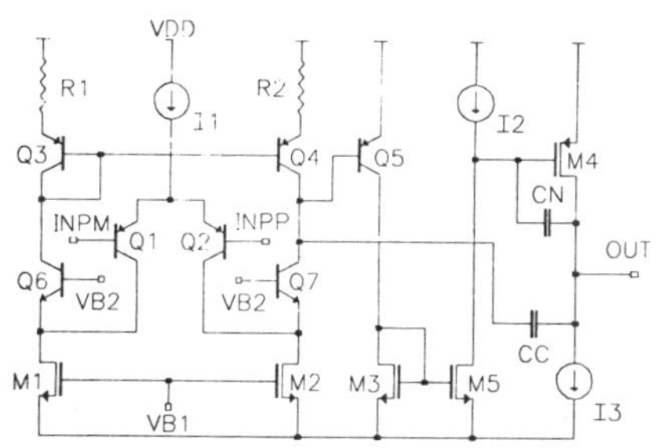

Fig. 6. Single-ended amplifier.

the single-ended output. Better results are expected at the differential output of the SC filter.

A plot of the filter frequency response for supply voltages of 1.5 and 2 V, in the linear frequency range from 0 to 4 kHz, is shown in Fig. 10. Minimization of the dependency of the frequency response on the supply voltage is a key objective for this design. This is because in line-fed systems a large variation of the supply voltage is generally encountered depending on the line length. For a 30% variation of the supply voltage in worst-case conditions (minimum supply voltage) a change of 0.1 dB occurs in the passband and approximately 0.2 dB in the stopband. This is adequate for our objectives, which call for 10% precision in the value of the synthesized impedance.

A plot of THD versus differential amplitude for a 1-kHz signal is shown in Fig. 11. The THD for a 2-V_{pp} differential signal at the SC filter output is −70 dB. The SC filter output has a THD better than −40 dB up to a 2.8-V_{pp} differential signal.

A summary of the achieved filter performance is shown in Table II. The noise density at the single-ended output of the chain is 300 nV/√Hz compared with a density of 270 nV/√Hz obtained with the simulation program SWAP for the SC part and with SPICE for the continuous-time part (smoothing and operational amplifier). The output noise is made up of the contribution of the smoothing

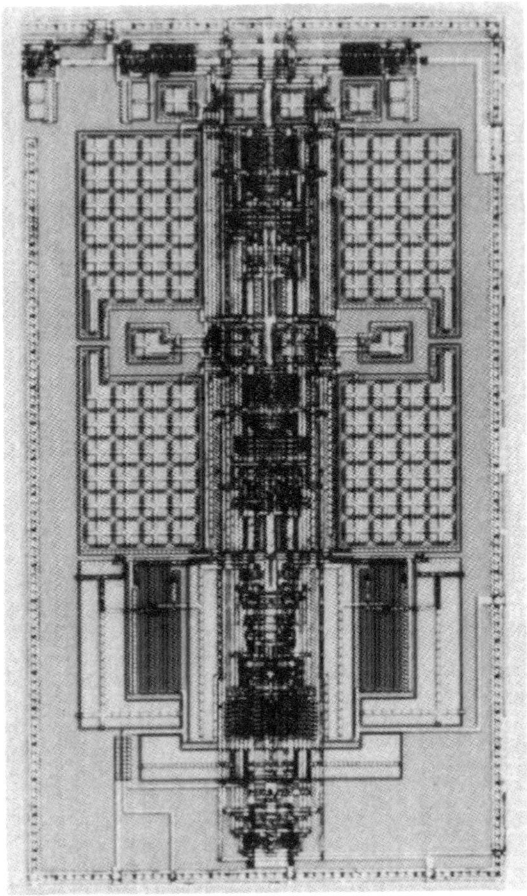

Fig. 7. Chip microphotograph.

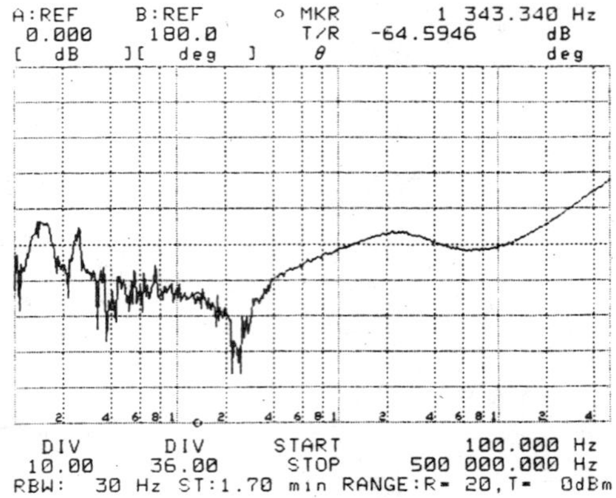

Fig. 9. PSR versus frequency.

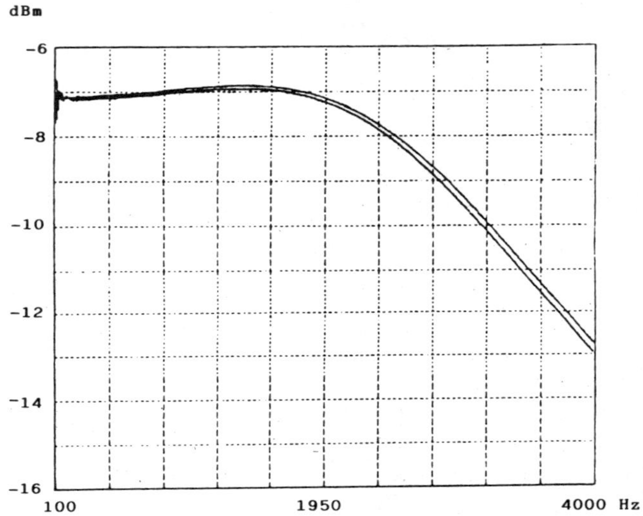

Fig. 10. Filter response at 1.5- and 2-V supply.

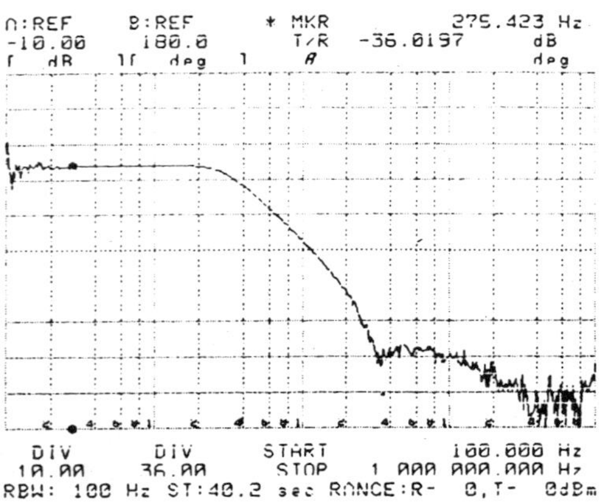

Fig. 8. Filter response.

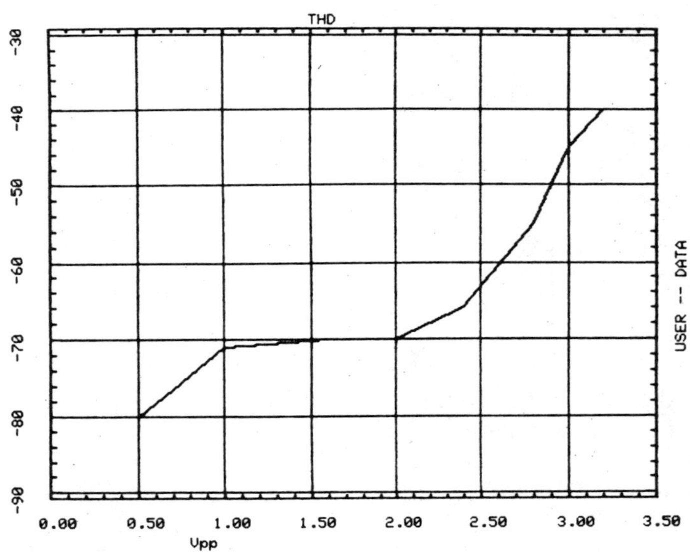

Fig. 11. THD versus amplitude.

filter with a gain of one plus that of the SC filter attenuated by 4 due to the attenuation of the differential-to-single-ended converter. The noise at the SC filter output is computed starting from the output noise measurement using simulation values for the contribution of the smoothing filter. This gives a worst-case estimation of 450 nV/$\sqrt{\text{Hz}}$. Integrating this noise density from 300 Hz to 3.4 kHz gives a total noise at the SC filter output of 25 μV_{rms}. Combining this result with the output swing at 1% distortion gives a dynamic range of approximately 92 dB.

TABLE II
FILTER ELECTRICAL PARAMETERS

Parameter	2-V Filter	2.8-V Filter
Current consumption	60 μA	70 μA
Minimum supply	1.5 V	2.4 V
Noise 300–3400 Hz	25 μV	40 μV
PSR 1 kHz	< 55 dB	< 50 dB
THD $F = 1$ kHz		
100 mV_{pp}	75 dB	76 dB
2 V_{pp}	71 dB	
2.4 V_{pp}	68 dB	
2.8 V_{pp}	52 dB	
Max. signal with 1% THD	3 V_{pp}	300 mV_{pp}
Residual clock	−70 dB	−75 dB
Dynamic range	92 dB	68 dB
Die area	600 mil^2	500 mil^2

The minimum operating voltage for the chain is 1.4 V. At this low supply a THD of −59 dB is achieved for a 2-V_{pp} differential signal at the SC filter output.

Some data on the second SC filter chain that works down to 2.5 V is also given in Table II. The main objective in this design was to achieve a very low residual clock energy. A −75-dB clock residual energy is obtained due in part to the use of a continuous-time CMFB circuit. On the other hand, a reduced dynamic range is obtained mainly due to the smaller output swing and not because of excess noise. Finally, the larger power consumption (70-μA supply current) is due to the different CMFB circuit used. This circuit, in fact, requires a relatively large current on the emitter degenerated differential pair in order to obtain the appropriate output swing.

VI. CONCLUSIONS

The ability to implement very low-supply-voltage (less than 2 V) high-performance SC filters in a standard BiCMOS technology was demonstrated. Performance comparable with 5–10-V implementations was achieved. The realized filters can be used as building blocks to implement more complex functions, like the active synthesis of a given impedance in line-fed telecom systems. Making use of these filters, a programmable telephone chip with a minimum number of external components is being designed.

ACKNOWLEDGMENT

The authors wish to thank P. Paradiso for the careful layout.

REFERENCES

[1] F. Callias, F. H. Salchli, and D. Girard, "A set of four IC's in CMOS technology for a programmable hearing aid," *IEEE J. Solid-State Circuits*, vol. SC-20, no. 2, pp. 301–312, Apr. 1989.

[2] P. Consiglio and M. Siligoni, "A programmable speech circuit suitable for telephone transducers," *IEEE J. Solid-State Circuits*, vol. SC-17, no. 6, pp. 1149–1157, Dec. 1982.

[3] F. Krummenacher, H. Pinier, and A. Guillaume, "Higher sampling rates in SC circuits by on chip clock-voltage multiplication," in *Proc. ESSCIRC* (Lausanne, Switzerland), Sept. 1983, pp. 123–126.

[4] J. H. Huijsing and D. Linebarger, "Low-voltage operational amplifier with rail-to-rail input and output ranges," *IEEE J. Solid-State Circuits*, vol. SC-20, no. 6, pp. 1144–1150, Dec. 1985.

[5] T. Adachi, A. Ishinawa, A. Barlow, and K. Takasuka, "A 1.4 V switched capacitor filter," in *Proc. CICC* (Boston), May 1990.

[6] R. Becker and J. Mulder, "SIGFRED: A low-power DTMF and signalling frequency detector," in *Proc. ESSCIRC* (Grenoble), Sept. 1990, pp. 5–8.

[7] K. A. I. Haloner, W. M. C. Sansen, and M. Steyaert, "A micropower fourth-order elliptical switch-capacitor low-pass filter," *IEEE J. Solid-State Circuits*, vol. SC-22, no. 2, pp. 164–173, Apr. 1987.

Paper 4.13

Tunable BiCMOS Continuous-Time Filter for High-Frequency Applications

Roberto Alini, Andrea Baschirotto, *Student Member, IEEE,* and Rinaldo Castello, *Senior Member, IEEE*

Abstract—**A new BiCMOS fully differential transconductor based on MOS transistors operated in the linear region is presented. The circuit has an equivalent nondominant pole located above 1.5 GHz. This makes it suitable for high-frequency continuous-time filters. A second-order low-pass filter using the new transconductor realized in a 2-μm BiCMOS technology is reported. The cutoff frequency f_0 of the cell is tunable in the range of 8–32 MHz and the quality factor is 2. The filter THD stays lower than −40 dB for an output signal up to 3.2 V_{p-p} at 5-MHz frequency. The area of the cell is 0.322 mm^2 and the power consumption (with f_0 = 25 MHz) is 30 mW with a single 5-V power supply.**

I. Introduction

DIGITAL signal processing (DSP) techniques are becoming more and more attractive as technology scaling progresses on. However, for some particular applications, as for example in very high-frequency filtering, analog signals processing often still remains advantageous, especially in terms of power consumption.

In the last few years monolithic analog filters in the megahertz range have been reported using both switched-capacitor (SC) [1]–[3] and continuous-time techniques [4]–[11]. Typical applications for these filters are in high-performance constant-density memory disks drives [5], [7], in video application [2], [4], or in demodulation systems [1].

In particular, for disk-drive applications low-pass filters with a passband of up to 50 MHz are required. These filters generally have only a moderate selectivity but they must give a precise phase response and they must also be programmable over a wide range. For this specific application continuous-time techniques are receiving increasing attention [5], [7]. In fact, even if SC filters present some advantages (like easy of programmability and especially the possibility of shifting in frequency the filter response by simply changing the clock frequency), they also have some drawbacks which make them less suitable for these high-frequency applications. In particular SC filters need pre- and post-filters (antialiasing and smoothing) which, due to the small clock-to-base-band frequency ratio, must have a sharp and precise cutoff characteristic not easily realized in integrated circuits. Furthermore, SC filters need a precise multiphase clock signal four times or more higher than the filter bandwidth.

Manuscript received January 17, 1992; revised July 9, 1992.

R. Alini was with the Department of Electronics, University of Pavia, 27100 Pavia, Italy. He is now with SGS-Thomson Microelectronics, 20010 Cornaredo (MI), Italy.

A. Baschirotto and R. Castello are with the Department of Electronics, University of Pavia, 27100 Pavia, Italy.

IEEE Log Number 9204138.

Because of this, continuous-time filters are often preferred for these high-frequency applications, especially when only moderate amplitude precision is required. It should be noticed that SC filters retain some important advantages as mentioned before. As a consequence, when programmability is important or high precision is required, SC filters may still be considered if the above-mentioned drawbacks can be overcome.

The two main techniques for the realization of continuous-time monolithic filters [4]–[13] are the MOSFET-C and the transconductance-C (G_m–C). In both cases the basic building block is the continuous-time integrator. The key performance of a high-frequency integrator is the phase shift at its unity-gain frequency. Deviation from the ideal 90° phase value is due to finite gain effects (resulting in a low-frequency dominant pole and so in a phase lead) and high-frequency effects (resulting in high-frequency non-dominant poles and so in a phase lag). For high-frequency applications the G_m–C approach is the most often used. A G_m–C integrator in fact operates in an open-loop way. This implies that the nondominant poles of the G_m–C integrator coincide with the nondominant poles of the transconductor. On the other hand, the open-loop operation of the G_m–C integrator makes it sensitive to parasitic capacitances. This is the major drawback of the G_m–C approach with respect to the MOSFET-C approach.

In this paper a novel BiCMOS fully differential transconductor is presented which uses MOS devices operating in the triode region to give a linear voltage-to-current conversion. Key characteristics of the proposed transconductor are: large linear range, reasonably low noise, wide bandwidth, easy tunability of the transconductance, low power consumption, and operation from a single 5-V supply.

A second-order cell implemented with the G_m–C approach using the proposed transconductor is reported. The biquadratic filter has a low-pass response with a quality factor Q of 2. The cutoff frequency can be tuned in the range 8–32 MHz. The THD for a 5-MHz input signal is lower than −40 dB up to 3.2-V_{p-p} output signal amplitude. The total integrated output noise is about 240 μV_{rms} (with f_0 = 25 MHz): this results in a dynamic range of 73

Reprinted from *IEEE J. Solid-State Circuits,* vol. 27, no. 12, pp. 1905–1915, Dec. 1992.

dB. This performance satisfies the requirements of the analog front end for a medium-speed disk-drive read-channel systems.

The paper is organized as follows. In Section II the novel BiCMOS transconductor is introduced, its principle of operation is explained, and simulated performances are given. In Section III the design of a second-order low-pass filter is developed. Finally, the measured performance are reported in Section IV with some details regarding the testing setup.

II. The BiCMOS Transconductor Stage

A. Basic Structure

The voltage-to-current converter or transconductor is the basic building block for G_m–C continuous-time filters. In fact, when it is loaded by a capacitor it realizes an integrator and, starting from an integrator, any transfer function can be realized, at least in principle, using active filtering techniques.

Key target specifications for a transconductor with several megahertz unity-gain frequency are: moderate power consumption, good linearity up to a large voltage swing, small excess of noise compared with a passive realization, and small phase errors due to high-frequency nonidealities and finite dc gain. Furthermore, in some cases programmability of the pole frequency may be required.

A simplified schematic of the proposed BiCMOS fully differential transconductor is shown in Fig. 1. The two input devices $M1$ and $M2$ are biased in the linear region of operation and determine the transconductance of the stage [8]. The value of the battery V_D determines the drain–source voltage of $M1$ and $M2$ in quiescent conditions (typically from 100 mV to several hundred millivolts). Since $M1$ and $M2$ operate in the linear region, the relationship between voltage and current is

$$I_0 = \mu C_{OX} \frac{W}{L} V_{DS} \left(V_{GS} - V_T - \frac{V_{DS}}{2} \right) \tag{1}$$

where I_0 is the bias current in $M1$ and $M2$, μ is the carrier mobility, C_{OX} is the gate oxide capacitance per unit area, and V_T is the threshold voltage. The transconductance g_m of the stage is given by

$$g_m = \frac{\partial I_{out}}{\partial V_{in}} = \mu C_{OX} \frac{W}{L} V_{DS} \approx \frac{I_0}{(V_{GS} - V_T)_0} \tag{2}$$

where V_{in} is the differential input voltage and I_{out} is the differential output current defined as $I_{out} = I_1 - I_2$. The last approximation in (2) is valid for $(V_{GS} - V_T) >> V_{DS}/2$.

Referring to (2), if μ and V_{DS} are constant over the entire operating range, there is a perfectly linear relation between I_{out} and V_{in}.

However, several causes for nonlinear behavior are present in the circuit and will be analyzed next. First, as the current through bipolar device $Q1$ ($Q2$) changes, the voltage at node A (B) changes as well. This produces a

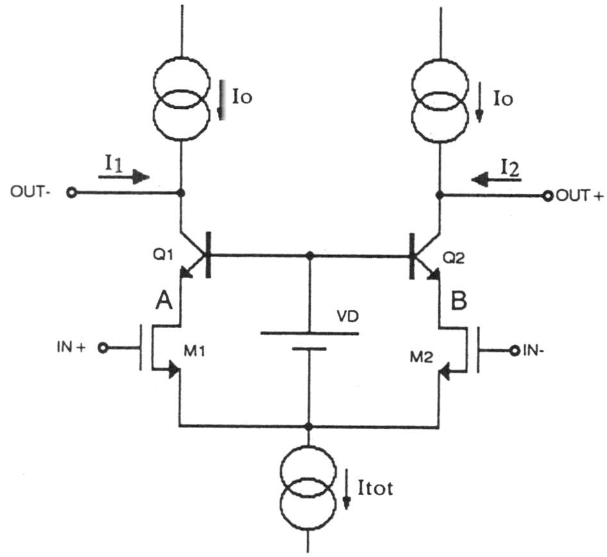

Fig. 1. Basic BiCMOS transconductor scheme.

change in the V_{DS} of the input devices, which in turn causes a nonlinear relationship between I_{out} and V_{in}. For this reason, to minimize the impedance at node A (B) while maintaining high speed of operation, bipolar transistors $Q1$ ($Q2$) are used as cascode devices. Using this solution the V_{DS} variation is reduced to the point of having a marginal effect on the overall transconductor linearity, provided that a sufficient V_{DS} for $M1$ and $M2$ is maintained.

A second cause of nonlinearity is the dependency of the carrier mobility μ of the input devices on their $V_{GS} - V_T$ as given by

$$\mu = \frac{\mu_0}{1 + \theta(V_{GS} - V_T)} \tag{3}$$

where θ ($\approx 0.1\ \mathrm{V}^{-1}$) is the parameter that accounts for the dependence of the mobility on the electrical field in the channel.

Considering only the effect of the mobility variation on the linearity of the circuit, the g_m can be expressed as follows:

$$g_m = \mu_0 C_{OX} \frac{W}{L} V_{DS} \frac{\left(1 + \theta \frac{V_{DS}}{2}(1 + \Delta)\right)}{(1 + \theta(V_{GS} - V_T)_0)^2} \cdot \left(1 - \frac{3}{4}\left(\frac{\theta V_{in}}{1 + \theta(V_{GS} - V_T)_0}\right)^2\right) \tag{4}$$

where Δ is the linearization coefficient of the body effect whose value is about 0.1.

In deriving (4) it has been assumed that $(\theta V_{in})^2 << 1$. From (4), the maximum differential input signal amplitude (peak to peak) V_{inMAX} for which the percentage variation of g_m remains always smaller than ϵ is given by

$$V_{\text{inMAX, pp}} = \frac{0.4}{\theta} (1 + \theta(V_{GS} - V_T)_0) \sqrt{\frac{\epsilon}{3}} \approx \frac{0.4}{\theta} \sqrt{\frac{\epsilon}{3}}. \tag{5}$$

From (5) a 1% g_m variation can be obtained up to a 2.3-$V_{\text{p-p}}$ differential input signal amplitude assuming $\theta = 0.1$. This is a very large linear range. Equations (4) and (5), however, are valid provided that the input MOS transistors remain in the triode region of operation, e.g., for $(V_{GS} - V_T) > V_{DS}$. This corresponds to an upper limit on the differential input signal amplitude given by

$$V_{\text{inMAX, pp}} = 4[(V_{GS} - V_T)_0 - V_{DS}]_{M1, M2}. \tag{6}$$

This value is often smaller than that given by (5).

Simulated performances of the basic transconductor are shown in Fig. 2(a) and (b). They refer to a circuit with the following characteristics: $V_{DS} = 0.25$ V, $V_{GS} - V_T = 0.9$ V, and $(W/L)_{M1, M2} = 90/2$. Fig. 2(a) shows the differential transconductance $g_m = \partial I_{\text{out}}/\partial V_{\text{in}}$ with the output nodes connected to ac ground. The g_m variation is less than $\pm 0.5\%$ for an input differential signal up to 2.3 $V_{\text{p-p}}$. For a higher input signal amplitude, one input transistor enters the saturation region and the linearity is degradated. Fig. 2(b) shows the total harmonic distortion of I_{out} as a function of the input signal amplitude for a 5-MHz sine wave. THD is lower than 55 dB up to 2.5-$V_{\text{p-p}}$ input signal amplitude.

No measurements of the transconductor as a stand-alone circuit were performed but they were inferred through the filter measurement. This is because the nonideality of the transconductor are more easily and accurately obtained from the complete filter measurement.

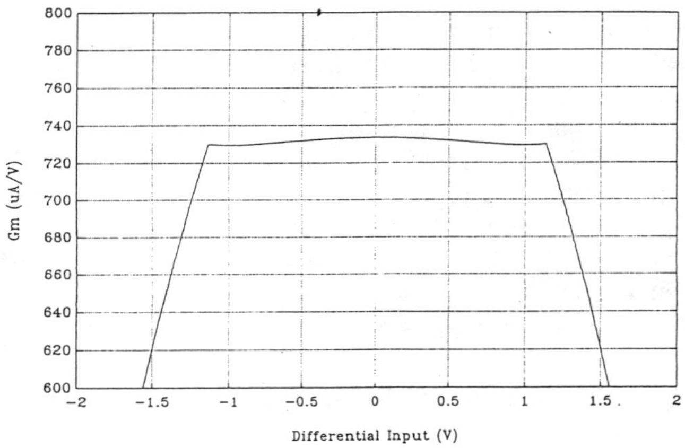

(a)

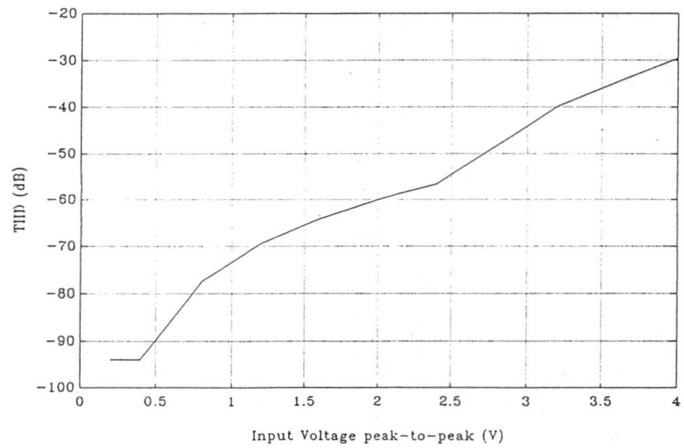

(b)

Fig. 2. (a) Transconductance value versus input differential amplitude (input linearity). (b) Differential output current THD versus peak-to-peak input voltage (at 5 MHz).

B. Complete Structure

A complete scheme of the proposed V–I converter is shown in Fig. 3. It includes the common-mode feedback (CMFB) circuit ($M3$ and $M4$) and the active loads realized with degenerated p-n-p current sources ($Q3$, $Q4$, $R3$, $R4$). Furthermore, the battery V_D of Fig. 1 is replaced by the diode-connected transistor Q_D and the resistance R_D. The current flowing through R_D is used to control V_{DS} of $M1$, $M2$ and through this to tune the g_m of the stage. The circuit has been designed with particular attention to minimize power consumption.

The main requirements of the CMFB circuit are speed of operation and linearity. The achieved CMFB bandwidth is larger than the integrator unity-gain bandwidth. This is because although the g_m of $M3$ and $M4$ is slightly smaller than that of the input devices, the common-mode output capacitance is only a fraction of the differential-mode output capacitance. In fact the common-mode circuit only sees the capacitances toward ground. Good linearity is achieved using large value for the $V_{GS} - V_T$ of $M3$ and $M4$, which are operating in the triode region. The

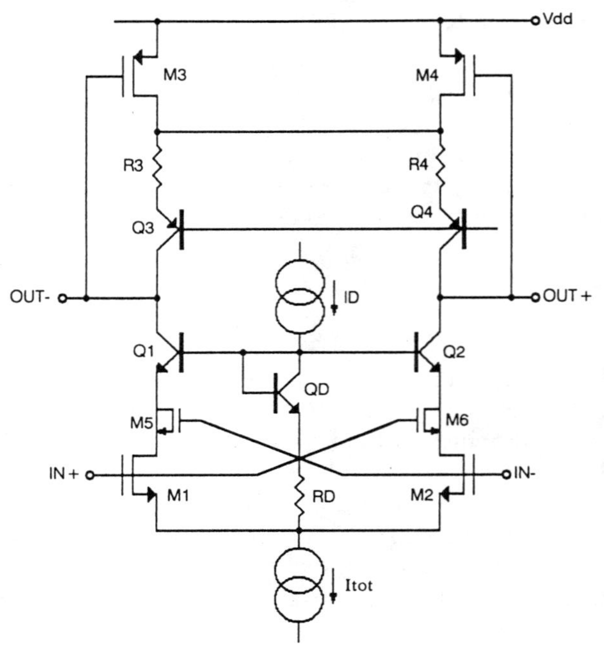

Fig. 3. Complete BiCMOS transconductor scheme.

adopted solution, however, also presents some drawbacks. In particular, substantial extra parasitic capacitance is added to the output node of the transconductor (due to the large size of $M3$ and $M4$). Furthermore, the maximum allowed output swing is reduced due to the voltage drop (about 300 mV) between the drain and the source of $M3$ and $M4$.

When several transconductors are interconnected to build filters, the input and the output nodes must have the same linear swing and the bias point of the input and the output nodes must be the same. Both of these requirements can be met by the new transconductor without any level shifter [7] or any folded structure [4]. This is possible because $M1$ and $M2$ are operated in the triode region with a V_{GS} that is at least 1 V larger than V_{DS}. The use of both level shifters and folded structures produces extra phase shift in the transconductor, which is very detrimental at high frequency. Furthermore, for a given required g_m and linear swing, both level shifters and folded structures require extra supply current not used to perform the basic voltage-to-current conversion, which increases the power consumption.

The frequency response of the transconductor is now considered. Assuming that the transconductor is loaded by a capacitor to form an integrator, the phase shift at the unity-gain frequency presents an error with respect to the ideal 90° due to the following effects.

The first effect is the finite voltage gain A_V of the stage, which results in a pole at f_0/A_V and causes a phase lead. For this reason the gain must be as high as possible. The gain of the circuit A_V is given approximately by

$$A_V = g_{m\,M1} \times \{[(r_0 \times g_m)_{Q1} \times r_{ds\,M1}]/[(r_0 \times g_m)_{Q3} \times R3]\}. \tag{7}$$

Using typical values for the transistor parameters and a voltage drop of 400 mV across $R3$ ($R4$), a gain of about 250 is obtained. Notice that the use of bipolar devices is also helpful in achieving sufficient gain.

The second effect is the presence of high-frequency singularities in the signal path, which result in a phase lag. The two main effects are the finite transit time in the input devices and the pole associated with the common-base device $Q1$ ($Q2$): they give rise to an equivalent nondominant pole located well above 1 GHz. In addition, the feedforward path around $M1$ ($M2$) through the gate-to-drain parasitic capacitance results in a right-hand side zero (at about 1.5 GHz), which also produces phase lag. However, by connecting two dummy devices $M5$ and $M6$ of area equal to 1/3 of that of $M1$ ($M2$) as shown in Fig. 3, a compensation of the feedforward signal occurs. Notice that if excess compensation is implemented, the right-hand side zero is moved to the left-hand side and this can partially compensate the effect of the high-frequency poles.

In this design great effort was made to move the nondominant singularities as far as possible away from the unity-gain frequency. No attempt was made to compensate the phase lag (due to the nondominant poles) with the phase lead (due to the finite gain) because this cannot give stable performance over technology and temperature variations. Furthermore, no Q-tuning technique is used in order to simplify the implementation and to save power.

The input-referred noise can be expressed in terms of the noise excess factor (NEF), which is the ratio between the actual noise and the noise of an ideal realization. Using this approach the input noise of a transconductor of value g_m is written as

$$V_n^2 = \text{NEF}\,\frac{2}{3}\frac{4kT}{g_m}. \tag{8}$$

For the ideal circuit of Fig. 1 the dominant noise contribution is due to the input devices. Since $M1$ and $M2$ are in the linear region, their input-referred noise can be written as follows [15]:

$$\begin{aligned} V_n^2 &= 2\,\frac{2}{3}\frac{4kT}{g_{m\,M1}}\left(1 + \frac{1}{(g_m r_{ds})_{M1}}\right) \\ &= 2\,\frac{2}{3}\frac{4kT}{g_{m\,M1}}\left(1 + \frac{V_{GS} - V_T - V_{DS}}{V_{DS}}\right). \end{aligned} \tag{9}$$

From this follows that

$$\begin{aligned} \text{NEF} &= 2\left(1 + \frac{V_{GS} - V_T - V_{DS}}{V_{DS}}\right) \\ &= 2\left(\frac{V_{GS} - V_T}{V_{DS}}\right). \end{aligned} \tag{10}$$

Using the nominal value of $V_{GS} - V_T$ of 900 mV, NEF varies from 4 to 16 as V_{DS} is adjusted from 450 to 110 mV to compensate for all possible parameter and temperature variations. This noise contribution is much less than that of a Gilbert-cell-based transconductor, when normalized to the voltage swing [7]. Compared with other MOS implementations [4]–[6], the new transconductor presents similar noise for the same voltage swing.

Including the noise due to the active load, the total input-referred noise for the new transconductor becomes

$$\begin{aligned} V_n^2 &= \left\{2\left[\frac{8}{3}\frac{kT}{g_{m\,M1}}\left(1 + \frac{1}{(g_m r_{ds})_{M1}}\right)\right]\right\}_{\text{input}} + \left\{2\left[\frac{4kT}{g_{m\,M1}}\frac{V_{GS} - V_T}{\Delta V_R}\right]\right\}_{\text{load}} \\ &= 2\left[\frac{8}{3}\frac{kT}{g_{m\,M1}}\left(\frac{V_{GS} - V_T}{V_{DS}} + \frac{3}{2}\,\frac{V_{GS} - V_T - \frac{V_{DS}}{2}}{\Delta V_R}\right)\right] \end{aligned} \tag{11}$$

where ΔV_R is the dc voltage drop across $R3$ and $R4$ and it is assumed to be much larger than kT/q. From (11) the load contribution can be reduced by increasing ΔV_R. A relatively small value of ΔV_R is, however, necessary to guarantee a sufficient large output swing. Assuming a worst-case supply voltage of 4.5 V, a typical ΔV_R of 400

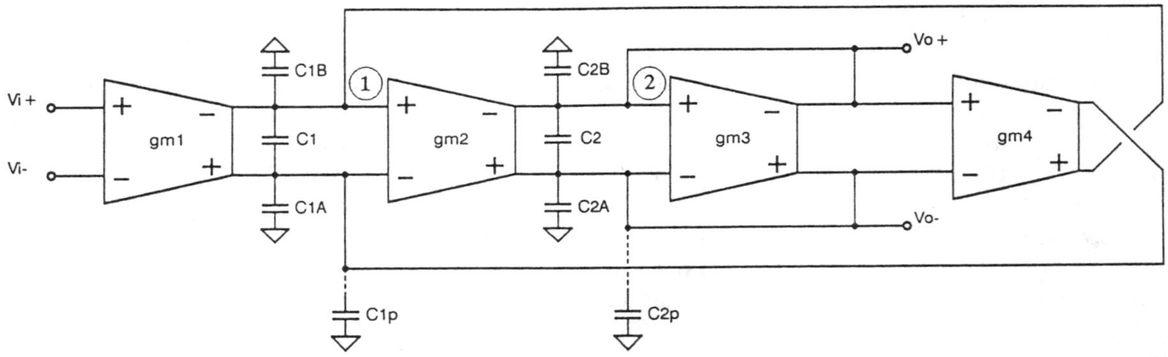

Fig. 4. Biquadratic filter scheme.

mV is chosen. In this case for $V_{GS} - V_T = 900$ mV and V_{DS} tuned in the range of 110–450 mV, an NEF in the range of 9–22 is obtained.

The offset of the stage is that of the input pair $M1$ and $M2$ plus that of $Q1$ and $Q2$ amplified by the ratio between the $V_{GS} - V_T$ and the V_{DS} of $M1$ ($M2$) plus that of $Q3$ and $Q4$ amplified by the ratio between the $V_{GS} - V_T$ of $M1$ ($M2$) and the voltage drop across $R3$ ($R4$). Since bipolar devices exhibit much less offset than MOS devices, the offset is dominated by $M1$ and $M2$.

The required area for a given g_m is small since the structure is very simple and $M1$ and $M2$ operated with large $V_{GS} - V_T$ and therefore they have smaller W/L ratio.

An important aspect in the design of continuous-time filters is the presence of parasitic capacitances at the input node (C_{PIN}) and at the output node (C_{POUT}) of the transconductor. C_{PIN} is due to the gate (oxide) capacitance of the input MOS $M1$ and $M2$. C_{POUT} is due to the junction capacitances of the bipolar transistor (C_{BC} and C_{CS}) and to the oxide capacitance of the MOS transistor in the CMFB stage.

III. Filter Design

The proposed transconductor has been used in the design of a second-order (biquadratic cell) low-pass filter. The quality factor of this filter is 2 and the cutoff frequency f_0 can be tuned externally in the range 8–32 MHz. This are typical specifications for a filter used as equalizer in disk-drive applications. The schematic of the filter is shown in Fig. 4. Table I gives the device sizes for the four transconductors used in the filter. The filter has the following transfer function [16]:

$$H(s) = \frac{\dfrac{g_{m1} g_{m2}}{C1\ C2}}{s^2 + s\dfrac{g_{m3}}{C2} + \dfrac{g_{m2} g_{m4}}{C1\ C2}}. \qquad (12)$$

The relative transconductance values are $g_{m1} = g_{m2} = g_{m4} = 2g_{m3}$. An integer value in the transconductance ratio is chosen to improve matching in order to get a precise phase response [4].

The total capacitance at the high-impedance nodes 1 and 2 of Fig. 4 is the sum of the nominal capacitor attached to the node and of the parasitic capacitance. A key design problem is to minimize the effects of the parasitic capacitances on the response (both phase and amplitude) of the filter. One approach is to have negligible parasitic capacitance with respect to the nominal one: this solution, however, requires very large capacitors and therefore large power consumption and is difficult to achieve for high-frequency applications. Another approach is to make the ratio between nominal and parasitic capacitance the same at all nodes. This could be achieved by adding dummy structures.

In this filter the ratio between parasitic and nominal capacitance has been kept as low as possible (less than 40%) for a given power dissipation. At the same time, the variation of the ratio between parasitic and nominal capacitance from node to node has been minimized (less than 10%). In addition to this optimization, the residual error due to the parasitic capacitance has been compensated by slightly adjusting the value of the nominal capacitance according to the results of simulations. No dummy structures were used in this implementation.

The capacitors at nodes 1 and 2 of Fig. 4 have been split into two parts: one connected from the output node to ground and one connected differentially between the output nodes. Since differential capacitors do not load the CMFB circuit, this allows to obtain a CMFB bandwidth comparable to the transconductor unity-gain frequency, as explained before.

TABLE I
Device Sizes

	$g_{m1} = g_{m2} = g_{m4}$	g_{m3}
$M1$, $M2$	90/2	45/2
Q_D	1	1
R_D	900	1800
$M3$, $M4$	200/2	100/2
$M5$, $M6$	30/2	15/2
$Q1$, $Q2$	2	2
$Q3$, $Q4$	2	1
$R3$, $R4$	400	800

A. Tuning Strategy

From (2) the transconductance g_m of the stage is linearly dependent on V_{DS}. The tuning strategy adopted for

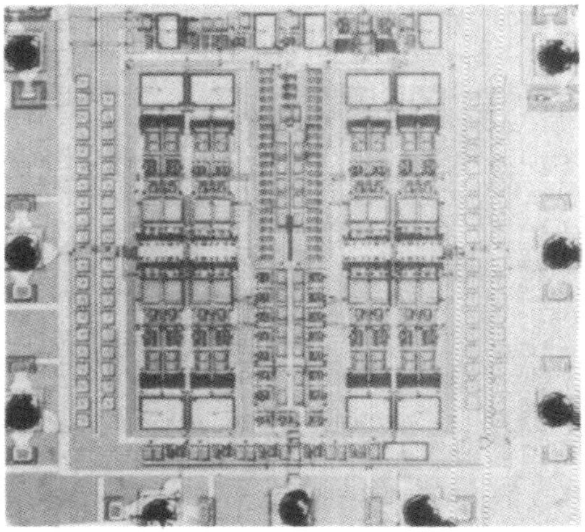

Fig. 5. Chip microphotograph.

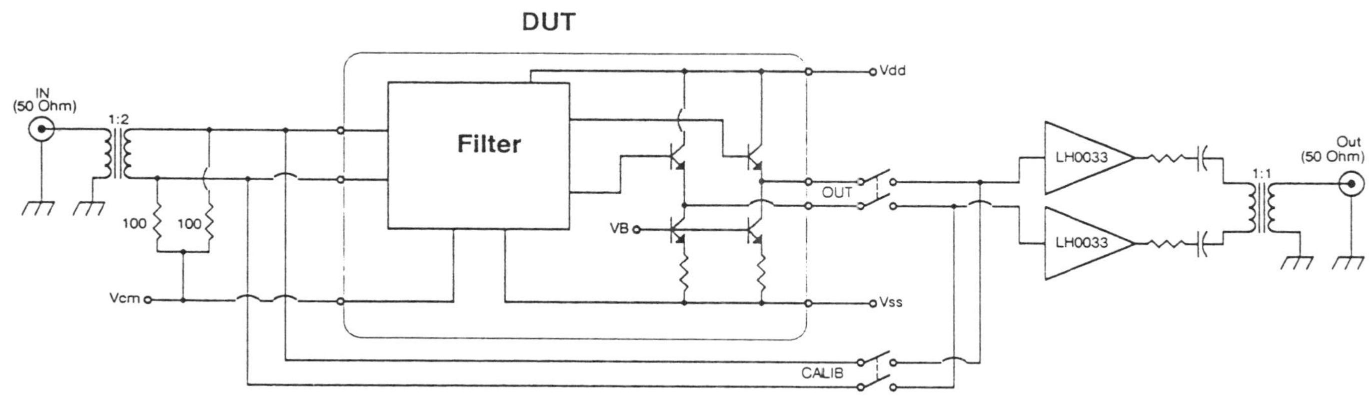

Fig. 6. Testing setup scheme.

the transconductance is to adjust the V_{DS} of $M1$ and $M2$ by changing I_D and the current I_{TOT} in such a way that the $V_{GS} - V_T$ of $M1$ and $M2$ remains constant. In the design the control of I_D and I_{TOT} is done externally in an open-loop fashion for simplicity. As an alternative a master–slave PLL-based tuning scheme can be implemented to provide the required I_D and I_{TOT} and is now in the process of being implemented on silicon.

IV. Experimental Results

The circuit has been fabricated in a 2-μm BiCMOS technology called HF2CMOS from SGS-Thomson Microelectronics. The microphotograph of the chip is shown in Fig. 5. Table I shows the devices size of the proposed realization.

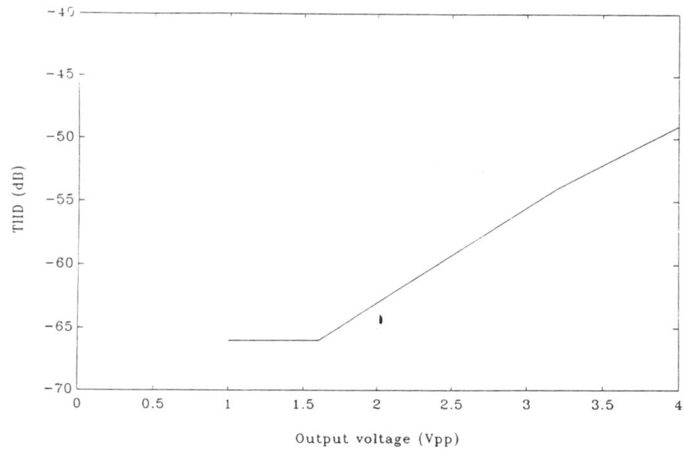

Fig. 7. Testing board distortion performance. F_{in} = 5 MHz.

A. Testing Setup

The block diagram of the testing setup is shown in Fig. 6. The key requirements of the testing setup are its frequency response in the band up to 100 MHz and its linearity, better than −50 dB for an output signal in the megahertz range up to 3.2 V_{p-p}. A single-ended-to-fully differential conversion at the input and a fully differential-to-single ended conversion at the output must be performed since the available instrumentations are single ended while the filter processes fully differential signals. This function must be realized without compromising the performance of the filter. Finally, the parasitic capacitance must be controlled. In particular bipolar buffers have been placed on chip at the filter output nodes and their input capacitance have been taken into account in the design of the

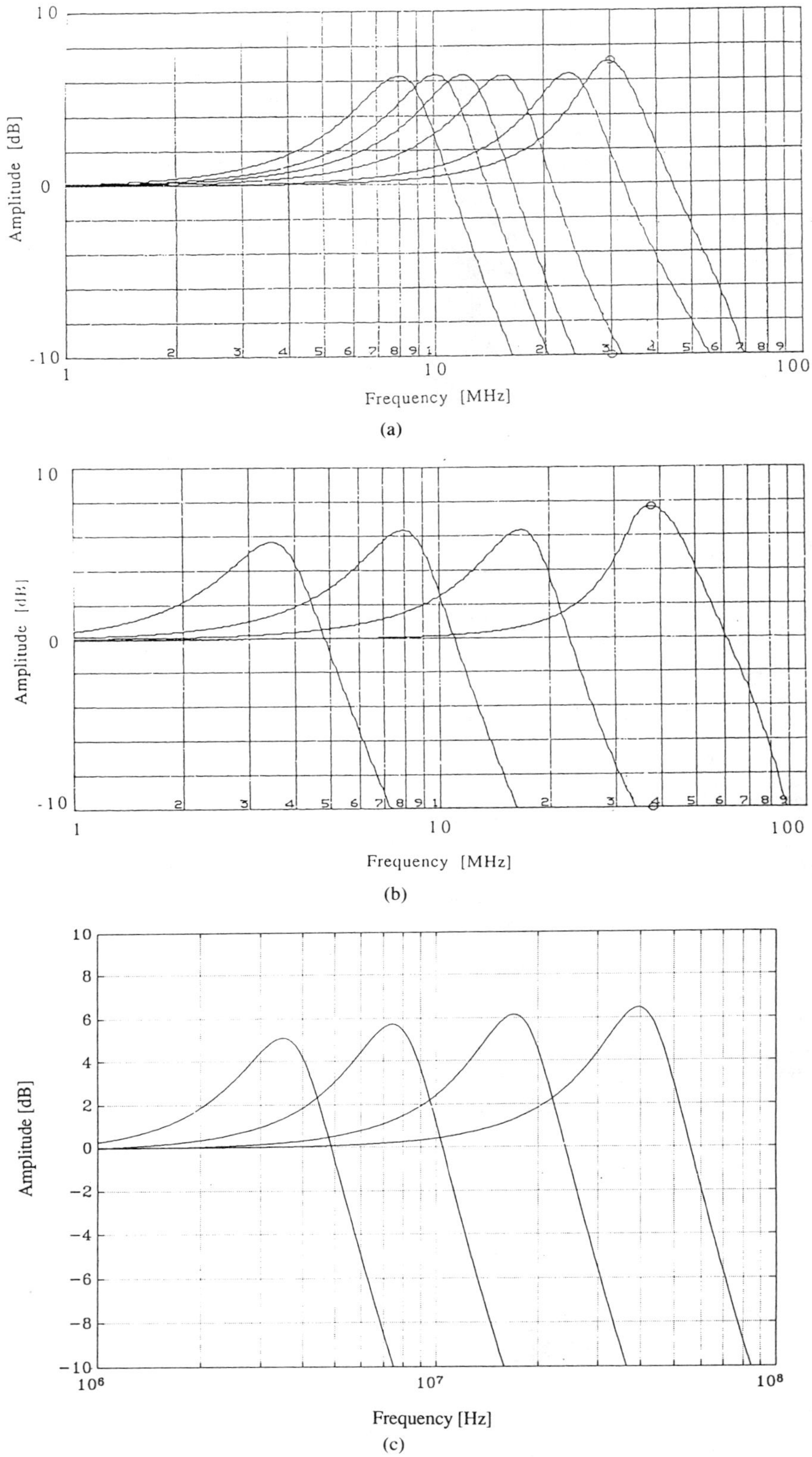

Fig. 8. (a) Measured amplitude performance with tuning. (b) Measured amplitude performance with extended tuning range. (c) Simulated amplitude performance with tuning.

capacitance of the filter. The input single-ended-to-fully differential conversion is performed by a transformer. The output of the filter, after the integrated bipolar buffers, feeds a high-performance buffer driving the output transformer which produces a single-ended signal. The linearity performance of this setup (without the device under

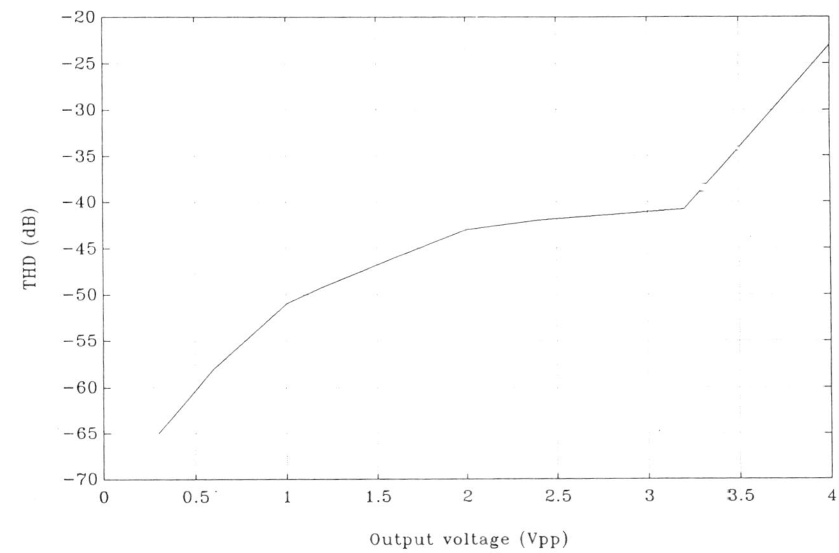

Fig. 9. Filter THD versus peak-to-peak output voltage (f_0 = 25 MHz, 5-MHz input signal).

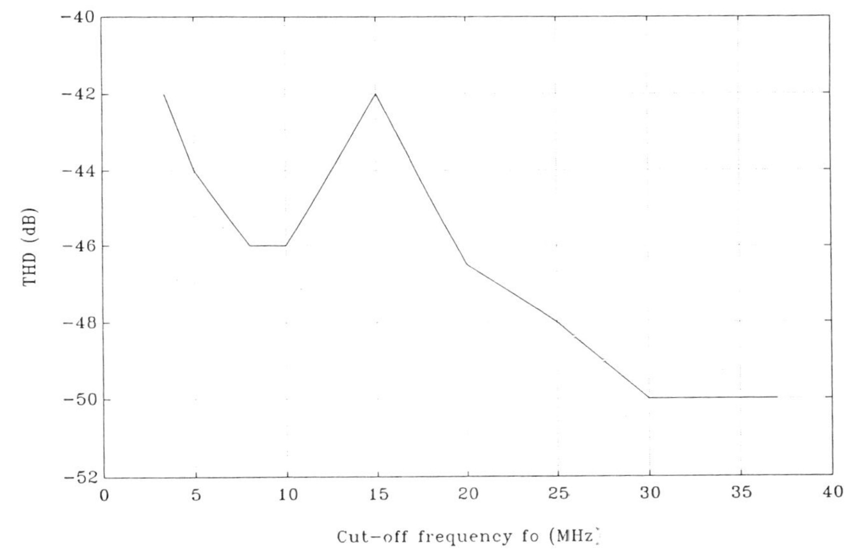

Fig. 10. Filter THD versus cutoff frequency f_0 (1.2-$V_{p\text{-}p}$ output voltage, 5-MHz input signal).

test) is shown in Fig. 7. Due to the transformers the setup circuit presents a high-pass cutoff frequency below 1 MHz. This is unimportant since the filter operates above 8 MHz and in disk-drive application the low-frequency signals are rejected before reaching the equalizer filter.

B. Measured Performance

The area of a single cell (without the common biasing circuit) is 0.322 mm^2 and the power consumption (with f_0 = 25 MHz) is 30 mW with a single 5-V power supply.

The amplitude response of the filter is shown in Fig. 8(a). The cutoff frequency f_0 is proportional to g_m, which is proportional to V_{DS}. Changing the external bias current and consequently the V_{DS} of the input MOS devices, the cutoff frequency is tuned in the range 8–32 MHz with a slight (<0.7 dB) variation of the peak value. Accepting a larger variation of the peak amplitude up to ± 1 dB, the range of tunability can be extended from 4 to 42 MHz as shown in Fig. 8(b).

In order to estimate the position in frequency of the equivalent nondominant pole that causes the measured peaking in the frequency response for higher f_0, a simulation was performed using SPICE. Fig. 8(c) shows the simulation results for the case of an equivalent nondominant pole frequency of 1.5 GHz. Comparison between Fig. 8(b) and (c) shows that indeed in the realized circuit the equivalent nondominant pole is around 1.5 GHz.

Over 14 measured chips, the quality factor mean value at $f_0 = 25$ MHz is 2.08 with a standard deviation of 2.5%. The deviation from the designed Q value of 2 is caused by an imprecision in the parasitic capacitance estimation.

The total harmonic distortion as a function of differ-

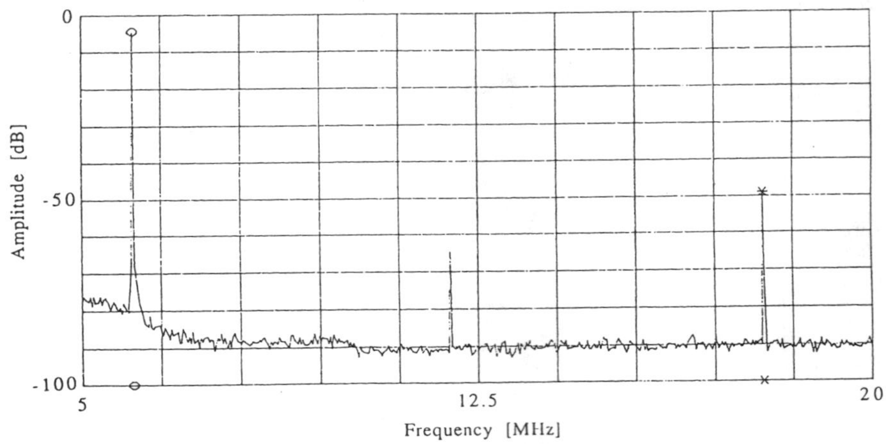

Fig. 11. Filter output spectrum for $f_0 = 25$ MHz and 2 $V_{p\text{-}p}$.

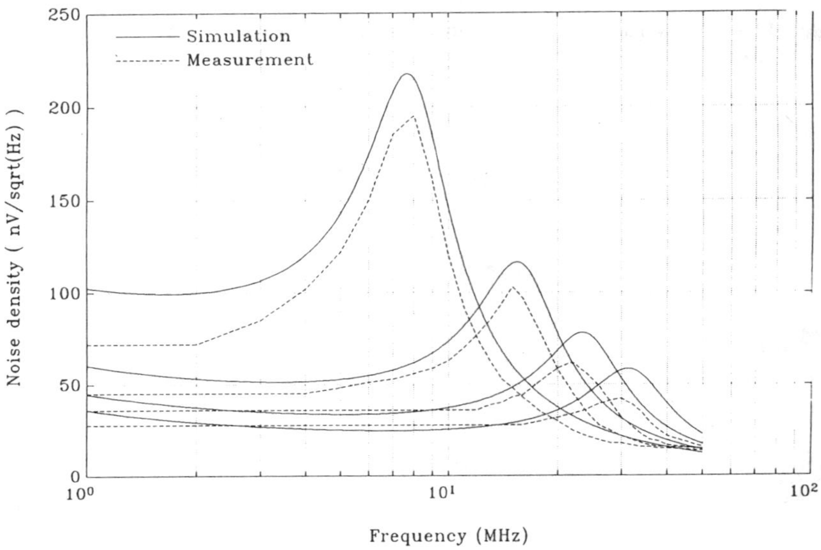

Fig. 12. Simulated and measured output noise spectrum for different filter cutoff frequency f_0.

ential output signal amplitude at 5 MHz is shown in Fig. 9 (with $f_0 = 25$ MHz). The THD stays below -40 dB up to 3.2 $V_{p\text{-}p}$. The THD for a 5-MHz output signal of constant amplitude of 1.2 $V_{p\text{-}p}$ as a function of the filter cutoff frequency f_0 is shown in Fig. 10. The degradation of the distortion performance at low cutoff frequency is due to the reduced V_{DS} of the input devices. In fact, as V_{DS} decreases, the effect of the finite impedance level at the drains of the input devices becomes more important. The degradation at $f_0 = 15$ MHz is due to the enhancement of the third harmonic for the 5-MHz input signal, due to the 6-dB peaking in the frequency response.

At higher frequency the THD improves due to the increase in the V_{DS} of the input devices. As seen in Fig. 10, linearity remains good up to the maximum f_0 that can be tuned.

The harmonic components of the output signal of 2 $V_{p\text{-}p}$ for a 5-MHz input sine wave are shown in Fig. 11.

The measured output noise spectrum is shown in Fig. 12 for different values of f_0. For the case of $f_0 = 25$ MHz, the integrated noise in the 1–50-MHz band is 240 μV_{rms}. This value, combined with the signal amplitude that produces less than 1% THD, results in a dynamic range of 73 dB. The measured results agree very well with the results of SPICE simulations shown in Fig. 12. The increase in output noise for lower values of f_0 is due to the fact that the noise power is proportional to $(V_{GS} - V_T)/V_{DS}$ as shown in (11). To reduce the cutoff fre-

TABLE II
MEASURED NOISE PERFORMANCE FOR DIFFERENT f_0

f_0	Total Output Noise (Bandwidth 1–50 MHz)	Equivalent Input Noise Density
8 MHz	460 μV_{rms}	65 nV/$\sqrt{Hz}$
16 MHz	330 μV_{rms}	47 nV/$\sqrt{Hz}$
24 MHz	250 μV_{rms}	36 nV/$\sqrt{Hz}$
32 MHz	200 μV_{rms}	30 nV/$\sqrt{Hz}$

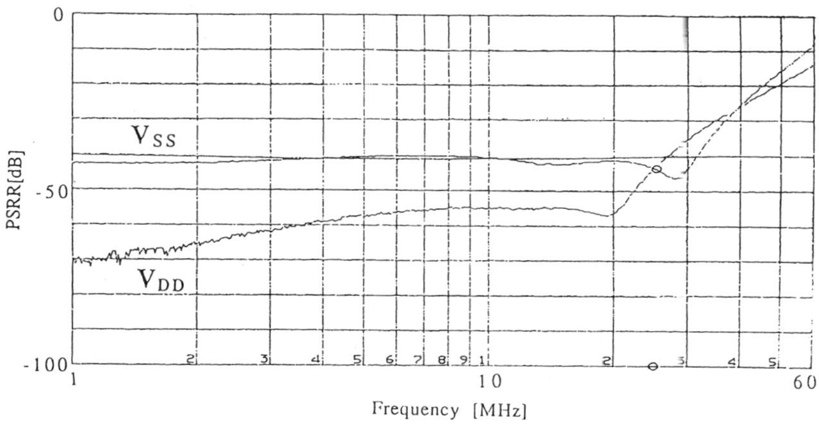

Fig. 13. PSRR for V_{DD} and V_{SS}.

TABLE III
SUMMARY OF THE MEASURED PERFORMANCE OF THE PROPOSED FILTER

Technology	2-μm BiCMOS
Power Supply Voltage	5 V
Power Consumption (@ f_0 = 25 MHz)	30 mW
Active Area	500 mil^2
THD (3.2 $V_{p\text{-}p}$ @ 5 MHz)	−40 dB
THD (1.2 $V_{p\text{-}p}$ @ 5 MHz)	−49 dB
Integrated Output Noise (1–50 MHz) f_0 = 25 MHz	240 μV_{rms}
Dynamic range (THD < −40 dB, @ 5 MHz)	73 dB
Dynamic range (THD < −60 dB, @ 5 MHz)	57 dB
Tuning Range (peak variation <0.7 dB)	8-32 MHz
Q measured mean value (14 samples)	2.08
Q measured standard deviation (14 samples)	2.5%
PSRR for V_{DD} @ 25 MHz	−43 dB
PSRR for V_{SS} @ 25 MHz	−43 dB

quency V_{DS} is decreased and therefore the noise power increases. Table II shows the measured total output noise and the equivalent input noise density for different values of the cutoff frequency.

The measured PSRR's for V_{DD} and V_{SS} are shown in Fig. 13. It is believed that the reduced PSRR for the V_{SS} supply with respect to the V_{DD} supply is caused by the n-p-n bipolar emitter followers at the output of the filter (Fig. 6), and not by the filter itself. These followers would be not used if the circuit were part of a mixed analog/digital systems in which the filter is not required to drive off-chip loads. In this case, where PSRR is very critical, it is expected to have the same good PSRR for both supplies.

A summary of the measured performance of the proposed filter is shown in Table III.

V. Conclusions

A new BiCMOS transconductor stage has been proposed. Its main characteristics are: high linearity, moderate noise, low excess phase at the unity-gain frequency, and small silicon area. The performance of the proposed transconductor has been demonstrated realizing a second-order low-pass continuous-time filter whose center frequency can be tuned in the range 8–32 MHz. The most important performances are a linearity of −40 dB up to 3.2-$V_{p\text{-}p}$ output signal amplitude and an in-band integrated noise of 240 μV_{rms}. The resulting dynamic range is 73 dB with THD < −40 dB, or 57 dB with THD < −60 dB.

This performance was achieved using a mature 2-μm BiCMOS technology and meets the specification required in disk-drive systems. The same results can be obtained using advanced and more expensive nonmixed technologies like submicrometer CMOS or oxide-isolated bipolar technologies. Future developments in a scaled-down 1.2-μm BiCMOS technology should allow the realization of a complete disk-drive analog front end suitable for next-generation products where the data transmission rate from the hard disk to the CPU is in the order of 50 Mb/s.

Acknowledgment

This research has been developed in cooperation with SGS-Thomson Microelectronics. In particular the authors wish to acknowledge Dr. M. Zuffada and his group at SGS-Thomson Microelectronics for useful discussion and for supporting the layout. The authors also wish to acknowledge Prof. F. Montecchi for suggestions and discussions and Prof. P. F. Manfredi and his group for assistance during testing.

References

[1] B. S. Song and P. R. Gray, "Switched-capacitor high-Q bandpass filter for IF applications," *IEEE J. Solid State Circuits*, vol. SC-21, pp. 924–933, Dec. 1986.
[2] M. S. Tawfik and P. Senn, "A 3.6-MHz cutoff frequency CMOS elliptic low-pass switched-capacitor ladder filter for video communication," *IEEE J. Solid-State Circuits*, vol. SC-22, pp. 378–384, June 1987.
[3] K. Matsui *et al.*, "CMOS video filters using switched-capacitor 14-MHz circuits," *IEEE J. Solid-State Circuits*, vol. SC-20, pp. 1096–1101, Dec. 1985.
[4] V. Gopinathan, Y. P. Tsividis, K.-S. Tan, and R. Hester, "Design considerations for high-frequency continuous-time filters and implementation of an antialiasing filter for digital video," *IEEE J. Solid State Circuits*, vol. 25, pp. 1368–1378, Dec. 1990.
[5] J. Khoury, "A 15 MHz CMOS continuous-time Bessel filter for disk drive," in *ISSCC Dig. Tech. Papers*, 1991.
[6] Y. T. Wang, F. Lu and A. A. Abidi, "A 12.5 MHz CMOS continuous-time bandpass filter," in *ISSCC Dig. Tech. Papers*, 1989, pp. 198–199.

[7] G. A. De Veirman and R. G. Yamasaki, "Fully integrated 5 to 15 MHz programmable bipolar Bessel lowpass filter," in *Proc. IEEE Int. Symp. Circuits Syst.* (New Orleans), 1990, pp. 1155–1158.

[8] J. L. Pennock, "CMOS triode transconductor for continuous-time active integrated filters," *Electron. Lett.*, vol. 21, no. 18, pp. 817–818, 1985.

[9] F. Krummenacher and N. Joehl, "A 4-MHz CMOS continuous-time filter with on-chip automatic tuning," *IEEE J. Solid-State Circuits*, vol. 23, pp. 750–758, June 1988.

[10] C. S. Park and R. Schaumann, "Design of a 4-MHz analog integrated CMOS transconductance-C bandpass filter," *IEEE J. Solid-State Circuits*, vol. 23, pp. 750–758, Aug. 1988.

[11] M. Banu and Y. Tsividis, "Fully integrated active *RC* filters in MOS technology," *IEEE J. Solid-State Circuits*, vol. SC-18, pp. 644–651, 1983.

[12] H. Khorammabadi and P. R. Gray, "High-frequency CMOS continuous-time filters," *IEEE J. Solid-State Circuits*, vol. SC-19, pp. 939–948, Dec. 1984.

[13] R. Castello, F. Montecchi, R. Alini, and A. Baschirotto, "A very linear BiCMOS transconductor for high-frequency filtering applications," in *Proc. IEEE Int. Symp. Circuits Syst.* (New Orleans), 1990, pp. 1364–1367.

[14] R. Alini, A. Baschirotto, R. Castello, and F. Montecchi, "Accurate MOS threshold voltage detector for bias circuitry," in *Proc. IEEE Int. Symp. Circuits Syst.* (San Diego), 1992.

[15] G. Nicollini, D. Pancini, and S. Pernici, "Simulation-oriented noise model for MOS devices," *IEEE J. Solid-State Circuits*, vol. SC-22, pp. 1209–1212, Dec. 1987.

[16] R. L. Geiger and E. Sanchez-Sinencio, "Active filter design using operational transconductance amplifiers: A tutorial," *IEEE Circ. Dev. Mag*, pp. 20–32, Mar. 1985.

[17] A. Van Bezooijen, N. Ramalho, and J. O. Voorman, "Balanced integrator filter at video frequencies," in *IEEE European Solid State Circuit Conf. Dig. Tech. Papers* (Milan), 1991, pp. 1–4.

Design Considerations of High-Dynamic-Range Wide-Band Amplifiers in BiCMOS Technology

Zhong-Yuan Chang, *Member, IEEE*, Willy M. C. Sansen, *Senior Member, IEEE*, and Michel S. J. Steyaert, *Associate Member, IEEE*

***Abstract*—Design techniques of high-dynamic-range wide-band amplifiers in BiCMOS technology for matching either a telescope (capacitive) or a wire (resistive) antenna are presented. The advantage of the BiCMOS process over bipolar and CMOS ones in designing such high-performance amplifiers is theoretically studied. It is analytically shown that by optimal combination of bipolar and MOS devices, BiCMOS amplifiers can be designed that show superior performance over the existing bipolar and CMOS realizations. The design principles are verified by the measurements on a bipolar test amplifier implemented in a linear gate array of a BiCMOS technology. A total equivalent input noise level as low as 0.25 μV_{rms} within 2.5-kHz IF band is achieved and a very high dynamic range of 135 dB is measured without the use of any AGC.**

I. Introduction

IT IS WIDELY expected that BiCMOS technologies will play an increasingly important role in implementing high-performance analog and digital VLSI's [1], [2]. The availability of both BJT and MOS transistors in a BiCMOS process provides circuit designers with the possibility of combining the desirable features of both kinds of devices in order to obtain best circuit performance. However, in comparison with bipolar and CMOS technologies, a BiCMOS technology requires more process steps and thus is more expensive. As a result, the use of a BiCMOS process should be justified by a superior circuit performance which is not possible to obtain using bipolar or CMOS processes. In this paper, the superiority of the BiCMOS process over BJT and CMOS processes is demonstrated by the design of low-noise, high-dynamic-range, wide-band amplifiers. Although, for the sake of convenience, the discussions will be limited to the design of wide-band preamplifiers for shortwave (SW) radio receivers, the obtained results apply to other applications as well.

Manuscript received December 31, 1990; revised July 10, 1991.

Z.-Y. Chang was with the Laboratory Electronika, Systemen, Automatisatie, Technologie (ESAT), Departement Elektrotechniek, Katholieke Universiteit Leuven, B-3001 Heverlee, Belgium. He is now with Alcatel Bell, 2000 Antwerpen, Belgium.

W. M. C. Sansen and M. S. J. Steyaert are with the Laboratory Electronika, Systemen, Automatisatie, Technologie (ESAT), Departement Elektrotechniek, Katholieke Universiteit Leuven, B-3001 Heverlee, Belgium.

IEEE Log Number 9103018.

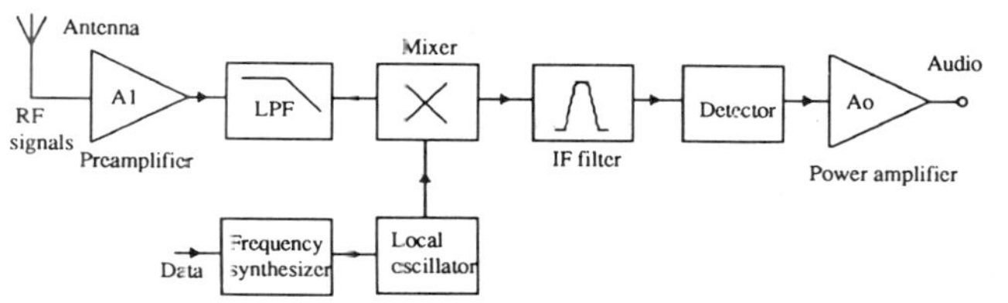

Fig. 1. Block schematic of radio receiver based on upconversion principle.

To reduce the cost of conventional radio receivers, receivers based on the so-called upconversion and AM synchronous detection principles appear to be more attractive. Fig. 1 shows a block schematic of a radio receiver based on the upconversion concept. The preamplifier realizes both amplification and noise matching functions for all incoming AM signals. After passing through a low-pass filter, signals are mixed with a local oscillator that shifts signal frequencies up to a fixed high intermediate frequency (e.g., IF = 10.7 MHz) instead of down to a low IF frequency, such as 450 kHz as used in conventional receivers. Because a high IF is used, a simple low-pass filter is sufficient to reject image frequencies. This kind of receiver differs from the conventional superheterodyne receivers [3] mainly in the absence of the front-end preselection *LC* tuning circuits. As a result, a significant cost reduction is obtained. However, the elimination of the RF input stage results in several new design problems for the wide-band front-end circuits. Firstly, the low-noise design of the amplifier becomes extremely important as the noise tuning mechanism in the conventional receivers is not more available. For practical portable radio receivers, a noise level of as low as 0.5 μV_{rms} is required to meet the 50-μV sensitivity specification. Secondly, due to the wide-band requirement, the preamplifier must be able to handle all signals coming from different transmitters simultaneously in the same manner. This requires the amplifier not only to be wide-band, but more importantly to show high linearity so as to avoid possible intermodulation distortions. Thirdly, in order to be able to handle the

Reprinted from *IEEE J. Solid-State Circuits,* vol. 26, no. 11, pp. 1681–1688, Nov. 1991.

wide range of input signals varying from 50 μV to 1 V with at least 40-dB S/N ratio, a dynamic range of up to 126 dB is required.

Such a high dynamic range has been realized by means of an AGC control of the preamplifier using a bipolar technology [4]. However, the AGC in [4] requires not only a complex control circuit, which increases design efforts and power consumption, but also some external components. The high power consumption and the increased cost due to the external components cannot be tolerated for portable radio receivers. In [5] a fully integrated CMOS amplifier based on a three-stage approach is presented which realizes a dynamic range of up to 130 dB without any AGC control. However, the three-stage approach consumes more power and requires a special frequency compensation technique.

In this paper, a new approach to such a preamplifier in BiCMOS technology is described. By taking advantage of the high-frequency characteristic of BJT transistors and the low input current noise feature of MOS devices, two-stage amplifiers are designed with a dynamic range of 135 dB without the use of any AGC control. Compared with the existing bipolar and CMOS approaches, the two-stage approach saves not only design efforts but also reduces power dissipation. In Section II, the feedback topologies of wide-band amplifiers for matching both telescope (capacitive) and wire (resistive) antennas are presented. The noise performance is analyzed in Section III for the two cases according to the type of input devices (BJT or MOS) to be used. A low-noise noninverting BiCMOS input stage will be presented in Section IV. Section V is devoted to the design of rail-to-rail class-AB output amplifiers. Based on the proposed BiCMOS amplifier configuration, a fully bipolar test amplifier has been implemented in a BiCMOS technology using a linear gate array approach. Measurement results on the test amplifier are summarized in Section VI. Finally, some concluding remarks will be given in Section VII.

II. Feedback Topologies

The feedback topologies for a telescope and a wire antenna are shown in Fig. 2. In Fig. 2(a) C_a is the capacitance of the telescope antenna, C_p is the parasitic cable capacitance, and C_f is the feedback capacitance. In the frequency range of interest where the loop gain is much larger than unity, the transfer function is given by C_a/C_f. A constant signal transfer is thus obtained independent of the parasitic input capacitance. In Fig. 2(b) the wire antenna is represented by a resistance R_s in series with a coupling capacitance C_s. The required constant signal transfer is realized by means of an overall resistive feedback with a resistor R_f. For both types of antennas, the voltage gain is set to be 6 dB in order to be able to handle a very wide input signal range from 50 μV to up to 1 V with a 5-V single power supply without use of any AGC control.

The most important feature of these feedback topologies appears in the core amplifier structure, which em-

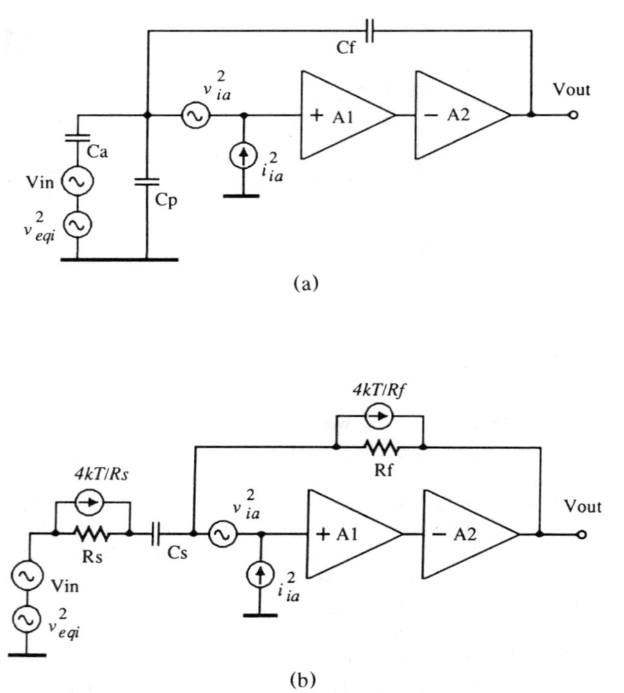

Fig. 2. Feedback topologies for (a) telescope antenna, and (b) wire antenna.

ploys a single-ended noninverting input amplifier $A1$ followed by a class-AB inverting output amplifier $A2$. A single-ended input is a must for achieving the best noise performance, while the noninverting character is required to allow the total negative feedback to be realized using a two-stage approach. Compared with the existing three-stage approach [5], the two-stage approach requires less design effort and saves power consumption. As will be shown in the next section, this two-stage approach yields only satisfactory results by using a BiCMOS technology.

It is easily seen that the total loop gains for both feedback configurations are similar so that one single-core amplifier is sufficient to match both antennas. Furthermore, in both cases the feedback does not introduce any phase shift, so the core amplifier must be designed to have an op-amp characteristic. The gain bandwidth (GBW) of the core amplifier must be sufficiently large so as to fulfill the following signal transfer requirements:

$$\mathrm{GBW}\frac{C_f}{C_f + C_a + C_p} > 26 \text{ MHz} \tag{1a}$$

$$\mathrm{GBW}\frac{R_s}{R_s + R_f} > 26 \text{ MHz} \tag{1b}$$

where 26 MHz represents the upper frequency of the SW band. For practical telescope antennas with $C_a = 10$ pF, $C_p = 5$ pF, and the feedback capacitance $C_f = 5$ pF, constraint (1a) becomes GBW > 104 MHz. For wire antennas with $R_s = 400\ \Omega$ and $R_f = 800\ \Omega$, condition (1b) becomes GBW > 78 MHz.

III. BJT Versus MOS Input Stage

A. MOS Input

Since both MOS and BJT transistors are available in the BiCMOS process, it is logical to examine which type of device yields the best noise performance. For the case of a MOS input, assuming the $1/f$ noise is insignificant in the SW band (3.8–26 MHz), the total equivalent input noise voltage in series with the signal source V_{in} for the telescope antenna amplifier is calculated as given by [6]

$$\nu_{\text{eqi}}^2 = \left[\frac{C_a + C_p + C_f + C_{in}}{C_a}\right]^2 \frac{8}{3} kT \frac{1}{g_m} \tag{2}$$

where g_m is the transconductance and $C_{in} = C_{GSI} + C_{GDI}$ is the input capacitance of the input MOS transistor. It can be shown [6] that (2) can be minimized by designing the input transistor to have a minimal channel length L and an optimal channel width W_{opt} and biasing it at the highest possible current level. Moreover, the optimal channel width is independent of frequencies and therefore the noise matching can be realized in the whole SW band. Under this optimal noise matching condition the equivalent input noise spectrum, taking into account both thermal and $1/f$ noise, is shown in Fig. 3. The assumption of neglecting the $1/f$ noise contribution is clearly justified. It is clearly seen that a total equivalent input noise level lower than 3 $\text{nV}/\sqrt{\text{Hz}}$ in the whole SW band can be obtained, which is sufficiently low to fulfill the sensitivity requirement.

B. BJT Input

Compared with MOS transistors, BJT transistors exhibit a lower equivalent input voltage noise due to their high transconductance, but a much higher equivalent input current noise due to the base shot noise. This high input current noise makes BJT transistors inadequate as input devices for matching high-impedance sources. Referring to Fig. 2(a), assuming a BJT input is used, the total equivalent input voltage noise is given by

$$\nu_{\text{eqi}}^2 = \left[\frac{C_a + C_p + C_f}{C_a}\right]^2 4kT\left(r_b + \frac{kT}{2qI_C}\right) + \left|\frac{1}{j\omega C_a}\right|^2 \frac{2qI_C}{\beta} \tag{3}$$

where r_b is the base resistance, I_C is the collector current, and β is the current gain of the input transistor. It is important to realize that the base shot noise results in an additional noise term which is inversely proportional to ω^2. To minimize the noise contributions of the base resistance and the base shot noise, the input transistor must be designed to have a small base resistance r_b and a large current gain β. Since for a given technology the

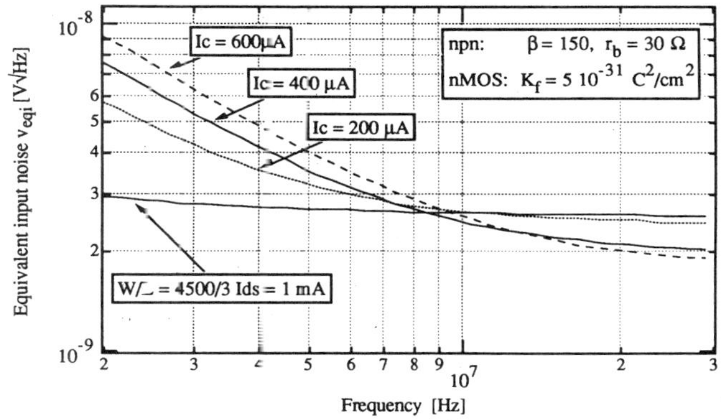

Fig. 3. Equivalent input noise for telescope antenna amplifier with MOS and BJT inputs.

base resistance is mainly concerned with layout techniques, a good layout such as a multibase configuration is of considerable importance. Under the normal operating conditions the current gain β is more or less constant and depends only on technological parameters so that little can be done about it by circuit designers. On the other hand, the collector bias current I_C is a free parameter for circuit designers. Therefore, the noise optimization is equivalent to the optimization of the collector current. By taking the derivative of (3) with respect to I_C the optimal collector current is obtained:

$$I_{C\,\text{opt}} = \frac{kT}{q}\omega(C_a + C_p + C_f)\sqrt{\beta}\,. \tag{4}$$

As (4) shows, the optimal collector current is directly proportional to the frequency. This makes it inadequate for realizing noise matching within the whole SW band. This is a considerable disadvantage compared to the MOS input where the optimal noise matching condition is totally independent of frequencies. Fig. 3 gives the equivalent input voltage noise density ν_{eqi} for three collector currents. Clearly, a lower noise density at the lower frequency end always corresponds with higher noise density at the high-frequency end and vice-versa. At low frequencies the noise comes mainly from the base shot noise, while at the high-frequency end the equivalent voltage noise generator turns out to be dominant. As a result, in order to minimize the equivalent input noise density in the whole SW band, a compromised I_C must be used.

It is important to note that in the case of the MOS input, the noise level in the whole SW band can always be lowered by increasing the dc bias current, while for the BJT input this is not possible. Thus, as far as the noise performance is the only concern, a MOS input should be used. However, the use of a MOS input has the disadvantage that a relatively high dc bias current I_{DS} is required.

Similar noise optimizations are carried out for the wire antenna. It is shown that in this case the choice of the

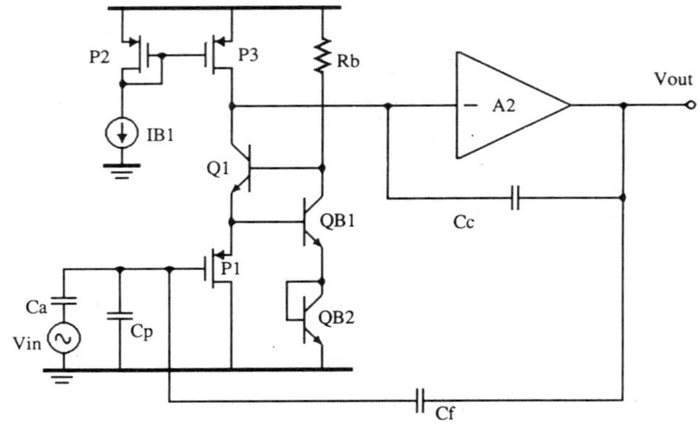

Fig. 4. Low-noise noninverting BiCMOS input structure.

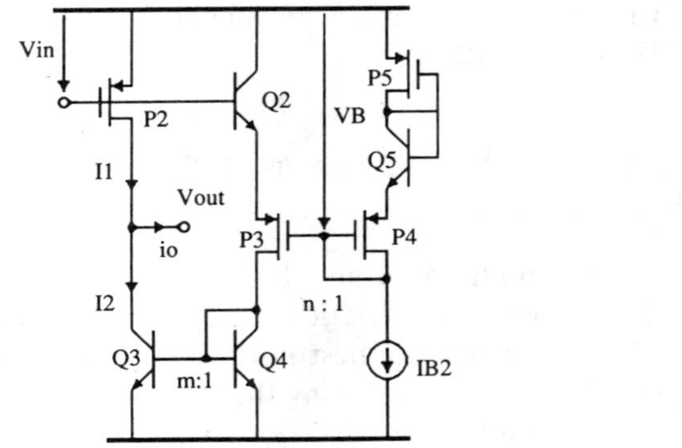

Fig. 5. BiCMOS class-AB output amplifier.

input device is less critical as the noise is by far dominated by the antenna and feedback resistances [7].

IV. Low-Noise Noninverting Input Stage

As Fig. 2 shows, the input amplifier is a single-ended noninverting amplifier, the circuit schematic of which is shown in Fig. 4. It is composed of a pMOS input transistor and a high-performance n-p-n transistor biased in the common-base configuration. The use of a MOS input transistor realizes the optimal noise matching, while a common-base n-p-n transistor takes care of noninverting amplification. It must be noted that satisfactory noise performance can only be obtained with this noninverting structure in a BiCMOS technology, because if a CMOS process is used, the common-base BJT transistor must be replaced by a nMOS transistor. The extra noise contribution of the nMOS transistor will double the noise level of the amplifier. In the case of a pure bipolar approach, the high input current noise hinders the noise performance from being optimized in the whole SW band.

The biasing of the common-base BJT transistor is realized by means of a bias network formed by $QB1$, $QB2$, and R_b. This biasing technique has several advantages due to the local negative feedback effect around the transistor $Q1$. First, the impedance looking into the emitter of $Q1$ is lowered by the feedback loop gain. As a result, the source degeneration due to $Q1$ is reduced by the same amount and therefore the equivalent input transconductance $g_{m\,\mathrm{eq}}$ is increased. High input transconductance is always desirable for low-noise design. Secondly, similar to the Wilson current mirror, the output impedance is increased by the loop gain, which results in a gain enhancement [8]. Finally, the local feedback loop brings about a dramatic reduction in the noise contribution of $Q1$. This can be easily understood physically as follows. Due to the virtual ground property of the local shunt feedback, the voltage noise source $4kT(r_{b1} + 1/2g_{m1})$ of $Q1$ is forced to manifest itself at the collector of $QB1$ instead of the internal base of $Q1$. Therefore, it does not contribute to the total output noise.

The extra noise contribution of the bias network comes mainly from the voltage noise of transistors $QB1$ and $QB2$. Because of the much higher transconductances of BJT's compared with MOS transistors, the voltage noise in a BJT transistor is generally much lower than that in a MOS one. Therefore, the extra noise contribution of the bias network can easily be made negligible compared with the input MOSFET.

Computer simulations show that by using this bias technique the noise contribution of $Q1$ is totally insignificant, and the extra noise contributions of the bias network R_b, $QB1$, and $QB2$ to the total equivalent input noise v^2_{eqi} are about 30% for the case where $QB1$ and $QB2$ have a base resistance of 20 Ω and are biased at 0.8-mA level. It should be noted that this small extra noise contribution can completely be eliminated if a dual power supply is used, since in this case the base of the common-base transistor $Q1$ can be tied to the noise-free analog ground.

V. Class-AB Output Stages

Considering the requirements of high dynamic range and low power consumption, a rail-to-rail class-AB approach is adopted for the output amplifier $A2$. The circuit schematic is shown in Fig. 5, where it is assumed that the BiCMOS process used does not have high-frequency vertical p-n-p transistors as is normally the case. This is a modified BiCMOS version of the CMOS source-coupled class-AB amplifiers proposed in [9] and [10]. The BiCMOS version has certain advantages over the CMOS one in that only one large MOS driving transistor $P2$ is required and much better high-frequency capability can be achieved, because the two parasitic poles at the source of $P3$ and the base of $Q3$ are much higher than in the CMOS approaches due to the much higher transconductance g_m of n-p-n transistors.

However, compared to the CMOS source-coupled class-AB amplifiers, the structure of the BiCMOS output stage is not symmetrical. It can thus be anticipated that the BiCMOS output stage is inferior to the CMOS one in terms of large signal distortion performance. However, it will be shown that by proper dimensioning of driving transistors, a symmetrical drive or output signal current can be obtained. Furthermore, a linear amplification can be achieved for the whole rail-to-rail output swing.

Under the condition that all transistors are in their normal working regions (i.e., MOSFET's in the saturation region and BJT's in the active forward region), the output push and pull currents are given by

$$I_1 = \frac{\beta_2}{2}(V_{in} - V_T)^2 \quad (5)$$

$$I_2 = \frac{m\beta_3}{2}(V_B - V_{BE} - V_{in} - V_T)^2 \quad (6)$$

where V_B is a constant dc bias voltage for the transistors $P3$ and $Q2$ and V_{BE} is the base–emitter voltage of $Q2$. It can be shown that in order to obtain a symmetrical output drive, the MOS transistor β's must be so chosen as to meet the condition $\beta_2 = m\ \beta_3 = \beta$. In this case, the output large signal current i_o which equals $I_1 - I_2$ is given by

$$i_o = \frac{\beta}{2}(V_B - V_{BE} - 2V_T)(V_B - V_{BE}) - \beta(V_B - V_{BE} - 2V_T)V_{in}. \quad (7)$$

In the first approximation one can neglect the changes in V_{BE} with respect to that of V_{GS} due to the exponential dependency of $I_C(V_{BE})$. In this case, the output current i_o is linearly related to the input signal voltage V_{in} despite the asymmetry in the structure.

The maximal linear output sinking current $i^-_{o\max}$ is determined by the condition $V_{in} = V_T$, under which the transistor $P2$ just turns off. For $V_{in} < V_T$ the output current is provided by the transistors $P3$ and $Q2$ only and therefore the sinking current i^-_o will increase more or less quadratically with V_{in}. The maximal linear sourcing current $i^+_{o\max}$ is determined by the point defined by $V_{in} = V_B - V_{BE} - V_T$, at which the driving transistors $P3$ and $Q2$ are OFF. Beyond this point, the sourcing current i^+_o equals the drain–source current of the driving transistor $P2$. It can be easily calculated that the maximal linear sourcing current equals the maximal sinking current and is given by

$$|i^+_{o\max}| = |i^-_{o\max}| = i_{o\max} = 2\beta\left(\frac{V_B - V_{BE}}{2} - V_T\right)^2 = 4I_0. \quad (8)$$

Thus, just as in the case of source-coupled symmetrical class-AB amplifiers, the maximal linear output signal current is exactly four times the output dc quiescent current [11]. This is a very important conclusion which allows circuit designers to determine the minimal dc quiescent current required for realizing a linear amplification for the whole range of the output signal swing.

The maximal linear output signal swing is determined by the saturation voltages of driving transistors $P2$ and $Q3$. The saturation voltage of the bipolar transistor $Q3$ is given by $V_{CE\,SAT} = kT/q \ln(1 + 1/\beta_R) + I_C r_c$ where β_R is the reverse current gain and r_C is the collector series resistance [12]. For practical bipolar transistors the first term depends only on the technological parameter β_R and is in the range of 100 mV. Since the second term is proportional to $I_C r_c$, it is important to design the output

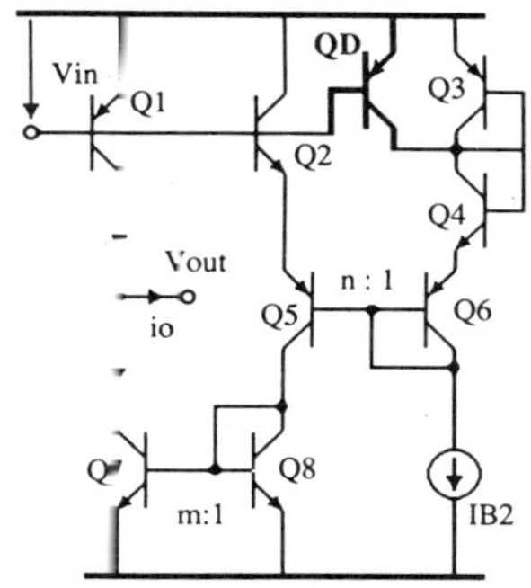

Fig. 6. Full bipolar version of BiCMOS class-AB amplifier of Fig. 4.

transistor $Q3$ with a minimal collector resistance r_c. The maximal linear positive voltage swing is determined by the saturation voltage of the driving transistor $P2$, which is given by $V_{SAT} = V_{GS} - V_T = \sqrt{2I_{DS}/\beta}$. Since the saturation voltage V_{SAT} is inversely proportional to the square root of the W/L ratio, a large W/L-ratioed pMOS driving transistor $P2$ is generally required.

The requirement of the large W/L-ratioed output MOS transistor can be eliminated if high-frequency p-n-p transistors are available in the BiCMOS process. In this case, the pMOS transistor $P2$ can be replaced by a p-n-p transistor as shown in Fig. 6. Compared with the BiCMOS output stage of Fig. 5, this full bipolar version requires much less chip area. In addition, the transconductance is much higher than that in Fig. 5 and therefore high gain and better high-frequency performance are obtained. On the other hand, the signal transfer can never be linear, as the difference between two exponential functions is not linear and the transconductance g_m of a BJT is totally independent of any geometrical parameters. To increase the symmetry, an extra driving stage is provided for $Q5$ by introducing an additional driving transistor QD. By matching QD to $Q3$ and biasing them at the same current level, the voltage gain of this driving stage is unity. As a result, the drive of $Q2$ and $Q6$ is now two times larger than that of $Q1$ and a more symmetrical drive is thus obtained. It should be noted that although the full bipolar version of the class-AB output stage is not linear under the open-loop condition, the suppression of its harmonic distortion under the closed-loop condition is stronger than in the CMOS version due to its higher loop gain.

VI. Experimental Results on a Test Amplifier

To verify the above design principle, a fully bipolar test amplifier has been realized in a linear gate array of an advanced BiCMOS technology where high-frequency vertical p-n-p transistors are available. The gate array used consists of n-p-n and vertical p-n-p transistors with various emitter areas, resistors, and capacitors of normal values and MOS transistors with limited W/L ratios. Since for optimal noise performance a relatively large MOS input transistor (i.e., $W/L = 4500/3$) is required, which is not available in the gate array, a low-noise

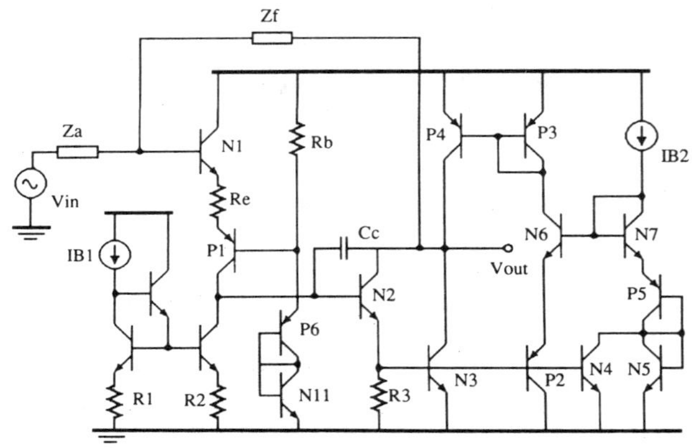

Fig. 7. Schematic of full bipolar core amplifier for telescope and wire antennas.

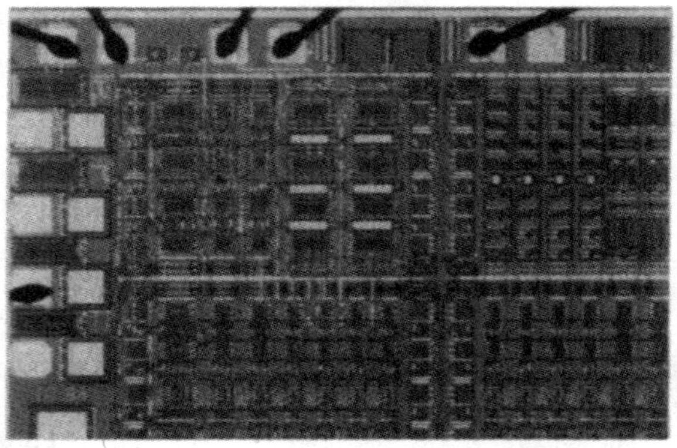

Fig. 8. Microphotograph of the test amplifier chip.

bipolar input transistor is used instead. Although in this case the noise performance is not optimal for the telescope antenna, it is still better than what has hitherto been realized. The circuit schematic of the bipolar test amplifier is shown in Fig. 7. Compared with Fig. 4, it is clearly seen that the test amplifier has exactly the same circuit structure as the proposed BiCMOS one. Consequently, the design principles can be verified by this bipolar test amplifier. The input stage is adapted in such a way that a low-noise n-p-n input transistor is used as the input device and a high-frequency vertical p-n-p transistor is employed for realizing the noninverting amplification function. As BJT transistors have a larger g_m/I ratio than their MOS counterparts, the slew rate (SR) of a bipolar amplifier is generally lower than that of a MOS one. To enhance the SR of the test amplifier, a small emitter resistance is added at the cost of small degradation in the S/N ratio. The active load of the input stage employs the emitter degeneration technique to limit its extra noise contribution.

The output class-AB stage uses the full bipolar approach as mentioned in Fig. 6. The Darlington configuration is used to enhance the impedance at the input stages' output and to reduce the capacitive loading of the second stage. Frequency compensation is provided by means of a simple Miller capacitance C_C. This is another advantage of the two-stage approach over the three-stage one where a special frequency compensation technique is necessary.

A microphotograph of the chip is shown in Fig. 8. A very regular structure of the gate array is easily recognized. The interconnections of different components are made by the second metal layer. The measured frequency transfer characteristics for both types of antennas are shown in Fig. 9 under the load condition of 500 Ω//10 pF. For both cases, a voltage gain of about 6 dB with −3-dB frequency of about 50 MHz is measured. Also, no peaking has been found in the transfer characteristics, which means that the phase margins for both cases are higher than 60°. The peaking phenomena occurring above 100 MHz are the result of high-frequency parasitic reflections. The main experimental results are summarized in

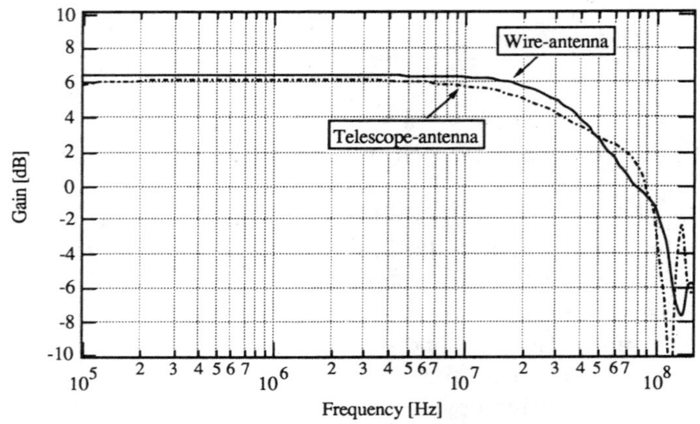

Fig. 9. Measured transfer function of the test amplifier.

TABLE I
MEASURED BiCMOS AMPLIFIER CHARACTERISTICS

	Telescope antenna	Wire antenna
V_{DD}	5 V	5 V
I_{total}	4.2 mA	4.2 mA
Gain	6 dB	6 dB
f_{3dB}	50 MHz	50 MHz
v_{eqi} (3–30 MHz)	4.2 nV/$\sqrt{Hz}$	4.5 nV/$\sqrt{Hz}$
$V_{o\,max}$	4 V_{pp}	4 V_{pp}
IMFDR3 at 5 MHz	100 dB	101 dB
at 10 MHz	94 dB	97 dB
SR	148 V/μs	151 V/μs
Load condition	500 Ω//10 pF	500 Ω//10 pF

Table I. The measured equivalent input noise within 2.5-kHz IF bandwidth is 0.25 and 0.21 μV_{rms} for the wire and the telescope antenna, respectively, which agrees very well with the theoretical calculated ones.

The amplifier is powered by a +5-V single supply. The measured 1-dB gain compression point is 2.1 V_{pp} corresponding with 4.2-V_{pp} output signal swing. The amplifier can drive a 4-V_{pp} signal into a 500-Ω//10-pF load with a measured total harmonic distortion of less than 1%. A very high dynamic range up to 135 dB is thus achieved without the use of any AGC. The intermodulation behavior is measured using a spectrum analyzer. Fig. 10 shows

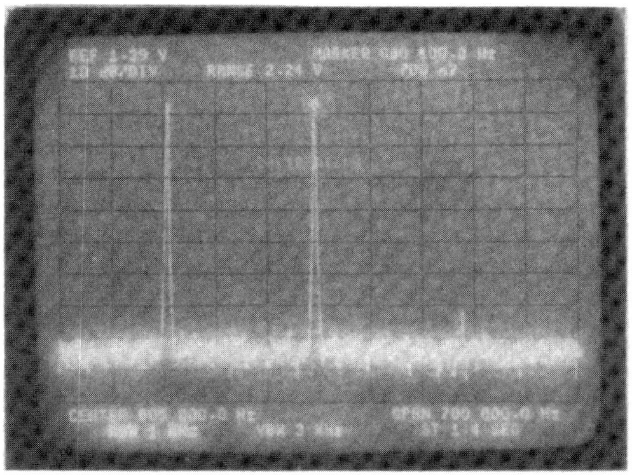

Fig. 10. Power spectrum of IM3 product at 5 MHz.

the spectra of two output signals and their IM3 product at $f = 5$ MHz. The intermodulation-free dynamic range (IMFDR3), defined as the ratio of the input signal level for which the IM3 product occurs above the noise level, is measured to be 100 dB at $f = 5$ MHz. The total dc quiescent current is 4.2 mA, of which 1 mA is used in the bias branch for the input stage in order to limit its extra noise contribution. As mentioned before, this dc bias current can be saved if a dual power supply is used. The class-AB common emitter output stage is biased at 2-mA level for large-signal linearity.

VII. Conclusions

The superiority of BiCMOS technology over bipolar and CMOS ones for analog applications is demonstrated by the design of low-noise high-dynamic-range amplifiers matching either a capacitive or a resistive antenna. Criteria of which type of active devices should be employed are derived analytically. It is theoretically shown that by optimal combination of bipolar and MOS devices, better circuit performance can be obtained with a simpler approach in BiCMOS technology than the existing bipolar and CMOS realizations. The analyses and design principles are verified by the measurement of a fully bipolar test amplifier implemented in a linear gate array of a BiCMOS technology. A total equivalent input noise level as low as 0.25 μV_{rms} within a 2.5-kHz IF band is achieved in the whole SW band and a very high dynamic range of 135 dB is measured without the use of any AGC.

Acknowledgment

The authors would like to thank Dr. P. De Wit and Dr. J. Lecocq from Philips Eindhoven for their stimulating interest and support.

References

[1] M. Kubo *et al.*, "Perspective on BiCMOS VLSI's," *IEEE J. Solid-State Circuits*, vol. 23, no. 1, pp. 5–11, Feb. 1988.

[2] P. J. Lim and B. A. Wooley, "An 8-b 200-MHz BiCMOS comparator," *IEEE J. Solid-State Circuits*, vol. 25, no. 1, pp. 192–199, Feb. 1990.

[3] G. Kennedy, *Electronic Communication Systems*, 3rd ed. New York: McGraw-Hill 1985.

[4] E. H. Nordholt, F. C. Nauta, and C. A. M. Boon, "A high-dynamic-range front end for an upconversion car-radio receiver," *IEEE J. Solid-State Circuits*, vol. SC-20, no. 3, pp. 688–696, June 1985.

[5] Z. Y. Chang and W. Sansen, "Low-noise, low-distortion CMOS AM wide-band amplifiers matching a capacitive source," *IEEE J. Solid-State Circuits*, vol. 25, no. 3, pp. 833–840, June 1990.

[6] Z. Y. Chang and W. Sansen, "Noise optimisation of CMOS wide-band amplifiers with capacitive sources," in *Proc ISCAS'89*, pp. 685–688.

[7] Z. Y. Chang and W. Sansen, *Low-Noise Wide-Band Amplifiers in Bipolar and CMOS Technologies*. Norwell, MA: Kluwer Academic, Dec. 1990.

[8] P. R. Gray and R. G. Meyer, *Analysis and Design of Analog Integrated Circuits*, 2nd ed. New York: Wiley, 1984. ch. 11.

[9] R. Castello and P. R. Gray, "A high-performance micropower switched-capacitor filter," *IEEE J. Solid-State Circuits*, vol. SC-20, no. 6, pp. 1122–1132, Dec. 1985.

[10] J. N. Babanezhad and R. Gregorian, "A programmable gain/loss circuit," *IEEE J. Solid-State Circuits*, vol. SC-22, no. 6, pp. 1082–1190, Dec. 1987.

[11] E. Seevinck and R. F. Wassenaar, "A versatile CMOS linear transconductor square law function circuit," *IEEE J. Solid-State Circuits*, vol. SC-22, no. 3, pp. 366–377, June 1987.

[12] S. M. Sze, *Physics of Semiconductor Devices*, 2nd ed. New York: Wiley, 1981.

Part 5
BiCMOS Smart Power Applications

SMART power ICs use devices that can handle currents between 0.2 A and 10 A and operated in the 30-V to 300-V range. The areas of application include motor speed control, lighting controls, automotive controls, and solid-state relays. BiCMOS technology allows the use of different types of power devices intended for divergent applications. Although the application of BiCMOS technology in the area of smart power is just beginning, the potential of using the technology in targeted custom-designed applications is quite high because of the design flexibility offered by the technology.

This part consists of two sections. Section 5.1 is an introduction and includes an invited overview by Huang. Section 5.2 covers the application of BiCMOS to the design of power and smart power circuits.

In Section 5.1, the papers by Baliga and Adler put into perspective the use of both MOS and bipolar in the design of high-power ICs. Section 5.2 addressed the design of high-voltage ICs and the application of BiCMOS to the design of a thermal-head intelligent driver. One paper is by Thomas et al. and the other by Tsumura.

Paper 5.1

An Overview of BiCMOS State-of-the-Art Smart Power Applications

J. S. T. HUANG
SOLID STATE ELECTRONICS CENTER
HONEYWELL, INC.
PLYMOUTH, MN 55441

Abstract—**The state-of-the-art power technology includes MOS-controlled bipolar structures such as the insulated gate bipolar transistor (IGBT). With the advent of BiCMOS process, "smart power," which integrates lateral power devices with signal conditioning and logic circuits on the same chip, becomes viable. With few process modifications, a wide variety of power devices having different current and speed characteristics can often be realized by varying cell designs and/or chip layouts. The lateral MOS-controlled thyristor (LMCT), having the lowest forward voltage drop among all lateral devices at high current levels, appears to be an attractive choice for future smart power applications.**

1. Advance in BiCMOS Power Devices

In the late seventies, the workhorse of MOS power technology was the double diffused MOS(DMOS). The lateral DMOS (LDMOS) is shown isolated in a typical BiCMOS integrated circuit in Fig. 1(a). Its vertical counterpart was obtained by moving the drain from the top to the bottom of the wafer, as shown in Fig. 1(b). Lateral power devices lend themselves more easily to integration with other circuitry on the same chip, thus making possible the so-called smart power ICs. DMOS conducts current and blocks voltage in one direction only. Its "on" resistance is rather high because of the lengthy high-resistance N^- drift region needed to support the breakdown voltage. In order to have low "on" resistance and still maintain high input impedance, one approach is to add an N^+ emitter in the P diffusion of the LDMOS, resulting in a lateral insulated base transistor (LIBT) [1], shown in Fig. 2(a). Its equivalent circuit as shown in Fig. 2(b) indicates a Darlington connection in which a bipolar transistor is driven by a power DMOS. Another way is to replace the N^+ drain in LDMOS with a P^+ anode, resulting in a lateral IGBT (LIGBT) structure shown in Fig. 3(a) [2]–[4]. In a LIGBT, the drain PN junction is used to inject holes to conductivity modulate the N^- drift region, thus reducing the "on" resistance. Although LIGBT, like LDMOS, conducts current unilaterally, it blocks voltage in both directions due to the presence of back-to-back

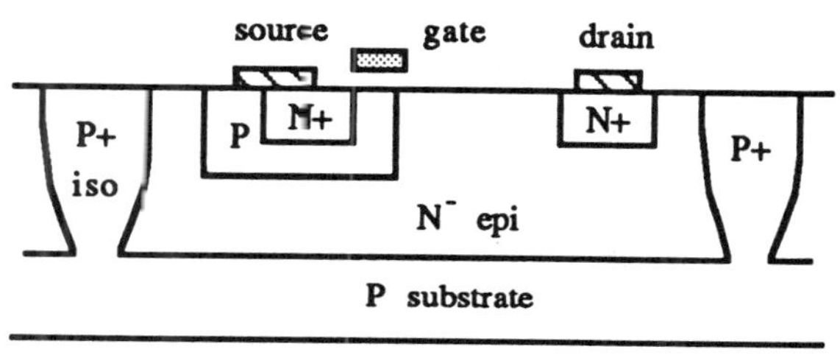

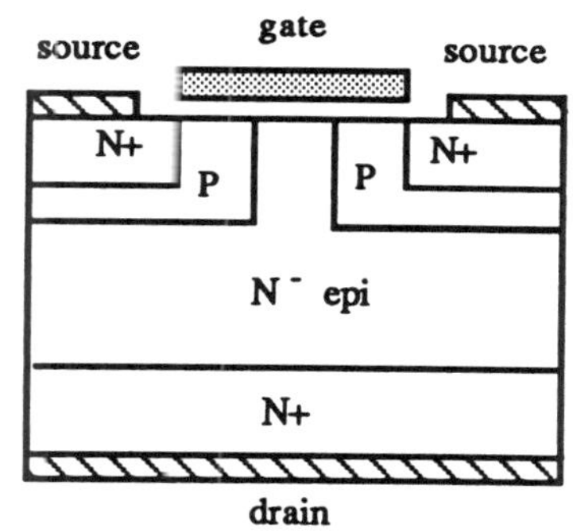

Fig. 1 Double-diffused MOS structures (DMOS); (a) cross section of a LDMOS for integrated circuits, (b) discrete vertical DMOS structure.

PN junctions. The equivalent circuit of a LIGBT including parasitics is shown in Fig. 3(b). At first sight, IGBT shows a built-in four-layer PNPN structure; but, in actual operation, latch-up is prevented by shorting the emitter-base junction of the NPN transistor. Thus, the vertical NPN transistor is not turned on. If the substrate PNP transistor is ignored, the equivalent circuit simplifies to that of Fig. 3(c), showing the current path that is predominantly MOS. This allows continuous gate-controlled output current with forces gate turn-off capability.

By proper material selection, cell design, and layout, LIGBT can be made to operate in the latch-up mode, provided that the latch can be terminated by a MOS gate. Such a device

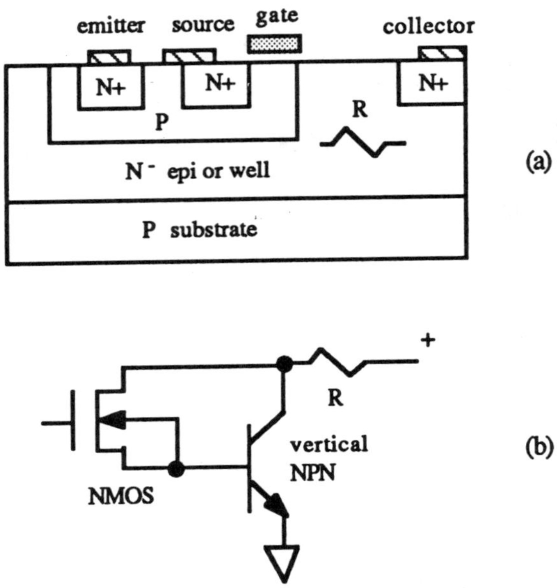

Fig. 2 Lateral insulated base transistor (LIBT); (a) cross section of LIBT (Darlington), (b) equivalent circuit of (a).

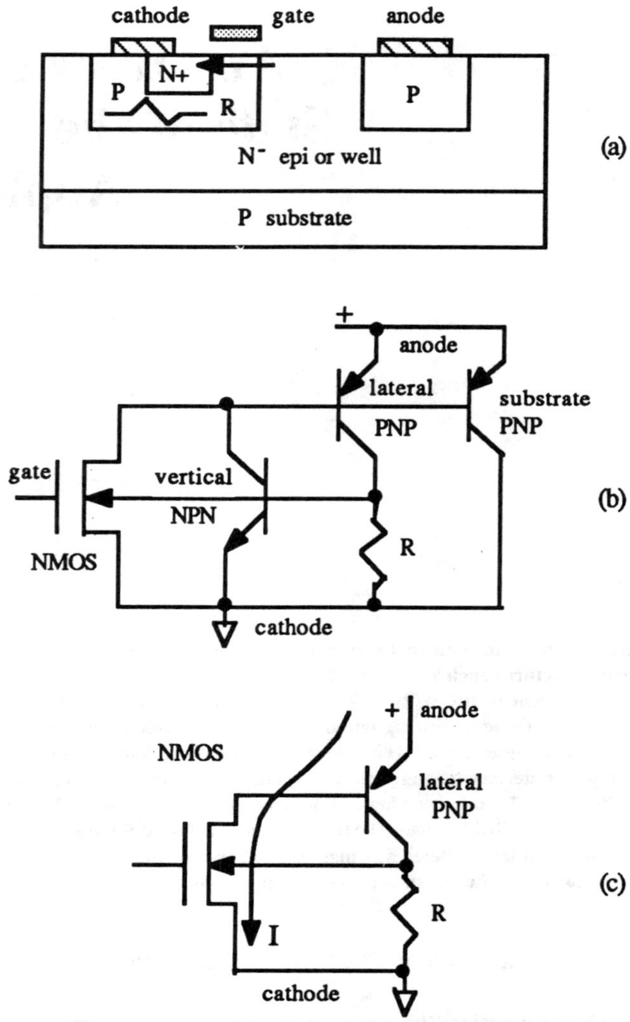

Fig. 3 Lateral IGBT structure; (a) cross section of LIGBT [2]–[4], (b) equivalent circuit of (a), (c) simplified equivalent circuit for LIGBT showing current flow.

is called a lateral MOS-controlled thyristor (LMCT) [5]–[7]. LMCTs exhibit the highest current handling capability per unit area or the lowest "on" resistance at high currents of all the lateral power devices available to date. A comparison of forward-conduction characteristics of all the devices discussed is given in Fig. 4. At a forward drop of 2 V, the LMCT has the highest forward current density of about 800 A/cm^2. For power devices operating in the MOS mode, the current-carrying capability is traded off for fast turn-off speed whereas in the bipolar mode, the speed is compromised because it takes time for minority carriers in the lightly doped N epi region to recombine. The "on" resistance versus speed trade-offs for various lateral power devices are compared in Fig. 5. These devices can all be made by essentially the same baseline process. Their characteristics can be tailored to a wide range of applications by merely making changes in device designs and/or layouts.

It is of interest to note that adding a N$^+$ diffusion in the surface P anode of Fig. 3(a) results in a symmetric structure, shown in Fig. 6(a), which is equivalent to two LMCTs connected back-to-back and is capable of ac or bilateral switching. An alternate version of the bilateral MCT (BMCT) with a single gate is shown in Fig. 6(b). The PMOS gate supplies holes to the base of the vertical NPN transistor. The turn-on sensitivity is expected to be better than the dual gate case in Fig. 6(a) because the base drive is now applied to the transistor with higher gain. The experimental switching characteristics of these two types of BMCT are described in Section 4.

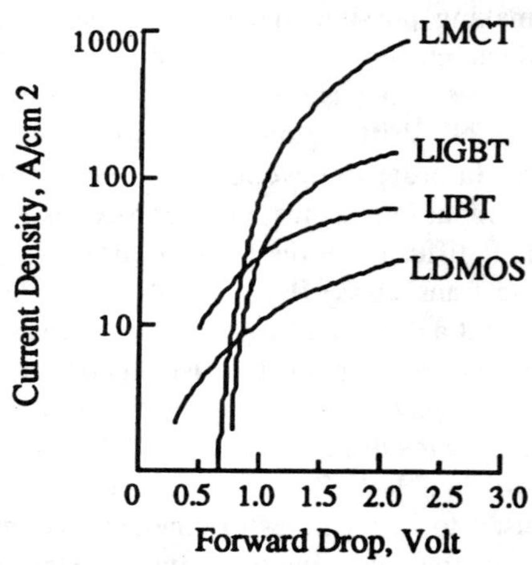

Fig. 4 Comparison of forward characteristics of devices with similar sizes and breakdown voltages.

2. BiCMOS and Smart Power

The processing of all the lateral devices that functionally merges MOS and bipolar transistors can usually be made compatible with either the modified CMOS or the BiCMOS

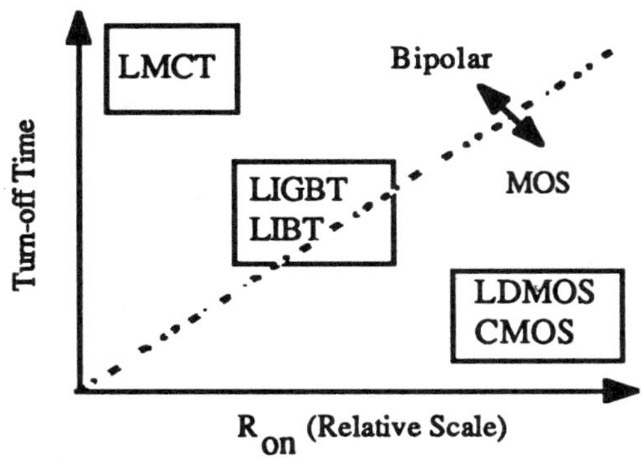

Fig. 5 Switching speed vs. "on" resistance trade-off.

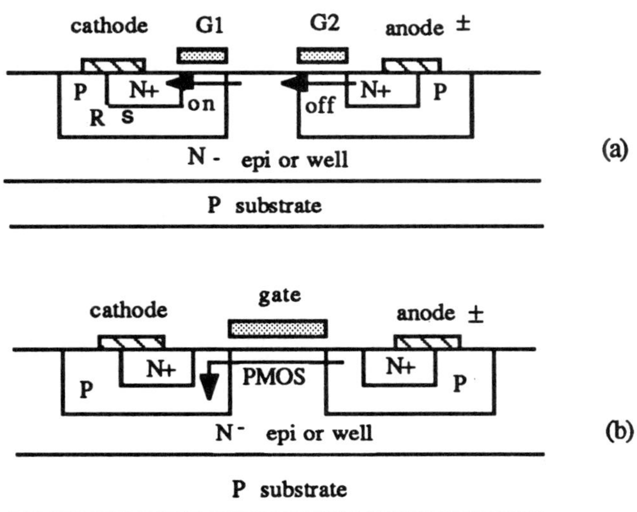

Fig. 6 Bilateral MCT (BMCT) structure; (a) cross section of a BMCT, (b) another version of BMCT with a single gate.

technologies. These devices built with BiCMOS process are either in the N-well or the N epi and are isolated from the rest of the circuits. In applications that require only one single power device for high-side switching, a vertical nonisolated IGBT or a MCT with P substrate acting as anode can offer higher current capability than its lateral counterpart. The P anode and the N^+ cathode of a lateral power device correspond to the base and the emitter diffusions of a bipolar transistor, respectively. Most of the MOS gates are self-aligned with respect to the source and the drain. Floating but shorted PN junctions at the surface, if needed, can be implemented with the same source/drain silicidation.

The BiCMOS technology used to implement smart power ICs is not likely to be the same as the state-of-the-art high-performance submicron technology for digital VLSI [8]. For example, polysilicon layers may not be needed for emitter and base contacts of the bipolar transistor, thus simplifying the process. Only one additional base implant step added to the conventional CMOS process could result in reasonably good linear bipolar transistors. In some cases, even the PMOS source/drain implants with relatively light dopings and deep junctions could serve as the base for NPN transistors. As a result, a straightforward CMOS process may be used to fabricate smart power ICs using lateral devices. Although the size of the device is usually large in order to handle the current, the channel length of the MOS control gate tends to be small so that sizable MOS currents are available to turn-on and turn-off the devices.

There are two important issues that need to be addressed in the implementation of smart power with BiCMOS technology. One is the parasitic substrate PNP transistor shown in the equivalent circuit of Fig. 3(b). The hole current flowing into the substrate detracts from the total injected hole current, thus lowering the lateral PNP gain, which happens to be one of the important parameters in determining the latch-ups in MCTs. Furthermore, the substrate current could potentially interfere with the operation of other circuits on the same chip. In fact, it is often contended that no LIGBT or LMCT can be truly isolated in junction-isolated ICs. However, this substrate current can be minimized by using a N^+ buried layer, which is readily available in most BiCMOS technologies to prevent CMOS latch-up and to serve as a barrier for substrate current. The silicon-on-insulator technology (SOI), however, has the potential to solve the isolation problem in power ICs. The limitation of the breakdown voltage is another concern for the lateral power devices. In vertical discrete devices, high-breakdown voltages are achieved through thick and lightly doped epi layers at no expense of the horizontal area. By contrast, to support high-breakdown voltage in a lateral device, large spacings between elements of the device are often required. This not only increases the device size, but could increase the forward voltage drop and decrease the current gain of the bipolar component. There are ways to optimize the breakdown voltage for lateral devices, as will be discussed in the next section. In general, lateral power devices built with BiCMOS process are often designed for low-voltage and high-current applications. With BiCMOS technology, choices of power devices are available. For forward drops less than 0.7 V, MOS or bipolar transistors are preferred. For forward drops greater than 0.7 V at high operating currents, the choice is LIGBT or LMCT.

3. Breakdown Voltage

It is possible to realize the higher bulk breakdown in lateral devices with relatively thin epi layers using the Resurf (reduced surface field) [9] or the LCC (lateral charge control) [10] technique. Figure 7(a) illustrates a typical geometry of a lateral DMOS. The epi thickness and doping concentration are such that the epi layer is completely depleted to the surface prior to the surface breakdown. As a result, the surface field is reduced and the bulk P substrate to N epi junction breakdown is realized. The Resurf concept is applicable to lateral IGBT [11] and to thin-film SOI devices where high-breakdown voltage can be obtained by controlling both the Si film and the buried oxide thicknesses [12], [13].

Another method to improve the bulk breakdown voltage is the use of multiple guard rings and field plates [14] as is sketched in Fig. 7(b). Qualitatively, field-limiting rings

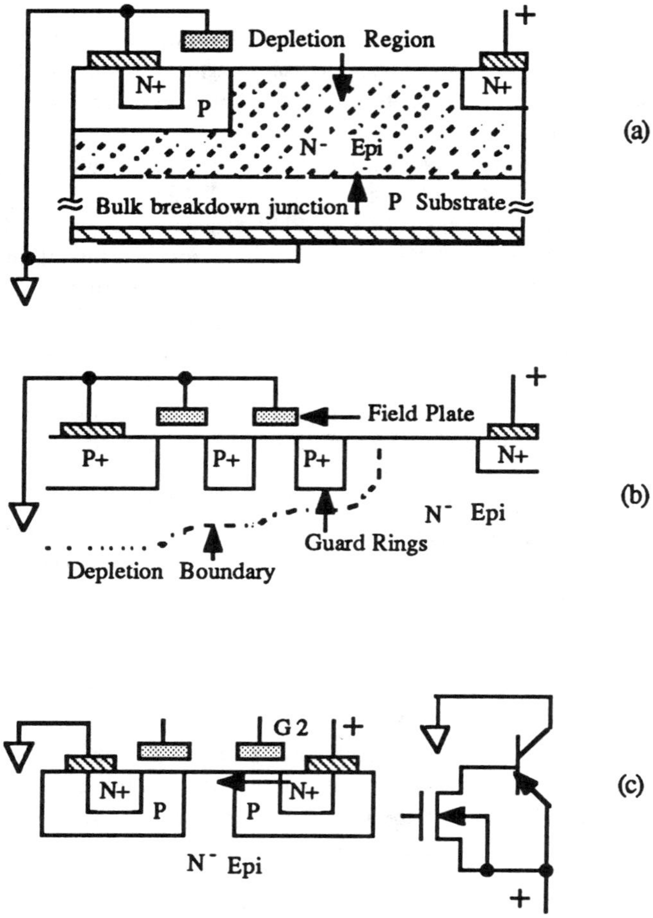

Fig. 7 Methods to enhance breakdown voltage, (a) resurf or LCC technique applied to LDMOS, (b) use of guard rings and field plates to diffuse surface field, (c) convert BV_{ceo} to BV_{ces}.

and plates, when properly placed, can help spread the depletion region at the surface and reduce the field. This way, a breakdown voltage close to the bulk value can be achieved. Because of the area occupied by the guard bands, the higher the breakdown voltage that is required, the larger the lateral device area becomes.

Yet another scheme to enhance the voltage capability uses a circuit technique. Figure 7(c) illustrates a symmetric LIGBT that is in the "off" state. A positive voltage applied to gate 2 turns on the NMOS transistors, shorting the emitter-base junction of the lateral PNP. This has the effect of converting a lower BV_{ceo} to a higher BV_{ces} of the lateral PNP transistor.

4. LMCT and BMCT

In the design of emitter-base shorts in a conventional gate turn-off (GTO) thyristor [15], there are conflicting requirements between large dV/dt capabilities and high gate sensitivities. This conflict is somewhat mitigated in the MOS-gated thyristor. The gate sensitivity, besides being dependent on the emitter-base short, also depends on the MOS geometry and its threshold voltage. This extra degree of freedom in design enables the use of a higher emitter-base short to increase the

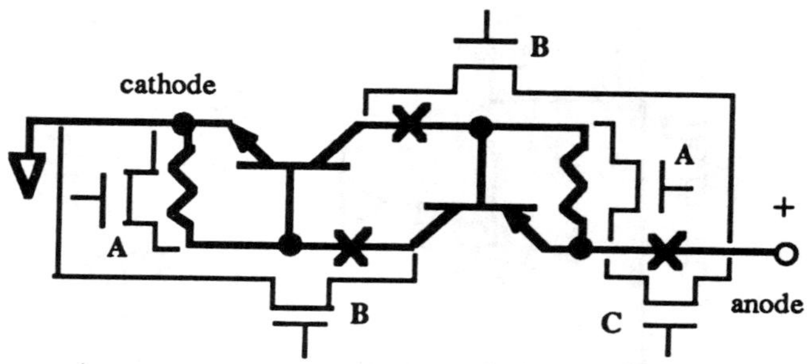

Fig. 8 Methods to turn off thyristor: A. emitter-base shorts, B. feedback loop interruption, and C. power path disruption.

dV/dt capability without adversely affecting device turn-on. Thus, the MCT, when compared with GTO, can be designed to operate at higher temperatures as well as in harsher radiation environments.

To increase the functional complexity of a smart power chip, the "on" resistance of the power device should be small, resulting in low power dissipation. The LMCT operating in the latch-up mode has the lowest "on" resistance and highest dV/dt capability of all lateral devices. It appears to be the most suitable choice for smart power applications in the future. The challenge in the design of LMCT is to develop methods to turn off the regenerative action using MOS gate structures. Figure 8 shows, in practice, three methods for the gate turn-off. In method A, the thyristor is turned off by introducing emitter-base shorts to raise the holding current above the operating current [16]. The base-current flow in method B is interrupted by pinching off one of the feedback paths. This technique has been applied to vertical discrete MCTs [17] using depletion-mode transistors for pinch-off. In method C, the operating current is disrupted by a MOS transistor causing the cathode (or anode) to float, as has been applied to the emitter switched thyristor (EST) [18]–[20].

Two examples of the bilateral MCT (BMCT) [21] are now described. One is an interrupted-gate-control BMCT shown in Fig. 6(a), in which the device is "on" regeneratively, so separate gate pulses are needed for turn-on and turn-off. The other is a continuous-gate-control BMCT, shown in Fig. 6(b). This device will be turned off when the same gate pulse that has turned it on before is removed. The symmetry of these bilateral devices allows designs of unique turn-off circuits that combine emitter-base short with emitter degeneration.

4.1. Interrupted-gate-control BMCT

Figure 9 shows the photomicrograph of the device that consists of 16 elemental cells occupying an area of 110 μm square. Each cathode cell is surrounded by four adjacent anode cells. As the device was made with the N-well CMOS process, the doping polarities are reversed from those shown in Fig. 6(a). Two polysilicon gates, one for turn-on and the other for turn-off, form interdigitated patterns that are underneath but orthogonal to the interdigitated metal lines of anode and cathode. The purpose of the design was to demonstrate turn-on and turn-off capabilities only. No attempt was made to optimize the breakdown voltage beyond what was offered by the

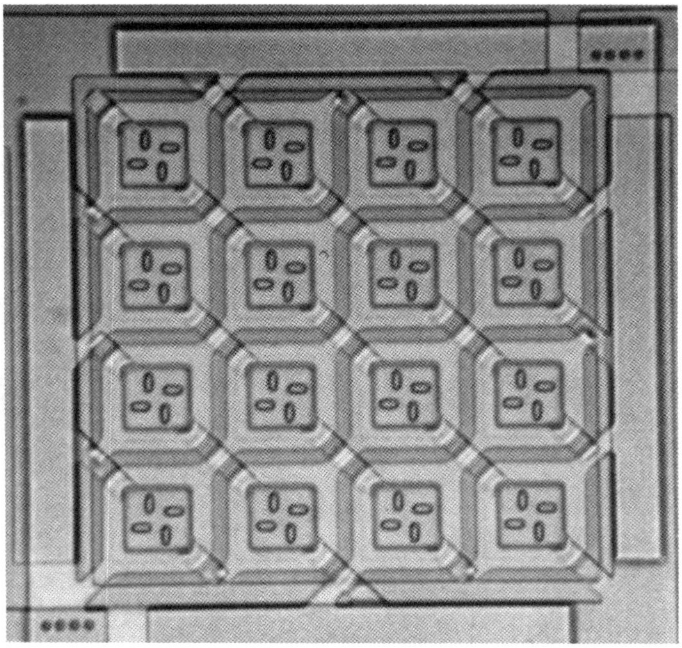

Fig. 9 Photomicrograph of interrupted-gate control BMCT.

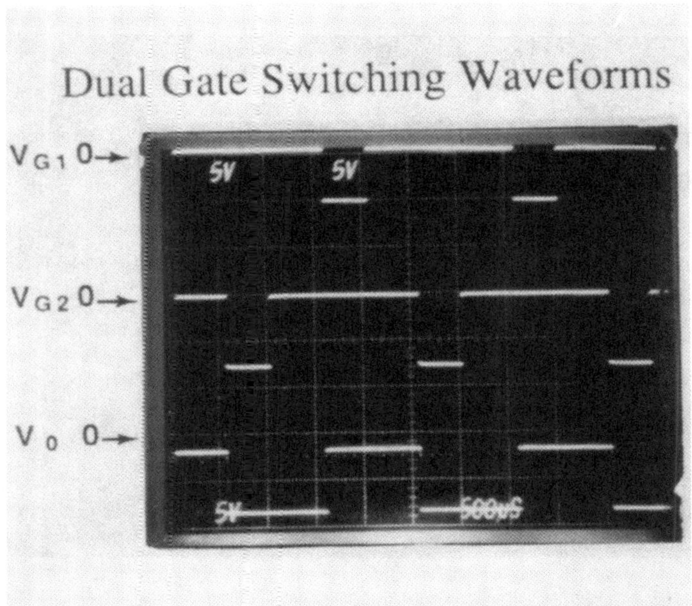

Fig. 10 Switching input and output waveforms of an interrupted-gate-control BMCT with doping polarities reversed from those in Fig. 6(a).

process. The current-handling capability, however, can be increased by simply paralleling more cells.

Assume initially the device in Fig. 6(a) is biased in the forward blocking mode. When the turn-on voltage exceeding the threshold voltage is applied to G1, electrons are injected into the base of the lateral PNP, which induce holes from the emitter of the PNP. As the potential drop from the hole current flowing through the shunt resistance R exceeds the PN junction threshold (~0.7 V), the vertical NPN starts to conduct, resulting in more electrons flowing into the N base. This regenerative process continues. At a sufficiently large device current called the trigger current, at which the effective current gain product of the bipolars exceeds unity, the device will be latched at a current essentially limited by the external resistor. Once the BMCT is on, it will remain on when the turn-on pulse is removed. To turn the device off, a separate gate pulse has to be applied to G2 to drain off the base current of the lateral PNP or equivalently to shunt the emitter-base junction of the PNP transistor. The on/off capability is clearly demonstrated by oscilloscope traces in Fig. 10, where the upper two traces are the turn-on and turn-off gate pulses and the bottom trace is the anode voltage in response to the input.

The maximum operating current density that can be switched off is experimentally found to exceed 800 A/cm^2. The switching speed is monitored on an oscilloscope by measuring the rise and the decay times of the anode voltage. The turn-on is fast, of the order of tens of nanoseconds. The turn-off time, as can be seen from Fig. 11, is about 0.7 µs. The longer turn-off time is due to excess carriers in the N epi region that have to be removed through the emitter-base short and the recombination process. It is shorter than those reported for the vertical device [22] partly because the charge storage volume in the low-voltage lateral device is smaller and partly because the current-flow path near the surface favors MOS turn-offs.

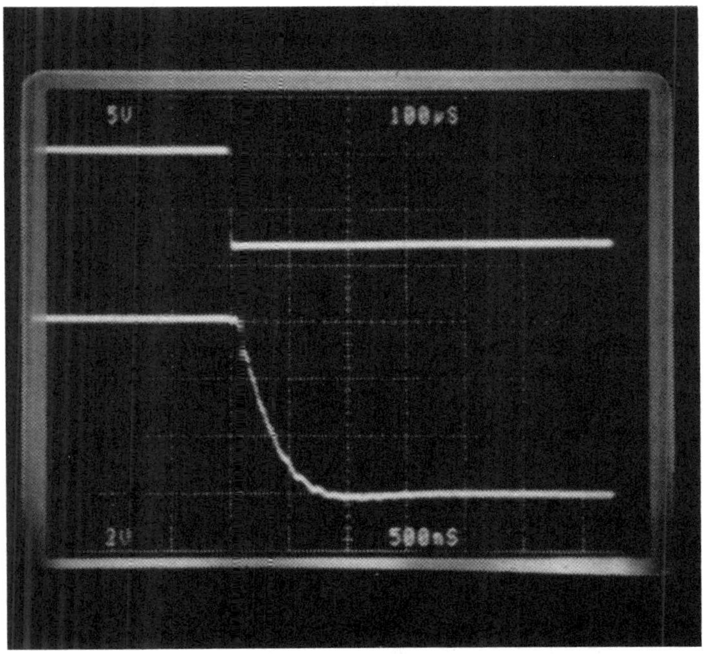

Fig. 11 The turn-off characteristics of an interrupted-gate-control BMCT.

4.2. Continuous-gate-control BMCT (Fig. 6(b))

The layout and the size of the continuous-gate-control BMCT are similar to those of the interrupted-gate-control BMCT shown in Fig. 9. It consists of distributed-cell elements

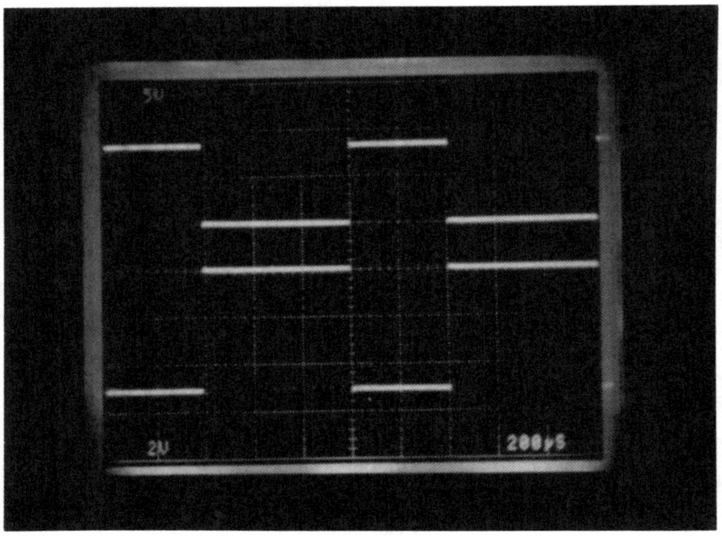

Fig. 12 Switching input (top) and output waveforms of a continuous-gate-control BMCT with doping polarities reversed from those in Fig. 6(b).

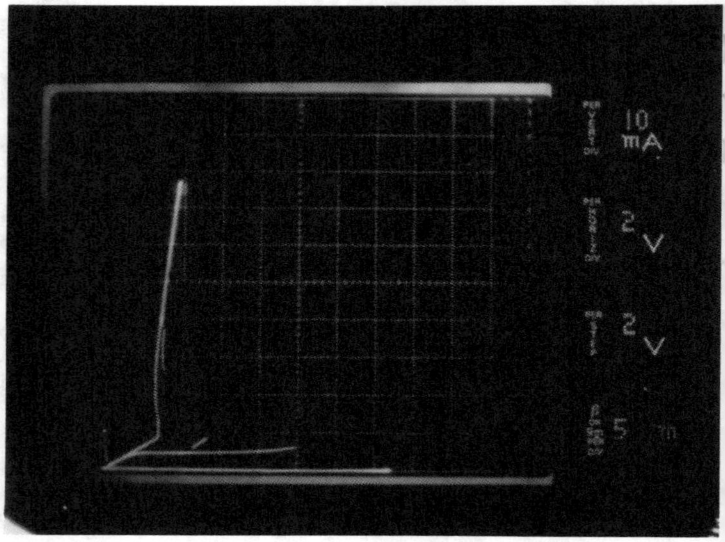

Fig. 13 Output I-V characteristics of a continuous-gate-control BMCT for V_g = 0 V, 2 V, and 4 V. Negative conductance regions result from positive feedback.

with polysilicon gates at the border. Application of a negative voltage to the gate turns on the PMOS transistor, initiating the regenerative process by injecting holes to the base of the vertical NPN transistor. When the device is latched, it conducts both the MOS and the bipolar currents. As the gate voltage is removed, the device will revert to its initial blocking state.

The switching waveforms of the continuous-gate-control BMCT are shown in Fig. 12, in which doping polarities of the device are reversed from those shown in Fig. 6(b). The output I-V characteristics for V_g equal to 0 V, 2 V, and 4 V are shown in Fig. 13, where negative conductance regions resulting from the regeneration process are apparent. The maximum operating current that can be controlled by the MOS is found to be about 650 A/cm^2.

5. Applications and Issues

Lateral devices used in smart power ICs could easily be designed to handle between 0.2 A to 10 A and operate in the 30-V to 300-V range. Potential applications include motor speed controls, lighting controls, automotive controls, switching power regulators, solid-state relays, etc. Possible benefits that can be accrued from smart power ICs are low cost, high reliability, improved performance, and reduced size and weight. But the most important benefit that can result from smart power is such value-added features as temperature sensing, over-current, and over-voltage protections not usually available with discrete devices.

Smart power chips tend to be custom-designed, targeted for specific applications. The fact that a baseline BiCMOS process can fabricate a whole gamut of power devices with different characteristics for a variety of applications appears particularly attractive. The ultimate decision on the type of power device to use, on whether or not to integrate, and on the extent of integration can only be made after careful trade-off studies on an application-by-application basis.

In the design of smart power chips with BiCMOS technology, some technical issues need to be investigated. Adequate voltage isolation and noise suppression between the high-voltage device and the low-voltage circuit will have to be provided. The internal heat generation and the resulting high chip temperature could be detrimental to circuit performance. Adequate packaging, offering good heat-sinking capabilities and high pin counts, must be developed. In this respect, the low ''on'' resistance of the LMCT devices should help mitigate the power-dissipation problem.

In ac switching, for example, the design of the biasing circuit becomes more complicated. Figure 14 shows a biasing scheme for the ac circuit using the continuous-gate-control BMCT shown in Fig. 6(b) [23]. When the power device is off, the gate is referenced to the N^- epi body, which is tied to the circuit's most positive potential at all times by a pair of PMOS transistors. A pair of NMOS transistors clamps the substrate to the most negative rails. Otherwise, the substrate could become forward-biased during the negative half of the ac cycle, causing the circuit to malfunction. When the device is switched on, both MOS clamps in Fig. 14 do not function, as there is little voltage across the power device. It is seen that both MOS clamps can easily be integrated with the ac power switch. In ac switching, turn-off poses no problem for the interrupted-gate-control BMCT, as it will turn itself off every time the ac wave-

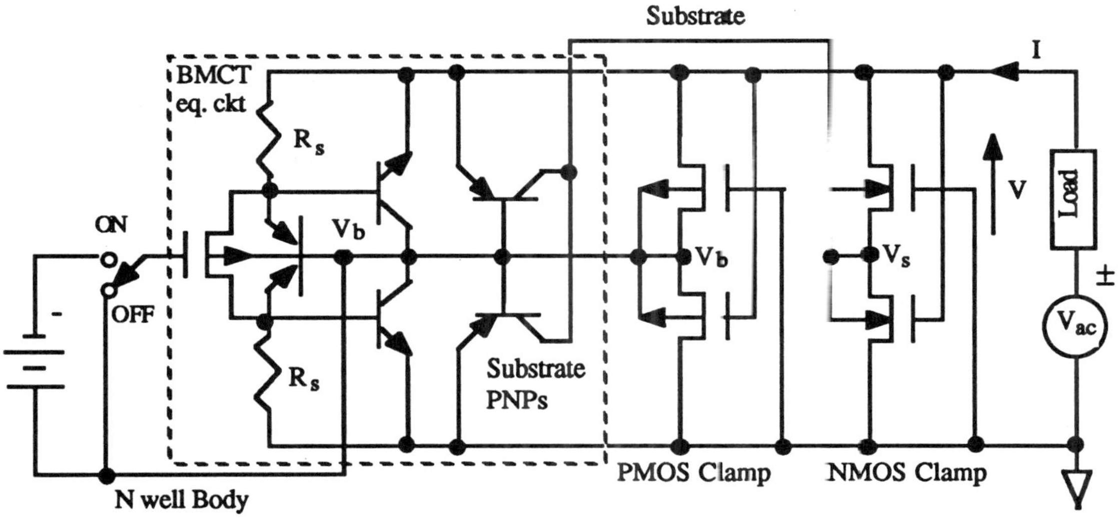

Fig. 14 Body and substrate biasing schemes for ac switching using continuous-gate-control BMCT.

form goes through zero, even when there is no turn-off voltage applied to the MOS gate.

6. Conclusions

It has been shown that the BiCMOS technology is suited for fabricating modern day smart power ICs. It allows the flexibility of designing different types of power devices intended for different applications. Lateral devices, in particular, provide the underpinning for power ICs with multiple power devices and ac-switching capabilities. The lateral MOS-controlled thyristor (LMCT) with MOS gate for turn-on and turn-off and bipolar for current-conduction appears to be the most attractive device for future smart power applications. Future BiCMOS technology built on fully or partially insulated substrates provides true isolation between various device types. It offers the potential for improved trade-offs in device performance and for enhanced flexibility in device design.

Finally, it should be emphasized that the smart power technology as a whole is market-and application-driven and will remain so in the future.

Acknowledgments

The author would like to thank J. M. Daughton, J. W. Schrankler, J. S. Kueng, and L. K. Cooper for helpful comments. He also would like to acknowledge T. Fabian and T. McAlpine for their assistance with experiments.

References

[1] Z. Parpia, J. G. Mena, and C. Salama, "A novel CMOS-compatible high-voltage transistor structure," *IEEE Trans. Electron Devices*, vol. ED-33, pp. 1948–1952, Dec. 1986.

[2] B. J. Baliga, M. Adler, P. Gray, R. Love, and N. Zommer, "The insulated gate rectifier (IGR): A new power switching device," in *IEDM Tech. Dig.*, pp. 264–267, Dec. 1982.

[3] J. Russell, A. M. Goodman, L. A. Goodman, and J. M. Neilson, "The COMFET—A new high conductance MOS-gated device," *IEEE Electron Device Lett.*, vol EDL-4, pp. 63–65, March 1983.

[4] D. N. Pattanayak, A L. Robinson, T. P. Chow, M. S. Adler, B. J. Baliga, and E. J. Wildi, "N-channel lateral insulated gate transistors: Part I—Steady state characteristics," *IEEE Trans. Electron Devices*, vol. ED-33, pp. 1956–1963, Dec. 1986.

[5] J. D. Plummer and B. W. Scharf, "Insulated-gate planar thyristors: I—Structure and basic operation," *IEEE Trans. Electron Devices*, vol. ED-27, pp. 380–387, Feb. 1980.

[6] J. S. T. Huang, J. S. Kueng, P. W. Green, and M. S. Holt, "Bidirectional lateral insulated gate transistors operated in controlled latch-up mode," Paper presented at Device Research Conference, University of Colorado, June 1988.

[7] M. Darwish, "A new lateral MOS-controlled Thyristor," *IEEE Electron Device Lett.* vol. 11, pp. 256–257, June 1990.

[8] R. H. Havemann and R. H. Eklund, "Overview of BiCMOS Device and process integration," in M. I. Elmasry (ed.), *Analysis and Design of BiCMOS Integrated Circuits*, New York: IEEE Press, 1994.

[9] J. A. Appels and H. M. J. Vaes, "High voltage thin layer devices (resurf devices)," in *IEDM Tech. Dig.*, pp. 238–241, Dec. 1979.

[10] E. J. Wildi, T. P. Chow, M. S. Adler, M. E. Cornell, and G. C. Pifer, "New high voltage IC technology," in *IEDM Tech. Dig.*, pp. 262–265, Dec. 1984.

[11] M. Darwish and K. Board, "Lateral resurf COMFET," *Electronics Lett.*, vol. 20, pp. 519–520, June 7, 1984.

[12] Y. S. Huang and B. J. Baliga, "Extension of RESURF principles to dielectrically isolated power devices," in *Proc. 3rd Int. Symp. on Power SC devices and ICs*, pp. 27–30, Baltimore MD, April 22–24, 1991.

[13] S. Merchant, E. Arnold, H. Baumgart, S. Mukherjee, H. Pein, and R. Pinker, "Realization of high breakdown voltage (>700 V) in thin SOI devices," in *Proc. 3rd Int. Symp. on Power SC devices and ICs*, pp. 31–35, Baltimore MD, April 22–24, 1991.

[14] K. P. Brieger, W. Gerlach, and J. Pelk, "Blocking capability of planar devices with field limiting rings," *Solid State Electronics*, vol. 26, pp. 739–745, Aug. 1983.

[15] M. Azuma and M. Kurata, "GTO thyristor," *Proc. IEEE*, vol. 76, pp. 419–427, April 1988.

[16] V. A. K. Temple, "MOST-controlled thyristor—A new class of power devices" *IEEE Trans. Electron Devices*, vol. ED-33, pp. 1609–1618, Oct. 1986.

[17] B. J. Baliga and H R. Chang, "Gate turn-off capability of depletion-mode thyristors," *IEEE Electron Device Lett.*, vol. 10, pp. 464–466, Oct. 1989.

[18] B. J. Baliga, "The MOS-gated emitter switched thyristor," *IEEE Electron Device Lett.*, vol. 11, pp. 75–77, Feb. 1990.

[19] M. S. Shekar, B. J. Baliga, M. Nandakumar, S. Tandon, and A. Reisman, "Characteristics of the emitter-switched thyristor," *IEEE Trans. Electron Devices*, vol. 38, pp. 1619–1623, July 1991.

[20] J. S. T. Huang, "The bilateral emitter switched thyristor (BEST)," in *IEDM Tech. Dig.*, pp. 249–252, Dec. 1992.

[21] J. S. T. Huang, "Characteristics of bilateral MOS-controlled thyristors (BMCT)," *IEEE Trans. Electron Devices*, vol. 40, pp. 1523–1529, Aug. 1993.

[22] Op. cit., V. A. K. Temple.

[23] J. S. T. Huang, "Applications of BiCMOS to smart power devices," Paper presented at the Int. Symp. on Circuits and System (ISCAS), New Orleans, LA, May 1990.

Evolution of MOS-Bipolar Power Semiconductor Technology

B. JAYANT BALIGA, FELLOW, IEEE

Invited Paper

A review of the innovations that have led to the evolution of a new power transistor technology based upon MOS gate control is provided. This device technology offers the advantage of very high input impedance which allows the control of the devices using low-cost integrated circuits. The physics of operation of the two types of devices in this category, power MOSFETs and power MOS-bipolar devices, are described. Trends in process technology and device ratings are analyzed. Based upon the superior performance of these devices, it is projected that they will completely displace the power bipolar transistor in the future.

I. Introduction

Ever since the invention of the transistor, the development of solid-state switches with increased power handling capability has been of interest in terms of expanding the application of these devices in systems. The development of a family of bipolar power devices in the 1960s can be related to this need to serve systems demanding higher power levels. In this regard, the thyristors have been found to lend themselves very well to scaling up voltage and current ratings. A survey of the development of these devices indicates a good correlation between their power ratings and the availability of large diameter, high resistivity, float-zone, silicon. In general, the availability of thyristors with higher power ratings tracks two years behind the availability of improved silicon wafers due to the time required to introduce the new silicon technology into the device fabrication lines and complete their characterization.

The reliance of the power electronics industry upon the bipolar devices has been shaken by the introduction of a new power device technology in the 1980s. This technology utilized MOS gate control to create a new family of power devices which can be driven with extremely low input power levels. In addition, with their inherently faster switching speed, these devices have created the opportunity to increase the operating frequency in power systems resulting in improvements in efficiency, size, and weight. In addition, the ability to integrate the control circuits for these devices and to provide protection against adverse circuit conditions has revolutionized the way in which power systems are being built.

This paper reviews the development of the MOS controlled device technology that has been a cornerstone for the integrated power electronics thrust in the industry. After a brief historical perspective, the paper describes the two key technologies that are being utilized for all the applications, namely, the power MOSFETs and the MOS-bipolar devices. Following this description of the devices, the trends in the evolution of the power ratings of these devices are provided including a projection of the market.

Manuscript received June 22, 1987; revised January 22, 1988.

The author is with the General Electric Company, Corporate Research and Development Center, Schenectady, NY 12301, USA.

IEEE Log Number 8820813.

II. Historical Perspective

A. Bipolar Devices

Although the focus of this paper is the MOS power device technology, the status of bipolar power devices prior to the introduction of these devices is worth reviewing in order to obtain a perspective upon the motivation for the development of the new technology. Bipolar power devices can be divided into two groups: the power thyristors and the power bipolar transistors. The thyristors are fabricated today with very large device area, and extremely high operating voltages have been achieved. Single devices fabricated from a 100-mm-diameter wafer of silicon are being manufactured with current handling capability of 1000 A and breakdown voltages of over 6500 V. Development effort is underway to increase the wafer diameter to 125 mm and to scale up the blocking voltage to 10 000 V. These devices being used in high voltage dc (HVDC) transmission systems. At present, no other device technology can match the performance of these devices and their application to very high power systems is expected to continue in the foreseeable future.

The bipolar power transistor and gate turn-off thyristor (GTO) have also been developed over the past 20 years to serve the needs of the power electronics industry. Their primary advantages over the thyristors have been the superior switching speed and the ability to interrupt current without reversal of the device voltage. These attributes have led to their application in higher frequency inverters which result in an increase in the efficiency and a reduction in the size of the snubber components [1].

Reprinted from *Proc. IEEE*, vol. 76, no. 4, pp. 409–418, April 1988.

One of the primary difficulties with the application of the bipolar power transistor is a degradation in the gain of the device with increasing blocking voltage capability. Since the bipolar transistor is fundamentally a current controlled device, a reduction in the gain results in larger control signals which must be delivered using expensive discrete circuits. The use of the Darlington configuration has greatly improved the performance of these devices. However, even in this case, typical devices with blocking voltages of 500 V exhibit current gains of 10 at high current levels with relatively high forward voltage drops in the on-state. Further, due to a limited safe-operating-area arising from current filamentation, these devices require snubber circuits which can raise system cost in their applications.

B. Power MOSFETs

The power MOS Field Effect Transistor (MOSFET) is a device that evolved from MOS integrated circuit technology. The motivation for the development of these devices arose from the large base drive current required by power bipolar transistors and their limited switching speed capability. The first attempts to develop high voltage MOSFETs were performed by redesigning lateral MOSFETs to increase their voltage blocking capability [2]. This work at Stanford University by Plummer was motivated by the interest in developing high speed switches for driving piezoelectric devices in medical electronics. The technology developed for these devices was lateral double-diffused MOS (DMOS) as illustrated in Fig. 1. In parallel with this effort, several

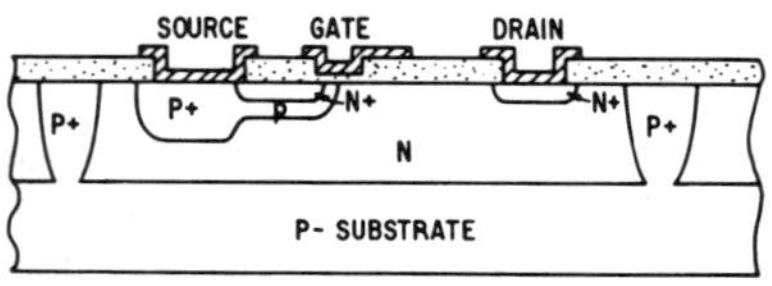

Fig. 1. Lateral DMOS high voltage driver for integrated circuits.

groups were working on the development of vertical power MOSFETs. This effort was motivated by the improved breakdown voltages and higher current ratings that could be obtained by moving the drain from the top to the bottom of the wafer as shown in Fig. 2(a) and (b). Although the vertical DMOS structure has been adapted by virtually all manufacturers of power MOSFETs today, the initial work in the early 1970s began with the use of the V-groove process. The

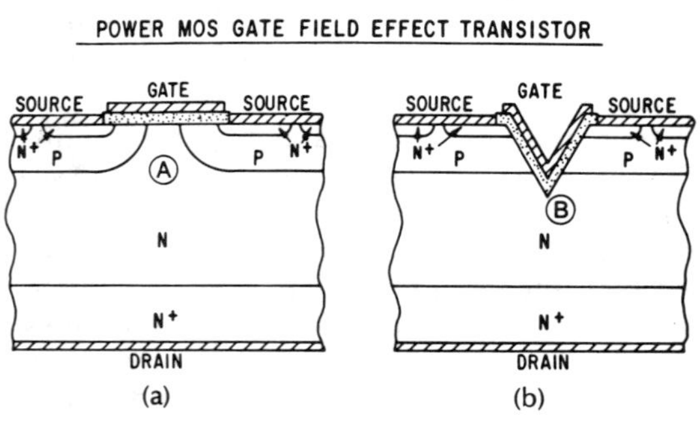

Fig. 2. Vertical power MOSFET structures for discrete power switching output devices. (a) Double diffused MOSFET (DMOS) structure. (b) V-groove MOS (VMOS) structure.

first high voltage devices were commercialized by Siliconix and International Rectifier in the early 1970s. The initial focus on the VMOS process was based upon the belief that the JFET pinch-off action between the p-base regions of adjacent cells in region *A* (see Fig. 2(a)), through which the drain current must flow, would result in a significantly higher resistance for the DMOS devices compared with the VMOS devices. However, detailed modeling studies of breakdown voltage of these devices pointed out the existence of very high electric fields at the bottom *B* of the V-groove (see Fig. 2(b)). This high field resulted in a significant reduction in the breakdown voltage compared to the DMOS geometry for the same drift layer doping and thickness. Since the on-resistance of the power MOSFET increases as the 2.5 power of the breakdown voltage rating, the VMOS structure was found to have a higher on-resistance than the DMOS structure for the same breakdown voltage [3], [4]. Further, the V-groove process exposed the gate oxide to ionic contamination resulting in unstable device threshold voltages which compromised production yield and device reliability.

The introduction of DMOS power FETs was originally regarded as a major threat to the power bipolar transistor. Initial claims of infinite current gain for the power MOSFETs were diluted by the need to design the gate drive circuit to account for the pulse currents required to charge and discharge the high input capacitance of these devices. This was especially true in high frequency applications where the power MOSFET was particularly valuable due to its inherently high switching speed (1 to 10 ns compared with 1 μs for bipolar transistors). The high speed capability was the result of current transport occurring solely via majority carriers. This eliminated the large storage and fall times observed in bipolar transistors due to minority carrier transport. The power MOSFETs were also found to exhibit superior safe-operating-area and output characteristics for paralleling when compared with the bipolar transistor. However, these merits of the power MOSFET were offset by a higher on-resistance per unit area and a much larger initial processing cost when compared to the bipolar transistor. In order to derive good characteristics, high resolution lithography with good registration between masking levels was crucial to power MOSFET fabrication and the yield was much lower than for bipolar transistors of the same size. Consequently, the high cost of power MOSFETs restricted their application to high-frequencies circuits (such as those used in switch-mode power supplies), and to low-voltage ($<$100 V) circuits where their on-resistance reached acceptable values.

With improvements made in process technology to obtain better yields and performance, the ratings of power MOSFETs have continued to grow over the last 10 years and their cost is now on par with bipolar transistors. At present, devices are available with breakdown voltages up to 1000 V at current levels of about 1 A and current handling capability of over 20 A at breakdown voltages below 100 V. The maximum pellet size that is in production today is about 250 mils $\times$ 250 mils in size, although larger devices (300 mils $\times$ 300 mils) have been reported in the literature [5]. These large-area low-voltage devices are being developed for use as synchronous rectifiers which are very low forward drop diodes that are aimed at replacing the Schottky output rectifiers in low-voltage ($<$5 V) switching power supplies.

C. MOS-Bipolar Hybrid Devices

With the discovery that power MOSFETs were not in a strong position to displace the power bipolar transistor, many researchers began to look at the possibility of combining these technologies to achieve a hybrid device which has a high input impedance and a low on-resistance. The most obvious step in this direction was to simply drive the bipolar transistor with a power MOSFET in the Darlington configuration shown in Fig. 3(a). This circuit element could

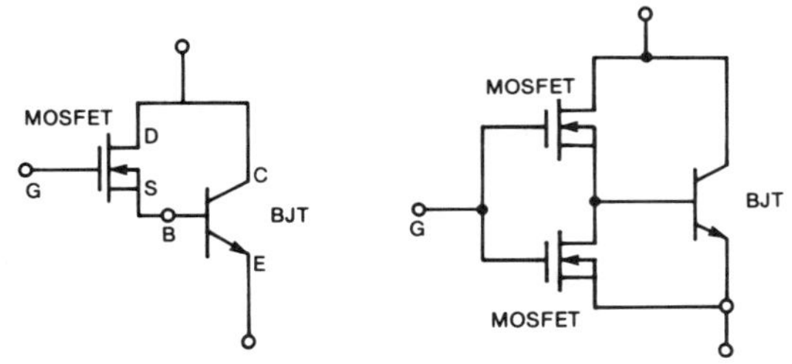

(a) (b)

Fig. 3. Hybrid MOSFET/bipolar transistor configuations for achieving a high input impedance and low on-resistance. (a) Darlington circuit with MOSFET providing base drive for bipolar transistor. (b) Circuit with MOSFETs for providing both base drive and emitter-base shorting for fast turn-off.

be gated-on like a power MOSFET and yet have a lower on-resistance than a MOSFET because most of the output current was handled by the bipolar transistor. One application for this configuration was in motor drives and processes to integrate these devices together on a monolithic chip were developed [6]. The drawback of this approach was, again, the poor performance of the power bipolar transistor. The low current gain (typically 10) of high voltage bipolar transistors made it essential to use a high-voltage MOSFET of equal size as a driver. In addition, the usual reverse base drive current applied to the bipolar transistor during turn-off could not be used without the addition of another expensive power MOSFET as illustrated in Fig. 3(b). It is also worth pointing out that this hybrid approach could only result in an average current density which was in-between that obtainable for the power MOSFET and bipolar transistors, while requiring process technology that must optimize the performance of both devices simultaneously [7].

An alternative hybrid approach to obtaining an MOS high input impedance device configuration with a high forward conduction current density was developed at the GE Research Center by Baliga in the late 1970s. In this approach, the power field controlled thyristor was used as the main current carrying high voltage device as illustrated in Fig. 4 and a power MOSFET was used to control the output current [8]. The field controlled thyristor is a device which operates like a p-i-n diode in its forward conduction mode. This results in a very high on-state current density for this device compared with the bipolar transistor even at high breakdown voltages. Before the development of this gating scheme, shown in Fig. 4, it had been necessary to apply a negative voltage to the gate of the FCT to hold it in the off-state and to provide large gate current pulses to achieve high speed turn-off [9]. These problems were eliminated by the circuit shown in Fig. 4 with the use of a low voltage power

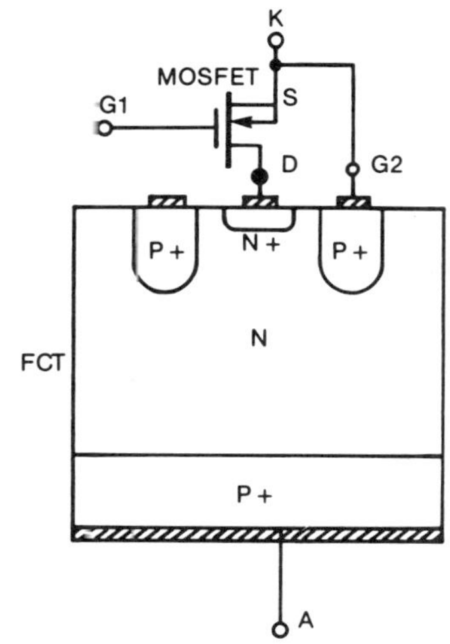

Fig. 4. Hybrid MOS/bipolar configuration for achieving high input impedance and high forward conduction current density by using a power MOSFET and field controller thyristor. Here, high speed forced gate turn-off is achieved by emitter switching the FCT by turning off the series MOSFET and diverting cathode current into the gate.

MOSFET connected in series with the FCT. These MOSFETs were considerably smaller than those required for the Darlington configuration in Fig. 3 due to their low voltage ratings. Yet it was highly desirable to integrate the MOSFET with the FCT to obtain a single monolithic device.

D. MOS-Bipolar Functional Integration

To obtain the maximum benefits of the MOS gate control and bipolar current conduction, it was necessary to discard the approach of using separate discrete power MOSFETs and bipolar devices and to conceptualize the merging of the physics within the unit cell structure. The first demonstration of the functional integration of MOS gate control and bipolar current transport was achieved in the late 1970s as a discrete vertical channel MOS gated thyristor by Baliga at the GE Research Center [10], and as a lateral MOS gated triac by Sharf and Plummer at Stanford University [11]. In these devices, an MOS gate structure was used to trigger the latch-up of a four-layer thyristor structure to obtain devices with much superior *di/dt* and *dv/dt* capability compared with conventional thyristors. These devices were later commercialized by Motorola for providing a protective crowbar function [12]. The widespread application of the devices was limited by the loss of gate control after turn-on, which made these devices unacceptable as replacements for bipolar transistors. In the early work on MOS-gated thyristors by Baliga at GE and Plummer at Stanford, it had been observed that gate controlled current flow could be observed at low current levels prior to the initiation of the latch-up of the thyristor. To create a truly MOS-controlled bipolar device it was necessary to extend this gate controlled regime to high current levels.

It took a conceptual breakthrough in the early 1980s to realize that the parasitic thyristor inherent in the above structure could be defeated by adequate emitter shorting, as illustrated in Fig. 5, to create the insulated gate transistor[1]

[1]The term Insulated Gate Bipolar Transistor (IGBT) has been accepted by JEDEC.

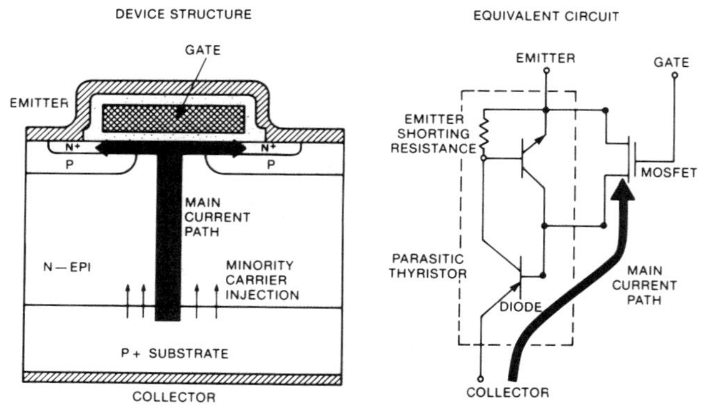

Fig. 5. The insulated gate transistor structure and its equivalent circuit. The bold arrows indicate the major current flow path. By effectively shorting the emitter-base junction of the upper *N-P-N* transistor, the circuit simplifies to that shown in Fig. 3(a).

[13]. This was achieved by Baliga and his colleagues at the General Electric Research Laboratory by a combination of strength in two-dimensional device modeling capability and extensive experience with process development for DMOSFETs and other novel device structures. A similar device called the COMFET was developed in the same time frame at the RCA Research Laboratory [14]. The modeling work at GE indicated that it was necessary to include a highly doped P^+ region in the center of each DMOS cell to reduce the lateral resistance in the p-base in order to suppress latch-up of the parasitic thyristor [15]. The ability to incorporate this P^+ region into a power MOSFET production process with a minimum of process modifications allowed device development and transition into production to occur rapidly. The first devices (IGT(A) in Fig. 6) exhibited excellent characteristics with forward conduction current densities of twenty times higher than for MOSFETs, and five times greater than for a bipolar transistor operating at a current gain of 10. These results demonstrated the value of functionally integrating the physics of an MOS structure for controlling the current and bipolar physics for conducting the current. However, the introduction of minority carriers for current transport resulted in a slow switching speed. The first IGTs had forced gate turn-off times of 10 to 20 μs. This relatively slow switching speed would have restricted the application of these devices to circuits operating at low frequencies such as directly off the ac power line. In order to use the device for a wider range of applications, including motor drives, it became necessary to develop a lifetime control process with which the turn-off speed could be reduced without degrading the MOS gate structure. This was achieved by using an electron irradiation process [16] originally developed for controlling the speed of p-i-n rectifiers [17]. This process created the ability to control the switching speed of the IGT in the range of turn-off times from 0.2 to 20 μs, which made the IGT a unique and versatile device which could be optimized to achieve the lowest combination of conduction and switching power losses for each application depending upon the operating frequency [18].

With the advent of these devices, the bipolar transistor is seriously threatened with replacement. The high operating current density of the IGT more than offsets the slightly higher cost of MOS fabrication. Its advantages of a much lower gate drive power which can be derived from low cost

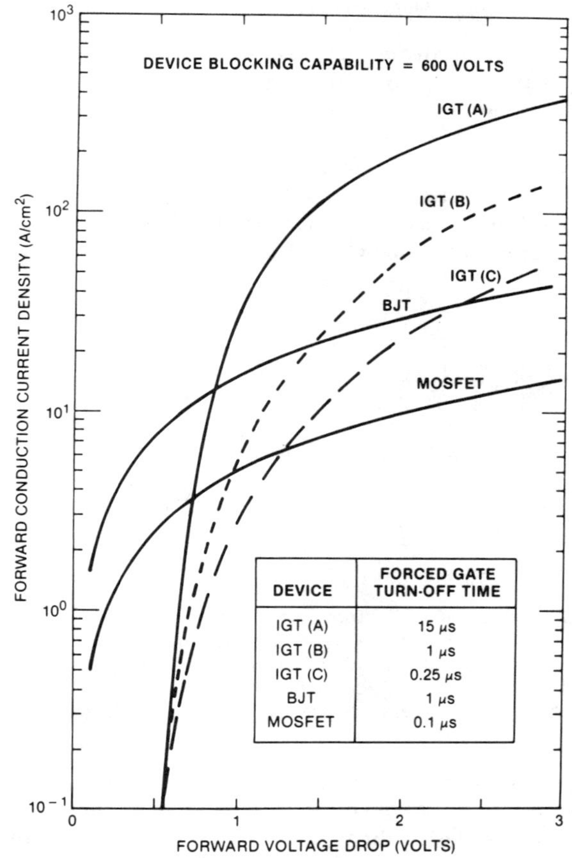

DEVICE	FORCED GATE TURN-OFF TIME
IGT (A)	15 μs
IGT (B)	1 μs
IGT (C)	0.25 μs
BJT	1 μs
MOSFET	0.1 μs

Fig. 6. Comparison of IGT forward conduction characteristics with power MOSFET and bipolar transistor. The three IGT curves correspond to different switching speeds given in the table. (Experimental data is used here.)

integrated electronics, its unique reverse blocking capability [13], superior safe-operating-area [19], [20], and high temperature performance [21] make it an ideal choice for high voltage (>100 V), lower frequency (<50 kHz) applications. Since the power MOSFET offers significantly superior performance compared to the bipolar transistor at breakdown voltages of less than 100 V and at frequencies above 50 kHz, it is predicted that the power bipolar transistor is likely to be completely replaced with power MOS devices in the near future.

III. Power MOSFET Technology

A. Power MOSFET Operation

The basic vertical DMOS structure is illustrated in Fig. 7. The device contains two important regions—the *N*-drift region that supports the high voltage, and the channel region which controls the on/off-state of the device. When the gate is connected to the source, the device blocks current flow for positive drain voltages. The *P-N* junction (*J*1) becomes reversed biased and a depletion layer extends into the *N*-drift region. The resistivity and thickness of this region determine the breakdown voltage. When a positive gate bias is applied, an n-channel (inversion layer) is formed at the surface of the *P*-base region. This allows electron transport from the N^+ emitter to the N^+ drain (substrate). Since increasing the gate voltage enhances the channel conduc-

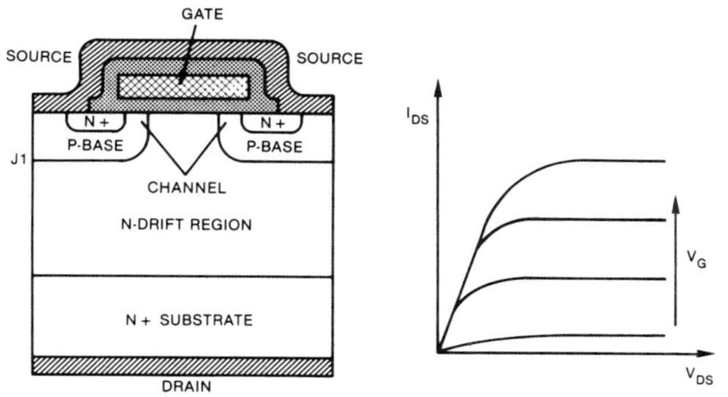

Fig. 7. Power MOSFET cross section and its output characteristics.

tivity, the device current increases with increasing gate voltage as illustrated on the right-hand side in Fig. 7.

B. Power MOSFET On-Resistance

The maximum current handling capability of the power MOSFET is determined by its on-resistance. The on-resistance is defined as the resistance of the device at low drain currents. It is primarily determined by a combination of the resistances of the channel and drift region. Other components that can impact the device are the contact and substrate resistances.

The channel and drift region resistance are in turn determined by the DMOS cell design. The cell design consists of many variables: 1) cell topology (circular, hexagonal, square, linear, etc.); 2) the polysilicon width; 3) the polysilicon window; 4) the channel length (determined by the difference in lateral diffusion between the *P*-base and the N^+ emitter); 5) gate oxide thickness; and 6) the gate drive voltage. The detailed design of the DMOS cell is beyond the scope of this paper and has been treated in depth elsewhere [4], [22]. An example of the impact of changing the polysilicon width upon the specific on-resistance (i.e., on-resistance per unit area) is provided in Fig. 8 for the case of a square cell design. In this figure, the impact of changing the cell window size is also shown. It is apparent that the on-resistance can be greatly reduced by decreasing the cell window size.

The minimum cell window size is determined by the device process technology, particularly the lithographic tolerances used in fabricating the cell structure. Over the years, there has been a continual advancement in this area resulting in the ability to reduce the cell size while maintaining a reasonable yield. The progressive decrease in cell size is illustrated in Fig. 9. At the start of this decade, the minimum cell size was about 50 microns. This corresponds to the upper curves in Fig. 8 which represent an optimum cell design with a cell window of 25 to 35 microns and a polysilicon width of 20 to 25 microns. With improvements in process technology (such as projection printing, better step coverage, dry etching, etc.), the minimum DMOS cell size has now been reduced to 20 microns (see Fig. 9). This corresponds to the lower curves in Fig. 8. As a consequence of this, the minimum specific on-resistance achievable for a 50-V DMOSFET has been reduced from about 7 milliohm-cm^2 to 1.7 milliohm-cm^2 as illustrated in Fig. 10. Simultaneously, the yield of larger chips has been improving and the largest power MOSFET chips that are manufacturable

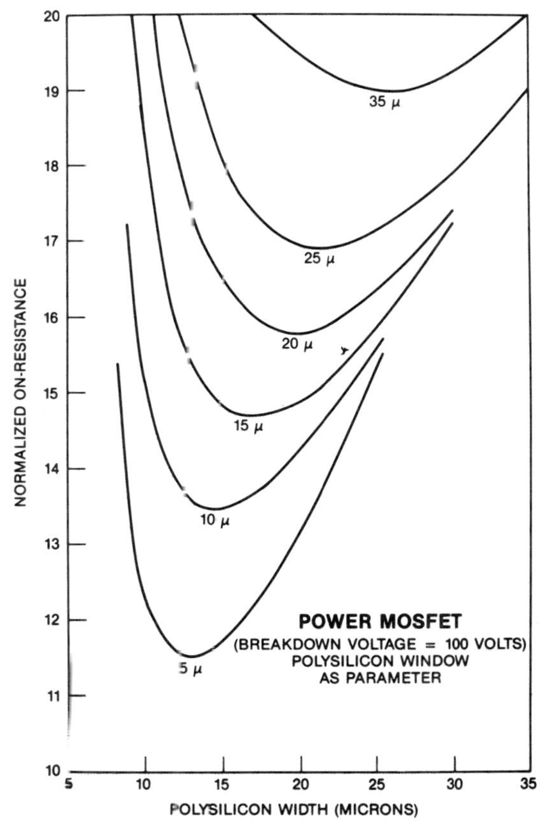

Fig. 8. Impact of changing the polysilicon width (gate overlap area) upon the on-resistance of a power MOSFET (theoretical calculations). All values are normalized to the ideal on-resistance of a 100-V device (5×10^{-4} ohm-cm^2). Each curve corresponds to a different polysilicon window size as determined by process technology limitations.

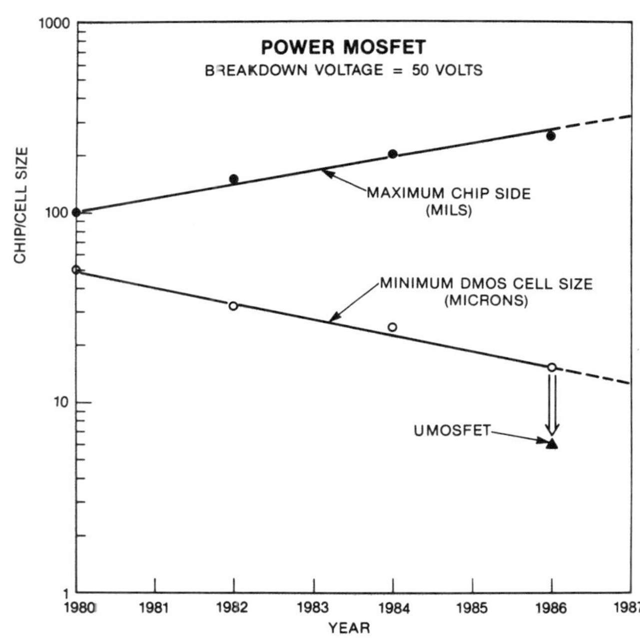

Fig. 9. Trends in DMOS cell size reduction and increase in maximum chip size in production. The minimum cell size is determined by process technology. The maximum die size is determined by defect densities which control yield. The UMOS data point represents a new technology that is not as yet in production in 1988.

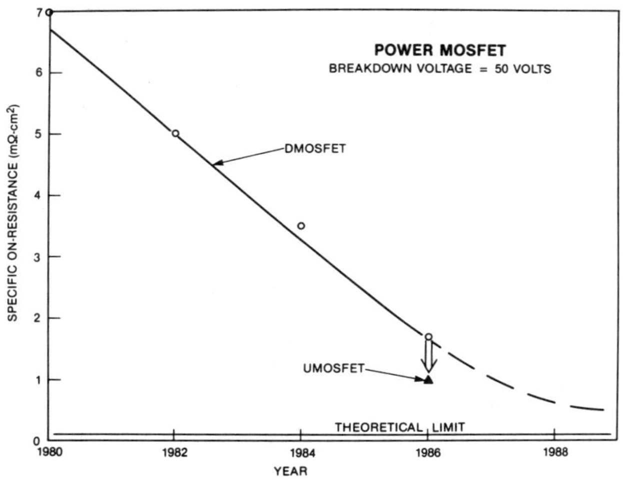

Fig. 10. Reduction in specific on-resistance (on-resistance for a chip with 1 cm^2 active area) of 50-V power MOSFETs with technological advances over the years. The UMOSFET data point represents a new technology that is not as yet in production in 1988. All the data are experimentally measured values.

with reasonable yield has grown from 100 mils in 1980 to 250 mils in 1986 (see Fig. 9).

The overall impact of the reduction in cell size and increase in the chip size is a rapid growth in maximum number of cells that can be integrated on a single chip. It can be seen from Fig. 11 that starting with only 2500 cells per chip in 1980, the power MOS industry has the capability to integrate nearly 200 000 cells per chip in 1986. This ability has resulted in a dramatic reduction in the lowest on-resistance available in a power MOSFET. Early devices (circa 1980) had on-resistances of over 100 milliohms. In 1988, devices with an on-resistance of less than 5 milliohms are becoming available. The resulting steady growth in power ratings of the DMOSFET implies greater penetration of this technology in systems.

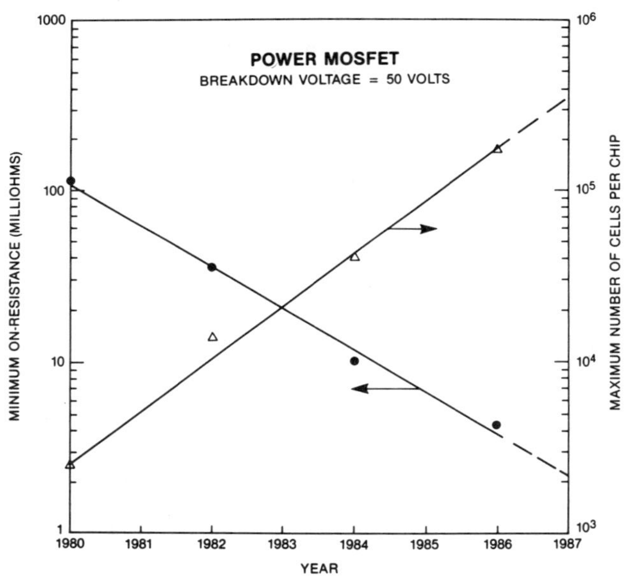

Fig. 11. Growth in the number of DMOS cells per chip for power MOSFETs and the corresponding reduction in the lowest available on-resistance.

It is worth pointing out that improvements discussed above for the low voltage power MOSFET are also applicable at higher voltages. A comparison of the specific on-resistance achievable in 1981 [23] with that obtained today in state-of-the-art devices is given in Fig. 12. In this figure, the ideal on-resistance limit for silicon is also provided because it defines the limit for this technology. An important conclusion that can be derived from this figure is that there is still large room for improvement of the specific on-resistance especially for the low voltage devices.

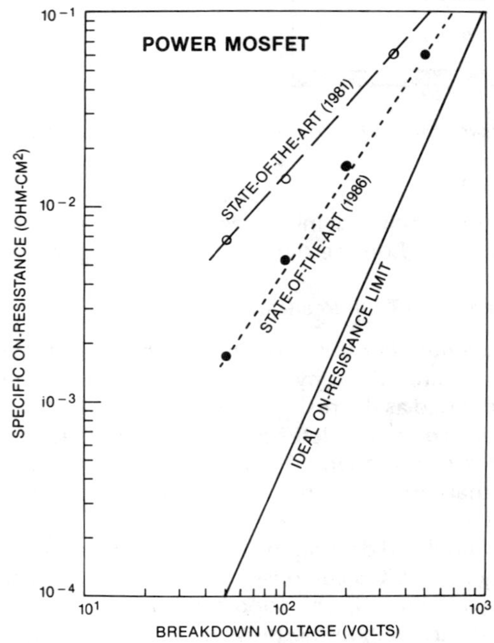

Fig. 12. Advances in specific on-resistance (experimental data) for power MOSFETs with various voltage ratings compared with the ideal silicon unipolar device limit. Progress in fabrication technology from 1981 to 1986 has resulted in a downward shift in this relationship, with larger improvements being achieved at lower voltage ratings where a larger room for improvement continues to exist.

1) UMOSFET Structure: The above discussion was confined to the DMOS cell structure. Some inherent disadvantages of that structure are the parasitic resistance of the region between the *P*-base diffusions (often called the JFET region) and the low channel density (channel width per cm^2 of active area). A big improvement in on-resistance can be obtained by using the UMOSFET structure illustrated in Fig. 13 [24], [25]. This structure is fabricated by forming vertically walled trenches which are refilled with the polysilicon gate after formation of the gate oxide. With this structure, the JFET region is eliminated. Further, a much smaller cell size can be achieved which results in an increase in the channel density.

The impact of the UMOSFET structure is illustrated in Figs. 9 and 10. By using a cell size of 6 microns, a specific on-resistance of 1 milliohm-cm^2 has been achieved [25]. Although this technology is in a development phase and no commercial devices are as yet available, it represents an important innovation that could drive the power ratings upwards in the future.

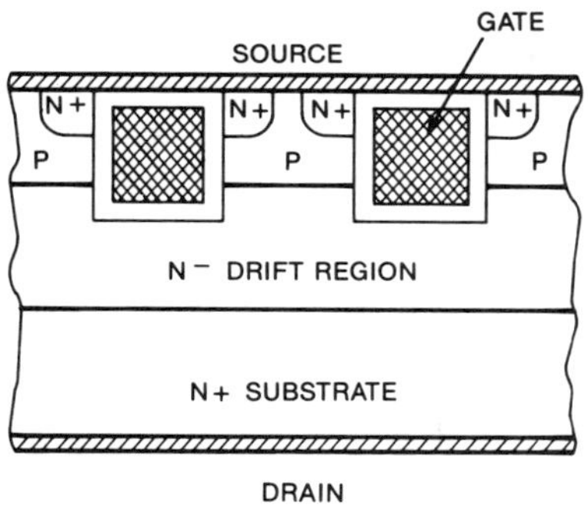

Fig. 13. Cross section of a vertical channel UMOSFET structure. This structure is fabricated using trench gate technology resulting in a large (~3*X*) increase in channel density.

IV. Power MOS-Bipolar Technology

Although power MOSFETs offer the advantage of a low control current, the sharp increase in their on-resistance with increasing voltage (see Fig. 12) results in poor power handling capability compared with the bipolar transistor. The impact of the power MOSFET is strongest in systems operating at below 100 V. A prime example of this is automotive electronics. To impact higher voltage systems, a new class of power devices based upon a combination of bipolar and MOS physics has been developed. The most mature example among these devices is the insulated-gate-transistor (IGT) [13].

This device can be fabricated by starting with a p^+ substrate and then growing the n-type drift layer on it. The DMOS gate structure is then formed on this drift layer by using a process similar to that described for power MOSFETs. Since this device contains two back-to-back high voltage junctions $J1$ and $J2$ (see Fig. 14), it is capable of symmetrical forward and reverse blocking capability. The forward blocking is achieved at junction $J1$ when the gate electrode is shorted to the emitter contact. To turn on the devices, a positive gate voltage must be applied to form an n-type channel at the surface of the p-base region under the gate. This channel then links the n^+ emitter to the n-drift layer as in the case of the power MOSFET. However, in the IGT, the positive bias on the collector causes the p^+ region to inject a high concentration of minority carriers into the n-drift region which severely reduces its resistance. Thus, in the on-state, the IGT behaves like a forward biased p-i-n rectifier and is capable of operating at high current densities equivalent to that of the GTO. At the same time, the current in the IGT can be controlled using the applied gate bias voltage. This feature of the IGT not only allows gate turn-off capability but is useful for limiting the collector current. Thus, the IGT exhibits the best feature of the power MOSFET, namely, voltage controlled operation with very high input impedance, as well as the best feature of bipolar devices, namely, high forward conduction current density. It has also been found that the on-state voltage drop of the IGT increases with increasing temperature [21]. This allows paralleling of these devices and ensures good current distribution within each device.

One of the outcomes of bipolar operation in the IGT is a reduced switching speed compared with the power MOSFET. The first IGT devices were reported to have a typical forced gate turn-off time ranging from 10 to 50 μs. This feature would have limited the IGT to very low frequency circuits. However, by using minority carrier lifetime control techniques, IGTs with gate turn-off times as low as 0.25 μs have been fabricated [16].

A comparison of the IGT forward conduction characteristics with the bipolar transistor and power MOSFET is provided in Fig. 6. The three IGT curves represent devices with different switching speeds [16]. From Fig. 6, it can be concluded that for high-voltage low-frequency applications, the IGT offers the advantages of a high input impedance and a chip size that is an order of magnitude smaller than the MOSFET. This advantage of the IGT has been found to be applicable for frequencies as high as 50 kHz. Since the process technology for IGTs closely parallels that for power MOSFET, the ability to increase the power handling capability per cm^2 has had a dramatic impact on the availability of power MOS gated chips with higher power ratings.

This ability to tailor the IGT characteristics over a broad range of frequencies to optimize its characteristics is another important feature of the device. Further due to the strong modulation of the n-drift layer resistance, the IGT current density varies relatively slowly as the square root of the blocking voltage capability [26]. This makes it an attractive device for high voltage applications. The combination of high input impedance and high conduction current density in the IGT makes it a superior power device in many circuits where bipolar transistors have been used in the past. Consequently, the IGT can be expected to displace the bipolar transistor in many of its applications. Power MOS-IGTs with continuous current ratings of 10 and 25 A at a blocking voltage capability of up to 1200 V have already become available. A rapid growth in both the current and voltage ratings of these devices is anticipated.

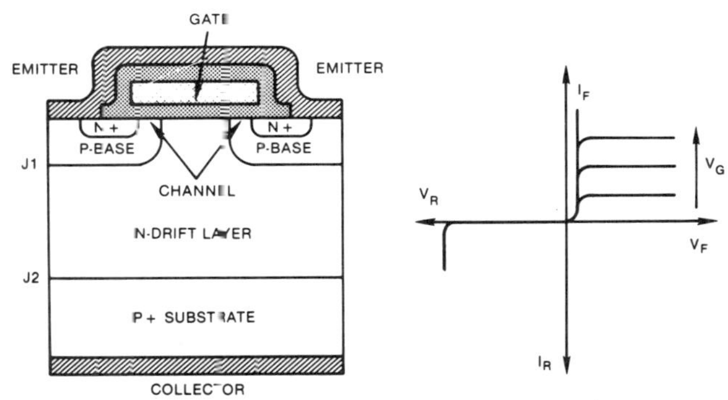

Fig. 14. Power MOS-IGT structure and its voltage controlled output characteristics. (Note the unique reverse blocking capability.)

A. MOS Turn-Off Thyristor

The MOS turn-off thyristor differs from the MOS gated thyristor described earlier in its ability to both turn on and turn off the current by using a signal applied to an MOS gate. These devices do not exhibit the fully gate controlled output characteristics that are exhibited by the bipolar transistor, power MOSFET, and IGT. Instead, they can be regeneratively turned on with the S-type negative resistance characteristic of thyristors and can be turned off like a gate turn-off (GTO) thyristor without the need to reverse the anode voltage. In comparison to a GTO, these devices have the advantage of a high input impedance MOS gate.

In comparison with the IGT, they can operate at an even higher forward conduction current density.

The basic structure of an MOS turn-off thyristor is shown in Fig. 15(a) and its equivalent circuit is provided in Fig. 15(b).

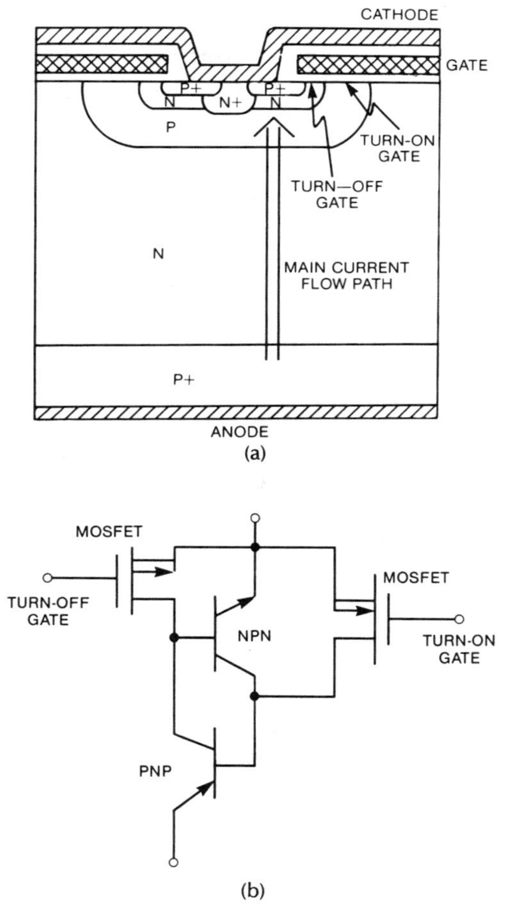

Fig. 15. (a) MOS turn-off thyristor structure. (b) Equivalent circuit of the MOS turn-off thyristor.

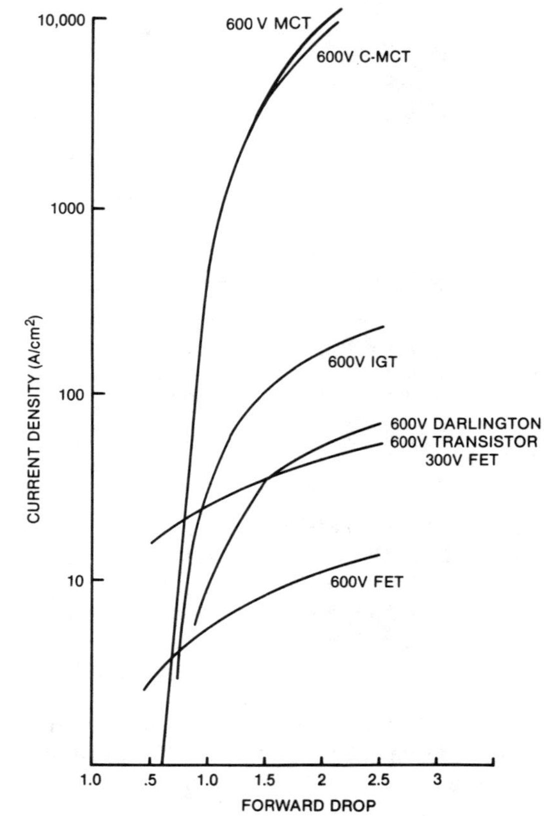

Fig. 16. Comparison of the forward conduction characteristics of the MOS turn-off thyristor with the IGT, MOSFET, and bipolar transistor. These curves were theoretically generated for the MCT and experimentally for the other devices.

The device consists of an FET structure integrated into the conventional P-N-P-N thyristor structure in such a manner that the emitter-base junction of the upper *N-P-N* transistor can be short-circuited by the application of a gate voltage to the FET. In the absence of the gate voltage, the device can be switched on either like a conventional thyristor or by utilizing the MOS gate as described before for the MOS gated thyristor. The current flow in the on-state occurs through the two coupled transistors as indicated by the arrow in Fig. 15(a). As a result of the strong injection of minority carriers into the *N*-drift region, its resistance becomes very small during current flow. In the on-state, the forward conduction characteristics approach those of a conventional thyristor, and these devices can operate at even (two to five times) greater current densities than an IGT as shown in Fig. 16.

To achieve forced gate turn-off, the p-channel FET within the device structure, which forms an active short circuit between the N^+-cathode and *P*-base regions, is turned on. Holes that enter the *P*-base now have an alternate path to the cathode contact: that is, they can bypass the N^+-*P* junction of the upper *N-P-N* transistor. To accomplish forced gate turn-off, the resistance in the hole current path must be so low that, when all the hole current is diverted to the p-channel FET, the forward bias of the N^+-*P* junction remains below 0.7 V. Computer modeling indicates that turn-off can be accomplished at very high current densities (>2000 A/cm²) with proper cell design [27]. Since the mobility for electrons is higher than that for holes in silicon, the complementary structure with an n-channel turn-off FET is expected to exhibit even higher turn-off current density.

Experimental demonstration of the operation of the MOS turn-off thyristor concept has been obtained on rudimentary n-channel devices fabricated with only small active areas. It has been found that, for a single-cell device with an active area of 20 × 20 μm, a current density of 6000 A/cm² can be turned off at 25°C and at low anode voltages for a resistive load by using a gate drive of 60 V. The maximum turn-off current density decreases to 2000 A/cm² for a more practical gate drive of 15 V. This indicates that the channel resistance is an important limitation to the turn-off process. A bigger problem that has been encountered is the severe reduction in the turn-off current density with an increasing number of cells. This phenomenon is believed to be due to nonuniform distribution of current during device turn-off. This factor, coupled with the additional derating of the maximum turn-off current density at high anode voltages [28] under inductive loads and at elevated temperatures, makes the MOS turn-off thyristor current handling capability too poor for circuit applications at the present time. It is also worth pointing out that the regenerative turn-on of these devices does not make them suitable as a substitute for the bipolar transistor, power MOSFET, or the IGT. The MOS turn-off thyristor is more suitable for circuits where GTOs are presently being applied.

V. Trends

Since the penetration of power MOS technology into an electronic system is dependent upon the availability of chips with higher power handling capability, it is interesting to track the growth in power ratings over the years. In the early 1980s, the only power MOS technology that was available was the MOSFET technology. As discussed earlier, the power handling capability of these devices is dependent upon the ability to reduce the specific on-resistance and to increase the maximum die size. From the data shown in Figs. 9 and 10, it can be concluded that:

a) the power MOS chip area is doubling every two years,
b) the specific on-resistance of power MOSFETs is decreasing by a factor of 1.4 every two years, and
c) the power MOSFET on-resistance is decreasing by a factor of 3 every two years.

These trends are expected to continue.

The cumulative impact of these developments upon the power handling capability of power MOSFETs is shown in Fig. 17 [29]. The power ratings have grown from 500 W in 1980 to 2500 W in 1987. The impact of the introduction of the IGT in 1983 can also be clearly seen in this figure. The power handling capability jumped by a factor of 5 at that time. Today, IGTs are available with ratings of 1200 V and 100 A. The IGT ratings are, therefore, growing at a much faster pace than for power MOSFETs. This is a direct consequence of the ease with which IGT voltage ratings can be increased compared with a power MOSFET. In the IGT, the current handling capability per unit area decreases as the square root of the voltage rating compared with a reduction proportional to the 2.5 power of the voltage for the power MOSFET [26]. Based upon the data in Fig. 17, the IGT ratings can be projected to grow by a factor of 3.4 every two years. This is twice as fast as the rate of increase in power MOSFET ratings.

The innovations made in power structures that have led to the observed rapid growth in power ratings are expected to continue. Another MOS-bipolar device that holds great promise is the MOS gated thyristor. The first MOS gated thyristors were reported in 1979 [10]. These devices had the advantages of an MOS controlled turn-on with improved *di/dt* and *dv/dt* capability. More recently, the concept has been extended to achieve MOS gated turn-off [27], [28]. These devices operate like thyristors in the on-state. This results in even superior modulation of the drift layer resistance as compared with the IGT. An increase in power handling capability per unit chip area by a factor of 2 to 3 over the IGT can be projected as illustrated in Fig. 16. The commercial introduction of these devices in the future can be expected to result in another boost to the power handling capability of the power MOS technology.

As the ratings of power MOS devices continue to grow and approach that of a larger and larger segment of the bipolar transistor ratings, a continual erosion of the bipolar transistor market can be foreseen. It has been projected

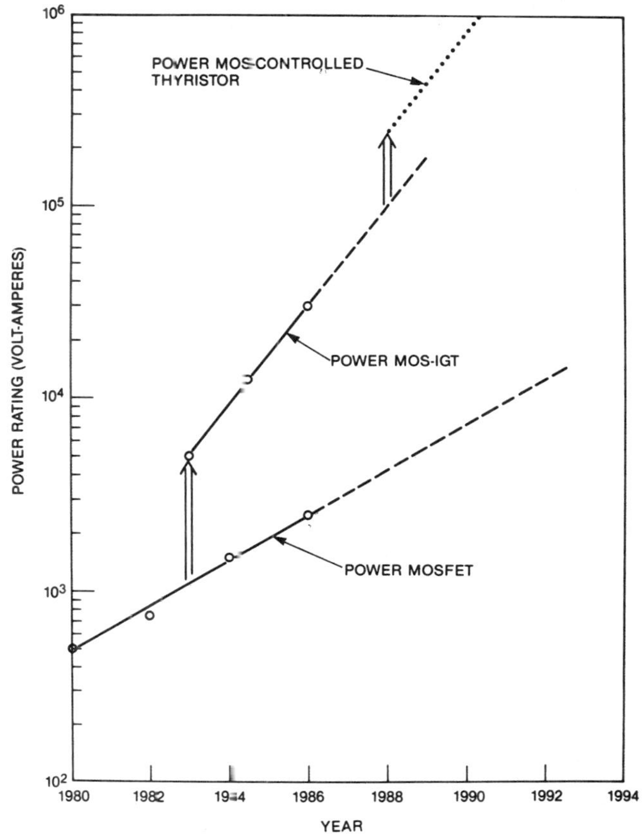

Fig. 17. Increase in the power handling capability of power MOS technology over the years. The power rating used here is the product of the blocking voltage rating and the continuous current rating. All the data are measured values.

that with the increasing application of power electronics in motor drives and factory automation, the power transistor market will grow from its present level of about $1 billion to $5 billion in 1995. As illustrated in Fig. 18, at present the

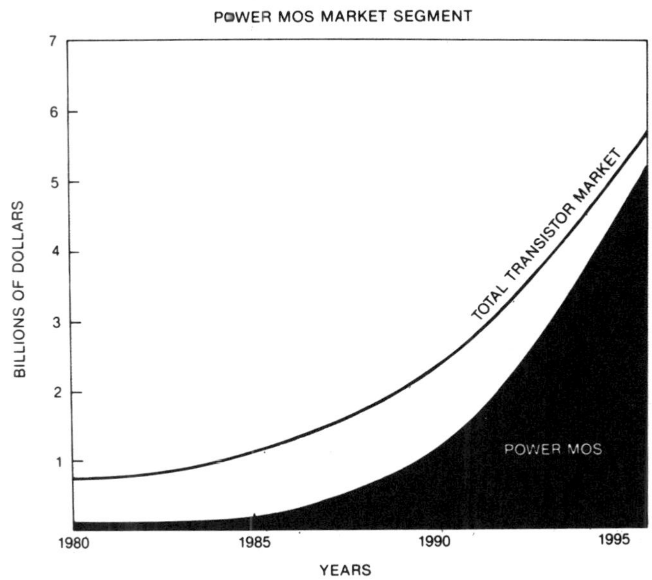

Fig. 18. Trends in the total transistor market. Note that an increasing proportion of the market will be taken over by the power MOS technology. Here the power MOS technology includes both FETs and MOS-bipolar devices.

market share for power MOS transistors represents about 10 percent of the total transistor market. However, by the mid-1990s, the power MOS transistor segment will grow to over 80 percent of the total market and the power MOS transistor technology is expected to become the dominant discrete device technology for the 1990s.

References

[1] B. J. Baliga and C. Chen, *Power Transistors*. New York, NY: IEEE Press, 1984.

[2] M. J. Declerq and J. D. Plummer, "Avalanche breakdown in high voltage D-MOS devices," *IEEE Trans. Electron Devices*, vol. ED-23, pp. 1-6, 1976.

[3] B. J. Baliga, *Modern Power Devices*. New York, NY: Wiley, 1987.

[4] S. C. Sun and J. D. Plummer, "Modeling of the on-resistance of LDMOS, VDMOS and VMOS power transistors," *IEEE Trans. Electron Devices*, vol. ED-27, pp. 356-367, 1980.

[5] R. P. Love, P. V. Gray, and M. S. Adler, "A large area power MOSFET designed for low conduction losses," *Int. Electron Devices Mtg.*, abstract 17.4, pp. 418-421, 1981.

[6] N. Zommer, "The monolithic HV BIPMOS," *Int. Electron Devices Mtg.*, abstract 11.5, pp. 263-266, 1981.

[7] T. Tanaka, Y. Yasuda, and M. Ohayashi, "A new MOS-gate bipolar transistor for power switches," *IEEE Trans. Electron Devices*, vol. ED-33, pp. 2041-2045, 1986.

[8] B. J. Baliga, "High gain power switching using field controlled thyristors," *Solid-State Electron.*, vol. 25, pp. 345-353, 1982.

[9] ——, "The asymmetrical field controlled thyristor," *IEEE Trans. Electron Devices*, vol. ED-27, pp. 1262-1268, 1980.

[10] ——, "Enhancement and depletion mode vertical channel MOS gated thyristors," *Electron. Lett.*, vol. 15, pp. 645-647, 1979.

[11] J. D. Plummer and B. W. Scharf, "Insulated gate planar thyristors," *IEEE Trans. Electron Devices*, vol. ED-27, pp. 380-394, 1980.

[12] A. Pshaenich, "The MOS-SCR, a new thyristor technology," *Motorola Engineering Bull.*, vol. ED-103, 1982.

[13] B. J. Baliga, M. S. Adler, P. V. Gray, R. P. Love, and N. Zommer, "The insulated gate rectifier (IGR): A new power switching device," *Int. Electron Devices Mtg.*, abstract 10.6, pp. 264-267, 1982.

[14] J. P. Russell, A. M. Goodman, L. A. Goodman, and J. M. Neilson, "The COMFET—A new high conductance MOS-gated device," *IEEE Electron Device Lett.*, vol. EDL-4, pp. 63-65, 1983.

[15] B. J. Baliga, M. S. Adler, P. V. Gray, and R. P. Love, "Suppressing latch-up in insulated gate transistors," *IEEE Electron Device Lett.*, vol. EDL-5, pp. 323-325, 1984.

[16] B. J. Baliga, "Fast switching insulated gate transistors," *IEEE Electron Device Lett.*, vol. EDL-4, pp. 452-454, 1983.

[17] B. J. Baliga and E. Sun, "Comparison of gold, platinum, and electron irradiation for controlling lifetime in power rectifiers," *IEEE Trans. Electron Devices*, vol. ED-24, pp. 685-688, 1977.

[18] B. J. Baliga, "The new generation of MOS power devices," in *Drive/Motors/ Controls Conf. Proc.*, pp. 139-141, 1983.

[19] A. Nakagawa, H. Ohashi, M. Kurata, H. Yamaguchi, and W. Watanabe, "Non-latchup, 1200 volt bipolar-mode MOSFET with large SOA," in *IEEE Int. Electron Devices Meeting*, abstract 16.8, pp. 860-861, 1984.

[20] A. Nakagawa, Y. Yamaguchi, K. Watanabe, H. Ohashi, and M. Kurata, "Experimental and numerical study of non-latch-up bipolar mode MOSFET characteristics," in *IEEE Int. Electron Device Meeting*, abstract 6.3, pp. 150-153, 1985.

[21] B. J. Baliga, "Temperature behavior of insulated gate transistor characteristics," *Solid-State Electron.*, vol. 28, pp. 289-297, 1985.

[22] C. Hu, M. H. Chi, and V. M. Patel, "Optimum design of power MOSFETs," *IEEE Trans. Electron Devices*, vol. ED-31, pp. 1693-1700, 1984.

[23] J. Baliga, "Switching lots of watts at high speed," *IEEE Spectrum*, vol. 18, no. 12, pp. 42-48, 1981.

[24] D. Ueda, T. Takagi, and G. Kano, "A new vertical power MOSFET structure with extremely reduced on-resistance," *IEEE Trans. Electron Devices*, vol. ED-32, pp. 2-8, 1985.

[25] H. R. Chang, R. D. Black, V. A. K. Temple, W. Tantraporn, and B. J. Baliga, "Ultra-low specific on-resistance UMOSFET," *Int. Electron Devices Mtg.*, abstract 28.3, pp. 642-645, 1986.

[26] T. P. Chow and B. J. Baliga, "Comparison of 300, 600, and 1200 V n-channel IGTs," *IEEE Electron Device Lett.*, vol. EDL-6, pp. 161-163, 1985.

[27] V. A. K. Temple, "MOS controlled thyristors (MCTs)," *Int. Electron Devices Mtg.*, abstract 10.7, pp. 282-285, 1984.

[28] M. S. Stoisek and H. Strack, "MOS-GTO—A turn-off thyristor with MOS-controlled emitter shorts," *Int. Electron Devices Mtg.*, abstract 6.5, pp. 158-161, 1985.

[29] B. J. Baliga, "Revolutionary innovations in power discrete devices," in *Int. Electron Devices Meeting*, abstract 5.1, pp. 102-105, 1986.

Paper 5.3

A Comparison Between BIMOS Device Types

MICHAEL S. ADLER, FELLOW, IEEE

***Abstract*—This paper deals with the design and comparison of three types of BIMOS transistors (cascade, cascode, and parallel combinations of MOSFET and bipolar transistors). The first section of the paper presents a technique for accurately determining the overall device *I-V* relations for the cascade (Darlington) combination based on the area ratio for the MOSFET and bipolar transistors. From this, one sees that the optimum area ratio varies from 1.5 for 700-V devices to 0.2 for 80-V devices. The second section of the paper deals with the design and comparison of the cascode and parallel device types. In these cases, no optimum area ratio based solely on device conduction exists and the device design and comparison is based on achieving the maximum current density limited by keeping the power dissipation to under 100 W/cm^2. Included in this analysis are losses due to conduction as well as switching. The results show that a 0.5 area ratio (MOSFET to bipolar transistor) is optimum for the cascode (series) combination and a 0.3 ratio is a good compromise for the parallel combination. The last section of the paper shows an overall comparison of all of the BIMOS device types together with the MOSFET and the bipolar transistor.**

I. Introduction

RECENTLY, there has been an emergence of new power devices and circuits that combine both MOS and bipolar devices. In some cases, this combination is done by monolithically combining the devices onto a single chip while in other cases, discrete devices are combined in a circuit. In most of these combinations, the MOS device is used to simplify the drive requirements of the combined device and/or to increase the switching efficiency due to the high speed of the MOS device. The bipolar device, whether monolithic or discrete, offers the advantage of low conduction losses. Presently, there are three combinations of MOS and bipolar transistors: the cascade combination (MOS-bipolar Darlington), the parallel combination where the MOS and bipolar devices are in parallel and across the load, and the cascode (emitter switches) combination where the MOS device is placed in the emitter of the bipolar device. The advantage of the cascade combination is drive simplicity since only a MOS gate need be driven, and in the second and third combinations the advantage is high speed and increased SOA for the bipolar transistor since the devices can be operated at their VCBO ratings rather than VCEO. One of the problems in designing these devices, designing circuits using the devices, or evaluating their cost effectiveness and performance is that it is not clear how big to make the MOS device relative to the bipolar device. The corollary to this problem is the question of the relationship between overall device performance (speed and conduction loss) and the relative sizes of the MOS and bipolar transistors. This is not simply an academic issue since concerns have been raised about the large size of the MOS driver device in the cascade combination that is required to equal the conduction losses of the all bipolar Darlington. In order to gain appreciation for these issues and inevitably the viability of these new device combinations a study will be presented in this paper as to the total device area for each of the device combinations (cascade, parallel, cascode) and the relation between MOS area and overall device performance. In addition, a comparison will be made between the various device combinations, the pure MOSFET, and the pure bipolar transistor as to switching efficiency (power loss on a device area basis and drive simplicity).

Manuscript received April 24, 1985; revised August 23, 1985.

The author is with General Electric Corporate Research and Development, Schenectady, NY 12345.

IEEE Log Number 8406509.

II. Cascade Device Combination

The first device to be analyzed is the cascade or Darlington combination where the power MOSFET drives the base of the bipolar transistor (BJT). This is shown in Fig. 1. The issue when designing devices of this type is the proper choice of the areas for the bipolar and the MOSFET devices. By making the MOSFET larger, one is providing increased base drive for the bipolar transistor, thus driving it further into saturation and reducing its forward drop. However, at some point the increased area of the MOSFET does not "pay off" in a sufficiently reduced forward drop or increased collector current to justify the increased area. While these arguments are obviously qualitative, the question arises as how to deal with them on a quantative basis. The technique that was used for doing this is given by the following.

The specific example here is for a Motorola BU205 transistor that has a 700-V sustaining voltage and a 2.5-A current rating. The Motorola *Power Device Data Handbook* [1] gives the collector to emitter voltage as a function of base current for different collector currents. From the same data sheet, the base to emitter voltage as a function of collector current can also be determined. The relevant data sheet is reproduced in Fig. 2 for illustration. From this data sheet, a table is produced that lists, for the three collector currents given in Fig. 2(a), the base current for collector to emitter voltages ranging from 5 to 1 V. Also taken from Fig. 2(c) are the base to emitter voltages for the same collector currents. Although not precisely true, the base to emitter voltage is assumed to be constant over the range of base currents given. Using this data, it is then possible to determine the collector to base voltage, and then, by knowing the base current, one can determine the

Reprinted from *IEEE Trans. Electron Devices*, vol. ED-33, no. 2, pp. 286–293, Feb. 1986.

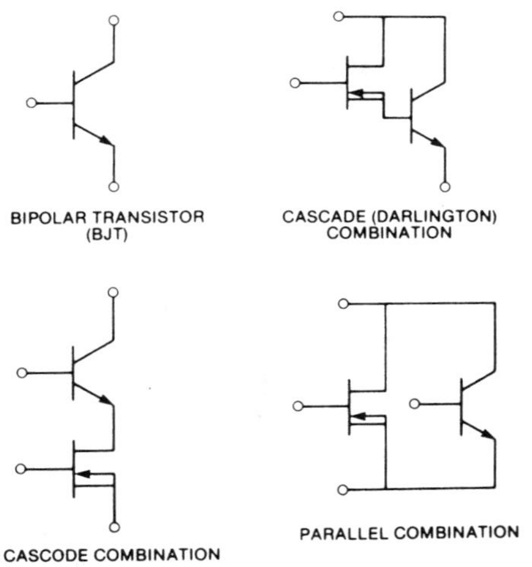

Fig. 1. Various BIMOS device combinations studied in this report paper.

value of the resistance that would have to be placed between the collector and base terminals of the bipolar transistor to generate the needed base current. This process is illustrated in Table I where the required MOS transistor resistance is listed as R_{MOS}.

The next step of determining the required area for a MOSFET transistor is done by noting that the resistance area product of available MOSFET's tends to follow a single functional relationship as a function of breakdown voltage. This is illustrated in Fig. 3 where the resistance-area product for a variety of available devices is plotted as a function of the device breakdown voltage. Also shown is the $V_{BD}^{2.5}$ power dependence of the device resistance (based on the linear scaling of the thickness of the drain region and resistivity with voltage and the decrease of the peak breakdown field with increasing doping.) Table II gives an average value for the resistance-area product for devices of various breakdown voltages based on the data in Fig. 3. Knowing this value one can then calculate the area of the required MOSFET (given as A_{MOS} in Table I). The last column in Table I gives the overall current density based on adding the areas of MOSFET and the bipolar transistor. It should be noted that the bipolar transistor base current must be added to the collector current in this computation since this current also flows from the device load. Note that the chip areas used in these calculations are for the active chip areas that are calculated by removing the area taken for the junction termination and the scribe lanes from the overall die areas [2].

Fig. 4 is the result of plotting the data in Table I where the ratio of the area of the MOSFET and the bipolar transistor is plotted as a function of the overall current density for five different values of collector-to-emitter voltage. As can be seen, there is an optimum area for the driver device. As explained earlier, this optimum results from the fact that once driven into hard saturation, there is no improvement in the bipolar current density or forward drop by further increasing the base drive. One should also note that in this context the concept of an optimum area ratio is based on valuing the area of the MOSFET and bipolar transistors on a uniform basis. In actuality, on an area basis, MOS transistors presently cost more than bipolar transistors so that this optimum, based on the total device area, may not be the optimum on an overall cost basis. However, lacking this detailed cost information, the optimum based on achieving the maximum overall current density will be used.

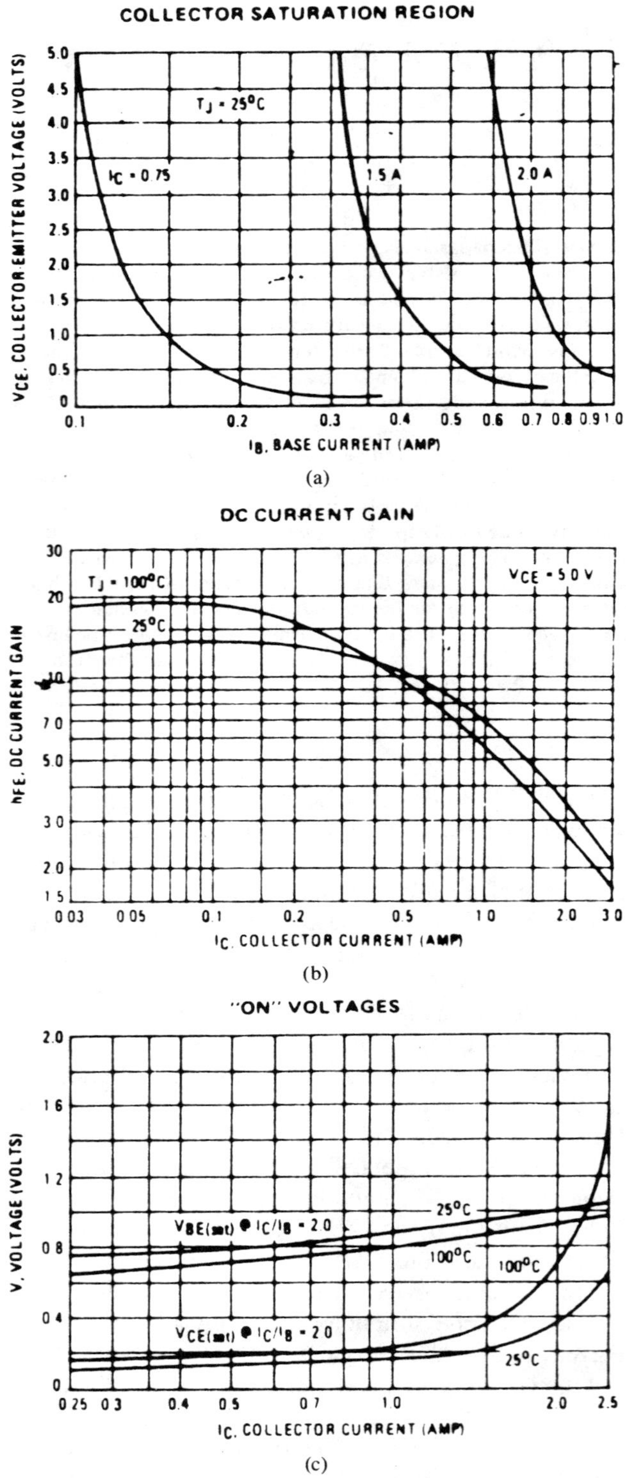

Fig. 2. Characteristics for the Motorola BU205 bipolar transistor.

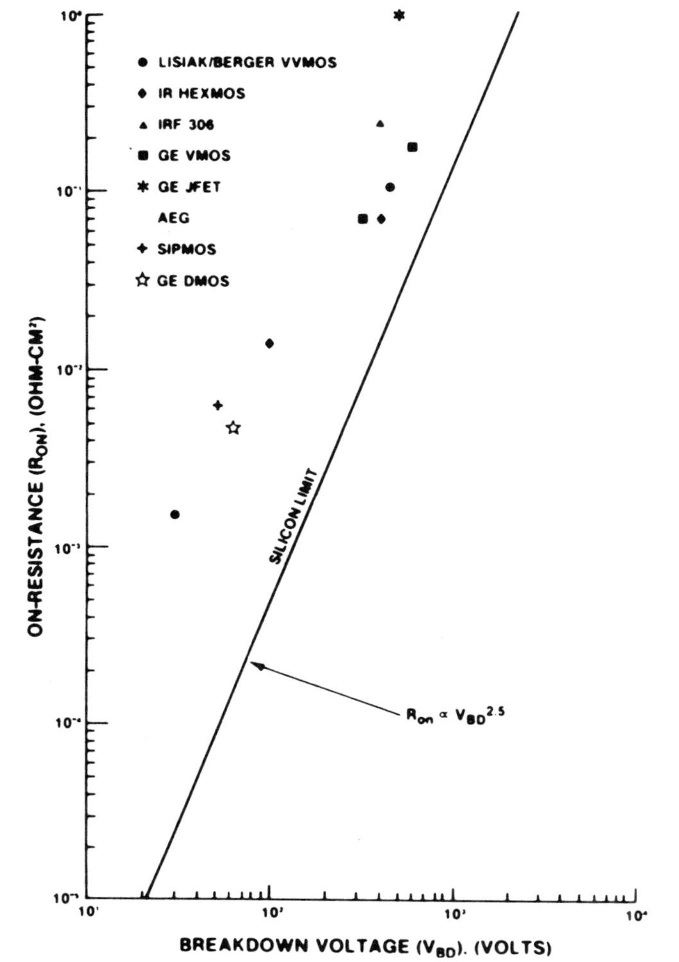

Fig. 3. Resistance-area product as a function of breakdown voltage for power MOSFET's.

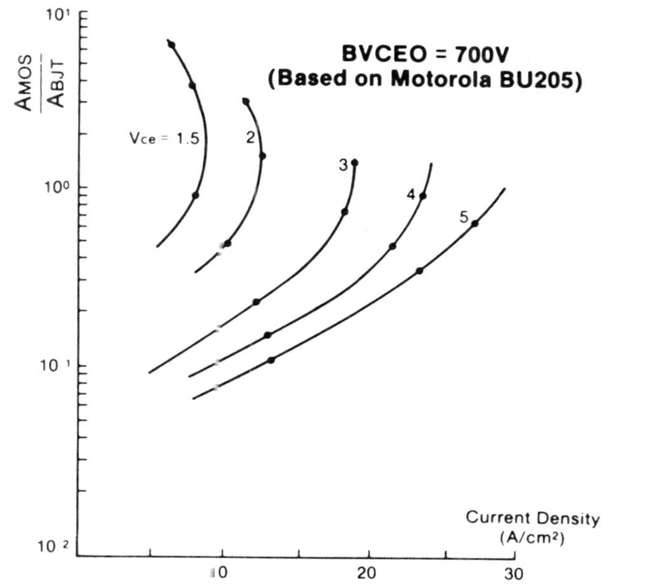

Fig. 4. Ratio of the areas of the driver MOSFET and the bipolar transistor as a function of the overall current denstiy for a 700-V cascade BIMOS device.

TABLE II
RESISTIVE AREA PRODUCT FOR POWER MOSFET DEVICES

Device Breakdown Voltage (V)	Resistance-Area Product ($\Omega \cdot cm^2$)
55	.01
80	.012
100	.014
150	.021
200	.03
300	.05
400	.075
500	.12
700	.26

TABLE I
MOTOROLA BU205 $BV_{CEO(SUS)}$ = 700 V, IC = 2.5 A, DIE AREA = 0.0904 cm^2, ACTIVE DEVICE AREA = 0.0582 cm^2

V_{ce}(V)	I_b (A)	V_{be} (V)	V_{cb}(V)	R_{MOS} (Ω)	A_{MOS} (cm)	J_T (A/cm²)
I_c = .75 A						
5	.1	.85	4.15	41.5	6.27×10^{-3}	13.2
4	.106	.85	3.15	29.7	8.76	12.8
3	.1107	.85	2.15	19.4	13.4	12.0
2	.1225	.85	1.15	9.39	27.7	10.2
1.5	.130	.85	.65	5.	52.	7.99
1	.147	.85	.15	1.02	255	2.86
I_c = 1.5 A						
5	.315	.95	4.05	12.9	2.0210^{-2}	23.2
4	.32	.95	3.05	9.53	2.73	21.3
3	.335	.95	2.05	6.12	4.25	18.2
2	.37	.95	1.05	2.84	9.17	12.5
1.5	.4	.95	.55	1.38	18.9	7.69
1	.46	.95	.05	.109	239.	6.6
I_c = 2.0 A						
5	.58	1.0	4	6.9	3.77×10^{-2}	26.9
4	.615	1.0	3	4.88	5.33	23.5
3	.64	1.0	2	3.13	8.33	18.7
2	.69	1.0	1	1.45	17.95	11.3
1.5	.725	1.0	.5	.690	37.7	6.26
1	.775	1.0	–			

Having generated a set of curves shown in Fig. 4, it then becomes possible to generate a set of overall current–voltage curves for a given area ratio. Such a set of curves are shown in Fig. 5 where it can be seen that the current density is uniformly higher for the area ratios at least up to unity. One can see from Fig. 4 that until one reaches an area ratio greater than two the current density increases monotonicaly for all of the voltage curves.

Figs. 6–8 are similar to Fig. 5 except that they are for devices with a 400-, 150-, and 80-V collector–emitter sustaining voltage rating, respectively. The specific devices shown are for the General Electric D44TE, the Motorola 2N6341, and the Motorola 2N5686 transistors, respectively. Qualitatively these curves are similar to those shown in Fig. 5 except that the optimum area ratio occurs at progressively lower values as the voltage is reduced. Although the corresponding curve to Fig. 4 is not shown, the optimum area ratio is approximately 1.5 for the 700-V device, 0.7 for the 400-V device, 0.5 for the 150-V device, and 0.25 for the 80-V device. Physically this can be understood because the current gains for the lower voltage devices tend to be higher than for the higher voltage devices. It is interesting to note that for bipolar transistors in hard saturation, the inverse of the current density-current gain product has the same $V_{BD}^{2.5}$ power law as the resistance area product in MOSFET's [3].

In order to gain an overall appreciation as to the tradeoff between the MOSFET, bipolar, and BIMOS cascade transistors, the current density at 3-V forward drop has been plotted as a function of breakdown voltage in Fig. 9. Two

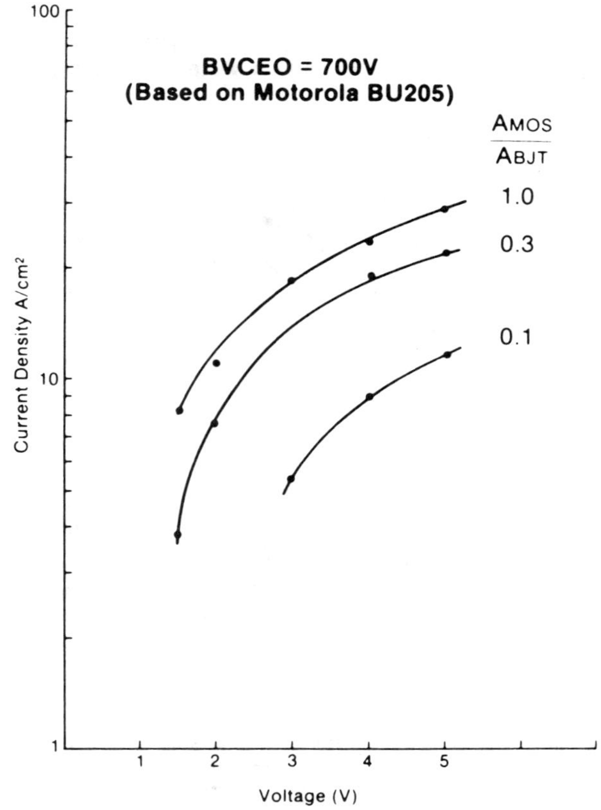

Fig. 5. Current density as a function of forward voltage for three different area ratios of MOS to bipolar transistor for a 700-V cascade transistor.

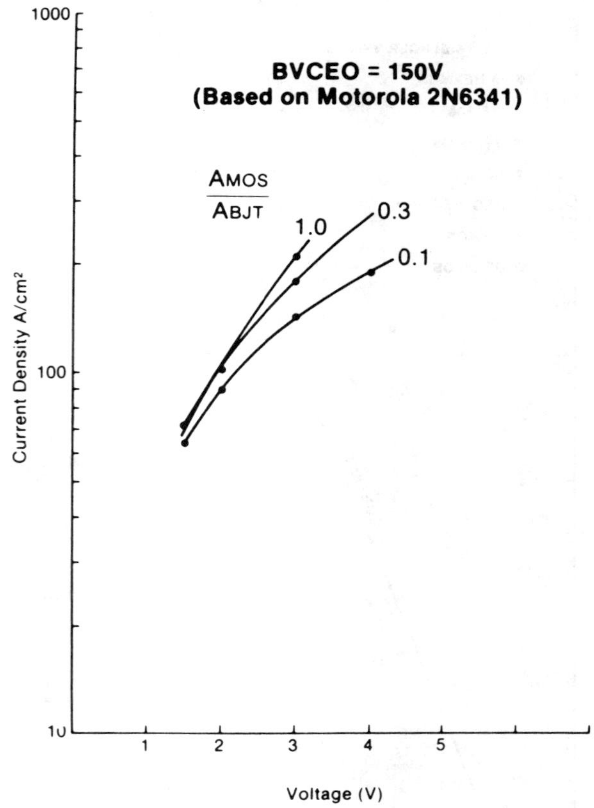

Fig. 7. Current density-forward voltage curves for a 150-V cascade BIMOS device.

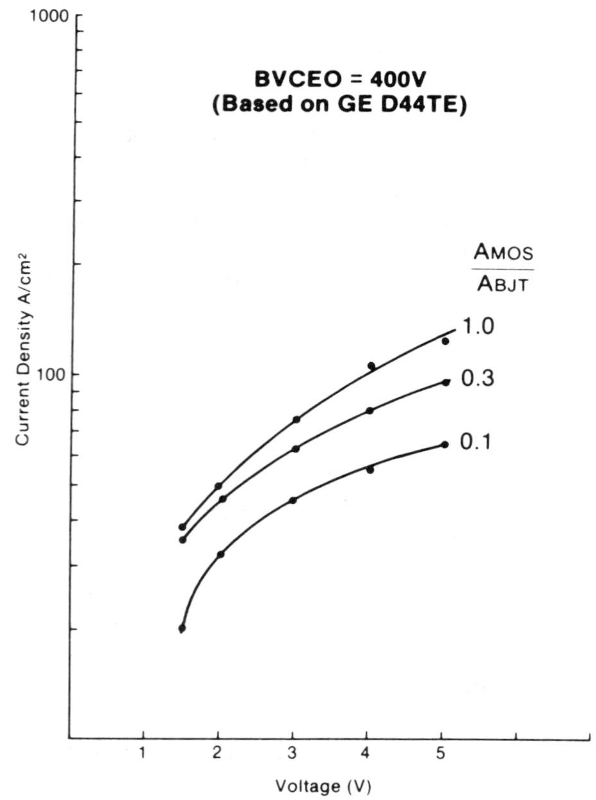

Fig. 6. Current density-forward voltage curves for a 400-V cascade BIMOS device.

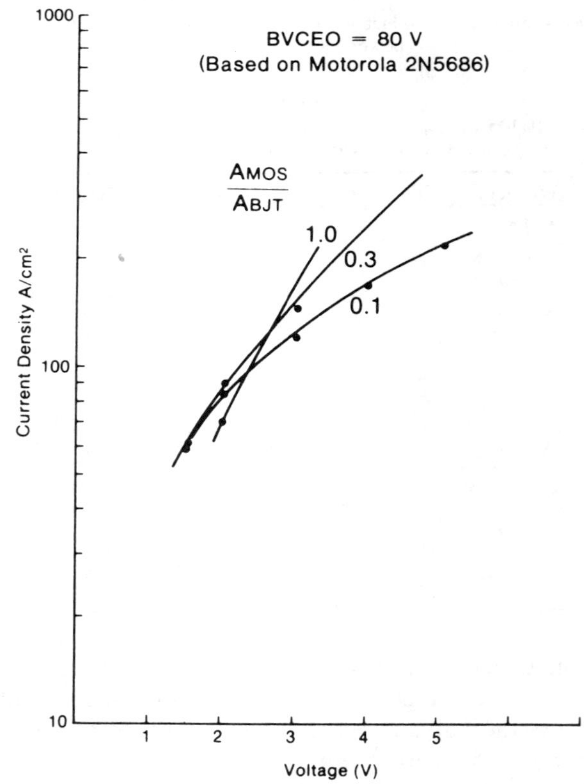

Fig. 8. Current density-forward voltage curves for an 80-V cascade BIMOS device.

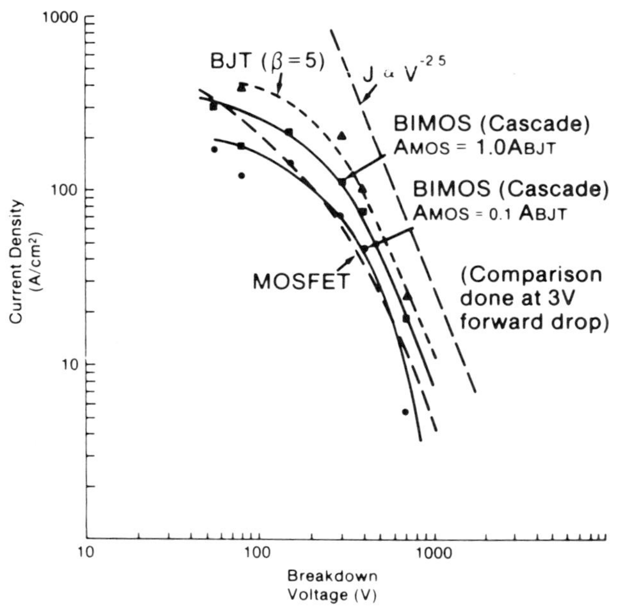

Fig. 9. Current density at a 3-V forward drop for MOS and bipolar transistors and for cascade type BIMOS device with two different area ratios.

curves are given for the BIMOS devices with the lower curve for a MOSFET to bipolar area ratio of 0.1 and the upper curve for an area ratio of unity. The pure MOSFET curve is based on the average performance for existing MOSFET devices given in Table II. The bipolar transistor curve is given for a forward current gain of five and is based on the same set of transistors discussed above in the context of Figs. 4–8.

Also shown is the theoretical curve where $J \alpha V^{-2.5}$. It is interesting that, as predicted above, the same overall dependence exists for MOSFET's, bipolars, and BIMOS's when measured on the basis of current density or current density at a fixed current gain. It is also interesting that all three device types follow the $V^{-2.5}$ dependence at high breakdown voltage but do not at the lower breakdown voltages. This is not surprising because for low-voltage bipolar transistors the current gain is mainly affected by the properties of the metallurgical base regions independently of the collector region and in the case of MOSFET's, the resistance is determined mainly by the spreading resistance from the MOS channels and not the bulk resistance of the extended drain regions. Because of this, the curves in Fig. 9 tend to flatten out and become independent of voltage at the lower breakdown voltages.

III. Comparisons Based on Maximum Power Densities

It is not possible to make meaningful comparisons of the cascode and parallel BIMOS device types together with the cascade, MOSFET, or simple bipolar transistors on simply the basis of current density at a fixed forward drop as done above. The problem is that no optimum overall device design for the area ratio of the MOSFET and bipolar devices exists strictly on the basis of device conduction. In fact, the main attributes of the cascode and parallel devices are associated with their switching performance. In the case of the cascode device, very high switching speeds are achieved because once the emitter of the bipolar transistor is open-circuited, the device is being turned off with a reverse gain of unity. Furthermore, because the emitter is open circuited, the gain of the bipolar transistor if broken so that the device can be operated at reverse voltages close to the BV_{CBO} rating rather than the $BV_{CEO(SUS)}$ rating. For many high-voltage devices this can amount to a factor of two in breakdown voltage. The price one pays for this is the need to drive and control two devices rather than one as well as the cost associated with the additional MOSFET device.

In the parallel combination, very high switching speeds result since the MOSFET device handles the load current when the combined device is switching [4]. To achieve this, of course, it is necessary to turn on the bipolar transistor after the MOSFET is on and turn it off before the MOSFET is turned off. Since the bipolar transistor is carrying the combined devices' conduction current when the device is not switching, the conduction losses are also low. A further benefit occurs because the bipolar transistor is not required to simultaneously have high voltage across it while carrying high current as would occur if it was doing the switching as well as carrying the conduction current. As a result, it should also be possible to operate the bipolar transistor near the BV_{CBO} rating rather than $BV_{CEO(SUS)}$. The price paid for these benefits is a significant increase in drive complexity since not only must drive currents be supplied to two devices but a specific nonsimultaneous ordering of the drive signals must be applied to the separate devices. In terms of ordering the drive complexity of the various devices discussed, the MOSFET and the cascade device are the simplest, followed by the bipolar transistor, the cascode device, and finally the parallel device.

In designing the cascode or parallel device types, the same problem of determining the ratio of the areas for the MOSFET and the bipolar transistor is a key issue. In doing this, however, simply looking at forward drop is not very instructive since, as noted above, there is no optimum value for these devices. A better basis for choosing the proper area ratio as well as for comparing these devices to the earlier device is to compare the maximum current densities at which the devices can be operated based on a limiting level of power dissipation.

To be meaningful, such a comparison should include conduction as well as switching losses. In doing this, the device combinations shown in Table III were used where the switching times for 700- and 400-V device types are given. In the case of the 700-V devices, the cascode and parallel combinations are based on the D44TE transistor. While this device only has a 400-V $BV_{CEO(SUS)}$ rating, it does have a BV_{CBO} rating of 700-V making it possible to use it for these device types (as discussed above). In the cascade and bipolar device types, the BU205 device must be used since it has a 700-V $BV_{CEO(SUS)}$ rating needed for switching at high votage. The 700-V MOSFET is based

TABLE III
SWITCHING PERFORMANCE OF BIMOS DEVICE TYPES

Device	Type	Bipolar Sustaining Voltage (V)	Device Breakdown Voltage (V)	t_r (μsec)	t_f (μsec)
MOSFET	Scaled version IRF330		700	.023	.023
BIMOS Cascade	700 V MOSFET & BU205 bipolar	700	700	.46	.46
BIMOS Cascode	50 V MOSFET	400	700	.042	.042
Bipolar	BU205	700	700	.6	.6
BIMOS Parallel	700 V MOSFET & D44TE bipolar	400	700	.023	.023
MOSFET	IRF330 (400 V)	400	400	.023	.023
BIMOS Cascade	400 V MOSFET & D44TE bipolar	400	400	.46	.46
BIMOS Cascode	50 V MOSFET & 2N6498 bipolar (300 V)	300	400	.042	.042
Bipolar	D44TE	400	400	.46	.46
BIMOS Parallel	400 V MOSFET & 2N64986 bipolar	300	400	.023	.023

on a scaled version of the International Rectifier IRF330, which is a 400-V device. For the 400-V devices, the 2N6498 devices are used for the cascode and parallel devices and the D44TE used for the cascade and bipolar devices. The switching times given in the table are based on measured values in inductive switching [5], [6] except for the BU205 and the 2N6498 which come from data sheet data [1].

To determine the maximum current density, the conduction losses based on a 50-percent duty cycle are added to the switching losses and the current density adjusted until a value of power density of 100 W/cm^2 is reached. This value is widely accepted as the maximum power density that can be dissipated in a device with single-sided cooling. An interesting problem arises in doing these calculations since three of the device types are actually combinations of two devices and there is a question of deciding what device area to base the power density upon. Arguing that the cascode and parallel combinations are likely to be made with either separate devices or devices located in different portions of the same chip, the power densities are calculated separately for the two devices making up the combined devices. The criteria then for the maximum current density is set by the device of the pair that turns out to have the highest power dissipation density. The current density as listed in the results, however, is the overall current density computed by adding up the areas of the two separate devices (as was described earlier for the cascade devices). In the case of the cascade device, the power dissipation is based on the total power dissipation of the two chips taken together. The reason for this is that the cascade device has the potential for being interwoven in such a way that the power is dissipated within the same area for each device type.

IV. CASCODE AND PARALLEL DEVICE TYPES

Before making the device comparison between all of the device types, it is appropriate to investigate the tradeoff in the design based on varying the ratios of the areas for the MOS and bipolar devices for the cases of the cascode

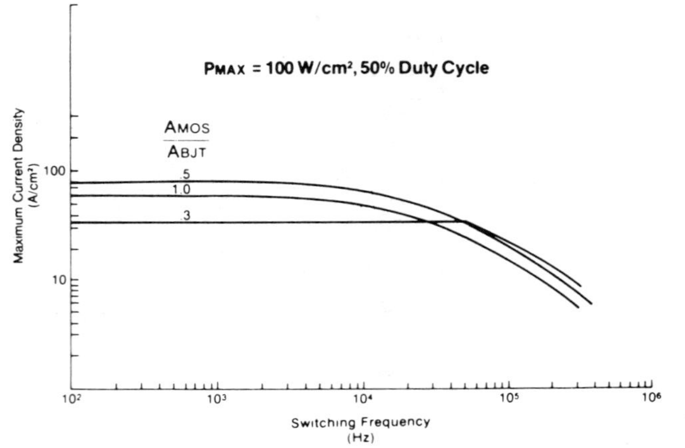

Fig. 10. Maximum current density as a function of switching frequency for three area ratios of MOS to bipolar transistors in a cascode combination.

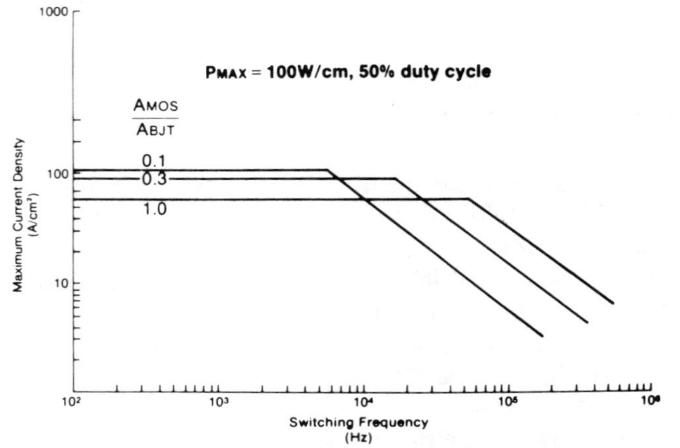

Fig. 11. Maximum current density as a function of switching frequency for three area ratios of MOS to bipolar transistors in a parallel combination.

and parallel device types. Fig. 10 shows the maximum current density for a 700-V cascode device for three different ratios of MOS and bipolar devices. The MOS transistor in this case is a 50-V device with a resistance-area product of 0.01 Ω/cm^2. As can be seen in Fig. 10, the 0.5-ratio device has the highest current density at all but the very highest switching frequencies. The issue here is that for area ratios greater than 0.5, the bipolar device is the limiting factor, but for smaller area ratios, the MOS device becomes the limiting factor. In the latter case, the overall maximum current density is reduced except at very high frequencies where the bipolar device switching losses again make it the limiting factor.

Fig. 11 shows the maximum current density for three area ratios of MOS and bipolar transistors connected in a parallel combination. As can be seen, the maximum current density at low frequencies increases as the area of the MOSFET device decreases. This occurs because the conduction losses in the bipolar transistor is the limiting factor and reducing the MOSFET area reduces the overall area of the device. At high frequencies, however, the switching losses in the MOSFET are the limiting factor and decreasing the area of the MOSFET decreases the maximum current density before reaching 100-W/cm^2 power density.

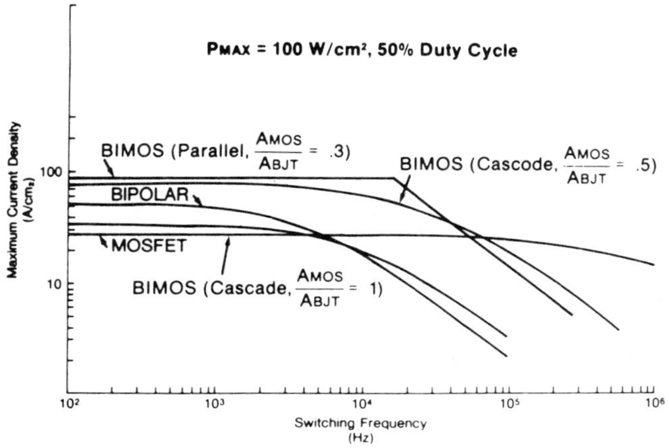

Fig. 12. Maximum current density as a function of switching frequency for the MOSFET, bipolar transistor, the cascade, cascode, and parallel BIMOS devices for devices rated at 700 V.

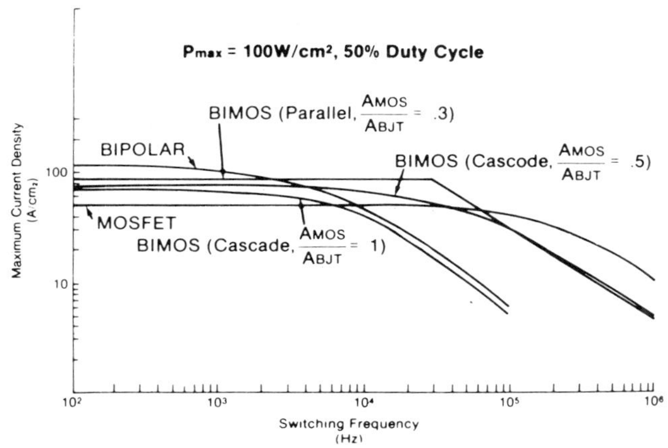

Fig. 13. Maximum current density as a function of switching frequency for the MOSFET, bipolar transistor, the cascade, cascode, and parallel BIMOS devices for devices rated at 400 V.

V. Comparison between the MOSFET, Bipolar Transistor, and the Cascade, Cascode, and Parallel BIMOS Devices

Figs. 12 and 13 show comparisons of the maximum allowable current densities of cascade, cascode, and parallel BIMOS device types together with the MOSFET and bipolar transistor. As before, the comparison is based on a maximum power density of 100 W/cm^2 in inductive switching for devices rated at 700-V (Fig. 12) and 400-V (Fig. 13). In both figures, the area ratios for the MOS and bipolar transistor in the BIMOS devices are 0.3, 0.5, and 1 for the parallel, cascode, and cascade device types, respectively. These ratios are the ratios that lead to the maximum overall current densities for the respective device types. For the devices requiring bipolar transistor drives, conduction losses associated with supplying base current from a 5-V source is included. In Fig. 12 for 700-V devices, one can see that at low frequencies the cascode and parallel devices are comparable and capable of the highest current densities of all of the device types. At frequencies above 70 kHz the MOSFET is the most efficient device. The bipolar transistor and the cascode devices are comparable at high frequencies with the bipolar transistor being more efficient at low frequencies.

For the 400-V device in Fig. 13, the situation at high frequencies is the same as in Fig. 12 where the MOSFET is the most efficient above 70 kHz. At low frequencies, however, the bipolar transistor is the most efficient. This relative change from the 700-V comparisons occurs because of the strong dependence of the current gain-current product with breakdown voltage (as discussed earlier) and the fact that there is less relative difference in the $BV_{CEO(SUS)}$ and BV_{CBO} ratings for 400-V devices as compared to 700-V devices.

VI. Summary

This paper deals with the design and comparison of three types of BIMOS transistors (cascade, cascode, and parallel combinations of MOSFET and bipolar transistors). The first section of the paper presents a technique for accurately determining the overall device I–V relations for the cascade (Darlington) combination based on the area ratio for the MOSFET and bipolar transistors. From this, one sees that there is a optimum area ratio and that it varies from 1.5 for 700-V devices to 0.2 for 80-V devices. These optima are based on the area ratio that achieves the overall highest current density for the combined device. Tacitly assumed in arriving at these optima is that there is a uniform tradeoff of the area of the MOSFET driver and the bipolar output device. In actuality, present costs are higher for MOSFET's than bipolar transistors of the same area so that the optimum on a cost basis will likely occur at lower area ratios than given.

The second section of the paper deals with the design and comparison of the cascode and parallel device types. In these cases, no optimum area ratio exists based solely on foward conduction drop. Instead, devices are compared based on the maximum current density at which the devices can be operated at 100-W/cm^2 power dissipation. While the specific choice depends on the frequency and voltage at which the device is operated, an area ratio of 0.5 for the cascode and an area ratio of 0.3 for the parallel combinations are good compromises.

On this same basis of a maximum power dissipation, all five device types are compared. At 700 V and at frequencies less than 70 kHz, the parallel and cascode device types are optimum. At 400-V and below, the bipolar transistor has the highest overall usable current density at low frequency. The change occurs because the current gain, current product of the bipolar transistor is increasing rapidly as the device blocking voltage is reduced, and the fact that at lower voltages, there is not as large a difference between the $BV_{CEO(SUS)}$ and the BV_{CBO} rating. The MOSFET continues to have the highest overall usable current density at high frequencies.

The one issue that the above comparisons (based on usable current density) ignore is that of base or gate driver simplicity. In this regard, the MOSFET and cascade BIMOS devices are the simplest, followed by the bipolar transistor, the BIMOS cascode, and the BIMOS parallel.

In the case of the latter, not only do separate signals need to be applied to the two devices but they must be phased properly in time. In many applications, the tradeoff between maximum current density (device area) will have to be made against simplicity and the cost of the gate/base drive.

A final issue is that of integrating the BIMOS devices on a single chip. For the cascade device, this can be easily done even to the point of interweaving the MOS driver device among the bipolar device fingers. The one problem that occurs is that the bipolar transistor in general will have a more resistive and thicker collector region than would be optimum for the MOSFET because of the need to operate the device below a $BV_{CEO(SUS)}$ rating. The MOSFET would optimally operate at the BV_{CBO} rating, and since there is a $V_{BD}^{2.5}$ power dependence to the area of the MOSFET with voltage, this could amount to a significant increase in the required MOSFET area. However, some device area could possibly be gained back if the MOS channels and emitter base fingers of the bipolar transistors were interwoven making use of area that is not actively used in bipolar transistors.

In the case of the cascode device, the MOSFET is a 50-V device and would necessarily have to be a lateral transistor rather than a vertical device if it were to be integrated onto the same chip as the bipolar. This will probably result in an increase in the required MOSFET area over what would occur if the device were made separately. For the parallel device combination, both devices share the same collector/drain contact and the emitter/source contact. As a result, it would be quite feasible to integrate the two devices onto a single chip. It is unlikely, however, that the devices could be interwoven. Since the devices operate separately, they would interact in an undesirable manner if interwoven.

References

[1] *Motorola Power Devices Data*, 2nd ed., Motorola, Inc., 1980.

[2] *Semiconductor Data Library, CHIPS*, vol. 8, series A, Motorola, Inc., 1976.

[3] M. S. Adler and S. R. Westbrook, "Power semiconductor switch devices—A comparison based on inductive switching," *IEEE Trans. Electron Devices*, to be published.

[4] N. Zommer, "The monolithic HV BIPMOS," in *IEDM Tech. Dig.*, pp. 263–266, 1981.

[5] M. S. Adler, S. W. Westbrook, and A. J. Yerman, "Power semiconductor devices—An assessment," in *IAS Conf. Rec.*, pp. 723–728, 1980.

[6] S. R. Westbrook, Masters thesis, Massachusetts Institute of Technology, June 1980.

Paper 5.4

High-Voltage Technology Offers New Solutions for Interface Integrated Circuits

GILLES THOMAS, GILLES TROUSSEL, AND FRANÇOISE VIALETTES

***Abstract*—A BIMOS IC technology improving the design of interface circuits that require either high-voltage (up to 120 V) of current (up to a few amperes per output) has been developed. Both bipolar and MOS complementary components are processed together on the same chip for low- and high-voltage applications. Various BIMOS power interface circuits are now in production, e.g., a motor driver, a high-voltage plasma display driver, and a printer head driver. This paper describes the BIMOS technology and the characteristics of its components. As applications, two circuits are presented: the UEB 4732 (plasma display driver) with complementary MOS push–pull output stages (120 V), and the UAA 2081 (stepper motor driver) with power bipolar transistors (1 A per output). Both circuits have a logical part designed with low-voltage CMOS (5–12 V).**

I. BIMOS Technology

THE NEWLY developed BIMOS technology is a semiconductor fabrication process mixing on the same silicon wafer bipolar components and complementary MOS devices in two voltage ranges, a low voltage of 12 V and a high voltage of 120 V for HVMOS and 80 V for bipolar. Fig. 1 shows a cross section of the process. One can identify the main features of a complementary bipolar process with n-epitaxy on p-substrate ⟨100⟩ and junction isolation.

To insure availability of a low-power and fair packing density logic, we chose to add to bipolar a CMOS with (1) silicon gate, (2) local oxidation and field implants, and (3) self-aligned source and drain. The "superposition" of both technologies can be used industrially and be profitable only if the number of photolithographic levels does not become too large (13 in our case). This situation leads to multifunction levels:

1) n-p-n base implant common to p-source and drain;
2) n-p-n emitter diffusion common to n-source and drain; and
3) p^- well of the NMOS common to isolation, p-channel for the LDMOS [12], and drift region around the HV PMOS drain [12].

Obviously, this situation leads to some compromises:

1) NMOS with drawn gate minimum of 6 μm;
2) PMOS with drawn gate minimum of 8 μm; and
3) HV MOS without self-alignment on the gate.

Manuscript received April 17, 1986; revised July 19, 1986.
The authors are with Thomson Semiconducteurs, Bipolar Division, BP 54, 38130 Saint Egreve, France.
IEEE Log Number 8610707.

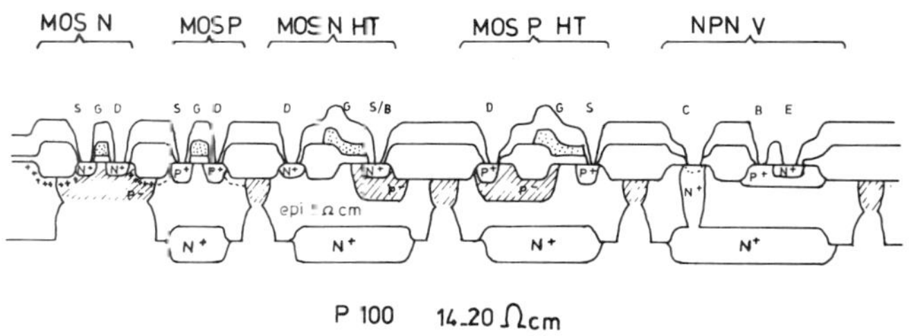

Fig. 1. BIMOS technology.

We have taken advantage of "up isolation" and buried-layer levels necessary to bipolar to give CMOS an excellent immunity to latch up. More precisely a buried layer lies under PMOS regions in order to reduce resistance R_{N^-} and a p^- well parasitic n-p-n is avoided by short circuiting the well to the substrate through the up isolation. Fig. 2 gives some physical and electrical characteristics of process and devices demonstrating their versatility, complexity and performances.

Fig. 3 presents the $I_{DS} = f(V_{DS})$ and $\sqrt{I_{DS}} = f(V_{GS})$ curves for complementary HVMOS devices. The effective length and breakdown voltage of these devices depend, by their construction, on alignment accuracy. Their interest lies in their 120-V blocking capacity and their ability to deliver current in the 250-mA range. However, a more efficient output device is the n-p-n with 0.45 mm^2 for 1-A output current. The n-p-n has $BV_{\mathrm{CEO}} = 80$ V; the Early voltage is in the range of 400 to 500 V and the $F_{t\max} \geqq 400$ MHz (Fig. 4).

The high-voltage p-n-p can be of the vertical or the lateral type. L-p-n-p have their base under local oxide and $\beta \geqq 80$. To improve further their Early voltage, the collector can be implanted into a p^- well acting as a drift region. The space-charge region extends then both in the n-base and p^- collector region. The design of L-p-n-p requires use of a polysilicon shield, properly connected to avoid parasitic MOS due to metal connections.

Thanks to the bipolar structure, it is possible to build good high-voltage diodes for use as protection against inductive overvoltages. These diodes (Fig. 5), due to their structure, occupy more space than an ordinary p^+-n diode with surrounding deep n^+ plug. But the gain of the parasitic substrate p-n-p associated with these diodes is reduced by a factor of 5 to 10. A low leakage current to the substrate is an interesting characteristic when one considers the deleterious effect of substrate current like parasitic

Reprinted from *IEEE Trans. Electron Devices*, vol. ED-33, no. 12, pp. 2016–2024, Dec. 1986.

	NPN	L PNP	UNITS
GAIN (β)	> 80	> 60	V
BVCEO	80	80	
$F_{T\ MAX}$	450	3	MHz
	NMOS	PMOS	
THRESHOLD VOLTAGE	+ 1.5	- 1.2	V
K' (μ Co/2)	10 - 11	2.2 - 2.6	μA/V²
BODY EFFECT	1.8 - 2	0.5 - 0.6	V^1/2
GATE OXIDE THICKNESS	800	800	Å
FIELD OXIDE MOS	> 12	< - 12	V

	VALUES	UNITS
SUBSTRAT P	10 - 20	Ωcm
EPITAXY N	5.5 - 7.5	Ωcm
R□ BASE	120 - 130	Ω□
R□ EMETTER	5 - 7	Ω□
R□ BURIED LAYER	9 - 12	Ω□
R□ POLY	20 - 30	Ω□

	BIPOLAR NPN	BIPOLAR PNP		CMOS BT N	CMOS BT P	CMOS HT N	CMOS HT P	
BV_{EB}	6,5	150 V	V_T	1,2-1,8	1,1-1,6	1,2-1,8	1,1-1,6	V
BV_{CEO}	80 V	80 V	BV_{DS}	24	26	120	120	V
H_{FE}	100 - 200	40 - 80	GM/W MAX	20	6,5	10	5	μA/V.μ
R_C/MM²	0,48	Ω/MM²	R_{ON}/MM²	0,24	2,76	13	68	Ω/MM²

BIMOS

- 13 LITHOGRAPHIC STEPS
- 5 PLASMA ETCHES
- 6 DEPOSITIONS (EPITAXY, 3 LPCVD, 1 CVD, 1 PECVD)
- 6 ION IMPLANTS (1 SB, 1 AS, 4 B)
- 6 ANNEALLINGS
- 6 OXIDATIONS (INCLUDING LOCAL OX AND GATE OX)
- 1 METALLIZATION (AL/SI)

BIMOS

DESIGN RULES

- MINIMUM SIZE CONTACT (4 μ x 8 μ OR 6 μ x 6 μ)
- POLY PITCH 12 μ (6 μ + 6 μ)
  . MOS N MINIMUM 6 μ
  . MOS P MINIMUM 8 μ
- ALUMINIUM PITCH 13 μ (9 μ + 4 μ)
- DESIGNED FOR 2 μ ALIGNMENT TOLERANCE
- ALL PERKIN ELMER 1/1 PROJECTION ALIGNS

Fig. 2. BIMOS electrical characteristics.

n-p-n action in between adjacent epi islands. For a 1 900-μm^2 n^+ cathode the forward voltage at 500 mA is 1.7 V while diode breakdown is 140 V, and 180 V to substrate is allowed.

The logic CMOS part is characterized by: 1) a density of 380 transistors/mm² (n- and p-types mixed); 2) a maximum frequency of 10 MHz (V_{cc} = 12 V) over the military temperature range (up to 125°C), for a static shift register; 3) choice of voltage supply (either 5 V to insure TTL compatibility or 12 V to improve immunity to external interference); 4) no threshold adjustment by implantation; 5) effective lengths and mobilities for minimum size MOS of: $LN = 2.8\ \mu m$ (μN_{max} − 700 cm²/V · s), $LP = 3.6\ \mu m$ (μP_{max} − 225 cm²/V · s); and (6) no latch up. Fig. 6 shows a plot of $I_{DS} = f(V_{DS})$ and $\sqrt{I_D/W} = f(V_{GS})$ for logic CMOS components.

We took advantage of the necessity of two field implants to implement a current limiting device with the p-field implant pinched by an arsenic n^+ field implant. This device, under the local oxidation, is characterized by V_p = 6 V, R_{ON} = 17.5 kΩ/□, BV = 26 V, I_{DSSO} = 250 μA/□ (Fig. 5).

To ease special analog design (amplifier, voltage reference), the low-voltage CMOS can be completed by low-voltage (12 V) bipolar components. These are deduced from HV bipolar components by adapting the design rules.

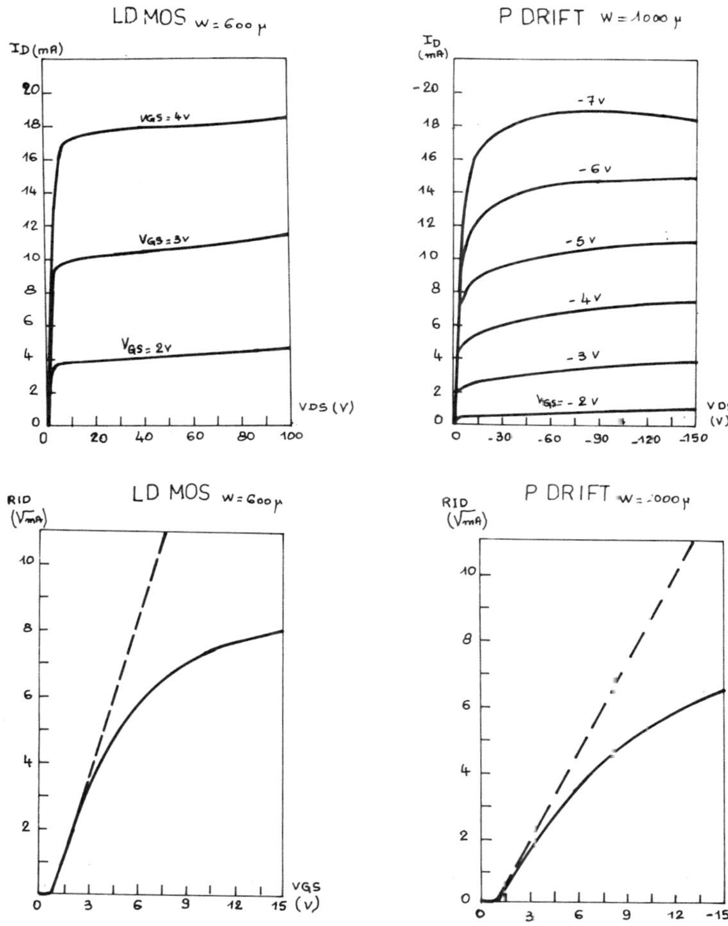

Fig. 3. High-voltage CMOS components.

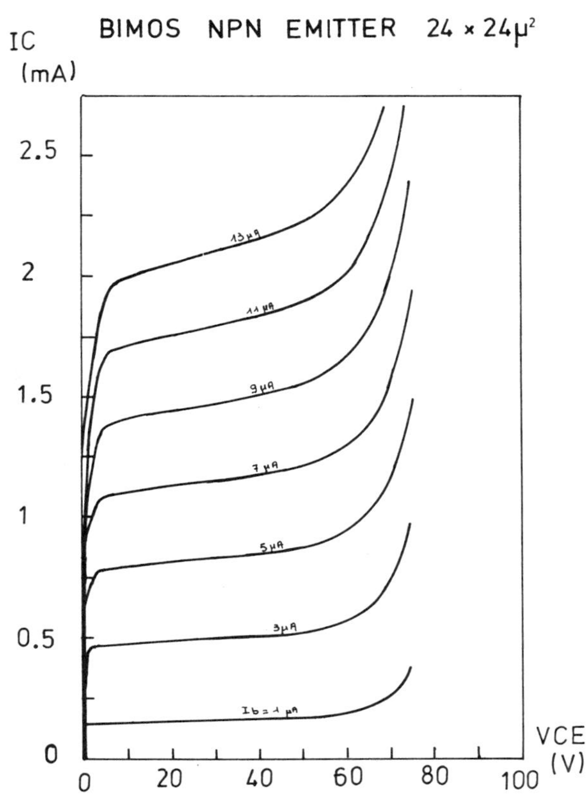

Fig. 4. High-voltage n-p-n.

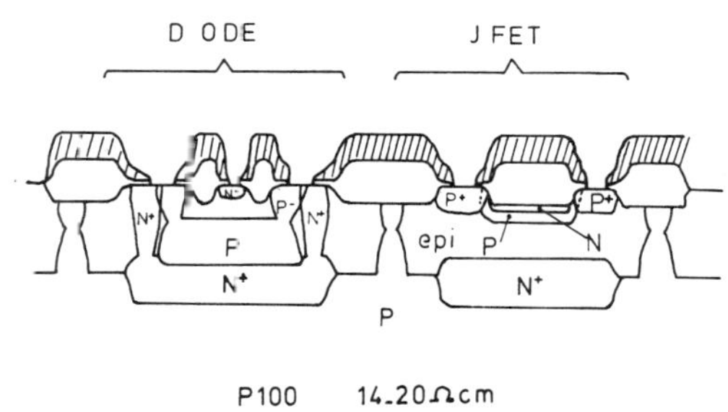

Fig. 5. Diode and JFET.

For an intermediate complexity process (13 masks), thus at acceptable industrial cost, the technology offers a broad variety of components and therefore flexibility for the designers. This flexibility is exemplified by the presentation of two circuits.

II. A Unipolar Stepper Motor Driving Circuit

As an application of BIMOS technology, Thomson Semiconducteurs and Triumph Adler have developed together the UAA 2081, which is a dual unipolar motor driver.

A. The System

A permanent magnet stepper motor consists of a permanent magnet rotor and stator poles with windings (Fig.

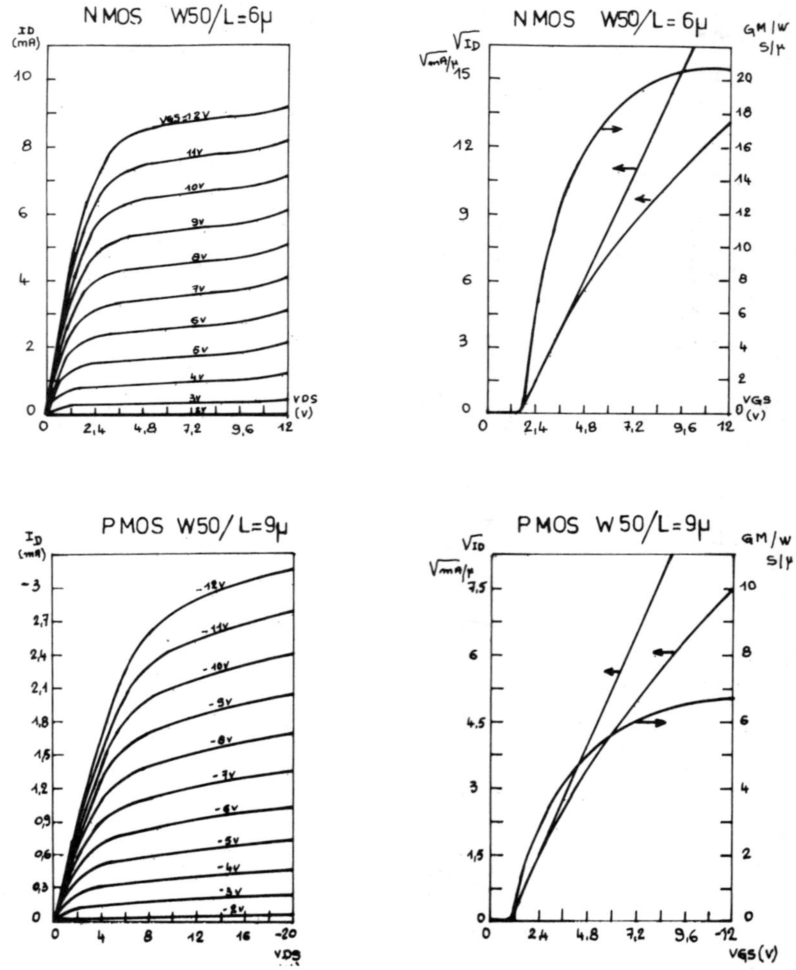

Fig. 6. Low-voltage CMOS.

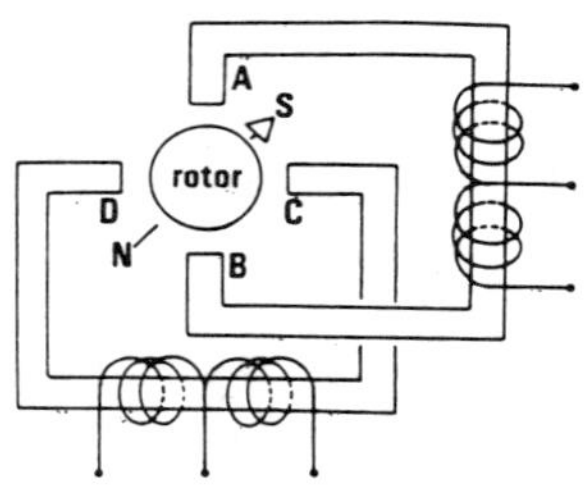

Fig. 7. Stepper motor schematic diagram.

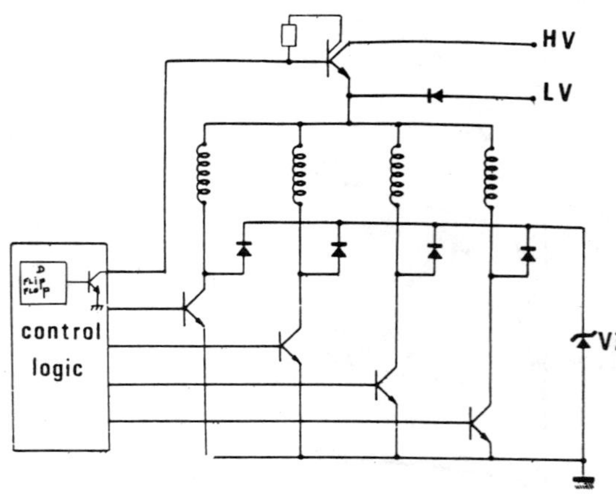

Fig. 8. Bilevel drive diagram.

7). When current flows through 1 or 2 windings, it creates a magnetic field to which the rotor will be aligned. Then, by switching off the current in one winding and by driving the next one, the field turns and the rotor rotates a step [1]. The one-phase-on or "wave drive" mode consumes the least power because only one motor phase is energized at one time; sequentially the *A*, *C*, *B*, and *D* winding is driven for 90° steps. The two-phase-on mode or "overlapping drive" mode offers less susceptibility to resonance and better single step damping; now *AC*, *BC*, *BD*, or *DA* windings are driven providing again 90° steps. By combining these two modes, the motor can be used in the half-step mode with 45° steps and an *A*, *AC*, *C*, *BC*, *B*, *BD*, *D*, *DA* sequence. This excitation mode produces a higher acceleration and the smoothest operation. According to these remarks, it is obvious that an extremely versatile integrated circuit solution is required to drive efficiently a stepper motor.

The combination of two basic drive approaches allows to get very short current rise and fall times in order to have good motor efficiency: the bilevel drive and the recirculation diodes (Fig. 8) [2], [3]. To get a short rise

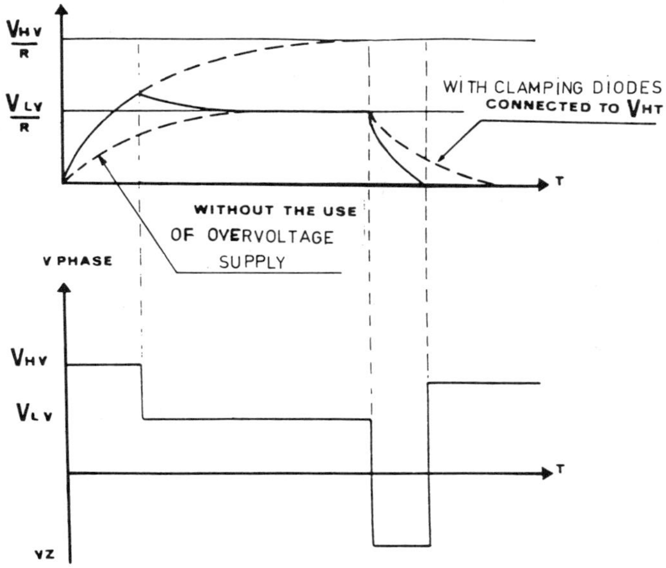

Fig. 9. Voltage and current for a phase.

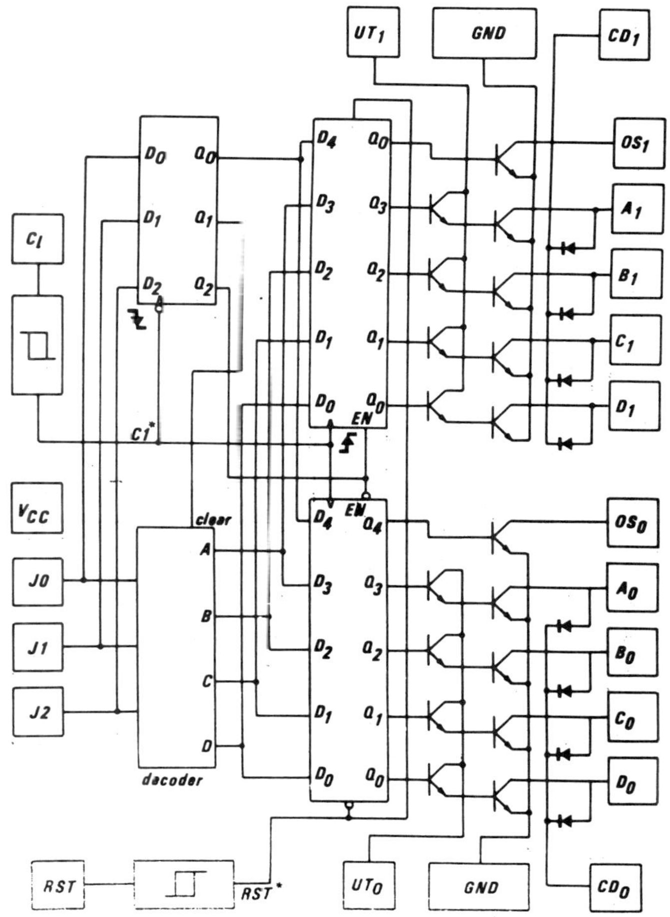

Fig. 10. UAA2081 block diagram.

time, a high voltage supply (20 to 40 V) is used for a rapid current increase at phase turn on. Then a low voltage (5 to 10 V) is used to sustain the current at the rated level. To get a short current fall time, a recirculation diode is used with a high Zener voltage (about 50 V) preventing high-voltage spikes from destroying the switching device, and making the current to decrease rapidly (Fig. 9). So a good circuit must integrate on the same chip both logical elements and power stages with rugged and low dissipation transistors.

B. *The Integrated Solution*

To get such a good circuit at a low cost, BIMOS technology was the only solution. The present circuit can drive independently two unipolar stepper motors with the bilevel solution [4]. For each motor, four n-p-n transistors deliver 1 A per phase with a low saturation voltage: 900 mV (outputs *A*, *B*, *C*, *D*) (Fig. 10). The transistors can withstand 60 V as BV_{CEO}. The area needed is about 0.45 mm^2 per transistor, but in return the saturation resistance is less than 0.8 Ω at $T_{amb} = 25°C$. The outputs are protected by recirculating clamp diodes (outputs *CD*) characterized by low drop voltage (1.5 V at 1 A) and very low leakage current to the substrate.

An n-p-n transistor (output *OS*) is used as a switch to select the power supply voltage applied on the motors' windings. It has been designed to a current of a few milliamperes (up to 8 mA) with a saturation voltage of 150 mV and drives an external Darlington transistor. As it switches high voltages, it has to withstand 60 V as BV_{CEO}. When transistor *OS* is off, the Darlington is on and the higher supply voltage gives a rapid current increase for improving speed and torque of the motor (Fig. 8). The driver current for switch saturation can be adjusted by external resistances (inputs *UT*) for minimum chip losses. Dissipation for an output of 2 × 0.5 A is 0.55 W, whereas it is 2 W at 2 × 1 A at I_{ut} of 50 mA and 100 mA, respectively, and $V_{cc} = 5$ V. A 24-pin dual in line package is used with an *R*th junction ambient of about 55°C/W.

For avoiding lateral n-p-n action between two adjacent epi islands due to current switching in the windings of the motors, we have included an extra n-epi island between them, short-circuited to ground. Thus, the n-p-n parasitic beta is reduced to less than 2×10^{-3} [5].

The logic part consists of 10 flip-flops with a reset signal and a 3- to 4-bit decoder that allows us to define independently all the different driving sequences for bilevel half/full steps. Thanks to the CMOS density, only a small area is necessary: 0.04 mm^2 per flip-flop and less than 0.15 mm^2 for the decoder. All the logic inputs (*J*0, *J*1, *J*2, *RST*, *Cl*) are LS TTL and CMOS compatible. Besides, *RST* and *Cl* inputs have been specially designed for high noise margin (thanks to a 400-mV hysteresis) and high protection against less than 200-ns disturbant signals. The device can operate over a wide range of logical supply voltage (4.5 to 7 V) and because of the use of the CMOS power, the logic current is 2 mA, +2 mA per activated power output.

The main interest is to require few external components to get a performant control device for two unipolar stepper motors (Fig. 11). A microcontroller with two 3-bit words defines acceleration pulse width, forward or re-

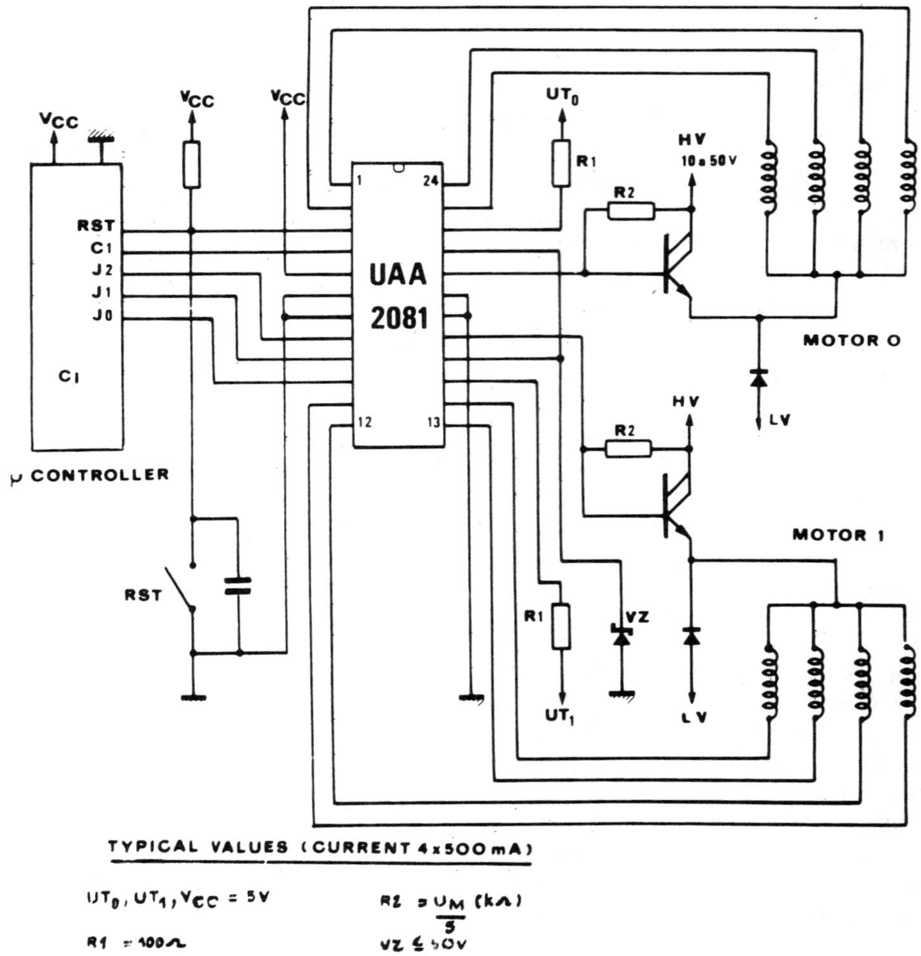

Fig. 11. UAA2081 Application synoptic.

verse rotation (CW or CCW), and half or full step mode. Dissipation is optimized by the choice of the resistance on the UT inputs. The adequacy of the driver transistors avoids the use of a too powerful supply. Fig. 12 shows a photomicrograph of the 14-mm^2 chip.

Fig. 12. Photomicrograph of UAA 2081.

III. A High-Voltage IC Driver for AC Plasma Display Panel (UEB 4732)

Thomson Semiconducteurs has designed the UEB 4732 in cooperation with the Thomson-CSF Electron Tube Division and with support from French Direction Générale de l'Armement (DGA) to answer the requirement of integrated drivers for ac plasma panels, and mainly for plasma panels of large dimensions (512 rows × 512 columns, 1024 × 1024, $\cdots$) [9]. These large display panels are now available from several suppliers in the United States, Japan, and Europe.

A. AC Plasma Display Panel

The ac PDP is an X–Y matrix gas discharge display with inherent memory. The pixel defined by the intersection of the line electrode Y and the column electrode X is driven by the voltage ($V_y - V_x$) created by the difference between the voltages V_y and V_x applied, respectively, to the line Y

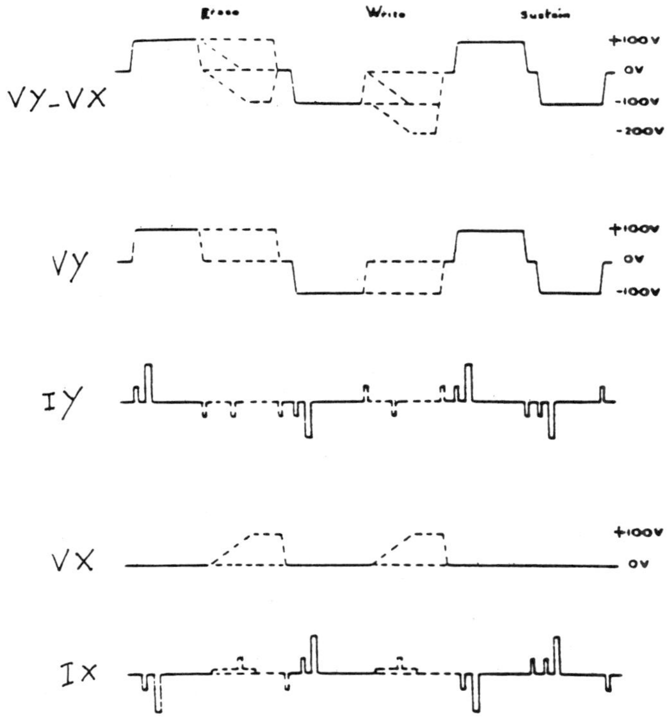

Fig. 13. Command signals for an ac PDP.

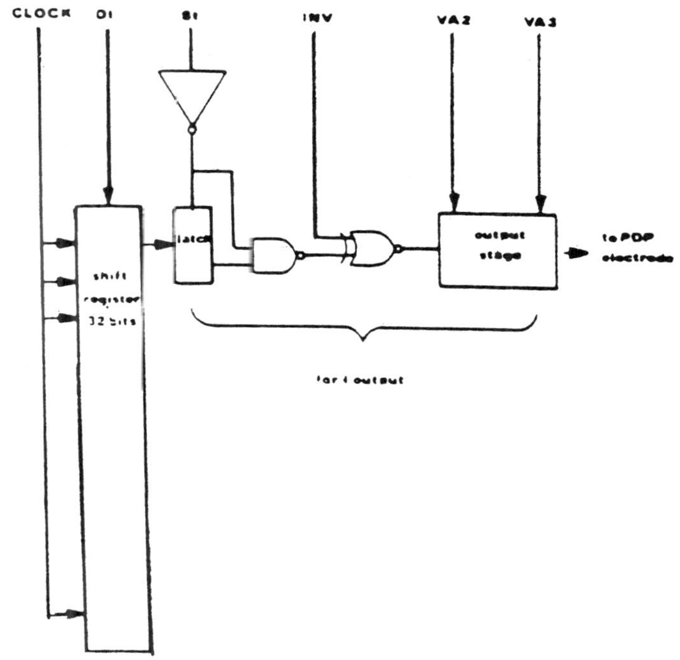

Fig. 14. UEB4732 block diagram.

and the column X. The WRITE, ERASE, and SUSTAIN signals ($V_y - V_x$) of a pixel are sketched on Fig. 12 with the voltages V_y and V_x of the corresponding electrodes, and with the current flow created by these signals.

Two types of signals are needed by the plasma panel (Fig. 12), [6], [7]: 1) A sustaining voltage, swinging between 0 and +100 V, and between 0 and −100 V simultaneously on all the electrodes. The switching time of the sustain waveform is short (<0.5 μs). The sustain waveform creates the luminous discharge current that can reach 100 mA for an electrode and lasts 0.2 μs. The drive circuit must have a low-voltage drop to maintain the luminous signal of the plasma. 2) Selective writing and erasure address signals [4], bringing the selected electrodes to 100 V above the others. These signals are slow (leading edge ≧1 μs). Their role is to charge the capacitor between a selected electrode and its neighboring electrodes, and to supply the WRITE and ERASE currents of the pixels. These currents appear during the raising slope of the signal and are always below 10 mA. They can be supplied across a high impedance (1 kΩ).

B. The Integrated Solution

The required signals are generated by the UEB 4732 from four CMOS logic inputs and one high-voltage signal [10]. The IC includes a CMOS logic part (decoding, 32-bit shift register with latches) and a low- to high-voltage interface (Fig. 13). The design choosen for the output stages (Figs. 14 and 15) takes advantage of the qualities of diodes and complementary HVMOS of HVBIMOS technology. Diodes with a low forward voltage ($\Delta V < 2$ V for $I_D = 100$ mA), transmit the sustaining signals [11]. During sustain, V_{A2} and V_{ref} are equal and connected to a sustain amplifier, common to all the IC drivers of one axis. The high-voltage stages are disabled during sustain, so the power consumption in the IC is lowered.

The selective signals are created by a complementary (120 V) LDMOSN and p-drift stage. The LDMOSN is directly driven by the logic part of the IC. The MOSP drift uses two reference current sources, common to the 32 output stages. This design presents the following advantages: a small number of high-voltage components per stage (2LDMOSN, 1 MOSPdrift, and 2 diodes), therefore reduced area needed; a low power consumption at a steady level, an easy control from the logic part of the IC (only one logic signal needed per stage).

The 64 diodes (two diodes per stage) are made of p^+-n with a deep n^+ plug. The area covered by the n-island of one diode is 0.038 mm^2. Taking into account the operating conditions of the circuit ($V_{A2} = V_{\text{ref}}$ during the peak currents through the diodes), low leakage diodes, demanding a bigger area, were not useful. The forward voltage is typically 1.2 V at $I_D = 100$ mA; BV is ≧ 130 V.

The PMOS drift transistor must have an $R_{DSON} < 1$ kΩ with $V_{GS} = -12$ V and $I_{DS} = 10$ mA on the whole military temperature range (−55°C, +125°C). Taking into account that this component is not self-aligned, and that its R_{DSON} doubles for a 100°C temperature rise, we have designed the MOSP drift with $W = 1000$ μm (area of one MOSP drift island = 0.08 mm^2). We obtained an R_{DSON} typically equal to 0.3 kΩ to ambient.

Two LDMOSN's are needed for each output stage. The biggest one, used with the MOSP drift in the push–pull output, must have $R_{DSON} < 1$ kΩ with $V_{GS} = 12$ V and $I_{DS} = 10$ mA on the whole military temperature range (−55°C, +125°C). The LDMOSN being favored com-

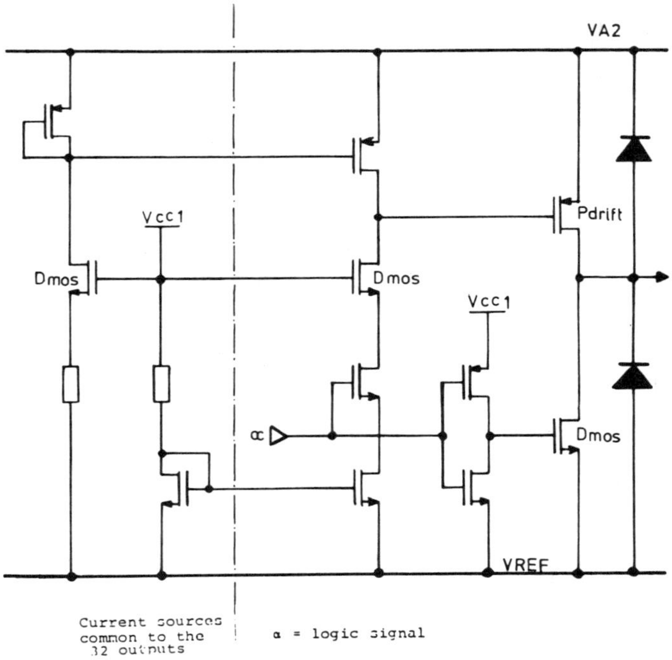

Fig. 15. UEB4732 schematic of one output stage.

Fig. 16. Photomicrograph of UEB4732.

pared to the MOSP, was designed with $W = 600\ \mu m$ (the area of one LDMOSN island is 0.045 mm^2). We obtained an R_{DSON} typically below 0.2 kΩ at 25°C ambient. The other LDMOSN, used in the command system of the MOSP drift, is mainly a link between high voltage and logic signals. It has a low level direct current and its on resistance is not a main parameter. We designed it with $W = 100\ \mu m$, so it's R_{DSON} is roughly 1 kΩ.

Fig. 16 shows a photomicrograph of the 20 mm^2 chip, with 14 mm^2 for the 32 output stages. It is packaged in a 40-pin DIL, in a version for commercial applications (0, 70°C) and in a version for military applications (−55°C, +125°C). A micropackage is on study.

IV. Conclusion

This article has presented the use of various HVBIMOS components in the IC design for two very different interface problems: the driving of two stepper motors (UAA2081) and the driving of an ac plasma display panel (UEB4732). BIMOS is also used in other applications such as thermal printer head driving (UCA4532) or in nonspecific IC (octuple latch UCN4801A). These various interface circuits are at present in full production. Diverse new circuits are on study and design, using BIMOS technology. With a production price only 20 percent higher than a standard high-voltage bipolar technology, HVBIMOS offers high-voltage bipolar components plus high-voltage complementary DMOS and low-voltage CMOS. This diversity allows a great reduction in the number of IC's needed for an application by the integration of logic and power functions on the same IC.

References

[1] H. Bollinger and D. Hirschi, "Le moteur pas à pas, organe d'entraînement du futur," *Tech. moderne*, vol. 3, p. 9, 1985.

[2] D. Jones, J. Anderson, and H. Manson, "The PM step motor, integrated combination, a low cost integrated step motor drive system," in *Proc. Motorcon*, p. 155, Sept. 1982.

[3] D. G. Wilson, "Step motor drivers: An analysis of various approaches," in *Proc. Motorcon*, p. 166, Sept. 1982.

[4] G. Troussel, "L'UAA2081: Une commande intégrée originale de deux moteurs pas à pas unipolaires réalisée en BIMOS," to be published.

[5] F. Van Zanten, "Dispostif de protection contre les courants de fuite dans les circuits intégrés," *INPI*, vol. 80, p. 10842, May 1980.

[6] J. Deschamps, "L'affichage de données graphiques et alphanumériques par panneau à plasma," *Rev. Technique Thomson-CSF*, vol. 10, no. 2, June 1978.

[7] Texas Instruments, *Display Driver Handbook*, 1984.

[8] T. N. Criscimagna, J. R. Beidl, and J. B. Trushell, "Write and erase waveform for high resolution ac plasma display panels," *IEEE*, 1980. 1980.

[9] F. Vialettes and G. Thomas, "BIMOS technology and its application to plasma panel IC driver," *Opto Dig.*, p. 245, 1985.

[10] L. Delgrange, M. Specty, and F. Vialettes, "Circuit de commande d'un panneau à plasma de type alternatif," *INPI*, vol. 85.

[11] L. Delgrange, J. Deschamps, and F. Vialettes, "A.C. plasma display panel control circuit," U.S. patent application 431 152.

[12] G. Thomas, "Structure de circuit intégré comportant des transistors CMOS à tenue en tension élevée et son procédé de fabrication," *INPI*, vol. 84.

[13] L. Degrange, F. Vialettes, J. Deschamps, and G. Thomas, "A high voltage IC driver for large-area A.C. plasma display panels," in *SID Dig.*, p. 103, 1984.

A BiCMOS Thermal Head Intelligent Driver with Density Controllers for Full-Tone Rendition Printers

MAKOTO TSUMURA, RYOZO TAKEUCHI, MEMBER, IEEE, AND ISAO SHIMIZU

Abstract **—A 3-μm BiCMOS thermal head driver employing pulse-width modulation dealing with 8-bit density input data (256 gray levels) is described. We have developed novel circuits, composed of 64×8-bit complex counters, which function as eight-line parallel 64-bit shift registers (shift mode) and as 64 counters which have 8 bits (count mode) by alteration of their mutual connections according to the mode signals. The complex counter controls the output pulse width according to the binary data and the clock intervals (minimum 100 ns). The shift registers can operate at 20-MHz clock. The driver consists of about 4500 CMOS gates and 128 bipolar transistors in a 2.8×8.8-mm² chip size. The breakdown voltage of the bipolar transistor BV_{cbo} is more than 35 V. The developed driver is especially suited for full-tone rendition printers. Applications of the driver include use in thermal print heads, LED print heads, LCD print heads, etc.**

I. Introduction

FULL-TONE rendition color printers, such as thermal transfer printers, are coming into great demand as video printers, CAE printers, and color copiers. These printers need high-quality full-tone printing characteristics which are obtained by controlling the heat energy of each microheater on the thermal print head through changes in the applied pulse width. The thermal head printing quality depends on the accuracy of the controlled pulse width, namely the number of gray levels. The conventional drivers [2], [4], which have been developed for black and white printers, cannot receive the multibit binary image signals directly, as they consist of only one-line shift registers. The converter is indispensable for full-tone rendition printers. The converter changes the multibit binary image signals to the lines of ON/OFF signals sliced in each gray level. Therefore, the amount of transmitted data increases in proportion to the number of gray levels. The printing ability of the conventional driver is inversely proportional to the increase in the number of gray levels, as shown in Fig. 1, because the number of data transmission times increases according to the gray level. Therefore, if the printing ability of the thermal head is higher than that of the driver as shown in the hatched area, the printing speed is limited by the number of gray levels and the operation speed of the driver.

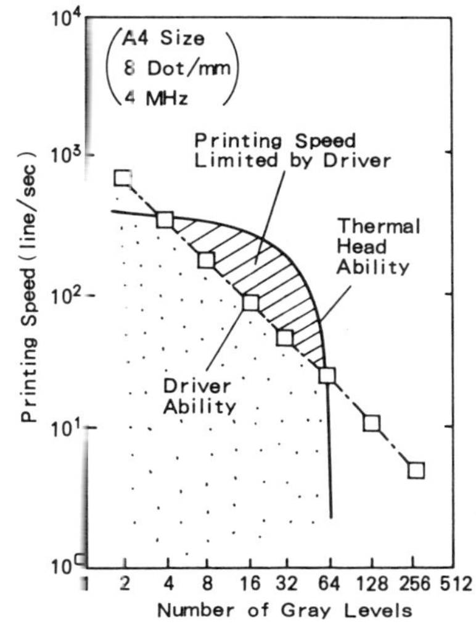

Fig. 1. Number of gray levels versus printing speed.

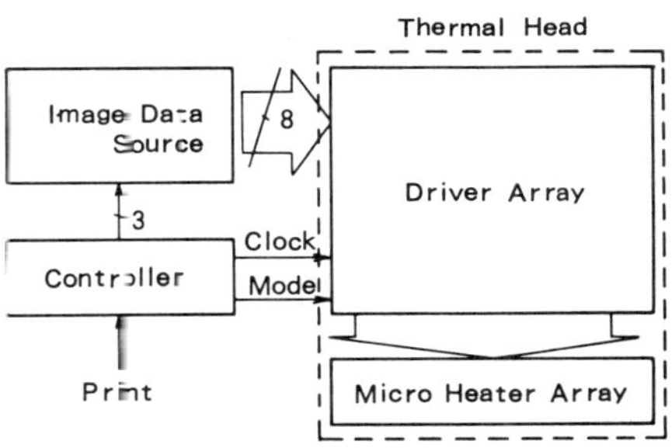

Fig. 2. Block diagram of full-color thermal transfer printer.

In order to resolve this contradiction, the driver has to receive binary image data directly, as shown in Fig. 2, and control the output pulse width according to these data. Since the number of times data are transmitted is only once per picture element cell, the driver achieves both high-speed and high-quality printing. Counter circuits are well suited for this kind of operation, because they can easily convert the binary data to the controlled output pulse width by addition of logic gates.

We have developed the first thermal head driver which is especially suited for full-tone rendition image printers. It

Manuscript received July 31, 1987; revised November 20, 1987.
M. Tsumura and R. Takeuchi are with Hitachi Research Laboratory, Hitachi Ltd., 4026 Kuji-cho, Hitachi-shi, Ibaraki-ken 319-12, Japan.
I. Shimizu is with Takasaki Works, Hitachi Ltd., 111 Nishiyokote-cho, Takasaki-shi, Gunma-ken 370-11, Japan.
IEEE Log Number 8719260.

Reprinted from *IEEE J. Solid-State Circuits*, vol. 23, no. 2, pp. 437–441, April 1988.

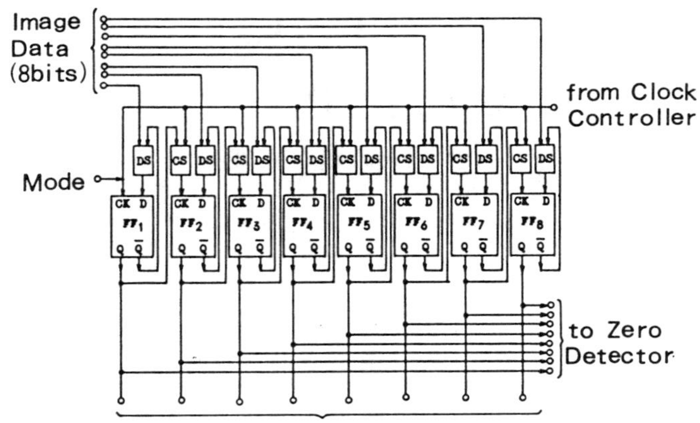

Fig. 3. Circuit organization of the 8-bit complex counter.

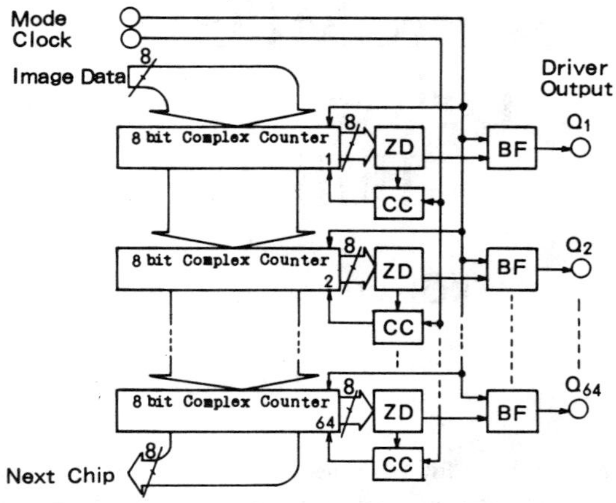

Fig. 4. Logic diagram of 64-bit thermal head driver.

is composed of 64×8-bit complex counters and 64-bit output buffers. This paper describes the architecture, circuit design, and device features of the 64-bit output, 256 gray level, BiCMOS thermal head intelligent driver.

II. Functional Description

Image data are usually transmitted from frame memory or A/D converter output of video signals in the form of 8-bit binary image data. In the case of a one-line shift-register-type driver, it is indispensable that the bit-sliced converter be outside the thermal head, as the size of the circuits is very large, and the converter has to operate more often with increases in the number of gray levels. Furthermore, it takes much time to transmit the huge number of data.

To decrease the size of the external circuits and the data propagation time and to get high-quality printing, the following functions are necessary: eight-line shift registers for transmitting image data to the line memory; 8-bit line memory for initializing the counters; and 8-bit counters for the pulse-width modulation. Flip-flop circuits are well suited for all of these operations. They can operate as shift registers, memories, and counters by altering their wire connections. Fig. 3 shows an 8-bit complex counter which can operate like a flip-flop to provide the three functions described next.

A. Shift Register and Memory Function

The 8-bit complex counter works as a component of the eight-line shift registers, by setting the data selectors *DS* to connect image data and data input *D*, and setting the clock selectors *CS* to connect the clock controller and clock input *CK* of the flip-flops. The clock signals are applied to each clock input *CK* through the clock controller, and then image data are transmitted to the next complex counter, synchronized by the clock signals (shift-register function). When the clock signals are stopped, the image data are memorized in each flip-flop (memory function).

B. Counter Function

By means of inputting each output $\bar{Q}$ to its own data input *D* and connecting the output *Q* and the clock *CK* by switching the data selector *DS* and the clock selector *CS*, respectively, the 8-bit complex counter works as an 8-bit down counter. The data selector *DS* and the clock selector *CS* are controlled by the mode signal. The data, which are stored just before alteration of the mode signal to the count mode, are used as the initial data of the counter. Therefore the loading procedures from the shift register to the counters are completed automatically.

Fig. 4 shows the logic diagram of a 64-bit intelligent thermal head driver including one 8-bit complex counter per one output buffer. The zero detector *ZD* and clock controller *CC* are combined in order to perform the pulse-width modulation. By setting the mode to the shift mode, 8-bit binary image data are transmitted by the clock signals. Then, by altering the mode to the count mode, if the initial data of the counter are not zero, the zero detector *ZD* turns the buffer on and the clock controller permits passing of a clock signal to the flip-flop. The clock signal continues to count down the output of the counter step stepwise until the *ZD* detects a zero value. If the counter output becomes zero, the clock controller inhibits the next clock signal. Therefore, the counter stops the operation by itself and the zero detector turns off the output buffer. As each complex counter has a zero detector and a clock controller, each output pulse width is controlled independently by the image data stored in the complex counter and the intervals of the clock signals.

III. Features of BiCMOS Devices

The developed thermal head driver demands higher integration and smaller chip size for mounting on a thermal head with high density, and it is necessary to operate the device with high speed and low power dissipation for high-speed and high-quality printing.

The BiCMOS process meets all these demands. It features a combination of bipolar and CMOS technologies,

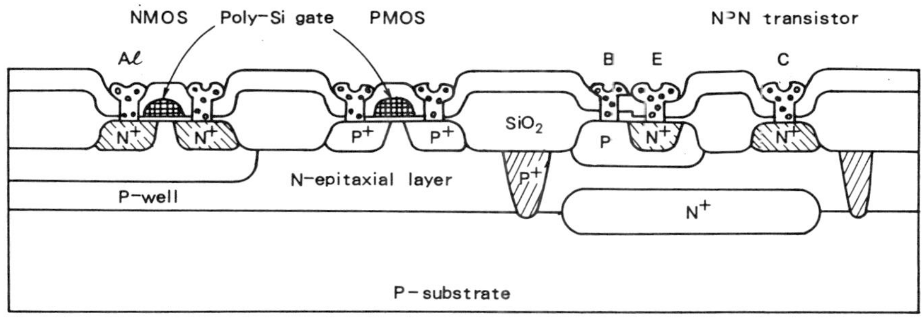

Fig. 5. Section structure of 3-μm BiCMOS device.

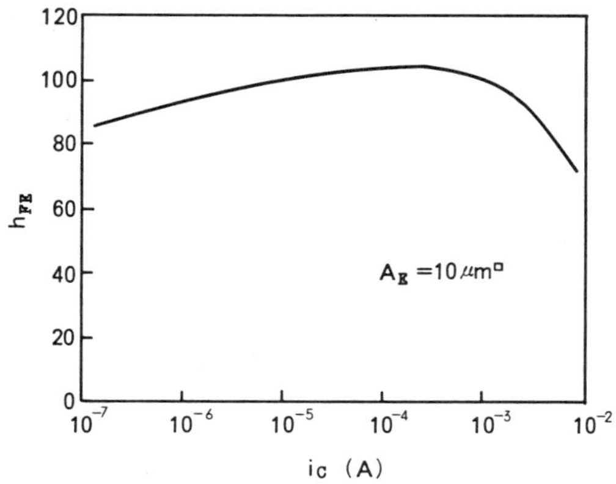

Fig. 6. Characteristics of n-p-n transistor.

TABLE I
CHARACTERISTICS OF 3-μm BiCMOS

Device		Item	Specification	Remarks
Bipolar Transistor	NPN	BV_{CEO}	15 min(V)	
		BV_{CBO}	35 typ(V)	Emitter size is 10 μm^2
		h_{FE}	100 typ	
		$f_{T\ max}$	600 MHz	
		r_{CS}	10 Ω	h_{FE}=10
	PNP	$\lvert BV_{CEO} \rvert$	15 min(V)	
		h_{FE}	50 typ	
		$f_{T\ max}$	2.5 MHz	
MOS Transistor	N channel	BV_{DS}	10 V	Gate Length 3 μm Thickness of Oxide 500 Å
		V_{TH}	0.55V	
	P channel	$\lvert BV_{DS} \rvert$	10 V	
		$\lvert V_{TH} \rvert$	0.55V	

and these two transistor types are integrated on the same chip. A cross section of the newly developed high-breakdown-voltage 3-μm BiCMOS device is shown in Fig. 5 and its characteristics are listed in Table I and Fig. 6. The base area and the source–drain area of the P-channel MOS transistor are formed with different diffusion conditions to improve precision of the bipolar transistor and diffusion resistor which determine performance of analog parts. The emitter area and the source–drain area of the N-channel MOS transistor are formed concurrently by the same diffusion. The CMOS part has the same performance as obtained by the 3-μm CMOS process. This BiCMOS device is made in the N-type epitaxial layer on a P-type substrate with a controlled surface impurity concentration, so V_{th} of the MOS transistor is precisely controlled and its deviations are small.

Fig. 7. Microphotograph of 64-bit thermal head driver.

IV. DEVICE CHARACTERISTICS

Fig. 7 is a microphotograph of the 64-bit thermal head driver fabricated by a high-breakdown-voltage 3-μm BiCMOS process [1], [3]. It consists of about 4500 CMOS gates and 128 bipolar transistors in a 2.80×8.80-mm^2 chip size. By using 64-bit thermal head drivers for line print heads with 8-dot/mm resolution, only 27 drivers are necessary for an A4 size line print head with 1728 microheaters. These driver chips can be arranged in a single line on a thermal print head.

The developed intelligent driver achieves the characteristics listed in Table II. It has a high breakdown voltage

TABLE II
FEATURES OF THE DRIVER

Item		Value
Number of Input Data		8 bits (256 Gray Levels)
Number of Controls		2
Operation Speed	Shift Mode	20 MHz
	Count Mode	10 MHz
I/O Interface		CMOS
Drivability		50 mA/bits

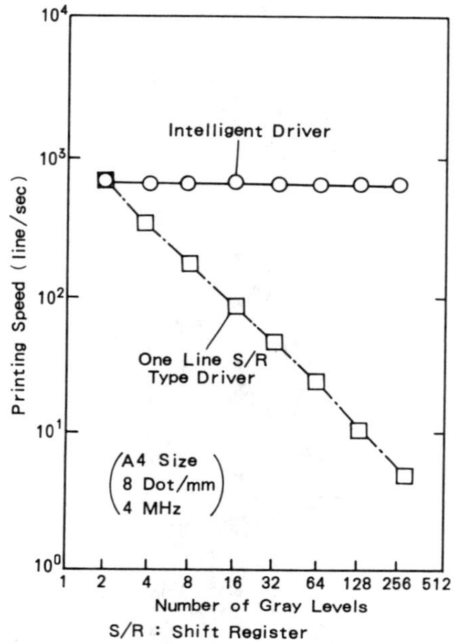

Fig. 8. Device application characteristics.

(BV_{cbo}) and low drop voltage (rcs) in the output buffers. Further, the low voltage dispersion between each buffer, low power dissipation, and small chip size are direct results of using the high-breakdown 3-μm BiCMOS process. These characteristics have the following advantages for high-quality full-tone rendition printers:

1) to decrease the power dissipation in power supply lines and buffers, through the high BV_{cbo} and low rcs;
2) to improve the accuracy of the energy applied to each microheater on the thermal head by lowering the voltage dispersion between each buffer; and
3) to realize a high-resolution print head with a small chip size.

V. APPLICATIONS

Fig. 8 shows the printing speed characteristics of a full-tone rendition printer which uses the developed intelligent driver (solid line) and the one-line shift-register-type driver [2], [4] (broken line). For the printer which employs the latter, the printing speed is the same as that of the intelligent driver for two gray levels such as needed in character printers, but it decreases inversely with the number of gray levels. The printing speed of the printer employing the intelligent driver is about 900 line/s for a 4-MHz shifting clock and it does not change up to 256 gray levels. Printing performance of 900 line/s for an A4 size full-tone rendition color image with 256 gray levels is obtained within 10 s.

VI. CONCLUSION

A BiCMOS thermal head intelligent driver has been developed. A complex counter which can function as shift register, memory, and counter was developed to decrease the size of external circuits, increase the printing speed, and improve the printing quality of a full-tone rendition printer. Only clock and mode signals are needed to control the thermal print heads which can print full-tone rendition images. Applications for the driver include not only thermal print heads, but also LED print heads, LCD print heads, etc.

ACKNOWLEDGMENT

The authors would like to thank T. Masuda and H. Fukuda for device designing and testing. They are also grateful to S. Iguchi, K. Hoya, and Dr. Y. Hori for helpful suggestions and constant encouragement.

REFERENCES

[1] T. Matsuda *et al.*, "A new Bi-CMOS structure for analog/digital systems," in *Europhysics Conf. Abstr. ESSDERC*, Sept. 1981, pp. 174–176.
[2] M. Tsumura *et al.*, "A 64 bit Bi-CMOS sink driver for thermal head," *IEEE Trans. Consumer Electron.*, vol. CE-30, pp. 348–352, Aug. 1984.
[3] K. Hoya, "Analog/digital compatible device technology and its applications," *Hitachi Rev.*, vol. 35, no. 5, pp. 231–236, 1986.
[4] Y. Tokuma *et al.*, "A powerful 32 bit thermal printing head driver LSI," *IEEE Trans. Consumer Electron.*, vol. CE-28, pp. 351–358, Aug. 1982.

Author Index

L

M

N

O

P

R

S

T

U

V

W

Y

Z

Subject Index

A

B

C

F

G

H

I

L

M

N

O

P

Editor's Biography

Mohamed I. Elmasry was born in Cairo, Egypt on December 24, 1943. He received the BSc. degree from Cairo University, Cairo, Egypt, and the MASc. and PhD. degrees from the University of Ottawa, Ottawa, Ontario, Canada, all in Electrical Engineering in 1965, 1970, and 1974 respectively. He has worked in the area of digital integrated circuits and system design for the last 30 years. Dr. Elmasry worked for Cairo University from 1965 to 1968 and for Bell-Northern Research, Ottawa, Canada, from 1972 to 1974. He has been with the Department of Electrical and Computer Engineering, University of Waterloo, Waterloo, Ontario, Canada, since 1974, where he is a Professor and founding Director of the VLSI Research Group. He also has a cross appointment with the Department of Computer Science where he is a Professor. Dr. Elmasry holds the NSERC/BNR Research Chair in VLSI design at the same University since 1986. He has served as a consultant to research laboratories in Canada and the United States, including AT&T Bell Labs, GE, CDC, Ford Microelectronics, Linear Technology, Xerox and BNR, in the area of LSI/VLSI digital circuit/subsystem design. During sabbatical leaves from Waterloo he was at the Micro Components Organization, Burroughs Corporation (Unisys), San Diego, California; Kuwait University, Kuwait; and Swiss Federal Institute of Technology, Lausanne, Switzerland. He has authored and co-authored over 200 papers on integrated circuit design and design automation and has several patents to his credit. He is editor of the following IEEE Press books: *Digital MOS Integrated Circuits,* 1981; *Digital VLSI Systems,* 1985; *Digital MOS Integrated Circuits II,* 1991 and *Analysis and Design of BiCMOS Integrated Circuits,* 1994. He is also author of the book *Digital Bipolar Integrated Circuits,* John Wiley, 1983; and a co-author of the books *Digital BiCMOS Integrated Circuits,* Kluwer, 1993; *Optimal VLSI Architectural Synthesis,* Kluwer, 1992, and *Digital Integrated Circuit Design,* Dur-ul-Qalam, 1993. Dr. Elmasry has served in many professional organizations in different positions and has received many Canadian and international awards. He is a founding member of the Canadian Conference on VLSI, the International Conference on Microelectronics, and the founding president of Pico Electronics Inc. Dr. Elmasry is a member of the Association of Professional Engineers of Ontario and is a Fellow of the IEEE for his contributions to digital integrated circuits.